Encyclopedia of Lasers and Optical Technology

ROBERT A. MEYERS, EDITOR
TRW, Inc.

Academic Press, Inc.
Harcourt Brace Jovanovich, Publishers
San Diego New York Boston
London Sydney Tokyo Toronto

This book is printed on acid-free paper. ∞

Academic Press, Inc.
San Diego, California 92101

United Kingdom Edition published by
Academic Press Limited
24–28 Oval Road, London NW1 7DX

Library of Congress Cataloging-in-Publication Data

Encyclopedia of lasers and optical technology / Robert A. Meyers, editor.
p. cm.
Includes bibliographical references and index.
ISBN 0-12-226693-5 (alk. paper)
1. Optics--Encyclopedias. 2. Lasers--Encyclopedias. I. Meyers, Robert A. (Robert Allen), Date.
TA1509.E53 1991
621.36'03--dc20 90-39014
CIP

Printed in the United States of America
90 91 92 93 9 8 7 6 5 4 3 2 1

Encyclopedia of Lasers and Optical Technology

Executive Advisory Board

Editorial Advisory Board

Contents

Contributors

Abraham, Neal. *Bryn Mawr College*. Laser Pulsations, Dynamical.

Albrecht, Georg F. *Lawrence Livermore National Laboratory*. Solid-State Lasers (in part).

Alfano, R. R. *City College of New York*. Ultrafast Laser Technology (in part).

Barr, L. D. *National Optical Astronomy Observatories*. Telescopes, Optical.

Bass, Michael. *University of Southern California*. Laser–Materials Interactions.

Bjorkholm, John E. *AT&T Bell Laboratories*. Laser Cooling and Trapping of Atoms.

Blyler, Jr., L. L. *AT&T Bell Laboratories*. Optical Fibers, Drawing and Coating (in part).

Boynton, Robert M. *University of California, San Diego*. Color, Science of.

Cherin, A. H. *AT&T Bell Laboratories*. Optical Fiber Communications (in part).

Delfyett, Peter J. *Bell Communications Research*. Semiconductor Injection Lasers (in part); Ultrafast Laser Technology (in part).

Demas, J. N. *University of Virginia*. Luminescence (in part).

Demas, S. E. *University of Virginia*. Luminescence (in part).

DiMarcello, F. V. *AT&T Bell Laboratories*. Optical Fibers, Drawing and Coating (in part).

Duley, W. W. *York University*. Gas Lasers.

Eberly, J. H. *University of Rochester*. Quantum Optics (in part).

Fischer, Robert E. *Ernst Leitz Canada Ltd*. Optical Systems Design.

Fisher, Robert A. *R. A. Fisher Associates*. Optical Phase Conjugation.

Freund, H. P. *Science Applications International Corporation*. Free-Electron Lasers (in part).

Gayen, S. K. *City College of New York*. Ultrafast Laser Technology (in part).

Gibbs, H. M. *University of Arizona*. Optical Circuitry (in part).

Gibson, U. J. *Dartmouth College*. Optical Circuitry (in part).

Gratze, S. C. *GEC Research UK*. Switching, Optical.

Guest, Clark C. *University of California, San Diego*. Holography.

Hariharan, P. *CSIRO Division of Applied Physics*. Interferometry, Optical.

Healy, W. P. *University College London and Royal Melbourne Institute of Technology*. Electrodynamics, Quantum.

Iga, Kenichi. *Tokyo Institute of Technology*. Microoptics.

Johnston, Jr., T. F. *Coherent, Inc*. Tunable Dye Lasers.

Kannari, Fumihiko. *Keio University*. Rare Gas-Halide Lasers (in part).

Katzir, Abraham. *Tel Aviv University*. Optical Fiber Techniques (Medicine).

Khitrova, G. *University of Arizona*. Optical Circuitry (in part).

Knee, Joseph L. *Wesleyan University*. Ultrashort Laser Pulse Chemistry and Spectroscopy.

Koch, S. *University of Arizona*. Optical Circuitry (in part).

Lee, Chang-Hee. *Bell Communications Research*. Semiconductor Injection Lasers (in part).

McAlister, Harold A. *Georgia State University*. Speckle Interferometry.

McArthur, David A. *Sandia National Laboratories*. Nuclear Pumped Lasers.

Measures, Raymond M. *University of Toronto Institute for Aerospace Studies*. Fiber-Optic Based Smart Structures.

Milonni, P. W. *Los Alamos National Laboratory*. Quantum Optics (in part).

Mirsalehi, Mir Mojtaba. *The University of Alabama in Huntsville*. Optical Information Processing.

Obara, Minoru. *Keio University*. Rare Gas-Halide Lasers (in part).

O'Gallagher, J. *University of Chicago*. Nonimaging Concentrators (Optics) (in part).

Parker, R. K. *Naval Research Laboratory*. Free-Electron Lasers (in part).

Payne, Stephen A. *Lawrence Livermore National Laboratory*. Solid-State Lasers (in part).

Peyghambarian, N. *University of Arizona*. Optical Circuitry (in part).

Pollock, Clifford L. *Cornell University*. Color Center Lasers.

Reintjes, John F. *Naval Research Laboratory*. Nonlinear Optical Processes.

Seaton C. T. *Coherent Radiation*. Optical Circuitry (in part).

Silfvast, William T. *AT&T Bell Laboratories*. Lasers.

Smith, P. W. *Bell Communications Research, Inc*. Mode-Locking of Lasers (in part).

Solimeno, Salvatore. *University of Naples*. Diffraction, Optical.

Solomon, Wayne C. *University of Illinois at Urbana–Champaign*. Chemical Lasers.

Stegeman, G. I. *University of Arizona*. Optical Circuitry (in part).

Tariyal, B. K. *AT&T Technologies, Inc*. Optical Fiber Communications (in part).

Warren, M. *Sandia National Laboratories*. Optical Circuitry (in part).

Waymouth, John F. *GTE Lighting Products*. Light Sources.

Weiner, A. M. *Bell Communications Research, Inc*. Mode-Locking of Lasers (in part).

Winston, R. *University of Chicago*. Nonimaging Concentrators (Optics) (in part).

Preface

The *Encyclopedia of Lasers and Optical Technology* provides a unique compendium resource for scientists, teachers, and engineers as well as university undergraduates and graduate students. The articles encompass basic theory, key information, and applications for those who either are expert in one of the subdisciplines and require additional information on another or have a basic education in science and wish to become familiar with modern lasers and optics.

Scientists and engineers who invent, design, evaluate, or test new applications for lasers and optical technology in medicine, information processing, communications, optical computing, manufacturing, sensing, robotics, energy, and defense will have at their fingertips basic theory, as well as descriptions of each laser or optical system with key parameters in mathematical, descriptive, illustrative, and tabular form. This information can be used for the assessment of test data, planning of experimentation, and preparation of technical reports, article manuscripts, proposals, and presentations. Technical supervisors, corporate managers, and government personnel will find this *Encyclopedia* useful as background information for project assessment and decision making.

This *Encyclopedia* is the first single work to cover all major facets of lasers and optics and to do so on a mathematically oriented, first-principle basis suitable to provide an understanding of the principles of this subject as well as its applications. It is essentially a condensation of key information from 38 textbooks that could have been written to detail the subject.

Each article is organized in a single, uniquely accessible format. The glossary section preceding each article allows the uninitiated researcher, developer, or evaluator to easily integrate the information presented in the subject article without reference to additional background information, while the cross-references allow access to related *Encyclopedia* articles containing supporting theory, techniques, and technologies. The carefully selected review article, reference, and textbook bibliographic entries at the conclusion of the articles provide additional background information. Access to information is provided by both article alphabetization and a detailed subject index. There is no other laser or optics compendium that presents information in this easily accessible and supportable form.

The *Encyclopedia* provides basic understanding of laser theory, laser types, and applications in an article entitled Lasers. Each major type of laser is presented in detailed articles on Gas Lasers, Rare Gas-Halide Lasers, Chemical Lasers, Free-Electron Lasers, Semiconductor Injection Lasers, Solid-State Lasers, Color Center Lasers, Tunable Dye Lasers, Ultrafast Laser Technology, and Nuclear Pumped Lasers. Laser phenomenology is described in articles on Mode-Locking of Lasers and Laser Pulsations, Dynamical. Processes for modifying laser radiation characteristics are presented in articles on Nonlinear Optical Processes and Optical Phase Conjugation.

Information crucial to experimentation and applications is presented in articles on Light Sources; Luminescence; and Color, Science of; while the theoretical basis for modern laser and optical science is presented in articles on Quantum Optics and Electrodynamics, Quantum.

Laser applications are covered in articles on Holography; Laser Cooling and Trapping of Atoms; Laser–Materials Interactions; Ultrashort Laser Pulse Chemistry and Spectroscopy; Optical Fiber Techniques (Medicine); Interferometry, Optical; Speckle Interferometry; and Fiber-Optic Based Smart Structures.

Important optical phenomena are presented in articles on Diffraction, Optical; Microoptics;

and Switching, Optical. Modern optical applications are described in Optical Systems Design; Optical Information Processing; Optical Fiber Communications; Nonimaging Concentrators (Optics); Telescopes, Optical; and Optical Circuitry. Also, fabrication of the basic materials of many modern optical systems applications is described in Optical Fibers, Drawing and Coating.

Robert A. Meyers

CHEMICAL LASERS

Wayne C. Solomon *University of Illinois at Urbana–Champaign*

GLOSSARY

Beam quality: Measure of the excellence of the beam, that is, how closely it approaches the theoretical limit of 1.0.

Chemical laser: Laser operating on a population inversion produced (directly or indirectly) in the course of an exothermic reaction.

Efficiency: Effectiveness of conversion of chemical energy into laser energy output.

Optical cavity: Portion of the flowing system containing the gain medium.

Specific power: Power/mass flow rate, given in kilojoules per kilogram.

Spectral transitions: Energy spacing for the lasing lines.

Unstable resonator: Optical extraction system composed of two convex mirrors, allowing for resonance within the optical cavity.

Chemical lasers are normally divided into two classes, that is continuous wave (cw) flowing systems and various types of pulsed devices. These can be subdivided into direct chemical pumping and transfer lasers, depending on how the lasing species receives its energy. The laser systems of greatest potential for scaling to high average powers are the cw class; we will, therefore, direct this discussion to cw systems offering the most promise. There has been dramatic progress in chemical laser research in the past 25 years. Since the first pulsed chemical laser was discovered by J. V. V. Kasper and G. C. Pimentel in 1965, a very large body of scientific research and engineering data have been produced. The most comprehensive survey of the chemical laser field was accomplished by A. A. Steponov, R. W. F. Gross, and J. F. Bott several years ago.

I. Continuous Wave, Direct Pumping Chemical Lasers

A. EARLY WORK AND OPERATING CHARACTERISTICS

In 1967, K. G. Anlauf reported the first experiments capable of producing a steady stream of population-inverted active molecules. This demonstrated the *potential* for operation of a class of continuous wave chemical lasers. The excitation is provided by the simple rearrangement of the chemical bonds of the initial species undergoing exothermic chemical reaction. The development of the first *cw chemical laser* devices quickly followed these early experiments. In 1969, two independent groups almost simultaneously observed lasing from a supersonic diffusion flowing laser. The principle of operation for the flowing devices is based on the mixing of reagents at reduced pressure, chemical reaction, and the subsequent nonequilibrium distribution of energy within the product particle. When an optical system is provided, these purely chemical processes are nearly simultaneously followed by stimulated emission of radiation from the active lasing medium.

Continuous wave lasing has been achieved by direct pumping from a wide variety of chemical reactions. The HF/DF or DF/HF lasers, around which most of our discussion will center, have been highly developed in the United States. Other less-well-known cw chemical lasers include HCl, HBr, CO, and HF overtone types. Many of these have been characterized by researchers in Europe and the Soviet Union. [*See* NUCLEAR PUMPED LASERS.]

The supersonic diffusion HF[DF] laser employing fuel combustion is the most important example. An oft-studied derivative of this is an electric arc-heated device. Both will be discussed. Typically, these laser variants employ a one-step excitation process known as the "cold" reaction mechanism.

$$F + H_2[D_2] \rightarrow HF^*[DF^*] + H[D], \quad \Delta H = -32 \text{ kcal/mole} \quad (1)$$

The existing lasers based on this chemistry are necessarily low-pressure devices due to the relatively short collisional lifetimes for the principal lasing species. The most efficient devices known have output energies (specific power) in the range 100–400 kJ/kg and corresponding chemical efficiencies of 5–20%. The higher chemical efficiencies are obtained at a lower range of optical cavity pressures. The HF laser radiates multiline at $\lambda = 2.5$–$3\ \mu m$, although for every process which we consider, there will be an analogous one governing the DF laser [$\lambda = 3.6$–$4.2\ \mu m$]. These wavelengths correspond to the vibration–rotation transitions in HF[DF].

Continuous wave lasers are designed with elements which can be best illustrated by considering a simple combustion-driven supersonic diffusion laser as an example. In this case, no external power supply is required; the laser operates purely on chemical energy. Figure 1 gives the main elements which make up this device. The laser combustor provides the appropriate thermodynamic conditions for the dissociation of the oxidizer (fluorine or fluorinated oxidizer). Diluent (helium) mixed with combustor fuel, D_2, is introduced into the combustor through a conventional gas–gas injector. Additional diluent plus an excess of F_2 enter through impinging jets, providing the complete mixture of combustion gases which undergo auto-ignition and rapidly achieve the desired equilibrium conditions. Combustion stagnation temperatures vary from 1500 to 2500 K, and stagnation pressures are in the range 30–300 psi in the subsonic flowing section. This short combustion zone has walls which are cooled with excess fuel or liquid coolant. The conditioned combustor gas containing the excess dissociated fluorine subsequently flows through an array of highly cooled small nozzles. These serve as the fine supersonic–supersonic mixing arrays which introduce the oxidizer stream into the optical cavity. These arrays of supersonic mixing nozzles are designed to be at their maximum efficiencies for particular operating conditions that produce flow at a temperature and pressure substantially lower than that in the combustor (optical cavity static temperatures are 200–400 K and pressures are 2–15 torr). Another function of the nozzle is to "freeze" chemically the F-atom mole fraction at or near its combustor value. The supersonic mixing nozzles thus establish appropriate pressure, temperature, and composition for the all-important chemical reactions which proceed rapidly in the laser cavity. The cavity mixing and chemical excitation reaction stage takes advantage of the large cavity inlet velocity provided by the supersonic nozzles, which act to extend the lasing-zone length and fill the optical cavity.

A wide variety of laser optical systems are employed to extract a beam from the cavity. Typical of these is the unstable resonator in Fig. 1. In this system, there is a central region near the optical axis (of such a radius that its Fresnel number is near unity) within which all the laser radiation is closely coupled together by diffraction. Therefore, the laser emission produced by this central region is coherent. However, because the convex mirrors magnify the beam continually as it propagates through the optical cavity, this central or uniphase mode spreads out to fill the entire laser medium, eventually overflowing around one mirror to form the laser output shown exiting the beam extraction mirror (scraper mirror). One way to think of this unstable resonator is as a low-Fresnel-number oscillator region surrounded by a multipass amplifier. Since phase control of the output beam occurs within the central portion of the resonator, a high-quality output beam is usually obtained.

The output beam has the form of an annulus with a zero-intensity central region produced by the part of the beam that is blocked by the extraction mirror. Such an output beam, when passed through a conventional telescope designed to accommodate an unobscured beam, has an annular intensity distribution which departs from the ideal uniform intensity (such would give the minimum far-field beam width). However, in high-power lasers, it is frequently found that the annular beam departs so little in beam quality from the uniform beam that other factors, such as medium homogeneity, are more important than uniform intensity when optimization of laser system performance is conducted.

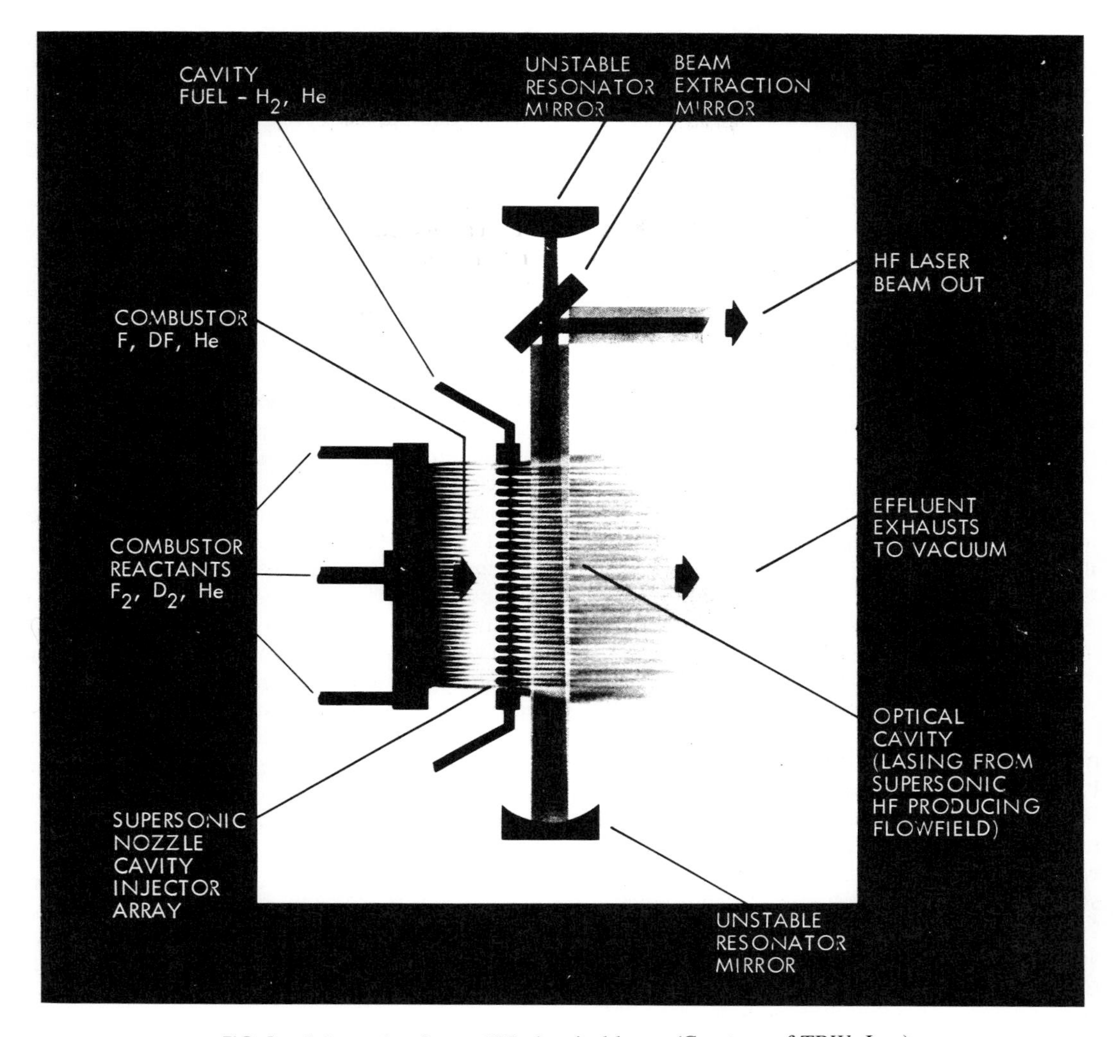

FIG. 1. Schematic of a cw HF chemical laser. (Courtesy of TRW, Inc.)

In the final stage of any flowing chemical laser, the processing of the exhaust effluent must be considered. This is usually accomplished by a smooth transition of the cavity shroud into a diffuser designed to raise the pressure and accommodate the low entrance Reynolds numbers. Diffuser efficiency can be considered an integral part of the overall design of the laser system. A common normal-shock diffuser model is usually adequate to optimize the performance of this section. Pumping of this effluent is subsequently accomplished after recovery of the available pressure within the system.

This completes our description of the generic cw chemical laser. The details and evolution of these systems will now be considered.

B. Experimental Results

Results obtained from arc-heated devices have been crucial in developing the required data base. Devices which can run for several hours, such as that illustrated in Fig. 2, have been used to guide the effort toward higher efficiencies. During these experiments, it was noted that the chemical reaction producing the excited HF is severely limited by diffusion and that fast mixing is the key to higher powers and better efficiencies. The gas generator unit illustrated here incorporates a successive arrangement of arc and plenum chambers to provide the source of atomic fluorine (eliminating the need for a subsonic combustor). A stream of helium mixed

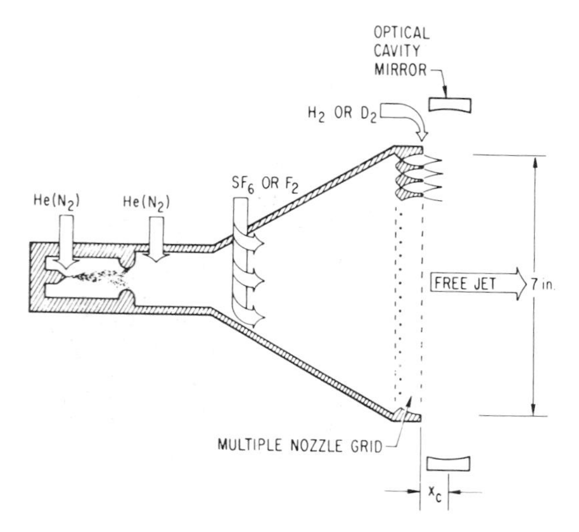

FIG. 2. Schematic of an arc-driven cw chemical laser. [Reprinted with permission from Gross, R. W. F., and Bott, J. F. (1976). "Handbook of Chemical Lasers." Wiley, New York.]

with fluorine atoms is accelerated through a supersonic nozzle array into the laser cavity. At the exit of the nozzle, there is a secondary fuel injector system used to supply the fuel H_2 (or D_2). It quickly becomes apparent that short, minimum-length nozzles are required to expand and accelerate the gas flow to provide supersonic mixing of the oxidizer–diluent mixture with the secondary fuel jet. These nozzles are decidedly different from wind-tunnel nozzles or rocket nozzles, in which the main object is to produce a gas flow that is in equilibrium and in which all the energy introduced by the heater is recovered as kinetic energy and not frozen in the internal energy modes of the molecules. In contrast, the primary nozzle of a chemical laser generates only a nonequilibrium chemical composition. The high velocity of the supersonic jet is also an important feature of the chemical laser. This is because the very rapid deactivation of the HF^* and, therefore, the very short effective lifetime of the lasing species result in short inversion zones even at supersonic velocities. Therefore, a premium is placed on the ability of a particular nozzle design to mix the fuel, H_2 (or D_2), rapidly into the supersonic flow. The special characteristics of arc devices have been particularly instructive in the effort to design proper supersonic mixing nozzles. Early experiments with these devices have provided the data base for rapid-mixing, two-dimensional, slit-nozzle arrays and large axisymmetric matrix nozzles. The latter led to the development of the fastest known supersonic mixing arrangement, which is the array of small circular axisymmetric nozzles in which the main fluorine jets are fully surrounded by fuel. In this design, the main jets are surrounded by hydrogen, and the distance the hydrogen has to diffuse is considerably shortened when compared to a slit-nozzle array of equal area ratio.

An analogous development with fine-scale axisymmetric nozzles has been conducted for combustion-driven low-pressure nozzles employing helium as a diluent (see Fig. 3). This has been followed by a number of sophisticated laboratory developments to provide integral two-dimensional, rapid-mixing nozzles for specific applications. This technology has been vastly expanded and demonstrated on a wide variety of combustion-driven lasers employing nitrogen diluent. However, experiments with a new technology eventually superceded these in what has proven to be a unique departure from these earlier integral nozzle arrays. The efficient fine-scale mixing obtained with the axisymmetric arrays in the presence of large boundary layers has been replaced by a clever hypersonic injection technique within the free stream. Several of these new arrays have been developed in such a way that highly efficient manufacturing methods are employed. The technology development referred to here is the employment of relatively large primary fluorine nozzles and injection of the H_2 (or D_2) by hypersonic wedges located across the nozzle exit plane. This method provides a well-established supersonic flow in the primary stream while allowing minimal boundary layer growth along the hypersonic wedges. Care is taken to ensure that the oblique shocks created by the hypersonic injectors do not block the large primary nozzle or create large optical path differences. This hypersonic wedge technology has been incrementally scaled up in the United States, utilizing modular devices such as the Multiple Purpose Chemical Laser and the Alpha verification module to provide the data base for the large engineering demonstration laser, Project Alpha, recently discussed by workers at TRW.

The Alpha cylindrical chemical laser is a high-power demonstration of several laser technologies. It provides the ability to wrap arrays of modules in the form of a large cylinder so that

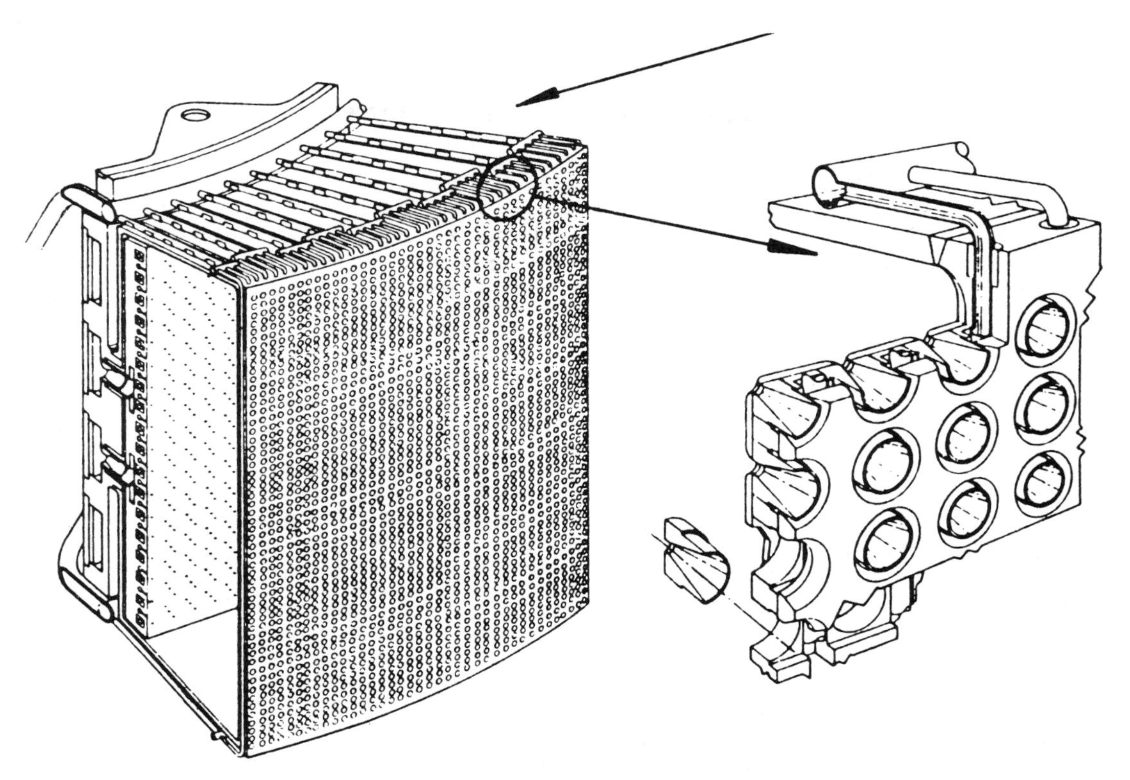

FIG. 3. Gain generator module with axisymmetric mixing nozzle. (Courtesy of Textron.)

the fluorine and diluent flow passes radially out around the annulus which holds the hypersonic hydrogen injectors (see Fig. 4). Mixing and chemical reaction occur in the annularly formed optical cavity. A specially designed annular resonator must be employed for mode control (see Fig. 5). Curvature in the radial direction on the annular mirrors allows mode control and intensity levels to be controlled in the compacted leg. The cylindrical configuration is preferred for very high powers because it allows for decreased gain length over that of conventional linear laser configurations. This provides many times the device output power without increasing the nozzle power flux, cavity injector length, or mode width. Such a resonator should also reduce effects from flow-field disturbances, diffraction losses, anomalous dispersion, and other optical loss mechanisms which might tend to produce a poor-quality beam.

Another trend has been toward improved mixing nozzles for the production of large linear lasers. Thus, a series of HF (or DF) chemical laser nozzle arrays have been developed for operation at high cavity pressure where mixing is likely to be slow and deactivation is rapid. These nozzles incorporate trips within the trailing edges of the arrays to enhance the stretching of the diffusion flame front. Several meters of individual banks of such nozzles can be coupled in a slightly tilted fashion (avoiding the alignment of repetitive flow disturbances) to achieve very high powers. Supersonic diffusers and ejectors have been developed for efficient recovery of pressure from such flows. The best example of a high power laser test device produced by scaling up the high-pressure trip nozzle technology is the Mid-Infrared Advanced Chemical Laser (MIRACL) system located at the White Sands Missile Range in New Mexico. The optical configuration for this system is a confocal unstable resonator similar to that illustrated in Fig. 1.

There have been a series of engineering demonstrations of large linear chemical lasers at TRW, leading to the development of the MIRACL. These demonstrations have been employed in exploring the beam characteristics, beam control technologies for pointing and

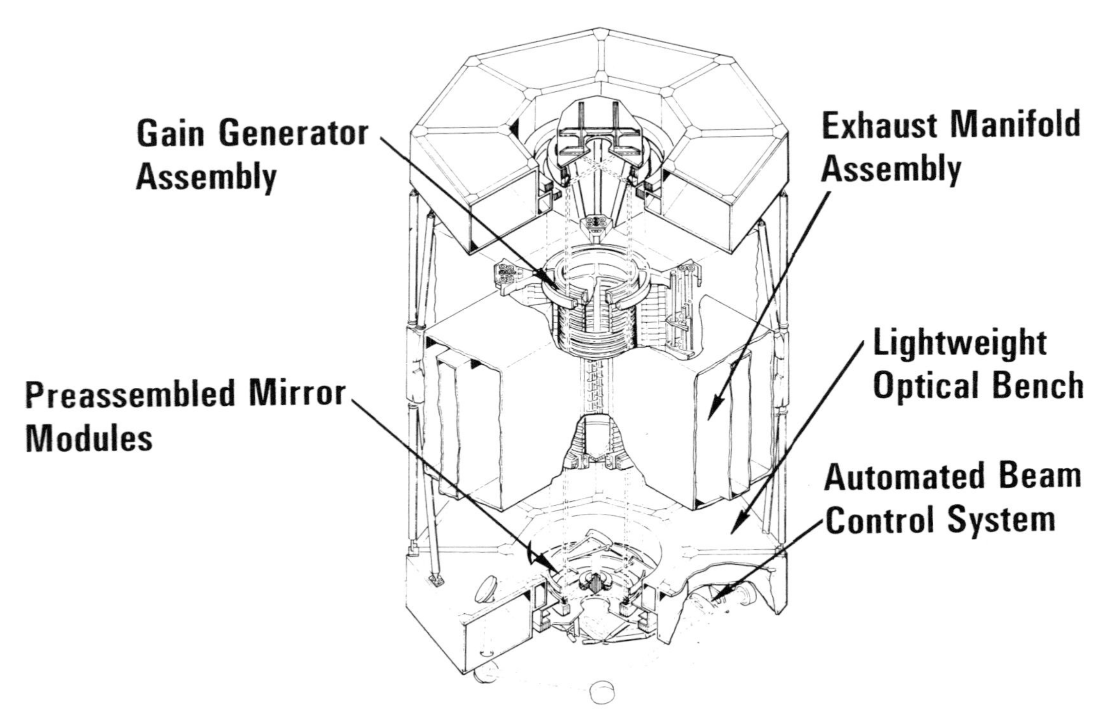

FIG. 4. Alpha HF chemical laser. (Courtesy of J. Miller, TRW, Inc.)

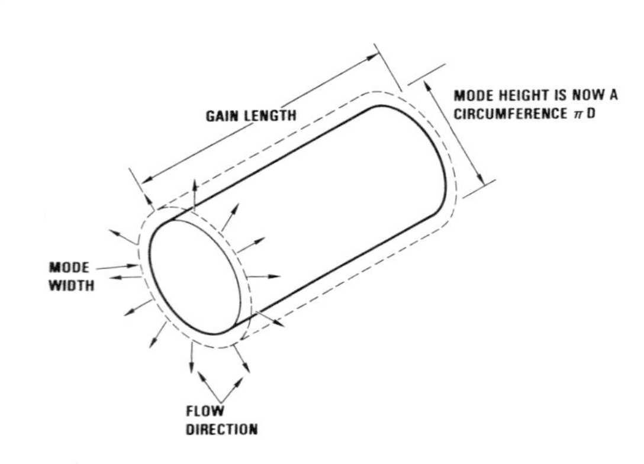

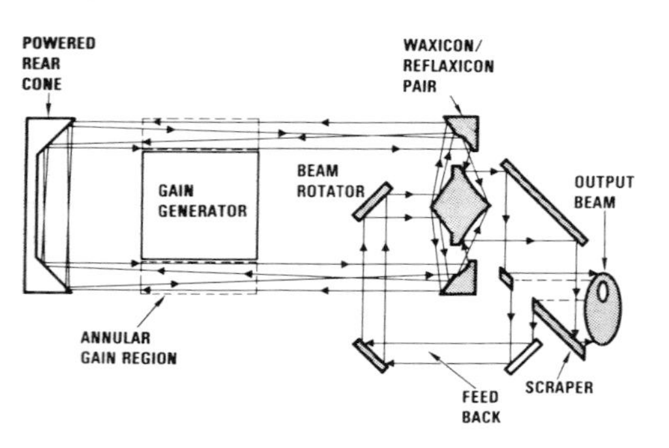

FIG. 5. Cylindrical laser resonator. (Courtesy of J. Miller, TRW, Inc.)

tracking, DF laser beam propagation, and beam effects testing. In addition to the pure power-scaling effects with linear lasers, which do not appear to be of consequence, there are media- and hardware-induced degradations on the beam quality of these lasers at higher powers. The principal effects are repetitive flow-field disturbances, mirror distortion, and diffraction. At laser powers achieved so far, the most serious contributions to poor beam quality appear to be those which result in optical path differences within the cavity. Some of these are evidently due to design compromises, but even with the most rigidly designed linear devices, there will be dramatic benefits from optical devices which are designed to correct phase front distortions. The methods currently available include phase sensing and correction with a deformable mirror. Advances in this area are coming rapidly.

C. Recent Advances

New phase conjugation methods such as stimulated Brillouin scattering (SBS) might well be applied to HF lasers to improve the beam to attain very high quality and brightness. With further developments in this area likely, one can easily conclude that practical consideration of cw chemical lasers to tens of megawatts may be possible.

A completely different approach to scaling and beam projection for chemical lasers involves the combination of beams from separate resonators. Thus, phased arrays of laser beams can provide a means of increasing the power and aperture size of laser transmitters. D. J. Bernard has provided a good example of this technique in experiments with coupled multiline cw HF lasers. Here the larger aperture is synthesized by combining the outputs of several phase-locked laser systems. In a coupled device, part of the output from one laser is injected into a second laser in order to achieve phase locking between the two outputs. Phase locking here is defined as the achievement of mutual coherence between two outputs from every spectral line. Complete phase locking has now been achieved between each spectral line of two lasers employing unstable confocal resonators. The demonstrations have been impressive, since the corresponding far-field peak intensity enhancement and reduction in far-field spot size have been very near the theoretical limit.

Many attempts have been made to produce a purely cw chemical laser that is capable of operating at wavelengths shorter than 2.8 μm. The techniques proposed involve lasing on electronic transitions and have not been successful. However, a short wavelength laser concept which offers potential for scaling to very high powers has proven its capabilities in W. Q. Jeffers's laboratory. The concept is to operate an HF chemical laser on the first overtone transitions, thus shifting the output to several bands centered at 1.33 μm. The overtone chemical laser is produced by providing for the optical resonator to remain below threshold for the fundamental transition while lasing on the overtone transitions. It has been found that overtone efficiencies exceeding 30% of the available power on the fundamental of an HF laser can be obtained when the cavity mirror reflectivity at 2.8 μm is near 1% while maintaining a relatively high reflectivity (>98%) at the overtone wavelength, 1.3 μm. The potential for an even more efficient overtone device for utilization with smaller optical telescopes which transmit well through the atmosphere appears to be quite high, according to recent work by L. H. Sentman (private communication).

II. Continuous Wave Transfer Chemical Lasers

The operation of a cw transfer chemical laser depends on the transfer of energy from excited reaction products, which are produced in a mixing and reacting flow system, to host molecules or atoms which can undergo stimulated emission. Most of the work has been concerned with systems utilizing vibrational–rotational transitions in CO_2 and, more recently, in the development of chemical oxygen-iodine (COIL) cw chemical lasing utilizing electronic transitions. Recent work has focused on the utilization of new reagents to extend the spectral range of the emission.

A. The DF–CO_2 Transfer Chemical Laser

The most highly developed cw transfer chemical laser is the subsonic and supersonic DF–CO_2 laser. This laser depends on the chain reaction of fluorine and deuterium followed by the transfer of vibrational quanta from excited DF molecules to CO_2 molecules as seen in reactions (1′–3).

$$F + D_2 \rightarrow DF^* + D, \qquad \Delta H = -32 \text{ kcal/mol} \tag{1'}$$

$$D + F_2 \rightarrow DF^* + F, \qquad \Delta H = -100 \text{ kcal/mol} \tag{2}$$

$$DF^* + CO_2 \rightarrow DF[v - 1] + \Delta CO_2[00^01] \tag{3}$$

The lasing occurs at 10.6 μm and corresponds to the usual P-20 transition, 00^01–100^0 band of the CO_2 molecule. This cw laser is undoubtedly the most versatile of all the known chemical lasers, having been operated successfully at multikilowatt powers over a range of cavity pressures of 10 to 250 torr, with extended lasing zones both subsonically and supersonically. Theory attributes this robustness to the large amount of chemical energy in storage, the rapid rate of reactions (1′–3), and the relatively low rate of deactivation of the upper state of CO_2. As a result of these factors, the chemical efficiency of these devices can be quite high.

B. The Chemical Oxygen–Iodine Laser

The chemical oxygen–iodine laser (COIL) is the only known chemically pumped cw laser utilizing electronic transitions. Overall processes involve the liquid phase generation and flow of excited metastable singlet oxygen, O_2 ($^1\Delta g$) into the gaseous phase. This is followed by a fast resonant electronic energy transfer to a metastable atomic iodine, $I(5^2P_{1/2})$ according to the following reaction set.

$$H_2O_2 + 2OH^- + Cl_2 \rightarrow O_2(^1\Delta g) + 2H_2O + 2Cl^- \quad (4)$$

$$2O_2(^1\Delta g) + I_2 \rightarrow 2O_2(^3\Sigma^- g) + 2I(5^2P_{3/2}) \quad (5)$$

$$O_2(^1\Delta g) + I(5^2P_{3/2}) \rightleftarrows O_2(^3\Sigma^- g) + I(5^2P_{1/2}) \quad (6)$$

$$I(5^2P_{1/2}) \rightarrow I(5^2P_{3/2}) + h\nu \; 1.315 \; \mu m \quad (7)$$

In the first reaction (4), the chemical energy originally contained in the hydrogen peroxide is stored in the singlet oxygen. An alkaline solution is necessary to provide sufficiently rapid generation of the gaseous oxygen. The sound reaction (5) (although speculative in nature) is needed to generate sufficient quantities of atomic iodine species. The final chemical reaction (6) is a reversible process needed to generate the excited-state iodine atom for the upper laser state. Since this process is reversible, the ratio of excited-state oxygen to ground electronic state must be sufficiently high to ensure that a population inversion in iodine atoms obtains.

The state of development of this COIL device is inherently tied to the development of the singlet oxygen chemical generator. The usual two-phase flow-excited oxygen generation methods tend to operate best at very low pressures (1–5 torr), and this inherently narrows the operational characteristics available in supersonic flowing systems. The adverse reaction of water and hydrogen peroxide vapor in the process of quenching O_2 ($^1\Delta g$) and excited iodine atoms makes it necessary to remove these deactivating agents. Thus, cooling of the gas stream after generation but prior to entering the laser cavity is necessary to remove the condensibles. A number of clever laboratory generation/cooling techniques have been employed in an effort to increase the power and efficiency of supersonic COIL lasers; however, no completely acceptable solution for scaling the generator component is currently at hand, and the potential remains to be realized.

This laser has the shortest wavelength of any cw chemical laser. Such a device is of interest because it has been operated to maximum efficiencies of as much as 40% and offers the potential to reduce the size of the large optics used for high-power systems. The single-line laser wavelength lends itself to good optical transmission through the atmosphere or silica fiber. In materials-processing applications, a coupling (absorption) of laser energy is also quite efficient at 1.3 μm. These same properties also permit some hope for application to the problem of controlled thermonuclear fusion.

Bibliography

Airey, J. R., and McKay, S. F. (1969). *Appl. Phys. Lett.* **15,** 401.

Anlauf, K. G., Maylotte, C. H., Pacey, P. D., and Polanyi, J. C. (1967). *Phys. Lett. A* **24,** 208.

Bernard, D. J., Chodzko, R. A., and Mirels, H. (1988). *AIAA J.* **26,** 1369–1372.

Blauer, J. A., Hager, G. D., and Solomon, W. C. (1979). *IEEE J. Quantum Electron.* **QE-15,** 602.

Cool, T. A., and Stephans, R. S. (1970). *Appl. Phys. Lett.* **16,** 55.

Driscoll, R. J., and Tregay, G. W. (1983). *AIAA J.* **21,** 241.

Duignan, M. T., Feldman, B. J., and Whitney, R. B. (1987). *Opt. Lett.* **12,** 111.

Giedt, R. R. (1973). The Aerospace Corporation Tech. Rep., TR-0073(3435)-1, The Aerospace Corporation, Los Angeles, California.

Gross, R. W. F., and Bott, J. F. (1976). "Handbook of Chemical Lasers." Wiley, New York.

Jeffers, W. Q. (1989). *AIAA J.* **27,** 64–66.

Kasper, J. V. V., and Pimentel, G. C. (1965). *Phys. Rev. Lett.* **14,** 3529.

McDermott, W. E., Pchelkin, N. R., Bernard, D. J., and Bousek, R. R. (1978). *Appl. Phys. Lett.* **32,** 469.

Meinzer, R. A. (1970). *Int. J. Chem. Kinet.* **2,** 335.

Miller, J. (1986). Advances in chemical lasers, *Proc. Int. Conf. Lasers, 1985*. STS Press, McLean, Virginia.

Miller, J. (1988). *Proc. Int. Conf. Lasers, 1987*. STS Press, McLean, Virginia.

Spencer, D. J., Jacobs, T. A., Mirels, H., and Gross, R. W. F. (1969). *Int. J. Chem. Kinet.* **1,** 295.

Steponov, A. A., and Schcheglov, V. A. (1982). *Sov. J. Quantum Electron.* (*Engl. Transl.*) **12,** 681–707.

Steponov, A. A., Schcheglov, V. A., and Yuryshev, N. N. (1985). *Sov. J. Quantum Electron.* (*Engl. Transl.*) **15,** 746–777.

Wilson, L. E., and Hook. D. L. (1976). *AIAA Pap.* **76–344.**

COLOR CENTER LASERS

Clifford R. Pollock *Cornell University*

Glossary

Cross section: Measure of the probability of a stimulated transition occurring. The larger the cross section, the more likely a photon will interact with the system.

Femtosecond: 10^{-15} second, abbreviated fsec.

Free spectral range (FSR): Frequency difference between adjacent resonances of an optical cavity.

Interstitial: Presence of an extra ion in the lattice at a site not normally occupied in a perfect crystal.

Quantum efficiency: Ratio of the number of emitted photons to the number of absorbed photons.

Stokes shift: Wavelength of the luminescence is generally greater than the wavelength of the stimulating radiation.

Threshold: Point where a laser has enough gain to begin to oscillate.

Vacancy: The absence of an ion at a lattice point that would normally be occupied.

Wave function: Distribution of the electron in space. This distribution is generally localized around a potential well.

Certain color centers in the alkali halides can be used to create broadly tunable, optically pumped lasers in the near infrared. These lasers may be operated in a continuous wave with output powers on the order of 1 W, or mode locked to yield pulses of less than 100 fsec duration. Laser-active color centers generally provide long-lived operation only when operated at cryogenic temperatures, although many of the centers can be stored at room temperature with no degradation. Operationally, color center lasers are analogous to dye lasers, except that instead of operating in the visible region, color center lasers cover the infrared tuning range from 0.8 to 4 μm. Their ability to generate tunable radiation in the near infrared has made color center lasers unique and indispensable in the study of guided wave optical devices, narrow bandgap semiconductors, and molecular spectroscopy.

I. Introduction to Color Center Lasers

Color centers are simple point defects in crystal lattices, consisting of one or more electrons trapped at an ionic vacancy in the lattice. These point defects are common in many crystalline solids and have been studied rather extensively in alkali-halide crystals (e.g., NaCl and KCl). Certain color centers have optical absorption and emission bands that make them suitable as laser gain media. The structure and physics of these centers and transitions will be reviewed in the next section. Lasers based on these color centers are closely analogous to organic dye lasers: they are optically pumped, can be broadly tuned, share a similar cavity design, and can generate ultrashort pulses. The significant difference between color center lasers and dye lasers is the tuning range: at present, through the use of several different types of centers and host lattices, the entire region between 0.8 and 4.0 μm can be covered. Tunable coherent radiation over this wavelength region is important for studies in optical communication, molecular and semiconductor spectroscopy, and various other specialized fields. It is this unique combination of broad wavelength tuning range and continuous wave power that makes the color center laser useful. [*See* LASERS.]

The laser-active color centers possess two characteristics ideal for efficient, tunable operation. First, their absorption and emission bands are homogeneously broadened: all the excited centers can contribute energy to a given laser mode, and all the centers will be equally well pumped by a single line laser operating within the pump absorption band. Homogeneous broadening allows efficient single-mode operation over the entire tuning range of the laser. Second, most of the transitions involved in laser emission are fully allowed. Such transition strengths when combined with broad homogeneous linewidths lead to large gains, quiet continuous wave (cw) operation, and the ability to generate ultrashort pulses ($\approx 10^{-13}$ sec duration). The single pass power in a 2-mm crystal containing a reasonable density of centers can exceed 100%. Such gains are in dramatic contrast to those obtained in solid state lasers that use transition metal ion impurities for the gain medium.

Color center lasers can be operated in pulsed, mode-locked, or cw fashion. In the cw mode, output powers exceeding 2 W have been achieved in certain crystals. Typical output powers and tuning ranges for the various color center lasers are summarized in the following sections of this article. In pulsed operation, output powers of 1 MW have been achieved, and it is reasonable to assume pulsed powers of many megawatts will be generated routinely in the future. [*See* MODE-LOCKING OF LASERS.]

The single frequency linewidth of the color center laser is truly exceptional in its purity. In single mode cw operation, linewidths below 4 kHz have been achieved. This performance is attributable to the solid state nature of the gain media: there are no moving parts in the laser cavity to perturb the phase of the laser. Such high frequency definition, coupled with broad tunability, has made the color center laser a powerful tool for spectroscopy and metrology.

It is perhaps the mode-locked operation of color center lasers that has been the most spectacular. Taking advantage of the broad emission bandwidth, the laser output can be transformed into a train of tunable picosecond and femtosecond duration pulses. The femtosecond color center laser was the first tunable subpicosecond pulse source, and is the only such source that has reasonable average output power (hundreds of milliwatts) and fast repetition rates (~100 MHz). Such short pulses are extremely useful for the investigation of ultrafast phenomena in semiconductor materials and optical fiber switches.

The laser active color centers are generally stable only when cooled to cryogenic temperatures. Although some center types allow for low-duty-cycle pulsed operation at room temperature, cw operation at room temperature usually leads to fading of the output power. At room temperature, the centers either thermally dissociate or become mobile and transform into nonlasing centers through attachment to other defects. Thermal degradation can be minimized or eliminated at reduced temperatures, hence the crystals are usually anchored to a cold finger maintained at liquid nitrogen temperature (77 K). Additionally, the radiative quantum efficiency for some of these centers increases significantly at reduced temperatures.

Due to the relatively short radiative lifetime of the color center, intense optical pumping from a laser source is necessary to achieve efficient laser operation. The pump laser depends on the crystal, but is usually a Nd:YAG laser operating at 1.06 or 1.32 μm, or an Ar- or Kr-ion laser operating in the visible. The color center laser cavity must contain dispersive elements such as a prism in order to facilitate tuning and line narrowing. Properly designed, a color center laser is usually capable of tuning over a range exceeding 25% of its central wavelength. Specific examples of the design and construction of the laser are described in this article.

II. Physics of Laser-Active Color Centers

A. Optical Emission Processes

While color centers exist in many different crystal lattices, most research to date has been done on point defects in alkali-halide crystals such as KCl. This review will concentrate on the alkali halide centers because they form the basis of practically all the useful color center lasers, and they are well understood. A representative sample of color centers in alkali-halide crystals is shown in Fig. 1. All laser-active color centers involve anion (halide ion) vacancies. The F center has the simplest structure, consisting of a

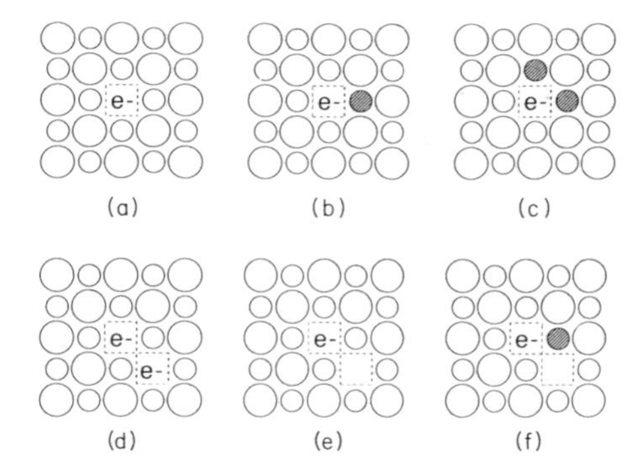

FIG. 1. A representative sample of color centers in alkali-halide crystals. The large and small circles represent negative and positive ions, respectively. Filled circles represent alkali impurities. (a) F center, (b) F_A center, (c) F_B center, (d) F_2 center, (e) F_2^+ center and (f) $(F_2^+)_A$ center.

single electron trapped at a vacancy surrounded by an essentially undisturbed lattice. If one of the neighboring alkali ions is a substitutional alkali impurity, say Li^+ in a KCl crystal, the center is called an F_A center. Similarly, the F_B center consists of an F center beside two substitutional impurities. Two adjacent F centers along a [110] axis of the crystal form the F_2 center. Its ionized counterpart is called the F_2^+ center, which is an F_2 center with only one electron. Analogous to the F_A center, the $(F_2^+)_A$ center consists of an F_2^+ center adjacent to a substitutional alkali ion. Of the various centers described in Fig. 1, the F_A, F_B, F_2^+, and $(F_2^+)_A$ defects form color center lasers. Note that the simple F center does not lase. The reason for this will be discussed below.

There are also several types of laser-active color center not shown in Fig. 1. These include the $(F_2^+)^*$ and $F_2^+ : O^{2-}$ centers, which consist of F_2^+ centers associated with an as yet unknown defect and with an oxygen ion, respectively. Also missing is the $Tl^\circ(1)$ center, which is schematically similar to the F_A center but is functionally more like a neutral Tl perturbed by an adjacent vacancy, and the N_2 center, which consists of three F-centers in a trigonal arrangement. Many other alkali-halide color centers are not shown, such as the larger aggregate centers (including F_4), but with few exceptions, these larger centers do not play a significant role in the physics of color center lasers. Laser-active color centers have also been reported in crystals other than the alkali halides, including the F^+ center in alkaline earth oxides and the H_3 and N_3 centers in diamond. These lasers have not been widely reproduced or studied, and there is still some doubt concerning the physics of their operation; hence they will not be discussed in this article.

The optical absorption and emission of the F center can be understood on a qualitative basis using a highly simplified quantum mechanical model, the particle-in-a-box. This model is based on the fact that F centers are essentially electrons trapped in a three-dimensional square well formed by the electrostatic potential of the surrounding positive ions. In this model, the energy between the ground state and first excited state is:

$$E_{2p} - E_{1s} = 3h^2/8ma^2$$

where $1s$ denotes the ground state, $2p$ the excited state, h is Planck's constant, a is the box dimension, and m is the mass of the electron. Note that the energy scales with the well dimension as a^{-2}. This model works amazingly well when applied to color centers. It is found experimentally that if distance a is taken as the nearest neighbor separation, the F-band energy can be related to a for most alkali halides as

$$E_F = 17.7a^{-1.84}$$

where a is in angstroms and E_F is in electron volts. This is known as the "Mollwo relation." As will be illustrated with examples below, many color centers follow a similar relation between the lattice constant and the energies of their absorption and emission bands. The model also predicts a strong coupling between the lattice dimensions and the transition energy. Lattice vibrations, called phonons, harmonically vary the actual dimensions of the square well in a period of less than 10^{-13} sec, causing the energy levels to vary on this time scale. Since this perturbation is random and occurs at all F centers at a rate faster than the excited state lifetime of the center, the absorption and emission bands of the center are homogeneously broadened.

B. F Centers

The F center is the most fundamental color center defect in the alkali-halide lattice. Although it is not laser-active, the optical properties of the F-center are important in understanding the laser physics of other color center lasers. The fundamental absorption band of the F center, called the "F" band, corresponds to a transition from the 1*s*-like ground state to the 2*p*-like first excited state of the square-well potential. The F band transition is very strong and dominates the optical spectrum of the alkali-halide crystal. In fact, the term *F center* comes from the German word *Farbe*, meaning color, and refers to the strong color imparted to the otherwise transparent alkali-halide crystals.

As noted above, the simple F center is not suitable for lasing. After excitation to the first excited state, the F electrons are very near the conduction band and can be easily ionized by thermal or optical energy. The strong possibility of self-absorption into the conduction band by photons emitted from other F centers also exists, destroying the potential gain mechanism as well as introducing loss. These deleterious effects result from an anomalous spatial relaxation following optical excitation, as illustrated in Fig. 2. After being excited, the electron is raised to the 2*p* state, where its spatial structure is more spread out than the 1*s* state. In response to this diffuse wave function, the lattice expands slightly, enlarging the dimensions of the potential well. This expanded lattice causes even further spreading of the excited-state wave function and lattice until lattice forces finally restrain the relaxed dimensions. The resultant wave function extends out several lattice constants and is called the relaxed excited state (RES). Due to the poor spatial overlap of the RES and the terminal state wave function, the emission dipole moment is relatively small, leading to a small gain cross section. No laser has ever been made based on simple F centers, and it appears unlikely that such a laser will ever exist. On the other hand, the F center is very useful for forming other varieties of color centers.

C. F_A(II) Centers and F_B(II) Centers

F_A centers are classified into two categories depending on their relaxation behavior after optical excitation. Type I centers, denoted F_A(I), behave almost identically to the simple F center just described. They display the same diffuse relaxed excited state, low gain cross section, and self-absorption. The other (and rarer) class of F_A centers are called type II. They are distinguished by a dramatically different relaxation process. After excitation, F_A(II) centers relax to a double-well configuration as shown in Fig. 3. A neighboring anion moves between the impurity and neighboring cation, creating a double-well potential. The resulting electronic wave functions remain localized and the transition moment is very strong. The resulting emission from this transition is Stokes shifted out to the 2–3 μm region.

The presence of the impurity ion in the F_A(II) system causes the P_z orbital to be distinguished from the $P_{x,y}$ orbitals (see Fig. 3). Transitions of the P_z orbital result in a longer wavelength absorption band that is often well resolved from the main band. In laser applications, this additional band considerably enhances the probability that a convenient pump source will overlap with the absorption.

Upon relaxation, the anion separating the two wells can move back into its original location or into the original vacancy. In the latter case, the F_A(II) center will have effectively rotated 90°. Such reorientation can lead to orientational bleaching of the band, where the center no longer absorbs light polarized along the original *z* axis. In practice, optimum power is achieved from F_A(II) crystals if a crystal ⟨100⟩ axis lies at 45° to the laser and pump polarization, so that

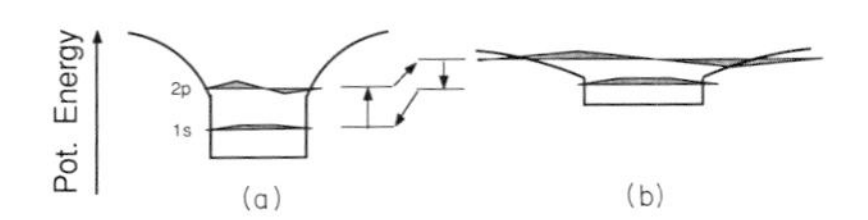

FIG. 2. Potential well and wave functions for the F center (a) before excitation (normal configuration) and (b) after excitation (relaxed configuration). The four-level energy scheme is shown between the wells. All energies are plotted with respect to the conduction band, hence the relaxed state appears to be at a higher energy than the first excited state.

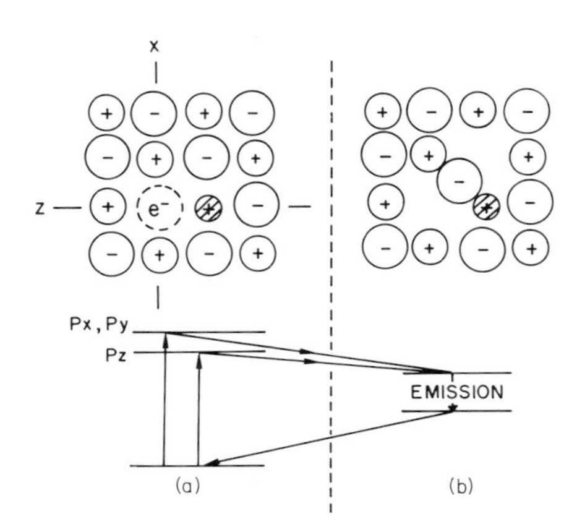

FIG. 3. Structure and energy-level diagram of the F_A(II) center (a) before optical excitation (normal configuration) and (b) after optical excitation (relaxed configuration).

either center orientation will have a component of the dipole moment aligned with the stimulating and pump fields.

$F_A(II)$ centers have been found only in lithium-doped alkali halides, specifically KF, KCl, and RbCl. The lithium is incorporated into the crystal during growth in concentrations approaching 0.02 mol %. Lasers have been built using the KCl and RbCl centers and, except for different tuning ranges, their characteristics are similar. The quantum efficiency η of the $F_A(II)$ luminescence in KCl : Li is about 40% at $T = 77$ K, and decreases slowly with increasing temperature until it approaches zero at $T \approx 300$ K. Efficient operation of $F_A(II)$ center lasers thus requires that the crystals be maintained at cryogenic temperatures during use. Liquid nitrogen ($T = 77$ K) is commonly used, being inexpensive and readily available. Coupled with the large Stokes shift, the low quantum efficiency leads to a maximum conversion efficiency of about 10% for the $F_A(II)$ laser.

F_B centers involve an F center beside two substitutional alkali impurities. Since the distribution of foreign alkali ions in an otherwise pure lattice is statistical, F_B centers are far fewer in number than F_A centers and are obtained in substantial quantities only when the impurity dopant concentration is fairly high (approximately 1%). Like the F_A center there are two types of F_B center, also classified by their relaxation behavior. The $F_B(I)$ center is formed when the two substitutional impurities lie along a common $\langle 100 \rangle$ axis. The $F_B(I)$ center has optical properties similar to the F center, which preclude the possibility of lasing action. The $F_B(II)$ center is formed when the two alkali impurities are adjacent to one another along a $\langle 110 \rangle$ axis of the crystal (see Fig. 1).

The $F_B(II)$ centers relax into a double-well configuration after excitation, similar to that shown in Fig. 3 for the $F_A(II)$ center. Their emission is almost entirely quenched at temperatures of 4 and 300 K, but reaches a maximum around 100 K. Optimum laser performance can be obtained from the $F_B(II)$ center with cw pumping, which raises the crystal temperature from 77 K to around 100 K. Pulsed operation is less effective in raising the crystal temperature, since the crystal has a chance to cool between pulses. One difficulty with $F_B(II)$ centers is that they are accompanied by substantial quantities of $F_A(I)$ and $F_B(I)$ centers, all of which have overlapping absorption bands with the $F_B(II)$ center. Consequently, the optical pump power suffers losses by these residual centers, diminishing the overall efficiency of the $F_B(II)$ center laser.

The $F_A(II)$ and $F_B(II)$ are among the most stable color center lasers presently known, providing stable tunable laser radiation in the 2.2 to 3.6 μm region. The operational lifetime of these lasers is almost entirely determined by secondary effects, such as crystal fogging due to small vacuum leaks or water desorption in the dewar. Figure 4 shows the power and tuning range of an optimized color center laser using $F_A(II)$ and $F_B(II)$ centers. Other relevant details of these and the other color center lasers are tabulated in Table I.

TABLE I. Performance of Common Color Center Lasers

Host lattice	Center	Pump wavelength (μm)	Tuning range (μm)	Maximum power (W)	Operational lifetime
LiF	F_2^+	0.647	0.82–1.05	1.8	Days
NaF	$(F_2^+)^*$	0.87	0.99–1.22	0.4	Weeks
KF	F_2^+	1.06	1.22–1.50	2.7	Days
NaCl	F_2^+	1.06	1.4–1.75	1	Days
NaCl : OH	$F_2^+ : O^{2-}$	1.06	1.42–1.85	3	Years
KCl : Tl	$Tl^{\circ}(1)$	1.06	1.4–1.64	1	Years
KCl : Na	$(F_2^+)_A$	1.32	1.62–1.95	0.05	Months+
KCl : K_2O	$F_2^+ : O^{2-}$	1.32	1.7–1.85	0.06	Months+
KCl : Li	$(F_2^+)_A$	1.32	2.0–2.5	0.4	Months+
KI : Li	$(F_2^+)_A$	1.7	3.0–4.0	0.006 (pulsed)	?
KCl : Na	$F_B(II)$	0.514	2.25–2.65	0.05	Years
KCl : Li	$F_A(II)$	0.514	2.3–3.0	0.2	Years
RbCl : Li	$F_A(II)$	0.647	2.6–3.6	0.1	Years
KCl	N_2	1.064	1.27–1.35	0.04 (pulsed)	Months+

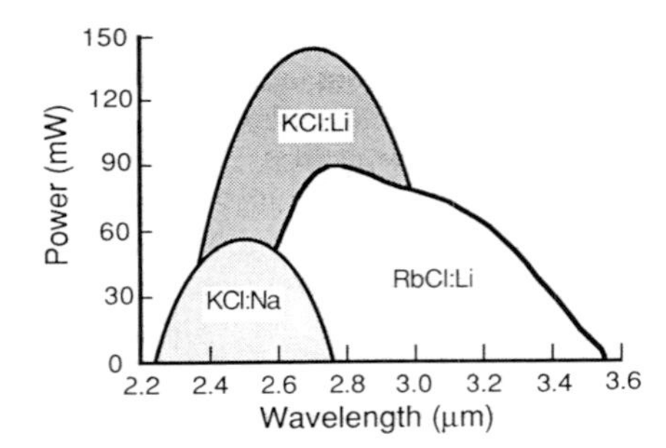

FIG. 4. Tuning range and output power from an optimized color center laser using $F_A(II)$ and $F_B(II)$ centers. (Adapted from German, 1986.)

D. F_2^+ Centers

A powerful but short-lived color center laser is based on the F_2^+ center. As indicated in Fig. 1, the F_2^+ center consists of two adjacent anion vacancies sharing one trapped electron. In contrast to the $F_{A,B}(II)$ centers, the F_2^+ relaxation following excitation entails only a slight enlargement of the surrounding lattice.

The configuration of the F_2^+ center is suggestive of an H_2^+ ion imbedded in a dielectric continuum. The two lattice vacancies play the role of the protons. The energy levels of such an ion are related to the free space case by

$$E_{F_2^+}(r, K_0) = (1/K_0^2)\, E_{H_2^+}(R)$$

where $R = r/K_0$ is the proton separation in free space, r is the distance between the lattice vacancies, K_0 is the dielectric constant of the lattice, and $E_{H_2^+}(R)$ is the energy function of the molecular hydrogen molecule in free space, which has been calculated for various states and separations R. Figure 5 shows an F_2^+ energy level diagram. The levels are named after their molecular ion counterparts. The left side of the energy diagram shows the RES levels. Comparison of measured ground and excited state levels of the F_2^+ center in several alkali-halide lattices with predicted levels from the imbedded H_2^+ model show excellent agreement. Dotted lines indicate nonradiative transitions.

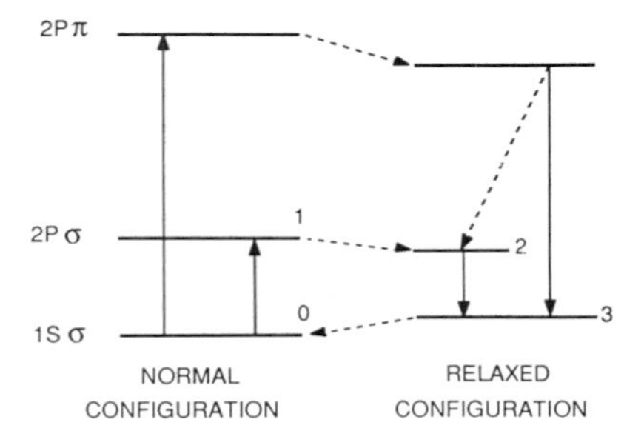

FIG. 5. Energy-level diagram of the F_2^+ center. The dotted lines represent nonradiative transitions.

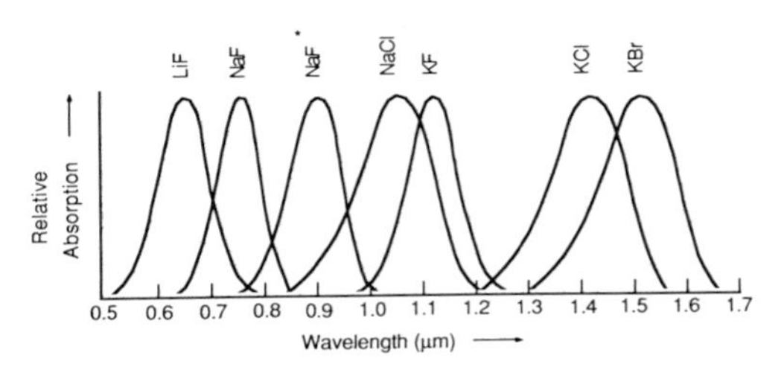

FIG. 6. Normalized absorption bands of the F_2^+ center in various alkali halides.

Since the alkali halides have a wide variety of lattice dimensions, r, and dielectric constants, K_0, one would expect the F_2^+ center to have a wide distribution of absorption energies in the various crystals. This is indeed the case, as was shown in Fig. 6, where the absorption bands of the F_2^+ center range from 0.67 μm in the tight LiF lattice up to 1.5 μm in the much larger KBr lattice. Similar to the F center, the broadening of individual absorption bands is caused by lattice phonons, and is homogeneous.

The transition of greatest interest for F_2^+ lasers is the $1S\sigma \rightarrow 2P\sigma$ transition between the ground state and first excited state. This transition has nearly ideal properties for laser operation: (1) the oscillator strength is large, allowing for a large gain cross section; (2) the quantum efficiency is 100%; (3) the Stokes shift is just enough to prevent overlap of the emission and absorption bands, forming an ideal four-level system; and (4) there are no excited-state absorptions that overlap with the emission. Lasers based on this transition are among the most powerful and efficient color center lasers ever made. Figure 7 shows normalized tuning curves

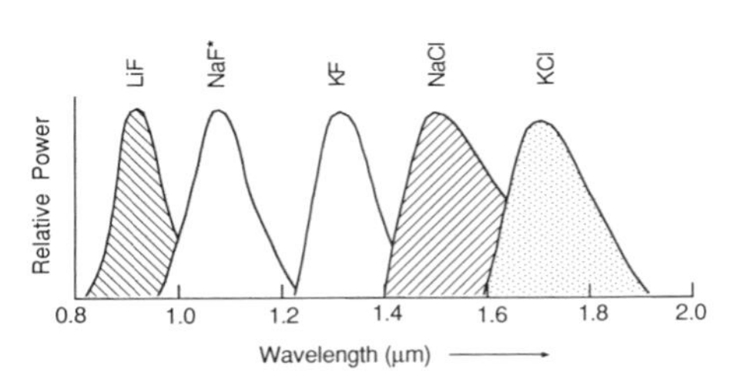

FIG. 7. Normalized tuning curves for some of the known F_2^+ lasers. The NaF* curve is from an $(F_2^+)^*$ laser.

of some common F_2^+ lasers. The tuning range extends from 0.8 to 1.9 μm, a range important for optical communication. [The NaF* band shown in Fig. 6 belongs to the $(F_2^+)^*$ center described below.]

F_2^+ center lasers have two drawbacks: (1) they display a slow fading of output power with extended operation and (2) the crystals must be stored at cryogenic temperatures at all times. The mechanism for the decay is not totally understood, but it is probably associated with reorientation of the center. Excitation to the $2P\pi$ state is known to lead to a reorientation of the center's axis through nonradiative relaxation to one of the other ⟨110⟩ directions in the lattice. Figure 8 illustrates the reorientation of an F_2^+ center. Under the intense optical fields of laser operation, multiphoton excitation from the laser pump source may excite the $2P\pi$ state and lead to reorientation. Since there is nothing in the lattice to pin the center to one location, repeated flipping of the F_2^+ centers cause them to take a random walk through the crystal. Eventually, it is likely they will run into other centers or defects, forming larger F-center aggregates.

The requirement of cold storage arises because the F_2^+ centers are formed through radiation damage. (Coloration techniques are described briefly in Section III.) At elevated temperatures the F centers and radiation byproducts (such as interstitials) tend to annihilate one another through thermal motion. Storage at reduced temperature ($T < 170$ K) stops this aggregation. Cryogenic specimen dewars are generally used for both long-term storage and transporting crystals from the radiation source.

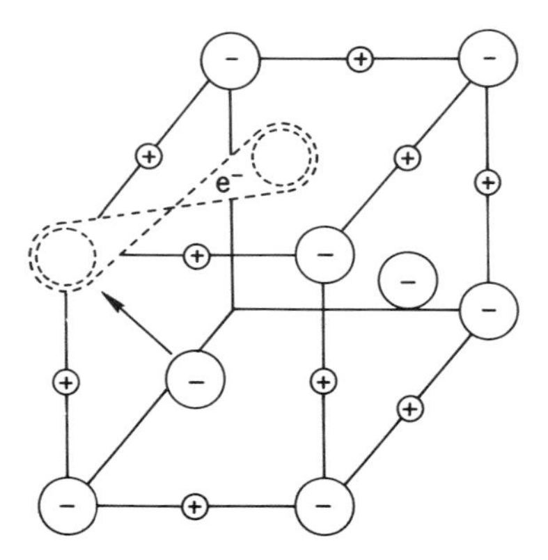

FIG. 8. Schematic representation of the reorientation of an F_2^+ center. A neighboring anion can move into one of the two vacancies, causing an effective rotation of the center.

The operational lifetime of an F_2^+ laser depends on how long the crystal has been pumped. Under the best conditions a single crystal can be made to operate for only several days. This decay, coupled with the awkwardness of creating and storing the active crystals, and the discovery of more stable color center lasers, has virtually eliminated the use of F_2^+ lasers. However, the F_2^+ center laser was the first powerful color center laser, and in the hands of skilled experimenters it has been used to generate tunable near-infrared radiation.

E. Stabilized F_2^+ Centers

The F_2^+ centers can be associated with certain defects in the crystal lattice to form more stable color centers with output characteristics similar to the F_2^+ center. To date, four types of stabilized centers have been reported: the $(F_2^+)_A$, $(F_2^+)^*$, $(F_2^+)^{**}$, and $F_2^+:O^{2-}$ centers. The $(F_2^+)_A$ center is an F_2^+ center located beside an alkali impurity (see Fig. 1). The $F_2^+:O^{2-}$ center is an F_2^+ center beside a doubly negative anion impurity ion. The $(F_2)^*$ and $(F_2^+)^{**}$ centers are F_2^+ centers associated with an as yet undetermined defect in the lattice, most likely an interstitial halogen defect.

These stabilized centers display most of the advantages of the F_2^+ center, such as large cross sections and unity quantum efficiency, but they offer a number of additional benefits as well. First, the centers are pinned at one point in the lattice, and as a result they display reduced or no fading. Second, new and useful tuning ranges can be created because the stabilizing defect slightly perturbs the energy levels. Finally, the room temperature shelf life of the centers is usually increased to essentially infinite length with no degradation of ultimate laser performance.

1. $(F_2^+)^*$ and $(F_2^+)^{**}$ Centers

The $(F_2^+)^*$ center appears in NaF after a radiation-colored F_2^+ center crystal is allowed to sit at room temperature for several days in the dark. The F_2^+ band disappears and is replaced by a new band at a longer wavelength. Pumping this new band produces emission that is shifted to a longer wavelength relative to the F_2^+ emission band, but otherwise displays all the characteristics of an F_2^+ center. Strong evidence has been found linking the density of $(F_2^+)^*$ centers to the dosage of radiation used to introduce color centers.

The $(F_2^+)^{**}$ center is similar to the $(F_2^+)^*$ but is even further shifted in wavelength for both the absorption and emission band. In contrast to the $(F_2^+)^*$, the $(F_2^+)^{**}$ center is only generated in NaF crystals which contain OH^-, thus it is thought that the stabilizing defect for the "double star" center is an artifact of radiation damaged OH^- in the crystal.

The "star" centers can be stored at room temperature, and in NaF they cover an important spectral window around 1.1 μm. Figure 7 includes a tuning curve of the NaF $(F_2^+)^*$ laser. Unfortunately, like the F_2^+ laser, the $(F_2^+)^*$ and $(F_2^+)^{**}$ centers also display a slow fading of output power with use. This fading is an order of magnitude slower than that of the F_2^+ laser, so they represent an improvement from the user point of view. To this date, the actual structure of the "star" centers is not known, although it is likely that they are nearly identical to the $F_2^+ : O^{2-}$ center described in Section II,E,3, with the difference that the star centers are created with radiation damage and the $F_2^+ : O^{2-}$ center is created through additive coloration.

2. $(F_2^+)_A$ Centers

A stable laser based on the $(F_2^+)_A$ center has been demonstrated in several lattices and currently is the only color center laser able to tune beyond 3.6 μm. $(F_2^+)_A$ center lasers combine the best characteristics of the F_A and F_2^+ laser: (1) they are operationally stable with no fading, (2) they are room temperature storable because they are additively colored (see Section III), and (3) they are reasonably powerful.

The stability of the center arises from the trapping of the otherwise mobile F_2^+ center to one location in the lattice, stopping the debilitating migration. The impurity also affects the energy levels of the F_2^+ center: the absorption bands are only slightly shifted, but the emission bands are always shifted to longer wavelengths. The size of the shift depends on the substitutional impurity: a small ion in a large lattice causes a large shift, while similar sized substitutional ions cause a smaller shift. This new degree of freedom is a boon to the laser engineer: by choosing a suitable dopant, the energy bands of the F_2^+ center can be shifted to new wavelength ranges. Examples of these new ranges are shown in Fig. 9, where the laser tuning curves of several $(F_2^+)_A$ centers in KCl, KI, and RbCl are displayed. To illustrate the wavelength shift caused by the impurity, the emission band of the F_2^+ center in pure KCl is also displayed, along with the two known $(F_2^+)_A$ centers in KCl.

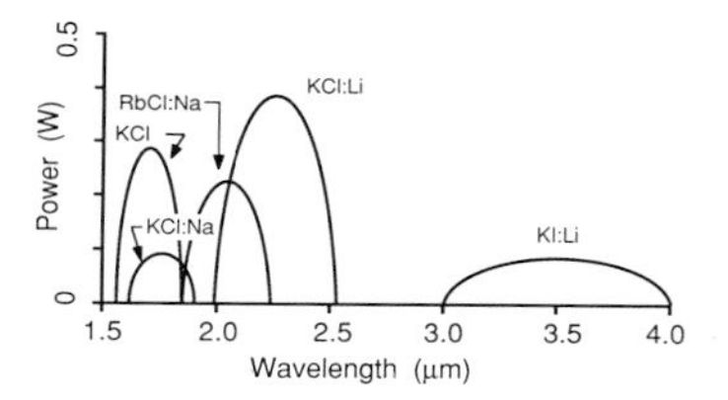

FIG. 9. Tuning curves of several $(F_2^+)_A$ lasers. The data from the KI : Li laser is extrapolated from pulsed data. The KCl tuning curve is for a pure F_2^+ laser and is shown for comparison to the other KCl systems.

To date, $(F_2^+)_A$ lasers have been able to produce output powers of only several hundred milliwatts. This is a fraction of the power available from a comparable F_2^+ laser, and is due to the low density of $(F_2^+)_A$ centers that can be created in a given crystal. Typical $(F_2^+)_A$ crystals only absorb about 30% of the pump power, limiting the overall conversion efficiency. The low absorption is due to a low concentration of centers, which in turn appears to be due to a lack of suitable electron traps. The F_2^+ and $(F_2^+)_A$ centers are each ionized, so there must be one stable electron trap in the crystal for each active center. In the additively colored $(F_2^+)_A$ center crystals, the traps are other color centers, such as the F_A center which traps two electrons to become an F_A^- center. (In the F_2^+ center crystal, artifacts from radiation damage such as interstitial ions serve this function.) Typical of most negatively charged color centers, the lowest lying energy level of this extra electron is energetically near the conduction band of the crystal, so that only a small amount of energy (thermal or optical) is needed to ionize the center, again creating a free electron. The F_A^- electron traps are subsequently slowly discharged by the intense pump light from the laser. To counter this, all $(F_2^+)_A$ lasers use an auxiliary ultraviolet light to continually reionize the $(F_2)_A$ centers to form $(F_2^+)_A$ centers. A dynamic equilibrium between $(F_2^+)_A$ and F_A^- centers thus exists in the crystal during lasing. In Section IV, an example of an $(F_2^+)_A$ laser is shown (see Fig. 12) that uses a mercury lamp to maintain lasing. Typically, lasing ceases within 1 sec of blocking the UV illumination onto the crystal. In spite of the comparatively low power of the $(F_2^+)_A$ laser, for many applications a stable, reliable laser source with 200 mW output power is sufficient.

3. The $F_2^+ : O^{2-}$ Center

The $F_2^+ : O^{2-}$ center forms one of the most stable and powerful color center lasers yet discovered. The center is an F_2^+ center adjacent to a doubly negative substitutional anion impurity, usually O^{2-} although S^{2-} has also been used. A schematic representation of the model is shown in Fig. 10. The structure is distinct from the $(F_2^+)_A$ center because the perturbing impurity is located at an anion site.

This new family of color center laser is operationally similar to the $(F_2^+)_A$ center in three ways: (1) the crystals are additively colored and can be stored at room temperature, (2) lasing is enhanced by an auxiliary light, and (3) the tuning range of each crystal is shifted to slightly longer wavelength than that of the F_2^+ center in the corresponding lattice. In addition, the center appears to be more robust than the $(F_2^+)_A$ center in terms of electron trapping and operating temperatures. Because the impurity ion is doubly negative (as opposed to the singly negative anion it replaces) the overall $F_2^+ : O^{2-}$ center is "charge neutral," so there is little tendency for a free electron to be attracted to and neutralize the laser-active center, as is the case with other F_2^+ systems. The $F_2^+ : O^{2-}$ center essentially has its own "built-in" electron trap.

The laser has been demonstrated in NaCl doped with O^{2-} and S^{2-}, in KCl and RbCl doped with O^{2-}, and will probably be found in other latttices as well. Best results to date have been obtained with the NaCl system, which conveniently operates over the 1.4 to 1.8 μm range. Figure 10 shows the absorption, emission, and laser tuning range of the NaCl system. Following additive coloration and room temperature exposure to UV light for 30 min, a stable absorption band forms near 1.09 μm, accompanied by a strong emission band centered at 1.6 μm. When cooled to 77 K and pumped by a cw Nd:YAG (neodymium-doped yttrium–aluminum–garnet) laser at 1.06 μm, the NaCl crystal has been made to lase with over 3 W of cw output power. Mode-locked operation of the NaCl laser has generated 4-psec pulses, which are at least a factor of two shorter than other stable color centers can produce. Using additive pulse mode-locking techniques (see Section V,C) 75-fsec pulses can be routinely generated with this laser.

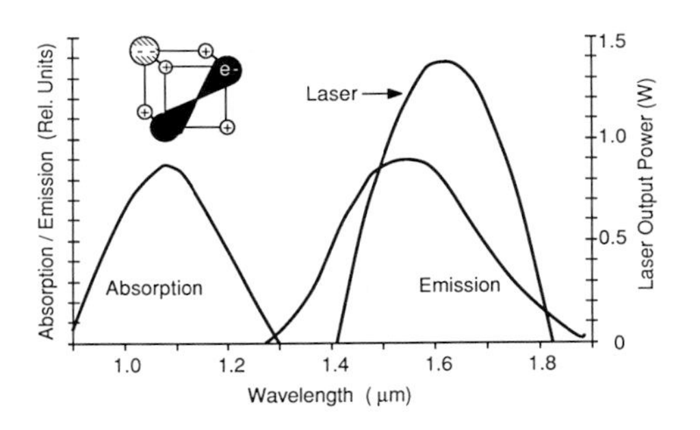

FIG. 10. Tuning curve for the NaCl $F_2^+ : O^{2-}$ center laser. The absorption and fluorescent emission bands are shown in relative units. The insert shows a schematic representation of the $F_2^+ : O^{2-}$ center. Note that the O^{2-} impurity is at an anion location, in contrast to the $(F_2^+)_A$ center.

Under the action of intense pumping, the $F_2^+ : O^{2-}$ center can reorient in the lattice. Only centers that are aligned with the pump polarization strongly interact with the radiation, so they are most likely to reorient. Unfortunately, once the center flips, it no longer is aligned with the driving polarization, so it no longer participates in the gain process. To counter this effect, an auxiliary light is simultaneously exposed onto the crystal during pumping. The auxiliary light acts to reorient the misaligned centers so that their dipole moment becomes aligned parallel to the pump- and laser-electric-field polarization. Commonly, the output of a HeNe laser or a frequency-doubled Nd:YAG laser is used. One milliwatt of auxiliary power is more than enough to counter the effects of intense pumping.

F. Tl°(1) Center

The Tl°(1) center consists of a neutral Tl atom perturbed by the field of an adjacent single anion vacancy. The superscript in the notation denotes that the Tl atom is neutral. The number in parentheses represents the number of vacancies that are adjacent to the atom. The Tl°(1) center has been demonstrated to lase in KCl and KBr lattices doped with Tl. The latter crystal has proven to be very difficult to grow, so most commonly the Tl°(1) center is used only in KCl. An F center trapped beside the Tl ion will find that its electron spends most of its time on the Tl^+ ion, effectively forming a neutral Tl atom that is perturbed by an adjacent positive vacancy.

The transition involved in the Tl°(1) laser is not related to other color center transitions discussed so far. This laser transition occurs between the perturbed $^2P_{1/2}$ and $^2P_{3/2}$ states of the free Tl atom. This transition is normally parity forbidden, but the strong odd symmetry perturbation caused by the positive ion vacancy mixes the ^2P states with higher lying states, and allows for a modest electric dipole to appear between these states. The absorption band caused by this transition is centered around 1.06 μm, and the emission band is centered at 1.5 μm. The Stokes shift comes about through lattice relaxation after excitation. The laser has a relatively narrow tuning curve, extending from 1.4 to 1.63 μm for the KCl host, and from 1.5 to 1.7 μm for the KBr host. This range is about half that expected for an F_2^+-type center. Output powers up to 1 W have been obtained from the KCl Tl°(1) laser. When mode locked, pulses as short as 9 psec have been directly generated.

The Tl°(1) center must be formed by radiation damage, usually using 2 MeV electron beams. Once formed, the center is operationally stable, but requires modest cooling ($T < 0°C$) for long-term storage. The laser properties of the crystal are destroyed if warmed above room temperature for short periods (days) or if pumped too hard in a laser. Due to these inconveniences, and since the NaCl $F_2^+ : O^{2-}$ offers more power, broader tuning, and easier handling, the Tl°(1) laser is slowly being displaced by NaCl.

G. N CENTERS

In most alkali halides a pair of absorption bands, called the N_1 and N_2 bands, exist at a wavelength slightly longer than the F_2 band. The structure of the N centers responsible for these bands has not been conclusively determined, even though the center has been the focus of many studies over the past two decades. The longer wavelength band, called the N_2 band, is usually attributed to the F_3 center in which the three point defects form a triangle in the lattice.

In KCl, the N_2 band overlaps the 1.06 μm line of the Nd : YAG laser, which forms a convenient pump for the center. The emission, shown in Fig. 11, spans from 1.1 to 1.5 μm. Lasing in a pulsed mode has been obtained from this center over the 1.23 to 1.35 μm range, making this the shortest wavelength stable color center laser yet reported. The laser only operates in a pulsed mode however. The pulsed operation is probably (but not yet conclusively) due to multiplet formation of the electrons in the excited state. Unlike all the other lasers described in this article, the N center has more than one electron, which leads to possible triplet or multiplet formation upon relaxation. Such multiplet formation could lead to deleterious absorptions in the lasing region and to a general depletion of laser-active centers. Nonetheless, the N-center laser represents the first of a new class of aggregate center lasers which may open the door to new systems and applications in the future.

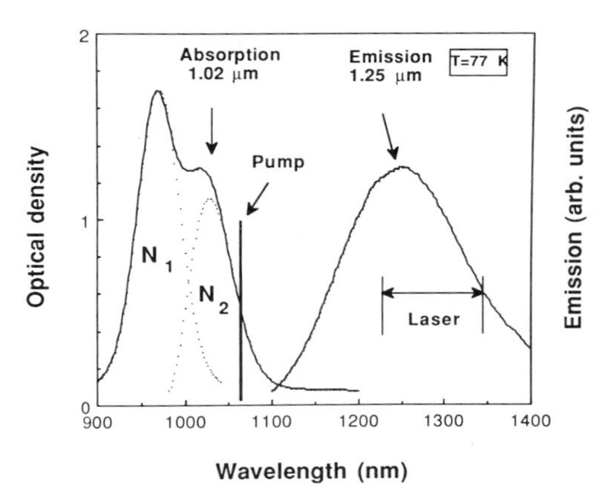

FIG. 11. Absorption and emission bands for the N center in KCl. When pumped by a Q-switched 1.06-μm laser, the N_2 center lases from 1.23 to 1.37 μm.

H. SUMMARY OF LASER PERFORMANCE

Table I lists relevant data concerning the performance of the more common color center lasers. Since the operational lifetime is sometimes an issue with color center lasers, the approximate useful period of a single crystal is also listed. Lifetimes listed as "Months+" represent centers for which no fading has been observed, but observation periods have only been for periods up to 3 months. Unless otherwise noted, the powers are for cw operation.

III. Formation of Color Centers

A question often asked is "How are these color centers formed?". In this section, several proven techniques for forming laser-active crystals will be described.

The first step in generating laser-active centers is to create a population of ordinary F centers, either through additive coloration or radiation damage. Additive coloration is the preferred coloration technique because the F centers produced are very stable, whereas radiation damaged crystals usually require cryogenic storage at all times. However some lasers, like the F_2^+ and the Tl°(1), can only be colored through radiation damage.

A. Additive Coloration

Additive coloration is achieved by diffusing a stoichiometric excess amount of alkali metal into an alkali-halide crystal. This diffusion process can be achieved by placing a crystal in a vapor bath of the alkali metal for a sufficient length of time. To get reasonable vapor pressures of metal requires temperatures on the order of 600°C. The equilibrium concentration for F centers is directly proportional to the metal vapor density.

An excellent apparatus for coloring crystals is the heat pipe. A detailed description of the heat pipe would be out of place in this article, but an excellent review of the method is given by Mollenauer (see the bibliography). Briefly, the heat pipe maintains a zone of pure alkali metal vapor at a precisely controlled pressure. An uncolored crystal is lowered into the alkali vapor for 30 to 60 min, during which time excess alkali ions diffuse into the crystal. To maintain charge neutrality, negative ion vacancies with electrons (F centers) must diffuse into the crystal in equal concentration. The ultimate density of F centers in the crystal is precisely controlled by adjusting the vapor pressure. The density of centers is often critical for optimized laser performance.

After coloration, the F centers are converted into the desired laser-active center through controlled aggregation. The F centers are made mobile by the illumination of the crystal with F-band light (light absorbed by the F center). This optical excitation is usually carried out with nothing more sophisticated than an unfiltered light bulb. The light ionizes some of the F centers, forming vacancies and free electrons. The free electrons become trapped elsewhere in the lattice, typically at another F center forming F^- centers. The vacancies wander through the lattice until they combine with another color center or with a foreign metal ion. Recapture of an electron (from one of the F^- centers) by the new center leads to the formation of an F_2 or F_A center, respectively.

The end product can be controlled through proper doping and temperature control during processing. If F_A or F_B centers are the desired end product, the host lattice should have a higher density of foreign ions than it has F centers. The wandering anion vacancy will then most likely run into a trapping impurity before it runs into another F center. Temperature control is critical: if $T < -50°C$, the vacancies become immobile, while at $T > 0°C$, the vacancies will not attach once they meet an impurity. If the crystal temperature is kept within these limits, an equilibrium population of F_A or F_B can be formed in about 30 min.

To create a more complicated center, such as an $(F_2^+)_A$, a multistep process, which must be empirically determined for each lattice and center, is usually followed. In KCl doped with Li for example, a high density of F centers is first put into the crystal through additive coloration. Aggregation at −20°C then leads to the formation of large populations of F_2 centers as well as F_A centers. The binding energy for F_{2A} centers (an F_2 center beside a substitutional impurity) is less than the thermal energy of the lattice at these temperatures, so few F_{2A} centers are formed in this first step. The crystal is then cooled to −70°C and illuminated with F light. At this temperature, the F centers are nearly immobile, but the F_2 centers are excited by the F light and relax by reorienting (see Fig. 7). Subsequent excitation–relaxation cycles lead to a random walk of the F_2 centers through the lattice, and eventually they are trapped at individual impurity ions, forming F_{2A} centers. Cooling to 77 K and exposing the crystal to UV light ionizes the F_{2A} centers, forming the desired $(F_2^+)_A$ centers. The free electrons are conveniently trapped by the residual F_A centers.

Both the F_A and $(F_2^+)_A$ centers dissociate if the crystal temperature is raised to ~300 K. In fact, if left at room temperature for a long time (days), the F centers will form large aggregate centers called colloids. The colloids are easily dispersed by briefly annealing the crystal at the coloration temperature for about 1 min. Due to this ability to be regenerated, additively colored crystals are considered to be room temperature storable for essentially infinite periods, with the proviso that the crystal may require annealing and reaggregation prior to use.

B. Radiation Damage

Radiation damage is a simple technique for creating color centers. No matter what source of radiation is used, whether it be X rays, γ rays, or high-energy electrons, the primary effect is to produce electron–hole pairs. In the alkali halides, the incident radiation strips the electron from a negatively charged halogen ion. Eventually the electron returns home, with 5–10 eV of kinetic energy. The released energy, in combination with thermal energy from the lattice, is usually enough to eject a halogen atom into an interstitial position in the lattice, leaving the electron behind in the vacancy to form an F center. If irradiation is carried out at low temperatures ($T < 170$ K), the vacancies and interstitials will remain frozen in place. But, if the temperature is raised sufficiently, the anion vacancies will migrate. Should it meet an interstitial, the two defects will annihilate each other, leaving a perfect lattice behind. Thus radiation damaged crystals must be stored at cold temperatures during and after coloration.

Radiation damage is required only for F_2^+ and Tl°(1) lasers. To form the F_2^+ center, the crystal is irradiated at about −100°C, then allowed to warm to room temperature for a few minutes, during which time thermally mobile anion vacancies aggregate with F centers. This warm aggregation process conveniently occurs during the mounting of the crystal in the laser: the crystal must be warmed to room temperature to prevent condensation or frost forming on the crystal surfaces. After mounting, the crystal is cooled to 77 K for use. It must remain below 170 K from then on.

The Tl°(1) center is processed in a similar manner, except with less severe temperature restrictions. After irradiation, the crystal must be kept below 0°C, which can easily be achieved in a home freezer. The steps involved in warming the crystal to mount it are sufficient to aggregate the F centers beside the Tl ions. White light at −10°C for 30 min completes the aggregation process. From then on, the crystal must be stored below 0°C to ensure reliable laser operation.

IV. Optical Gain from Color Centers

A. Stimulated Emission Cross Section

Color centers form an ideal four-level system for laser operation. The energy level diagram of the F_2^+ center in Fig. 4 shows a good example of this. The relevant energy levels for laser operation are labeled 0 through 3, representing the ground state (0), the first excited state (1), the relaxed excited state (RES) (2), and the lower laser level (3). The laser depends on stimulated transitions between levels 2 and 3.

Optical gain is usually defined in terms of a stimulated emission cross section σ, which can be calculated from readily measurable quantities:

$$\sigma = \lambda^2 \eta/(8\pi n^2 \delta\nu \tau_{obs})$$

where τ_{obs} is the measured radiative lifetime of the center, η the quantum efficiency of the emission process, n the index of refraction, λ the wavelength of emission, and $\delta\nu$ the full width at half maximum (FWHM) of the luminescence. For color centers, η is defined as the number of quanta emitted per quanta absorbed. A value less than unity implies that a nonradiative process is occurring in addition to spontaneous emission. Table II lists typical values of τ, η, $\delta\nu$, and σ for a few laser-active color centers.

The gain coefficient γ (cm^{-1}) for a laser medium is related to the cross section by

$$\gamma = \sigma(N_2 - (g_2/g_3)N_3)$$

where N_2 and N_3 are the population densities of the relaxed-excited and relaxed-ground state of the transition, respectively, and g_2/g_3 is the ratio of degeneracies for the two states. The population of the upper laser level N_2 depends on the pumping rate. For color center lasers, virtually all of the centers pumped from the ground state arrive in the relaxed-excited state. If we assume cw pumping and no saturation of the ground state (i.e., $N_0 \gg N_2$ at all times), simple rate equations show that the unsaturated population of N_2 is

$$N_2 = P\tau/h\nu$$

where P is the power per unit volume absorbed by the crystal, $h\nu$ the pump photon energy, and τ the lifetime of the upper state. Due to the relatively short radiative lifetime of the color center, the pump intensity required to achieve a useful population in level 2 must be in the 10^5 W/cm^2 range. Such large intensities can most readily be achieved using a tightly focused pump laser beam. Most color center lasers are pumped by the Ar^+ ion or Nd : YAG laser.

The population of the lower laser level, N_3, decays rapidly into the normal ground state, 0, through lattice contraction. Like the $1 \rightarrow 2$ transition, the $3 \rightarrow 0$ transition is nonradiative and occurs within 10^{-12} sec, so the residual population N_3 is essentially zero. The gain in a color center system thus depends primarily on the upper state population.

Energy extraction from a laser crystal involves the key process of stimulated emission. High intensity is needed to induce stimulated transitions and overwhelm the spontaneous emission from level 2. The characteristic inten-

TABLE II. Laser Design Parameters

Center	λ_0 (μm)	τ_{obs} (nsec)	η (%)	$\Delta\nu$ (THz)	σ (10^{-16} cm^2)	I_{sat} (kW/cm^2)
$F_A(II)$	2.7	200	40	15	1.7	9.4
F_2^+	1.5	80	100	30	1.6	45
$(F_2^+)_A$	2.3	170	100	20	2.7	8.2
$F_2^+ : O^{2-}$	1.6	160	100	45	0.9	9.0
$Tl^{\circ}(1)$	1.5	1600	100	15	0.2	21
N_2	1.3	210	~100	20	0.4	18

sity for a stimulated emission process is called the saturation intensity I_{sat}, and is given by the equation

$$I_{sat} = h\nu/\sigma\tau$$

At $I = I_{sat}$, the stimulated emission rate just equals the spontaneous emission rate. I_{sat} values for several types of color center are given in Table II. A laser is usually designed to operate with an intracavity intensity I in the range $I_{sat} < I < 10\ I_{sat}$. The lower limit of I is obviously for efficiency. The upper limit comes about from the reducing marginal increase in extraction efficiency with increasing intensity.

A linear color center laser cavity is schematically shown in Fig. 12. The cavity has two arms, one with a tight beamwaist at which the color center crystal is placed and one with a nominally collimated beam where tuning elements may be inserted. The crystal is oriented at Brewster's angle to minimize reflective losses. The tight focus at the crystal establishes the high intensity needed for efficient saturation of the population inversion. The spot size at the beam focus, designated the beamwaist parameter ω_0, is typically in the range of 20 to 35 μm for color center lasers.

The beam size expands due to diffraction as the light propagates away from the waist. The characteristic distance over which the beam radius remains smaller than $\sqrt{2}\omega_0$ is called the

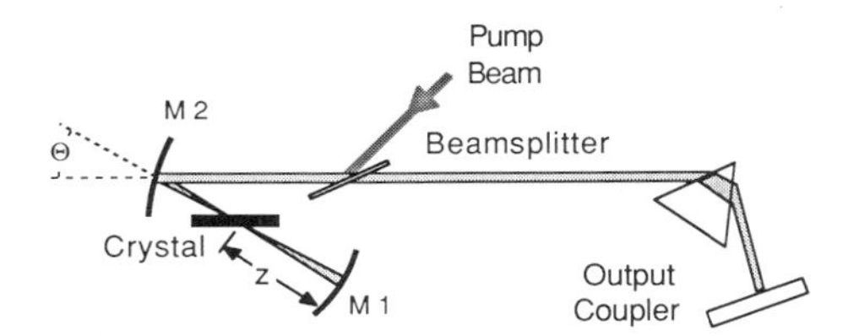

FIG. 12. Schematic of a linear color center laser. The beamsplitter is dichroic, reflecting the pump wavelength yet transmitting the color center wavelength. Tuning is accomplished by rotating the output coupler.

confocal parameter b and is given by

$$b = 2\pi n\omega_0^2/\lambda$$

where n is the index of refraction and λ the wavelength of the beam. Because efficient stimulated emission requires high intensity, the gain media outside of the confocal parameter makes little contribution to the overall power. Therefore color center laser crystals are usually made no thicker than b, with typical values of b in the 1 to 3 mm range.

Pump absorption is determined by the Beer's law expression

$$P_{abs} = P_{in}(1 - e^{-\sigma_g N_0 l})$$

where σ_g is the ground state absorption cross section (which is approximately the same size as the emission cross section for the laser-active color centers), N_0 the ground state population, l the crystal thickness, and P_{in} and P_{abs} represent the input and absorbed pump power, respectively. Efficient pumping demands that most of the pump power be deposited in the crystal, namely $e^{-\sigma_g N_0 l} < 0.1$. However, if $N_0 l$ is too high, the pump power will be absorbed in a thin region, causing severe local heating problems. Good results have been obtained with $N_0 l$ set to give 80–90% absorption per pass.

A clever aspect of the folded cavity shown in Fig. 12 is that it is astigmatically compensated. A Brewster angle crystal introduces a certain amount of astigmatism that is exactly opposite in sign to the astigmatism introduced by the off-axis folding mirror. The total astigmatism can be eliminated by choosing the proper angle of incidence θ of the beam on the folding mirror.

B. Color Center Laser Cavity

Since the color center laser is operated at cryogenic temperatures, the crystal must be enclosed in a vacuum chamber to provide thermal insulation and to prevent condensation on the crystal surfaces. The focusing optics, M1 and M2 of Fig. 12, are usually located inside the vacuum chamber to avoid the astigmatism effects of windows. These mirrors must be carefully prealigned prior to operation. The crystal is usually mounted on a translatable cold finger so that the position of the beamwaist on the crystal can be adjusted. In this way, any local defects or scratches in the crystal can be avoided.

The collimated arm, sometimes called the "tuning arm," is directed out of the main vacuum chamber through a Brewster's angle window. Tuning elements can then be easily placed and adjusted in the tuning arm without disrupting the vacuum surrounding the crystal. The tuning arm is often separately evacuated as well in order to avoid absorptive losses from H_2O and CO_2 bands in the 2.7-μm region.

C. Optical Pumping Techniques

Due to the thickness of the gain media, it is necessary to pump the crystal coaxially with the cavity mode to achieve reasonable efficiency. Several schemes have been employed for injecting the pump beam into the cavity. One method involves the use of dichroic mirrors, where one of the cavity mirrors is given a coating which is transparent to the shorter wavelength pump laser, but reflective for the color center laser wavelength. An example of this technique is shown in the ring laser of Fig. 13. In this example, the pump beam enters the cavity through an end mirror, and the cavity lens focuses both the laser mode and pump beam onto the crystal.

A second popular method is to use a Brewster's angle beamsplitter with a special coating which transmits nearly 100% of the color center laser radiation, yet reflects nearly all of the pump radiation. This technique is shown in Fig. 12. The injected pump beam is focused onto the crystal by the curved mirror M1, and power that is not absorbed on the first pass through the crystal is reflected back by mirror M2 for a second pass. One advantage of this scheme is that the mirrors can be coated with broadband en-

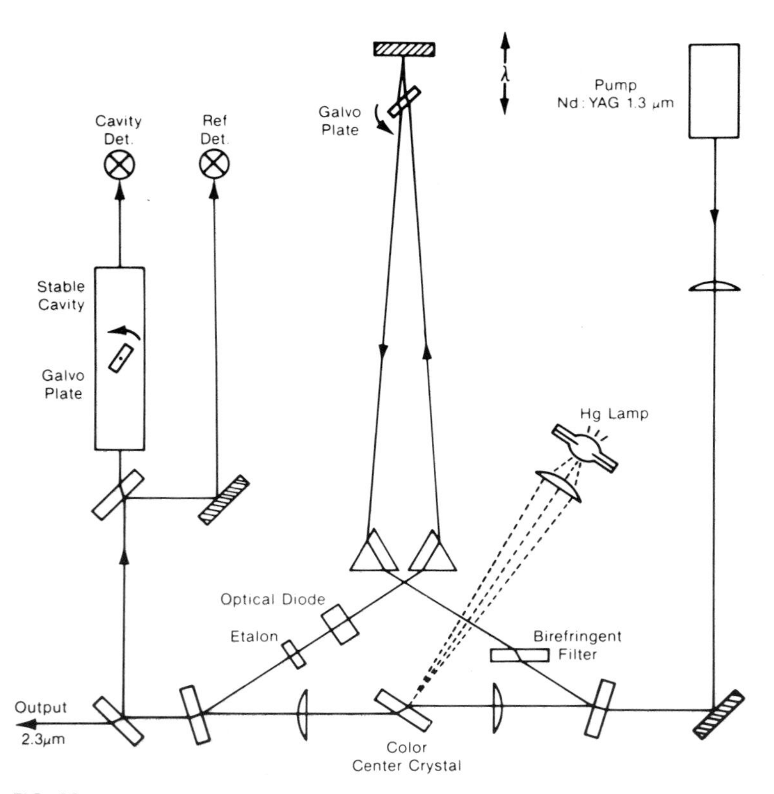

FIG. 13. Actively stabilized color center ring laser. The wavelength is adjusted by translating the top mirror. The mercury lamp is required for $(F_2^+)_A$ center operation.

hanced silver, which permits one set of mirrors to operate over a broad wavelength range.

V. Color Center Laser Operation

A. Broadband Tuning Techniques

Color center emission bands are homogeneously broadened by lattice vibrations. This broadening leads to the wide tuning range of the color center laser, and its homogeneous nature allows for straightforward single-mode operation. However, this same broad tuning range makes it nearly impossible to control the output wavelength with only one tuning element. Generally, a hierarchy of dispersive elements are required for single-mode operation. These elements include a coarse tuner for general wavelength selection, an intermediate dispersive element such as an etalon for individual mode selection, and a cavity length adjustment for continuous frequency scanning of the individual mode. For mode-locked operation, however, it is only necessary to use a coarse tuner.

A Brewster angle prism is often used as a simple and inexpensive tuning element. The prism has the advantage of being low loss and monotonic in wavelength transmission; however, it has the disadvantages of having relatively low dispersion and deviating the direction of the beam as the wavelength is changed. The prism is generally not used in favor of other elements such as the grating or birefringent tuner.

The Littrow mounted diffraction grating has very high dispersion. The Littrow orientation causes the refracted beam to retroreflect along the incident beam. If the blaze angles are chosen correctly, gratings can exhibit first-order retroreflections of >95%. A practical method of using a grating tuner is shown in Fig. 14. The first-order reflection feeds back into the laser and tunes it according to $\lambda = 2d \sin \theta$, where d is the groove spacing and θ is defined in the figure. The zeroth order reflection serves as the output coupling. The zeroth order reflection is bounced off a second mirror adjusted such that a constant deviation of 180° is achieved with respect to the input beam. A small pick-off mirror then directs the beam away from the cavity. The laser wavelength can be linearly scanned using a "sine bar," where motion in the x direction causes a linear change in $\sin \theta = x/h$. The dispersion of the grating is given by

$$\delta\lambda = 2d \cos \theta \, \delta\theta$$

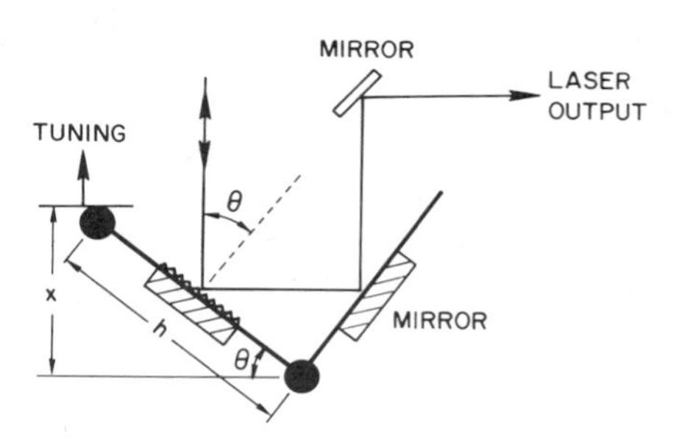

FIG. 14. Littrow-oriented diffraction grafting for tuning a laser. The zeroth order reflection is used as the output coupling, while the first order reflection is fed back into the laser.

where $\delta\theta$ is the laser beam diffraction angle. It has been found that a laser ceases to operate when the returning beam is misaligned by the angle $\Delta\theta = \lambda/40\omega$, where ω is the spot size of the beam. Assuming a 1-mm beam spot size, $\lambda = 1.5$ μm, and $d = 600$ mm^{-1}, we find $\delta\lambda = 0.6$ Å, which is about 30 times narrower than the bandwidth of a prism.

An intermediate selectivity tuning element can be constructed using birefringent plates in the cavity placed at Brewster's angle to the beam. Tuning is accomplished by rotating the plates about their normal axis. A linearly polarized incident beam emerges in general as an elliptically polarized beam. This exit beam suffers substantial reflection losses at the Brewster surfaces of the cavity. However, for certain wavelengths the output polarization is linear and unrotated, and for these there is no loss. These eigenwavelengths vary with plate rotation angle θ as

$$\lambda = (\lambda_p/m)[1 - \cos^2\phi \sin^2\theta]$$

where $\lambda_p = (n_0 - n_e)t/\sin \theta$, m is an integer, t the plate thickness, and ϕ Brewster's angle. The bandwidth $\delta\lambda$ is a complicated function of the number of polarizing surfaces in the cavity but has a magnitude that falls somewhere between the dispersion of the prism and the grating. An advantage of the birefringent tuner is that it requires no displacement of the beam during tuning, in contrast to the prism or grating. One disadvantage of the birefringent tuner arises from the multiple orders that exist in a plate, making it possible for the laser to hop back toward line center when tuned to the extremes of the tuning curve.

B. Single Mode Operation

It is possible to operate a color center laser in a narrowline single mode that can be tuned to any wavelength within the tuning range of the crystal. Such capability has proven useful for molecular spectroscopists, where the excellent frequency of definition and tunability of the color center laser have allowed detailed studies of many important molecular species.

1. Standing Wave Linear Laser

Using a coarse tuning element, the homogeneously broadened standing wave laser oscillates in one primary mode and one or two secondary modes caused by spatial hole burning. The frequency spacing of the secondary modes from the primary mode is given by

$$\Delta \nu_{\text{hole}} = c/4z$$

where z is the distance between the end mirror and the active media (see Fig. 12). The physical mechanism causing oscillation at the hole burning frequencies is illustrated in Fig. 15. The standing wave of the primary mode "burns" holes into the population inversion of the gain media, saturating the gain to the level where gain equals cavity loss. At the nodes of the standing wave, the gain is not saturated and remains at values well above that needed to exceed threshold. A second mode which is spatially $\pi/4$ out of phase with the primary mode will perfectly overlap this periodic gain and will see the excess gain. Unless mode selective losses are included in the cavity, this second "hole burning" mode will oscillate.

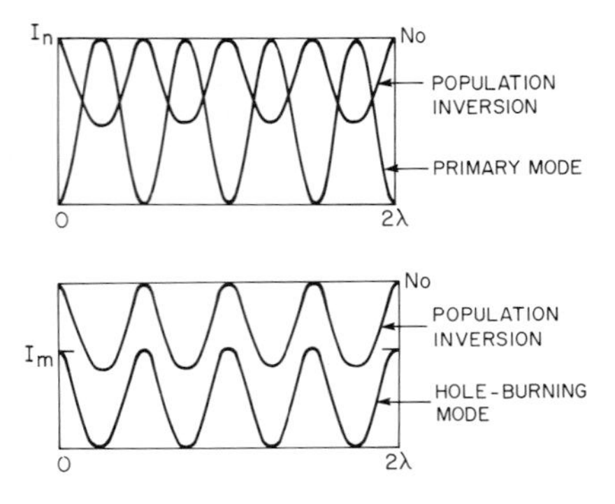

FIG. 15. Plot of the spatial gain profile in the grain medium of a standing wave laser. The antinodes of the standing wave saturate the gain, while at the nodes the gain remains high. A second mode can find gain at the nodes.

Single-mode operation in a linear laser is usually achieved by making the hole burning spacing large (small z) and using a combination grating and etalon. An etalon is a Fabry–Perot optical resonator consisting of two plane parallel mirrors separated by a distance l. The etalon has a "picket fence" of transmission peaks, equally spaced in frequency by the free spectral range (FSR)

$$\text{FSR} = c/2nl$$

where n is the refractive index between the mirrors. The etalon selects one cavity mode of the laser, and the grating selects one order of the etalon. The optimum FSR is chosen to be twice the hole burning mode spacing $\Delta\nu_{\text{hole}}$. The single-mode output power is $\approx 70\%$ that of the multimode laser. This power drop is due to the loss of the hole burning mode.

Scanning the etalon and coarse tuner in tandem, the laser will tune in steps of the cavity FSR, $c/2l_c$, where l_c is the cavity length. Typical cavity FSRs are 300 MHz. Tuning the laser frequency between adjacent cavity modes can be achieved by smoothly increasing the cavity length by $\lambda/2$. This is usually accomplished with a PZT driven cavity mirror or a tilted Brewster angle plate. To achieve smooth continuous scans, the three tuning elements must be synchronized.

2. Traveling Wave Ring Laser

The ring laser overcomes the hole burning limitations of the standing wave laser by operating in a traveling wave mode, allowing uniform saturation of the gain media. This has two advantages: first, single mode operation is more efficient, since one mode can extract all the power, and second, because there is no gain for other modes, low finesse optics can be used for tuning.

Figure 13 shows a schematic of a frequency stabilized ring laser. Instead of curved mirrors, two antireflection coated lenses are used for focusing the cavity mode into the color center crystal. In this scheme, the lenses serve both as focusing elements and as vacuum barriers for the evacuated region surrounding the crystal. Astigmatism is compensated by tilting the lenses at a small angle to the beam axis.

A unique property of ring lasers is that the direction of the traveling wave is not determined unless some form of biasing is introduced. Commonly, a combination of a "Faraday effect" rotator and an "optically active" plate is employed to force oscillation in only one direction. This combination is called an "optical diode." The Faraday effect is due to magnetically induced birefringence which nonreciprocally rotates the polarization of transmitted light through a small angle

$$\theta = VlH$$

where V is the Verdet constant of the material, l the length of the device, and H the strength of the magnetic field inside the material. At wavelengths longer than 1.5 μm, YIG (yttrium–iron–garnet) is used for the Faraday rotator because of its large Verdet constant. YIG becomes lossy below 1.5 μm, so SF-6 glass is used for the shorter wavelength applications. Following the Faraday device, an optically active plate is used as a reciprocal rotator. Depending on the direction of travel, the reciprocal device will either rotate the polarization to a further degree, or for oppositely directed waves, rotate the polarization back to its original state. Light whose polarization has been rotated will suffer loss at polarizing Brewster surfaces, while unrotated light will see no additional loss. This small direction dependent loss is enough to force the laser to oscillate in one direction only. Typical rotation values are $\sim$2%.

The linewidth of the single-mode laser is determined by fluctuations in the optical path length of the resonator. These fluctuations are caused primarily by mechanical instabilities of the optical mounts and by temperature variations of the color center crystal. The mechanical system is perturbed by floor vibrations and acoustics, both of which can be minimized with proper design to yield a laser linewidth of $\approx$10 kHz. Pump power amplitude noise is a major source of frequency instability. Changes in pump power cause temperature variations which alter the index of refraction of the crystal, effectively modulating the cavity length. The magnitude of this thermal tuning has been measured to be around 20 kHz per milliwatt change of pump power. For example, a pump power of 1 W with 1/2% amplitude noise will have a minimum linewidth of 100 kHz.

Active frequency stabilization can be applied to color center lasers to yield spectacularly narrow linewidths. The ring laser in Fig. 13 shows one scheme for active frequency stabilization. A portion of the output is directed through a thermally and acoustically isolated Fabry–Perot optical resonator. Comparing the frequency of the laser with the stable transmission passband of the Fabry–Perot yields an error signal that is proportional to the frequency excursion. This error signal is electronically amplified and used to alter the laser cavity length through a PZT driven mirror and a Brewster angle plate. The PZT mirror compensates for small high-frequency excursions, while the Brewster plate is galvonometrically tilted to control low-frequency noise and drift. The laser in Fig. 13 achieved a stabilized linewidth of less than 4 kHz.

C. Mode-Locked Operation

Color center lasers are relatively easy to mode-lock by synchronous pumping. In view of the large homogeneous luminescence bandwidth, the ultimate limits on pulse width with mode-locked color center lasers are under 1 psec. Mode-locking refers to the phase-locking of hundreds of adjacent longitudinal cavity modes. In synchronous pumping, the cavity length L of the color center laser is adjusted such that the cavity round trip time $2L/c$ corresponds exactly to the period between pulses of the mode-locked pump laser. The dramatic temporal gain modulation produced by such pumping excites many sidebands, whose frequency separation corresponds to the color center cavity mode spacing. The sidebands build up in intensity through the homogeneous gain of the medium.

In practice, synchronous mode-locking is achieved by pumping with a mode-locked laser, such as an ion laser or Nd : YAG laser, and adjusting the cavity length of the color center laser until the above length condition is satisfied. It is generally necessary to add a low-dispersion tuning element to the mode-locked cavity in order to obtain transform limited pulses; (i.e., the product $\Delta\nu\, \Delta\tau$, where $\Delta\tau$ and $\Delta\nu$ represent the FWHM of the temporal width and bandwith of the pulse, respectively, is a minimum for the pulse shape employed by the laser). With no tuning element, the pulses will have excess frequency bandwidth but will not necessarily be any shorter in duration. Such non-transform-limited pulses can lead to anomalous results and excess dispersion in applications. A prism or single plate birefringent tuner is thus often employed in mode-locked color center lasers. Typically, synchronous pumping mode-locking has generated 5–15 psec pulses from the various color center lasers.

Several new mode-locking techniques have recently been introduced which provide subpicosecond pulses from color center lasers. The first method uses passive mode-locking through the introduction of a suitable semiconductor saturable absorber in the laser cavity. To date, 200-fsec pulses have been generated at fixed wavelengths in the 1.5 and 2.7 μm region.

The second technique is called additive pulse mode-locking (APM). Figure 16 shows a schematic representation of an additively mode-locked laser. The laser consists of two coupled cavities, one with the color center gain media and the other containing a single mode optical fiber. A portion of the mode-locked output is coupled onto the optical fiber, where it experiences a nonlinear effect called self phase modulation (SPM). SPM effectively adds bandwidth to the pulse by "red-shifting" the leading edge and "blue-shifting" the trailing edge of the pulse. The two cavities are made equal length so that when the broadened pulse is coupled back into the laser it interferes with the cavity pulse. The two pulses destructively interfere in the wings, leading to a shorter pulse being reinjected back into the gain media. This shortening process repeats with every round trip until a bandwidth limitation is reached, such as due to a tuning element. To date, 75-fsec pulses have been generated that are transform-limited and tunable from 1.48 to 1.75 μm, making this the first tunable femtosecond source. APM techniques are now being extended to other sources, such as Ti : sapphire and Nd : YAG lasers.

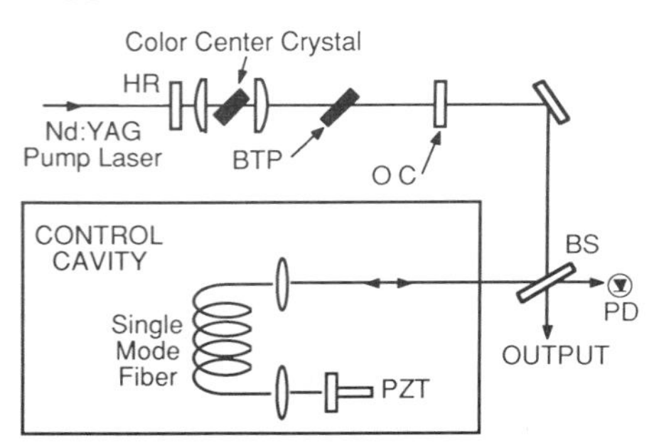

FIG. 16. Additive pulse mode-locked laser. The output of the color center laser is coupled to a nonlinear fiber. The combination produces tunable femtosecond duration pulses. (OC, output coupler; BS, beam splitter; BTP, birefringent tuner plate; HR, high reflector; PZT, Pb-doped zirconium titanate piezotransducer.)

VI. Future Developments

The color center's unique ability to generate broadly tunable light in the infrared ensures its continued usefulness, especially as a research tool for semiconductors, optical communication devices, and optical fibers. The recent development of tunable femtosecond pulses in the spectral region where optical fibers operate has made possible many new experiments that are likely to influence the communication systems of the future. It is likely that new color center sources will be developed which extend the coverage of stable lasers to new wavelength regions. Currently, there are no stable continuous wave color center lasers that operate below 1.4 μm, which is an important region for fiber optics research. Undoubtedly, new sources will be developed that provide convenient, useful tuning bands.

Bibliography

Farge, Y., and Fontana, M. P. (1979). "Electronic and Vibrational Properties of Point Defects in Ionic Solids," North-Holland Publ., Amsterdam.

Geogiou, E. T., Pinto, J. F., and Pollock, C. R. (1987). Optical properties and formation of oxygen-perturbed F_2^+ color center in NaCl, *Phys. Rev. B* **35,** 7636.

Geogiou, E. T., Carrig, T. J., and Pollock, C.R. (1988). Stable, pulsed, color center laser in pure KCl tunable from 1.23 to 1.35 μm, *Opt. Lett.* **13,** 987.

German, K. R. (1986). Optimization of F_A(II) and F_B(II) color center lasers, *J. Opt. Soc. Am. B: Opt. Phys.* **3,** 149.

Ippen, E. P., Haus, H. A., and Liu, L. Y. (1989). Additive pulse mode locking, *J. Opt. Soc. Am. B: Opt. Phys.* **6,** 1736.

Mollenauer, L. F. (1979). Color center lasers, *in* "Methods of Experimental Physics," Vol. 15B (C. L. Tang, ed.). Academic Press, New York.

Mollenauer, L. F. (1985). Color center lasers, *in* "Laser Handbook," Vol. 4 (M. Stitch, ed.). North-Holland Publ., Amsterdam.

Mollenauer, L. F., and Stolen, R. H. (1984). The soliton laser, *Opt. Lett.* **29,** 13.

Mollenauer, L. F., Vieira, N. D., and Szeto, L. (1983). Optical properties of the Tl°(1) center in KCl, *Phys. Rev. B* **27,** 5332.

Pinto, J. F., Georgiou, E. T., and Pollock, C. R. (1986). Stable color center laser in OH-doped NaCl tunable from 1.41 to 1.81 μm, *Opt. Lett.* **11,** 519.

COLOR, SCIENCE OF

Robert M. Boynton *University of California at San Diego*

GLOSSARY

Chromaticity: Ratios x, y, z of each of the tristimulus values of a light to the sum of the three tristimulus values X, Y, Z, these being the amounts of three primaries required to match the color of the light.

Chromaticity diagram: Plane diagram formed by plotting one of the three chromaticity coordinates against another (usually y versus x).

Color: Characteristics of sensations elicited by light by which a human observer can distinguish between two structure-free patches of light of the same size and shape.

Colorant: A substance, such as a dye or pigment, that modifies the color of objects or imparts color to otherwise achromatic objects.

Colorimetry: Measurement and specification of color.

Color matching: Action of making a test color appear the same as a reference color.

Color order: System of reference whereby the relation of one color to another can be perceived and the position of that color can be established with respect to the universe of all colors.

Color rendering: General expression for the effect of a light source on the color appearance of objects in comparison with their color appearance under a reference light source.

Color temperature: Absolute temperature of a blackbody radiator having a chromaticity closest to that of a light source being specified.

Metamerism: 1. Phenomenon whereby lights of different spectral power distributions appear to have the same color. 2. Degree to which a material appears to change color when viewed under different illuminants.

Optimal colors: Stimuli that for a given chromaticity have the greatest luminous reflectance.

Primaries: 1. Additive: Any one of three lights in terms of which a color is specified by giving the amount of each required to match it by combining the lights. 2. Subtractive: Set of dyes or pigments that, when mixed in various proportions, provides a gamut of colors.

Radiance: Radiant flux per unit solid angle (intensity) per unit area of an element of an extended source or reflecting surface in a specified direction.

Reflectance: Ratio of reflected to incident light.

Reflection: Process by which incident flux leaves a surface or medium from the incident side, without change in wavelength.

Color science examines a fundamental aspect of human perception. It is based on experimental study under controlled conditions susceptible to physical measurement. For a difference in color to be perceived between two surfaces, three conditions must be satisfied: (1) There must be an appropriate source of illumination, (2) the two surfaces must not have identical spectral reflectances, and (3) an observer must be present to view them. This article is concerned with the relevant characteristics of lights, surfaces, and human vision that conjoin to allow the perception of object color.

I. Physical Basis of Perceived Color

The physical basis of color exists in the interaction of light with matter, both outside and inside the eye. The sensation of color depends on physiological activity in the visual system that begins with the absorption of light in photoreceptors located in the retina of the eye and ends with patterns of biochemical activity in the brain. Perceived color can be described by the color names white, gray, black, yellow, orange, brown, red, green, blue, purple, and pink; these 11 basic color terms have unambiguous referents in all fully developed languages. All of these names (as well as combinations of these and many other less precisely used nonbasic color terms) describe colors, but white, gray, and black are excluded from the list of those called hues. Colors with hue are called chromatic colors; those without are called achromatic colors.

Although color terms are frequently used in reference to all three aspects of color (e.g., one may speak of a sensation of red, a red surface, or a red light), such usage is scientifically appropriate only when applied to the sensation; descriptions of lights and surfaces should be provided in physical and geometrical language.

II. CIE System of Color Specification

A. Basic Color-Matching Experiment

The most fundamental experiment in color science entails the determination of whether two fields of light, such as those that might be produced on a screen with two slide projectors, appear the same or different. If such fields are abutted and the division between them disappears to form a single, homogeneous field, the fields are said to match. A match will, of course, occur if there is no physical difference between the fields, and in special cases color matches are also possible when substantial physical differences exist between the fields. An understanding of how this can happen provides an opening to a scientific understanding of this subject.

Given an initial physical match, a difference in color can be introduced by either of two procedures, which are often carried out in combination. In the first instance, the radiance of one part of a homogeneous field is altered without any change in its relative spectral distribution. This produces an achromatic color difference. In the second case, the relative spectral distribution of one field is changed such that, for all possible relative radiances of the two fields, no match is possible. This is called a chromatic color difference.

When fields of different spectral distributions can be adjusted in relative radiance to eliminate all color difference, the result is termed a metameric color match. In a color-matching experiment, a test field is presented next to a comparison field and the observer causes the two fields to match exactly by manipulating the radiances of so-called primaries provided to the comparison field. Such primaries are said to be added; this can be accomplished by superposition with a half-silvered mirror, by superimposed images projected onto a screen, by very rapid temporal alternation of fields at a rate above the fusion frequency for vision, or by the use of pixels too small and closely packed to be discriminated (as in color television). If the primaries are suitably chosen (no one of them should be matched by any possible mixture of the other two), a human observer with normal color vision can uniquely match any test color by adjusting the radiances of three monochromatic primaries. To accomplish this, it sometimes proves necessary to shift one of the primaries so that it is added to the color being matched; it is useful to treat this as a negative radiance of that primary in the test field. The choice of exactly three primaries is by no means arbitrary: If only one or two primaries are used, matches are generally impossible, whereas if four or more primaries are allowed, matches are not uniquely determined.

The result of the color-matching experiment can be represented mathematically as $t(T) = r(R) + g(G) + b(B)$, meaning that t units of test field T produce a color that is matched by an additive combination of r units of primary R, g units of primary G, and b units of primary B, where one or two of the quantities r, g, or b may be negative. Thus any color can be represented as a vector in R, G, B space. For small, centrally fixated fields, experiment shows that the transitive, reflexive, linear, and associative properties of algebra apply also to their empirical counterparts, so that color-matching equations can be manipulated to predict matches that would be made with a change in the choice of primaries. These simple relations break down for very low levels of illumination and also with higher levels if the fields are large enough to permit significant contributions by rod photoreceptors or if the fields are so bright as to bleach a significant fraction of cone photopigments, thus altering their action spectra.

Matches are usually made by a method of adjustment, an iterative, trial-and-error procedure whereby the observer manipulates three controls, each of which monotonically varies the radiance of one primary. Although such settings at the match point may be somewhat more variable than most purely physical measurements, reliable data result from the means of several settings for each condition tested. A more serious problem, which will not be treated in this article, results from differences among observers. Although not great among those with normal color vision, such differences are by no means negligible. (For those with abnormal color vision, they can be very large.) To achieve a useful standardization—one that is unlikely to apply exactly to any particular individual—averages of normal observers are used, leading to the concept of a standard observer.

In the color-matching experiment, an observer is in effect acting as an analog computer, solving three simultaneous equations by iteration, using his or her sensations as a guide. Although activity in the brain underlies the experience of color, the initial encoding of information related to wavelength is in terms of the ratios of excitations of three different classes of cone photoreceptors in the retina of the eye, whose spectral sensitivities overlap. Any two physical fields, whether of the same or different spectral composition, whose images on the retina excite each of the three classes of cones in the same way will be indiscriminable. The action spectra of the three classes of cones in the normal eye are such that no two wavelengths in the spectrum produce exactly the same ratios of excitations among them.

B. Imaginary Primaries

Depending on the choice of primaries, many different sets of color-matching functions are possible, all of which describe the same color-matching behavior. Figure 1 shows experimen-

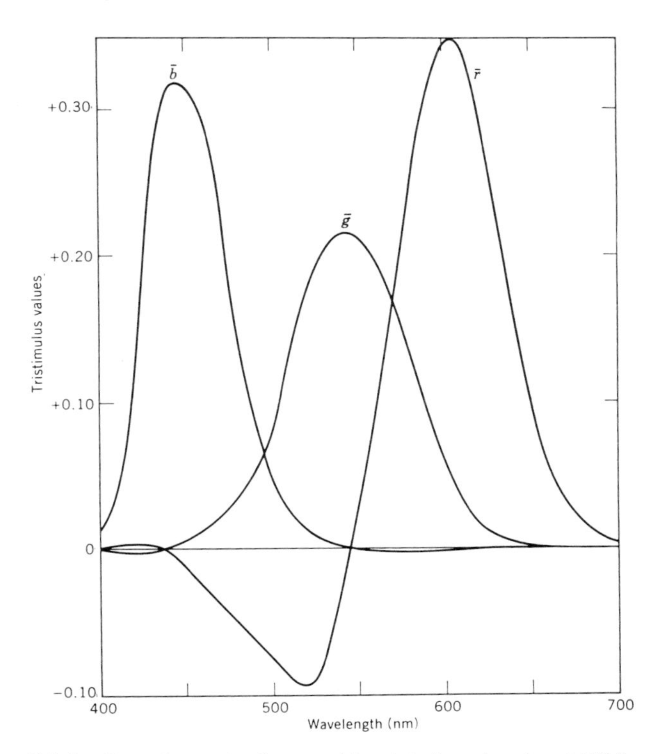

FIG. 1. Experimental color-matching data for primaries at 435.8, 546.1, and 700.0 nm. [From Billmeyer, F. W., Jr., and Saltzmann, M. (1981). "Principles of Color Technology," 2nd Ed. Copyright © 1981 John Wiley & Sons, Inc. Reprinted by permission of John Wiley & Sons, Inc.]

tal data for the primaries 435.8, 546.1, and 700.0 nm. Depicted in Fig. 2 are current estimates of the spectral sensitivities of the three types of cone photoreceptors. These functions, which have been inferred from the data of psychophysical experiments of various kinds, agree reasonably well with direct microspectrophotometric measurements of the absorption spectra of outer segments of human cone photoreceptors containing the photopigments that are the principal determinants of the spectral sensitivity of the cones.

The cone spectral sensitivities may be regarded as color-matching functions based on primaries that are said to be imaginary in the sense that, although calculations of color matches based on them are possible, they are not physically realizable. To exist physically, each such primary would uniquely excite only one type of cone, whereas real primaries always excite at least two types.

Another set of all-positive color-matching functions, based on a different set of imaginary primaries, is given in Fig. 3. This set, which makes very similar predictions about color

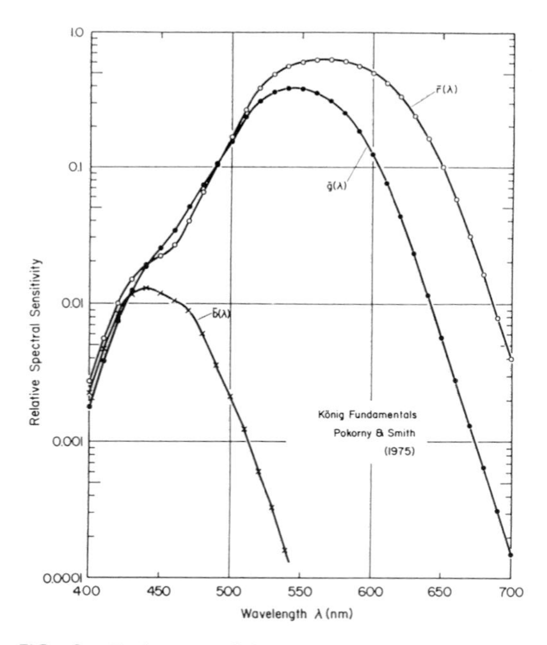

FIG. 2. Estimates of human cone action spectra (König fundamentals) derived by V. Smith and J. Pokorny. [From Wyszecki, G., and Stiles, W. S. (1982). "Color Science: Concepts and Methods, Quantitative Data and Formulae," 2nd Ed. Copyright © 1982 John Wiley & Sons, Inc. Reprinted by permission of John Wiley & Sons, Inc.]

matches as the cone sensitivity curves, was adopted as a standard by the International Commission on Illumination (CIE) in 1931.

By simulating any of these sets of sensitivity functions in three optically filtered photocells, it is possible to remove the human observer from the system of color measurement (colorimetry) and develop a purely physical (though necessarily very limited) description of color, one that can be implemented in automated colorimeters.

C. Chromaticity Diagram

A useful separation between the achromatic and chromatic aspects of color was achieved in a system of colorimetry adopted by the CIE in 1931. This was the first specification of color to achieve international agreement; it remains today the principal system used internationally for specifying colors quantitatively, without reference to a set of actual samples.

The color-matching functions $\bar{x}(\lambda)$, $\bar{y}(\lambda)$, and $\bar{z}(\lambda)$ are based on primaries selected and smoothed to force the $\bar{y}(\lambda)$ function to be proportional to the spectral luminous efficiency function $V(\lambda)$, which had been standardized a decade earlier to define the quantity of "luminous flux" in lumens per watt of radiant power. The $\bar{x}(\lambda)$, $\bar{y}(\lambda)$, and $\bar{z}(\lambda)$ functions were then scaled to equate the areas under the curves, an operation that does not alter the predictions they make about color matches.

To specify the color of a patch of light, one begins by integrating its spectral radiance distribution $S(\lambda)$ in turn with the three color-matching functions:

$$X = k \int S(\lambda)\bar{x}(\lambda)\, d\lambda$$

$$Y = k \int S(\lambda)\bar{y}(\lambda)\, d\lambda$$

$$Z = k \int S(\lambda)\bar{z}(\lambda)\, d\lambda$$

The values X, Y, and Z are called relative tristimulus values; these are equal for any light having an equal-radiance spectrum. Tristimulus values permit the specification of color in terms of three variables that are related to cone sensitivities rather than by continuous spectral radiance distributions, which do not. Like R, G, and B, the tristimulus values represent the coordinates of a three-dimensional vector whose angle specifies chromatic color and whose length characterizes the amount of that color.

Chromaticity coordinates, which do not depend on the amount of a color, specify each of

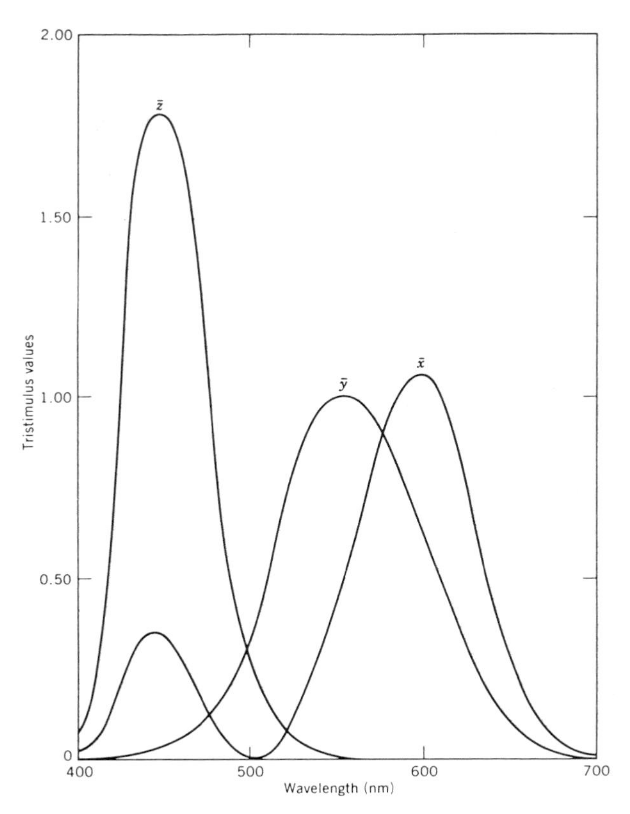

FIG. 3. Tristimulus values of the equal-energy spectrum of the 1931 CIE system of colorimetry. [From Billmeyer, F. W., Jr., and Saltzmann, M. (1981). "Principles of Color Technology," 2nd Ed. Copyright © 1981 John Wiley & Sons, Inc. Reprinted by permission of John Wiley & Sons, Inc.]

the tristimulus values relative to their sum:

$$x = X/(X + Y + Z); \qquad y = Y/(X + Y + Z);$$
$$z = Z/(X + Y + Z)$$

Given any two of these, the third is determined (e.g., $z = 1 - x - y$). Therefore, full information about chromaticity can be conveniently represented in a two-dimensional diagram, with y versus x having been chosen by the CIE for this purpose. The resulting chromaticity diagram is shown in Fig. 4. If one wishes to specify the quantity of light as well, the Y tristimulus value can be given, allowing a color to be fully specified as x, y, and Y, instead of X, Y, and Z. The manner in which the quantity of light Y is specified is determined by the normalization constant k.

Depending on the choice of primaries for determining color-matching functions, many other chromaticity diagrams are possible. For example, the set of color-matching functions of Fig. 1 leads to the chromaticity diagram of Fig. 5. This so-called *RGB* system is seldom used.

The affine geometry of chromaticity diagrams endows all of them with a number of useful properties. Most fundamental is that an additive mixture of any two lights will fall along a straight line connecting the chromaticities of the mixture components. Another is that straight lines on one such diagram translate into straight lines on any other related to it by a change of assumed primaries. The locations of the imaginary primaries X, Y, and Z are shown in Fig. 5, where one sees that the triangle formed by them completely encloses the domain of realizable colors. The lines $X–Y$ and $X–Z$ of Fig. 5 form the coordinate axes of the CIE chromaticity diagram of Fig. 4. Conversely, the lines $B–G$ and $B–R$ in Fig. 4 form the coordinate axes of the chromaticity diagram of Fig. 5. The uneven grid of nonorthogonal lines in Fig. 4, forming various an-

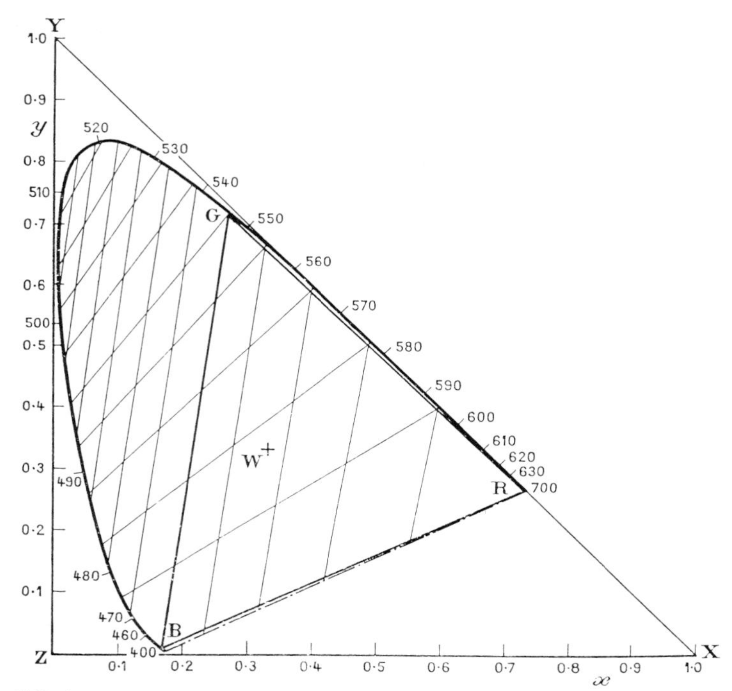

FIG. 4. The *XYZ* chromaticity diagram showing the locations of the *RGB* primaries of Fig. 5 and the projection of the rectilinear grid of that figure onto this one. [From LeGrand, Y. (1957). "Light, Colour, and Vision," 2nd Ed. Wiley (Interscience), New York.]

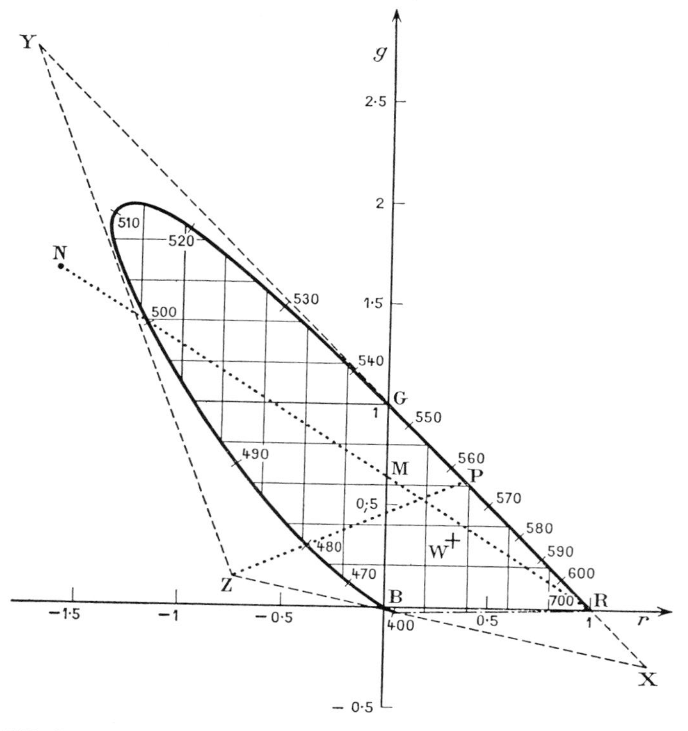

FIG. 5. The *RGB* chromaticity diagram showing the locations of the *XYZ* primaries of Fig. 4. [From LeGrand, Y. (1957). "Light, Colour, and Vision," 2nd Ed. Wiley (Interscience), New York.]

gles at their intersections, translates into the regular grid of evenly spaced, orthogonal lines in Fig. 5. This illustrates that angles and areas have no instrinsic meaning in chromaticity diagrams.

The CIE in 1964 adopted an alternative set of color-matching functions based on experiments with large (10°) fields. Their use is recommended for making predictions about color matches for fields subtending more than 4° at the eye.

Table I lists values of the CIE color-matching functions for 2° and 10° fields at 10-nm wavelength values. Tables for 1-nm wavelength values for 2° and 10° fields are available in "Color Measurement," the second volume in the series Optical Radiation Measurements, edited by F. Grum and C. J. Bartleson.

III. Color Rendering

From an evolutionary viewpoint, it is not surprising that sunlight is an excellent source for color rendering. Its strong, gap-free spectral irradiance distribution (Fig. 6) allows the discrimination of a very large number of surface-color differences. Color appearance in sunlight provides the standard against which the adequacy of other sources for color rendering is often judged.

A. Best and Worst Artificial Sources for Color Rendering

Of the light sources in common use today, low-pressure sodium is one of the poorest for

TABLE I. Spectral Tristimulus Values for Equal Spectral Power Source

a. CIE 1931 Standard Observer

Wavelength (nanometer)	$\bar{x}(\lambda)$	$\bar{y}(\lambda)$	$\bar{z}(\lambda)$	Wavelength (nanometer)	$\bar{x}(\lambda)$	$\bar{y}(\lambda)$	$\bar{z}(\lambda)$
380	0.0014	0.0000	0.0065	580	0.9163	0.8700	0.0017
385	0.0022	0.0001	0.0105	585	0.9786	0.8163	0.0014
390	0.0042	0.0001	0.0201	590	1.0263	0.7570	0.0011
395	0.0076	0.0002	0.0362	595	1.0567	0.6949	0.0010
400	0.0143	0.0004	0.0679	600	1.0622	0.6310	0.0008
405	0.0232	0.0006	0.1102	605	1.0456	0.5668	0.0006
410	0.0435	0.0012	0.2074	610	1.0026	0.5030	0.0003
415	0.0776	0.0022	0.3713	615	0.9384	0.4412	0.0002
420	0.1344	0.0040	0.6456	620	0.8544	0.3810	0.0002
425	0.2148	0.0073	1.0391	625	0.7514	0.3210	0.0001
430	0.2839	0.0116	1.3856	630	0.6424	0.2650	0.0000
435	0.3285	0.0168	1.6230	635	0.5419	0.2170	0.0000
440	0.3483	0.0230	1.7471	640	0.4479	0.1750	0.0000
445	0.3481	0.0298	1.7826	645	0.3608	0.1382	0.0000
450	0.3362	0.0380	1.7721	650	0.2835	0.1070	0.0000
455	0.3187	0.0480	1.7441	655	0.2187	0.0816	0.0000
460	0.2908	0.0600	1.6692	660	0.1649	0.0610	0.0000
465	0.2511	0.0739	1.5281	665	0.1212	0.0446	0.0000
470	0.1954	0.0910	1.2876	670	0.0874	0.0320	0.0000
475	0.1421	0.1126	1.0419	675	0.0636	0.0232	0.0000
480	0.0956	0.1390	0.8130	680	0.0468	0.0170	0.0000
485	0.0580	0.1693	0.6162	685	0.0329	0.0119	0.0000
490	0.0320	0.2080	0.4652	690	0.0227	0.0082	0.0000
495	0.0147	0.2586	0.3533	695	0.0158	0.0057	0.0000
500	0.0049	0.3230	0.2720	700	0.0114	0.0041	0.0000
505	0.0024	0.4073	0.2123	705	0.0081	0.0029	0.0000
510	0.0093	0.5030	0.1582	710	0.0058	0.0021	0.0000
515	0.0291	0.6082	0.1117	715	0.0041	0.0015	0.0000
520	0.0633	0.7100	0.0782	720	0.0029	0.0010	0.0000
525	0.1096	0.7932	0.0573	725	0.0020	0.0007	0.0000
530	0.1655	0.8620	0.0422	730	0.0014	0.0005	0.0000
535	0.2257	0.9149	0.0298	735	0.0010	0.0004	0.0000
540	0.2904	0.9540	0.0203	740	0.0007	0.0002	0.0000
545	0.3597	0.9803	0.0134	745	0.0005	0.0002	0.0000
550	0.4334	0.9950	0.0087	750	0.0003	0.0001	0.0000
555	0.5121	1.0000	0.0057	755	0.0002	0.0001	0.0000
560	0.5945	0.9950	0.0039	760	0.0002	0.0001	0.0000
565	0.6784	0.9786	0.0027	765	0.0001	0.0000	0.0000
570	0.7621	0.9520	0.0021	770	0.0001	0.0000	0.0000
575	0.8425	0.9154	0.0018	775	0.0001	0.0000	0.0000
580	0.9163	0.8700	0.0017	780	0.0000	0.0000	0.0000
Totals					21.3714	21.3711	21.3715

b. CIE 1964 Supplementary Observer

Wavelength (nanometer)	$\bar{x}_{10}(\lambda)$	$\bar{y}_{10}(\lambda)$	$\bar{z}_{10}(\lambda)$	Wavelength (nanometer)	$\bar{x}_{10}(\lambda)$	$\bar{y}_{10}(\lambda)$	$\bar{z}_{10}(\lambda)$
380	0.0002	0.0000	0.0007	580	1.0142	0.8689	0.0000
385	0.0007	0.0001	0.0029	585	1.0743	0.8256	0.0000
390	0.0024	0.0003	0.0105	590	1.1185	0.7774	0.0000
395	0.0072	0.0008	0.0323	595	1.1343	0.7204	0.0000
400	0.0191	0.0020	0.0860	600	1.1240	0.6583	0.0000
405	0.0434	0.0045	0.1971	605	1.0891	0.5939	0.0000
410	0.0847	0.0088	0.3894	610	1.0305	0.5280	0.0000
415	0.1406	0.0145	0.6568	615	0.9507	0.4618	0.0000
420	0.2045	0.0214	0.9725	620	0.8563	0.3981	0.0000
425	0.2647	0.0295	1.2825	625	0.7549	0.3396	0.0000
430	0.3147	0.0387	1.5535	630	0.6475	0.2835	0.0000
435	0.3577	0.0496	1.7985	635	0.5351	0.2283	0.0000
440	0.3837	0.0621	1.9673	640	0.4316	0.1798	0.0000
445	0.3867	0.0747	2.0273	645	0.3437	0.1402	0.0000
450	0.3707	0.0895	1.9948	650	0.2683	0.1076	0.0000
455	0.3430	0.1063	1.9007	655	0.2043	0.0812	0.0000
460	0.3023	0.1282	1.7454	660	0.1526	0.0603	0.0000
465	0.2541	0.1528	1.5549	665	0.1122	0.0441	0.0000
470	0.1956	0.1852	1.3176	670	0.0813	0.0318	0.0000
475	0.1323	0.2199	1.0302	675	0.0579	0.0226	0.0000
480	0.0805	0.2536	0.7721	680	0.0409	0.0159	0.0000
485	0.0411	0.2977	0.5701	685	0.0286	0.0111	0.0000
490	0.0162	0.3391	0.4153	690	0.0199	0.0077	0.0000
495	0.0051	0.3954	0.3024	695	0.0138	0.0054	0.0000
500	0.0038	0.4608	0.2185	700	0.0096	0.0037	0.0000
505	0.0154	0.5314	0.1592	705	0.0066	0.0026	0.0000
510	0.0375	0.6067	0.1120	710	0.0046	0.0018	0.0000
515	0.0714	0.6857	0.0822	715	0.0031	0.0012	0.0000
520	0.1177	0.7618	0.0607	720	0.0022	0.0008	0.0000
525	0.1730	0.8233	0.0431	725	0.0015	0.0006	0.0000
530	0.2365	0.8752	0.0305	730	0.0010	0.0004	0.0000
535	0.3042	0.9238	0.0206	735	0.0007	0.0003	0.0000
540	0.3768	0.9620	0.0137	740	0.0005	0.0002	0.0000
545	0.4516	0.9822	0.0079	745	0.0004	0.0001	0.0000
550	0.5298	0.9918	0.0040	750	0.0003	0.0001	0.0000
555	0.6161	0.9991	0.0011	755	0.0002	0.0001	0.0000
560	0.7052	0.9973	0.0000	760	0.0001	0.0000	0.0000
565	0.7938	0.9824	0.0000	765	0.0001	0.0000	0.0000
570	0.8787	0.9556	0.0000	770	0.0001	0.0000	0.0000
575	0.9512	0.9152	0.0000	775	0.0000	0.0000	0.0000
580	1.0142	0.8689	0.0000	780	0.0000	0.0000	0.0000
Totals					23.3294	23.3324	23.3343

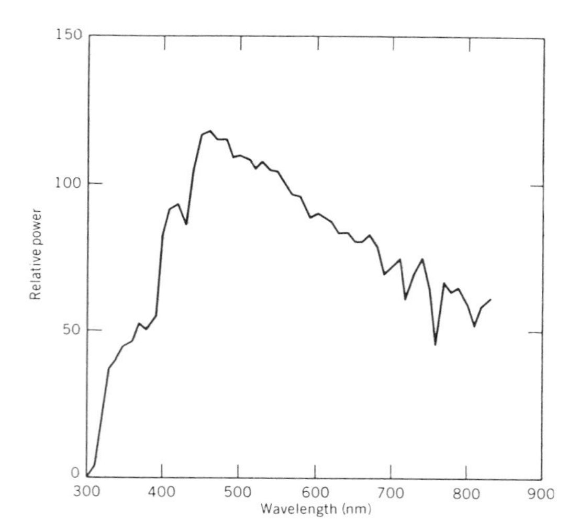

FIG. 6. Spectral power distribution of typical daylight. [From Billmeyer, F. W., Jr., and Saltzmann, M. (1981). "Principles of Color Technology," 2nd Ed. Copyright © 1981 John Wiley & Sons, Inc. Reprinted by permission of John Wiley & Sons, Inc.]

color rendering, coming very close to being one of the worst possible. This illuminant consists mainly of the paired sodium lines that lie very close together (at 589.0 and 589.6 nm) in the "yellow" region of the spectrum; although some other spectral lines are also represented, these are present at such low relative radiances that low-pressure sodium lighting is for practical purposes monochromatic. [*See* LIGHT SOURCES.]

For a surface that does not fluoresce, its spectral reflectance characteristics can modify the quantity and geometry of incident monochromatic light, but not its wavelength. Viewed under separate monochromatic light sources of the same wavelength, any two surfaces with arbitrarily chosen spectral distributions can be made to match, both physically and visually, by adjusting the relative radiances of incident lights. Therefore, no chromatic color differences can exist under monochromatic illumination.

The best sources for color rendering emit continuous spectra throughout the visible region. Blackbody radiation, which meets this criterion, is shown for three temperatures in Fig. 7. These curves approximate those for tungsten sources at these temperatures.

B. Intermediate Quality of Fluorescent Lighting

Much of the radiant flux produced by incandescence emerges as infrared radiation at wavelengths longer than those visible; this is not true of fluorescent light, which is more efficiently produced, accounting for its widespread use. This light results from the electrical energizing of mercury vapor, which emits ultraviolet radiation. Although itself invisible, this radiation elicits visible light by causing the fluorescence of phosphors suspended in a layer coating the inside of a transparent tube.

Fluorescent lamps emit energy at all visible wavelengths, which is a good feature for color rendering, but their spectra are punctuated by regions of much higher radiance whose spectral locations depend on the phosphors chosen and the visible radiations of mercury vapor. Radiant power distributions of six types of fluorescent lamps are shown in Fig. 8.

C. Efficacy

The amount of visible light emitted by a source is measured in lumens, determined by integrating its radiant power output $S(\lambda)$ with the spectral luminous efficiency function $V(\lambda)$. The latter, which is proportional to $\bar{y}(\lambda)$, peaks at ~555 nm. Therefore, the theoretically most efficient light source would be monochromatic at this wavelength, with the associated inability to render chromatic color differences. Efficacy does not include power lost in the conversion from electrical input to radiant output, which may vary independently of the efficacy of the light finally produced.

D. Correlated Color Temperature

A blackbody, or Planckian radiator, is a cavity within a heated material from which heat cannot escape. No matter what the material, the walls of the cavity exhibit a characteristic spectral emission, which is a function of its temperature. The locus of the chromaticity coordinates corresponding to blackbody radiation, as a function of temperature, plots in the chromaticity diagram as a curved line known as the Planckian locus (see Fig. 4). The spectral distribution of light from sources with complex spectra does not approximate that of a Planckian radiator. Nevertheless, it is convenient to have a single index by which to characterize these other sources of artificial light. For this purpose the CIE has defined a correlated color temperature, determined by calculating the chromaticity coordinates of the source and then locating the point on the blackbody locus perceptually closest to these coordinates.

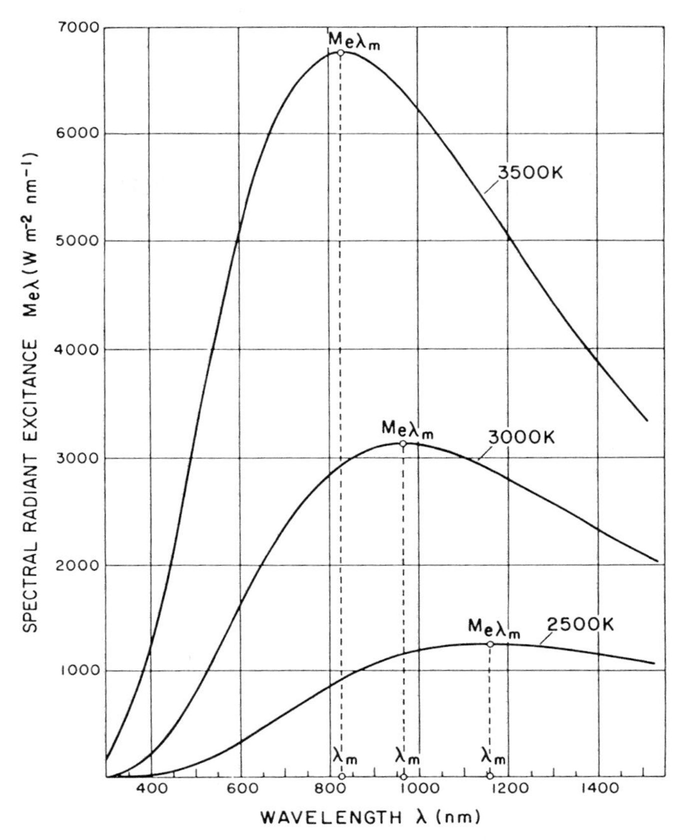

FIG. 7. Spectral radiance distributions of a blackbody radiator at three temperatures. [From Wyszecki, G., and Stiles, W. S. (1982). "Color Science: Concepts and Methods, Quantitative Data and Formulae," 2nd Ed. Copyright © 1982 John Wiley & Sons, Inc. Reprinted by permission of John Wiley & Sons, Inc.]

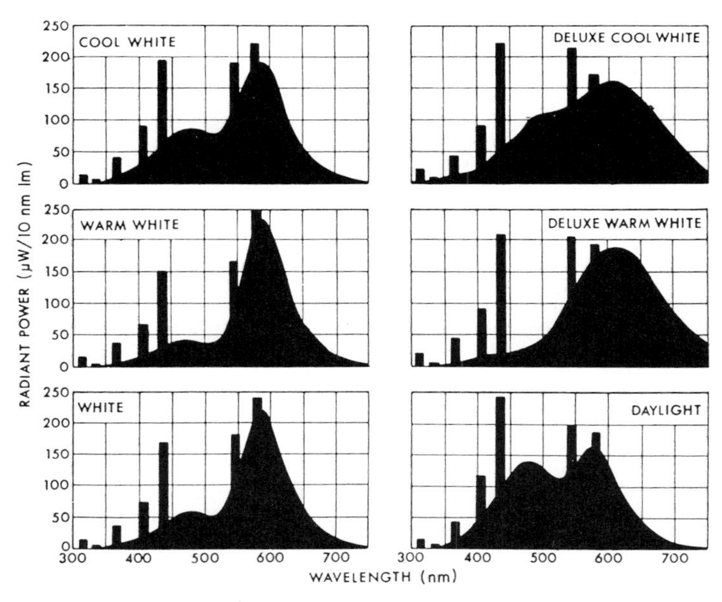

FIG. 8. Spectral radiance distributions of six typical fluorescent lamps. [From Kaufman, J. E., ed. (1981). "IES Lighting Handbook; Reference Volume." © 1981 Illuminating Engineering Society of North America.]

E. Color-Rendering Index

The CIE has developed a system for attempting to specify the quality of color rendering supplied by any light source. The calculations are based on a set of reflecting samples specified in terms of their reflectance functions. The calculations begin with the choice of a reference illuminant specified as a blackbody (or daylight) radiator having a color temperature (or correlated color temperature) as close as possible to the correlated color temperature of the test illuminant; the choice of reference illuminant depends on the correlated color temperature of the test illuminant (daylight is used as a reference above 5000 K). For each of the samples, defined by their spectral reflectance functions, the amount of color shift ΔE introduced in going from reference to test illuminant is determined using the CIELUV formula described in Section VI,B. There are 14 reference samples in all. A special color-rendering index R_i, peculiar to each sample, is calculated as $100 - 4.6\ \Delta E$. Most commonly a single-number index is calculated from the mean of a subset of eight special color-rendering indices to provide a final value known as the general color-rendering index R_a. The factor 4.6 was chosen so that a standard warm white fluorescent lamp would have an R_a of ~50; tungsten-incandescent sources score very close to 100. Table II gives R_a values for several commonly used artificial sources. Despite its official status, R_a is of limited value because of its many arbitrary features, especially its dependence on so limited a set of color samples. It is most useful for distinguishing large differences in color rendering, but not so useful for discriminating among sources of very high color-rendering properties. Individual values of R_i can be useful for determining the manner in which light sources differ in their color-rendering properties.

The intermediate color-rendering properties of most fluorescent light sources are closer to the best than to the worst. Mercury vapor and high-pressure sodium sources, widely used for street lighting, have poor color-rendering properties that fall between those of fluorescent and low-pressure sodium illumination.

IV. Global Surface Properties

The term *reflection* characterizes any of a variety of physical processes by which less than 100% of the radiant energy incident on a body at

TABLE II. Color and Color-Rendering Characteristics of Common Light Sources[a]

Test Lamp Designation	CIE Chromaticity Coordinates		Correlated Color Temperature (Kelvins)	CIE General Color Rendering Index	CIE Special Color Rendering Indices, R_i													
	x	y		R_a	R_1	R_2	R_3	R_4	R_5	R_6	R_7	R_8	R_9	R_{10}	R_{11}	R_{12}	R_{13}	R_{14}
Fluorescent Lamps																		
Warm White	.436	.406	3020	52	43	70	90	40	42	55	66	13	−111	31	21	27	48	94
Warm White Deluxe	.440	.403	2940	73	72	80	81	71	69	67	83	64	14	49	60	43	73	88
White	.410	.398	3450	57	48	72	90	47	49	61	68	20	−104	36	32	38	52	94
Cool White	.373	.385	4250	62	52	74	90	54	56	64	74	31	−94	39	42	48	57	93
Cool White Deluxe	.376	.368	4050	89	91	91	85	89	90	86	90	88	70	74	88	78	91	90
Daylight	.316	.345	6250	74	67	82	92	70	72	78	82	51	−56	59	64	72	71	95
Three-Component A	.376	.374	4100	83	98	94	48	89	89	78	88	82	32	46	73	53	95	65
Three-Component B	.370	.381	4310	82	84	93	66	65	28	94	83	85	44	69	62	68	90	76
Simulated D_{50}	.342	.359	5150	95	93	96	98	95	94	95	98	92	76	91	94	93	94	99
Simulated D_{55}	.333	.352	5480	98	99	98	96	99	99	98	98	96	91	95	98	97	98	98
Simulated D_{65}	.313	.325	6520	91	93	91	85	91	93	88	90	92	89	76	91	86	92	91
Simulated D_{70}	.307	.314	6980	93	97	93	87	92	97	91	91	94	95	82	95	93	94	93
Simulated D_{75}	.299	.315	7500	93	93	94	91	93	93	91	94	91	73	83	92	90	93	95
Mercury, clear	.326	.390	5710	15	−15	32	59	2	3	7	45	−15	−327	−55	−22	−25	−3	75
Mercury, improved color	.373	.415	4430	32	10	43	60	20	18	14	60	31	−108	−32	−7	−23	17	77
Metal Halide, clear	.396	.390	3720	60	52	84	81	54	60	83	59	5	−142	68	55	78	62	88
Xenon, high pressure arc	.324	.324	5920	94	94	91	90	96	95	92	95	96	81	81	97	93	92	95
High pressure sodium	.519	.418	2100	21	11	65	52	−9	10	55	32	−52	−212	45	−34	32	18	69
Low pressure sodium	.569	.421	1740	−44	−68	44	−2	−101	−67	29	−23	−165	−492	20	−128	−21	−39	31
DXW Tungsten Halogen	.424	.399	3190	100					All 100 except for $R_6 = R_{10} = 99$									

[a] Lamps representative of the industry are listed. Variations from manufacturer to manufacturer are likely, especially for the D series of fluorescent lamps and the high-intensity discharge lamps. A high positive value of R_i indicates a small color difference for sample i. A low value of R_i indicates a large color difference.

each wavelength is returned without change of wavelength. Reflection is too complicated for detailed specification at a molecular level for most surfaces and wavelengths of light. For this reason and because the molecular details are unimportant for many practical purposes, methods have been devised for measuring the spectral reflectance of a surface—the spectral distribution of returned light relative to that which is incident. Reflectance depends on the wavelength and angle of incidence of the light, as well as the angle(s) at which reflected light is measured.

A. Specular and Diffuse Reflectance

A familiar example of specular reflectance is provided by a plane mirror, in which the angles of light incidence and reflectance are equal. An ideal mirror reflects all incident light nonselectively with wavelength. If free of dust and suitably framed, the surface of an even less than ideal real mirror is not perceived at all; instead, the virtual image of an object located physically in front of the mirror is seen as if positioned behind.

Although specular reflectance seldom provides information about the color of a surface, there are exceptions. In particular, highly polished surfaces of metals such as gold, steel, silver, and copper reflect specularly. They also reflect diffusely from within but do so selectively with wavelength so that the specular reflection is seen to be tinged with the color of the diffuse component. More often, because highlights from most surfaces do not alter the spectral distribution of incident light, specular reflection provides information about the color of the source of light rather than that of the surface.

Diffuse reflectance, on the other hand, is typically selective with wavelength, and for the normal observer under typical conditions of illumination it is the principal determinant of the perceived color of a surface. A surface exhibiting perfectly diffuse reflectance returns all of the incident light with the distribution shown in Fig. 9, where the luminance (intensity per unit area) of the reflected light decreases a cosine function of the angle of reflection relative to normal. As such a surface is viewed more and more obliquely through an aperture, a progressively larger area of the surface fills the aperture—also a cosine function. The two effects cancel, causing the luminance of the surface and its subjective counterpart, lightness, to be independent of the angle of view.

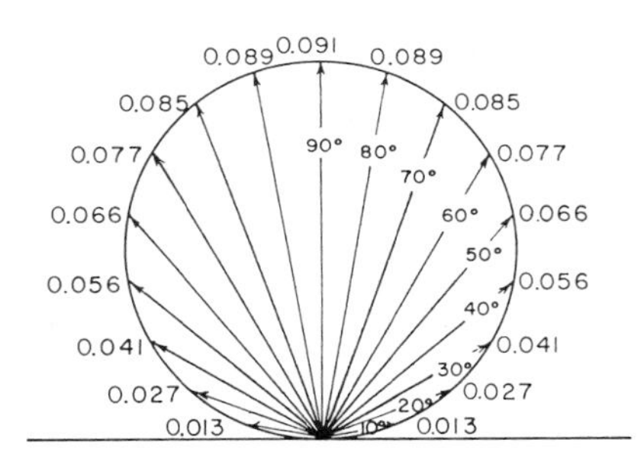

FIG. 9. Intensity distribution of light reflected from a perfectly diffuse (Lambertian) surface, showing proportion of reflected light within 5° of each indicated direction. [From Boynton, R. M. (1974). *In* "Handbook of Perception" (E. C. Carterette and M. P. Friedman, eds.), Vol. 1. Copyright 1974 Academic Press.]

No real surface behaves in exactly this way, although some surfaces approach it. Some simultaneously exhibit specular and diffuse reflectance; that of a new automobile provides a familiar example. The hard, highly polished outer surface exhibits specular reflectance of some of the incident light. The remainder is refracted into the layers below, which contain diffusely reflecting, spectrally selective absorptive pigments suspended in a binding matrix. Light not absorbed is scattered within this layer with an intensity pattern that may approximate that of a perfectly diffuse reflector. Because of the absorptive properties of the pigments, some wavelengths reflect more copiously than others, providing the physical basis for the perceived color of the object.

Many intermediate geometries are possible, which give rise to sensations of sheen and gloss; these usually enable one to predict the felt hardness or smoothness of surfaces without actually touching them.

B. Measuring Diffuse Surface Reflectance

The diffuse spectral reflectance of a surface depends on the exact conditions of measurement. To some extent these are arbitrary, so that in order for valid comparisons of measurements to be made among different laboratories and manufacturers, standard procedures are necessary. To agree on and specify such procedures has been one of the functions of the CIE, which has recommended four procedures for measuring diffuse spectral reflectance, the most sophisticated of which is illustrated at the bottom left in Fig. 10. It makes use of an integrating

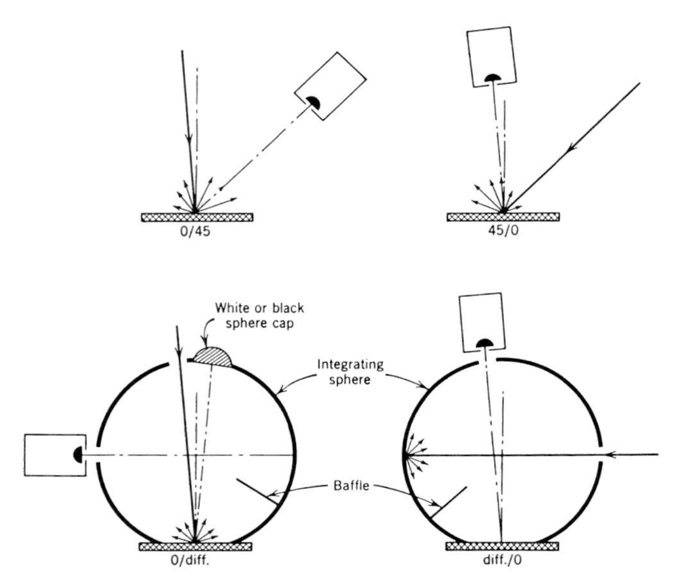

FIG. 10. Schematic diagram showing the four CIE standard illuminating and viewing geometries for reflectance measurements. [From Wyszecki, G., and Stiles, W. S. (1982). "Color Science: Concepts and Methods, Quantitative Data and Formulae," 2nd Ed. Copyright © 1982 John Wiley & Sons, Inc. Reprinted by permission of John Wiley & Sons, Inc.]

sphere painted inside with a highly reflecting and spectrally nonselective paint made from barium sulfate. When light is admitted into an ideal integrating sphere, the sphere "lights up" uniformly as a result of multiple diffuse internal reflections. The size of the sphere does not matter so long as the ports cut into it, which permit entry and exit of the incident and reflected light, do not exceed 10% of the total surface area.

The surface to be measured fills an opening at the bottom, oriented horizontally. The incident light, which should be of limited cross section in order to be confined to the sample being measured, enters at an angle of 5° to normal. To eliminate the specular component of reflection from the measurement, a light trap is introduced, centered at an angle of 5° on the other side of normal; the remaining diffusely reflected component illuminates the sphere. Ideally, the exit port could be located almost anywhere. In practice, it is located as shown, so that it "sees" only a small opposite section of the sphere. As an added precaution, a small baffle is introduced to block the initially reflected component of light, which otherwise would strike the sphere in the area being measured. When the cap, shown in the lower left of Fig. 10 is black, it eliminates the specular (or direct) component and only the diffuse reflectance is measured. When the cap is white (or when the sphere's surface is continuous, as at the bottom right of the figure), both specular and diffuse components contribute and the measurement is called total reflectance. Measurements are made, wavelength by wavelength, relative to the reflectance of a calibrated standard of known spectral reflectance. The spectral sensitivity of the detector does not matter so long as it is sufficiently sensitive to allow reliable measurements.

The arrangement of Fig. 10 ensures that all components of diffusely reflected light are equally weighted and that the specular component can be included in or eliminated from the measurement. Often, however, there is no true specular component, but rather a high-intensity lobe with a definite spread. This renders somewhat arbitrary the distinction between the specular and diffuse components. Operationally, the distinction depends on the size of exit port chosen for the specular light trap. Reflectance measurements are usually scaled relative to what a perfectly diffuse, totally reflecting surface would produce if located in the position of the sample. Figure 11 shows the diffuse spectral reflectance curves of a set of enamel paints that are sometimes used as calibration standards.

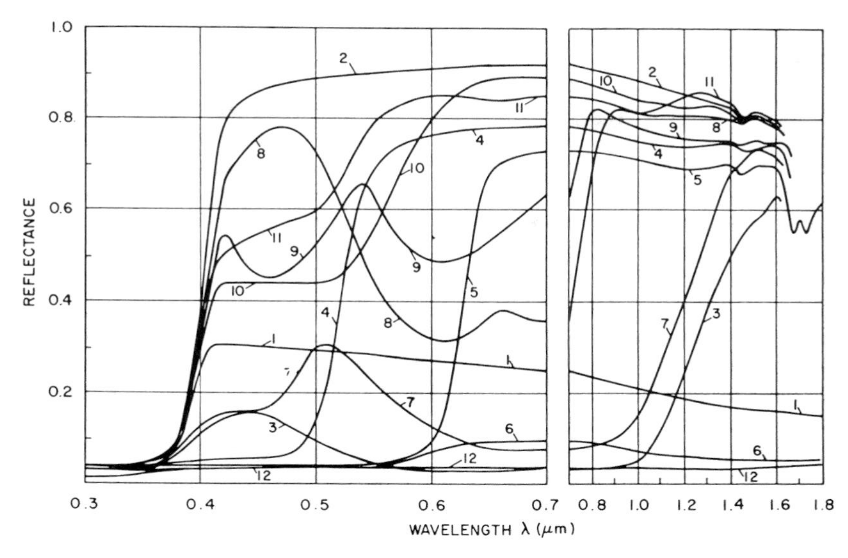

FIG. 11. Diffuse spectral reflectance curves of a set of enamel paints having the following color appearances: (1) medium gray; (2) white; (3) deep blue; (4) yellow; (5) red; (6) brown; (7) medium green; (8) light blue; (9) light green; (10) peach; (11) ivory; (12) black. [From Wyszecki, G., and Stiles, W. S. (1982). "Color Science: Concepts and Methods, Quantitative Data and Formulae," 2nd Ed. Copyright © 1982 John Wiley & Sons, Inc. Reprinted by permission of John Wiley & Sons, Inc.]

C. Chromaticity of an Object

The chromaticity of an object depends on the spectral properties of the illuminant as well as those of the object. A quantity $\phi(\lambda)$ is defined as $\rho(\lambda)S(\lambda)$ or $\tau(\lambda)S(\lambda)$, where $\rho(\lambda)$ symbolizes reflectance and $\tau(\lambda)$ symbolizes transmittance. Whereas reflectance ρ is the fraction of incident light returned from a surface, transmittance τ is the fraction of incident light transmitted by an object. Tristimulus values are then calculated as follows:

$$X = k \int \phi_\lambda \bar{x}(\lambda)\, d\lambda$$
$$Y = k \int \phi_\lambda \bar{y}(\lambda)\, d\lambda$$
$$Z = k \int \phi_\lambda \bar{z}(\lambda)\, d\lambda$$

Calculation of chromaticity coordinates then proceeds as described above for sources. If an equal-energy spectrum is assumed, the source term $S(\lambda)$ can be dropped from the definition of $\phi(\lambda)$. When the chromaticity of a surface is specified without specification of the source, an equal-energy spectrum is usually implied.

D. Fluorescence

The practice of colorimetry so far described becomes considerably more complicated if the measured surface exhibits fluorescence. Materials with fluorescing surfaces, when excited by incident light, generally both emit light at a longer wavelength and reflect a portion of the incident light. When making reflectance measurements of nonfluorescent materials, there is no reason to use incident radiation in the ultraviolet, to which the visual mechanism is nearly insensitive. However, for fluorescent materials these incident wavelengths can stimulate substantial radiation at *visible* wavelengths. The full specification of the relative radiance (reflection plus radiation) properties of such surfaces requires the determination, for *each* incident wavelength (including those wavelengths in the ultraviolet known to produce fluorescence), of relative radiance at *all* visible wavelengths, leading to a huge matrix of measurement conditions. As a meaningful and practical shortcut, daylight or a suitable daylight substitute can be used to irradiate the sample and a spectrophotometer can be located at the exit port to register the spectral distribution of the reflected light, which will include the component introduced by fluorescence. Daylight substitutes are so difficult to obtain that the most recent standard illuminant sanctioned by the CIE, called D-65, has been specified only mathematically but has never been perfectly realized. (A suitably filtered, high-pressure xenon arc source comes close.)

E. Optimal Colors

Because of the broadband characteristics of the cone spectral sensitivity functions, most of the spectrum locus in the chromaticity diagram is very well approximated by wave bands as broad as 5 nm. A reflecting surface that completely absorbed all wavelengths of incident broadband (white) light and reflected only a 5-nm wave band would have a chromaticity approximating the midpoint of that wave band along the spectrum locus. Such a surface would also have a very low reflectance because almost all of the incident light would be absorbed. Extending the wave band would increase reflectance, but at the cost of moving the chromaticity inward toward the center of the diagram, with the limit for a nonselective surface being the chromaticity of the illuminant. For any particular reflectance, the domain of possible chromaticities can be calculated; the outer limit of this domain represents the locus of optimal surface colors for that chromaticity.

The all-or-nothing and stepwise reflectance properties required for optimal surface colors do not exist either in nature or in artificially created pigments (see Fig. 11), which tend to exhibit instead gently sloped spectral reflectance functions. For any given reflectance, therefore, the domain of real colors is always much more restricted than the ideal one. Figure 12 shows the CIE chromaticity diagram and the relations between the spectrum locus, the optimal colors of

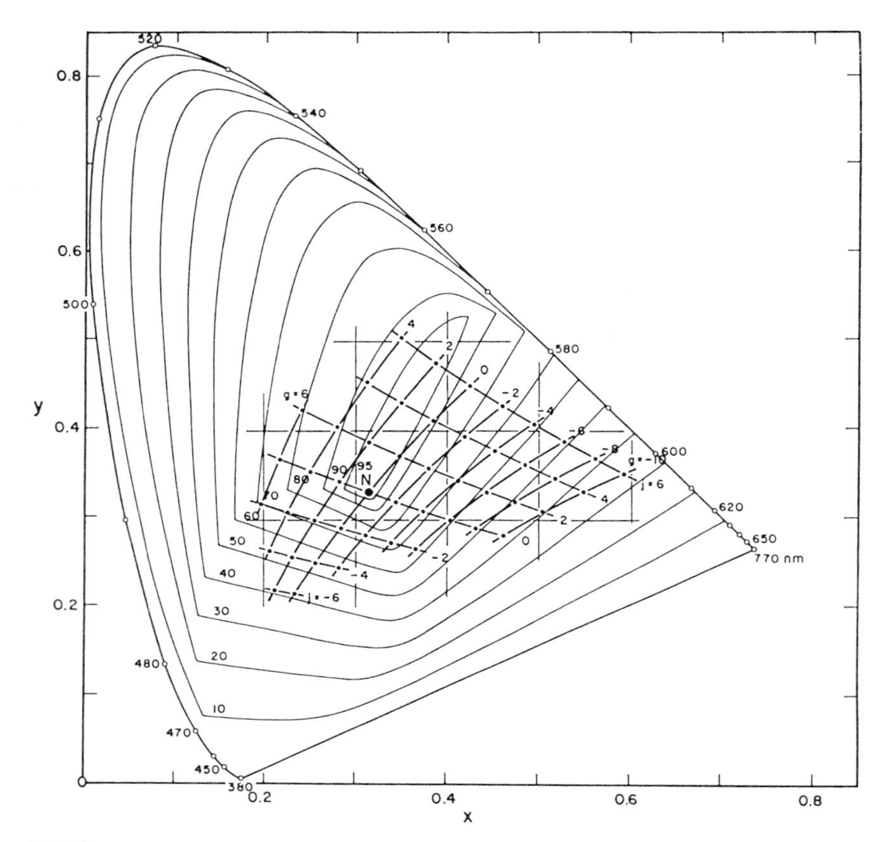

FIG. 12. Locus of optimal colors of reflectances indicated, also showing the locations of 47 surface colors of ~30% reflectance developed by the Optical Society of America to be equally spaced perceptually. [From Wyszecki, G., and Stiles, W. S. (1982). "Color Science: Concepts and Methods, Quantitative Data and Formulae," 2nd Ed. Copyright © 1982 John Wiley & Sons, Inc. Reprinted by permission of John Wiley & Sons, Inc.]

several reflectances, and the real surface colors of the Optical Society of America Uniform Color Scales set.

F. Metamerism Index

As already noted, *metamerism* refers to the phenomenon whereby a color match can occur between stimuli that differ in their spectral distributions. In the domain of reflecting samples the term carries a related, but very different connotation; here the degree of metamerism specifies the tendency of surfaces to change in perceived color, or to resist doing so, as the spectral characteristics of the illuminant are altered. Surfaces greatly exhibiting such changes are said to exhibit a high degree of metamerism, which from a commercial standpoint is undesirable.

Most indices of metamerism that have been proposed depend either on the assessed change in color of specific surfaces with change in illuminant, calculated by procedures similar to those used for the color-rendering index of illuminants, or on the number of intersections of the spectral reflectance functions of the samples being assessed. For two samples to be metameric, these functions must intersect at least three times; in general, the more intersections, the lower is the degree of metamerism that results, implying more resistance to color change with a change in illuminant. In the limiting case, where the curves are identical, there is no metamerism and the match holds for all illuminants.

Except for monochromatic lights on the curved portion of the spectrum locus in the chromaticity diagram, the number of possible metamers is mathematically infinite. Taking into account the limits of sensitivity of the visual system for the perception of color differences, the number of possible metamers increases as one approaches the white point on the chromaticity diagram, moving inward from the outer limits of realizable reflecting colors.

V. Physical Basis of Surface Color

The physical basis of the color of a surface is related to processes that alter the spectral distribution of the returned light in the direction of an observer, relative to that of the incident illumination.

A. Color from Organic Molecules

The action of organic molecules, which provide the basis for much of the color seen in nature, has been interpreted with accelerating precision since about 1950 in the context of molecular orbital theory. The interaction of light with small dye molecules can be completely specified, and although such detailed interpretation remains impractical for large dye molecules, even with currently available supercomputers, the origin of color in organic molecules is considered to be understood in principle at a molecular level.

Most organic dyes contain an extended conjugated chromophore system to which are attached electron donor and electron acceptor groups. Although the wavelength of a "reflected" photon is usually the same as that of the incident one, the reflected photon is not the same particle of light as the incident one. Instead, the surface is more properly regarded as a potential emitter of light, where (in the absence of incandescence, fluorescence, or phosphorescence) incident light is required to trigger the molecular reaction that produces the emitted radiation. Except for fluorescent materials, the number of emitted photons cannot exceed *at any wavelength* the number that are incident, and the frequency of each photon is unchanged, as is its wavelength if the medium is homogeneous. In considering nonfluorescent materials, the subtle exchange of reflected for incident photons is of no practical importance, and the term *reflection* is often used to describe the process as if some percentage of photons were merely bouncing off the surface.

B. Colorants

A colorant is any substance employed to produce reflection that is selective with wavelength. Colorants exist in two broad categories: dyes and pigments. In general, dyes are soluble, whereas pigments require a substrate called a binder. Not all colorants fall into either of these categories. (Exceptions include colorants used in enamels, glasses, and glazes.)

C. Scatter

Because pigments do not exist as dissociated, individual molecules, but instead are bound within particles whose size and distribution may vary, the spectral distribution of the reflected light depends only partly on the reaction of the dye or pigment molecules to absorbed light. In addition to the possibility of being absorbed, reflected, or transmitted, light may also be scattered. Particles that are very small relative to the wavelength of light produce Rayleigh scattering, which varies inversely as the fourth power of wavelength. (Rayleigh scatter causes the sky to appear blue on a clear day; without scatter from atmospheric particles, the sky would be black, as seen from the moon.) As scattering particles become larger, Mie scattering results. Wavelength dependence, which is minimal for large-particle scatter, becomes a factor for particles of intermediate size. The directionality of scattered light is complex and can be compounded by multiple scattering. There are two components of Rayleigh scattering, which are differentially polarized. Mie scattering is even more complicated than the Rayleigh variety, and calculations pertaining to it are possible to carry out only with very large computers.

Scatter also occurs at object surfaces. For example, in "blue-eyed" people and animals, the eye color results mainly from scatter within a lightly pigmented iris. As a powder containing a colorant is ground more finely or is compressed into a solid block, its scattering characteristics change and so does the spectral distribution of the light reflected from it, despite an unchanging molecular configuration of the colorant. Often such scatter is nonselective with wavelength and tends to dilute the selective effects of the colorant. A compressed block of calcium carbonate is interesting in this respect because it comes very close to being a perfectly diffuse, totally reflecting, spectrally nonselective reflector.

D. Other Causes of Spectrally Selective Reflection

The spectral distribution of returned light can also be altered by interference and diffraction. Interference colors are commonly seen in thin films of oil resting on water; digital recording disks now provide a common example of spectral dispersion by diffraction. [*See* DIFFRACTION, OPTICAL.]

Light is often transmitted partially through a material before being scattered or reflected. Various phenomena related to transmitted light per se also give rise to spectrally selective effects. In a transmitting substance, such as glass, light is repeatedly absorbed and reradiated, and in the process its speed is differentially reduced as a function of wavelength. This leads to wavelength-selective refraction and the prismatic dispersion of white light into its spectral components.

The most common colorants in glass are oxides of transition metals. Glass may be regarded as a solid fluid, in the sense of being a disordered, noncrystalline system. The metal oxides enter the molton glass in true solution and maintain that essential character after the glass has cooled and hardened. Whereas much colored glass is used for decorative purposes, color filters for scientific use are deliberately produced with specific densities and spectral distributions caused by selective absorption (which usually also produces some scatter), by reflection from coated surfaces, or by interference.

The visual characteristics of metals result from specular reflection (usually somewhat diffused) which, unlike that from other polished surfaces, is spectrally selective. If the regular periodicity of their atoms is taken into account, the reflectance characteristics of metals can also be understood in terms of the same molecular orbital theory that applies to organic colorants. In this case it serves as a more fundamental basis for band theory, in terms of which the optical and electrical conductance properties of metals and semiconductors have classically been characterized.

E. Subtractive Color Mixture

The addition of primaries, as described in Section I,A, is an example of what is often termed additive color mixture. Four methods of addition were described, all of which have in common the fact that photons of different wavelengths enter the eye from the same, or nearly the same, part of the visual field. There are no significant interactions between photons external to the eye; their integration occurs entirely within the photoreceptors, where photons of different wavelengths are absorbed in separate molecules of photopigments, which for a given photoreceptor are all of the same kind, housed within the cone outer segments.

Subtractive color mixing, on the other hand, is concerned with the modification of spectral light distributions external to the eye by the action of absorptive colorants, which, in the simplest case, can be considered to act in successive layers. Here it is dyes or pigments, not lights, that are mixed. The simplest case, approximated in some color photography processes, consists of layers of nonscattering, selectively absorptive filters. Consider the spectral transmittance functions of the subtractive primaries called cyan and yellow in Fig. 13 and the

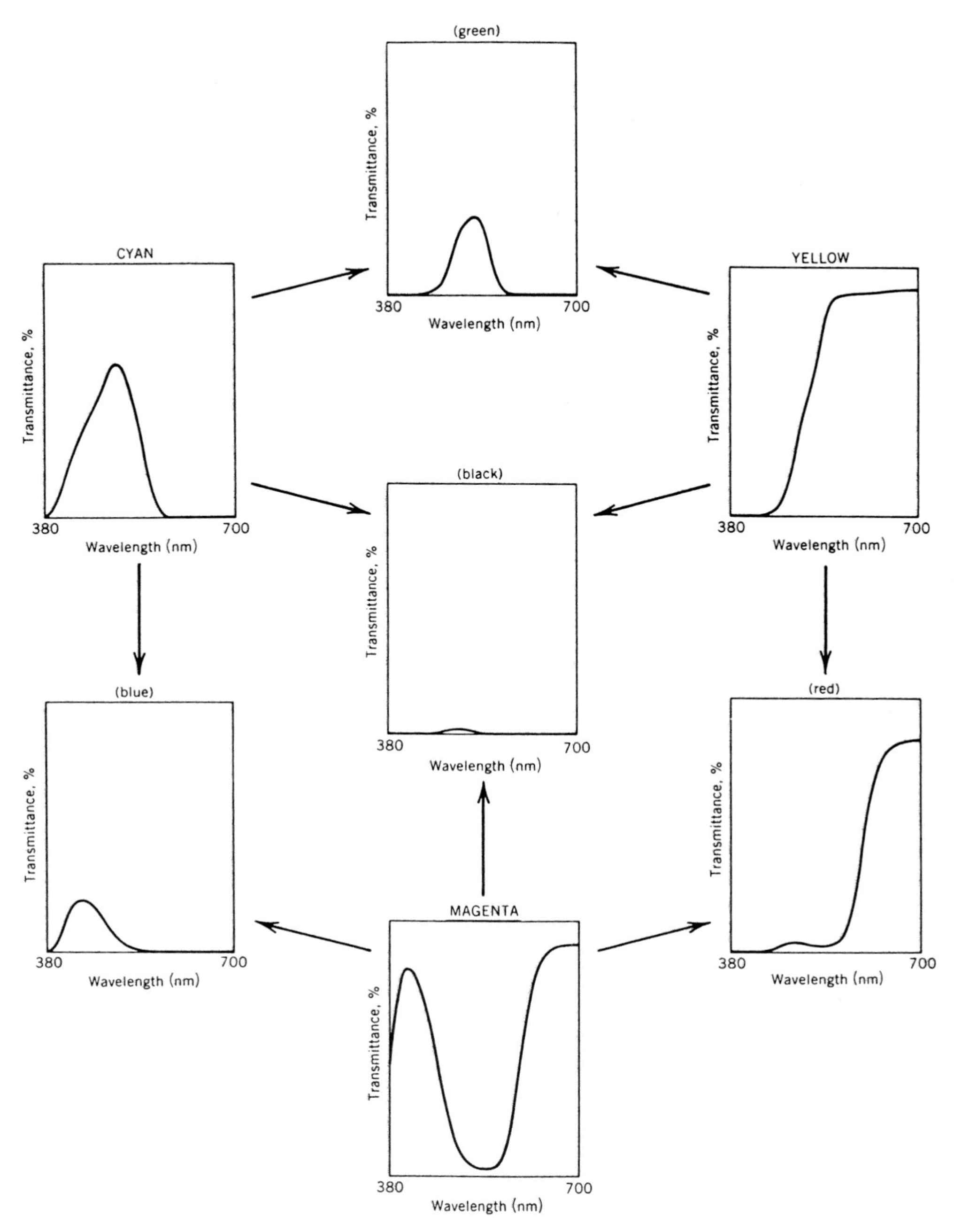

FIG. 13. Spectrophotometric curves of a set of subtractive primary filters and their mixtures, superimposed in various combinations. [From Billmeyer, F. W., Jr., and Saltzmann, M. (1981). "Principles of Color Technology," 2nd Ed. Copyright © 1981 John Wiley & Sons, Inc. Reprinted by permission of John Wiley & Sons, Inc.]

result of their combination: green. The transmittance function for the resulting green is simply the product, wavelength by wavelength, of the transmittance functions of cyan and yellow. When a third subtractive primary (magenta) is included in the system, blue and red can also be produced by the combinations shown. If all three subtractive primaries are used, very little light can pass through the combination, and the result is black.

If the filters are replaced by dyes in an ideal nonscattering solution, transmittance functions of the cyan, yellow, and magenta primaries can be varied quantitatively, depending on their concentration, with little change of "shape"; that is, each can be multiplied by a constant at each wavelength. By varying the relative concentrations of three dyes, a wide range of colors can be produced, as shown by the line segments on the CIE chromaticity diagram of Fig. 14. Subtractive color mixtures do not fall along straight lines in the chromaticity diagram.

Dichroic filters, such as those used in color television cameras, ideally do not absorb, but instead reflect the component of light not transmitted, so that the two components are complementary in color. By contrast, examination of an ordinary red gelatin filter reveals that the appearance of light reflected from it, as well as that transmitted through it, is red. The explanation for the reflected component is similar to that for colors produced by the application of pigments to a surface.

Consider elemental layers within the filter and a painted surface, each oriented horizontally, with light incident downward. In both cases, the light incident at each elemental layer consists of that not already absorbed or backscattered in the layers above. Within the elemental layer, some fraction of the incident light will be scattered upward, to suffer further absorption and scatter before some of it emerges from the surface at the top. Another fraction will be absorbed within the elemental layer, and the remainder will be transmitted downward, some specularly and some by scatter. In the case of the painted surface, a fraction of the initially incident light will reach the backing. In the case of the red gelatin filter, light emerging at the bottom constitutes the component transmitted through the filter. For the painted surface, the spectral reflectance of the backing will, unless perfectly neutral, alter the spectral distribution of the light reflected upward, with further attenuation and scattering at each elemental layer, until some of the light emerges at the top.

The prediction of color matches involving mixtures of pigments in scattering media is, as

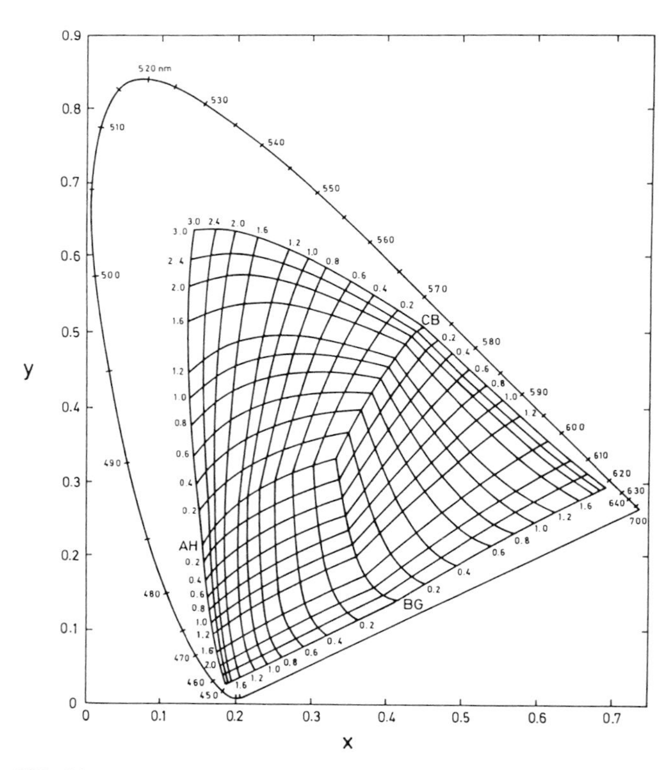

FIG. 14. Chromaticities in a white light (~6500 K) of indicated combinations of dyes AH, BG, and CB in various concentrations. (Courtesy D. L. MacAdam and Springer-Verlag.)

the example above suggests, not a simple matter. For this purpose, a theory developed by Kubelka and Munk in 1931, and named after them, is (with variations) most often used. Complex as it is, the theory nevertheless requires so many simplifying assumptions that predictions based on it are only approximate. Sometimes Mie scattering theory is applied to the problem of predicting pigment formulations to match a color specification, but more often empirical methods are used for this purpose.

VI. Color Difference and Color Order

Paradoxically, exact color matches are at the same time very common and very rare. They are common in the sense that any two sections of the same uniformly colored surface or material will usually match physically and therefore also visually. Otherwise, exact color matches are rare. For example, samples of paints of the same name and specification, intended to match, seldom do so exactly if drawn from separate batches. Pigments of identical chemical specification generally do not cause surfaces to match if they are ground to different particle sizes or suspended within different binders. Physical matches of different materials, such as plastics and fabrics, are usually impossible because differing binders or colorants must be used. In such cases—for example, matching a plastic dashboard with the fabric of an automobile seat—metameric matches must suffice; these cannot be perfect for all viewing conditions and observers. Given the difficulty or impossibility of producing perfect matches, it is important to be able to specify tolerances within which imperfect matches will be acceptable.

The issue of color differences on a more global scale will also be considered. Here concern is with the arrangement of colors in a conceptual space that will be helpful for visualizing the relations among colors of all possible kinds—the issue of color order.

A. Color Difference Data

Figure 15 shows the so-called MacAdam discrimination ellipses plotted in the CIE chromaticity diagram. These were produced more than 40 years ago by an experimental subject who repeatedly attempted to make perfect color matches to samples located at 25 points in chromaticity space. The apparatus provided projected rather than surface colors, but with a

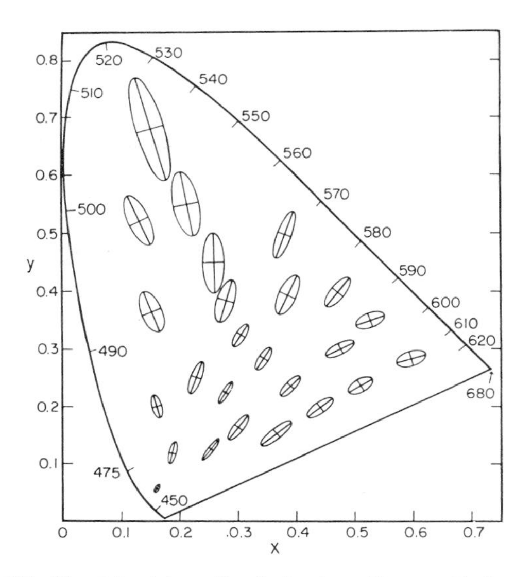

FIG. 15. MacAdam discrimination ellipses, 10 times actual size. (Courtesy D. L. MacAdam and Springer-Verlag.)

specified achromatic surround. For a set of settings at a given reference chromaticity, the apparatus was arranged so that the manipulation of a single control caused chromaticity to change linearly through the physical match point in a specified direction while automatically keeping luminance constant. Many attempted matches were made for each of several directions, as an index of a criterion sensory difference. The standard deviations of which were plotted on both sides of each reference chromaticity. Each of the ellipses of Fig. 15 was fitted to collections of such experimental points. MacAdam developed a system for interpolating between the measured chromaticities, and further research extended the effort to include luminance differences as well, leading to discrimination ellipsoids represented in x–y–Y space. By this criterion of discrimination, there are several million discriminable colors.

Early calculational methods required the use of graphical aids, some of which are still in widespread use for commercial purposes. Very soon it was recognized that, if a formula could be developed for the prediction of just-discriminable color differences, measurements of color differences could be made with photoelectric colorimeters. Many such formulas have been proposed. To promote uniformity of use, the CIE in 1976 sanctioned two systems called CIELAB and CIELUV, the second of which will be described here.

B. CIELUV Color Difference Formulas

It has long been recognized that the 1931 CIE chromaticity diagram is perceptually nonuniform, as revealed by the different sizes and orientations of the MacAdam ellipses plotted thereon. For the evaluation of chromatic differences, the ideal chromaticity space would be isotropic, and discrimination ellipsoids would everywhere be spheres whose cross sections in a constant-luminance plane would plot as circles of equal size.

Many different projections of the chromaticity diagram are possible; these correspond to changes in the assumed primaries, all of which convey the same basic information about color matches. The projection of Fig. 16 is based on the following equations:

$$u' = 4X/(X + 15Y + 3Z);$$

$$v' = 9Y/(X + 15Y + 3Z)$$

The CIELUV formula is based on the chromatic scaling defined by this transformation combined with a scale of lightness. The system is related to a white reference object having tristimulus values X_n, Y_n, Z_n, with Y_n usually taken as 100; these are taken to be the tristimulus values of a perfectly reflecting diffuser under a specified white illuminant.

Three quantities are then defined as

$$L^* = 116Y/Y_n - 16$$

$$u^* = 13L^*(u' - u'_n)$$

$$v^* = 13L^*(v' - v'_n)$$

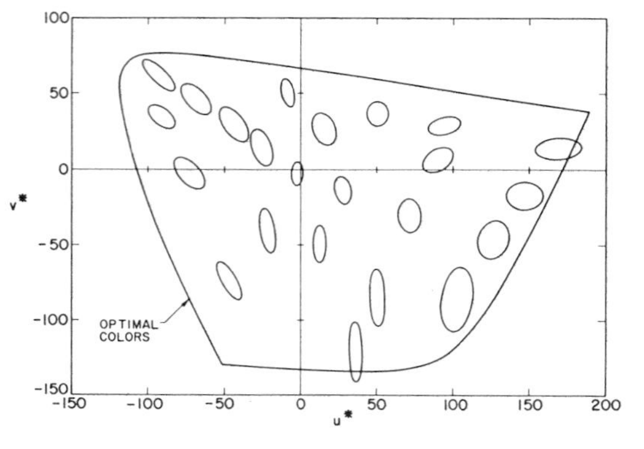

FIG. 16. MacAdam ellipses ($L^* = 50$) shown in the CIE u^*, v^* diagram. [From Wyszecki, G., and Stiles, W. S. (1982). "Color Science: Concepts and Methods, Quantitative Data and Formulae," 2nd Ed. Copyright © 1982 John Wiley & Sons, Inc. Reprinted by permission of John Wiley & Sons, Inc.]

These attempt to define an isotropic three-dimensional space having axes L^*, u^*, and v^*, such that a color difference ΔE^*_{uv} is defined as

$$\Delta E^*_{uv} = \Delta L^* + \Delta u^* + \Delta v^*$$

The MacAdam ellipses, plotted on the u^*, v^* diagram, are more uniform in size and orientation than in the CIE diagram. Without recourse to nonlinear transformations, this is about the greatest degree of uniformity possible. These data and the CIE's color difference equations are recommended for use under conditions in which the observer is assumed to be adapted to average daylight; they are not recommended by the CIE for other conditions of adaptation.

It is not difficult to write a computer program that will calculate the ΔE^*_{uv} values appropriate to each member of a pair of physical samples. Starting with knowledge of the spectral reflectance distributions of the samples and the spectral irradiance distribution of the illuminant, one calculates the tristimulus values X, Y, and Z. From these, the L^*, u^*, and v^* values for each sample are calculated and inserted into the final formula. Given that voltages proportional to tristimulus values can be approximated using suitably filtered photocells and electronics, it is a short step to the development of fully automated devices that, when aimed in turn at each of two surfaces, will register a color difference value.

The CIELUV formula is only one of more than a dozen schemes that have been suggested for calculating color differences, some simpler but most more elaborate. None of these performs as well as would be desired. Correlations of direct visual tests with predictions made by the best of these systems, including CIELUV, account for only about half the experimental variance. Different formulas make predictions that correlate no better than this with one another. In using CIELUV to predict color differences in self-luminous displays, agreement is lacking concerning the appropriate choice of reference white. For industrial applications in which reflecting materials are being evaluated, differential weighting of the three components entering into the CIELUV equation may be helpful, and to meet the demands of specific situations, doing so can significantly improve the predictive power of the system. For example, when samples of fabrics are being compared, tolerance for luminance mismatches tends to be greater than for mismatches along the chromatic dimensions.

Despite these problems and limitations, calculation of chromatic differences by formula has proved useful, automated colorimeters for doing so exist, and the practice may be regarded as established, perhaps more firmly than it should be. Room for improvement at a practical level and for a better theoretical understanding of the problem certainly exists.

C. Arrangement of Colors

In the years before the development of the CIE system of color specification, which is based on radiometric measurement, colors could be described only by appeal to labeled physical samples. Any two people possessing a common collection of samples could then specify a color by reference to its label. Whereas in principle such sets of colors could be randomly arranged, an orderly arrangement is clearly preferable in which adjacent colors differ by only a small amount.

For more than 100 years, it has been thought that colors can be continuously arranged in a domain composed of two cones sharing a common base, as shown in Fig. 17. A line connecting the cone apices defines the axis of achromatic colors, ranging from white at the top to black at the bottom. A horizontal plane intersecting one of the cones, or their common base, defines a set of colors of equal lightness. Within such a plane colors can be represented in an orderly way, with gray at the center and colors of maximum saturation on the circumference. The hues on the circumference are arranged as they are in the spectrum, in the order red, orange, yellow, green, blue, and violet, with the addition of a range of purples shading back to red. In moving from gray at the center toward a saturated hue, the hue remains constant while the white content of the color gradually diminishes as its chromatic content increases, all at constant lightness. As the intersecting horizontal plane is moved upward, the represented colors are more highly reflecting and correspondingly lighter in appearance, but their gamut is more restricted as must be so if only white is seen at the top. As the intersecting horizontal plane is moved downward, colors become less reflecting and darker in appearance, with progressively more restricted gamuts, until at the bottom only a pure black is seen.

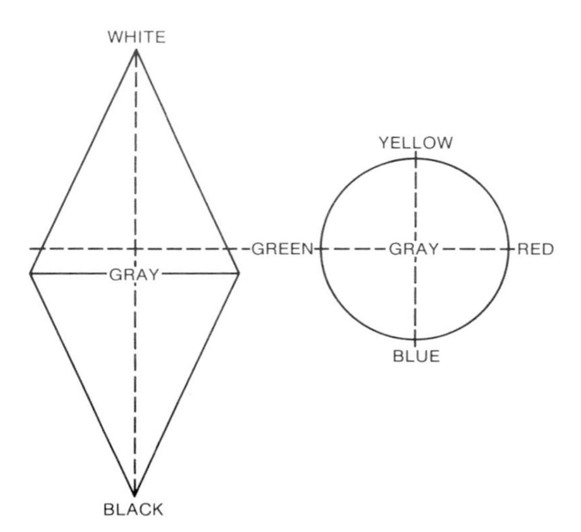

FIG. 17. Representation of three-dimensional domain of surface colors.

Considering a typical cross section, a chromatic gray is at its center. Significant features, first observed by Newton, pertain to the circumferential colors. First, they represent the most highly saturated colors conceivable at that level of lightness. Second, adjacent hues, if additively mixed, form legitimate blends; for example, there is a continuous range of blue-greens between blue and green. Third, opposite hues cannot blend. For example, yellow and blue, when additively mixed, produce a white that contains no trace of either component, and the sensations of yellow and blue are never simultaneously experienced in the same spatial location. Two colors that when additively mixed yield a white are called complementary colors, and in this kind of color-order system they plot on opposite sides of the hue circle.

There are several such systems in common use, of which only one, the Munsell system, will be described here. In this system, the vertical lightness axis is said to vary in *value* (equivalent to lightness) from 0 (black) to 10 (white). At any given value level, colors are arranged as described above but labeled circumferentially according to the hues blue, green, yellow, red, purple (and adjacent blends) and radially according to their saturation (called chroma). Figure 18 illustrates the system.

CIE chromaticity coordinates and Y values have been determined for each of the Munsell samples—the so-called Munsell renotations. A rubber-sheet type of transformation exists between the locations of Munsell colors at a given lightness level and their locations the CIE diagram, as shown in Fig. 19 for Munsell value 5. This figure illustrates that color order can also be visualized on the CIE diagram. A limitation

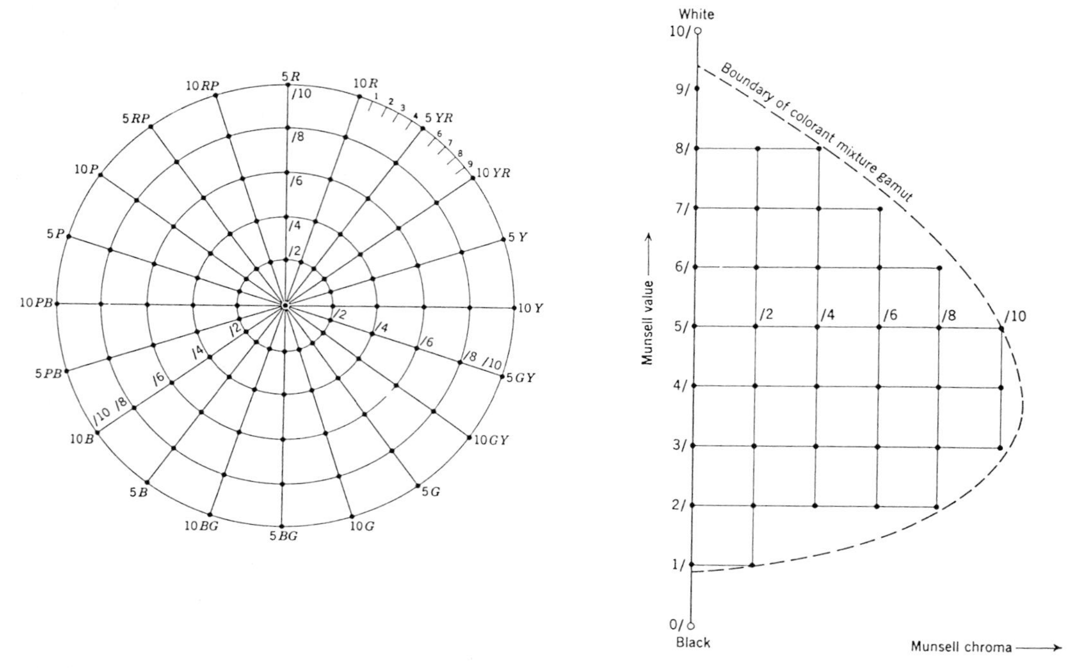

FIG. 18. Organization of colors in the Munsell system. [From Wyszecki, G., and Stiles, W. S. (1982). "Color Science: Concepts and Methods, Quantitative Data and Formulae," 2nd Ed. Copyright © 1982 John Wiley & Sons, Inc. Reprinted by permission of John Wiley & Sons, Inc.]

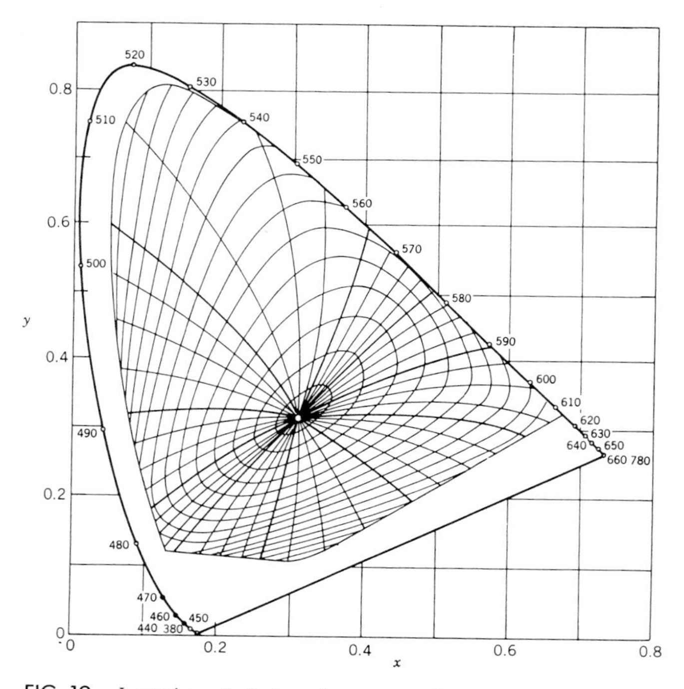

FIG. 19. Location of circles of constant chroma and lines of constant hue from the Munsell system plotted on the CIE chromaticity diagram. [From Billmeyer, F. W., Jr., and Saltzmann, M. (1981). "Principles of Color Technology," 2nd Ed. Copyright © 1981 John Wiley & Sons, Inc. Reprinted by permission of John Wiley & Sons, Inc.]

of the CIE diagram for this purpose, in addition to its perceptually nonuniform property, is that it refers to no particular lightness level. The dark surface colors, including brown and black, do not exist in isolated patches of light. These colors are seen only in relation to a lighter surround; in general, surface colors are seen in a complex context of surrounding colors and are sometimes called related colors for this reason.

Although surrounding colors can profoundly influence the appearance of a test color in the laboratory situation, these effects are seldom obvious in natural environments and probably represent an influence of the same processes responsible for color constancy. This concept refers to the fact that colors appear to change remarkably little despite changes in the illuminant that materially alter the spectral distribution of the light reaching the retina. In other words, the perceived color of an object tends, very adaptively, to be correlated with its relative spectral reflectance, so that within limits color seems to be an unchanging characteristic of the object rather than triply dependent, as it actually is, on the characteristics of the illuminant, object, and observer.

VII. Physiological Basis of Color Vision

Progress in modern neuroscience, including a progressively better understanding of sensory systems in physical and chemical terms, has been especially rapid since about 1950 and continues to accelerate. The following is a very brief summary of some highlights and areas of ignorance related to color vision.

The optical system of the eye receives light reflected from external objects and images it on the retina, with a spectral distribution that is altered by absorption in the eye media. By movements of the eyes that are ordinarily unconsciously programmed, the images of objects of interest are brought to the very center of a specialized region of the retina, known as the fovea centralis, where we enjoy our most detailed spatial vision. The color of objects in the peripheral visual field plays an important role in this process.

The spectral sensitivities of the three classes of cone photoreceptors shown in Fig. 2 depend on the action spectra of three classes of photopigments, each uniquely housed in one type of cone. The cones are very nonuniformly distributed in the retina, being present in highest density in the fovea and falling off rapidly to lower density levels across the remainder of the retina.

The colors of small stimuli seen in the periphery are not registered so clearly as in foveal vision, but if fields are enlarged sufficiently, the periphery is capable of conveying a great deal of information about color. In the fovea there are few if any short-wavelength-sensitive (S) cones, and the long-wavelength-sensitive (L) and middle-wavelength-sensitive (M) cones are present in roughly equal numbers and very high density.

The L and M cones subserve spatial vision and also provide chromatic information concerned with the balance between red and green. Outside the fovea, the proportion of S cones increases but is always very small. The coarseness of the S-cone mosaic makes it impossible for them to contribute very much to detailed spatial vision; instead, they provide information concerned almost exclusively with the second dimension of chromatic vision.

As noted earlier, color is coded initially in terms of the ratios of excitation of the three kinds of cones. A single cone class in isolation is colorblind, because any two spectral distributions can excite such cones equally if appropriate relative intensities are used. The same is true of the much more numerous rod photoreceptors, which can lead to total colorblindness in night vision, where the amount of light available is often insufficient to be effective for cones. At intermediate radiance levels, rods influence both color appearance and, to a degree, color matches. Interestingly, vision at these levels remains trichromatic in the sense that color matches can be made using three primaries and three controls. This suggests that rods feed their signals into pathways shared by cones, a fact documented by direct electrophysiological experiment.

Light absorption in the cones generates an electrical signal in each receptor and modulates the rate of release of a neurotransmitter at the cone pedicles, where they synapse with horizontal and bipolar cells in the retina. The latter deliver their signals to the ganglion cells, whose long, slender axons leave the eye at the optic disk as a sheathed bundle, the optic nerve. Within this nerve are the patterns of impulses, distributed in about a million fibers from each eye, by means of which the brain is exclusively informed about the interaction of light with objects in the external world, on the basis of which form and color are perceived.

Lateral interactions are especially important

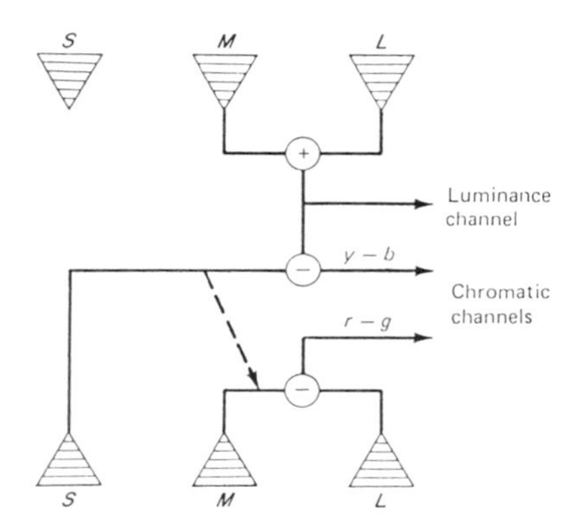

FIG. 20. Opponent-color model of human color vision at the retinal level.

for color vision. The color of a particular area of the visual scene depends not only on the spectral distribution of the light coming from that area, but also on the spectral distribution (and quantity) of light coming from other regions of the visual field. In both retina and brain, elaborate lateral interconnections are sufficient to provide the basis for these interactions.

Figure 20 summarizes, in simplified fashion, a current model of retinal function. The initial trichromacy represented by L, M, and S cones is transformed within the retina to a different trichromatic code. The outputs of the L and M cones are summed to provide a luminance signal, which is equivalent to the quantity of light as defined by flicker photometry. The L and M cone outputs are also differenced to form a red–green signal, which carries information about the relative excitations of the L and M cones. An external object that reflects long-wavelength light selectively excites the L cones more than the M, causing the red–green difference signal to swing in the red direction. A yellow–blue signal is derived from the difference between the luminance signal and that from the S cones.

The color appearance of isolated fields of color, whether monochromatic or spectrally mixed, can be reasonably well understood in terms of the relative strengths of the red–green, yellow–blue, and luminance signals as these are affected in turn by the strength of the initial signals generated in the three types of cones. In addition, increasing S-cone excitation appears to influence the red–green opponent color signal in the red direction.

There has been a great deal of investigation of the anatomy and physiology of the brain as it relates to color perception, based on an array of techniques that continues to expand. The primary visual input arrives in the striate cortex at the back of the head; in addition there are several other brain centers that receive visual input, some of which seem specifically concerned with chromatic vision. The meaning of this activity for visual perception is not yet clear. In particular, it is not yet known exactly which kinds or patterns of activity immediately underly the sensation of color or exactly where they are located in the brain.

Bibliography

Billmeyer, F. W., Jr., and Saltzman, M. (1981). "Principles of Color Technology," 2nd Ed. Wiley, New York.

Boynton, R. M. (1979). "Human Color Vision." Holt, New York.

Grum, F., and Bartleson, C. J. (1980). "Optical Radiation Measurements," Vol. 2. Academic Press, New York.

Kaufman, J. E., ed. (1981). "IES Lighting Handbook: Reference Volume," Illuminating Engineering Society of North America, New York.

MacAdam, D. L. (1981). "Color Measurement: Theme and Variations." Springer-Verlag, Berlin and New York.

Mollon, J. D., and Sharpe, L. T., eds. (1983). "Colour Vision: Physiology and Psychophysics." Academic Press, New York.

Nassau, K. (1983). "The Physics and Chemistry of Color." Wiley, New York.

Wyszecki, G., and Stiles, W. S. (1982). "Color Science: Concepts and Methods, Quantitative Data and Formulae," 2nd Ed. Wiley, New York.

Zrenner, E. (1983). "Neurophysiological Aspects of Color Vision in Primates," Springer-Verlag, Berlin and New York.

DIFFRACTION, OPTICAL

Salvatore Solimeno *University of Naples*

GLOSSARY

Airy pattern: Far field diffracted by a circular aperture illuminated by a plane wave or field on the focal plane of an axial symmetric imaging system.

Babinet's principle: The sum of the field diffracted by two complementary screens (corresponding to a situation in which the second screen is obtained from the first one by interchanging apertures and opaque portions) coincides with the field in the absence of any screen.

Boundary diffraction wave: Representation introduced by Rubinowicz of the deviation of the diffracted field from the geometrical optics one as a wave originating from the screen boundary.

Diffraction-limited system: Optical system in which aberrations are negligible with respect to diffraction effects.

Far-field pattern: Diatribution of the field diffracted at an infinite distance from an obstruction.

Fourier methods: Analysis of diffraction and aberration effects introduced by optical imaging systems based on the decomposition of the field in a superposition of sinusoids.

Fraunhofer diffraction: Diffraction at a very large distance from an aperture.

Fresnel diffraction: Diffracted in proximity to an aperture.

Fresnel number of an aperture: Parameter related to the deviation of the field behind an aperture from the geometrical optics one.

Geometric theory of diffraction: Method introduced by J. B. Keller for calculating the field diffracted from an obstacle based on a suitable combination of geometrical optics and diffraction formulas for particular obstacles (wedges, half-planes, cylinders, spheres, etc.).

Green's theorem (Helmholtz–Kirchhoff integral theorem): Representation of the field by means of an integral extending over a closed surface containing the field point.

Huygens–Fresnel principle: Every point on a primary wave front serves as the source of secondary wavelets such that the wave front at some later time is the envelope of these wavelets.

Kirchhoff method: Expression of the field amplitude at any point behind a screen by means of an integral extending only over the open areas in the screen and containing the unobstructed values of the field amplitude and its normal derivative.

Modulation transfer function: Function measuring the reduction in contrast from object to image, that is, the ratio of image-to-object modulation for sinusoids of varying spatial frequencies.

Optical transfer function: Function measuring the complex amplitude of the image transmitted by an optical system illuminated by a unit amplitude sinusoidal pattern, versus the spatial frequency.

Spatial frequency: Representation of an object or an image as a superposition of sinusoids (Fourier components).

Spread function: Image of a point source object; its deviation from an impulse function is a measure of the aberration diffraction effects.

Any deviation of light rays from rectilinear paths which cannot be interpreted as reflection or refraction is called diffraction. As it stands today diffraction optics (DO) deals mainly with the description of fields in proximity to caustics, foci, and shadow boundaries of wave fronts delimited by apertures and with the far fields. Imaging systems, diffraction gratings, optical resonators, and holographic and image processing systems are examples of devices that depend for their ultimate performance on DO. Rudimentary solutions to DO problems can be obtained by using the Huygens–Fresnel principle. More accurate solutions are obtained by solving the wave equation with the Helmholtz–Kirchhoff integral formula. The modern geometric theory of diffraction combines ray optical techniques with the rigorous description of the field diffracted by typical obstacles (wedges, cylinders, spheres) to represent with great accuracy the light field diffracted by a generic obstruction.

I. History

A. The First Ideas

The phenomenon of diffraction was discovered by the Jesuit Father Francesco Maria Grimaldi and described in the treatise *Physico Mathesis de Lumine, Coloribus et Iride,* published in 1665, two years after his death. As it stands today, diffraction optics (DO) is the result of a long and interesting evolution originating from the ideas illustrated in 1690 by Christian Huygens in his celebrated "Traité de la Lumière." Thomas Young had the great merit of introducing the ideas of wave propagation to explain, following Newton, the corpuscular theory of optical phenomena. He introduced the principle of interference, illustrated in his three Bakerian lectures read at the Royal Society in 1801–1803. By virtue of this principle, Young was able to compute for the first time the wavelengths of different colors. Later, in 1875, Augustine Jean Fresnel presented to the French Academy the famous treatise "La diffraction de la Lumière," in which he presented a systematic description of the fringes observed on the dark side of an obstacle where he was able to show agreement between the measured spacings of the fringes observed and those calculated by means of the wave theory. [*See* COLOR, SCIENCE OF.]

B. The Wave Theory and Maxwell Equations

A remarkable success of the wave theory of light was recorded in 1835 with the publication in the Transactions of the Cambridge Philosophical Society of a fundamental paper by Sir George Biddell Airy, director of the Cambridge Observatory, in which he derived his famous expression for the image of a star seen through a well-corrected telescope. The image consists of a bright nucleus, known since then as *Airy's disk,* surrounded by a number of fainter rings, of which only the first is usually bright enough to be visible to the eye. Successive developments, until Maxwell's publication in 1873 of the "Treatise on Electricity and Magnetism," took advantage of Fresnel's ideas to solve a host of diffraction problems by using propagation through an elastic medium as a physical model. In particular, in 1861 Glebsch obtained the analytic solution for the diffraction of a plane wave by a spherical object.

C. Fourier Optics

Since the initial success of the Airy formula, the theory of diffraction has enjoyed increasing popularity, providing the fundamental tools for quantitatively assessing the quality of images and measuring the ability of optical systems to provide well-resolved images. To deal with this complex situation, Duffieux proposed in 1946 that the imaging of sinusoidal intensity patterns be examined as a function of their period. Then the optical system becomes known through an optical transfer function (OTF) that gives the system response versus the number of lines of the object per unit length. More recently, optical systems have been characterized by means of the numerable set of object fields that are faithfully reproduced. This approach, based on the solution of Fredholm's integral equations derived from the standard diffraction integrals, has allowed the ideas of information theory to be applied to optical instruments.

II. Mathematical Techniques

Diffraction optics makes use of a number of analytical tools:

1. Spectral representations of the fields (plane, cylindrical, spherical wave expansions; Hermite–Gaussian and Laguerre–Gaussian beams; prolate spheroidal harmonics).
2. Diffraction integrals.

3. Integral equations.
4. Integral transforms (Lebedev–Kontorovich transform, Watson transform).
5. Separation of variables.
6. Wiener–Hopf–Fock functional method.
7. Wentzel–Kramers–Brillouin (WKB) asymptotic solutions of the wave equations.
8. Variational methods.

In many cases the solutions are expressed by complex integrals and series, which can be evaluated either asymptotically or numerically by resorting to a number of techniques:

1. Stationary-phase and saddle-point methods.
2. Boundary-layer theory.
3. Two-dimensional fast Fourier transform (FFT) algorithm.

III. Helmholtz–Kirchhoff Integral Theorem

A. Green's Theorem

Let $u(\mathbf{r})e^{i\omega t}$ be a scalar function representing a time-harmonic ($e^{i\omega t}$) electromagnetic field which satisfies the Helmholtz wave equation

$$\nabla^2 u + \omega^2 \mu \varepsilon u = 0 \tag{1}$$

where $\nabla^2 = \partial^2/\partial x^2 + \partial^2/\partial y^2 + \partial^2/\partial z^2$ is the Laplacian operator and μ and ε are the magnetic permeability (expressed in henry per meter in mksa units) and the dielectric constant (faradays per meter), respectively, of the medium. It can be shown that the field $u(\mathbf{r}_0)$ at the point $\mathbf{r}_0$ depends linearly on the values taken by $u(\mathbf{r})$ and the gradient ∇u on a generic closed surface S containing $\mathbf{r}_0$ (*Green's theorem*):

$$u(\mathbf{r}) = \oiint_S \left[G(\mathbf{r}, \mathbf{r}') \frac{\partial u(\mathbf{r}')}{\partial n_0} - u(\mathbf{r}') \frac{\partial G(\mathbf{r}, \mathbf{r}')}{\partial n_0} \right] dS' \tag{2}$$

where $G(\mathbf{r}_0, \mathbf{r}) = \exp(-ikR)/4\pi R$, $R = |\mathbf{R}| = |\mathbf{r} - \mathbf{r}_0|$, and $\partial/\partial n_0$ represents the derivative along the outward normal $\hat{\mathbf{n}}_0$ to S.

For a field admitting the ray optical representation $u(\mathbf{r}) = A(\mathbf{r})e^{-ikS(\mathbf{r})}$ the integral in Eq. (2) reduces to

$$u(\mathbf{r}) = \frac{i}{2\lambda} \oiint_S \frac{A(\mathbf{r}')}{R} e^{-ik(R+S)} (\cos\theta_i + \cos\theta_d)\, dS' \tag{3}$$

where θ_i and θ_d are, respectively, the angle between the ray passing through $\mathbf{r}'$ and the inward normal $\hat{\mathbf{n}}_0$ to S and the angle between the direction $-\hat{\mathbf{R}}$ along which the diffracted field is calculated and $\hat{\mathbf{n}}_0$. In particular, when the surface S coincides with a wave front, $\theta_i = 0$ and $\cos\theta_i + \cos\theta_d$ reduces to the *obliquity factor* $1 + \cos\theta_d$.

B. Huygens–Fresnel Principle

The integral of Eq. (3) allows us to consider the field as the superposition of many elementary wavelets of the form

$$du(\mathbf{r}) = \frac{e^{-ikR}}{4\pi R} \left[\frac{\partial u(\mathbf{r}')}{\partial n_0} + u(\mathbf{r}')\hat{\mathbf{n}}_0 \cdot \mathbf{R} \left(ik + \frac{1}{R} \right) \right] dS' \tag{4}$$

According to the above equations every point on a wave front serves as the source of spherical wavelets. The field amplitude at any point is the superposition of the complex amplitudes of all these wavelets. The representation of a field as a superposition of many elementary wavelets is known as the *Huygens principle,* since it was formulated in his "Traité de la Lumière" in 1690. Fresnel completed the description of this principle with the addition of the concept of interference.

C. Kirchhoff Formulation of the Diffraction by a Plane Screen

Let us consider the situation in which a plane Π, of equation $z' = \text{const}$, separates the region I containing the sources from the homogeneous region II, where the field must be calculated. In this case, replacing the function G in Eq. (2) by

$$G_{\pm}(\mathbf{r}, \mathbf{r}') = \frac{\exp(-ik|\mathbf{r} - \mathbf{r}'|)}{4\pi|\mathbf{r} - \mathbf{r}'|} \pm \frac{\exp(-ik|\mathbf{r}_s - \mathbf{r}'|)}{4\pi|\mathbf{r}_s - \mathbf{r}'|} \tag{5}$$

with $\mathbf{r}_s$ referring to the specular image of $\mathbf{r}$ with respect to Π, it is easy to verify that Eq. (2) reduces to

$$\begin{aligned} u(\mathbf{r}) &= -2 \iint_\Pi \frac{\partial u(x', y', z')}{\partial z'} \frac{e^{-ikR}}{4\pi R} dx'\, dy' \\ &= 2 \iint_\Pi u(x', y', z') \left(ik + \frac{1}{R} \right) \frac{e^{-ikR}}{4\pi R} \\ &\quad \times \cos\theta_d dx'\, dy', \end{aligned} \tag{6}$$

use having been made of the relation $\partial G_{+}(\mathbf{r}', \mathbf{t})/\partial n_0 = G_{-}(\mathbf{r}', \mathbf{r}) = 0$ for $\mathbf{r}'$ on Π. In particular, for an aperture Σ on a plane screen illuminated by the incident field $u_i(\mathbf{r})$, Eq. (6) can be approximated by

$$u(\mathbf{r}) = 2 \iint_{-\infty}^{+\infty} P(x', y') u_i(x', y', z')\left(ik + \frac{1}{R}\right) \times \frac{e^{-ikR}}{4\pi R} \cos \theta_d \, dx' \, dy' \tag{7}$$

where $P(x', y')$ (the *pupil function*) takes on the value one if (x', y') belongs to the aperture and vanishes otherwise. In writing Eq. (7) we have tacitly assumed that the field on the exit aperture coincides with that existing in the absence of the aperture; this approximation, known as the *Kirchhoff principle,* is equivalent to the assumption that a finite exit pupil does not perturb the field on the pupil plane. Since presumably the actual perturbation is significant only near the pupil edge, we expect the error related to the application of Kirchhoff's principle to be neglected provided the aperture is sufficiently large. Exact analysis of the effects produced by some simple apertures (e.g., half-plane) confirms the validity of Kirchhoff's hypothesis for calculating the field near the *shadow boundaries* separating the lit region from the shady side; the error becomes relevent only for field points lying deep in either the lit or the dark regions.

IV. Diffraction Integrals

A. Fresnel and Fraunhofer Diffraction Formulas

Let us consider a field different from zero only on a finite plane aperture Σ. If we indicate by a the radius of the smallest circumference encircling the aperture and assume that $|z - z'| \gg a$, we can approximate the distance R with $|z - z'| + \frac{1}{2}[(x - x')^2 + (y - y')^2]/|z - z'|$, so that the diffraction integral of Eq. (7) reduces to the *Fresnel formula*

$$u(x, y, z) = \frac{i}{\lambda d} e^{-ikd} \iint_{\Sigma} u(x', y', z') \exp\left\{-i \frac{k}{2d} \times [(x - x')^2 + (y - y')^2]\right\} dx' \, dy' \tag{8}$$

where $d = |z - z'|$.

When $d \gg a^2\pi/\lambda$, where a is the radius of the aperture, we can neglect the terms of the exponential in the integrand of Eq. (8) proportional to $x'^2 + y'^2$, so that

$$u(x, y, z) = \frac{i}{\lambda d} e^{-ikR_0} \iint_{\Sigma} u(x', y', z') \times \exp\left(ik \frac{xx' + yy'}{d}\right) dx' \, dy' \tag{9}$$

where $R_0 = d + (x^2 + y^2)/2d$. The above equation, referred to as the *Fraunhofer diffraction formula,* allows us to express the far field in terms of the two-dimensional Fourier transform of u on the aperture plane, evaluated for $k_x = kx/d$ and $k_y = ky/d$. The Fraunhofer fields of some typical apertures illuminated by plane waves are plotted in Fig. 1.

B. Rotationally Invariant Fields and Airy Pattern

The far field radiated by a circular aperture can be obtained by assuming A [see Eq. (3)] in-

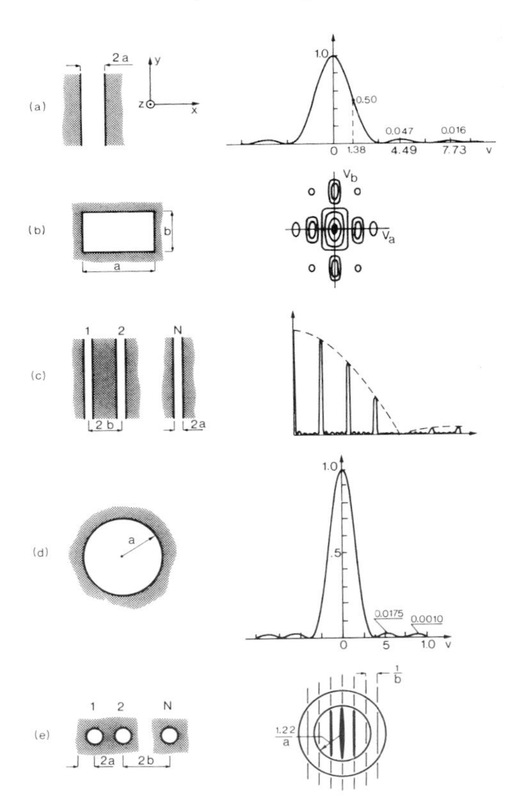

FIG. 1. Diffraction patterns of typical apertures. The pattern functions $|G(\theta, \phi)|^2$ are respectively proportional to (a) $[\sin(v)/v]^2$, $v = ka \sin \theta \cos \phi$; (b) $\{[\sin(v_a)/v_a][\sin(v_b)/v_b]\}^2$, $v_a = ka \sin \theta \cos \phi$, $v_b = kb \sin \theta \sin \phi$; (c) $\{[\sin(v_a)/v_a][\sin(Nv_b)/\sin v_b]\}^2$, $v_a = ka \sin \theta \cos \phi$, $v_b = kb \sin \theta \sin \phi$; (d) $(2J_1(v)/v)^2$, $v = ka \sin \theta$; (e) $\{[2J_1(v_a)/v_a][\sin(Nv_b)/\sin v_b]\}^2$, $v_a = ka \sin \theta$, $v_b = kb \sin \theta \cos \phi$.

dependent of the angle ϕ and integrating the integral on the right-hand side of Eq. (9) with respect to ϕ, thus obtaining

$$u(\rho, z) = -i\,\mathrm{NA}\,ka\,\exp(-ikd - ikd\theta^2/2) \times \int_0^1 A(ax)J_0(ka\theta x) \times \exp[-ikS(ax)]x\,dx \qquad (10)$$

where $\rho = \sqrt{x^2 + y^2}$, $\theta = \rho/d$, NA is the numerical aperture of the lens; and $u = A \exp(-ikS)$. In particular, for $A = 1$, $S = 0$, Eq. (10) gives

$$u(\rho, z) \propto 2J_1(ka\theta)/ka\theta \qquad (11)$$

J_1 being the Bessel function of first order. The intensity distribution associated with the above field is known as the *Airy pattern* (see Fig. 1d). Because of the axial symmetry, the central maximum corresponds to a high-irradiance central spot, known as the *Airy disk*. Since $J_1(v) = 0$ for $v = 3.83$, the angular radius of the Airy disk is equal to $\theta = 3.83/ka$.

The central spot is surrounded by a series of rings corresponding to the secondary maxima of the function $J_1(v)/v$, which occur when v equals 5.14, 8.42, 11.6, etc. On integrating the irradiance over a pattern region, one finds that 84% of the light arrives within the Airy disk and 91% within the bounds of the second dark ring.

The Airy pattern can be observed at a finite distance by focusing a uniform spherical wave with a lens delimited by a circular pupil. In this case the quantity v is replaced by $k\,\mathrm{NA}\,\rho$, where $\rho = \sqrt{x^2 + y^2}$.

C. Resolving Power

If we consider the diffraction images of two plane waves, it is customary to assume as a resolution limit the angular separation at which the center of one Airy disk falls on the first dark ring of the other (*Rayleigh's criterion of resolution*). This gives for the angular resolution

$$\theta_{\min} \simeq 1.22\lambda/D \qquad (12)$$

D representing the diameter of the exit pupil.

D. Fields in the Focal Region of a Lens

Imaging systems are designed with the aim of converging a finite conical ray congruence, radiated by a point source on the object plane, toward a focal point on the image plane. In most cases, the field relative to the region between the source and the exit pupil can be calculated by geometrical optics methods, that is, by evaluating the trajectories of the rays propagating through the sequence of refracting surfaces. However, downstream from the exit pupil, we are faced with the unphysical result of a field vanishing abruptly across the shadow boundary surface formed by the envelope of the rays passing through the edge of the *exit pupil*. To eliminate this discontinuity it is necessary to resort to the diffraction integral representation. In particular, the field on the exit aperture can be assumed to coincide with that existing in the absence of the aperture and can be calculated by geometrical optics methods.

When the numerical aperture of the beam entering or leaving the lens is quite large, it is necessary to account for the vector character of the electric field **E**. This occurs, for example, in microscope imaging, where the aperture of the beam entering the objective can be very large. As a consequence, for rotationally symmetric lenses, the focal spot for linearly polarized light is not radially symmetric, a fact that affects the resolving power of the instrument.

If we choose a Cartesian system with the z axis parallel to the optic axis and the plane $z = 0$ coinciding with the Gaussian image of the object plane $z = z_0$ (<0) (see Fig. 2), it can be shown for a field point very close to the Gaussian image that Eq. (3) generalizes to the *Luneburg–Debye integral*

$$\mathbf{E}(\mathbf{r}) = i\frac{e^{-ikV_0}}{\lambda}\iint_{A} \mathbf{E}'(\hat{\mathbf{n}}_0)\exp[-ik(p(x-\bar{x}) + q(y-\bar{y}) + rz)]\,d\Omega \qquad (13)$$

where $\mathbf{E}' = Re^{ikR}\mathbf{E}$, $p = -n_{0x}$, $q = -n_{0y}$, and $r = -n_{0z}$ are the *direction cosines* of the ray passing through (ξ, η, ζ) (see Fig. 3) while $d\Omega =$

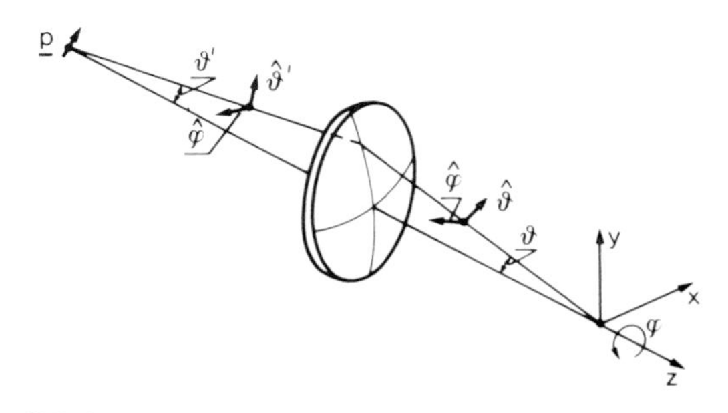

FIG. 2. Mutual orientation of the spherical coordinate systems relative to the source and the image formed by a lens.

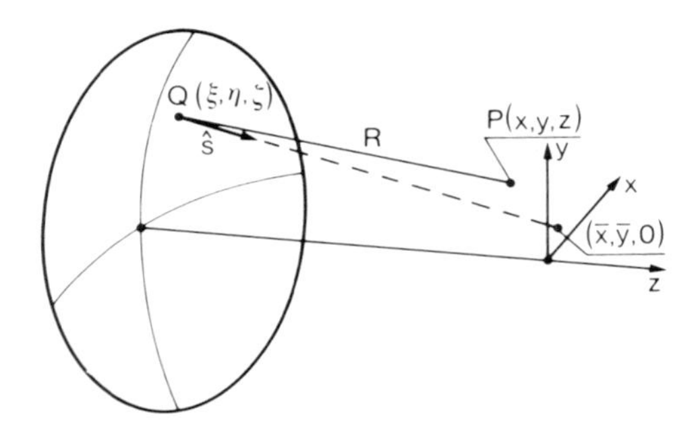

FIG. 3. Schematic representation of field point $P(x, y, z)$, wave front $Q(\xi, \eta, \zeta)$, and Gaussian image $(\bar{x}, \bar{y}, 0)$.

$d\bar{A}/R^2$ is the solid angle that the surface element $d\bar{A}$ of the exit pupil subtends at the focus.

It is now useful to introduce *optical coordinates* v, $\bar{u}$ defined by

$$v = k\rho \text{ NA}, \qquad \bar{u} = kz \text{ NA}^2 \tag{14}$$

where NA = $\sin \theta_{max}$ represents the *numerical aperture*, θ_{max} indicating the half-aperture in the *image space*. In this way, the Luneburg–Debye integral, for NA sufficiently small and a field linearly polarized along the x axis, reduces to

$$\begin{aligned} \mathbf{E}(\mathbf{r}) &\propto \hat{\mathbf{x}} \exp(-i\bar{u}/\theta_{max}^2) \\ &\times \int_0^{2\pi} d\phi \int_0^1 d\Theta\, \Theta\, \mathbf{E}_i(\Theta, \phi) \\ &\times \exp[iv\Theta \cos(\phi - \psi) + i\bar{u}\Theta^2/2] \end{aligned} \tag{15}$$

A three-dimensional plot of the field amplitude on a focal plane is shown in Fig. 4, and the streamlines of the Poynting vector in a focal region are represented in Fig. 5.

E. Field near a Caustic

When the field point approaches a *caustic*, then two or more rays having almost equal directions pass through P (see Fig. 6). In this case the diffraction integral I for a two-dimensional field takes the form

$$\begin{aligned} I &\propto \frac{1}{\lambda^{1/2}} \int_{-\infty}^{+\infty} \exp[-ik(as + bs^3)]\, ds \\ &\propto \frac{R^{1/2}}{\lambda^{1/2}} \left(\frac{2}{k\rho_c}\right)^{1/3} \text{Ai}\left[-\left(\frac{k^2\rho'^6}{4\rho_c^4}\right)^{1/3}\right] \end{aligned} \tag{16}$$

Ai(·) being the Airy function. A plot of the field in proximity to the caustic is shown in Fig. 7.

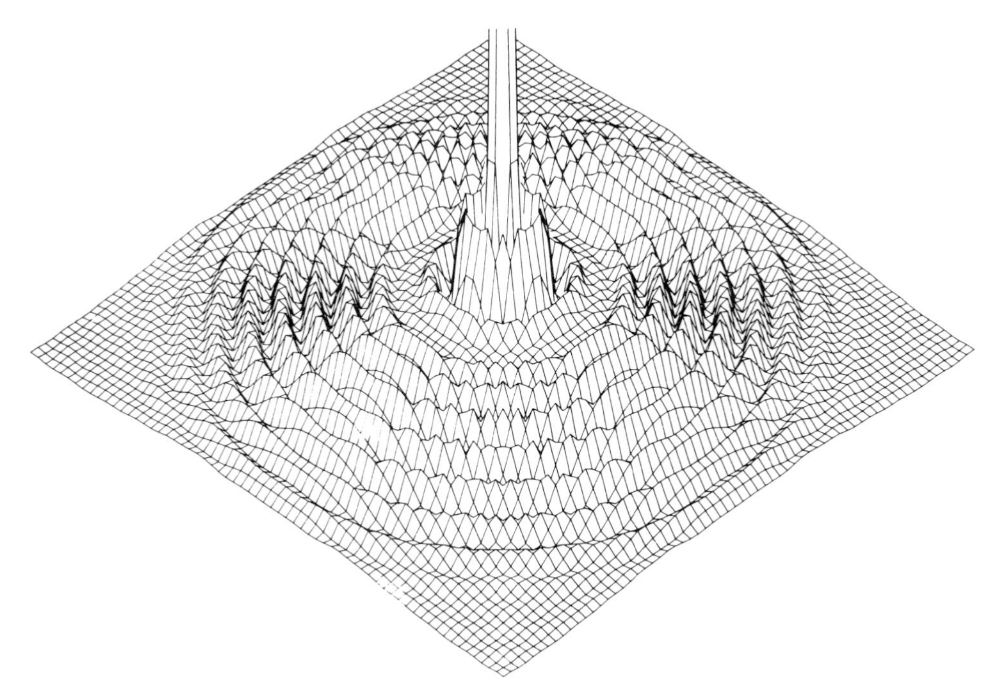

FIG. 4. Diffraction pattern of a ring-shaped aperture with internal diameter 0.8 times the external one, illuminated by a plane wave at a distance corresponding to a Fresnel number of 15. The three-dimensional plot was obtained by using an improved fast Fourier transform algorithm. [From Luchini, P. (1984). *Comp. Phys. Commun.* **31,** 303.]

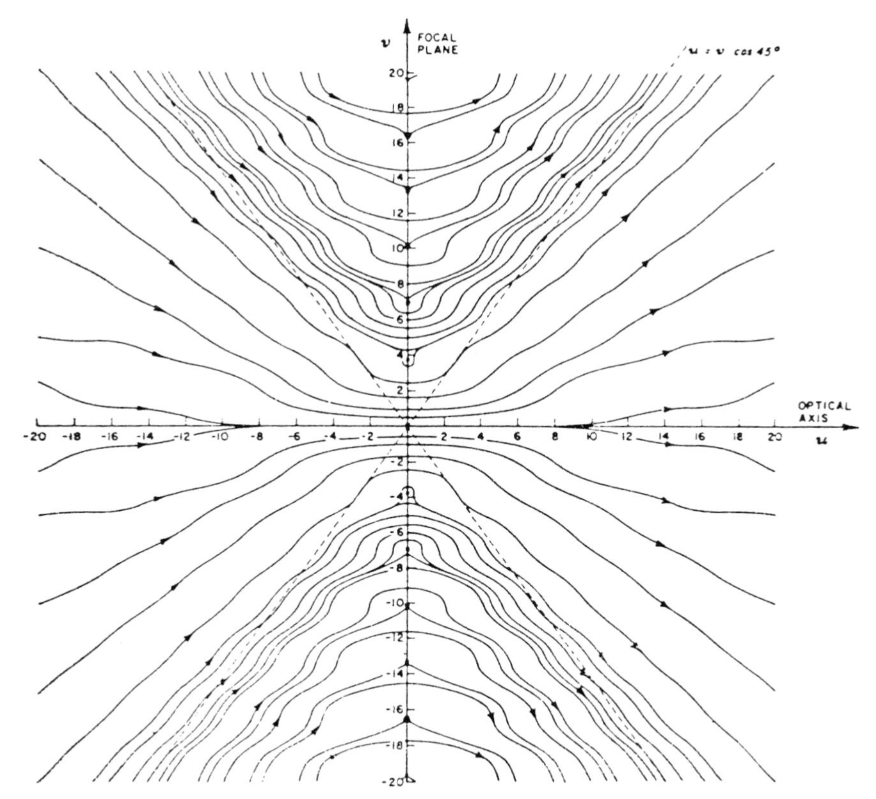

FIG. 5. Flow lines of the Poynting vector near the focus of an aplanatic system with angular semiaperture of 45. [From Boivin, A., Dow, J., and Wolf, E. (1977). *J. Opt. Soc. Am.* **57,** 1171.]

F. Aberrations

In order to describe the diffraction effects of the deviation of an optical system K from ideal behavior, it is convenient to describe the aberrations as a departure of the wave fronts from some ideal surfaces. In particular, we consider a wave front W passing through the center of the exit pupil of K and a reference sphere W_s having the same curvature radius R of W. The distance W_0 between these two surfaces is called the ab-

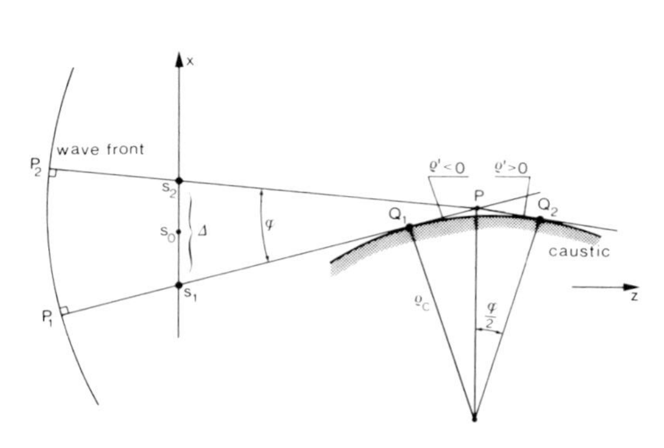

FIG. 6. Geometry for calculation of the field in proximity to a caustic.

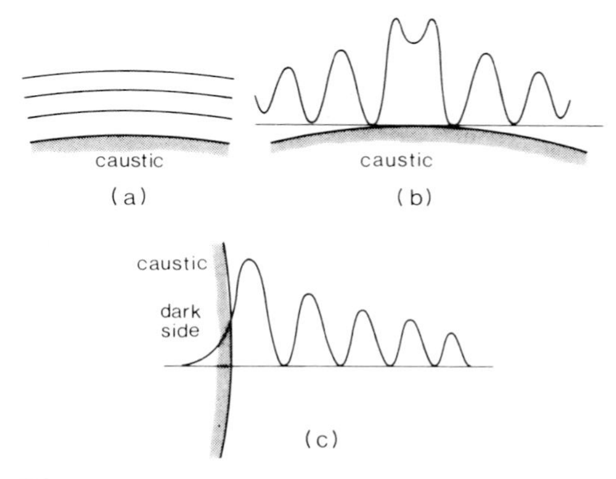

FIG. 7. Fringes in proximity to a caustic (a) and field amplitude along a ray (b) and a normal to a caustic (c).

erration function. As a consequence the Fresnel formula (8) can be rewritten as

$$u(x, y, z) = \frac{i}{\lambda d} e^{-ikkd} \times \iint_{\Sigma} u_{\text{ideal}}(x'y'z') \exp\left\{-i \frac{k}{2d}[(x - x')^2 + (y - y')^2] - ikW_0(x' y')\right\} \quad (17)$$

In particular, the spherical aberration is represented by the function

$$W_0 = B(x^2 + y^2)^2 \quad (18)$$

For small circular apertures the aberration function can be represented at the 4th lowest order in the radial coordinate ρ by the Seidel primary aberrations $W^{(4)}$, that is

$$kW^{(4)}(\rho, \phi) = BR_4^{(0)}(\rho) + CR_2^{(2)}(\rho) \cos 2\theta - DR_2^{(0)}(\rho) + ER_1^{(1)}(\rho) \cos \theta + FR_3^{(1)}(\rho) \cos \theta \quad (19)$$

The radial coordinate ρ is normalized to the radius of the output pupil so that it spans the interval 0–1, and θ stands for the angular coordinate. The five coefficients B, C, D, E, and F refer in order to the spherical aberration, the astigmatism, the field curvature, the distortion, and the coma. The functions $R_m^{(n)}$ are the so-called Zernike polynomials,

$$R_n^{(m)}(z) = \sum_{s=0}^{(n-|m|)/2} (-1)^s \times \frac{(n-s)!}{s![(n+m)/2 - s]![(n-m)/2 - s]!} z^{n-2s} \quad (20)$$

V. Fock–Leontovich Parabolic Wave Equation

Let us represent the field in the form

$$u(\mathbf{r}) = e^{-ik_0 z} A(\mathbf{r}) \quad (21)$$

where z is an arbitrary direction and $A(\mathbf{r})$ satisfies the equation

$$\frac{\partial}{\partial z} A(\mathbf{r}) = -\frac{i}{2k_0}\left(\nabla_t^2 + \frac{\partial^2}{\partial z^2}\right) A \quad (22)$$

where ∇_t^2 stands for the transverse Laplacian.

If A is a slowly increasing function of z, we can neglect the second order derivative of it with respect to z, thus obtaining the Fock–Leontovich parabolic wave equation

$$\frac{\partial}{\partial z} A(\mathbf{r}) = -\frac{i}{2k_0} \nabla_t^2 A(\mathbf{r}) \quad (23)$$

Accordingly, the electromagnetic propagation is reduced to an irreversible diffusive process similar to those associated with wave function evolution in quantum mechanics. It is straightforward to show that the Green's function associated with the above equation coincides with the Fresnel kernel characteristic of paraxial propagation.

A. Hermite–Gauss Beams

We look for a trial solution of the above equation by putting

$$A(\boldsymbol{\rho}, z) = \frac{1}{w(z)} f\left(\sqrt{2}\, \frac{\rho}{w(z)}\right) \times \exp\left[-i((l + m + 1)\psi(z) + \frac{k}{2q(z)} \rho^2)\right] \quad (24)$$

with $\boldsymbol{\rho} = \hat{\mathbf{x}}x + \hat{\mathbf{y}}y$. The terms $w(z)$, $q(z)$, $\psi(z)$, and f are functions to be determined by plugging the above function into the parabolic wave equation. In particular, we choose

$$f = \frac{1}{\sqrt{\pi\, 2^{l+m+1}\, l!m!}} \times H_l\left(\sqrt{2}\, \frac{x}{w}\right) H_m\left(\sqrt{2}\, \frac{y}{w}\right) \quad (25)$$

where

$$\psi(z) = \arctan \frac{2z}{kw_0^2}$$

$$w^2(z) = w_0^2(1 + z^2/z_R^2)$$

with z_R being the so-called Rayleigh length (equivalent to the focal depth) of the beam, which is related to the spot size w_0 in the waist by the relation

$$z_R = \frac{\pi w_0^2}{\lambda} \quad (26)$$

The complex curvature radius q is given by

$$\frac{1}{q} = \frac{1}{z + iz_R} = \frac{1}{R(z)} - i\frac{\lambda}{\pi w^2(z)} \quad (27)$$

with R the curvature radius of the wave front,

$$R(z) = z + \frac{z_R^2}{z} \quad (28)$$

Finally, H_m is the Hermite polynomial of order m.

B. Laguerre–Gauss Beams

If we use polar coordinates, it may be preferable to represent the field as a superposition of Gauss–Laguerre modes,

$$f = \sqrt{\frac{2}{\pi}\frac{p!}{(l+p)!}}\left(\sqrt{2}\,\frac{\rho}{w}\right)^l \times L_p^l\left(2\,\frac{\rho^2}{w^2}\right)\cos\left(l\theta + \varepsilon\,\frac{\pi}{2}\right) \quad (29)$$

with $\varepsilon = 0, 1$ and L_p^l the Laguerre polynomial.

These modes can be used for representing a generic field in the paraxial approximation. In particular, if we expand the field on an aperture we can immediately obtain the transmitted field by using the above expressions for propagating the single modes. For example, we can use gaussian beams having the same curvature of the wave front of the incoming field and a spot size depending on the illumination profile.

C. Expansion in Gauss–Laguerre Beams

If we represent the field on the output pupil of an optical system as a superposition of Gauss–Laguerre modes, we can describe the effects on the diffracted field of the limited size of the pupil and of the aberrations by introducing a matrix. That is,

$$\mathbf{E}_\mathrm{d} = \mathbf{K}\cdot\mathbf{E}_\mathrm{i} \quad (30)$$

where $\mathbf{E}_{\mathrm{i,d}}$ is a vector whose components are the generally complex amplitudes of the Gauss–Laguerre modes composing the incident (diffracted) field. $\mathbf{K}$ is a matrix which can be represented as a product of $\mathbf{K}_\mathrm{p}$, which represents the finite size of the pupil,

$$K_{Ppp'}^{ll'\varepsilon\varepsilon'} = \delta_{pp'}\delta_{ll'}\delta_{\varepsilon\varepsilon'} - \delta_{ll'}\delta_{\varepsilon\varepsilon'}\sqrt{\frac{p!p'!}{(l+p)!(l'+p')!}} \times e^{-u}\sum_{k=0}^{p+p'+l}\frac{u^k}{k!}\sum_{m+m'+l>k}(-1)^{m+m'} \frac{(m+m'+l)!}{m!m'!}\binom{p+l}{p-m}\binom{p'+l}{p'-m'} \quad (31)$$

where $u = 2a^2/w^2$, with a the pupil radius and w the spot size of the gaussian beam in correspondence of it. $\mathbf{K}_\mathrm{a}$ accounts for the aberrations,

$$\mathbf{K}_\mathrm{a} = \exp(ik\mathbf{W})$$

with $\mathbf{W}$ being the aberration matrix. In particular, for spherical aberrations $\mathbf{W}$ takes the form

$$W_{pp'\varepsilon\varepsilon'}^{ll'} \propto \delta_{ll'}\delta_{pp'}\delta_{\varepsilon\varepsilon'} - \delta_{ll'}\delta_{\varepsilon\varepsilon'}\sqrt{\frac{p!p'!}{(l+p)!(l+p')!}} \times \sum_{mm'}\frac{(-1)^{m+m'}}{m!m'!}\binom{p+l}{p-m} \times\binom{p'+l}{p'-m'}(m+m'+l+2)! \quad (32)$$

VI. Geometric Theory of Diffraction

As we know from experience, the edges of an illuminated aperture shine when observed from the shadow region. This fact was analyzed by Newton, who explained it in terms of repulsion of light corpuscles by the edges. In 1896 Arnold Sommerfeld obtained the rigorous electromagnetic solution of the half-plane diffraction problem. Using this result, it can be shown that the total field splits into a geometrical optics wave and a diffracted wave originating from the edge. Subsequently (1917), Rubinowicz recast the (scalar) diffraction integral for a generic aperture illuminated by a spherical wave in the form of a line integral plus a geometrical optics field. Parallel to this development, J. B. Keller (1957) successfully generalized the concept of ray by including those diffracted by the edges of an aperture.

In order to emphasize the geometric character of his approach, Keller called it the *geometric theory of diffraction* (GTD).

A. Diffraction Matrix

The asymptotic construction of diffracted fields will be illustrated by using as example the metallic wedge (see Fig. 8) illuminated by a plane wave having the magnetic field parallel to the edge (*s-wave or TM wave*). It can be shown that for $\rho \to \infty$ the diffracted field is given by

$$u(\rho,\phi) \sim D_\mathrm{s}(\phi,\phi')\frac{e^{-ik\rho}}{\rho^{1/2}} + \sum_q \exp[-ik\rho\cos(\phi-\phi')] \times [U(\phi-\beta_q+2\pi) - U(\phi-\beta_q)] \quad (33)$$

where $U(x)$ is the unit step function and

$$D_\mathrm{s}(\phi,\phi') = -\frac{e^{-i\pi/4}\lambda^{1/2}}{N}\left\{\frac{1}{\cos(\pi/N) - \cos[(\phi-\phi')/N]} + \frac{1}{\cos(\pi/N) - \cos[(\phi+\phi')/N]}\right\} \quad (34)$$

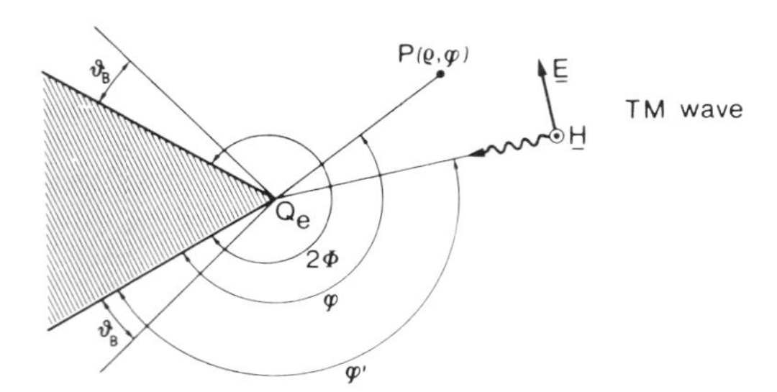

FIG. 8. Geometry of a wedge-shaped region.

where $N = 2\pi/\Phi$ and D_s is the *diffraction coefficient* of the wedge illuminated by an s-wave. For a wave having the electric field parallel to the edge (p-wave) the coefficient D_p can be obtained from D_s by changing to minus the plus sign before the second term on the right side of Eq. (34).

The above results have been extended to a metallic wedge illuminated by rays forming an angle β with the tangent $\hat{\mathbf{e}}$ to the edge at the diffraction point Q_e (see Figs. 9 and 10). It can be shown that the diffraction rays form a half-cone with axis parallel to $\hat{\mathbf{e}}$ and aperture equal to the angle β formed by the incident ray with the edge. If we consider the projection of the incident and diffracted rays on a plane perpendicular to the edge at Q_e, the position of the diffracted rays, forming a conical surface, is given by the angle ϕ_e, while the direction of the incident ray is defined by ϕ'_e. The electric component of the edge-diffracted ray can be expressed in the form

$$\mathbf{E}_d(\mathbf{r}) = [\rho_1/[r(\rho_1 + r)]]^{1/2} e^{-ikr} \times \mathbf{D}(\phi, \phi'; \beta) \cdot \mathbf{E}_i(Q_e) \qquad (35)$$

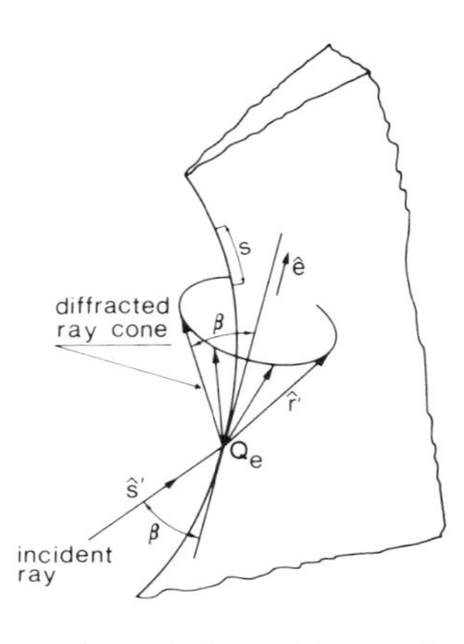

FIG. 9. Cone of rays diffracted by an edge point Q_e.

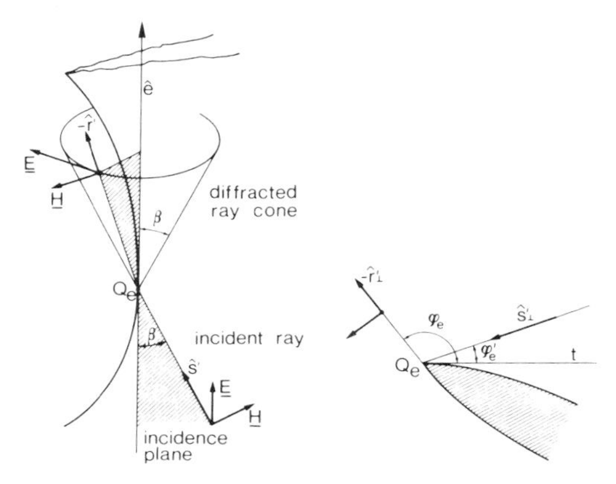

FIG. 10. Geometry for an edge-diffracted field.

where $\mathbf{r}$ represents the distance of the field point P from Q_e, and ρ_1 stands for the distance of P from the focal point along the diffracted ray. The *diffraction matrix* $\mathbf{D}$ has been derived by Kouyoumjian and Pathak and put in the form

$$\mathbf{D}(\phi, \phi'; \beta) = \hat{\boldsymbol{\beta}}_d \hat{\boldsymbol{\beta}}_i D_p + \hat{\boldsymbol{\phi}} \hat{\boldsymbol{\phi}}' D_s \qquad (36)$$

where $\hat{\boldsymbol{\beta}}_d = \hat{\boldsymbol{\phi}} \times \hat{\mathbf{r}}$, $\hat{\boldsymbol{\beta}}_i = \hat{\boldsymbol{\phi}}' \times \hat{\mathbf{s}}'$, and D_p and D_s, which are a generalization of the diffraction coefficient (19) for $\beta \neq \pi/2$, are given by

$$D_{\substack{p\\s}} = \frac{e^{-i\pi/4} \sin(\pi/N) \lambda^{1/2}}{N \sin \beta} \times \left[-\frac{1}{\cos(\pi/N) - \cos[(\phi - \phi')/N]} \pm \frac{1}{\cos(\pi/N) - \cos[(\phi + \phi')/N]} \right] \qquad (37)$$

B. Diffraction from a Slit

The GTD formalism can be applied conveniently to the calculation of the field diffracted by a slit of width $2a$ and infinite length. For simplicity, we assume a plane incident wave normal to the edges. As a first approximation we take the field on the aperture coincident with the incident field (Kirchhoff approximation). Then, following J. B. Keller, we can say that the field point P at a finite distance is reached by two different rays departing from the two edges and by a geometrical optics ray, if any (see Fig. 11a). The contribution of the diffracted rays can be expressed in the form

$$\mathbf{E}_d(\theta) = \frac{e^{-ik\rho}}{\lambda^{1/2}} \left[\mathbf{D}\left(\frac{3}{2}\pi + \theta, \frac{\pi}{2}; \frac{\pi}{2}\right) + \mathbf{D}\left(\frac{3}{2}\pi - \theta; \frac{\pi}{2}; \frac{\pi}{2}\right) \right] \cdot \mathbf{E}_i \qquad (38)$$

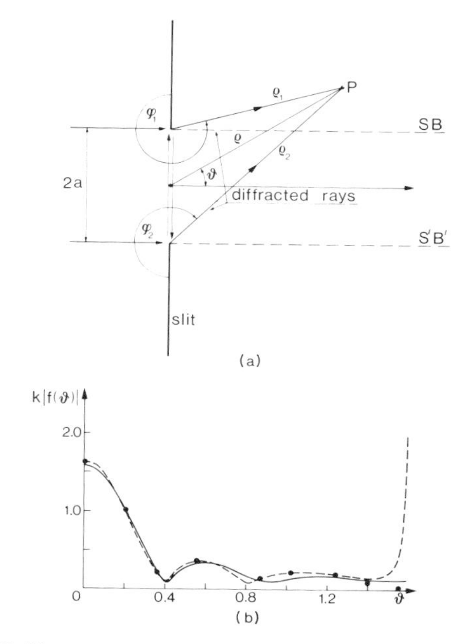

FIG. 11. Diffraction by a slit of width $2a$ (a) and relative far-field pattern (b) for a p-wave and $ka = 8$. The solid curve results from single diffraction; the dashed curve includes the effects of multiple diffraction. The dots represent the exact solution. [From Keller, J. B. (1957). *J. Appl. Phys.* **28,** 426.]

A point that deserves comment is the use of the Kirchhoff approximation, which can be considered valid when the slit width is much larger than the field wavelength. This approximation can be improved, as shown by Keller, by taking into account the multiple diffraction undergone by the rays departing from each edge and diffracted from the opposite one (see dotted line in Fig. 11b).

VII. Coherent and Incoherent Diffraction Optics

An ideal imaging system can be described mathematically as a mapping of the points of the object plane Π_o into those of the image plane Π_i. For a finite wavelength and delimited pupil a unit point source located at (x_0, y_0) produces a field distribution $K(x, y; x_0, y_0)$ called the *impulse response,* which differs from a delta function $\delta^{(2)}(x - \bar{x}, y - \bar{y})$ centered on the Gaussian image of the object having coordinates $(\bar{x}, \bar{y})$. As a consequence, diffraction destroys the one-to-one correspondence between the object and the image.

The departure of K from a delta function introduces an amount of uncertainty in the reconstruction of an object through its image. This is indicated by the fact that two point sources are seen through an optical instrument as clearly separate only if their distance is larger than a quantity W roughly coincident with the dimension of the region on Π_i where K is substantially different from zero. The parameter W, which measures the smallest dimension resolved by an instrument, is proportional to the wavelength. This explains the increasing popularity of UV sources, which have permitted the implementation of imaging systems capable of resolutions better than 0.1 μm, a characteristic exploited in microelectronics for the photolithographic production of VLSI circuits.

A. Impulse Response and Point Spread Function

Let us consider a unit point source in (x_0, y_0, z_0) producing a spherical wave front transformed by a composite lens into a wave converging toward the paraxial image point $(\bar{x}, \bar{y}, z)$. Using the Luneburg–Debye integral, we can express the impulse response $K(x, y; x_0, y_0)$ on Π_i as an integral extended to the wave front of the converging wave:

$$\begin{aligned} K(x, y; x_0, y_0) &= \exp\left(-ik\frac{x_0^2 + y_0^2}{2d_o} - ik\frac{\bar{x}^2 + \bar{y}^2}{2d}\right. \\ &\quad \left. + iv_x\bar{p} + iv_y\bar{q}\right) \times \frac{\Omega}{2\pi} \\ &\quad \times \iint P(p, q)A(p, q) \\ &\quad \exp[i(v_x p + v_y q)]\, dp\, dq \\ &\equiv e^{-i\Phi}\bar{K}(x - \bar{x}, y - \bar{y}) \end{aligned} \qquad (39)$$

where $v_x = k(x - \bar{x})$ NA and $v_y = k(y - \bar{y})$ NA are the optical coordinates of the point (x, y) referred to the paraxial image $(\bar{x}, \bar{y})$, NA is the numerical aperture of the lens in the image space, P the pupil function, p and q are proportional to the optical direction cosines of the normal to the converging wave front in the image space in such a way that $p^2 + q^2 = 1$ on the largest circle contained in the exit pupil, and d_o and d are the distances of the object and the image from the respective principal planes.

Physically we are interested in the intensity of the field, so it is convenient to define a *point spread function* $t(x, y; x_0, y_0)$ given by

$$t(x, y; x_0, y_0) = |K(x, y; x_0, y_0)/K(\bar{x}, \bar{y}; x_0, y_0)|^2 \qquad (40)$$

In particular, for *diffraction-limited instruments* with circular and square pupils, respectively, $\bar{K}$ is

$$\bar{K} \propto 2J_1(v)/v \qquad \bar{K} \propto \frac{\sin v_x}{v_x}\frac{\sin v_y}{v_y} \tag{41}$$

with

$$v_x = k\text{NA}(x + Mx_0), \qquad v_y = k\text{NA}(y + My_0)$$
$$v = (v_x^2 + v_y^2)^{1/2} \tag{42}$$

M being the *magnification* of the lens and J_1 the Bessel function of first order. [*See* OPTICAL SYSTEMS DESIGN.]

B. Coherent Imaging of Extended Sources

We can apply the superposition principle to calculate the image field $i(x, y, z)$ corresponding to the extended object field $o(x_0, y_0, z_0)$ defined on the plane region Σ_0, obtaining

$$i(x, y, z) = \iint_{\Sigma_0} K(x, y, z; x_0, y_0, z_0)o(x_0, y_0, z_0) \times dx_0\, dy_0 \tag{43}$$

If we are interested in the intensity, we have

$$|i(x, y, z)|^2 = \iint_{\Sigma_0} dx_0\, dy_0 \iint_{\Sigma_0} dx_0'\, dy_0' \times K(x, y, z; x_0, y_0, z_0) \times K^*(x, y, z; x_0', y_0', z_0) \times o(x_0, y_0, z_0) \tag{44}$$

C. Optical Transfer Function

If we are interested in the intensity distribution $I(x, y) \propto \langle i(x, y)i^*(x, y)\rangle$ Eq. (44) gives for isoplanatic systems

$$I(x, y) = \frac{1}{M^2}\iint_{\Sigma_0} I_0\left(-\frac{x_0}{M}, -\frac{y_0}{M}\right) E(x - x_0, y - y_0)\, dx_0\, dy_0 \tag{45}$$

where M is the magnification. With Fourier transformation, this relation becomes

$$\tilde{I}(\alpha, \beta) \propto T(\alpha, \beta)\tilde{I}_0(\alpha, \beta) \tag{46}$$

where $\tilde{I}$ is the two-dimensional Fourier transform of $I(x, y)$ and $T(\alpha, \beta)$ the *optical transfer function* (OTF) of the systems (see Fig. 12). For

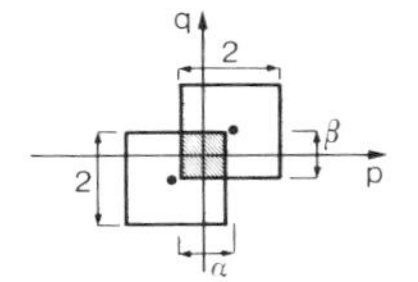

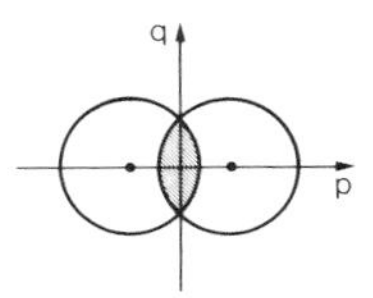

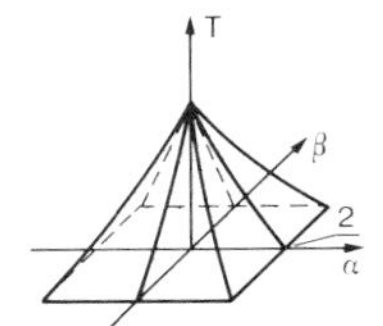

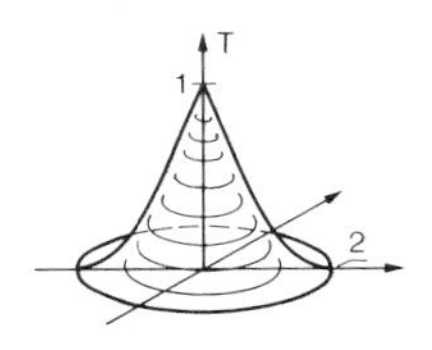

FIG. 12. Optical transfer functions related to incoherent illumination of square (left) and round (right) pupils. T is proportional to the convolution of the pupil function [see top diagram and Eq. (47)].

an isoplanatic system, $T(\alpha, \beta)$ is proportional to the Fourier transform of the point-spread function t, so in absence of aberrations

$$T(\alpha, \beta) = \frac{\iint_{-\infty}^{+\infty} \bar{t}(v_x, v_y)\exp(i\alpha v_x + i\beta v_y)\, dv_x\, dv_y}{\iint_{-\infty}^{+\infty} \bar{t}(v_x, v_y)\, dv_x\, dv_y} = \frac{\iint_{-\infty}^{+\infty} P(p + \alpha/2, q + \beta/2)\, P(p - \alpha/2, q - \beta/2)\, dp\, dq}{\exp[-ik \iint_{-\infty}^{+\infty} P(p, q)\, dp\, dq} \tag{47}$$

by virtue of convolution and Parseval's theorems. The modulus of T is called the *modulation transfer function* (MTF).

For a circular pupil, the OTF reads

$$T(\omega) = \begin{cases} (2/\pi)[\arccos(\omega/2) - \omega(1 - \omega^2/4)^{1/2}] & 0 \le \omega \le 2 \\ 0 & \text{otherwise} \end{cases} \tag{48}$$

where $\omega = (\alpha^2 + \beta^2)^{1/2}$. Then for a circular pupil the spatial frequency ω is limited to the interval (0, 2). The dimensionless frequency ω, conjugate of the optical coordinate v, is related to the *spatial frequency* f, expressed in cycles per unit length, by $f = k\omega$ NA. The expression just given for the OTF indicates that the diffraction sets an upper limit to the ability of the system to resolve a bar target with a normalized spatial frequency greater than 2.

BIBLIOGRAPHY

Barakat, R. (1980). *In* "The Computer in Optical Research" (B. R. Frieden, ed.) p. 35–80. Springer-Verlag, Berlin and New York.

Born, M. and Wolf, E. (1970). "Principles of Optics." Pergamon, Oxford, 1970.

Duffieux, P. M. (1983). "The Fourier Transform and its Application to Optics." Wiley, New York.

Gaskill, J. D. (1978). "Linear Systems, Fourier Transforms and Optics," Wiley, New York.

Hansen, R. C., ed. (1981). "Geometric Theory of Diffraction." IEEE Press, New York.

Kouyoumjan, R. G., and Pathak, P. H., *Proc. IEEE* **62,** 1448 (1974).

Northover, F. H., "Applied Diffraction Theory." Amer. Elsevier, New York, 1971.

S. Solimeno, B. Crosignani and P. Di Porto, "Guiding, Diffraction and Confinement of Optical Radiation." Academic Press, Orlando, 1986.

ELECTRODYNAMICS, QUANTUM

W. P. Healy *University College London and Royal Melbourne Institute of Technology*

GLOSSARY

Anomalous magnetic moment: Difference between the intrinsic magnetic moment of a charged spin-$\frac{1}{2}$ particle and that predicted by the single-particle Dirac theory.

Coherent state: State of the quantized radiation field in which the average electric and magnetic fields and the average energy are the same as the corresponding quantities for a state of the classical electromagnetic field.

***C, P,* and *T* symmetries:** Invariance of quantum electrodynamics under the operations of charge conjugation (or matter–antimatter interchange), parity (or left–right interchange), and time reversal, respectively.

Dirac equation: Four-component relativistic quantum mechanical wave equation for a spin-$\frac{1}{2}$ particle.

Einstein's *A* and *B* coefficients: Factors determining the rates of spontaneous emission, induced emission, and absorption of radiation by atoms.

Feynman diagram: Pictorial representation of a process in quantum electrodynamics in which states of particles or atoms are depicted as lines and their interactions as vertices where two or more lines meet.

Gauge invariance: Independence of a quantity of the choice of potentials used to represent the electromagnetic field.

Lamb shift: Change in atomic energy levels (from the values predicted by the single-particle Dirac theory) caused by electromagnetic interactions, or the splitting of spectral lines due to this change.

Leptons: Spin-$\frac{1}{2}$ particles subject to the weak and, if charged, the electromagnetic force, but not subject to the strong nuclear force, and including electrons, muons, tauons, neutrinos and their antiparticles.

Maxwell's equations: General fundamental equations for the electromagnetic field, summarizing the basic laws of electromagnetism.

Occupation-number state: State of a quantized field (such as the Maxwell or Dirac field) that has a definite number of particles or quanta in each field mode.

Photon: Particle or quantum of the electromagnetic field that travels at the speed of light, has no charge or rest mass, and has intrinsic spin 1.

Renormalization: Elimination of unobservable mass and charge of bare particles in favor of observed mass and charge of physical particles.

***S*-matrix element:** Probability amplitude for a scattering process in which the incoming and outgoing particles are specified by their momenta and polarization or spin states.

Quantum electrodynamics is the fundamental theory of electromagnetic radiation and its interaction with microscopic charged particles, particularly electrons and positrons. In its most accurate form, the theory combines the methods of quantum mechanics with the principles of special relativity; often, however, it is sufficient to treat the charged particles in nonrelativistic approximation. Each part of the complete dynamical system of radiation and charges displays a characteristic wave–particle duality. Thus, electrons behave in many circumstances as particles, but they can also exhibit wave properties such as interference and diffraction. Similarly electromagnetic radiation, which was considered classically as a wave field, may have particle properties ascribed to it under suitable conditions, e.g., in scattering experiments. The particles or quanta associated with the electromagnetic field are called photons. Quantum electrodynamics is a highly successful theory, despite certain mathematical and interpretational difficulties inherent in its formulation. Its success is due in part to the weakness of the coupling between the radiation and the charges, which makes possible a perturbative treatment of the interaction of the two parts of the system. The theory accounts for many phenomena, including the emission or absorption of radiation by atoms or molecules, the scattering of photons or electrons, and the creation or annihilation of electron–positron pairs. Its most famous predictions concern the electromagnetic shift of energy levels observed in atomic spectra and the anomalous magnetic moment of the electron; both of these predictions are in good agreement with experimental results. Quantum electrodynamics also includes the interaction of photons with muons and tauons (which differ from electrons only in mass) and their antiparticles. The validity of the theory has been tested in high-energy collision experiments involving these particles down to distances less than 10^{-15} cm.

I. Introduction

A. Early Theories of Light

Since quantum electrodynamics is the modern theory of electromagnetic radiation, including visible light, it is instructive to begin with a brief historical review of previous theories. The nature of light has long been a subject of interest to philosophers and scientists. In the fifth century B.C., Empedocles of Acragas held that light takes time to travel from one place to another but that we cannot perceive its motion. He knew that the moon shines by light reflected from the sun and was also aware of the cause of solar eclipses. Heron of Alexandria, who is thought to have lived in the first or second century A.D., discussed the rectilinear propagation properties of light. In his book *Catoptrica* he derived the law of reflection using a principle of minimal distance. The law of refraction was not formulated until 1621, when it was discovered experimentally by Snell. Snell's law was later derived theoretically from Fermat's celebrated principle of least time.

From about the middle of the seventeenth century to the end of the nineteenth century there were two competing, and mutually contradictory, theories of light. The wave theory was initiated by Hooke and Huygens following the first observations of interference and diffraction. Huygens enunciated a principle, based on the wave theory, from which he derived the laws of reflection and refraction. He also discovered the polarization properties of light. These properties, as well as the law of rectilinear propagation, were difficult to explain by the wave theory, which at that time dealt only with longitudinal waves in a hypothetical "aether," analogous to sound waves in air. These difficulties led Newton to propose a corpuscular theory, according to which light is emitted from luminous bodies in a stream of small particles or corpuscles. Newton's views inhibited any further advances in the wave theory until about the beginning of the nineteenth century. In the meantime, the fact that light has a finite speed was confirmed by Römer through observations of eclipses of the moons of Jupiter. This occurred in 1675, more than two millenia after the time of Empedocles. (The speed of light in empty space is denoted by c and is approximately 2.998×10^{10} cm/sec in cgs units.)

B. Classical Electrodynamics

The revival of the wave theory began with Young's interpretation of interference experiments. In particular, the destructive interference of two light beams at certain points in space seemed totally inexplicable on the corpuscular hypothesis but was readily accounted for by the wave theory. Young also suggested that light waves execute transverse rather than longitudinal vibrations, as this could then explain the observed polarization properties. The wave theory was further developed by Fresnel, who applied it to phenomena involving diffraction, interference of polarized light, and crystal optics. An important test of the theory was provided by the comparison of the speeds of light in media with different refractive indexes. According to the wave theory light travels slower in an optically denser medium, but according to the corpuscular theory it travels faster. The results of experiments carried out in 1850 agreed with the predictions of the wave theory.

The wave theory was in a certain sense completed when Maxwell established his equations for the electromagnetic field and showed that they have solutions corresponding to transverse electromagnetic waves in which both the electric and magnetic induction field vectors oscillate perpendicularly to the direction of propagation. The speed of these waves in empty space could be calculated from constants (the permittivity and permeability of the vacuum) obtained by purely electric and magnetic measurements and was found to be the speed of light. This conclusion became the basis of the electromagnetic theory of light. It was subsequently found that the frequencies of visible light form only a small part of the complete spectrum of electromagnetic radiation, which also includes radio waves, microwaves and infrared radiation on the low-frequency side, and ultraviolet radiation, X-rays, and gamma rays on the high-frequency side.

C. Photons

Despite the success of classical electromagnetic theory in dealing with the propagation, interference, and scattering of light, experiments carried out about the end of the nineteenth century and the beginning of the twentieth century led to the reintroduction of the corpuscular theory, though in a form different to that proposed by Newton. The departure from classical concepts began in 1900 when Planck published his law of black-body radiation. In this law the quantum of action h (approximately 6.626×10^{-27} erg sec), now known as Planck's constant, made its first appearance in physics. Planck's law for the variation with frequency of the energy in black-body radiation at a given temperature is closely related to the existence of discrete energy levels for the electromagnetic field, even though Planck, in his original derivation of the law, did not consider the field itself to be quantized. A black body is one that absorbs all the electromagnetic energy incident on it. It was shown by Kirchoff in 1860 that when such a body is heated, the emitted radiation does not depend on the detailed composition of the body but only on its absolute temperature. Radiation confined in a state of thermal equilibrium in a cavity with perfectly reflecting walls behaves as black-body radiation. According to classical electromagnetic theory, the cavity radiation can undergo simple harmonic motion at a number of certain allowed or characteristic frequencies ν, the values of which depend on the shape and size of the enclosure. These so-called radiation oscillators may be quantized, as in ordinary quantum mechanics. Then for each nonnegative integer n, an oscillator with frequency ν has a nondegenerate stationary state with energy $nh\nu$ above the ground-state energy (see Fig. 1). The possible values of the energy at this frequency thus form a discrete set $0, h\nu, 2h\nu, 3h\nu, \ldots$ instead of a continuum. It can be shown that quantization of the oscillators in this way for all the allowed frequencies leads directly to Planck's law.

In 1905, Einstein made use of the idea of light quanta in order to explain the photoelectric effect and later applied it to the emission as well as the absorption of radiation by atoms. The light quantum hypothesis states not only that the energy of monochromatic radiation of frequency ν is made up of integral multiples of the quantum $h\nu$, but also that the momentum is made up of integral multiples of the quantum h/λ, where λ is the wavelength of the radiation (ν and λ are related by the equation $\nu\lambda = c$). This contrasts sharply with the classical picture in which the energy and momentum are regarded as continuously variable. The existence of discrete light quanta, or photons, is not immediately evident on a macroscopic scale, however. Due to the smallness of Planck's constant, even in a weak electromagnetic field there is an enormous number of photons, provided the frequency is not too high. For example, black-body radiation at a

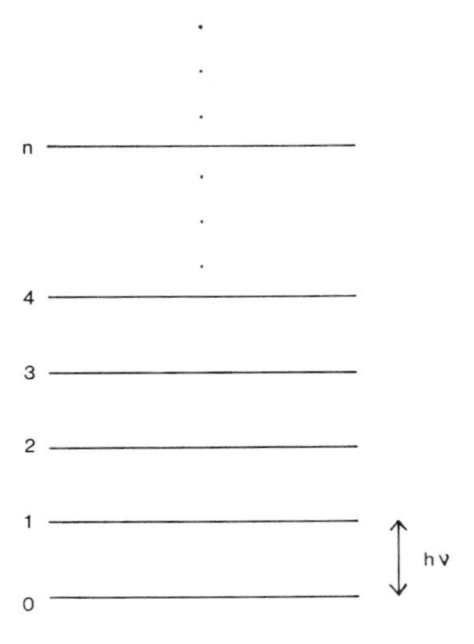

FIG. 1. The horizontal lines represent the discrete energy levels of a quantized harmonic oscillator of frequency ν. The energy of the ground state is taken to be 0 and the equally spaced levels $0, h\nu, 2h\nu, \ldots, nh\nu, \ldots$ are labelled by the quantum numbers $0, 1, 2, \ldots, n, \ldots$, respectively. Excitation of the radiation field at frequency ν to level $nh\nu$ corresponds to the addition of n photons, each with energy $h\nu$, to the field.

temperature of 300 K (room temperature) contains about 5.5×10^8 photons/cm^3, most of which correspond to frequencies in the infrared part of the electromagnetic spectrum. At a temperature of 6000 K (roughly that at the surface of the sun), the bulk of the radiation has frequencies in the visible spectrum, and there are about 4.4×10^{12} photons/cm^3. (The total number of photons in black-body radiation is proportional to the cube of the absolute temperature.)

Individual photons manifest themselves only through their interaction with atomic systems. According to Einstein's treatment of the absorption and emission of radiation, for example, an atom in a stationary state can make a transition to a lower or a higher energy level accompanied by the creation or annihilation, respectively, of a photon. If the atomic energies are E_r and E_s, where $E_r > E_s$, then the energy $h\nu$ of the photon must equal the difference $E_r - E_s$ (see Fig. 2). This is called Bohr's frequency condition and is equivalent to the law of conservation of energy applied to the complete system of atom and radiation; any energy lost or gained by the atom is given up to or abstracted from the radiation field in the form of photons. It should be noted that the number of photons in the radiation field need not be constant—photons can be created or annihilated through the interaction of the field with atoms.

The scattering of X rays by free electrons also furnishes direct evidence for the corpuscular properties of radiation. In 1922, Compton discovered that when X rays of wavelength λ are incident on a graphite target, the scattered X rays have intensity peaks at two wavelengths, λ and λ', where $\lambda' > \lambda$. The shift in wavelength given by $\Delta\lambda = \lambda' - \lambda$ is a function of the angle of scattering (i.e., the angle between the direction of the incident and scattered X rays) but is independent of wavelength and the target material. The X rays with unchanged wavelength λ were understood to have been elastically scattered by atoms, which suffer no appreciable recoil, and they could readily be accounted for on the basis of classical electrodynamics. The scattered X rays with shifted wavelength λ', however, required a new interpretation. If it is assumed that the incident X rays consist of photons, then these may collide with essentially free electrons in the target. In this case a photon gives up some of its energy $h\nu$ to an electron and is scattered with a lower frequency ν' and a longer wavelength λ', where $\nu'\lambda' = c$.

The wavelength shift $\Delta\lambda$ can be calculated as a function of scattering angle by using the laws of conservation of energy and momentum. By treating the problem relativistically and taking the electron to be at rest initially (see Fig. 3), it is not difficult to show that $\Delta\lambda$ depends on the scattering angle θ alone through the formula

$$\Delta\lambda = \lambda_c(1 - \cos\theta)$$

where the constant λ_c is the Compton wavelength given by

$$\lambda_c = h/mc \simeq 2.43 \times 10^{-10} \text{ cm}$$

Here m is the rest mass of the electron (approximately 9.11×10^{-28} g). This formula was verified experimentally. The energy and distribution of the recoil electrons and scattered X rays were also in accord with the predictions of the photon theory. [*See* DIFFRACTION, OPTICAL.]

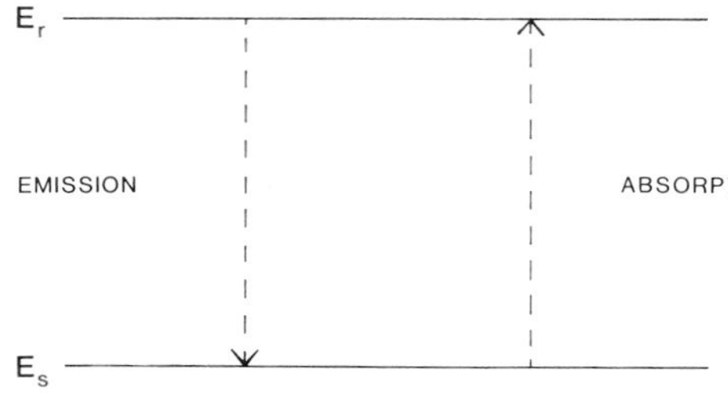

E_r

EMISSION

ABSORPTION

E_s

FIG. 2. An atom can make a transition from a higher energy level E_r to a lower level E_s while emitting a photon of frequency ν, where $h\nu = E_r - E_s$. The emission may be spontaneous or induced by radiation. The atom can make the upward transition from level E_s to level E_r by absorbing a photon of frequency ν.

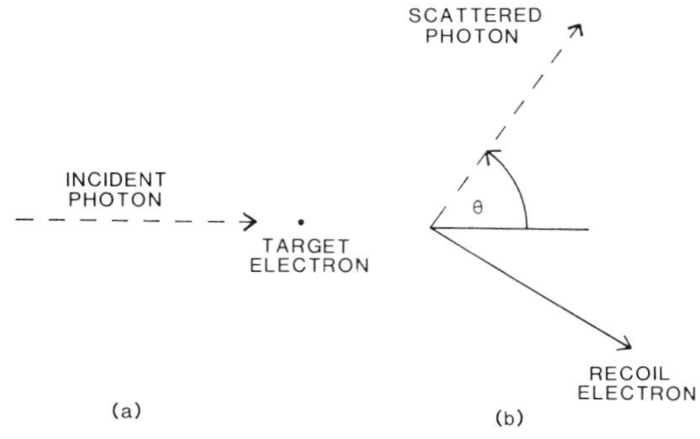

FIG. 3. Photon and electron in the Compton effect (a) before collision and (b) after collision. The scattering angle θ is the angle between the initial and final directions of the photon.

D. Quantum Electrodynamics

The use of the photon concept to explain certain phenomena does not imply a return to a naive classical particle view of light and other forms of electromagnetic radiation. A proper account must also be given of the wave properties of radiation, such as interference and diffraction. Indeed, the formulas for the energy and momentum of the photons—$h\nu$ and h/λ—are based on the assumption that the photons are associated with waves of definite frequency and wavelength. The nature of electromagnetic radiation is such that it appears, *under different experimental conditions*, sometimes to have particle properties and sometimes to have wave properties—these two aspects are said to be complementary. A single coherent theory that encompassed the dual nature of radiation, and with it settled the age-old controversy between the wave theory and the corpuscular theory, was made possible only by the development of quantum mechanics in the mid-1920s. In 1927, Dirac used the new methods of quantization, which had been successfully applied to atomic systems, to quantize the radiation field enclosed in a cavity, and was thus able to give a fully dynamical treatment of the emission and absorption of light by atoms. The beginning of quantum electrodynamics may be taken to date from this time.

The wave properties of radiation can be adequately described by using Maxwell's equations for the electromagnetic field, and these are retained as operator equations in the quantum theory of radiation. Suppose, for example, that the electromagnetic field has a node (i.e., a point where the field amplitudes always vanish) due to interference at P. Then an atom placed at P has, in so far as it can be regarded as a geometrical point, zero probability of absorbing a photon from the field. The field amplitudes are, however, subject to uncertainty relations, involving Planck's constant h, which are analogous to the Heisenberg uncertainty relations for the position and momentum of a particle in ordinary quantum mechanics. The origin of the uncertainty relations for the fields may be understood by considering a simple example.

Let $\bar{\mathscr{E}}$ denote the average value of a component of the electric field over a volume V and a time interval T. (Since a field component at a definite point in space and a definite instant of time appears an abstraction from physical reality, only such average values need be considered.) Now $\bar{\mathscr{E}}$ may be found by measuring the change produced by the field in the momentum of a charged test body occupying the volume during this time. Although the position and momentum of the test body are uncertain by amounts Δq and Δp that satisfy the Heisenberg uncertainty relation $\Delta q\, \Delta p \sim h$, it can be shown that this does not impair the accuracy of the field measurement, provided a sufficiently massive and highly charged body (which is therefore part of a macroscopic measuring instrument) is used. The charge Q on the body must be such that the product $Q\Delta q$ is large; $\bar{\mathscr{E}}$ can then be measured to any desired accuracy $\Delta\bar{\mathscr{E}}$. However, in the measurement of two average field strengths $\bar{\mathscr{E}}$ and $\bar{\mathscr{E}}'$ (taken over two space regions V and V' during two times intervals T and T', respectively), it may not be possible to make both $\Delta\bar{\mathscr{E}}$ and $\Delta\bar{\mathscr{E}}'$ as small as is desired. If the separation distance L between V and V' (see Fig. 4) is such that most light signals emitted from V during the time interval T will reach V' during the time interval T', then the measurement of $\bar{\mathscr{E}}$ will influence that of $\bar{\mathscr{E}}'$ in a way that is to some extent unknown. For the field produced by the test body used to measure $\bar{\mathscr{E}}$ is superimposed on $\bar{\mathscr{E}}'$ and cannot be fully subtracted out, as its value is somewhat uncertain (due to the uncertainty Δq in the position of the test body). This field, and hence $\Delta\bar{\mathscr{E}}'$, can indeed be made as small as is desired by making the product $Q\Delta q$ sufficiently small, but then $\Delta\bar{\mathscr{E}}$ becomes relatively large. The experimental conditions for measurements of $\bar{\mathscr{E}}$ and $\bar{\mathscr{E}}'$ are complementary—those that serve to measure $\bar{\mathscr{E}}$ more precisely will measure $\bar{\mathscr{E}}'$ less precisely, and vice versa. The order of magnitude of the uncertainly product is given by

$$\Delta\bar{\mathscr{E}}\, \Delta\bar{\mathscr{E}}' \sim h/L^3T$$

and is independent of both Q and Δq. Thus, only for well-separated regions or over long intervals of time can both averages be measured with unlimited accuracy.

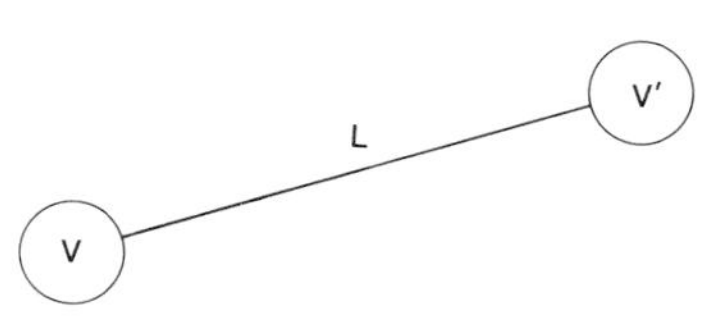

FIG. 4. Regions V and V' in which the average electric fields $\bar{\mathscr{E}}$ and $\bar{\mathscr{E}}'$ are measured during time intervals T and T', respectively. The separation distance L is such that most light signals emitted from V during the interval T will reach V' during the interval T'.

E. Electrons and Positrons

Dirac's original radiation theory had to be modified to bring it into line with the special theory of relativity. This was true particularly of the treatment of the charged particles with which the electromagnetic field interacts. In 1928, Dirac had developed a one-particle relativistic wave equation for the electron that automatically accounted for the observed electron spin and predicted values for the fine structure of the energy levels of the hydrogen atom and of hydrogen-like ions that were in agreement with the experimental data of that time. The Dirac equation, however, also has extraneous solutions corresponding to negative-energy states. To eliminate these, Dirac introduced in 1930 the so-called hole theory, according to which most of the negative-energy states are occupied, each having one electron. Any unoccupied states, or holes, may be interpreted as particles with positive energy and positive charge. These particles were at first thought by Dirac to be protons, but were later identified with positrons, or antiparticles of electrons.

The experimental discovery of the positron by Anderson in 1932 lent support to Dirac's hole theory. Nevertheless, difficulties remained, such as the infinite (but unobservable!) charge density associated with the "sea" of negative-energy electrons. These difficulties can be removed, however, by treating Dirac's one-particle wave equation for the electron as a field equation and subjecting it to a process of quantization, similar in some respects to the quantization of the classical electromagnetic field. This method, which is often referred to as second quantization, was applied to the Dirac equation by Heisenberg and others and resulted in the appearance of electrons and positrons, on an equal footing, as quanta of the Dirac field, just as photons appear as quanta of the Maxwell field. There are, however, some differences between the methods of second quantization used for the Dirac and Maxwell fields, which stem from the different characteristics of the associated particles or quanta. Photons have zero rest mass (but have nonzero momentum because they travel at the speed of light), are electrically neutral, have spin 1 (in units of $\hbar$, which is Planck's constant divided by 2π), and are bosons (i.e., any number of photons can occupy a given state). Electrons and positrons have the same nonzero rest mass, carry equal but oppositely signed charges (by convention, this is negative for the electron and positive for the positron), have spin $\frac{1}{2}$, and are fermions (i.e., not more than one electron or positron can occupy a given state). It was shown by Pauli that there is a connection between the spin of a particle and its so-called statistics—particles with integer spin are bosons and are not subject to the exclusion principle, whereas particles with half odd-integer spin are fermions and are subject to the exclusion principle. It is necessary for photons to be bosons in order that the quantized electromagnetic field may have a classical counterpart, which is realized in the limit of large photon occupation numbers. The quantized Dirac field, on the other hand, does not have a physically realizable classical limit.

F. Divergences and Renormalization

In quantum electrodynamics, as in classical electrodynamics, there are no known exact solutions to the equations for the complete dynamical system of radiation and charges. Indeed, from a purely mathematical viewpoint, the question of even the existence of such solutions is still an open one. Approximate solutions may be found by assuming that the coupling between the two parts of the system is weak and using perturbation theory. This is justified by the smallness of the fine-structure constant α, which gives a measure of the strength of the coupling:

$$\alpha = e^2/4\pi\hbar c \approx \tfrac{1}{137}$$

where e is the magnitude of the charge on the electron and rationalized cgs units are being used ($e \approx 1.355 \times 10^{-10}$ g$^{1/2}$ cm$^{3/2}$/sec).

It was found in the 1930s and 1940s that the calculations for many processes, when taken beyond the first approximation, gave divergent results. Some divergences (the so-called infrared divergences) were due to deficiencies in the approximation method itself. Others (ultraviolet divergences) were associated with the problem of the structure and self-energy of the electron and other elementary particles. This problem had also arisen in classical electrodynamics, where the electron was assumed to have a structure-dependent electromagnetic contribution included in its inertial mass. In quantum electrodynamics, however, there occurred additional divergences of a radically different nature, due to effects that have no classical analogues. For example, the possibility of electron–positron pair creation gave rise to an infinite vacuum polarization in an external field and also implied an infinite self-energy for the photon.

The need to extract finite results from the formalism became acute when refinements in experimental technique revealed small discrepancies between the observed fine structure of the energy levels of atomic hydrogen and that given by Dirac's one-particle relativistic wave equation. These differences, whose existence had been suspected for some time, were measured accurately by Lamb and Retherford in 1947. In the same year, Kusch and Foley found that the value of the intrinsic magnetic moment of an electron in an atom also differs slightly from that predicted by the Dirac theory. To show that these discrepancies could be explained as radiative effects in quantum electrodynamics, it was necessary first to recognize that the mass and charge of the bare electrons and positrons that appear in the formalism cannot have their experimentally measured values. Since the electromagnetic field that accompanies an electron, for example, can never be "switched off," the inertia associated with this field contributes to the observed mass of the electron; the bare mechanical mass itself is unobservable. Similarly, an electromagnetic field is always accompanied by a current of electrons and positrons whose influence on the field contributes to the measured values of charges. The parameters of mass and charge, therefore, had to be renormalized to express the theory in terms of observable quantities. The results for the shift of energy levels (now known as the Lamb shift) and the anomalous magnetic moment of the electron then turned out to be finite and were, moreover, in good agreement with experiment. The use of explicitly relativistic methods of calculation, developed by Tomonaga and Schwinger, was essential in avoiding possible ambiguities in this procedure. Further important contributions were made by Dyson, who showed that the renormalized theory gave finite results for interaction processes of arbitrary order, corresponding to arbitrary powers of the coupling constant e, and by Feynman, who introduced a diagrammatic representation of the mathematical expressions for these processes, which are often of considerable complexity.

The Feynman-diagram technique and Dyson's perturbation theory are now part of the standard formulation of quantum electrodynamics. This formulation and some of its applications will be outlined in Sections II and III. In this article only the electromagnetic interactions of electrons and positrons (or, more generally, of charged leptons, which include muons and tauons and their antiparticles) are considered. These particles also participate in the so-called weak interaction (and, of course, in the much weaker gravitational interaction). A unified theory of the electromagnetic and weak interactions has been developed in recent years. Many elementary processes, however, are dominated by electromagnetic effects, and these alone form the subject matter of quantum electrodynamics.

II. Nonrelativistic Quantum Electrodynamics

A. Approximations

Any treatment of the pure radiation field based on Maxwell's equations in empty space must satisfy the principles of the special theory of relativity, even though it might not be expressed in a form that makes this evident. Quantum electrodynamics has, however, a well-defined nonrelativistic limit in so far as the motion of the charged particles with which the electromagnetic field interacts is concerned. The nonrelativistic theory is of an approximate character, but it involves a much simpler mathematical formalism than its more exact relativistic counterpart. Moreover, it can be applied to a wide range of problems in physics and chemistry, particularly in the areas of atomic spectroscopy, intermolecular forces, laser physics, and quantum optics. [*See* QUANTUM OPTICS.]

Nonrelativistic quantum electrodynamics provides an accurate description of phenomena when the following two conditions are satisfied:

1. The charged particles move at such slow speeds (in the inertial frame of a given observer) that their masses can be considered constant and equal to their rest masses. Since the relativistic mass of a particle with speed v and rest mass m is $m/\sqrt{(1 - v^2/c^2)}$, this requires that $v/c \ll 1$. Now this inequality generally holds for the constituent particles of atoms under normal laboratory conditions. For example, the root mean square speed $\bar{v}$ (relative to the supposedly slowly moving nucleus) of the electron of a hydrogen-like ion in a state with principal quantum number n is $Ze^2/(2nh)$ where Ze is the nuclear charge. If $Z = 1$ and $n = 1$ (the hydrogen atom in its ground state), then $\bar{v}/c$ equals the fine-structure constant α (approximately $\frac{1}{137}$) and the corresponding fractional increase in mass (over and above the rest mass) is only about 3 parts in 10^5.

This ratio is larger for higher values of Z but smaller for higher values of n. The variation of mass with velocity is, therefore, expected to be appreciable only for the inner-shell electrons of the heavier elements.

2. The number of each type of charged particle (electron, proton, etc.) is conserved; that is, such particles are neither created nor destroyed in any process. This assumption imposes a restriction on the frequency ν of the radiation with which the particles may interact, since photons of sufficiently high energy are capable of creating particle–antiparticle pairs. This possibility requires an energy of order mc^2 (where m is the rest mass of the lightest charged particle, namely the electron) and will therefore be excluded if $\nu \ll \nu_c$, where ν_c is defined by $h\nu_c = mc^2$ and is about 10^{20} Hz. (Here ν_c is the frequency associated with the Compton wavelength of the electron given by $\lambda_c = h/mc$.) It follows that hard X rays and high-energy gamma rays are to be omitted from consideration in this section.

B. An Assembly of Photons

The classical electromagnetic field in empty space is equivalent to an infinite number of one-dimensional simple harmonic oscillators. One oscillator is associated with each plane wave component of the field, specified by its frequency ν, wave vector $\mathbf{k}$ (where $|\mathbf{k}| = 2\pi\nu/c$), and unit polarization vector $\hat{\mathbf{e}}$. The waves are transverse waves, which implies that the polarization vector is perpendicular to the direction of propagation $\hat{\mathbf{k}}$ (see Fig. 5). Hence, for each propagation direction, there are two independent polarization vectors $\hat{\mathbf{e}}^{(\lambda)}$ ($\lambda = 1, 2$). A radiation oscillator may therefore be labelled by the pair ($\mathbf{k}, \lambda$), which specifies the frequency, propagation direction, and polarization for the corresponding mode of the field.

A mathematical description of an assembly of noninteracting photons is obtained when each of the radiation oscillators is treated as a quantum mechanical system. This involves little more than the use of the matrix theory of the harmonic oscillator developed in elementary quantum mechanics but extended to cover the case of a set of independent oscillators. The result of this quantization of the electromagnetic field can be briefly summarized. States of the complete system are represented by vectors in a generalized (in fact, infinite-dimensional) vector space and dynamical variables (such as energy and momentum) by linear operators, which act on the vectors to produce other vectors of the same kind. The vacuum state is that for which every oscillator has its lowest energy. It can be assumed, for convenience, that the energy of the vacuum state is zero. The so-called zero-point energy $\frac{1}{2}h\nu$ of an oscillator with frequency ν is therefore discarded, but this amounts merely to a shift in the datum point for measuring energies. (Nevertheless, changes in the zero-point

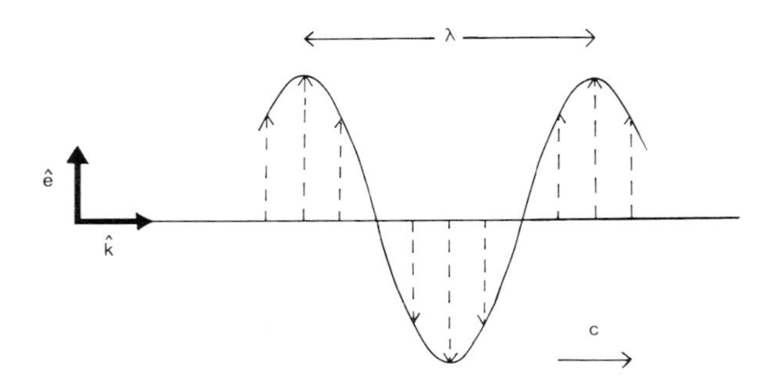

FIG. 5. A linearly polarized electromagnetic wave of wavelength λ propagating in empty space with speed c in the direction $\hat{\mathbf{k}}$. The (real) unit polarization vector $\hat{\mathbf{e}}$ and $\hat{\mathbf{k}}$ together determine the plane of polarization, at any point of which the magnetic induction vector $\mathcal{B}$ (broken arrows) is parallel to $\hat{\mathbf{e}}$ and oscillates in simple harmonic motion of frequency ν, where $\nu\lambda = c$. The electric vector $\mathcal{E}$ (not shown) also oscillates with frequency ν and in phase with $\mathcal{B}$ but is perpendicular to the plane of polarization.

energy can give rise to measureable forces, for example, between conducting plates. This is called the Casimir effect.)

If a radiation oscillator of mode $(\mathbf{k}, \lambda)$ is excited to its nth stationary state, with energy $nh\nu$, then this is taken to correspond physically to the presence of n photons, each with energy $h\nu$, for that mode of the field. An occupation-number state is one with a specified number of photons in each mode. The number $n_{\mathbf{k}\lambda}$ of photons in mode $(\mathbf{k}, \lambda)$ is then called the occupation number for that mode. Only a finite number of occupation numbers can be nonzero and the total numbers of photons is $\sum n_{\mathbf{k}\lambda}$ with the summation extending over occupied modes. Similarly, the total energy E of the photons in an occupation-number state is $\sum(n_{\mathbf{k}\lambda} h\nu)$. The vacuum state can be thought of as an occupation-number state for which every occupation number is zero.

The operator that represents the total energy of the radiation field is called the Hamiltonian operator and is denoted by H_{RAD}. It can be expressed in terms of photon annihilation and creation operators. The annihilation operator for mode $(\mathbf{k}, \lambda)$, when acting on an occupation-number state vector, reduces the number of photons for that mode by one, and when acting on the vacuum-state vector gives the zero vector. Similarly the creation operator increases the number of photons by one. (In the context of the elementary theory of the harmonic oscillator, these operators are usually called lowering and raising operators, respectively.)

A general state of the radiation field at time t is represented by a state vector Ψ of unit length that is a linear combination of the occupation-number state vectors, with coefficients $c(\ldots, n_{\mathbf{k}\lambda}, \ldots; t)$ depending on the occupation numbers and the time. The square of the magnitude of $c(\ldots, n_{\mathbf{k}\lambda}, \ldots; t)$ is the probability that if a measurement of the occupation numbers is carried out at time t, then these will be found to have precisely the values $\ldots, n_{\mathbf{k}\lambda}, \ldots$. So long as no measurements are made on the system, the time evolution of the state vector is governed by Schrödinger's equation

$$i\hbar \frac{\partial \Psi}{\partial t} = H_{\text{RAD}} \Psi$$

The (unit) length of the state vector does not change with time and so the dynamical behavior of the system may be said to correspond to a pure rotation in the generalized vector space. Indeed, the individual probabilities $|c|^2$ do not change with time, since the probability amplitudes c change only through a phase factor $\exp(-iEt/\hbar)$. This is consistent with the fact that for the free field, photons are neither created nor destroyed.

C. The Quantized Electromagnetic Field

While the treatment of the radiation field as an assembly of photons may seem to emphasize its corpuscular aspects, the wave properties are, nevertheless, also contained in the formalism. In particular, Maxwell's equations in empty space remain valid, although they now appear as operator equations rather than as equations for classical fields. Both the electric field $\mathcal{E}$ and the magnetic induction field $\mathcal{B}$ become operators that can be expressed as linear combinations of the photon annihilation and creation operators. If these expressions are inserted into the classical formula for the field energy, then the expansion of the Hamiltonian operator in terms of the annihilation and creation operators is recovered. Thus,

$$H_{\text{RAD}} = \frac{1}{2} \iiint (\mathcal{E}^2 + \mathcal{B}^2)\, dV$$

(It is true that an infinite zero-point energy also appears. This energy may, however, be discarded, as were the zero-point energies of the individual oscillators.) Similarly, the classical expression

$$\frac{1}{c} \iiint \mathcal{E} \times \mathcal{B}\, dV$$

for the field momentum, obtained from Poynting's theorem, implies, when reinterpreted in terms of annihilation and creation operators, that a momentum $\hbar\mathbf{k}$ is to be ascribed to each photon of mode $(\mathbf{k}, \lambda)$. This is in agreement with Einstein's hypothesis, since $\hbar|\mathbf{k}| = h/\lambda$.

Another consequence of the quantization of the field is the occurrence of uncertainties or fluctuations that have no counterpart in the classical theory. In the vacuum state, for example, the mean value of the electric field is zero but its root mean square deviation from the mean, $\Delta\mathcal{E}$, is nonzero. The fluctuation $\Delta\mathcal{E}$ arises from the collective zero-point motions of the radiation oscillators and, if calculated for a nonzero volume with linear dimensions of order L and a nonzero time interval with length of order T, assumes a value whose order of magnitude is given by

$$\Delta\mathcal{E} \sim \begin{cases} \dfrac{\sqrt{\hbar c}}{L^2}, & \text{if } L \geq cT \\[2ex] \dfrac{\sqrt{\hbar c}}{L(cT)}, & \text{if } L \leq cT \end{cases}$$

In any other of the occupation-number states, for which the mean values are also zero, the field fluctuations are of greater magnitude than in the vacuum state. Now the occupation-number states resemble incoherent superpositions of classical plane-wave states, since they are associated with definite wave vectors and polarization vectors but do not have well-defined phases (in the classical sense), whereas a classical plane-wave field has a simple harmonic time dependence. There exist other states of the quantized field, however, called coherent or quasi-classical states, in which the phase is more well defined but the number of photons, and hence the energy and momentum, is less sharp than for the occupation-number states. To each state of the classical field there corresponds a unique coherent state of the quantized field such that (a) the mean values of the quantized field components are equal to the classical field components and (b) the mean value of the quantized energy is equal to the classical energy. The coherent states are also remarkable in the following respect: the field fluctuations for these states are exactly the same as those for the vacuum state.

The operators representing the components of the electric field $\mathcal{E}$ and the magnetic induction field $\mathcal{B}$ satisfy certain commutation relations, which may be derived from those for the photon annihilation and creation operators. Just as in ordinary quantum mechanics the commutation relation

$$qp - pq = i\hbar$$

between the position operator q and the momentum operator p of a particle leads to the Heisenberg uncertainty relation

$$\Delta q \, \Delta p \sim \hbar$$

so the commutation relations between the components of $\mathcal{E}$ and $\mathcal{B}$ lead to uncertainty relations for the electromagnetic field strengths. These uncertainty relations are in agreement with the way in which the field strengths can, at least in principle, be measured by means of macroscopic test bodies. This was shown in detail by Bohr and Rosenfeld in 1933.

D. Interactions of Photons and Atoms

The quantized radiation field has so far been considered as a system by itself. A set of nonrelativistic charged particles, interacting through instantaneous Coulomb forces and also, perhaps, acted on by prescribed external static electric or magnetic fields, can also be considered as a system by itself, as in ordinary quantum mechanics. This system will, for convenience, be referred to as an atom, although it may really be a molecule, an ion, or a collection of atoms, molecules, or ions. It is assumed that there are N charged particles with masses m_1, $m_2, \ldots, m_N$, charges $e_1, e_2, \ldots, e_N$, position operators $\mathbf{q}_1, \mathbf{q}_2, \ldots, \mathbf{q}_N$, and momentum operators $\mathbf{p}_1, \mathbf{p}_2, \ldots, \mathbf{p}_N$. The Hamiltonian operator for this system is given by

$$H_{\text{ATOM}} = \sum_{\alpha=1}^{N} \frac{1}{2m_\alpha} \mathbf{p}_\alpha^2 + U$$

where the first term represents the kinetic energy and the second the potential energy. The potential energy U depends on the positions and momenta of the particles, their charges, and the external fields, if any are present.

It is often a good approximation to treat the nuclei as fixed and to regard the coordinates and momenta of the electrons alone as dynamical variables. This is possible because of the large mass of the protons and neutrons compared with that of the electrons (proton mass $\approx$ 1836 $\times$ electron mass). The fixed-nuclei approximation involves, among other things, the neglect of the recoil of the atoms which should accompany the absorption or emission of photons. The recoil velocity is, however, normally very small. For example, the speed imparted to a hydrogen atom by a photon with a frequency in the visible spectrum is of the order of 10^{-8} times the speed of light *in vacuo*. Such a speed results in only a very slight Doppler shift in the frequency of the emitted radiation. In the fixed-nuclei approximation, the Hamiltonian operator H_{ATOM} has, in general, a discrete set of energy levels $E_r, E_s, \ldots$ corresponding to bound states as well as a continuous set of energy levels E corresponding to ionized states. Here $r, s, \ldots$ are shorthand notations for sets of quantum numbers sufficient to specify the states completely.

A state vector for the complete system consisting of the atom and the radiation field is obtained by multiplying a state vector for the field directly into a state vector for the atom. For example, there are states for which the photon occupation numbers have definite values and the atom is in a stationary state with a definite energy. The general state of the complete system is, at any instant, a superposition of such product states.

The Hamiltonian H for the complete system is not simply the sum of the radiation and atomic

Hamiltonians given previously. This sum must be supplemented by an interaction term H_{INT}:

$$H = H_{\text{RAD}} + H_{\text{ATOM}} + H_{\text{INT}}$$

The inclusion of the interaction term is essential if the operator equations of motion are to reproduce (a) Maxwell's equations for $\mathcal{E}$ and $\mathcal{B}$ with the charges and currents as sources and (b) the Lorentz-force law for the charged particles when acted on by $\mathcal{E}$ and $\mathcal{B}$, that is, the expected equations of motion for the interacting systems. The interaction Hamiltonian H_{INT} contains some operators that refer to the field and some that refer to the particles and, hence, is responsible for the coupling between the two parts of the complete system. In the absence of H_{INT}, the product vectors of the type mentioned above represent stationary states in which the photon occupation numbers are constant and the atom has a fixed energy. Due to the presence of H_{INT}, however, transitions between these states can occur, in which, for example, the atom loses or gains energy and the number of photons is correspondingly increased or decreased. The interaction Hamiltonian may be expressed as the sum of two parts, one proportional to e and the other to e^2:

$$H_{\text{INT}} = eH_1 + e^2H_2$$

where H_1 is linear and H_2 is quadratic in the photon annihilation and creaton operators. As a consequence, these two terms give rise to processes in which the number of photons changes by one or two, respectively.

The use of the so-called Coulomb gauge is very convenient in nonrelativistic theory. In this gauge only the transverse electromagnetic field, which is a superposition of modes with transverse polarization vectors ($\hat{\mathbf{e}}^{(\lambda)} \cdot \hat{\mathbf{k}} = 0$), is quantized. The effect of the longitudinal field, responsible for the instantaneous Coulomb interaction between the charges, is treated as a potential as in ordinary quantum mechanics and is included in the expression for H_{ATOM}.

The time evolution of the complete system is governed by Schrödinger's equation

$$i\hbar \frac{\partial \Psi}{\partial t} = H\Psi$$

where now H is the total Hamiltonian and Ψ represents the state of both the field and the atom at time t. No exact solutions of this equation are known. Fortunately, however, H_{INT} is of order e and, hence, can be regarded as a small perturbation to the unperturbed Hamiltonian $H_{\text{RAD}} + H_{\text{ATOM}}$. Time-dependent perturbation theory can then be used to calculate approximately the probabilities for transitions between unperturbed states. The total energy is always exactly conserved in transitions between initial and final states. Since the perturbation is small, the unperturbed energy is approximately conserved in such transitions.

E. Applications

Applications of the theory to the emission and absorption of photons by atoms and the scattering of photons by free electrons will now be considered. [*See* MICROOPTICS.]

1. Spontaneous Emission—Einstein's *A* Coefficient

If initially (a) the atom is in an excited state r with energy E_r and (b) the radiation field is in the vacuum state, then there is a probability that after a time t a photon of mode ($\mathbf{k}$, λ) has been created and the atom has made a transition to a state s with lower energy E_s, where

$$h\nu \approx E_r - E_s$$

Since there are no photons at all present initially, this process is known as spontaneous emission. It is represented graphically by the Feynman diagram in Fig. 6. Single-photon spontaneous emission involves, in the lowest order of perturbation theory, only that term in the interaction Hamiltonian that is proportional to e.

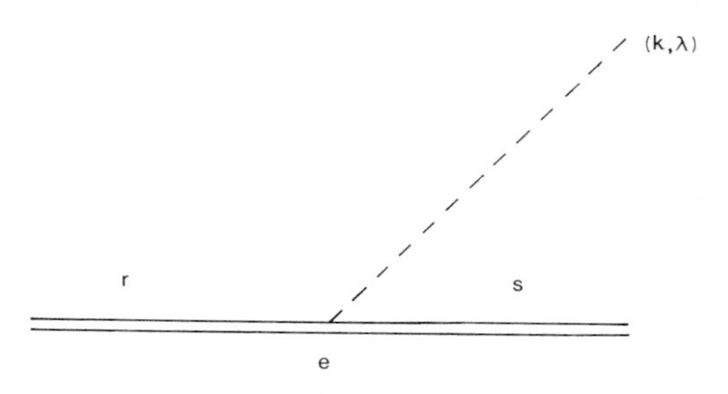

FIG. 6. Feynman diagram for spontaneous emission. The left-hand and right-hand portions of the parallel horizontal lines represent the initial and final atomic states r and s, respectively. (Double lines are used to indicate that the electrons are not free but are bound to the atomic nucleus.) The dotted line represents the emitted photon of mode ($\mathbf{k}$, λ). This is created when the atom undergoes the transition $r \rightarrow s$. The vertex labelled e corresponds to the first-order term in the interaction Hamiltonian, which is responsible for this process in the lowest order of perturbation theory.

Furthermore, the so-called dipole approximation can be used for optical or lower frequencies and bound states of atoms or small molecules, since then the wavelength of the emitted photon is much larger than the dimensions of the region in which the atomic wave functions differ significantly from zero. The emission probability can sometimes be expressed in terms of a constant transition rate (that is, a probability per unit time for the transition to occur) known as Einstein's A coefficient. The total transition rate for emission of the photon in any direction and with any polarization is given in dipole approximation by

$$A_s^r = (16\pi^3\nu^3/3hc^3)|\boldsymbol{\mu}^{rs}|^2$$

where $\boldsymbol{\mu}^{rs}$ denotes the dipole transition moment, which can be calculated once the wave functions for the atomic states r and s are known. Thus, in dipole approximation, Einstein's A coefficient is proportional to the cube of the transition frequency and the square of the length of the dipole transition moment.

The reciprocal of A_s^r is the average lifetime of the upper state r with respect to the lower state s. For example, for optical transitions with a photon wavelength of order 5000 Å (1 Å $= 10^{-8}$ cm) and a dipole transition moment of order ea_0 (where a_0 is the Bohr radius of hydrogen, approximately 0.53 Å), the lifetime is of order 10^{-8} sec. The transition probability is proportional to the time t so long as t is large compared with the atomic period $1/\nu$ and small compared with the lifetime. Since, with the above assumptions, the period is of order 10^{-15} sec, there is indeed a range of values of t that satisfy both conditions. The detection of the emitted photons must take place at times t lying in this range, or else the emission rate is not approximately constant.

2. Absorption and Stimulated Emission—Einstein's *B* Coefficients

If the atom is initially at the lower level E_s, but there is radiation already present, it may make a transition to the higher level E_r by absorbing a photon with energy approximately equal to $E_r - E_s$. The Feynman diagram for absorption is shown in Fig. 7. The transition rate for this process is proportional to the photon occupation number $n_{\mathbf{k}\lambda}$ and hence to the intensity I (erg cm^{-3} Hz^{-1}) of the incident radiation in the spectral region from which the photon is absorbed. If the atom is bathed in isotropic unpolarized radiation (so that I is independent of $\hat{\mathbf{k}}$ and λ), the total absorption rate is $B_r^s I$ where B_r^s is Einstein's B coefficient for absorption, given in dipole approximation as

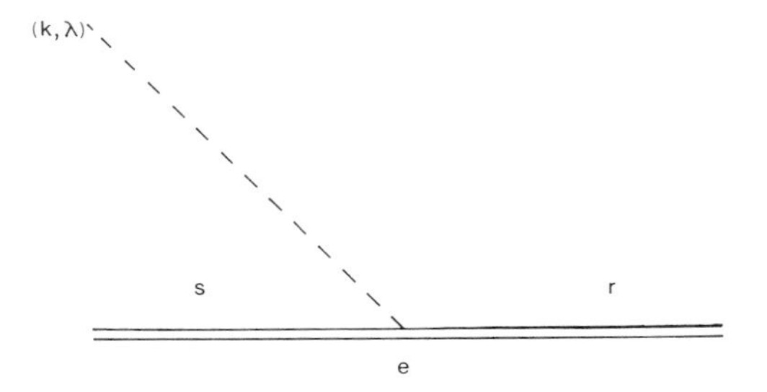

FIG. 7. Feynman diagram for absorption. In the initial state (left of diagram), the atom has energy E_s and there is a photon of mode ($\mathbf{k}$, λ) present, whereas in the final state (right of diagram), the atom has higher energy E_r and the photon has been annihilated.

$$B_r^s = (2\pi^2/3h^2)|\boldsymbol{\mu}^{rs}|^2$$

The upper limit on the time for the validity of this transition rate is now much less than the reciprocal of $B_r^s I$. Times less than this upper limit but much greater than the period can be found, provided the intensity of the radiation is not too high.

For an atom initially at the upper level E_r with radiation present as before, there is a probability for a transition to the lower level E_s accompanied by the emission of a photon with the same characteristics as some of those in the incident beam. This emission may, of course, occur spontaneously, that is, even when all the photon occupation numbers are zero. There is in addition, however, emission that is stimulated or induced by the incident radiation, at a rate proportional to its intensity. For isotropic unpolarized radiation, the stimulated emission rate is $B_s^r I$, where the B coefficient for emission $r \rightarrow s$ is the same as that for absorption $s \rightarrow r$, that is, $B_s^r = B_r^s$.

3. Thomson Scattering

The nonrelativistic limit of the Compton scattering of photons by free electrons is known as Thomson scattering. This limit applies when both the electron and photon momenta have magnitudes small compared to mc. If $\mathbf{p}$ and $\mathbf{p}'$ denote the initial and final momenta of the electron and ($\mathbf{k}$, λ) and ($\mathbf{k}'$, λ') denote the wave vectors and polarizations of the incident and scattered photons, then it follows from the laws of conservation of energy and momentum that $k' \approx k$ and hence that $p' \approx p$. Thus, the magnitudes of the momenta are effectively unaltered, although, in general, their directions change. In particular, for the limit considered, there is no shift in the frequency of the scattered photon.

The Feynman diagrams that give the leading contributions to Thomson scattering are shown in Fig. 8. Each diagram depicts the incident photon and initial electron arriving from the left, and the scattered photon and recoil electron disappearing to the right. The contribution of Fig. 8a arises from the e^2 term in the interaction Hamiltonian; it may be said that here the incident photon is annihilated and the scattered photon simultaneously created. In Fig. 8b and c, on the other hand, the annihilation and creation are represented by two different one-photon vertices, each arising from the e term in the interaction Hamiltonian. These diagrams differ in the order in which the creation and annihilation take place. Overall momentum is conserved at every vertex.

It must be emphasized, however, that the contributions of Fig. 8a–c cannot be physically separated, since only the initial and final states are observed. For this reason, the intermediate states are often referred to as virtual states. Indeed, it may be shown that the contributions of Fig. 8b and c effectively cancel, as their sum is of order $\hbar k/(mc)$ times that of Fig. 8a.

In scattering experiments the measured quantity is the cross section (having dimensions cm^2), defined as the number of scattered particles per unit time divided by the number of incident particles per unit area per unit time. For Thomson scattering, the differential cross section per unit solid angle Ω for scattering the photon with polarization λ' and direction within $d\Omega$ of $\hat{\mathbf{k}}'$ is given, from the contribution of Fig. 8a, by

$$\frac{d\sigma}{d\Omega} = r_0^2|\hat{\mathbf{e}}^{(\lambda)} \cdot \hat{\mathbf{e}}^{(\lambda')}|^2 = r_0^2 \cos^2 \Theta$$

where (a) it is assumed that the polarization vectors are real, (b) Θ is the angle between the directions of polarization of the incoming and outgoing photons, and (c)

$$r_0 = e^2/4\pi mc^2 \approx 2.82 \times 10^{-13} \text{ cm}$$

and is the so-called classical electron radius. (This is the radius that an electron of uniformly distributed charge must have if its electrostatic energy is to equal its rest energy.) It should be noted in particular that, with the approximations indicated, the cross section is independent of frequency and vanishes if the incident and scattered polarization vectors are perpendicular.

If the incident photons are randomly polarized and the polarization of the scattered photons is not observed, then the cross section should be averaged over initial polarization indexes and summed over final polarization indexes. The resulting differential cross section per unit solid angle depends only on the scattering angle θ (where $\cos\theta = \hat{\mathbf{k}} \cdot \hat{\mathbf{k}}'$):

$$d\sigma/d\Omega = \tfrac{1}{2}r_0^2(1 + \cos^2\theta).$$

The total unpolarized cross section, obtained by integrating this over all solid angles, is given by

$$\sigma_{\text{tot}} = (8\pi/3)r_0^2 \approx 6.65 \times 10^{-25} \text{ cm}^2.$$

4. Other Applications

The field of application of nonrelativistic quantum electrodynamics has expanded considerably in recent years due to the development of lasers and their use as spectroscopic tools for investigating a variety of physical and chemical systems. Lasers are sources of highly coherent and very intense beams of light. In the subject of quantum optics, the quantum statistical properties, such as the degree of coherence, of the light beam itself may be the object of investigation. For example, the quasi-classical states of the radiation field referred to earlier exhibit a higher

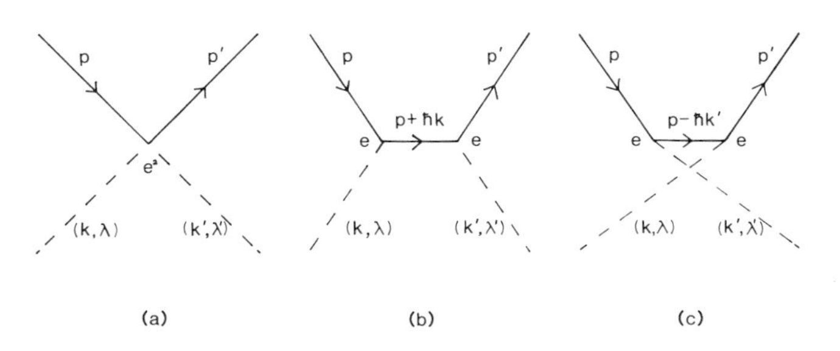

FIG. 8. Feynman diagrams for Thomson scattering. Dotted lines represent photons and solid lines represent free electrons.

degree of coherence than the occupation number states, and these in turn are less chaotic than fields in thermal equilibrium, in which the distribution of photons follows Planck's radiation law. [*See* QUANTUM OPTICS.]

The high intensity of the light beams that may be achieved by using laser sources can also give rise to nonlinear effects that are unobservable at lower intensities. A typical example of a nonlinear process is third-harmonic generation, that is, the absorption of three photons of frequency ν by an atom and the emission of a single photon of frequency 3ν (the third harmonic of the incident frequency). The rate for this process (after which the atom returns to its initial state and overall energy is consequently conserved) is proportional to the cube of the intensity of the incident beam. This should be contrasted with a linear process such as single-photon absorption for which the transition rate is proportional to the intensity itself, the factor of proportionality being Einstein's B coefficient.

III. Relativistic Quantum Electrodynamics

A. Relativistic Theory

Relativistic quantum electrodynamics is formed by the union of the special theory of relativity, characterized by the speed of light, and quantum mechanics, characterized by Planck's constant. In discussing the relativistic theory it is useful and customary to employ the natural system of units, in which speeds are measured as multiples of c and angular momenta are measured as multiples of $\hbar$. Since no natural length appears in the theory, lengths continue to be measured in centimeters. The expression for any quantity in (rationalized) natural units is obtained from the corresponding expression in (rationalized) cgs units simply by setting $c = 1$ and $\hbar = 1$. For example, the cgs expressions $\hbar\mathbf{k}$, mc^2, and $e^2/(4\pi\hbar c)$ for the momentum of a photon, the rest energy of an electron and the fine structure constant, respectively, become $\mathbf{k}$, m and $e^2/(4\pi)$, respectively, in natural units. It is also easy to convert from natural units to cgs units, by inserting appropriate factors of $\hbar$ and c.

If quantum electrodynamics is to satisfy the principles of the special theory of relativity, its equations must be covariant under Lorentz transformations. Lorentz transformations relate the space–time coordinates x, y, z, and t of events as seen by observers using inertial frames of reference moving with uniform velocity relative to each other. (The coordinates x, y, z, and t are the components of a four-dimensional vector, to be denoted simply by x.) A covariant equation has the same form for two such observers. The fact that physical laws are expressible as covariant equations means that these laws are the same for all observers using inertial reference frames.

The state of a quantum-mechanical system (for example, the electromagnetic field *in vacuo*) is specified in relativistic theory on a three-dimensional spacelike hyperplane. In a given inertial frame, this consists of either all the points in three-dimensional space at a particular instant of time or all the events on a two-dimensional plane moving for all time perpendicularly to itself with constant speed greater than that of light. Two distinct events on a spacelike hyperplane cannot be connected by signals travelling with speed less than or equal to the speed of light, and so two measurements made in the vicinity of the corresponding space–time points will not interfere. This is known as microscopic causality. The whole of the four-dimensional space–time manifold is filled with a set of parallel spacelike hyperplanes, which may be labelled by an invariant timelike parameter τ, with τ ranging from $-\infty$ to ∞. The evolution of the system is described by specifying the state for each hyperplane τ and is determined dynamically, through Schrödinger's equation, on the interval $[\tau_1, \tau_2]$, if the state is specified at τ_1 and no measurements are made until τ_2.

B. Electrons and Positrons

The one-particle relativistic theory of the electron is based on the Dirac equation. This is a differential equation, with matrix coefficients, for a spinor wave function $\psi(x)$ having four components $\psi^\mu(x)$ ($\mu = 0, 1, 2, 3$). The requirement that the Dirac equation be covariant determines the behavior of ψ under Lorentz transformations. Since the electron is now described by a four-component spinor rather than by a one-component scalar wave function, it has extra degrees of freedom, over and above those allowed by the Schrödinger theory. These correspond to the spin or intrinsic angular momentum of magnitude $\frac{1}{2}$ (in natural units). Hence the spin appears automatically in the Dirac theory of the electron and does not have to be added on in an *ad hoc* fashion, as it does in nonrelativistic quantum mechanics.

The difficulties of interpretation associated with the negative-energy solutions of the Dirac equation have already been mentioned. These difficulties disappear in the second-quantized version of the theory, in which electrons and positrons are treated on an equal footing and all have positive energy. Moreover, this version provides a calculus for processes involving annihilation and creation of electrons and positrons—in the high-energy regime, the number of these particles is no longer conserved.

The spinor ψ and its related adjoint spinor $\bar{\psi}$ are first expressed in terms of plane-wave solutions of the free-particle equation. These solutions correspond to particles with energy E, momentum $\mathbf{p}$, and rest mass m satisfying the relativistic energy–momentum relation

$$E^2 = |\mathbf{p}|^2 + m^2$$

The coefficients in the expansions of ψ and $\bar{\psi}$ may then be interpreted as annihilation and creation operators for electrons and positrons in definite momentum and spin states. The spin states can be chosen to be helicity states of the electrons and positrons, that is, states in which the component of spin in the direction of motion is either $\frac{1}{2}$ (right-hand helicity) or $-\frac{1}{2}$ (left-hand helicity). It is only the component of spin in the direction of motion that is invariant under Lorentz transformations.

The algebra of the creation and annihilation operators for electrons and positrons differs from that of the creation and annihilation operators for photons. (Technically, it involves anticommutation instead of commutation relations.) The difference arises from the fact that, whereas photons are bosons, electrons and positrons are fermions, and are subject to the exclusion principle. The only possible occupation numbers for electrons or positrons are therefore 0 or 1. This is in agreement with Pauli's spin-statistics theorem, since the spin of an electron or positron is half an odd integer. It can be shown that quantizing the Dirac field by using commutation relations instead of anticommutation relations leads to a Hamiltonian operator with energy levels that are not bounded below and, hence, one for which no stable vacuum (ground) state exists. (Similarly, quantizing the Maxwell field, which has intrinsic spin 1, by using anticommutation relations instead of commutation relations leads to a breakdown of microscopic causality.)

C. Covariant Quantization of the Electromagnetic Field

The quantization of the electromagnetic field in the Coulomb gauge, though very useful for dealing with bound systems in nonrelativistic approximation, is not a manifestly covariant procedure. In the Coulomb-gauge formalism, only the transverse field is quantized, while the longitudinal field gives rise to an instantaneous interaction between the charges. The division of the field into transverse and longitudinal components, however, is not Lorentz covariant—these components do not transform separately on going from one inertial frame to another. To exhibit the covariance of the theory, it is necessary to use a gauge condition on the electromagnetic potentials that is itself covariant. The most convenient such condition is the so-called Lorentz condition. This condition also leads to certain difficulties which are, however, overcome in the formalism developed by Gupta and Bleuler.

The electromagnetic field is, in the first instance, quantized in a covariant way without reference to the Lorentz gauge condition. In contrast to the noncovariant treatment, there are now, for each wave vector $\mathbf{k}$, four types of photon, corresponding to timelike and longitudinal as well as two transverse polarization vectors, which are, in addition, four-dimensional rather than three-dimensional vectors. Moreover, the inner (or scalar) product of the infinite-dimensional vector space on which the photon creation and annihilation operators act is not positive definite; that is, there exist nonzero vectors in this space the square of whose length is zero or negative. (This is due to the metric of the space–time continuum, which distinguishes timelike from spacelike directions. Thus, the four-dimensional vector x is spacelike, lightlike, or timelike, relative to the origin, according as $x^2 + y^2 + z^2 - t^2$ is positive, zero, or negative.) This constitutes a serious difficulty, since the quantum mechanical statistical interpretation requires a positive-definite inner product. For the resolution of this problem, the use of the Lorentz gauge condition, which has yet to be imposed, is of decisive importance.

It may be shown that neither the Lorentz condition nor Maxwell's equations are satisfied as operator equations in the covariant theory, because they are incompatible with the commutation relations. They are, however, satisfied as equations for expectation values (and hence are satisfied in the classical limit), provided a subsidiary condition is imposed on those state vectors that are to represent physically realizable states. The effect of the subsidiary condition is to make the timelike and longitudinal photons unobservable in real states of the system. These states have either no timelike or longitudinal photons at all or only certain allowed admixtures of them. Moreover, changing the allowed admixtures is merely equivalent to carrying out a gauge transformation that maintains the Lorentz condition. The allowed admixtures are always such that the contributions of timelike and longitudinal photons to, for example, the energy and momentum, cancel out, and only the contributions of the transverse, observable photons remain. Similarly, the statistical interpretation of the theory is consistent, when this is restricted to the calculation of probabilities for physically realizable states.

Despite the fact that timelike and longitudinal photons are unobservable in real states of the system, their presence is important and cannot be neglected in intermediate or virtual states. For example, the Coulomb interaction may be described in terms of the virtual exchange of timelike and longitudinal photons by charged particles. The appearance of these photons in the formalism is also required, of course, if the theory is to be manifestly Lorentz covariant.

D. Symmetries and Conservation Laws

The coupling between the quantized Maxwell and Dirac fields is represented by a Lorentz-invariant interaction Hamiltonian density $\mathcal{H}_{\mathrm{INT}}$ that links the scalar and vector potentials to the charge and current densities. Here $\mathcal{H}_{\mathrm{INT}}$ is linear in the electromagnetic potentials and bilinear in the spinor fields ψ and $\bar{\psi}$ and is also of order e—there is no e^2 term as in nonrelativistic theory. The interaction Hamiltonian density may be derived from a Lagrangian density known as the minimal-coupling Lagrangian density.

It is interesting to note that certain continuous symmetries of the coupled systems are reflected in the structure of the complete Lagrangian density $\mathcal{L}$, which is Lorentz invariant and gauge invariant. According to a theorem of Noether, these symmetries must lead to conservation laws. For example, the invariance of $\mathcal{L}$ under time displacements implies the conservation of energy; its invariance under space displacements and rotations implies the conservation of linear and angular momentum, respectively; and its invariance under gauge transformations implies the conservation of charge.

The complete system also has three discrete symmetries. It is invariant under (a) charge conjugation C, that is, the interchange of particles and antiparticles (which affects only electrons and positrons, since the photon is its own antiparticle); (b) the parity operation P, that is, space inversion or the interchange of left and right; and (c) time reversal T. This invariance under C, P and T is not shared by all the laws of nature. The nonconservation of parity in the weak interaction, which is responsible for the dynamics of beta emission, was suggested by Lee and Yang in 1956 and subsequently confirmed experimentally. That the combined transformation of charge conjugation and parity is also not a symmetry follows from the decay of the long-lived neutral K meson into two charged pions, a decay that is forbidden by CP conservation. Invariance under the combined CPT transformation, established on very general assumptions (Lorentz covariance and locality), then implies that time reversal is also not a symmetry of the physical world. Hence, the separate conservation of C, P, and T is only an approximation which is, however, valid for phenomena that are adequately described by electrodynamics alone.

E. The *S* Matrix and Feynman Diagrams

The S matrix in quantum electrodynamics is used to calculate probability amplitudes for processes in which particles (electrons, positrons or photons) that are initially free are allowed to interact and scatter. In the so-called interaction picture of the motion, the state vector Ψ for the complete system evolves under the influence of the interaction Hamiltonian $\mathcal{H}_{\mathrm{INT}}$ alone, and the S operator (or scattering operator) maps the state vector on the hyperplane $\tau = -\infty$ (that is, long before the interaction takes place) onto the state vector on the hyperplane $\tau = \infty$ (that is, long after the interaction has ceased):

$$\Psi(\infty) = S\Psi(-\infty)$$

The S operator can be developed as a power series in the coupling constant e. With the help of a theorem due to Wick, the structure of the nth-order contribution, corresponding to the nth power of e in the expansion, may be systematically analysed and represented by Feynman diagrams. It is usually convenient to use Feynman diagrams in energy–momentum space. These represent all possible virtual processes that can take place for given initial and final momentum and polarization or spin states of the particles. The Feynman rules enable expressions for the probability amplitude or S-matrix element S_{fi} for the process $i \rightarrow f$ to be written down directly from the diagrams. From this the cross section for the process may be calculated to a given order in e and compared with the experimentally obtained value.

The lowest order of perturbation theory ($n = 1$) involves only the first power of the interaction Hamiltonian $\mathcal{H}_{\mathrm{INT}}$, which is linear in the photon annihilation and creation operators and bilinear in the fermion (electron or positron) annihilation and creation operators. This gives rise to processes such as those depicted in the Feynman diagrams of Fig. 9. These diagrams are called basic vertex diagrams. There are in all eight such diagrams, corresponding to processes in which a photon is either annihilated or created and two fermions are annihilated or created or one is annihilated and the other created.

Every Feynman diagram is a combination of some or all of the eight basic vertex diagrams—an nth-order diagram contains n vertices. Energy and momentum (which together form a four-dimensional vector) are conserved at every vertex. (This is in contrast to the nonrelativistic theory, in which momentum but not energy is conserved in virtual processes.) However, the relativistic relation between energy and momentum need not be satisfied for virtual particles. Now this relation cannot be satisfied by all the particles participating in a basic vertex process, which must therefore be a virtual rather than a real process. For example, electron–positron annihilation with the production of a single photon is forbidden by energy–momentum conservation, even though it is allowed by charge conservation. Hence the basic vertex diagrams can appear only as parts of larger Feynman diagrams depicting processes for which overall energy and momentum are conserved and the relativistic energy–momentum relation is satisfied by the (real) particles in the initial and final states.

As an example of a real process, consider the Compton scattering of photons by electrons. This is allowed in the second order of perturbation theory, and the Feynman diagrams, each containing two vertices, are shown in Fig. 10. The corresponding polarized cross section for the laboratory reference system, in which the target electron is initially at rest, is given by the Klein–Nishina formula:

$$\frac{d\sigma}{d\Omega} = \frac{\alpha^2}{4m^2}\left(\frac{\nu'}{\nu}\right)^2\left[\frac{\nu}{\nu'} + \frac{\nu'}{\nu} + 4(\varepsilon \cdot \varepsilon)^2 - 2\right]$$

Here ν and ν' are the frequencies of the incident and scattered photons, respectively, and ε and ε' are their (four-dimensional) transverse polarization vectors, which in this formula are assumed to be real (so that the photons are linearly polarized). In the low-energy limit ($\nu \ll m$ and $\nu' \approx \nu$), this reduces to the Thomson cross section derived from the nonrelativistic theory. (Note that $r_0 = \alpha/m$.) The unpolarized cross section, obtained by averaging over initial and summing over final polarizations, is given by

$$\frac{d\sigma}{d\Omega} = \frac{\alpha^2}{2m^2}\left(\frac{\nu'}{\nu}\right)^2\left[\frac{\nu}{\nu'} + \frac{\nu'}{\nu} - \sin^2\theta\right]$$

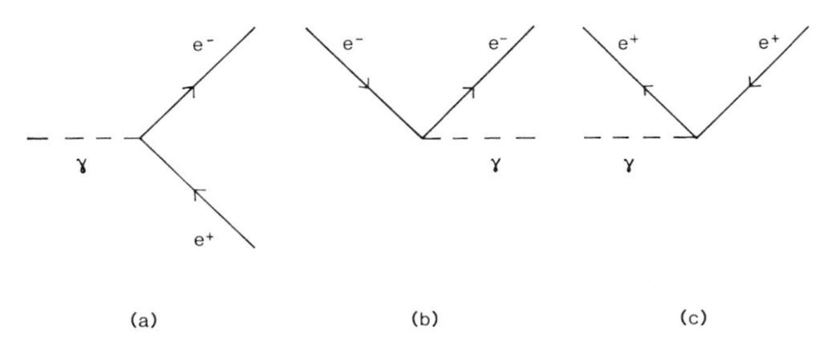

FIG. 9. Virtual processes depicted by basic vertex diagrams (to be viewed from left to right). (a) Photon (γ) annihilation and electron–positron (e^-e^+) pair production. (b) Electron scattering and photon creation. (c) Photon annihilation and positron scattering. Note the convention for the sense of the arrows used on the fermion lines to distinguish electrons and positrons.

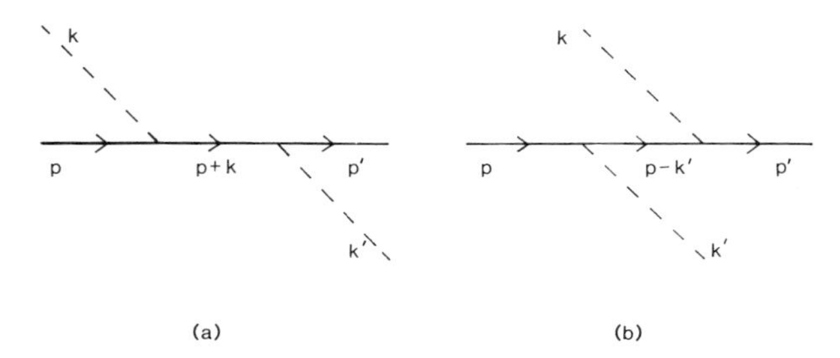

FIG. 10. Feynman diagrams for Compton scattering. The lines are labelled by the four-dimensional energy–momentum vectors of the particles. Energy and momentum are conserved overall and at every vertex. Polarization and spin labels have been suppressed. Diagrams (a) and (b) differ in the order in which the incident photon is annihilated and the scattered photon created.

where θ is the angle of scattering, as in Fig. 3. This reduces to the unpolarized Thomson cross section in the low-energy limit.

F. Radiative Corrections

The first approximation to the S-matrix element for a given process may be improved by adding contributions from higher-order perturbation theory. These contributions, known as radiative corrections, often, though not always, involve integrals with ultraviolet divergences (that is, the integrals tend to infinity as the upper limits on the momenta of the virtual photons or fermions involved tend to infinity). For example, radiative corrections of second order in e (or of first order in α) relative to the lowest-order term are expected when one of the modifications shown in Fig. 11 is made in a Feynman diagram. Each of the integrals corresponding to the modified diagrams, however, has an ultraviolet divergence.

The divergence difficulties of relativistic quantum electrodynamics may be overcome by first regularizing the theory, that is, by altering it so that all the integrals converge. In the method of dimensional regularization, for example, this is achieved by replacing (in a well-defined sense) divergent four-dimensional expressions by convergent $(4 - \varepsilon)$-dimensional expressions, where $\varepsilon > 0$. This may be described as reducing the dimensions of energy-momentum space from 4 to $4 - \varepsilon$. The regularized theory is not equivalent to quantum electrodynamics, which is restored only in the limit as $\varepsilon \to 0$, in which limit the divergences reappear.

The mass and charge of the fermions are then renormalized; that is, the predictions of the

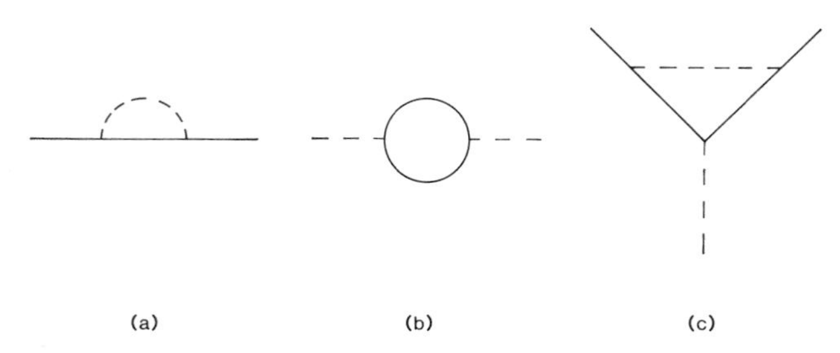

FIG. 11. Modifications of a fermion line, a photon line, and a basic vertex part leading to second-order radiative corrections. (a) Fermion self-energy arising from emission and reabsorption of virtual photons. (b) Photon self-energy (or vacuum polarization) arising from virtual pair creation and annihilation. (c) Vertex modification arising from virtual photon exchange.

regularized theory are expressed in terms of the observed mass and charge of the physical particles rather than the unobservable mass and charge of the bare particles. In this connection, certain relations between fermion self-energy and vertex-modification contributions (see Fig. 11a and c), known as Ward's identities, allow a great simplification to be made. In particular, they imply that charge renormalization arises solely from vacuum polarization effects (see Fig. 11b). It is important to note also that mass and charge renormalization would have to be carried out even if no divergences appeared in the formalism.

Finally, quantum electrondynamics is recovered by removing the regularization. If the method of dimensional regularization is used, this means taking the limit as $\varepsilon \to 0$. In this limit, infinities reappear in the relations between the observed and bare masses and charges. These relations, however, are not susceptible to experimental verification, as the bare masses and charges themselves are unobservable. Moreover, as $\varepsilon \to 0$, the physical predictions of the theory (for example, radiative corrections to scattering cross sections or electromagnetic shifts of energy levels) are finite in all orders of perturbation theory and are expressed in terms of the observed masses and charges. (For this reason quantum electrodynamics is said to be a renormalizable quantum field theory.) These predictions can therefore be tested against experimental results.

1. The Lamb Shift

The nonrelativistic Lamb shift for atomic hydrogen was first calculated by Bethe in 1947, following the experiments of Lamb and Retherford. Whereas Dirac's one-particle relativistic theory predicts that the $2S_{1/2}$ and $2P_{1/2}$ states of the hydrogen atom have the same energy, Lamb and Retherford showed that the $2S_{1/2}$ level is actually higher and found a difference in energy corresponding to a frequency of about 1000 MHz. In Bethe's treatment, the effect was interpreted as a difference between the electron's self-energy when free (Fig. 11a) and when bound to the proton (Fig. 12a); a cutoff of order mc was used for the momenta of the virtual photons emitted in each case. The calculation gave no shift for the $2P_{1/2}$ level but did give an upward shift of about 1040 MHz for the $2S_{1/2}$ level and was therefore, in view of the nonrelativistic treatment and the approximations made, in good agreement with the observed value of the separation. This agreement has been enhanced by subsequent refinements in both theory and experiment.

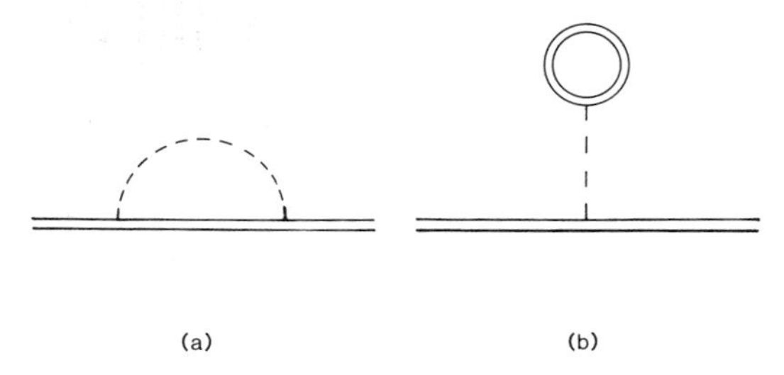

FIG. 12. Feynman diagrams for second-order radiative corrections to atomic energy levels (Lamb shift). (a) Electron self-energy and (b) vacuum polarization.

The relativistic treatment of the Lamb shift, which is a bound-state problem, requires a more elaborate formalism (involving the so-called bound interaction picture) than the S-matrix theory outlined above. In addition to electron self-energy effects, there are also vacuum polarization effects (see Fig. 12) and effects due to the finite mass and nonzero size of the nucleus. (The proton is not a pointlike object but has an effective radius of about 0.86×10^{-13} cm.) Vacuum polarization gives a downward shift of about 27 MHz to the $2S_{1/2}$ level. There are still difficulties in calculating higher-order binding corrections to the Lamb shift and these have resulted in discrepancies of the order of 0.045 MHz between different calculated values. The discrepancies are small but nevertheless larger than the experimental uncertainties.

Recent theoretical values for the $2S_{1/2} - 2P_{1/2}$ level splitting in atomic hydrogen given by Borie are 1057.843(15) MHz and 1057.888(15) MHz where the figures in brackets represent uncertainties in the last two digits quoted. These values are based on earlier calculations by Mohr and Erickson, respectively, but take previously neglected nuclear-size effects into account. Recent experimental values are 1057.826(20) MHz, obtained by Andrews and Newton in 1976, and 1057.845(9) MHz, obtained by Lundeen and Pipkin in 1981. Despite the discrepancies between the theoretical values, the agreement between theory and experiment is impressive. The uncertainties, both theoretical and experimental, in the Lamb-shift frequency are now of the order of 10^4 Hz. This should be compared with a frequency of order 10^9 Hz for the Lamb shift itself and a frequency of order 10^{15} Hz for an optical transition.

2. The Anomalous Magnetic Moment of the Electron

The comparison of the measured and calculated values of the anomalous magnetic moment of the electron is regarded as an important test of quantum electrodynamics. The anomalous moment arises from small deviations of the electron's gyromagnetic ratio from 2, which is the value predicted by the Dirac theory. The gyromagnetic ratio g_{e^-} is defined through the relation between the intrinsic magnetic moment **M** of the electron and its spin angular momentum **S**, namely

$$\mathbf{M} = -g_{e^-}\left(\frac{e}{2mc}\right)\mathbf{S}$$

The directly measured quantity is not the gyromagnetic ratio (or g-factor, as it is also called) itself but the electron anomaly a_{e^-}, which is the difference between this ratio and 2, all divided by 2. Thus,

$$a_{e^-} = \frac{g_{e^-} - 2}{2}$$

The electron anomaly is, like the g-factor, a dimensionless constant. Its value is approximately one-tenth of one percent and has been both measured and calculated with great precision. The experimental and theoretical values of a_{e^-} agree to nine places of decimals:

$$a_{e^-} = 0.001159652$$

The value of g_{e^-} is obtained from this by simple arithmetic:

$$g_{e^-} = 2.002319304$$

In experiments carried out at the University of Washington in Seattle, the accuracy of the measurement has been greatly increased. In these experiments, electrons are held in a configuration of static electric and magnetic fields in a cavity with linear dimensions of order 1 cm. The arrangement is known as a Penning trap. Even single electrons can be held in it for weeks at a time. The electrons in a Penning trap have a discrete set of energy levels and are sometimes considered as part of an atom with a nucleus of macroscopic size, namely the experimental apparatus or, indeed, the earth on which it rests; the atom is called geonium.

The most recent measured value of a_{e^-} was reported by Van Dyck, Schwinberg, and Dehmelt in 1985 as

$$a_{e^-}^{\text{exp}} = 0.001159652193(4)$$

where again the figure in brackets represents the probable uncertainty in the last digit. The electron anomaly, or the g-factor, is the most accurately known of all physical constants. Its measurement does not depend on a knowledge of either the values of other physical constants or the strength of the magnetic field involved in the experiment.

The theoretical value of a_{e^-} is obtained by considering the scattering of electrons by an external (prescribed) field. Feynman diagrams for the lowest-order contribution to this process and a radiative correction of order α are shown in Fig. 13. The change in the momentum of the scattered electron is supplied by the external field. The lowest-order contribution to the electron anomaly is known exactly (its value $\alpha/2\pi$ was calculated by Schwinger in 1948), as is the contribution of order α^2. Further contributions of order α^3 and α^4 have also been calculated, partly analytically and partly numerically. (The contribution of order α^4 arises from 891 different Feynman diagrams!) A recent theoretical value, due to Kinoshita and Lindquist, is given by

$$a_{e^-}^{\text{th}} = 0.001159652460(127)\ (075)$$

where the first error stems from experimental uncertainties in the value of the fine-structure constant α and the second from computational and other theoretical uncertainties. The slight inconsistency, in the last three digits, between the experimental and theoretical values has not yet been resolved.

The positron anomaly a_{e^+} was measured by Schwinberg, Van Dyck, and Dehmelt in 1981 using the geonium experiment. They concluded

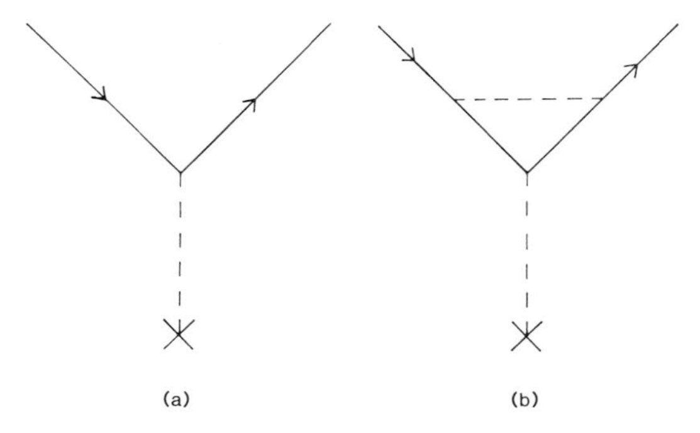

FIG. 13. Feynman diagrams for electron scattering by an external field (Denoted by X). (a) Zeroth-order contribution yielding a g-factor of 2 or an electron anomaly of 0. (b) Radiative correction of order α yielding a g-factor of $2 + (\alpha/\pi)$ or an electron anomaly of $\alpha/2\pi$.

that any difference between the ratio of the positron g-factor to the electron g-factor and unity must be less than 10^{-10}. This conclusion is strong evidence for the validity of the *CPT* theorem. Any departure of g_{e^+}/g_{e^-} from 1 would signal a breakdown of the combined charge conjugation, parity and time-reversal transformation as a symmetry of nature.

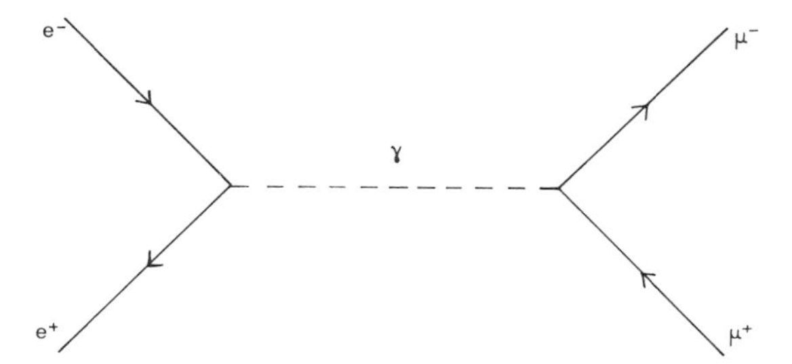

FIG. 14. Muon pair production through electron–positron annihilation. The lepton number (in this case 0) is conserved at each vertex for each kind of lepton separately.

G. Interaction of Photons and Leptons

Relativistic quantum electrodynamics may readily be extended to include the interaction of photons with certain other charged particles besides electrons and positrons. These are the muon (symbol μ^-) and the tauon (symbol τ^-) and their antiparticles μ^+ and τ^+. Muons and tauons have, within experimental accuracy, the same charge ($-e$) and spin ($\frac{1}{2}$) as electrons, but different masses. While the rest energy (measured in electron volts) of the electron is about 0.511 MeV, that of the muon is about 105.659 MeV and that of the tauon is (with a possible error of about 3 MeV) about 1784 MeV. The fact that the electron, muon, and tauon seem to have identical characteristics (apart from mass) is known as e–μ–τ universality. The muon and the tauon have lifetimes of order 10^{-6} sec and 10^{-13} sec, respectively. Electrons, muons, and tauons are all called leptons (as are neutrinos, which are, however, uncharged); they do not, in contrast to hadrons, experience the strong (nuclear) force. On the other hand, they do participate in the weak and gravitational interactions as well as in the electromagnetic interaction.

An example of a scattering process that involves more than one kind of lepton is muon pair production in electron–positron collisions. The Feynman diagram for the lowest-order contribution to this is shown in Fig. 14. An electron and a positron are annihilated and a virtual photon created; this in turn is annihilated and a muon and an antimuon created. For the process to occur, the electron and positron together must have at least the threshold energy equal to twice the rest energy of the muon (about 211 MeV). It should be noted that in Fig. 14 the lepton number, defined as the number of leptons minus the number of antileptons, is conserved at each vertex for both electrons and muons. This is true generally (and for tauons as well) and arises from the form of the interaction Hamiltonian. Each basic vertex involves only one type of lepton or antilepton. There are no vertices involving, for example, the annihilation of an electron and the simultaneous creation of a muon.

In the center-of-mass reference system (in which the electrons and positrons collide head-on with the same energy E), the threshold energy is reached when the particles have been accelerated to speeds about 0.999988 times the speed of light. For energies much greater than the threshold energy (or speeds even closer to that of light), the unpolarized differential and total cross sections for muon pair production in the center-of-mass system reduce to

$$\frac{d\sigma}{d\Omega} = \frac{\alpha^2}{16E^2}(1 + \cos^2\theta)$$

and

$$\sigma_{\text{tot}} = \frac{\pi\alpha^2}{3E^2}$$

where θ is the angle between the incoming electron and outgoing muon (or between the incoming positron and outgoing antimuon). The second of these formulas has been verified in experiments using total center-of-mass energies of the order of 30 GeV.

The results of the high-energy experiments can be used to set bounds on possible deviations from exact quantum electrodynamics. The existence of heavy photons, for example, would modify the structure of the theory. (Thus, in the static limit, the Coulomb potential would no longer have a simple inverse distance dependence). To good approximation, the cross section for muon pair production would be altered according to

$$\sigma_{\text{tot}} \to \sigma_{\text{tot}}\left(1 \pm \frac{4E^2}{\Lambda_\pm^2}\right)$$

where $\Lambda_\pm$ are cutoff parameters with the dimensions of energy. For consistency with the experimental results $\Lambda_{\pm}$ must be at least of the order of 150 GeV. (Recent results obtained at the Stanford Linear Accelerator Center suggest that $\Lambda_{\pm} > 172$ GeV.) This corresponds to a test of the pointlike nature of photon–lepton interactions down to distances of the order of $1/\Lambda_{\pm}$, that is, less than 10^{-15} cm.

Bibliography

Barut, A. O. (Ed.) (1984). "Quantum Electrodynamics and Quantum Optics," N.A.T.O. A.S.I. Series Vol. 110. Plenum, New York.

Berestetskii, V. B., Lifshitz, E. M., and Pitaevskii, L. P. (1982). "Quantum Electrodynamics," Course of Theoretical Physics Vol. 4, 2nd ed. Translated from the Russian by J. B. Sykes and J. S. Bell. Pergamon, Oxford.

Craig, D. P., and Thirunamachandran, T. (1984). "Molecular Quantum Electrodynamics: an Introduction to Radiation–Molecule Interactions." Academic Press, Orlando.

Feynman, R. P. (1985). "QED: The Strange Theory of Light and Matter." Princeton Univ. Press, Princeton, New Jersey.

Gräff, G., Klempt, E., and Werth, G. (Eds.) (1981). "Present Status and Aims of Quantum Electrodynamics," Lecture Notes in Physics Vol. 143, Springer-Verlag, Berlin and New York.

Healy, W. P. (1982). Non-Relativistic Quantum Electrodynamics." Academic Press, New York.

Mandl, F., and Shaw, G. (1984). "Quantum Field Theory." Wiley, New York.

FIBER-OPTIC BASED SMART STRUCTURES

Raymond M. Measures *University of Toronto Institute for Aerospace Studies*

GLOSSARY

Composite material: A fiber-reinforced polymer. Generally divided into thermosets and thermoplastics. Graphite or Kevlar fibers in an epoxy matrix are examples of a thermoset; one of the most popular thermoplastics involve graphite fibers in a PEEK matrix.

Fiber-optic sensor: A length of optical fiber in which the properties of light are influenced by some transduction mechanism. For example, if an interferometric fiber-optic sensor is subjected to changes in strain or temperature, there is a corresponding change in the phase of light propagating through the optical fiber.

Fly-by-light: The use of light signals transmitted through optical fibers to convey sensor and control information between the cockpit and the various systems of an aircraft.

Intelligent structure: A smart structure that is capable of adaptive learning.

Optical fiber: A transparent fiber comprising a central core and an outer cladding. Light launched into the core at one end will be confined to propagate through the core with very little attenuation.

Passive smart structure: A structure that possesses a built-in optical microsensor system for evaluating its state in real time.

Reactive smart structure: A structure that possesses a built-in optical microsensor system for evaluating its state in real time and an actuator control system for modifying that state when necessary.

Structurally integrated fiber-optic sensor: A fiber-optic sensor that is either bonded to the surface of a structure or, in the case of a composite material, embedded within the structure.

A smart structure possesses a structurally integrated optical microsensor system for determining its state. This built-in sensor system can be thought of as an optical nervous system which can, in real time, evaluate the strain or deformation of the structure, monitor if it's vibrating or subject to some external pressure, check its temperature, and warn of the appearance of any hot spots. Most important of all a smart structure should maintain a constant vigilance over its structural integrity and indicate if damage is inflicted or fatigue is developing. In more advanced smart structures the information provided by the sensor system could be used for controlling some aspect of the structure, such as its stiffness, shape, position, orientation, rotation, or acceleration; such systems might be called "reactive smart structures" to distinguish them from the simpler "passive smart structures," which only sense their state. Eventually, smart structures will be developed that will be capable of adaptive learning, and these can be termed "intelligent structures." The combination of structures, sensor systems, actuator control systems, and neural networks could lead to the broad class of structures indicated in Fig. 1.

The successful development of smart structures technology could lead to aircraft that are safer, lighter, more efficient, easier to maintain and to service; pipelines, pressure vessels, and storage tanks that constantly monitor their structural integrity and immediately issue an alert if any problem is detected; space platforms that check for pressure leaks, unwanted vibration, excess thermal buildup, and devia-

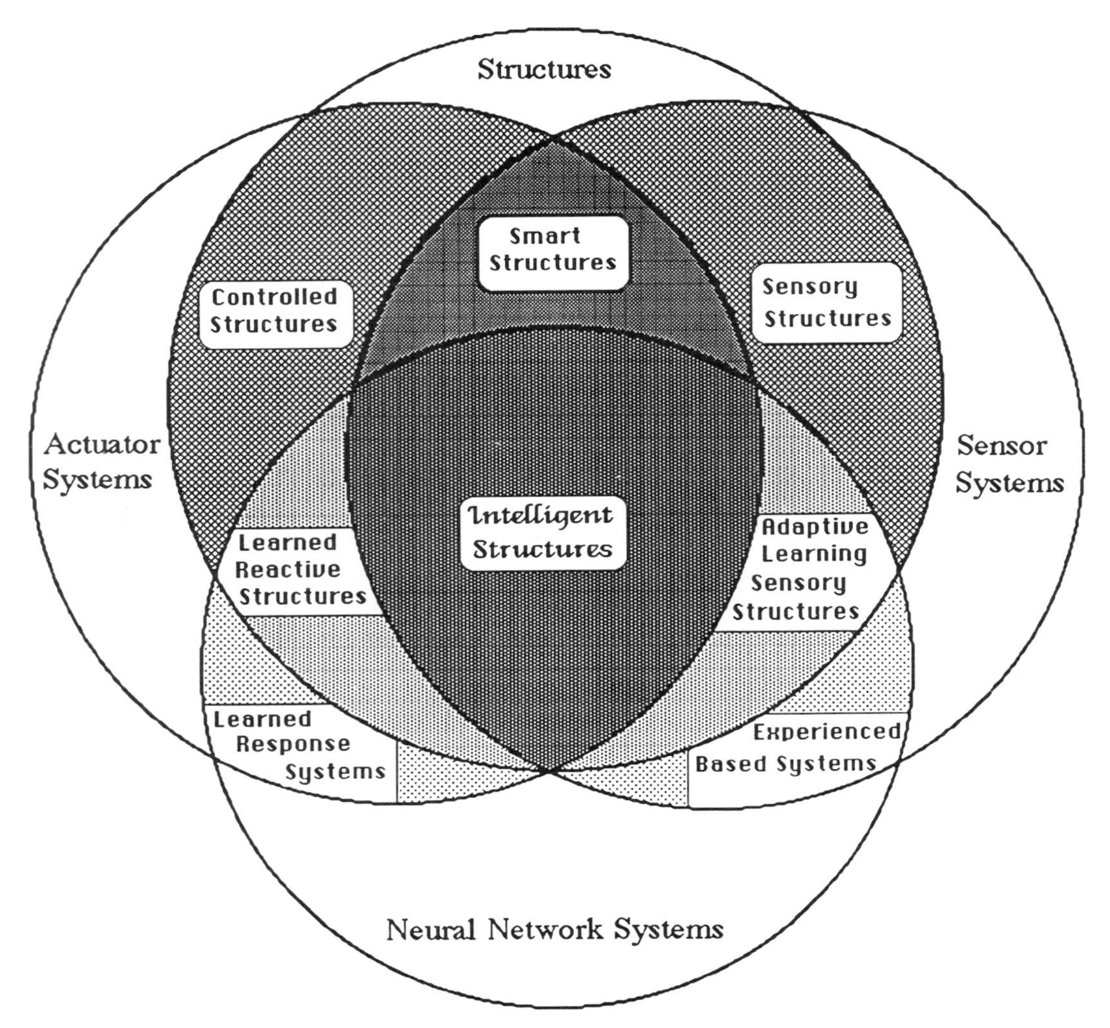

FIG. 1. Types of structures possible by the confluence of structures, sensor systems, actuator control systems, and neural networks.

tion from some preassigned shape. In many situations this optical sensing network would be part of a control system that could take some appropriate action. Such structures might also learn by means of a neural network the correct response for achieving a desired movement. In the twenty-first century this could lead to aircraft without flaps or ailerons that would instantly adjust their shape to conform with the flight conditions, and large flexible space structures that would actively damp out unwanted vibrations and maintain some preassigned shape. The development of structurally integrated optical microsensors combined with distributed actuator systems could also lead to the manufacture of flexible structures that are more adaptable and provide much finer motion control than their more rigid counterparts. Flexible manipulators built on these principles would endow robotic systems with human-like control capabilities.

I. Introduction

Optical fibers can either be bonded to the surface of metal, plastic, ceramic, or other structures or embedded within composite material structures during their manufacture. These sensors could be used to detect and locate impacts, anomalous hot spots, damage, and structural deformation. In the case of composite materials, internal damage generated by impacts, manufacturing flaws, excessive loading or fatigue could be detected and assessed and, most important, the growth in the extent of these damage zones could also be monitored. Structurally embedded optical fibers sensors could undertake "real-time" monitoring of the cure state in thermosets or the degree of consolidation in thermoplastics. The residual stress during manufacture could be determined, and subsequently the structure could be examined for internal defects both prior to handling and after installation (see Table I). This could lead to improvements in the quality control of composite products.

In-service monitoring of structural loads, especially in high-load regions like wing roots of aircraft, could be of considerable benefit in helping to avoid structural overdesign. Monitoring the dynamic response of structural components during operation might also provide insight into the structural integrity of the system. Indeed, smart structures technology could change engineering ethics, making it "*unacceptable engineering to permit mechanical failures that lead to death, injury, or major environmental catastrophes*." In the twenty-first century accidents due to mechanical failure will be attributed to HUMAN ERROR for not building in sufficient strength redundancy or a sensor system to warn of impending failure.

Two of the most important considerations for aircraft are system weight and life-cycle costs. Protection against lightning and various forms of electromagnetic interference also figure prominently in these deliberations. Unfortunately, where composite materials are being used to replace metal, fly-by-wire systems require extensive shielding. Even in the absence of a direct lightning strike, magnetic field coupling can generate voltages that can burn out various components and fragile circuits used in many electronic systems. Although these problems can be overcome, even in the hostile environment in which a military aircraft might find itself, a considerable weight penalty is associated with the appropriate shielding. The use of optical fibers for the transmission of sensor, encoder, and flight control signals would alleviate this problem. Indeed, in many instances optical fibers can serve as the sensors, encoders, and the data and control signal conduits (Fig. 2). These remarkable virtues of optical fibers could see fly-by-light supersede fly-by-wire in the next decade.

Military aircraft are forever pushing at the boundaries of high performance, and today that comes at the cost of aerodynamic stability. Some of these aircraft are almost impossible to fly without sophisticated flight controls that continuously sense the flight en-

TABLE I. Types of Fiber-Optic Sensors

Category	Single optical fiber	Localized sensing	Single ended	Sensitivity
Interferometric				
Mach–Zehnder	No	Yes	No	High
Michelson	No	Yes	Yes	High
Fabry–Perot	Yes	Yes	Yes	High
Bragg grating	Yes	Yes	Yes	High
Polarimetric				
Low birefringence	Yes	No	Yes	Low
High birefringence with 45-degree splices	Yes	Yes	Yes	Low
Intensiometric				
Bistable	Yes	No	No	Poor
Microbend	Yes	Yes	No	Low
Modified cladding/core	Yes	Yes	Yes	Low
Modalmetric				
Few mode	Yes	Yes	No	Low
Elliptic core	Yes	Yes	No	Low
Twin core	No	Yes	No	Low

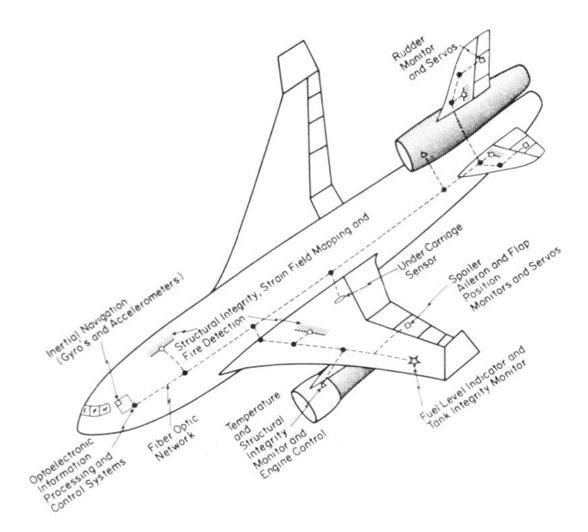

FIG. 2. Potential fiber-optic systems within future aircraft.

vironment and act to keep it stable. Another attractive feature of fiber-optic sensors that are embedded within the structure (or skin) of an aircraft is that they can perform their sensing function without affecting the external aerodynamic shape. This would be particularly important at the supersonic and hypersonic regimes presently being seriously considered even for civilian vehicles and probably represents an extreme example of where vigilant monitoring will be required. [For a broad overview see Measures (1989a).]

II. The Need for Built-in Fiber-Optic Structural Integrity Systems

A. Composite Materials in the Aerospace Field

The special properties of composite materials, such as their high strength and stiffness-to-weight ratio, are leading to their increasing use by the aerospace industry. These attributes not only provide a direct incentive but have a "multiplier" effect that operates to reduce fuel consumption and engine weight. In addition these materials possess other desirable properties, such as low thermal expansion and resistance to fatigue and corrosion. Furthermore, their elastic and thermal properties can be tailored. Although initially composite materials were used only for secondary structures and control surfaces, recently some military and civilian aircraft have used them for primary structures.

Unfortunately, the very nature of these fiber-reinforced polymers makes them prone to certain kinds of damage. In particular, fiber-reinforced composite materials are very susceptible to impact loading, because the epoxy matrix is at least an order of magnitude weaker than the embedded fibers of carbon or Kevlar. The damage characteristics of these materials when subjected to subperforation impact can be classified as indentation, fiber breakage, matrix cracking, fiber-matrix debonding, and interply disbonding, better known as "delamination." Among these damage modes, *delamination* has been found to be the most serious. This form of damage can be induced by relatively low energy impacts and is particularly insidious, as it can reduce both the tensile and the compressive strength with very little visible surface damage. This has given rise to the commonly used acronym, BVID, for barely visible impact damage. Such impacts can result from a carelessly dropped tool, runway stones, hail, or a collision with a bird, and once started the region of damage can grow with load cycling incurred in normal operation.

Although rigorous structural testing is used to ensure that any aircraft structure put into service is safe and can withstand the service loads and environments—degradation of composites do occur due to moisture, chemical attack, thermal spikes, fatigue, overloading, erosion, impact damage, and lightning strikes. Accidental damage can also arise during transportation, installation, and operation. Consequently, a large proportion of aircraft maintenance time is currently spent inspecting and testing composite parts for manufacturing and service-induced defects and cracks. Because visual inspection is of such limited value in these materials, a variety of nondestructive evaluation techniques have been developed for the purpose of assessing damage to composite structural components. Ultrasonic probing is the most widely used NDE technique for composites and bonded joints.

B. Virtues of a Built-in Structural Integrity System

Unfortunately, all conventional damage assessment techniques are limited to ground use and require the structure to be taken out of service. Most techniques, including the more reliable ultrasound C-scan and X-ray techniques, are not practical in the field and often require disassembly and isolation of the component involved. When hand-held scanning systems are employed, the inspections are time consuming and require highly skilled technicians.

Clearly, the development of a *built-in* damage assessment system that could constantly check the structural integrity of each important or critical component could greatly alleviate concerns over the introduction of composite materials, as well as substantially reduce the cost of their use in aircraft due to savings in down time for inspection. For composite materials a built-in monitoring system could be based on a grid of optical fibers embedded within the structure at the time of manufacture. The simplest form of such a system would involve the interrogation of the optical fibers by a source of light, and disruption in their transmission would be used to determine the location and extent of damage. This could certainly be used in the case of damage resulting from impacts, but could also be used more generally. Growth of a damage zone resulting from fatigue, mishandling, excessive load, or flaws created during manufacture could also be detected by this system. Structural integrity monitoring by an embedded optical fiber array could also be of considerable benefit to many other structures that rely heavily on composite materials, such as storage tanks, pressure vessels, and pipelines.

III. Structurally Integrated Fiber-Optic Sensors

Optical fibers are usually a little thicker than a human hair and basically comprise three cyclindrical regions: a *core*, where the refractive index is a little higher than in the surrounding *cladding*, and a *buffer* or protective jacket (Fig. 3). Light launched into the core of an optical fiber is confined and guided over considerable distances. This has led the communication industry to gradually replace electrical cables with optical fibers, with the substitution of photons for electrons as the information media. Multimode optical fibers have a core radius that is large enough for light to propagate in many different patterns or modes, and an appreciable amount of power can be transmitted. Single-mode optical fibers, on the other hand, have a very small core radius (comparable to the wavelength of light) and transmit only the lowest order mode.

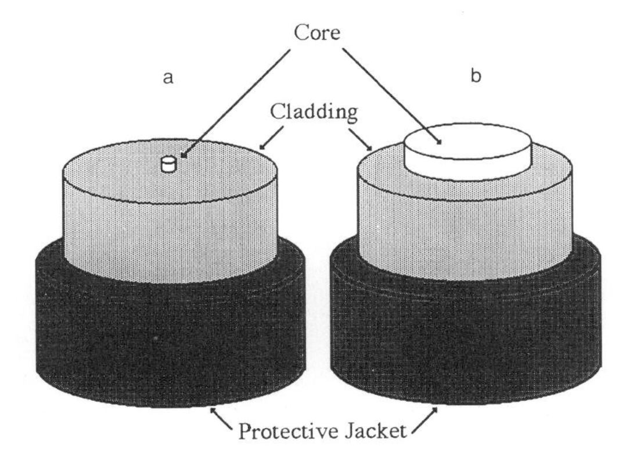

FIG. 3. Single and multimode optical fibers.

A. Merits of Fiber-Optic Sensors

Optical fibers possess many attractive features that make them eminently suitable to serve as the sensory system for a smart structure. In essence they are

light in weight
very small
immune to electromagnetic interference
inert and corrosion resistant
safe and rarely initiate fires
capable of being very sensitive sensors
embeddable within composite materials
nonperturbing in regard to structural properties
multifunctional
compatible with optical bus or network
amenable to multiplexing and signal processing with integrated optics

Although *structurally integrated fiber-optic reticulate sensors* (SIFORS) technology is in its infancy, its elements are well defined and are illustrated in the generic view presented as Fig. 4. Light is launched into a grid or network of optical fiber sensors that are embedded within a composite material structure through some input interface (connector) and is subsequently modified by its passage through the structure. Information about the state of the structure is impressed upon the transmitted (or in certain instances the reflected) light by a number of mechanisms. These include interactions that change the intensity, phase, frequency, polarization, wavelength, or modal distribution of the radiation propagating along the optical fiber. What makes the fibers particularly attractive for embedding within composite structures is that for the most part they serve the dual role of sensor and pathway for the signal, and since they are dielectric in nature they are compatible with the composite material, avoiding the creation of electrical pathways within the structure. This will become very important as composites are used increasingly in aircraft that have to fly through thunder storms.

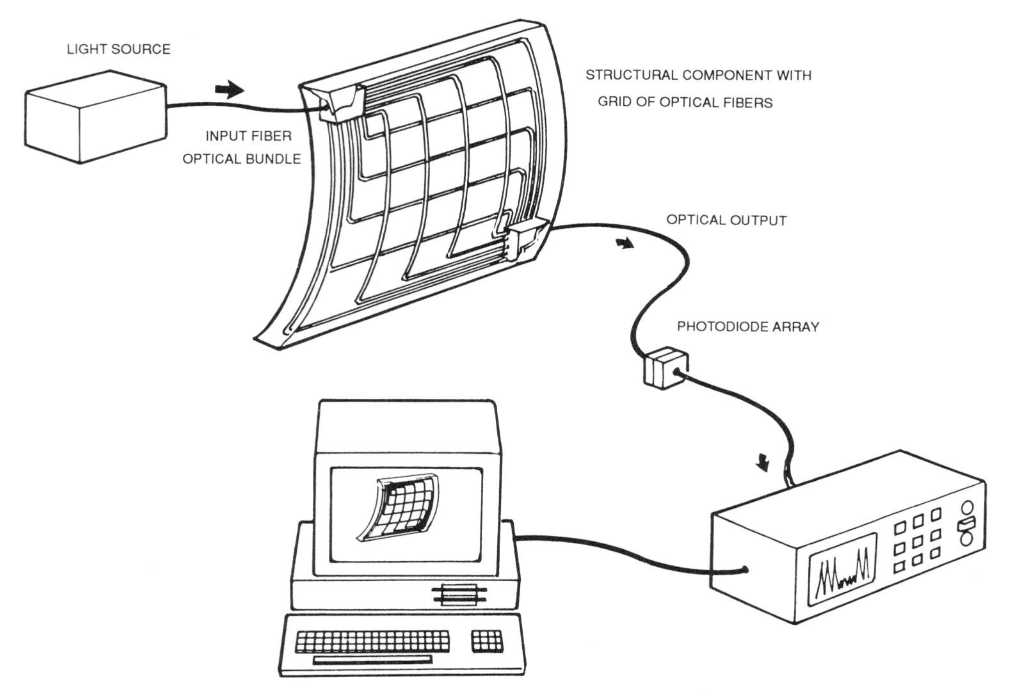

FIG. 4. Generic fiber-optic smart structure involving a light source, an input optical fiber (or bundle), the optical grid or sensor array, the photodetection system, and the interrogation system with a computer and monitor.

B. Types of Fiber-Optic Sensor

Fiber-optic sensors can be classified into four major categories: intensiometric, interferometric, polarimetric, and modalmetric. The type of fiber-optic sensor selected will primarily depend on the application. If more than one type of optical fiber sensor could be used for a given measurement then secondary considerations, such as the ease of integrating the sensor into the structure, might serve to determine the selection. Three questions pertinent to this selection are whether only a single optical fiber is required per sensor; whether the sensing region can be localized (i.e., made with insensitive lead-in and lead-out optical fibers); and whether the sensor can be single-ended. The sensitivity is also very important. Some idea of the relative sensitivity of the various sensors can be gauged by judging them in terms of their sensitivity to strain. Table I provides a finer subdivision of fiber-optic sensors and a comparison of their relative attributes.

Intensiometric sensors depend on a variation of the radiant power transmitted through a multimode optical fiber. The simplest form of this sensor involves the presence or absence of light (a bistable approach). For example, fracture of an optical fiber can serve as the basis of a damage-sensing system. It has long been recognized that small bends in an optical fiber lead to a loss of its light, as the curvature permits core radiation to leak into the cladding, from which it escapes the fiber. The smaller the radius of curvature the greater the loss. Microbend fiber-optic sensors rely on this principle, using a small device to convert an applied force into a microbend in the optical fiber and thereby reduce its transmission of light. Optical time domain reflectometry (OTDR) uses a time-of-flight measurement to not only detect but also locate breaks in an optical fiber by means of Fresnel reflection from the fracture. Other types of intensiometric fiber-optic sensors are based on fluorescence, Raman scattering, and blackbody emission. Optical fibers with modified segments of either the core or the cladding can also serve as sensors. These are used often to measure temperature; an index-matching effect can be used for composite cure monitoring.

Interferometric sensors represent a large class of extremely sensitive optical fiber sensors. These primarily function by sensing the phase change induced in light that has propagated along a single-mode optical fiber. There are five basic types of interferometric fiber-optic sensor: Mach–Zehnder, Michelson, Fabry–Perot, Bragg, and the Sagnac. The latter is primarily used in the development of fiber-optic gyros and for this reason is not included in Table I. The Michelson uses two closely spaced single-mode optical fibers, where one optical fiber serves as a reference. The sensing region is localized between the mirrored ends of the two optical fibers (Fig. 5), and interference associated with changes in the strain or temperature of the structure modulates the intensity of light incident on the detector. The coupler acts to mix the light from the two optical fibers. The differential Mach–Zehnder fiber-optic sensor is primarily a transmissive-based interferometer that also involves two closely spaced optical fibers, where localization is achieved by the introduction of an additional element of length in the sensing optical fiber. If both the input and output interfaces are on the same side of a structure, the optical fibers form two closely spaced loops and the sensing region becomes the small difference in the length between these two loops.

A Fabry–Perot fiber-optic sensor involves a single monomode optical fiber with a sensing region defined by a cavity comprising two mirror surfaces that are parallel to each other and perpendicular to the axis of the optical fiber. A change in the optical path length between the mirrors leads to a shift in the frequencies of the cavity modes. In some ways the Fabry–Perot cavity represents the simplest interferometric sensor and has several very attractive features: the reference and sensing optical fiber are one and the same, up to the first mirror that constitutes the sensing region; it can form a single-ended configuration (see Fig. 6); it can be wavelength-multiplexed; there is no need for phase preservation in the connector to a structure; it is a very sensitive sensor and is capable of excellent spatial resolution with a well-defined sensing region.

Bragg reflection from a region of periodic variation in the core index of refraction can also form the basis of a fiber-optic sensor. In this system the spectrum of light reflected from the Bragg grating is in the form of a narrow spike with a center wavelength that is linearly dependent on the product of the wavelength of the periodic variation and the mean core

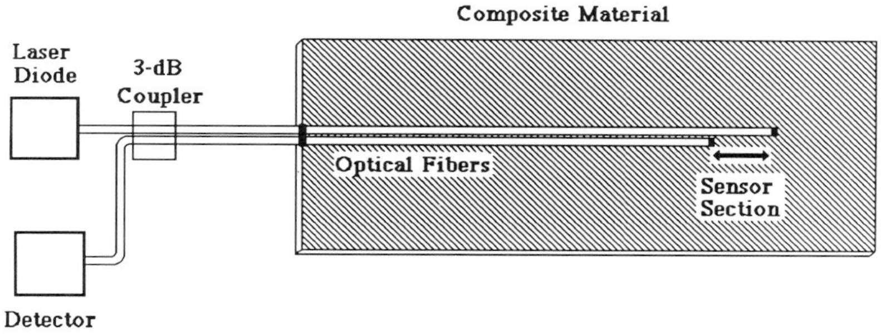

FIG. 5. Michelson interferometric fiber-optic sensor embedded within a composite panel.

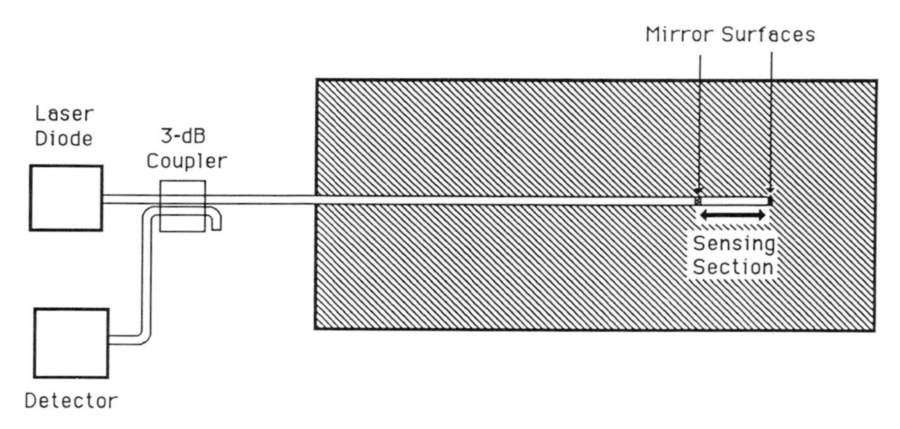

FIG. 6. Fiber-optic sensor based on a Fabry–Perot interferometer.

index of refraction. Consequently, changes in strain or temperature to which the optical fiber is subjected will shift the reflected spectrum center wavelength and constitute the mode of measurement.

Modalmetric sensors involve changes in the distribution of light within the optical fiber. These involve single optical fibers in which modal interferometry is exploited in circular or elliptic core, single-mode optical fibers excited with light below the cutoff wavelength (the wavelength above which light propagates in only a single spatial mode). In general the sensitivity of this type of sensor is far less than the interferometric sensors discussed above, and localization of the sensing region is difficult. Fiber-optic sensors that rely on evanescent coupling of light can also be included under this heading, since these devices may also be thought to involve a redistribution of modal energy. In a twin-core fiber the evanescent field in the cladding permits the light to be exchanged between the two cores in concert with changes in the strain or temperature of the optical fiber.

Polarimetric sensors rely on evaluating the change in the phase of the two orthogonal polarization eigenmodes of a single-mode optical fiber. Consequently, a measurement of the change in the state of polarization of the light emanating from the optical fiber can be employed to determine the change in the strain or temperature to which the optical fiber is subjected. The polarimetric fiber-optic sensor can also be single-ended by employing a mirrored end, and the sensing region can be localized by fusion-splicing it to the lead-in/out optical fiber with its axes rotated through 45° (Fig. 7).

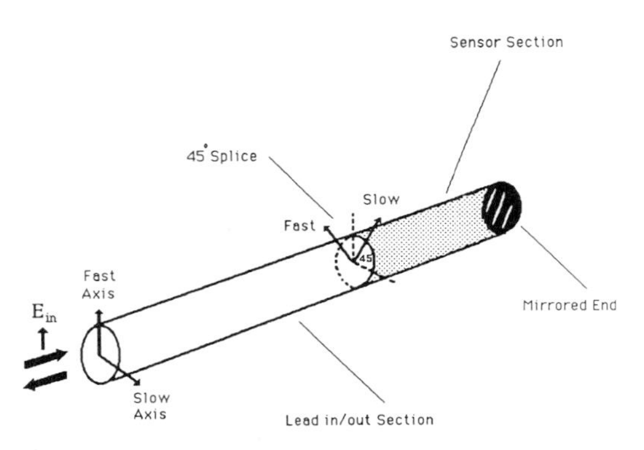

FIG. 7. Polarimetric fiber-optic sensor in which the sensing section is *localized* between the mirrored end and the 45-degree rotation of the polarization eigenaxes. Light launched into the lead-in section of optical fiber is linearly polarized and aligned with one of the fiber's polarization eigenaxes.

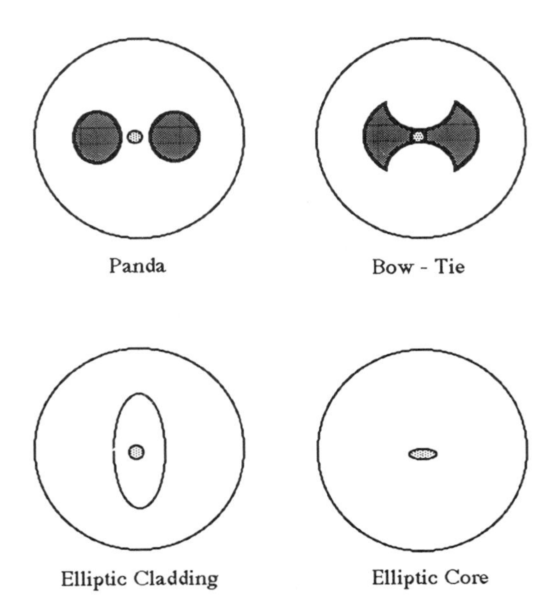

FIG. 8. Four types of polarization-maintaining optical fibers.

Unfortunately, the nonlinear (cosine) relationship between the signal and the environmental factors, such as strain and temperature, in a structure is common to all of the sensors (see the theoretical section), with the exception of intensiometric and Bragg sensors. This gives rise to serious problems of signal fading and ambiguity in regard to the direction of phase change. There are, however, a number of quadrature detection schemes for overcoming this difficulty (Jackson and Jones, 1986).

For sensitive fiber-optic interferometric sensors, the circular symmetry of the optical fiber may not be adequate to stabilize the two orthogonal linear polarization eigenmodes, and this can lead to *polarization-fading* and the possibility of cross coupling between the modes. These effects represent a source of noise in interferometric measurements based on small changes (of the order of a few fringes). One of the most common methods of avoiding this problem is to use polarization-maintaining single-mode optical fiber. The high birefringence designed into this kind of optical fiber stabilizes the polarization because the two eigenmodes are forced to have an appreciable difference in their respective propagation constants. This also limits the exchange of energy between the two modes. This high birefringence is created in one of two ways. The optical fiber is fabricated with a deliberate asymmetry, such as an elliptic core, or by stress-loading the core. Several designs have been introduced to this end (Fig. 8).

C. System Considerations

1. Configuration and Structural Considerations

Designing a smart structure will require questions of *configuration* to be addressed. For example, should the fiber-optic sensors be embedded within the structure or bonded to its surface; can the fiber-optic sensors be double-ended or are they required to be single-ended (i.e., the signal exits through the same optical fiber as the exciting light entered the structure); are the fiber-optic sensors restricted to a single plane (which would be the case if surface bonded, since optical fibers should not cross each other), or can an orthogonal grid be incorporated in the design; should the fiber-optic sensors make point, line, or distributed measurements.

Other important system design considerations would include the density of the optical fiber sensors, their arrangement, and if embedded, their depth and orientation with respect to the reinforcing fibers. Extensive studies at the University of Toronto Institute for Aerospace Studies on damage assessment for composite materials using embedded optical fibers has provided answers to some of these questions. This work demonstrated that where fracture of an embedded optical fiber is used to sense damage the optimum orientation is perpendicular to the adjacent reinforcing fibers. Conversely, if the optical fibers are required to have the greatest chance of surviving in the presence of damage (as would be the case if they were serving as data conduits or sensors), they should be oriented parallel to the adjacent reinforcing fibers (see Fig. 9). If damage-sensing optical fibers cannot be placed between collinear plies, then their optimum orientation would be dictated by the direction from which damage is likely to originate (Fig. 10).

This research also indicated that for damage detection the embedded optical fibers should be located as close to the surface of maximum tensile strain as possible. In the case of impacts on thin laminated structures that are able to flex, this would be the rear surface, (i.e., the surface farthest from the impact). In the case of thick structures or ones that are more rigidly supported, contact stresses can lead to high tensile strain close to the impact surface.

Careful consideration has therefore to be given to the placement of optical fibers intended to remain functional in the presence of impact damage. Research at the University of Toronto Institute for Aerospace Studies has also demonstrated that it may be necessary to intermittently etch the optical fibers to

Optimum Configuration for Damage Sensing Based on Fracture of the Optical Fibers

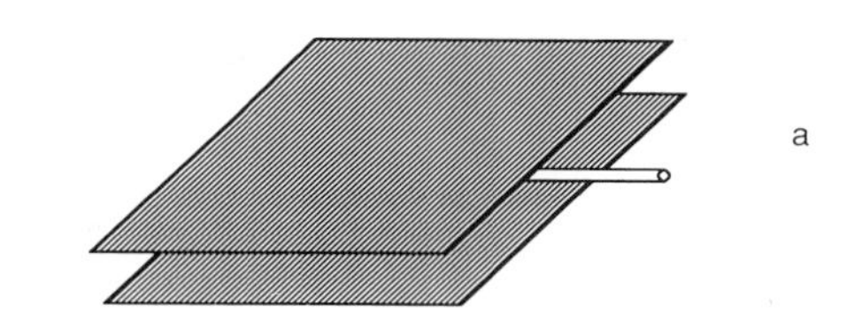

Optimum Configuration for Survivability of Optical Fibers Used as Sensors or Data Links

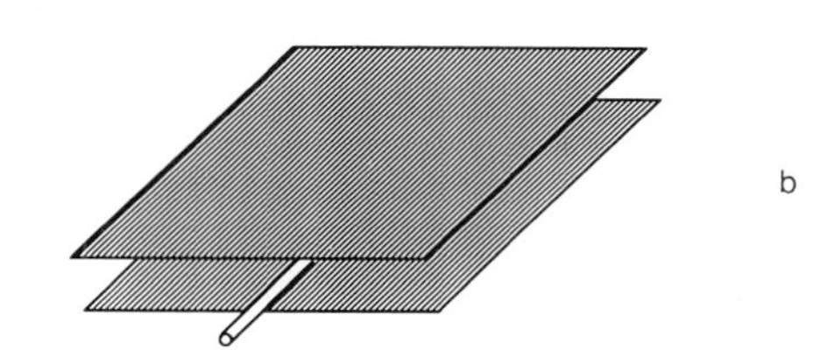

FIG. 9. Optimum orientation for optical fibers embedded between collinear plies within composite materials for the purpose of (a) evaluating damage based on fracture of the optical fibers and (b) serving as data links or sensors.

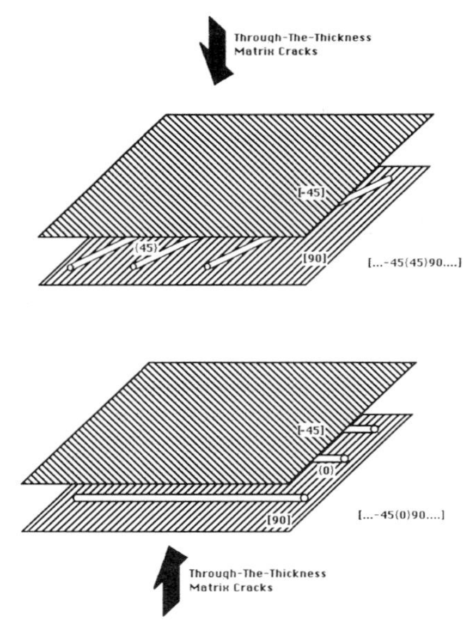

FIG. 10. Illustration of the rule used to determine the best orientation for optical fibers embedded between two noncollinear plies of composite material for the purpose of evaluating damage based on fracture of the optical fibers. This rule states that the optical fibers should be oriented orthogonal to the ply (i.e., the direction of the reinforcing material fibers) that lies between the optical fibers and the most likely site of damage.

control their damage sensitivity. A list of some of the fundamentals of composite damage assessment based on the fracture of embedded optical fibers is presented in Table II. This includes identification of through-the-thickness matrix cracks as the primary mechanisms for fracture of optical fibers embedded within composite materials. In the case of damage assessment systems based on fracture of an embedded grid of optical fibers, three basic configurations are possible. These are illustrated in Fig. 11.

Other factors that need to be addressed when designing fiber-optic sensors to be embedded within a composite structure include the material of the optical fibers in terms of its temperature survivability during the cure and compatibility with the material of the structure; the size of the optical fibers in order to minimize their perturbation; the type of coating needed to protect the optical fibers and to provide the appropriate coupling to the structure for a given application.

2. Influence of Embedded Optical Fibers on Material Properties

The influence of optical fibers on the properties of the structure in which they are embedded can also be regarded as an important issue under system considerations. Some work has already been undertaken in an attempt to address this concern, and preliminary results from the University of Toronto Institute for Aerospace Studies suggest that no noticeable deterioration of the mechanical properties occurs. In particular, it has been shown that neither the tensile nor the compressive strengths of Kevlar/epoxy panels are compromised by the presence of embedded optical fibers of 125-μm diameter (see Fig. 12). It has also been shown that, if anything, the critical energy release rate is increased by the presence of embedded optical fibers (see Table III). This suggests that embedded optical fibers can actually enhance the fracture toughness of the composite material in regard to delamination. However, much more definitive research will be needed before optical fibers can be embedded with confidence within structures, such as aircraft components, intended to have a 15- to 20-year working life.

In the case of carbon/epoxy material, embedded 125-μm optical fibers, with their buffer coating removed, have been shown photographically to cause a minimum perturbation to the microstructure when oriented parallel with the reinforcing carbon fibers. However, when similar optical fibers are oriented perpendicular to the reinforcing carbon fibers, sub-

TABLE II. Fundamentals of Composite Damage Assessment Based on the Fracture of Embedded Optical Fibers

- *Fracture mechanism* for embedded optical fibers is through-the-thickness *matrix cracks*.
- *Damage sensitivity* of embedded optical fibers can be *tailored* by an intermittent etching process.
- *Maximum damage sensitivity* is attained when the optical fibers are embedded close to the surface of maximum tensile load.
- *Threshold* impact energy for fracture of embedded optical fibers depends on composite layup and orientation and placement of the optical fibers.
- Damage sensitivity and reliability are *optimized* when the optical fibers are embedded between *collinear* plies and oriented *orthogonal* to ply direction.
- Optical fibers embedded between *noncollinear* plies should be oriented *orthogonal to the ply* separating them from the region of damage.
- If this aligns the optical fibers with the other ply, then the orientation should bisect the larger angle between the plies.

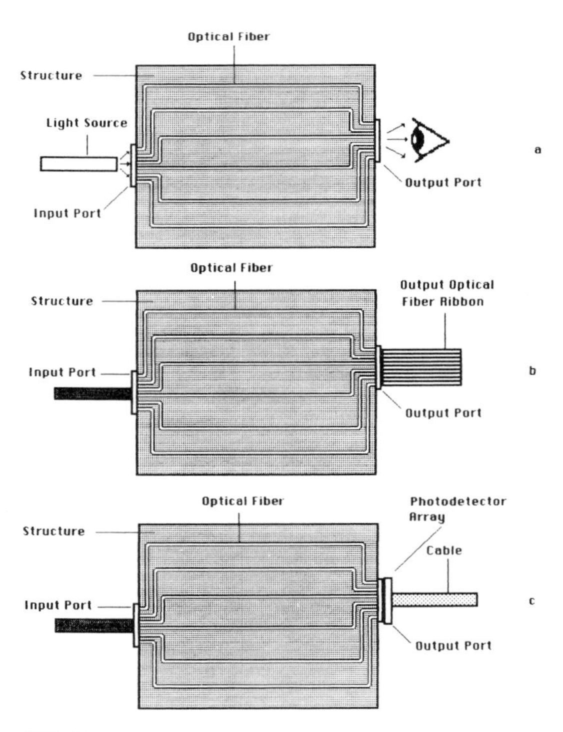

FIG. 11. Three possible optical interface arrangements for a transmissive based optical-fiber sensor system.

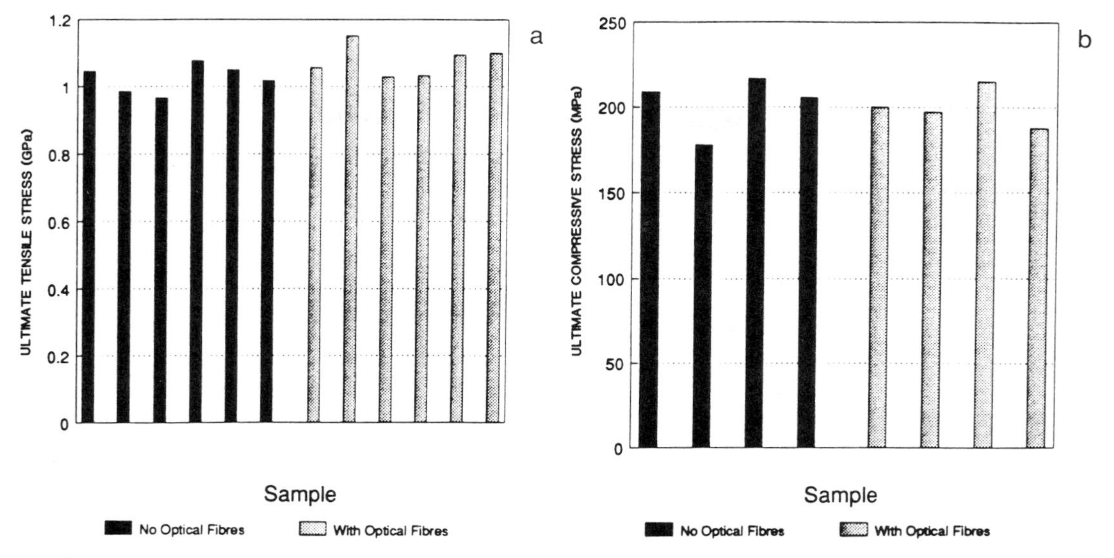

FIG. 12. (a) Ultimate tensile strength of instrumented [0, 0{90}0, 0] and uninstrumented $[0_4]$ 4-ply Kevlar/epoxy laminates. (b) Ultimate compressive strength of instrumented $[0_4\{90\}0_4]$ and $[0_8]$ 8-ply Kevlar/epoxy laminates.

TABLE III. Critical Energy Release Rate, G_c (100 mm) for Kevlar/Epoxy Panel Samples with and without Embedded Optical Fibers

Composite sample configuration	Critical energy release rate, G_c
Reference sample with no optical fibers	
$[O_8/O_8]$	1.09 (kJ/m^2)
$[O_8\{0\}O_8]$	1.12 (kJ/m^2)
$[O_8\{45\}O_8)$	1.14 (kJ/m^2)
$[O_8\{90\}O_8]$	1.24 (kJ/m^2)
Reference sample with no optical fibers	
$[O_2/90/O_5/90/O_5/90/O_2]$	1.43 (kJ/m^2)
$[O_2/90/O_5\{0\}90/O_5/90/O_2]$	1.44 (kJ/m^2)
$[O_2/90/O_5\{45\}90/O_5/90/O_2]$	1.46 (kJ/m^2)
$[O_2/90/O_5\{90\}90/O_5/90/O_2]$	1.48 (kJ/m^2)

stantial resin voids are created around each optical fiber. This perturbation to the microstructure is more severe when the acrylate buffer coating is left on the embedded optical fibers, since their diameter is then about 250 μm and the coating melts if it is acrylate. Clearly, careful consideration will have to be given to the diameter of optical fibers and the nature of their coatings if the fibers are to be embedded within composite structures and function correctly for the useful life of the structure.

3. Demultiplexing/Multiplexing

Multiplexing is the merging of data from a number of different channels into one channel, while demultiplexing is the inverse (Fig. 13). Both are expected to play an important role in the development of smart structures. The primary parameters used in multiplexing schemes are wavelengths, time, frequency, phase, and space. Wavelength division multiplexing (WDM) involves each sensor operating at a different wavelength and looks the most promising for certain types of fiber-optic sensors to be used in smart structures. The relevant wavelength encoding can be performed by the light source, spectral filters, resonant cavities, or dispersive elements. Demultiplexing is achieved by means of a spectral analyzer.

If the sensors have to be single ended, then, as shown in Fig. 14(a), couplers would be required to divert some fraction of the returning light signal to the photodetection system. It has been assumed that the interface to the structure would include a device for optical demultiplexing/multiplexing so that only one (armored) optical fiber is needed to convey the interrogating light in and the signals out of the structure. If this is not the case and each optical fiber sensor has its own input fiber, it would also require its own coupler, light source, and detector, and the system would be bulky, heavy, expensive, and less practical.

In the case of a double-ended system there are three possible detection modes. The output light from

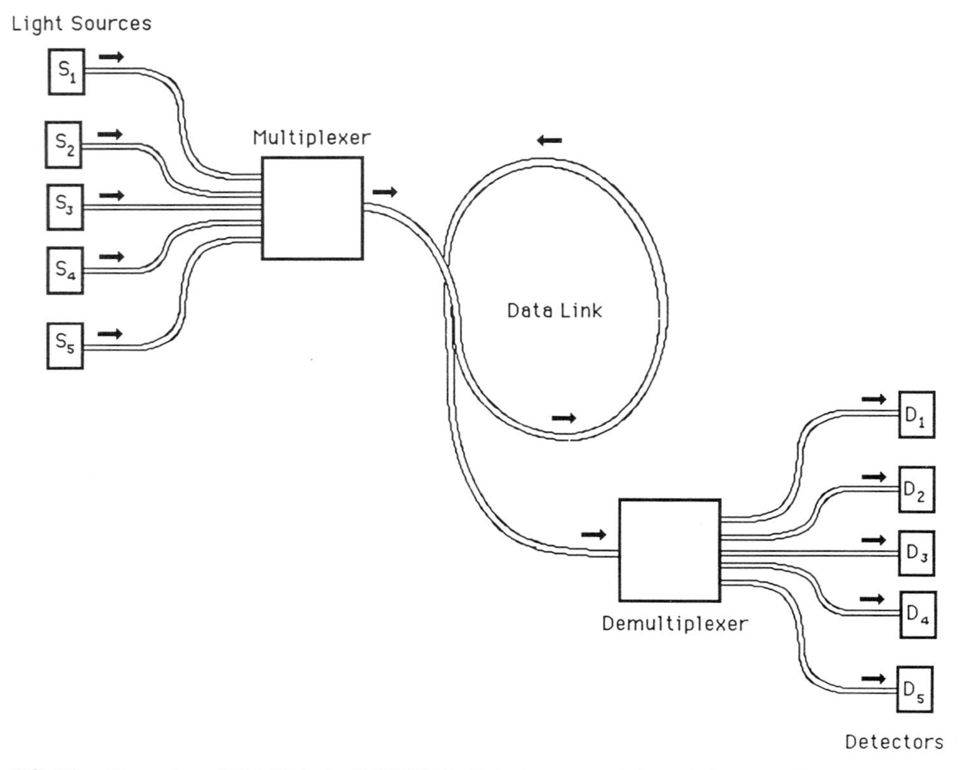

FIG. 13. Illustration of "Multiplexing" (MUX) the light from several channels into one channel and "Demultiplexing" (DEMUX) the light from one channel into several channels.

the network of optical fiber sensors could be communicated to an external photodetection system by means of an optical ribbon or fiber-optic bundle; if spatial coherency is not maintained, it would be necessary to identify the detector viewing each optical fiber sensor. An alternative arrangement would involve mounting the photodetection system directly on the output port; this may be practical in many situations because of the small size of photodiode arrays. The third possibility would rely on a small multiplexing output device to recombine the optical signals and send them all down one armored optical fiber to the photodetection system [see Fig. 14(b)]. We have assumed in each of these double-ended arrangements that the interrogating light is delivered to the structure through one armored optical fiber and is then distributed to each of the fiber-optic sensors by means of an optical demultiplexer. For many situations it may be just as convenient to use a fiber-optic bundle, or an optical ribbon, to deliver the light to each sensor on a one-to-one basis.

In terms of system components, the factors that would determine the choice of light source include wavelength, brightness, size, whether pigtailed to an optical fiber, and cost. The first decision in regard to the choice of photodetector would hinge on whether a single detector or an array of detectors is needed; then the questions of wavelength response, sensitivity, and size would have to be addressed. If couplers are required it would then be necessary to decide between 3-dB (50:50) couplers or asymmetric couplers and to consider if they need to be polarization-splitting couplers.

IV. Single-Mode Optical Fiber Sensor Theory

Strain, deformation, and structural loading can be evaluated by phase or polarization transduction. Figure 15 represents a generic system that can be used to discuss either interferometric or polarimetric fiber-optic sensors. In principle, the *signal*, as influenced by the measurand field through the transduction mechanism, is mixed with a *reference* wave and the resulting interference signal is detected. If the signal and reference wave have the same optical frequency

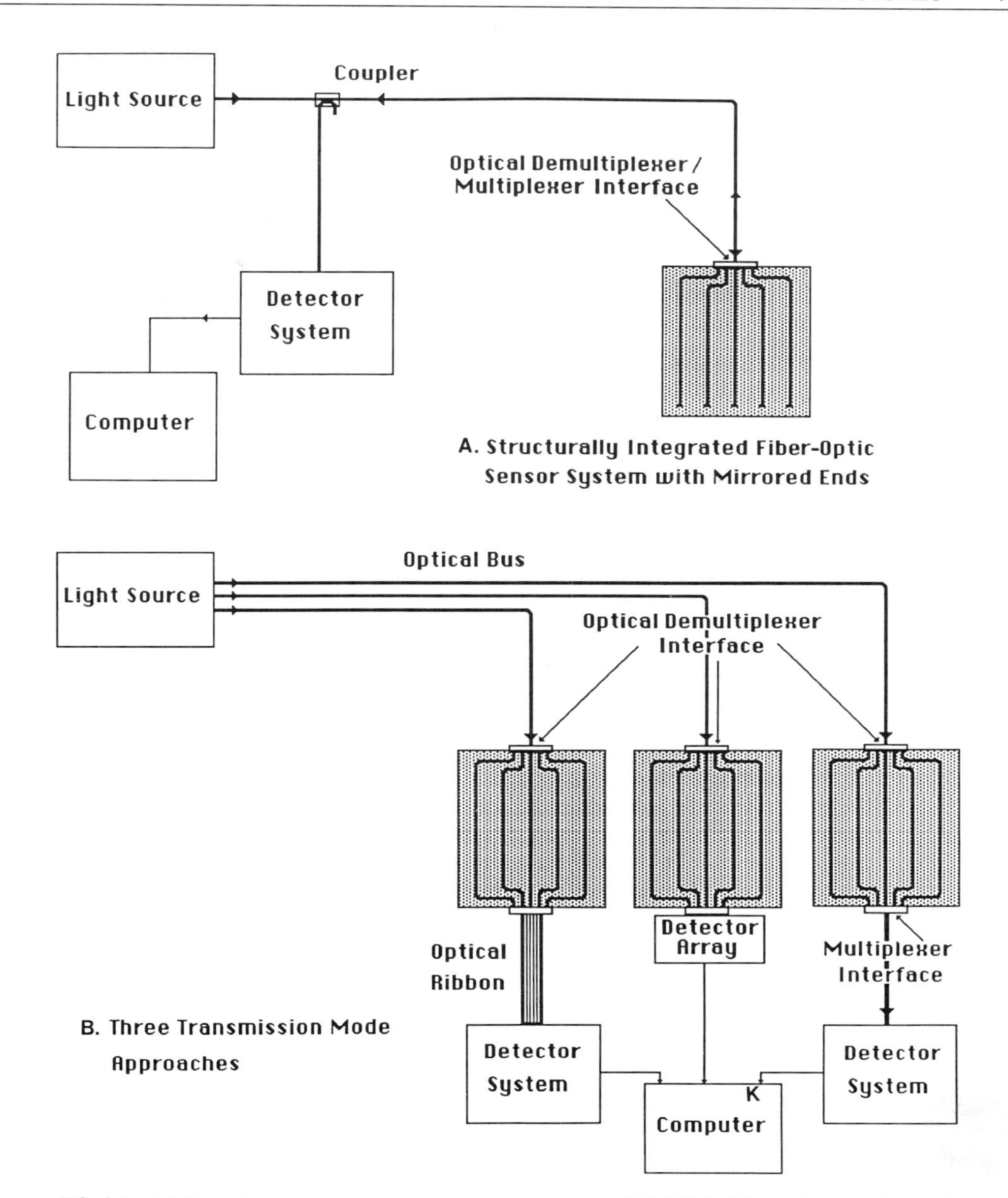

FIG. 14. (a) Illustrates how the combination of a coupler and a DEMUX/MUX interface would permit an array of structurally integrated single-ended fiber-optic sensors to be interrogated by a single input/output optical fiber.(b) Illustrates three detection modes possible with double-ended, transmission based structurally integrated fiber-optic sensor arrays. Excitation of the array of optical fibers in each structure is assumed to be accomplished by only one input optical fiber and an appropriate demultiplexer interface.

we refer to "homodyne" detection; if their frequencies are different we speak of "heterodyne" detection. We assume no attenuation.

In both a Michelson and a Mach–Zehnder interferometer the signal and reference paths are spatially separate, and we assume that none of the components are birefringent. In the case of polarimetry the two waves are orthogonal polarization states that are spatially equivalent. The couplers in Fig. 14 are described by matrices K_1 and K_2, while the two paths

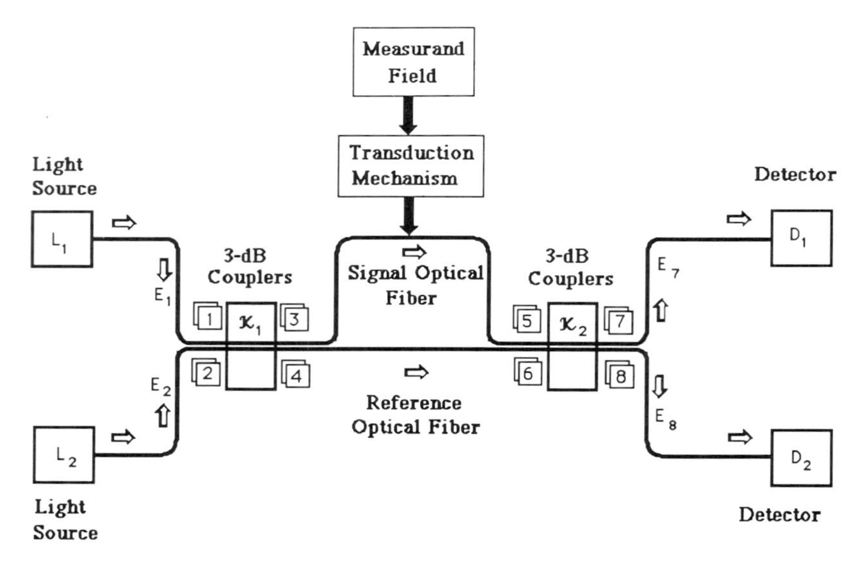

FIG. 15. Generic theoretical model of a single-mode optical fiber sensor.

(a) and (b) are together described by the transduction matrix T. In general the detected (output) field $\{E_7, E_8\}$ is related to the incident (input) field $\{E_1, E_2\}$ by the relation:

$$E_{\text{out}} = K_2 T K_1 E_{\text{in}} \tag{1}$$

If we assume 3-dB couplers so that the power is shared equally between the two outputs, then

$$K_1 = K_2 = \frac{1}{\sqrt{2}} \begin{bmatrix} 1 & i \\ i & 1 \end{bmatrix} \tag{2}$$

and the transduction matrix can be expressed in the form

$$T = \begin{bmatrix} \exp(i\phi_a) & 0 \\ 0 & \exp(i\phi_b) \end{bmatrix} \tag{3}$$

where ϕ_a and ϕ_b represent the phase change experienced by the waves propagating along paths (a) and (b), respectively.

Substitution of Eqs. (2) and (3) into (1) yields

$$E_7 = E_1 \{\exp(i\phi_a) - \exp(i\phi_b)\}/2 \tag{4}$$

and

$$E_8 = iE_1 \{\exp(i\phi_a) + \exp(i\phi_b)\}/2 \tag{5}$$

The corresponding photodetector signals for output ports 7 and 8 are

$$I_7 = I_1[1 - \cos(\phi_a - \phi_b)]/2 \tag{6}$$

and

$$I_8 = I_1[1 + \cos(\phi_a - \phi_b)]/2 \tag{7}$$

In general we can write

$$\Delta\phi = \phi_a - \phi_b \tag{8}$$

We see that each of the outputs are cosine in nature, which means that they are multivalued, clearly nonlinear. Furthermore, the signals cannot be used to determine the direction of change; in other words, $\cos(\Delta\phi) = \cos(-\Delta\phi)$, and the response becomes very insensitive whenever $\Delta\phi = n\pi$, $\{n = 0, 1, 2, \ldots\}$. This latter condition is termed *signal fading*.

Both of these problems can be avoided if we have two *quadrature* outputs, that is, two signals that differ by $\pi/2$ and can be expressed in the form

$$X = A \cos\{\Delta\phi\} \tag{9}$$

and

$$Y = B \sin\{\Delta\phi\} \tag{10}$$

for then one would always be at the point of maximum sensitivity when the other was at the point of minimum sensitivity. In fact, if the following operation is performed,

$$\int \{X\, dY/dt - Y\, dX/dt\}\, dt = -AB\, \Delta\phi \tag{11}$$

we see that a signal directly proportional to $\Delta\phi$ can be obtained. Unfortunately, we see from Eqs. (6) and (7) that the outputs from the interferometer are antiphase (i.e., different by π). If the typical measurements involve many cycles (or fringes) then signal fading is not an issue; however, the *ambiguity* in regard to the direction of change remains a problem.

In the presence of attenuation the signals can be expressed in the form

$$I_7 = I_1[1 - U\cos(\Delta\phi)]/2 \quad (12)$$

and

$$I_8 = I_1[1 + U\cos(\Delta\phi)]/2 \quad (13)$$

where $0 < U < 1$ is termed the *visibility* constant. In the case of a Mach–Zehnder interferometer the signal and reference paths are spatially separate, and ϕ_a and ϕ_b represent the phase change experienced by the waves propagating along the physically different paths (a) and (b), respectively.

Two techniques that can be used to produce quadrature signals and are suitable for smart structures involve

1. Replacing the 4-port (2 × 2) output direction coupler of Fig. 14 with a 6-port (3 × 3) direction coupler.
2. Switching the absolute frequency of the laser (in a time that is short compared to the times of interest) between two frequencies f_1 and f_2, which correspond to an effective phase difference for the unbalanced interferometer of $\pi/2$, that is,

$$2\pi nL\{f_1 - f_2\}/c = \pi/2 \quad (14)$$

One of the simplest detection modes for a fiber-optic Mach–Zehnder interferometer involves mixing the outputs from the two optical fibers to form a fringe pattern and detecting any phase shift from the movement of this fringe pattern. This is accomplished with two strategically positioned detectors, designed to sample a small fraction of a fringe. Although this can provide quadrature detection, a superior technique that utilizes a much larger fraction of the light involves the use of two gratings, each shifted by an appropriate amount, to provide the desired quadrature demodulation. The advantages of this approach are that the interrogation system can be compact, requires no feedback loops or couplers, and there is no requirement for generation of a carrier frequency nor the introduction of any device in the reference arm. Furthermore, low noise is achieved, and a frequency response of 50 MHz is attainable.

In general the change in phase $\Delta\phi$ can be related to the physical change in length ΔL and the change in the index of refraction Δn, by the relation

$$\Delta\phi = k[n\Delta L + L\Delta n] \quad (15)$$

The second term, which involves the change in the optical waveguide properties, is the more significant. Of the two phenomena, the strain–optic effect and mode dispersion, that constitute this second term the latter is usually negligible. Consequently, we can write

$$\Delta n = -n^3[(P_{11} + P_{12})e_r + P_{12}e_z]/2 \quad (16)$$

where P_{11} and P_{12} represent the strain–optic (photoelastic) coefficients, and e_r and e_z represent the radial and longitudinal strain, respectively. Values of these strain–optic coefficients, P_{11} and P_{12}, for single-mode optical fibers with a pure silica core and B_2O_3 doped cladding, are $P_{11} = 0.113$ and $P_{12} = 0.252$, values that are about 7% lower than for bulk silica.

In the case of pure axial strain,

$$e_r = -\mu e_z \quad (17)$$

where μ is Poisson's ratio (typically 0.17 for glass). In which case Eq. (17) can take the form

$$\Delta\phi = kLne_z[1 - n^2\{(1 - \mu)P_{12} - \mu P_{11}\}/2] \quad (18)$$

For a given wavelength this can be expressed in terms of a *strain gauge factor* S, that is to say,

$$\Delta\phi = SLe_z \quad (19)$$

In the case of silica optical fibers (with $n = 1.458$),

$$S = 1.13 \times 10^7 \text{ rad strain}^{-1}\text{ m}^{-1}$$

When considering the phase change of an optical fiber bonded to, or embedded within, a structure the issue of *apparent strain* has to be considered. From Eq. (15) we can write

$$\Delta\phi/\Delta T = k[n\Delta L/\Delta T + L\Delta n/\Delta T] \quad (20)$$

When the thermal expansion coefficient for the structure differs from that of the optical fiber, a change of temperature leads to the creation of differential strain between the optical fiber and the structure. A similar problem arises with electrical resistive strain gauges. In this case the effect is minimized by tailoring the strain gauge material to have a thermal expansion coefficient as close as possible to that of the structure. The problem is complicated further in the case of composites because their anisotropic properties lead to an *apparent strain* that depends quite strongly on the orientation of the optical fiber relative to the reinforcing fibers. It should be noted that this effect always needs to be taken into account and is not restricted to interferometric sensors. Indeed, a comparison between polarimetric and interferometric fiber-optic sensors has found that the former has an "apparent strain" sensitivity of 20 μstrain/°C compared to a value of 7 μstrain/°C for the Michelson fiber-optic sensor.

In polarimetry the waves travel the same physical

path, but the two orthogonal polarization states travel with different velocities due to the birefringent properties of the optical fiber. Under these circumstances,

$$\Delta\phi = kLB \quad (21)$$

where k is the free-space propagation constant, L is the length of the optical fiber, and

$$B = [n_f - n_s] \quad (22)$$

is the *birefringence* of the optical fiber, with n_f and n_s representing the index of refraction for the fast and slow eigenmodes, respectively. The *beat length* L_p of an optical fiber is given by the simple relation

$$L_p = 2\pi/kB \quad (23)$$

and we see that

$$\Delta\phi = 2\pi L/L_p \quad (24)$$

which means that the beat length corresponds to the length of optical fiber needed for the two orthogonal polarization modes to change their relative phase by 2π.

Polarimetric-based fiber-optic sensors can be viewed as a form of differential interferometer in which radiation in the two polarization eigenmodes propagate at different speeds within a single-mode optical fiber. The state of polarization, SOP, consequently evolves along the optical fiber in the manner indicated in Fig. 16. Optimized results are obtained if the optical fiber is highly birefringent and both polarization eigenmodes are equally excited. This latter condition can be achieved if the light launched into the optical fiber is either circularly polarized or linearly polarized at 45° to the polarization eigenaxis of the fiber. In general the sensitivity of these sensors is significantly (about 100 times) less than that of interferometric fiber-optic sensors.

The usual method of localizing the sensing region of a polarimetric-based fiber-optic sensor is to orient the polarization eigenaxes of this section at 45° to that of the lead-in and lead-out sections of the optical fiber (Fig. 7). It is also important to ensure that the light launched into the lead-in optical fiber is linearly polarized and aligned with one of the polarization eigenaxes. This is most easily accomplished in an all-fiber system by using polarizing optical fiber in the lead-in and lead-out sections. An alternative, and more readily available, method of launching light correctly into the optical fiber would use a small thin-film polarizer in the connector to the structure. The transmission axis of this polarizer would have to be accurately aligned with one of the eigenaxes of the polarization-maintaining lead-in optical fiber. This type of sensor can be operated in either a transmission or reflection mode. The latter approach, in which the sensing section is terminated by a mirrored end, is more attractive from the smart structures perspective in that it constitutes a single-ended device, is twice as sensitive for the same sensing length, and requires only one 45° splice. Recently it has been shown possible to achieve a 45° rotation by means of local heating and twisting of the optical fiber. This eliminates the need for the fusion splices.

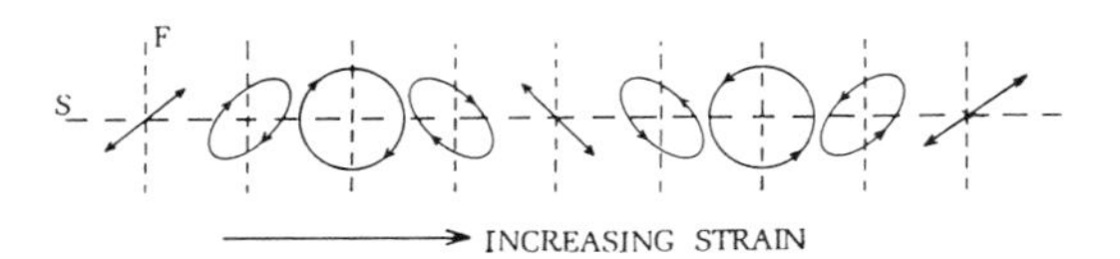

FIG. 16. Variation in the state of polarization of light emerging from a birefringent medium subject to increasing strain.

The state of polarization of linearly polarized light launched into one of the polarization eigenmodes of the lead-in optical fiber is unchanged by environmental effects as it propagates along this section of optical fiber. On arriving at the 45° splice it will equally populate the two polarization eigenstates of the sensing section of the optical fiber. After traversing the sensing region (once for the transmission-based system or twice for the case of the mirrored end) these orthogonally polarized beams are incident upon the lead-out optical fiber, which in effect samples the new state of polarization determined by the birefringence of the sensing fiber. Clearly, as the birefringence of the sensing region responds to the changes in the environment (e.g., strain or temperature), this is reflected in the change in the state of polarization of the light leaving the sensing region. A photodetector viewing the intensity of light in one of the polarization eigenmodes through the use of a polarizing element of the lead-out optical fiber will respond to this variation in the state of polarization of the light that exits the sensing region. The output intensity for the two polarization eigenmodes can in general be expressed in the form

$$I_1 = I_0[1 + \cos(\phi_f - \phi_s)] \quad (25)$$

and

$$I_2 = I_0[1 - \cos(\phi_f - \phi_s)] \quad (26)$$

where $\phi_f = kn_f L$, and $\phi_s = kn_s L$, k is again the free-space propagation constant, and L is the sensing length of the optical fiber.

The problems of signal fading and fringe ambiguity for polarimetric fiber-optic sensors can be avoided by also operating at the quadrature point. Two methods of producing quadrature signals with such sensors that are also suitable for use with smart structures are

1. using a 3 × 3 coupler and necessary electronics for the required signal processing, and
2. switching between two lasers in order to shift the wavelength by an amount that corresponds to an effective phase difference for the two polarization eigenmodes of $\pi/2$.

It should be noted that the wavelength shift corresponding to a $\pi/2$ phase difference in the case of polarimetry is much larger than normally required for interferometry. For that reason it may be necessary to switch between two lasers rather than to shift the output wavelength of one laser (by switching the injection current, see Fig. 23).

V. Structural Measurements Possible with Fiber-Optic Sensors

The potential breadth of applications that can be undertaken with optical fiber sensors embedded within composites can be gauged by reference to Table IV. These can be divided into four groups: fabrication control; structural integrity; loading, shape, and vibration; and thermal state. Structural integrity monitoring (or damage assessment) will be considered in detail in the next section. Here we shall review the more general kinds of structural measurements possible with fiber-optic sensors.

TABLE IV. Three Groups of Applications for Fiber-Optic Sensors Embedded in Advanced Composite Materials

Manufacturing	Installation	In-service operation
Cure mapping Residual stress measurement Defect detection and location	Detect bad handling Sense excessive force Warn of incorrect riveting Check bond integrity Detect impacts Damage evaluation	Damage assessment Structural integrity monitoring Real time load evaluation Structural dynamic response monitoring Opto-control monitoring Thermal mapping Strain and deformation mapping Remaining service lifetime evaluation

Spatial resolution with optical-fiber sensors can be attained by using

distributed sensors
a grid of line integrated sensors
a grid of point sensors

Distributed sensors are, in principle, very attractive for they would permit the use of fewer sensors and represent a more effective use of the optical fibers, in that each element of the optical fiber would be used for both measurement and data transmission. In practice the spatial resolution attainable is somewhat limited. This is particularly true if the measurements have to be made in real time. Furthermore, the need for high reliability and graceful degradation tends to detract somewhat from the advantages of distributed sensors, since a single break in the optical fiber effectively takes out of commission a large number of sensors.

A grid of line-integrated fiber-optic sensors can be employed to map a scalar field, such as pressure or temperature, by use of the inverse Radon transform. Indeed, line-integrated measurements of position or angular orientation may even have specific value in certain forms of structural control situations. However, a grid of line-integrated optical fiber sensors cannot, in general, be used to map a vector or tensor field, such as strain. Exceptions can arise where the field is constrained in such a way that it is not truly two-dimensional in nature.

Where the field is two-dimensional, three independent measurements are required at each point to uniquely specify the field. This can be accomplished by using a set of optical rosette sensors, each comprising three small identical sensors set at a different orientation. The first practical optical analog of the electrical strain rosette was built and tested in our laboratory and was based on three Michelson fiber-optic sensors (Fig. 17).

A. Strain Measurements

In terms of smart structures, strain and deformation represent two of the most important measurements. Furthermore, real-time measurements of the strain in a structure normally permit its state of vibration to also be evaluated. Although a variety of fiber-optic sensing techniques (intensiometric, interferometric, polarimetric, and modalmetric) have been proposed for determining strain, in general, the interferometric based approaches offer the highest

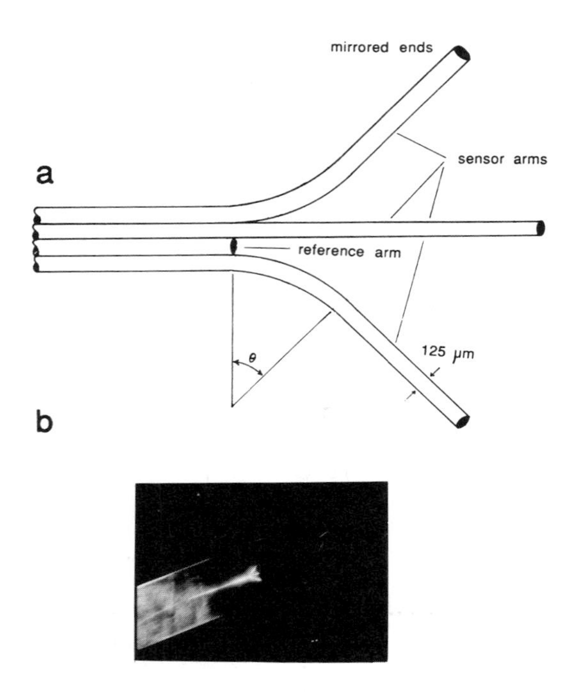

FIG. 17. (a) Physical arrangement of the optical strain rosette based on three Michelson fiber-optic sensors. (b) Illustration of both optical strain rosette illuminated by helium-neon laser and electrical strain rosette.

sensitivity. Furthermore, it is expected that for most smart structure applications local measurements of strain will be required. Thus, in considering any sensing technique the ease of ensuring lead-in, lead-out insensitivity will constitute an important factor.

The potential properties of an interferometric fiber-optic strain gauge (the term "gauge" is used to mean localized) can be as good, if not better, than a conventional electrical strain gauge in respect to temperature range, range resolution, and maximum strain. However, the optical fiber strain gauge is likely to have superior fatigue life and be more conducive to embedding within composite materials, due to its small size and absence of the electrical leads and solder joints required by conventional resistive strain gauges.

Although it is possible to build a Mach–Zehnder fiber-optic strain sensor, a gauge based on a Michelson fiber-optic interferometer offers several advantages, including a better defined sensing region; a single-ended configuration; twice the sensitivity (as it is a double-pass system); and a much higher spatial resolution, because the Michelson is not limited by the minimum radius of curvature of the optical fiber, as is the Mach–Zehnder interferometric sensor. Unfortunately, both of these sensors require two closely spaced optical fibers. This represents a perturbation to any structure within which they are embedded, and the requirement of exact phase cancellation in their common lengths may not easily be achieved. Furthermore, application of this type of dual fiber sensor to smart structures could be hindered by the difficulty of producing a connector that could provide phase preservation between both optical fibers at all times and under all conditions of vibration and temperature. However, in the case of the Mach–Zehnder fiber-optic sensor, the outputs from the two optical fibers can be permitted to overlap and interfere directly and the resulting fringe pattern sampled by a thin slit (or grating mask) and a detector. As the strain or temperature in the structure changes, the phase difference between the two beams is modulated and the fringe pattern sweeps across the detector.

We have undertaken the first strain measurements with a Michelson optical strain gauge embedded within a cantilever beam made of thermoplastic (carbon/PEEK) and found an excellent linear relation between the number of fringes and the strain produced by end-loading of this cantilever beam (Fig. 18). Although considerable care was taken to position the two optical fibers very close together to ensure lead-in and lead-out insensitivity, some movement during the consolidation led to the primary source of noise. Use of a 3 × 3 output coupler permits this all-fiber Michelson optical strain gauge to have passive quadrature detection. The strain gauge factor for this device when embedded within graphite/PEEK was determined to be 13.9 (± 1.1) degrees per μstrain per cm.

Recently, we have also shown that a Michelson optical strain gauge embedded within a Kevlar/epoxy

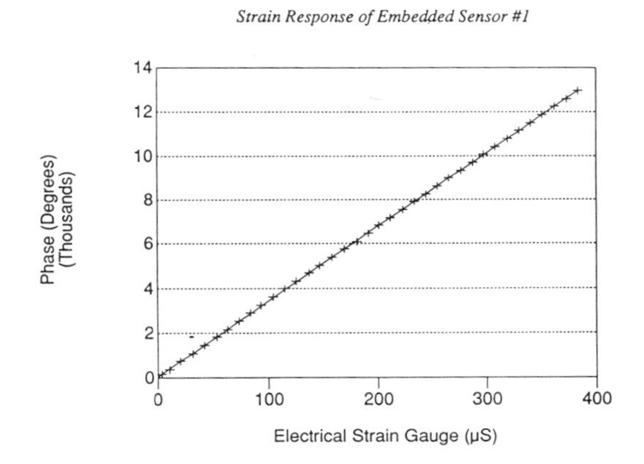

FIG. 18. (a) Strain response of a Michelson fiber-optic strain sensor embedded within a carbon/PEEK thermoplastic cantilever beam. (b) Schematic of the experimental arrangement.

panel can detect acoustic emission arising from a delamination generated by out-of-plane loading of the panel. Figure 19 indicates the experimental arrangement, including the image-enhanced backlighting system used to observe delaminations within the loaded panels. Figure 20 reveals a delamination recorded by this means and the concomitant acoustic emission signal sensed by the embedded Michelson fiber-optic sensor. In this instance, only signals with a frequency greater than 70 kHz were permitted to be detected, and a piezoelectric phase modulator was employed to eliminate the drift and dc strain (Fig. 21).

In the event that an arbitrary state of strain in two dimensions needs to be determined, three independent measurements of the strain field have to be undertaken. This requires the development of a fiber-optic strain rosette. We have fabricated the first

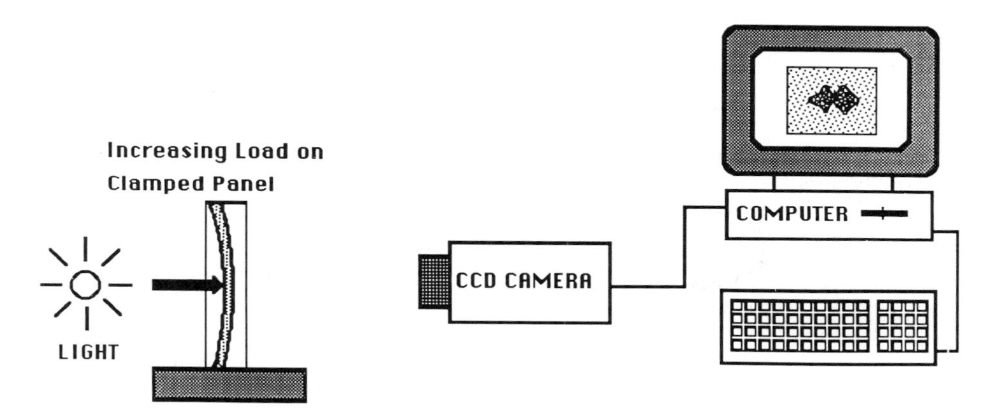

FIG. 19. Schematic representation of the experimental arrangement used to study the formation and growth of delaminations produced within Kevlar/epoxy composite panels by out-of-plane loading. Image-enhanced backlighting is employed to map the load-induced delaminations.

Growth of Delamination Zone
in [0/0/90/90/{90}/90/90/0/0] Sample

a

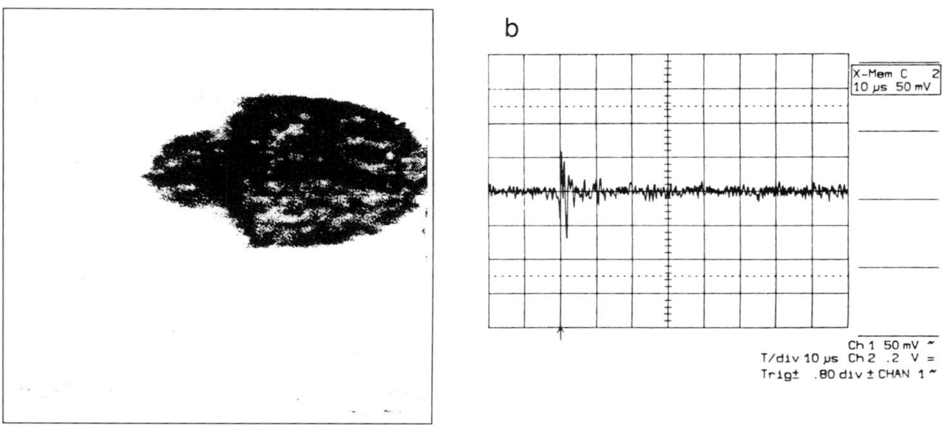

Enhanced backlit image

Acoustic emission signal from embedded optical fibre sensor

FIG. 20. (a) Example of a delamination produced within a Kevlar/epoxy panel by out-of-plane loading and recorded by image-enhanced backlighting. (b) Corresponding acoustic emission signal detected by an embedded Michelson fiber-optic sensor.

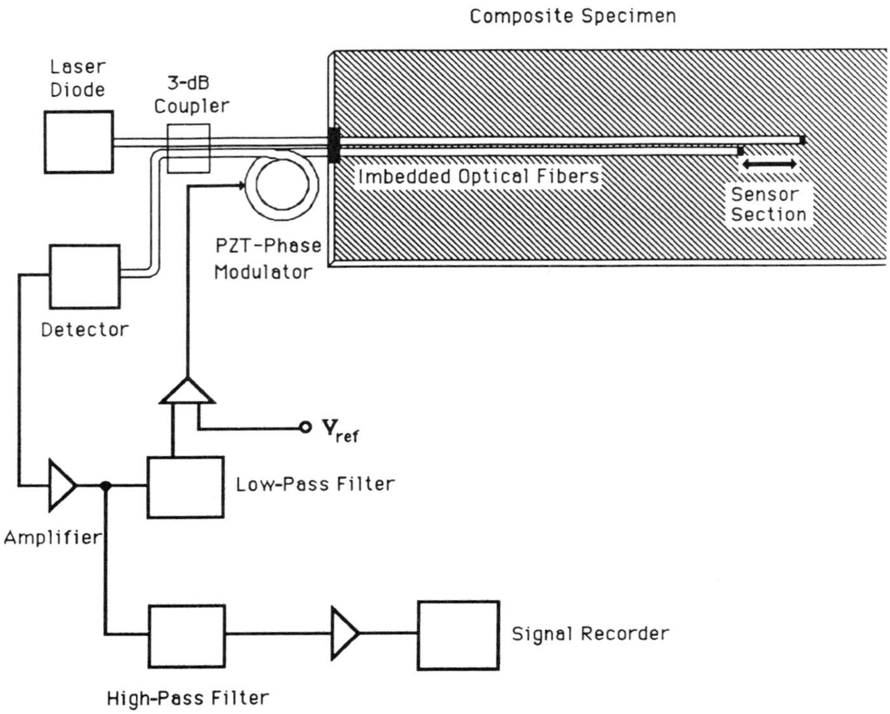

FIG. 21. Schematic of the embedded Michelson fiber-optic sensor used to detect acoustic emission through a high-pass filter. Also shown is the phase modulation technique employed to ensure quadrature detection.

optical strain rosette based on three Michelson fiber-optic sensors (Fig. 17). This prototype device has been tested for use in smart structures by embedding it within a graphite/PEEK cantilever beam and comparing its measurement of the strain with that of a conventional resistive strain rosette adhered to the surface of the beam. A representative example of this comparison is displayed as Fig. 22. The agreement is excellent for a prototype device and clearly indicates the viability of using embedded optical strain rosettes for mapping strain fields. More recently, we have fabricated optical strain rosettes based on three Fabry–Perot fiber-optic sensors and tested them within composite structures.

When optical fibers are embedded within composite materials, due consideration must be given to the effects of the cure, or consolidation, process on their coating. For example, acrylic coatings have been observed to melt during the cure process in an autoclave; this can lead to poor bonding, asymmetrical loading, and the creation of large, resin-rich perturbations of the composite around the optical fiber. Indeed, even the much larger diameter of the optical fiber with its coating will, in general, represent a significant perturbation. Improved coatings that are thin, capable of surviving the cure or consolidation process, and able to provide excellent bonding remain to be developed. Currently, metal or polyimide coatings are under consideration.

An important consideration when considering any fiber-optic sensor system for a smart structure is the relative ease of developing a practical interface. In the case of a set of interferometric fiber-optic strain sensors, this involves a rugged and reliable alignment of a number of single-mode optical fibers. This is no minor problem, since the core diameter of each optical fiber is of the order of one micrometer. Although this task has been accomplished in the telecommunications field, smart structures impose a number of different constraints on the connector: it must be small (so as not to perturb the structure); it must not introduce any arbitrary phase, even when the structure is subject to all forms of vibration and motion; and it must accurately maintain the alignment of many optical fibers simultaneously over a wide range of temperatures.

The interface problems would be somewhat relaxed if each sensor involved only one optical fiber, as both the sensing and reference paths coincide in the lead-in and lead-out sections of the sensor. A fiber-optic strain sensor based on a Fabry–Perot interferometer operated in the back-reflection mode involves only one optical fiber, has a well-defined sensing region, is a single-ended system with double-pass enhanced sensitivity, and is capable of high spatial resolution. The difficulty of creating the first mirror surface within the optical fiber (Fig. 6) has represented a formidable barrier to the development of this sensor for smart structures. Recent success in the fabrication a Fabry–Perot fiber-optic sensor based on fusion splicing offers considerable promise in terms of smart structures.

Another form of single optical fiber sensor that is under consideration is based on the creation of a Bragg grating along a designated length of an optical fiber with a germanosilicate (GeO_2 doped silica) core by means of laser-induced photorefraction. Step-like changes in the core refractive index, $\Delta n/n$, of about 0.01% have been produced by this means. The resulting periodic index perturbation acts as a narrow-band reflection filter (typically 25 GHz) with a center-line frequency that shifts with a change of temperature or strain. Reflectivities of about 60% have been achieved with a 1-cm long grating that has a sensitivity of about 0.05 nm per Ksi. In principle, this approach could be adapted to produce a quasi-distributed fiber-optic sensor by impressing a number of such gratings, each with a different spacing and therefore different center-line frequency, along the length of an optical fiber. Scanning the laser frequency would allow each of the sensing elements to be interrogated in turn. Additional virtues of this sensing concept are that it is single ended, it does not require quadrature detection, and it lends itself to wavelength-division demultiplexing/multiplexing.

Several other potential strain sensors are based on the use of a single optical fiber. A number of these rely on the difference in the propagation speed of two or more transverse modes within the same optical fiber. The term *modal interferometry* has been coined to designate sensors that involve this concept. The number of such modes permitted to propagate along an optical fiber is controlled by the V-number of the radiation. This V-number is given by the relation

$$V = ka\{n_1^2 - n_2^2\}^{1/2} \tag{27}$$

where a is the core radius; k is the propagation constant (inversely proportional to the wavelength); and n_1, n_2 represent the index of refraction of the core and the cladding, respectively. If the wavelength of the radiation is such that $V < 2.405$, then only a single transverse mode will propagate unattenuated. A single-mode optical fiber operated at a wavelength less than this cutoff wavelength (i.e., at a wavelength such that V is a little larger than 2.405) will permit two transverse modes to propagate. Although radiation exchanged between the modes of such a few mode optical fiber has the potential to constitute a strain sensor, the sensitivity is considerably less than that of interferometric sensors. Localized sensing can be achieved by employing optical fibers (for the lead-in and lead-out segments of the system) with a cutoff wavelength below that of the radiation used.

An example of an all-fiber, single-ended, polarimetric strain sensor with a well-defined sensing section and a dual-wavelength quadrature system is dis-

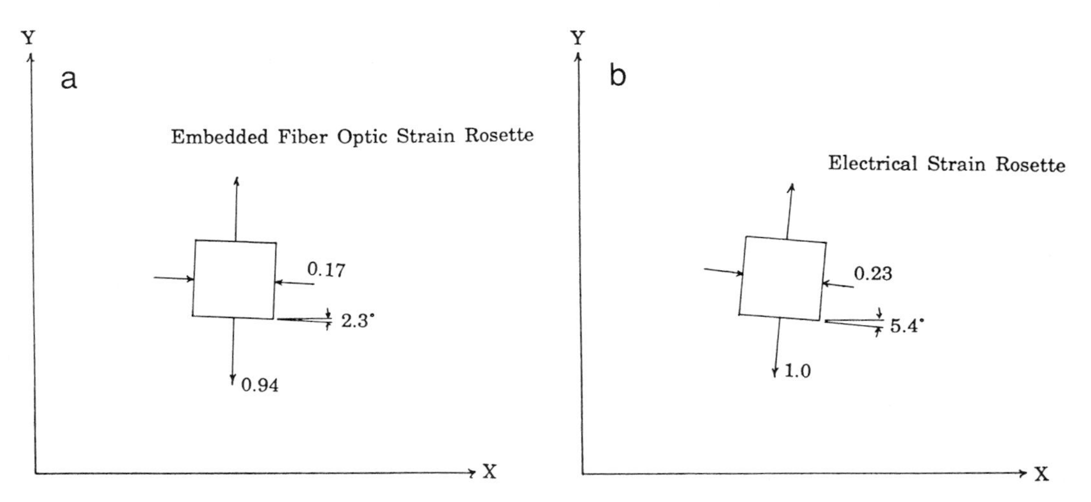

FIG. 22. Comparison of the "principal strains" and the inclination of the principal axes measured (a) with a Michelson based fiber-optic strain rosette (see Fig. 16) embedded within a thermoplastic beam and (b) a standard resistive strain rosette mounted on the surface of the beam.

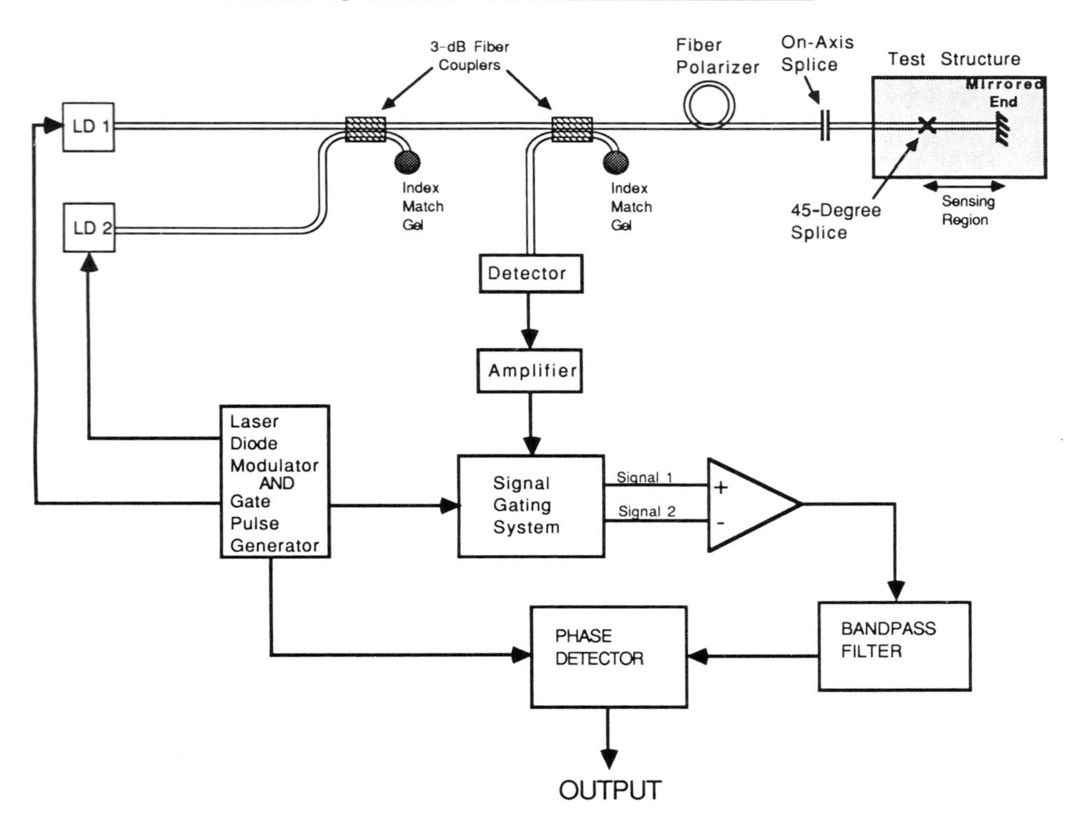

FIG. 23. A schematic of a dual wavelength, single-ended, all-fiber polarimetric strain fiber-optic sensor.

played as Fig. 23. Highly birefringent, bow-tie, optical fiber was used in this sensor, and lead-in, lead-out insensitivity was achieved by using a 45° splice and a mirrored end to isolate the sensing region. Linearly polarized light from a laser is launched into one of polarization axes of the lead-in optical fiber. This is accomplished by means of an in-line fiber polarizer. The polarization state of the light is not changed while it propagates with its polarization aligned with one of the polarization axes of the lead-in optical fiber. At the 45° splice the light equally excites both of the polarization modes, and the state of polarization of the light is modified as it propagates along this sensing section of the optical fiber. The fiber polarizer serves as the analyzer during the return path. The strain gauge factor for this sensor was evaluated to be approximately 0.048 degrees per μstrain per cm when the sensor was adhered to the surface of an aluminum cantilever beam. It can be seen that polarimetric sensors have a strain sensitivity that is about two orders of magnitude less than that of their interferometric counterparts. Polarimetric sensors are thus well suited to measuring the integrated strain over fairly large distances (i.e., anything greater than 10 cm).

Although phase preservation through a connector to a structure is not a problem for this sensor, polarization preservation has to be considered. Another attractive feature of a polarimetric strain sensor is the possibility of attaining a degree of temperature compensation by dividing the sensing section in two with a 90° splice. Polarimetric fiber-optic sensors have been used to monitor the variation of strain in a woven glass-fiber reinforced plastic specimen loaded to failure, and to detect stress waves created by focusing a 40-joule CO_2 laser pulse on the surface of a composite material panel.

B. Fabrication and Installation Monitoring

Fabrication monitoring of composite material structures could lead to both optimization and quality control of these products. Inspecting (or, more accurately, interrogating) the structure after installation would represent a bonus that could be very important with future aircraft and space structures, where errors leading to premature failures could be very costly.

Three examples will be considered:

1. Cure monitoring of thermoset composite materials
2. Consolidation monitoring of thermoplastic composite materials
3. Rivit and screw installation examination

Thermoset composite materials comprise strength-reinforcing fibers (such as carbon, aramid, or glass) embedded within a resin matrix (like epoxy). The optimum mechanical properties of these materials are only achieved if their fabrication includes the correct curing cycle at elevated temperatures and pressures. Unfortunately, the degree of cure during fabrication can vary with position for a large structure. Quality control, especially for aerospace structures, and cost demand a more controlled cure, and this can only be achieved if the cure state can be mapped by an *in-situ* sensor.

Localized cure monitoring has been demonstrated by making a small section of an optical fiber from a sample of the fully cured resin used in the composite material. When this optical fiber is embedded within the uncured structure and illuminated, the segment made of the fully cured resin will guide the light, as its index of refraction is greater than that of the surrounding uncured material. As the structure cures (i.e., polymerizes), its index of refraction increases and the guiding properties of this sensing section of the optical fiber diminish. When the surrounding material is fully cured, the transmission of the optical fiber falls to zero. The cure point is, as a consequence, unequivocally determined by a complete loss of light; this approach is seen to be indifferent to the temperature or pressure of the material and requires no calibration. Its major drawbacks are that it requires the sensing section of the optical fiber to be made of the same resin as the structure, and when cured these optical fibers can no longer be part of the ongoing optical neurosystem.

A totally different approach toward cure monitoring is based on observations of viscosity-dependent fluorescence of nonreactive probe dye molecules (a fluorophor) added to the resin. For example, diglycidolether of bisphenol-A has been shown to experience a 16-fold increase in its fluorescence during a room-temperature cure. The eventual outcome of this approach might be the development of a nonreactive fluorophor that can be used in conjunction with optical fibers, embedded to monitor the structural properties, to reliably monitor the cure state of the structure.

The presence of embedded fiber-optic sensors capable of measuring both the pressure and temperature distributions within a composite structure during the cure process might provide enough information to ensure optimization and quality control. The same may also be said for *thermoplastic* where it is necessary to monitor the degree of consolidation. Indeed, we have demonstrated that a Michelson interferometric optical fiber sensor can be embedded within carbon/PEEK specimens and can be used to measure both temperature and strain within such structures.

Another important diagnostic role for which optical fibers may be quite suitable is checking various installation procedures. For example, the compressive load applied to a structure by bolts or rivets could be measured. In this context, either interferometric, polarimetric, modalmetric, and even microbend fiber-optic sensors may be suitable for checking the compressive loads associated with installation procedures. This is an important application because overtightening can lead to surface cracks and excessive subsurface stress concentrations, which can initiate matrix damage that would greatly shorten the fatigue life or, worse, initiate a major failure.

Although in the laboratory it is often easy to demonstrate the ability of a given fiber-optic sensor to monitor a specific event, this is a far cry from performing the same function in the very noisy world of operational systems. Localization is essential for fiber-optic sensors operating in an operational environment if they are to discern the desired signal from any extraneous noise.

C. Temperature Measurements

Temperature measurements can be required in their own right or to correct strain measurements. Most of the interferometric, polarimetric, and modalmetric strain sensors discussed above could also be used to evaluate the change in temperature, if it is known *a priori* that the applied strain remains constant. The temperature of a structure can also be evaluated by optical fibers in a variety of other ways. These include blackbody emission, differential absorption, fluorescence decay time, and Raman backscattering.

Localized hot spots can be detected by the blackbody emission arising from the section of optical fiber with an elevated temperature, or with special optical fibers possessing sections of modified cladding for which the index of refraction becomes equal to or greater than that of the core at temperatures above a critical value. The optical fiber consequently becomes very lossy at temperatures above this value. Differential absorption has been used to measure temperatures with rare earth (Nd_2O_3-doped aluminosilicate glass core) optical fibers up to about 1000 K, or with a chip of semiconductor (GaAs or CdTe) spliced into an optical fiber from 260 to 570 K. Thermooptical fiber-optic sensors can also be employed for the low-temperature range (77–150 K) using Europium (Eu^{3+})-doped fluoride glass.

The dependence of the fluorescence decay time of neodymium (Nd^{3+})-doped glass with temperature when excited with infrared radiation has been shown to be capable of evaluating the temperature over the range 270–620 K. A great virtue of this approach is its independence from variations in the intensity or wavelength of the light-emitting diode source or the sensitivity of the detector. Although a distributed fiber-optic temperature sensor has been developed based on the ratio of Raman Stokes to anti-Stokes backscattering, its poor spatial resolution would limit its use to very large structures.

VI. Damage Assessment with Structurally Integrated Optical Fibers

In the short term the most important aspect of smart structure technology will be the development of structural components with "built-in" damage assessment systems. These will permit the structural integrity of any major or critical component to be monitored at all times. A network of suitable optical fibers could be used for a variety of damage assessment tasks. Current methods of aircraft structural integrity monitoring often involves both disassembling and component-by-component (sometimes rivet by rivet) inspection. This represents a very time-consuming and labor-intensive procedure that is often prone to human error. Considerable savings in time, cost, and human resources could be achieved if real-time, automatic, integrity-monitoring systems could be developed. One of the current approaches involves bonding crack-detecting wires onto components. Cracks encountering these wires interrupt their current and reveal their presence. Unfortunately, this technique has a number of serious drawbacks and limitations:

Corrosion can affect wire crack-sensor performance.
The sensitivity for fine cracks is poor.
The system is subject to, and can generate, electrical interference.
It cannot reliably be used within composites.

A. Damage Assessment for Composite Materials Based on Optical Fiber Fracture

Aircraft wing leading edges, submarine sonar domes, and propellers are examples of composite structures with complex geometries involving compound curvatures and a thickness that varies in two or more directions. These structures are very difficult to diagnose for damage by conventional techniques like ultrasound and X-radiography. Fracture of optical fibers, with the attendant disruption in their transmitted light, represents the simplest technique for damage detection. The first attempt at detecting damage within composite material with embedded optical fibers was based on the use of regular buffered multimode optical fiber (Corning 1516 with a diameter of 250 μm), and was only capable of detecting quite severe damage. Another factor contributing to the relative insensitivity of the optical fibers to damage was the alignment of the optical fibers parallel to the reinforcing fibers.

Optical fibers have been shown to have particular relevance to the aircraft industry. They have been shown to be capable of detecting cracks in metal surfaces, bond failure in lap joints, airframe delaminations and fractures in the bonds between stringers and the airframe skin, and delaminations in composite structures.

1. Treatment, Orientation, and Depth Optimization

In working with Kevlar/epoxy we found it necessary to devise a special treatment for the optical fibers that enabled us to tailor their fracture strength to the point where they would detect barely visible impact damage (BVID). Control of the damage sensitivity was achieved by periodic etching of the sensing length of the optical fiber with a saturated solution of ammonium hydrogen difluoride. An example of the variation in the threshold impact energy for fracture of the optical fibers embedded in a specific manner within a given layup as a function of the etch time is presented as Fig. 24.

An extensive analysis of the sensitivity to fracture of embedded, treated optical fibers with respect to their orientation relative to the material-reinforcing fibers in the adjacent plies and their depth within a structure has been undertaken. The result of this

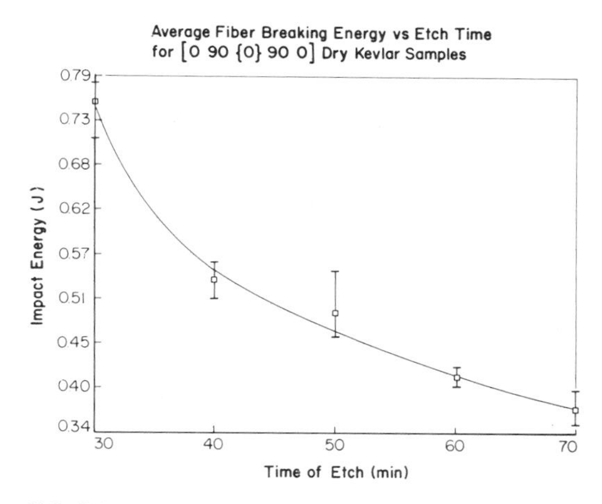

FIG. 24. Variation with etch time of threshold impact energy for fracture (TIEF) of optical fibers embedded in the midplane of a 4-ply $[0_4]$ Kevlar/epoxy laminate.

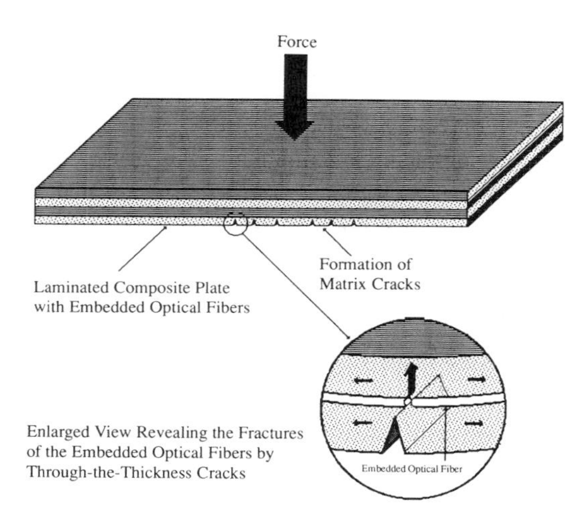

FIG. 25. Fracture mechanism for optical fibers embedded within composite materials is primarily through-the-thickness matrix cracks.

work has allowed us to formulate the following general rules:

1. Damage sensitivity and reliability is optimized when the optical fibers are embedded between collinear plies and oriented orthogonal to the ply direction.
2. Optical fibers embedded between noncollinear plies should be oriented orthogonal to the ply separating them from the expected region of damage.
3. If this arrangement aligns the optical fibers with the other ply, then their orientation should bisect the larger angle between the plies.

We have also determined that for optimum sensitivity the optical fibers should be located as close to the surface of maximum tensile load as possible. This corresponds to the rear surface in the case of thin laminates. For thicker laminates, or thin laminates that are rigidly reinforced, this is no longer the case, due to the reduced flexure and high contact stresses. Under these circumstances the optical fibers have to be embedded closer to the impact surface for optimum sensitivity.

2. Fracture Mechanism for Embedded Optical Fibers

We have established that each degree of treatment leads to a fairly well-defined *threshold impact energy for fracture* (TIEF) of the embedded optical fibers in any given composite layup and that this TIEF also depends on the orientation and location of the embedded optical fibers within the composite. Furthermore, these threshold energies can be closely matched to the threshold impact energy for delamination of the material by an appropriate choice of etch time for the optical fiber. The TIEF for optical fibers embedded collinear with the material-reinforcing fibers in the adjacent plies can be six times the TIEF of a similar optical fiber placed in the same location within the same composite but oriented perpendicular to the reinforcing fibers. Results of this nature and other observations have led us to suggest that the mechanism responsible for fracturing embedded optical fibers is primarily through-the-thickness matrix cracks (see Fig. 25).

3. Detection of Load-Induced Growth of a Damaged Area

In a series of experiments that combined light transmission measurements with light bleeding we have been able to demonstrate that load-induced *growth* of a region of delamination within a composite panel could be detected by an embedded grid of suitably damage-sensitized optical fibers. These experiments involved out-of-plane loading of Kevlar/epoxy panels of various layups and thicknesses. A schematic of the arrangement is presented as Fig. 19. The delamination damage produced by out-of-plane loading was determined with image-enhanced backlighting, even in the case of complex composite structures such as the thick panel, comprising 13 different plies and a honeycomb reinforcement, used to simulate a planar section of the leading edge of an aircraft wing.

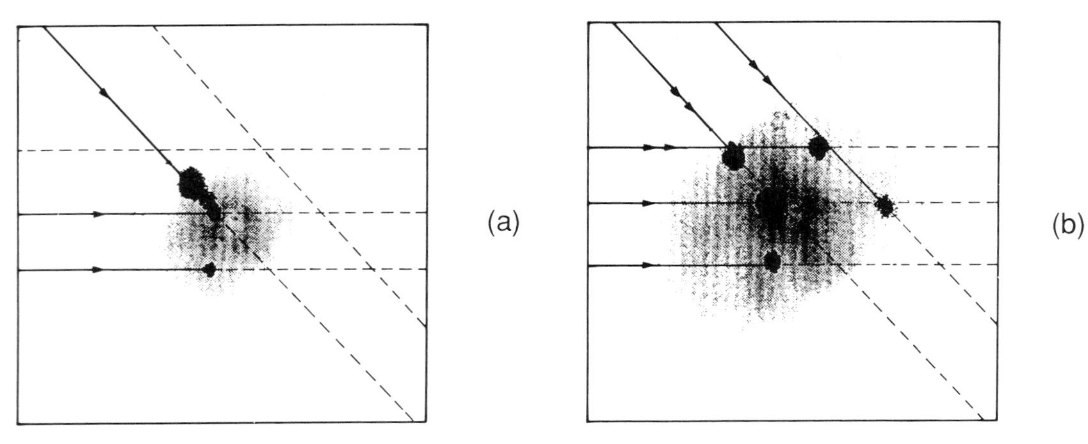

FIG. 26. (a) A video image of a backlit Kevlar/epoxy laminate with a layup $[0_w, 0_2, 45, 0_2\{90\}0, -45\{45\}90, 0_w$, -Honeycomb, $0_w, 90, 0_w]$ corresponding to a planar version of an aircraft wing leading edge, subject to a central out-of-plane force. The curly brackets $\{x\}$ indicating the location and x the orientation in degrees of the embedded optical fibers. The shadow (grey area) reveals the region of delamination while the light-bleeding (dark) spots indicate the fracture locations of the embedded optical fibers orientated at 90 and 45 degrees. (b) The growth in the area of delamination resulting from a larger out-of-plane force is clearly seen in the second video image in terms of both the larger shadow and the additional light-bleed spots.

Figure 26 provides a representative set of results that reveal (a) fracture of embedded optical fibers (as seen by light bleeding—the dark spots) within the region of delamination (grey shadow produced by image-enhanced backlighting); and (b) as the load is increased and this region of delamination grows, more fractures of the optical fibers are observed. Although inaccessibility or paint will, in general, preclude observation of light bleeding, the associated changes in the transmission characteristics of the embedded optical fibers will be used for damage assessment. Our results to date are encouraging and strongly support the viability of using embedded optical fibers to assess the growth of a region of delamination under load conditions. The original cause of this delamination could have been an impact or a manufacturing flaw.

4. Fiber-Optic System for an Aircraft Wing Leading Edge

We have recently collaborated with Boeing (De Havilland Division) of Canada to fabricate a prototype aircraft wing leading edge instrumented with a structurally integrated fiber-optic damage assessment system (Fig. 27). This system uses three arrays of damage-sensitized optical fibers embedded within three separate lamina (Fig. 28). These optical-fiber arrays were embedded within the leading edge after being mounted in the appropriate configuration on thin adhesive films. This is illustrated for the optical

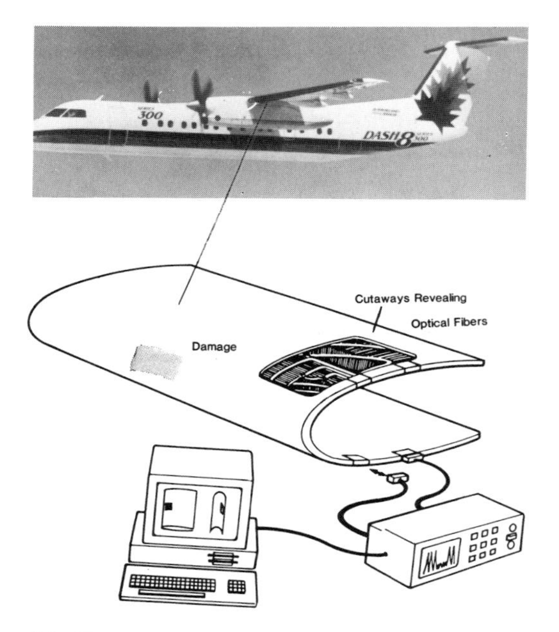

FIG. 27. Schematic of aircraft wing leading edge instrumented with three arrays of damage-sensing optical fibers.

fibers oriented at 90° in Fig. 29. Although visual inspection of the optical-fiber integrity is quite feasible, the human eye is not good at discerning modest changes (say a 30% drop) in intensity associated with a partial fracture. A photodiode array that is butt-fitted against the output port for the embedded optical

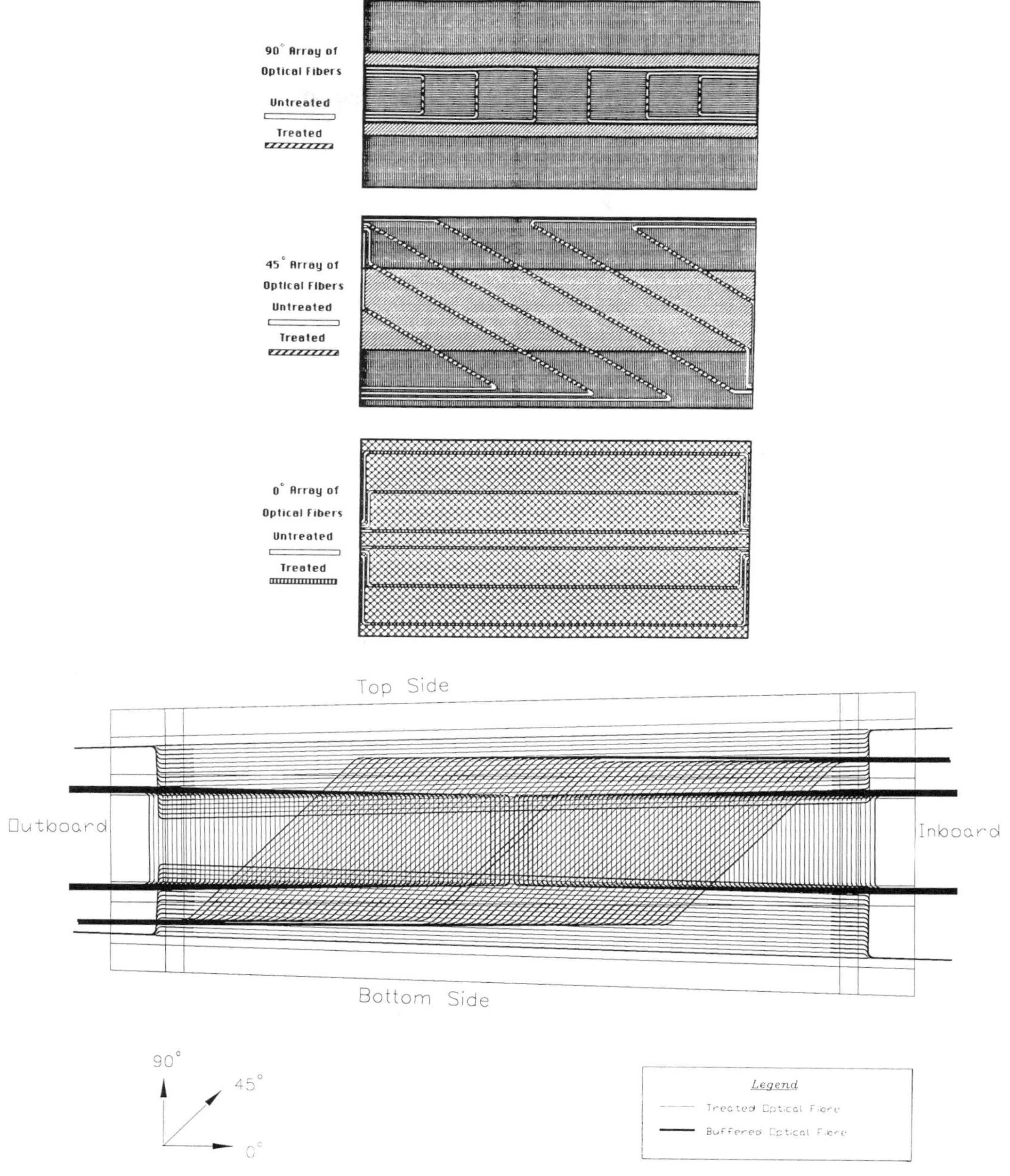

FIG. 28. (a) Ideal arrangement of the three arrays of embedded optical fibers that constitute the damage-sensing system in the aircraft wing leading edge. (b) Actual arrangement used in the first leading edge with built-in fiber-optic damage detection system.

fibers will thus be used in evaluating this system. The output from this photodiode array will be interpreted by a computer, and a display of the predicted region of damage will be presented on a monitor. This type of system could be used for periodic interrogation when the leading edge is dismounted from the wing or adapted for real-time monitoring during flight using an onboard computer. Fig. 30 is a photograph of the first aircraft leading edge with a built-in fiber-optic damage detection system undergoing tests at the University of Toronto Institute for Aerospace Studies. The bright laser light bleed spots indicate

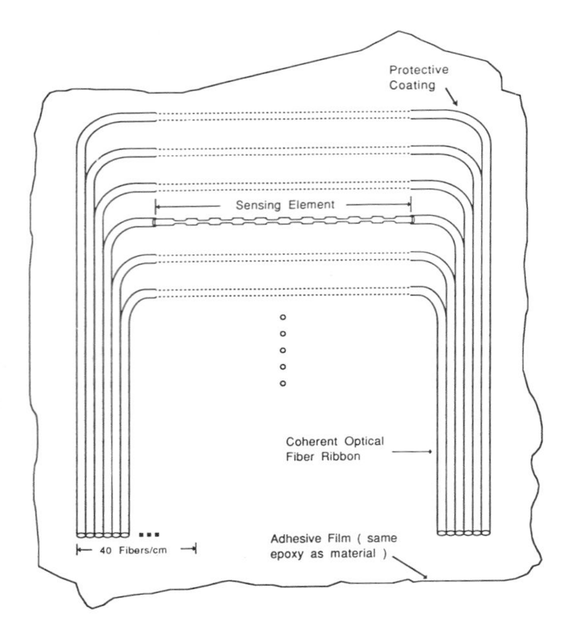

FIG. 29. A schematic illustration of the damage-sensitized 90-degree optical fibers as laid in the resin film for mounting within the leading edge.

location of impact damage inflicted on the leading edge with the pendulum hammer shown to the left of the photograph.

VII. Future Prospects

Smart structure technology could lead to a revolution in engineering concepts and ethics. In the twenty-first century it might be quite unacceptable engineering to build any structure or vehicle (that could endanger the lives of people or the environment) without including an integrated sensor system to warn of impending mechanical failure. For pipelines, pressure vessels, storage tanks, etc., simple "built-in" fiber-optic structural monitoring systems could provide an early warning of structural weaknesses and prevent many of the failures that today are responsible for a number of environmental problems. In major structures such as dams, nuclear reactors, or large buildings, this kind of system could warn of the development of structural faults or, after events like earthquakes, provide important information regarding structural soundness.

In aircraft, where composite materials are finding increasing application, structurally integrated fiber-optic sensor systems capable of detecting damage and assessing its growth in time could find acceptance as part of the new fly-by-light technology (Fig. 2). Through the use of artificial intelligence such a system might also determine the effect of this damage in terms of limiting flight maneuvers or evaluating the reduction in the useful life of duly instrumented components. This could also greatly improve maintenance efficiency and effectiveness.

FIG. 30. First composite aircraft leading edge fabricated with a structurally integrated fiber-optic damage detection system. One array of embedded, damage sensitized optical fibers is seen illuminated by a Helium-Neon laser. The optical fibers can be seen within the leading edge by the low level of red laser light leaking along their length (recorded in the dark). The much brighter spots of red laser light (recorded under normal lighting conditions) indicates the sites of delamination where the embedded optical fibers have been fractured by low energy impacts. In operational systems, the accompanying drop in the transmission of light in the optical fibers would indicate the creation of damage.

Beyond structural integrity, the concept of using light in conjunction with optical fibers for sensing the shape, deformation, temperature, state of vibration, etc., also looks very promising. Indeed, it may eventually be possible to build aircraft with wings that sense their load distribution with arrays of integrated fiber-optic sensors and adapt their shape directly with distributed actuators to optimize flight control. Such wings would have no ailerons, flaps, motors, or cables. Everything would be integrated into the structure, effecting a considerable weight saving, a major simplification in the construction, a significant reduction in maintenance and down time, and a commensurate improvement in reliability and safety. Furthermore, if this can be accomplished with opto-control systems there would be additional improvements in safety. This technology might also find application to space platforms, especially those large, flexible space structures where active shape control is required, and to robotic systems where human-like dexterity could be of value.

Structures with a fiber-optic nervous system and a structurally integrated distributed array of actuators (or muscles) might best be controlled by neural networks that would be capable of adaptive learning. Such entities might learn from their own experiences and from their environment and evolve into truly intelligent structures.

BIBLIOGRAPHY

Afromowitz, M. A., and Lam, K. Y. (1988). Fiber optic cure sensor for thermoset composites. *Fiber Opt. Smart Struct. Skins* **986,** 135–139.

Culshaw, B., and Dakin, J., eds. (1988). "Optical Fiber Sensors: Principles and Components." Artech House, Boston and London.

Culshaw, B., and Dakin, J., eds. (1989). "Optical Fiber Sensors: Systems and Applications." Artech House, Boston and London.

Dakin, J. P., and Wade, C. A. (1984). Compensated polarimetric sensor using polarization-maintaining optical fiber in a differential configuration. *Electron. Lett.* **20,** 51–53.

Freidah, J. T., Cahill, R. F., Turner, A. A., Holmes, S. A., and Wagoner, R. E. (1985). "Passive Homodyne Optical Grating Demodulator: Principles and Performance." *Fiber Optic and Laser Sensor III,* SPIE Vol. **566,** 114–121.

Giallorenzi, T. G., Bucaro, J. A., Dandridge, A., Sigel Jr., G. H., Cole, J. H., Rashleigh, S. C., and Priest, R. G. Optical fiber sensor technology. *IEEE J. Quant. Elect.* QU-**18,** 626–665.

Hofer, G. B. (1987). Fiber optic damage detection in composite structures. In *Composites* **18,** 309–316.

Hogg, W. D., Turner, R. D., and Measures, R. M., Polarimetric fiber optic structural strain sensor characterisation. In *Fiber Opt. Smart Struct. Skins* **1170,** 60.

Jackson, D. A., and Jones, J. D. C. (1986). "Fiber Optic Sensors." In *Optica Acta* **33,** 1469–1503.

Koo, K. P., Tveten, A. B., and Dandridge, A. (1982). Passive stabilization scheme for fiber interferometers using (3X3) fiber directional couplers. *Appl. Phys. Lett.* **41,** 616–618.

Le Blance, M., Dubois, S., McEwen, K., Hogg, D., Park, B., Tsaw, W., and Measures, R. M. (1989). Development of a fiber optic damage detection system for an aircraft leading edge. *Fiber Opt. Smart Struct. Skins* **1170,** 25.

Liu, K., Ferguson, S. M., and Measures, R. M. (1989). Damage detection in composites with embedded fiber optic interferometric sensors. *Fiber Opt. Smart Struct. Skins* **1170,** 22.

Measures, R. M. (1989a). Smart structures with nerves of glass. *Progress in Aerospace Sciences* **26,** 289–351.

Measures, R. M. (1989b). "Fiber optics smart structures program at UTIAS." *Fiber Opt. Smart Struct. Skins* **1170,** 11.

Morey, W. W., Meltz, G., Glenn, W. H. (1989). Fiber optic Bragg grating sensors. *Fiber Opt. Laser Sens.* **1169,** 12.

Udd, Eric, ed. (1988). *Fiber Opt. Smart Struct. Skins* **986.**

Valis, T., Tapanes, E., and Measures, R. M. (1989). Localized fiber optic strain sensors embedded in composite materials. *Fiber Opt. Smart Struct. Skins* **1170,** 53.

FREE-ELECTRON LASERS

H. P. Freund *Science Applications International Corp.*
R. K. Parker *Naval Research Laboratory*

GLOSSARY

Amplifier: A device used to increase the amplitude of some input signal. The term *superradiant amplifier* is often used to describe a free-electron laser that is configured to amplify noise (i.e., random fluctuations) in the beam rather than an externally supplied signal. In contrast to oscillator configurations, there is no reflection of the signal, and the superradiant amplifier is a single-pass device.

Beat wave: A composite wave formed by the superposition of two waves having different angular frequencies (ω_1, ω_2) and wavenumbers (k_1, k_2). Beat waves form at the sum and difference frequencies ($\omega_1 \pm \omega_2$) and wavenumbers ($k_1 \pm k_2$).

Compton regime: This denotes operation in a free-electron laser in which the dominant mechanism is stimulated Compton scattering. In this process, the wiggler field (which appears to be a backwards-propagating electromagnetic wave in the rest frame of the electrons) produces secondary electromagnetic waves by scattering off the electrons.

Electrostatic accelerators: Include Van de Graaf and Cockcroft–Walton configurations. Van de Graaf accelerators accelerate charges by passing a moving belt through the coronal discharge from an array of points. The charge is then carried away by the belt to a field-free region where it can be extracted from the accelerator. The maximum voltages that can be achieved with electrostatic accelerators are in the range of from 10–30 MV.

Free-streaming: The free and unrestrained propagation of particles without hindrance by external forces. This is often used in a context that denotes trajectories that are unbounded.

Gain: A measure of the amplification of the input signal in an amplifier. The gain of an amplifier is often measured in decibels (dB), which is ten times the common logarithm of the ratio of the output power of the amplifier to the input drive power.

Linear induction accelerators (induction linacs): Linear accelerators that operate by inducing an electromotive force in a cavity through a rapid change in the magnetic field strength. In effect, the electron beam acts as the analog of the secondary winding in a transformer.

Larmor rotation: The circular rotation of charged particles in a uniform magnetic field. Also called *cyclotron* rotation.

Master oscillator power amplifier (MOPA): A shorthand notation for a power amplifier in which the source of the input drive signal is an external, or *master,* oscillator.

Microtrons: Cyclic accelerators in which the particles execute circular motion in a uniform magnetic field that carries the beam through an rf accelerating cavity, one in each cycle.

Modulators: A pulsed-voltage source that includes an energy storage element (such as a capacitor bank) and a switching system to discharge the energy through some load. [See *Pulse-line accelerators.*]

Oscillator: A device used to generate periodic (i.e., oscillatory) signals without the necessity of an external drive signal. Free-electron laser oscillators are constructed by the insertion of the wiggler/electron-beam system within a reflecting cavity in which the signal makes many passes through the interaction region. As a result, the signal grows from noise and is amplified during each pass. As a result, the signal can grow to high intensities without the need for a strong input (driving) signal.

Phase space: The multidimensional space formed by the position and momentum of a particle (or of an ensemble of particles).

Phase velocity: The speed of propagation for a single wave of angular frequency ω and wavenumber k. The phase velocity is determined by the ratio ω/k and defines the speed of a point of constant phase of the wave.

Ponderomotive wave: A slowly-varying wave formed by the beating of two waves.

Pulse-line accelerators: Draw an intense current from a diode by means of a transmission line from a high-voltage capacitor bank. The capacitor bank and transmission line form a distributed capacitance network. The diode acts as a load through which the energy is discharged. [See *Modulators*.]

Radio-frequency linear accelerators (rf linacs): A linear accelerator that employs radio frequency (rf) cavities for electron acceleration. The particles are accelerated in cylindrical cavities that require a high-power source for the rf fields. The rf fields in these cavities may be either traveling or standing waves. In the case of traveling-wave configurations, which are most often employed for electron accelerators, the phase velocity of the rf fields must be synchronized with the desired electron velocity.

Raman regime: This denotes operation in a free-electron laser in which the dominant mechanism is a three-wave scattering process. It occurs when the electron-beam density is high enough that the longitudinal (i.e., electrostatic) waves driven by the beam exert a greater force than the ponderomotive wave. In this process, as opposed to stimulated Compton scattering, the wiggler field scatters off the beam-driven longitudinal waves to produce the secondary electromagnetic waves.

Separatrix: The line or surface in the phase space of a particle that distinguishes between two classes of trajectories as, for example, between bounded and unbounded motion.

Storage rings: A toroidal configuration in which beams of electrons and/or positrons circulate for periods of the order of several hours. Short bunches of electrons and positrons are injected into the torus and guided around the ring by a system of bending and focusing magnets. Due to the curved trajectories of the beam, the particles lose energy during each circuit by means of synchrotron radiation. To compensate for this loss, an rf accelerating cavity is included in the ring. The ring itself may be circular or polygonal; however, the straight sections in polygonal rings facilitate the insertion of the wiggler magnets required for free-electron lasers.

Undulator: See *Wiggler*.

Untrapped trajectories: Unbounded trajectories of particles under the influence of some external force. Examples include parabolic and hyperbolic trajectories of bodies subject to a central gravitational force or the unrestrained motion of a circular pendulum.

Wiggler: The periodic magnet used in free-electron lasers to generate an undulatory motion in the electron beam. The term *wiggler* is specifically used for the magnets employed to generate coherent radiation in free-electron lasers, as opposed to the similar periodic magnets (referred to as *undulators*) employed in synchrotron light sources to generate spontaneous, or incoherent, radiation.

In its fundamental concept, the free-electron laser is an extremely adaptable light source that can produce high-power coherent radiation across virtually the entire electromagnetic spectrum. In contrast, gas and solid-state lasers generate light at well-defined wavelengths corresponding to discrete energy transitions within atoms or molecules in the lasing media. Dye lasers are tunable over a narrow spectral range but require a gas laser for optical pumping and operate at relatively low power levels. Further, while conventional lasers are typically characterized by energy conversion efficiencies of only a few percent, theoretical calculations indicate that the free-electron laser is capable of efficiencies as high as 65%, while efficiencies of 40% have been demonstrated in the laboratory.

Applications of free-electron lasers to date range from experiments in solid-state physics to molecular biology, and novel designs are under development for such diverse purposes as communications, radar, and ballistic missile defense. At the present time, however, free-electron lasers have been largely confined to the laboratory. Most have been built around available electron accelerators, and although they have the potential to emit light anywhere from the microwave to the ultraviolet, researchers have encountered difficulties in getting them to lase at visible and shorter wavelengths. Only recently have free-electron lasers begun to come into their own, as accelerators are designed for their specific needs, and user facilities are set up so that researchers in other disciplines can take advantage of this new source of intense and tunable light.

I. Introduction

The free-electron laser was first conceived almost four decades ago and has since operated over a spectrum ranging from microwaves through visible light. There are plans to extend this range to the ultraviolet. In a free-electron laser, high-energy electrons emit coherent radiation, as in a conventional laser, but the electrons travel in a beam through a vacuum instead of remaining in bound atomic states within the lasing medium. Because the electrons are free-streaming, the radiation wavelength is not constrained by a particular transition between two discrete energy levels. In quantum mechanical terms, the electrons radiate by transitions between energy levels in the continuum and, therefore, radiation is possible over a much larger range of frequencies than is found in a conventional laser. However, the process can be described by classical electromagnetic theory alone.

The radiation is produced by an interaction among three elements: the electron beam, an electromagnetic wave traveling in the same direction as the electrons, and an undulatory magnetic field produced by an assembly of magnets known as a *wiggler* or *undulator.* The wiggler magnetic field acts on the electrons in such a way that they acquire an undulatory motion. The acceleration associated with this curvilinear trajectory is what makes radiation possible. In this process, the electrons lose energy to the electromagnetic wave, which is amplified and emitted by the laser. The tunability of the free-electron laser arises because the wavelength of light required for the interaction between these three elements is determined by both the periodicity of the wiggler field and the energy of the electron beam.

Although the basic principle underlying the free-electron laser is relatively simple, the practical application of the concept can be difficult. In 1951 Hans Motz of Stanford University first calculated the emission spectrum from an electron beam in an undulatory magnetic field. At the time, coherent optical emission was not expected due to the difficulty of bunching the electron beam at short wavelengths; however, it was recognized that maser (microwave amplification through stimulated emission of radiation) operation was possible. Experiments performed by Motz and co-workers shortly thereafter produced both incoherent radiation in the blue-green part of the spectrum and coherent emission at millimeter wavelengths. The application of undulatory magnetic fields to the maser was independently invented by Robert Phillips in 1957 in search of higher power than was currently available from microwave tubes. The term ubitron was used at this time as an acronym for undulating beam interaction. Over the succeeding seven years, Phillips performed an extensive study of the interaction and pioneered many innovative design concepts in use today. Whereas the original microwave experiment at Stanford observed an output of 1–10 W, Phillips achieved 150 kW at a 5-mm wavelength. However, the full potential of the free-electron laser was unrecognized, and the ubitron program was terminated in 1964 due to a general shift in interest from vacuum electronics to solid-state physics and quantum electronics.

A resurgence of interest in the concept began in the mid-1970s, when the term free-electron laser was coined in 1975 by John Madey to describe an experiment at Stanford University. This experiment produced stimulated emission in the infrared spectrum at a wavelength of 10.6 μm using an electron beam from a radio-frequency linear accelerator (rf linac). The first optical free-electron laser was built using the ACO storage ring at the Université de Paris Sud, and has been tuned over a broad spectrum. More recently, stimulated emission at visible and ultraviolet wavelengths has been reported using the VEPP storage ring at Novosibirsk in the Soviet Union. Visible wavelength free-electron lasers have also been built both at Stanford University and by a Boeing Aerospace/Los Alamos National Laboratory collaboration based on rf linacs, and there is interest in the use of an rf linac to drive a free-electron laser in the ultraviolet at Los Alamos National Laboratory. The rf linac has also been the basis for a longstanding infrared free-electron laser experimental program at Los Alamos.

In parallel with the work at Stanford, experimenters at several laboratories began work on microwave free-electron lasers, successors to the ubitron. Those projects, at the Naval Research Laboratory, Columbia University, the Massachusetts Institute of Technology, Lawrence Livermore National Laboratory, TRW, and the Ecole Polytechnique in France, differed from the original work of Phillips by using intense relativistic electron beams with currents of the order of a kiloampere and voltages in excess of a megavolt. The principal goal of this effort was the production of high absolute powers, and the results ranged from a peak power of the order of 2 MW at a wavelength of 2.5 mm at Columbia, through 70 MW at a 4-mm wavelength at the Naval Research Laboratory, to a maximum power figure of 1 GW obtained by Livermore at an 8-mm wavelength. This latter result represents an efficiency (defined as the ratio of the output radiation power to the initial electron-beam power) of 35% and was made possible by the use of a nonuniform wiggler field.

At the present time, free-electron lasers have been constructed over the entire electromagnetic spectrum. This spectral range is summarized in Fig. 1, in which we plot the peak power of a sample of conventional lasers, microwave tubes, and free-electron lasers as a function of wavelength. At wavelengths above 0.1 mm, free-electron lasers already either match or exceed power levels obtainable from conventional technology. At shorter wavelengths, conventional lasers can be found with higher power than is currently available from free-electron lasers. However, free-electron laser technology is rapidly maturing, and this situation is likely to change in the future.

II. General Principles

An electron beam that traverses an undulatory magnetic field emits incoherent radiation. Indeed, this is the mechanism employed in synchrotron light sources. In conventional terminology, the periodic magnetic field in synchrotron light sources is referred to as an *undulator*, whereas that used in free-electron lasers is called a *wiggler*, although there is no fundamental difference between them. It is necessary for the electron beam to form coherent bunches in order to give rise to the stimulated emission required for a free-electron laser. This can occur when a light wave traverses an undulatory magnetic field such as a wiggler because the spatial variations of the wiggler and the electromagnetic wave combine to produce a beat wave, which is essentially an interference pattern. It is the interaction between the electrons and this beat wave that gives rise to the stimulated emission in free-electron lasers.

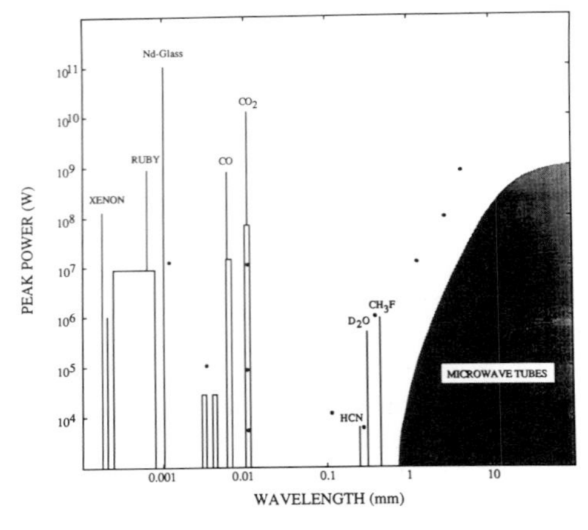

FIG. 1. An important criterion in the judgement of radiation sources is the peak power available at specific wavelengths. A comparison of free-electron lasers, represented by dots, with conventional lasers and microwave sources is shown.

This beat wave has the same frequency as the light wave, but its wavenumber is the sum of the wavenumbers of the electromagnetic and wiggler fields. With the same frequency, but a larger wavenumber (and thus a shorter wavelength), the beat wave travels more slowly than the light wave; for this reason it is called a *ponderomotive* wave. Since the ponderomotive wave is the combination of the light wave and the stationary (or magnetostatic) field of the wiggler, it is the effective field experienced by an electron as it passes through the free-electron laser. In addition, since the ponderomotive wave propagates at less than the speed of light *in vacuo* it can be synchronous with the electrons limited by that velocity. Electrons moving in synchronism with the wave are said to be in *resonance* with it and will experience a constant field—that of the portion of the wave with which it is traveling. In such cases, the interaction between the electrons and the ponderomotive wave can be extremely strong.

A good analogy to the interaction between the electrons and the ponderomotive wave is that of a group of surfers and a wave approaching a beach. If the surfers remain stationary in the water the velocity difference between the wave and the surfers is large, and an incoming wave will merely lift them up and down briefly and then return them to their previous level. There is no bulk, or average, translational motion or exchange of energy between the surfers and the wave. But if the surfers "catch the wave" by paddling so as to match the speed of the wave, then they can gain significant momentum from the wave and be carried inshore. This is the physical basis underlying the resonant interaction in a free-electron laser. However, in a free-electron laser, the electrons amplify the wave, so the situation is more analogous to the surfers "pushing" on the wave and increasing its amplitude.

The frequency of the electromagnetic wave required for this resonant interaction can be determined by matching the velocities of the ponderomotive wave and the electron beam. This is referred to as the phase-matching condition. The interaction is one in which an electromagnetic wave characterized by an angular frequency ω and wavenumber k and the magnetostatic wiggler with a wavenumber k_w produce a beat wave with the same frequency as the electromagnetic wave but with a wavenumber equal to the sum of the wavenumbers of the wiggler and electromagnetic waves (i.e., $k + k_w$). The velocity of the ponderomotive wave is given by the ratio of the frequency of the wave to its wavenumber. As a result, matching this velocity to that of the electron beam gives the resonance condition in a free-electron laser

$$\frac{\omega}{k + k_w} \cong v_z$$

for a beam with a bulk streaming velocity v_z in the z-direction. The z-direction is used throughout this article to denote both the bulk streaming direction of the electron beam and the symmetry axis of the wiggler field. The dispersion relation between the frequency and wavenumber for waves propagating in free space is $\omega \cong ck$, where c denotes the speed of light *in vacuo*. Combination of the free-space dispersion relation and the free-electron laser resonance condition gives the standard relation for the wavelength as a function of both the electron-beam energy and the wiggler period

$$\lambda \cong \frac{\lambda_w}{2\gamma_z^2}$$

where $\gamma_z = (1 - v_z^2/c^2)^{-1/2}$ is the relativistic time-dilation factor related to the electron-streaming energy, and $\lambda_w = 2\pi/k_w$ is the wiggler wavelength. The wavelength, therefore, is directly proportional to the wiggler period and inversely proportional to the square of the streaming energy. This results in a broad tunability that permits the free-electron laser to operate across virtually the entire electromagnetic spectrum.

How does a magnetostatic wiggler and a forward-propagating electromagnetic wave, both of whose electric and magnetic fields are directed transversely to the direction of propagation, give rise to an axial ponderomotive force that can extract energy from the electron beam? The wiggler is the predominant influence on the electron's motion. To understand the dynamical relationships between the electrons and the fields, consider the motion of an electron subject to a helically symmetric wiggler field. An electron propagating through a magnetic field experiences a force that acts at right angles to both the direction of the field and to its own velocity. The wiggler field is directed transversely to the direction of bulk motion of the electron beam and rotates through 360° in one wiggler period. An electron streaming in the axial direction, therefore, experiences a transverse force and acquires a transverse velocity component on entry into the wiggler. The resulting trajectory is helical and describes a bulk streaming along the axis of symmetry as well as a transverse circular rotation that lags 180° behind the phase of the wiggler field. The magnitude of the transverse wiggle velocity, denoted by v_w, is proportional to the product of the wiggler amplitude and period. This relationship may be expressed in the form

$$\frac{v_w}{c} \cong 0.934 \frac{B_w\lambda_w}{\gamma_b}$$

where the wiggler period is expressed in units of centimeters, B_w denotes the wiggler amplitude in tesla, and

$$\gamma_b = 1 + \frac{E_b}{m_e c^2}$$

denotes the relativistic time-dilation factor associated with the total kinetic energy E_b of the electron beam (where m_e denotes the rest mass of the electron, and $m_e c^2$ denotes the electron rest energy).

Since the motion is circular, both axial and transverse velocities have a constant magnitude. This is important because the resonant interaction depends on the axial velocity of the beam. In addition, since the wiggler induces a constant-magnitude transverse velocity, the relation between the total electron energy and the streaming energy can be expressed in terms of the time-dilation factors in the form

$$\gamma_z \cong \frac{\gamma_b}{\sqrt{1 + 0.872\, B_w^2\lambda_w^2}}$$

As a result, the resonant wavelength depends on the total beam energy and the wiggler amplitude and period through

$$\lambda \cong (1 + 0.872\, B_w^2\lambda_w^2)\,\frac{\lambda_w}{2\gamma_b^2}$$

It is the interaction between the transverse wiggler-induced velocity with the transverse magnetic field of an electromagnetic wave that induces a force normal to both in the axial direction. This is the ponderomotive force. The transverse velocity and the radiation magnetic field are directed at right angles to each other and undergo a simple rotation about the axis of symmetry. A resonant wave must be circularly polarized with a polarization vector that is normal to both the transverse velocity and the wiggler field and that

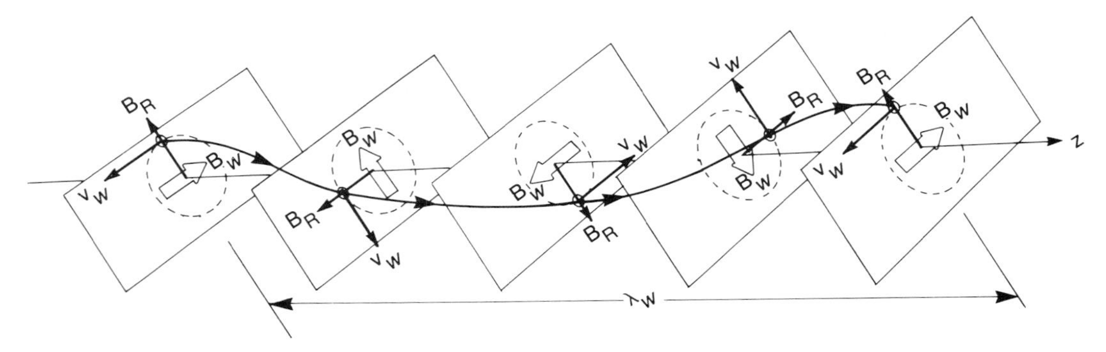

FIG. 2. The electron trajectory in a helical wiggler includes bulk streaming parallel to the axis of symmetry as well as a helical gyration. The vector relationships between the wiggler field B_w, the transverse velocity v_w, and the radiation field B_R of a resonant wave are shown in the figure projected onto planes transverse to the symmetry axis at intervals of one quarter of a wiggler period. This projection of the orbit is circular, and the transverse velocity is directed opposite to that of the wiggler. A resonant wave must be circularly polarized, with a polarization vector that is normal to both the transverse velocity and the wiggler field and which rotates in synchronism with the electrons. The electrons then experience a slowly varying wave amplitude. In addition, the transverse velocity and the radiation magnetic field are directed at right angles to each other and undergo a simple rotation about the axis of symmetry. The interaction between the transverse velocity and the radiation field induces a force in the direction normal to both that coincides with the symmetry axis.

rotates in synchronism with the electrons. This synchronism is illustrated in Fig. 2 and is maintained by the aforementioned resonance condition.

To understand the energy transfer, we return to the surfer analogy and consider a group of surfers attempting to catch a series of waves. In the attempt to match velocities with the waves, some will catch a wave ahead of the crest and slide forward while others will catch a wave behind the crest and slide backward. As a result, clumps of surfers will collect in the troughs of the waves. Those surfers who slide forward ahead of the wave are accelerated and gain energy at the expense of the wave, while those who slide backward are decelerated and lose energy to the wave. The wave grows if more surfers are decelerated than accelerated, and there is a net transfer of energy to the wave. The free-electron laser operates by an analogous process. Electrons in near resonance with the ponderomotive wave lose energy to the wave if their velocity is slightly greater than the phase velocity of the wave, and gain energy at the expense of the wave in the opposite case. As a result, wave amplification occurs if the wave lags behind the electron beam.

This process in a free-electron laser is described by a nonlinear pendulum equation. The *ponderomotive phase* Ψ $[= (k + k_w)z - \omega t]$ is a measure of the position of an electron in both space and time with respect to the ponderomotive wave. The ponderomotive phase satisfies the circular pendulum equation

$$\frac{d^2}{dz^2}\Psi = K^2 \sin \Psi$$

where the pendulum constant is proportional to the square root of the product of the wiggler and radiation fields

$$K \approx 8.29 \frac{\sqrt{B_w B_R}}{\gamma_b}$$

Here K is expressed in units of inverse centimeters and the magnetic fields are expressed in tesla. The pendulum equation can be reduced to

$$\frac{1}{2}\left(\frac{d\Psi}{dz}\right)^2 + U(\Psi) = H$$

where H has the form of the Hamiltonian or total energy of the system, and

$$U(\Psi) = K^2 \cos \Psi$$

is the ponderomotive potential. The electron trajectories through the wiggler, therefore, may be expressed as

$$\frac{d\Psi}{dz} = \pm\sqrt{2H - 2K^2 \cos \Psi}$$

Observe that the first derivative of the phase [i.e., $d\Psi/dz = (k + k_w) - \omega/v_z$] is a measure of the electron-streaming velocity; hence, this equation effectively describes the electron velocity as a function of the phase of the ponderomotive wave.

There are two classes of trajectory: trapped and

untrapped. The *separatrix* describes the transition between trapped and untrapped orbits and occurs when $H = K^2$. Hence, the separatrix is defined by a pair of curves in the phase space for $(d\Psi/dz, \Psi)$

$$\frac{d\Psi}{dz} = \pm 2K \sin\left(\frac{\Psi}{2}\right)$$

Free-streaming, untrapped orbits are characterized by $H > K^2$ and occupy that region of the phase space outside of the separatrix,

$$\frac{d\Psi}{dz} > \left| 2K \sin\left(\frac{\Psi}{2}\right) \right|$$

The trapped orbits are those for which $H < K^2$ within the bounds of the separatrix. The free-streaming orbits correspond to the case in which the pendulum swings through the full 360° cycle. The electrons pass over the crests of many waves, traveling fastest at the bottom of the troughs and slowest at the crests of the ponderomotive wave. In contrast, the electrons are confined within the trough of a single wave in the trapped orbits. This corresponds to the motion of a pendulum that does not rotate full circle, but is confined to oscillate about the lower equilibrium point. In the dynamical process, illustrated in Fig. 3, the pendulum constant evolves during the course of the interaction. Electrons lose energy as the wave is amplified; hence, the electrons decelerate and both the pendulum constant and separatrix grow. Ultimately, the electrons cross the growing separatrix from untrapped to trapped orbits.

In typical operation, electrons entering the free-electron laser are free-streaming on untrapped trajectories. The gain in power during this phase of the interaction increases as the cube of the distance z in the case of relatively low-current operation, in which the total gain is less than unity. This case is often

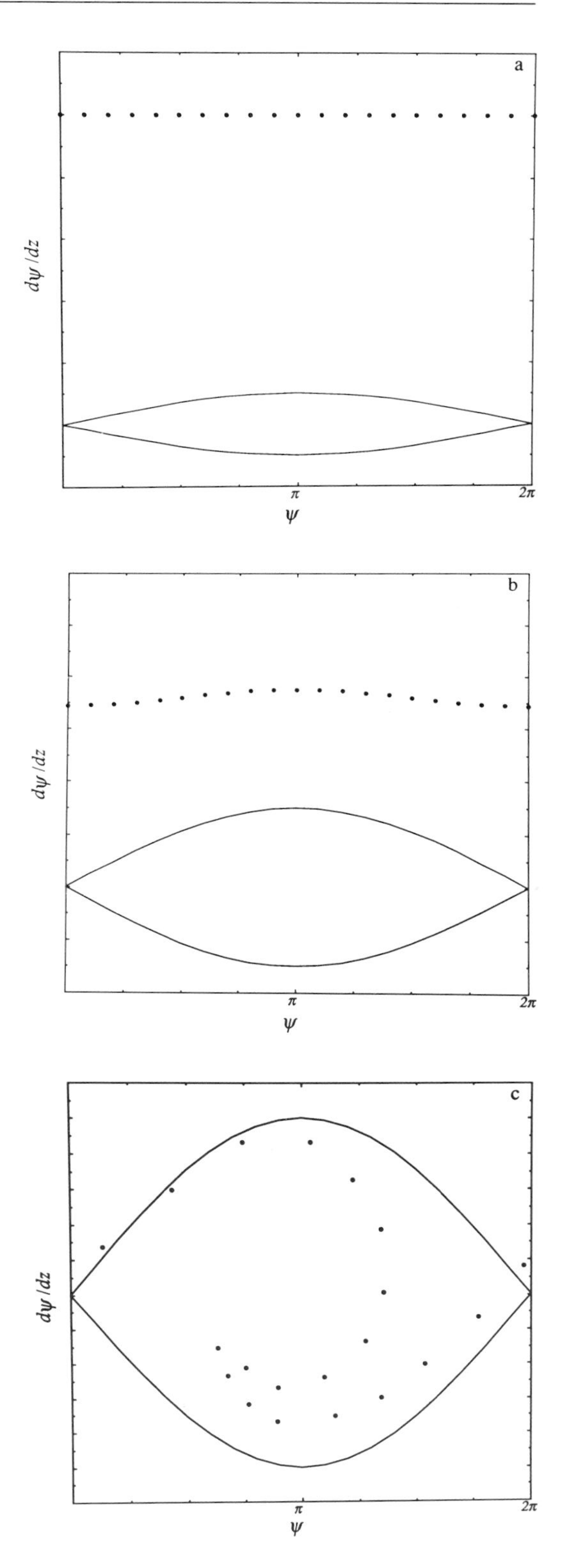

FIG. 3. The evolution of the phase space distribution of an electron beam is illustrated schematically for three general stages of the interaction. In the initial state (a), a monoenergetic electron population is located far from the separatrix. As the interaction progresses (b), the electrons lose energy to the wave, and the separatrix grows. Observe that the electrons are still on untrapped trajectories; however, the coherent bunching mechanism characteristic of the interaction has begun. In the final stage (c), the bulk of electrons have crossed the separatrix and become trapped. Electron trajectories after this point circulate about within the separatrix, alternately gaining from and losing energy to the wave. Saturation occurs when the net rate of energy transfer vanishes. Observe that all electrons do not become trapped.

referred to as the low-gain Compton regime, and the power gain in decibels is given by

$$G(\mathrm{dB}) \approx 1.09 \frac{\omega_b^2}{c^2 k_w^2} \frac{v_w^2}{c^2} (k_w z)^3 F(\theta)$$

where

$$\omega_b = \sqrt{\frac{4\pi e^2 n_b}{\gamma_b m_e}}$$

is the so-called *electron plasma frequency*. Here e represents the charge of the electron, and n_b denotes the bulk density of the electron beam. The spectral shape of this gain is determined by

$$F(\theta) = \frac{d}{d\theta}\left(\frac{\sin \theta}{\theta}\right)^2$$

where $\theta = (\omega/v_z - k - k_w)z/2$. This spectral function is shown in Fig. 4 and exhibits a maximum at $\theta \approx -1.3$, for which $F \approx 0.54$; hence, the peak gain occurs at a wavelength

$$\lambda \approx \frac{\lambda_w}{2\gamma_z^2\left(1 + \frac{2.6}{k_w z}\right)}$$

This regime is relevant to operation at short wavelengths in the infrared and optical spectra. These experiments typically employ electron beams generated by radio-frequency linear accelerators, microtrons, storage rings, and electrostatic accelerators in which the total current is small.

In contrast, experiments operating at microwave and millimeter wavelengths often employ high-current accelerators, and the wave amplification is exponential. In these cases, two distinct regimes are found. The high-gain Compton (sometimes called the *strong-pump*) regime is found when

$$\frac{\omega_b}{ck_w} \ll \frac{1}{16} \gamma_z^2 \left(\frac{v_w}{c}\right)^2$$

In this regime, the maximum gain in the signal power over a wiggler period is given by

$$G(\mathrm{dB}) \approx 37.5 \left(\frac{v_w^2}{c^2} \frac{\omega_b^2}{c^2 k_w^2}\right)^{1/3}$$

and is found at the resonant wavelength.

The opposite limit, referred to as the collective Raman regime, is fundamentally different from either the high- or low-gain Compton regimes. It occurs when the current density of the beam is high enough that the space-charge force exceeds that exerted by the ponderomotive wave. The gain in power over a wiggler period in this regime varies as

$$G(\mathrm{dB}) \approx 27.3 \frac{v_w}{c} \sqrt{\gamma_z \frac{\omega_b}{ck_w}}$$

In this regime, the space-charge forces result in electrostatic waves that co-propagate with the beam and are characterized by the dispersion relations

$$\omega = k_{sc} v_z \pm \frac{\omega_b}{\gamma_z}$$

which describe the relation between the frequency ω and wavenumber k_{sc}. These dispersion relations describe positive- and negative-energy waves corresponding to the " + " and " − " signs, respectively. The interaction results from a stimulated three-wave scattering process. This is best visualized from the perspective of the electrons, in which the wiggler field appears to be a backwards propagating electromagnetic wave called a *pump* wave. This pump wave can scatter off the negative-energy electrostatic wave (the *idler*) to produce a forward-propagating electromagnetic wave (the *signal*). The interaction occurs when the wavenumbers of the pump, idler, and signal satisfy the condition $k_{sc} = k + k_w$, which causes a shift in the wavelength of the signal to

$$\lambda \approx \frac{\lambda_w}{2\gamma_z^2\left(1 - \frac{\omega_b}{\gamma_z c k_w}\right)}$$

Observe that the interaction in the Raman regime is shifted to a somewhat longer wavelength than occurs in the high-gain Compton regime.

Wave amplification can saturate by several processes. The highest efficiency occurs when the elec-

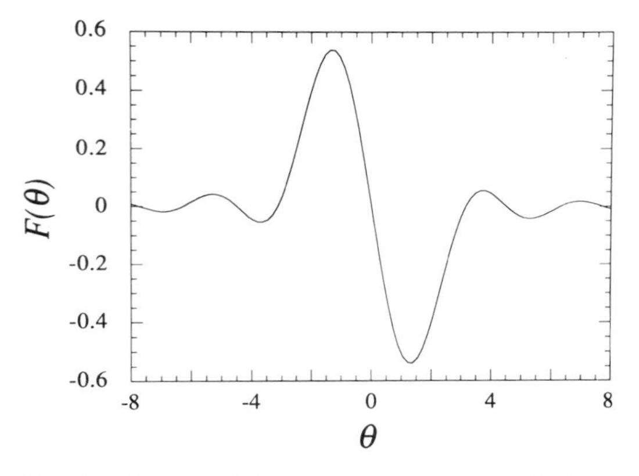

FIG. 4. The spectral function measures the strength of the interaction as a function of the detuning of the wave from resonance with the electron beam. The maximum gain is found at $\theta \approx -1.3$. In contrast, the wave is damped for $\theta \approx 1.3$.

trons are trapped in the ever-deepening ponderomotive wave and undergo oscillations within the troughs. In essence, the electrons are initially free-streaming over the crests of the ponderomotive wave. Since they are traveling at a velocity faster than the wave speed, they come upon the wave crests from behind. However, the ponderomotive wave grows together with the radiation field, and the electrons ultimately will come upon a wave that is too high to cross. When this happens, they rebound and become trapped within the trough of the wave. In analogy to the oscillation of a pendulum, the trapped electrons lose energy as they rise and gain energy as they fall toward the bottom of the trough. As a result, the energy transfer between the wave and the electrons is cyclic, and the wave amplitude ceases to grow and oscillates with the electron motion in the trough. The ultimate saturation efficiency for this mechanism can be estimated from the requirement that the net change in electron velocity at saturation is equal to twice the velocity difference between the electron beam and the ponderomotive wave. This results in a saturation efficiency of

$$\eta \approx \frac{\gamma_b}{2(\gamma_b - 1)} \left(\frac{v_w^2}{2c^2} \frac{\omega_b^2}{c^2 k_w^2} \right)^{1/3}$$

in the high-gain Compton regime, and

$$\eta \approx \frac{\gamma_b}{\gamma_b - 1} \frac{\omega_b}{\gamma_z c k_w}$$

in the collective Raman regime.

However, this process places stringent requirements on the quality of the electron beam. The preceding formulas apply to the idealized case of a monoenergetic (or cold) beam. This represents a theoretical maximum for the gain and efficiency, since each electron has the same axial velocity and interacts with the wave in an identical manner. A monoenergetic beam is physically unrealizable, however, and all beams exhibit a velocity spread that determines a characteristic temperature. Electrons with axial velocities different from the optimal resonant velocity are unable to participate fully in the interaction. If this axial velocity spread is sufficiently large that the entire beam cannot be in simultaneous resonance with the wave, then the fraction of the electron beam that becomes trapped must fall. Ultimately, the trapping fraction falls to the point where the trapping mechanism becomes ineffective, and saturation occurs through the thermalization of the beam. Thus, there are two distinct operating regimes: the *cold* beam limit characterized by a narrow bandwidth and relatively high efficiencies, and the *thermal* regime characterized by a relatively broader bandwidth and sharply lower efficiencies.

The question of electron-beam quality is the most important single issue facing the development of the free-electron laser at the present time. To operate in the cold beam regime, the axial velocity spread of the beam must be small. It is convenient to relate the axial velocity spread to an energy spread to obtain an invariant measure of the beam quality suitable for a wide range of electron beams. In the case of the low-gain limit, this constraint on the beam thermal spread is

$$\frac{\Delta E_b}{E_b} \ll \frac{1}{N_w}$$

where ΔE_b represents the beam thermal spread, and N_w is the number of periods in the wiggler. In the high-gain regimes, the maximum permissible energy spread for saturation by particle trapping is determined by the depth of the ponderomotive or space-charge waves, which is measured by twice the difference between the streaming velocity of the beam and the wave speed. The maximum permissible thermal spread corresponds to this velocity difference, and is one-half the saturation efficiency for either the high-gain Compton or collective Raman regimes, that is,

$$\frac{\Delta E_b}{E_b} \ll \frac{1}{2} \eta$$

Typically, this energy spread must be approximately 1% or less of the total beam energy for the trapping mechanism to operate at millimeter wavelengths, and decreases approximately with the radiation wavelength. Hence, the requirement on beam quality becomes more restrictive at shorter wavelengths and places greater emphasis on accelerator design.

Two related quantities often used as measures of beam quality are the *emittance* and the *brightness*. The emittance measures the collimation of the electron beam and may be defined in terms of the product of the beam radius and the average pitch angle (i.e., the angle between the velocity and the symmetry axis). It describes a random pitch-angle distribution of the beam that, when the velocities are projected onto the symmetry axis, is equivalent to an axial velocity spread. In general, therefore, even a monoenergetic beam with a nonvanishing emittance displays an axial velocity spread. The electron beam brightness is an analog of the brightness of optical beams and is directly proportional to the current and inversely proportional to the square of the emittance. As such, it describes the average current density per unit pitch angle and measures both the beam intensity and the degree of collimation of the electron trajectories. Since the gain and efficiency increases with increasing beam current for fixed emittance, the brightness is a complementary measure of the beam quality. Although it is important to minimize the emittance and maximize the brightness in order to optimize performance, both of these measures relate to the free-electron laser only insofar as they describe the axial velocity spread of the beam.

Typical free-electron laser efficiencies range up to approximately 12%; however, significant enhancements are possible when either the wiggler amplitude or period are systematically tapered. The free-electron laser amplifier at Livermore, which achieved a 35% extraction efficiency, employed a wiggler with an amplitude that decreased along the axis of symmetry, and contrasts with an observed efficiency of about 6% in the case of a uniform wiggler. The use of a tapered wiggler was pioneered by Phillips in 1960. The technique has received intensive study of late. Tapered-wiggler designs have also been shown to be effective at infrared wavelengths in experiments at Los Alamos National Laboratory and, using a superconducting rf linac, at Stanford University.

The effect of a tapered wiggler is to alter both the transverse and axial velocities. Since the transverse velocity is directly proportional to the product of the amplitude and period, the effect of gradually decreasing either of these quantities is to decrease the transverse velocity and, in turn, increase the axial velocity. The energy extracted during the interaction results in an axial deceleration that drives the beam out of resonance with the wave; hence, efficiency enhancement occurs because the tapered wiggler maintains a relatively constant axial velocity (and phase relationship between the electrons and the wave) over an extended interaction length. The pendulum equation is modified in the presence of a tapered wiggler, and has the form

$$\frac{d^2}{dz^2}\Psi = K^2(\sin\Psi - \sin\Psi_{res})$$

where the resonant phase Ψ_{res} is determined by the wiggler taper. The wiggler taper can be accomplished either through the amplitude or period. However, it is technically simpler to taper the amplitude, and most tapered-wiggler experiments have employed this approach. In this case, the resonant phase varies as

$$\sin\Psi_{res} = \left(\frac{v_w}{v_z}\right)^2 \frac{k + k_w}{K^2} \frac{1}{B_w} \frac{d}{dz} B_w$$

Integration of the tapered-wiggler pendulum equation results in an equation similar to that found in the case of a uniform wiggler, with a ponderomotive potential

$$U(\Psi) = K^2(\cos\Psi + \Psi\sin\Psi_{res})$$

If the wiggler amplitude is a decreasing function of axial position, then $\sin\Psi_{res} < 0$ and the average potential decreases linearly with the ponderomotive phase. The difference between the ponderomotive potential for a uniform and a tapered wiggler is illustrated in Fig. 5. As a result, the motion is similar to that of a ball rolling down a *bumpy* hill and accelerating as it falls.

The enhancement in the tapered-wiggler interaction efficiency is proportional to the decrement in the wiggler field of ΔB_w, and satisfies

$$\Delta\eta \approx \frac{0.872 B_w^2 \lambda_w^2}{1 + 0.872 B_w^2 \lambda_w^2} \frac{\Delta B_w}{B_w}$$

In practice, a tapered wiggler is effective only after the bulk of the beam has become trapped in the ponderomotive wave. In single-pass amplifier configurations, therefore, the taper is not begun until the signal has reached saturation in a section of uniform wiggler, and the total extraction efficiency is the sum of the uniform wiggler efficiency and the tapered wiggler increment. Numerical simulations indicate that

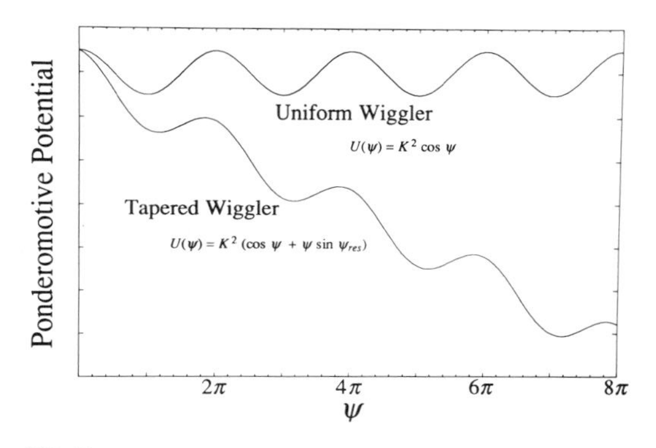

FIG. 5. A comparison of the ponderomotive potentials for uniform and tapered wigglers. The extraction efficiency of the free-electron laser can be enhanced by introducing a taper in the wiggler field. In particular a gradual reduction in either the amplitude or period of the wiggler will reduce the magnitude of the transverse velocity and accelerate the electrons in the axial direction. This is illustrated by the effect of the taper, for $\sin\Psi_{res} < 0$, on the ponderomotive potential. In such a potential, an electron behaves in a manner analogous to a ball rolling down a *bumpy hill* and accelerating as it falls.

total efficiencies as high as 65% are possible under the right conditions.

Once particles have been trapped in the ponderomotive wave and begin executing a bounce motion between the troughs of the wave, the potential exists for exciting secondary emission referred to as sideband waves. These sidebands are caused by the beating of the primary signal with the ponderomotive bounce motion. This bounce motion is at the pendulum period that depends weakly on the depth of the ponderomotive well and the amplitude of the bounce motion. For deeply trapped particles that oscillate near the bottom of the well, the pendulum equation can be approximated as a harmonic oscillator with a wavenumber equal to the pendulum constant K. The sidebands, therefore, occur at wavenumbers shifted both upward and downward from the radiation wavenumber by this value (i.e., $k_{\pm} = k \pm K$, where $k_{\pm}$ denotes the sideband wavenumber). Note that the bounce period is, typically, much longer than the radiation wavelength (i.e., $K \ll k$), and these sidebands are found at wavelengths close to the wavelength of the primary signal.

The difficulties imposed by the presence of sidebands is that they may compete with and drain energy from the primary signal. This is particularly crucial in long tapered-wiggler systems designed to trap the beam at an early stage of the wiggler and then extract a great deal more energy from the beam over an extended interaction length. In these systems, unrestrained sideband growth can be an important limiting factor. As a result, a great deal of effort has been expended on techniques of sideband suppression. One method of sideband suppression was employed in a free-electron laser oscillator at Columbia University. This experiment operated at a 2-mm wavelength in which the dispersion due to the waveguide significantly affected the resonance condition. As a consequence, it was found to be possible by proper choice of the size of the waveguide to shift the sideband frequencies out of resonance with the beam. Experiments on an infrared free-electron laser oscillator at Los Alamos National Laboratory indicate that it is also possible to suppress sidebands by (1) using a Littrow grating to deflect the sidebands out of the optical cavity or (2) changing the cavity length.

The foregoing description of the principles and theory of the free-electron laser is necessarily restricted to the idealized case in which the transverse inhomogeneities of both the electron beam and wiggler field are unimportant. This is sufficient for an exposition of the fundamental physics of the free-electron laser. In practice, however, these gradients can have important consequences on the performance of the free-electron laser. It is beyond the scope of this article to delve into these subjects in depth, and we will present only a simple sketch of the types of effects to be encountered. The most important effect that is found if the wiggler field varies substantially across the diameter of the electron beam is that the electron response to the wiggler will vary as well. In practice, this means that an electron at the center of the beam will experience a different field than an electron at the edge of the beam, and the two electrons will follow different trajectories with different velocities. As a result of this, the wave-particle resonance that drives the interaction will be broadened, and the gain and efficiency will decline. In essence, therefore, the transverse wiggler inhomogeneity is manifested as an effective beam thermal spread. The bounded nature of the electron beam also affects the interaction, since wave growth will occur only in the presence of the beam. Because of this, it is important in amplifier configurations to ensure good overlap between the injected signal and the electron beam. Once such overlap has been accomplished, however, the dielectric response of the electron beam in the presence of the wiggler can act in much the same way as an optical fiber to refractively guide the light through the wiggler.

III. Experiments and Applications

There are three basic experimental configurations for free-electron lasers: amplifiers, oscillators, and superradiant amplifiers. In an amplifier, the electron beam is injected into the wiggler in synchronism with the signal to be amplified. The external radiation source that drives the amplifier is referred to as the master oscillator and can be any convenient radiation source, such as a conventional laser or microwave tube. As a consequence, this configuration is often referred to as a master oscillator power amplifier (MOPA). Because amplification occurs during one pass through the wiggler, MOPAs require intense electron beam sources that can operate in the high-gain regime. Oscillators differ from amplifiers in that some degree of reflection is introduced at the ends of the wiggler so that a signal will make multiple passes through the system. The signal is amplified during that part of each pass in which the radiation copropagates with the electrons and allows for a large cumulative amplification over many passes, even in the event of a low gain per pass. Oscillators are typically constructed to amplify the spontaneous (i.e., shot) noise within the beam, and no outside signal is necessary for their operation. However, a long-pulse accelerator is required because a relatively long time

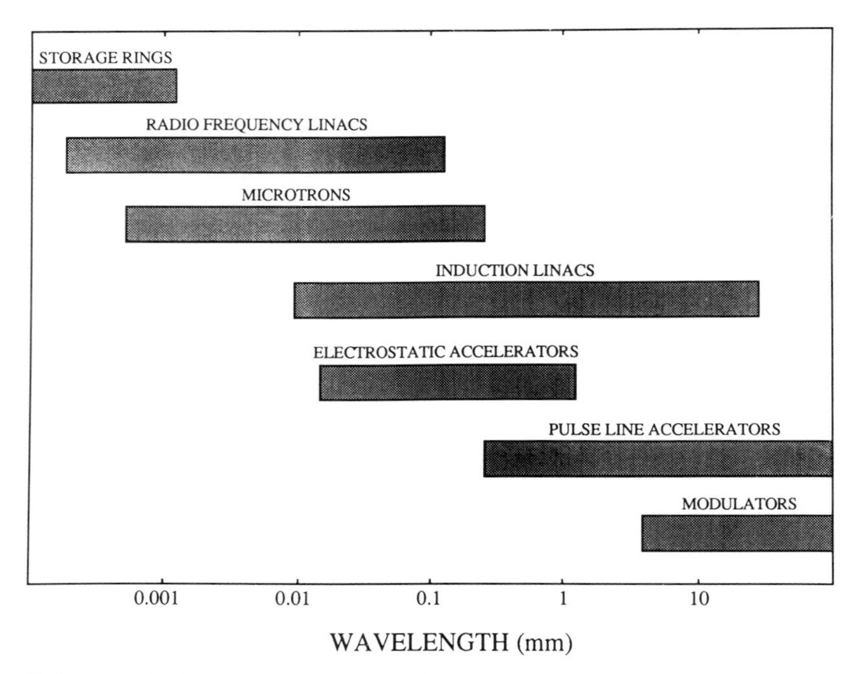

FIG. 6. The limits on the wavelengths possible with different accelerators depend both on the electron beam energies that may be achieved as well as on the state of wiggler development. The wavelength ranges that may be accessed by accelerators given the present state of accelerator and wiggler technology are shown.

may be required to build up to saturation. Superradiant amplifiers are devices in which the shot noise in the beam is amplified over the course of a single pass through the wiggler and, like amplifiers, require high-current accelerators to drive them. Since the shot noise present in the beam is generally broadband, the radiation from a superradiant amplifier is typically characterized by a broader bandwidth than a MOPA.

The optimal configuration and type of accelerator used in a free-electron laser design depends on the specific application, and issues such as the electron-beam quality, energy, and current are important considerations in determining both the wavelength and power. In general, however, each accelerator type is suited to the production of a limited range of wavelengths, as shown in Fig. 6. In addition, the temporal structure of the output light from a free-electron laser corresponds to that of the electron beam. Thus, either a pulsed or continuous electron-beam source will give rise to a pulsed or continuous free-electron laser. Free-electron lasers have been constructed using virtually every type of electron source, including storage rings, radio-frequency linear accelerators, microtrons, induction linacs, electrostatic accelerators, pulse-line accelerators, and modulators. Since the gain in a free-electron laser increases with current but decreases with energy, accelerators producing low-current/high-energy beams are generally restricted to the low-gain regime. Accelerator types that fall into this category include rf linacs, microtrons, storage rings, and electrostatic accelerators. In contrast, intense beam accelerators such as induction linacs, pulse-line accelerators, and modulators are suitable electron-beam sources for high-gain systems.

Storage rings are typically characterized by multiple electron pulses continuously circulating through the ring. Each pulse is several nanoseconds in duration, and the output light from a free-electron laser driven by a storage ring is a continuous stream of picosecond bursts. In addition, while storage rings produce high-quality and high-energy beams of low to moderate currents, the electron pulses are recirculated through both the ring and the wiggler, and the stability of the ring is disrupted by the extraction of too much energy. Hence, storage rings are feasible for applications that require uniform and continuous short-wavelength radiation sources but do not demand high output powers.

The first successful operation of a storage ring free-electron laser was at the Université de Paris Sud at Orsay. This experimental configuration was that of an oscillator and made use of the ACO storage ring, which operates at energies and average currents in the range of 160–224 MV and 16–100 mA. The laser was tuned across a broad band of the visible spectrum, but was first operated at wavelengths in the neighborhood of approximately 0.65 μm. The peak output power from the oscillator was 60 mW over the 1-ns duration of the micropulses, which corresponds to an intracavity power level of 2 kW. The average power extracted from the system was typically of the order of 75 mW. Higher harmonic emission was also detected in the ultraviolet, however, which posed a problem since radiation at these wavelengths resulted in the ultimate degradation of the optical system.

An ultraviolet free-electron laser oscillator has also been achieved using the VEPP storage ring at Novosibirsk. This experiment employed an optical klystron configuration in which the wiggler was composed of two distinct sections separated by a diffusive drift space. In this configuration, the first section operates as a prebuncher for the electron beam, which subsequently enhances the gain in the second wiggler section. Operating the storage ring at 350 MV and a peak current of 6 A, experimenters were able to obtain coherent emission at wavelengths as short as 0.3 μm and average output power as high as 6 mW.

Radio-frequency linacs use a series of cavities that contain rapidly varying electromagnetic (rf) fields to accelerate streams of electrons. The beams they produce are composed of a sequence of macropulses (typically of microseconds in duration) each of which consists of a train of shorter picosecond pulses. Microtrons produce beams with a temporal structure similar to that of rf linacs but, unlike the rf linac, are composed of a single accelerating cavity coupled to a magnet that causes the electron beam to recirculate through the cavity many times. The output light from a free-electron laser built with these accelerators, therefore, is similar to the output from a storage ring. Recent experiments at Stanford University and Boeing Aerospace have also demonstrated the feasibility of the rf linac to produce visible light. In addition, rf linacs and microtrons are suitable for high-power free-electron lasers, since the electrons are not recirculated and the energy extraction is not limited by the disruption of the beam.

Free-electron lasers based on rf linacs have demonstrated operation over a broad spectrum extending from the infrared through the visible. The initial experiments conducted by Madey and co-workers at Stanford University resulted in (1) an amplifier that operated at a wavelength of 10.6 μm with an overall gain of 7%, and (2) a 3.4-μm oscillator that produced peak and average output power of 7 kW and 0.1 mW, respectively. In collaboration with TRW, the superconducting rf linac (SCA) at Stanford University has been used to drive a free-electron laser oscillator that has demonstrated efficiency enhancement with a tapered wiggler, operation at visible wavelengths, and beam recirculation. The significance of the latter point is that the overall wall-plug efficiency can be enhanced by recovery and reuse of the spent electron beam subsequent to its passage through the free-electron laser. The tapered-wiggler experiment operated in the infrared at a wavelength of 1.6 μm and peak power levels of 1.3 MW. This yields a peak extraction efficiency of approximately 1.2%, which constitutes an enhancement by a factor of three over the efficiency in the case of an untapered wiggler. Operation at visible wavelengths was also found at 0.52 μm and peak power levels of 21 kW. The superconducting technology embodied in the SCA can enable the rf linac to further compete with storage rings by operating in a near steady-state mode. Both energy recovery and enhancement of the extraction efficiency by means of tapered wiggler were also demonstrated at Los Alamos National Laboratory. Starting in 1981 with a tapered-wiggler free-electron laser amplifier that obtained an extraction efficiency of 4% at a wavelength of 10.6 μm, researchers went on to (1) extend that to a 5% extraction efficiency in an oscillator configuration, and (2) demonstrate a 70% energy recovery rate with beam recirculation.

The limitations that storage rings, rf linacs, and microtrons impose on free-electron laser design stems from restrictions on the peak (or instantaneous) currents that may be obtained and thus limits the peak power from a free-electron laser. High peak powers may be obtained by using induction linacs, pulse-line accelerators, or modulators that produce electron beams with currents ranging from several amperes through several thousands of amperes, and with pulse times ranging from several tens of nanoseconds through several microseconds. Induction linacs operate by inducing an electromotive force in a cavity through a rapid change in the magnetic field strength. In effect, the electron beam acts as the secondary winding in a transformer. For example, the Advanced Test Accelerator (ATA) at Livermore is an accelerator of this type and can achieve energies and currents as high as 50 MV and 10 kA over a duration of 50 ns. At lower energies, pulse-line accelerators and conventional microwave tube modulators are available. Pulse-line accelerators produce beams with energies up to several tens of megavolts, currents of several tens of kiloamperes, and pulse times up to 50–100 ns. As a result, pulse-line accelerators and modulators have been applied exclusively to microwave generation.

Amplifier experiments employing induction linacs have been performed at the Naval Research Laboratory, Lawrence Livermore National Laboratory, and at the Institute for Laser Engineering at Osaka University in Japan and have demonstrated operation from the microwave through the infrared spectra. A superradiant amplifier at the Naval Research Laboratory employed a 650-kV/200-A electron beam and produced 4 MW at a wavelength of 1 cm. The Livermore experiments have employed both the Experimental Test Accelerator (ETA) and the ATA. The free-electron laser amplifier experiment at an 8-mm wavelength was conducted with the ETA operating at approximately 3.5 MV and a current of 10 kA. However, due to beam quality requirements, only about 10% of the beam was found to be usable in the free-electron laser. This ETA-based MOPA has been operated in both uniform and tapered-wiggler configurations. In the case of a uniform wiggler, the measured output power was in the neighborhood of 180 MW, which corresponds to an extraction efficiency of about 6%. A dramatic improvement in the efficiency was achieved, however, using a tapered wiggler. In this case, the total output power rose to 1 GW for an efficiency of 35%. The ATA has been used for a high-power MOPA design at a wavelength of 10.6 μm. Experiments are currently in progress at both Livermore and the Institute for Laser Engineering at Osaka on the use of induction linacs in high-power MOPAs at frequencies above 140 GHz for the purposes of radio-frequency heating of magnetically confined plasmas for controlled thermonuclear fusion.

An important consideration in the construction of free-electron lasers with the intense beams generated by these accelerators is that an additional source of focusing is required to confine the electrons against the self-repulsive forces generated by the beam. This can be accomplished by the use of additional magnetic fields generated by either solenoid or quadrupole current windings. Quadrupole field windings were employed in the 8-mm amplifier experiment at the Lawrence Livermore National Laboratory; the interaction mechanism in a free-electron laser is largely, though not entirely, transparent to the effect of the quadrupole field. In contrast, the solenoidal field has a deep and subtle effect on the interaction mechanism. This arises because a solenoidal field results in a precession, called Larmor rotation, about the magnetic field lines that can resonantly enhance the helical motion induced by the wiggler. This enhancement in the transverse velocity associated with the helical trajectory occurs when the Larmor period is comparable to the wiggler period. Since the Larmor period varies with beam energy, this relation can be expressed as

$$B_0 \approx 1.07 \frac{\gamma_b}{\lambda_w}$$

where the solenoidal field B_0 is expressed in tesla and the wiggler period is in centimeters. For fixed-beam energies, therefore, this resonant enhancement in the wiggler-induced velocity requires progressively higher solenoidal fields as the wiggler period is reduced.

The effect of a resonant solenoidal magnetic field is to enhance both the gain and saturation efficiency of the interaction. This was demonstrated in a superradiant amplifier experiment using the VEBA pulse-line accelerator at the Naval Research Laboratory. In this experiment, the output power from the free-electron laser was measured as a function of the solenoidal field as it varied over the range of 0.6–1.6 T. The beam energy and current in the experiment were 1.35 MV and 1.5 kA, respectively, and the wiggler period was 3.0 cm. It should be remarked that, due to the high current in the experiment, the beam was unable to propagate through the wiggler for solenoidal fields below 0.6 T. The magnetic resonance was expected for a solenoidal field in the neighborhood of 1.3 T, and the experiment showed a dramatic increase in the output power for fields in this range. Other experiments that have demonstrated the effect of a solenoidal field have been performed at the Massachusetts Institute of Technology, Columbia University, and the Ecole Polytechnique in France.

The maximum enhancement in the output power in the experiment at the Naval Research Laboratory was observed for solenoidal fields slightly above the resonance. In this case, the nature of the interaction mechanism undergoes a fundamental change. In the absence of a solenoidal field (or for fields below the magnetic resonance), the axial velocity of the electrons decreases as energy is lost to the wave while the transverse velocity remains relatively constant. In contrast, the result of the strong solenoidal field is to cause a negative-mass effect in which the electrons accelerate in the axial direction as they lose energy to the wave. The bulk of the energy used to amplify the wave is extracted from the transverse motion of the electrons. Computer simulations of free-electron lasers operating in this strong solenoidal regime indicate that extremely high extraction efficiencies (in

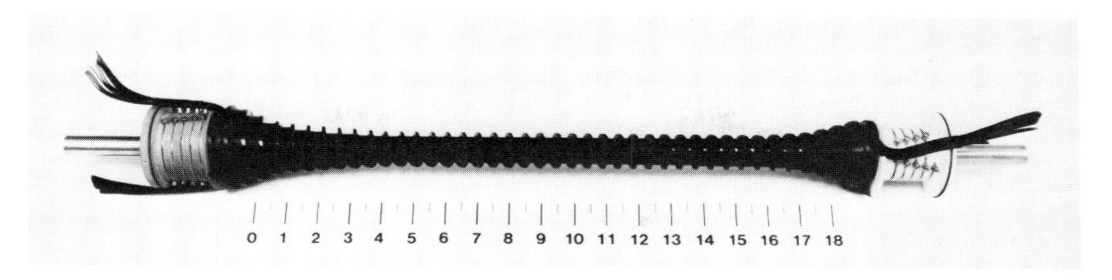

FIG. 7. One of the principal wiggler designs in use is the bifilar helical current winding. The example shown here exhibits flares at both the entrance and the exit to the wiggler, to facilitate the transition of the electron beam both into and out of the wiggler field. The scale shown is in centimeters.

the neighborhood of 50%) are possible without recourse to a tapered-wiggler field. However, operation in this regime is precluded below submillimeter wavelengths. This is because the solenoidal field required to achieve this magnetic resonance varies directly with the beam energy and inversely with the wiggler period. Because of this, impractically high fields are required for wavelengths in the far infrared and below.

It is important to bear in mind that the high peak-versus-average power and oscillator-versus-MOPA distinctions between the different aforementioned accelerator technologies are becoming blurred by advances in the design of both rf and induction linacs. On the one hand, the development of laser-driven photocathodes at Los Alamos National Laboratory have dramatically increased both the peak (of the order of 400 A) and average currents achievable with rf linacs. As a result, a high-gain MOPA experiment is under design at Los Alamos National Laboratory. It is significant in this regard that a Stanford University collaboration with Rocketdyne Inc. has already achieved MOPA operation using the Mark III rf linac. On the other hand, new induction linacs are under development that may be fired repetitively at the rate of up to several thousand times per second. Successful completion of these development programs will enable high average power free-electron lasers to be constructed using this technology.

Although their average power is lower than that of linacs, electrostatic accelerators can produce continuous electron beams using charge recovery techniques. In such a process, the electron beam is recirculated through the wiggler and back into the accelerator in a continuous stream. Using this technology, the electrostatic accelerator holds promise as an electron beam source for a continuous or long-pulse free-electron laser. However, given practical restrictions on the size of such accelerators, which limit energies and currents, electrostatic accelerators have been restricted to the construction of free-electron laser oscillators that operate from the microwave regime through the infrared spectrum.

The breadth of free-electron laser experiments includes many different wiggler configurations and virtually every type of accelerator in use today. The wiggler has been produced in planar, helical, and cylindrical forms by means of permanent magnets, current-carrying coils, and hybrid electromagnets with ferrite cores. An example of a bifilar helical wiggler currently in use at the Naval Research Laboratory is shown in Fig. 7. Helical wiggler fields are produced by a current-carrying bifilar helical coil in which the field increases radially outward from the symmetry axis and provides magnetic focusing to confine the beam against the mutually repulsive forces between the electrons. In a planar wiggler, both the transverse and axial components of the velocity oscillate in synchronism with the wiggler. As such, the interaction is determined by the average, or root-mean-square, wiggler field. Because of this, planar wigglers require a stronger field to produce the same effect as helical wigglers. This is compensated for, however, by the ease of adjustment allowed by a planar design, in which the strengths or positions of the individual magnets can be altered to provide either a uniform or a tapered field. In contrast, the only adjustment possible for a bifilar helix is the strength of the field.

One practical constraint on the development of free-electron laser oscillators at the present time is mirror technology for infrared and shorter wavelengths; this constraint relates to both reflectivity and durability. The reflectivity is important since the net gain of an oscillator decreases as the mirror losses increase, and oscillation is possible only if the amplification due to the free-electron laser interaction exceeds the losses at the mirrors. The reflectivity is a measure of this loss rate, and must be kept sufficiently high that the energy losses at the mirrors do not overwhelm the gain. The issue of durability relates to the power level that any specific mirror material can endure without suffering optical damage. In this sense, optical damage refers to a decrease in the reflectivity. Note that the extreme case of the complete burning out of the mirrors might be described as a catastrophic drop in the reflectivity. Problems exist in finding materials with a high enough reflectivity and durability to operate in the infrared and ultraviolet spectra; even the visible presents problems. For example, the visible free-electron laser oscillator at the Université de Paris Sud experienced mirror degradation due to harmonic emission in the ultraviolet. At extremely high power levels, solutions can be found through such techniques as the grazing-incidence mirrors used by Boeing Aerospace in which the optical beam is allowed to expand to the point where the power density on the mirrors is low. In the infrared, oscillator experiments at Los Alamos National Laboratory originally employed a dielectric mirror material with a high reflectivity at low power levels. However, recent observations indicate that nonlinear phenomena occur in this material at high power levels that effectively reduce the reflectivity, and that the use of copper mirrors substantially improves performance. An additional problem occurs at high power levels due to thermal distortion of the optical surface. To combat this problem, actively cooled mirrors are under development.

The principal biomedical applications of the free-electron laser are surgery, photocoagulation and cauterization, photodynamic therapy, and the *in vivo* thermal destruction of tissue through a process called photothermolysis. The most common surgical technique is the thermal ablation of tissue, which requires a laser producing 10–100 W at a wavelength of approximately 3 μm. This corresponds to a strong absorption resonance of the water molecule characterized by relatively little scattering of the light by the tissues. In contrast, photocoagulation requires a shorter wavelength of approximately 1–1.5 μm, which is also strongly absorbed by the water molecule but exhibits a higher degree of scattering throughout the surrounding tissue. It is important to observe that the tunability of the free-electron laser holds the potential for a surgical laser that may be tuned in a single sequential process from 3 μm down to 1 μm to give both clean surgical incisions and cauterization. Another advantage is that the optical pulse can be tailored to meet specific requirements by control of the temporal structure of the electron beam. For example, short pulses are useful in ophthalmic therapy for the surgical disruption of pigmented tissue, while longer pulses are required for retinal photocoagulation.

Photodynamic therapies rely on the injection of photosensitive dyes that are preferentially concentrated in malignant tissue. Subsequent irradiation excites a photooxidation process that is toxic to the tumorous tissue. The principal dyes are photosensitive at wavelengths between 0.6–1.7 μm and are used to treat tumors of the lung, bladder, and gastrointestinal tract at early stages in their development and also as a palliative treatment at later stages of growth. At relatively high average powers (up to 100 W), a free-electron laser makes possible the simultaneous treatment of relatively large masses of tissue. However, the high-energy electron beams needed to produce these wavelengths also produce relatively large X-ray fluxes, and the entire facility including power supply, accelerator, wiggler, optical system, and X-ray shielding is likely to be rather bulky and complex. Hence, in consideration of the rapid development of conventional laser sources at these wavelengths, the long-term biomedical applications are envisioned to be in (1) the initial refinement of these therapeutic techniques, (2) large centralized facilities for tumor treatment, and (3) experimental research tools.

Applications to research are unimpaired by considerations of the bulk and complexity of a free-electron laser facility; the first user facility was established by Luis Elias at the University of California at Santa Barbara in 1984. This facility, shown in Fig. 8, employs a long-pulse 3-MV electrostatic accelerator. The free-electron laser produces a peak power of as much as 10 kW over a range in wavelengths of 390–1000 μm and is suitable for a wide range of experiments in the biomedical, solid-state, and surface sciences. In the field of photobiology, since the DNA molecule is sensitive to infrared wavelengths, the free-electron laser can study such behavior as the variation in the DNA mutation rate with wavelength. In addition, experiments have been conceived in the linear and nonlinear excitations of phonons and magnons, ground and excited-state Stark splitting, the generation of coherent phonons, phonon amplification by stimulated emission, induced phase transitions, and semiconductor band-gap structure. The latter application may prove relevant to the study of high critical-temperature superconductors.

More recently, two user facilities have been constructed at Stanford University based on both the Mark III rf linac and the SCA. The Mark III facility, established by John Madey and co-workers, is tunable over the range 0.5–10 μm and produces 60 kW over a pulse time of several microseconds. Due to the electron energies available from the Mark III, however, the fundamental operation occurs at wavelengths in the neighborhood of 1–3 μm. Shorter

FIG. 8. The first free-electron user facility was established at the University of California at Santa Barbara. This device operates in the far infrared and employs an electrostatic accelerator, as shown in the background. A magnetically focused system of electron-beam optics is used to transport the beam from the accelerator through the wiggler (in foreground) and back to the accelerator. (Photo courtesy of Luis Elias.)

wavelengths are achieved through the use of frequency-doubling techniques common to laser engineering. In this case, the tunability, high power, and temporal structure offer a unique opportunity to study surgical applications. In particular, experiments have been conducted in the cutting of both bone and soft tissue with encouraging results. Since spot sizes of the order of 100–1000 μm and power densities as high as several megawatts per square centimeter are possible, the cutting mechanism is not thermal ablation but direct plasma formation of the irradiated tissue with extremely clean and localized incisions. In contrast to conventional lasers (which produce lower power levels over longer pulse times), the combination of high power and short pulses results in less scar formation and more rapid healing. Indeed, the power densities available with this free-electron laser have raised concern that current optical fiber technology may ultimately prove inadequate to the task of directing the radiation, and research has begun in the development of optical fibers capable of handling higher intensities. In addition, experiments have been conducted to study semiconductor band-gap structure as well as the multiphoton spectroscopy of germanium and polyacetylene. The Mark III free-electron laser has been relocated at Duke University.

The SCA free-electron laser facility commonly operates in a continuous mode in which the macropulses are generated at a frequency of 20 Hz (that is, 20 macropulses per second) over a timescale of several hours. A typical macropulse is of 5-ms duration and is composed of a train of 3-ps micropulses separated by 84.6 ns. Average electron-beam currents over a macropulse can reach 200 μA with a peak current of 6 A over a micropulse. The peak voltage of the superconducting linac is 75 MV, but the recirculation system allows this to rise as high as 150 MV. The free-electron laser has operated over wavelengths ranging from 0.5–3.5 μm and is capable of operation in the ultraviolet. This range will soon be extended to 15 μm, using a wiggler supplied by Spectra Technologies, Inc.

Experiments to date have been concerned with dynamical processes on picosecond time scales in materials such as, for example, photon echoes in dye molecule/glass systems. To this end, the picosecond micropulses and the relatively large macropulse separation (which is long enough for conventional optical pulse selection techniques to be used) produced by the SCA are crucial. For a recent experiment, the free-electron laser was operated at a wavelength of 1.54 μm with a linewidth of 0.08% and was stable over a timescale of several hours. The output power was approximately 300 kW over a micropulse, 12 W averaged over a macropulse, or 1 W over the longer timescale.

The operation of the SCA is soon to be upgraded by the implementation of a novel electron injector/acceleration scheme. In a conventional rf linac, the electron beam is accelerated by a single-frequency rf signal that varies sinusoidally over time. Since the amount of electron acceleration depends on the rf power, the electron pulse must be synchronized to the peak of the rf signal and its duration kept short. In general, the longer the electron pulse, the larger the variation in rf power over the pulse and the greater the energy spread of the electron beam. To minimize the beam energy spread, which degrades the performance of the free-electron laser, researchers at the SCA plan to use a composite rf signal composed of waves at multiple frequencies. In a manner analogous to the way in which a *square-wave* signal can be built up from a large composite spectrum of waves, this process will extend the duration of the peak rf signal. As a consequence, both the duration and power in the electron beam will increase at little or no cost to the beam quality. It is estimated that the peak and average beam currents will increase to approximately 50 A and 1 mA, respectively, with corresponding increases in the output of the free-electron laser.

As a measure of the impact of advances in accelerator technology on the free-electron laser, it should be noted that the Mark III linac was the accelerator used by Hans Motz in 1951, and that the present range of experiments was made possible by the continual improvement of the original design. Indeed, a user facility was under construction at Vanderbilt University late in 1989 based on a further improved Mark III linac. This free-electron laser was designed to operate over a spectral range of 0.2–10 μm, and experiments are currently envisioned to study wound healing and scarification after short-pulse laser surgery, neurosurgical shrinkage of irradiated tumors, the differential transport of molecules across membranes, Raman active normal modes of DNA, and high-intensity laser-induced damage in glasses and crystals.

Another user facility, under construction by a collaboration between the National Institute of Standards and Technology and the Naval Research Laboratory, is scheduled for completion in 1991. This free-electron laser is based on a steady-state racetrack microtron designed to produce an electron beam ranging from 17–185 MV and having a peak current of 2 A. The variation in electron energy is accomplished by extraction of the beam after a variable number of passes through the accelerating cavity. The electron beam, and by extension the optical beam, will consist of a continuous train of 3-ps pulses with a repetition rate of 66 MHz. The fundamental wavelength will be tunable from 200 nm through 10 μm with average and peak powers ranging from 10–200 W and 40–1000 kW, respectively. The facility will be usable for research and development in the fields of medicine, materials science, chemistry, and physics.

In general, the principal advantages of the free-electron laser as a research tool are high intensity (relative to currently available sources), tunability, and a temporal structure controlled by the characteristics of the accelerator. Tunability permits the selection of specific energy states for study, while an appropriate tailoring of the temporal structure of the pulses allows the time evolution and decay of excited states to be investigated. In addition, it should be remarked that there are no good infrared sources available with wavelengths ranging from 2–5 μm. Important areas of investigation are the bulk and surface properties of semiconductors—in particular, the band-gap structure. Some of the most commercially important semiconductors exhibit band gaps in the range of 0.25–2.5 μm, and it is important to extend the spectral range of study to within 0.1 μm about this range. A high-intensity, tunable free-electron laser producing pulses of approximately 10–100-ps duration at a repetition rate of several megahertz would permit the study of the dynamic excitation and subsequent decay of electrons into unoccupied energy states.

Other applications include laser photochemistry and photophysics that require sources in the visible and ultraviolet spectra. Initial development of an ultraviolet free-electron laser is being conducted at Los Alamos National Laboratory using rf linac technology. Since the electron beam is composed of short bursts at high repetition rates, rf linacs produce radiation with the desired temporal structure. The principal competition in the ultraviolet comes from incoherent synchrotron light sources that also make use of undulator magnets. The advantage of a free-electron laser over synchrotron sources is the increased counting rates resulting from a larger photon flux, which would make practical a large number of currently marginal experiments. A nonexhaustive list of experiments possible with a coherent visible through ultraviolet source includes the multiphoton ionization of liquids, chemistry of combustion and of molecular ions, high-resolution polyatomic and fluorescence spectroscopy, time-resolved resonance Raman spectroscopy, spin-polarized photoemission and photoemission microscopy, magnetooptical studies of rare-earth elements, and studies of optical damage from high-intensity ultraviolet radiation.

Industrial applications are envisioned in materials production and photolithography. The requirements for materials production are sources in the near infrared (2.5–100 μm) and the ultraviolet through soft X-ray spectra, and involve the pyrolytic production of powders for catalysts, the near-stoichiometric production of high-value chemicals and pharmaceuticals, and the pyrolytic and photolytic deposition of thin films on substrates. In photolithography, the surface of a wafer is coated with a layer of a photoresisting substance of which a part is illuminated. The wafer is then processed by the removal of either the exposed or unexposed portions of the resist. Photolithography is primarily done with the optical lines of the mercury arc that occur over the spectral range of 436–357 μm; the excimer laser is under consideration for this purpose as well. Three principal lithographic techniques are either in use or under consideration. Contact printing is performed by bringing a mask and wafer into contact, whereas in proximity printing a small space separates these two components. The preferred technique is that of projection printing in which the image of the mask is projected onto the wafer from a greater distance than in contact or proximity printing, which allows the use of larger masks and improves usable mask lifetime. Each of these techniques requires uniform and stable sources of illumination. The sources may be either continuous or pulsed; however, possible difficulties with pulsed sources (e.g., rf linac driven free-electron lasers) are that (1) the instantaneous pulse power necessary to give a sufficiently high average power may also be high enough to damage the wafer, and (2) extremely high uniformity from pulse to pulse is required. In addition, short wavelength sources may render the nonlithographic direct printing of wafers possible and eliminate the need for photoresists.

Applications also exist for high-power microwaves and submillimeter waves in the fields of communications, radar, and plasma heating. We confine the discussion to the heating of a magnetically confined plasma for controlled thermonuclear fusion. The reactor design of greatest interest is the Tokamak, which confines a high-temperature plasma within a toroidal magnetic *bottle*. A thermonuclear reactor must confine a plasma at high density and temperature for a sufficiently long time to ignite a sustained fusion reaction. In its original conception, the Tokamak was to be heated to ignition by an ohmic heating technique whereby a current is induced by means of a coil threading the torus. As such, the Tokamak acts like the secondary winding in a transformer. However, recent developments indicate the need for auxiliary heating, and the resonant absorption of submillimeter radiation has been proposed for this purpose. The frequencies of interest are the harmonics of the electron cyclotron frequency and range from about 280 GHz through 560 GHz. In the case of the Compact Ignition Torus, which is the proposed design of the next major Tokamak experiment in the United States, the applied magnetic field is 10 T and the resonant wavelengths are in the neighborhood of 1 mm and less. The estimates of the power requirements are, of necessity, relatively crude but indicate the need for an average power of approximately 20 MW over a pulse time of nearly 3 s. At present there are no sources capable of meeting this requirement. However, the free-electron laser has operated in this spectral region at power levels of this order, but over a much shorter pulse time. As such, it represents one of several competing concepts.

A more controversial application of high-power, long-pulse free-electron lasers is strategic defense against intercontinental ballistic missiles. In this regard, planners envision a large-scale ground-based laser that would direct light toward a target by means of both ground-based and orbiting mirrors. Designs based on both amplifier and oscillator free-electron lasers are being pursued.

In experiments during the late 1980s, a free-electron laser amplifier at Livermore amplified a 14-kW input signal from a carbon dioxide laser at 10.6 μm to a level of approximately 7 MW, a gain of 500 times. Boosting the input beam to 5 MW yielded a saturated power of 50 MW. In initial experiments researchers used a 15-m planar wiggler with a uniform period and amplitude; they have since lengthened the wiggler to 25 m, and a tapered version to further increase the extraction efficiency has been considered.

Boeing Aerospace has built an experimental free-electron laser oscillator in collaboration with Los Alamos National Laboratory based on a 5-meter-long planar wiggler and an advanced radio-frequency linac. The linac produces electron beams with energies as high as 120 MV. The oscillator has lased in the red region of the visible spectrum at a wavelength of 0.62 μm and at power levels a billion times that of the normal spontaneous emission within the cavity. The average power over the course of a 100-μs pulse is about 2 kW. The corresponding conversion efficiency is about 1%, but the peak power is a more respectable 40 MW. Even though oscillators typically generate short-wavelength harmonics that can damage the cavity mirrors, no degradation has been observed thus far. Work is underway to convert the uniform wiggler to a tapered configuration to increase the oscillator's efficiency.

Unclassified figures indicate that pulses of visible or near-infrared light at an average power of about 10–100 MW over a duration of approximately one second are required to destroy a missile during its boost phase. This means lengthening the pulses or increasing the peak power levels of existing free-electron lasers by a factor of a million or more. Depending on laser efficiency and target hardness, a collection of ground-based free-electron lasers would require somewhere between 400 MW and 20 GW of power for several minutes during an attack. (For comparison, a large power plant generates about 1000 MW.) For these and other reasons, it is not clear whether it will be practical to scale up free-electron lasers to the power levels required. Current and future experiments will help to resolve this question.

IV. Summary and Discussion

The aforementioned discussion includes a necessarily abridged list of recently conceived applications of the free-electron laser. The fundamental principles of the free-electron laser are understood at the present time, and the future direction of research is toward evolutionary improvements in electron-beam sources (in terms of beam quality and reliability) and wiggler designs. The issues, therefore, are technological rather than physical, and the free-electron laser can be expected ultimately to cover the entire spectral range discussed. In this regard, it is important to recognize that the bulk of the experiments to date have been performed with accelerators not originally designed for use in a free-electron laser, and issues of the beam quality important to free-electron laser operation were not adequately addressed in the initial designs. As a consequence, the results shown do not represent the full potential of the free-electron laser, although many of the experiments have produced record power levels. The only accelerators specifically designed for use with a free-electron laser are the rf linac at Boeing Aerospace, an induction linac at Lawrence Livermore National Laboratory, and the electrostatic accelerator at the University of California at Santa Barbara. At the present time, an intensive program at Los Alamos National Laboratory is directed toward adapting the technology of rf linacs to the design of short-wavelength free-electron laser oscillators. In particular, it is found that wakefields induced by pulses of high peak currents in the accelerator and beam transport system can result in serious degradation in beam quality, and that the minimization of these effects is an important consideration in the overall design of the beam line in free-electron lasers.

Another important direction for future research is the design of short-period wigglers that permit short-wavelength operation with relatively low-voltage electron beams. An alternate approach to the production of short-wavelength radiation with moderate-energy electron beams is the use of higher harmonic interactions. For example, if the third harmonic interaction is employed [i.e., $\omega = (k + 3k_w)v_z$], then the energy requirement for a fixed-wavelength output is reduced by a factor of approximately $\sqrt{3}$. Indeed, lasing has been achieved at the third harmonic in experiments at (1) Stanford University using the Mark III linac at a wavelength of 1.4–1.8 μm and (2) Los Alamos National Laboratory at a wavelength of 4 μm. The reduction in the voltage requirement has important practical implications in the simplification of accelerator design problems and the reduction in the production of secondary X-rays and neutrons. The combination of improved accelerators and short-period wigglers will reduce both the size and complexity of free-electron laser systems and open doors to a host of new practical applications of the technology.

Free-electron laser research and development at the present time is international in scope with a wide range of experiments either in operation or in the planning stages throughout the United States, Europe, Japan, China, and the Soviet Union. These projects include a wide range of designs based on all the accelerator types and cover a spectral range extending from microwaves through the ultraviolet. The technology can therefore be expected to make the transition from development in the laboratory to active exploitation over a range of applications during the next decade.

Bibliography

Deacon, D. A. G., Elias, L. R., Madey, J. M. J., Ramian, G. J., Schwettman, H. A., and Smith, T. I. (1977). First operation of a free-electron laser. *Phys. Rev. Lett.* **38**(16), 892.

Freund, H. P., and Parker, R. K. (1989). Free-electron lasers. *Sci. Am.* **260**, 84.

Jacobs, S. F., Pilloff, H. S., Sargent, M., Scully, M. O., and Spitzer, R., eds. (1980). "Physics of Quantum

Electronics: Free-Electron Generators of Coherant Radiation," Vol. 7. Addison-Wesley, Reading, Massachusetts.

Jacobs, S. F., Pilloff, H. S., Sargent, M., Scully, M. O., and Spitzer, R., eds. (1982). "Physics of Quantum Electronics: Free-Electron Generators of Coherant Radiation," Vols. 8, 9. Addison-Wesley, Reading, Massachusetts.

Marshall, T. C. (1985). "Free-Electron Lasers." Macmillan, New York.

Martellucci, S., and Chester, A. N., eds. (1983). "Free-Electron Lasers." Plenum Press, New York.

Motz, H. (1951). Applications of the radiation from fast electron beams. *J. Appl. Phys.* **22**(5), 527.

Phillips, R. M. (1988). History of the ubitron. *Nucl. Instrum. Methods Phys. Rev. Sect. A* **272**(1), 1.

Roberson, C. W., and Sprangle, P. (1989). A review of free-electron lasers. *Phys. Fluids B* **1**(1), 3.

Sessler, A. M., and Douglas, V. (1987). Free-electron lasers. *Am. Sci.* **75,** 34.

GAS LASERS

W. W. Duley *York University*

GLOSSARY

Band: Grouping of spectral lines in the emission or absorption spectrum of a molecule.

Continuous wave: Refers to dc operation of laser; typically a steady (dc) output for 1 sec or more.

Excimer: Molecule that is strongly bound in an excited state but normally has a dissociative ground state. The term excimer comes from excited dimer. Excimers also include exciplexes (excited complexes).

Mode: Output characteristic of laser; normally refers to the intensity distribution of output light intensity.

Population: Subset of atoms or molecules in a particular energy level.

Pumping: Means of producing an inversion in a laser system.

Transverse excitation–atmospheric pressure (TEA): Excitation geometry where the discharge is applied transverse to the optical path; gas pressure is close to 1 atm.

Tunable: Having an output wavelength that can be adjusted over a range of 1 nm or more.

Vacuum ultraviolet: Region of the spectrum with wavelengths less than 200 nm.

Laser emission has been obtained from most elements of the periodic table in gaseous form as well as from a host of molecular species. Output wavelengths available from gas lasers span the spectrum from the far infrared to the X ray. Output powers are a function of the mode of excitation—either continuous wave or pulsed—as well as of the laser gas. Gas lasers may be pumped electrically, chemically, thermodynamically, or optically by other laser or light sources. Few gas lasers are tunable over an appreciable wavelength range.

I. Introduction

The first gas laser was operated in 1961, only one year or so after Maiman's demonstration of the feasibility of obtaining maser-type emission at optical frequencies. Gas lasers had been envisioned in fundamental patents for quantum electronic systems but difficulties associated with pumping of candidate systems inhibited the development of practical gas laser devices. The operation of an He–Ne laser at a wavelength of 1152.27 nm showed that these problems could be overcome and opened the door to the rapid development of the gas laser field. [*See* LASERS.]

The five or so years following the announcement of laser oscillation in He–Ne saw the development of gas lasers based on ionized rare gas atoms, a variety of neutral atoms, and simple molecules such as N_2 and CO_2. In recent years this pace of discovery has slackened somewhat, although the mid-1970s saw the development of excimer and gold lasers, and attention has focused instead on the commercialization of these and other gas laser systems (Table I). At the same time, interest has shifted to the use of gases as nonlinear media for the generation of tunable vacuum ultraviolet radiation. The free electron laser, which, in a sense, is also a gas laser, has been the subject of much recent development and offers the possibility of obtain-

TABLE I. History of Commercial Gas Lasers

Laser	First reported	Date commercialized
He–Ne	1961	1962
Far infrared	1963	1969
Iodine	1964	1983
CO_2	1964	1966
Rare gas ion	1964	1966
Nitrogen	1966	1969
Copper vapor	1966	1981
HF/DF	1967	1977
He–Cd	1968	1970
Excimer	1975	1976
Gold vapor	1978	1982

ing high-power laser radiation tunable over a wide range of wavelengths.

Today laser emission has been obtained at well over 1000 wavelengths from gaseous elements in atomic or ionic form. Most elements in the periodic table have been induced to lase when in gaseous form, generally via pulsed or continuous-wave (cw) discharge pumping. A multitude of molecular gases as well as radical and short-lived transient species have also exhibited laser emission. These molecular lasers emit at wavelengths extending from the vacuum ultraviolet (H_2 laser, excimer lasers) to the far infrared (HCN laser). The most important of these from the point of view of industrial applications is the CO_2 laser operating at 10.6 μm.

II. Helium–Neon Laser

Laser oscillation in a mixture of He and Ne gas was first reported in 1961 by Ali Javan and colleagues working at Bell Laboratories. Helium–neon has the distinction of being the first gaseous system to exhibit laser oscillation. Since 1961 the He–Ne laser has become widely available as a compact, reliable commercial product used in many industrial applications, primarily in the measurement and metrology fields.

The active medium in the He–Ne laser is a gaseous mixture of He and Ne with proportions of about 10 : 1 in a sealed Pyrex tube at a pressure of several torr. This gas is excited in a positive column discharge with a current of some 100 mA. Optimum operating parameters (e.g., discharge current, He : Ne ratio, total gas pressure) are functions of tube diameter and also depend on the output wavelength selected.

TABLE II. Wavelengths λ (in Air and Vacuum) for Important He–Ne Laser Lines

λ (air) (nm)	λ (vacuum) (nm)
543.364	543.515
632.816	632.991
1152.27	1152.59
3391.32	3392.24

The He–Ne laser can be made to emit at a variety of discrete wavelengths in the visible and infrared regions of the spectrum (Table II); however, dominant spectral lines lie at 632.8 nm (the He–Ne red line) and 3391.3 nm (see Fig. 1). For operation at 632.8 nm, emission at 3391.3 nm must be suppressed since the two lines share the same upper level. Oscillation on the high-gain 3391.3 nm line depletes the excited state population for 632.8 nm emission. Discrimination against the infrared line can be accomplished by adjusting the reflectivity of the laser mirrors so as to have peak reflection at 632.8 nm together with low reflectance at 3391.3 nm. Alternatively, application of a longitudinal magnetic field has been shown to suppress oscillation at 3391.3 nm by Zeeman splitting. Magnetic fields of ~200 G are typically required.

The low gain of the 632.8 nm laser line in He–Ne makes for a relatively inefficient system with low output powers. While cw output powers of ~1 W are attainable from laboratory devices, standard commercially available He–Ne lasers offer powers of between 1 and 50 mW at 632.8 nm. Low-power He–Ne lasers are compact, with tube lengths of 20 cm or so, and emit either linearly polarized or randomly polarized light in a TEM_{00} mode. Higher-power He–Ne lasers may be 2 m or so in length. A summary of output characteristics is given in Table III.

TABLE III. Output Characteristics of He–Ne Lasers

Wavelength	632.82 nm
Power	0.5–50 mW
Beam diameter	0.5–2.0 mm
Beam divergence	0.5–2 mrad
Stability	5%/hr
Coherence length	0.1–0.3 m
Lifetime	10–20,000 hr

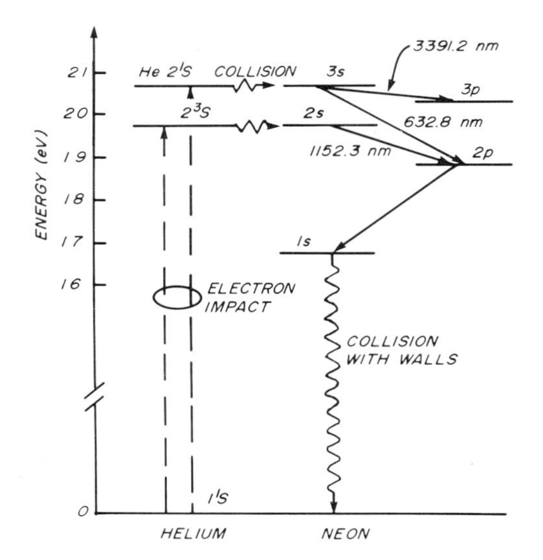

FIG. 1. Simplified energy level diagram for He–Ne laser showing laser transitions.

III. Rare Gas Ion Lasers

Laser emission has been observed under pulsed and cw excitation of the rare gases Ne, Ar, Kr, and Xe (see Table IV) in various ionization states. Rare gas ion lasers when operated in a cw mode provide up to 20 W of power at wavelengths in the spectral range 450–700 nm and up to several watts of power when oscillating in the 350 nm region. Lasers of the rare gas ion type therefore provide some of the highest cw powers available over the wavelength range between 350 and 700 nm (Table V). This is achieved at the expense of inefficient operation; the ArII laser, for example, has an efficiency of only about 0.1%. Thus 10 kW of electrical power is required to obtain 10 W of laser output.

An energy level diagram showing some of the atomic and ionic states involved in the ArII laser is shown in Fig. 2. It is apparent that each 3-eV photon emitted by an excited ArII ion is the result of the ionization of a neutral Ar atom (ionization potential 15.76 eV) together with excitation of the resulting ion to states nearly 20 eV above the ground state of the ion. A number of mechanisms have been proposed for the excitation of the upper ArII levels in the laser discharge. Electron collisions may excite these states directly via a single step from the ArI level or indirectly by means of a two-step transition involving the ArII ground state as an intermediate level. Collisional excitation of higher ArII or ArIII states followed by radiative cascade back to the upper laser levels is probably also important.

In general, the power P per unit length L emitted by the cw ArII laser is described by the expression

$$P/L = K(JR)^2$$

TABLE IV. Rare Gas Ion Lasers—Primary Laser Wavelengths (cw Operation)

Ion	Wavelength in air (nm)	Ion	Wavelength in air (nm)
NeII	332.376	KrII	520.831
	334.552		530.865
	337.830		568.188
	339.286		647.088
	371.310		676.442
ArIII	351.112		752.546
	363.789	XeIII	345.424
ArII	454.504	XeIV	364.551
	457.935	XeIII	378.097
	465.789		406.041
	472.686		421.401
	476.486		424.024
	487.989		427.259
	496.508		460.302
	501.716		495.418
	514.531		500.778
KrIII	350.742		515.906
	356.422		523.893
	406.737		526.017
	413.132		526.189
KrII	468.041		535.288
	476.243		539.459
	482.517		541.915
			627.081

TABLE V. Output Characteristics of Ar and Kr Rare Gas Ion Lasers

Characteristic	Low power	High power
Wavelength[a] (nm)	457–530	457–530
Power[b] (W)	0.25	20
Beam diameter (mm)	2.0	1.9
Beam divergence (mrad)	1.5	0.40
Stability (%/day)	5	0.5
Beam amplitude noise[c] (% rms)	2	0.5
Lifetime	2000 hr	18 month

[a] 647–676 nm for Kr.
[b] Maximum 5 W for Kr operation.
[c] 0–100 kHz.

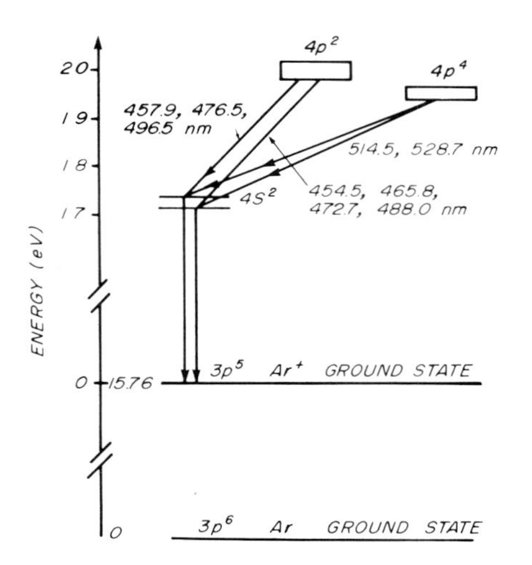

FIG. 2. Energy levels involved in the blue-green emission from the ArII laser.

where J is the current density in the laser discharge, R the radius of the discharge column, and K a constant which takes the value $2 \times 10^{-3} \leq K \leq 5 \times 10^{-3}$ when JR has units of amperes per reciprocal centimeter and P/L is in watts per meter. Typically, for a commercial ArII laser oscillating in the blue-green one has $P/L \sim 5$–10 W m^{-1} while $JR \sim 50$ A cm^{-1}. Optimum tube gas pressure depends on discharge current and discharge bore diameter. Pressures in the range 150–300 mTorr are usual, however.

The Doppler-limited linewidth of rare gas ion laser emission is generally in the range 5–10 GHz. With cavity stabilization the laser linewidth can be $\simeq$500 MHz with a drift of ~100 MHz hr^{-1}. With a linewidth of 10 GHz, mode locking can produce pulse lengths of 100 psec. In ArII and KrII lasers mode locking yields pulses with a peak power of ~1 kW at a pulse repetition frequency (prf) of $\approx$150 MHz. Cavity dumping can produce narrow pulses at a prf of $\geq$1 MHz with peak powers ~10^2 times the cw power. [*See* MODE-LOCKING OF LASERS.]

The large discharge current required for operation of Ar and Kr ion lasers places strong demands on bore materials. Current tube technology involves the use of segmented tungsten disk bores for high-power systems. Earlier technology centered on the use of graphite or beryllium oxide discharge channels. Beryllium oxide technology is currently used in several low-power air-cooled ArII lasers.

IV. Metal Vapor Lasers

Laser emission has been observed from vapors of a variety of metal atoms and their ions (see Fig. 3) under either pulsed or cw excitation. Most of these lasers have yet to be commercialized for one reason or another (low efficiency, competitive simpler system, etc.). For this reason, we limit the present discussion to two of the more important metal vapor systems—the He–Cd and Cu vapor lasers.

A. Helium–Cadmium Lasers

The two strong laser lines emitted by CdII in the He–Cd laser have wavelengths of 325.029 and 441.565 nm (in air). The cw output powers are typically 5–10 mW on the 325 nm line and 10–50 mW on the 441 nm line, both for fundamental mode (TEM_{00}) operation. Multimode powers are about 50% larger. Primary applications of the He–Cd laser have been in the electronic printing industry. The Xerox Company, for example, has used the He–Cd laser for high-speed writing on a selenium photoconductor.

The He–Cd laser operates with Cd vapor at a pressure of ~2 mTorr maintained by evaporating metallic Cd. The He pressure is typically 4–6 Torr and a dc discharge is passed axially through the laser tube. Reliability problems are related to the consumption of Cd as the tube ages and to plating out of metallic Cd on tube windows. The overall lifetime of commercial He–Cd laser tubes is now in the 4000-hr range.

B. Copper Vapor Laser

The neutral Cu atom has two very high gain transitions at 510.554 and 578.213 nm that can be

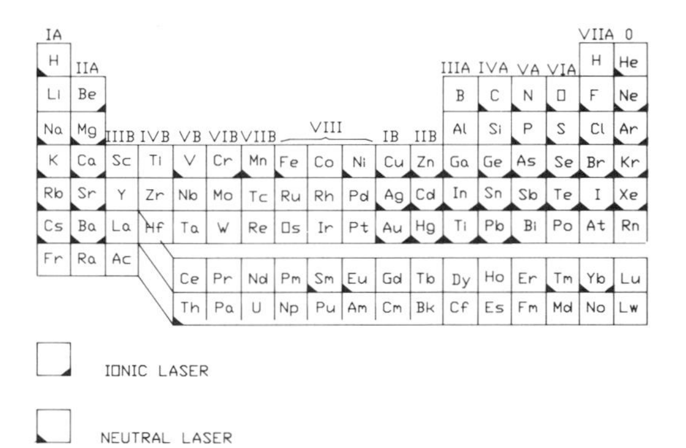

FIG. 3. Periodic table showing the elements that have exhibited laser emission in gaseous form.

pumped efficiently in pulsed discharges at very high repetition rates. Average powers of 40–50 W are available from commercial Cu vapor laser at prf's of up to 20 kHz. Laboratory systems have been pulsed at rates up to 150 kHz. Applications have centered on the use of these lasers to pump dye lasers, primarily for uranium enrichment.

The low vapor pressure of metallic Cu, together with the need to maintain a Cu pressure of ~0.5 Torr in the laser discharge, has hindered the development of these systems. Thermal evaporation of Cu metal requires that the metal be heated to 1400–1600°C, a severe constraint to operation. Recently, interest has focused on the use of excess heat from the laser discharge to maintain the requisite Cu vapor pressure.

V. Carbon Dioxide Laser

Laser emission from the CO_2 molecule was first observed by Patel in 1964. In Patel's early experiments, a pulsed discharge was passed through pure CO_2 and laser output powers were small. It was soon realized, however, that the overall efficiency of the CO_2 laser could be greatly increased by the incorporation of N_2. Nitrogen molecules that are vibrationally excited in the laser discharge collisionally excite the upper CO_2 laser level as follows:

$$N_2(X^1\Sigma_g^+, v'' = 1) + CO_2(00^00) \rightarrow$$

$$N_2(X^1\Sigma_g^+, v'' = 0) + CO_2(00^01) - 18\ \mathrm{cm}^{-1}$$

where standard spectroscopic notation is used to designate molecular states. Infrared laser emission occurs from rotational levels of the CO_2 (00^01) excited vibrational state to rotational levels of either the $(10^00, 02^00)_{\mathrm{I}}$ or $(10^00, 02^00)_{\mathrm{II}}$ vibrational state (see Fig. 4). These transitions

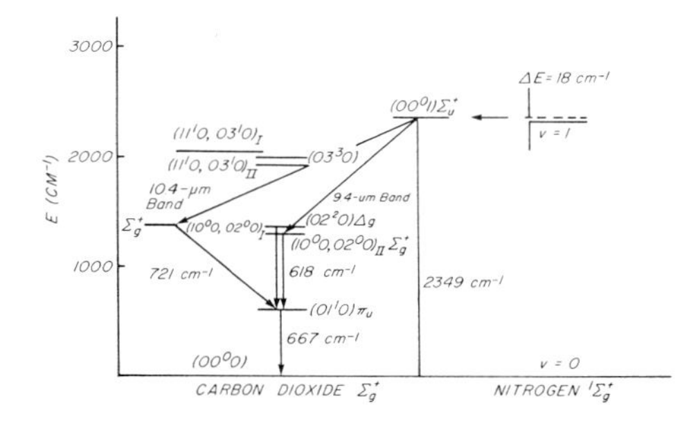

FIG. 4. Energy level diagram for CO_2 and N_2 molecules showing 10.4- and 9.4-μm bands.

form bands at 10.4 and 9.4 μm, respectively. Because rotational transitions are involved and many rotational levels are populated under discharge conditions, the CO_2 laser can be tuned to oscillate at a large number of discrete wavelengths within the 10.4- and 9.4-μm bands. In fact, at pressures of nearly 10 atm these lines broaden into a continuum. Under these conditions pulsed CO_2 lasers can be continuously tuned over a wavelength range of several micrometers near 10 μm.

Unless the CO_2 laser cavity contains a means by which individual wavelengths may be selected from within the 10.4- and 9.4-μm bands, oscillation occurs primarily on those lines that have the highest gain. Under most excitation conditions these are the $P(18)$, $P(20)$, and $P(22)$ lines of the 10.4-μm band. Their wavelengths (in vacuum) are $P(18)$, 10.5713 μm; $P(20)$, 10.5912 μm; and $P(22)$, 10.6118 μm. Since these three lines are near 10.6 μm, this is the emission wavelength that is usually quoted for the CO_2 laser.

Laser emission from CO_2 has also been observed at wavelengths in three bands near 4.3 μm as well as numerous bands between 11 and 18 μm. All are seen only under pulsed conditions.

The population inversion required to obtain laser emission from CO_2 may be established in a variety of ways. Excitation of the upper CO_2 laser level (00^01) can occur by inelastic collision with low-energy electrons or by a nearly resonant energy transfer from vibrationally excited N_2. Both mechanisms have been exploited in practical laser devices.

The CO_2 laser usually operate on a mixture of CO_2, N_2, and He gases. The optimum proportions of the three components depend on the mode of excitation, but generally for flowing cw systems the $CO_2:N_2:$He ratio is in the range 1 : 1 : 8 with a total gas pressure of 5–15 Torr. The role of He in the laser mixture is, in part, to quench the population that builds up in (01^10) and other low-lying levels of the CO_2 molecule after laser emission has occurred (see Fig. 4) and to stabilize the glow discharge. Helium atoms are translationally excited during such processes, and this heat must be removed from the laser gas for efficient operation. In practice, in slow-flow devices heat is removed by collisions with the walls of the laser tube. By increasing gas flow to the point where convective cooling occurs, the overall size of the laser may be reduced while maintaining high output power. Typical values for the power per unit discharge length generated in a slow axial flow CO_2 laser is 75 W m^{-1}, while for fast-flow convectively cooled devices this ratio increases to 200–600 W m^{-1}.

Heat can also be removed by flowing the laser gas mixture transverse to the discharge direction (Fig. 5). This gas flow and excitation mode are now utilized in high-power (5–20 kW) commercial cw CO_2 lasers. Some properties of these and other cw CO_2 lasers are summarized in Table VI. Many cw CO_2 lasers can also be pulsed electrically or in *Q*-switching mode to obtain output pulses with peak power $\sim 10^2$ larger than the cw power. Pulse duration can extend from less than one microsecond to milliseconds. With intracavity modulation or mode locking pulse technique, pulse lengths as short as a few nanoseconds are obtainable in a high-pressure CO_2 laser.

The transverse excitation–atmospheric pressure (TEA) laser exploits a transverse discharge geometry to induce a uniform self-sustained avalanche discharge in the laser gas at high pressure. By employing a preionization technique to initially ionize CO_2 laser mixtures, arcing can be eliminated. The result is stable operation at high energy input and at high prf. Such TEA lasers exhibit a phenomenon known as gain switching. Here rapid pumping induces a large population in the upper laser level. When the electromagnetic field builds up within the cavity the population inversion is already large, so rapid amplification occurs. The result is the emission of an intense gain-switched pulse lasting about 100–200 nsec (Fig. 6). After this pulse, populations in the upper and lower laser levels have been equalized. However, vibrationally excited N_2 molecules continue to pump the upper laser level, leading to the subsequent emission of a broad (1–10 μsec) pulse following the gain-switched pulse.

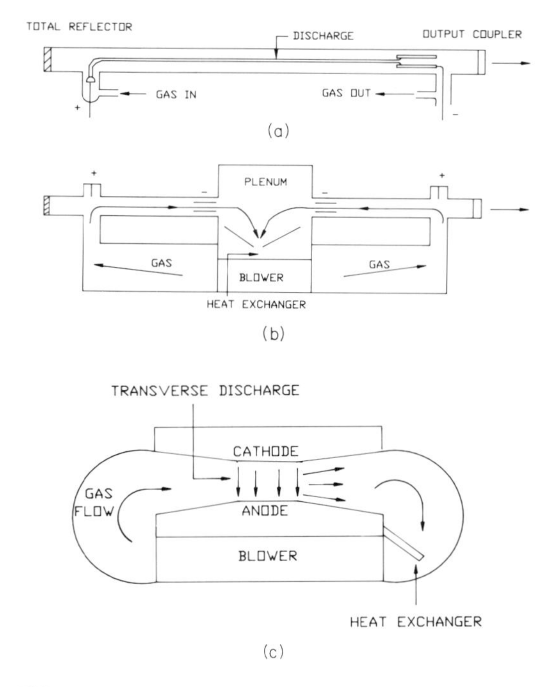

FIG. 5. Discharge and gas flow configurations for the CO_2 laser: (a) axial flow, (b) fast axial flow, and (c) transverse flow and discharge.

Commercial TEA lasers, of the sort used in laser machining applications, typically emit up to 3 J per pulse at prf's up to 50 Hz. The output beam of these lasers is rectangular in cross section with dimensions of $\sim 2 \times 3$ cm. High-energy TEA lasers are available that emit up to $2–3 \times 10^3$ J per pulse at low (0.05 Hz) prf.

Mode locking of the TEA laser output occasionally occurs spontaneously. Active mode locking yields ~1-nsec pulses at atmospheric pressure. Operation at higher pressure can yield mode-locked pulses with durations as short as 100 psec.

TABLE VI. Continuous Wave (cw) CO_2 Lasers

Configuration	Output power (W)	Beam diameter (mm)	Beam divergence (mrad)	Features
Waveguide	≤ 25	1.5–3	10	Sealed tube
Longitudinal excitation	≤ 75	5	3	Sealed tube
Slow axial flow longitudinal excitation	100–1000	5–30	1.5–3	Pulsed, line-tunable
Fast axial flow longitudinal excitation	500–3000	10–20	1.5–3	Compact size
Transverse flow transverse discharge	5000–15,000	>15	>3	High power

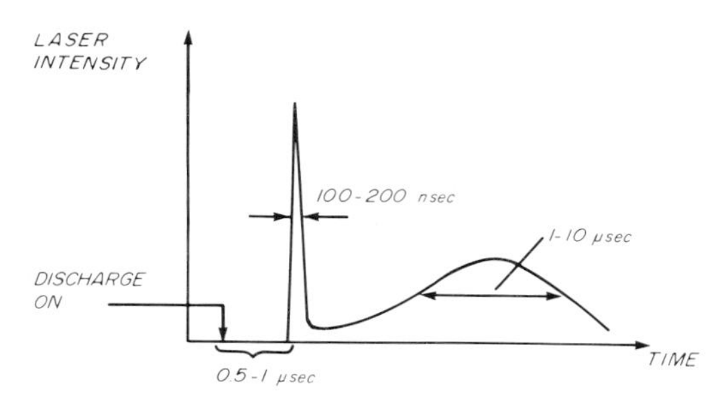

FIG. 6. Time dependence of TEA laser output. Both the gain-switched and broad pulses can contain comparable amounts of energy.

Large pulse energies can be extracted from electron beam-controlled CO_2 lasers. In these devices a high current of ~100-keV electrons is used to preionize the laser gas. When the preionization charge density is sufficiently large, a subsequent main electrical discharge can be operated at a lower field than that required for a self-sustained discharge. Hence the discharge field strength may be adjusted to optimize the CO_2 laser excitation and this laser output. In this way CO_2 laser pulse energies of 2×10^3 J with microsecond duration can be generated.

In the gasdynamic laser (GDL) a hot CO_2 gas mixed with hot N_2 and He with perhaps some added H_2O is allowed to expand supersonically through a nozzle. When the gas contains the right combination of quenching species (e.g., H_2O) the population of CO_2 molecules in the lower laser level can be reduced rapidly while still maintaining much of the population in the upper laser level. The overall result is the establishment of a population inversion between the two states. Large amounts of laser power can be extracted from such a system. Quasi-cw outputs of ~400 kW have been recorded from GDLs.

The GDL can also be operated on a pulsed basis with gas flow generated by the combustion of a mixture of CO, O_2, N_2, and H_2. Output pulses of millisecond duration and containing ~20 J have been reported.

VI. Excimer Lasers

Certain molecules such as ArF, KrF, and XeCl are strongly bound only in excited states. The ground state of these molecules, which are called excimers, are characterized by small dissociation energies or in some cases by repulsive potential energy curves. In the excimer laser electrical pumping creates atoms and ions that combine to form excimer molecules. Since the ground state of such molecules is essentially empty because of rapid dissociation, a population inversion between the excimer state and the ground state is easily obtained. Transitions between these states occur in most instances in the ultraviolet or vacuum ultraviolet. Thus excimer lasers emit primarily at wavelengths shorter than 350 nm (see Table VII). The high gain of excimer laser transitions leads to efficient operation—the ratio of average laser output power to electrical power into an excimer laser can be as large as 0.04. Furthermore, the short radiative lifetime of the excimer level (~10 nsec) yields short pulse operation at high prf's. Some excimer lasers have been operated at prf's up to 2000 Hz. Average powers up to 200 W can be obtained from commercial excimer lasers.

In excimer lasers operating on transitions involving molecules containing halogen atoms the halogen is obtained by electrical dissociation of a parent compound (e.g., HCl or F_2) that can be relatively easily introduced into the laser cavity by standard gas handling techniques. However, the reactive nature of these precursor gases and their dissociation products means that care must be taken to ensure that metal and glass surfaces in contact with the laser mixture are coated with materials that are chemically resistant. Nickel and brass have been found to be effective as electrode materials, while a coating of Teflon

TABLE VII. Output Characteristics of Excimer Lasers—Values for Typical Commercial Lasers

Gas	F_2	ArF	KrCl	KrF	XeCl	XeF
Wavelength[a] (nm)	157	193	222	249	308	350
Pulse energy (mJ)	10	400	45	550	200	275
prf (Hz)	40	90	130	100	160	100
Average power (W)	0.5	30	6	45	30	20
Pulse length (nsec)	6	10	8	12	8	15

[a] Wavelength refers to the B → X transition.

protects the laser tube itself from corrosion. Despite these precautions, systems must be passivated by extended contact with halogen before stable operation can be obtained.

During operation of the laser, particulate matter and chemical impurities build up in the laser cavity. Particulate matter may deposit on the system optics and degrade output power, while molecular impurities and the reduction of the concentration of precursor halogen-bearing molecules reduce system gain. The overall result is a loss of output power. The limited lifetime of excimer gas mixtures was a major problem with early lasers. However, improvements in cavity design and electrical excitation have greatly extended the lifetime of excimer gas mixtures and reduced operating costs. In addition, the use of external gas processors that continually remove particulate matter and low vapor pressure impurities has extended operating times to up to several days on a single fill of laser gas. This makes the excimer laser one of the more economical lasers to operate, even though the cost of primary gases (e.g., Xe and Kr) can be high. The use of neon as a buffer gas in some systems (e.g., the KrF laser) can, however, be a major cost consideration.

VII. Other Molecular Lasers

A. Nitrogen Laser

Laser emission from the N_2 molecule can be excited at pressures of ~100 Torr in a transverse discharge geometry. The output wavelength of the N_2 laser lies at 337.1 nm and involves vibronic transitions within the $C^3\Pi_u \rightarrow B^3\Pi_g$ second positive band system. Nitrogen laser radiation consists of a single pulse of amplified spontaneous emission with a duration of ~10 nsec. Typical output powers are up to 1 MW with pulse energies of ≈10 mJ. Pulse repetition rates can be 100 Hz. Picosecond pulse widths are also possible with high-pressure, short-pulse excitation.

The short wavelength and narrow pulse width of the N_2 laser, together with an excitation geometry that yields an output beam with a rectangular cross section, have led to extensive use of these lasers as pumps for pulsed dye lasers. A major limitation, however, is the low energy per pulse—almost one to two orders of magnitude less than that from the excimer laser.

B. Carbon Monoxide Laser

The CO laser can be operated as a cw axial flow device, as a gas dynamic laser, and under TEA-type excitation conditions. It can also be run as a chemical laser. Laser emission occurs between high vibrational levels of ground state CO molecules. The output wavelength is line-tunable between 5 and 7 μm. Highly excited vibrational levels are populated in a discharge via a collisional process known as anharmonic pumping:

$$\mathrm{CO}\ (v = i) + \mathrm{CO}\ (v = j) \rightarrow \mathrm{CO}\ (i + n) + \mathrm{CO}\ (j - n)$$

where $n = 1, 2, \ldots, i \geq j$. In the anharmonic process CO molecules can be raised to vibrational states with v as high as 35. The laser emission involves the radiative cascade process

$$\mathrm{CO}\ (v, J) \rightarrow \mathrm{CO}\ (v - 1, J \pm 1) + h\nu$$

where J is the rotational quantum number. Output power of the cw CO laser is greatly enhanced by cooling the gas to 77 K. Output powers of 20 W are available from commercial cw devices. TEA-type excitation yields 10-mJ pulses of ~1-μsec duration at a prf of 10 Hz.

C. The HF/DF Laser

The exothermic chain reaction between H_2 and F_2 to yield vibrationally excited HF is the basis for an efficient chemical laser requiring little input of electrical power. Continuous-wave output powers of 2.2 MW at 2.7 μm (HF) and 3.8 μm (DF) have been produced in this way. In a pulsed mode, pulse energies of 5 kJ have been reported. The reaction between F atoms and H_2

$$\mathrm{F} + \mathrm{H_2} \rightarrow \mathrm{HF}\ (v) + \mathrm{H}$$

yields HF molecules in vibrationally excited states peaking at $v = 2$ for HF and $v = 3$ for DF. A population inversion with respect to lower vibrational levels can be established as the result of this selective excitation. The nonchain HF laser, for example, oscillates on lines of the $2 \rightarrow 1$, $1 \rightarrow 0$, and $3 \rightarrow 2$ vibrational bands, while the chain reaction HF laser typically oscillates on lines up to the $6 \rightarrow 5$ band.

To promote the reaction and then freeze in the high-temperature equilibrium, a mixture of H_2 and F_2 gases is heated in a high-temperature precombustor (Fig. 7). Extra F_2 is added downstream before the gaseous mixture is allowed to

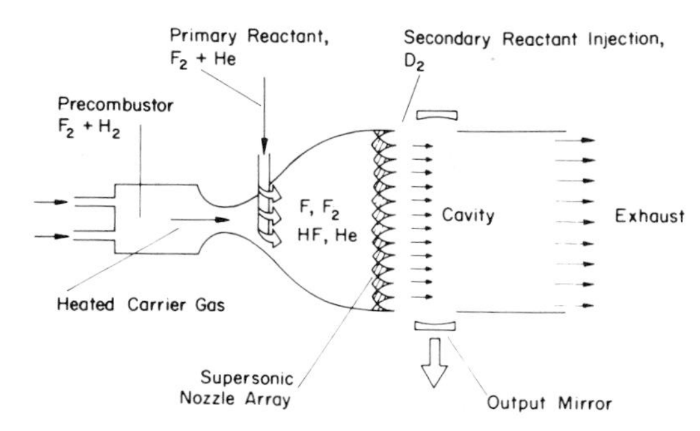

FIG. 7. Schematic diagram of the elements of a cw supersonic chemical laser. [From Cool, T. A. (1979). In "Methods of Experimental Physics," Vol. 158 (L. Marton and C. Marton, eds.). Academic Press, New York.]

expand through an array of supersonic nozzles. Supersonic expansion together with quenching of low vibrational levels results in the establishment of a population inversion downstream. Laser oscillation then occurs transverse to the direction of gas flow.

While more emphasis has been placed on the development of cw HF/DF chemical lasers for military applications, smaller versions are available as commercial products. Output powers of ~150 W can be obtained from such cw devices. High-power pulsed HF lasers initiated by intense electron beams have been developed as candidates for the inertial confinement fusion energy driver.

D. Far-Infrared Lasers

Molecular emission at wavelengths in excess of 10 μm are dominated by pure rotational transitions. Such transitions can be excited thermally, in glow discharges or via optical pumping. The latter technique has proved to be the most versatile in view of the availability of the CO_2 laser as a line-tunable pump. To date, more than 50 molecules have been pumped optically with the CO_2 or other mid-infrared lasers. Emissions occurs in the form of discrete lines at wavelengths extending to the millimeter region.

Optical pumping of selected rotational levels in a molecule such as NH_3 or CH_3OH leads to the establishment of a population inversion between the excited level and one or more rotational levels at lower energy. With the high powers available from pulsed and cw CO_2 (or N_2O) lasers, the pump transition can be relatively easily saturated, yielding significant excited state populations. Where an exact resonance does not exist between the pump wavelength and a transition of the excited molecule, Stark shifting may be employed to bring the molecular transition into resonance.

VIII. Laser Radiation via Nonlinear Effects

The nonlinear electrical susceptibility of many gases when exposed to intense laser radiation has been exploited for some time in the generation of tunable vacuum ultraviolet (VUV) radiation. The polarization $\bar{P}(w)$ of a medium when exposed to an electrical field containing components at frequencies w_1, w_2, and w_3 contains the term

$$\bar{P}(w) = \sum_{123} X^{(3)}(w = w_1 + w_2 + w_3)\bar{E}(w_1) \cdot \bar{E}(w_2) \cdot \bar{E}(w_3)$$

where $\bar{E}(w_1)$ is the electric field at frequency w_1 and so forth. As a result, the system can generate radiation of frequency w which is the sum of individual applied components. Energy levels involved in such transitions are shown schematically in Fig. 8.

The third-order susceptibility can be written

$$X^{(3)}(w = w_1 + w_2 + w_3) = \frac{3e^4}{4\hbar^3} \times \frac{\mu_{01}\mu_{12}\mu_{23}\mu_{03}}{(\Omega_{30} - w_1 - w_2 - w_3)(\Omega_{20} - w_1 - w_2)(\Omega_{10} - w_1)}$$

where e is electron charge, $\hbar$ is Planck's constant/2Π, the μ_{ij}'s are dipole matrix elements, and Ω_{30} etc. are frequency differences between real states and the ground state. It is apparent that $X^{(3)}$ has resonances when $\Omega_{20} - w_1 = 0$ and so forth; however, those involving one- or three-photon processes are accompanied by strong ab-

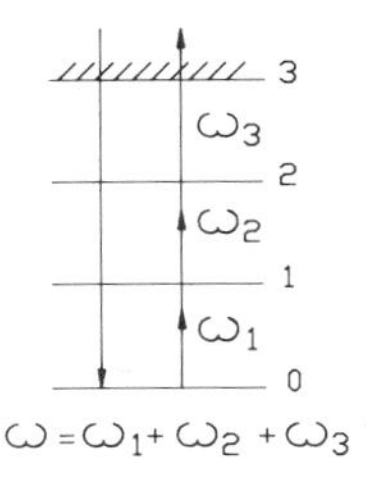

FIG. 8. Energy levels and transitions involved in four-wave sum mixing to generate radiation of frequency w.

TABLE VIII. Atomic Vapor Used for Generation of Tunable VUV Laser Radiation via Four-Wave Sum Mixing

Vapor	Tunable wavelength range (nm)	Laser system
Sr	195–178	N_2–dye
Mg	174–140	N_2–dye
	129–121	KrF–dye
Zn	140–106	XeCl/KrF–dye
Hg	125–93	Nd : YAG–dye
Xe	206–160	Nd : YAG–dye
Ke : Kr	147–140	Nd : YAG–dye
Kr	130–110	Nd : YAG–dye
		Excimer–dye
Ar	102.7[a]	XeCl
Xe	83[a]	cw–dye, Kr–dye
H_2	79[a]	ArF–dye
Ar, H_2, Kr	64[a]	cw–dye, ArF–dye
Ar	57[a]	Xe_2

[a] Small tuning range.

sorption in the nonlinear medium. Hence, tunable radiation at frequency $w = 2w_1 \pm w_2$ is usually obtained by tuning $2w_1$ to resonate with a two-photon transition, preferably chosen such that $2w_1 + w_2$ resonates with an autoionizing state above the ionization limit of the absorber.

Fortunately, many atomic vapors satisfy the requirement of accessible two-photon states together with a strong autoionization resonance at high energy (Table VIII). Experimentally, the beam from a fixed-frequency dye laser (w_1) is combined with that of a tunable dye laser in a Glan prism to form a collinear beam that enters a cell containing the nonlinear vapor. When $2w_1$ is tuned to a two-photon resonance in the vapor, a beam at $w = 3w_1$ can be generated. However, when the laser emitting at w_2 is tuned into resonance or near resonance with an autoionizing level, then an enormous enhancement of coherent output at $w = 2w_1 + w_2$ is observed. Phase matching of output and input beams is required for optimum results. Conversion efficiencies for visible power to VUV power as high as 10^{-3} have been reported under pulsed excitation. Output (VUV) linewidths of 0.02 cm^{-1} have been obtained. Molecular gases can also be used for four-wave sum mixing.

BIBLIOGRAPHY

Bennett, W. R. (1979). "Atomic Gas Laser Transition Data." Plenum Press, New York.

Duley, W. W. (1976). "CO_2 Lasers. Effects and Applications." Academic Press, New York.

McIlrath, T. J., and Freeman, R. R. (1982). "Laser Techniques for Extreme Ultraviolet Spectroscopy" *American Institute of Physics*, **90.** New York.

Tang, C. L. (1979). "Quantum Electronics." **15** A,B. Academic Press, New York.

Weber, M. J. (1982). "CRC Handbook of Laser Science and Technology" CRC Press Inc. **II,** Boca Raton, Florida.

HOLOGRAPHY

Clark C. Guest *University of California, San Diego*

GLOSSARY

Diffraction: Property exhibited by waves, including optical waves. When part of a wavefront is blocked, the remaining portion spreads to fill in the space behind the obstacle.

Diffraction orders: When a wave passes through regularly spaced openings, such as a recorded holographic interference pattern, the diffracted waves combine to form several beams at different angles. These beams are called diffraction orders.

Fringes: Regular pattern of bright and dark lines produced by the interference of optical waves.

Hologram: Physical record of an interference pattern. It contains phase and amplitude information about the wavefronts that produced it.

Holograph: Although this is an English word, meaning signature, it is often improperly used as a synonym for hologram.

Holography: Process of recording holograms and reproducing wavefronts from them.

Index of refraction: Property of transparent materials related to their polarizability at optical frequencies. The speed of light in a vacuum divided by the speed of light in a material gives the index of refraction for the material.

Interference: When two waves are brought together they may be in phase, in which case their amplitudes add, or they may be out of phase, in which case their amplitudes cancel. The reinforcement and cancellation of wavefronts due to their relative phase is called interference. Waves that do not have the same phase structure will add in some regions of space and cancel in others. The resulting regions of high and low amplitude form an interference pattern.

Parallax: Difference, due to perspective, in a scene viewed from different locations.

Planewave: Wave configuration in which surfaces of constant phase form parallel flat planes. All the light in a planewave is travelling the same direction, perpendicular to the surface of the planes.

Reconstructed beam: Light diffracted by a hologram that reproduces a recorded wavefront.

Reconstructing beam: Beam that is incident on a hologram to provide light for the reconstructed beam.

Reconstruction: Either the process of reading out a recorded hologram, or the wavefront produced by reading out a hologram.

Refractive index: See ''index of refraction.''

Spherical wave: Wave configuration in which surfaces of constant phase form concentric spherical shells or segments of shells. Light in an expanding spherical wave is propagating radially outward from a point, and light in an converging spherical wave is propagating radially inward toward a point.

Surface relief pattern: Ridges and valleys on the surface of a material.

Wavefront: Surface of constant phase in a propagating wave.

Holography is the technology of recording wavefront information and producing reconstructed wavefronts from those recordings. The record of the wavefront information is called a hologram. Any propagating wave phenomenon such as microwaves or acoustic waves is a candidate for application of the principles of holography, but most interest in this field has centered on waves in the visible portion of the electromagnetic spectrum. Therefore, this article will concentrate on optical holography.

I. Introduction

Although holography has many applications, it is best known for its ability to produce three-dimensional images. A hologram captures the perspective of a scene in a way that no simple photograph can. For instance, when viewing a hologram it is possible by moving one's head to look around objects in the foreground and see what is behind them. Yet holograms can be recorded on the same photographic film used for photographs.

Two questions naturally occur: What makes holograms different from photographs, and how can a flat piece of film store a three-dimensional scene? Answering these questions must begin with a review of the properties of light. As an electromagnetic wave, light possesses several characteristics: amplitude, phase, polarization, color, and direction of travel. When the conditions for recording a hologram are met, there is a very close relationship between phase and direction of travel. The key to the answers to both our questions is that photographs record only amplitude information (actually, they record intensity, which is proportional to the square of the amplitude), and holograms record both amplitude and phase information. How ordinary photographic film can be used to record both amplitude and phase information is described in Section II of this article.

Another interesting property of holograms is that they look nothing like the scene they have recorded. Usually a hologram appears to be a fairly uniform gray blur, with perhaps a few visible ring and line patterns randomly placed on it. In fact, all the visible patterns on a hologram are useless information, or noise. The useful information in a hologram is recorded in patterns that are too small to see with the unaided eye; features in these patterns are about the size of a wavelength of light, one two-thousandth of a millimeter.

One useful way to think of a hologram is as a special kind of window. Light is reflected off the objects behind the window. Some of the light passes through the window, and with that light we see the objects. At the moment we record the hologram, the window "remembers" the amplitude and direction of all the light that is passing through it. When the hologram is used to play back (reconstruct) the three-dimensional scene, it uses this recorded information to reproduce the the original pattern of light amplitude and direction that was passing through it. The light reaching our eye from the holographic window is the same as when we were viewing the objects themselves. We can move our heads around and view different parts of the scene just as if we were viewing the objects through the window. If part of the hologram is covered up, or cut off, the entire scene can still be viewed, but through a restricted part of the window.

There are actually many different types of holograms. Although photographic film is the most widely used recording material, several other recording materials are available. The properties of a hologram are governed by the thickness of the recording material and the configuration of the recording beams. The various classifications of holograms will be discussed in Section III. Holograms can be produced in materials that record the light intensity through alterations in their optical absorption, their index of refraction, or both. Materials commonly used for recording holograms are discussed in Section IV.

Holograms have many uses besides the display of three-dimensional images. Applications include industrial testing, precise measurements, optical data storage, and pattern recognition. A presentation of the applications of holography is given in Section V.

II. Basic Principles

Photographs record light-intensity information. When a photograph is made, precautions must be taken to ensure that the intensities in the scene are suitable; they must be neither too dim nor too bright. Holograms record intensity and phase information. In addition to the limits placed on the intensity, the phase of light used to record holograms must meet certain conditions as well. These phase conditions require that the light is coherent. There are two types of coherence, temporal and spatial; the light used for recording holograms must have both types of coherence. Temporal coherence is related to the colors in the light. Temporally coherent light contains only one color: it is monochromatic. Each color of light has a phase associated with it; multicolored light cannot be used to record a hologram because there is no one specific phase to record. Spatial coherence is related to the direction of light. At any given point in space, spatially coherent light is always travelling in one direction, and that direction does not change with time. Light that constantly changes its direction also constantly changes its relative phase in space, and so is unsuitable for recording holograms. [*See* COLOR, SCIENCE OF.]

Temporal and spatial coherence are graded quantities. We cannot say that light is definitely coherent or definitely incoherent; we can only say that a given source of light is temporally and spatially coherent by a certain amount. Ordinary light, from a light bulb for example, is temporally and spatially very incoherent. It contains many different colors, and at any point in space it is changing directions so rapidly that our eyes cannot keep up; we just see that on average it appears to come from many different directions. The temporal coherence of ordinary light can be improved by passing it through a filter that lets only a narrow band of colors pass. Spatial coherence can be improved by using light coming from a very small source, such as a pinhole in an opaque mask. Then we know that light at any point in space has to be coming from the direction of the pinhole. Ordinary light that has been properly filtered for color and passed through a small aperture can be used to record holograms. However, light from a laser is naturally very temporally and spatially coherent. For this reason, practically all holograms are recorded using laser light. [*See* LASERS.]

The simplest possible hologram results from bringing together two linearly polarized optical planewaves. Imagine that a planewave is incident at an angle θ_1 on a flat surface. At a particular moment in time we can plot the electric field of the planewave at positions on that surface. This is done in Fig. 1(a). If a second planewave is incident on the same surface at a different angle θ_2, its electric field can also be plotted. This, along with the combined field from both planewaves is plotted in Fig. 1(b). Both planewaves incident on the surface are, of course, travelling forward at the speed of light. In Fig. 1, parts (c)–(e) show the electric field at the observation surface for each planewave and for the combined field when the waves have each travelled forward by one-fourth, one-half, and three-quarters of a wavelength, respectively. The interesting thing to notice is that the locations on the observation plane where the total electric field is zero remain fixed as the waves travel. Locations midway between zero electric field locations experience an oscillating electric field. To an observer, locations with constant zero electric field appear dark, and locations with oscillating electric field appear bright. These alternating bright and dark lines, called fringes, form the interference pattern produced by the planewaves. Likewise, locations with zero electric field will leave photographic film unexposed, and locations with oscillating electric field will expose film. Thus, the interference pattern can be recorded.

The interference fringes resulting from two planewaves will appear as straight lines. Wavefronts do not have to be planewaves to produce an interference pattern, nor do the wavefronts have to match each other in shape. Interference between arbitrary wavefronts can appear as concentric circles or ellipses, or as wavy lines.

The distance L from one dark interference fringe to the next (or from one bright fringe to the next) depends on the wavelength λ of the light and the angle θ between the directions of propagation for the wavefronts,

$$L = \lambda/[2 \sin(\theta/2)] \qquad (1)$$

For reasonable angles, this fringe spacing is about the size of a wavelength of light, around one two-thousandth of a millimeter. This explains why holograms appear to be a rather uniform gray blur: the useful information is recorded in interference fringes that are too small for the eye to see.

Variations in the amplitudes of the recording beams are also recorded in the hologram and contribute to accurate reproduction of the reconstructed wavefront. During hologram recording, locations of zero electric field will occur only if the two beams are equal in amplitude. If one beam is stronger than the other, complete cancelation of their electric fields is impossible, and the depth of modulation, or contrast, of the recorded fringes is decreased. In practical terms, this means that all of the hologram is exposed to some extent.

There are two steps to the use of ordinary photographs, taking (and developing) the photograph, and viewing the photograph. Similarly, there are two steps to using a hologram: recording (and developing) the hologram, and reconstructing the holographic image. As we have just seen, recording a hologram amounts to recording the interference pattern produced by two coherent beams of light. Reconstructing the holographic image is usually accomplished by shining one of those two beams through the de-

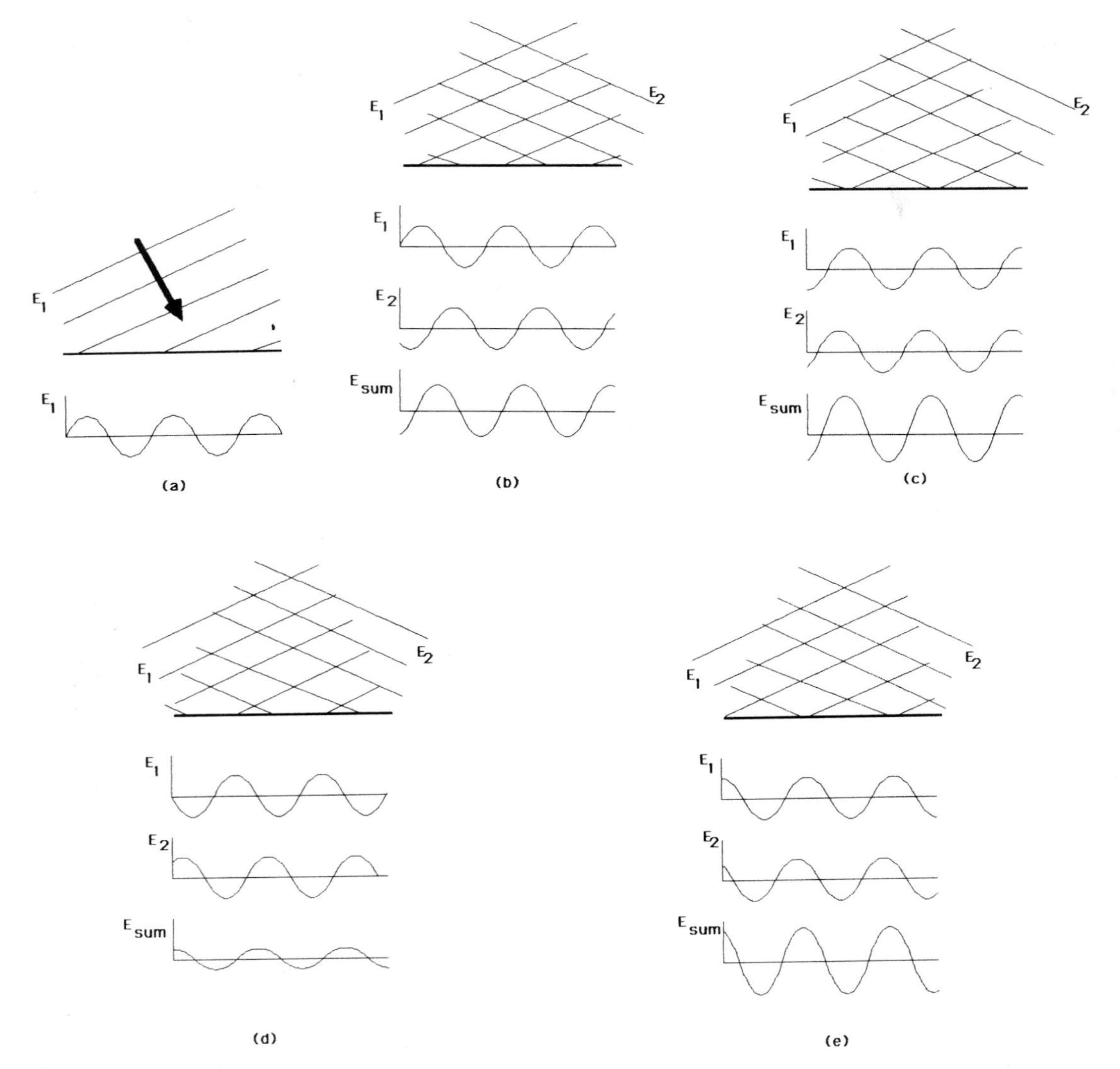

FIG. 1. (a) The electric field of a planewave incident on a surface. The electric field due to two planewaves incident on a surface at different angles as the waves are (b) initially, (c) after one-quarter wavelength of propagation, (d) after one-half wavelength of propagation, and (e) after three-quarters wavelength of propagation.

veloped hologram. Through a wave phenomenon known as diffraction, the recorded interference fringes redirect some of the light in the reconstructing beam to form a replica of the second recording beam. This replica, or reconstructed, beam travels away from the hologram with the same variation in phase and amplitude that the original beam had. Thus, for the eye, or for common image recording instruments such as a photographic camera or a video camera, the reconstructed wavefront is indistinguishable from the original wavefront and therefore possesses all the visual properties of the original wavefront, including the three-dimensional aspects of the scene. [*See* DIFFRACTION, OPTICAL.]

The interference fringe spacing that is recorded depends on the angle between the recorded beams. During reconstruction, this fringe spacing is translated, through diffraction, back into the proper angle between the reconstructing and the reconstructed beams. During reconstruction, locations on the hologram with high fringe contrast divert more optical energy into the reconstructed beam than locations with low fringe contrast, thereby reproducing the amplitude distribution of the original beam.

For most holograms, even if the contrast of the recorded fringes is large, not all of the light in the reconstructing beam can be diverted into the reconstructed image. The ratio of the amount of optical power in the reconstructed image to the total optical power in the reconstructing beam is the diffraction efficiency of the hologram. For common holograms recorded on photographic film, the maximum diffraction efficiency is limited to 6.25%. The rest of the light power, 93.75%, passes through the hologram unaffected, or ends up in beams that are called higher diffraction orders. These higher diffraction orders leave the hologram at different angles and are generally not useful. As will be noted in Section III, certain types of holograms, notably thick-phase holograms, are able to eliminate the undiffracted light and the higher diffraction orders; they can produce a diffraction efficiency close to 100%.

Certain assumptions that are made in the explanation given above should now be discussed. First, it is assumed that both beams used for recording are the same optical wavelength, that is, the same color. This is necessary to produce a stationary interference pattern to record. It is also assumed that the wavelength of the reconstructing beam is the same as that of the recording beams. This is often the case, but is not necessary. Using a reconstructing beam of a different wavelength changes the size of the reconstructed image: a shorter wavelength produces a smaller image, and a longer wavelength produces a larger image. Oddly, the dimensions of the image parallel to the plane of the hologram scale linearly with wavelength, but the depth dimension of the image scales proportional to the square of the wavelength, so three-dimensional images reconstructed with a different wavelength will appear distorted. Also, if the change in wavelength is large, distortions will occur in the other dimensions of the image, and the diffraction efficiency will decrease.

Another assumption is that the two recording beams have the same polarization. If beams with different polarizations or unpolarized beams are used, the contrast of the fringes, and therefore the diffraction efficiency of the hologram, is decreased. Beams that are linearly polarized in perpendicular directions cannot interfere and so cannot record a hologram. The polarization of the reconstructing beam usually matters very little in the quality of the reconstruction. An exception to this is when the hologram is recorded as an induced birefringence pattern in a material. Then the polarization of the reconstructing beam should be aligned with the maximum variation in the recording material index of refraction.

It is also possible that the reconstructing beam is not the same as one of the recording beams, either in its phase variation, its amplitude variation, or both. If the structure of the reconstructing beam differs significantly from both of the recording beams, the reconstructed image is usually garbled. One particular case where the image is not garbled is if the reconstructing beam has the same relative phase structure as one of the recording beams, but approaches the hologram at a different angle. In this case, provided the angle is not too large and that the recording behaves as a thin hologram (see Section III for an explanation of thin and thick holograms), the reconstructed image is produced.

Photographic film is assumed to be the recording material used in the example above. Many other materials can be used to record holograms, and these are the subject of Section IV. Photographic film gives a particular type of recording, classified as a thin absorption hologram. The meaning and properties of different classifications of holograms are dealt with in Section III.

III. Classification of Holograms

Many different types of holograms are possible. Holograms are classified according to the material property in which the interference pattern is recorded, the diffraction characteristics of the hologram, the orientation of the recording beams with respect to the hologram, and the optical system configuration used for recording and reconstructing the hologram.

Holograms are recorded by exposing an optically sensitive material to light in the interference pattern produced by optical beams. For example, exposing photographic film to light triggers a chemical reaction that, after development, produces a variation in the optical absorption of the film. Portions of the film exposed to high optical intensity become absorbing, and unexposed portions of the film remain transparent. Other materials also exhibit this characteristic of changing their optical absorption in response to exposure to light. Holograms that result from interference patterns recorded as variations in material absorption are known as amplitude holograms.

There are also materials whose index of refraction changes in response to exposure to light. These materials are usually quite transparent, but the index of refraction of the material increases or decreases slightly where it is exposed to light. Holograms that result from interference patterns recorded as index-of-refraction variations are known as phase holograms. During reconstruction, light encountering regions with a higher index of refraction travels more slowly than light passing through lower-index regions. Thus the phase of the light is modified in relation to the recorded interference pattern.

It is not correct to assume that amplitude holograms can reconstruct wavefronts with only amplitude variations, and phase holograms can reconstruct wavefronts with only phase variations. In Section II it was explained that wavefront direction (i.e., phase) is recorded by interference fringe spacing, and wavefront amplitude is recorded by interference fringe contrast. In reality, both amplitude and phase types of holograms are capable of recording wavefront amplitude and phase information.

Holograms are also classified as being "thin" or "thick." These terms are related to the diffraction characteristics of the hologram. A thin hologram is expected to produce multiple diffraction orders. That is, although only two beams may have been used for recording, a single reconstructing beam will give rise to several reconstructed beams, called diffraction orders. Another property associated with thin holograms is that if the angle at which the reconstructing beam approaches the hologram is changed, the hologram continues to diffract light, with little change in diffraction efficiency. The diffraction orders will rotate in angle as the reconstructing beam is rotated. Thick holograms, on the other hand, produce only a single diffracted beam; a portion of the reconstructing beam may continue through the hologram in its original direction as well. Also, noticeable diffraction efficiency for thick holograms occurs only if the reconstructing beam is incident on the hologram from one of a discrete set of directions, called the Bragg angles. If the beam is not at a Bragg angle, it passes through the hologram and no diffracted beam is produced. The property of thick holograms that diffraction efficiency falls off if the reconstructing beam is not at a Bragg angle is called angular selectivity. Many thick holograms can be recorded in the same material and reconstructed separately by arranging for their Bragg angles to be different.

The terms thin and thick were originally applied to holograms based solely on the thickness of the recording material. The situation is, in fact, more complicated. Whether a particular hologram displays the characteristics associated with being thick or thin depends not only on the thickness of the recording material, but also on the relative sizes of the optical wavelength and the interference fringe spacing, and on the strength of the change produced in the absorption or refractive index of the material.

The next category of hologram classification has to do with the arrangement of the recording beams (and therefore the reconstructing and reconstructed beams) with respect to the recording material. When two planewave beams produce interference fringes, the fringes form a set of planes in space. The planes lie parallel to the bisector of the angle between the beams, as shown in Fig. 2(a). If the recording material is arranged so that both recording beams approach it from the same side, fringes are generally perpendicular to the material surfaces, as shown in Fig. 2(b), and a transmission-type hologram is formed. During readout of the transmission hologram, the reconstructing and the recon-

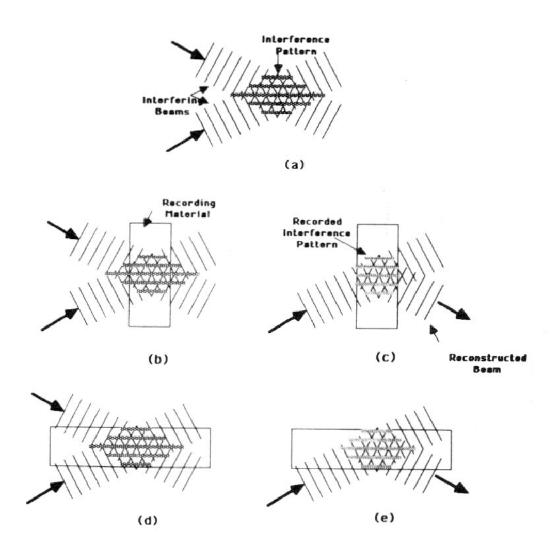

FIG. 2. (a) The interference produced by two planewave beams. (b) The orientation of the recording material for a transmission hologram. (c) The configuration used for reconstruction with a transmission hologram. (d) The orientation of the recording material for a reflection hologram. (e) The configuration used for reconstruction with a reflection hologram.

structed beams lie on opposite sides of the hologram, as in Fig. 2(c). Alternatively, the recording material can be arranged so that the recording beams approach it from opposite sides. In this case, the fringes lie parallel to the surfaces of the material, as shown in Fig. 2(d), and a reflection-type hologram is formed. For a reflection hologram, the reconstructing and the reconstructed beams lie on the same side of the hologram, portrayed in Fig. 2(e)

The three hologram classification criteria discussed so far—phase or absorption, thick or thin, and transmission or reflection—play a role in determining the maximum possible diffraction efficiency of the hologram. Table I summarizes the diffraction efficiencies for holograms possessing various combinations of these characteristics. Because the fringes for reflection holograms lie parallel to the surface, there must be an appreciable material thickness to record the fringes; therefore, thin reflection holograms are not possible and are absent from the table. (Often, holograms use reflected light but have fringes perpendicular to the material surface and are properly classified as transmission holograms.) Notice that phase holograms are generally more efficient than absorption holograms, and thick phase holograms are able to achieve perfect efficiency; all of the reconstructing beam power is coupled into the reconstructed beam. Keep in mind that the figures in the table represent the absolute highest diffraction efficiencies that can be achieved. In practice, hologram diffraction efficiencies are often substantially lower.

The final classification criterion to be discussed has more to do with the configuration of the optical system used for recording than with the hologram itself. For recording a hologram, coherent light is reflected from or transmitted through an object and propagates to the position of the recording material. A second beam of coherent light produces interference with light from the object, and the interference pattern is recorded in the material. If the object is very close to the recording material or is imaged onto the recording material, then an image-plane hologram is formed. If the separation between the object and the recording material is a few times larger than the size of the object or the material, a Fresnel hologram is formed. If the object is very far from the recording material, a Fraunhofer hologram is recorded. Another possible configuration is to place a lens between the object and the recording material such that the distance between the lens and the material is equal to the focal length of the lens. This arrangement produces a Fourier transform hologram, so called because during reconstruction, the Fourier transform of the reconstructed wave must be taken (with a lens) to reproduce the recorded image. These previous configurations use a planewave as the second recording beam. If an expanding spherical wave is used instead, with the source of the spherical wave near the location of the object, a quasi-Fourier-transform hologram, or lensless Fourier transform hologram, is formed. This type of hologram shares many of the properties of the true Fourier transform hologram.

TABLE I. Maximum Diffraction Efficiencies of Hologram Classes

Thickness	Modulation	Configuration	Maximum efficiency (%)
Thin	Absorption	Transmission	6.25
Thin	Phase	Transmission	33.90
Thick	Absorption	Transmission	3.70
Thick	Absorption	Reflection	7.20
Thick	Phase	Transmission	100.00
Thick	Phase	Reflection	100.00

Other classes of holograms also exist, such as polarization holograms, rainbow holograms, synthetic holograms, and computer generated holograms. However, these are specialized topics, best dealt with separately from the central concepts of holography.

IV. Recording Materials

There are many materials that can be used for recording holographic interference patterns. Some materials record the pattern as a change in their optical absorption; this yields absorption holograms. Other materials record the patterns as changes in their index of refraction or as a relief pattern on their surface; these materials produce phase holograms. The thickness of the recording layer is also important to the characteristics of the hologram, as discussed in Section III.

Practical concerns related to holographic recording materials include the recording resolution, material sensitivity as a function of optical wavelength, and the processing steps required to develop the hologram. The spacing of interference fringes can be adjusted by changing the angle between the beams: a small angle gives large fringes, and a large angle gives fringes as small as one-half the wavelength of the light used. An ideal recording material would have a resolution of at least 5000 fringes per millimeter.

Some materials are sensitive to all visible wavelengths, and others to only a portion of the spectrum. The wavelength sensitivity of the recording material must be matched to the light source used. Sensitive recording materials are generally desirable, since high sensitivity reduces the recording exposure time and the amount of optical power required of the source. Long exposures are undesirable because of the increased chance that a random disturbance will disrupt the coherence of the recording beams.

Many recording materials require some chemical processing after exposure to develop the holographic pattern. Some materials develop when heat is applied, and a few materials require no processing at all: the hologram is immediately available. Of course, the need for developing complicates the system and introduces a delay between recording the hologram and being able to use it. Important characteristics of the most common hologram recording materials are summarized in Table II.

Silver halide photographic emulsions are the most common recording material used for holograms. They are a mature and commercially available technology. The emulsion may be on a flexible acetate film or, for greater precision, a flat glass plate. Photographic emulsions have a very high sensitivity and respond to a broad spectral range. The ordinary development procedure for photographic emulsions causes them to become absorptive at the locations that have been exposed to light. Thus, an absorption hologram is produced. Alternate developing procedures employing bleaches leave the emulsion transparent but modulate the index of refraction or the surface relief of the emulsion. These processes lead to phase holograms. Many photographic emulsions are thick enough to produce holograms with some characteristics of a thick grating.

Dichromated gelatin is one of the most popular materials for recording thick phase holograms. Exposure to light causes the gelatin molecules to crosslink. The gelatin is then washed, followed by dehydration in alcohol. Dehydration causes the gelatin to shrink, causing cracks and tears to occur in the regions of the gelatin that are not crosslinked. The cracks and tears produce a phase change in light passing through those regions. Phase holograms recorded in dichromated gelatin are capable of achieving diffraction efficiencies of 90% or better with very little optical noise. The primary limitations of dichromated gelatin are its very low sensitivity and the undesirable effects of shrinkage during development.

Photoresists are commonly used in lithography for fabrication of integrated circuits, but can be utilized for holography too. Negative and positive photoresists are available. During developing, negative photoresists dissolve away in locations that have not been exposed to light, and positive photoresists dissolve where there has been exposure. In either case, a surface relief recording of the interference pattern is produced. This surface relief pattern can be used as a phase hologram either by passing light through it or by coating its surface with metal and reflecting light off it. The photoresist can also be electroplated with nickel, which is then used as the master for embossing plastic softened by heat. The embossing process can be done rapidly and inexpensively, and so is useful for mass produc-

TABLE II. Holographic Recording Materials

Material	Modulation	Sensitivity (J/cm^2)	Resolution (line pairs/mm)	Thickness (μm)
Photographic emulsion	Absorption or phase	$\sim 5 \times 10^{-5}$	~5000	<17
Dichromated gelatin	Phase	$\sim 7 \times 10^{-2}$	>3000	12
Photoresist	Phase	$\sim 1 \times 10^{-2}$	~1000	>1
Photopolymer	Phase	$\sim 1 \times 10^{-2}$	3000	3–150
Photoplastic	Phase	$\sim 5 \times 10^{-5}$	>4100	1–3
Photochromic	Absorption	~2	>2000	100–1000
Photorefractive	Phase	~3	>1000	5000

ing holograms for use in magazines and other large-quantity applications.

Photopolymers behave in a fashion similar to photoresists, but instead of dissolving away during development, exposure of a photopolymer to light induces a chemical reaction in the material that changes its index of refraction or modulates its surface relief. Some photopolymers require no development processing, and others must be heated or exposed to ultraviolet light.

Photoplastics are noted for their ability to record and erase different holographic patterns through many cycles. The photoplastics are actually multilayer structures. A glass substrate plate is coated with a conductive metal film. On top of this is deposited a photoconductor. The final layer is a thermoplastic material. For recording, a uniform static electric charge is applied to the surface of the thermoplastic. The voltage drop due to the charge is divided between the photoconductor and the thermoplastic. The structure is then exposed to the holographic interference pattern. The voltage across the illuminated portions of the photoconductor is discharged. Charge is then applied a second time to the surface of the device. This time excess charge accumulates in the regions of lowered voltage. The device is now heated until the thermoplastic softens. The electrostatic attraction between the charge distribution on the surface of the thermoplastic and the conductive metal film deforms the plastic surface into a surface relief phase hologram. Cooling the plastic then fixes it in this pattern. The hologram may be erased by heating the plastic to a higher temperature so that it becomes conductive and discharges its surface.

Photochromics are materials that change their color when exposed to light. For example, the material may change from transparent to absorbing for a certain wavelength. This effect can be used to record absorption holograms. Furthermore, the recording process can be reversed by heating or exposure to a different wavelength. This allows patterns to be recorded and erased. These materials, however, have very low sensitivity.

Photorefractive materials alter their refractive index in response to light. These materials can be used to record thick phase holograms with very high diffraction efficiency. This recording process can also be reversed, either by uniform exposure to light or by heating. These materials, too, have rather low sensitivity, but research is continuing to produce improvements.

V. Applications

Holography is best known for its ability to reproduce three-dimensional images, but it has many other applications as well. Holographic nondestructive testing is the largest commercial application of holography. Holography can also be used for storage of digital data and images, precise interferometric measurements, pattern recognition, image processing, and holographic optical elements. These applications are treated in detail in this section.

An image reconstructed from a hologram possesses all the three-dimensional characteristics of the original scene. The hologram can be considered a window through which the scene is viewed. As described in Section II, a hologram records information about the intensity and direction of the light that forms it. These two quantities (along with color) are all that the eye uses to perceive a scene. The light in the image reconstructed by the hologram has the same intensity and direction properties as the light from the original scene, so the eye sees an image that is nearly indistinguishable from the original. There are two important aspects in which the holographic reconstruction of a scene differs from the original: color and speckle. Most holograms are recorded using a single wavelength of light. The image is reconstructed with this same wavelength, so all objects in the scene have the same color. Also, because of the coherence properties of laser light, holographic images do not appear smooth: they are grainy, consisting of many closely spaced random spots of light called speckles. Attempts to produce full-color holograms and eliminate speckle will be described in this section.

The simplest method of recording an image hologram is shown in Fig. 3(a) Light reflecting off the object forms one recording beam, and light incident on the film directly from the laser is the other beam needed to produce the interference pattern. The exposed film is developed and then placed back in its original position in the system. The object that has been recorded is removed from the system, and the laser is turned on. Light falling on the developed hologram reconstructs a virtual image of the object that can be viewed by looking through the hologram toward the original position of the object, as shown in Fig. 3(b).

Many variations on the arrangement described above are possible. Often, the portion of the beam directed toward the object is split into

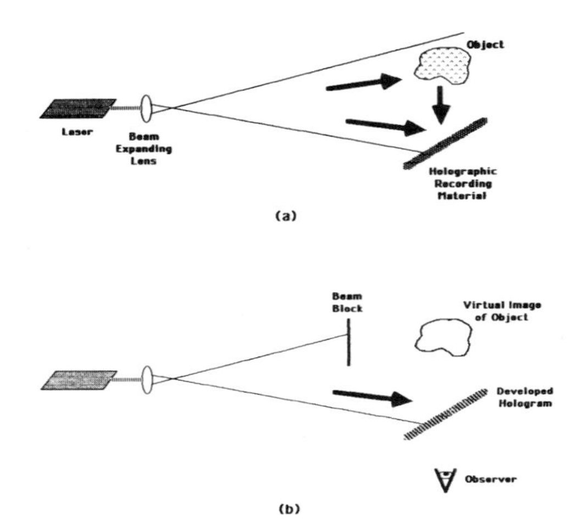

FIG. 3. A simple optical system for (a) recording and (b) viewing a transmission hologram.

several beams that are arranged to fall on the object from different directions so that it is uniformly illuminated. Attention to lighting is especially important when the scene to be recorded consists of several separate objects that are different distances from the film. Another recording arrangement is to have the laser beam pass through the film plane before reflecting from the object, as shown in Fig. 4(a). A reflection hologram is formed with this arrangement, and can be viewed as shown in Fig. 4(b). An important advantage of the reflection hologram is that a laser is not needed to view it: a point source of white light will work quite well. Use of white light to reconstruct a hologram is not only more convenient, but eliminates speckle too. This is possible because the reflection hologram also acts as a color filter, efficiently reflecting light of only the proper wavelength. It is important, however, that source is very small (as compared to its distance from the hologram), otherwise the reconstructed image will be blurred.

A hologram that reconstructs an image containing all the colors of the original scene would, obviously, be a desirable accomplishment. The principle obstacle in achieving this goal is that holographic interference patterns result only when very monochromatic light is used. Recording a hologram with a single color and reconstructing the image with several colors does not work. Reconstructing a hologram with a wavelength other than the one used to record it changes the size and direction of the reconstructed image, so the variously colored images produced with white-light illumination are not aligned. It is possible to record three different holograms of the same object on one piece of film, using a red, a blue, and a green laser. The three holograms can be reconstructed simultaneously with the appropriate colors, and a reasonable representation of the colors of the original object is produced. The problem, however, is that each reconstructing beam illuminates not only the hologram it is intended for, but the other two holograms as well. This leads to numerous false-color images mingled with the desired image. The false images can be eliminated with a thick hologram recording material by using the angular selectivity of thick holograms.

Another approach to color holography is the stereogram. Sets of three separate holograms (one each for red, green, and blue) are recorded for light coming from the scene at various angles. A projection system using a special screen is used for viewing. Light from different angular perspectives is directed at each eye of the viewer, thus providing the parallax information needed to yield a three dimensional image. This system requires that the viewer is positioned rather exactly with respect to the screen.

Interferometry is a means of making precise measurements by observing the interference of optical wavefronts. Since holograms record phase information about the light from an object, they are useful in making before and after comparisons of the deformation of objects in response to some change in their environment. A typical arrangement for holographic interferometry is shown in Fig. 5. The first step is to record and develop a hologram of the object. The hologram is replaced in the system, and the image of the object is reconstructed from it. The object

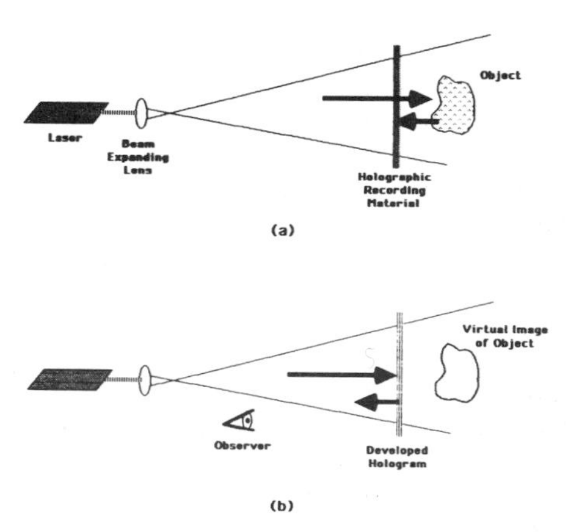

FIG. 4. A simple optical system for (a) recording and (b) viewing a reflection hologram.

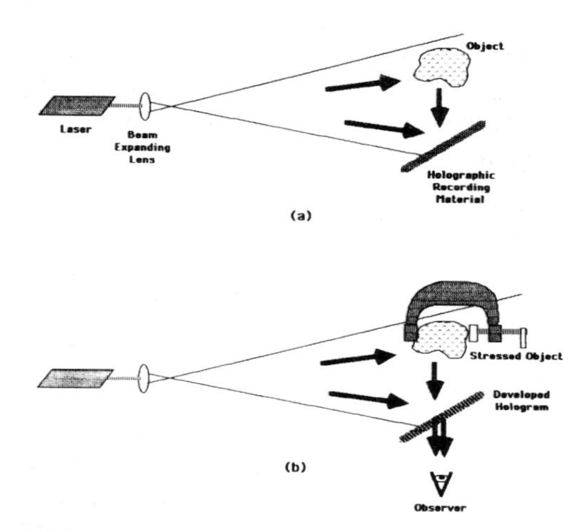

FIG. 5. (a) For real-time holographic interferometry, a hologram of the object is recorded. (b) Then the stressed object is viewed through the developed hologram.

itself is subjected to some change and illuminated as it was to record the hologram. When viewing the object through the hologram, light from the object and from the holographic reconstruction of the object will interfere. If the surface of the object has deformed, bright and dark fringes will appear on the surface of the object. Each fringe corresponds to a displacement in the surface of the object by one-quarter of the optical wavelength. Thus, this is a very sensitive measurement technique.

The same principles apply to transparent objects that undergo an index of refraction change. Changes in index of refraction alter the optical path length for light passing through the object. This leads to interference fringes when the object is viewed through a holographic recording of itself. Such index of refraction changes can occur in air, for example, in response to heating or aerodynamic flows.

A slight modification to the technique described above can provide a permanent record of the interference pattern. A hologram of the original object is recorded but not developed. The object is subjected to some change and a second recording is made on the same film. The film is now developed and contains superimposed images of both the original and the changed object. During reconstruction, wavefronts from the two images will interfere and show the desired fringe pattern.

Another variation of holographic interferometry is useful for visualizing periodic vibrations of objects. Typically, the object is subjected to some excitation that causes its surface to vibrate. A hologram of the vibrating object is made with an exposure time that is longer than the period of the vibration. A time integration of the lightwave phases from locations on the object is recorded on the film. Portions of the object that do not move, vibrational nodes, contribute the same lightwave phase throughout the exposure. This produces constructive interference, and these locations appear bright in the reconstruction. Bright and dark fringes occur elsewhere on the object, with the number of fringes between a particular location and a vibrational node indicating the amplitude of the vibration.

Deformations on the surface of an object in response to applied pressure or heating can often be used to make determinations concerning changes within the volume of the object under the surface. For this reason, holographic interferometry is useful for holographic nondestructive testing; characteristics of the interior of an object can be determined without cutting the object apart. If interference fringes are concentrated on one area of a stressed object, that portion of the object is probably bearing a greater portion of the stress than other locations. Or a pattern of fringes may indicate a void in the interior of the object, or a location where layers in a laminated structure are not adhering.

Holography plays an important role in the field of optical data processing. A common optical system using a holographic filter is shown in Fig. 6. One application of this system is pattern recognition through image correlation. This is useful for detecting the presence of a reference object in a larger scene. The first step is to record a Fourier hologram of the object that is to be detected. A transparency of the object is placed in the input plane and is holographically recorded onto photographic film in the filter

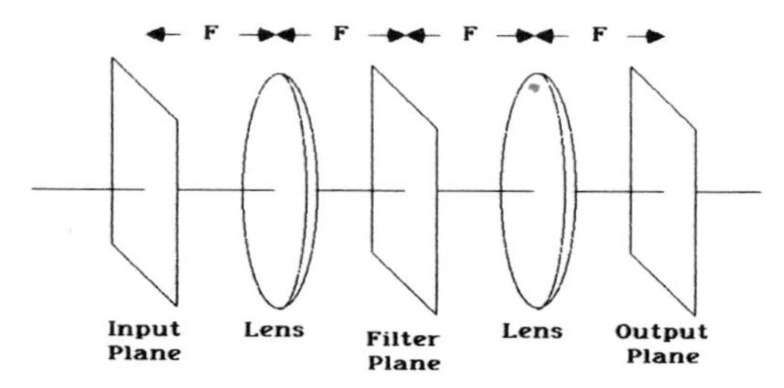

FIG. 6. A simple, yet versatile, optical system for using holographic filters for image processing.

plane; a reference beam is also incident on the film. After exposure, the film is removed from the system, developed, and then replaced in the filter plane. The scene believed to contain occurrences of the reference object is placed in the input plane and illuminated. A portion of the light distribution in the output plane will represent the correlation of the input image with the reference image. This means that wherever the reference image occurs in the input scene, a bright spot of light will be present at the corresponding position on the output plane. As a specific example, the reference object might be a printed word. A hologram of the word would be placed in the filter plane. Input to the system could be pages of printed text. For each occurrence of the selected word within the text, a bright spot would be present at the output plane. In the simple form described, the system can detect occurrences of the reference object that are very close to the same size and rotational orientation as the object used to record the hologram. However, research is being conducted to produce recognition results without regard to scaling or rotation of the object, and has already met with considerable success.

Other image-processing operations can be performed using the same optical system. The hologram is recorded to represent a frequency-domain filter function rather than a reference object. For example, blurring of an image produced by an unfocused optical system can be modeled as a filtering of the spatial frequencies in the image. Much of the detail can be restored to a blurred image by placing a hologram representing the inverse of the blurring function in the filter plane and a transparency of the blurred image in the input plane. Holographic filters can also be used to produce other useful image-processing operations, such as edge enhancement.

Holography is also attractive for data storage. Because holograms record information in a distributed fashion, with no one point on the hologram corresponding to a particular part of the image, data recorded in holographic form are resistant to errors caused by defects in the recording material. Returning to the analogy of the hologram as a window through which to view an image, if part of the hologram is obscured or destroyed, it is still possible to recover all the data simply by looking through the remaining usable portion of the window. Obviously, if part of an ordinary photograph containing data is destroyed, the data on that portion of the photograph is irrevocably lost. The data stored in a holographic system can be pages of text, images, or digitally encoded computer data. If thick recording materials are used, many pages of data can be stored in one piece of material by utilizing the angular selectivity of thick holograms.

Holograms can be recorded to provide the same functions as refractive optical elements such as lenses and prisms. For example, if a hologram is recorded as the interference of a planewave and a converging spherical wave, when the developed hologram is illuminated with a planewave it will produce a converging spherical wave, just as a positive lens would. Holographic optical elements (HOEs), as they are called, have two principle disadvantages with respect to the more common refractive elements. HOEs work as designed only for one wavelength of light, and because they usually have a diffraction efficiency less than 100%, not all of the light is redirected as desired. However, HOEs also have advantages over refractive elements in certain applications. First, the optical function of a HOE is not linked to its physical shape. A HOE may be placed on a curved or angled surface. For example, HOEs are used on the surface of the visor on pilots' helmets to serve as a projection lens for instrumentation displays. Also, HOEs can be created that provide the function of refractive elements that would be very difficult to fabricate, such as an off-axis segment of a lens. Several HOEs can be recorded on the same piece of material to give the function of multiple refractive elements located at the same spatial position. Another popular configuration is to record a HOE on the surface of an existing lens. The lens takes care of most of the refraction required by an optical system, and the HOE provides small additional corrections to reduce aberrations. A second advantage of HOEs is their small physical volume and weight. A HOE recorded on film can provide the same function as a thick, heavy piece of glass. In particular, lenses with a large aperture and short focal length can be quite massive when fabricated in glass, while those characteristics are readily produced using a HOE. Finally, there are applications in which the limited range of working wavelengths of HOEs is desirable. Reflective holograms are often used in these applications. Information presented in the intended color can be reflectively imaged toward a viewer while the scene behind the hologram, containing mainly other colors, is also visible through the hologram without distortion.

The applications just cited are only a sampling of the uses of holography. It is a field that is the subject of ongoing research and that enjoys the continuing discovery of new forms and applications.

VI. History of Holography

The fundamental principle of holography, recording the phase of a wavefront as an intensity pattern by using interference, was devised by Dennis Gabor in 1948 to solve a problem with aberrations in electron microscopy. He also coined the word hologram from Greek roots meaning whole record. The results of Gabor's work of translating electron diffraction patterns into optical diffraction patterns and then removing aberrations from the resulting image were less than satisfactory. Other researchers following his lead were not significantly more successful, so this first application of holography soon faded from the scene.

Other researchers experimented with optical holography for its own sake during the early and mid-1950s, but results were generally disappointing. At that time, holography suffered from two important disadvantages. First, no truly interesting application had been found for it. The rather poor-quality images it produced were laboratory curiosities. Second, coherent light to record and view holograms was not readily available. The laser was not available until the early 1960s. Coherent light had to be produced through spectral and spatial filtering of incoherent light.

In 1955 an interesting, and eventually very successful, application for holography was uncovered. Leith and Upatnieks, working at the University of Michigan's Radar Laboratory, discovered that holography could be used to reconstruct images obtained from radar signals. Their analysis of this technique led them to experiment with holography in 1960, and by 1962 they had introduced an important improvement to the field. Gabor's holograms had used a single beam to produce holograms, recording the interference of light coming only from various locations on the object. This led to reconstructed images that were seriously degraded by the presence of higher diffraction orders and undiffracted light. Leith and Upatnieks introduced a second beam for recording. The presence of this second beam separated the reconstructed image from other undesired light, thus significantly improving the image quality.

Also in the early 1960s, Denisyuk was producing thick reflection holograms in photographic emulsions. These holograms have the advantage that they can be viewed with a point source of white (incoherent) light, since the reflection hologram acts also as a color filter.

Advances in practical applications of holography began to occur in the mid 1960s. In 1963, Vander Lugt introduced the form of holographic matched filter that is still used today for pattern recognition. Powell and Stetson discovered holographic interferometry in 1965. Other groups working independently introduced other useful forms of this important technique.

The first form of thin transmission hologram that can be viewed in white light, the rainbow hologram, was developed by Benton in 1969. Another form of hologram that can be viewed in white light, the multiplex hologram, was produced by Cross in 1977.

Bibliography

Caulfield, H. J., (ed.) (1979). "Handbook of Optical Holography." Academic Press, New York.

Goodman, J. W. (1968). "Introduction to Fourier Optics." McGraw-Hill, New York.

INTERFEROMETRY, OPTICAL

P. Hariharan *CSIRO Division of Applied Physics*

GLOSSARY

Coherence: Statistical measure of the similarity of the fields produced by a light wave at two points separated in space and/or time.

Doppler effect: Shift in the frequency of a wave observed with a moving source or detector, or when it is reflected or scattered by a moving object.

Incoherent source: Source consisting of a large number of individual emitters that radiate independently of each other.

Laser: Light source that radiates by stimulated emission. The output from a laser is highly directional and monochromatic.

Localized fringes: With an incoherent light source the contrast of the fringes produced by an interferometer is usually a maximum in a particular plane. The fringes are then said to be localized in this plane.

Moiré fringes: Relatively coarse fringes produced by the superposition of two fine fringe patterns with slightly different spacings.

Optical fiber: Glass fiber surrounded by a transparent sheath with a lower refractive index. If the diameter of the fiber is comparable to the wavelength, light propagates along it in a single guided mode.

Optical path: Product of the geometrical path traversed by a light wave and the refractive index of the medium.

Polarizing beamsplitter: Optical element that separates light waves polarized in two orthogonal planes.

Quarter-wave plate: Device that introduces an optical path difference of a quarter wavelength between two orthogonally polarized waves.

Speckle: Granular appearance of a rough surface illuminated by a laser caused by the superposition of scattered light waves with random phase differences.

Visibility: Visibility of interference fringes is defined by the relation $V = (I_{max} - I_{min}) / (I_{max} + I_{min})$, where I_{max} and I_{min} are the irradiances at adjacent maxima and minima, respectively.

Optical interferometry comprises a range of techniques that use light waves to make extremely accurate measurements. For many years optical interferometry remained a laboratory technique. However, as a result of several recent innovations such as lasers, optical fibers, holography, and the use of digital computers for image processing, optical interferometry has emerged as a very practical tool with many applications.

I. Interference of Light

If two or more waves are superposed, the resultant displacement at any point is the sum of the displacements due to the individual waves. This is the well-known phenomenon of interference. With two waves of equal amplitude it is possible for their effects to cancel each other at some points so that the resultant amplitude at these points is zero.

The colors of an oil slick or a thin film of air enclosed between two glass plates are due to the interference of light waves. To observe such interference patterns (fringes), the interfering waves must have exactly the same frequency. This normally implies that they must be derived from the same light source. In addition, to maximize the visibility of the fringes, the polarization of the interfering light waves must be the same.

Only a few interference fringes can be seen with white light because as the optical path difference increases, the phase difference between the interfering light waves differs for different wavelengths. However, interference fringes can be seen with much larger optical path differences if light with a very narrow spectral bandwidth is used.

Optical interferometers split the light from a suitable light source into two or more parts that transverse separate paths before they are recombined. Common types of interferometers are the Michelson, the Mach–Zehnder, and the Sagnac, which use two-beam interference, and the Fabry–Perot, which uses multiple-beam interference. Applications of optical interferometry include accurate measurements of distances, displacements, and vibrations, tests of optical systems, studies of gas flows and plasmas, microscopy, measurements of temperature, pressure, electrical, and magnetic fields, rotation sensing, and even the determination of the angular diameters of stars.

II. Measurements of Length

One of the earliest applications of optical interferometry was in length measurements where it led to the replacement of the meter bar by an optical wavelength as the practical standard of length. Several lasers are now available that emit highly monochromatic light whose wavelength has been measured extremely accurately. With such a laser, optical interferometry can be used for very accurate measurements of distances of a hundred meters or more.

A. Absolute Measurements of Length

Electronic fringe counting is now widely used for length interferometry. In the Hewlett–Packard interferometer a helium–neon laser is forced to oscillate simultaneously at two frequencies separated by a constant difference of about 2 MHz, by applying an axial magnetic field. As shown in Fig. 1, these two waves, which are circularly polarized in opposite senses, are converted to orthogonal linear polarizations by a quarter-wave plate. A polarizing beamsplitter reflects one wave to a fixed corner reflector C_1, while the other is transmitted to a movable corner reflector C_2. The returning waves pass through a polarizer, so that the transmitted components interfere and are incident on the detector D_S.

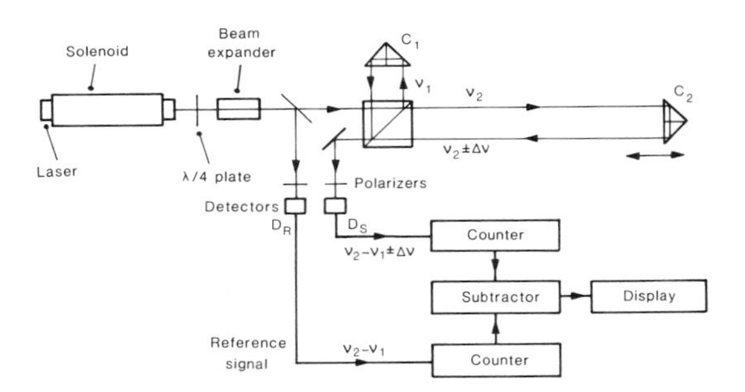

FIG. 1. Fringe-counting interferometer using a two-frequency laser. [After J. N. Dukes and G. B. Gordon (1970). *Hewlett-Packard J.* **21**(12), 2–8. © Copyright 1986 Hewlett–Packard Company. Reproduced with permission.]

The outputs from D_S and a reference detector D_R go to a differential counter. If the two reflectors are stationary, the frequencies of the two outputs are the same, and no net count accumulates. If one of the reflectors is moved, the change in optical path in wavelengths is given by the net count.

Another technique, which can be used if the distance to be measured is known approximately, involves synthetic long-wavelength signals. This technique is based on the fact that if two wavelengths λ_1 and λ_2 are simultaneously incident on a two-beam interferometer, the envelope of the fringes corresponds to the interference pattern that would be obtained with a synthetic wavelength $\lambda_s = \lambda_1\lambda_2/|\lambda_1\lambda_2|$.

The carbon dioxide laser can operate at several wavelengths, which have been measured accurately, and is, therefore, well suited to such measurements. The laser is switched rapidly between two of these wavelengths and the output signal obtained from a detector as one of the interferometer mirrors is moved is squared, low-pass filtered, and processed in a computer to obtain the phase difference. Distances up to 100 m can be measured with an accuracy of one part in 10^7.

Yet another method is to use a semiconductor laser whose frequency is swept linearly with time by controlling the injection current. For an optical path difference D, the two beams reach the detector with a time delay D/c, where c is the speed of light, and they interfere to yield a beat signal with a frequency

$$f = (D/c)(df/dt) \tag{1}$$

where df/dt is the rate at which the laser frequency is varying with time.

B. Measurements of Very Small Changes in Length

A number of interferometric techniques are also available for accurate measurements of very small changes in length. One method is based on phase compensation. Changes in the output intensity from the interferometer are detected and fed back to a phase modulator in the measurement path so as to hold the output constant. The drive signal to the modulator is then a measure of the changes in the optical path.

Another method involves sinusoidally modulating the phase of the reference beam. Under these conditions the average phase difference between the interfering beams can be determined from a comparison of the amplitudes of the components in the output of the detector at the modulation frequency and at its second harmonic.

A third group of methods is based on heterodyning (light beats). For this a frequency difference is introduced between the two beams in the interferometer, usually by means of a pair of acoustooptic modulators operated at slightly different frequencies. The output from a detctor then contains an oscillatory component at the difference frequency whose phase corresponds to the phase difference between the two interfering wave fronts.

Light beats can also be produced by superposing the beams from two lasers operating on the same transition and can be used to measure changes in length very accurately. For this purpose two mirrors are attached to the ends of the specimen to form a Fabry–Perot interferometer. The frequency of a laser is then locked to a transmission peak of the interferometer, so that the wavelength of this slave laser is an integral submultiple of the optical path difference in the interferometer. A displacement of one of the mirrors results in a change in the wavelength of the slave laser and hence in its frequency. These changes are measured to better than one part in 10^8 by mixing the beam from the slave laser at a fast photodiode with the beam from a frequency-stabilized reference laser and measuring the beat frequency.

III. Measurements of Velocity and Vibration Amplitude

Light scattered from a moving particle has its frequency shifted by the Doppler effect by an amount proportional to the component of the velocity of the particle along the bisector of the angle between the directions of illumination and observation. With a laser source this frequency shift can be detected by the beats produced either by the scattered light and a reference beam or by the scattered light from two illuminating beams incident at different angles. An initial frequency offset can be used to distinguish between positive and negative flow directions. Laser–Doppler interferometry is now used widely to measure flow velocities. Another industrial application has been for noncontact measurements of the velocity of moving material.

Laser–Doppler techniques can also be used to analyse surface vibrations using an interferometer in which a frequency offset is introduced between the beams by an acoustooptic modulator. The output from a detector then consists of a component at the offset frequency (the carrier) and two sidebands. Vibration amplitudes down to a few thousandths of a nanometer can be determined by a comparison of the amplitudes of the carrier and the sidebands, while the phase of the vibration can be obtained by comparison of the carrier with a reference signal.

IV. Optical Testing

Another major application of interferometry is in testing optical components and optical systems. The instruments commonly used for this purpose are the Fizeau and the Twyman–Green interferometers. The Fizeau interferometer is widely used to compare flat surfaces. However, with a laser source it can carry out a much wider range of tests, including tests on concave and convex surfaces.

The output of such an interferometer is a fringe pattern that can be interpreted by an observer quite readily; unfortunately, the process of extracting quantitative data from it is tedious and time consuming. This has led to the use of digital computers for analyzing such fringe patterns.

A. Digital Techniques

A typical digital system for fringe analysis uses a television camera in conjunction with a video frame memory and a minicomputer. Since the fringes only give the magnitude of the errors and not their sign, a tilt is introduced between the interfering wave fronts so that a linear phase gradient is added to the actual phase differences that are being measured.

Much higher accuracy can be obtained by directly measuring the optical path difference between the two interfering wave fronts at an array of points covering the interference pattern. A number of electronic techniques are now available for this purpose.

In one method (phase shifting) the optical path difference between the interfering beams is varied linearly with time, and the output current from a detector located at a point on the fringe pattern is integrated over a number of equal segments covering one period of the sinusoidal output signal. Alternatively, the optical path difference between the interfering wave fronts is changed in equal steps and the corresponding values of the intensity are measured. Three measurements at each point provide enough data to calculate the original phase difference between the wave fronts. Since a photodiode array or a charge-coupled detector array can be used to implement this technique, measurements can be made simultaneously at a very large number of points covering the interference pattern.

The simplest way to generate the phase shifts or phase steps is by mounting one of the mirrors of the interferometer on a piezoelectric transducer to which appropriate voltages are applied. Another way is to use a semiconductor laser whose output wavelength can be changed by varying the injection current. If the optical path difference between the two arms of the interferometer is D, a wavelength change $\Delta\lambda$ results in the introduction of an additional phase difference between the beams,

$$\Delta\varphi \approx 2\pi D \Delta\lambda / \lambda^2 \qquad (2)$$

Figure 2 shows a three-dimensional plot of the errors of a surface produced by an interferometer with a digital phase measurement system. Because of their speed and accuracy such interferometers are now used extensively in the production of high-precision optical components.

B. Tests of Aspheric Surfaces

Many optical systems now use aspheric surfaces. The simplest way of testing such a surface is to generate a table of wave front data giving the theoretical deviations of the wave front from the best-fit sphere and to subtract these values from the corresponding measurements. Surfaces with large deviations from a sphere can be tested either by using long-wavelength (infrared) light or by recording phase data with two wavelengths. These are used to calculate the phase differences between adjacent data points corresponding to a longer synthetic wavelength. The profile of the surface can then be obtained by integrating these differences.

Surfaces with large deviations from a sphere can also be tested with a shearing interferometer in which the interference pattern is produced by superposing different portions of the test wave front. In a lateral shearing interferometer two images of the test wave front are superposed with a small mutual lateral displacement. For a small shear the interference pattern corresponds to the derivative of the wave-front errors and the deviations to be measured are considerably smaller than the errors themselves. Evaluation of the wave-front aberrations is easier with a radial shearing interferometer in which interference takes place between two images of the test wave front of slightly different sizes.

Surfaces with very large deviations from a sphere are best tested with a suitably designed null lens, which converts the wave front leaving the surface under test into an approximately spherical wave front, or with a computer-generated hologram (CGH). Figure 3 is a schematic of a setup using a CGH in conjunction with a Twyman–Green interferometer to test an aspherical mirror. The CGH resembles the interference pattern formed by the wave front from an aspheric surface with the specified profile and a tilted plane wave front and is positioned so that the mirror under test is imaged on to it. The deviation of the surface under test from its specified shape is then given by the moiré pattern formed by the actual interference fringes and the CGH, which is isolated by means of a small aperture placed in the focal plane of the imaging lens.

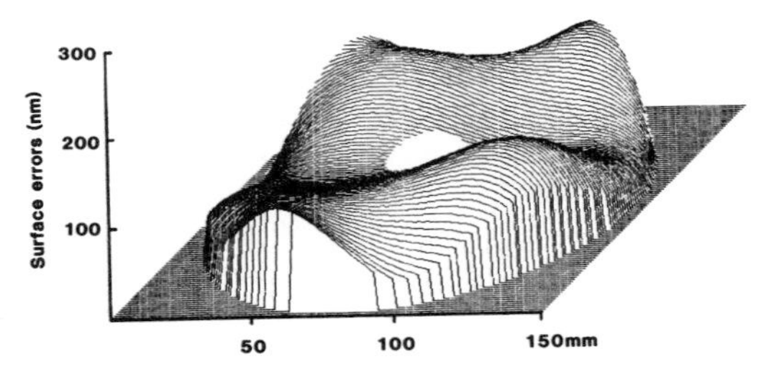

FIG. 2. Three-dimensional plot of the residual errors of a concave mirror obtained with a digital phase-measuring interferometer.

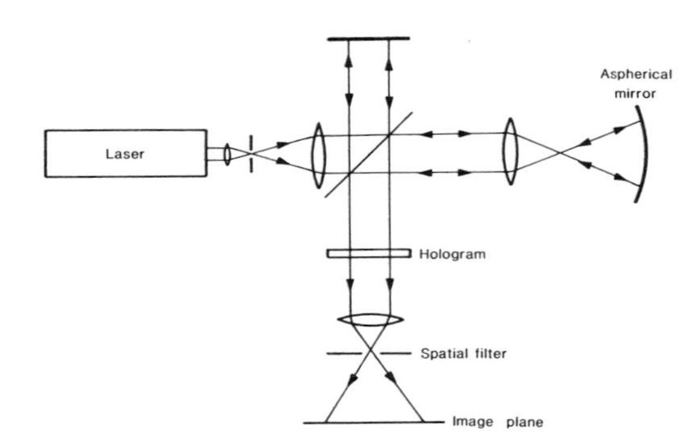

FIG. 3. Interferometer using a computer-generated hologram to test an aspheric mirror. [From J. C. Wyant and V. P. Bennett (1972). *Appl. Opt.* **11**(12), 2833–2839.]

V. Studies of Refractive Index Fields

A significant field of application of optical interferometry has been in studies of diffusion, fluid flow, combustion, and plasmas, where changes in the refractive index can be related to changes in pressure, temperature, or relative concentration of the different components.

The Mach–Zehnder interferometer is commonly used for such studies. It has several advantages for such work: the separation of the two beams can be made as large as desired, the test section is traversed only once, and fringes localized in a plane in the test section can be obtained with an extended incoherent source such as a flash lamp.

Measurements of changes in the optical path difference can now be made extremely rapidly to better than 0.01 wavelength by heterodyne techniques, using either an image–dissector camera to scan the interference pattern, or an array of detectors coupled to individual phase-to-voltage converters.

VI. Holographic and Speckle Interferometry

Holography makes it possible to use interferometry for measurements on objects with rough surfaces. Holographic interferometry is now a powerful tool for nondestructive testing and strain analysis.

Initially, a hologram of the object is recorded by illuminating it with a laser and allowing the light reflected by the object to fall on a high-resolution photographic plate along with a reference beam from the same laser. When the processed hologram is replaced in exactly the same position and illuminated with the same reference beam it reconstructs an image that is superimposed exactly on the object. If a stress is applied to the object, interference between the wave front reconstructed by the hologram and the wave front from the deformed object gives rise to fringes that contour the changes in shape of the object. Weak spots and defects are revealed by local changes in the fringe pattern.

Very accurate measurements of the optical path differences in the interference pattern can be made by the digital phase-stepping technique. The data from three or more such measurements made with different directions of illumination can then be processed to obtain the surface displacements and the principal strains.

Holographic interferometry can also be used for measurements on vibrating objects. One method (time-average holographic interferometry) involves recording a hologram with an exposure long compared to the period of vibration. The reconstructed image is then covered with fringes that can be used to map the vibration amplitude. More accurate measurements can be made using stroboscopic illumination in conjunction with the phase-stepping technique.

A faster, though less accurate, technique for such measurements is speckle interferometry, which involves recording the interference pattern formed between the speckled image of the object when it is illuminated with a laser and a reference beam from the same source. Any change in the shape of the object results in a change in the optical path difference between the two wave fronts and a consequent change in the intensity distribution in the speckled image. Two such speckled images can be recorded electronically and their difference extracted to give fringes similar to those obtained by holographic interferometry.

The digital phase-stepping technique can also be used with electronic speckle pattern interferometry. In this case each speckle is treated as an individual interference pattern, the light from the object having a particular amplitude and phase. If the optical path difference at each such point with respect to the reference beam is measured by the phase-stepping technique before and after the surface moves, the change gives a direct measure of the surface displacement at that point.

VII. Interference Microscopy

An important application of optical interferometry is in microscopy. Interference microscopy provides a noncontact method for studies of surface structure when stylus profiling cannot be used because of the risk of damage. In the Mirau interferometer shown in Fig. 4, light from the microscope illuminator is incident, through the objective, on a semitransparent

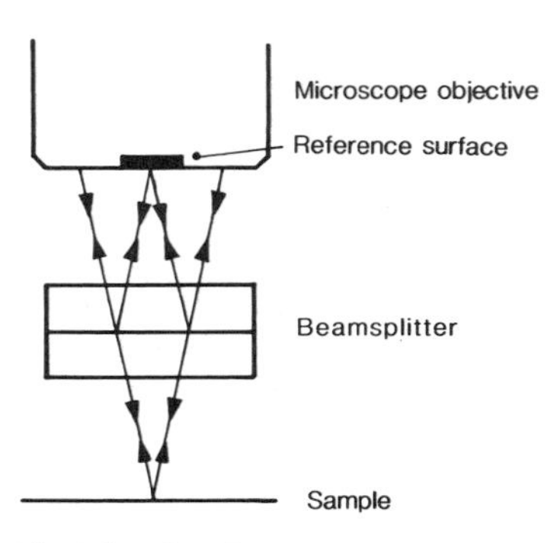

FIG. 4. The Mirau interferometer.

mirror. The transmitted beam goes to the test surface, while the reflected beam goes to a reference surface. These two beams are recombined at the same semitransparent mirror and return through the objective. The interference pattern formed in the image plane contours the deviations from flatness of the test surface. Very accurate measurements of surface profiles and estimates of roughness can be made using the digital phase-shifting technique.

Another application of interference microscopy is for studies of transparent living cells that cannot be stained without damaging them. The Nomarski interferometer, which is commonly used for such work, is a shearing interferometer that uses two Wollaston (polarizing) prisms to split and recombine the beams. Two methods of observation are possible. With small isolated objects it is convenient to use a lateral shear larger than the dimensions of the object. Two images of the object are then seen, covered with fringes that contour the phase changes due to the object. With an extended object the shear is made much smaller than the dimensions of the object. The interference pattern then shows the phase gradient.

VIII. Interferometric Sensors

It is possible to set up interferometers in which the two paths are single-mode optical fibers. Since the optical path length in a fiber changes when it is stretched and is also affected by its temperature, fiber interferometers can be used as sensors for several physical quantities. It is possible to have very long noise-free paths in a small space, so that high sensitivity can be obtained. Figure 5 shows a typical optical setup using optical fiber couplers to divide and recombine the beams and a fiber stretcher to modulate the phase of the reference beam. The output is picked up by a photodetector and measurements are made with either a heterodyne system or a phase-compensating system. Detection schemes involving either a modulated laser source or laser frequency switching have also been used.

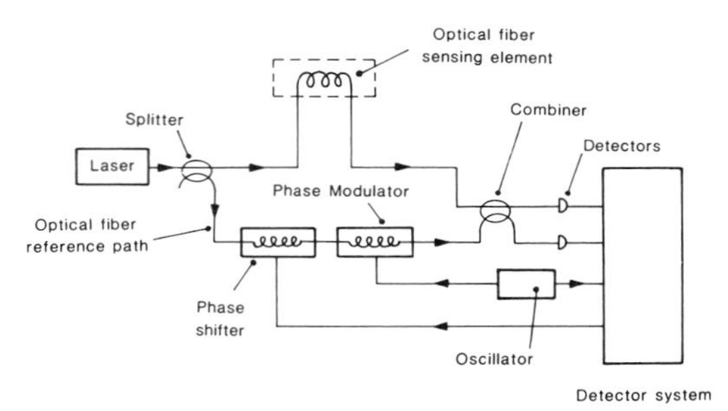

FIG. 5. Fiber-optic interferometric sensor. [From T. G. Giallorenzi, J. A. Bucaro, A. Dandridge, G. H. Sigel Jr., J. H. Cole, S. C. Rashleigh, and R. G. Priest (1982). *IEEE J. Quant. Electron.* **QE-18**(4), 626–665. © 1982 IEEE. Reproduced with permission.]

Fiber interferometers have been used as sensors for mechanical strains and changes in pressure and temperature. They can also be used for measurements of magnetic and electric fields by bonding the fiber to a suitable magnetostrictive or piezoelectric element.

Another application of fiber interferometers has been in rotation sensing where they have the advantages over gyroscopes of instantaneous response, very small size, and relatively low cost. In this case the two waves traverse a closed multiturn loop made of a single optical fiber in opposite directions. If the loop is rotating with an angular velocity ω about an axis making an angle θ with the normal to the plane of the loop, the phase difference introduced between the two waves is

$$\Delta\varphi = (4\pi\omega LR \cos\theta)/\lambda c \qquad (3)$$

where L is the length of the fiber, R is the radius of the loop, λ is the wavelength, and c is the speed of light.

IX. Stellar Interferometers

Even the largest stars have angular diameters of about 0.01 arcsec which is well below the resolution limit of the largest telescopes. However, since a star can be considered as an incoherent circular source, its angular diameter can be calculated from the coherence of the light received from it, which in turn can be obtained from measurements of the visibility of the interference fringes formed by light collected from two points at the ends of a long horizontal base line. For a uniform circular source of angular diameter α, the visibility of the fringes is

$$V = 2J_1 \frac{(\pi\alpha B/\lambda)}{(\pi\alpha B/\lambda)} \qquad (4)$$

where B is the length of the base line and λ is the wavelength; the visibility of the fringes falls to zero when $B = 1.22\ \lambda/\alpha$.

The first stellar interferometer, which was built by Michelson in 1921, used two mirrors whose spacing could be varied, mounted on a 6-m-long support on the 2.5-m telescope at Mt. Wilson, California. The beams reflected by these mirrors were reflected by two other mirrors to the main telescope mirror that brought them to a common focus, at which the interference fringes were formed.

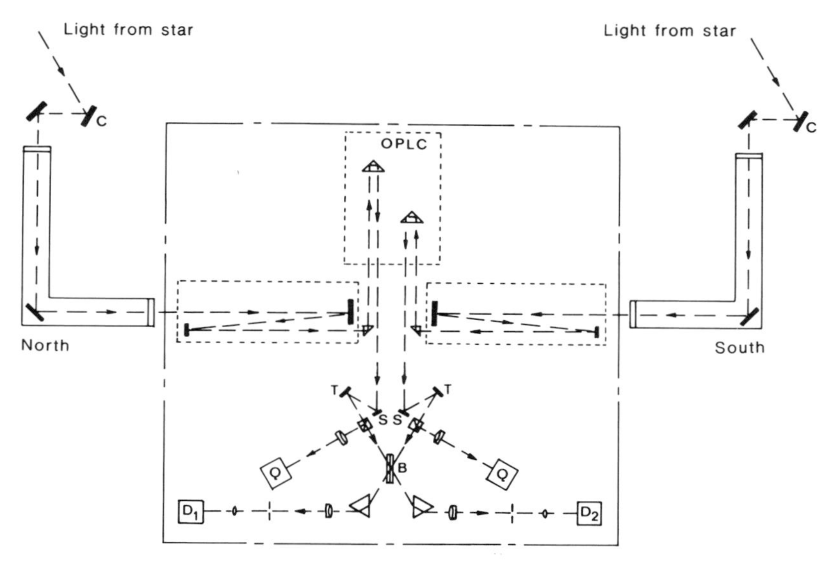

FIG. 6. A modern stellar interferometer. (Courtesy J. Davis, University of Sydney.)

Several practical problems limited the length of the base line that could be used with this interferometer, the most important being lack of stability of the fringes due to rapid random changes in the two optical paths caused by atmospheric turbulence. Modern electronic techniques have now overcome these problems.

Figure 6 shows the optical system of a stellar interferometer designed to make measurements over base lines up to 1 km. Two coelostats (C) at the ends of the north–south base line send light via a system (OPLC) that equalizes the two optical paths to the beamsplitter (B) where they are combined. Two piezoelectric-actuated tilting mirrors (T) controlled by two quadrant detectors (Q) ensure that the two images of the star are exactly superimposed. Two detectors (D_1 and D_2) measure the total flux in a narrow spectral band in the two interference patterns over a sampling interval of a few seconds. During the next sampling interval an additional phase difference of 90° is introduced between the two beams by two mirrors (S) mounted on piezoelectric translators. The visibility of the fringes can then be obtained from the average value of the square of the difference between the signals from the two detectors.

Bibliography

Born, M., and Wolf, E. (1980). "Principles of Optics." Pergamon, Oxford.

Culshaw, B. (1984). "Optical Fiber Sensing and Signal Processing." Peregrinus, London.

Durst, F., Melling, A., and Whitelaw, J. H. (1976). "Principles and Practice of Laser–Doppler Anemometry." Academic Press, London.

Françon, M., and Mallick, S. (1971). "Polarization Interferometers: Applications in Microscopy and Macroscopy." Wiley (Interscience), New York.

Hariharan, P. (1985). "Optical Interferometry." Academic Press, San Diego.

Hariharan, P. (1987). Interferometry with lasers, *in* "Progress in Optics" (E. Wolf, ed.) Vol. 24. North-Holland, Amsterdam.

Hecht, E., and Zajac, A. (1987). "Optics." Addison-Wesley, Reading, Massachusetts.

Malacara, D. (1978). "Optical Shop Testing." Wiley, New York.

Steel, W. H. (1983). "Interferometry." Cambridge Univ. Press, London and New York.

Vest, C. M. (1979). "Holographic Interferometry." Wiley, New York.

LASER COOLING AND TRAPPING OF ATOMS

John E. Bjorkholm *AT&T Bell Laboratories*

Glossary

Dipole force: Force exerted on an atom due to the stimulated scattering of light by the atom. Also called the stimulated force or the gradient force.

Laser cooling: Using laser radiation pressure forces to extract kinetic energy from atomic motion.

Laser radiation pressure: Radiation pressure exerted by a laser beam. Because laser beams are very bright, laser radiation pressure forces are generally much larger than the radiation pressure forces exerted by beams of incoherent light.

Optical molasses: Three-dimensional configuration of laser beams used to cool atoms to very low temperatures (on the order of 10^{-4} K).

Optical trapping: Using laser radiation pressure forces to confine an atom, or a group of atoms, within a small region of space.

Quantum heating: Heating of atomic motion caused by the fluctuations of the radiation pressure forces around their average values.

Radiation pressure: Forces exerted on a body due to the generalized scattering of light by that body.

Resonance-radiation pressure: Radiation pressure exerted on an atom by light tuned close to the frequency of one of the atom's resonance transitions.

Spontaneous force: Force exerted on an atom due to the spontaneous scattering of light by the atom. Also called the scattering force.

Laser light can exert significant forces on a free atom when the light frequency is tuned to be close to, or equal to, the frequency of one of the atom's resonance-absorption lines. These resonance-radiation pressure forces have found unique application with atomic beams. Most noteworthy, gaseous collections of atoms in vacuum have been cooled to ultralow temperatures (as low as 10^{-4} K); also, such ultracold atoms have been optically trapped within a small volume of space for appreciable periods of time (many seconds). When viewed in hindsight, 1987 may well be seen as the year when the study of laser radiation pressure on atoms came of age. In previous years efforts were mainly devoted to understanding the forces in detail and to learning how to use them. In the years ahead most of the effort will be focused on using these techniques to help in carrying out measurements of a more fundamental nature. The purpose of this article is to discuss the basic forces exerted on atoms by resonance-radiation pressure and to describe some of the new optical techniques that can be used to manipulate atoms in ways not heretofore possible.

I. Introduction

It is well known that a beam of light carries momentum and that it will exert forces on objects that it illuminates. These forces arise out of the generalized scattering of the light by the object. Before the invention of the laser, however, applications of radiation pressure were virtually nonexistent. This was because the light emitted by conventional incoherent light sources is neither intense enough nor spectrally narrow enough to cause large effects. Nonetheless, in 1933 O. Frisch was able to demonstrate the transverse deflection of some of the atoms in an atomic beam of sodium caused by the light from a resonance lamp. The deflection observed was very small, about 3×10^{-5} rad, and was caused by the absorption and re-emission of a single photon by each deflected atom! The invention of the laser, with its intense and highly directional light beams, dramatically changed the situation. Now it was possible for an atom or other body to interact with a large number of photons in a short period of time. This was made clear in 1970 when A. Ashkin used a laser beam to accelerate, trap, and manipulate micrometer-sized, transparent dielectric spheres suspended in water. He further pointed out that significant radiation pressure forces similarly could be exerted on an atom if the frequency of the light was tuned near the frequency of one of the atomic resonance transitions. This was an attractive idea since neutral atoms are not easily manipulated using conventional techniques. Ashkin's seminal paper started worldwide thinking about laser radiation pressure on atoms. Experiments demonstrating these forces followed less rapidly. As laser radiation pressure became better understood throughout the 1970s, new applications became apparent. In 1975 it was realized that light pressure could be used to cool atoms to very low temperatures and in 1978 Ashkin proposed a particularly simple optical trap for atoms. It was not until the mid-1980s that some of these interesting possibilities were actually demonstrated experimentally. The delay between initial conception and eventual realization was caused by the need to develop the complex techniques and equipment required for the experiments. The first demonstrations of the basic forces and of the optical cooling and trapping of atoms utilized sodium atoms and precisely tuneable, cw dye lasers operating near the 589-nm resonance line of sodium. The lasers and the associated apparatus were quite complex and expensive. Recent advances have led to great simplification. In particular, since 1986 relatively simple and inexpensive cw GaAs diode lasers have been used to cool and trap cesium atoms, using the cesium resonance line at 852 nm.

II. Radiation Pressure Forces

The forces of laser radiation pressure are of two types. The first is usually referred to as the spontaneous force, but is sometimes called the scattering force. The second is variously referred to as the dipole force, the stimulated force, or the gradient force. A complete description of these forces is complex since it must account for the quantized nature of light. That is, an atom scatters only one photon at a time and this fact leads to statistical fluctuations of the forces in direction and in time. These quantum fluctuations tend to heat the atomic motion and are a limiting factor in many applications of radiation pressure. A complete description of these fluctuations is beyond the scope of this article, but some of the consequences of these fluctuations will be discussed.

A. Spontaneous Force

The spontaneous force of radiation pressure arises because of the spontaneous scattering of nearly resonant light by the atom. It is most easily understood using the photon picture of light. Consider Fig. 1 which shows an atom illuminated by a traveling-wave light beam of frequency $\nu = c/\lambda$, where c is the speed of light and λ is the light wavelength. Because the light is tuned to be nearly resonant with the atom, the atom occasionally absorbs a photon from the beam and makes the transition to its excited state. In absorbing the photon the atom picks up the photon momentum h/λ. After a short time the atom decays back to its ground state and, in the process, re-emits a photon of frequency ν in some random direction, as shown by the out-going wavey lines in the figure. When averaged over many scattering events, the out-going photons carry away no momentum since the scattering distribution is symmetric. Thus F_s, the average force exerted on the atom due to spontaneous scattering, is in the direction of the light propagation. Its magnitude is the rate at which momentum is absorbed from the incoming photons and is given by

$$F_s = (h/\lambda)(f/\tau)$$

where f is the probability that the atom is in its excited state and τ is the excited-state lifetime. To proceed further, we make the simplifying assumption that the atom can be described by the two-level model. This model assumes that the atom has only two energy levels, those of the ground and excited states. While this model is usually overly simplistic, its use makes a discussion of the basic physics much

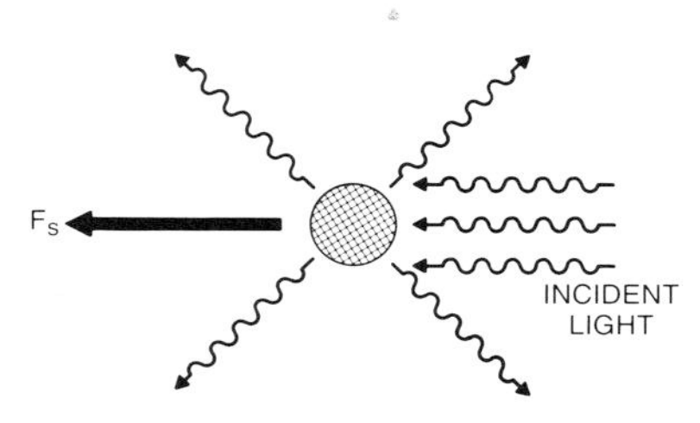

FIG. 1. Schematic diagram showing an atom being illuminated by a beam of nearly resonant light. The out-going wavey lines indicate photons spontaneously scattered by the atom and the vector F_S denotes the resulting average spontaneous force exerted on the atom.

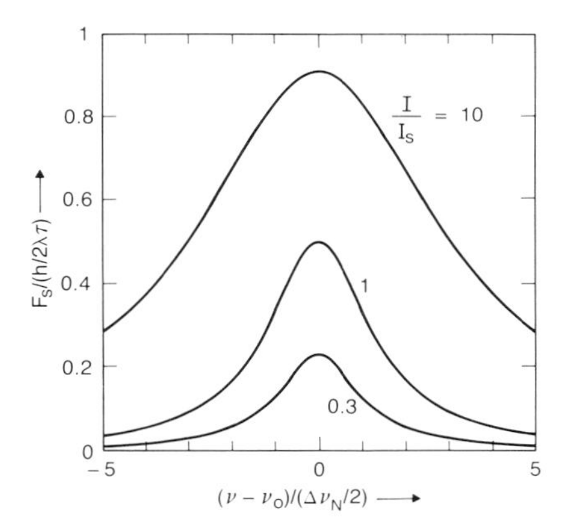

FIG. 2. The normalized spontaneous force as a function of the normalized light frequency for three values of I/I_S.

easier. The model is widely used, but rarely can it be applied to actual situations without at least some modification. For the two-level atom,

$$f = p/2(1 + p)$$

where p is the saturation parameter and is given by

$$p = (I/I_s)/(1 + q^2)$$

In this equation, I is the light intensity, I_s the saturation intensity for the transition, and q the normalized detuning given by $q = 2(\nu - \nu_0)/\Delta\nu_N$ where ν_0 is the resonant frequency of the transition and $\Delta\nu_N = 1/2\pi\tau$ its natural linewidth (FWHM). Figure 2 shows the dependence of F_S on detuning of the light from resonance for several values of I/I_s. The frequency dependence of the force reflects the atomic absorption line shape. For $p << 1$ the line shape is Lorentzian. The natural linewidth is usually quite narrow, $\Delta\nu_N = 10$ MHz for the sodium atom and 5.2 MHz for cesium; this is why precisely tuneable lasers are required to exert significant spontaneous forces on atoms. When $p >> 1$ the line exhibits broadening and the maximum force saturates. The maximum force is small but nonetheless significant; for sodium and cesium, the corresponding accelerations amount to roughly 10^5 and 6×10^4 times the acceleration of gravity, respectively. [*See* LASERS.]

A number of experiments have demonstrated or have used the spontaneous force on neutral atoms of sodium and cesium, which are alkali metals. Alkali atoms are particularly useful since they have a single electron in their outermost shell and a correspondingly simple atomic energy level structure. Atoms with more than one outermost electron have more complicated level structure and this leads to experimental difficulties. Nonetheless, spontaneous forces have been exerted on excited neon atoms in which a metastable excited level functioned as an effective ground state. Spontaneous forces have also been utilized on singly ionized atoms of magnesium and barium contained in ion traps. These atoms have two electrons in their outermost shell, so the singly ionized species have a reasonably simple "alkali-metal-like" level structure.

B. Fluctuations of the Spontaneous Force

The force F_s is the net force averaged over many spontaneous scattering events. Because an atom scatters only one photon at a time, and because each photon is scattered into a random direction, there are fluctuations of the instantaneous force around the average. These fluctuations cause the atomic motion to contain a significant random component, which grows with time. This heating of the atomic motion is described by a so-called momentum diffusion coefficient, D_p. The rate at which W, the kinetic energy associated with the random motion, grows is given by $dW/dt = D_p/m$, where m is the atomic mass. For a uniform, traveling, plane wave and in the limit $p << 1$, this expression becomes

$$dW/dt = (m/2)(h/m\lambda)^2(p/\tau)$$

The quantity $h/m\lambda$ is the speed of atomic recoil due to the absorption of a single photon; for the sodium atom it is about 3 cm/sec and for cesium it is 0.35 cm/sec. For situations in which p is large or when the light field contains intensity gradients the expression for D_p becomes much more complex. Counteracting quantum heating is crucial for the cooling and trapping of atoms.

C. Dipole Force

The dipole force of resonance-radiation pressure is most easily understood using the wave picture of light. In this picture it is simply the force exerted on an induced dipole situated in an electric field gradient. It can also be viewed as arising from the stimulated scattering of light by the atom. The average dipole force can be written as

$$\overline{F}_d = (4\pi/c)\alpha\,\overline{\nabla}I$$

where α is the atomic polarizability and I is the light intensity. For the idealized two-level atom,

$$\alpha = -\frac{1}{2}\frac{(\lambda/2\pi)^3 q}{(1 + q^2)(1 + p)}$$

Several characteristics of this force should be stressed. First, it exists only when there is a gradient of the light intensity. Second, the dipole force has no

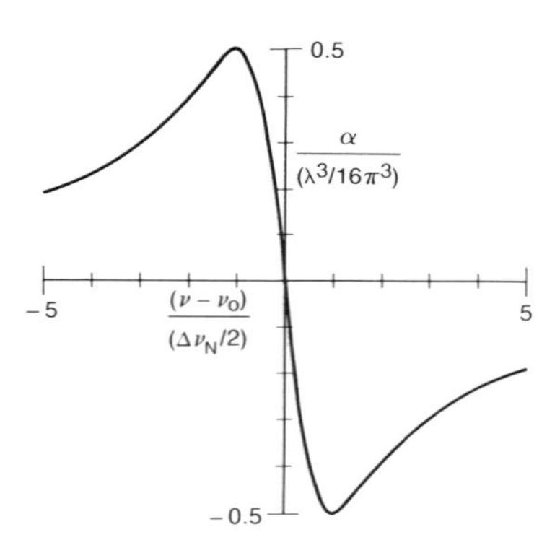

FIG. 3. The normalized atomic polarizability as a function of the normalized light frequency for $I/I_S << 1$.

upper limit; importantly, it can be much larger than the spontaneous force. Finally, the frequency dependence of this force is dispersive in character, as shown in Fig. 3. The force is zero for $\nu = \nu_0$; for $\nu < \nu_0$, the direction of the force is such as to pull the atom into the high-intensity regions of the light beam; for $\nu > \nu_0$, the atom is pushed away from the intense regions. Notice also that the force can be large even for $q >> 1$.

It is often convenient to think of the dipole force as being derivable from a conservative optical potential $U(\bar{r})$, which is given by

$$U(\bar{r}) = (h\,\Delta\nu_N/4)q \ln[1 + p(\bar{r})]$$

This potential energy is the same as the shift in energy of the atomic ground state caused by the optical Stark effect. Thus, whenever dipole forces are exerted on an atom, there will also be optical Stark shifts of the atomic energy levels. Because these level shifts can often be large compared with $\Delta\nu_N$, it is usually difficult to effectively apply the spontaneous force to an atom that is simultaneously subjected to a dipole force.

D. Fluctuations of the Dipole Force

Due to the quantum nature of light there are also random fluctuations of the dipole force about its average. These fluctuations are much more difficult to describe and understand than are those of the spontaneous force. We will not consider them here, but it must be realized that the quantum heating caused by these fluctuations can be very large, often very much greater than that due to spontaneous fluctuations. The dipole force and its fluctuations are proportional to the gradient of the light intensity. Thus dipole force quantum heating can be exceptionally large in light fields that have large intensity gradients, as in a standing-wave field.

E. Early Demonstrations of the Forces

Early demonstrations of the spontaneous force were made by illuminating a sodium atomic beam at normal incidence with the light from a cw dye laser tuned onto the atomic resonance. The spontaneous force caused substantial deflection of the atoms. Deflection angles as large as 5×10^{-3} rad were observed, corresponding to the scattering of about 200 photons and an acquired transverse velocity of 600 cm/sec. The maximum deflections obtained were limited by the Doppler shifts associated with the transverse speed. That is, as the atoms acquired transverse speed, they were Doppler shifted out of resonance with the light and the force was greatly diminished. The transverse speed of 600 cm/sec corresponds to a Doppler shift of 10 MHz, the full natural linewidth of sodium.

The first demonstration of the dipole force was made around 1980 by superimposing a copropagating cw dye laser beam on top of an atomic beam of sodium atoms. The laser beam had a Gaussian intensity profile (TEM_{00} mode) and was tuned several gigahertz away from the atomic resonance. Because of the intensity gradients and the cylindrical symmetry of the illumination, transverse dipole forces were exerted on the atoms. For tunings below resonance the forces were such as to pull the atoms to the axis of the laser beam; in other words, the light exerted focusing, confining forces on the atoms. For tunings above resonance the light forces were opposite and brought about defocusing effects on the atoms. The dramatic changes in the atomic beam profile caused by these forces are shown in Fig. 4. Because the light was tuned far from resonance, the effects caused by the average spontaneous force were small. Significantly, however, it was demonstrated that the size of the spot to which the atomic beam could be focused was determined by the transverse heating of the atomic motion caused by the fluctuations of the spontaneous force. While not yet demonstrated, it should be possible to focus atomic beams to very small spot sizes by using a TEM_{01}, or donut-mode, laser beam tuned above the atomic resonance. In this case the atoms would tend to be concentrated on the laser beam axis where the light intensity is lowest and where the spontaneous heating is minimized.

In a generalized sense these experiments demonstrated that laser beams can be used to manipulate atomic motion in useful ways. It is expected that the use of lasers to modify and control atomic motion

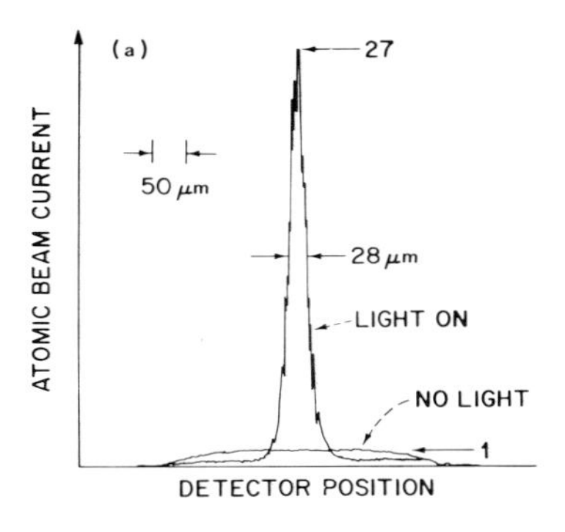

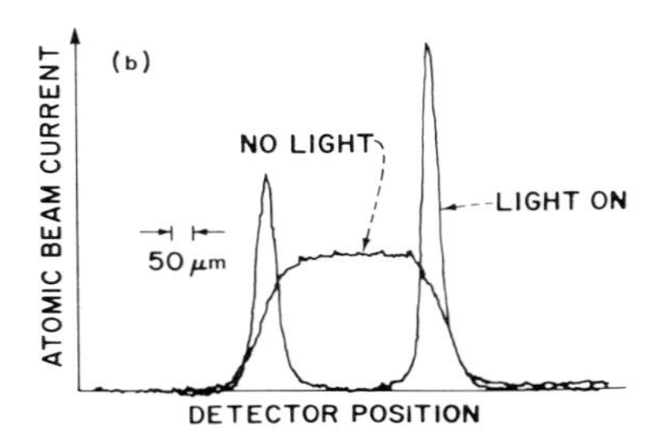

FIG. 4. Focusing and defocusing of an atomic beam by a superimposed, Gaussian laser beam. The figures show the atomic beam current density as a function of position, with the light on and off. In (a) the light frequency is less than the atomic resonance frequency and the atomic beam is focused by the light. In (b) the light is tuned above the atomic resonance and defocusing takes place.

will become a useful technique in atomic beams work.

III. The Single-Beam, Dipole-Force Optical Trap

The first optical trap to be demonstrated is deceptively simple to describe. As shown in Fig. 5, this trap is formed by a sharply focused Gaussian-mode

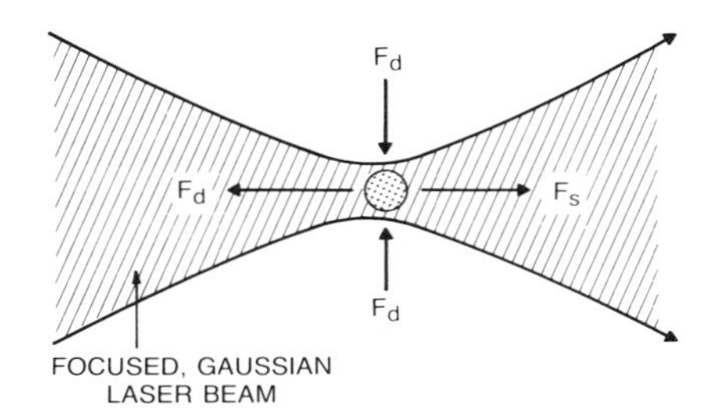

FIG. 5. Schematic diagram of the single-beam, dipole-force optical trap. There is a point of stable equilibrium for the atom just beyond the focus of the light beam. The dipole and spontaneous forces exerted on the atom are shown.

laser beam tuned far below the atomic resonance ($q << -1$). The intensity of a Gaussian (or TEM_{00}-mode) laser beam propagating along the z axis (the longitudinal direction) and focused at $z = 0$ is given by

$$I(r, z) = (2P/\pi w^2) \exp(-2r^2/w^2)$$

where $w(z) = w_0[1 + (\lambda z/\pi w_0^2)^2]^{1/2}$, w_0 is the focal spot size at $z = 0$, and P is the power in the laser beam. As shown in Fig. 5, there is a point of stable equilibrium for the atom just beyond the focus. Transverse confinement of the atom is provided by the transverse dipole forces. Longitudinal confinement is brought about by balancing the longitudinal spontaneous force with the longitudinal dipole force that exists because of the strong longitudinal gradients of the intensity (strong focusing). It is useful to consider the depth of the trap for parameters comparable to those used in the first demonstration of optical trapping. In those experiments, which trapped the sodium atom, the laser power P was 200 mW, the focal spot size w_0 was 10 μm, and the longitudinal trap depth was maximized by detuning the laser -150 GHz from the resonance (corresponding to $q = -3 \times 10^4$). For these parameters, and approximating the sodium atom with the two-level model, the transverse well depth is found to be equivalent to an atomic temperature of 25 mK and the longitudinal trap depth equivalent to 15 mK. For reasons to be explained later, the trap depths for the actual experiment were only about 40% of the above values.

From this example it now can be readily understood why it took a number of years to develop the techniques required to demonstrate an optical trap. The problems that needed to be confronted and solved were as follows. First, since the trap is very shallow, a collection of ultracold atoms ($T <<$ 1 mK) is needed to load it. No source of such cold atoms existed prior to 1985. Second, the trap volume is very small, on the order of 10^{-7} cm^3. Thus, to efficiently load the trap, the collection of cold atoms must also be dense. Finally, even if cold atoms could be loaded into the trap, the quantum fluctuations of the forces would rapidly heat them. Without an effective means for counteracting this heating, the initially cold atoms would "boil" out of the trap about 10 msec after being loaded. Thus, before an optical trap could be demonstrated it was necessary to develop techniques for cooling atoms to ultralow temperatures and for keeping them cold in the presence of quantum heating. As will shortly be described, the solution to these problems is a technique called "optical molasses," which was first demonstrated in 1985.

IV. Laser Cooling of Atoms

In this section we will discuss the techniques that have been used to cool atoms to temperatures low enough for placing them in optical traps.

A. Atomic Beam Slowing

The seemingly most straightforward way to slow, or cool, the atoms in an atomic beam is to let them propagate against a light beam tuned to the atomic resonance. The deceleration corresponding to the maximum spontaneous force is $h/2m\lambda\tau$; for sodium and cesium, this amounts to -9.15×10^7 cm/sec^2 and -5.7×10^6 cm/sec^2, respectively. Consider a typical thermal atomic beam of sodium atoms, for which the longitudinal velocity distribution peaks at roughly 10^5 cm/sec. In principle, atoms moving at this speed could be brought to rest in about 1 msec, during which time the atoms would travel over a distance of about 46 cm while scattering about 3×10^4 photons. Unfortunately, such efficient deceleration is not straightforward to achieve because of the Doppler shifts of the light frequency that occur as the atom slows down. For instance, a velocity change of only 10^3 cm/sec causes a Doppler shift of about 17 MHz, which is larger than the 10-MHz linewidth of the sodium absorption line. In other words, unless something is done, the atoms quickly shift out of resonance with the light and the deceleration is greatly reduced.

Several experimental techniques have been devised to counteract these Doppler shifts and to keep the atoms in resonance with the light as they slow down. The first technique uses spatially dependent Zeeman shifts of the atomic energy levels to compensate for the changing Doppler shifts. In this method, the atomic beam is directed down the bore of a solenoid having a tapered longitudinal magnetic field. The field varies in such a way that the Zeeman shifts of the atomic energy levels compensate for the changing Doppler shifts as the atoms slow down. In the second technique the frequency of the laser is directly "chirped," or swept in frequency, at a rate appropriate for keeping the atoms in resonance with the light. Both techniques work well and both have been used to bring the mean longitudinal velocity of some of the atoms in an atomic beam to zero.

These "stopped" atoms, however, are not cold enough for use in optical trap experiments because of the quantum heating that occurs as they are slowed down. Considering sodium once again, an atom starting with a longitudinal velocity of 10^5 cm/sec scatters 3×10^4 photons in being stopped. Because of the random scattering of the emitted photons, there is a spread in the velocity distribution along each axis given approximately by $(h/m\lambda)\sqrt{N/3}$, where N is the number of photons scattered in bringing the atom to rest. For sodium this amounts to roughly 300 cm/sec; the corresponding atomic temperature is roughly 75 mK. In order to achieve optical trapping, even colder atoms are required.

Atoms having temperatures of this magnitude have found other applications. For example, the first trapping of neutral atoms was demonstrated in 1985 using a magnetic trap that had a potential well depth on the order of 5 K. This trap was filled using cold atoms obtained using the Zeeman slowing technique. Techniques similar to the chirping technique have also been used to cool hot atoms confined in ion traps that have well depths on the order of 10 eV (corresponding to a temperature of about 2×10^5 K). Ions in traps have been cooled to temperatures of about 10 mK.

B. Optical Molasses

"Optical molasses" is a technique that uses the spontaneous force to rapidly cool already cold atoms to much lower temperatures.

The basic idea behind optical molasses is easily described for one dimension. Consider an atom illuminated from opposite directions by two laser beams of equal intensity traveling along the x axis. The light frequency is tuned to be slightly below the atomic resonance. When the light intensity is low, the net force acting on the atom is simply the sum of the forces of each beam acting alone. When the atom is at rest it sees no net spontaneous force since the forces exerted by the two light beams are equal and opposite. Now let the atom have a velocity component along the x axis. In this situation the frequency of the beam propagating against the atomic motion is Doppler shifted closer to resonance and the force it exerts on the atom increases. The opposite holds for the copropagating light beam. As a result, there is a net average force that opposes the atomic velocity and that is proportional to it. For small velocities it is given by

$$F_x = (8\pi h/\lambda^2)(I/I_s)(q/1 + q^2)v_x = -\beta v_x$$

where I is the intensity of each beam. This expression is valid as long as the Doppler shifts are small compared with $\Delta\nu_N$; for sodium this corresponds to velocities less than 150 cm/sec. This force appears as a viscous damping force to the atom. Damping is maximized for $q = -1$, which corresponds to a detuning a $\Delta\nu_N/2$ below the atomic resonance. An initial velocity exponentially damps to zero with a decay time of m/β. For the sodium atom, $q = -1$, and $I/I_s = 0.1$, this decay time is 16 μsec; thus the damping is seen to be quite strong.

Optical molasses is the extension of the above ideas to three dimensions, using three pairs of oppositely propagating laser beams along the three orthogonal axes. A slowly moving atom situated in the mutual intersection of the six laser beams experiences strong three-dimensional viscous damping and its average velocity is rapidly reduced to zero. Because of the quantum fluctuations of the optical forces, the atoms do not actually come to rest. As will be discussed, the atoms execute a random walk motion. Consequently, atoms find it difficult to escape from the optical molasses and they are confined within it for a long time. In spite of containment times approaching 1 sec that have been achieved with sodium, optical molasses is not an optical trap since there are no restoring forces exerted on the atoms.

The long time required for an atom to escape from optical molasses is easily understood in terms of the quantum fluctuations. An atom in optical molasses with an average velocity of zero experiences a velocity "kick" of $h/m\lambda$ in a random direction each time it scatters a photon. Each velocity kick is damped out and the net result is that the atom executes a random walk motion in three dimensions with a step size of $(\lambda/2\pi)(I/I_s)^{-1}$. After a time t, the mean-square deviation of the atom from its starting point is

$$\langle r^2 \rangle = (\lambda/2\pi)^2(I/I_s)^{-2}N$$

where N is the number of scattered photons and is roughly given by $(6I/I_s)(t/\tau)$, for $I/I_s << 1$. Confinement times of about 0.5 sec are obtained using optical molasses with a diameter of 1 cm. More careful analysis yields somewhat shorter confinement times.

While the average atomic velocity in optical molasses is zero, the mean-square atomic velocity is determined by the interplay of the heating caused by the fluctuations of the optical forces and the cooling caused by the average spontaneous force. Equilibrium is established when the rate of heating equals the average cooling rate. A careful analysis, which includes the dipole heating rate due to the standing-wave nature of the light beams, yields an equilibrium temperature for the collection of atoms given by

$$kT_{eq} = m\langle v^2 \rangle = h\Delta\nu_N/2$$

For the sodium atom $T_{eq} = 240$ μK and for cesium it is 100 μK.

Optical molasses was first demonstrated in 1985 using the sodium atom and cw dye lasers. The experiments were carried out using a pulsed atomic beam. Atoms initially traveling with a speed of 2×10^4 cm/sec were decelerated to 2×10^3 cm/sec using a chirped slowing laser beam. These slow atoms were then allowed to drift into the optical molasses region, which was roughly spherical in shape and about 0.5 cm^3 in volume, where final cooling and retention took place. Atomic densities of 10^6 cm^{-3} and retention times of several tenths of a second were obtained. A direct measurement of the temperature of the atoms in optical molasses was made using a time-of-flight technique. It was found that T_{eq} was 240 μK, in agreement with the quantum heating prediction. During 1987 similar, but experimentally simpler, optical molasses experiments were carried out using cw laser diodes and the cesium atom; the atomic temperature achieved was the predicted limit of 100 μK.

The ideas discussed in this section apply only in the low intensity limit, $I < I_s$. In the high intensity limit, $I >> I_s$, the behavior of the optical forces becomes much more complicated and, in some situations, can confound our physical understanding. As an example, for low intensities optical molasses provides damping of the atomic motion for tunings below the atomic resonance and heating for tunings above resonance. In the high intensity limit, the situation is reversed! In this case it is the dipole forces, which exist because of the standing waves in the optical molasses configuration, that do the cooling. This situation is usually referred to as "stimulated molasses." The damping rate can far exceed that for the usual optical molasses, but so does the heating. The equilibrium temperature achieved with stimulated molasses is not as low as that achieved with spontaneous molasses.

V. Optical Trapping of Atoms

A. Single-Beam, Dipole-Force Trap

The successful demonstration of optical molasses in 1985 provided experimentalists with the remaining tools needed to accomplish optical trapping. Optical trapping of atoms was first carried out in 1986 using sodium atoms, cw dye lasers, and the single-beam, dipole-force optical trap described in Section III.

The experiment was carried out by first injecting sodium atoms into optical molasses. After a delay of several milliseconds to allow the atoms to reach equilibrium, the optical trap beam was introduced into the interior of the optical molasses, as shown sche-

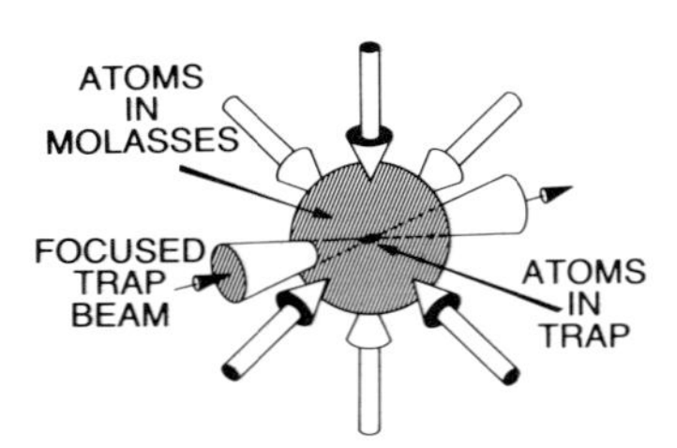

FIG. 6. Schematic diagram of the interaction region used for trapping atoms. The broad arrows represent the collimated laser beams that intersect to form "optical molasses." The shaded sphere represents the fluorescence emitted by the collection of ultracold atoms contained and executing random-walk motion within the optical molasses. The optical trap is formed just beyond the focus of the trap laser beam, which is also shown. The black spot represents the intense fluorescence emitted by the dense collection of atoms confined within the optical trap.

matically in Fig. 6. Trapping was observed as the buildup of a small, but intense, spot of fluorescence situated within the much larger and much weaker cloud of fluorescence from the atoms in the optical molasses (see Fig. 7). The brightness of the small spot indicated that the density of trapped atoms was much higher than the density of atoms in optical molasses. Recall that optical molasses is needed to keep the trapped atoms cool. However, optical molasses is not effective when the trapping light is present because of the large optical Stark shifts associated with the potential well of the trap. Thus, in this experiment it was necessary to alternatively chop on and off the trap and optical molasses beams. Furthermore, it was important that the chopping be done rapidly enough; if too slow, the oscillation of the atoms within the optical potential well becomes unstable and trapping does not occur.

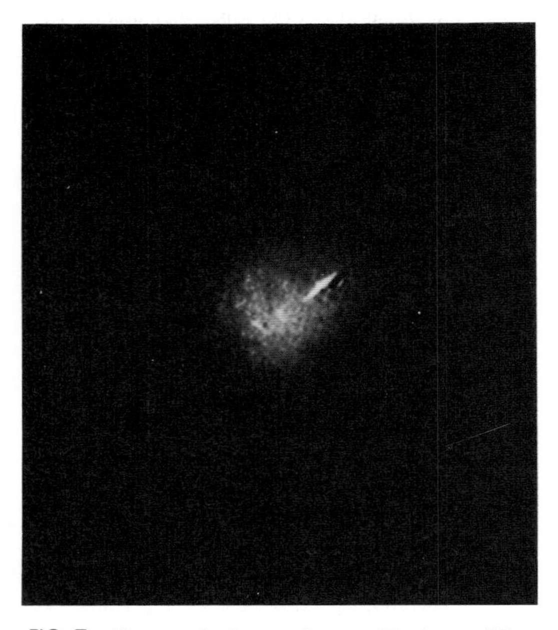

FIG. 7. Photograph of trapped atoms. The large, diffuse cloud is fluorescence emitted by atoms contained within optical molasses. The small, bright spot is the fluorescence from the much higher density of atoms confined within the optical trap.

The optical trap was observed to behave as expected from calculations in all respects but one. The one discrepancy was the observed trap lifetime; it was found only to be about 1 sec, about four orders if magnitude less than expected from simple calculations. It was surmised, and later confirmed, that the short lifetime was determined by the imperfect vacuum in the vacuum chamber (about 10^{-9} Torr). Trapped atoms were ejected from the trap by even very weak collisions with the residual gas atoms, which were at a temperature of about 300 K.

A summary of the operating parameters for the original optical trap and the results obtained with it are as follows. The trap beam was focused within optical molasses to a spot size of 10 μm, its power was 200 mW, and it was tuned approximately 150 GHz below the sodium D1 resonance line. Good results were obtained for chopping periods ranging from about 0.4 to 10 μsec. Under best conditions about 10^3 atoms were trapped within a volume of 10^{-9} cm^3 at a density of 10^{12} cm^{-3}. The atomic temperature was inferred to be about 400 μK.

B. Spontaneous Force Optical Traps

In the early 1980s there was a good deal of speculation about the various forms that optical traps might take. This speculation led to what has come to be called the optical Earnshaw's theorem. Simply put, the theorem states that it is impossible to form a dc optical trap using only forces that are directly proportional to the light intensity. Since the spontaneous force is proportional to I in the low intensity limit, this theorem initially was interpreted as stating that optical traps could not be constructed using only the spontaneous force. However, it is now realized that there are at least two ways to avoid this conclusion. The first is to allow for time dependence of the forces. Thus spontaneous force traps can be constructed in which at least some of the optical beams exhibit time dependence. These traps are sometimes referred to as "optodynamic" traps. Another way around the theorem is to break the assumption that the forces are linearly related to the optical intensity.

Thus saturation of the forces due to high intensities, the optical pumping of real multilevel atoms, and the application of other force fields can all be used to allow dc, spontaneous force traps to be constructed. Similar techniques have been used in ion traps to avoid the consequences of the electrical Earnshaw's theorem.

The first spontaneous force optical trap was demonstrated during 1987 and is referred to as the "magnetic molasses" trap. This trap is constituted by superimposing a simple spherical quadrupole magnetic field on an optical molasses formed using circularly polarized light beams. The magnetic field causes spatially dependent Zeeman shifts of the atomic energy levels. For sodium these Zeeman shifts result in restoring forces that are proportional to the displacement of an atom from the origin (defined by the magnetic field). This trap is much larger (several millimeters diameter) and much deeper (about 500 mK) than the single-beam, dipole-force trap. As a result, it is easier to fill and many more atoms could be trapped. Trap lifetimes were increased by reducing the chamber background pressure. Figure 8 shows an example of the exponential decay of the trap population exhibiting a 1/e-lifetime of 65 sec. Trap lifetimes as long as 100 sec were obtained at a background pressure of 2×10^{-10} Torr. As many as 10^7 sodium atoms were confined in the magnetic molasses trap at densities as high as 10^{11} cm^{-3}.

When the magnetic molasses trap was loaded with a high initial atomic density, the decay with time of the number of trapped atoms was observed to be nonexponential and faster than the exponential decay observed with low initial fills. An example of such an observation is shown in Fig. 9. It was found that the departure from exponential decay was proportional to the square of the density of trapped atoms, indicating that the additional losses were caused by collisions between the ultracold trapped atoms. This finding is of interest for several reasons. First, density-dependent losses such as these are detrimental since they limit the maximum atomic densities that can be achieved in an optical atom trap. This will make it difficult to push a great deal further into a new regime of gas physics in which high atomic densities and ultralow temperatures are simultaneously achieved. On the other hand, it is now possible to observe the effects of the relatively infrequent collisions between slow atoms. In work with thermal atomic beams or gas cells such collisions are swamped out by the much more frequent collisions between fast atoms. Collisions between slow atoms are not well understood and are only beginning to be studied. Recent observations of associative ionization between slow atoms in an optical trap have indicated that the cross section for the process is several orders of magnitude larger than for atoms traveling at typical thermal velocities.

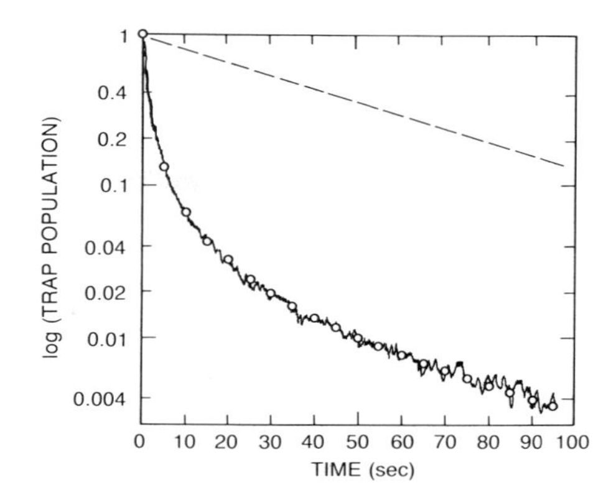

FIG. 9. A logarithmic plot of the nonexponential decay of the atomic population in a magnetic molasses trap observed for a high initial atomic density (about 10^{11} cm^{-3}). The initial decay is 150 times faster than the decay that would be obtained in the absence of density-dependent losses, as shown by the dashed straight line.

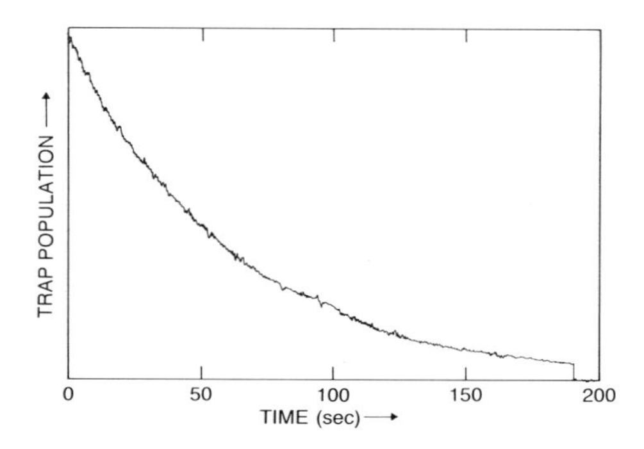

FIG. 8. Time dependence of the atomic population in a "magnetic molasses" optical trap as a function of time. The decay is well described by an exponential curve having a 1/e-lifetime of 65 sec.

VI. Conclusion

The year 1987 is somewhat of a "watershed" year in that the nature of work on laser radiation pressure on atoms is undergoing a subtle change. In the past, efforts were directed towards demonstrating and understanding the basic forces and the ways in which they can be utilized to affect atomic trajectories and to cool and trap atoms. In the future, the emphasis will be placed on using these forces and techniques as experimental tools that will make it possible to carry out measurements of a fundamental nature that had not heretofore been possible. Areas of interest include the study of atomic collisions at slow speeds, collisions between slow atoms and surfaces, the possible observation of collective quantum effects between cold atoms, possible extension of laser radiation pressure to molecules, and precision spectroscopic measurements. Because of these tantalizing prospects, the number of scientists interested in applying laser radiation pressure on atoms is undergoing rapid growth.

Bibliography

Theoretical discussions:

Cook, R. J. (1979). *Phys. Rev. A* **20,** 224.

Dalibard, J., and Cohen-Tannoudji, C. (1985). *J. Opt. Soc. Am. B* **2,** 1707.

Gordon, J. P., and Ashkin, A. (1980). *Phys. Rev. A* **21,** 1606.

Experimental demonstrations:

Bjorkholm, J. E., Freeman, R. R., Ashkin, A., and Pearson, D. B. (1980). *Opt. Lett.* **5,** 111.

Bjorkholm, J. E., Chu, S., Ashkin, A., and Cable, A. (1987). *In* "Advances in Laser Science—II" (Lapp, M., Stwalley, W. C., and Kenney-Wallace, G. A., eds.), *Am. Inst. Phys. Conf. Proc.* **160,** 319. Am. Inst. Phys. New York.

Chu, S., and Wieman, C. (feature eds.) (1989). *J. Opt. Soc. Am. B, Opt. Phys.* **6**(11), 2020.

Chu, S., Hollberg, L. W., Bjorkholm, J. E., Cable, A., and Ashkin, A. (1985) *Phys. Rev. Lett.* **55,** 48.

Chu, S., Bjorkholm, J. E., Ashkin, A., and Cable, A. (1986). *Phys. Rev. Lett.* **57,** 314.

Ertmer, W., Blatt, R., Hall, J. L., and Zhu, M. (1985). *Phys. Rev. Lett.* **54,** 996.

Migdall, A., Prodan, J., Phillips, W. D., Bergeman, T., and Metcalf, H. (1985). *Phys. Rev. Lett.* **54,** 2596.

Phillips, W. D., and Metcalf, H. J. (1982). *Phys. Rev. Lett.* **48,** 596.

Phillips, W. D., and Metcalf, H. J. (1987). *Sci. Am.*, March, p. 50.

Prodan, J., Migdall, A., Phillips, W. D., So, I., Metcalf, H., and Dalibard, J. (1985). *Phys. Rev. Lett.* **54,** 992.

Raab, E. L., Prentiss, M. G., Cable, A., Chu, S., and Pritchard, D. E. (1987). *Phys. Rev. Lett.* **59,** 2631.

Watts, R. N., and Wieman, C. E. (1986). *Opt. Lett.* **11,** 291.

Wineland, D. J., and Itano, W. M. (1987). *Phys. Today*, June, p. 34.

LASER–MATERIALS INTERACTIONS

Michael Bass *University of Southern California*

GLOSSARY

Complex propagation constant: Quantity γ (= $\alpha + j\beta$) used to describe the propagation of light in a general medium; α gives the attenuation and β the wave vector magnitude, both in reciprocal centimeters.

Laser-driven plasma: Optically thick plasma found near a surface heated to the point of vaporization by an intense laser beam; the beam ionizes some of the vapor and through the process of avalanche breakdown creates the plasma.

Nonlinear absorption: Process in which the attenuation coefficient itself is a function of the light intensity.

Photoablation: Process by which short-wavelength laser light breaks molecular bonds in matter and causes material to be ejected from the surface with very little heating of the sample.

Surface electromagnetic wave: Wave formed on an irradiated surface when an irregularity on the surface scatters the light energy; the amount of energy coupled can be significant.

The interaction of laser light with matter is in many ways like the interaction of ordinary light with matter. The difference between them and the interest in laser light arises from the fact that laser light can be much more intense than light from other sources. This property of laser light has opened new research areas in nonlinear optics, laser plasma generation, and new types of materials processing. It has also led to applications such as laser cutting, drilling, welding, and heat treating. The high fluxes possible allow consideration of such futuristic applications as power generation through inertial confinement fusion, X-ray generation for microelectronics, and laser weaponry.

I. Deposition of Light Energy in Matter

A. ABSORPTION

The interaction between light, a form of electromagnetic energy, and materials takes place between the optical electric field and the most nearly free electrons in the medium. In quantum mechanical terms, one speaks of interactions between photons and electrons in the presence of the lattice of nuclei. The lattice must be present to ensure momentum conservation. For the purposes of this article, the classical viewpoint of an electric field causing electrons to move in the presence of damping or frictional forces is sufficient to describe the process of optical absorption.

Conduction band electrons in metals are able to interact with an optical electric field. Their motion is damped by collisions with the vibrating lattice and so some of the light energy is transferred to the lattice. In this manner the material is heated. In semiconductors the motions of both electrons in the conduction band and holes in the valence band must be considered. In dielectrics the electrons are effectively bound to the atoms or molecules that compose the material. The applied optical field induces a polarization in the material. Upon relaxation some of the energy in the polarization is coupled to the lattice and the material is heated. These processes of absorption of energy from an optical field can be treated by classical electromagnetics. That is, Maxwell's equations, the constitutive equations of matter, and the boundary conditions for each material can be solved in each case.

1. Absorption from an Electromagnetic Point of View

a. Propagation in a General Dielectric Material. Let the propagation of the optical electromagnetic wave in a medium be given by the complex propagation constant γ where

$$\gamma = \alpha + j\beta \qquad (1)$$

so that a plane wave electric field is described by

$$E(z, t) = \mathrm{Re}\{E(0) \exp(-\gamma z) \exp[j(\omega t - \beta z)]\} \qquad (2)$$

Here ω is the radian frequency of the field, α the attenuation coefficient, and β the propagation coefficient.

A dielectric is described by the complex dielectric constant

$$\varepsilon = \varepsilon_1 + j\varepsilon_2 \qquad (3)$$

and a magnetic susceptibility μ. This enables us to define a complex index of refraction

$$\eta = n + jk \qquad (4)$$

in which n determines the propagation vector and k determines the attenuation of the wave in the medium. It is easy to show that

$$n = \{[(\varepsilon_1^2 + \varepsilon_2^2)^{1/2} + \varepsilon_1]/2\}^{1/2} \qquad (5)$$

and

$$k = \{[(\varepsilon_1^2 + \varepsilon_2^2)^{1/2} - \varepsilon_1]/2\}^{1/2} \qquad (6)$$

By applying Maxwell's equations and the dielectric boundary conditions as to the case of a lossy dielectric we find that

$$\gamma = j\omega\{(\mu\varepsilon_1)[1 - j(\varepsilon_2/\varepsilon_1)]\}^{1/2} \qquad (7)$$

Note that if we had treated the lossy dielectric as a medium having a finite conductivity σ, we would have obtained

$$\gamma = j\omega\{(\mu\varepsilon)[1 - j(\sigma/\omega\varepsilon)]\}^{1/2} \qquad (8)$$

where ε is the dielectric permittivity. It is clear that Eqs. (7) and (8) are identical with ε replaced by ε_1 and σ/ω by $-\varepsilon_2$.

Lossless dielectrics are media having σ identically zero. The complex propagation constant is then purely imaginary and there is no absorption in the medium.

b. Perfect Metals. A perfect metal is a material having an infinite conductivity or a material in which

$$\alpha/\omega\varepsilon \gg 1$$

In this case we find

$$\alpha = \beta = (\omega\mu\sigma/2)^{1/2} \qquad (9)$$

which means that a field propagating in a metal will be attenuated by a factor of $1/e$ when it has traveled a distance

$$\delta = (2/\omega\mu\sigma)^{1/2} \qquad (10)$$

The quantity δ is called the skin depth and at optical frequencies for most metals it is ~50 nm or about one-tenth of the wavelength of green light. After a light beam has propagated one skin depth into a metal, its intensity is reduced to $1/e^2$ or 0.135 of its value on the surface.

The energy that is no longer in the electromagnetic wave when light is attenuated in a metal drives conduction currents in the so-called skin layer. These in turn produce heating through ohmic losses and the light energy thereby appears in the metal as heat. Heating of the metal at depths below the skin layer takes place by means of thermal conduction.

c. Lossy, Nonconductive Dielectrics. As mentioned above, the results derived for a general dielectric are equivalent to those for a metal when the proper substitutions are made for ε_1 and ε_2. This allows us to consider the case where

$$\varepsilon_2/\varepsilon_1 \gg 1$$

and write the attenuation coefficient for the lossy dielectric as

$$\alpha = \omega(\mu\varepsilon_2/2)^{1/2}$$

$$= (2\pi/\lambda_0)(\mu_R\varepsilon_{2R}/2)^{1/2} \qquad (11)$$

where $(\mu_0\varepsilon_0)^{1/2}$ is recognized as the speed of light in free space, $\omega = 2\pi c/\lambda_0$, and the subscript "R" indicates the relative permittivity and susceptibility. For electromagnetic radiation with a free space wavelength of 1000 nm propagating in such a dielectric we see that

$$\alpha > 10^4\ \mathrm{cm}^{-1}$$

This means that the electromagnetic energy will be reduced to $1/e^2$ of its value within 10^{-4} cm.

The nonconductive dielectric is a medium with very few free or conduction band electrons at room temperature and so we cannot consider ohmic losses as the mechanism for heating. Its interactions with electromagnetic radiation involve valence band electrons. When such electrons in this type of material absorb light energy they can be raised either to the conduction band or to some impurity state lying within the band gap. They then relax back to the valence band. The process of relaxation can be radiative (fluorescence and phosphorescence) or nonradiative. In either case phonons (lattice vibrations) are generated and the material is heated. The heat generation is localized to the region in which the absorption occurs and, as in metals, heating beyond this region takes place by thermal heat conduction.

2. Lorentz and Drude Models

The classical theory of absorption in dielectric materials is due to H. A. Lorentz and in metals it is the result of the work of P. K. L. Drude. Both models treat the optically active electrons in a material as classical oscillators. In the Lorentz model the electron is considered to be bound to the nucleus by a harmonic restoring force. In this manner, Lorentz's picture is that of the nonconductive dielectric. Drude considered the electrons to be free and set the restoring force in the Lorentz model equal to zero. Both models include a damping term in the electron's equation of motion which in more modern terms is recognized as a result of electron–phonon collisions

These models solve for the electron's motion in the presence of the electromagnetic field as a driving force. From this, it is possible to write an expression for the polarization induced in the medium and from that to derive the dielectric constant. The Lorentz model for dielectrics gives the relative real and imaginary parts of the dielectric constant as

$$\varepsilon_{1R} = 1 + (Ne^2/\varepsilon_0 m)\frac{\omega_0^2 - \omega^2}{(\omega_0^2 - \omega^2)^2 + \Gamma^2\omega^2} \quad (12)$$

and

$$\varepsilon_{2R} = (Ne^2/\varepsilon_0 m)\frac{\Gamma\omega}{(\omega_0^2 - \omega^2)^2 + \Gamma^2\omega^2} \quad (13)$$

In these expressions N is the number of dipoles per unit volume, e the electron charge, m the electron mass, Γ the damping constant, ω_0 the resonance radian frequency of the harmonically bound electron, ω the radian frequency of the field, and ε_0 the permittivity of free space. Equations (12) and (13) are sketched in Fig. 1. The range of frequencies where ε_1 increases with frequency is referred to as the range of normal dispersion, and the region near $\omega = \omega_0$ where it decreases with frequency is called the range of anomalous dispersion.

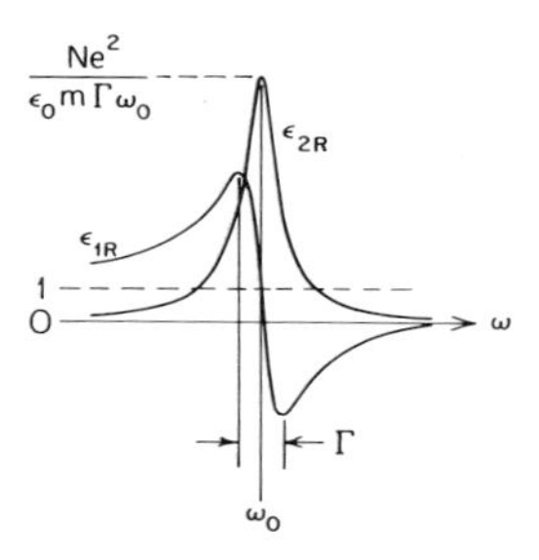

FIG. 1. Frequency dependences of ϵ_{1R} and ϵ_{2R}.

In the preceding discussion the contributions of the electronic polarizability to the dielectric constant were considered. Other contributions occur such as stimulation of vibrations of ions in ionic crystals. This type of contribution is very small at optical frequencies because of the large mass of the ions compared to that of the electrons. In other words, the ionic polarizability is much smaller than the electronic polarizability at optical frequencies. We can therefore consider only the electronic terms when evaluating optical absorption. By inserting Eq. (13) into Eq. (11) we obtain the absorption coefficient of the lossy, nonconductive dielectric in terms of the properties of the electron's kinematics assumed by Lorentz. That is,

$$\alpha = (2\pi/\lambda)(\mu_R/2)(Ne^2/\varepsilon_0 m) \times \left(\frac{\Gamma\omega}{(\omega_0^2 - \omega^2)^2 + \Gamma^2\omega^2}\right)^{1/2} \quad (14)$$

The Drude model for metals assumes that the electrons are free to move. This means that it is identical to the Lorentz model except that ω_0 is set equal to zero. The real and imaginary parts of the metal's dielectric constant are then given by

$$\varepsilon_{1R} = 1 - (Ne^2\varepsilon_0 m)\frac{1}{\omega^2 + \Gamma^2} \quad (15)$$

$$\varepsilon_{2R} = (Ne^2\varepsilon_0 m)\frac{\Gamma}{\omega(\omega^2 + \Gamma^2)} \quad (16)$$

The quantity Γ is related to the mean time between electron collisions with lattice vibrations. T (i.e., to the problem of electron–phonon scattering). By considering the motion of electrons able to make collisions with lattice vibrations in an electric field E having radian frequency ω, it is straightforward to show that the average velocity is

$$\nu = -\frac{eE}{m}\frac{\mathrm{T}}{1 - j\omega\mathrm{T}} \quad (17)$$

The conductivity at this frequency is then

$$\sigma = \frac{Ne^2\mathrm{T}}{m}\frac{1}{1 - j\omega\mathrm{T}} \quad (18)$$

where the dc conductivity is given by

$$\sigma_{\mathrm{dc}} = Ne^2\mathrm{T}/m \quad (19)$$

From Eqs. (15) and (16) we see that if we allow $\Gamma = 1/\mathrm{T}$, then

$$\varepsilon_1 = 1 - \left(\frac{\sigma_{\mathrm{dc}}}{\varepsilon_0}\right)\mathrm{T}\frac{1}{\omega^2\mathrm{T}^2 + 1} \quad (20a)$$

and

$$\varepsilon_2 = \left(\frac{\sigma_{\mathrm{dc}}}{\varepsilon_0}\right)\left(\frac{1}{\omega}\right)\frac{1}{\omega^2\mathrm{T}^2 + 1} \quad (20b)$$

At electromagnetic field frequencies that are low, that is, when $\omega\mathrm{T} \ll 1$, we have

$$\varepsilon_1 = 1 - \sigma_{\mathrm{dc}}\mathrm{T}/\varepsilon_0 \quad (21)$$

and

$$\varepsilon_2 = \sigma_{\mathrm{dc}}/\varepsilon_0\omega \quad (22)$$

At such frequencies $\varepsilon_2 \gg \varepsilon_1$, and since $\Gamma = 1/\mathrm{T}$ Eq. (14) gives

$$\alpha = (\omega\mu\sigma/2)^{1/2} \quad (23)$$

which is exactly the result we obtained earlier when treating absorption from an electromagnetic point of view. In other words, the optical properties and the dc conductivity of a perfect metal are related through the fact that each is determined by the motion of free electrons. At high frequencies transitions involving bound or valence band electrons are possible and there will be a noticeable deviation from this simple result of the Drude model. However, the experimental data reported for most metals are in good agreement with the Drude prediction at wavelengths as short as 1 μm.

3. Temperature Dependence

Another aspect of the absorption of light energy by metals that should be noted is the fact that it increases with temperature. This is important because during laser irradiation the temperature of a metal will increase and so will the absorption. The coupling of energy into the metal is therefore dependent on the temperature dependence of the absorption. This property is easy to understand if we remember that all the light that gets into a metal is absorbed in it. The question that must be addressed is how the amount of incident optical energy that is not reflected from the metal's surface depends on temperature. Recalling the Fresnel expression for electric field reflectance and applying to it the real and imaginary parts of the complex index of refraction for a metal–air interface, we can write the field reflectivity. Multiplying this by its complex conjugate, we find the intensity reflection coefficient for a metal

$$R_{\mathrm{I}} = 1 - 2\mu\varepsilon_0\omega/\sigma \quad (24)$$

Since the conductivity σ decreases with increasing temperature, R_{I} decreases with increasing temperature. As a result, more incident energy actually gets into the metal and is absorbed when the temperature is raised. This is true even though the absorption at high temperatures takes place in a deeper skin depth.

B. Nonlinear Absorption

Since the advent of lasers made possible high-intensity optical fields it has been possible to explore interactions of light with matter in which the response of the material is not linear with the optical electric field. This subject, known as nonlinear optics, has become a major field of research and has led to such useful optical devices as frequency converters, tunable parametric devices, and optically bistable elements. This section presents a semiclassical discussion of nonlinear optics and how it gives rise to nonlinear absorptions, in particular two-photon absorption (TPA).

The polarization of a medium and the optical electric field applied to it are linked by the material's susceptibility X, a tensor quantity. In the previous section we considered the limit of small optical fields, where the susceptibility is a function of the dielectric constants only and is independent of the field. In this case the polarization vector **P** is related to the optical electric field **E** by the expression

$$\mathbf{P} = \mathrm{X} * \mathbf{E} \quad (25)$$

Equation (25) is the relationship on which optics was built prior to 1961. As a result of lasers and the high fields they produce it is now necessary to allow that X can be a function of the optical field. This is accomplished by writing the field-dependent susceptibility as a Taylor series expansion in powers of the optical field and thus expressing the polarization as

$$\mathbf{P} = \mathrm{X}_1 * \mathbf{E} + \mathrm{X}_2 * \mathbf{E} * \mathbf{E} + \mathrm{X}_3 * \mathbf{E} * \mathbf{E} * \mathbf{E} + \cdots \quad (26)$$

Maxwell considered this form of relation in his classical treatise on electricity and magnetism but, for simplicity, retained only the first-order term.

The first term on the right of Eq. (26) gives rise to the "linear" optics discussed previously. The second term gives rise to optical second-harmonic generation or frequency doubling and optical rectification. The third term results in third-harmonic generation and self-focusing.

Absorption processes involving one or more photons are also described by the susceptibilities X_n employed in Eq. (26). By considering the continuity equation for the flow of energy and the fact that the susceptibilities are complex quantities,

$$X_n = X_n' + jX_n''$$

the imaginary part of the third-order term can be shown to give rise to an altered form of Beer's law. That is,

$$I(z) = \frac{I(0)\exp(-\alpha_0 z)}{1 + (\alpha_{\mathrm{TPA}}/\alpha_0)I(0)[1 - \exp(-\alpha_0 z)]} \qquad (27)$$

where $\alpha_0 = 4\pi k X_1$ is the conventional linear attenuation coefficient, k the propagation vector in the medium, and α_{TPA} the nonlinear attenuation coefficient, given by

$$\alpha_{\mathrm{TPA}} = \frac{32\pi^2}{c^2}\,\omega X_3'' I \qquad (28)$$

This is the part of the absorption that is linearly dependent on the intensity. From a quantum mechanical point of view, such a property is the result of processes in which two photons are absorbed simultaneously. In other words, a material with energy levels separated by U must be considered able to absorb simultaneously two photons, each having energy $U/2$. As a result, when studying the absorption of light in materials it is no longer sufficient to measure it at one low intensity. Contributions from such higher-order processes as TPA can occur and must be measured. For example, TPA is sufficient in some semiconductors to enable their excitation to lasing inversions. It is also a powerful tool for studying the presence of deep-level dopants in semiconductors, since they contribute energy levels that enhance TPA.

Higher-order multiphoton absorption processes are possible. There is some evidence that three-photon processes have been detected. However, nonlinear processes higher than TPA require optical fields that are very high. These fields may result in TPA, and the electrons that are thereby freed may be accelerated by the optical field to form a catastrophic electron avalanche breakdown. When this happens the material's effective absorptivity becomes 100%, too much energy is deposited in the irradiated volume, and severe mechanical damage known as intrinsic laser-induced damage follows. This sequence of events is responsible for setting intensity limits for materials and, consequently, for fixing the minimum size of optical components used in high-power lasing systems.

C. Laser-Driven Plasma Coupling and Decoupling

One of the most spectacular features of laser–materials interactions is the formation of a laser-driven plasma in the irradiated region (see Fig. 2). This phenomenon occurs in almost all types of intense laser irradiation, continuous wave (cw) or pulsed, on surfaces, and in the bulk of solids, liquids, and gases. It can prevent the laser light from further coupling to the sample or it can enhance the coupling. Its most important applications are in driving the target compression used to produce fusion energy by inertial confinement and in creating point sources of soft X-rays for X-ray lithography in microelectronics.

When an intense laser beam is focused inside a transparent medium a laser-induced break-

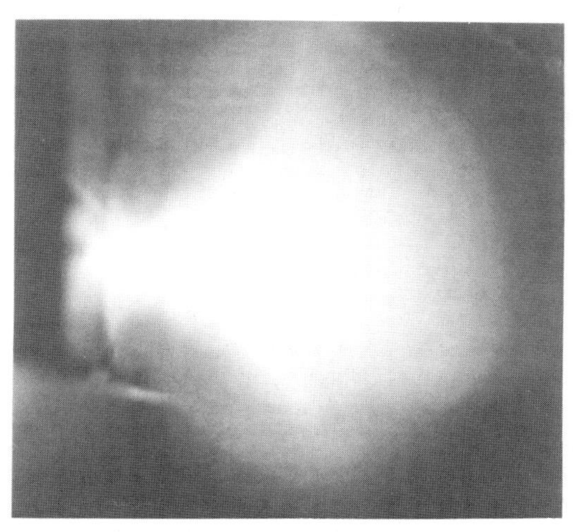

FIG. 2. Plasma generated by a pulsed CO_2 transverse electrical discharge at atmospheric pressure (TEA) laser incident from the right on a copper target in vacuum. The outer region contains hot copper atoms. The central region is a very bright blue-white color, indicating highly ionized, very hot copper vapor.

down and plasma can be formed in as short a time as 6 psec. This very fast formation is the result of the very highly nonlinear process of plasma formation. The process is so complex that it remains a major subject of current research. As a result, only a qualitative discussion is presented in this article.

If the target reaches a high enough temperature to induce mass loss, it is essential to account for the energy carried away by the removed material in order to determine the energy absorbed. Mass can be removed by such processes as vaporization, melt removal, and pyrolysis or burning. When vaporization occurs some of the ejected material can be ionized. Then, through the process of inverse bremsstrahlung, the vapor can be broken down by the optical field to produce an optically thick plasma. This plasma is a very hot blackbody radiator and emits an intense blue-white light called a laser-induced spark.

As most irradiations take place in air, we consider that case. At laser intensities slightly greater than the plasma formation threshold intensity, a laser-supported combustion (LSC) wave can be ignited. These waves occur at intensities from 2×10^4 to 10^7 W/cm^2 with both pulsed and cw lasers. The ignition of an LSC wave takes place in the ejected target vapor and the heated vapor transfers energy to the surrounding air. The LSC wave thus generated propagates away from the target surface along the beam path and drives a shock wave ahead of itself in the air (see Fig. 3).

The nature of the coupling of the light to the material when a plasma is formed depends on the beam parameters of energy, spot size, and pulse duration as well as air pressure and target material. For long pulse times and low laser intensities the formation of an LSC wave usually results in decreased coupling as the wave propagates away from the surface. This is the case in most laser materials processing applications requiring melting or material removal (i.e., welding, drilling, and cutting).

At high intensities a laser-supported detonation (LSD) wave is ignited. In this case absorption takes place in a thin zone of hot, high-pressure air behind the detonation wave (see Fig. 4). Since this wave takes air away from the surface, expansion fans form to satisfy the boundary conditions at the target surface. The plasma remains nearly one-dimensional until the expansion fans from the edge reach the center. This time is given approximately by the beam radius divided by the speed of sound in the plasma. In the vicinity of the surface there is no laser absorption and the plasma properties are determined by its isentropic expansion. The LSD wave plasma expansion away from the surface is very rapid. As a result, though it intercepts and absorbs incoming laser energy, the LSD wave plasma does not reradiate this energy as strongly absorbed ultraviolet light into the area originally irradiated.

Computations for two-dimensional LSD wave plasmas show that for short laser pulses (i.e., shorter than 1 μsec) the overall coupling of light to metal targets can be as large as 25%. Since most metals have absorptions near 1–2% in the infrared, this represents a substantial enhancement of the coupling. However, the coupling remains the same for increasing intensities but is spread out over increasing areas compared to that of the irradiated spot. Therefore, for short pulses, while the total coupling coefficient may be large, the energy deposited in the irradiated area may be smaller than if the LSD wave plasma had never been formed.

When an LSC wave plasma is formed, it is possible to find enhanced coupling in the irradi-

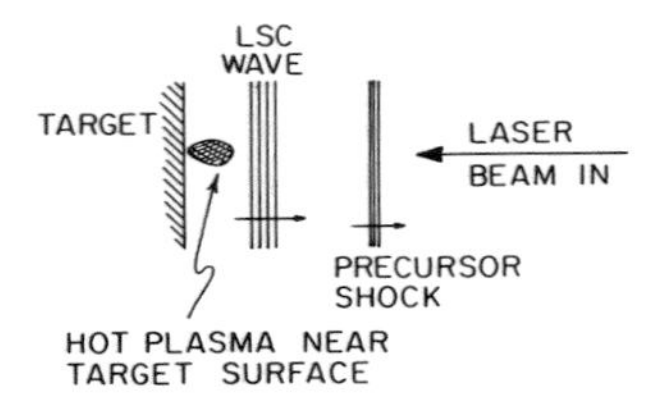

FIG. 3. Sketch of a one-dimensional laser-supported combustion (LSC) wave. The hot plasma near the target can radiate ultraviolet light, which is strongly absorbed by the target material. The LSC wave is of low enough density to allow the laser light to reach and heat the plasma.

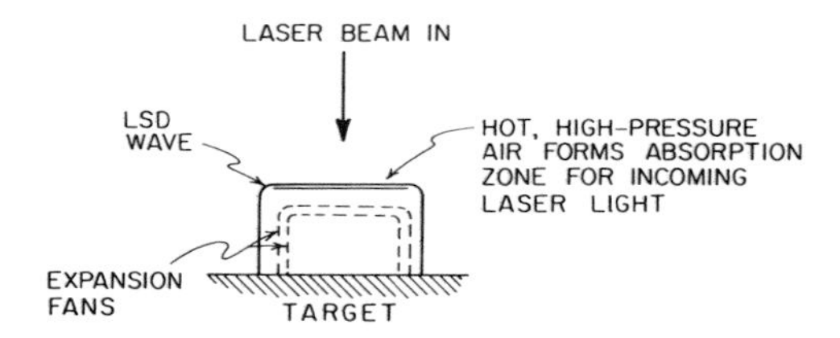

FIG. 4. Sketch of a laser-supported detonation (LSD) wave. The incoming laser light is absorbed in the LSD wave front as it propagates away from the target. This spreads any reradiated energy in the ultraviolet over a large area and reduces the local energy deposition.

ated region. This results from energy transfer by radiation from the hot, high-pressure plasma adjacent to the surface. The LSC wave propagates into the air at a low speed, and a large fraction of the incoming laser energy is used to heat the plasma to a temperature in excess of 20,000 K. This plasma radiates very efficiently in the ultraviolet part of the spectrum, where most materials, in particular, metals, absorb more strongly than in the visible and infrared. As long as the LSC wave expansion remains nearly one-dimensional the plasma-radiated ultraviolet light remains localized in the originally irradiated area and the local heating is enhanced.

Efficient local coupling of incident energy to the sample requires that the plasma be ignited near the surface, an LSC wave be generated, and the plasma expansion remain one-dimensional during the irradiation. These requirements define a range of intensities over which enhanced coupling will be observed. The minimum intensity is clearly that which ignites a plasma at the surface in the vaporous ejecta. The maximum intensity is that which generates an LSD wave instead of an LSC wave. Pulse durations or dwell times of cw sources must be less than that required for radial expansion of the plasma to set in over the dimension of the irradiated spot. Thus, enhanced coupling will occur for pulses shorter than the spot radius divided by the speed of sound in the plasma ($\sim 5 \times 10^5$ cm/sec). Once radial expansion begins, the pressure and the plasma temperature drop rapidly. Concomitantly, so does the efficiency of the plasma as an ultraviolet light radiator.

As mentioned above, the irradiation conditions used in many laser materials processing applications correspond to the two-dimensional LSC wave plasma case. Therefore, plasma formation reduces the coupling of laser energy into the irradiated region, and further increases in the laser intensity, either by increasing the laser power or decreasing the spot size, are futile. To deal with this problem, He cover gas is often used to suppress plasma formation. In some cases strong gas jets are used to blow the plasma away and allow the laser light to enter the irradiated material.

Another aspect of laser-driven vaporization and plasma formation that should be considered is the momentum transferred to the surface. In leaving the surface at high speed the ejected material carries away a substantial amount of momentum. This appears as a recoil momentum of the surface and, since it occurs during the short duration of the laser irradiation, it results in a substantial impulse to the surface. This process has been used to shock-harden certain materials. An ablative coating is placed on the area to be shock-hardened and it is struck by a pulsed laser beam. The impulse due to the removal of the ablative coating produces the desired hardening. The transfer of an impulse by laser irradiation is another subject of major research interest, in part due to its potential value as a laser weapons effect.

D. Surface Electromagnetic Waves

Researchers studying such diverse topics as laser processing of semiconductors, laser-induced damage, laser materials processing, and laser-driven deposition processes have observed the formation of "ripples" in the irradiated region (see Fig 5). All of these observations had in common the facts that linearly polarized lasers were used and that the material in the irradiated region had reached its melting point. The ripples were spaced by the light wavelength when the light was at normal incidence. These facts and

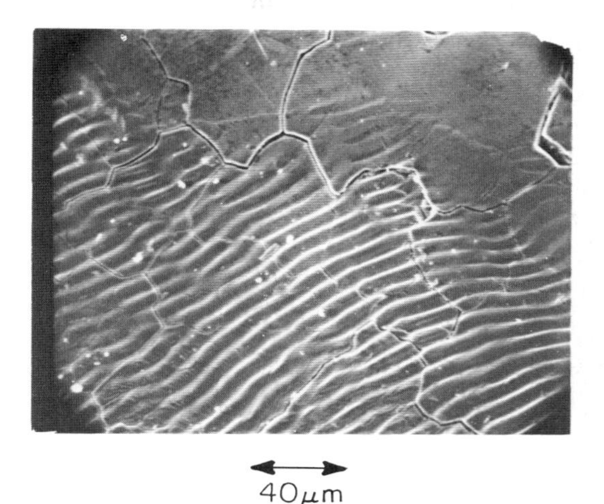

FIG. 5. Scanning electron micrograph of ripples spaced by $\sim 10\ \mu m$ produced in 304 stainless steel exposed to a cw CO_2 laser able to melt metal. The sample was stationary under a normally incident 10.6-μm beam polarized perpendicular to the ripples. The residual ripples are found near the edges of the melt pool, where freezing is sufficiently rapid to preserve the ripples. In the central part of the melt pool convective motion in the melt obscures the ripples. The ripples are the result of surface electromagnetic waves (SEW) launched by scattering the incident light and subsequent interference between two electromagnetic waves.

additional data demonstrated that the ripples were the result of the following sequence of events:

1. The incident light was scattered by some surface irregularity and a surface electromagnetic wave was launched.
2. In the region irradiated this wave was able to interfere with the incident light.
3. The interference pattern produced localized differences in the surface temperature, and when the material became molten it flowed due to surface tension effects or differential vapor pressures to conform with the varying light intensity.
4. The molten material froze quickly enough following irradiation that the ripples were detectable afterward.

The key element in the formation of these ripples is the fact that surface electromagnetic waves (SEW) were generated. This is another means by which light energy can be coupled into a material and one that was not obvious from prelaser knowledge. The fact that a gratinglike interference pattern is set up allows one to apply the electromagnetic theory of gratings to understand the process. From this theory, it is clear that for a particular line spacing, there is a particular grating height that can dramatically enhance the absorption of a material over that of a smooth surface. The generation of SEW can therefore result in greatly enhanced coupling of laser light to materials.

On the other hand, if the surface waves propagate, they may carry energy away from the irradiated area. If the dimension of the irradiated area is small compared to the SEW attenuation length, this may be a major factor in redistributing the absorbed energy. The irradiated region will not become as hot as it might if no SEW were generated. It should be noted that this effect would give rise to a beam spot size dependence of laser heating when the beam diameters used are in the range of one SEW attenuation length.

E. Photoablation

A new and exciting light–matter interaction called photoablation has been demonstrated with the advent of high peak intensity ultraviolet lasers. This occurs with the excimer lasers (i.e., ArF, KrF, XeF, or XeCl) operating at wavelengths from ~250 to 350 nm. The lasers provide high energies (~1 J or more per pulse) in pulses of ~30 nsec duration. Most organic materials absorb very strongly at these wavelengths. The energies of photons in the ultraviolet are sufficient to efficiently break bonds in such materials. When this happens the fragments are subjected to very high pressures and are ejected from the surface (see Fig. 6). The process takes place entirely within one absorption depth (~20 nm) and so provides a very controllable means of material removal. In addition, the energy deposited in the material is removed as kinetic energy of the ejecta. As a result, there is very little thermal heating of the sample and the process is confined to the area irradiated. With lasers operating at reasonably high repetition rates, this process can be used to remove significant amounts of material.

The use of excimer lasers and photoablation is still under consideration in research and development laboratories. However, it is clearly suitable for removing protective coatings or insulation and for certain medical applications where great delicacy is required.

II. Laser Heating of Materials

A. Thermal Diffusion Problem and Thermal Diffusivity

We are concerned with the heating of materials that results from several of the laser light absorption processes discussed in the previous sections. This can be approached in a number of nearly equivalent ways, all of which require a

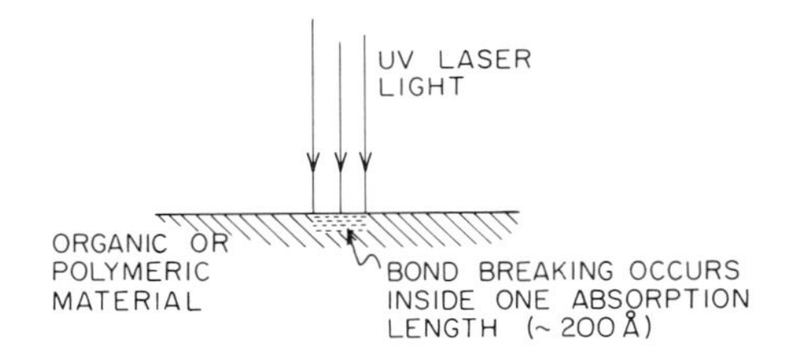

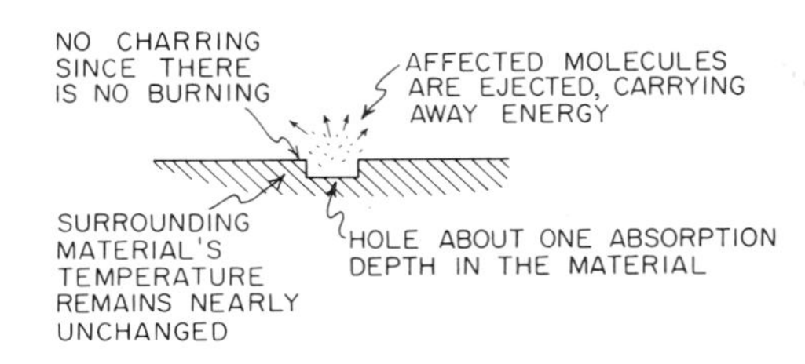

FIG. 6. Sketch of the process of ultraviolet laser irradiation-induced photoablation.

knowledge of (1) the target's optical and thermal properties during the irradiation, (2) the laser beam distribution, (3) the dynamics of the irradiation process, and (4) the processes of phase change in the target material. As would be expected, it is extremely difficult to solve the heat flow problem exactly in the general case and so reasonable approximations are used. In addition, only problems that are easy to solve are attempted. These are then useful as guides to the solution of other problems.

Consider the equation of heat conduction in a solid with the laser energy absorbed on the irradiated surface as a heat source. By judiciously selecting the beam geometry, dwell time, and sample configuration, the problem may be reduced to solvable one- and two-dimensional heat flow analyses. Phase transitions can be included and the temperature distributions that are produced can be calculated. Examples will be selected to provide specific guidance in the choice of lasers and materials. The result of all this will be an idea of the effects that one may produce by laser heating of solids.

The equation for heat flow in a three-dimensional solid is

$$\rho C \frac{\partial T}{\partial t} = \frac{\partial}{\partial x}\left(K \frac{\partial T}{\partial x}\right) + \frac{\partial}{\partial y}\left(K \frac{\partial T}{\partial y}\right) + \frac{\partial}{\partial z}\left(K \frac{\partial T}{\partial z}\right) + A(x, y, z, t) \quad (29)$$

where ρ is the material density in grams per cubic centimeter, K the thermal conductivity in watts per centimeter per degree Celsius, and C the heat capacity in joules per centimeter per degree Celsius; these are material properties that will depend on temperature and position. The quantity $A(x, y, z, t)$ is the rate at which heat is supplied to the solid per unit time per unit volume in joules per second per cubic centimeter and $T = T(x, y, z, t)$ is the resulting temperature distribution in the material. Figure 7 defines the coordinate system used.

The temperature dependence of the properties results in a nonlinear equation that is very difficult to solve exactly. Where the functional dependence of these quantities on temperature is known, it is sometimes possible to use numerical integration techniques to obtain a solution. A further complication arises from the temperature dependence of $A(x, y, z, t)$ through that of the material's absorptivity. When phase transitions occur one can attempt a solution of the problem by solving for each phase separately

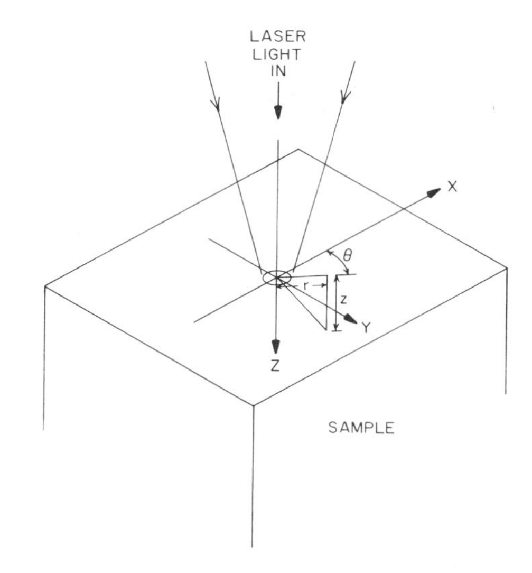

FIG. 7. Sketch of coordinate system used in evaluating responses of materials to laser irradiation.

and including the heat required for the transition where appropriate.

For most materials the thermal properties do not vary greatly with temperature and can be assigned an average value for the temperature range to be studied. In this case it is possible to solve the heat flow problem. A further simplification is obtained by assuming that the material is homogeneous and isotropic. Under these conditions Eq. (29) reduces to

$$\nabla^2 T - \frac{1}{\kappa}\frac{\partial T}{\partial t} = -\frac{A(x, y, z, t)}{K} \quad (30)$$

where $\kappa = K/\rho C$ is the thermal diffusivity. In the steady state $\partial T/\partial t = 0$, resulting in

$$\nabla^2 T = -\frac{A(x, y, z, t)}{K} \quad (31)$$

If there is no heat source, as in the case of cooling of heated material, the temperature distribution will be given by

$$\nabla^2 T = \frac{1}{\kappa}\frac{\partial T}{\partial t} \quad (32)$$

in the time-dependent case and

$$\nabla^2 T = 0 \quad (33)$$

in the steady state.

A physical insight into the meaning of the quantity κ is useful. A simple analysis of the units show that

$$(\kappa t)^{1/2} = \text{distance}$$

This distance or a multiple thereof is generally known as the thermal diffusion distance for the particular problem and is a handy quantity to use when scaling the effects of laser heating. It is often useful to know how the optical absorption depth compares with the thermal diffusion distance during a laser irradiation. To see the meaning of the thermal diffusion depth more clearly, consider the following particular solutions of Eq. (32):

$$T = T_0 e^{t/\mathrm{T} \pm z/z_D}$$

and so

$$\frac{\partial T}{\partial t} = \frac{1}{\mathrm{T}} T$$

and

$$\frac{\partial^2 T}{\partial z^2} = \left(\frac{1}{z_D}\right)^2 T$$

Thus

$$\left[\left(\frac{1}{z_D}\right)^2 - \frac{1}{\kappa \mathrm{T}}\right] T = 0$$

and we have a solution if

$$z_D = (\kappa \mathrm{T})^{1/2}$$

Therefore a characteristic distance is related to a characteristic time by $(\kappa \mathrm{T})^{1/2}$.

For a more realistic case consider

$$T = T_0 t^{-1/2} e^{-z^2/4\kappa t} \qquad \text{for } t > 0$$

Then we have

$$\frac{\partial T}{\partial t} = T_0 \left(-\frac{1}{2t^{3/2}} + \frac{z^2}{4\kappa t^{5/2}}\right) e^{-z^2/4\kappa t}$$

and

$$\frac{\partial^2 T}{\partial z^2} = T_0 \left(-\frac{1}{2\kappa t^{3/2}} + \frac{z^2}{4\kappa t^{5/2}}\right) e^{-z^2/4\kappa t}$$

which satisfy

$$\frac{\partial^2 T}{\partial z^2} - \frac{1}{\kappa}\frac{\partial T}{\partial z} = 0$$

At $t = 0$ we will set $T = T_0$ for $z = 0$ and $T = 0$ for $z > 0$. Now consider that at any time t_p the temperature at

$$z_D = 2(\kappa t_p)^{1/2}$$

is $1/e$ times that at $z = 0$.

The second case corresponds to the release of the quantity of heat $2\rho C T_0(\pi\kappa)^{1/2}$ per unit area over the plane $z = 0$ at time $t = 0$. This could be a decent approximation to the case of laser heating of a semi-infinite metal irradiated by a short pulse [defined in terms of the relative values of beam radius and $(\kappa t_p)^{1/2}$] in a uniform beam where we wish to know a temperature on the beam axis. The quantity of heat per unit area would be the laser intensity I times the pulse duration T_p times the fraction absorbed or

$$I\mathrm{T}_p\alpha = 2\rho C T_0(\pi\kappa)^{1/2}$$

where α is the absorptance of the metal. Then

$$T_0 = \frac{\alpha I \mathrm{T}_p}{2\rho C(\pi\kappa)^{1/2}}$$

and

$$T(z, t) = \frac{\alpha I \mathrm{T}_p}{2\rho C(\pi\kappa)^{1/2}} t^{-1/2} e^{-z^2/4\kappa t}$$

or

$$T(z, t) = \frac{\alpha I \mathrm{T}_p}{\rho C \pi^{1/2}} \frac{e^{-z^2/4\kappa t}}{(4\kappa t)^{1/2}} \qquad \text{for } t > 0$$

The important role of the quantity $z_D = 2(\kappa t)^{1/2}$ in describing the process of laser heating is clear; z_D defined above is the thermal diffusion distance.

B. Absorption in a Very Thin Surface Layer

These are cases where

$$\alpha^{-1} \ll (\kappa \mathrm{T}_{\text{laser}})^{1/2}$$

where $\mathrm{T}_{\text{laser}}$ is the duration of the irradiation.

1. Uniformly Illuminated Surface

The incident intensity is given by

$$I_0(t) = \begin{cases} 0, & t < 0 \\ I_0, & t \geq 0 \end{cases}$$

and is assumed to be uniformly distributed. The temperature distribution on-axis is sought by assuming no heat diffusion in the x and y directions, even though this may be important in certain applications. The results of this special case are very useful since most often the on-axis temperature is the desired quantity.

The thermal diffusion equation with this heat source and the boundary condition

$$T(z, t) = 0 \qquad \text{for } t < 0 \text{ and all } z$$

has the solution

$$T(z, t) = \left(\frac{2\alpha I_0}{K}\right) (\kappa t)^{1/2} \operatorname{ierfc}\left[\frac{z}{2(\kappa t)^{1/2}}\right] \tag{34}$$

where

$$\operatorname{ierfc}(X) = \int_X^\infty \operatorname{erfc}(X')\, dX'$$

and

$$\text{erfc}(X) = 1 - \text{erf}(X) = \frac{2}{\pi}\int_X^\infty e^{-(X')^2}\, dX'$$

and

$$\text{erf}(X) = \frac{2}{\pi}\int_0^X e^{-(X')^2}\, dX'$$

The error function erf(X) has the following properties:

$$\text{erf}(0) = 0 \qquad \text{erf}(\infty) = 1 \qquad \text{erf}(-X) = -\text{erf}(X)$$

$$\text{erfc}(0) = 1 \qquad \text{erfc}(\infty) = 0$$

and

$$\text{ierfc}(0) = 1/\sqrt{\pi}$$

Thus the surface temperature is given by

$$T(0, t) = \frac{2\alpha I_0}{K}(\kappa t)^{1/2}\left(\frac{1}{\pi}\right)^{1/2} \tag{35}$$

It is proportional to $t^{1/2}$, and if there were no phase changes (i.e., melting and vaporization) the temperature would continue to increase. Note that the energy flux absorbed by the surface is

$$E = \int_0^t \alpha I\, dt = I_0 t \alpha_0$$

but because of conduction the surface temperature increases more slowly, that is, as $t^{1/2}$. This means that a greater surface temperature can be achieved for a given laser pulse energy by shortening the pulse and increasing I_0.

Since we have mentioned pulses, let us model a simple pulse as one where

$$I = \begin{cases} 0, & t < 0 \\ I_0, & 0 \le t \le \text{T} \\ 0, & \text{T} < t \end{cases}$$

During the interval 0 to T the temperature is as given above with the maximum temperature obtained at $t = \text{T}$. For $t > \text{T}$

$$T(z, t) = \frac{2\alpha I_0}{K}\left\{(\kappa t)^{1/2}\,\text{ierfc}\left[\frac{z}{2(\kappa t)^{1/2}}\right] - [\kappa(t - \text{T})]^{1/2}\,\text{ierfc}\left[\frac{z}{2\kappa(t - \text{T})}\right]^{1/2}\right\} \tag{36}$$

Remembering that this problem is only a one-dimensional approximation to the actual case and that the assumption of an infinitely thick medium requires that the slab thickness L be

$$L > 2(\kappa \text{T})^{1/2}$$

allows one to use these results to calculate the temperature distribution achieved. As pointed out above, the result will only be correct near the beam axis. The cooling rate following the irradiation in this example is obtained by taking the time derivative of $T(z, t)$ in Eq. (36).

2. Uniformly Illuminated Circle of Radius A

The power is

$$P = \begin{cases} 0, & t < 0 \\ P_0, & t \ge 0 \end{cases}$$

Using cylindrical coordinates as shown in Fig. 7, one finds

$$T(r, z, t) = \frac{\alpha P_0}{2\pi A K}\int_0^\infty \frac{d\lambda}{\lambda} J_0(\lambda r) J_1(\lambda A) \times \left\{e^{-\lambda z}\,\text{erfc}\left[\frac{z}{2(\kappa t)^{1/2}} - \lambda(\kappa t)^{1/2}\right] - e^{\lambda z}\,\text{erfc}\left[\frac{z}{2(\kappa t)^{1/2}} + \lambda(\kappa t)^{1/2}\right]\right\} \tag{37}$$

where J_0 and J_1 are Bessell functions of the first kind. This expression may be evaluated numerically. It is far more useful in planning experiments to evaluate the result directly under the beam since this will be the hottest place. In other words, set $r = 0$ and $z = 0$, to find

$$T(0, 0, t) = \frac{2\alpha P_0(\kappa t)^{1/2}}{\pi A^2 K} \times \left\{\frac{1}{\pi^{1/2}} - \text{ierfc}\left[\frac{A}{2(\kappa t)^{1/2}}\right]\right\} \tag{38}$$

Note that if $A \gg 2(\kappa t)^{1/2}$, then at those values of t one can treat the problem as the simple one in Section II,B,1. In other words, while the beam radius is greater than the thermal diffusion distance, the problem can be reduced to the simple case for the on-axis temperature. However, as time marches on this will break down and there will be differences. Most obviously, T does not become infinite but instead as $t \to \infty$

$$T(0, 0, \infty) = \alpha P_0/\pi A K = T_{max} \tag{39}$$

This reveals a curious point—the quantity P_0/A determines the maximum achievable temperature. If, for example, we wish to heat the material without melting the surface, then we must choose P_0/A such that $T_{max} < T_{melt}$.

3. Uniformly Illuminated Rectangle

It is often the case that the focused laser beam is not circular. In fact, most excimer laser beams are rectangular. Assuming that the light distribution is uniform, we can examine the rectangular distribution. The solution on-axis and on the surface is

$$T(0, 0, \infty) = [\alpha P/2K\pi la] [a \sinh^{-1}(l/a) + \sinh^{-1}(a/l))] \quad (40)$$

at late times. Here a is the width and l the length of the irradiated rectangle.

For example, if the width of the rectangle a and the radius in the circular case A were set equal and we asked what power would be needed to reach a specific temperature in each case we would have

$$\frac{P_{\text{circ}}}{P_{\text{rect}}} = \frac{1}{2l}\left[a \sinh^{-1}\left(\frac{l}{a}\right) + l \sinh^{-1}\left(\frac{a}{l}\right)\right] \quad (41)$$

Even a square beam (l = a) requires more power to produce a specific temperature than a circular beam. This is obvious since a larger area is irradiated and the absorbed energy must heat up more material. To achieve the same maximum temperature, more input must be provided. If $l/a = 3$, then we see from Eq. (41) that the power must be increased by a factor of ~2. This simple example demonstrates the importance of good beam quality and focusing optics in order to obtain maximum heating from laser irradiation.

4. Gaussian Beam Illumination

The Gaussian beam intensity distribution is

$$I = I_0 e^{-r^2/w^2} \quad (42)$$

[Note: It is common to describe a Gaussian mode of a laser by the electric field distribution

$$E = E_0 e^{-r^2/w_0^2}$$

and then give

$$I = I_0 e^{-2r^2/w_0^2} = I_0 e^{-r^2/(w_0/\sqrt{2})^2}$$

The beam parameter w used in this treatment is the radius at which the *intensity* has fallen to $1/e$ of its on-axis value. It is the radius at which the electric field has fallen to $e^{-1/2}$ of *its* on-axis value. If we used w_0 to describe the Gaussian, then it corresponds to the radius at which the intensity has fallen to e^{-2} of its on-axis value.]

The following expression can be obtained for the temperature distribution in a semi-infinite solid due to irradiation at the surface by an instantaneous ring source of radius r' and total deposited energy Q:

$$T_{\substack{\text{inst.}\\\text{ring}}}(r, z, t) = \frac{Q}{4\rho C(\pi\kappa t)^{3/2}} \times \exp\left(\frac{-r^2 - r'^2 - z^2}{4\kappa t}\right) \mathcal{I}_0\left(\frac{rr'}{2\kappa t}\right) \quad (43)$$

In this expression $\mathcal{I}_0$ is the modified Bessel function of order zero.

For a Gaussian source we have

$$Q = q_0 e^{-r'^2 w^2} 2\pi r' \, dr'$$

where q_0 is the energy per unit area at the origin. Inserting this into Eq. (43), we integrate to obtain

$$T_{\substack{\text{inst.}\\\text{Gaussian}}}(r, z, t) = \frac{q_0 w^2}{\rho C(\pi\kappa t)^{1/2}(4\kappa t + w^2)} \times \exp\left(-\frac{z^2}{4\kappa t} - \frac{r^2}{4\kappa t + w^2}\right) \quad (44)$$

If q_0 is replaced by $\alpha I_0 \mathrm{T_p}$, where $\mathrm{T_p}$ is the duration of a pulse, and

$$w \gg (4\kappa \mathrm{T_p})^{1/2}$$

we can use Eq. (44) for the temperature distribution following irradiation by a short pulse. (*Note: Short pulse is defined in terms of whether heat diffuses significantly with respect to the beam dimension during the pulse. If it does not, then the pulse is "short."*)

Rewriting Eq. (44) gives

$$T_{\substack{\text{inst.}\\\text{Gaussian}}}(r, z, t) = \frac{\alpha I_0 \mathrm{T_p}}{\rho C(\pi\kappa t)^{1/2}}\left(\frac{1}{4\kappa t/w^2 + 1}\right) \times \exp\left[-\frac{(z/w)^2}{4\kappa t/w^2} - \frac{(r/w)^2}{4\kappa t/w^2 + 1}\right] \quad (45)$$

Defining the unitless quantities

$$T' = w\rho C\pi^{1/2} T/2\alpha I_0 \mathrm{T_p} \qquad t' = 4\kappa t/w^2$$

$$z' = z/w \qquad r' = r/w$$

enables writing

$$T' = \left(\frac{1}{t'}\right)^{1/2} \frac{1}{t' + 1} \exp[-z'^2/t' - r'^2/(t' + 1)] \quad (46)$$

For any z' and t', we see that T' decreases with

r' as $e^{-r'/t'+1}$. Similarly for any t' and r', T' decreases as $e^{-z'^2/t'}$. Thus, having found the on-axis surface temperature, one can easily find the temperature at any point in the material!

For copper $\kappa \sim 1$ cm^2/sec and when $w = 0.5$ mm we would require

$$T_p \ll 6 \times 10^{-4} \text{ sec}$$

For steel $\kappa \sim 0.15$ cm^2/sec and for $w = 0.5$ mm we would require

$$T_p \ll 4 \times 10^{-3} \text{ sec}$$

In the case of pulsed Nd : YAG lasers where $T_p \sim 10^{-5}$ sec, Eq. (44) is quite adequate. For CO_2 TEA laser pulses with $T_p \sim 10^{-6}$ sec, Eq. (44) is again acceptable. However, for discharge pulsed CO_2 lasers with $T_p \sim 10^{-3}$ sec or for cw lasers (or for smaller w) the noninstantaneous form must be used.

The case of cw Gaussian illumination can be treated by setting $q_0 = \alpha I_0(\mu)\, d\mu$ (where μ is a dummy variable for time) and integrating $T_{\substack{\text{inst.}\\ \text{Gaussian}}}$ from 0 to t. This gives

$$T_{\substack{\text{noninst.}\\ \text{Gaussian}}}(r, z, t) = \frac{\alpha I_0^{\max} w^2}{K} \frac{\kappa^{1/2}}{\pi} \times \int_0^t \frac{p(t-\mu)\, d\mu}{(\mu)^{1/2}(4\kappa\mu + w^2)} \times \exp\left(-\frac{z^2}{4\kappa\mu} - \frac{r^2}{4\kappa\mu + w^2}\right)$$

where $I_0(\mu) = I_0^{\max} p(\mu)$. Setting

$$\mu' = 4\kappa\mu/w^2, \qquad z' = z/w, \qquad r' = r/w$$

gives

$$T_{\substack{\text{noninst.}\\ \text{Gaussian}}}(r', z', t') = \frac{\alpha I^{\max} w}{K 2\pi^{1/2}} \times \int_0^{t'} \frac{p(t'-\mu')\, d\mu'}{(\mu')^{1/2}(\mu'+1)} \times \exp\left(-\frac{z'^2}{\mu} - \frac{r'^2}{\mu+1}\right)$$

and so for a dimensionless $T' = 2\pi^{1/2} KT/\alpha I_0^{\max} w$ we have

$$T' = \int_0^{t'} \frac{p(t'-\mu')\, d\mu'}{(\mu')^{1/2}(\mu'+1)} \times \exp\left(\frac{z'^2}{\mu} - \frac{r'^2}{\mu+1}\right) \tag{47}$$

At the surface $z' = 0$ and at the center $r' = 0$ and so

$$T' = \int_0^{t'} \frac{p(t'-\mu')\, d\mu'}{(\mu')^{1/2}(\mu'+1)}$$

For a cw laser or one where $T_p \gg w^2/4\kappa$

$$P(t'-\mu') = 1$$

and

$$T' = 2 \tan^{-1}(t')^{1/2}$$

or

$$T(0, 0, t) = \frac{\alpha I_0^{\max} w}{K\pi^{1/2}} \tan^{-1}\left(\frac{4\kappa t}{w^2}\right)^{1/2} \tag{48}$$

5. Comparison of Results

The several expressions for the induced surface temperature are summarized below.

For the uniformly irradiated surface of a semi-infinite sample (Section II,B,1)

$$T_A(0, 0, t) = \frac{2\alpha I_0}{K}\left(\frac{\kappa t}{\pi}\right)^{1/2}$$

For the uniformly irradiated beam of radius A (Section II,B,2)

$$T_B(0, 0, t) = \frac{2\alpha P_0}{\pi A^2 K}(\kappa t)^{1/2} \times \left[\frac{1}{\pi_{1/2}} - \operatorname{ierfc}\frac{A}{2(\kappa t)^{1/2}}\right]$$

and

$$T_B(0, 0, \infty) = \frac{\alpha P_0}{\pi A K}$$

For the noncircular uniform beam (Section II,B,3)

$$T_C(0, 0, \infty) = \frac{\alpha P_0}{2K\pi l a}\left[a \sinh^{-1}\left(\frac{l}{a}\right) + l \sinh^{-1}\left(\frac{a}{l}\right)\right]$$

For the Gaussian beam (Section II,B,4) with $I = I_0 e^{-r^2/w^2}$

$$T_D(0, 0, t) = \frac{\alpha I_0 w}{K\pi^{1/2}} \tan^{-1}\left(\frac{4\kappa t}{w^2}\right)^{1/2}$$

and

$$T_D(0, 0, \infty) = \frac{\alpha I_0 w \pi^{1/2}}{2K}$$

It is appropriate to ask how much difference there is between these results. For example, if a Gaussian beam were used but treated as if it were a circular beam of radius w with $I = I_0$, what would happen?

By comparing cases II,B,4 and II,B,2 at equilibrium it is clear that

$$\frac{T_D(0, 0, \infty)}{T_B(0, 0, \infty)} = \frac{\alpha I_0 w \pi^{1/2}/2K}{\alpha P_0/\pi w K} = \frac{\pi^{1/2}}{2} = 0.886$$

The error in approximating the Gaussian by a uniform circle with average intensity equal to the peak on-axis intensity of the Gaussian is only 11%. Considering the approximations involved in setting the thermal and optical properties equal to some average value, this is acceptable. A similar conclusion could be reached when comparing the other on-axis surface temperatures at other times. Thus, an acceptable estimate of the on-axis surface temperature produced by a beam with $I(r = 0) = 0$ is obtained from the simplest case with a finite beam radius, that is, case II,B,1.

6. Some Numerical Examples

Now consider some numbers to get a feeling for the scale of things and for when to include phase changes. The properties of three interesting materials are listed in Table I.

The other property needed to estimate the temperatures that can be obtained is the material's absorptivity for the laser in question. The carbon phenolic can be assumed with reasonable accuracy to absorb 100% of incident 1.06- and 10.6-μm laser light. The Al and 204 stainless steel absorptivities can be questioned as they will depend on surface finish and the presence of any molten material. Also, the absorptivity of a metal is expected to increase with temperature. Recent measurements for Al and 1016 steel show that at 10.5 μm the Drude model gives accurate values for both the absorptivity and its temperature coefficient. At 1.06 μm the experimental results do not agree with the Drude model. The absorptivity is higher than predicted. Furthermore, whenever even a small amount of any metal is melted ~20% of the incident laser light seems to be absorbed. This may be due to an unexpected increase in absorptivity at the melting temperature or to geometric considerations on melting. However, a fair estimate of the absorptivity of almost any metal exposed to 1.06- and 10.6-μm laser beams is ~10%. (You cannot go too far wrong with this estimate, particularly if a little melting occurs.)

For a Gaussian beam at the surface and on-axis

$$T(0, 0, t) = \frac{\alpha I_0 w}{K\pi^{1/2}} \tan^{-1}\left(\frac{4\kappa t}{w^2}\right)^{1/2}$$

$$= \frac{\alpha I_0}{K\pi^{1/2}} 2(\kappa t)^{1/2}$$

for small t or for $w \gg 2(\kappa t)^{1/2}$

$$= \frac{\alpha P_0}{K\pi w^2} 2 \left(\frac{\kappa t}{\pi}\right)^{1/2}$$

and

$$T(0, 0, \infty) = \frac{\alpha I_0 w \pi^{1/2}}{2K}$$

$$= \frac{\alpha P_0}{2K\pi^{1/2} w}$$

TABLE I. Material Thermal Properties

Material	K (W/cm °C) Solid	K (W/cm °C) Liquid	C (J/g °C)	ρ (g/cm³)	κ (cm²/sec)
Al	2.0	1.0	1.0	2.7	0.74
304 S.S.	0.26	0.26	0.6	8.0	0.054
Carbon phenolic	0.01	—	1.7	1.45	0.004
Material	L_m (kJ/g)	L_v (kJ/g)	T_m[a] (°C)	T_v[a] (°C)	
Al	0.4	11	640	2430	
304 S.S.	0.27	4.65	1430	2980	
Carbon phenolic		14.6		4000	

[a] T_m and T_v are measured as the increase from 20°C need to melt or vaporize, respectively.

Let us assume a 500-W beam focused to a spot with $w = 0.02$ cm (200 μm). Then we have $T(0, 0, \infty) = 353$°C for Al, 2712°C for 304 S.S., and 705,237°C for carbon phenolic. Obviously, Al gets warm, 304 stainless steel melts, and carbon phenolic vaporizes. We can see from this the role of K and α. Note that more power or smaller w enters the answer linearly.

Now

$$\frac{T(0, 0, t)}{T(0, 0, \infty)} = \frac{4(\kappa t)^{1/2}}{w\pi}$$

Let us assume that the approximation $w \gg 2(\kappa T)^{1/2}$ holds if

$$w/2(\kappa t)^{1/2} = 5$$

For times such that this inequality holds we have

$$\frac{T(0, 0, t)}{T(0, 0, \infty)} \sim 0.3$$

In other words, we reached one-third of the final temperature in the following times: Al, 5.4×10^{-6} sec; 304 S.S., 7.4×10^{-5} sec; and carbon phenolic, 1×10^{-3} sec.

C. Time Required to Achieve Melting or Vaporization

It is interesting to consider the time required to achieve melting or vaporization for a given intensity. Assuming that this can be accomplished while we are in the "short time" or diffusion-free range, we have

$$T(0, 0, t) = \frac{2\alpha P_0}{K\pi^{3/2}w^2}(\kappa t)^{1/2}]$$

and so

$$t_m = \pi^3 w^4 K^2 T_m^2 / 4\alpha^2 P_0^2 \kappa \qquad (49)$$

or

$$v = \pi^3 w^4 K^2 v / 4\alpha^2 P_0^2 \kappa$$

This expression is quite useful in designing a process for a given laser. For example, if the laser is fixed in output capability and a certain material is to be treated, the only variable that can be adjusted to obtain melting is w. Thus it is crucial to prepare proper beam-handling optics. (Of course, it helps to have t_m or $v \propto w^4$.)

1. Rates of Heating and Cooling

Some elementary considerations are presented concerning pulsed heating and cooling. (These matter a great deal when considering studies of rapid resolidification, special alloying, annealing, and metastable phases.) We are concerned with order-of-magnitude estimates of the heating and cooling rates and for this purpose will assume no latent heat due to phase transitions. There are two limiting cases.

1. When α is very large or the optical absorption depth α^{-1} is very small compared to the thermal diffusion depth $(\kappa t_p)^{1/2}$. Here t_p is the duration of the irradiation.

In this case the energy absorbed per unit area (assuming a uniform beam) is $I_0 t_p$ and it is used to heat a layer $(\kappa t_p)^{1/2}$ thick. Thus

$$\Delta T = \alpha I_0 t_p / C\rho(\kappa t_p)^{1/2}$$

and the heating rate is

$$\Delta T / t_p = \alpha I_0 / C\rho(\kappa t_p)^{1/2}$$

After the laser pulse it takes about the same time, t_p, for the heat from the layer to diffuse a distance $(\kappa t_p)^{1/2}$ into the material. During this time the temperature at the surface drops by an amount that is of order of magnitude equal to ΔT. Thus the cooling rate is also

$$\Delta T / t_p = \alpha I_0 / C\rho(\kappa t_p)^{1/2}$$

2. If α^{-1} is greater than $(\kappa t)^{1/2}$, the light is absorbed in the medium according to

$$I = I_0 e^{-\alpha z}$$

and a temperature distribution is created that is roughly given as

$$T(z) = (1 - R) I_0 (e^{-\alpha z}) t_p / C$$

where R is the reflectivity at the surface. After the pulse, cooling will occur if heat diffuses a distance of $\sim 1/\alpha$, and so the cooling time is given approximately as

$$t_c = (\alpha^{-1})^2 / \kappa$$

Thus the cooling rate at the surface is

$$\Delta T(z) / t_c = (1 - R)\alpha^3 I_0 \kappa t_p / \rho C$$

In many applications it is necessary to obtain $T = T_m$. By comparison it is clear that the cooling rate for case 1 will always be very high. In fact,

$$dT/dt = T_m / t_p$$

and if one has sufficient energy to obtain T_m in, say, a 10^{-9} sec pulse, one will have

$$dT/dt \approx 10^{12} \text{ °C/sec}$$

III. Responses of Materials

When laser light is absorbed by a material it results in localized heating and sometimes in a physical change in the heated matter. The change may be desired, as in the many materials processing applications that have been demonstrated, or undesired, as in the formation of laser-induced damage in optical components.

The variety of materials responses to laser heating is so great that not all can be covered here. The key to understanding this phenomenon is to recognize that laser heating is often very localized and very rapid. As estimated in Section II,C, the cooling rate for very localized surface heating can be very large. The result is an ability to rapidly heat and cool materials to produce desired changes. Such desired changes as hardening, annealing, melting, and vaporization are obtained routinely in laser materials processing.

Hardening steels and cast irons requires heating above the martensitic transition point and quenching the material. The laser does this by heating a surface layer (the material in which either the absorption occurred or the heat diffused during the irradiation) and then allowing it to quench by cooling conductively into the bulk. This process has a significant advantage over bulk heating and more conventional quenching methods in that there is very little workpiece distortion. That is, a laser-hardened sample is ready for use with little or no remachining.

In hardening large areas the laser beam can be spread out by defocusing or by scanning a focused beam across the surface. This is possible because in the hardening process one wishes to heat the metal and not to melt it. The overlapped trails of hardened material that are produced contain annealed metal in the regions where they intersect. However, it is often possible to design the beam and scan parameters to minimize this effect.

Sometimes it is necessary to create alloys on the surface by local rapid heating and cooling. Again, the laser is an ideal tool for this type of controlled melting process. Use of this type of alloying can reduce the amount of expensive hard alloy required for a given purpose. One step beyond alloying is forming metastable forms of alloyed metals by very rapid melt resolidification. This technique, which owes its existence to the very high cooling rates discussed in Section II,C, is used to create and study certain glassy forms of metals. It is hoped that these corrosion- and wear-resistant forms can be reliably created by laser heating and cooling techniques.

The process of melting and vaporizing materials is the way in which lasers weld and drill or cut. These require more tightly focused light to achieve the necessary temperatures. When matter becomes molten, the vapor pressure above it can depress the surface and form a hole. The hole can be supported by the vapor pressure and, because light is more strongly absorbed in the hole, a deep hole is generated. This is the process of forming a "keyhole" to obtain deep-penetration drilling, cutting, or welding. In the welding process the atmosphere and irradiation conditions are chosen to encourage proper mixing of the materials to be welded. In drilling or cutting the conditions are chosen to enhance material removal from the irradiated zone; thus, auxiliary gas jets are used to blow away the molten material and sometimes to react with the heated matter to speed the reaction.

While laser treatment of metals was the first area of laser materials processing to receive major attention, it is by no means the only activity in which lasers are used to process materials. A great deal of work has gone into developing laser techniques for processing organic materials, composites, and ceramics. One type of processing, readily visible in many homes, is the use of lasers in carving very detailed images in wood. This type of artwork is made possible by the selective nature of laser heating and the ease with which optical imaging techniques and computer-controlled workpiece handling can be integrated.

Laser processing of semiconductors, in particular silicon, has generated very wide interest in the electronics industry. The application arose from the need to find efficient and effective ways to anneal ion-implanted silicon. The process of ion implantation creates structural damage in semiconductor material. Annealing an ion-implanted silicon wafer by conventional means requires heating in a convection furnace at temperatures over 1000°C for times of the order of $\frac{1}{2}$ hr. As a result, wafers are often warped and undergo some degree of chemical decomposition. With microelectronic devices dependent on precise electronic and physical properties these present serious problems. In the mid-1970s researchers in the Soviet Union, and later in the United States and Europe, began to study the use of lasers to anneal the implanted surfaces of silicon wafers. While the work began as a way to treat damage due to ion implantation, it has been applied to such other problems as the removal of dislocation networks introduced by high-temperature diffusion and the reduction of misfit defects in epitaxial silicon on sapphire. Lasers have also been used to provide localized, controllable heat sources for epitaxial growth of evaporated silicon on crystalline silicon, for the growth of large-grain polysilicon from small-grain polysilicon, and for the formation of metallic silicides and metal overlayers on silicon.

The interaction that makes this type of processing possible is again that of localized rapid heating and cooling. When sufficient heating has taken place there can be a solid phase rearrangement of atoms in the silicon that was damaged by the implantation or, if melting has been achieved, recrystallization can occur with the undamaged material serving as the seed. There is no debate that the processes of interaction that allow these effects to take place are thermal when cw lasers or lasers with pulse durations greater than 100 nsec are used. If pulses less than 100 nsec in duration are employed, solid phase rearrangements cannot take place quickly enough and the material must be melted by the irradiation. The process is similar to that of laser glazing of metals by very rapid melting and resolidification.

Initially there was some concern over the process of semiconductor annealing using pulses with durations in the subpicosecond range. This centered on the issue of whether there was time for thermal processes to operate or whether some other interaction had to be invoked to explain the observed phenomenon. Elegant experiments were performed to which the process of melting and resolidification with ultrashort pulse irradiation was monitored by Raman scattering. The interaction process was confirmed to be thermal in nature. This is a very important scientific result in laser–materials interactions research in that it confirms the fact that thermalization takes place in times as short as several femtoseconds.

Other laser–materials interactions applied in the electronics industry involve the unique ability of laser light to be focused to very small areas and to heat only the areas irradiated. This allows lasers to serve a wide range of resistor and capacitor trimming applications. Also, when properly focused, lasers are able to remove unwanted bridges in microcircuits. If the atmosphere above the substrate is selected to form desired deposits when the substrate or the gas is heated, then lasers can be used to personalize microcircuit masks and to repair damaged circuits. Work with ultraviolet lasers may make it possible to process microcircuit components with submicrometer dimensions.

Lasers of sufficient intensity can cause very highly reflective or transparent materials to be damaged as a result of very high order processes that come into play when very high optical intensities are employed. In Section I,B, the process of nonlinear absorption and breakdown was described. These processes result in very rapid conversion of the nonabsorbing matter into matter that absorbs nearly 100% of the incident light. The absorption takes place locally in the irradiated region and the heated material melts and may vaporize. When melting or vaporization occurs in in the bulk of a transparent substance severe cracking also takes place. In addition, some materials display slip banding after intense laser irradiation, indicating that they were heated to the point where plastic deformation could occur.

While breakdown due to nonlinear processes represents the intrinsic failure mechanism of nonansorbing materials, laser-induced damage is most often determined by the presence of some discrete inclusion or irregularity in the irradiated volume. Damage due to absorption at such defects can be detected as arrays of discrete, randomly located damage sites within the irradiated region. In thin films and on most bare surfaces this type of defect-determined damage dominates the interaction and response of the matter with laser light.

Bibliography

Bass, M. (1983). *In* "Physical Processes in Laser Materials Interactions," pp. 77–116. Plenum, New York.

Bertolotti, M., ed. (1983). "Physical Processes in Laser Materials Interactions," pp. 175–220. Plenum, New York.

Brown, W. L. (1983). *In* "Laser Materials Processing" (M. Bass, ed.), pp. 337–406. North-Holland Publ., Amsterdam.

Duley, W. W. (1976). "CO_2 Lasers: Effects and Applications." Academic Press, New York.

Johnson, A. W. *et al.* (eds.) (1989). "Laser and Particle Beam Chemical Processes on Surface Materials." Research Society Symposium Proceedings, Vol. 129.

Ready, J. F. (1971). "Effects of High Power Laser Radiation." Academic Press, New York.

Wooten, F. (1972). "Optical Properties of Solids." Academic Press, New York.

Yardley, J. T. (1985). *In* "Laser Handbook" (M. Bass and M. L. Stitch, eds.), Vol. 5, pp. 405–454. North-Holland Publ., Amsterdam.

LASER PULSATIONS, DYNAMICAL

Neal Abraham *Bryn Mawr College*

GLOSSARY

Beat frequency: Modulation frequency of the laser intensity, which is the difference between two optical frequencies found for the optical field.

Chaos: Irregular fluctuations resulting from deterministic nonlinear interactions.

Dynamics: Deterministic evolution.

Laser: Combination of an optical resonator and a medium that can store energy and amplify light over some range of wavelengths.

Mode: Particular solution for a resonant wave pattern in the laser resonator that corresponds to a particular optical frequency.

Mode locked: Operating condition for which modes have fixed amplitudes, equal spacings in frequency, and fixed phase differences.

Nonlinear: Proportional to some power of the variable or to the product of two variables. Nonlinear laser medium—the amplification by the medium is not simply a constant times the intensity of the light. Nonlinear equations: each variable depends on powers or products of variables.

Period doubling: Change of a periodic pulsing pattern to one that takes twice as long (twice as many pulses) to repeat.

Population inversion: Stored energy in a laser medium by having more atoms (molecules) in the relevant high-energy state than in the relevant low-energy state. This ratio of the two populations is inverted with respect to the usual thermodynamic equilibrium conditions.

Q switching: Modulation of the amount of light energy lost on transiting the laser cavity. A common way to excite mode locking or giant pulses.

Quasiperiodic pulsing: Pulsing governed by two independent frequencies.

Rabi oscillation: Oscillating extraction and storage of energy in the laser medium by a constant intensity light field.

Relaxation oscillation: Oscillatory exchange of energy between the laser medium and the light field in the laser cavity.

Slowly varying amplitude approximation: It is assumed the light can be described as a wave of nearly fixed frequency and wavelength, which has an amplitude that changes slowly with respect to the period of the optical frequency of the wave.

Spontaneous emission: Random incoherent light emitted by the laser medium.

Stimulated emission: Light emitted by the laser medium in the presence of another field. The stimulated emission matches the present field in wavelength, phase, and direction.

Stochastic: Random fluctuations such as resulting from a large number of independent effects.

Lasers may emit their intense, highly directional and coherent beams either at a constant rate [sometimes called continuous wave (cw)] or in a pulsed fashion. Pulsations are often of greater use than constant intensity, so the history of lasers is replete with efforts to both understand spontaneous pulsations and regulate them. In the last few years, it has become clear that many laser pulsations originate in internal dynamical processes and cannot be attributed to internal or external noise or simply to the constructive interference of emissions at the resonant frequencies of the laser cavity. In this respect, the recent discoveries and studies of laser pulsations draw on the language of the relatively new field of nonlinear dynamics and descriptions of behavior in terms of dynamical chaos or turbulence. This provides an alternative perspective to the more conventional engineering view of lasers as devices and gives new insight into how laser pulsations can be generated or controlled. [*See* LASERS; ULTRAFAST LASER TECHNOLOGY.]

The key features of a laser are a nonlinear medium (in which energy can be stored and from which it can be extracted in the form of light) and a resonant cavity formed by mirrors that recirculate a portion of the emitted light for further interaction with the medium. As the recirculated light is an electromagnetic wave, it will build up only if it constructively interferes with the field from previous round trips in the cavity and only if the medium can be stimulated to emit at the wavelength of the circulating field. Usually, the range of possible emission wavelengths of the medium is only a small fraction of the most resonant wavelength. Typical values range from 10^{-5} for gas lasers and 10^{-3} for solid-state lasers to as much as 10^{-2} for dye lasers. Nevertheless, because optical wavelengths are small (of order 10^{-6} m), there is the possibility that more than one wavelength within the resonance bandwidth of the medium is also resonant with the laser cavity. The simplest cavity resonance condition is given by the requirement that the length of the laser cavity equal a multiple of the wavelength. If the laser cavity is about 1 m in length then the cavity resonance frequencies are spaced by only 10^{-6} of their absolute value. The number of cavity resonance frequencies interacting with the medium provides the basis for a fundamental classification of laser operation as being either single mode or multimode.

I. Characteristic Times for Laser Processes

A laser cannot be imagined as involving an instantaneous process of turning a control signal into light. Instead, because emission involves energy transfer from a source to an atomic or molecular medium and then to light and because the light energy can be reabsorbed by the emitting medium, there is a dynamical response with associated delays that can reshape or distort any control signal or lead to spontaneous pulsations.

Four characteristic times exist for laser pulsations. The first is the round-trip time in the laser cavity (the inverse of the "mode-beating" frequency between adjacent cavity modes), which is most important if the light energy is compressed in time to a pulse that is much shorter than the time for it to travel once around the path defined by the laser mirrors. Such pulsations are characteristic of multimode operation. The second time is the period of oscillatory energy exchange, called relaxation oscillations, between the energy of the light field in the laser cavity and the energy stored in the medium. The third is the period of coherent modulation of the emission of light by the medium that occurs in the presence of a strong and constant intensity light field. These oscillations are known as *Rabi oscillations*. The fourth is the characteristic time of external perturbations of the laser resulting from modulation of the rate of excitation of the medium, the length of the laser cavity, the energy loss from the laser, or some other parameter of the system. The possibility of having sustained pulsations is enhanced whenever two or more of the characteristic processes have nearly equal frequencies.

The general focus here will be on dynamical processes in the laser or on its response to the time dependence of the external control of the laser. The first sections deal with single-mode pulsations, the phenomena under which a single resonant mode of the laser field displays slow modulation of its amplitude. The later sections consider how the intermode beat frequencies lead to complex modulation of the laser or how they can be created by external controls. A few simple examples are discussed first.

A. Mode Beating

When the laser operation includes modes of different wavelengths, the total intensity is modulated in time as illustrated by the following example. Suppose frequencies f_1 and f_2 correspond to two modes given by

$$f_1 = c/\lambda_1, \qquad f_2 = c/\lambda_2$$

where λ_1 and λ_2 are the wavelengths and c is the speed of light. Then, the total electric field at a fixed point in space in the output of the laser can be written

$$E(t) = E_1 \cos(2\pi f_1 t + \phi_1) + E_2 \cos(2\pi f_2 t + \phi_2)$$

where E_1 and E_2 are the amplitudes of the two modes. The total intensity at this point is given by the square of the electric field and can be written as

$$\begin{aligned} I(t) = {} & E_1^2 \left[\tfrac{1}{2} + \tfrac{1}{2}\cos(4\pi f_1 t + 2\phi_1)\right] \\ & + E_2^2 \left[\tfrac{1}{2} + \tfrac{1}{2}\cos(4\pi f_2 t + 2\phi_2)\right] \\ & + E_1 E_2 \cos[2\pi(f_1 - f_2)t + \phi_1 - \phi_2] \\ & + E_1 E_2 \cos[2\pi(f_1 + f_2)t + \phi_1 + \phi_2] \end{aligned}$$

Since the optical frequencies are very high (of order 10^{14} Hz), no detector can provide an electronic output at these frequencies. Instead, it must average over oscillations at $2f_1$, $2f_2$, and $f_1 + f_2$. The possible detected signal denoted by I is then

$$I(t) = \tfrac{1}{2}E_1^2 + \tfrac{1}{2}E_2^2 + E_1E_2 \cos[2\pi(f_1 - f_2)t + \phi_1 - \phi_2]$$

The result shows intensity oscillations at the frequency that is the difference between the frequencies of the two modes. If the modes have the same transverse spatial pattern, then the period of these pulsations corresponds to the time for the light in the laser cavity to make one round trip. Much of the multimode pulsation phenomena can be understood from simple generalizations of this example, as will be discussed in Section III.

B. Relaxation Oscillations

The relaxation oscillation of energy exchange between the optical field and the laser material is damped in most laser systems and thus can typically be observed only if the laser is switched in some way. Then the laser restores itself to a steady, time-independent operation after an oscillatory transient. An example for a NdP_5O_{14} laser is shown in Fig. 1. These pulsations occur only if the average rate of decay of the field in the empty laser cavity is greater than the decay rate for the energy stored in the medium. Many lasers exhibit this property, including other solid-state lasers (ruby, Nd: glass, Nd: YAG); many low-pressure molecular lasers, such as (CO_2, N_2O, CO, and NO); and semiconductor lasers.

II. Single-Mode Pulsations

A. Homogeneously Broadened Lasers

The classic model studied by laser theorists for spontaneous pulsations in a single-mode laser is that

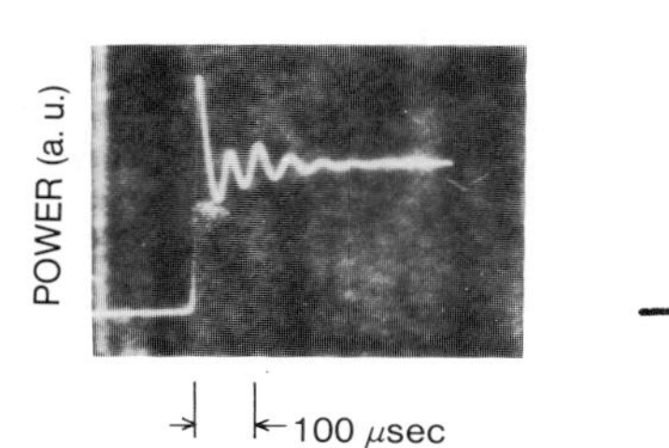

FIG. 1. Damped oscillatory transient in the output of a NdP_5O_{14} laser at 1.3 μm after an abrupt switch in the otherwise constant excitation of the laser. [Reprinted with permission from W. Klische, H. R. Telle, and C. O. Weiss (1984). *Opt. Lett.* **9**, 561.]

for a simple medium made of a collection of identical two-level atoms. First introduced by Uspenskiy and Oraevsky in 1963 and popularized in the Western literature by Haken, it has been extensively studied and has been shown to be practically equivalent to a model developed by Lorenz to describe convective turbulence in fluids. Hence, the Haken–Lorenz model is of special interest because it is one of the simplest models rooted in physical reality that has irregular pulsation solutions of the type known as dynamical chaos. In this model, the laser operates with a constant intensity output for a range of excitation levels reaching from the threshold value needed for laser operation to a value 10 or 20 times higher. Above this second threshold, the laser output spontaneously changes to chaotic pulsations. If the laser cavity is slightly detuned from resonance with the medium, the pulsations become more periodic, showing an inverse period-doubling cascade before becoming periodic and then disappearing in favor of stable laser intensity for large enough detunings. The origin of this spontaneous modulation can be thought of as a matching between the relaxation oscillations and the coherent Rabi oscillations. Despite (or because of) the simplicity of this model, the minimum conditions for reaching the second laser threshold have been difficult to achieve experimentally. The decay rate of the laser field in the cavity must be larger than the sum of the decay rates of the population inversion and the atomic polarization—called "the bad cavity condition." This typically requires a relatively lossy cavity because the material polarization usually relaxes very quickly, and then one must have sufficient excitation of the medium to have optical amplification to offset these losses. The stringency of these conditions made it seem for more than 20 yr after the model was developed that it could not be achieved for any laser material. Quite recently, however, Weiss and co-workers at the PTB in the Federal Republic of Germany have demonstrated that optically pumped FIR lasers (discussed further below) can closely approximate the predicted behavior of the Lorenz–Haken model as shown in Fig. 2.

B. Inhomogeneously Broadened Lasers

Perhaps the easiest spontaneous single-mode pulsations to observe experimentally are those in inhomogeneously broadened lasers as first reported by Casperson and Yariv in 1974. In 1978, Casperson outlined a clever intuitive picture for the origin of these pulsations. The single mode of the laser interacts with only those atoms whose resonant frequencies are close to the laser frequency while other at-

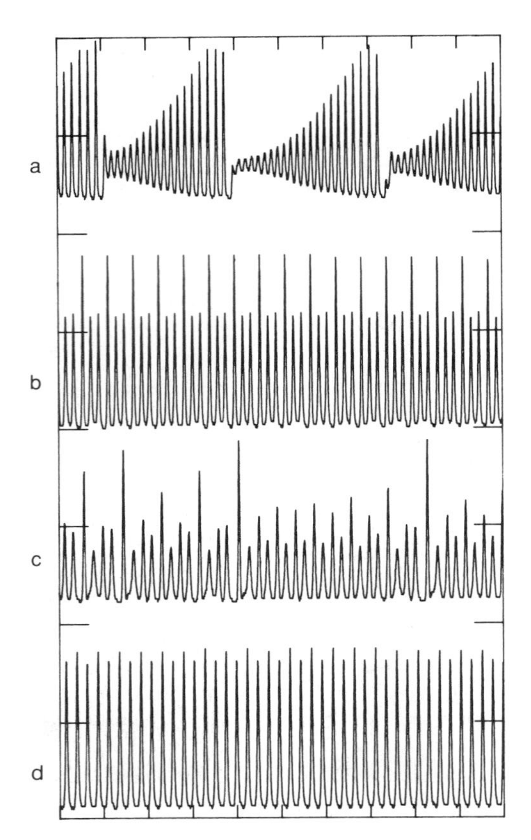

FIG. 2. Intensity pulsations from an optically pumped NH_3 laser at 81 μm showing behavior emulating that predicted for the single-mode homogeneously broadened laser. (a) Irregular Lorenz-like spiral chaotic pulsations for a resonant cavity; (b) period-three pulsations for a small detuning; (c) period-doubling chaos; and (d) period-two pulsations with increasing detuning. (Courtesy of C. O. Weiss.)

oms are largely unaffected. The number of these atoms is depleted by the strong field. The missing atoms in a narrow frequency range near the laser frequency cause the index of refraction (and hence the speed of light and the wavelength of light in the medium) to vary rapidly with frequency. The result is that weak fields at nearby frequencies can have the same wavelength as the laser mode and hence they are also resonant, and, having gain from undepleted atoms, they grow. The resulting beat frequencies between the sidebands and the original frequency cause intensity pulsations. Complex patterns of periodic and chaotic intensity patterns are observed for excitations only a few times above the lasing threshold. Samples of experimental results and numerical solutions of a theoretical model are shown in Fig. 3. More recently more formal numerical and analytical solutions of models and detailed experiments have

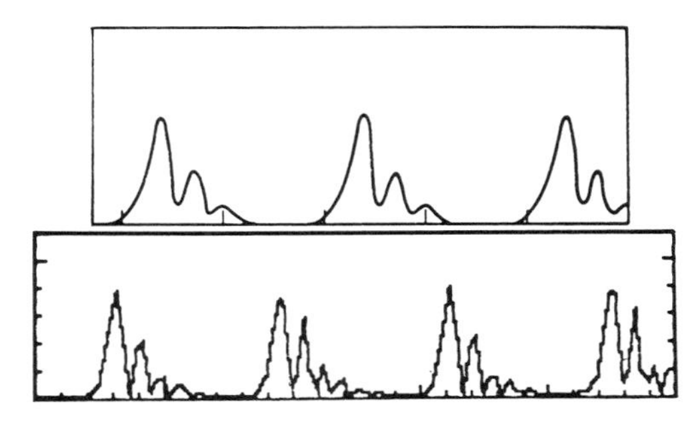

FIG. 3. Intensity pulsation patterns from experiments and numerical modeling for a Xenon laser at 3.51 μm. [Reprinted with permission from L. W. Casperson (1978). *IEEE J. Quantum Electron.* **QE-8,** 756. Copyright © 1978 IEEE.]

expanded the understanding of spontaneous pulsations in these lasers in both ring and standing wave geometries.

C. Optically Pumped Lasers

The dynamics of optically pumped lasers have been intensively studied since 1984 when Weiss and Klische proposed that they might be used to achieve the Haken–Lorenz model of a laser with a homogeneously broadened two-level material. Theoretical work shows that it is hard to justify a two-level model when the upper lasing level is not incoherently excited but rather is pumped by a constant laser intensity from a lower (nonlasing) level. Three-level models are inherently more complex than two-level ones and the various types of dynamical pulsations that are predicted include many more periodic pulsations and types of pulsations not found in the Haken–Lorenz model. Many of these features have been found in work with far infrared and mid-infrared lasers in the range of 12 to 400 μm for which the upper level is pumped by such molecular lasers as CO_2. Periodic pulsing is frequently observed, and the transition with increasing excitation from stable laser operation is noted for the appearance of periodic pulsations (at about the Rabi and relaxation oscillation frequencies) followed by period doubling to chaos. Surprisingly, despite the difficulty of theoretical justification, it does appear that experimental behavior remarkably similar to the predictions of the simpler two-level model can be achieved in some of these lasers (see Section II, A).

D. Lasers with Saturable Absorbers

For the wide class of lasers that do not spontaneously generate single-mode pulsations, it is possible

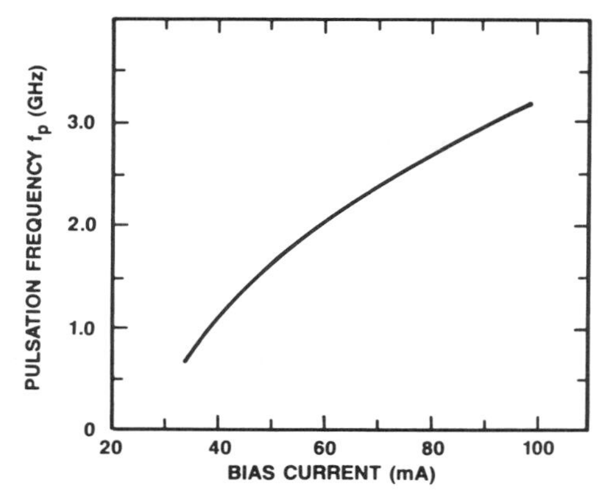

FIG. 4. Frequency of sustained pulsations at the relaxation oscillation frequency, induced in a semiconductor laser by the presence of saturable absorbers from proton-bombardment damage. [Reprinted from M. Kuznetsov, D. Z. Tsang, J. N. Walpole, Z. L. Lau, and E. P. Ippen (1987). *Appl. Phys. Lett.* **51,** 895.]

to induce such behavior by adding an optically absorbing medium inside the laser cavity. The key feature of the absorber is that it must be bleachable, that is, it must become more transparent upon illumination with sufficient light. In this case, it is easy to show that lasers which would otherwise exhibit only transient damped relaxation oscillations will, instead, provide sustained trains of pulses. The pulsing frequency can be approximately that of the relaxation oscillations (if the absorber has a fast recovery time) or at some mean between the relaxation oscillation rates of the laser medium and absorbing medium separately (if the absorber relaxes more slowly). Pulsing of this type has been found in ruby lasers, molecular gas lasers with other absorbing gases, and semiconductor lasers with impurities, damaged regions or separate subsections of the material acting as the absorber. Samples of this behavior are shown in Fig. 4. This method of inducing pulsations is sometimes called *passive Q switching;* passive because the dynamics occur within the laser itself and are not driven from the outside, and Q switching as the overall quality of the laser cavity, as governed by its losses, is changed by the saturable opacity of the absorber.

E. Modulation

Modulation of the excitation or loss of a laser at a frequency close to the relaxation oscillation frequency is an *active* way to induce sustained pulsations. Long used since the work of Helleman in the early 1960s as a way to generate regular Q-switched pulses, it has more recently been shown by Arecchi

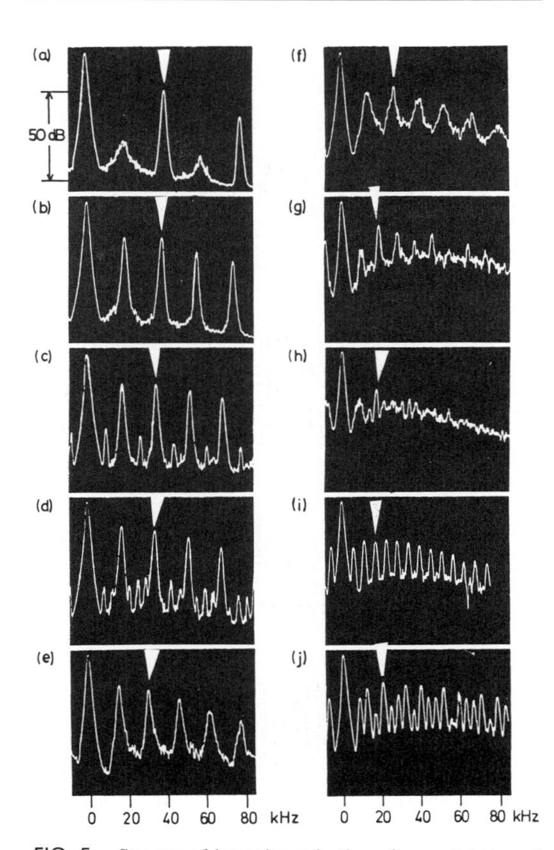

FIG. 5. Spectra of intensity pulsations for modulation of the excitation of the solid-state laser in Fig. 1 showing period doubling (subharmonics) to chaos (h) and period-3 and period-5 windows. The modulation frequency is marked by an arrow. [Reprinted with permission from W. Klische, H. R. Telle, and C. O. Weiss (1984). *Opt. Lett.* **9,** 561.]

and others that it is a convenient way to achieve period doublings in the pulsations or chaos. The specific response of the laser is quite sensitive to detuning of the modulation frequency from the relaxation frequency or to changes in the degree of modulation. Sample results are shown in Figs. 5 and 6 for modulation of CO_2 and solid-state lasers. Generally, it is found that modulation of the loss of the laser (with an electro-optic modulator, for example) is much more effective in generating complicated pulsation patterns than is modulation of the laser excitation.

Because the response of the laser and the interactions of the field and the lasing material are nonlinear, the laser will also resonantly respond when the modulation frequency is a harmonic or subharmonic of the relaxation oscillation frequency. This is obvious from some of the results of Fig. 5. The optimum frequency for resonant response also shifts somewhat depending on the amplitude of the modu-

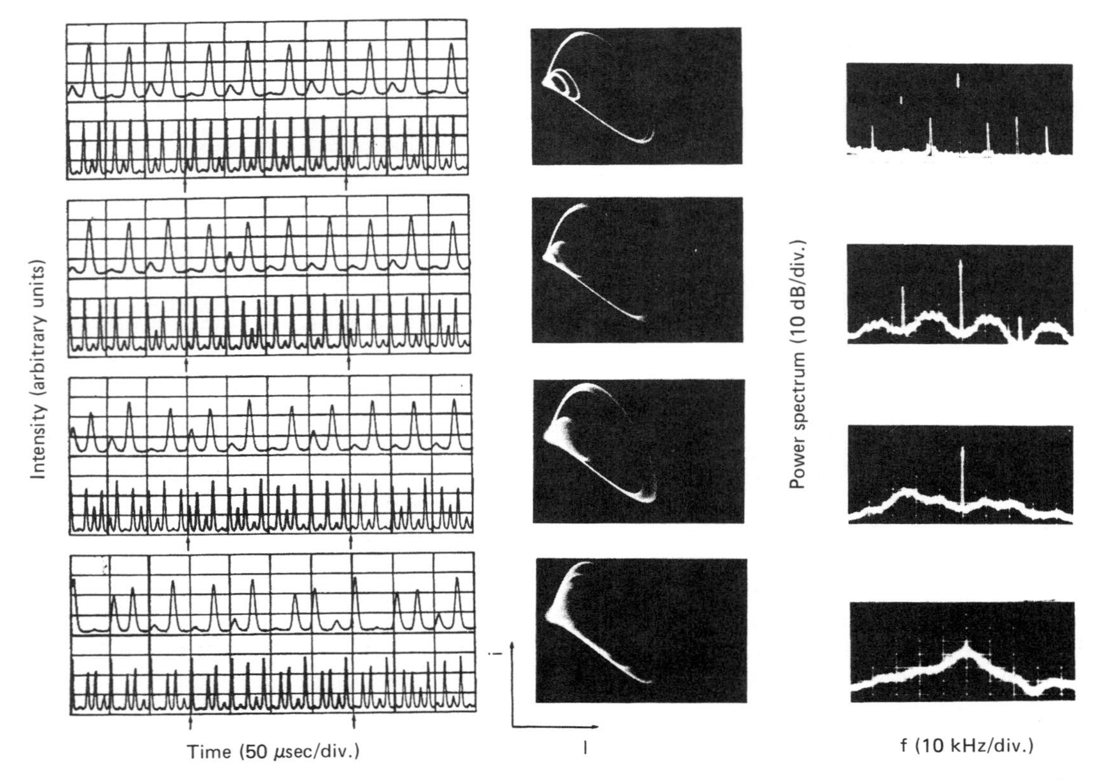

FIG. 6. Periodic and chaotic pulsations from the modulation of the loss of a CO_2 laser (time traces, phase portraits, spectra). [Reprinted with permission from R. Meucci, A. Poggi, F. T. Arecchi, and J. R. Tredicce (1988). *Opt. Commun.* **65,** 151.]

lation, another common effect in nonlinear systems. Winful and Chen in 1984 combined the intracavity absorber (making a free-running laser that pulsed) and modulation to find a variety of quasiperiodic and locked phenomena in a semiconductor laser with locking occurring for many different rational ratios of the two frequencies.

F. Injection Locking

An important way to stabilize the amplitude, frequency, and phase of a laser is to *seed* it by injecting a portion of light from another laser. When successful in slaving the injected oscillator to the master oscillator, this method is useful for achieving stronger signals or high spatial and temporal coherence in arrays of lasers. However, if the injected signal is not large enough compared to that of the free-running slave laser, locking will not occur. Instead the free-running slave laser field and the injected signal combine to form a modulated intensity which then drives the nonlinear dynamics of the slave laser. Sometimes this results in no more than the simple beat frequency (effectively no interaction). However, especially if the beat frequency is close to the relaxation oscillation frequency, the pulsations that result are much like the Q switching induced spiking in the intensity. On closer examination, one can observe that the frequency and phase of the slave laser nearly lock to the master oscillator, and then rather suddenly the phase of the slave laser slips by 2π and nearly locked conditions resume. While not yet of great technological interest, the case of not-quite-locked interaction of two laser oscillators has provided rich patterns of periodic and chaotic pulsations.

III. Multimode Pulsation

A. Bidirectional Ring Laser

In a standing wave laser, the forward and backward going waves are not independent because they are connected by reflection at the mirrors. In contrast, the forward and backward waves in a ring laser are less strongly coupled, interacting only through their mutual drawing of energy from the single lasing medium. If the laser is rotated about an axis perpendicular to the plane of the ring cavity, the forward and backward waves have different resonant frequencies for the same mode number (same number of wavelengths). This frequency difference can be detected by combining the outputs of two waves. The beat frequency provides the basis of laser gyroscopes now used in advanced guidance systems. In practical devices, it is found that if the frequency difference is too small the modes couple strongly enough to lead either to frequency locking or to irregular dynamical pulsations. Usually this is attributed to scattered light from one mode that acts as an injected signal for the other mode.

Even in the absence of backscattering, the two beams are coupled strongly through interaction in the medium. Not only do the modes compete for the available energy stored in the medium, but the standing wave interference pattern of the two waves saturates the medium more strongly at points of constructive interference and less strongly where there is destructive interference. The resulting corrugated pattern in the medium acts like a coherent scatterer and reflects each wave into the other.

In some cases, the grating is not important because the population either diffuses quickly or naturally relaxes rapidly. In these cases, usually one mode dominates the other, although the relative dominance can be inverted, and switches can be induced by noise or other disturbances. When the grating lasts a relatively long time, the modes can dynamically interact leading to regular and chaotically irregular pulsations. Samples of pulsations predicted and observed for such lasers are shown in Fig. 7. The fast modulation is typically the residual of relaxation oscillations, and the slower switching depends on the population inversion recovery rate and on the cavity detuning.

B. Mode Locking

Short pulses are most easily generated by the constructive interference of many equally spaced modes of the same spatial pattern. Then, the height of each pulse depends on the square of the number of modes, and its time duration is approximately the time for light to make one round trip in the laser cavity divided by the number of modes. The interval between output pulses is the time for the pulse formed in the laser to circulate once in the cavity. However, because the modes interact with the same medium, they compete with each other to some degree and one cannot directly obtain simultaneous free-running operations on many modes. In addition, the dispersive ef-

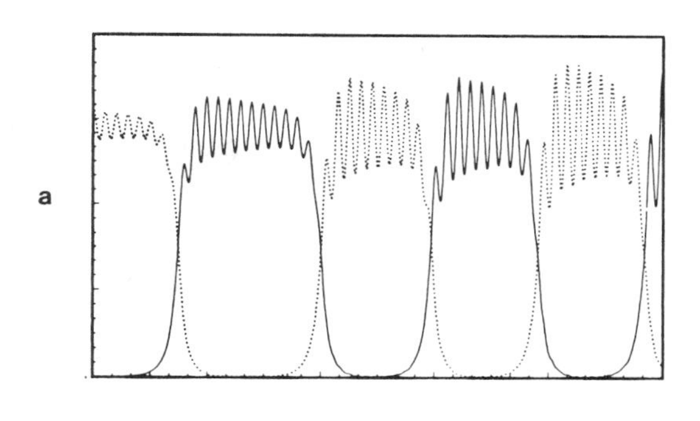

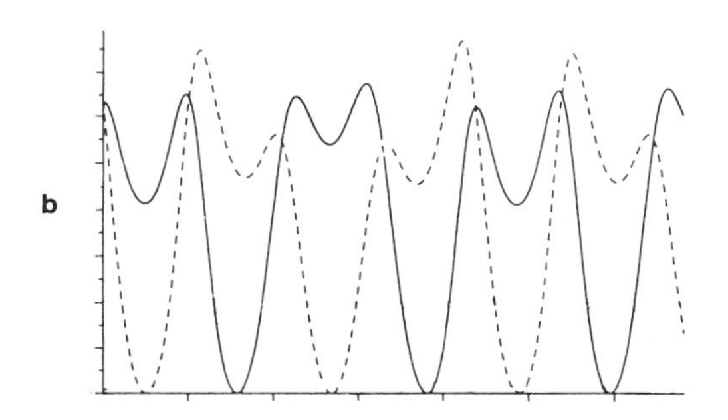

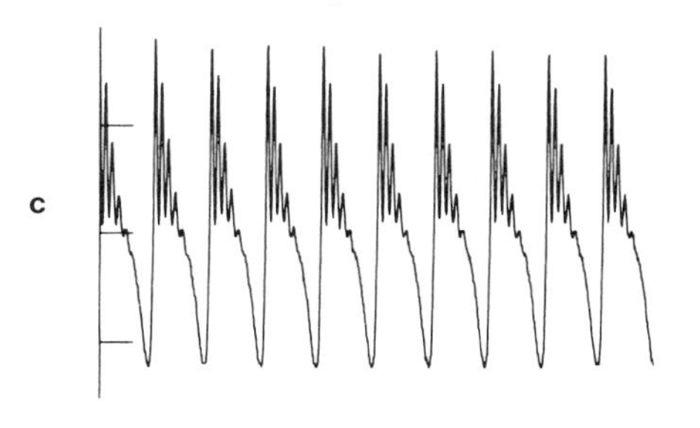

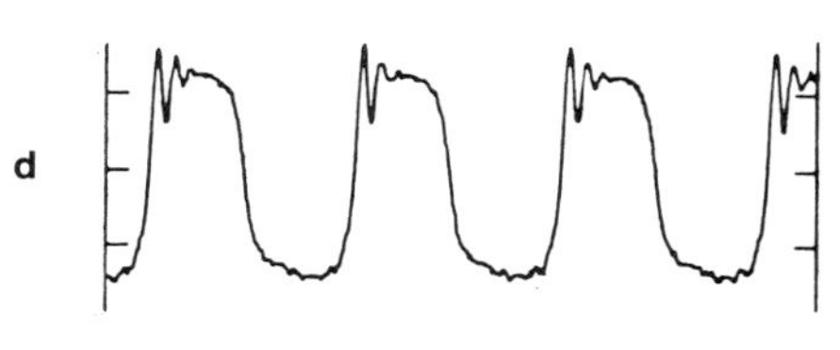

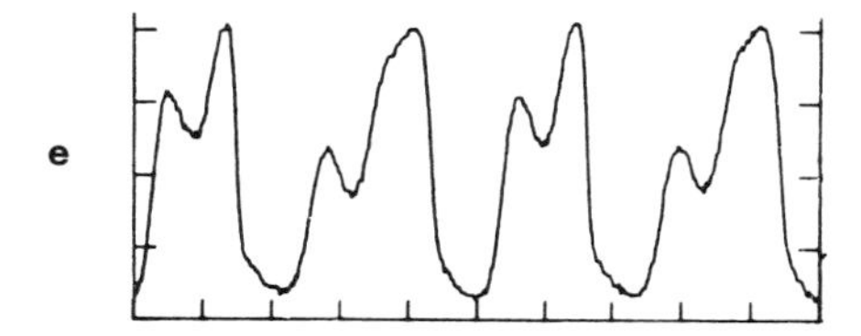

FIG. 7. (a) and (b) Predictions for the intensity pulsations in the two modes of a bidirectional CO_2 laser. (Courtesy of L. M. Hoffer, G. L. Lippi, and N. B. Abraham.) Also, observations of the pulsations of one mode of a bidirectional NH_3 FIR laser at 81.5 (c) and 153 μm (d) and (e). (Courtesy of C. O. Weiss and N. B. Abraham.)

fects of the medium cause the mode frequencies to be unequally spaced to some degree, making mode-locked operations more difficult. A mechanism for spontaneous mode locking was proposed in 1964 by Lamb, who noted that any two modes at different frequencies interacting with the medium can combine to generate a third optical frequency. If the modes are nearly equally spaced, the third frequency lies near a third mode and the *combination frequency* may act as an injected signal to pull the third mode into a locked condition. By this mutual interaction, some many-mode lasers spontaneously enter the phase-locked and equal frequency spacing conditions necessary for short pulse generation.

Mode locking can also be accomplished if the laser is modulated at the intermode spacing frequency (active mode locking) or if a saturable absorber is placed within the laser (passive mode locking) to strengthen mode interactions and to stimulate the circulation of spatially localized pulses in the laser cavity.

The relaxation oscillation frequency and the Rabi frequency can also play strong roles in mode locking. Coupling between modes is enhanced if their frequency spacing matches either of these two dynamical oscillation frequencies of the field-matter interactions. Examples of these results for semiconductor and dye lasers are shown in Figs. 8 and 9, respectively.

The foregoing discussion has been limited to the interaction of longitudinal modes, those having the same transverse field pattern and differing only in

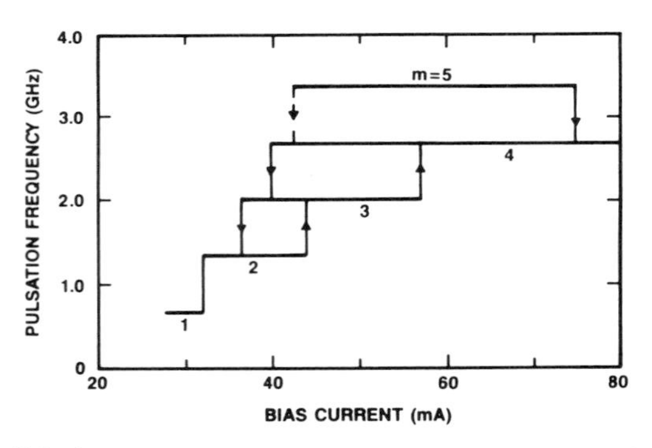

FIG. 8. Mode-locked operation involving modes spaced by different numbers of quiescent modes as indicated for a semiconductor laser for which the relaxation oscillation frequency is adjusted by changing the laser excitation. The laser mode spacings were reduced by using antireflection coating of the laser diode and a 22-cm-long external cavity. Proton bombardment of the laser also provided fast internal saturable absorbers. [Reprinted with permission from M. Kuznetsov, D. Z. Tsang, J. N. Walpole, Z. L. Lau and E. P. Ippen (1987). *Appl. Phys. Lett.* **51**, 895.]

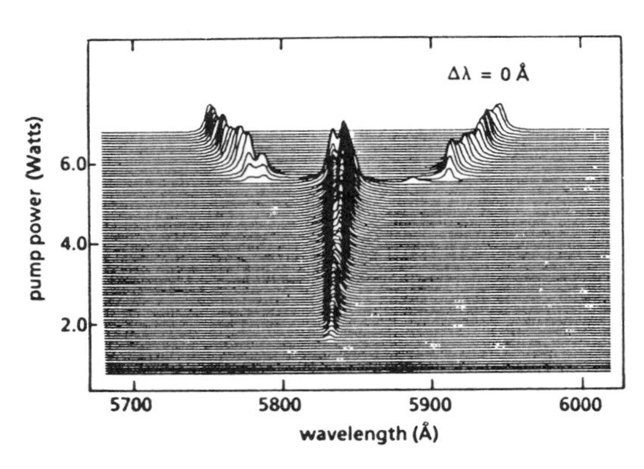

FIG. 9. Optical spectrum of dye laser operation showing splitting of single-mode operation into two-mode operation where the mode spacing grows with laser excitation according to the growth in the Rabi oscillation frequency. [Reprinted with permission from C. R. Stroud, Jr., K. Koch, S. Chakmajian, and L. W. Hillman (1986). *In* "Optical Chaos" (J. Chrostowski and N. B. Abraham, eds.), *Proc. SPIE* **667**, 47.]

the number of wavelengths in the longitudinal direction needed to complete a round trip in the cavity. Modes of different transverse field patterns have different intermode frequency spacings, and in some circumstances, they too can be locked together. The resulting pulsation rate is not directly related to the round-trip time in the cavity but rather to the frequency spacing. The laser output pulsations in this case also appear in a spatially varying transverse pattern as the modes of different field profiles interfere constructively at different places at different times.

C. Combination Tones

If three or more modes coexist in a laser without locking, they can generate overall modulations that are at frequencies much lower than the mode spacing frequencies. These arise from a beating between the combination frequencies and the adjacent modes. The resulting beat frequency is given by the difference between the adjacent intermode spacings and is often in the range of tens of kilohertz to a few megahertz. This low frequency noise is typical of unlocked multimode operation and vanishes completely when mode locking is accomplished. Systematic studies of this phenomena by Weiss and co-workers and by Halas and co-workers in 1983 revealed that period doubling and quasiperiodic routes to chaotic low-frequency modulation are possible in these systems. In both cases, multiple transverse mode operation in inhomogeneously broadened lasers was studied. Inhomogeneous broadening of the laser medium

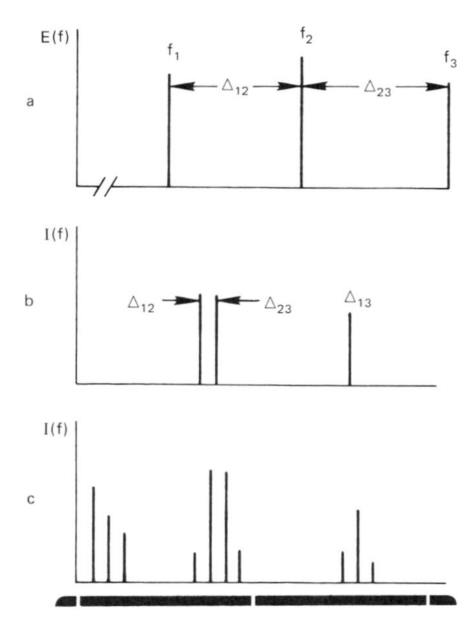

FIG. 10. Schematic origin of combination tone pulsing. (a) Optical spectrum of a three-mode laser, (b) intensity pulsation spectrum of uncoupled three-mode operation, and (c) intensity pulsation spectrum from combination tone pulsing with low-frequency pulsing generated by the second-order differences in the mode frequencies.

reduces the mode interaction and provides intensity dependent dispersive effects that tend to favor the combination tone pulsing over mode locking. A sequence of optical and intensity spectra illustrating this effect is shown in Figs. 10 and 11.

When the combining modes are transverse modes, then one can also study the spatial patterns of the laser output. When there is simple mode-beating, the pattern blinks on and off rapidly at different points in the pattern in a kind of "machine-gun" pattern. However, if the modes are very close together in frequency, they may lock to a common frequency with the result being a time-independent though spatially distorted pattern. When the modes are far apart and when many modes are involved, the spatio-temporal dynamics has complexity similar to that of two-dimensional fluid turbulence.

D. External Cavities

In some lasers, particularly semiconductor lasers used for coherent applications, one wishes to suppress multimode operation in favor of stable single-mode behavior. A design to accomplish this in long laser cavities (which have modes closely spaced in frequency) is to insert an extra pair of reflecting sur-

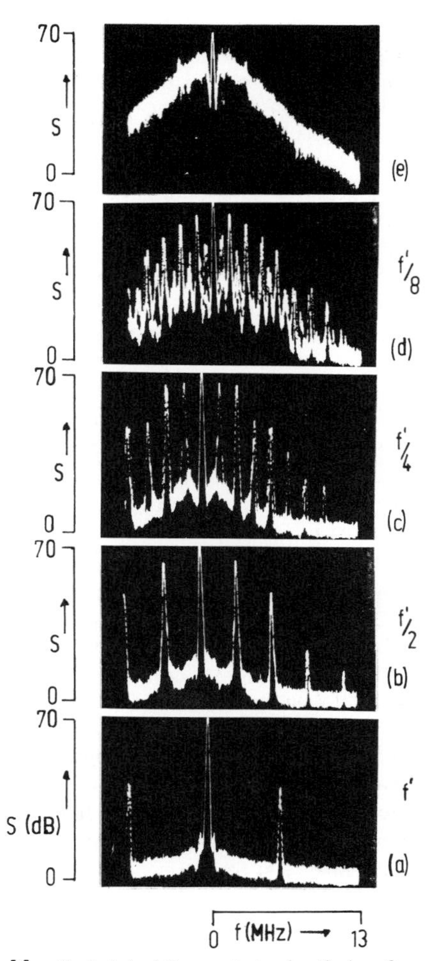

FIG. 11. Period-doubling route to chaotic low-frequency combination tone pulsing in the HeNe 6328-μm laser. [Reprinted with permission from C. O. Weiss, A. Godone, and A. Olafsson (1983). *Phys. Rev. A* **28,** 892.]

faces (called an etalon) within the round-trip path in the laser cavity. The resonant frequencies can be widely spaced because of a short distance between the surfaces. Oscillation on a mode of the laser cavity, which also is resonant with the etalon, is selected while other modes of the cavity are suppressed by their lack of resonance with the etalon.

For lasers that use the reflection from the end surfaces of the laser material as the laser mirrors, as is done in most semiconductor lasers, it is not possible to insert another etalon in the cavity. Tsang at AT&T Bell Laboratories overcame this by splitting the semiconductor in two pieces (called "c^3" for *cleaved coupled cavity*), creating an etalon between the pieces and also providing two regions for differing excitation.

Another approach for semiconductor lasers has been the addition of an external third mirror with the idea that optical feedback might suppress noise and provide selective enhancement of one of the laser modes. Often this has been found to lead to spontaneous pulsations. This occurs when the external path length provides delay of the optical feedback roughly corresponding to the period of the relaxation oscillations in the laser. As such frequencies are of order 10^9 Hz, the corresponding external cavity round-trip length is 30 cm. Müller and Glas in 1984 and Otsuka have demonstrated that a wide range of periodic and chaotic oscillations can be excited by even very weak feedback from such external reflectors.

IV. Noise

Lasers carry with them their own intrinsic noise, that of spontaneously emitted incoherent light from the laser medium. In addition, thermal, mechanical, acoustical, electrical, and excitation fluctuations can all be present in practical laser devices. Noise disrupts the coherent dynamical interaction of the optical field and the laser medium. This can provide modulation of otherwise stable, constant intensity lasers, and it can also disrupt periodic or chaotic oscillations making them either more or less regular depending on the particular operating conditions. For example, when systems that respond to abrupt disturbances with damped relaxation oscillations are continuously disrupted by noise, they tend to retain a relatively large amplitude of noisy modulation at the relaxation oscillation frequency. Noise-driven irregular pulsations can be distinguished from irregular chaotic pulsations because of the higher degree of coherence and correlation inherent in the dynamical processes that generate the chaos.

Bibliography

Abraham, N. B. (1983). A new focus on laser instabilities and chaos, *Laser Focus* **19,** 73 (May).

Abraham, N. B., and Firth, W. E. (1990). Transverse effects in nonlinear optics, *J. Optical Society of America* B **7,** June and July.

Abraham, N. B., Lugiato, L. A., and Narducci, L. M., eds. (1985). Special issue on instabilities in active media, *J. Optical Society of America* B **2,** 5–264.

Abraham, N. B., Mandel, P., and Narducci, L. M. (1988). Dynamical instabilities and pulsations in lasers, *in* "Progress in Optics," Vol. XXV (E. Wolf, ed.). Elsevier, Amsterdam.

Abraham, N. B., Arecchi, F. T., and Lugiato, L. A., eds. (1988). "Instabilities and Chaos in Quantum Optics II" (lecture notes of a summer school). Plenum, New York.

Arecchi, F. T., and Harrison, R. G., eds. (1987). "Instabilities and Chaos in Quantum Optics" (collection of research reviews). Springer-Verlag, Berlin and New York.

Bandy, D. K., Oraevsky, A. N., and Tredicce, J. R. (1988). Laser instabilities, *J. Optical Society of America* B **5,** May.

Boyd, R. W., Raymer, M. G., and Narducci, L. M., eds. (1986). "Optical Instabilities" (proceedings of an international conference with tutorial reviews). Cambridge Univ. Press, London and New York.

Harrison, R. G. (1987). Dynamical instabilities and chaos in lasers, *Contemporary Physics*.

Harrison, R. G., and Biswas, D. J. (1986). Chaos in light, *Nature*, **321,** 394–401.

Harrison, R. G., and Biswas, D. J. (1985). Pulsating instabilities and chaos in lasers, *Progress in Quantum Electronics* **10,** 147–228.

Narducci, L. M., and Abraham, N. B. (1988). "Lecture Notes on Lasers and Laser Dynamics." World Scientific, Singapore.

Weiss, C. O. (1988). Chaotic laser dynamics, *Optics and Quantum Electronics* **20,** 1–22.

LASERS

William T. Silfvast *AT&T Bell Laboratories*

GLOSSARY

Absorption: Extinction of a photon of light when it collides with an atom and excites an internal energy state of that atom.

Emission: Radiation produced by an atomic species when an electron moves from a higher energy level to a lower one.

Frequency: Reciprocal of the time it takes for a lightwave to oscillate through a full cycle.

Gain: Condition that causes a beam of light to be intensified when it passes through a specially prepared medium.

Linewidth: Frequency or wavelength spread over which emission, absorption, and gain occur in the amplifier.

Mode: Single frequency beam of light that follows a unique path as it grows in the amplifier and emerges as a beam.

Photon: Discrete quantum of light of an exact energy or wavelength.

Population inversion: Condition in which more atoms exist in a higher energy state than a lower one, leading to amplification, or gain, at a wavelength determined by electron transitions between those states.

Wavelength: Distance over which light travels during a complete cycle of oscillation.

A laser is a device that amplifies, or increases, the intensity of light, producing a strong, highly directional or parallel beam of light of a specific wavelength. The word "laser" is an acronym for "light amplification by stimulated emission of radiation." Stimulated emission is a natural process, first recognized by Einstein, that occurs when a beam of light passes through a medium and initiates or stimulates atoms of that medium to radiate more light in the same direction and at the same wavelength as that of the original beam. A specific laser device (see Fig. 1) consists of (1) an amplifying, or gain, medium that produces an increase in the intensity of a light beam and (2) an optical resonator or mirror arrangement that provides feedback of the amplified beam into the gain medium thereby producing the beamlike and ultrapure frequency or coherent properties of the laser. The optical cavity, or resonator, which typically consists of two highly reflecting mirrors arranged at opposite ends of an elongated amplifier, allows a strong beam of light to develop due to multiple reflections that produce growth of the beam as it bounces back and forth through the amplifier. A useful beam emerges from a laser either by making one of the laser mirrors partially transmitting or by using a mirror with a small hole in it. Some lasers have such a high gain that the intensity increase is large enough, after only a single pass of the beam through the amplifier, that mirrors are not necessary to achieve a strong beam. Such amplifiers produce a directional beam by having the gain medium arranged in a very elongated shape, causing the beam to grow and emerge from the amplifier in only the elongated direction.

Some of the unique properties of lasers include high beam power for welding or cutting, ultrapure frequency for communications and holography, and an ultraparallel beam for long-distance propagation and extremely tight focusing. Laser wavelengths cover the far infrared to the near infrared, the visible to the ultraviolet, and the vacuum ultraviolet to the soft X-ray region. Laser sizes range from small semiconductor lasers the size of a grain of salt, for use in optical

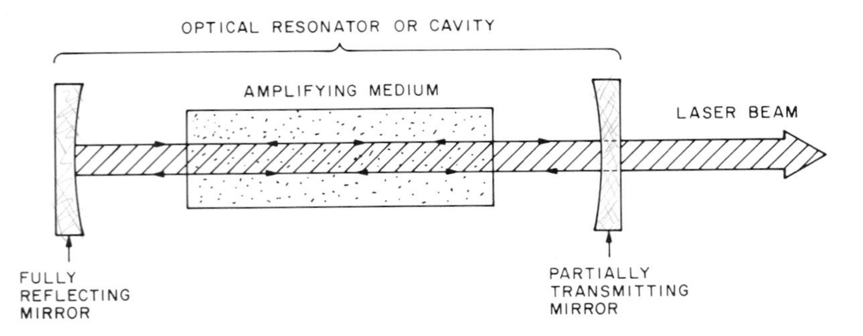

FIG. 1. Laser components including amplifying medium, optical cavity, and laser beam.

communications, to large solid-state and gas lasers the size of a large building, for use in laser fusion programs.

I. Laser History

Charles Townes was the first person to take advantage of the stimulated emission process as an amplifier by conceiving and constructing the first maser (an acronym for "microwave amplification by stimulated emission of radiation"). The maser produced a pure beam of microwaves that were anticipated to be useful for communications in a similar way to that of a klystron or a traveling-wave tube. The first maser was produced in ammonia vapor and the inversion occurred between two energy levels that produced gain at a wavelength of 1.25 cm. In the maser, the radiation wavelengths are comparable to the size of the device and therefore the means of producing and extracting the beam could not obviously be extrapolated to the optical spectrum in which the wavelengths of light are of the order of 100th the size of a human hair.

In 1958, Townes along with Arthur Schawlow began thinking about extending the maser principle to optical wavelengths. At that time they developed the concept of a laser amplifier and an optical mirror cavity to provide the multiple reflections thought to be necessary for rapid growth of the light signal into an intense visible beam. Townes later (1964) shared the Nobel prize in physics with A. Prokhorov and N. Basov of the Soviet Union for the development of the maser–laser principle.

In 1960, Theodore Maiman of the Hughes Research Laboratories produced the first laser using a ruby crystal as the amplifier and a flash lamp as the energy source. The helical flash lamp surrounded a rod-shaped ruby crystal and the optical cavity was formed by coating the flattened ends of the ruby rod with a highly reflecting material. In operation, an intense red beam emerged from the ends of the rod when the flash lamp was initiated.

Shortly after the ruby laser came the first gas laser, developed in 1961 in a mixture of helium and neon gases by A. Javan, W. Bennett, and D. Herriott of Bell Laboratories. At the same laboratories, L. F. Johnson and K. Nassau first demonstrated the now well-known and high-power neodymium laser. This was followed in 1962 by the first semiconductor laser demonstrated by R. Hall at the General Electric Research Laboratories. In 1963, C. K. N. Patel of Bell Laboratories discovered the infrared carbon dioxide laser which later became one of the most powerful lasers. Later that year A. Bloom and E. Bell of Spectra-Physics discovered the first ion laser, in mercury vapor. This was followed in 1964 by the argon ion laser developed by W. Bridges of Hughes Research Laboratories and in 1966 the blue helium–cadmium metal vapor ion laser discovered by W. T. Silfvast, G. R. Fowles, and B. D. Hopkins at the University of Utah. The first liquid laser in the form of a fluorescent dye was discovered that same year by P. P. Sorokin and J. R. Lankard of the IBM Research Laboratories leading to the development of broadly tunable lasers. The first of the now well-known rare-gas–halide excimer lasers was first observed in xenon fluoride by J. J. Ewing and C. Brau of the Avco–Everett Research Laboratory in 1975. In 1976 J. M. J. Madey and co-workers at Stanford University developed the first free-electron laser amplifier operating at the infrared carbon dioxide laser wavelength. In 1985 the first soft X-ray laser was successfully demonstrated in a highly ionized selenium plasma by D. Matthews and a large number of co-workers at the Lawrence Livermore Laboratories.

II. Laser Gain Media

A. Energy Levels and the Emission and Absorption of Light

All lasers (except the free-electron laser) result from electron energy changes among discrete energy levels of atomic species, including: (1) individual atoms or ions, (2) small uniquely bonded groups of atoms (molecules), (3) periodically arranged groups of atoms (semiconductors or crystalline solids), or (4) randomly arranged groups of atoms (liquids and amorphous solid structures). All of these species contain a lowest energy level (ground state) in which the electrons reside at low temperatures, and a spectrum of higher lying levels that are occupied when energy is pumped into the species either by heating or irradiating it with light or energetic particles such as fast electrons or fast atomic particles. Light originates from these species when electrons jump or decay from some of these high lying, or excited, energy levels to lower lying energy levels. The energy that is lost by the particular atomic material when the electron decays is given up in the form of a photon, a discrete particle of light. In order to satisfy the law of conservation of energy, the emitted photon must be of the exact energy corresponding to the energy difference between the higher lying level and the lower lying level, and the wavelength or frequency of the emitted photon is associated with that energy difference. Many photons together can form a beam or a wave of light.

Light, emitted when an electron decays, can occur either spontaneously, due to inherent interactions of the atomic structure, or by stimulated emission whereby the electron is forced or driven to radiate by an approaching photon of the appropriate energy or wavelength (see Fig. 2). Absorption, the opposite process of stimulated emission, occurs when an atom having an electron in a low lying energy level absorbs and thereby eliminates an approaching photon, using the absorbed energy to boost that electron to a higher lying energy level.

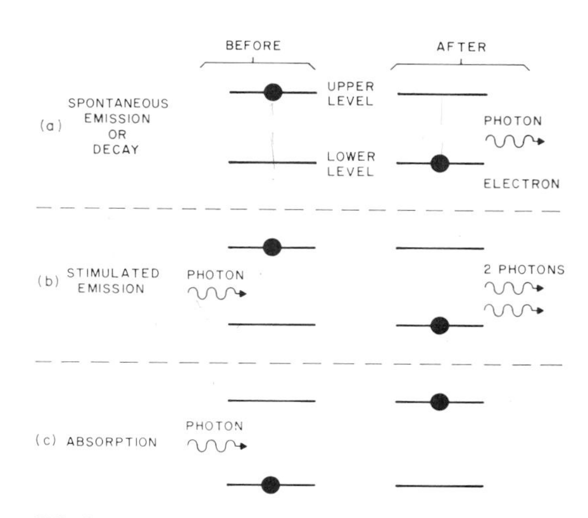

FIG. 2. Electronic transitions between energy levels depicting (a) spontaneous emission, (b) stimulated emission, and (c) absorption.

B. Population Inversions

A laser amplifier is produced when conditions are created within the amplifying medium such that there are more atoms having electrons at a higher energy level than at a lower energy level for a specific pair of levels. This condition is known as a population inversion since it is the opposite or inverse of almost all physical situations at or near thermal equilibrium in which there are more atoms with electrons in lower energy levels than at higher levels. Under the conditions of a population inversion, when a beam of light passes through the amplifier more photons will be stimulated than absorbed thereby resulting in a net increase in the number of photons or an amplification of the beam.

C. Excitation Mechanisms in Lasers (Energy Sources)

Since all laser emission involves radiation from excited states of atoms, the energy must be fed to those atoms to produce the excited states. This energy is provided in the form of highly energetic electrons (moving at rapid speeds of the order of 10^8 to 10^9 cm/sec), energetic heavier particles such as protons, neutrons, or even other atoms, or electromagnetic radiation (light) in the form of (1) a broad frequency spectrum of emission such as a flash lamp or (2) a narrow frequency spectrum provided by another laser. The most common excitation source is that provided by energetic electrons since they are easily accelerated by applying an electric field or voltage drop to an amplifier. Electrons are typically used in most gas and semiconductor lasers whereas light is most often used in liquid (dye) lasers and crystalline solid-state lasers. Electron excitation sources tend to be the most efficient since flash lamps are themselves generally excited by electrons before they are used to pump lasers. Heavier particles are generally less efficient as pumping sources since they are much more difficult to energize than either electrons or lamps.

D. Inversions in Gases, Metal Vapors, and Plasmas

Inversions in gases, metal vapors, and plasmas are generally produced by applying a voltage drop across the elongated gain region thereby producing an electric field that accelerates the electrons. These rapidly moving electrons then collide with the gas atoms and excite them to a wide range of excited energy levels. Some of those levels decay faster than others (primarily by spontaneous emission) leaving population inversions with higher levels. If the populations in the inverted levels are high enough, then the gain may be sufficient to make a laser. Typically if 1 in 10^5 or 10^6 atoms is in a specific upper laser level, that will be a sufficient fraction to produce enough gain to make a laser. Most gas lasers have relatively low gains and therefore amplifier lengths of the order of 25 to 100 cm are necessary. Since spontaneous emission rates are much faster for shorter wavelength transitions, power input for short-wavelength lasers is significantly higher than for visible and infrared lasers. Typical gas pressures for gas lasers range from 1/1000th to 1/100th of an atmosphere although there are some gas lasers that operate at atmospheric pressure and above and require very closely spaced electrodes (transverse excitation) in order to produce the necessary excited state populations.

E. Inversions in Liquids (Dyes)

Most excited states of liquids decay so rapidly by collisions with surrounding atoms or molecules (10^{-13} sec) that it is difficult to accumulate enough population in an upper laser level to make significant gain. Also, since it is difficult to use electron excitation in liquids, the primary energy source is optical excitation, either by flash lamps or by other lasers. Fluorescing dyes are the best liquid media for lasers. The fact that dyes fluoresce suggests that their excited energy levels stay populated long enough to lose their energy by radiation of light rather than by collisions with surrounding atoms or electrons. To help establish the population inversion, the lower laser level of a dye decays very rapidly by collisions.

F. Inversions in Semiconductors

Inversions in semiconductors are produced when a *p–n* junction is created by joining two slightly different semiconducting materials (in a similar way to that of a transistor). The *n*-type material has an excess of electrons and the *p*-type material has an excess of holes (missing electrons). When they are joined, the excess electrons of the *n*-type material are pulled over into the *p* region and vice versa by their charge attraction, causing the electrons and holes in that region to recombine and emit recombination radiation. This neutralizes the junction region, leaving a small, inherent electric field to prevent further recombination of the remaining electrons and holes near the junction. If an external electric field is applied in the appropriate direction, by applying a voltage across the junction, more electrons and holes can be pulled together, causing them to recombine and emit more radiation but also to produce an inversion. This inversion occurs on transitions originating from energy levels located just above the inherent energy band gap of the material. Semiconductor lasers operate in a similar way to light-emitting diodes except that the requirements for constructing the lasers are much more restrictive than the diodes, due primarily to the necessity for higher electric current densities that are essential to produce large gains, and also due to the need for better heat-dissipation capabilities to remove the heat produced by the higher current densities.

G. Inversions in Solids

Inversions in most solid-state lasers are obtained by implanting impurities (the laser species) within a host material such as a crystal or a glass in a proportion ranging from approximately 1 part in 100 to 1 part in 10,000. In most solid-state lasers the impurities are the form of ions in which the energy states are screened from the surrounding atoms so the energy levels are narrow, like those of isolated atoms or ions, rather than broad like those of liquids. In color center lasers the impurities are crystal defects produced by irradiating the crystal with X rays. In solids, as in liquids, electrons cannot easily be accelerated by electric fields to excite the laser energy levels of the impurity species so the energy must be fed to the medium via flash lamps or other lasers. The input lamp energy occurs over a broad wavelength region to a large number of excited energy levels. These levels then decay to the upper laser level which acts much like a temporary storage reservoir, collecting enough population to make a large inversion with respect to lower lying levels that have a rapid decay to the ground state.

H. Bandwidth of Gain Media

The frequency or wavelength spectrum (gain bandwidth) over which gain occurs in a laser amplifier is determined by a number of factors. The minimum width is the combined width of the energy levels involved in the laser transition. This width is due primarily to the uncertainty of the natural radiative decay time of the laser levels. This width can be increased by collisions of electrons with the laser states in high-pressure gas lasers, by interaction of nearby atoms and bonding electrons in liquids and solids, and by Doppler shifted frequencies in most gas lasers. This gain linewidth of the laser amplifier is only part of the contribution to the linewidth of the laser beam. Significant line narrowing due to optical cavity effects will be described in the next section.

III. Laser Beam Properties

A. Beam Properties

The beam properties of a laser, such as the direction and divergence of the beam and the wavelength or frequency characteristics that are not related to the bandwidth of the laser gain medium, are determined largely by the laser structure. The features of the structure affecting the beam properties include the width and length of the gain medium, the location, separation and reflectivity of the mirrors of the optical cavity (if the gain is low and the gain duration long enough to make use of mirrors), and the presence of losses in the beam path within the cavity. These features determine unique properties of the laser beam referred to as laser modes.

B. Shape of the Gain Medium

If a laser gain medium were in the shape of a round ball, then stimulated emission would occur equally in all directions and the only result might be a slight increase in the intensity of the light, for a slightly shorter duration than would occur if gain were not present, but the effect would probably not be noticeable to an observer. The goal of a laser designer is to cause most of the laser photons to be stimulated in a specific direction in order to produce a highly directional beam, at the expense of allowing those same photons to radiate in random directions by either stimulated or spontaneous emission (as was the case for the round ball). This is achieved by making the gain medium significantly longer in one dimension than in the other two.

C. Growth of the Beam and Saturation

When gain is produced in the amplifier and spontaneously emitted photons begin to be amplified by stimulated emission, photons that are emitted in directions other than the elongated direction of the amplifier soon reach the walls of the medium and die out. The photons that are emitted in the elongated direction continue to grow by stimulating other atoms to emit additional photons in the same direction until all of those photons reach the end of the amplifier. They then arrive at the mirror and are reflected back through the amplifier where they continue to grow. Finally, after a number of round trips, a beam begins to evolve. If the duration of the gain is long enough, amplification will lead to more photons in the beam than there are atoms available to stimulate, and growth can therefore no longer occur. The beam is then said to be saturated and the beam power reaches a steady value, determined by the amount of energy being fed into the upper laser level by the pumping source. If the population inversion can be maintained on a continuous basis in the amplifier, the laser-beam output becomes steady and the laser is referred to as a continuous wave, or cw, laser. If the gain only lasts for a short duration, the laser output occurs as a burst of light and the laser is referred to as a pulsed laser.

D. Optical Cavity (Optical Resonator)

The optical cavity or resonator, typically comprised of a mirror at each end of the elongated gain region (see Fig. 1), allows the rapidly growing beam to bounce back and forth or resonate between the mirrors. Although the first laser used flat mirrors, as suggested in the original Schalow–Townes paper, in 1961 Fox and Li suggested the use of slightly curved mirrors, especially for cavities where the amplifier consisted of a long narrow tube, in order to reconcentrate the beam in the center of the gain medium after each reflection from the mirrors thereby reducing the diffraction losses of the narrow tube. Stable modes of lower loss are possible for curved mirrors than for flat mirrors if the separation between the mirrors is less than twice their radius of curvature. Mirror reflectivities of 99.9% at the laser wavelength, using dielectric layered coatings, make possible laser operation under conditions of very low gain.

E. Stable and Unstable Resonators

The term resonator implies a wave that is in harmony or resonance with the device that is generating the wave, whether it be an organ pipe, a flute, or a microwave cavity. An optical wave can also have this property. The term "resonance" suggests that an exact integral number of wavelengths (a mode) of the wave fit between the mirrors of the resonator. A stable resonator refers to a mirror arrangement (usually one with a mirror at each end of the elongated cavity), producing modes that are continually reproducible during the duration that gain occurs in the amplifier (which could be a thousandth of a second or many days or longer). An unstable resonator is a mirror arrangement that is used to obtain modes when the amplifier gain is high and has a short duration (less than a millionth of a second) such that a normal mode would not have time to evolve. In that situation, the energy is extracted by using a mirror arrangement in which the beam begins to resonate in a small unstable region of the amplifier. Part of this beam is leaked into the larger portion of the amplifier where it rapidly grows and extracts most of the amplifier energy in a few passes. [*See* MODE-LOCKING OF LASERS.]

F. Laser Modes

Laser modes are wavelike properties relating to the oscillating character of a light beam as the beam passes back and forth through the amplifier and grows at the expense of existing losses. The development of modes involves an attempt by competing light beams of similar wavelengths to fit an exact number of their waves into the optical cavity with the constraint that the oscillating electric field of the light beam is zero at each of the mirrors. This is much like a vibrating guitar string which is constrained at each end by the bridge and a fret, but is free to vibrate with as many nodes and antinodes in the region in between as it chooses. As an example, a laser mode of green light having a wavelength of exactly 5.0×10^{-5} cm will fit exactly 1,000,000 full cycles of oscillation between laser cavity mirrors separated by a distance of exactly 50 cm. Most lasers have a number of modes operating simultaneously, in the form of both longitudinal and transverse modes, which give rise to a complex frequency and spatial structure within the beam in what might otherwise appear as a relatively simple, pencil-like beam of light.

G. Longitudinal Modes

Each longitudinal mode is a separate light beam traveling along a distinct path between the mirrors and having an exact integral number of wavelengths along that path. In the example of green light mentioned previously, three different longitudinal modes would have very slightly different wavelengths of green light (indistinguishable in color to the eye) undergoing respectively 1,000,000, 1,000,001, and 1,000,002 full cycles of oscillation between the mirrors while traveling exactly the same path back and forth through the amplifier (see Fig. 3). In this situation each mode would differ in frequency by exactly 300 MHz as determined by the velocity of light (3 ×

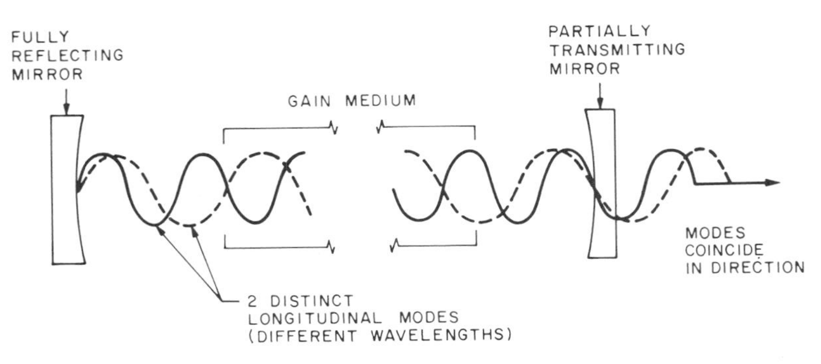

FIG. 3. Two distinct longitudinal modes occupying the same spatial region of the laser optical cavity.

10^{10} cm/sec) divided by twice the cavity length (2 × 50 cm). A gas laser amplifier having a relatively narrow gain width of 3 GHz could fit 10 longitudinal modes within the gain bandwidth whereas a liquid (dye) laser having a bandwidth covering up to one-fifth of the visible spectrum of light could have as many as 100,000 longitudinal modes all oscillating simultaneously.

H. Transverse Modes

Whereas longitudinal modes involve many light beams traveling exactly the same path through the amplifier, but differing in wavelength by an amount determined by the total number of wave cycles that fit between the mirrors, different transverse modes are represented by slightly different optical paths as they travel through the amplifier (Fig. 4). Thus each transverse mode traveling over its unique path could consist of several longitudinal modes oscillating along that path. In most instances, closely located transverse modes differ in frequency by a smaller value than do adjacent longitudinal modes that follow the same path through the amplifier.

IV. Laser Linewidth

A. Homogeneous Linewidth of Gain Media

All laser amplifiers have a finite frequency width or wavelength width over which gain can occur. This width is related to the widths of the energy levels involved in the population inversion. Single atoms or ions have the narrowest widths, determined by the very narrow energy levels inherent in the atomic structure. The gain linewidth for these amplifiers is the sum of the linewidths of the upper and lower laser levels. The linewidth is determined by both a homogeneous component and an inhomogeneous component. The homogeneous component includes all mechanisms that involve every atom in both the upper and lower laser levels of the gain medium in identically the same way. These mechanisms include the natural radiative lifetime due to spontaneous emission, broadening due to collisions with other particles such as free electrons, protons, neutrons, or other atoms or ions, or power broadening in which a high intensity laser beam rapidly cycles an atom between the upper and lower levels at a rate faster than the normal lifetime of the state. In each of these effects, the rate at which the process occurs determines the number of frequency components or the bandwidth required to describe the process, with faster processes needing broader linewidths. These rates can vary anywhere from 10^6 Hz in the infrared to 10^{10} Hz in the soft X-ray spectral region for spontaneous emission and up to 10^{12} Hz or more or rapid collisional broadening.

B. Inhomogeneous Linewidth of Gain Media

The inhomogeneous component of the broadening results from processes that affect different atoms in the upper and lower laser levels in different ways depending on unique characteristics of those atoms. The most common example of inhomogeneous broadening is Doppler broadening or motional broadening that results from the random thermal motion of the atoms due to their finite temperature. In this effect, atoms traveling in one direction would see an approaching lightwave as having a specific frequency and atoms traveling in the opposite direction would see that same lightwave as having a lower fre-

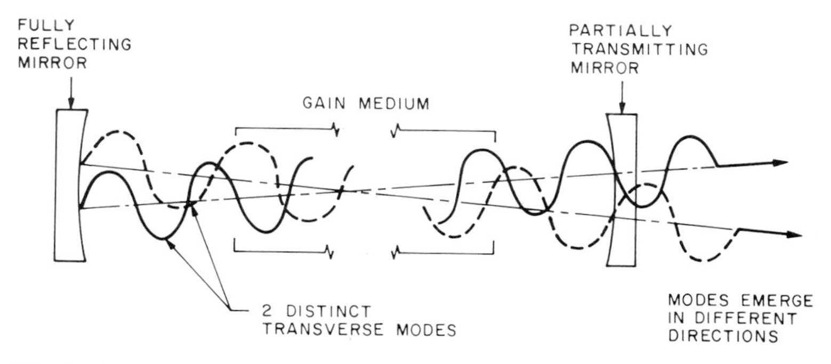

FIG. 4. Two distinct transverse modes oscillating over different spatial regions of the laser optical cavity.

quency. In a similar way, light emitted from each of those atoms would be seen by an observer as having different frequencies or wavelengths even though each atom emitted the same frequency. This Doppler effect is similar to that when a train approaches with its horn blowing and the pitch of the sound changes from a high tone to a lower tone as the train first approaches the listener and then moves away, even though the frequencies emitted from the horn remain the same.

C. Linewidth in Laser Amplifiers

The dominant broadening effect for most gas laser amplifiers is Doppler broadening which produces a linewidth of the order of 10^9 to 10^{10} Hz for visible and shorter wavelength amplifiers. Molecules have linewidths that are determined not only by the jumping of electrons between energy levels, as is the case of individual atoms, but also by the rotation of the atoms around a common center of gravity and by the vibrations of the atoms as though they were tied together by various springs. These rotations and vibrations produce a series of emissions of light over a relatively narrow range of wavelengths and in some instances these emissions can overlap in frequency to produce broad emission lines. If no overlap occurs, then the emission consists of an equally spaced series of narrow lines.

Liquids have their energy levels broadened due to the interaction of the closely packed atoms of the liquid. Such dense packing leads to a very rapid collisional interruption of the electronic states causing them to smear out into very broad energy bands thereby producing emission spectrum widths of the order of 10^{13} to 10^{14} Hz.

Solids, in addition to having closely located atoms (only slightly more dense than liquids) in many cases have a periodic structure due to the regularity of the atomic distribution (much as in the way eggs are lined up in an egg carton). This produces in many species a large energy gap between the lowest energy level and the first excited energy level. Emission from this excited level can have a very broad frequency spectrum similar to that of liquids; however, there also exist discrete levels within the bandgap, known as exciton states, that have a much narrower emission spectrum (of the order of 10^{12} Hz), due to electron–hole pairs bonding together to form atomiclike particles. Such exciton states are involved in some semiconductor lasers. Another class of emission and broadening in solids occurs when an atom like chromium is embedded in another solid material such as aluminum oxide (sapphire) to make a ruby crystal. In such a crystal, the outer electrons of the chromium atoms are shared with the surrounding aluminum oxide atoms while the inner electrons are screened from the neighboring atoms causing the atom to behave more like an isolated atom having narrow energy levels. These crystalline solids, containing specific impurities, can have narrow emission lines of the order of 10^{11} to 10^{12} Hz. The first laser was made in ruby using such levels.

D. Line Narrowing Due to Cavity Effects

A laser cavity tends to select specific wavelengths (Fig. 5) within the normal gain bandwidth of the gain medium as determined by the wavelengths of light that have an exact integral number of waves that fit between the mirrors (modes). These modes, which are equally spaced in frequency or wavelength, tend to be amplified at the expense of other wavelengths that suffer losses by not exactly "fitting" between the mirrors. In principal these modes can "narrow up" to widths of the order of a few hertz, but cavity stability problems tend to keep them from going much below 1000 Hz unless extremely stable environments and rigid cavity structures are available.

V. Laser Wavelengths

A. Range of Wavelengths

Wavelengths of electromagnetic radiation covering the spectral region where lasers occur are referred to in terms of fractions of meters. The two most common units are the micrometer (μm) or 10^{-6} m and the nanometer (nm) 10^{-9} m. Micrometers are used to designate the infrared region ranging from 0.7 μm to approximately 1,000 μm. Nanometers are used to cover the spectral range from the visible at 700 nm (0.7 μm) down to approximately 10 nm in the soft X-ray region. Each spectral region covers a specific wavelength range although the boundaries are not always exact. The infrared is broken up into three regions. The far infrared ranges from about 15 to approximately 1,000 μm (approaching the microwave region). The middle infrared covers from 2 to 15 μm and the near infrared ranges from 0.7 to 2 μm. The visible region in-

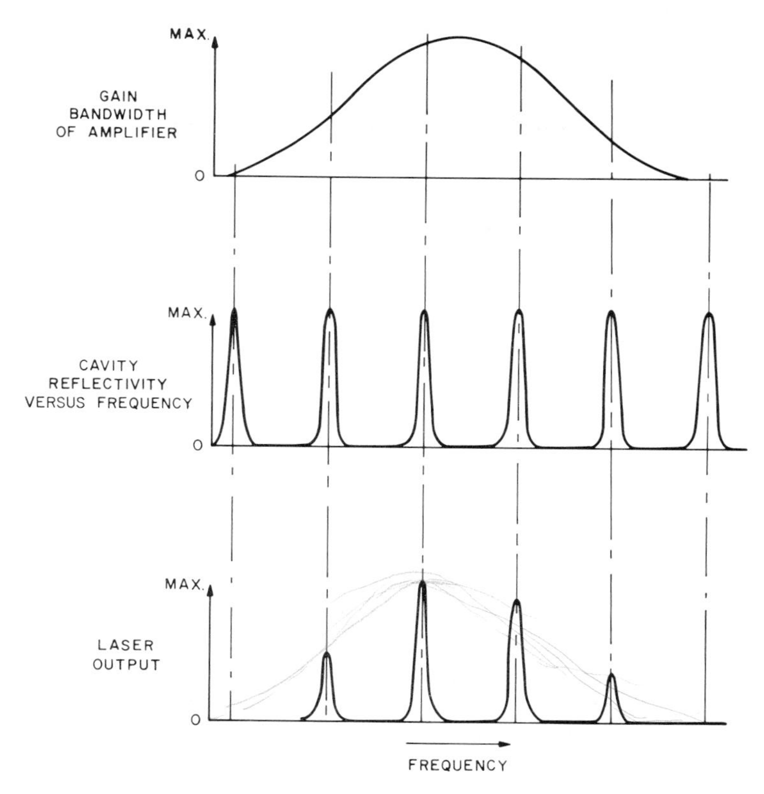

FIG. 5. Laser output resulting from effects of laser amplifier gain bandwidth and laser optical cavity modes.

cludes the rainbow spectrum ranging from the violet at 400 to 440 nm, the blue from 440 to 490 nm, the green from 490 to 550 nm, the yellow from 550 to 580 nm, the orange from 580 to 610 nm, and the red from 610 to 700 nm. The ultraviolet ranges from 400 to 200 nm with the 400- to 300-nm region termed the near ultraviolet and the 200- to 300-nm region the far ultraviolet. The vacuum ultraviolet covers the range from 100 to 200 nm, since it is the region where radiation can no longer transmit through the air (thus requiring a vacuum) because of absorption due to air molecules (mostly oxygen) and because transmission optics can still be used in this region. The shortest region is the extreme ultraviolet, which extends into the soft X-ray region, covering the range from 10 to 100 nm. In this region only reflection optics can be used and even then the reflectivities of the best materials are quite low (especially below 40 nm) when compared to those available in the visible. Free-electron lasers have the potential for operating at all wavelengths and will therefore not be singled out for any specific wavelength in this section.

B. Infrared Lasers

The most powerful lasers occur in the middle and near infrared, but a large number of lasers have been produced in the far infrared. These include the discharge excited water-vapor lasers with wavelengths ranging from 17–200 μm and the cyanide laser operating at 337 μm. The other principle far infrared lasers are the laser-pumped gases of methyl fluoride (450–550 μm) and ammonia (81 μm). In the middle infrared, the carbon dioxide laser is the most significant laser operating primarily at 9.6 and 10.6 μm. The other strong laser is the carbon monoxide laser which emits in the 5–6-μm region. There is also an important chemical laser, the hydrogen–fluoride laser, which operates at 3.7 μm. In the near infrared, many more lasers are available. The most significant laser is the neodymium laser, usually available in a yttrium–aluminum–garnet (YAG) host crystal referred to as the Nd:YAG laser, or in a glass host called a Nd:glass laser, both operating at 1.06 μm. Semiconductor lasers operate over the range from 0.7 to approximately 1.70 μm; their wavelengths being extended in both directions with continuing research. Color center lasers are tunable from 0.8 to 4.0 μm and the alexandrite laser is tunable from 0.7 to 0.8 μm. Tunable dye lasers also extend into the infrared to about 1.5 μm.

C. Visible Lasers

Visible lasers are primarily dominated by gas lasers and tunable dye lasers. The helium–neon laser at 633 nm is probably the most commonly seen laser. The argon ion laser covers the blue and green spectrum with the most prominent wavelengths at 488 and 515 nm. The krypton ion laser covers the spectrum from green to red with some of the most prominent wavelengths at 521, 568, and 647 nm. The helium–cadmium laser operates in the blue at 442 nm. Tunable dye lasers operate over the entire visible spectrum, generally pumped by a slightly shorter wavelengh laser such as the argon ion laser or the frequency doubled or tripled Nd:YAG laser. The rhodamine 6G dye has the lowest pump threshold and operates as a laser over a wavelength range from 570 to 630 nm.

D. Ultraviolet Lasers

There are fewer lasers available in the ultraviolet primarily because pump thresholds become much higher than for visible lasers and because cavity mirror reflectivities are lower and less durable. The two prominent continuous lasers are the argon ion laser, operating primarily at 351 nm and the helium cadmium laser at 325 nm. The rare-gas–halide excimer lasers produce high pulsed powers at 351 nm for xenon–fluoride, 308 nm for xenon chloride, and 248 nm for krypton fluoride. The pulsed nitrogen laser operating at 331 nm was extensively used for pumping dye lasers but this has now largely been replaced with the frequency doubled and tripled Nd:YAG laser and the excimer lasers. Tunable dye lasers as short as 310 nm are available but they are not easily pumped by commercial lasers and therefore frequency mixing and doubling of visible and infrared dye lasers is the most common technique for producing tunable laser radiation in this region.

E. Vacuum Ultraviolet Lasers

The molecular hydrogen laser was the first laser developed in this spectral region, operating in the 120- and 160-nm ranges but it never became a useful device. There are now only two readily available vacuum ultraviolet lasers. These are the argon–fluoride excimer laser operating at 193 nm and the fluorine excimer laser emitting at 153 nm. Much of the research requiring lasers in this spectral region is accomplished by frequency summing and mixing of various visible and ultraviolet lasers to produce coherent output, but such techniques are not yet readily available to the general laser community.

F. Extreme Ultraviolet and Soft X-Ray Lasers

Until 1984 there were no lasers to report in this spectral region. At that time stimulated emission was reported in krypton at 93 nm. In 1985 several highly ionized atoms provided laser output in the soft X-ray end of this spectrum, including twenty-fourth ionized selenium at 21 nm, fifth ionized carbon at 18 nm, twenty-ninth ionized yttrium at 15 nm and thirty-second ionized molybdenum at 13 nm. These lasers are so new that no experiments have yet been carried out with them. They are initiated with powerful neodymium and carbon dioxide lasers the size of large buildings and are therefore not easily duplicated in other laboratories.

VI. Types of Lasers

A. Gas Lasers

The most common types of gas lasers are the helium–neon laser, the argon and krypton ion lasers, the carbon dioxide laser, the rare-gas–halide excimer lasers, and the chemical lasers, most notably the hydrogen–flouride laser. Metal vapor lasers also fit into this category but are treated separately in Section V. With few exceptions these lasers receive their energy input via collisions of gas atoms with high-energy electrons. This energy is provided by applying a high voltage between electrodes located within the gaseous medium in order to accelerate the electrons to the necessary high energies. In some instances the electrons first excite a storage level in a separate species within the gaseous medium rather than directly pumping the laser state. The energy is subsequently transferred from that storage level to the laser level of the lasing species by direct collisional exchange of energy. [*See* GAS LASERS.]

1. Helium–Neon Laser

The helium–neon laser was the first gas laser. The original laser transitions were in the near infrared but the most commonly used transition is the red laser at a wavelength of 632.8 nm. This laser, which has more units in use than any other laser, is available in sizes ranging from approximately 10 cm in length to over 100. It has continuous power outputs ranging from less than a milliwatt to over 100 mW and has a lifetime approaching 50,000 h for some commercial units. The excitation mechanism involves electrons colliding with helium atoms to produce helium metastable atoms, which then transfer their energy to neon laser levels. This laser is used in surveying, construction, supermarket checkout scanners, printers, and many other applications.

2. Argon and Krypton Ion Lasers

Argon and krypton ion lasers were discovered shortly after helium–neon lasers. They were the first lasers to operate in the green and blue regions of the spectrum and some versions provide ultraviolet output. These lasers have the capability of producing more than 20 W of continuous power for the largest versions. The size of the laser tubes range from 50 to 200 cm in length with a separate power supply. They are relatively inefficient and consequently require high input power and water cooling for most units. The high power requirements, which put great demands on the strength of the laser-discharge region, limit the lifetime of the high-power versions of these lasers. Some smaller, lower power versions of the argon ion laser are air cooled and offer lifetimes of 5000 h. The excitation mechanisms for these lasers involve electrons collisions first populating the ion ground states of the argon and krypton species with subsequent electron excitation to the upper laser level. Applications include phototherapy of the eye, pumping dye lasers, printing, and lithography.

3. Carbon Dioxide Laser

The carbon dioxide lasers are some of the most powerful lasers, operating primarily in the middle infrared spectral region at a wavelength of 10.6 μm. They range from small versions with a few milliwatts of continuous power to large pulsed versions the size of large buildings producing over 10,000 J of energy. They are among the most efficient lasers (up to 30%) and can produce continuous output powers of over 100 kW in room-size versions. Small versions of these lasers are referred to as waveguide lasers because the excitation region is of a cylindrical shape small enough to guide the beam down the bore in a waveguide type of mode. They can produce continuous power outputs of up to 100 W from a device smaller than a shoe box (with a separate power supply). The lasers typically operate in a mixture of carbon dioxide, nitrogen, and helium gases. Electron collisions excite metastable (storage) levels in nitrogen molecules with subsequent transfer of that energy to carbon dioxide laser levels. The helium gas acts to keep the average electron energy high in the gas-discharge region. This laser is used for a wide variety of applications including eye and tissue surgery, welding, cutting, and heat treatment of materials, laser fusion, and beam weapons. Figure 6 shows a carbon dioxide laser being used to drill a hole in a turbine blade. The white streamers are hot metal particles being ablated from the region of the hole.

4. Rare-Gas–Halide Excimer Lasers

The rare-gas–halide excimer lasers, relative newcomers to the laser industry, operate primarily in the ultraviolet spectral region in mixtures of rare gases, such as argon, krypton, or xenon with halide molecules such as chlorine and fluorine. They include the argon–fluoride laser at 193 nm, the krypton–fluoride laser at 248 nm, the xenon–chloride laser at 308 nm, and the xenon–flouride laser at 351 nm. These lasers typically produce short pulses of energy ranging from tens of millijoules to thousands of joules in pulse durations of 10 to 50 nsec and repetition rates of up to 1000 pulses per second. They range in size from an enclosure that would fit on a kitchen table top to lasers the size of a very large room. They are relatively efficient (1–5%) and provide useful energy in a wavelength region that has not had powerful lasers available previously. Their operating lifetimes are related to the development of discharge tubes, storage regions, and gas pumps that can tolerate the corrosive halogen molecules that are circulated rapidly through the gain region. Typical lifetimes are of the order of the time it takes to produce 10^6 to 10^7 pulses. The laser species are mixed with helium gas to provide a total pressure of 2 atms. Excitation occurs by electron dissociation

FIG. 6. Carbon dioxide laser used to drill holes in a turbine blade. White streamers are hot metal particles being ejected from the hole region.

and ionization of the rare gas molecule to produce Ar^+, Kr^+, or Xe^+ ions. These ions then react with the halide molecules pulling off one of the atoms of that molecule to create an excited-state dimer (abbreviated as excimer) molecule. This excimer molecule then radiates rapidly to an unstable (rapidly dissociating) lower laser level. The very high operating pressures cause the molecules to react rapidly in order to produce the upper laser levels at a rate that can compete with the rapid decay of those levels. Applications include laser surgery, pumping of dye lasers, and lithography. [*See* RARE GAS-HALIDE LASERS.]

5. Chemical Lasers

In these lasers the molecules undergo a chemical reaction that leaves the molecule in an excited state that has a population inversion with respect to a lower lying state. An example of this type of laser is the hydrogen–fluoride laser in which molecular hydrogen and molecular fluorine react to produce hydrogen–fluoride molecules in their excited state resulting in stimulated emission primarily at 2.8 μm. There are no commercially available chemical lasers but they have undergone extensive development for military applications.

B. Metal Vapor Lasers

Metal vapor lasers are actually a type of gaseous laser since the laser action occurs in the atomic or molecular vapor phase of the species at relatively low pressures, but the lasers have peculiar problems associated with vapors, such as having to vaporize a solid or liquid into the gaseous state either before or during the excitation and lasing process. These problems, along with the problems associated with controlling the condensed vapors after they diffuse out of the hot region in some designs, and with minimizing the corrosive effects of hot metal atoms and ions in other designs, have caused them to be classified in a separate category. The two most well-known types of metal vapors lasers are the helium–cadmium ion laser and the pulsed copper vapor laser.

1. Helium–Cadmium Laser

The helium–cadmium laser operates primarily at two wavelengths, in the blue at 441.6 nm and in the ultraviolet at 325.0 nm. It produces continuous power outputs of the order of 5–100 mW for the blue and 1–25 mW in the ultraviolet, in sizes ranging from 50–200-cm long. The cadmium vapor is obtained by heating the cadium metal in a reservoir located near the helium discharge. The cadmium vapor diffuses into the excited helium gas where it is ionized and the cataphoresis force on the cadmium ions causes them to move towards the negative potential of the cathode thereby distributing the metal relatively uniformly in the discharge region to produce a uniform gain. Typical operating life for the laser is of the order of 5000 h with the limiting factors being the control of the vapor and the loss of helium gas via diffusion through the glass tube walls. The laser levels are excited by collisions with helium metastable atoms, by electron collisions, and by photoionization resulting from radiating helium atoms. Applications include printing, lithography, and fluorescence analysis.

2. Copper Vapor Laser

This laser operates in the green at 510.5 nm and in the yellow at 578.2 nm. It efficiently (2%) produces short laser pulses (10–20-nsec duration) of 1 mJ of energy at repetition rates of up to 20,000 times per second yielding average powers of up to 20 W. The size is similar to that of an excimer laser. Commercial versions of this laser are designed to heat the metallic copper up to temperatures of the order of 1600°C in order to provide enough copper vapor to produce laser action. The lifetime associated with operating these lasers at such high temperatures has been limited to a few hundred hours before servicing is required. The excitation mechanism is primarily by electron collisions with ground-state copper atoms to produce upper laser states. Applications of these lasers include uranium isotope enrichment, large-screen optical imaging, and pumping of dye lasers. A gold vapor laser that is similar to the copper laser, but emits red light at a wavelength of 624.0 nm, is used for cancer phototherapy.

C. Solid-State Lasers

These lasers generally consist of transparent crystals or glasses as "hosts" within which ionic species of laser atoms are interspersed or "doped." Typical host materials include aluminum oxide (sapphire), garnets, and various forms of glasses with the most common lasing species being neodymium ions and ruby ions (the first laser). In color center lasers the host is typically an alkali–halide crystal and the laser species is an electron trapping defect in the crystal. The energy input in all of these lasers is provided by a light source that is focused into the crystal to excite the upper laser levels. The light source is typically a pulsed or continuously operating flash lamp, but efficient diode lasers are also being used to pump small versions of neodymium lasers and argon ion lasers are used to pump color center lasers.

1. Neodymium Lasers

Neodymium atoms are implanted primarily in host materials such as yttrium–aluminum–garnet (YAG) crystals or various forms of glasses in quantities of approximately one part per hundred. When they are implanted in YAG crystals, the laser emits in the near infrared at 1.06 μm with continuous powers of up to 250 W and with pulsed powers as high as several megawatts. The YAG crystal growth difficulties limit the size of the laser rods to approximately one centimeter in diameter. The YAG host material, however, has the advantage of having a relatively high thermal conductivity to remove wasted heat, thus allowing these crystals to be operated at high repetition rates of the order of many pulses per second. Glass hosts also produce Nd lasers in the 1.06-μm wavelength region but with a somewhat broader bandwidth than YAG. They can also be grown in much larger sizes than YAG, thereby allowing the construction of very large amplifiers, but glasses have a much lower thermal conductivity, thus requiring operation at much lower repetition rates (of the order of one pulse every few minutes or less). Thus Nd:YAG is used for continuous lasers and relatively low-energy pulsed lasers (1 J per pulse) operating at up to 10 pulses per second whereas glass lasers exist in sizes up to hundreds of centimeters in diameter, occupy large buildings and are capable of energies as high as 100 kJ per pulse for laser fusion applications. Neodymium lasers typically have very long lifetimes before servicing is required, with the typical failure mode being the replacement of flash lamps. Neodymium lasers are used for surgery applications, for pumping dye lasers (after doubling and tripling their frequencies with nonlinear optical techniques), as military range finders, for drilling holes in solid materials, and for producing X-ray plasmas for X-ray light sources, and for laser fusion and for making X-ray lasers.

2. Ruby Laser

The ruby laser, the first laser discovered, is produced by implanting chromium ions into an aluminum oxide crystal host and then irradiating the crystal with a flash lamp to excite the laser levels. Although ruby lasers were frequently used during the early days of the laser, the difficulties associated with growing the crystals, compared with the ease of making neodymium lasers, has led to their being used much less often in recent times.

3. Color Center Lasers

Color center lasers use a different form of impurity species implanted in a host material in quantities of one part per ten thousand. In such lasers the laser species is generally produced by irradiating the crystal with X rays to produce defects that attract electrons. These defect centers produce energy levels that absorb and emit light and are capable of being inverted to produce gain. Color center lasers typically operate in the infrared from 0.8–4 μm and are tunable within that range by using different crystals having different emission wavelengths. Their tunability makes them attractive lasers for doing spectroscopy. [*See* COLOR CENTER LASERS; ULTRAFAST LASER TECHNOLOGY.]

D. Semiconductor Lasers

Semiconductor or diode lasers, typically about the size of a grain of salt, are the smallest lasers yet devised. They consist of a *p–n* junction formed in an elongated gain region, typically in a gallium–arsenide crystal, with parallel faces at the ends to serve as partially reflecting mirrors. They operate with milliamps of current at a voltage of only a few volts. The entire laser package is very small and could be incorporated into an integrated circuit board if required. Heterostructure lasers, a more recently developed type of diode laser, include additional layers of different materials of similar electronic configuration, such as aluminum, indium and phosphorous on the sides of the junction to help confine the electronic current to the junction region in order to minimize current and heat dissipation requirements. Semiconductor lasers range in wavelengh from 0.7 to 1.8 μm with typical continuous output powers of up to 10 mW. By constructing a row of *p–n* junctions next to each other, all of the separate gain media can be forced to emit together in a phased array to produce an effective combined power output of over one watt. Applications for semiconductor lasers are primarily in the communications field in which the near-infrared beams can be transmitted over long distances through low-loss fibers. In addition, they have recently found a large market as the reading device for compact disc players. Figure 7 shows a diode laser array, consisting of 10 diode lasers, recording at a high data rate onto a multitrack optical disk.

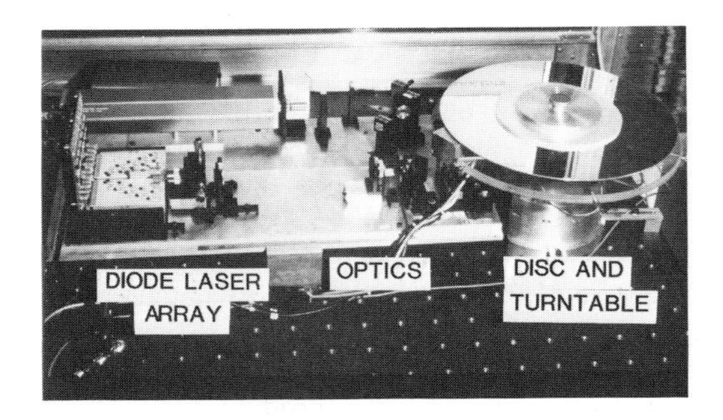

FIG. 7. High-data-rate multichannel optical recording using an array of addressable diode lasers. [Courtesy of RCA Laboratories.]

E. Liquid (Dye) Lasers

Dye lasers are similar to solid-state lasers in that they consist of a host material (in this case a solvent such as alcohol) in which the laser (dye) molecules are dissolved at a concentration of the order of one part in ten thousand. Different dyes have different emission spectra or colors thus allowing dye lasers to cover a broad wavelength range from the ultraviolet (320 nm) to the infrared at about 1500 nm. A unique property of dye lasers is the broad emission spectrum (typically 30–60 nm) over which gain occurs. When this broad gain spectrum is combined with a diffraction grating or a prism as one of the cavity mirrors, the dye laser output can be a very narrow frequency beam (10 GHz or smaller) tunable over a frequency range of 10^{13} Hz. Frequency tuning over even larger ranges is accomplished by inserting different dyes into the laser cavity. Dye lasers are available in either pulsed (up to 50–100 mJ) or continuous output (up to a few watts) in table-top systems that are pumped by either flash lamps or by other lasers such as frequency-doubled or -tripled YAG lasers or argon ion lasers. Most dye lasers are arranged to have the dye and its solvent circulated by a pump into the gain region from a much larger reservoir, since the dye degrades slightly during the excitation process. Dyes typically last for 3 to 6 months in systems where they are circulated. Dye lasers are used mostly for applications where tunability of the laser frequency is required, either for selecting a specific frequency that is not available from one of the solid-state or gas lasers, or for studying the properties of a material when the laser frequency is varied over a wide range. Most of these applications are in the area of scientific experiments. Another large application of dye lasers is for producing ultrashort optical pulses by a technique known as mode locking. In this process, all of the longitudinal modes of a dye laser (as many as 10,000) are made to oscillate together (in phase), causing individual pulses as short as 50 fsec (5×10^{-14} sec) to emerge from the laser, spaced at intervals of the order of 20 nsec. These short pulses are of interest in studying very fast processes in solids and liquids and may have applications for optical communications. [*See* TUNABLE DYE LASERS.]

F. Free-Electron Lasers

Free-electron lasers are significantly different from any other type of laser in that the laser output does not occur from discrete transitions in atoms or molecules of gases, liquids, or solids. Instead, a high-energy (of the order of one million electron volts) beam of electrons is directed to pass through a spatially varying magnetic field that causes the electrons to oscillate back and forth in a direction transverse to their beam direction, at a frequency related to the magnet separation in the transverse direction and also to the energy of the electron beam. This oscillation causes the electrons to radiate at the oscillation frequency and to stimulate other electrons to oscillate and thereby radiate at the same frequency, in phase with the original oscillating electrons, thereby producing an intense beam of light emerging from the end of the device. Mirrors can be placed at the ends of the magnet region to feed the optical beam back through the amplifier to stimulate more radiation and cause the beam to grow. The free-electron laser, although still more of a laboratory curiosity than a useful device, offers to produce very high-power output over a wide range of wavelengths from the far infrared to the vacuum ultraviolet.

VII. Applications of Lasers

A. Laser Properties Associated with Applications

Laser applications are now so varied and widespread that it is difficult to describe the field as a single subject. Topics such as surgery, welding, surveying, communications, printing, pollution detection, isotope enrichment, heat treatment of metals, eye treatment, drilling, and laser art involve many different disciplines. Before these individual topics are reviewed, the various characteristics or features of lasers that make them so versatile will be summarized. These properties include beam quality, focusing capabilities, laser energy, wavelength properties, mode quality, brightness properties, and pulse duration.

1. Beam Quality

When a laser beam emerges from a laser cavity, after having evolved while reflecting back and forth between carefully aligned mirrors, it is highly directional, which means that all of the rays are nearly parallel to each other. This directional property, if considered as originating from a single longitudinal mode emerging from a laser cavity, would have such a low beam divergence that a laser aimed at the moon would produce a spot on the moon only 5 miles in diameter after having travelled a distance of 239,000 miles! That is a very small divergence when compared to a flashlight beam which expands to five times its original size while traveling across a room.

2. Focusing Capabilities

Because of the parallel nature of the beam when it passes through a lens (of good quality) all of its rays are concentrated to a point at the focus of the lens. Since diffraction properties of the light must be taken into account, the beam cannot be focused to an infinitely small point but instead to a spot of a dimension comparable to the wavelength of the light. Thus a green light beam could be focused to a spot 5.0×10^{-5} cm in diameter, a size significantly smaller than that produced by focusing other light sources.

3. Laser Energy

The primary limitations on laser energy and power are the restrictions on the size of the amplifier and the damage thresholds when the beam arrives at the laser mirrors and windows. A 1 W continuous carbon dioxide laser beam, focused on a fire brick, will cause the brick to glow white hot and begin to disintegrate. A carbon dioxide laser has been made to continuously operate at a level 10^5 times that powerful! Similarly, a pulsed laser of 5.0×10^{-4} J, when focused into the eye, can cause severe retinal damage. A pulsed neodymium laser has been made that produces 10^8 times that energy!

4. Wavelength Properties

The capability of generating a very pure wavelength or frequency leads to many uses. It allows specific chemical reactions to be activated, wavelength sensitive reflective and transmissive effects of materials to be exploited, and certain atomic processes to be preferentially selected. Some of these processes can use the fixed frequencies of gas and solid-state lasers. Others require the broad tunability of dye lasers.

5. Mode Quality

The ultrapure frequency available from a single-mode output of a laser not only makes possible the concentrated focusing capabilities mentioned but it provides a very pure frequency that is capable of being modulated in a controlled way to carry information. Since the amount of information that is carried over an electromagnetic (light or radio) wave is proportional to the frequency of the radiation, a single-mode optical wave can carry over 10^4 times the information of a microwave and 10^9 times that of a radio wave. This realization is possible whether the optical wave is carried through space, through the atmosphere, or through a tiny glass fiber the size of a human hair.

6. Brightness Properties

The brightness properties are represented by the amount of light concentrated in a specific area during a definite amount of time in a specific wavelength or frequency interval. Such properties are applicable for use in holography and other coherent processes. The effective brightness temperature of a laser can be 10^{20} to 10^{30}°C, much higher than any other man-made light source.

7. Pulse Duration

The duration of a laser beam can vary anywhere from a continuous (cw) beam that will last as long as the power is supplied to the amplifier and as long as the amplifier keeps producing gain (50,000 h for some gas lasers and potentially many tens of years for some semiconductor lasers) to pulses as short as 6.0×10^{-15} sec for the shortest pulses from a pulse-compressed, mode-locked dye laser. Many applications require very short pulses, including high-speed digital signal transmission for communications and also surgical applications that would result in damage to surrounding tissue via heat conduction if longer pulses were used. Other applications require very long or continuous light fluxes to produce effects in materials that are relatively insensitive to the light beam but would be destroyed by shorter duration, higher-intensity beams.

B. Communications

One of the earliest recognized applications of the laser, because of its capability of producing a very pure frequency, was in the field of communications. Earlier developments using radio waves and then microwaves naturally led to thoughts of using optical frequencies in a similar manner to take advantage of the increased information carrying bandwidth that would be essential for the information age. This concept has now progressed to the point where the semiconductor laser with its small size, low power consumption, and high reliability, in conjunction with optical fibers as transmission media and rapidly developing product lines of optical connectors, couplers, modulating devices, etc., is a major component in communication systems. Using very short, discrete pulses of light to transmit digitally encoded signals is a more recent advancement that will most likely become the ultimate technique for both long and short range communications.

C. Medical

Medical applications are often the most publicized uses of lasers primarily because they affect the general public more directly than other applications. The first applications were in the field of ophthalmology where the laser was used to weld detached retinas and photocoagulate blood vessels that had grown into the region in front of the retina, thereby blocking vision. In such instances the laser beam easily passes through the transparent portions of the eye, including the cornea and lens, to the region of its intended use where the laser energy is absorbed for treatment. Lasers were also developed as diagnostic tools in cell sorting devices for blood treatment and in fluorescent analysis of Pap smears. The more recent applications in laser surgery are perhaps the most far reaching applications of lasers in medicine. The ability of a laser beam to perform very localized cutting along with cauterization of the incision has produced a wide range of surgical procedures including eye operations (see Fig. 8), gynecologi-

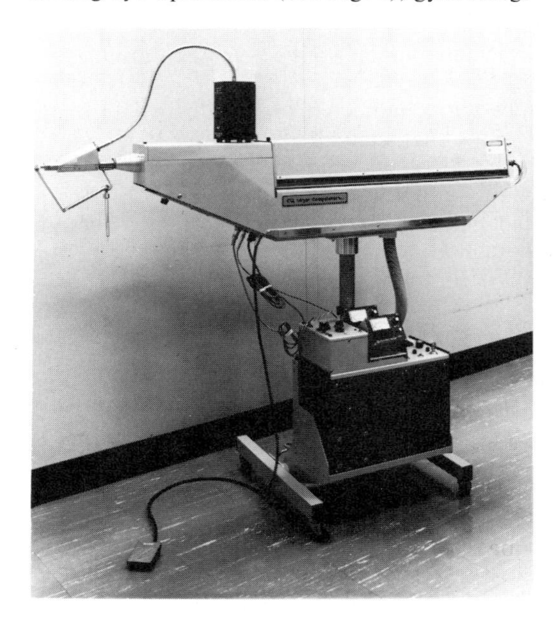

FIG. 8. Carbon dioxide laser eye surgery instrument showing articulated laser beam delivery arm on the left. [Courtesy AT&T Bell Laboratories.]

cal operations, throat and ear surgery, removal of birthmarks, and the most recent prospect of reaming out clogged arteries.

D. Materials Processing

The accurate focusing capability of the laser, allowing the concentration of high power into a small area, makes it a natural device for working with materials. Drilling accurate, tiny holes using a computerized control is now routine in many manufacturing centers, ranging from drilling holes in nipples of baby bottles to drilling holes through high-strength steel. Cutting and welding of high-temperature materials with a laser beam are also effective procedures due to the laser's ability to concentrate energy and power. Another use of lasers is in the heat treatment of metallic surfaces. Pistons of automobile engines, for example, were previously treated in ovens thereby requiring the heating of the entire piston, whereas now a high-power laser can be quickly scanned over specific locations of the piston, locally melting and resolidifying the surface in a short time without having to treat the entire piston.

E. Construction

The use of the laser in the construction industry primarily involves its effectiveness as a straight line reference beam. It is used in the construction of buildings by using a device that has three mutually perpendicular laser beams emerging to provide the reference lines for the sides and vertical alignment of the buildings. It is used as a reference beam for leveling and grading. In this procedure, a laser beam is arranged to continually scan a region under development so that the beam regularly provides a reference level for a grading tractor. Lasers are also used in surveying, as a horizontal reference level.

F. Information Processing

Information processing includes many applications involving reading information in one form and transmitting and converting it to another form. Supermarket bar code readers are probably the most well-known application of this type in which a laser beam scans the digitally encoded "bar code" on grocery items. The reflected, digitally encoded beam is then transmitted to a computer where the product is identified and the information is sent to the checkout register printout. In compact disc players, a diode laser is focused into the grooves on the disc and the reflected, modulated light is detected, converted to an electrical signal, and sent to an amplifier and speaker system. In holograms, the light that is reflected from an object is recorded on film in a way that denotes not only the intensity of the light from various portions of the object but also the relative phase of each of those portions. Reconstruction of the image then provides a three-dimensional image of the original object. Use of the laser in projection is accomplished by scanning various colors of lasers across a screen much as an electron beam is scanned in a television set. The light of the various laser colors is adjusted in intensity at each point on the screen to produce a realistic image of the original object.

G. Remote Sensing

A laser beam provides a unique opportunity to access a distant region without having to install sensing devices in that region. The region could be a hazardous or polluted area, a mine where poisonous gases are potentially located, or an ozone layer in the upper atmosphere. By various techniques the laser can scatter a small portion of its intense beam off of the impurity species, giving off characteristic radiation that is detected back at the source of the laser without requiring human access to the remote region. This technique makes it possible to do quantitative measurements in real time to provide valuable information that is often not available by other techniques.

H. Military

Military applications of lasers most often fall in the category of radar or ranging devices which determine accurate distances to specific targets or map out regions for future access. Lasers also can be directed onto specific targets (usually using an invisible infrared laser) to serve as an illuminating source that will guide infrared sensitive bombs or provide aiming of gunfire. More recently they have begun to be developed as a directed energy weapons; however, the size of such devices prevents them from becoming relatively portable field weapons.

I. Laser Fusion and Isotope Enrichment

Both of these processes involve the use of lasers in developing alternative energy sources for the future when oil supplies begin to dwindle. The laser fusion program is attempting to produce miniature hydrogen fusion reactions in which the fusion energy will be captured as heat that can be used to drive electric power generators. In a conceptual power plant, powerful lasers (1 MJ of energy) will be focused on tiny pellets of special hydrogen isotopes, compressing and heating them to temperatures as high as 10^8 degrees, densities 10^3 times that of solid densities and durations of the order of one billionth of a second (the conditions under which the fusion reactions occur). Laser isotope enrichment has already been shown to be an effective process for enriching concentrations of special isotopes of uranium for use in atomic fission reactors, in a way that is much less costly than the gaseous diffusion and centrifuge processes that have been used since World War II. This enrichment process is accomplished by using selective laser wavelengths that react with the desired rare isotope but not with the more common isotope of uranium.

J. Lithography and Printing

Lasers can be used as sources to expose paper in printing systems, to expose printed circuits for electronic circuit design, or to expose photoresist material to make electronic microchips. They can also be used to deposit circuit material in specific locations on microchips. This procedure is done by focusing the laser onto locations where deposition is required, thereby producing a localized chemical reaction that precipitates the solid material from a molecular gas or vapor containing that material.

K. Laser Art

Lasers are used in many artistic media. The laser light show is probably the most well known. In such a show, laser beams of various colors are directed around a dark room or on a screen in synchronization with a musical presentation. The complicated but artistic images provide a spectacular visual display. Laser are also used in etching or burning artistic patterns on various media.

L. Miscellaneous Applications

Several other applications that do not fall into the other categories are worth mentioning. Laser marking devices are used to put product labels, serial numbers, and other information on items that are difficult to permanently mark by other means, such as high-strength metal and ceramic materials. The laser beam actually melts or ablates the surface in the region where the marking is desired. In a similar way a laser is used to remove material from weights on automobile wheels as determined by computerized balancing instruments, while the wheels are still rotating, thus saving time and improving balancing precision. Lasers also serve as ultrasensitive reference beams in earthquake detection instruments.

Bibliography

Ready, J. F. (1978). "Industrial Applications of Lasers." Academic Press, New York.

Thyagarajan, K., and Ghatak, A. K. (1981). "Lasers—Theory and Applications." Plenum, New York.

Yariv, A. (1985). "Optical Electronics," 3rd ed. Holt, New York.

LIGHT SOURCES

John F. Waymouth *GTE Lighting Products*

GLOSSARY

Arc: Name applied to a high-pressure electric discharge, originally derived from the bowed shape assumed by a horizontal unconfined discharge as a result of upward convective displacement of the center while ends remained anchored to electrodes.

Ballast: Electric circuit device for operation of a discharge lamp, combining a high open-circuit voltage for reliable initiation (ignition or starting) of the discharge, together with internal series impedance to regulate current flow to the design value and decrease the output voltage at ballast terminals to the steady-state operating voltage of the discharge at that current.

Color Rendering Index (CRI): Measure of the capability of a light source to illuminate colored objects in such a way that the colors are perceived as "normal" (see Section II).

Color temperature: Measure of the apparent color of the light emitted from a light source, being the temperature of the blackbody radiator having the closest color match. Designated as correlated color temperature if exact match is not possible (this distinction is frequently ignored in practice).

Dumet: Composite material used for lead-in wires in glass-to-metal seals for lamps using soda–lime–glass envelopes, comprising a core of low thermal expansion nickel-iron and a cladding of high-expansion copper in such proportions that the net expansion matches that of the glass.

Efficacy: Measure of performance of light sources, equal to the ratio of light output to electrical power input, normally given in lumens per watt. Since this ratio is not dimensionless, the word *efficiency* is avoided for this quantity.

Hard glass: Generic jargon term for glasses having softening and working temperatures greater than soda–lime glass but less than fused silica (quartz).

Lead-in wires: Component of a lamp through which electric current is introduced, through a glass-to-metal seal, into the interior of the hermetically sealed envelope. In industry jargon, these are termed lead wires, but that term is not used here to avoid confusion with the element lead.

Luminance: Measure of the intensity of light, equal numerically to the light flux in lumens per unit area per unit solid angle. (formerly referred to as brightness).

Polycrystalline alumina (PCA): Translucent, ceramic material used for discharge tubes of high-pressure sodium lamps (see Section V).

Press seal: Glass-to-metal seal in a lamp fabricated from fused silica (quartz), used in tungsten–halogen lamps (see Section IV) and high-intensity discharge lamps (see Section V). Fabricated by mechanically pressed heated, softened silica glass to and around thin foils of molybdenum, which serve as lead-in wires.

Resonance radiation: Radiation emitted in an atomic transition, in which the lower state of the transition pair is the lowest energy state of the atom, occupied in a large fraction of the atoms present.

Stem: Component of a lamp incorporating the glass-to-metal seals of the lead-in wires together with mating surfaces of glass ready for flame-fusion sealing to the glass envelope. The stem usually includes a tube through which air from the envelope interior can be exhausted after the stem is sealed to the envelope.

Wall-loading: Measure of the power input to a discharge lamp, in watts per unit area of wall surface. An important design parameter for most discharge lamps.

Next to the invention of the wheel, humanity's oldest and most important invention is the artificial light source, a means or device for converting some other form of energy into light. From the dawn of human history until ~200 years ago, such devices could convert only chemical energy into light, through the medium of fire. For ~100 years, commercially useful electric light sources have been available—a technological accomplishment of stupendous magnitude that has transformed our lives.

I. Introduction

Approximately 5 billion incandescent lamps, 2 billion fluorescent lamps, and 200 million high-pressure discharge lamps are in service around the world. They consume ~25% of the worldwide electric energy production. Electric energy production, in turn, accounts for ~25% of worldwide consumption of energy.

Electric lamps produce the light that makes it possible to carry on all forms of human activity far outside the sunrise-to-sunset time limitations faced by earlier societies, and thus greatly increasing our productivity. They make our streets and highways safer after dark, and they decorate and illuminate all manner of public places. Finally, they do all this at a cost to us that is one-quarter percent of our gross personal product, individually or gross national product, collectively.

This article describes the scientific and technological bases of the major electric light sources. Each section discusses the mechanisms of radiation, the energy balance and the factors that set the limits for performance and efficiency; the fabrication technology and the principal deviations from ideality; representative examples of commercially available products; and the direction of research and development toward improved future products.

II. Electromagnetic Radiation

A. Emission and Absorption of Radiation by Matter

A ray of electromagnetic radiation transiting a material substance may be characterized in terms of the number of photons of frequency ν crossing unit area per second, per unit solid angle, per hertz. As it traverses the substance, it may be diminished as a result of absorption, or it may be augmented as a result of spontaneous and stimulated emission. Stimulated emission results in photons being added to the beam, in phase and in the identical direction as the stimulating photon (i.e., within the same solid angle as the incident ray). Spontaneously emitted photons are emitted (in the absence of nuclear orientation effects not usually encountered in light sources) with equal probability in all directions. That is, the fraction of the spontaneously emitted photons within a given solid angle is equal to $d\Omega/4\pi$.

1. Radiation Transport Differential Equation

These considerations may be quantified in the radiation transport equation, Eq. (1), in which $n(u, x)$ and $n(l, x)$ are the number densities at position x of upper and lower states of the transition pair resulting in emission of frequency ν, and $A(\nu)$ is the spontaneous transition rate per second, per hertz of bandwidth, at frequency ν.

$$\mathbf{i}\frac{d\Gamma(\nu, x)}{dx} = \mathbf{i}\frac{A(\nu)}{4\pi} + k(\nu)n(u, x)\Gamma(\nu, x) - \alpha(\nu, x)\Gamma(\nu, x) \quad (1)$$

Note that $A(\nu)$ is dimensionless. The expression $\Gamma(\nu, x)$ is the vector spectral radiance in photons per unit-area-second per hertz of bandpass per steradian, assumed for simplicity to have only an x component, and $\mathbf{i}$ is a unit vector in the x direction. The expression $\alpha(\nu)$ is the absorption coefficient of radiation of frequency (ν), and $k(\nu)$ is the stimulated emission coefficient of frequency (ν), both per unit of path length through the radiating medium.

Einstein first demonstrated that absorption, spontaneous emission, and stimulated emission coefficients are related. That is,

$$k(\nu) = \alpha(\nu, x)/(n(l, x)g_u/g_l)$$

$$\alpha(\nu, x) = \frac{A(\nu)}{8\pi\nu^2}c^2 n(l, x)g_u/g_l \quad (2)$$

The expressions g_u, g_l are the statistical weights

of upper and lower states, respectively, and c represents the velocity of light.

Incorporating these relationships in the radiation transport equation gives the following result:

$$\mathbf{i}\,\frac{d\Gamma(\nu, x)}{dx} = \alpha(\nu, x)\left\{\frac{2\nu^2}{c^2}\left[\frac{n(u, x)g_l}{n(l, x)g_u}\right]\mathbf{i} + \left[\frac{n(u, x)g_l}{n(l, x)g_u} - 1\right]\Gamma(\nu, x)\right\} \quad (3)$$

The source of the energy input to the material that maintains a population of upper states in the face of depletion by radiation does not concern us in this section.

2. Local Thermal Equilibrium

For a system locally in thermal equilibrium, at a local temperature $T(x)$, the number densities of upper and lower states are related through a Boltzmann factor:

$$n(u, x) = n(l, x)\,\frac{g_u}{g_l}\exp(-h\nu/kT(x)) \quad (4)$$

For systems not in local thermal equilibrium (LTE) we may still define a radiation temperature from Eq. (4) for the transition in question, with the stipulation that the radiation temperature will not necessarily be the same for all possible transition pairs.

3. Blackbody Radiation

In an infinite material medium in which the temperature is independent of position, global thermal equilibrium exists, and the value of $d\Gamma(\nu, x)/dx$ becomes zero. Setting it equal to zero in Eq. (3), and solving for $\Gamma(\nu)$ yields the blackbody, or ideal radiator law, expressed in the units of photons per second per unit area per steradian per hertz of bandpass.

$$\Gamma_{BB}(\nu) = \mathbf{i}\,\frac{2\nu^2}{c^2}\Big/(e^{h\nu/kT} - 1) \quad (5)$$

Therefore, the radiation transport equation may be rewritten for systems in LTE in terms of the local temperature and the local blackbody spectral radiance:

$$\mathbf{i}\,\frac{d\Gamma(\nu, x)}{dx} = \alpha(\nu, x)(1 - e^{-h\nu/kT})\{\Gamma_{BB}(\nu, x) - \Gamma(\nu, x)\} \quad (6)$$

For most light sources, the value of $h\nu/kT$ is greater than or equal to 3, so that the first term $[1 - e(-h\nu/kT)]$ is essentially equal to unity and is usually neglected.

Note that in Eq. (6) the temperature may vary with position and therefore the corresponding blackbody spectral radiance as well; note also that the units of spectral radiance are those chosen for the blackbody spectral radiance. Although Eq. (6) is derived in terms of photons per second, it is much more general and can be used with whatever units for blackbody radiance are convenient and familiar. Note also that the growth and decay of a ray of radiation in transiting a material medium is proportional to the absorption coefficient. Where it is less than the blackbody level, spectral radiance will increase only at those frequencies for which absorption coefficient is greater than zero. Material substances cannot emit electromagnetic radiation at any frequency at which they do not also absorb.

Therefore, determination of the radiant emission spectrum of a material substance requires a determination of its absorption spectrum.

B. Sources of Absorption

The processes that contribute to absorption and emission of radiation in the UV, visible, and IR regions of the electromagnetic spectrum are principally electronic transitions. These may include free-free and free-bound transitions of free electrons, which result in absorption and emission of radiation over wide ranges of frequency (continuum emission). This is the most characteristic feature of emission from heated metals. Many electric discharge lamps take advantage of emission and absorption by electronic transitions in isolated atoms. Absorption and emission "lines" in the radiation spectrum from such transitions are usually broadened by a variety of perturbing influences, such as (1) collisional or pressure broadening due to collisions between radiating atoms and perturbing atoms; (2) Stark broadening from the electric fields of nearby charged particles, electrons and ions, in the plasma of a discharge lamp; (3) Doppler broadening as a result of thermal motions of the radiating atoms; and (4) resonance broadening as a result of perturbation by identical atoms in the lower state of the transition pair.

In addition to atomic line emission, emission as a result of electronic transitions in molecules, as well as vibrational and rotational transitions, contributes absorption and emission in some media.

It is beyond the scope of this article to deal with the *a priori* calculation of the frequency-dependent absorption coefficient required for the use of Eq. (6). However, it is clear from the foregoing that the totality of absorption processes must be analyzed with full knowledge of necessary transition probabilities and broadening parameters applicable to the system in question. For those cases in which absorption coefficients can be calculated, and radiation temperature profile determined, the integration of Eq. (6), repeated for the desired range of frequencies or wavelengths and for selected angular orientation to the surface, permits calculation of emitted spectral radiance.

In many cases of interest in the science and technology of electric discharge light sources, however, necessary basic data are unknown; they must be either estimated or used as variable parameters in an empirical fitting to observational data. Research and development of light sources therefore remains an empirical science in many cases but with continual improvement in calculation models to guide the empirical program.

C. Spectral Radiance versus Position and Wavelength

Once the value of absorption coefficient as a function of frequency and position is known and temperature is known as a function of position for a material medium in local thermal equilibrium, radiance may be calculated as a function of position by integration of Eq. (6). For systems not in LTE, integration of Eq. (3) is required.

The optical density of the medium is defined as

$$\tau(\nu) = \int_0^{\bar{x}} \alpha(\nu, x)\, dx$$

In two limiting cases, optically thick, $\tau(\nu) \gg 1$, and optically thin, $\tau(\nu) \ll 1$, particularly simple results can be obtained.

1. Constant Temperature

For the optically thin case, $\Gamma(\nu, x)$ remains everywhere $\ll \Gamma_{BB}(\nu)$, and it may be neglected in comparison. Radiance increases monotonically with distance through the medium:

$$\mathbf{i}\,\frac{d\Gamma(\nu, x)}{dx} = \alpha(\nu, x)\Gamma_{BB}(\nu) = \mathbf{i}\,\frac{A(\nu)n(u, x)}{4\pi} \qquad \left(\frac{h\nu}{kT} \gtrsim 3\right) \quad (7)$$

For the nonoptically thin case of constant temperature, integration of Eq. (6) gives:

$$\Gamma(\nu, x) = \Gamma_{BB}(\nu)\{1 - e^{-\alpha(\nu)\cdot x}\} \quad (8)$$

For $\tau \gtrsim 3$, radiance approaches the blackbody level. Wherever absorption coefficients are sufficiently large that the substance is optically thick, it is not necessary to know the absorption coefficient exactly; radiance is everywhere equal to local blackbody value. Because of the relatively large absorption coefficient for free-free transitions of electrons in metals as a result of large free electron density, and because temperature inside a heated wire is essentially independent of position, this case corresponds reasonably well to the situation inside the metal filament of an incandescent lamp. The internal spectral radiance essentially equals the blackbody level from the UV to the IR.

2. Nonconstant Temperature: Self-Reversed Lines

A common situation in electric discharge light sources is optically thick emission of atomic line radiation from a discharge in LTE with a position-dependent temperature, as shown in Fig. 1. Although the absorption coefficient at the line's center may be high, leading to a large optical depth, absorption coefficient decreases with increasing frequency shift from line center because of the broadening processes. Dependent on the wavelength shift from line center, τ may be either $\gg 1$, 1, or $\ll 1$. Integration of Eq. (6) for these three cases leads to the result shown in Fig. 1. At the line center where absorption coefficient is high, spectral radiance is essentially everywhere equal to the local blackbody value, rising from a low level at $x = 0$ to a maximum coincident with maximum temperature and decreasing to a very low value at $x = D$. In the far wings of the line, where optical depth is low, there is essentially no absorption of any radiation emitted into the ray, and radiance increases monotonically to a maximum at $x = D$ but reaches only a relatively low level. In the vicinity of $\tau = 1$, however, spectral radiance continues to increase for all those values of x for which it is less than local blackbody radiance and decreases for those values of x for which the converse is true. The distance scale over which these changes can take place, however, is $1/\alpha$.

Therefore, the value of emergent radiance at $x = D$ is much greater than blackbody at $x = D$. It typically reaches 10–20% of blackbody spectral radiance at maximum temperature.

The frequency or wavelength distribution in

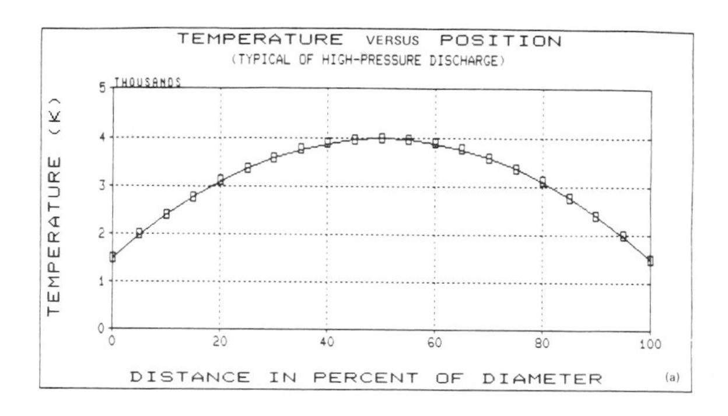

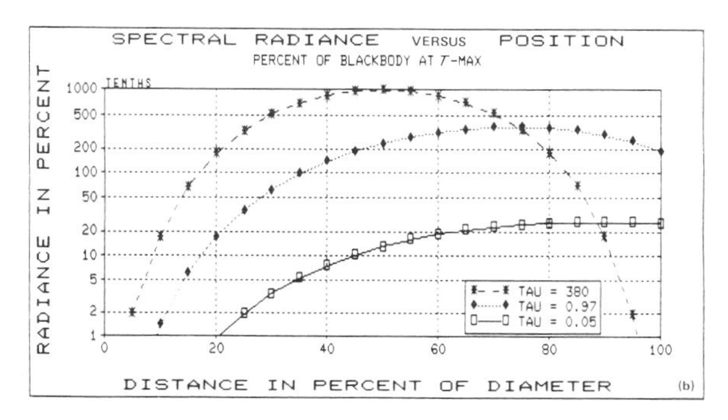

FIG. 1. (a) Profile of temperature along a diameter through an electric discharge in local thermal equilibrium (LTE) (typical). (b) Positively directed spectral radiance along a diameter through the LTE discharge of (a) for three different values of optical depth between temperature maximum and discharge tube wall. Radiance for $\tau = 380$ is indistinguishable from local blackbody radiance at the temperature corresponding to each position.

the emitted line (see Fig. 2) shows a minimum (reversal) at the center wavelength of the atomic line, with maxima in emission on either side. Such emission lines are commonly referred to as self-reversed. The optical depth between temperature maximum and escape point is typically about unity at wavelength of maximum emission.

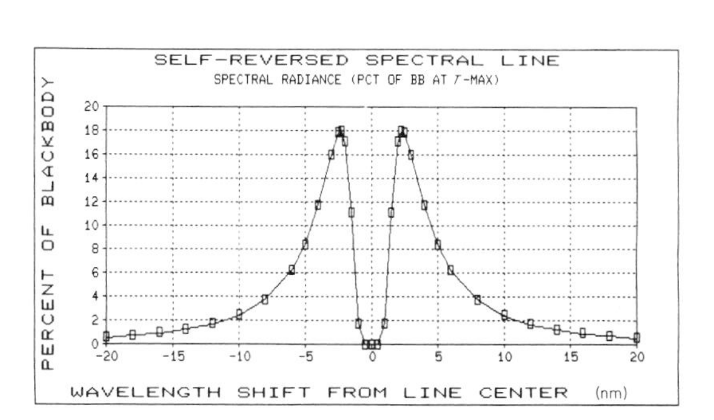

FIG. 2. Spectral distribution of emitted atomic resonance line radiation for temperature profile of Fig. 1(a), illustrating self-reversal, calculated for sodium resonance line (589.0–589.6 nm) for 0.1 atm pressure in a tube 1.0 cm in diameter. Line center (reversal minimum) corresponds to $\tau = 380$ in Fig. 1(b); maxima correspond to $\tau = 0.97$; 10-nm shift from line center (either side) corresponds to $\tau = 0.05$.

D. Reflection at Bounding Surfaces

The integration of the radiation transport equation does not of itself permit calculation of the radiation emitted by a bounded material medium. There is an interface at the boundary where the radiating medium meets the external environment and in general a mismatch of indices of refraction. As a result of the difference in the indices of reflection, there is a reflection loss so that a portion of the internally generated radiation does not escape. For discharge lamps in which the radiating plasma is contained in a transparent vessel, the reflection losses are relatively small approximately 4–5% at each of several glass surfaces. It is more severe in discharges contained in translucent vessels, which embody many scattering centers within the vessel wall that result in significant diffuse reflection of escaping radiation. For incandescent metal sources—such as tungsten-filament lamps—however, it is extremely significant because of the large index of refraction of the metal, especially in the far IR range.

The escaping flux is equal to $(1 - r)\Gamma_{BB}$, which is usually expressed as $\varepsilon\Gamma_{BB}$, where $\varepsilon = (1 - r)$ is called the emittance. The reflectance, r, of a metal surface can be approximately calculated by treating the free electrons of the metal as a free-electron gas, having a plasma frequency greater than the radiation frequency. Provided collision frequency of electrons in the metal is small in comparison to radiation frequency, a simplified Drude formula gives for $1 - r$, where ρ is electric resistivity in ohm centimeters, and λ is wavelength in centimeters:

$$\varepsilon(\lambda) = 0.365\left(\frac{\rho}{\lambda}\right)^{1/2} - 0.0464\left(\frac{\rho}{\lambda}\right) \qquad (9)$$

Figure 3 shows the observed spectral emittance of tungsten as a function of wavelength, together with the values calculated from Eq. (9). The simplified model is a fairly accurate representation of the IR emittance. It is clear that reflection at the surface drastically reduces IR radiation escape, by as much as 90% at 10 μm, for example. Therefore, even though spectral radiance internal to a heated tungsten body reaches the blackbody level over the entire range of optical frequencies, escaping radiance is at maximum less than 50% of the blackbody value in the visible, declining toward 10% at 10 μm. As is shown in Section IV, this has important consequences for the efficacy of tungsten-filament lamps.

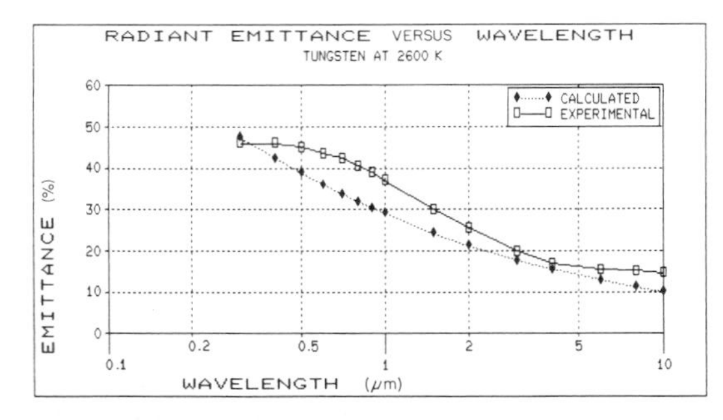

FIG. 3. Radiant emittance of tungsten as a function of wavelength, experimentally measured, and calculated from Eq. (9).

III. Light

A. Electromagnetic Radiation and the Eye

The proper measure of the "goodness" of artificial light sources is not how well they emit electromagnetic radiation, but how well they emit electromagnetic radiation that can be perceived and detected by the eye—that is, light. Therefore, of concern in light source development is not only radiant power, but also the radiant power weighted by the biological action spectra of vision. These correspond to two distinct perceptions of light, luminance (formerly called brightness) and color. There prove to be two distinct action spectra for luminance: scotopic, or dark-adapted, and photopic, or light-adapted. Under the vast majority of situations involved in applications of artificial light sources, the luminance level is high enough that the eye is light-adapted, and the photopic action spectrum is the proper one to use. In photopic vision, there are three sets of photoreceptors, called cones, sensitive to red, green, and blue radiations, respectively. In combination, these are sensitive to the wavelength band of the electromagnetic spectrum between approximately 380 nm and 780 nm.

Signal processing in the eye–brain system is such that luminance information comes from the sum of the responses of red (long wavelength) cones and green (intermediate wavelength) cones. Color information comes from a combination of the luminance signal and the differences between the responses of the red, green, and blue (short wavelength) cones.

The relative spectral sensitivity functions of these responses have been determined from an extended series of luminance and color matching experiments by a number of standard observers who have been selected to eliminate various pathologies from consideration. The average values of these functions have been agreed on by deliberations of the *Commission Internationale de L'Eclairage* (International Commission on Illumination), known as the CIE.

1. Luminance and Lumens

The luminance function is designated as $\overline{y(\lambda)}$, and the total light output of a light source in lumens is determined from the convolution integral over the wavelength interval (or range) of the radiated spectral power distribution $P(\lambda)$ weighted by the $\overline{y(\lambda)}$ function:

$$L = k \int \overline{y(\lambda)} P(\lambda)\, d\lambda \qquad (10)$$

The coefficient k has been arbitrarily set at 683 lm/W so that the lumen output value determined according to Eq. (10) should agree with comparison photometry based on older so-called standard candles. In this older definition, one standard candle emits 1 lm/sr. The values of the Y function as a function of wavelength are shown in Fig. 4. The luminous efficacy of a light source is determined by dividing its total lumen output by the input power required to maintain it in a constant light-emitting state and is measured in lumens per watt.

2. Color

The color of light can be expressed in terms of two indices, x and y, determined from three other functions, X, Y, and Z, called tristimulus values:

$$x = \frac{X}{X + Y + Z}$$
$$y = \frac{Y}{X + Y + Z} \qquad (11)$$

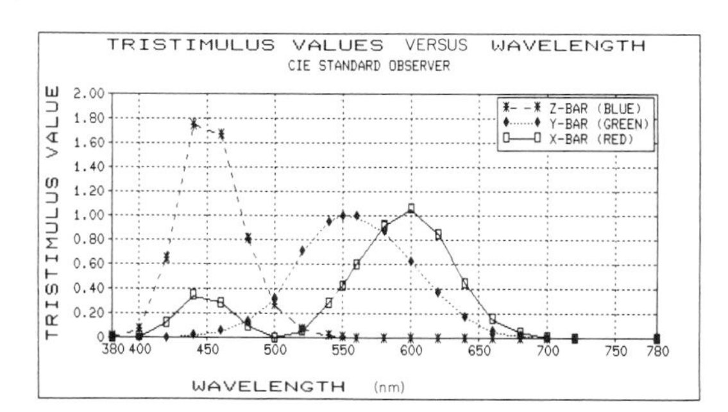

FIG. 4. Values at 20-nm intervals of the red, green, and blue tristimulus functions for computing color properties of light sources. Note that although these are ultimately dependent on the spectral responses of the three different types of retinal cones, they are not identically equal to those response functions. The Y-bar (green) function is the same as the photopic action spectrum for luminance.

The functions X, Y, Z are themselves determined from convolution integrals over the spectral power distribution, weighted according to the tristimulus functions $\overline{x(\lambda)}$, $\overline{y(\lambda)}$, and $\overline{z(\lambda)}$. The $\overline{y(\lambda)}$ function is the same one used for luminance calculations. All three functions are shown in Fig. 4. The x and y values may be measured directly by photometers with suitably calibrated filters, which are in effect analog computers for calculating the convolution integrals.

Figure 5 shows the locus of monochromatic spectrum colors on the x–y plane, forming the CIE 1931 chromaticity diagram (x–y diagram). Color points x,y for nonmonochromatic radiation sources lie in the space enclosed by the locus of spectrum colors. The x,y value for the sum of two sources of color x',y' and x'',y'' are found along a line joining the two x,y coordinates, which is a useful additive property. The curved line extending through the center of the diagram is the locus of colors of blackbody sources of the indicated temperatures. The color points for daylight also fall on or near the blackbody locus, toward the high-temperature end. For reasons probably rooted in evolution, the preferred color domain for most artificial light sources is also on or near the blackbody locus; x,y points above the locus are perceived as too

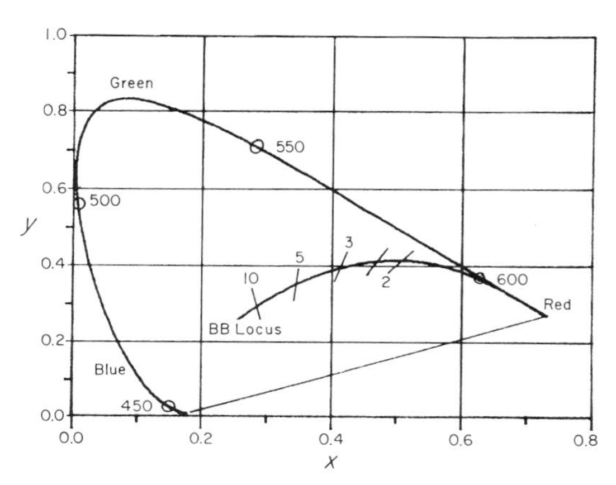

FIG. 5. CIE chromaticity (x–y) diagram: outer curve is the locus of monochromatic spectrum colors, labeled at selected wavelengths (nm); all real colors are contained within the bounded region. The inner curve is the locus of colors of blackbody radiators ranging from low temperatures at the right to high temperatures at the left; selected temperatures are labeled (in 1000 K). Light sources perceived as white have x–y values near the blackbody (BB) locus between 2500 K and 5000 K.

green, those below as gloomy and purple. Therefore a common, although far from exact, representation of the color of a light source is its correlated color temperature (or just color temperature): the temperature of the blackbody source having the closest color match. The x–y diagram, although useful for representing colors of light sources, suffers from the fact that equal perceptible color differences are not represented by equal vector distances between x,y points in different parts of the diagram. Therefore, other systems for color representation are also in use, but these do not concern us here.

3. Color Rendering

Color rendition is another property of light sources. Color rendering is a measure of the degree of distortion in the *apparent* colors of colored pigments perceived when they are illuminated by the source in question, in comparison to their apparent colors when they are illuminated by a standard source. For example, a source emitting monochromatic blue light and monochromatic red light in such proportions that the composite color lies on the blackbody locus at a color temperature of 3000 K will cause a green pigment to appear black or purple in comparison to the blackbody of the same color temperature—although the colors of the light emitted from the two sources are indistinguishable. The tristimulus values of the light reflected from a pigment can be calculated using formulas identical with those of Eq. (10), except that they also include the spectral reflectance of the pigment as a weighting factor.

The color rendering index (CRI) of a light source is computed by calculating the shifts in a uniform color space of the color coordinates in a series of standard pigments illuminated by the light source from the computed color coordinates for the same pigments illuminated by a standard source of the same color temperature. For color temperatures less than 5000 K, the reference source is a blackbody; for color temperatures greater than 5000 K, it is a standard daylight. Continuous spectrum sources, such as incandescent lamps, have high CRIs, near 100, since their spectra are nearly blackbody. Multiline spectrum sources, such as some high-pressure discharge lamps, have CRIs ranging from 50 to 80. Fluorescent lamps also have CRIs of 60 and greater. CRI values less than 40 are generally not desirable for illuminating spaces occupied by humans.

IV. Incandescent Lamps

A. Introduction

The most common form of electric light source is the incandescent lamp, which consists of a solid filament, heated by the passage of electric current, in a hermetically sealed glass envelope from which all traces of oxidizing gases have been removed. Figure 6 is a schematic diagram illustrating the principal components of such a lamp. Present-day incandescent lamps use a tungsten filament in a coiled or multiply-coiled helix. Most types use an inert gas filling in the bulb to retard evaporation of the filament, thereby permitting higher filament temperatures for a given design life.

B. Radiation Mechanism

The radiation mechanism in incandescent lamps is the continuous spectrum resulting from deflection and unquantized radiation of valence electrons (traveling freely through the conduction band of the solid) by collisions with nuclei, each other, and lattice phonons. The optical depth is great enough that the internal radiation flux reaches the blackbody level throughout the UV–visible–IR spectrum. A large index of refraction causes reflection losses in the escape of this radiation from the surface, yielding an effective emittance of ~0.45 in the visible, decreasing to approximately 0.10 to 0.15 at 10-μm wavelength in the IR. Thus, tungsten emits a larger fraction of total radiation into the visible (giving approximately 30–40% higher luminous efficacy) than a blackbody at the same temperature, with a comparable increase in this visible frac-

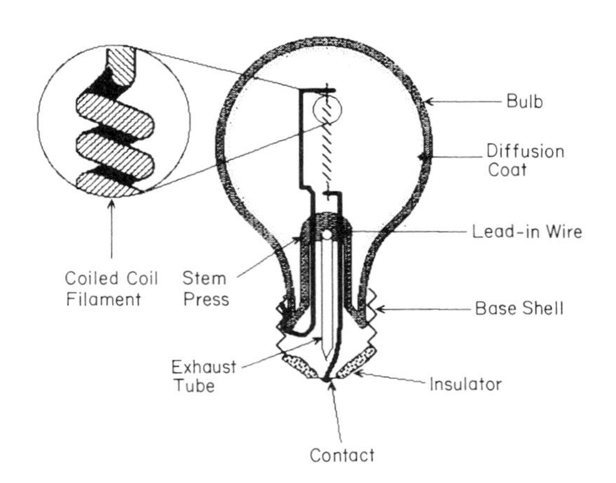

FIG. 6. Schematic diagram, in section, illustrating components of a typical incandescent lamp.

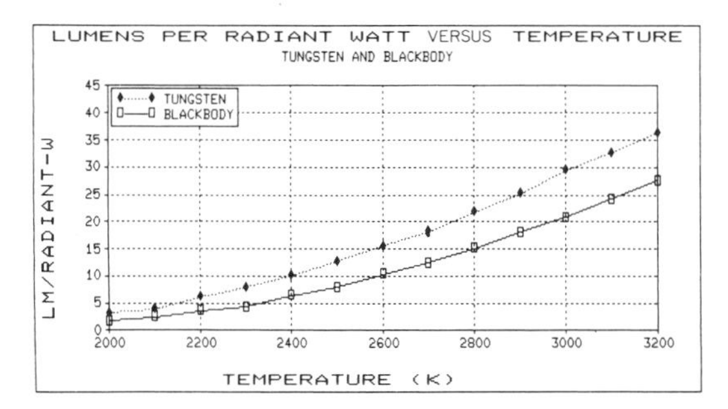

FIG. 7. Plots of luminous efficacy versus temperature for tungsten and blackbody incandescent radiators. The higher luminous efficacy for tungsten is the result of its higher radiant emittance in the visible in comparison to the IR range (see Fig. 3).

tion (and luminous efficacy) with increasing temperature (see Fig. 7). For efficient incandescent lamps, the temperature of the radiator must be as high as possible, yet consistent with constraints on service life. Visible spectral power distribution is blackbodylike, with a color temperature of 50 to 100 K higher than true; the CRI is in the high 90s.

C. Energy Balance

Ohmic heating by the energizing current raises filament temperature to dissipate power by radiation, convective cooling by the gas filling, and solid conduction to filament supports and electric lead-ins. The power consumption of the lamp at a given input voltage is determined by the filament resistance, a function of the length and cross-sectional area of the wire and resistivity of the wire material. Resistivity of tungsten increases strongly with temperature, and is 15-fold less at room temperature than at operating temperature; inrush current at initial switch-on is correspondingly higher than operating current. At normal operating temperatures (~2750 K), resistivity is ~80 $\mu\Omega$ cm. For a 100-W lamp operating from a 120-V line and requiring a resistance of 144 Ω, the ratio of cross-sectional area to length is ~5 nm. This dictates wires a few hundredths of a millimeter in diameter for total filament lengths of ~0.5 m. Even more slender filaments are required for operation at lower wattages or from 240-V power lines. For a given filament design, electric power input and filament temperature increase with increasing line voltage.

The gaseous convection loss from the filament is minimized by coiling the filament into a helix or multiple helix because of the existence at the hot surface of a stationary boundary layer of gas, with thickness comparable to the diameter of the helices. Thus, the effective surface area cooled by the gaseous convection is that of the cylindrical envelope of the helix, much less than that of the filament wire itself. The effective radiating surface is also smaller than that of the filament wire surface as a result of the coiling, but by not so great a factor. This is because radiation emitted into the interior of the helix does have a reasonable probability of escape between the turns, after one to several reflections, absorptions, and reemissions. In typical household incandescent lamps, 10–15% of the input power is lost in gaseous convection, 80–85% in radiation, and the balance in conduction to the filament supports and electric lead-ins. The reduction in evaporation rate resulting from the gas fill permits ~300°C higher operating temperature for equal life as in vacuum, which results in as much as 80% increase in luminous efficacy of the emitted radiation, more than justifying a 10–15% reduction due to gaseous convection loss.

D. Performance Limitations Resulting from Material Limitations

Maximizing luminous efficacy of incandescent lamps requires as high a radiator temperature as possible. Evaporation of the filament material and the maintenance of mechanical strength at elevated temperatures limit the permissible temperature so that tungsten is the material of choice.

1. Vaporization

Vaporization of tungsten varies as $\exp(-105000/T)$, which at 2750 K equates to a fourfold increase per 100 degrees and is the principal determinant of lamp life. Failure occurs when less than 1% of the filament mass has evaporated. A thin spot along the length of the filament has higher resistance and higher I-squared-R heating because the same total current must pass through every cross-sectional area. It consequently operates at a higher temperature and vaporizes faster than adjacent spots, becoming thinner faster. Failure generally occurs when the thinnest point along the length of the filament is reduced in diameter by ~50%. High inrush current at switch-on frequently causes failure at the weakest spot. However, the same mechanism could cause failure under steady burning conditions within a few hours.

The lives of incandescent lamps are not significantly dependent on the number of switch-ons per operating hour.

Thin spots, or hot spots, may originate from nonuniformities in wire diameter or nonuniformities in coil turn spacing (caused by the coiling process itself), distortions in mounting of the filament, or shifts during operation at the elevated temperatures ("squirm" and "sag").

Both luminous efficacy and evaporation rate depend on filament temperature, reflected in an empirical "LPW/DL" law: design life varies as the inverse seventh or eighth power of design lumens-per-watt. Consequently, the sum of lamp plus electricity cost to produce a given illumination is the minimum at an optimum design life. If the design life is too short, although efficacy will be the greatest and energy cost the least, the cost of bulbs and replacement bulbs is too high. And if design life is too long, although lamp cost is minimized, energy cost increases. In household use (no labor cost for bulb replacement) the bulb life of ~1000 hr gives the minimum total cost of light. In commercial or industrial use, replacement labor cost may exceed the cost of the bulb itself; thus, longer design lives are needed.

Life and luminous efficacy of an incandescent lamp are strongly affected by operation at service voltages other than the design value. Operation at 10% greater than design voltage will shorten lamp life by a factor of 3, but it will give 35% more light. Operation at 10% less than design voltage will lengthen life 400% at the cost of 28% less light.

Incorporating an inert gas filling in the bulb obstructs the flow of vapor from the filament, converting the rate-limiting step from primary vaporization to diffusion. The dependence on filament temperature remains, however, since the density gradient is determined by the equilibrium pressure of tungsten vapor at the filament surface, which has the exact same temperature dependence as the primary evaporation rate. The loss rate of filament material is dependent on the diffusion coefficient of tungsten in the gas; life increases approximately linearly with pressure, and increases with atomic mass number of the gas. Household lamps are generally filled with argon plus a few percent of nitrogen at ~1 atm of operating pressure. The nitrogen prevents the establishment of an electric discharge in the gas between the lead-in wires. Krypton fill is used in certain premium types for higher efficiency; the lower diffusion coefficient of tungsten in krypton permits slightly higher filament temperature for equal life, and the convective losses in krypton are smaller.

2. Mechanical Strength

Wire for lamp filaments must initially be ductile enough to coil to a mandrel diameter comparable with wire diameter, without splitting or "green-stick" fracture. But the filaments must become rigid and nonductile in operation at 75% of melting temperature so that the turns of the coil do not shift, and the wire does not stretch. Control of the microstructure, not only of the as-drawn wire but also after recrystallization at elevated temperature, produces this astonishing result.

Non-sag tungsten for lamp use incorporates approximately 30–50 ppm of elemental potassium introduced in compound form in the powder preparation stage of a powder-metallurgy process. Controlled wire-drawing in multiple stages, with intermediate anneals, results in a fine-grained fibrous-looking microstructure of elongated grains parallel to the wire axis. The potassium, which is insoluble in tungsten, is dispersed as uniformly as possible throughout the wire and is concentrated as minute globules along grain boundaries. The fibrous microstructure imparts to the wire a flexibility analogous to rope so that the wire can be bent easily without fracture.

At temperatures between 1500 K and 2000 K (much less than the service temperature), the tungsten recrystallizes, and the small fibrous grains grow to sizes comparable with the wire diameter. Without potassium dopant, the preferred crystal habit would be an equiaxed hexagonal structure (see Fig. 8) with many grain boundaries approximately perpendicular to the wire axis, along which slippage or sliding occurs at elevated temperatures so that the wire "sags" or "squirms." With potassium-doped wire, however, the recrystallization is quite different. Insoluble and volatile at the recrystallization temperature, the potassium globules vaporize and create voids distributed in long rows parallel to the wire axis. The rows of voids act as barriers to grain boundary migration so that propagation of grain boundaries across the wire is greatly inhibited in comparison to propagation along the wire axis. The resulting recrystallized structure, also shown in Fig. 8, then comprises elongated interlocking grains (with greater than 10 : 1 aspect ratio in high-quality wire) oriented

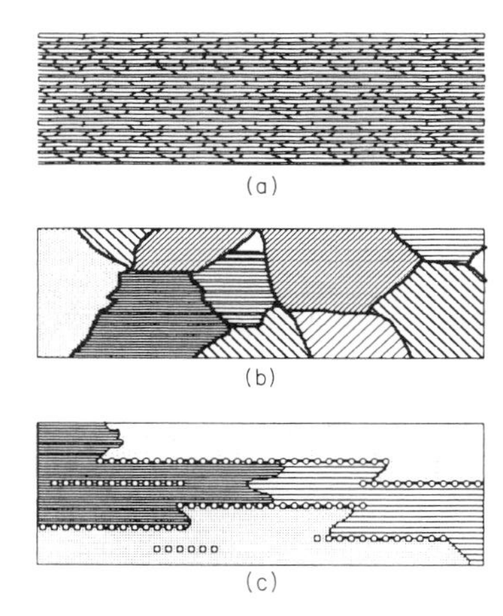

FIG. 8. Schematic diagram illustrating tungsten metal microstructure: (a) as-drawn, fibrous; (b) recrystallized, undoped wire, equiaxed; (c) recrystallized, potassium-doped wire, interlocking.

along the wire axis, with few if any perpendicular grain boundaries. The long-grain boundaries exhibit much greater resistance to sliding than short perpendicular boundaries, which greatly increases the hot strength of the wire.

E. Fabrication Technology

Assembly and processing of incandescent lamps takes place on highly automated machine groups operating at speeds between one and two finished lamps per second, essentially without operator intervention. Such manufacturing methods are not only necessary to maintain low costs, but also to cope with the sheer volume of production required to fill the demand, ~1 billion lamps per year in the United States alone.

Lamp manufacture begins with fabrication of tungsten wire coils, typically wound on molybdenum or steel wire mandrels, which are acid-dissolved after stress-relieving. Primary coil-winding is done fully automatically at speeds in excess of 10,000 rpm, with precision of pitch better than 1%. Secondary coil-winding is slower but equally automatic and precise. Glass bulbs are made of soda–lime glass, blown into rotating molds at speeds of a thousand parts per minute.

Stem presses are fabricated by automatically sealing lead-in wires into a press, fusing glass to metal with gas-oxygen and gas-air fires. The lead-in wires themselves are a controlled-thermal-expansion composite comprising a low-expansion nickel-iron core and a high-expansion cladding of copper (Dumet) in such proportions as to provide an expansion match to the glass. After glassworking, the filament supports are automatically bent, and the coiled filaments are automatically fed and mounted. Meanwhile, a light-diffusing particulate coating is applied electrostatically to the inside of the bulb, and both bulb and mount are automatically transferred to the sealing machine where the glass of the bulb is fused to the glass of the stem with gas-oxygen fires. Automatically transferred to an exhaust machine, the sealed bulbs are exhausted and backfilled with gas through an exhaust tube integral with the stem, after which the exhaust tube is fused closed and cut off by gas fires. The evacuated bulb is transferred to a basing machine where the external leads are threaded to and affixed to the base, and the filaments are "flashed" to recrystallize the wire in its high-temperature microstructure. Automatic testing and packaging follow.

1. Influence of Impurities

Certain impurities substantially degrade the life and lumen maintenance performance of incandescent lamps, even when their filament temperatures are exactly as specified. An important object of the process and material control is to minimize these impurities. The elements hydrogen, oxygen, and carbon and the transition metals iron, nickel, and chromium have significant deleterious effects.

Carbon can cause formation of tungsten carbides, resulting in brittleness of the filament and susceptibility to fracture under mechanical shock. Since graphite is the preferred lubricant for the wire-drawing process, it is plain that cleaning processes to remove carbon must be under very careful control. Transition metals result in exaggerated grain growth and tend to defeat the purpose of the potassium doping in the recrystallization process. Many of the tools and jigs and fixtures involved in the assembly process are made of steel, and the lead-in wires themselves are typically made of nickel-plated steel. Contamination of filaments and coils from these sources must be monitored and controlled to a minimum.

Oxygen in gaseous form is a particularly damaging impurity since it reacts with the tungsten filament, thus forming volatile tungsten oxides. As little as 10 parts per billion of oxygen in the gas will remove tungsten from the surface of the filament as fast as it vaporizes at 2750 K. That is, the oxygen will cut the life of the lamp in half. Water vapor is a particularly troublesome form of volatile oxygen since the reaction to form tungsten oxide at the filament surface releases hydrogen. The tungsten oxide evaporates to the lower temperature bulb wall, at which temperature the reaction reverses and the evolved hydrogen reduces the tungsten oxide to tungsten, thus reforming water to return to the filament to react again.

Largely to control the water content, present day incandescent lamps contain one or more getters—active chemical compounds that react with and remove water, oxygen, hydrogen, or all three, from the gaseous atmosphere of the lamp to minimize the degree of attack on the tungsten filament. Common getters include phosphorus and its compounds and active metals such as zirconium, aluminum, and their alloys. Specific formulations are proprietary to the several manufacturers. Most manufacturers use proprietary automated on-line spectroscopic detection of oxygen and water vapor to cull out leakers and contaminated lamps that would be very short-lived in service.

F. Commercially Available Products and Performance Characteristics

Table I enumerates the performance characteristics of a limited number of commercially available products, while Fig. 9 is a photograph illustrating a sampling of the variety of incandescent lamp types that are available.

G. Areas of Research and Development

The major deficiency of incandescent lamps is an unfavorable fraction of the emitted radiation in the visible spectrum, less than 10%, resulting in poorer luminous efficacy than other sources. The search for materials with greater selectivity of emittance, or of selective emitter surface coatings compatible with high-temperature tungsten, is a low-level concern in many laboratories. A more promising avenue may be the use of selective reflective materials on the envelope that reflect IR rays back to the filament while transmitting visible rays, thereby reducing the power input required to keep the filament at a given temperature and produce a given amount of light. There are a variety of different reflective materials having suitable optical properties. But a major problem remains: the returning radiation must in fact hit and be absorbed by the filament. This requires significant improvement in mechanical precision of the envelope geometry and of coil location in the center, at manufacture, as well as similar improvements in subsequent movement of the coil—both due to mechanical shock in shipping, and due to sag or squirm in operation. Research and development

TABLE I. Representative Incandescent Lamps (120 V)

Power (W)	Bulb type[a]	Current (A)	Rated life (hr)	Light output (lm)	Efficacy (lm/W)	Color temperature[b] (K)
6	S14	0.050	1500	40	6.7	2370
10	S14	0.083	1500	86	8.6	2450
25	A19	0.208	2500	235	9.4	2550
40[c]	A19	0.333	1500	445	11.1	2770
60[c]	A19	0.500	1000	870	14.5	2800
100[c]	A19	0.833	750	1,710	17.1	2870
150[c]	A21	1.250	750	2,780	18.5	2900
200[c]	A23	1.670	750	3,830	19.2	2930
300	PS30	2.500	1000	5,960	19.9	2940
500	PS35	4.170	1000	10,600	21.2	2960
1000	PS52	8.330	1000	23,100	23.1	3030
1500	PS52	12.500	1000	33,620	22.4	3070

[a] S = spherical, A = common household lamp shape, PS = pear-shaped. Number gives maximum diameter in $\frac{1}{8}$ inches.

[b] Color rendering index for all types is essentially 100.

[c] Most common household types; ratings are for lamp with diffusing powder-coated bulbs (soft white). For all other types, ratings are for inside-frost bulbs.

FIG. 9. Sample of the variety of types of incandescent lamps.

aimed at solution of these problems in a cost-effective way is an active area of concern in several organizations.

V. Tungsten–Halogen Lamps

A. Similarities to Incandescent Lamps

Tungsten–halogen (T–H) lamps, comprising a tungsten-filament radiator in a transparent glass envelope with a gas filling, are substantially the same as incandescent lamps in all aspects of their radiation generation and energy balance. Shared characteristics include dependence of efficacy and life on filament temperature and many of the limitations on performance resulting from material limitations.

B. Unique Features of Tungsten–Halogen Lamps

Tungsten–halogen (T–H) lamps use much higher gas filling pressures than do their incandescent counterparts, typically 5–10 atm in operation, depending on type. Consequently, the life of the filament against evaporation in T–H lamps at the same temperature would be several times longer; the gain is taken instead by operating the filament at higher temperatures to result in about 20% higher efficacy for equal life. To safely contain internal pressures greater than 1 atm requires small-diameter, heavy-walled vessels fabricated of refractory glasses operating at much higher temperatures than the bulb of an incandescent lamp. To prevent early blackening of the bulb by evaporated tungsten due to its close proximity to the filament, the T–H lamp makes use of a chemical cleaning cycle in which gaseous halogens react with evaporated tungsten, forming volatile halides that do not condense on the hot bulb walls, thereby keeping the walls clean. The tungsten halides decompose on the hot filament, redepositing the tungsten on them. Unfortunately, the temperature dependence of the redeposition is not the same as that of evaporation, and tungsten is not redeposited on the hottest spots from which it primarily evaporates. Thus the halogen cycle is not a regenerative one that of itself lengthens the life of the filament.

1. Principal Construction Features

Figure 10 shows the principal construction features of the two principal types of T–H

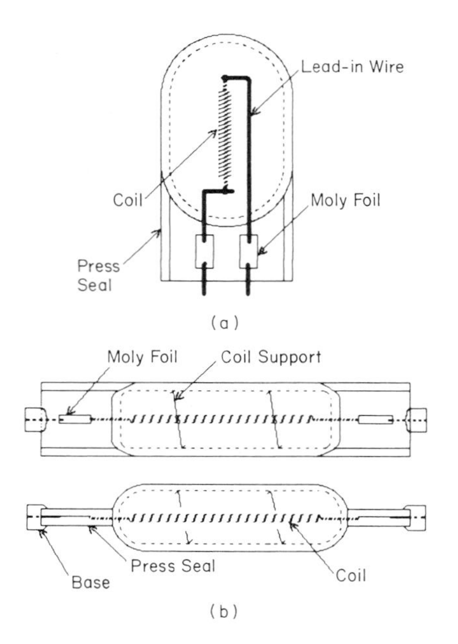

FIG. 10. Schematic diagram, illustrating components of typical T–H lamps. (a) Single-ended type, with press seal incorporating two molybdenum-foil feedthroughs. (b) Double-ended, shown in top and side views to illustrate geometry of press-seal; the coil support is usually a spiral tantalum or molybdenum wire loop affixed to the coil and loosely contacting the quartz wall, to center and support the elongated coiled filament.

lamps: single-ended and double-ended. The most common glass used for such lamps is either pure fused silica (quartz) or Vycor brand of high-silica glass (Corning Glass Works), which have softening temperatures near 1500°C and safe service temperatures of 900 to 1000°C. Glass-to-metal seals for electric lead-ins for these glasses use a very thin molybdenum foil with thickness to width ratio of 1/80 or less and having a very shallow taper at the edge to less than 25% of the thickness. This configuration permits the molybdenum to yield like plastic to accommodate the factor of 10 to 12 difference in thermal expansion between quartz and molybdenum without developing sufficient stress in the surrounding glass to cause it to fail. Such seals are susceptible to oxidation failure when operated at high temperature in air. For each type of T–H lamp, a maximum seal temperature is specified to ensure that seal failure does not occur before normal filament burnout.

Quartz and Vycor envelopes require extremely elevated temperatures for glassworking operations, which make lamp sealing slow and therefore expensive. For some types of T–H lamps, in which relatively high-production volumes are needed, aluminosilicate glasses (hard glass) are used. The sealing temperature of these glasses is 300–400°C less than quartz or Vycor, and thermal expansion is comparable to that of molybdenum. Thus rigid molybdenum wire lead-ins can be used and rapid sealing cycles used.

2. Chemical Cleaning Cycle

Tungsten–halogen lamps use reversible reactions of the form

$$W + X \leftrightarrow WX(g)$$

$$W + X + O \leftrightarrow WOX(g)$$

in which X is a halogen (usually iodine or bromine). The reactions proceed to the right at low temperatures to react with evaporated tungsten, forming volatile products that do not condense on the wall of the bulb, which therefore remains clean despite its close proximity to the filament. The maintenance of light output during the life of the lamp is therefore essentially 100%. In contact with the tungsten filament at elevated temperatures, the halides and oxyhalides decompose, redepositing tungsten on the filament, and releasing halogen. Careful control of the cycle is needed to ensure that clean-up is sufficiently rapid to prevent wall-blackening, without being so rapid as to erode the low-temperature tungsten-filament legs before normal burnout occurs. This is accomplished by (1) selection of halogen and halogen pressure and (2) control of the oxygen pressure.

C. Practical Problems

1. Fabrication Technology

The fabrication technology of T–H lamps is different from that for incandescent lamps in three respects. Tungsten–halogen lamps require (1) higher glassworking temperatures, (2) longer sealing times, and (3) preflushing of the vessel with inert gas before sealing to prevent oxidation of the filament during sealing. Filaments are recrystallized (stabilized) by firing in atmosphere furnaces before assembly to improve accuracy of filament positioning, especially in lamps for use in projection optical systems. Filling the vessel to pressures greater than atmospheric poses another unique problem in T–H lamp fabrication, solved by condensing out the fill gas into liquid form in the bulb with liquid nitrogen coolant before sealing so that flame sealing of the exhaust tube can be done in the conventional way, with less-than-atmospheric internal pressure. A final major technological difference is the necessity to dispense accurate amounts of corrosive halogens automatically on the exhaust machines.

As a result of these differences, the manufacture of many types of T–H lamps is not nearly as automated as it is for incandescent lamps. Only in the manufacture of types produced in quantities of millions per year, such as automotive headlamps, are fully integrated automatic machine groups for mount assembly, sealing, exhausting and filling, and basing used. For other types, of lower production volume, automated and semiautomated, highly flexible machines (capable of rapid type change) are used for the individual processes, and the parts are manually transferred from machine to machine.

2. Influence of Accidental Impurities on Chemistry

The cyclic halogen cleanup reactions in a T–H lamp are critically dependent on oxygen content since the oxyhalide reaction is in parallel with the halide reaction. Oxygen is present as an impurity in the tungsten itself, as well as in lead-in wires and supports. Moreover, variations in the degree of oxidation protection at sealing lead to variations in oxygen content of the lamp. Carbon, again an unavoidable impurity in the tungsten, serves to getter oxygen as carbon monoxide, in which form it is substantially inert. Hydrogen forms water from oxygen, providing water-cycle bulb blackening in parallel with evaporation. Thus, depending on the relative amounts of carbon, hydrogen, and oxygen, the role of the oxygen may be accelerated cleanup (and filament-leg erosion), neutral, or accelerated blackening.

The fact that substantial fractions of the inputs of all of these quantities result from the normal impurity contents of the lamp materials requires extremely close material control and compensatory adjustment of process to produce uniform product.

3. Mechanical Strength Problems Exaggerated

All the mechanical strength problems of incandescent lamps are present in T–H lamps to an exaggerated degree because of the higher operating temperatures of T–H filaments. Most manufacturers establish more stringent specifications for wire in T–H lamps than in incandescent lamps to avoid sag and squirm in operation. The problem of filament strength, as well as leg erosion, has in the past prevented development of low-wattage 120-V types because of the extremely fine filament wire they require. Improved materials, processes, and process control now permit manufacture of 120-V lamps having as little as 45 W.

D. Commercially Available Products and Performance Characteristics

The principal applications of T–H lamps have been in two areas: outdoor floodlighting (using high-wattage double-ended types and projection-type applications, photographic or audiovisual (using single-ended types). Single-ended quartz or Vycor T–H lamps have been widely used in Europe for automotive headlamps. Hard glass halogen capsules in sealed-beam-type outer envelopes have recently been introduced for this application in the United States, where they provide equal low-beam performance at two-thirds the electric power, and brighter high-beam performance at equal power in comparison to the incandescent product. Use in studio, theater, and television (STTV) applications, although small in absolute numbers, has revolutionized that illumination technology. Tables II and III summarize the characteristics of some double-ended and single-ended types, respectively. Recent application of aluminosilicate (hard glass) T–H capsules in sealed reflector (PAR) lamps has resulted in cost-effective products that compete with traditional incandescent applications, providing equal illumination and life, with very substantial energy savings. Table IV provides a comparison of performance of hard glass T–H PAR lamps with their incandescent predecessors. Figure 11 is a photograph illustrating a sampling of some of the various types.

E. Principal Areas of Research and Development

The development of low-wattage, 120-V, hard glass T–H capsules that can be inexpensively manufactured at high production speeds has opened new opportunities for T–H lamps in traditional incandescent applications, with energy savings of 20% or more. Many organizations are actively investigating the extension of this ad-

TABLE II. Representative Double-Ended Tungsten–Halogen Lamps

Power (W)	Type[a]	Length (in.)	Line (V)	Current (A)	Rated life (hr)	Light output (lm)	Efficacy (lm/W)	Color temperature[b] (K)
500	500T3Q/CL	$4\frac{1}{16}$	120	4.17	2000	11,100	22.2	3000
1000	1000T6Q/CL	$5\frac{5}{8}$	120	8.33	2000	22,000	22.2	3000
1000	1000T3Q/CL	$10\frac{1}{16}$	240	4.17	2000	21,500	21.5	3000
1500	1500T3Q/CL	$10\frac{1}{16}$	240	6.25	2000	35,800	23.9	3000
1500	1500T3Q/CL	$10\frac{1}{16}$	277	5.42	2000	34,400	22.9	3000

[a] T = Tubular, Number is diameter in $\frac{1}{8}$ in., Q = quartz, CL = clear.
[b] Color rendering index for all types is essentially 100.

vantage to residential illumination. Such a product would be indistinguishable in function and operation from the present-day incandescent lamp but would provide the same illumination with less electric power consumption. Application of the selective-reflector principle to T–H lamps is also being actively pursued.

VI. High-Intensity Discharge Lamps

High-intensity discharge (HID) lamps convert into radiation the power dissipation of an electric current passing through a gaseous medium at a pressure greater than or equal to 1 atm. Much higher radiating temperatures can be achieved than in any incandescent solid; appropriate selection of the gaseous medium results in favorable spectral distributions of radiated power, with a much smaller fraction of IR rays. As a result, such sources are very bright, and are 3–10 times as efficient as incandescent lamps. Figure 12 shows a schematic diagram illustrating the principal components of such a lamp.

Three principal types of HID lamps are distinguished by their radiating species: high-pressure mercury (mercury), high-pressure sodium (sodium, or HPS), and metal halide (M-H). Each comprises an inner discharge tube (arc tube) containing the high-pressure gas or vapor enclosed in a hermetically sealed outer envelope (jacket). The outer jacket is required for thermal insulation, protection of the arc tube seals from oxidation, and absorption of any short wave-

TABLE III. Representative Single-Ended Tungsten–Halogen Lamps

Power (W)	Type[a]	Length (in.)	Line (V)	Current (A)	Rated life (hr)	Light output (lm)	Efficacy (lm/W)	Color temperature[b] (K)
100	100Q/CL	$2\frac{3}{4}$	120	0.833	1000	1,800	18.0	3000
250	250Q/CL	$3\frac{1}{8}$	120	2.080	2000	5,000	20.0	3000
300	ELH[c]	$1\frac{3}{4}$	120	2.500	35	c	c	3350
600	DYV[d]	$2\frac{1}{2}$	120	5.000	75	17,500	29.1	3200
500	BTL[e]	$4\frac{1}{4}$	120	4.170	500	11,000	22.0	3050
750	BTN[e]	$4\frac{3}{8}$	120	6.250	500	17,000	22.7	3050
1,000	BTR[e]	$4\frac{3}{8}$	120	8.330	200	28,000	28.0	3200
1,500	CXZ[f]	8	120	12.500	325	38,500	25.7	3200
5,000	DPY[f]	11	120	41.700	500	143,000	28.6	3200
10,000	DTY[f]	15	120	83.300	300	291,000	29.1	3200

[a] Three-letter designations have no dimensional significance.
[b] Color rendering index for all types is essentially 100.
[c] Single-ended T–H capsule prefocused in ellipsoidal reflector for film projection applications.
[d] Principal application is in overhead projectors.
[e] Medium prefocus base for replacement of incandescent types in stage lighting.
[f] Prefocus bi-post base; primary application in studio and TV lighting.

TABLE IV. Comparison of Performance of Parabolic-Aluminized-Reflector (PAR) Lamps: Standard Incandescent versus PAR with Internal Hard-Glass Halogen (HGH) Capsule

	Center beam intensity (candela)[a]	Beam ½-angle[b] (degrees)	Field ½-angle[c] (degrees)
90-W HGH-narrow spot	22,000	4	9
150-W incandescent PAR spot	17,000	4	18
45-W HGH spot	5,000	6–7	13–15
75-W incandescent PAR spot	4,800	6–7	13–15
45-W HGH flood	1,800	14	20
75-W Incandescent PAR flood	1,800	15	24

[a] One Candela = 1 lm/sr.
[b] Angular deviation between center of beam and ½-maximum intensity.
[c] Angular deviation between center of beam and $\frac{1}{10}$-maximum intensity.

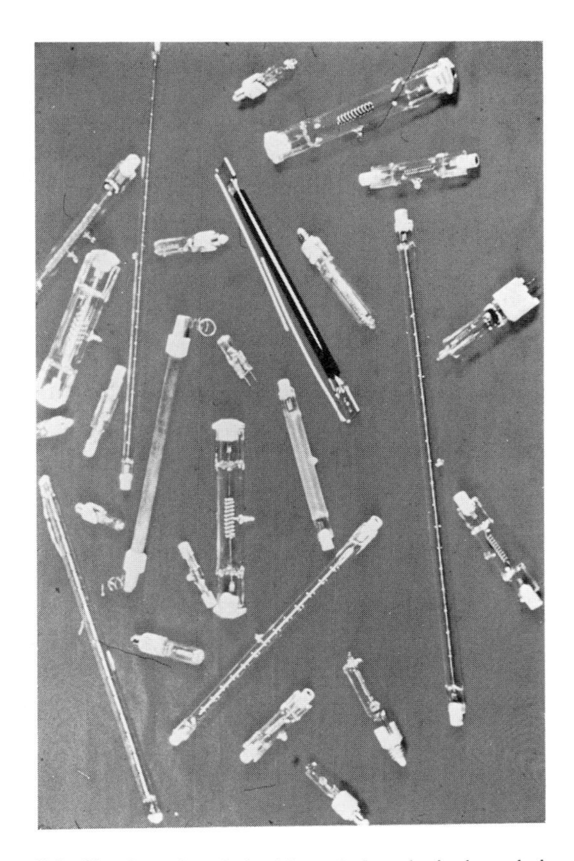

FIG. 11. Sample of double-ended and single-ended T–H lamp types.

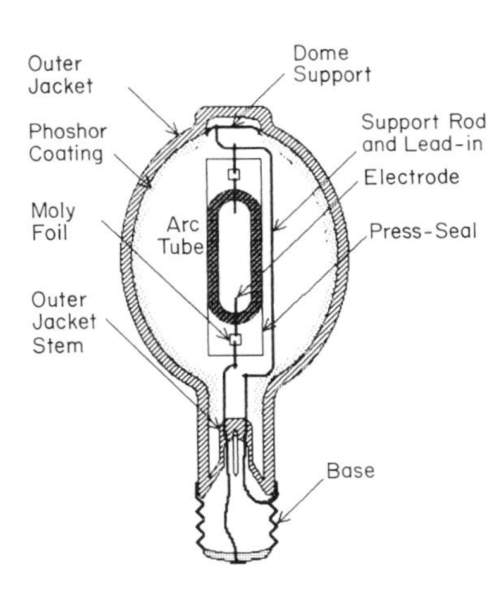

FIG. 12. Schematic diagram illustrating principal components of HID lamps, shown in section. Additional components (not shown) in many types are included to facilitate ignition.

length UV rays that may be emitted from the arc tube. Arc tubes for mercury and M-H lamps are quartz, similar to T–H lamp capsules; arc tubes for HPS lamps are fabricated from translucent polycrystalline alumina (PCA) to withstand corrosion by hot molten and gaseous sodium.

A. Radiation Mechanism

The passage of current ionizes the gaseous medium, converting it to a "plasma" containing high densities of free electrons plus positive ions in approximately equal concentrations but with little or no net charge density. The gas offers resistance to the passage of current because electrons collide with gas atoms. The I-squared-R power dissipation heats the gas to very high temperatures. Unlike a metallic conductor, the gas has very low thermal conductivity; there is a large temperature gradient between the center axis and the walls. The gas at the center reaches temperatures far greater than the boiling point of any material, while the walls of the arc tube remain at modest temperatures (1000–1500 K, depending on type). Rapid collisions sharing energy among electrons, ions, gas atoms, and excited gas atoms promote local thermal equilibrium (LTE), and the radiating properties are described by Eq. (6) of Section I, with absorption coefficients determined by the optical transitions of the gaseous medium. Such a discharge is called a high-pressure arc discharge, or arc.

The emitted radiation consists primarily of the line spectra of the elements present in the gas:

mercury in mercury lamps, and sodium in HPS lamps. Although mercury vapor is also present in the gas in HPS lamps, the temperature is insufficient to excite mercury to states that contribute to absorption in the visible wavelengths, and mercury lines are essentially absent from the spectrum.

Metal halide lamps contain atomic metal radiators that either do not vaporize at wall temperatures accessible to quartz or react vigorously with quartz at those temperatures. They are introduced into the arc tube in the form of metal iodides, which are quite volatile at ~1000 K and relatively nonreactive to quartz. The metal iodides participate in a halogen cycle, similar to that in T–H lamps. Iodide molecules evaporate from the tube wall and enter the high-temperature core of the arc where they dissociate, freeing metal atoms to contribute their optical transitions to the radiation of the gas mixture. Free metal and iodine atoms diffuse, or are convected, back into the low temperature regions of the discharge and chemically recombine to metal iodide, preventing condensation on, or reaction with, the quartz wall by metal atoms. The emitted radiation is dominated by the line spectrum of the added metals, with contributions from mercury (which is also present), and minor radiation from metal iodide molecules. Arc temperatures at the core are too low to excite iodine significantly, and its emission lines are barely detectable.

In all three types of HID lamps, the absorption spectra exhibit broadening of the natural line widths by a variety of processes: (1) Stark broadening due to perturbation of energy levels by electric fields from electrons and ions; (2) resonance broadening by the interaction of excited atoms with neighboring identical nonexcited atoms; and (3) van der Waals broadening caused by interaction of excited atoms with nearby foreign atoms. Calculation of the absorption coefficient as a function of wavelength requires knowledge of the broadening mechanisms, as well as of the density of absorbing atoms, and of the probabilities of their optical transitions. Calculation of the spectral radiance at the arc tube wall requires knowledge of the radial temperature profile. Nevertheless, accurate radiation transport calculations have been available for some time for HPS lamps and are becoming available for M-H lamps. In all three types, optical depths range from very thin to very thick. Because of the large temperature gradients, self-reversed lines are common. Spectral radiances at emission line maxima approach 50% or more of the blackbody level at the arc axis temperature for strong nonresonance lines and 25% or more of that level for resonance lines. Infrared transitions in most atoms, between upper energy levels near the ionization limit, are not strongly excited at the arc temperatures. Their contribution to the absorption spectrum is weak, as are corresponding emission lines. By design, electron density is maintained at a sufficiently low level that free-free transitions of electrons colliding with each other and with ions, leading predominantly to emission in the IR, are minimized. The consequent weak emission of HID lamps in the IR contributes strongly to their high efficacy. Proper choice of radiating atoms can favor emission in the visible over the UV wavelengths as well. M-H lamps permit the choice of radiating atoms that provide for emission of all colors, so that the CRI is also high.

Figure 13 shows spectral power distributions of the radiation emitted from the three types of HID lamps. These spectra are taken on instruments with relatively wide spectral bandwidth and do not reveal the self-reversed nature of the spectral lines except in the case of the sodium resonance lines at 589 nm in the HPS lamp.

B. ENERGY BALANCE

1. Arc Discharge

Electric power is dissipated in the arc tube in three modes: heating electrodes to electron-emitting temperatures, heat conduction through the radial temperature gradient in the gas to the walls, and radiation from the high temperature gaseous medium. Table V shows the approximate percentages of power dissipation in these modes for 400-W mercury, HPS, and M-H lamps. Electrode loss is minimized by designing the lamp to operate at as low a current, and as high a voltage drop, for a given power as is practical. Voltage drop between electrodes increases with increasing interelectrode spacing (arc length) and with increasing gas pressure. Heat conduction loss is approximately a constant loss per unit length of arc. The percentage lost in heat conduction is minimized by operation at as high a power input per unit length as possible, that is, operation with as short an arc length for a given power input as possible. Maximizing voltage drop while minimizing arc length dic-

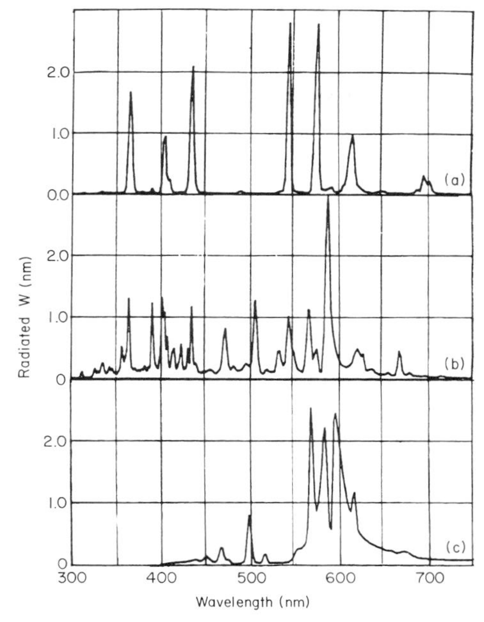

FIG. 13. Spectral distribution of radiated power (W/nm) of HID lamp types in use in the United States, all at 400-W input power. (A) Mercury lamp with phosphor-coated outer jacket; strong peaks at 365, 405, 436, 546, and 578 nm are atomic lines of mercury, while the peak at 615 nm is red emission from the phosphor, excited by UV from the arc tube; color temperature = 4000 K, CRI = 45. (B) Sodium-scandium M-H lamp; the mercury lines are visible in the spectrum, the strong line at 590 nm is due to sodium, and the balance is due to groups of lines emitted by scandium atoms (unresolved at the spectrometer bandpass used here); color temperature = 4000 K, CRI = 65. (C) HPS lamp; all radiation is from sodium, the major yellow peak with a minimum at 590 being the self-reversed resonance line of sodium (see Fig. 2); color temperature = 2100 K, CRI = 22.

tates high gas pressures, 1–10 atm, depending on type.

2. Lamp–Electrical-Circuit System

Gases are normally insulators; to establish current flow through a gas requires generation of free electrons and ions. This is accomplished in discharge lamps such as HID lamps by application of sufficiently high potential across the discharge tube terminals so that a single initial electron may gain enough energy to ionize a gas atom. This creates an ion–electron pair, the electron of which gains enough energy to create a new ion–electron pair, and so on, in an avalanche, which eventually results in sufficient free electrons in the gas to carry the required discharge current at the desired potential drop. The necessary starting, (ignition) and breakdown potential is typically several times higher than the ultimate operating voltage. Therefore, the electrical circuit must provide a high open-circuit voltage for ignition but must include current-limiting impedance to drop the voltage across the lamp terminals to the operating value, which is relatively independent of current, once the discharge is ignited. If the circuit had no current-limiting impedance, but continuously applied the ignition voltage, the electron avalanche would continue to grow unabated, and current would rapidly rise to destructive levels.

The electric circuit device that performs these functions is called a ballast. In principle, the current-limiting impedance could be resistive; however, I-squared-R dissipation in the resistance would add to power consumption of the circuit without contributing any light, thus reducing system efficacy. Since HID lamps operate on

TABLE V. Approximate Percentages of Power Dissipation in High-Intensity Discharge Lamps

Lamp type	Electrode loss	Heat conduction plus wall absorption	Radiation IR	Radiation UV	Radiation Visible
Mercury[a]	12	38	10	24	16
Metal halide	12	25	17	19	31
High-pressure sodium[b]	4	37	32	0	27

[a] Low wall-loading in mercury lamps results in large conduction loss, which together with unfavorable fraction of radiation in UV results in low efficiency of visible radiation in comparison to other types.

[b] High thermal conductance of sodium vapor results in large heat conduction loss; strong IR emission lines of sodium also reduce fraction radiated in the visible rays. Visible radiation is predominantly near the maximum of the photopic response curve; therefore, luminous efficacy is high.

alternating current, the necessary impedance in their ballasts is reactive: either inductive or inductive plus capacitive, minimizing power dissipation in the ballast. Many HID lamps require ignition voltages greater than 120 V; the ballasts also include step-up transformers to increase the open-circuit voltage to the ignition value.

High-pressure sodium lamp ballasts and some M-H ballasts provide the ignition voltage in the form of very brief pulses of several kilovolts peak from a circuit that is disabled once the lamp is ignited. On alternating current, the discharge must also be reignited each half cycle; the peak instantaneous voltage required to accomplish this is typically 30% more than the instantaneous arc voltage during the majority of the half cycle. The ballast must provide the necessary reignition voltage under operating conditions with the pulse-voltage circuit inactive.

The ignition voltage required is a strongly increasing function of gas pressure. Therefore, all HID lamps use a low pressure of a rare gas as a starting gas, typically 20–100 Torr, and provide the required high operating pressure by mercury vapor. At ignition, the mercury is cold and condensed out, with a vapor pressure of a millitorr, and does not result in a high-ignition voltage. In operation, the arc tube wall is hot, all the mercury content of the capsule is in the gas phase, and the pressure is at the high level necessary to provide the design value of operating voltage. Use of mercury vapor to provide the high pressure has the added advantage that heat conductance of mercury is very low, minimizing heat conduction loss.

Less than ideal transformers and reactances result in power dissipation in the ballast, essentially proportional to volt-amperes handled, the product of open-circuit voltage times the operating current. The open-circuit voltage required is determined by ignition requirements of the lamp. Minimizing lamp current for a given power requires that operating voltage of the lamp be as high as possible, that is, as close to ignition voltage as possible. Typical commercial ballast systems for HID lamps have losses of 10 to 15% of lamp power.

C. Performance Limitations Resulting from Material Limitations

Unlike incandescent lamps, HID lamps do not embody a single mechanism that is the primary determinant of lamp life; there is not a one-to-one inverse correspondence between efficacy and life. Moreover, lives of HID lamps are prodigiously long; most mercury and HPS types have rated lives of 24,000 hr, while M-H lamps have rated lives from 7500 to 20,000 hr. Since typical operating hours in dusk-to-dawn outdoor service or two-shift indoor commercial service are 4000 hr per year, these rates represent operating lifetimes of 2 to 6 years. Nevertheless, design choices that lead to higher efficacy are constrained by material limitations that result in shorter life.

Failure mechanisms in mercury and HPS lamps include (1) exhaustion of electron-emitting activator material from the electrodes, resulting in failure to ignite or reignite each half cycle; (2) gas contamination, resulting in ignition failure; (3) arc tube seal failure, resulting in loss of hermeticity; (4) excessive arc tube blackening by evaporants from the electrode; or (5) wall material degradation, causing decrease in output to unacceptable levels and requiring replacement before failure. Metal halide and HPS lamps also may suffer from loss of radiating atomic species by unwanted chemical reactions that consume them irreversibly, therefore causing loss of light output as well as failure to reignite.

These considerations dictate materials of the arc tube, electrodes, and electrode activator compounds to minimize rates of adverse processes. All of the processes described, however, occur more rapidly as temperature is increased. While efficiency increases as input power per unit length is increased, so does arc tube temperature, and with it the rate of the adverse reactions discussed in the preceding paragraph. These constraints are generally embodied in the form of rule-of-thumb design rules regarding acceptable wall-loading (arc power input per unit area of arc tube wall surface). Experience indicates that acceptable lamp lives in mercury lamps are obtained at wall-loadings of 10 to 12 W/cm^2; M-H designs are commonly found at 13 to 17 W/cm^2; and HPS can use designs at 15 to 20 W/cm^2 by virtue of the more refractory nature of PCA. For all types, efficacy increases as wall-loading increases, and most designs are at or near these limits. In some cases in which somewhat shorter life is permitted, higher wall-loadings than the values cited may be used.

Molybdenum foil seals in both mercury and M-H lamps have adequate life at much higher temperatures than in T–H lamps because they are protected from oxidation by an inert atmosphere in the outer jacket. The electrical lead-in in HPS lamps involves a niobium (also known as columbium) metal member, chosen for expansion match to PCA and sealed with a polycrystalline oxide mixture, fusible without melting either niobium or PCA. The niobium is protected from oxidation by a vacuum in the outer jacket. In HPS lamps, resistance of sealing compounds to attack by sodium sets a limit to maximum seal temperature, thereby limiting the cold spot temperature that determines maximum pressure of sodium vapor inside the arc tube.

D. Practical Considerations

1. Fabrication Technology

Fabrication of mercury and M-H lamps involves aspects similar to a combination of T–H and incandescent lamp fabrication since two hermetically sealed vessels are required: one of fused quartz, incorporating molybdenum foil seals, and an outer jacket of glass using a stem seal similar to an incandescent lamp. The major qualitative difference is that processing of mercury arc tubes also involves dispensing a precisely determined quantity of mercury into a precisely determined internal volume of arc tube since those two factors determine operating pressure when all the mercury is vaporized. To that difference in M-H lamps is added a requirement for also dispensing extremely hygroscopic iodide salts without any contact with the atmosphere. The salts themselves are dehydrated in advance to only a few parts per million of water content, and they must be handled in atmosphere glove boxes free of moisture and oxygen to a few parts per million.

Wide differences exist among manufacturers and types in the level of automation of manufacture of mercury lamps. Some popular types are manufactured on fully integrated production lines that seal and process the arc tube automatically, mount it on the outer jacket stem press automatically, and seal and process the outer jacket automatically at speeds of dozens per minute. For others, one or more stages of the process involve jigs-and-fixtures manual operations.

Metal halide lamp manufacturing is much less automated because of lower production volumes and more difficult and demanding process requirements. With most manufacturers, arc tube fabrication is primarily a scaled-up laboratory type of process. Arc tube processing takes place on automatic exhaust machines. Final assembly and outer jacket processing usually proceeds through the mercury lamp line, sharing equipment and processes.

High-pressure sodium lamp manufacture is a different ceramic-based technology. Polycrystalline alumina tubes themselves are manufactured either by a press-and-sinter ceramic process, or an extrusion-sinter process, using as raw material a very pure alumina powder of precisely specified surface area, to which controlled quantities of sintering aid compounds have been added. The resultant translucent tubing has total diffuse transmittance of 95% or more, and single-wall inline transmittance of 10%. Arc tube sealing, exhausting, and filling may be either a two-step process or single step, depending on manufacturer. In the two-step process, both niobium end caps are cemented at once with a fusible oxide mixture to the tube ends in a vacuum or atmosphere furnace; one of the end caps is equipped with an exhaust tube. The arc tube is evacuated and mercury and sodium fillings are dispensed through the exhaust tube in a separate process, after which the niobium exhaust tube is pinched shut in a cold weld. In the one-step process, each end of the arc tube is sealed separately. The second seal is made in an atmosphere furnace containing the desired fill pressure of starting gas, with mercury-sodium amalgam already dispensed into the tube, residing at the cool first sealed-end; there is no exhaust tube, and no separate exhaust and filling process. The final assembly into an outer jacket and outer jacket processing are similar to those for mercury lamps, and they may share equipment.

2. Unfavorable Reactions and the Effects of Impurities

In all three classes of HIDS lamps, gaseous impurities that interfere with ignition or reignition must be minimized. Hydrogen is a particularly troublesome species since it is a principal component of water vapor, a common and difficult impurity to remove. In addition to surface contaminations, dissolved hydroxyl in quartz is a major problem. This is dissociated photolytically by UV rays from the discharge, and the liberated hydrogen diffuses rapidly into the arc tube interior. Manufacturing processes for quartz tubes are tailored to reduce the hydroxyl content to a parts-per-million level; optical absorption in an IR absorption band of OH is used as a quality-control tool. Because of rapid diffusion of hydrogen through hot quartz, moisture in the outer jacket must also be minimized since it may also be dissociated by UV light, and the liberated hydrogen may diffuse through the arc tube wall into the interior. Hydrogen is especially harmful in M-H lamps because of reaction with iodide salts, forming HI and liberating metal. The metal reacts with quartz and is lost to the halogen cycle; the hydrogen iodide is gaseous and interferes with electron avalanche growth in ignition by capturing free electrons.

Oxygen is also deleterious, especially in M-H and HPS lamps, because it promotes reaction of metal atoms with arc tube wall material. Hydrocarbon vapors in the outer jacket are also decomposed thermally and photolytically, liberating damaging hydrogen as well as depositing amorphous carbon on the arc tube wall surface, reducing optical transmission.

In addition to the foregoing, two adverse reactions in M-H lamps must be minimized. The inadvertent reaction of free metals with the quartz wall of the arc tube—as a result of some failure to complete the halogen cycle—causes a loss of metal, reduction in partial pressure of metal vapor ultimately entering the arc, and consequently of radiation emitted by the metal. In addition, such reactions also liberate silicon metal, which reacts with iodine to form volatile silicon tetraiodide. This compound decomposes at the electrode temperature, and deposits molten silicon on the electrode. The molten silicon fluxes recrystallization and regrowth of the electrode, drastically distorting its shape and degrading its performance. This becomes one of the life-limiting processes in M-H lamps by adversely affecting the reignition process every half cycle, ultimately to the point that the ballast capability to reignite does not suffice.

The second adverse reaction is the electrolytic loss of sodium from sodium iodide in the arc tube fill. In operation, there are always a few parts per million of sodium ions dissolved in quartz in contact with sodium iodide as a result of reaching thermochemical equilibrium in the reversible reaction between the quartz and the iodide. The quantity is not harmful to the quartz, nor does it represent a significant depletion of sodium from the initial dose dispensed. However, sodium ions are mobile in quartz, and it has been found that negative charging of the outer surface of the quartz arc tube by photoelectrons emitted from various parts of the outer jacket will attract the sodium ions to the outer surface to be neutralized and evaporate. The depletion of the ionic concentration in the inner surface then permits the forward reaction to proceed to provide more sodium ions, which are electrolyzed in turn, until eventually a very substantial fraction of the original sodium dose has been lost. Outer jacket designs providing minimum photoelectric-emitting surfaces are used to mitigate this problem.

E. Commercially Available Products and Performance Characteristics

HID lamps are primarily used in outdoor floodlighting, street and highway lighting, and indoor industrial and commercial applications in which their exceptionally long service lives and high efficacies outweigh their high initial cost and the added expense of the required ballasts. In such service, relatively high mounting heights have been traditional so that large areas could be illuminated from relatively few fixtures. This has dictated in the past that the more popular types were from 8000 to 100,000 lm in the United States and from 5000 to 50,000 lm in Europe. However, the increasing application of HPS and M-H lamps in lower ceiling height indoor commercial service, requiring smaller lumen packages—4000 to 10,000 lm—has resulted in many new lower wattage types being introduced within the last few years.

Because of the relatively small fraction of visible radiation output by mercury lamps, they provide only 50% of the efficacy of HPS and M-H lamps, or, conversely, they require twice the electrical power consumption to produce a given level of illumination. In addition, the absence of any red emission lines in the visible spectrum of mercury means that the CRI of mercury lamps is poor; the vast majority of such lamps in service today use a red-emitting phosphor coating applied to the inside surface of the outer jacket. Ultraviolet light emission from the arc is converted to red light by the phosphor, providing acceptable color rendering for most purposes. As a result of lower electrical power consumption by HPS and M-H lamps, the total cost of light obtained with these sources is less than that of mercury lamps. High-pressure sodium and M-H applications are expanding at a much more rapid rate, and are supplanting mercury lamps in many installations.

The most common type of HPS lamp is designed to operate with sodium pressure that gives maximum efficacy, with a CRI of ~22. At lower sodium pressures, there are insufficient sodium atoms in the vapor to provide all the sodium radiation that might be obtained. At higher sodium pressures, there is excessive resonance broadening of the sodium resonance line; this has the effect of shifting radiation away from the yellow region of the spectrum and into orange-red, where the eye is less sensitive. As a result, although the CRI may be improved to 60 to 80, and total efficiency of radiation slightly increased, luminous efficacy decreases ~25% for a CRI of 50 to 60 and about 60% for a CRI of 70 to 80. The higher sodium pressure also causes more rapid rates of adverse reactions with tube wall and sealing cement; life ratings are typically reduced by 50% or more. Although these lamps are commercially available from several manufacturers, they have not won wide acceptance in the marketplace.

Several different metal-iodide combinations are available in M-H lamps from different manufacturers. The preferred combination in the United States and in Japan uses a blend of sodium and scandium iodides. Scandium provides a multiline spectrum throughout the visible spectral range, similar to the rare earth elements, and results in a CRI of ~65, suitable for most commercial applications. Sodium increases the efficacy and permits adjustment to color temperatures from 3000 to 5000 K, which is the preferred range in the United States. In Europe, a mixture of indium, thallium, and sodium iodides has been used: indium for blue emission, thallium for green, and sodium for yellow and orange. A gamut of color temperatures is possible, depending on mixture ratios.

Also common in Europe is a dysprosium–holmium–thallium–iodide mixture, with or without sodium. The rare earth elements have multiline emission spectra giving good color rendition. Thallium adds green, improving efficacy. In the absence of sodium, the color temperature of this mixture is ~6000 K; with sodium added, it can be lowered to the 3500 to 4000 K range.

Table VI gives representative data for various examples of commercially available products in the mercury, HPS, and M-H families. Figure 14 shows some of the variety of types and sizes available.

F. Research and Development

Since efficacy of HID lamps increases with increasing wall-loading, there is active research for materials capable of operation at higher temperatures in more corrosive environments. A major focus of development is extending the range of lamp sizes to lower wattages and lumens output, with minimal loss of efficacy, in both the HPS and the M-H families. A practical deficiency of all types of M-H lamps is variation in color, both initially and in the course of the operating lifetime, as a result of variations in ratios of radiating constituents and variations in arc tube temperature. Considerable development is being expended to reduce these variations. Since M-H lamps may use any of ~50 metals as iodides, singly or in combination, a semi-infinite set of possibilities is available, and low-level exploratory activities to examine these continues. Research into M-H lamps focuses on developing more accurate arc models, which include not only radiation transport but also deal with the thermochemistry of molecules, atoms, and radicals, together with the detailed radial and axial energy balances and transport, all in a self-consistent way. Research and development activity to improve HPS lamp color without such large penalties in efficacy or life is a concern in many laboratories.

VII. Fluorescent Lamps

The fluorescent lamp is a discharge lamp in which UV radiation emitted by a low-pressure discharge in a mixture of mercury vapor and a noble gas is converted to visible light by a fluorescent phosphor on the inside of the discharge tube wall. Fluorescent lamps operate at wall-loadings of 0.04–0.20 W/cm^2—much lower than those for HID lamps. However, since fluorescent lamps require no outer jacket, overall they yield 2000–4000 lm/L of lamp volume, ~10% as much as HID lamps. Figure 15 is a schematic diagram illustrating the major components of a fluorescent lamp. Since there are a wide variety of fluorescent phosphors emitting nearly every color of the rainbow, and these may be used in blends or multilayered coatings, a very large range of colors and CRIs is available from this family of lamps, without significant change in the discharge medium itself.

A. Radiation Mechanisms

1. Discharge Plasma

The UV radiation is created by the passage of electric current through a mixture of mercury vapor at a few millitorr pressure, and a noble gas (usually argon, krypton, or neon, or some combination of them) at a few torr pressure. The power density is adjusted to result in a coldest

TABLE VI. Representative High-Intensity Discharge Lamps

Power (W)	Bulb[a] (type/MOL)	Operating[b] (V)	Current (A)	Rated life (hr)	Light output (lm)	Efficacy (lm/W)	Color temperature (K)	Color rendering index
Mercury lamps[c]								
175	BT28/8$\frac{5}{16}$	130	1.5	24,000	8,500	49	4,000	45
400	BT37/11$\frac{1}{2}$	135	3.2	24,000	23,000	58	4,000	45
1000	BT56/15$\frac{3}{8}$	265	4.0	16,000	55,000	55	4,000	45
Metal-halide lamps[d]								
100	ED17/5$\frac{7}{16}$	100	1.1	5,000	8,500	85	3,200	65
175	BT28/8$\frac{5}{16}$	130	1.5	7,500	14,000	80	4,400	65
400	BT37/11$\frac{1}{2}$	135	3.2	20,000	34,000	85	4,000	65
1000	BT56/15$\frac{3}{8}$	265	4.3	12,000	110,000	110	3,900	65
1500	BT56/15$\frac{3}{8}$	265	6.3	3,000	155,000	103	3,700	65
High-pressure sodium								
35	B17/5$\frac{1}{2}$	52	0.9	16,000	2,250	64	1,900	22
50	B17/5$\frac{1}{2}$	52	1.25	24,000	4,000	80	2,100	22
70	B17/5$\frac{1}{2}$	52	1.70	24,000	6,300	90	2,100	22
150	ED23$\frac{1}{2}$/7$\frac{1}{2}$	55	3.25	24,000	16,000	107	2,100	22
250	E18/9$\frac{3}{4}$	100	2.75	24,000	26,100	104	2,100	22
400	E18/9$\frac{3}{4}$	100	4.25	24,000	50,000	125	2,100	22

[a] Letters designate bulb shape; digits are diameter in $\frac{1}{8}$ in. MOL = maximum overall length.
[b] All lamps must be operated with appropriate ballasts rated for the line-supply voltage to regulate current and operating voltage of the discharge of these values.
[c] Data for phosphor-coated mercury lamps only.
[d] Ratings are for clear scandium-sodium types only.

FIG. 14. Sample of HID lamp types.

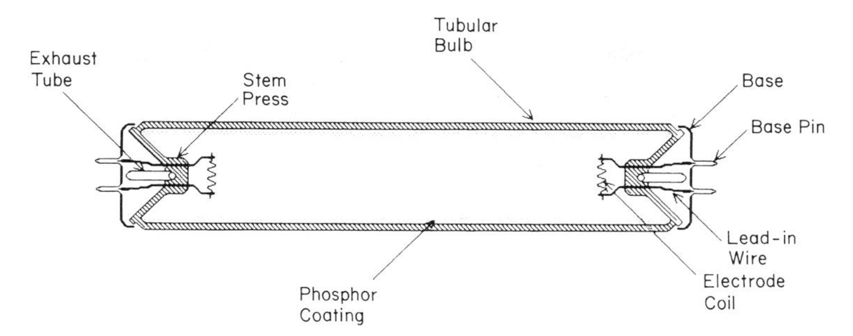

FIG. 15. Schematic diagram of the principal components of a typical fluorescent lamp, shown in section.

spot temperature, ~40°C, at which the saturated mercury vapor pressure is ~6 mtorr. At this pressure, the maximum efficiency of generation for the 254-nm resonance radiation of mercury is obtained, corresponding to the $^3P_1$-$^1S_0$ transition. The discharge is very far from LTE; electron temperature is greater than radiation temperature (which is itself different for the several energy levels of the mercury atom), and both are very much greater than the gas-kinetic temperature. The energy inputs and flows that maintain these temperature differences are discussed in Section VII,B. The equation describing radiation transport is Eq. 3 of Section I. The discharge has an optical depth of 50 to 100 at the center of the resonance line, about unity for some of the stronger visible and near-UV lines, and much less than unity for other emission lines. There is essentially no excitation of the radiation of the noble gas. The large optical depth of the resonance line results from the fact that the lower level of this transition is the ground state of the mercury atom, occupied in ~99% of the atoms present. Escape of this radiation from the plasma is therefore by a number of sequential absorptions and reemissions. This process, referred to in the literature as "imprisonment of resonance radiation," has the effect of prolonging the effective radiative lifetime of the excited state, thus increasing the chance for dissipation of excitation energy by some nonradiative transition. Efficiency of generation of resonance radiation therefore depends on mercury atom number density (and hence vapor pressure) as follows: at low pressures since there are insufficient mercury atoms to be excited by electrons, which consequently dissipate energy in other, nonuseful, ways, increasing pressure results in efficiency increase; at high mercury pressures since imprisonment time is too long, and too great a fraction of energy input to creating excited mercury atoms is dissipated in nonradiative transitions, increasing pressure results in efficiency decrease. Optimum mercury pressure results from balancing these two competing trends.

In this non-LTE case, calculation of radiation flux density via Eq. 3 of Section I is extremely complex, so much so that it has not been solved exactly. To integrate Eq. (3) for radiation flux requires knowledge of $n(u, r)$, the radial density distribution of mercury atoms in the $^3P_1$ upper state of the transition pair. However, instead of being constrained to an LTE value, this number density must be determined by solving coupled rate equations involving electron-collision transitions to and from the state, as well as radiation emission and absorption transitions. The rates of electron-collision transitions are proportional to local electron density, which in turn depends on the local rate of production and ambipolar diffusion loss. Much of the production of new electrons comes from ionization of mercury atoms in the very $^3P_1$ state whose concentration is to be calculated. Thus, the total problem involves solution of coupled integrodifferential equations.

So far, model calculations in such discharges have arbitrarily decoupled the equations to obtain a reasonably tractable problem. The radial distribution of excited atoms has been calculated (and with it the imprisonment time), assuming it was determined only by emission and absorption processes in a decaying ensemble of excited atoms. The imprisonment time is then used as a constant in the solution of the radial electron density and collision rate equations. Despite the arbitrary nature of the assumptions, this route to solution has given excellent correspondence to experiment. This can only mean that the actual radial distribution of excited atoms in the discharge is not far removed from the one calculated on the basis of radiation transport alone. Adding to the complexity of radiation transport in the system is that the 185-nm resonance transition (between $^1P_1$ and $^1S_0$ states) of mercury must be considered along with the 254-nm resonance transition.

2. Phosphors

Ultraviolet radiation reaching the walls of the discharge tube is absorbed by and converted to visible light by coatings of luminescent phosphor powders applied to them. Such phosphors are composed of inorganic crystalline materials, optically transparent in the pure state but synthesized to incorporate specific luminescent centers to absorb the desired wavelengths of UV light and emit visible rays. Such luminescent centers are typically strongly coupled to ligand fields of the solid; therefore, excitation by a UV photon creates excited state complexes in high vibrational energy states. These interact with lattice phonons to relax states of the excited manifold to lower vibrational energy before radiating. Consequent to loss of energy by phonon relaxation, the energy of the photon emitted on reradiation is substantially less than that of the absorbed one.

For all but the rare earth activators, coupling to the crystal fields results in emission over a broad band of wavelengths instead of narrow lines. Skillful empirical selection of activators and host crystals has over the years led to a library of phosphors that absorb well at 254 and 185 nm UV wavelengths and emit photons at longer wavelengths. Many peak-emission wavelengths from 300 to 800 nm are available, all with quantum efficiencies that approach unity.

Note that although near-unity quantum efficiency may be obtained, energy efficiency is only ~50% since a UV photon carries 4.86 eV of energy, whereas a visible photon at the maximum of the eye sensitivity curve only carries 2.23 eV. Note that radiation output from the phosphor surface far exceeds blackbody radiation at the phosphor temperature (~40°C), in apparent violation of the fundamental conclusion of Section I. The luminescent centers, however, are radiatively coupled by UV photons to excited mercury atoms, which are in turn collisionally coupled to the electron gas in the plasma; the latter have an effective temperature of ~11,000 K. The blackbody radiance at that temperature could set the maximum thermodynamic limit to radiation output; the radiance of the phosphor surface is much less than that level.

Figure 16 illustrates spectral power distributions of (a) a fluorescent lamp intended as a substitute for an incandescent lamp, (b) the conventional "cool white" fluorescent color used in many commercial installations, and (c) a fluorescent lamp used in photocopy applications.

B. Energy Balance

1. Discharge Plasma

Passage of electric current through the gas imparts energy to it, producing (1) a continuing supply of new electron–ion pairs by ionization of the mercury constituent of the gas to maintain

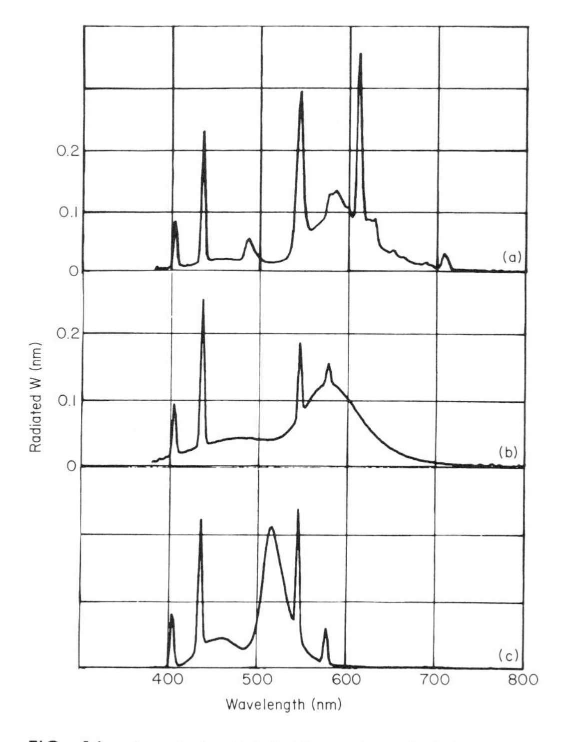

FIG. 16. Spectral distribution of radiated power (W/nm) of several fluorescent lamp types, all at 40-W input power. (a) Phosphor system including premium rare earth phosphors, intended for service as incandescent lamp substitute; color temperature = 3000 K, CRI = 75. (b) Standard cool white, the most common color for commercial applications; color temperature = 4200 K, CRI = 62. (c) Fluorescent lamp for photocopy applications, scale in arbitrary units. Since photoreceptor is not the human eye, color temperature and CRI are inappropriate designations and are not given. Visible mercury lines at 405, 436, 546, and 578 nm are evident in these spectral power distributions.

it in the plasma state, and (2) excited mercury atoms that radiate their characteristic spectrum. Because gas pressures are a small fraction of an atmosphere, the principal loss process for ions and electrons is by diffusion to the walls. Since radial electric fields in the plasma retard diffusion of electrons and enhance diffusion of ions, both species are constrained to diffuse to the wall at the same rate (ambipolar diffusion). Once at the wall, they recombine to reform neutral mercury atoms.

The energy input and dissipation process may be considered to take place in three stages:

1. Free electrons in the plasma are accelerated by the axial gradient of potential between anode and cathode.
2. As a result of frequent collisions, electrons share with each other energy gained in acceleration, establishing a steady-state electron energy distribution which is approximately maxwellian in character, with a "temperature" of ~11,000 K, ~1 eV.
3. The electrons dissipate their energy in collisions with atoms in the gas.

The high energy fraction of this distribution, greater than a few eV, may make inelastic collisions with mercury atoms, exciting them to states from which they may radiate on relaxation to the ground state. In a fraction of the cases, the mercury atom may receive enough energy to detach a valence electron, resulting in a new electron/positive-ion pair. The low energy fraction of electrons have insufficient energy to make exciting collisions (for which a threshold energy is required). The electrons can, however, make so-called collisions of the second kind, colliding with an excited mercury atom and extracting the excitation energy as kinetic energy, leaving the formerly excited atom in a state of lower excitation. Collisions of this kind by electrons substitute radiationless transitions for radiative transitions of excited atoms, reducing the amount of radiation emitted.

All electrons, low energy as well as high, make elastic collisions with gas atoms that transfer a small fraction of electron kinetic energy to the gas atom. Although the fraction of energy lost in any one collision is small, elastic collisions are very frequent; therefore, cumulative loss by this process represents a significant power dissipation.

Steady-state electron temperature is determined by the equality between rate of gain of energy by the entire ensemble of electrons from acceleration in the axial potential gradient and by the total rate of energy loss by the ensemble of electrons by all the energy loss processes that occur. The rate of ionization of mercury by high-energy electrons is proportional to the fraction of electrons having the threshold ionization energy, that is, exponentially dependent on electron temperature. There is a unique value of electron temperature at which rate of ionization exactly equals rate of loss and electron concentration in the plasma remains constant. Electron temperature increases monotonically with increasing axial potential gradient, and there is consequently a unique value of axial potential gradient which will result in the exact value of the electron temperature required to maintain the discharge in the steady state. The ballast circuit then must include internal impedance that decreases potential applied to the lamp terminals when current through the lamp increases so that the discharge can operate stably. Transient changes in electron density and current result in the alteration of discharge potential and potential gradient that produces changes of the opposite sign in electron density.

The efficiency of the discharge is determined by the ratio of power dissipated by electrons which is radiated as 254-nm resonance radiation to total power dissipated by electrons in all processes. Typically, for a 40-W fluorescent lamp, 10–15% of the input power is dissipated in electrode losses, 23% as elastic collision loss, 2% as visible and near-UV radiation, 8–9% as 185-nm resonance radiation, and 55% as 254-nm UV radiation. The astonishing efficiency of conversion of power in the plasma into 254-nm resonance radiation is the reason for the technological importance of this type of light source. Even after reduction by a factor 2 in conversion at near unity quantum efficiency to visible rays by a phosphor, it still permits high luminous efficacy (80 lm/W in commodity types, and up to 100 lm/W in premium types).

This section demonstrates the importance of the fraction of high energy electrons in the distribution (i.e., the electron temperature) in determining efficiency of the device. Electron temperature is amenable to control by the lamp designer, through adjustment of the ambipolar diffusion loss rate of electrons and ions. If diffusion rate is increased, so does the electron temperature required for steady-state operation. Diffusion rate is determined by the tube diameter and by pressure and atomic mass number of the noble gas. The heavier the gas or the higher its pressure, the lower the diffusion rate and the lower the electron temperature. The smaller the tube diameter, the faster the diffusion loss rate and the higher the electron temperature. However, noble gas pressure and type also determine elastic scattering loss by an ensemble of electrons having a given temperature. Equally, tube diameter is a determining factor in the imprisonment time for resonance radiation. Moreover, at a given current, current density increases with decreasing diameter, and with it, electron density increases. The higher the electron density, the greater is the probability of a collision of the second kind before radiation; as a result, efficiency decreases with increasing current density. Thus, ultimate performance of the device depends in a very complex way on a multiply connected set of variables.

The development of reasonably accurate discharge models incorporating rate constants for ~25 different collision processes as well as radiation transport was a major step forward in the understanding of the physics of this unique device. However, it did not result in significant improvement in efficiency of standard fluorescent lamps, which had already been optimized well by patient, empirical Edisonian techniques.

2. Lamp–Electrical-Circuit System

As in the case of HID lamps, fluorescent lamps require a higher applied potential for ignition than for steady-state operation. They too are operated on alternating current so that the necessary current limiting impedance can be reactive rather than resistive. Ignition potential is greatly reduced by heating the cathodes to electron-emitting temperatures to provide many free electrons to initiate avalanches in the breakdown process. The cathodes themselves are multiply-coiled helices of tungsten, the interstices of which are impregnated with alkaline-earth oxides for enhanced electron emission. They are heated by the passage of current through the tungsten wire of the helices.

There are two principal methods of accomplishing this during ignition. The simplest is the switch-start system, preferred in most of the world except in the United States. At starting, the two filaments are momentarily connected in series across the output terminals of the ballast so that the short circuit current of the ballast (limited by the internal impedance) flows through them, heating them to emission temperatures. After a brief interval the switch opens, applying the open-circuit potential of the ballast (together with any transient pulses of potential resulting from interruption of current in a circuit containing inductance) to the terminals of a lamp with heated cathodes, which then ignites. In the rapid-start ballast system, preferred in the United States, individual filament transformers wound on the ballast core continually supply 3.75 *V* to the cathodes, and the resulting current flow heats them to a sufficient temperature to emit electrons. Although this heating is not required during operation, since power received from the discharge end losses is adequate to keep the cathodes hot, the heater power has in the past been left connected during operation.

Recent concern with energy savings has resulted in introduction of lamps incorporating internal bimetal switches that disconnect one of the electrode terminals in operation, interrupting the flow of heater current. Ballast-lamp combinations in which the interruption is accomplished by the ballast have also been introduced.

As is the case in HID lamps, there are power losses in the ballast that must be added to lamp power to determine system power and efficacy. Commodity-type ballasts typically have ballast loss of 15% of lamp power. Again, recent concern with energy saving has resulted in the introduction of premium ballasts in which ballast loss is reduced by 50%. A great deal of engineering activity has recently been expended in development of electronic ballasts using semiconductor components and operating the lamps at hf, ~25 kHz; less lossy impedance can be easily provided at such frequencies, and the lamp itself is ~10–15% more efficient (primarily because of reduced electrode loss). Accordingly, up to 30% energy savings for a given illumination level over commodity lamps operated on commodity ballasts can be achieved with such a system. Unfortunately, electronic ballasts are much more expensive than copper and iron ones. In addition, premium lamp types and premium magnetic ballast types achieve equally impressive energy savings at lower cost. Consequently, market acceptance of electronic ballasts has been limited.

C. Performance Limitations Resulting from Materials Limitations

Fluorescent lamps are also extremely long-lived devices, with rated lives up to 20,000 hr at 3 burning hours per start on rapid-start circuits. There is not a fundamental relationship between efficiency and lamp life, as in incandescent lamps. Lamp life is primarily determined by the erosion of electron emission material from the cathodes. Since the process of ignition results in momentarily enhanced erosion, lamp life is adversely affected by increasing numbers of ignitions per thousand burning hours. A lamp with 20,000 hr life rating at 3 hr per start will last 40,000 hr if simply ignited once and never shut off. A lamp which is started once per minute of operation, will last 2000–5000 starts, only a few hundred hours. Lamps that are intended for service requiring many ignitions per hour are operated on ballasts supplying continuous filament heat, even when the lamp is not operating, to minimize damage to the electrodes at starting.

Electron emission materials that are more durable, but have equivalent electron emission as the alkaline earth oxides, would increase the flexibility of operation of fluorescent lamps in intermittent service. But no such materials have been found.

A major material limitation in fluorescent lamp operation is in the durability of fluorescent phosphors. These deteriorate in operation as a result of photolytic decomposition and color center formation, ion bombardment, and chemical reactions with mercury, glass, and impurity gases. The rate of deterioration of phosphors under these influences increases with the wall-loading (i.e., the flux density of UV rays and ions) and with wall temperature. The degree of this deterioration depends on the particular type of phosphor. Standard cool white fluorescent lamps, the most common type, lose ~15–20% of their output over the rated life of the lamp. Doubling the power input to achieve higher light output from a given size lamp, results in more than twice as much lumen depreciation, despite many years of empirical development to improve the phosphor durability.

Since a major obstacle to application of fluorescent lamps as substitutes for incandescent lamps is the size of the fluorescent lamp to achieve a given light output, it is plain that this material limitation has been a significant one. Recently developed rare earth–based phosphors have been found to be much more durable under high-loading conditions and to be hardly affected by operation at double and triple the customary wall-loading. Thus, most compact fluorescent lamps intended for incandescent replacement service have used these phosphors. The cost of these materials, however, is many times that of conventional phosphors, and for the present the cost appears to restrict these applications to commercial rather than residential service.

D. Practical Considerations

1. Fabrication Technology

Fabrication technology of fluorescent lamps has much in common with high-speed incandescent lamp manufacturing. It is highly automated with automatic transfer of parts in process from one station to another and little or no manual handling. There are two principal differences: first, the necessity to apply a uniform coating of powdered fluorescent materials to the inside wall of the tubular bulb, and, second, the constraints imposed by processing of electrode materials and the evacuation of very large internal lamp volumes through exhaust tubes of small diameter (limited by the capability of flame-seal rapidly and reliably).

The phosphor itself is manufactured off the site by blending raw materials of controlled particle size and firing at elevated temperatures to synthesize the proper compounds. The necessity for control of particle size in firing results from the fact that these materials are very susceptible to mechanical damage in the usual types of milling processes common to formulating paints. The application of the phosphor to the bulb is basically to make a paint with the phosphor as a pigment, flow-coating over the inside surface of the bulb to a thickness controlled by viscosity and drying. The remaining binder is removed by pyrolysis, passing the coated bulbs through furnaces at nearly the melting temperature of the glass, with axial air flow to react with the carbon and hydrogen constituents.

After coating and binder burnout, the stems with lead-in wires, electrodes, and exhaust tubes, are flame-sealed to the ends of the bulbs, and the assembly is transferred automatically to the exhaust machine. Because alkaline earth oxides used for the electron emission mix are extremely reactive to moisture, these materials are applied to the electrode coils as alkaline earth carbonates, and the carbonates are decomposed to oxides by heating as part of the evacuation process. Because of the large volume of gas to be exhausted and the small size of the exhaust tube, conventional vacuum evacuation is not fast enough. Therefore, the air and carbon dioxide are removed from the bulb by flushing inert gas in one exhaust tube and out the other; the entire process takes place in the viscous flow regimen instead of in the free-molecule flow. In this way, several liter-atmospheres of air can be removed and replaced by the noble gas filling in a fraction of a minute. Basing, ageing, testing, and packing are automatic, at production speeds as high as 50 to 100 lamps per minute.

In common with all other light sources, fluorescent lamps are sensitive to certain impurities, which must be effectively removed even at the extremely high processing speeds dictated by necessity for low-cost manufacture. Water vapor, the bête noir of all lamps, is troublesome in fluorescent lamps as well. It dissociates in the discharge, liberating hydrogen that attacks the phosphor and liberating oxygen that reacts with the electron emission coating, which "poisons" the cathode, and with mercury vapor, which forms solid deposits of dark mercury oxides on the surface of the phosphor. Residual hydrocarbons from the phosphor coating application result in carbon deposition on the phosphor as well as the liberation of hydrogen that will further damage the phosphor. If the bakeout temperature is increased too much, alkali from the glass will react with the phosphor, damaging it initially. Thus the baking process must be very critically controlled. Hydrogen and water vapor will also act to increase the ignition potential of the lamp in early life; this is usually a transient effect, however, because of cleanup during operation by the reactions noted. Such gas contamination may in extreme cases prevent the lamp from igniting the first time that it is placed in service, thus causing premature lamp failure. Testing for this effect, as well as cleanup of small amounts of impurities, is the purpose of the ageing step in the manufacturing process.

E. Commercially Available Products and Performance Characteristics

Fluorescent lamps are used primarily in indoor commercial lighting, where their high efficacy and low brightness make pleasantly diffused illumination of large areas possible at low total cost of light. Only a modest CRI, ~60, is required for this service, and most fluorescent lamps use phosphors with these characteristics. Retail merchandising places more stringent requirements on color rendition, and applications as incandescent substitutes requires not only a high CRI, but also color temperatures comparable with incandescent lamps, ~3000 K. These are obtained, together with improved resistance to degradation at high loadings, by recently developed phosphors based on rare earth compounds.

Table VII presents characteristics of fluorescent lamps intended for commercial and merchandising illumination, while Table VIII gives the characteristics of compact fluorescent lamps intended as substitutes for incandescent lamps. Figure 17 is a photograph illustrating various types of fluorescent lamps.

F. Research and Development

Research and Development in fluorescent lamp technology is focused on three areas: development of more efficient fluorescent lamps for commercial and merchandising service; development of improved phosphors; and development of compact fluorescent lamps for incandescent replacement, with the eventual hope that they can supplant incandescent lighting in many residential applications.

The first of these areas involves a range of activities from changes of diameter, fill-gas mixture and pressure, and phosphor improvements,

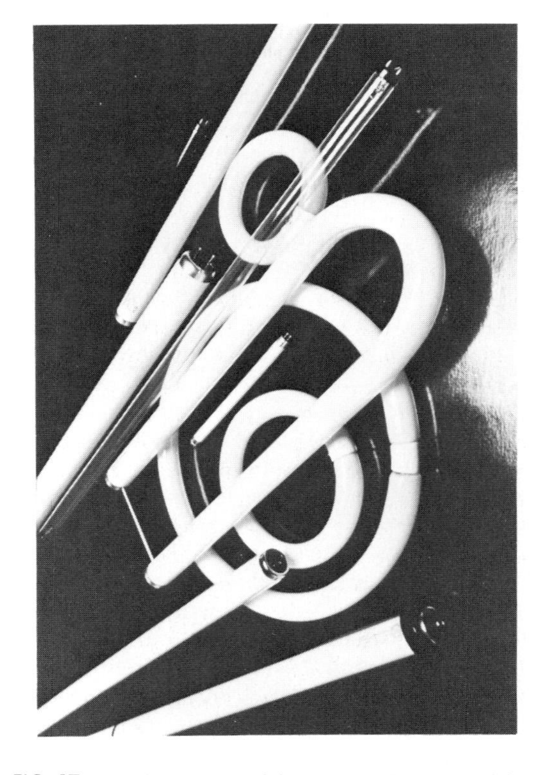

FIG. 17. Typical types of fluorescent lamps used in industrial and commercial applications.

TABLE VII. Representative Fluorescent Lamp Types for Industrial and Commercial Applications (United States Only)

Power (W)	Bulb[a]	Type	Operating current (V)	(A) (Nominal)	Rated life (hr)	Light output (lm)	Efficacy (lm/W)	Color temperature (K)	Color rendering index
Standard types									
40	48T12	CW/RS[b]	105	0.425	20,000	3150	79	4200	62
75	96T12	CW/IS	195	0.425	12,000	6300	84	4200	62
110	96T12	CW/HO	150	0.800	12,000	9200	84	4200	62
40	22.5T12U	CW/RS	105	0.425	18,000	3000	75	4200	62
Energy-saving retrofit types[c]									
34	48T12	Standard[d]/RS	85	0.425	20,000	2925	86	4150	48
34	48T12	Deluxe[d]/RS	85	0.425	20,000	2925	86	4100	67
60	96T12	/IS	155	0.425	12,000	5850	98	Available in standard & deluxe	
95	96T12	/HO	130	0.800	12,000	8800	104	Available in standard & deluxe	
34	22.5T12U	Standard[d]	85	0.425	18,000	2800	82	4150	48
Energy-saving new equipment types[e]									
32	48T8	Cool	135	0.265	20,000	2900	91	4100	75
32	48T8	Warm	135	0.265	20,000	2900	91	3100	75

[a] Nominal length in inches—tubular—diameter in ⅛ in.
[b] Phosphor color/ballast type: CW = cool white, see Fig. 16 for spectral power distribution. RS = rapid start, IS = instant start, HO = high output.
[c] Retrofit types are physically interchangeable with standard types and operate at lower power on same ballasts by virture of lower design operating voltages.
[d] Generic nomenclature to represent two principal high-efficacy phosphor systems.
[e] Use premium high-efficacy phosphor system combining high efficacy and high CRI; require special ballast.

to fundamental studies aimed at increasing the primary efficiency of generation of resonance radiation. It has been discovered, for example, that there is a slight isotopic effect in the imprisonment time. Increasing the concentration of the 196 isotope to ~4%, (normally present to only 0.15% in natural mercury, which has six stable isotopes), reduces imprisonment time by ~10%, which can be translated into 3 to 5% higher efficiency of generation of resonance radiation. Although this may not seem like a large reduction, it translates into a 1–2 W reduction per lamp for equal light output. Over the life of the lamp, this amounts to a reduction of 20 to 40 kW hr of energy, a substantial amount, worth approximately $2–4. If mercury enriched to the required amount can be obtained at a cost of $1–10 per gram, the value of energy savings can justify the increased cost of lamp manufacture. Investigation of the possibility of such an isotopic enrichment, either by photochemical or other techniques developed for uranium, is proceeding under the sponsorship of the U.S. Department of Energy.

Phosphor development is an activity that has been ongoing since the introduction of the fluorescent lamp more than 40 years ago. In the past, phosphor development has focused on extending the range of colors available and on evolutionary improvements in quantum efficiency and lumen maintenance of existing commercial phosphors. The primary focus at the present time is on discovering less expensive substitutes for the rare earth phosphors that nevertheless retain the high efficacy, good color, and extreme durability under high loading of those materials.

The real ferment in the field is in the development of a compact fluorescent for incandescent replacement. The chief obstacle is that radiation transport and electric characteristics of fluorescent lamps dictate that the discharge component is long and slender and requires a ballast. Incandescent lamps, however, are short and fat, and the fixtures that hold them allow little or no room for a ballast. To date, most of the approaches are the topological equivalent to spaghetti: a long slender tube is bent into a contorted continuous bundle, sometimes with a ballast tucked in among its segments. A cover is then placed over the entire bundle, which is then

TABLE VIII. Representative Compact Single-Ended Fluorescent Lamp Types for Incandescent Lamp Substitutes

	Retrofittable[a]		New equipment only	
Type	Circular with central ballast	Folded-U in light-diffusing cover	Linked parallel tubes (twin tube)	Double-D folded
Shape	Ring-shaped	Jar-shaped or globular	Two-finger-shaped	Square
Ballast	Integral	Integral	Separate	Separate
Lamp/ballast/system wattage[b]	22/5.3/27.3	13/5/18	13/5/18	16/5/21
Light output (lm)[b]	1080	900	900	1050
Efficacy (system)[b]	40	50	50	50
Equivalent incandescent lamp	75 W/16 lm/W	60 W/14.5 lm/W	60 W/14.5 lm/W	75 W/16 lm/W
Color temperature/color rendering index	3000/67	2700/80	2700/80	2800/80
Life (hr)	7500–12,000	5000	10,000	5000
Size (mm)	203	72 diameter × 175 long	26 wide × 188 long	134 square
Weight (g)	450[c]	520[c]	45 (lamp only)	60 (lamp only)
Principal manufacturers	All major U.S.	Philips, Toshiba, Hitachi, Matsushita, Mitsubishi[d]	All major U.S. and European	Thorn (England)

[a] Retrofittable types will fit in some incandescent lamp sockets as direct replacements. Retail prices of $7–20 make economics uncertain for residential user. Long life minimizes lamp replacement costs in commercial applications, making economics more attractive.
[b] Most common or popular type.
[c] Integral ballast contributes 80% of the weight.
[d] Philips, Toshiba, and Matsushita manufacture this type with the folded-U tubular lamp hermetically sealed with nonhermetic cover. Hitachi and Mitsubishi manufacture it with a hermetically sealed evacuated outer cover, and nonhermetically sealed internal folded-U lamp.

sold as a compact fluorescent lamp for $10 or more. A major problem with this approach is that bending glass tubing is slow, dictating production at no more than 10 to 20 pieces per minute instead of 50 to 100. Also, ballasts are costly and heavy and will outlive several lamps (so it is wasteful to attach them integrally and throw them away when the lamps fail). Furthermore, the ballasts will not fit in most incandescent lamp sockets and are too costly for all but the most dedicated devotees of life-cycle costing.

Consequently, although there are hundreds, or perhaps thousands of designs patented or disclosed around the world, only one or two have achieved even a modest measure of success in the marketplace (see Fig. 18). A prime example is the SL lamp (TM - N.V. Philips Gloeilampenfabrieken), which has achieved a secure market niche in Europe, principally in incandescent downlight applications in indoor commercial lighting. The long life and minimized labor cost to replace the SL lamp is more important than its high efficacy in the relative success of this lamp (See column 2 of Table VIII).

A great deal more success is expected in the future from new designs of single-ended low-wattage fluorescent lamps intended for new fixture applications in interior commercial lighting that would otherwise be served by long-life incandescent lamps: hotel corridor applications,

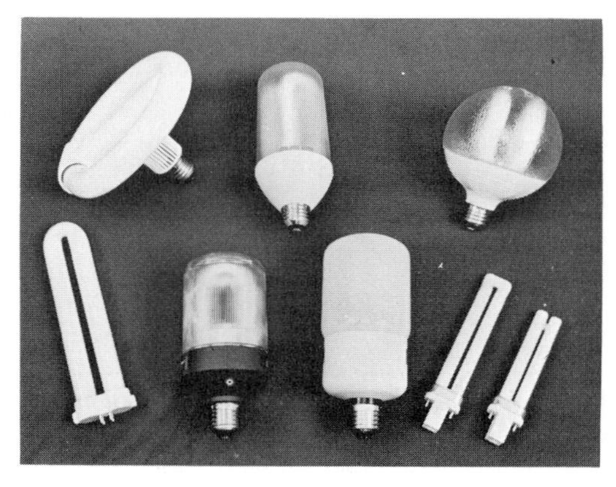

FIG. 18. Sample of fluorescent lamp types used as substitutes for incandescent lamps.

with 8000 burning hours per year, for example. A prime example of this type of lamp is the twin-tube U-shaped lamp noted in column 3 of Table VIII. Long life, high efficiency, and relatively low cost for the lamp combine to make the product highly competitive in the marketplace.

Meanwhile, the search for the ultimate replacement for the residential incandescent lamp continues.

VIII. Summary

A. Cost Comparison

Table IX compares several measures of cost for the different families of lamps described in this article. Included are costs of the lamps and the electric power to operate them; ballast, fixture, installation, and maintenance costs are not included. Costs indicated are based on 1985 prices for lamps and electricity. Changes since then have been about equal to inflation, and have not altered relative rankings.

Based on lamp cost alone per thousand lumens of light output, incandescent lamps are the least expensive. Allowing for the fact that discharge lamps outlast incandescent manyfold, a somewhat more realistic measure of lamp cost is the cost per million lumen-hours, for which the fluorescent lamp is least expensive. Operation of lamps and delivery of light requires consumption of electrical power, at a cost per million lumen-hours indicated in column 4 of Table IX, assuming $0.06/kW hr electricity cost. The least expensive light source in this regard is the HPS lamp, by virtue of its very high efficacy. Life cycle cost per million lumen-hours is the sum of lamp and power cost, and is given in column 5 of Table IX. Note that cost of electric power is 70–95% of the total so that HPS lamps are again the least expensive, even though their lamp cost per million lumen hours 33% more than that of the fluorescent.

For this reason, HPS lamps have become the light source of choice for nearly all kinds of outdoor lighting, where their relatively poor CRI and undesirably low color temperature are not fatal flaws. For indoor commercial applications, fluorescent and M-H are preferred, despite their higher cost of light, because of more favorable color temperature and CRI in comparison to HPS. Despite the preference for the other sources, however, there are increasing numbers

TABLE IX. Comparison of Lamp Costs, Operating Costs, and Total Life-Cycle Costs for Major Lamp Families[a]

Lamp family	Lamp cost $/thousand lm	Lamp cost $/million lm hr	Electrical power cost $/million lm hr	Life-cycle cost $/million lm hr (%)[b]	
Incandescent	0.69[c]	0.92	3.81	4.73	(80.5%)
Tungsten–halogen	2.34	1.17	2.70	3.87	(69.8%)
Fluorescent	1.25	0.052	0.84	0.89	(94.2%)
Mercury	1.67	0.069	1.19	1.26	(94.5%)
Metal halide	2.32	0.116	0.937	1.05	(89.0%)
High-pressure sodium	1.66	0.069	0.53	0.60	(88.5%)

[a] 1985 lamp prices and electrical costs ($.06/kWh); relative rankings not changed appreciably since then. Price does not include ballast, fixture, or installation cost.
[b] Figures in parentheses are percentage of life-cycle costs represented by electric energy.
[c] Least expensive types are set in boxes.

of indoor HPS applications (driven by cost considerations); for many industrial applications, color is relatively unimportant, and for indoor commercial applications, careful choice of decor emphasizing reds, oranges, yellows, and browns minimizes unfavorable color distortions. Mercury lamps appear to be marked for obsolescence; in fact, in the United States, sales of HPS lamps have exceeded sales of mercury lamps in every year since 1983, and sales of HPS lamps are growing faster.

Bibliography

Elenbaas, W. (1972). "Light Sources." Crane Russak and Co., New York.

Kaufman, J. E., and Haynes, H. (Eds.) (1981). "Ies Lighting Handbook," Illuminating Engineering Society, New York.

Waymouth, J. F. (1971). "Electric Discharge Lamps," (1971). MIT Press, Cambridge, Massachusetts.

Waymouth, J. F., and Levin, R. E. eds. (1980). "Designer's Handbook, Light Source Applications," GTE Marketing Services, West Seneca, New York.

LUMINESCENCE

J. N. Demas
S. E. Demas *University of Virginia*

GLOSSARY

Color centers: Absorbing sites in solids caused by lattice defects, trapped electrons or holes, or the formation of new chemical species.

Excimer: Excited complex that does not exist in the ground state and is formed between one excited and one ground state molecule of the same type.

Exciplex: Excimer formed between two molecules of different types.

Fluorescence: Luminescence characterized by very short lifetimes; typically a spin-allowed process.

Hole: In solids, an electron-deficient center that frequently can move through the lattice.

Internal conversion: Relaxation of a system from an upper state to a lower one of the same spin multiplicity.

Intersystem crossing: Conversion of a system from a state of one spin multiplicity to another.

Laser: Acronym for light amplification by stimulated emission of radiation—a stimulated emission device that produces intense, highly directional, coherent, monochromatic optical radiation.

Luminescence: Emission of ultraviolet (UV), visible, or infrared (IR) radiation of excited materials.

Phosphorescence: Luminescence characterized by a long lifetime; frequently a spin-forbidden process.

Quenching: Deactivation of an excited state by a nonemissive pathway.

Stimulated emission: Photon emission from an excited species promoted by the presence of other photons.

Trap: Lattice defect or chemical center in solids that can trap an electron or a hole.

Luminescence is the emission of ultraviolet (UV), visible, or infrared (IR) radiation from materials and arises from a radiative transition between an excited state and a lower state. The classification of the luminescence depends on how the excited state was derived. Photoluminescence arises following excitation by the absorption of a photon of light. Electroluminescence and cathodoluminescence arise from electric current flow in solids or solutions or in gases during an electrical discharge. Chemiluminescence arises during chemical reactions, and bioluminescence is chemiluminescence in biological systems. Radioluminescence arises from the passage of ionizing radiation or particles through matter. Thermoluminescence occurs during gentle sample heating.

I. Introduction

Since the beginning of recorded history, and undoubtedly much earlier, individuals have been fascinated by luminescence. The cold bioluminescences of glowworms, rotting wood, and sea creatures and the spectacular light shows of the aurora borealis have been particularly intriguing, and a great deal of effort has been made to understand their origins. Until the advent of quantum mechanics, however, the fundamental origins of these emissions could not be satisfactorily explained.

There were numerous ingenious attempts to quantify luminescence phenomena using photographic and manual recording of emission behavior; however, especially for broad molecular emissions, the major breakthroughs in luminescence studies tended to parallel instrumental developments. In particular, the high-sensitivity commercial photomultiplier tube marketed in the 1940s and the low-cost spectrofluorimeters of the 1950s can be credited with much of the modern information, theories, and applications of luminescence. More recently, lasers and nanosecond and subnanosecond decay time instruments have revolutionized the types of information that can be extracted.

Any study of luminescence should address the following key questions: (1) What is the molecular and atomic nature of the origin of the luminescence? (2) What are the detailed paths of molecular excitation and deactivation? (3) What are the structures of excited states? (4) Can one rationally design systems with specific and useful properties or exploit existing properties?

This article is concerned primarily with the phenomenological aspects of each type of luminescence rather than the theoretical underpinnings of the subject. The origins and factors affecting luminescence are described. Some experimental methodologies for studying luminescences are examined. Finally, applications of the various luminescences are described. Emissions of very high energy photons from nuclear or inner-electron-shell transitions or from the nonspecific incandescences of hot solids or plasmas are excluded.

II. Origins of Luminescences

Specifically considered are emissions that arise by radiative transitions between two states of atomic, molecular, or extended molecular systems. A radiative transition is one in which the energy is released as a photon. The nature of the emission depends on the nature of the initial and final states and the route to the excited state.

First, types of excited states are categorized, the factors that influence excited state emission are described, and then the methods of excited state population that define the nature of the emission are discussed.

Figures 1 and 2 show some of the wealth and complexities of atomic and molecular emissions. Figure 1 shows the absorption and emission (relative intensity as a function of wavelength) spec-

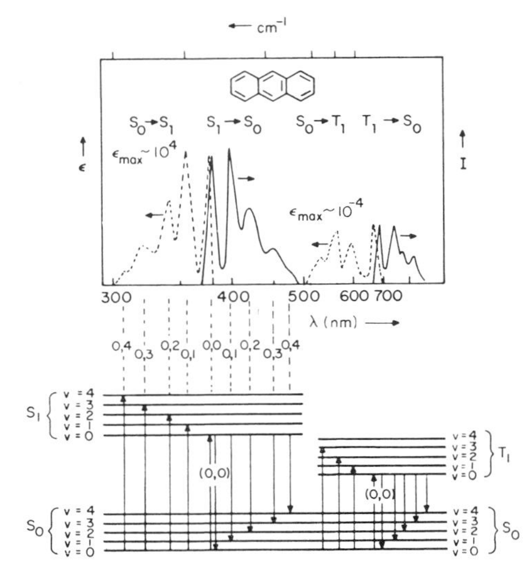

FIG. 1. Absorption (dashed lines) and emission (solid lines) of anthracene. The lower portion displays the electronic and vibrational assignments of the absorption and emission bands. [Reprinted with permission from Turro, N. (1978). "Modern Molecular Photochemistry," Benjamin/Cummings, Menlo Park, California.]

tra of anthracene. There are two distinct emissions. The high-energy band at 400 nm is characterized by a short luminescence lifetime of a few nanoseconds, while the lower-energy emission at 700 nm can be characterized by millisecond lifetimes. The overlap between the lowest-energy absorption and the high-energy emission is characteristic of this type of system. The regular progression of peaks on both emission systems is also common to many molecular systems.

Figure 2 shows the cathodoluminescence of atomic mercury in a low-pressure discharge. Particularly noteworthy is the exceptional nar-

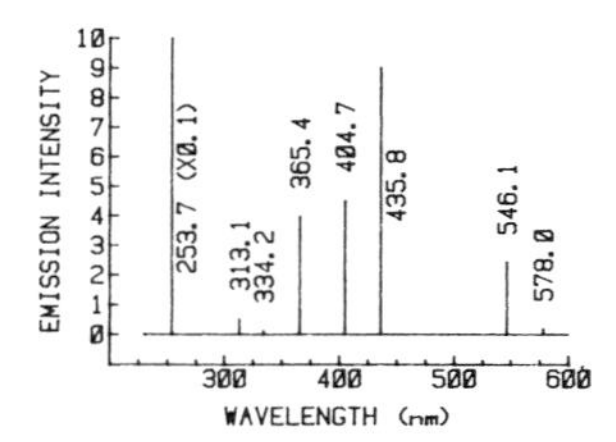

FIG. 2. Emission spectrum of low-pressure mercury vapor discharge. Wavelengths in nanometers are adjacent to each line. Note that the 253.7 nm has been attenuated by a factor of 10 to get it on scale.

rowness of the atomic versus the molecular emissions. The characteristics and differences of these emissions are discussed in Section III.

III. Excited State Types

A. Spin Multiplicity

A simplified excited state diagram is pictured in Fig. 3. Details of the quantum mechanical origins and nature of excited states are not presented here. The system is characterized by a singlet ground state, denoted by S_0, and singlet excited states, denoted by S_i ($i = 1, 2, \ldots$), are indicated. Singlet states arise when all the electrons are spin-paired. Also shown in the excited state manifold is the lowest triplet state, denoted by T_1. Triplets arise when there are two unpaired spins. This type of system corresponds to the vast majority of organic molecular species and occurs when the lowest energy configuration of the system is due to all of the electrons being spin-paired.

Excited states of such a system generally arise when a paired electron is promoted from a filled to an unoccupied orbital. The electron can remain paired with the electron left behind to form excited singlet states, or it can undergo a "spin flip" and become unpaired; this results in a triplet state. The triplet state derived from a specific orbital promotion is of lower energy than the corresponding singlet state (Figs. 1 and 3).

Oxygen and metal ions are the common stable exceptions to this type of excited state diagram. Atomic species in flames or discharges are also frequent exceptions. Oxygen has a triplet ground state with singlet and triplet excited

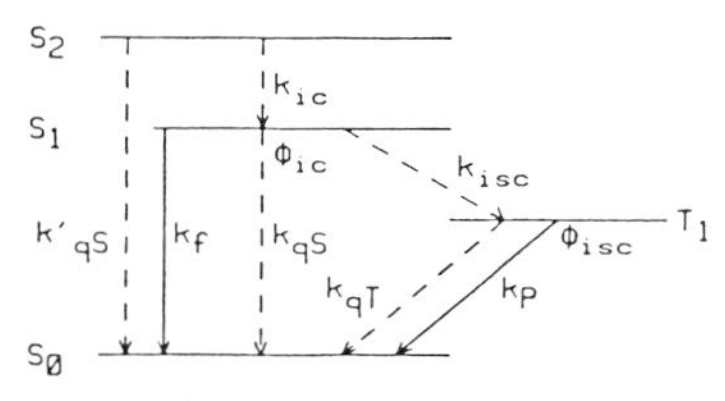

FIG. 3. Schematic energy level diagram, or Jablonski diagram, for a molecule showing the possible paths of energy degradation. Solid lines represent radiative emissive processes, and dashed lines represent nonradiative processes. Rate constants and efficiencies of the indicated constants are denoted by k's and ϕ's. [Reprinted with permission from Demas, J. N. (1983). *J. Chem. Ed.* **60,** 803. Copyright 1983 Division of Chemical Education, American Chemical Society.]

states. The ordering of excited singlets and triplets are inverted over those of Fig. 1, however, with the singlets being below their corresponding triplets. Metal ions can exhibit a multitude of excited state multiplicities, which can range from doublets (one unpaired electron) for Cu^{2+} and Na, quartets (three unpaired electrons) for Cr^{3+}, and octets (seven unpaired electrons) for Eu^{2+}.

Regardless of the nature of the ground state, however, the excited states can have spin multiplicities that are the same as, or different from, the ground state. Spin selection rules control whether a transition between states is allowed or forbidden. Transitions between states of the same multiplicity are spin-allowed, while all others are forbidden. Forbiddenness does not mean that a transition will not occur at all, but that it will not occur as readily as an allowed one. Allowed transitions are characterized by strong absorptions, large rate constants, and short lifetimes. Spin-forbidden transitions exhibit weak absorptions, long lifetimes, and low rate constants. Compare the allowed 400-nm absorption with the forbidden 650-nm absorption of anthracene (Fig. 1), where the allowed higher-energy transition is 10^8 times more intense.

Figure 3 is based on the assumption that spin is a good quantum number. This assumption is not always correct, especially for species of high atomic number. Spin–orbit coupling can mix orbital and spin angular momentum, and then the concept of electron spin fails. It is necessary to discuss the states of the system in terms of the good quantum number J. Pragmatically, spin–orbit coupling scrambles the singlet and triplet states and gives a large component of the other spin character to the state. Thus, the mixing of singlet character into a triplet state can greatly increase the allowedness of spin-forbidden transitions.

B. Fluorescence and Phosphorescence

Traditionally there has been a phenomenological characterization of emission type. Short-lived emissions have been considered fluorescences and long-lived emissions, phosphorescences. One would then infer that phosphorescence arises from spin-forbidden processes and fluorescence from spin-allowed processes.

In the case of many discrete molecular systems this categorization is correct. Figure 1 demonstrates the simultaneous presence of fluorescence and phosphorescence. The short-lived emission is the fluorescence and the long-lived emission the phosphorescence.

This simple model can break down when one considers the more complex case of fluorescent lamp and television phosphors. After use, a television screen glows in a dark room for minutes to hours. In keeping with the empirical classification these glows are called phosphorescences. As it turns out, however, these very long lifetimes are generally associated with slow secondary trapping processes that have nothing at all to do with the fundamental luminescence processes. Indeed, the fundamental luminescence step in many phosphorescences is a spin-allowed process.

One also sees numerous incorrect or misleading references in the literature. The emissions of rare earth elements (e.g., Tb^{3+} and Eu^{3+}) and of uranyl are frequently referred to as fluorescences even though their lifetimes are hundreds of microseconds to milliseconds. Furthermore, quantum mechanically these emissions are best described as spin-forbidden processes. Thus, by the criteria of both lifetime and quantum mechanics, these emissions are actually phosphorescences. If there is doubt about the origins of an emission, it is best referred to as a luminescence.

C. Energy Degradation Pathways (Nonradiative Pathways)

It is impossible to talk about luminescence without considering additional nonradiative processes. The anthracene emission spectrum (Fig. 1) is made up of both a fluorescence and a phosphorescence. These emissions occur with the same efficiencies regardless of whether S_1 or an upper singlet state is directly excited. Furthermore, because of the weakness of the $S_0 \rightarrow T_1$ absorption, it is usually extremely difficult to excite the triplet state directly. However, efficient phosphorescences on excitation into the singlet states are common. Finally, there are very few molecules that have emission efficiencies (photons emitted per photon absorbed) of close to 100%. These results imply the existence of both efficient nonradiative deactivation pathways and radiationless interconversions between states of the same, and of different, multiplicities.

Figure 3 shows a simplified representation of these additional processes. Relaxation within a manifold of the same multiplicity is called internal conversion. In condensed media, internal conversion is very fast compared with the rates of radiative emission from upper singlet states and accounts for the rarity of efficient upper-level emission. This rapidity arises because of the closeness of the lower levels, the absence of spin restrictions, and the availability of vibrational levels of the lower states that provide an efficient vibrational cascade mechanism for relaxation. In condensed media the best known example of an upper excited state emission is the $S_2 \rightarrow S_0$ fluorescence of azulenes.

Radiationless deactivation to the ground state from S_1 is also a special case of internal conversion. However, since it competes directly with the main emission process, it is given the separate term *quenching*. The decreased rate of quenching from S_1 to the ground state is attributable to the much larger energy gap between these two levels compared with the spacing between upper singlets.

Crossing between states of different multiplicities (e.g., singlet; to triplets) is also possible even though the process is spin-forbidden. Conversion between states of different multiplicities is called intersystem crossing. Indeed, in some systems with small energy gaps between the singlet and triplet states, and with reduction of the forbiddenness because of spin–orbit coupling, intersystem crossing can be so fast compared with radiative coupling to the ground state that only phosphorescence is observed.

In the triplet manifold as well as in the singlet manifold internal conversion usually causes rapid relaxation to the lowest triplet state before emission occurs. The emitting triplet state is also susceptible to direct quenching to the ground state. Indeed, because of the forbiddenness of phosphorescence, the long-lived triplet state is very susceptible to quenching; room-temperature phosphorescences are relatively rare and generally not very efficient.

In the gas phase, especially at low pressures, where collisions are infrequent, internal conversion is much less rapid due to the absence of solvent or other molecular vibrations to help carry away the excess energy. This reduced efficiency of internal conversion makes upper excited state emissions much more prevalent.

D. Atomic and Molecular Excited States

The states of discrete atoms are described by first determining the one-electron atomic orbitals, then adding the total number of electrons by filling the lowest-energy orbitals with two electrons per orbital. The ground state is derived from this configuration. Excited states are then generally derived by considering the configurations arising from promotion of an electron from an occupied to an unoccupied orbital.

For example, the ground state configuration of atomic mercury is $[Xe](4f)^{14}(5d)^{10}(6s)^2$, where [Xe] stands for the closed-shell xenon core. If everything but the 6*s* electrons are denoted as core, the lowest excited states of atomic mercury are given by $(core)(6s)^1(7s)^1$ and $(core)(6s)^1(6p)^1$.

The state diagram for atomic mercury and some of the radiative transitions responsible for emission lines of Fig. 2 are shown in Fig. 4. The other transitions can be derived from energy differences between states. The state designations based on the quantum numbers *S*, *L*, and *J* are shown above each set of states. The superscript denotes the spin multiplicity of the state, *M*, and is related to the spin angular momentum quantum number *S* by $M = 2S + 1$. The orbital symmetry of the state is determined by the orbital angular momentum quantum number *L* and is given by the capital letter. The *J* quantum number, which arises from coupling of spin and orbital angular momentum and represents the total angular momentum, is the subscript. For example, the 253.65-nm emission line arises from a transition from $^3P_1$ to the $^1S_0$ ground state ($S = 1 \rightarrow S = 0$; $L = 1 \rightarrow L = 0$; $J = 1 \rightarrow J = 0$).

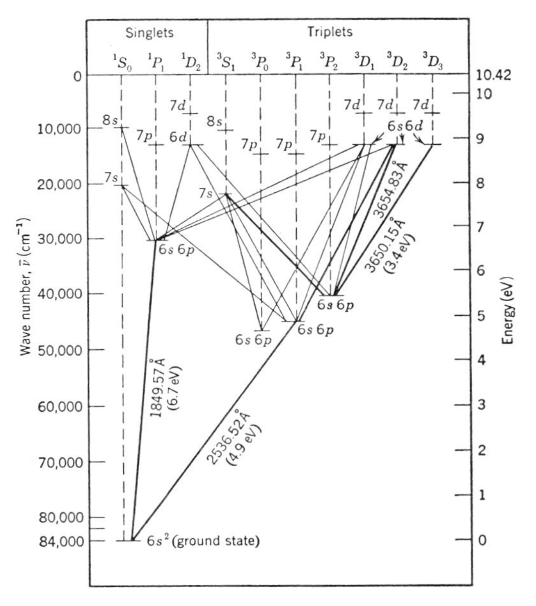

FIG. 4. Energy level and state diagram for atomic mercury. The term symbols for the states are indicated across the top. Some of the radiative transitions are indicated by solid lines. The orbital configuration is indicated on each state. For example, 6*s* 6*p* denotes a $(6s)^1(7p)^1$ outer-shell configuration, and 7*d* denotes a $(6s)^1(7d)^1$. The core is omitted for clarity. [Reprinted with permission from Leverenz, H. W. (1950). "An Introduction to Luminescence of Solids," John Wiley & Sons: New York.]

Molecular excited states are derived in much the same way, except that the orbitals of the system are described by molecular orbital theory. The single-electron molecular orbitals are made up of combinations of atomic orbitals derived from the different atoms in the molecule. Thus, the molecular orbitals extend over the entire molecule and are not localized on a single atom. This delocalization makes for very rich bonding and spectroscopy. As with the atomic case the electrons are added to fill up the lowest-energy orbitals in order to derive the ground state configuration. Excited states usually arise from orbital promotions of electrons from occupied to unoccupied orbitals.

Excited states of molecular systems are derived from a variety of electron configurations. In organic systems the configurations responsible for the low-energy states generally involve π–π^* and n–π^* states. The π–π^* states are derived from the promotion of an electron from a π-bonding to a π-antibonding orbital (e.g., anthracene). The n–π^* excited states are derived from the promotion of an electron in a nonbonding orbital to a π^* antibonding orbital; an example is ketones, where an electron in one of the nonbonding oxygen orbitals is promoted to the antibonding π orbital between the carbon and oxygen atoms.

Metal complexes introduce more new states. The coordinating ligands can contribute low-lying π–π^* or n–π^* states. Splittings of the degenerate *d* orbitals by a nonspherical ligand environment can give rise to metal-localized *d*–*d* transitions in metal complexes with *d* electrons. In addition, there are charge-transfer transitions derived from the promotion of an electron from a metal-localized orbital to a ligand-localized orbital or from ligand to metal orbitals.

A comparison of Figs. 1 and 2 shows a remarkable difference between the molecular and the atomic emission spectra. The atomic spectrum is incredibly sharp; while the molecular spectrum is very broad and exhibits regular progressions. The atomic states are simple because of the absence of any other vibrational or rotational states. In contrast, large molecules have a large number of vibrational and hindered rotational states superimposed on the simple energy level diagrams of Fig. 3 and 4. Furthermore, the molecule can exist in a large number of conformations in the solvent matrix, each with a characteristic absorption and emission. These factors result in a broadening of the molecular transitions.

A more complete energy level diagram is given in Fig. 1, where a dominant molecular vibration has its energy levels superimposed on each electronic state. This figure shows why the absorption and emission tend to overlap with, and be mirror images of, one another. A well-defined vibrational progression is characteristic of systems in which there is little distortion on going from the ground to the excited state. Where distortions occur, the vibrational structure is smeared out and the emission band is broadened and red-shifted.

An interesting type of hybrid atomic molecular system is exemplified by rare earth ions in crystal lattices or in molecular complexes. The electronic configuration of rare earth ions is (core)$(f)^n$ ($n = 0$ to 14). The lowest excited states are derived, not by orbital promotions, but by rearrangement of the electrons within the f shell. Furthermore, these f electrons are so well shielded within the atom that the excited state transitions are very insensitive to the environment around the atom. Thus, the transitions of rare earth elements look more like atomic transitions than molecular ones. Atomic state classifications are used because of the small perturbations on the atomic transitions.

Figure 5 shows emission spectra for a neodymium(III)-doped glass at room- and liquid-nitrogen temperatures. The quasi-atomic line spectra are very clear, especially at 77 K; compare these spectra with Figs. 1 and 2. Emission narrowing on cooling is common and one of the reasons why emissions are frequently studied at low temperatures. In this case the 77 K emission is only 0.16 nm wide.

E. Excimers and Exciplexes

Even if one fully understands ground state chemistry, one may find surprises in the excited state manifold, where totally unexpected species suddenly appear. A classic example of this is encountered in electrically excited mixtures of Ar and F_2. There are no known stable Ar compounds. However, one sees an intense *ArF emission derived from

$$^*\mathrm{Ar} + \mathrm{F}_2 \rightarrow {}^*\mathrm{ArF} + \mathrm{F} \quad (1)$$

where asterisks denote excited species. The reason for the existence of *ArF but not of ArF is that *Ar is not the same chemical species as Ar;

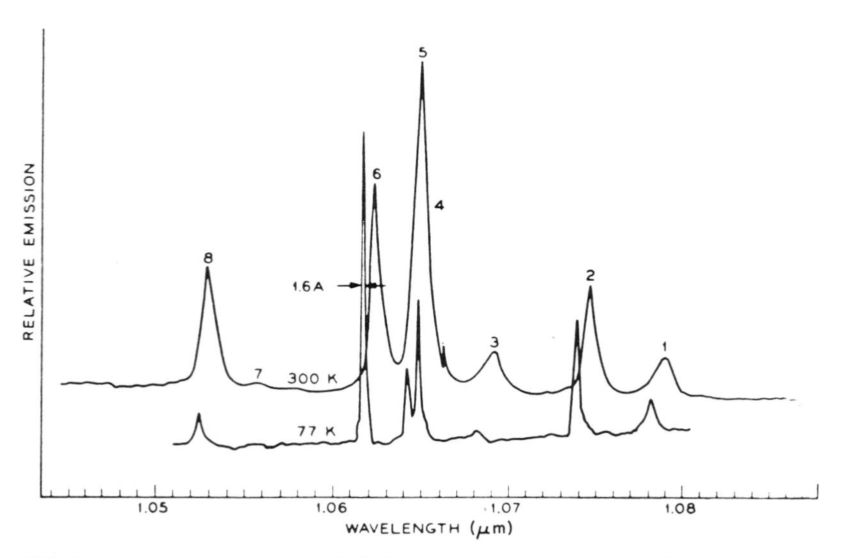

FIG. 5. Emission spectra of Nd^{3+} in $Y_3Al_5O_{12}$ at room and liquid-nitrogen temperatures. [Reprinted with permission from Van Uitert, L. G. (1966). *In* "Luminescence of Inorganic Solids" (P. Goldberg, ed.), p. 516, Academic Press, New York.]

it has a completely different electronic configuration. Ar has a closed-shell [Ne]$(3s)^2(3p)^6$ electronic configuration with no free bonding electrons and so does not form ArF. The lowest excited state of *Ar, however, is [Ne]$(3s)^2(3p)^5(4s)^1$, which has unpaired *s* and *p* electrons. Chemically this configuration is very similar to that of potassium metal; a free *s* electron is bound to a singly charged core. Not surprisingly the bonding in *ArF is ionic and very much like that of KF.

Rare gas chemistry can be even more complex. At high pressures the rare gas halide can react to give triatomic species

$$^*\text{RgX} + \text{Rg} \rightarrow {}^*\text{Rg}_2\text{X} \quad (2)$$

where Rg stands for a rare gas and X for a halogen. Figure 6 shows the emission spectra of several triatomic rare gas compounds.

Excimers and exciplexes are chemically stable excited state species that can exist only in the excited state and do not have a corresponding ground state form. Excimers are *exc*ited state d*imers* formed by the association of two identical subunits. Exciplexes are *exci*ted state com*plexes* formed of two distinctly different subunits. *ArF is an exciplex. If the two reactants are chemically similar, the complex is a mixed excimer.

The classic excimer, and first to be discovered, was pyrene. Pyrene exhibits no tendency to associate with itself in the ground state. At higher concentrations, however, excited state pyrene associates strongly with a ground state pyrene to form the pyrene excimer, which exhibits an intense emission that is shifted to the red of the monomer emission.

Figure 7 shows the exciplex emission of pyrene on silica gel. The high-energy structured emission is the pyrene monomer, while the broad low-energy emission derives from exciplexes formed from closely located adsorbed pyrenes.

An interesting dimeric emission arises in the chemiluminescent reactions of excited state singlet oxygen. Under the chemical conditions of generation, high concentrations of $^1O_2$ exist. Dimerlike species pool energy to produce higher-energy emissions:

$$^1\text{O}_2 + {}^1\text{O}_2 \rightarrow ({}^1\text{O}_2)_2 \rightarrow 2\ \text{O}_2 + h\nu \quad (3)$$

While the emission of $^1O_2$ is in the IR, the "dimol" emission is a spectacular red. Combination bands arising from states derived by simultaneous excitation on both oxygen molecules are observed.

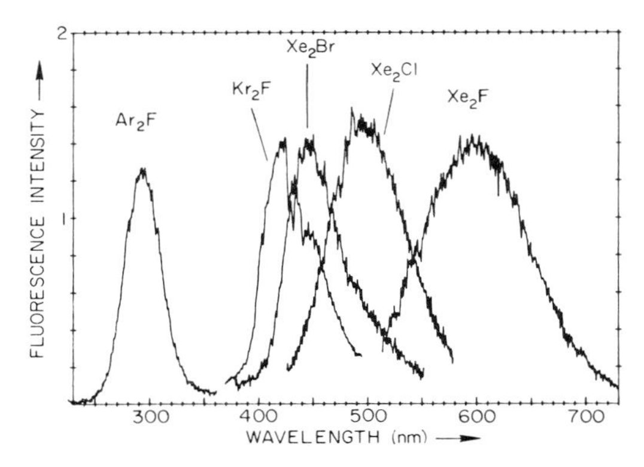

FIG. 6. Fluorescence spectra of rare gas trimers in electron-beam-excited high-pressure rare gas–halogen mixtures. [Reprinted with permission from Huestis, D. L., Marowsky, G., and Tittel, F. K. (1983). *In* "Excimer Lasers—1983," AIP Conference Proceedings No. 100, Subseries on Optical Science and Engineering No. 3 (C. K. Rhodes, H. Egger, and H. Pummer, eds.), p. 240, American Institute of Physics, New York.]

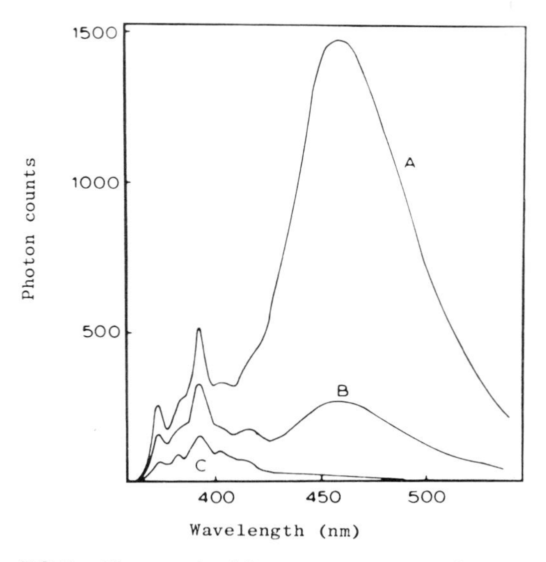

FIG. 7. Time-resolved fluorescence spectra of pyrene adsorbed on solid silica. The spectra are for the following delays after a short excitation pulse: (A) 7–52 nsec, (B) 108–162 nsec, and (C) 347–404 nsec. The structureless 460-nm band is the pyrene excimer, and the structured high-energy emission is the monomer. The excimer has a short lifetime, which enhances its emission at short times. [Reprinted with permission from Ware, W. R. (1983). *In* "Time-Resolved Fluorescence Spectroscopy in Biochemistry and Biology" (R. B. Cundall and R. E. Dale, eds.), p. 53, Plenum Press, New York.]

Although not an exciplex, a related area of excited state behavior is acid–base reactions. Again because of the differences in electronic configurations of the ground and excited states, the ground and excited states can have greatly different pK values. That is, the reaction

$$H^+ + {}^*A^- \leftarrow pK_a{}^* \rightarrow {}^*HA \tag{4}$$

has a different pK than the ground state pK_a. This excited state pK_a, or pK_a*, frequently differs from the ground state pK_a by 5 to 10 pK units. Thus, a strong acid may become a very weak acid in the excited state, or a weak acid may become a super acid. Bases can behave similarly. As protonated and unprotonated forms of a species can exhibit very different properties, emissions of species exhibiting excited state acid-base chemistry show remarkable and, to the uninitiated, unexpected variations in spectra with pH.

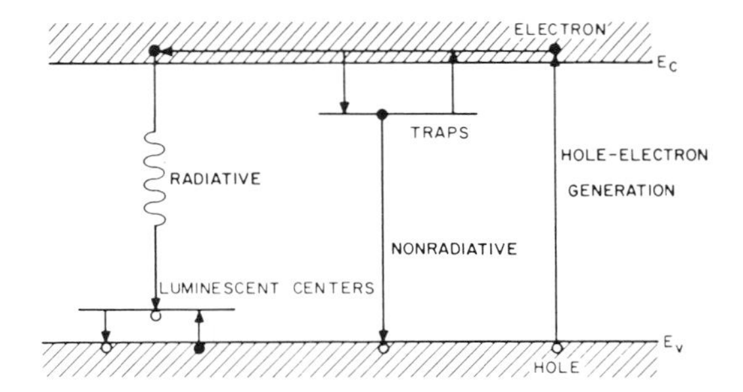

FIG. 8. Representation of radiative and nonradiative processes in solids. The lower striped area is the valence band, and the upper is the conduction band. [Reprinted with permission from Sze, S. M. (1981). "Physics of Semiconductor Devices," 2nd ed., Wiley, New York.]

F. Band States

In strongly interacting solids, the concept of localized atomic or molecular states fails. Orbitals on adjacent atoms or molecules can interact so strongly that the molecular orbitals of the composite system must be described as extending over the entire lattice rather than being localized on a specific atom or molecule. A consequence of this is that there are no longer discrete states of the system, but a series of very closely spaced levels that make up bands. The lower levels are valence bands, and the upper unoccupied levels are conduction bands. Conduction occurs when the electrons in the valence band are promoted to the conduction band, where they are free to move through the lattice. Depending on the energy gap (forbidden band) between the filled valence band and the conduction band, the solid is an insulator, semiconductor, or conductor. A small gap permits electrons to be thermally excited to the conduction band at room temperature for conductors or semiconductors. In insulators, the gap is too large to yield any appreciable concentration of charge carriers. Luminescence seems to be restricted to semiconductors and insulators.

Excitation of an electron from the valence to the conduction band produces an excited state of the system, which can be treated as any other type of excited state and can give rise to luminescence. Electron promotion leaves behind a positively charged center, or hole. Both the electron and the hole can freely move through the solid and are responsible for photoconductivity. This system is shown schematically in Fig. 8.

The electron and the hole can undergo secondary processes that influence emission. Both can be trapped at sites in the lattice. Traps may be defects in the lattice that arise from missing ion sites, interstitial ions, or replacement of normal lattice ions with impurities that may introduce additional lattice defects. Furthermore, other species may function directly as traps if they are easily oxidized or reduced. Figure 8 also shows a schematic representation of electron and hole trapping. An electron is trapped by dropping from the conduction band into a potential energy well. A hole is trapped by pulling an electron from an oxidizable site.

Trapped holes or electrons can be well-defined species with their own spectroscopy, including characteristic absorption and emission spectra. Such systems are called *color centers* because of their characteristic colors. For example, an electron trapped in a halide ion vacancy in an alkaline halide lattice is called an F center. Sodium chloride has a yellow F center, potassium chloride a magenta one, and potassium bromide a blue one. These F centers can undergo reasonably efficient low-temperature emission. Doping of halide matrices with the activator can produce a number of new types of centers involving such species as Ag^{2+}, Ag^0, Ag_2^+, and Ag_2^0.

Emission can result from direct recombination of the conduction electron with the hole. More commonly, luminescence in band solids arises from impurities. Electrons in the conduction band can relax back to a hole close to the activator; the energy released excites the activator, which luminesces. The transition to the trap itself may be radiative. Alternatively, if the hole is trapped by oxidizing the activator, the recombination is a reduction of the center; the chemical energy released by this reaction can lead to excitation of the center. This is a form of chemiluminescence.

IV. Methods of Studying and Characterizing Excited States

Excited state processes are usually studied and characterized by the following general approaches: emission and excitation spectra, luminescence efficiencies, polarization, temporal behavior, temperature effects, interactions with other species, double-resonance methods, fluorescence line narrowing, spin echos, transient gratings, and site-selective spectroscopy. This information is correlated with absorption processes. Several of the most common approaches are discussed.

An emission spectrum is the relative intensity of emission as a function of wavelength. Data are generally acquired by scanning through the emission with a monochromator. The relative intensity is measured with an optical detector such as a photomultiplier tube or semiconductor detector. These directly obtained data are not corrected for the transmission characteristics of the optics or for the variations in the detector's sensitivity with wavelength. Uncorrected spectra may bear little resemblance to true luminescence spectra and must be corrected, usually by calibrating the response of the system with a source of known spectral distribution such as a standard lamp.

Excitation spectra are obtained by measuring the relative emission intensity at a fixed wavelength while scanning the excitation source. For weakly absorbing solutions, the amount of light absorbed will be directly proportional to the sample absorbance. If the emission efficiency is independent of excitation wavelength, then the excitation spectrum will match the absorption spectrum. As with emission spectra the directly obtained spectra are distorted by the variations in light output of the source versus wavelength. Data are corrected by measuring the excitation source intensity as a function of wavelength.

Excited state lifetime measurements are extremely useful diagnostic tools of excited state processes. The standard method is to excite the sample with a pulse that is shorter in duration than the decay phenomena and then watch the relaxation by monitoring the luminescence. It is also possible to monitor the decay by following the excited state absorption spectrum or the electron spin resonance spectrum. Using mathematical tricks, one can also measure lifetimes appreciably shorter than the excitation. For extremely short decays, picosecond pulse probe techniques are used. Here, a sample is probed using an optical delay line where time between excitation and monitoring is set by adjusting the distance the probe plus travels before striking the sample. Lifetimes in the low nanosecond range are readily measured using emission relaxation methods, while picosecond methods can measure subpicosecond decays.

In terms of the rate constants and paths indicated in Fig. 3, the fluorescence and phosphorescence lifetimes are given by

$$\tau_f = 1/(k_f + k_{qS} + k_{isc}) \tag{5a}$$

$$\tau_p = 1/(k_p + k_{qT}) \tag{5b}$$

where the subscripts f and p denote the fluorescence and phosphorescence processes respectively, q denotes a quenching path, and S and T denote processes from the singlet and triplet states, respectively; k_{isc} is the rate constant for intersystem crossing between S_1 and the triplet manifold.

Luminescence quantum efficiencies (photons emitted per photon absorbed) are given by

$$\phi_f(S_1) = k_f\tau_f \tag{6a}$$

$$\phi_p(S_1) = \phi_{isc}k_p\tau_p \tag{6b}$$

$$\phi_p(T_1) = k_p\tau_p \tag{6c}$$

$$\phi_{ic} = k_{ic}/(k_{ic} + k'_{qs}) \tag{6d}$$

$$\phi_{isc} = \phi_p(S_1)/\phi_p(T_1) \tag{6e}$$

where ϕ_{ic} is the efficiency of internal conversion from the upper excited singlet to S_1 and ϕ_{isc} is the efficiency of intersystem crossing between S_1 and T_1. The parenthetical S_1 and T_1 denote the state into which the photons are absorbed.

Note that, by measuring the fluorescence efficiency on excitation into the emitting and any upper state, one can determine the internal conversion efficiency. By measuring the phosphorescence efficiency on excitation to S_1 and T_1, one can determine the intersystem crossing efficiency. Furthermore, from the luminescence yields, the lifetimes of each state, and the intersystem crossing efficiency, one can determine k_f, k_p, k_{isc}, and the k_q's, which largely define the dynamics of the lower excited state process.

An example of these methods is illustrated in Fig. 9, which shows the corrected excitation and absorption spectra as well as the relative luminescence yield of *trans*-dibromotetra(pyridine) rhodium(III) bromide.

The broad-band red emission exhibits a lifetime of 500 μsec, which clearly indicates a spin-forbidden phosphorescence. There is no fluorescence; therefore, $k_{qS} + k_{isc} \gg k_f$. The relatively intense bands at 25 and 26 kK (1 K = 1 cm^{-1}) correspond to an $S_0 \rightarrow {}^1(d\text{–}d)$ transition, where *d–d* indicates an excited state derived within the metal-localized *d*-orbitals. The much weaker 20-kK band is the spin-forbidden S_0 to $^3(d\text{–}d)$ excited state transition and is the inverse of the 15 kK emission. The intense band starting at 29 kK is another metal-localized state. If relaxation from all levels to the emitting level proceeds with 100% efficiency (i.e., $\phi_{ic} = \phi_{isc} = 1$), then the excitation spectrum should match the absorption spectrum. In this example, the invariance with wavelength of the emission yield on excitation into different singlet states and the emitting triplet level leaves no doubt as to the unity efficiency relaxation of all upper levels to the emitting level. Radiationless rate constants for deactivation of the emitting level are also easily calculated and can be correlated with theories.

Polarized emission spectra can be obtained in either crystals or randomly oriented samples. The direction and degree of polarization of the emission relative to the polarization of the excitation beam are recorded. From the variations of this polarization as a function of the absorption bands excited, one can frequently infer the molecular axis along which the emission originates or the absorptions arise.

An exceptionally powerful tool for unraveling the dynamics of excited state geometry changes is time-resolved polarization anisotropy. Basically, one looks at the degree of emission polarization following excitation by a short polarized excitation pulse. If the molecules stay in a fixed orientation during the emission, the polarization will remain constant during the decay. If the molecule rotates during the emission, the degree of polarization will fall as the originally ordered system becomes randomized. From the kinetics of the depolarization one can map out the nature and the rates of such depolarization processes as energy transfer and localized or whole molecule rotations. This method is invaluable for studying the dynamics of motion of large biomolecules.

Temperature effects on luminescence efficiencies, lifetimes, and spectral distributions are valuable diagnostic tools for finding the energies of excited states and for exploring excited state relaxation processes. For example, Fig. 10 shows the excited state lifetime and emission yield for [Ru(bpy)$_3$]$^{2+}$ (2,2′-bipyridine). The odd temperature dependence of the emission can be ascribed to the existence of three states, which are all in thermal equilibrium with one another. Each state has a characteristic radiative and ra-

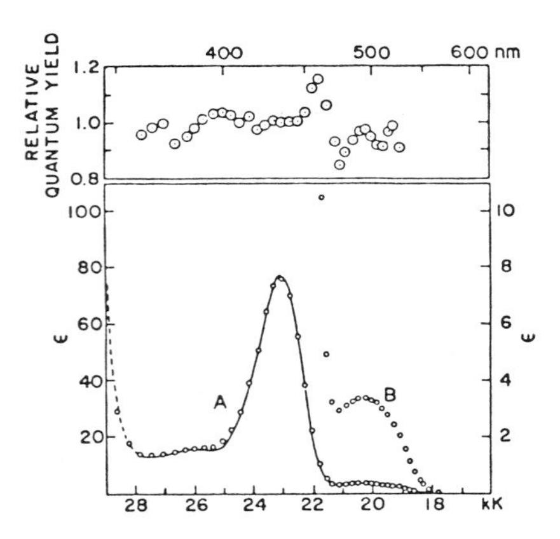

FIG. 9. Absorption (solid line) and excitation spectrum (○) of *trans*-[RhBr$_2$py$_4$]$^+$ (py = pyridine) at 77 K. Curve A is for the scale on the left and curve B for the scale on the right. The excitation spectrum is normalized to the absorption maximum. The relative emission efficiency as a function of excitation energy is in the upper graph. [Reprinted with permission from Demas, J. N., and Crosby, G. A. (1970). *J. Am. Chem. Soc.* **92,** 7626. Copyright 1970 American Chemical Society.]

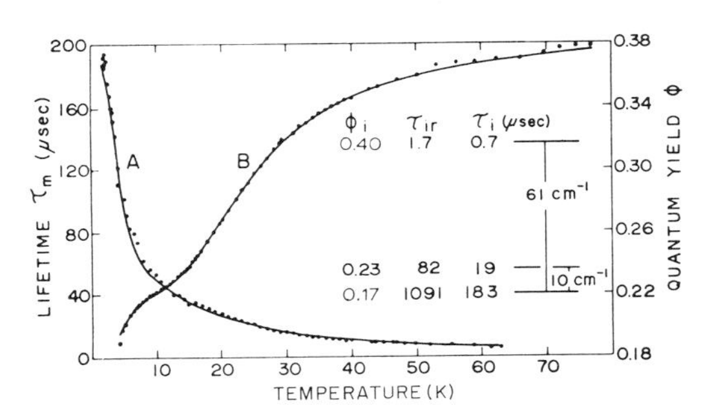

FIG. 10. Effects of temperature on the lifetime (A) and luminescence efficiency (B) of [Ru(bpy)$_3$]$^{2+}$ (2,2-bipyridine) in poly(methyl methacrylate). The solid lines are the calculated curves using the three-level energy diagram shown. For each sublevel, the efficiency as well as the radiative and nonradiative lifetimes used to fit the curves are shown. [Reprinted with permission from Hagrrigan, R. W., Hager, G. D., and Crosby, G. A. (1973). *Chem. Phys. Lett.* **21,** 487.]

diationless lifetime as well as different emission yields. The variation in lifetime with temperature arises from the variation in Boltzmann population of the three levels. The fitting of these temperature curves permitted determination of the energy spacing of the levels and their lifetimes. State assignments were then based on these results.

Another interesting temperature-related effect is E-type delayed fluorescence. This is a long-lived emission the spectrum of which is indistinquishable from the prompt fluorescence. Delayed fluorescence arises by thermal back-population of the emitting state. Thus, the triplet functions as a storage state, and the fluorescence tracks the triplet concentration.

V. Processes Affecting Luminescence

A. Intramolecular Processes

Intramolecular excited state processes and methods of studying them have been discussed in Sections III and IV. A few additional aspects need be mentioned only briefly. Intramolecular processes are very sensitive to the environment and to excited state types and ordering. For example, n–π^* excited states exhibit very efficient intersystem crossing, while π–π^* states do not. Thus, ketones such as benzophenone, the lowest excited states of which are n–π^*, exhibit almost pure phosphorescences, while biphenyl, the lowest excited states of which are π–π^*, exhibits predominantly fluorescence because of less efficient intersystem crossing. Similarly, molecules that phosphoresce strongly because the lowest excited state is n–π^* can be converted to strongly fluorescent materials by raising the n–π^* state above the fluorescent π–π^* state. This can frequently be done by using a hydrogen-bonding solvent, which ties up the nonbonding electrons and makes their removal energetically more difficult.

Rare earth chelate complexes with organic ligands exhibit a form of intramolecular energy transfer. Rare earth compounds generally absorb weakly, but strongly absorbing compounds can be made by coupling a strongly absorbing ligand to the metal ion. If the energy levels on the ligand are above the emitting rare earth level, then efficient energy transfer from the ligand-localized excited state to the metal state can occur. Indeed, in many of these systems, only metal emissions are observed, even though all of the excitation is into the organic ligand.

Another example is radiationless coupling of energy from a species to adjacent or bound solvent molecules. H_2O is a much more efficient deactivator of many molecular systems than D_2O, because of the higher vibrational frequencies that more easily bridge the gap between the ground and excited states. For example, crystalline Eu^{3-} and Tb^{3+} hydrates are more than 10 times less luminescent than the corresponding deuterates. This behavior has been used for counting the number of water molecules around metal ions in proteins or the average solvent exposure of metal complex sensitizers in organized media such as micelles.

B. Bimolecular Processes

Excited states that persist for any appreciable time are also susceptible to quenching by other chemical species. Excited state deactivators are called quenchers. Quenching can occur by a variety of processes,

$$^*D + Q - k_2 \rightarrow D + Q + \Delta \qquad (7a)$$

$$D + {}^*Q \qquad (7b)$$

$$D^+ + Q^- \qquad (7c)$$

$$D^- + Q^+ \qquad (7d)$$

$$\text{reaction products} \qquad (7e)$$

where k_2, is the bimolecular rate constant for deactivation of the excited state. If more than one deactivation pathway is present, k_2 is the sum of all processes affecting the state.

Equation (7a) denotes catalytic deactivation of the excited state. For example, external heavy atoms can increase spin–orbit coupling in a luminescent molecule during a collision and enhance intersystem crossing from a luminescent singlet to a nonemissive triplet or quench a triplet state back to the ground state.

Equation (7b) indicates energy transfer from a donor to an acceptor molecule. Efficient energy transfer quenching must generally be an exothermic reaction. The photodynamic effect is an example of bimolecular energy transfer. A dye in the presence of both light and oxygen will kill organisms. Chemically very reactive singlet oxygen is formed by energy transfer deactivation of dye triplet states. The chemically reactive singlet oxygen then kills the organism.

Electron transfer quenching denoted by Eqs. (7c) and (7d) can be either oxidative or reductive depending on whether the excited species is oxidized or reduced. Electron transfer forms the basis of a number of solar energy conversion schemes and can occur in the excited state even though the ground state species are thermodynamically stable. This again points out the great differences in chemistry that can arise between ground and excited state reactions, and stresses the chemical uniqueness of the excited state relative to the ground state.

Finally, Eq. (7e) indicates any other type of chemical reaction. Examples include the addition of singlet oxygen to an olefin or the hydrogen abstraction of a proton from a protic solvent by a triplet ketone.

The kinetics of Eqs. (7) yield two Stern–Volmer equations relating emission intensity or lifetime to the concentration of the quencher,

$$(\phi_0/\phi) - 1 = K_{SV}[Q] \quad (8a)$$

$$(\tau_0/\tau) - 1 = K_{SV}[Q] \quad (8b)$$

$$K_{SV} = k_2\tau_0 \quad (8c)$$

where the ϕ's are emission intensities and τ's are excited state lifetimes. The subscript 0 denotes the value in the absence of quencher. The K_{SV}'s and k_2's are Stern–Volmer and bimolecular quenching constants, respectively. Thus, plots of the experimentally determined left-hand side versus [Q] provide K_{SV}'s or k_2's if the unquenched lifetime is known. The k_2's, in particular, provide a great deal of fundamental information concerning the interaction of excited states with quenchers. Figure 11 shows typical lifetime and intensity Stern–Volmer plots. The two plots are the same within experimental error if all quenching is diffusional. Ground state or associational deactivation provides different plots.

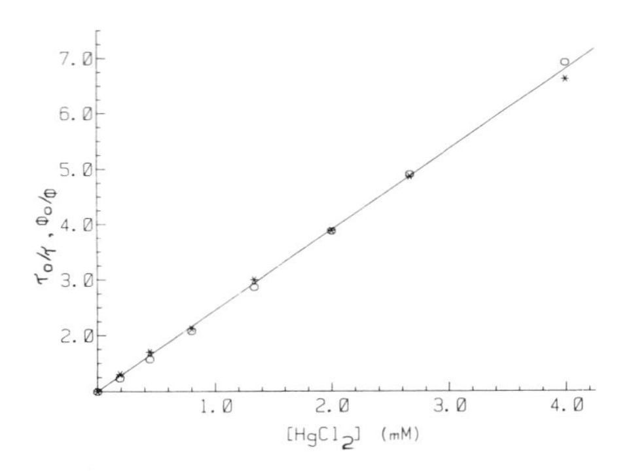

FIG. 11. Lifetime (*) and intensity (○) Stern–Volmer quenching plots for deactivation of $[Ru(phen)_3]^{2+}$ (phen = 1,10-phenanthroline) by $HgCl_2$. [Reprinted with permission from Hauenstein, Jr., B. L., *et al.* (1984). *Inorg. Chem.* **23,** 1101. Copyright 1984 American Chemical Society.]

If two partners in a collision are excited, new excited state reactions are observed. For example, if two excited triplets collide, one can have an energy-pooling, spin-conserving reaction that yields an excited singlet state. Also, two excited singlets can annihilate one another to form ground state species:

$$T + T \rightarrow {}^*S + S \quad (9a)$$

$${}^*S + {}^*S \rightarrow S + S \quad (9b)$$

Triplet–triplet annihilation is a common form of triplet decay under intense excitation conditions where high triplet concentrations exist. Because of the long lifetimes of the triplet state, triple–triplet annihilation can also yield a *P*-type delayed fluorescence, which can persist for many microseconds after the cessation of irradiation. Singlet–singlet annihilation is less common because of the difficulty of achieving sufficiently high concentrations of the short-lived species. However, with the advent of modern high-flux lasers it is readily observed and can be a significant nonradiative pathway.

Energy transfer does not require that the molecules be in contact with one another. Resonance energy transfer can occur at distances far exceeding the physical contact distance if the emission spectrum of the donor overlaps the absorption of the acceptor and the acceptor absorption spectrum is intense or highly allowed. Dipole–dipole resonance energy transfer, which is also known as Förster transfer, can occur at distances approaching 100 nm in favorable cases, and 30 to 50-nm transfers are common. The energy collection in the photosynthetic unit generally consists of hundreds of antenna chlorophylls that collect the energy and then transfer it by a resonance mechanism to the active chlorophyll.

In crystals composed of a single component, long-range energy transfer can occur by a contact mechanism. The close proximity of the molecules permits energy to hop from one molecule to its neighbor. By a series of hops the excitation can sample a very large volume. If there are any quenchers in this volume, the energy can be transferred to the quencher, and no luminescence of the major component is observed. The classic example of this is the phosphorescence of naphthalene, where it was necessary to reduce the concentration of β-methylnaphthalene to below 10^{-7} mol fraction to see the host emission.

C. Stimulated Emission

Light absorption occurs if the photon energy matches the energy gap between the ground and excited states. The photon, in effect, induces a transition between the ground and excited states with the loss of a photon from the radiation field and the production of an excited state. However, the inverse process is also possible. If the excited molecule is exposed to photons the energies of which exactly match the energy gap between the excited state and a lower energy state, the photon induces a transition between the upper and lower states. The net result is the addition of a photon to the radiation field rather than a loss. This process is called stimulated emission. Not only is the photon of the same energy as the stimulating photon, but it is also of the same phasing and is traveling in the same direction as the original photon.

Stimulated emission is not generally observed because, for it to occur efficiently, a significant fraction of the molecules must be in the excited state. Furthermore, the population of the upper state must have a higher-population than the terminating state (population inversion). Because of the symmetry of the absorption and the stimulated emission process, absorption will always be more efficient than stimulated emission if the terminal state concentration exceeds that of the upper state.

If stimulated emission occurs in a resonant cavity with a high degree of optical feedback, very intense highly directional monochromatic radiation results. Such systems are called lasers (light amplification by stimulated emission of radiation) and have countless practical and fundamental applications including surveying, weaponry, excited state lifetime determinations, and luminescence studies.

D. Multiphoton Processes

At the low fluxes obtained with most conventional light sources, the only absorption processes generally noted involve single photons. With the high fluxes available from lasers, however, nonlinear multiphoton processes occur. At high fluxes the absorption criterion is satisfied if the excited state transition energy equals the sum of the energy of two photons. Thus, absorption of two photons produces an excited state that is inaccessible by a single photon absorption. Indeed, efficient two-photon excitations can arise with light having energies less than the lowest excited state of the system.

Generally, because multiphoton processes have much lower cross sections than single-photon processes it is not possible to monitor the depletion of the beam as in an absorption experiment. However, luminescence from the excited species formed provides a powerful and extremely sensitive tool for monitoring the generation of excited states.

Figure 12 shows a two-photon excitation spectrum of naphthalene using visible excitation. The emission is monitored in the UV. The molecular absorptions are transitions occurring at half the indicated wavelength (twice the energy).

In addition to having different energy requirements from single-photon processes, the selection rules for two photon absorptions also differ. Thus, multiphoton absorptions provide a valuable tool for locating and studying states otherwise invisible by conventional one-photon spectroscopy.

Multiphoton absorptions are not limited to two photons of the same wavelength. If a sample is subjected to intense irradiation by two beams of different colors, absorption can also occur if the sum of the energies of the two different photons matches the energy of a transition. Even more complex schemes can arise if three or more photons of different wavelengths are required to induce transitions between states to arrive at the final monitored state.

E. Photon-Stimulated Emission and Quenching

Quite unrelated to simultaneous two-photon excited luminescence is a form of sequential

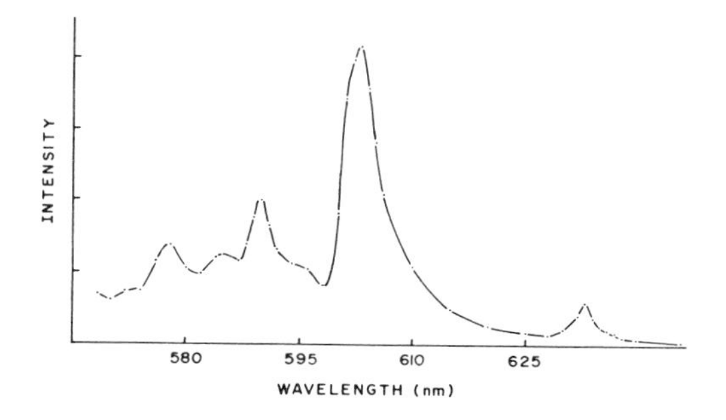

FIG. 12. Two-photon excitation spectrum for naphthalene. The absorption is into states that occur at half the indicated wavelength. Emission is monitored in the ultraviolet. [Reprinted with permission from Wirth, M. J., and Lytle, F. E. (1978). "New Applications of Lasers," ACS Symposium Series 85 (G. M. Hieftje, ed.), p. 24. Copyright 1978 American Chemical Society.]

multiphoton excited luminescence or quenching. This type of behavior is especially prevalent in solid phosphors. Initially, a high-energy photon or charged particle excites the system to a metastable state. This excited system may be a true excited state, but more often it is an electron trapped in a high-energy site. Optical excitation with different colored light can lead to a stimulated emission if the electron is ejected from the trap by light of an energy matching the gap between the trap and the conduction band (Fig. 8). Once in the conduction band the electron is free to recombine with a suitable center to produce luminescence. Generally, because of the small gaps between the traps and the conduction band, emission of this type is IR-stimulated.

Similarly, photons of suitable energy can induce transitions of the excited electron to lower nonemissive excited states. This results in quenching of subsequent emission processes. Because of the larger energy gaps to quenching levels, photon-stimulated quenching usually requires higher-energy photons than does photon-stimulated emission. Indeed, it is not uncommon for the source that excites the phosphor also to be capable of quenching the luminescence.

IR-stimulated emission has potential application in IR detectors. The most sensitive optical detectors are photomultipliers, which do not respond to farther IR photons. However, if a suitable phosphor is pumped up and then exposed to the IR source, stimulated visible photons can be detected by a photomultiplier. A related application is in light-emitting diodes, where an IR-emitting diode is used to pump up a visible-emitting phosphor by sequential photon absorptions.

VI. Types of Luminescence

Luminescence is categorized by the method of initiation. Although each type of luminescence is described and one or more applications given, the subject is so broad that an exhaustive discussion is impossible.

A. PHOTOLUMINESCENCE

Photoluminescence arises when the emitting excited state is generated by the absorption of a photon. Excitation can be either a single or multiphoton process.

The applications of photoluminescence are legion. They can be as mundane as black-light posters or as esoteric as multiphoton processes for sophisticated state analyses of molecules or ions. A particularly powerful application of photoluminescence is in quantitative analysis. The excitation and emission spectra of materials vary greatly. For example, the characteristic excitation and emission spectra provide "fingerprints" of different crude oils and have been used to trace the sources of oil spills or illegal dumping. Furthermore, many nonluminescent materials can be made luminescent by suitable, and frequently highly selective, chemical reactions. Concentrations can then be determined from photoluminescence intensities.

A major advantage of luminescence over absorption methods of analysis is the frequently much greater sensitivity of the emission methods. This sensitivity enhancement relies in part on the nature of the measurement. Absorption measurements of weakly absorbing samples depend on being able to measure small differences in the intensities of large transmitted signals, which is an intrinsically difficult problem. On the other hand, in emission measurements one is looking for small signals on essentially zero backgrounds.

Emission measurements are so sensitive that it is possible to detect single atoms in the gas phase. Furthermore, advances in laser-based flow cytometry promise to make single-molecule detection in condensed media possible.

The selectivity of luminescence methods is best exemplified by fluoroimmunoassay (FIA), by which specific antigens can be detected even in a messy medium such as a serum. Radioimmunoassay (RIA) has dominated the field when the ultimate in sensitivity is required (pico- to femtomoles). The problems of working with and disposing of radioactive tracers have given impetus to the use of the somewhat less sensitive FIA. Furthermore, advances in flow cytometry and laser-based detection schemes promise to make FIA fully competitive with RIA.

B. CHEMILUMINESCENCE

In chemiluminescence (CL) the energy necessary for excited state generation is derived from the energy released in a chemical reaction. Excluding flames, probably the first man-made CL was the air oxidation of phosphorus, discovered by Brand in 1669. This discovery is the subject of a classic and beautifully detailed engraving (Fig. 13).

Three basic processes can initiate CL: (1) decomposition of a high-energy species to lower-

FIG. 13. Chemiluminescence on air oxidation of phosphorus. A portion of *The Discovery of Phosphorus* by the alchemist Brand (1669), engraving by William Pether (1775) after the painting by Joseph Wright, Fisher Collection, Pittsburgh. [Reproduction courtesy of the Fisher Collection, Fisher Scientific Co.]

energy ones, (2) exothermic reaction of two or more components, and (3) electron transfer reactions. These processes are represented by

$$A \rightarrow {}^*B \quad \text{or} \quad {}^*B + C \tag{10a}$$

$$A + B \rightarrow {}^*C + D \quad \text{or} \quad {}^*C \tag{10b}$$

$$A^+ + B^- \rightarrow {}^*A + B \quad \text{or} \quad A + {}^*B \tag{10c}$$

Atom transfer reactions are included in Eq. (10b). While this classification is convenient, it is frequently simplistic. Ternary reactions are possible, and electron transfer steps can also be involved in the first two types.

In all cases the energy released by the reaction must be adequate to generate the excited state. Although a small energy shortfall can be made up by drawing energy from the thermal pool, efficient CLs usually have large excess driving energies. Furthermore, there must be an efficient pathway for the ground state reactants to pass over to the surface corresponding to excited state products.

It is not necessary for one of the chemical products to be the luminescent species or even to have an available excited state. If the energy is released in a system in which a suitable acceptor is readily available, the energy can be transferred to the acceptor, which then luminesces.

Examples of Eq. (10a) CLs are the reactions of 1,2-dioxetenes. Dioxetenes can give rise to efficient CLs and are implicated in a variety of bioluminescences. Dioxetenes can be formed biochemically by oxidations of olefins or aromatic molecules:

$$>C{=}C< \; + \text{ oxidizer} \longrightarrow -\underset{|}{\overset{|}{C}}-\underset{|}{\overset{|}{C}}- \;\; (\text{O—O ring}) \tag{11}$$

The CL arises on cleavage to ketones, with one being excited and the other being in the ground state:

$$-\underset{|}{\overset{|}{C}}-\underset{|}{\overset{|}{C}}- \;\; (\text{O—O ring}) \longrightarrow \; >C{=}O^* + \; >C{=}O \tag{12}$$

The excited ketone can then luminesce or transfer energy to another luminescent species.

The long-lived afterglow in electrical discharges in nitrogen is an example of recombination CL [Eq. (10b)],

$$N + N + M \rightarrow {}^*N_2 + M \tag{13}$$

where M is a third body required to carry away excess energy to prevent the hot product from promptly redissociating. This reaction can give rise to pink, yellow, and blue afterglows. In spite of enormous effort, this system is far from being fully understood.

Flames are rich sources of CL. For example, the blue glow of hydrocarbon flames arises from C_2 as well as CN and OH radicals. At least some of these emissions are CLs.

Even the luminescence of many trace elements in flames can be attributed in part to chemical excitation of the metal by such processes as

$$H + OH + M \rightarrow {}^*M + H_2O \tag{14a}$$

$$H + H + M \rightarrow {}^*M + H_2 \tag{14b}$$

where M represents a metal atom.

Hydrazide CL, typified by the oxidation of Luminol, is given by the following reaction sequence,

$$Fe^{2+} + H_2O_2 \rightarrow Fe^{3+} + OH^- + \cdot OH$$

[R-substituted phthalhydrazide (C=O, NH, NH, C=O ring)] $\xrightarrow{OH^-}$ [R-substituted phthalhydrazide anion (C=O, N^-, NH, C=O ring)] $\xrightarrow{\cdot OH}$

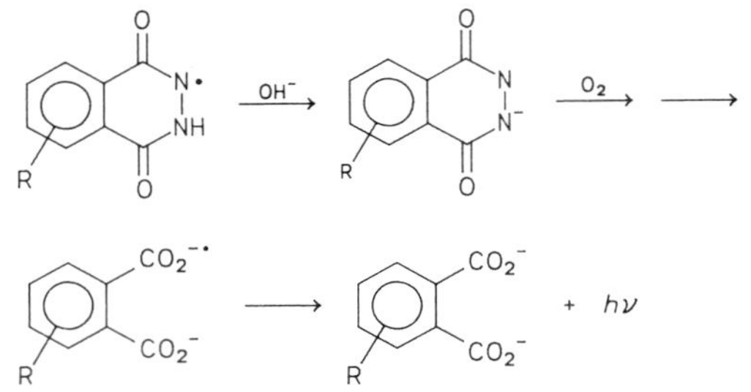

(15)

where R = NH_2 for Luminol. That the emissive species is the phthalate dianion is shown by the agreement between the CL and photoluminescence spectra of the product. Thus, although paths to the excited state are quite different, the final excited state is the same.

Only a few milligrams of Luminol produce an impressive light show, but the CL efficiency (photons emitted per number of molecules reacted) is, in fact, only about 1%. This result emphasizes the impressive quantity of light that a mole of photons represents.

The third class of CLs is initiated by electron transfer processes. This type is exemplified by the one-electron transfer reduction of $[Ru(bpy)_3]^{3+}$ (bpy = 2,2′-bipyridine), which yields a highly visible orange CL with hydrazine, water, and a variety of other reductants. Indeed, with $NaBH_4$ as the reductant, the spectacular emission is highly visible in a well-lighted room.

The complexity of simple CL systems is demonstrated by the simple $[Ru(bpy)_3]^{3+}$/oxalate system:

$$[Ru(bpy)_3]^{3+} + C_2O_4^{2-} \rightarrow [Ru(bpy)_3]^{2+} + C_2O_4^{-} \quad (16a)$$

$$[Ru(bpy)_3]^{3+} + C_2O_4^{-} \rightarrow {}^*[Ru(bpy)_3]^{2+} + 2\,CO_2 \quad (16b)$$

The initial reduction step does not appear to be energetic enough to excite the complex, and a secondary reaction with the energetic $C_2O_4^-$ free radical is required for excitation. This type of behavior, in which the initial step is not sufficiently energetic to initiate CL, is fairly common, and thus reactive radicals are common in the actual CL step.

A special case of electron transfer CL is that in which the reductant is the ultimate reductant, an electron. $[Ru(bpy)_3]^{3+}$ exhibits an intense CL on reaction with hydrated electrons:

$$[Ru(bpy)_3]^{3+} + e^- \rightarrow {}^*[Ru(bpy)_3]^{2+} \quad (17)$$

The efficiency of excited state production approaches 100%.

Applications of CL are diverse. There are commercial emergency lights that require only the breaking of a seal to permit the mixing of solutions. The long-lived CL is bright enough to read by.

One of the simplest and most sensitive NO analyzers utilizes the very efficient CL of the reaction

$$NO + O_3 \rightarrow {}^*NO_2 + O_2 \quad (18)$$

With very simple instrumentation, part-per-billion levels of NO can be accurately and precisely measured.

Another useful analytical application uses metastable $^*N_2$ to transfer energy to luminescent metals or molecules. Concentrations as low as 10^6 atoms per cubic centimeters are readily measured.

Chemical lasers use CLs. In a chemical laser the excited state population inversion is produced directly by chemical energy. One of the most efficient chemical lasers uses the hydrogen–fluorine reaction, which is initiated by the dissociation of F_2,

$$F_2 + h\nu \text{ (or electrical energy)} \rightarrow 2\,F \quad (19a)$$

$$H_2 + F \rightarrow H + {}^*HF \quad (19b)$$

$$F_2 + H \rightarrow F + {}^*HF \quad (19c)$$

where *HF is vibrationally excited HF produced with population inversion. The laser transitions are near-IR emissions arising between vibrational levels of HF.

C. Bioluminescence

Bioluminescence (BL) is a CL arising from living matter. Fireflies and the glow of disturbed microorganisms in a ship's wake are probably the best known. Although grouped separately, BLs are actually CLs in which the reactants are produced by and organized in living organisms.

Not surprisingly, one of the best known BLs is the firefly reaction, which involves the enzymatic oxidation of a luciferin. This reaction can have an incredible efficiency approaching 100%. The luciferin molecule and a number of synthetic analogs have been studied to elucidate the mechanism. The mechanism appears to involve a peroxide decomposition with free radical intervention.

$$\text{Photinus pyralis} \xrightarrow[\text{Enzymes}]{O_2} \text{(oxyluciferin radical)} + CO_2 + h\nu \quad (20)$$

Bioluminescence is much more pervasive than originally suspected. Low-level BLs are common to many fundamental and essential biological processes such as lipid peroxidation, intracellular redox processes, and catalase decomposition of poisonous H_2O_2.

Bioluminescence has had a crucial role in direct studies of cellular and biochemical processes. For example, there is a CL associated with the formation of the ultimate carcinogen benzo[*a*]pyrene-7,8-dihydrodiol 9,10-epoxide from benzo[*a*]pyrene. Also, a very sensitive CL assay for the benzo[*a*]pyrene-7,8-diol has been developed. Bioluminescence also continues to play a pivotal role in the development of the fundamental concepts of CL.

D. Electroluminescence

Electroluminescence (EL) is luminescence occurring on electron flow in a solid-state device or electrochemical cell. The mechanisms of these two ELs are quite different.

In the electrochemical cell, electrochemical oxidations and reductions occur. It is thus, not surprising that most ELs are electron transfer CLs, which are also called electrochemiluminescences (ECLs). The latter are usually produced by rapidly reversing the potential on an electrochemical cell. In this manner both oxidized and reduced species are generated near the same electrode. Electron transfer reactions between oxidizer and reductant are responsible for the ECL:

$$A + e^- \rightarrow A^- \quad (21a)$$

$$B - e^- \rightarrow B^+ \quad (21b)$$

$$A^- + B^+ \rightarrow {}^*A + B \quad \text{or} \quad A + {}^*B \quad (21c)$$

Except for chemical side reactions these systems do not run down and thus provide light indefinitely. Many such reactions have been studied as light sources, but light-emitting diodes have proved too simple, stable, and efficient to compete against. Among the ELs that have been studied the most thoroughly and for the longest time are the polycylic aromatic hydrocarbons in aprotic solvents. These systems are difficult to work with, however, because of the great reactivity of the high-energy organic radicals.

Metal complexes also provide efficient ECLs. For example, $[Ru(bpy)_3]^{2+}$ is readily oxidized to the +3 and reduced to the +1 oxidation states. Reaction of the +1 and +3 species gives a very efficient CL. Because of the robustness and relatively low reactivity of the +3 and +1 states, which are stable chemical species rather than free radicals, these inorganic systems are much easier to work with than the highly reactive free radical aromatic hydrocarbons.

An ECL need not be reversible. If the solution contains an electroinactive reactant that exhibits CL with a product of the electrode reaction, then one has an EL. The system will eventually run down when the reactant is consumed. Figure 14 shows the ECL of $[Ru(bpy)_3]^{3+}$ with oxalate. The Ru(III) is generated electrochemically at the electrode and yields a CL with the electroinactive oxalate. The spiking is caused by the pulsing of the electrode potential to form the Ru(III) with periods of low potential to permit more oxalate to diffuse up to the electrode before the next pulse. Also shown is the photoluminescence of the Ru(II) complex. The agreement between the ECL and the photoluminescence spectra leaves no doubt that the Ru(II) complex is the active species.

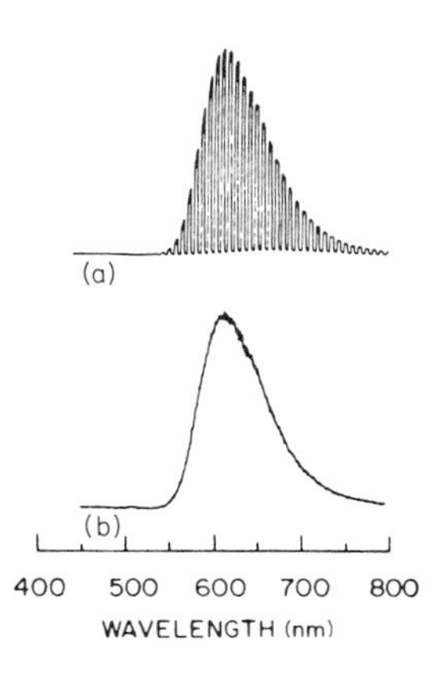

FIG. 14. Electrochemiluminescence (a) and photoluminescence (b) spectra of $[Ru(bpy)_3]^{2+}$. The ECL is from electrogenerated $[Ru(bpy)_3]^{3+}$ in the presence of oxalate. The spiking in A is caused by pulsing of the electrode potential. [Reprinted with permission from Rubinstein, I., and Bard, A. J. (1981). *J. Am. Chem. Soc.* **103,** 512. Copyright 1981 American Chemical Society.]

Solid-state EL devices fall into two categories: (1) film-type semiconductor devices and (2) solid-state diodes.

The earliest EL was obtained by subjecting a ZnS phosphor to an ac potential. The resultant current flow in the semiconductor excited the electrons to the conduction band. Recombination of the electrons and holes either directly or at traps gave rise to the luminescence.

Electroluminescent panels are a common source of low-level night lighting. Their efficiencies are much too low for general lighting.

Solid-state diode emitters are called light-emitting diodes (LEDs). An LED consists of a broad band-gap semiconductor diode formed by bonding a *p*-type (hole-rich) semiconductor material to an *n*-type (electron-rich) semiconductor. The resultant *pn* junction has both rectification and emitting properties.

LEDs exhibit light emission when forward-biased, which causes current flow through the junction. Current flow arises by the flow of holes from the hole-rich *p* material to the *n* material and by the flow of electrons from the *n*- to the *p*-type semiconductor. The electrons, for example, are now in the conduction band of the *p* semiconductor material, which has recombination sites suitable for radiative transitions from the conduction to the valence band. The light emission processes arise from direct band-gap radiative recombination with these holes or with suitable sensitizer centers. A similar fate potentially awaits the holes that traverse the junction and find themselves in an environment rich in conduction electrons. However, radiative recombination in the *p* semiconductor is more efficient, and LEDs are physically built so that emitted radiation can most freely escape from the *p* material.

Commercially available LEDs are green, yellow, red, and different ranges of IR. Blue-emitting devices have been built. Figure 15 shows typical emission spectra for different LEDs. The emission wavelength is controlled by the band gap of the semiconductor material. By supplying adequate optical feedback, solid-state LED lasers can be made. At this time only IR lasers are commercially available.

LEDs are almost universally used as state indicators and numeric displays on consumer and commercial electronics. The lower power consumption of liquid crystal displays, however, is resulting in their superseding LEDs except where self-luminosity is required.

Electrochemiluminescence devices are not commercialized. They do, however, have applications in some analytical work. The ECL of $[Ru(bpy)_3]^{3+}$ with oxalate is a sensitive analytical method for oxalate.

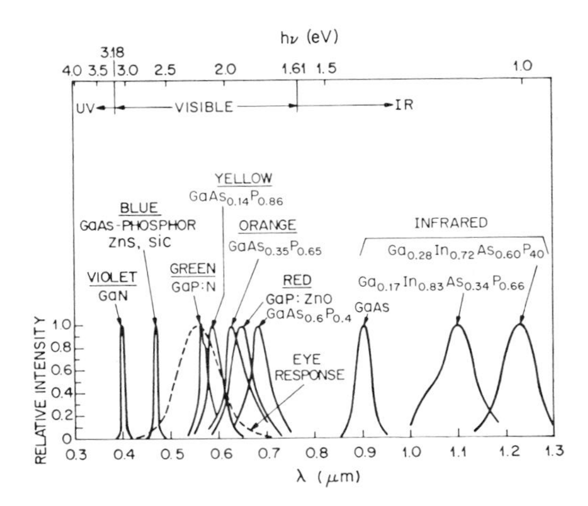

FIG. 15. Emission spectra of light-emitting diodes. The spectral sensitivity of the human eye is shown for comparison. [Reprinted with permission from Sze, S. M. (1981). "Physics of Semiconductor Devices," 2nd ed., John Wiley & Sons, New York.]

E. Cathodoluminescence

Cathodoluminescence is luminescence arising in gas or low-pressure electrical discharges. The luminescence can arise either in the electrical discharge itself or on the electrode or target.

Luminescences from electrodes result from excitation of electrode material from the impact of electrons or charged particles on the surfaces. Electronic excitation can result directly from inelastic collisions of particles with atoms or molecules. Chemical bonds can be broken, and high-energy electrons can be ejected. These high-energy secondary electrons can produce other electronic excitations, chemistry, or ionizations. The recombination of electrons with oxidized sites can release enough energy to excite either the product of the electron transfer reaction or excitable species in the neighborhood. Chemical reactions of decomposition products can produce CL.

Gas-phase excitation arises from the same processes as surface excitation. Inelastic collisions can excite molecules or atoms, and species can be decomposed or ionized. Reactions between products can yield CL or new species that can be excited, in turn giving rise to new emissions.

Gas-phase emission need not be limited to species originally in the gas phase. The energy transferred to the surface by electron or ion impact can eject or sputter surface material. This sputtering material, once introduced into the discharge, can be excited by the processes just described.

Secondary processes such as energy transfer and excited-state–excited-state processes can also occur. These affect the nature and efficiency of the emission.

Cathodoluminescence forms the basis of innumerable practical devices. Color and black-and-white televisions are examples of phosphors excited by electron impact on solids. The ubiquitous hollow cathode source of elemental emission lines uses electrode sputtering to introduce the desired element into the discharge. Advertising signs and discharge lighting are very common.

The fluorescent lamp represents an interesting application of both cathodoluminescence and photoluminescence. The discharge is in a low-pressure rare gas with a trace of mercury vapor. The 254-nm atomic mercury emission dominates the optical output (Fig. 2). The inside of the discharge tube is coated with a UV-excitable, visible-emitting phosphor, which transforms the invisible UV into usable visible light. By changing the phosphor, the emission color tint can be varied, as in the reddish greenhouse lamp. The long-wavelength black light works similarly except that the phosphor emits in the 300- to 400-nm region.

F. Radioluminescence

Radioluminescence arises from interactions between ionizing radiation and matter. This radiation may be particles, such as α or β particles, and cosmic rays or photons, such as X rays or γ rays. Materials exhibiting radioluminescence are called scintillators, named after the scintillations, or bursts of light, that arise when radiation strikes the material.

The mechanisms of excitation by ionizing radiation are very similar to those listed under cathodoluminescence. The X- and γ-ray excitations are initiated by energy loss of the photons as they interact with matter. For higher energies, energy loss occurs by photoionization. Substrates of high atomic number absorb energy more efficiently than those of low atomic number because of the high concentration of electrons.

Charged particles and the secondary electrons arising from γ rays behave like the charged particles of gas discharges striking a surface, and the processes of excited species generation are the same. Due to the penetrating power of radiation, however, excitation can penetrate much more deeply.

An important by-product of ionizing radiation in solids is that electrons are frequently trapped in lattice sites (Fig. 8). Depending on the depth of the trap sites, the electron may either recombine rapidly or be trapped for an extended period of time, even for millions of years. Release of these electrons can then yield a delayed recombination luminescence. This delayed luminescence in thermoluminescence is discussed in Section VI,G.

Radioluminescence forms the heart of many radiation detectors. The traditional γ detector is thallium-activated sodium iodide. The high atomic number of iodide increases the efficiency of energy absorption, and the energy is used to excite the Tl^+ luminescent center.

β-Particle detectors are crystals of organic molecules (e.g., anthracene) or solutions of highly fluorescent organic scintillators dissolved in solid aromatic polymers (e.g., polystyrene) or aromatic solutions. The substrate of an aromatic polymer or solvent is an essential component. The fluorescent indicator is generally present in relatively low concentrations, and much of the excitation occurs away from the scintillator. The bulk of the initial excited states formed are thus the easily excited bulk aromatic solvents. Due to the closeness of the aromatic species, the excitation can migrate or hop from molecule to molecule until the energy comes close enough to the scintillator molecule for efficient energy transfer. The scintillator molecule must have an excited singlet state that is below the energy levels of the solvent molecules. Thus, the scintillator traps the energy and emits.

Organic scintillators form the basis of the important liquid scintillation counter. The penetrating power of β particles is very limited. Therefore, counting weak β emitters such as tritium is very difficult with windowed detectors because the window or the sample itself absorbs most of the radiation. The penetration problem is solved by homogeneously dissolving the material to be counted in a "cocktail" consisting of the solvent and the scintillator. Since the particles are emitted directly in the presence of the scintillator, the penetrating power becomes irrelevant.

G. Thermoluminescence

Thermoluminescence (TL) is luminescence that arises on gentle warming of a material and usually occurs below incandescence. Reading the sample by heating it destroys the activation, so the readout is a single-shot experiment. Most TL materials can, however, be reactivated and used repeatedly.

The mechanism of TL is shown in Fig. 8. The activation process consists of trapping an electron in a trap site. The trap site must be deep enough to prevent rapid removal of the electron at the storage temperature. Activation can be accomplished by ionizing radiation or by light. Once the sample is activated, the readout is performed by heating the sample. When the temperature becomes sufficiently high, the electrons are thermally excited into the conduction band. These conduction electrons can migrate through the lattice until they find a hole to relax into. If the hole is an emissive center, or is near one, the center is excited and emits. Alternatively, the relaxation to the hole can be emissive.

The term thermoluminescence is misleading since it does not mean thermal generation of the excited system. *Thermal-stimulated luminescence* much more accurately describes the phenomenon.

In TL measurements or development, readout is performed by heating the sample at a uniform rate and viewing the sample emission. A filter may be added over the detector to restrict the viewing wavelength. Figure 16 shows a typical TL glow curve. The development supplies information on the distribution and depth of traps. Deeper traps require higher temperatures to boil the electrons out. For example, a 375°C glow curve corresponds to a trap lifetime of several million years. Indeed, elaborate models have been developed to extract quantitative information about trap depth and distribution from glow curves.

In most examples of TL, one is examining a delayed photoluminescence. Particularly in solids subjected to radiation chemistry, the nature of the traps and of the emissive centers is modified by irradiation. Therefore, the TL emission may look quite different from the photoluminescence of the unactivated solid.

The principal use of TL is in radiation dosimetry. The TL signal is proportional to dose over a very wide range. Modern TL dosimeters are sensitive in the microrad region. Thermoluminescence dosimeters have a number of advantages. They are inexpensive, robust, reusable, and sensitive, and they intrinsically integrate the dose. Furthermore, they require no power source or connections. This makes them especially useful for *in vivo* monitoring during radiation therapy since they can be easily implanted or swallowed.

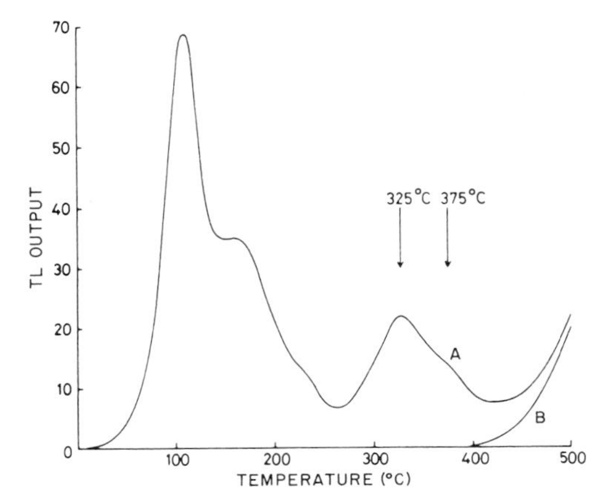

FIG. 16. Thermoluminescence glow curve (a) of quartz extracted from pottery for use in TL dating. The sample has been artificially irradiated with 550 rads of β radiation as part of the calibration process. Curve (b) is the red-hot glow measured on reheating the sample. The TL is composed of both the stored and irradiated components. [Reprinted with permission from Aitken, M. J., and Fleming, S. J. (1972). *In* "Topics in Radiation Dosimetry," Supplement 1 (F. H. Attix, ed.), Academic Press, New York, p. 1.]

Lithium fluoride is one of the most popular personal dosimeters. It is sensitive down to 10^{-2} rad and linear to $>10^3$ rads. The average atomic number of LiF is similar to that of tissue, which gives the dosimeter a response indicative of tissue irradiation. The traps are deep enough to give long storage times without appreciable fading.

The early work on TL demonstrates the difficulty of studying emissions that intrinsically rest on impurity sites. The first attempts at TL dosimetry date back to the early 1950s, when Farrington Daniels attempted to use LiF as a dosimeter. The original work was abandoned owing to problems of sensitivity and storage time. When the work was resumed in the early 1960s, it was discovered that all the earlier TLs of LiF arose from impurity centers that were no longer present in commercially purer LiF. This led to a considerable effort to elucidate the impurity problems and eventually led to usable systems. Also, other useful systems were discovered by other groups.

Another type of sensitive, and at one time widely used, dosimeter utilized radiophotoluminescence. Irradiation of a crystal produces color centers. When photoexcited, many of these color centers emit. Since the emission intensity is directly proportional to the concentration of color centers, a dosimeter is available. An advantage of photoluminescent dosimeters is that the readout is not destructive and can be repeated.

Another important use of TL is in archaeological and geological dating when radiocarbon dating is unsuitable (Fig. 16). Heating destroys any TL and initializes the phosphor. As all rocks have some degree of radioactivity from ^{238}U, ^{235}Th, and ^{40}K, the TL can be used to measure the length of time that the sample was irradiated since cooling. For igneous rocks this is the time since formation. For pottery it is the time since the pottery was fired. Such dating can give results accurate to better than 10%.

H. Flame Emissions

With the exception of hydrogen–oxygen flames, virtually all flames exhibit pronounced visual emissions. We shall discuss briefly the origin of some of these emissions. Luminescence can arise from the major components of the reaction or from trace materials. The orange glow of candle flames and oxidant-starved gas flames arises, not from luminescence, but from the incandescence of carbon particles.

Luminescence from a state is independent of how the state was populated. Many flames are hot enough and the excited state of elements and compounds low enough that a significant excited state population can be achieved thermally. For example, the emitting state of atomic sodium is at 589 nm. For flames of various temperatures the percentages of molecules in the excited state are as follows: 2000 K, 1×10^{-3}%; 3000 K, 6×10^{-2}; and 4000 K, 0.4%. While these excited state populations may seem very small, they are in fact very large in comparison with the concentrations that could be achieved by all but the most intense laser sources in photoluminescence experiments. This efficiency is readily seen by the intense yellow sodium fluorescence when even tap water is introduced into a relatively cool Bunsen burner flame.

An interesting aspect of flame spectroscopy is that a flame can be too hot to yield good elemental emissions. Too hot a flame can thermally ionize many of the atoms and reduce the population of the emissive element. This propensity of luminescent species to ionize can be suppressed by increasing the concentration of free electrons in this flame. For example, ionization of calcium can be suppressed by spiking the flame with cesium or potassium, which have lower ionization potentials and increase the concentration of free electrons. Many of the products of the chemical reactions contributing to the flame are produced in excited states and emit (see Section VI,B).

While not strictly speaking flames, inductively coupled plasma torches produce plumes that appear to be flames. Furthermore, these plasmas behave like extremely high temperature flames, which are capable of exciting all but the most recalcitrant elements.

One of the most useful applications of flames is for elemental quantitative analysis. High-temperature flames reduce most complex matrices to their elemental components. Many elements, especially metals, then emit in the flames, and the emission can be used for analytical quantification. Even many elements that do not emit directly can still be analyzed in flames by photoluminescence or by atomic absorption.

Because of the chemical generation of many of the excited states in flames, population inversion over the ground state can result. This population inversion can form the basis of a laser. The best known chemical laser is the HF or DF laser made by burning hydrogen or deuterium in fluorine. Visible or UV chemical lasers are still being sought.

I. Miscellaneous Emissions

Other types of luminescence include triboluminescence, sonoluminescence, and crystalloluminescence. To date none of these intriguing forms of luminescence appears to have any practical applications.

Triboluminescence or piezoluminescence is the emission by crystals when they are broken or ground up. Amusing, if not practical, is the parlor demonstration of grinding a wintergreen-flavored Life Saver® between one's teeth in a darkened room to produce bright blue-green flashes. Triboluminescence is, in most cases, due to the production of strong electric fields along fracture lines with a concomitant electrical breakdown. The observed luminescence is then a mechanically induced cathodoluminescence.

Crystalloluminescence is emission during the

growth of crystals and may arise from cleavage. It is thus a form of triboluminescence from internally generated breakage.

Sonoluminescence is the emission of liquids subjected to intense ultrasound. This is probably due to cavitation with the production of a variety of chemical species that subsequently show chemiluminescence.

Acknowledgment

We gratefully acknowledge support of the National Science Foundation (CHE 82–06279 and 86-00012) and the donors of the Petroleum Research Fund, administered by the American Chemical Society.

Bibliography

Adam, W., and Cilento, G. (eds.) (1982). Chemical and Biological Generation of Excited States. Academic Press, New York.

Alkemade, C. Th. J., Hollander, Th., Snellman, W., Seeger, P. J., D. ter Harr (eds.) (1982). Metal vapours in flames, *International Series in Natural Philosophy,* **103**. Pergamon Press, New York.

Cundall, R. B., and Dale, R. E. (1983). Time–Resolved Fluorescence Spectroscopy in Biochemistry and Biology. Plenum Press, New York.

Demas, J. N. (1983). Excited State Lifetime Measurements. Academic Press, New York.

Harvey, E. Newton (1957). A History of Luminescence. American Philosophical Society. Philadelphia.

Horowitz, Yigal S. (ed.) (1984). Thermoluminescence and Thermoluminescent Dosimetry, **1–3**. CRC Press. Boca Raton, Florida.

Lakowicz, Joseph R. (1983). Principles of Fluorescence Spectroscopy. Plenum Press, New York.

Sze, S. M. (1981). Physics of Semiconductor Devices, second edition, John Wiley & Sons. New York.

Weber, Marvin J. (1982). CRC Handbook of Laser Science and Technology, **1, 3**. CRC Press. Boca Raton, Florida.

Yen, W. M., and Selzer, P. M. (eds.) (1981). Laser spectroscopy of solids, *Topics in Applied Physics,* **49**. Springer–Verlag, Berlin.

MICROOPTICS

Kenichi Iga *Tokyo Institute of Technology*

GLOSSARY

Distributed-index microlens: Microlenses which utilize the lens effect due to refractive index distribution and this provides a flat surface.

Microlens: Lens for microoptic components with dimensions small compared to classical optics; large numerical aperture (NA) and small aberration are required.

Microoptic components: Optical components used in microoptics such as microlenses, prisms, filters, and gratings, for use in constructing microoptic systems.

Microoptic systems: Optical systems for lightwave communications, laser disks, copy machines, lightwave sensors, etc. which utilize a concept of microoptics.

Planar microlens: Distributed-index microlens with a three-dimensional index profile inside the planar substrate; planar technology is applied to its construction and two-dimensional arrays can be formed by the photolithographic process.

Stacked planar optics: Microoptic configuration composed of two-dimensional optical devices such as planar microlenses and other passive as well as active devices.

Microoptics utilizes a number of tiny optical components for use with electrooptics such as lightwave communications, laser disks, and copying machines.

Since optical fiber communication began to be considered as a real communication system which provides many possibilities, a new class of optical components has become necessary. It must be compact and lightweight, have high performance, and contain various functions that are different from classical optics. Integrated optics, which utilizes a concept of a planar dielectric waveguide, is thought to be the optics of the future, but presently it is difficult to find systems in use that consist of components with the *guided-wave* configuration. On the other hand, microoptic components made of *microlenses* and other tiny optical elements are pragmatically used in real optical systems such as lightwave communications and *laser disk* systems. They make use of all concepts of classical optics as well as of beam optics and partly even of guided-wave optics. One of the new important devices introduced in the course of research is a *distributed index (DI) lens*. It uses the refraction of light rays by the index gradient coming from the nonuniform index existing inside the medium. The point is that we can make a lens with flat surfaces, which is essential for optical fiber applications since we can put fibers directly in contact with the lens, while the classical lens immediately loses its lens action when some other materials touch its surface. [*See* OPTICAL FIBER COMMUNICATIONS.]

Therefore, the distributed index lens plays an important role in microoptics. We can thus define microoptics as an optics which utilizes microlenses and other tiny optical elements, with the high performance and reliability required in heavy-duty electrooptic systems.

I. Microoptics: Its Roles in Optoelectronics

A. COMPARISON OF OPTICAL COMPONENTS

Great progress in optical fiber communication has been made and many working systems have been installed. Three types of optical components used in optical fiber communication systems have been considered:

1. microoptics, which consists of microlenses such as gradient-index lenses or tiny spherical lenses;
2. optical fiber circuits made from manufactured fibers; and
3. integrated optics.

There have been many problems in the first two schemes, such as optical alignment and manufacturing process, and integrated optics devices are still far from the usable level, as shown in Table I. [*See* OPTICAL CIRCUITRY.]

B. Roles of Microoptics

The role of microoptics is then believed to be not as a substitute for other components such as guided-wave optic components or fiber optic circuits but for the purpose of fully utilizing these optical systems more effectively by cooperating with them. It is hoped that some more modern concepts will evolve to integrate microoptic devices without leaving them as old discrete optics. One of the ideas may be *stacked planar optics,* which will be discussed in detail in Section VII.

II. Basic Theory for Microoptics

A. Ray Theory

Light propagation is described simply by a ray that indicates a path of light energy. Actually, sometimes a very thin beam from a laser or collimated light through a pinhole appears as a "ray" that certainly indicates the path of light energy. In a conventional optical system, which is constructed with lenses, mirrors, prisms, and so on, a ray is represented by a straight line that is refracted or reflected at a surface where the refractive index changes. On the other hand, in a distributed index medium, a ray does not follow a straight line; rather it takes a curved trajectory as if it were affected by a force toward the higher refractive index. The ray trajectory in a distributed index medium is calculated from ray equations that are second-order partial differential in nature. Here, the ray is mathematically defined as a curve perpendicular to the wavefront being calculated from the limited case $\lambda \rightarrow 0$. According to this definition, the tangential direction of the ray corresponds to the direction of the time-averaged Poynting vector. This definition is consistent with Fermat's principle (i.e., the ray trajectory takes the shortest optical path length). We derive ray equations from Fermat's principle because the variational method can be used from which we can derive ray equations easily in any desired coordinate system. The optical path length is defined by

$$\alpha = \int_{P_1}^{P_2} n \, ds \tag{1}$$

where n is the refractive index of the medium of interest (see Fig. 1). By using coordinate variables $x(\tau)$, $y(\tau)$, and $z(\tau)$ as functions of the independent parameter τ, we have the equation of optical path length,

$$\alpha = \int_{\tau_1}^{\tau_2} n(x, y, z)\sqrt{x'^2 + y'^2 + z'^2} \, d\tau \tag{2}$$

TABLE I. Advantages of Stacked Planar Optics[a]

Classification of encountered problem	Optical system: Discrete	Waveguide	Stacked
Fabrication process	Yes	No	No
Surface preparation	Yes	Yes	No
Optical alignment	Yes	Yes	No
Coupling	No	Yes	No
Integration of different materials	No	Yes	No
Single–multi-compatibility	No	Yes	No
Polarization preference	No	Yes	No
Mass-production	Yes	No	No
2-D array	Yes	Yes	No
	large-scale optics ←		

[a] From Iga, K., Kokubun, Y., and Oikawa, M. (1984). "Fundamentals of Microoptics." Academic Press, Orlando.

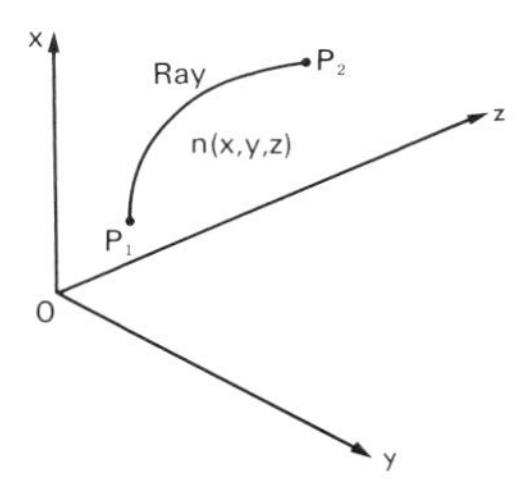

FIG. 1. Ray propagation in a distributed-index medium. [From Iga, K., Kokubun, Y., and Oikawa, M. (1984). "Fundamentals of Microoptics." Academic Press, Orlando.]

Here, ds in Eq. (1) is rewritten as $\sqrt{x'^2 + y'^2 + z'^2}$, and the prime means differentiation with respect to τ, where τ_1 and τ_2 are, respectively, the initial and final values of τ. To find the ray trajectory, we use Euler's equation; that is, when the functional F is a function of x, y, z, x', y', z', and τ;

$$F = F(x, y, z; x', y', z'; \tau) \tag{3}$$

and integration of F takes a stationary value, relations among x, y, and z are described by

$$\begin{aligned}(d/d\tau)(\partial F/\partial x') &= \partial F/\partial x\\ (d/d\tau(\partial F/\partial y') &= \partial F/\partial y\\ (d/d\tau)(\partial F/\partial z') &= \partial F/\partial z\end{aligned} \tag{4}$$

Since the functional of Eq. (4) is

$$\begin{aligned}&F(x, y, z; x', y', z'; \tau)\\ &\quad = n(x, y, z)\sqrt{x'^2 + y'^2 + z'^2}\end{aligned} \tag{5}$$

Euler's equations, satisfying Eq. (4), are written (with $n_x = \partial n/\partial x$) as

$$\begin{aligned}\frac{d}{d\tau}\frac{nx'}{\sqrt{x'^2 + y'^2 + z'^2}} &= n_x\sqrt{x'^2 + y'^2 + z'^2}\\ \frac{d}{d\tau}\frac{ny'}{\sqrt{x'^2 + y'^2 + z'^2}} &= n_y\sqrt{x'^2 + y'^2 + z'^2}\\ \frac{d}{d\tau}\frac{nz'}{\sqrt{x'^2 + y'^2 + z'^2}} &= n_z\sqrt{x'^2 + y'^2 + z'^2}\end{aligned} \tag{6}$$

If we give certain meaning to the parameter $d\tau = ds/n$, we obtain differential ray equations in Cartesian coordinates:

$$d^2x/d\tau^2 = \partial(\tfrac{1}{2}n^2)/\partial x \tag{7a}$$

$$d^2y/d\tau^2 = \partial(\tfrac{1}{2}n^2)/\partial y \tag{7b}$$

$$d^2z/d\tau^2 = \partial(\tfrac{1}{2}n^2)/\partial z \tag{7c}$$

When we treat a DI rod lens with the index expressed by Eq. (2), the partial derivative of n^2 with respect to z is zero when we choose the z axis as the optical axis. Equation (7c) is integrated and we have

$$dz/d\tau = C_i \tag{8}$$

For the ray incident at $x = x_i$, $y = y_i$, and $z_i = 0$, we have

$$C_i = n(x_i, y_i, 0)\cos\gamma_i \tag{9}$$

Since $d\tau = ds/n$, here $\cos\gamma_i$ is a directional cosine of the ray at $z = 0$ with respect to the z axis.

Elimination of τ from Eq. (7) using Eqs. (8) and (9) gives

$$\frac{d^2x}{dz^2} = \frac{1}{2n^2(r_i)\cos^2\gamma_i}\frac{\partial n^2(x, y)}{\partial x} \tag{10a}$$

$$\frac{d^2y}{dz^2} = \frac{1}{2n^2(r_i)\cos^2\gamma_i}\frac{\partial n^2(x, y)}{\partial y} \tag{10b}$$

where z denotes the axial distance, r_i the incident ray position, and i the direction cosine of the incident ray.

As will be discussed later, optical components with a distributed index may play an important role in the microoptics area. There have been various ways of expressing the index distribution of such devices. One method is a power series expansion of the refractive index with respect to coordinates. The expression of the refractive index medium that we present is of the form

$$\begin{aligned}n^2(r) = n^2(0)[1 &- (gr)^2 + h_4(gr)^4\\ &+ h_6(gr)^6 + \ldots]\end{aligned} \tag{11}$$

for a circularly symmetric fiber or rod, where g is a focusing constant expressing the gradient index and h_4 and h_6 represent higher-order terms of the index distribution and are closely related to aberration. In Eq. (1) $n(0)$ expresses the index at the center axis when $r = 0$. Advantages of this expression can be summarized as follows:

1. We can substitute Eq. (11) in the wave equation, which contains an inhomogeneous dielectric constant

$$\nabla^2E + k_0^2n^2(r)E + [\nabla n^2(r)] = 0$$

where $k_0 = 2\pi/\lambda$. This type of wave equation, which can usually be reduced to a scalar form, is commonly applied to an optical fiber that has a distributed index at the core.

2. Equation (11) can also be used to ascertain the ray trajectory (x, y) by solving the ray equation (10).

A solution to the meridional ray has been obtained in closed form.

3. It is convenient for contrasting with the α-class expression, which is common in optical multimode fibers; that is,

$$n^2(r) = \begin{cases} n^2(0)[1 - 2\,\Delta(r/a)^\alpha] & (r \leqq a) \\ n_2^2 & (r > a) \end{cases} \tag{12}$$

where $\Delta = [n(0) - n_2]/n(0)$.

4. It is convenient for relating to the density of particles contributing to the refractive index (i.e., the Clausius–Mossoti relation or the Lorenz–Lorentz relation including n^2 terms rather than n terms).

Recently we proposed and demonstrated the feasibility of a planar microlens having three-dimensional index distribution. To solve the ray equation in the three-dimensional problem, the form of Eq. (1) is extended in matrix form as

$$n^2(r, z) = n^2(0, 0)[1, gz, (gz)^2, (gz)^3, \ldots]$$

$$(N) = \begin{bmatrix} 1 \\ (gr)^2 \\ (gr)^4 \end{bmatrix} \tag{13}$$

where the index matrix is given by

$$N = \begin{bmatrix} 1 & -1 & \nu_{04} & \nu_{06} \\ \nu_{10} & \nu_{12} & \nu_{14} & \nu_{16} \\ \nu_{20} & \nu_{22} & \nu_{24} & \nu_{26} \\ \vdots & \vdots & \vdots & \vdots \end{bmatrix} \tag{14}$$

To equate this with Eq. (11) for the axially symmetric case, we can set $\nu_{04} = h_4$, $\nu_{06} = h_6$, and so on.

B. Wave Theory

The light ray we have introduced is convenient and easy to understand. When we attempt to solve a diffraction problem, however, we should treat the light as a wave. By using a wavefront having a certain relation to the light ray, we can accomplish this.

The wavefront is defined as a surface constructed with a set of equioptical path-length points from a light source. On a wavefront the light phase is constant. The wavefront is described by the eikonal equation

$$(\partial\phi/\partial x)^2 + (\partial\phi/\partial y)^2 + (\partial\phi/\partial z)^2 = n^2(x, y, z) \tag{15}$$

where ϕ is the light phase and corresponds to the optical pass. The wavefront is defined as $\phi =$ const.

When the light rays are focused at one point O, a set of the equioptical lengths from the origin O is a sphere. The sphere is orthogonal to the convergent rays, and a lens wave aberration is defined from this principle. The lens is used to focus the light rays on one point, but when the lens has an aberration, the rays cannot focus. In this case, the wave front is nonspherical and the displacement of the actual wave front from a spherical one is the wave aberration. Since ray aberration is also reduced from a wave aberration, wave aberration is measured to evaluate lenses.

C. Diffraction Theory

Spatial information, which is described by an electric field $f(x, y, 0)$ at $z = 0$ in Cartesian coordinates, is transformed while propagating in free space by a relation obeying the well-known Fresnel–Kirchhoff integral

$$f(x, y, z) = \frac{j}{kz}\exp(-jkz)\int\int f(x', y', 0) \times \exp\left[-\frac{jk}{2z}((x - x')^2 + (y - y')^2)\right] dx'\, dy' \tag{16}$$

where z is the propagation distance and k the propagation constant defined by $2\pi/\lambda$. In polar coordinates the expression is given in the same way as

$$f(r, \theta, z) = \frac{j}{kz}\exp(-jkz)\int\int f(r', \theta', 0) \times \exp\left[-\frac{jk}{2z}(r^2 - 2rr'\cos(\theta - \theta') + r'^2)\right] r'\, dr'\, d\theta' \tag{17}$$

A simple example is diffraction from a circular aperture, where $f(r, \theta, 0)$ is given by a constant within the region $r < A$ with A the radius of the aperture. The result is given by:

$$|f(r, \theta, z)/f(0, \theta, z)|^2 = [2J_1(\rho)/\rho]^2$$

where $J_1(\rho)$ is the first-order Bessel function and $\rho = ka(r/z)$. The profile is known as an Airy pattern as shown in Fig. 2.

D. Beam Theory

The Fresnel–Kirchhoff (FK) integral for transformation of a light beam through free space and a DI medium is discussed. In this sec-

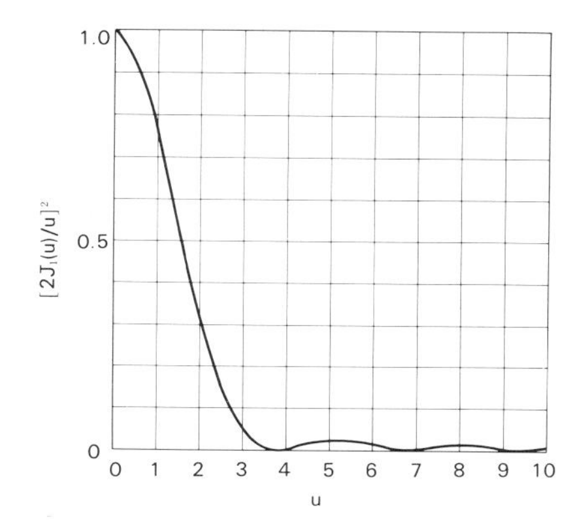

FIG. 2. Airy function. [From Iga, K., Kokubun, Y., and Oikawa, M. (1984). "Fundamentals of Microoptics." Academic Press, Orlando.]

tion we shall describe how a Gaussian beam changes when propagating in free space and in a DI medium.

We shall assume a Gaussian beam at $z = 0$ given by

$$f(x', y', 0) = E_0 \exp[-\tfrac{1}{2}((x'^2 + y'^2)/s^2)] \quad (18)$$

Positional change of the beam can be calculated by Eq. (16) and is expressed as

$$f(x, y, z) = E_0 \exp(-jkz)(s/w) \times \exp[-\tfrac{1}{2}p(x^2 + y^2) + j\phi] \quad (19)$$

Here the spot size w and radius R of the phase front are given by

$$w = s\sqrt{1 + (z/ks^2)^2}$$
$$R = z[1 + (ks^2/z)^2] \quad (20)$$

Parameters P and ϕ are defined by

$$P = 1/w^2 + j(k/R)$$
$$\phi = \tan^{-1}(z/ks^2) \quad (21)$$

From Eq. (19) it can be seen that the transformed beam is still Gaussian, although there are spot-size and phase-front changes. It is clear that R expresses the phase front if we take into consideration the phase condition

$$kz + (k/2R)r^2 = \text{const} \quad (22)$$

With this equation, the functional dependence of the phase front $z = -(1/2R)r^2$ can be reduced. When R is positive, the phase front is convex, as seen from $z = +\infty$.

Let us next examine the parameter z/ks^2 that appears in Eqs. (20) and (21). When this parameter is rewritten as

$$z/ks^2 = (1/2\pi)(s^2/\lambda z)^{-1} \quad (23)$$

and the Fresnel number N is defined as

$$N = s^2/\lambda z \quad (24)$$

the *Fresnel number* is a function of wavelength, distance, and spot size and expresses normalized distance. Regions can be characterized according to N such that

$N \ll 1$ (Fraunhofer region)

$N > 1$ (Fresnel region)

When the point of observation is located at a point some distance from the origin ($N \ll 1$), the spotsize w can be approximated from Eq. (21) as

$$w \cong z/ks \quad (25)$$

The spreading angle of the beam is, therefore,

$$2\,\Delta\theta = 2w/z = 0.64(\lambda/2s) \quad (26)$$

This is analogous to the spreading angle of a main lobe of a diffracted plane wave from a circular aperture, given by

$$2\,\Delta\theta = 1.22(\lambda/D) \quad (27)$$

Figure 3 presents the waveform coefficients P_0, P_1, and P_2 at $z = 0$, z_1, and z_2, respectively. If the spot sizes and curvature radii of the wavefront are given by s, w_1, and w_2 and ∞, R_1, and R_2, respectively, the coefficients can be expressed as

$$P_0 = 1/s^2$$
$$P_1 = 1/w_1^2 + jk/R_1 \quad (28)$$
$$P_2 = 1/w_2^2 + jk/R_2$$

From Eqs. (18)–(20),

$$1/P_0 = 1/P_1 + jz_1/k$$
$$1/P_0 = 1/P_2 + jz_2/k \quad (29)$$

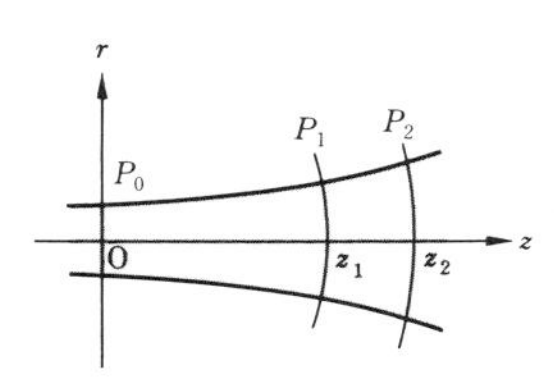

FIG. 3. Transformation of waveform coefficients. [From Iga, K., Kokubun, Y., and Oikawa, M. (1984). "Fundamentals of Microoptics." Academic Press, Orlando.]

When P_0 is eliminated, the relationship between P_1 and P_2 is reduced to

$$P_1 = P_2/[1 + (j/k)(z_2 - z_1)P_2] \tag{30}$$

This is a special case of the linear transform

$$P_1 = (AP_2 + B)/(CP_2 + D) \tag{31}$$

It is very convenient to use the matrix form

$$\tilde{F} = \begin{bmatrix} A & B \\ C & D \end{bmatrix} \tag{32}$$

to calculate the transform for a system composed of many tandem components. It is then possible to obtain a total F matrix with the product of the matrices expressed as

$$\tilde{F} = \tilde{F}_1 \times \tilde{F}_2 \times \tilde{F}_3\, x \cdots \tag{33}$$

Table II presents a tabulation of the waveform matrices associated with some optical components. It is not difficult to obtain these matrix forms by calculating the change of a Gaussian beam when it passes through these optical components. Kogelnik also proposed a matrix form for the same purpose, but it is somewhat different from the definition introduced here.

Figure 4 shows that ray position x_1 and ray slope $\dot{x}_1$ at the incident position are related to x_2 and $\dot{x}_2$ by the same matrix representation; that is,

$$\begin{bmatrix} jk\dot{x}_1 \\ x_1 \end{bmatrix} = \begin{bmatrix} A & B \\ C & D \end{bmatrix} \begin{bmatrix} jk\dot{x}_2 \\ x_2 \end{bmatrix} \tag{34}$$

TABLE II. Waveform Matrices for Various Optical Systems[a]

Optical system	Waveform matrix, F
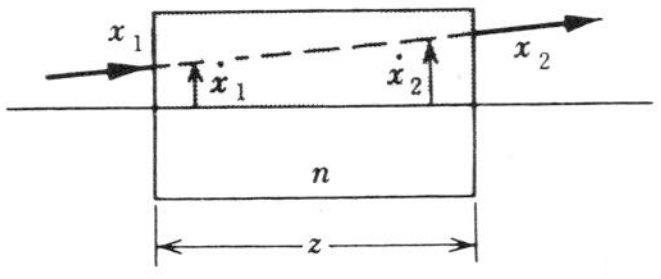 FREE SPACE	$\begin{bmatrix} 1 & 0 \\ j\frac{z}{k} & 1 \end{bmatrix}$, $k = k_0 n$
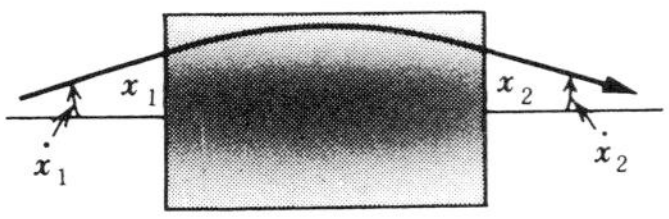 POSITIVE DI LENS	$\begin{bmatrix} \cos gz & jkg \sin gz \\ j\frac{1}{kg}\sin gz & \cos gz \end{bmatrix}$, $k = k_0 n(0)$
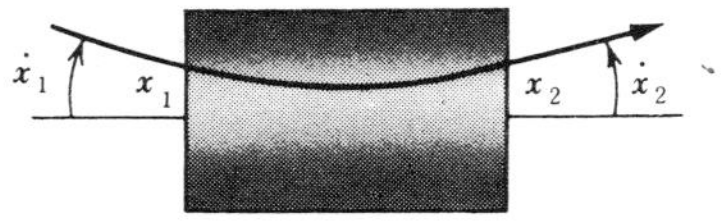 NEGATIVE DI LENS	$\begin{bmatrix} \cosh gz & -jkg \sinh gz \\ j\frac{1}{kg}\sinh gz & \cosh gz \end{bmatrix}$, $k = k_0 n(0)$
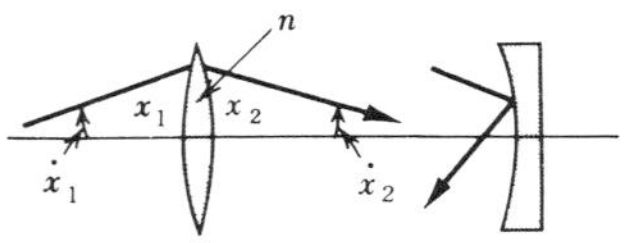 CONVEX LENS, CONCAVE MIRROR	$\begin{bmatrix} 1 & \frac{jk}{f} \\ 0 & 1 \end{bmatrix}$, $k = k_0 n$
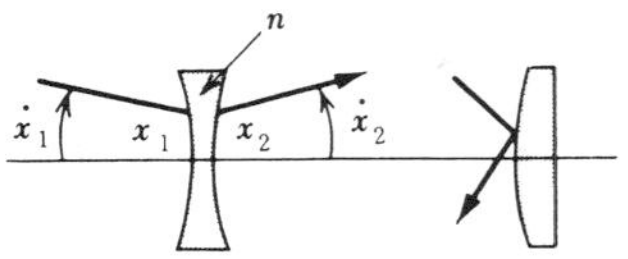 CONCAVE LENS, CONVEX MIRROR	$\begin{bmatrix} 1 & -\frac{jk}{f} \\ 0 & 1 \end{bmatrix}$, $k = k_0 n$

[a] From Iga, K., Kokubun, Y., and Oikawa, M. (1984). "Fundamentals of Microoptics." Academic Press, Orlando.

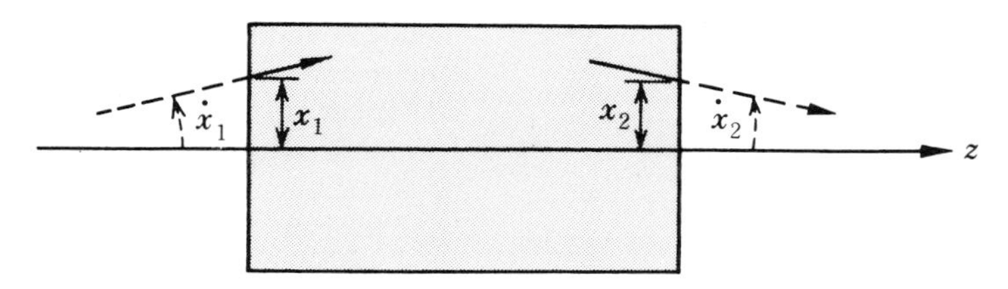

FIG. 4. Relationship of ray positions and ray slopes. [From Iga, K., Kokubun, Y., and Oikawa, M. (1984). "Fundamentals of Microoptics." Academic Press, Orlando.]

The propagation constant k is included in Eq. (34) to make it possible to treat a tandem connection of optical components having different refractive indices.

E. Guided-Wave Theory

In order to obtain a basic concept of guided-wave components including optical fibers and planar dielectric waveguides, we summarize the treatment of a simple planar waveguide consisting of three layers, as shown in Fig. 5. The tangential field components of TE modes are E_y and H_z. Since these two components are directly related to the boundary condition, it is convenient to deal with the E_y component instead of the H_z component. From Maxwell's equation, E_y must satisfy

$$d^2E_y/dx^2 + (k_0^2n^2 - \beta^2)E_y = 0 \qquad (35)$$

Special solutions of Eq. (35) in the core are cosine and sine functions. In the cladding, the solutions are classified into two types, namely, the evanescent (exponentially decaying) solution for $n_2k < \beta < n_1k$ and the sinusoidal oscillating solution for $\beta < n_2k$. The former is called the guided mode. Some amount of optical power of a guided mode is confined in the core, and the remainder permeates from the cladding. The latter is called a set of radiation modes and the power is not confined in the core. The group of all guided and radiation modes constitutes a complete orthogonal set and any field can be expanded in terms of these guided and radiation modes. First, we discuss guided modes.

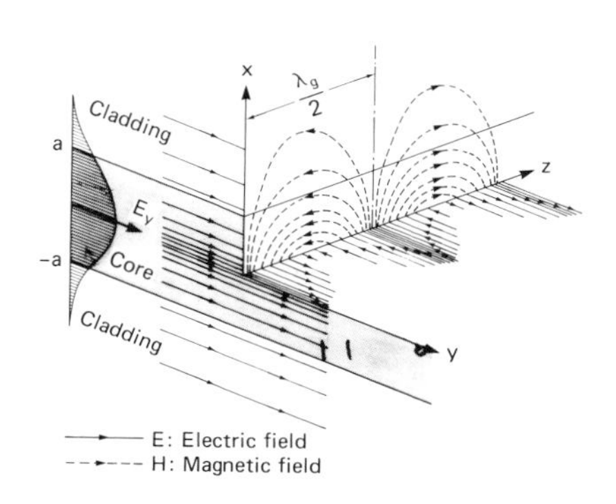

FIG. 5. Planar waveguide. [From Iga, K., and Kokubun, Y. (1986). "Optical Fibers." Ohm, Tokyo.]

The solution of a guided mode must satisfy the boundary conditions stating that the tangential components of the electric field must be continuous at the core–cladding boundary and approach zero at $x \to \infty$. From these conditions, we obtain the mode distributions

$$\begin{aligned} E_y(x) &= A_e \cos(\kappa x), \qquad |x| \leqq a \\ &= A_e \cos(\kappa a)\ \exp(-\gamma|x| - a) \\ &\qquad |x| \geqq a \quad \text{(even)} \end{aligned} \qquad (36)$$

$$\begin{aligned} E_y(x) &= A_o \sin(\kappa x), \qquad |x| \leqq a \\ &= A_o \sin(\kappa a)\ \exp(-\gamma|x| - a) \\ &\qquad |x| \geqq a \quad \text{(odd)} \end{aligned} \qquad (37)$$

where we have defined κ and γ as

$$\kappa^2 = k_0^2n_1^2 - \beta^2 \qquad (38)$$

$$\gamma^2 = \beta^2 - k_0^2n_2^2 \qquad (39)$$

Then H_z is another tangential field component continuous at the core–cladding boundary. From this boundary condition, the eigenvalue equations for TE even and odd modes are derived as Eqs. (37) and (38), respectively,

$$\tan \kappa a = (\kappa a)/\gamma a) \qquad \text{(even)} \qquad (40)$$

$$\tan(\kappa a - \pi/2) = (\kappa a)/\gamma a) \qquad \text{(odd)} \qquad (41)$$

The solutions of these eigenvalue equations can be normalized by introducing new parameters b and V defined by

$$b = (\beta/k_0 - n_2)/(n_1 - n_2) \qquad (42)$$

where

$$V = k_0an_1\sqrt{2\,\Delta} = \sqrt{(\kappa a)^2 + (\gamma a)^2} \qquad (43)$$

and giving a simple closed form,

$$V = (\pi/2)/\sqrt{(1 - b)} \times [(2/\pi)\tan^{-1}(\sqrt{b/(1 - b)} + N] \quad (44)$$

From this expression we obtain a dispersion curve that relates V and b, as shown in Fig. 6. When the waveguide parameters n_1, n_2, and a and wavelength λ are given, the propagation constant β_N of any mode can be obtained from Eq. (44). The mode number is labeled in the order of increasing N, and TE_0 is the fundamental mode. This mode number N corresponds to the number of nodes in the field distribution.

When the propagation constant of one guided mode reaches $n_2k(b \rightarrow 0)$, the mode is cut off, and the V value is then called the cutoff V value. By putting $\gamma = 0$ and $\kappa a = V$, the cutoff V value of TE modes is easily obtained from Eqs. (42)–(44) as

$$V = (\pi/2)N \qquad (N = 0, 1, 2, \ldots) \quad (45)$$

The cutoff V value of TE_1 gives a single mode condition, because, when V is smaller than $\pi/2$, only the TE mode can propagate. The single mode condition is important for designing single mode waveguides and single mode optical fibers with an arbitrary refractive index profile.

From Eq. (44) we can obtain the group velocity ν_g which is the velocity of a light pulse through the waveguide.

III. Fundamental Microoptic Elements

A. Microlens

In microoptics, several types of microlenses have been developed. A spherical microlens is used mostly to gather light from a laser diode by taking advantage of the high numerical aperture and the small Fresnel reflection of light from its spherical surface. Moreover, the precise construction of the sphere is very easy but its mount to a definite position must be considered specially. Spherical aberration is of course large.

A distributed-index (DI) rod lens with a nearly parabolic radial index gradient related to light focusing fiber was developed in 1968. The ray trace in the DI rod lens is expressed by

$$x = x_i \cos(gz) + (\dot{x}_i/g)\sin(gz) \quad (46)$$

where x_i and $\dot{x}_i$ are the incident ray position and slope.

It is readily known that the ray is transmitted with a sinusoidal trace, as shown in Fig. 7, and the periodicity pitch of the trace is

$$L_p = 2\pi/g \quad (47)$$

If we cut the rod into lengths $L_p/4$, the lens acts as a single piece of positive (focusing) lens. On the other hand, if the rod length is $\frac{3}{4}L_p$ an elect image can be formed by a single piece of lens. In addition, it has the merit of having a flat surface, and other optical elements, such as dielectric multilayer mirrors, gratings, and optical fibers, can be cemented directly without any space between them. Another feature is its ability to constitute the conjugate image device (i.e., real images with unity magnification can be formed by a single piece of rod microlens). This is illustrated in Fig. 8.

A planar microlens was invented for the purpose of constructing two-dimensional lightwave components. A huge number of microlenses

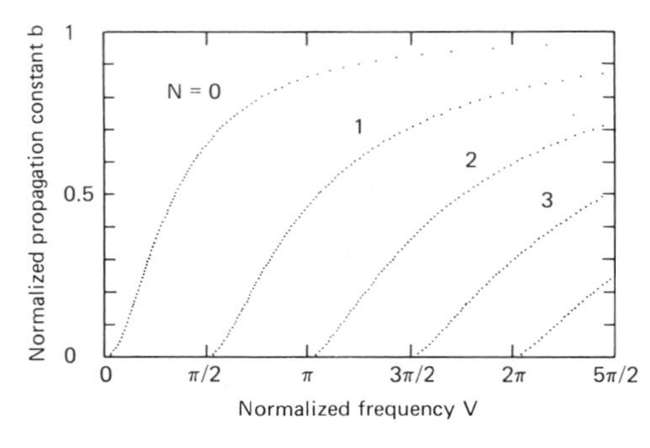

FIG. 6. Dispersion curve for a TE mode of a planar waveguide.

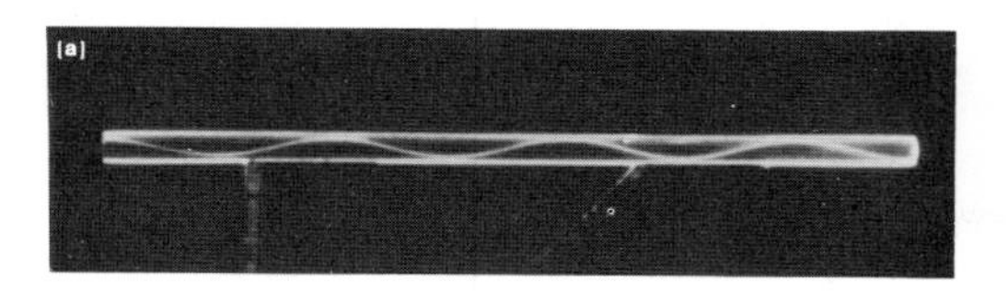

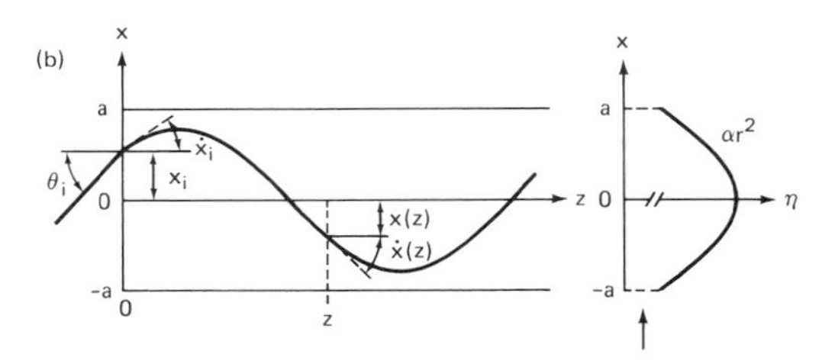

FIG. 7. Sinusoidal light ray trajectory in a DI rod. [From Iga, K., Kokubun, Y., and Oikawa, M. (1984). "Fundamentals of Microoptics." Academic Press, Orlando.]

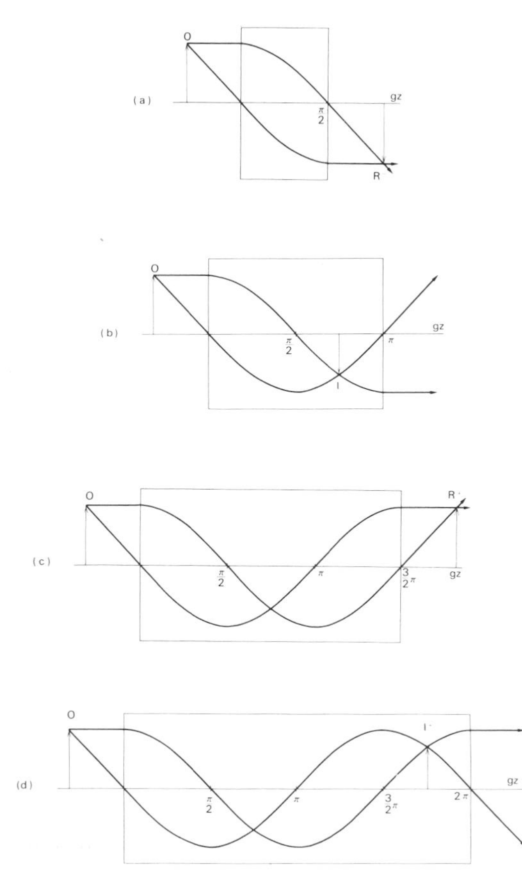

FIG. 8. Imaging configurations with various length DI rods. [From Iga, K., Kokubun, Y., and Oikawa, M. (1984). "Fundamentals of Microoptics." Academic Press, Orlando.]

of 0.1–2 mm in diameter can be arranged two-dimensionally in one substrate, and their position is determined by the precision of the photomasks. The applications to lightwave components and multi-image forming devices are considered.

The importance of a microlens for focusing the light from a diode laser onto a laser disk is increasing in potential. Some new types of microlenses are being developed. One of them is a mold lens with aspheric surfaces that can be permitted for diffraction-limited focusing.

B. Grating

A grating normally used in monochromators is used for multiplexing or demultiplexing different wavelengths. The grating fundamentals are described in standard optics textbooks. The band elimination property is sharp, and reliable components can be constructed. One slight problem is that the angle of the exit ray is changed if the wavelength varies. This is serious in wavelength demultiplexers because the wavelength of currently used semiconductor lasers varies considerably with temperature.

C. Multilayer Mirror and Filter

The dielectric-coated multilayer mirror or filter does not have the problem that grating filters have. The basic theory of multilayer mirrors and filters can be found in classical optics references. The reflectivity and transmittance can be designed. The change in optical properties with age must be especially examined in conventional applications (e.g., resistivity versus moisture and temperature change).

D. Aperture Stop and Spatial Frequency Filter

An aperture stop is used to eliminate light from an unwanted direction. In special cases various types of spatial filters are inserted in order to cut out unwanted modes.

E. Otical Fiber

A fiber component consists of manufactured optical fibers. A branch is obtained by polishing fibers to bare the core region for coupling. A directional coupler, polarizer, Faraday rotator, and so on can be constituted only of optical fibers. The merit of the fiber component is that the mode of the component is identical to that of the fibers used.

IV. Microoptic Components

A. Focuser

A light-focusing component (focuser) is the simplest and still-important component. We first discuss a simple focuser composed of a single lens. The diffraction limit of the focused spot D_s is given by

$$D_s = 1.22\lambda/\mathrm{NA} \cong 1.22 f\lambda/a \qquad (48)$$

where λ is the wavelength, $2a$ the aperture stop diameter, f the focal length of the employed lens, and NA the numerical aperture of the lens. This formula can be applied to a DI lens.

The focuser for the laser disk system must have the smallest aberration since the lens is used in the diffraction limit. Another important point is the working distance of the lens. In the laser disk application some space must be considered between the objective lens and the focal point because we have to have a clearance of about 2 mm and must also consider the disk thickness. Therefore, we need a lens with a large diameter (~5 mm) as well as a large NA (~0.5). First, the microscope objective was considered and several types of microlenses have been developed, such as the mold plastic aspheric lens and the DI lens with a spherical surface.

A microlens is employed for the purpose of focusing light from a laser into an optical fiber. A spherical lens is a simple one with an NA large enough to gather light from a semiconductor laser emitting light with 0.5–0.6 NA. In the single mode fiber application the combination of spherical and DI lenses is considered.

B. Branch

High-grade optical communication systems and electrooptic systems need a component that serves to divide light from one port to two or more ports. This is a branching component not necessarily used for separating power equally into two branches. A simple branch consists of a lens and prisms. In Fig. 9 we show a branch made of a manufactured distribution index lens. A microoptic branch is used inherently both for single mode and multimode fibers and no polarization preference exists. Another possibility is to utilize a planar waveguide as shown in Fig. 10. We have to design separately for single or multimodel fibers.

C. Coupler

A power combiner, or simply a coupler, is a component for combining light from many ports into one port. In general, we can use a branch as a combiner if it is illuminated from the rear. A

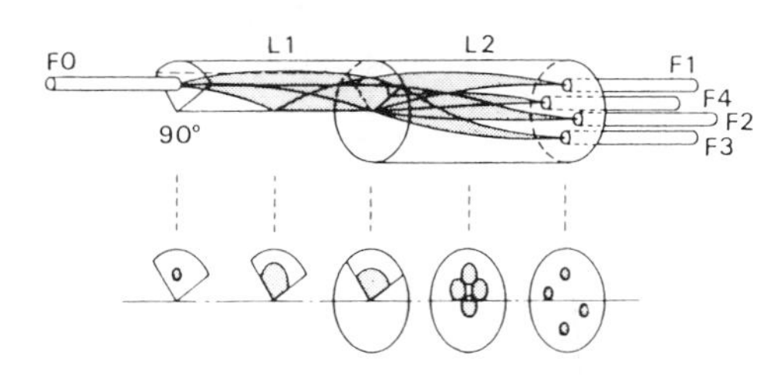

FIG. 9. Optical branch made of DI rod lenses. [From Kobayashi, K., Ishikawa, R., Minemura, K., and Sugimoto, S. (1979). *Fibers Integr. Opt.* **2,** 1.]

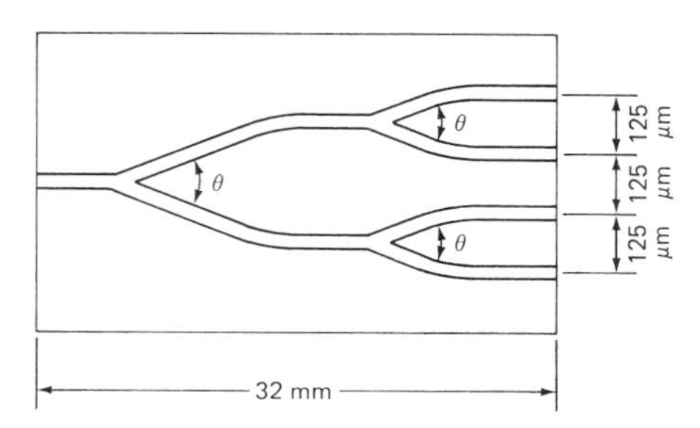

FIG. 10. Waveguide branch. [From Okuda, E. Tanaka, I. and Yamasaki, T. (1984). *Appl. Opt.* **23,** 1745.]

directional coupler is a component consisting of a branch and coupler, as shown in Fig. 11. The light from port 1 is divided into ports 2 and 3, and the light from port 3 exits from ports 1 and 4. A component consisting of a DI lens and a half-mirror is shown in Fig. 11. There is one component made of coupled fibers in which two fibers are placed so that the light in the two fibers can be coupled with each other (i.e., light propagating in one waveguide couples into the other while propagating along the guide).

A star coupler or optical mixer branches m ports into n ports, which serves to send light to many customers as in data highway or local area networks (LAN). Figure 12 shows a mixer made of manufactured fibers and Fig. 13 utilizes a planar waveguide configuration.

D. Wavelength Multiplexer/Demultiplexer

A wavelength *multiplexer* (MX)/*demultiplexer* (DMX) is a device that combines/separates light of different wavelengths at the transmitter receiver, which is needed inevitably for a communication system using many wavelengths at the same time. There exists a device consisting of a multilayer filter and DI lenses, as shown in Fig. 14, which is good for several wavelengths and one with a grating and lenses as in Fig. 15. The grating DMX is available for many wavelengths but the remaining problem is that the beam direction changes when the wavelength varies (e.g., as a result of the change in source

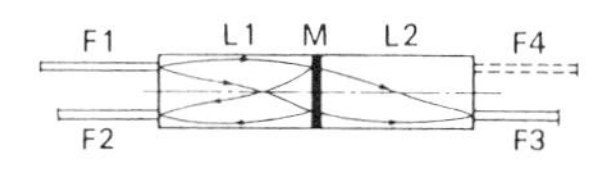

FIG. 11. Directional coupler. [From Uchida, T., and Kobayashi, K. (1982). *Jpn. Annu. Rev. Electron. Comput. Telecommun. Opt. Devices Fibers* **4,** 179.]

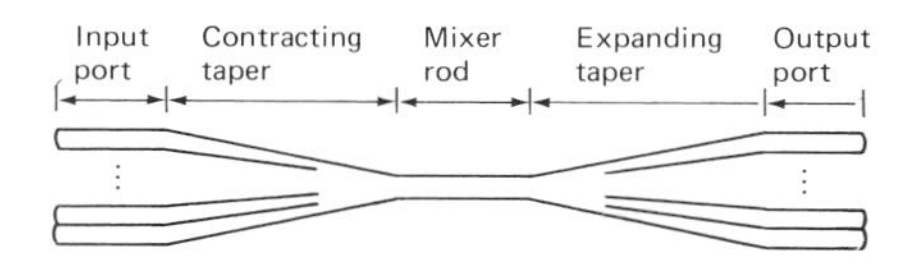

FIG. 12. Optical mixer made of fabricated fibers. [From Ohshima, S., Ito, T., Donuma, K., and Fujii, Y. (1984). *Electron. Lett.* **20,** 976.]

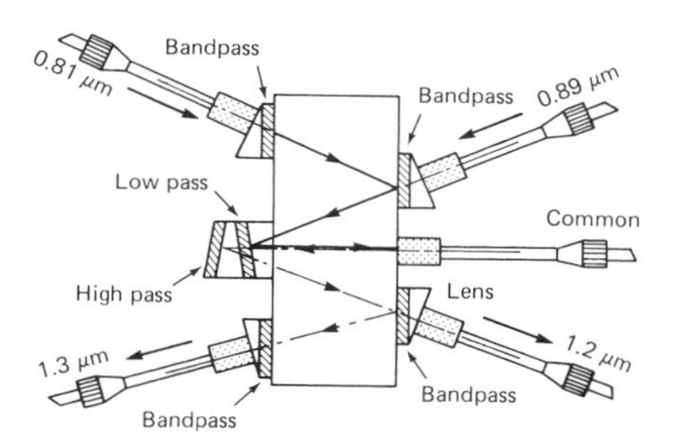

FIG. 14. DMX using multilayer filter. [From Watanabe, R., Sano, K., and Minowa, J. (1983). *IOCC*'83, 30C1-2.]

temperature). Therefore in this case the wavelength of the utilized light must be stabilized.

E. Optical Isolator

An optical isolator is used in sophisticated lightwave systems where one cannot allow light reflection that might perturb the laser oscillator. Of course, to maintain system performance one wishes to introduce an optical isolator, but the cost sometimes prevents this. The principle of the isolator is illustrated in Fig. 16. The Faraday rotator rotates the polarization of incident light by 45°. Then the reflected light can be cut off by the polarizer at the input end while the transmitted light passes through the analyzer at the exit end. Lead glass is used for the short-wavelength region and YIG ($Y_2Fe_5O_{12}$) for the long-wavelength region. A device with 1 dB of insertion loss and 30 dB of isolation is developed for wavelength multiplexing communications.

F. Functional Components

Functional components such as a switch and a light modulator are important for electrooptic systems. Several types of optical switches have been considered:

1. mechanical switch.
2. electrooptic switch, and
3. magnetooptic switch.

A switch as shown in Fig. 17, using the same idea as that of the isolator, will be introduced to switch laser sources in undersea communications for the purpose of maintaining the system when one of the laser devices fails.

A beam deflector is important in the field of laser printers. A rotating mirror is used now, but some kind of electrooptically controlled device is required to simplify the system.

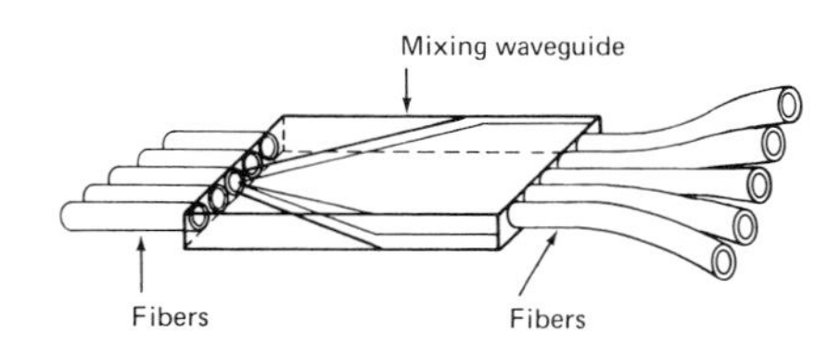

FIG. 13. Waveguide star coupler. [From Minowa, J., Tokura, N., and Nosu, K. (1985). *IEEE J. Lightwave Technology* **LT-3,** 3, 438.]

G. Imager

In this section we deal with some imaging components (imagers), especially those consisting of distribution-index (DI) lenses, including a simple imager, conjugate imager, and multiple imager. In the early history of lightwave transmission, a DI medium was considered a promising device among continuously focusing light guides having geometries of fibers (e.g., by Kapany) or slabs (e.g., by Suematsu).

Some types of lenslike and square-law media have been proposed and studied for light beam waveguides for laser communication systems. Various types of gas lenses and focusing glass fibers provide examples of such media, whose dielectric constants are gradually graded according to a square law with regard to the distance

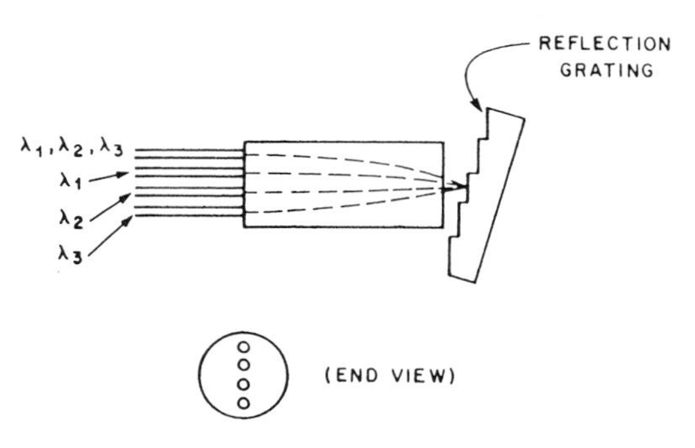

FIG. 15. DMX using a grating. [From Tomlinson, W. J. (1980). *Appl. Opt.* **19,** 1117.]

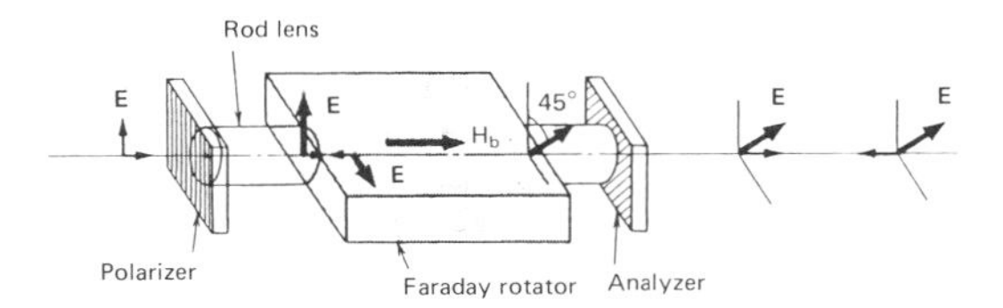

FIG. 16. Optical isolator. [From Suematsu, Y., and Iga, K. (1976). "Introduction to Optical Fiber Communications." Ohm, Tokyo (Wiley, New York, 1980).]

from the center axis. As for the optical characteristics of such media, it is known that a Hermite–Gaussian light beam is guided along the axis of the lenslike medium and, moreover, images are transformed according to a definite transform law that not only maintains the information concerning their intensity distribution but also that concerning their phase relation. This is thought to be one of the significant characteristics of gas lenses and focusing glass fibers, a characteristic different from that of a step-index glass fiber.

Various authors have reported on imaging properties of DI media. The imaging property of a gas lens was investigated, in which the lens formula and optical transfer function were obtained on the basis of geometrical optics, and some imaging experiments were made using a flow-type gas lens. In a paper describing the optical characteristics of a light-focusing fiber guide (SELFOC), Uchida *et al.* mentioned the experimental imaging property and measured the resolving power of the SELFOC lens. In each of these papers, imaging properties of DI media are interpreted in terms of geometrical optics.

When a DI medium is applied to coherent optics such as in interferometry and holography, however, it is important that a two-dimensional system theory based on wave optics be introduced into the treatment of transforms by an optical system with a DI medium. In this article we introduce such a theory, which applies an integral transform associated with a DI medium into the system theory of optics. This will enable us to learn not only about the imaging condition but also about some types of transform representations.

We express the index profile of a DI medium by Eq. (11). If we express the transverse field component by $\exp(j\omega t - j\beta z)$, the function ψ for the index profile is given approximately by the scalar wave equation

$$\frac{1}{r}\frac{d}{dr}\left(r\frac{d\psi}{dr}\right) + \frac{d^2\psi}{d\theta^2} + k_0^2 n^2(r)\psi = \beta^2\psi \quad (49)$$

where $k_0 = 2\pi/\lambda$.

The normal modes associated with a square-law medium, using the first two terms of Eq. (49), are known to be Hermite–Gaussian functions, as shown in Fig. 18.

The characteristic spot size w_0 of the fundamental mode is given by

$$w_0 = a/\sqrt{V} \quad (50)$$

where the normalized frequency V is written as

$$V = kn(0)a\sqrt{2\,\Delta} \quad (51)$$

with $\Delta = [n(0) - n(a)]/n(0)$. In the usual DI lenses, V is larger than 3000, since we have $\lambda = 0.5\ \mu\text{m}$, $n(0) = 1.5$, $a = 0.5$ mm, and $\Delta = 5\%$. The characteristic spot size w_0 is therefore smaller than the core radius a by a factor of 50 to 100. The ratio $w_0/a = \sqrt{V}$ is a measure to indicate whether we can use this gradient medium as an imaging lens, because a sinusoidal ray trace is distorted if w_0/a is not small. The propagation constant associated with the index profile given

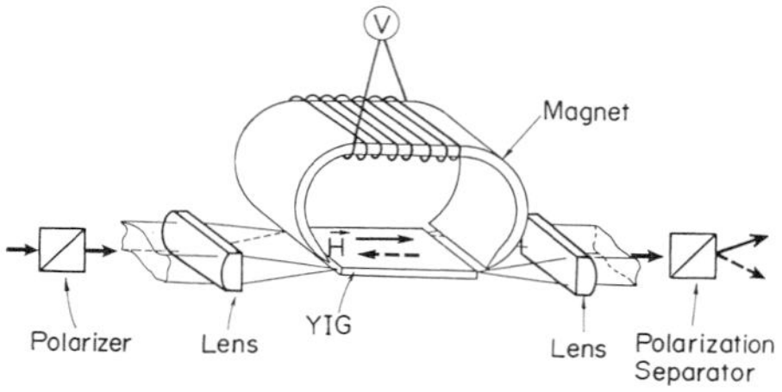

FIG. 17. Optical switch. Arrow indicates a vector quantity. [From Shirasaki, M. (1984). *Jpn. Annu. Rev. Electron. Comput. Telecommun. Opt. Devices Fibers* **11,** 152.]

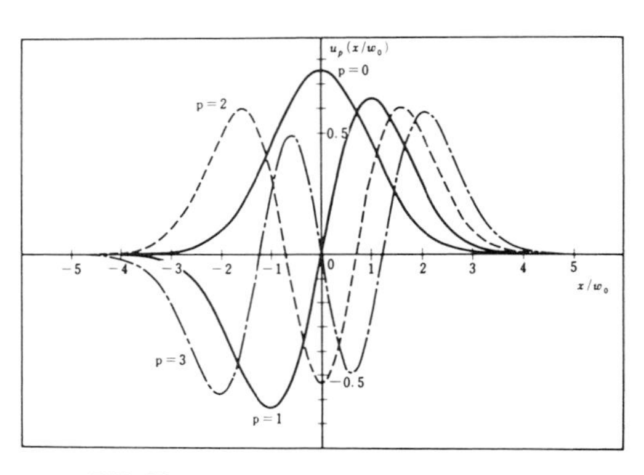

FIG. 18. Hermite–Gaussian functions.

by Eq. (49), including higher-order terms, is obtained by a perturbation method and is expressed in terms of a series expansion in powers of $g/k(0)$, where $k(0) = k_0 n(0)$, as

$$\begin{aligned}\beta_{lm}/k_0 = {} & 1 - (2l + m + 1)(g/k(0)) \\ & + \tfrac{1}{2}\{h_4[\tfrac{3}{2}(2l + m + 1)^2 \\ & + \tfrac{1}{2}(1 - m^2)] - (2l + m + 1)^2\} \\ & \times (g/k(0))^2 + O[(g/k(0))^3] + \cdots \end{aligned} \tag{52}$$

We should note that $g/k(0)$ is of the order of 10^{-4} to 10^{-3}. If $m = 0$, Eq. (52) is associated with the radially symmetric mode that corresponds to meridional rays.

If we calculate the group velocity v_g by differentiating $n(0)$ with respect to ω, we see that the minimum dispersion condition is $h_4 = \frac{2}{3}$, the same result obtained from the WKB method.

V. Microoptic Systems

A. Lightwave Communications

There is a wide variety of microoptic components used in lightwave communication systems. The simplest system consists of a light source such as a laser diode, an optical fiber, and a detector at the receiving end, as shown in Fig. 19. The component employed is said to focus the laser light into the fiber. The wavelength multiplexing (WDM) system needs a more sophisticated combination of devices. At the output end the multiplexer (MX) combines multiple wavelengths into a single piece of fiber, and on the contrary the demultiplexer (DMX) is utilized for the purpose of dividing different signals on different wavelengths. The important points have been introduced in the previous section, and we have to pay attention to the near-end reflection which affects the perturbation of the laser oscillator to obtain stable operation. Many long-haul transmission systems have been or are about to be installed in many countries (e.g., in the U.S. the Northeast corridor system is working between Washington, D.C., and Boston, and in Japan a 2000-km transmission system has been developed as part of INS (integrated network system). By 1988, transatlantic and transpacific underseas cables will be in use for international communications having a very wide bandwidth of about several thousand voice channels. The local area network (LAN) will be a most popular lightwave communication. A video transmission system is an attractive medium for education and commercials. At the Tokyo Institute of Technology television classrooms were introduced to connect two campuses separated by 27 km by eight-piece single mode fibers. Every minute 400-Mbit/sec signals are transmitted back and forth. The quantity of the microoptic components employed will increase more and more as higher speeds and more complex designs are considered.

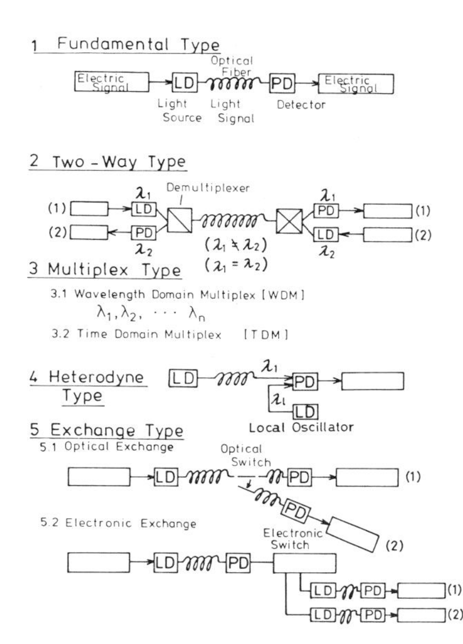

FIG. 19. Some lightwave communication systems. [From Suematsu, Y. (1983). *Proc. IEEE* **71**, 692–721.]

B. Laser Disks

One of the most popular systems is a laser disk for audio known as a compact disk (CD), in which PCM signals are recorded on the rear side of a plastic transparent disk and laser light reads them as shown in Fig. 20. A video disk and optical disk file for computer memory will be the successor to the CD. The optics used there consist of a combination of microoptic components. The light from a semiconductor laser is collimated by a collimating lens with large NA, passes through a beam splitter, and is focused on the rear surface of the disk by a focusing objective lens. The light reflected from the disk changes its radiation pattern as a result of the depth of the pits recorded on the disk, and the variation of light intensity is detected by a matrix light detector. The most important element is the focusing lens, since it must be designed

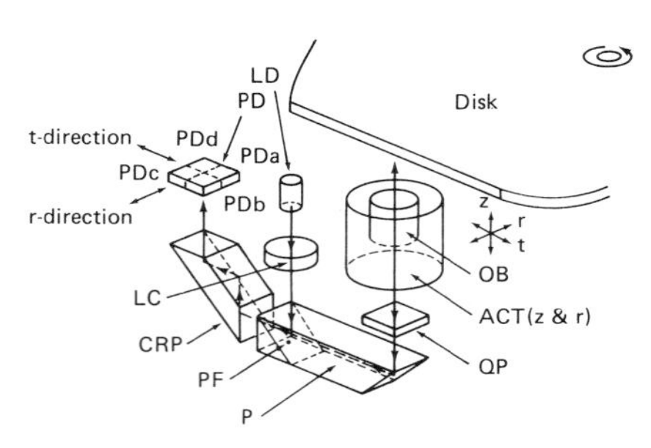

FIG. 20. Compact audio disk system. [From Musha, T., and Morokuma, T. (1984). *Jpn. Annu. Rev. Electron. Comput. Telecommun. Opt. Devices Fibers* **11,** 108.]

with a diffraction limit while remaining lightweight because motional feedback is employed to compass the disk deformation.

C. Copiers

A lens array consisting of distributed index lenses is introduced in a desk-top copying machine in which conjugate images (erect images with unit magnification) of a document can be transferred to the image plane as illustrated in Fig. 21. This configuration is effective for eliminating a long focal length lens and the total volume of the copying machine can be drastically reduced. The number of lenses produced is approaching more than 10^8.

D. Autofocusers

An autofocusing element to provide easy focusing for a steel camera is now becoming popular. Multiple images are formed by a lens array,

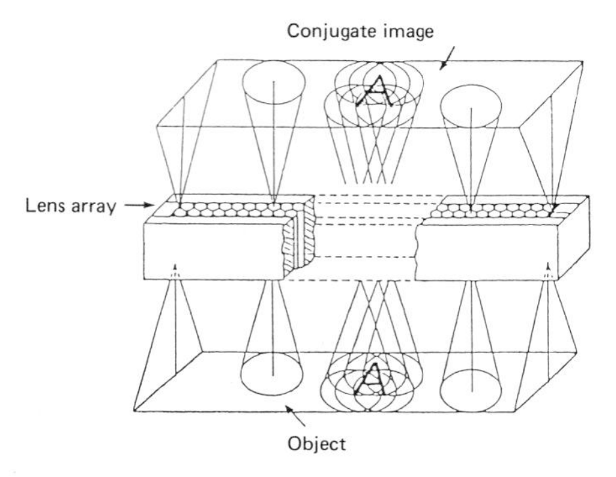

FIG. 21. Conjugate image made by DI lens array for a copying machine. [From Kitano, I. (1983). *Jpn. Annu. Rev. Electron. Comput. Telecommun. Opt. Devices Fibers* **5,** 151.]

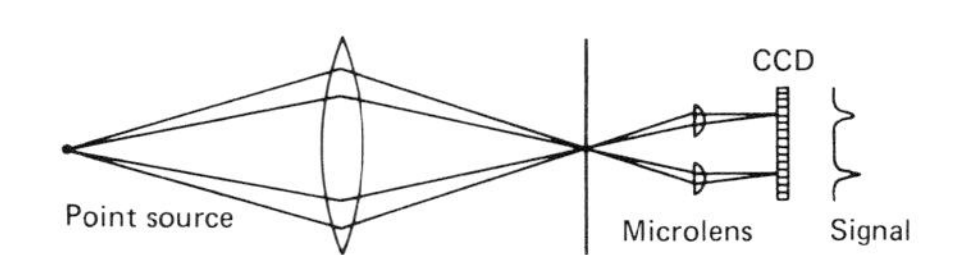

FIG. 22. Autofocus microoptic system. [From the Minolta catalogue, Nagoya, Japan.]

and a CCD detector generates an error-defocus signal until the main lens is automatically moved to the correct position, as in Fig. 22. Some types of microlens arrays have now been developed. A planar microlens array will be one of such arrays since it has the merit of being easy to mask by a photolithographic technique.

E. Fiber Sensors

A fiber gyro and other lightwave sensing systems are considered in various measurement demands. The microoptic element employed is some type of interferometer consisting of a beam splitter and half-mirrors, as shown in Fig. 23. Single mode components that match the mode of the single mode employed and polarization-maintaining fiber are necessary. An optical circuit made of manufactured fiber is an interesting method.

F. Optical Computers

A future technique may be an optical parallel processor such as a TSE computer; this idea is illustrated in Fig. 24. Some functional devices such as optical AND and OR elements based on semiconductor materials must be developed to obtain this sophisticated system. The two-dimensional configuration made of planar microlenses and a surface-emitting laser array will be very helpful.

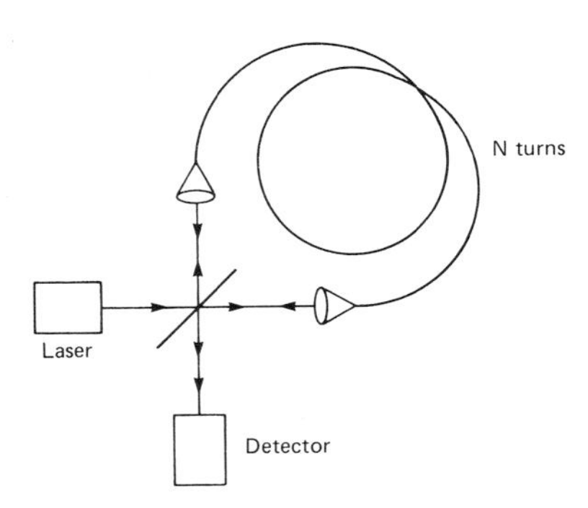

FIG. 23. Fiber-optic gyro. [From Ezekiel, S., and Arditty, H. J. (1982). "Fiber-Optic Rotation Sensors." Springer-Verlag, Berlin and New York.]

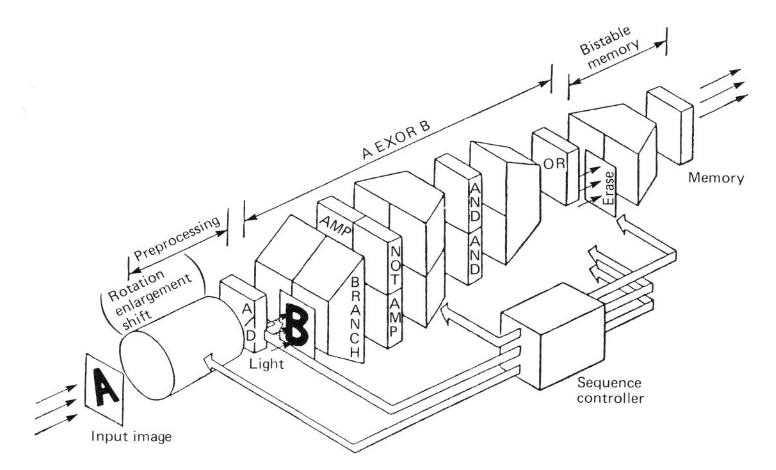

FIG. 24. Idea of an optical computer. [From Seko, J. (1984). *Oyo Butsuri* **53**, 409.]

VI. Characterization of Microoptic Components and Systems

A. Measurement of Refractive Index Distribution

1. Longitudinal Interference Method

The sample (fiber or preform rod sample) is cut into a thin round slice and its surfaces are polished to be optically flat. This thin sample is examined under an interference microscope, as shown in Table III. The refractive index profile is calculated from the fringe shift. An automatic measuring system has been developed by incorporating an image processing apparatus, such as a vidicon camera, and a computer. The spatial resolution limit is about 0.7 μm. If the sliced sample is too thick, the incident ray is refracted through the sample and causes an error. Therefore, the sample must usually be polished to a thickness of less than 100 μm. This takes much time, which prevents this method from being introduced into the fiber manufacturing process as a testing system. Accuracy of index is limited to about 0.0005 because of the roughness of the polished surfaces.

2. Transverse Interference Method

The sample is immersed in index-matching oil and is observed in its transverse direction by using an interference microscope (Table III). The index profile is calculated from the fringe shift. Before the author began this study, analysis based on the straight-ray trajectory had always been used to calculate the index profile. However, it is now known that accuracy can be increased by using an analysis that includes ray refraction. There also exists another method that uses the ray refraction angle to calculate the index profile, but the accuracy is not very good.

3. Transverse Differential Interference Method

The transverse differential method is an interference method modified to apply to thick samples, such as focusing rod lenses and optical fiber preform rods. Instead of a transverse interference pattern, a transverse differential interference pattern, differentiated with respect to the transverse distance, is used to calculate the index profile.

4. Focusing Method

When an optical fiber with an axillary symmetric index distribution is illuminated in its transverse direction by an incoherent light source, the fiber acts as a lens so that the incident light is focused on a plane placed behind the fiber as seen from Table III. If the light intensity distribution is uniform with respect to the incident plane, the index profile can be calculated from the focused light intensity distribution. This method can be applied to preform rods as well as to fibers and is one of the promising methods, along with the transverse and transverse differential interference methods.

TABLE III. Various Measuring Methods of Index Profile[a]

	Measurement time	Accuracy	Sample preparation	Correction of elliptical deformation
Longitudinal interference method	long	good	mirror polish	easy
Near-field pattern method	short	fairly good	cleave	possible
Reflection method	medium	fairly good	cleave	possible
Scattering pattern method	short	fairly good	not necessary	difficult
Transverse interference method	short	good	not necessary	possible
Focusing method	short	good	not necessary	not practical
Spatial filtering method	short	good	not necessary	possible

[a] From Iga, K., Kokubun, Y., and Oikawa, M. (1984). "Fundamentals of Microoptics." Academic Press, Orlando.

This method can be applied to axially nonsymmetric preforms.

B. Measurement of Composition

1. X-Ray Microanalyzer (XMA) Method

The XMA method measures the dopant concentration profile, which is related to the index profile, by means of an XMA (X-ray microanalyzer). The contribution of dopants such as P_2O_5, GeO_2, and B_2O_5 to the refractive index can be obtained separately, but accuracy is not good because of the low signal to noise ratio.

2. Scanning Electron-Beam Microscope (Etching) Method

When the end surface of a distributed index sample is chemically etched, the etching speed depends on the dopant concentration. Therefore, the unevenness can be observed by a scanning electron-beam microscope (SEM).

C. Reflection Pattern

The reflection coefficient of a dielectric material is related to the refractive index at the incident surface. The refractive index profile of an optical fiber can be measured by utilizing this principle. A laser light beam with a small spot size is focused into the end surface of a sample, and the reflection coefficient is measured by comparing the incident and reflected light intensity, as shown in Table III. The refractive index profile is obtained from the reflection coefficient profile by shifting the reference point. Accuracy is strongly affected by the flatness of the end surface. A fractured end surface gives better results than does a polished end surface. For borosilicate fibers, the result changes rapidly with time because of atmospheric exposure of the dopant. Spatial resolution is usually limited to about 1 to 2 μm by the spot size of the incident beam. This effect of the finite beam spot size can be corrected by numerical calculation.

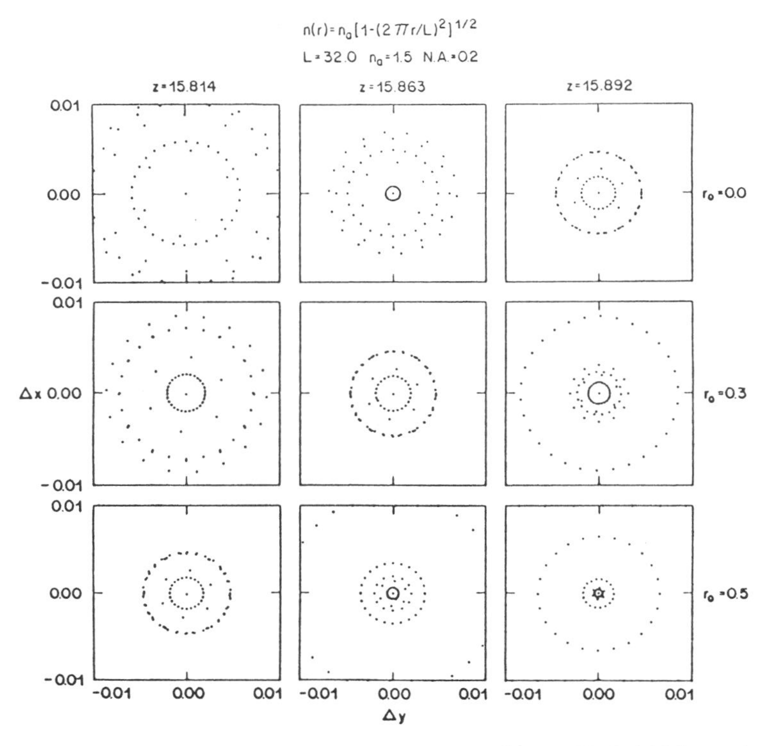

FIG. 25. Spot diagram for a DI lens. [From Tomlinson, W. J. (1980). *Appl. Opt.* **19,** 1117.]

A 0.3-μm spatial resolution and 5% total accuracy of the refractive index has been obtained by this correction.

D. Scattering Pattern

The scattering pattern is classified into both a forward scattering pattern method and a backward scattering pattern method. In the case of the forward scattering pattern method, the sample is immersed in an index-matching oil and the forward scattering pattern is observed. The refractive index profile is calculated from the scattering pattern by using a computer. The error in this method is determined from the product of the core radius a and the index difference Δn, and it increases with an increase of $a\ \Delta n$; as a numerical example, when $a\ \Delta n = 0.04$ mm, the error is 5%. Therefore, this method is applicable only to single mode fibers. Since this method requires many sampling points (500–1000), it is necessary to collect the data automatically.

On the other hand, the index profile can also be obtained from the backward scattering pattern. This method does not require index-matching oil and is applicable to thick samples such as preform rods. However, since the backward scattering pattern is tainted by externally reflected light, it is not suitable for precise measurements. Furthermore, the accuracy of this method is very sensitive to the elliptical deformation of the core cross section.

E. Near-Field Pattern

When all the guided modes of a multimode waveguide are excited uniformly by using an incoherent light source, the near-field pattern of output optical power is similar to the refractive index profile. Since it is difficult to satisfy the incident condition strictly and the near-field pattern is affected by leaky modes and the absorption loss difference of guided modes, this method cannot provide accurate measurements. Although several improvements, such as a correction factor for leaky modes, a refracting ray method that is free from the leaky mode effect, and a spot scanning method, have been made to increase accuracy, this method is being used only as an auxiliary technique.

F. Far-Field Pattern

This method utilizes the far-field pattern of output optical power instead of the near-field pattern. This method is applicable only to single mode waveguides. The former method is not very accurate because of modal interference within the far field. The latter requires an optical detector with a large dynamic range and the error is more than 5%.

G. Spot Diagram

The spot diagram is a standard characterization of classical optical systems. This is also applied to dielectric optical waveguides and fibers as well as microoptic components. In Fig. 25, we show a typical spot diagram for a DI rod lens which exhibits some aberrations due to the index distribution.

F. Optical Transfer Function

The optical transfer function (OTF) presents the fineness with which we can transmit spatial information in the spatial frequency domain. The OTF $H(s)$ is defined with s the spatial frequency,

$$H(s) = \frac{1}{N} \sum_{i=1}^{N} \exp(jsx_i) \tag{53}$$

where x_i is the ray position given by the spot diagram and N is the total number of spots. Figure 26 gives one example of OTF obtained from the spot diagram of a DI lens.

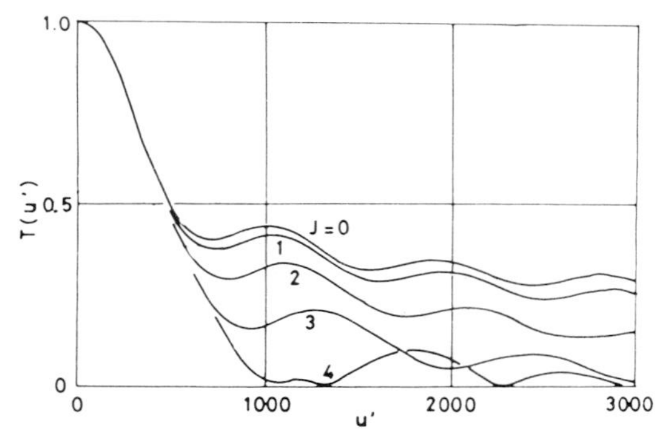

FIG. 26. OTF for a DI lens. [From Iga, K., Hata, S., Kato, Y., and Fukuyo, H. (1974). *Jpn. Appl. Phys.* **13**, 79.]

G. Phase Space Expression

It is very convenient to utilize a phase space consisting of the ray position and ray slope to express a mode spectrum of a multimode waveguide. Figure 27 shows a phase space plot of the output end of a branching waveguide.

VII. Stacked Planar Optics

A. Concept of Stacked Planar Optics

Stacked planar optics consists of planar optical components in a stack, as shown in Fig. 28. All components must have the same two-dimensional spatial relationship, which can be achieved from planar technology with the help of photolithographic fabrication, as used in electronics. Once we align the optical axis and adhere all of the stacked components, two-dimensionally arrayed components are realized; the mass production of axially aligned discrete components is also possible if we separate individual components. This is the fundamental concept of stacked planar optics, which may be a new type of integrated optics.

B. Planar Microlens Array

To have stacked planar optics, all optical devices must have a planar structure. The array of microlenses on a planar substrate is required in order to focus and collimate the light in optical circuits. A planar microlens is fabricated by selective diffusion of a dopant into a planar substrate through a mask, as shown in Fig. 29. We can have an array of planar microlenses with a 1.6- to 2.0-mm focal length and a numerical aperture (NA) of 0.34. We have confirmed that the substrate NA can be increased to as high as 0.54 by stacking two microlenses. This value is considered to be large enough for use as a focusing light from laser diodes.

A planar microlens is fabricated by using an electromigration technique, which was described by Iga, Kokubun, and Oikawa. The substrate is a planar glass 40 × 40 × 3 mm, where planar microlenses were formed as a 40 × 40-matrix with a 1-mm pitch. The radius of the mask is about 50 μm and the radius of the resultant lens is 0.45 mm. The focused spot of the collimated He–Ne laser beam ($\lambda = 0.63\ \mu$m) was measured with the planar microlens. We could observe an Airy-like disk originating from dif-

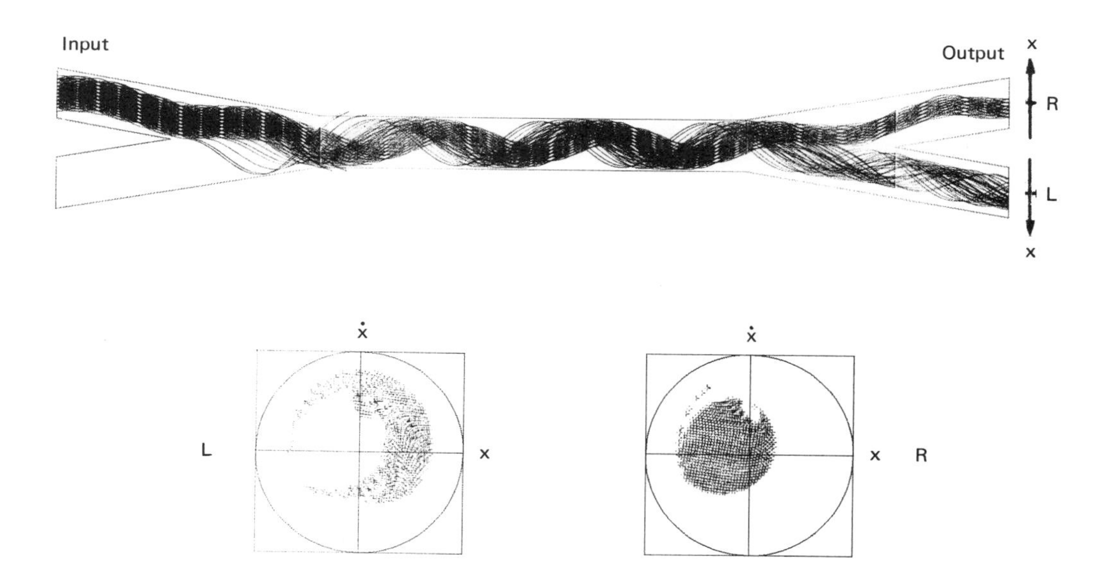

FIG. 27. Phase space plot of a waveguide branch. [From Kokubun, Y., Suzuki, S., Fuse, T. and Iga, K. (1986). *Appl. Opt.*]

fraction and aberration, as shown in Fig. 30. The spot diameter is 3.8 μm, small enough in comparison with the 50-μm core diameter of a multimode fiber, even when we use it in the long-wavelength region 1.3–1.6 μm. The data for available planar microlenses are tabulated in Table IV.

C. Design Rule of Stacked Planar Optics

A proposed possible fabrication procedure for stacked planar optics is as follows:

1. design of planar optical devices (determination of thickness, design of mask shape, etc.);

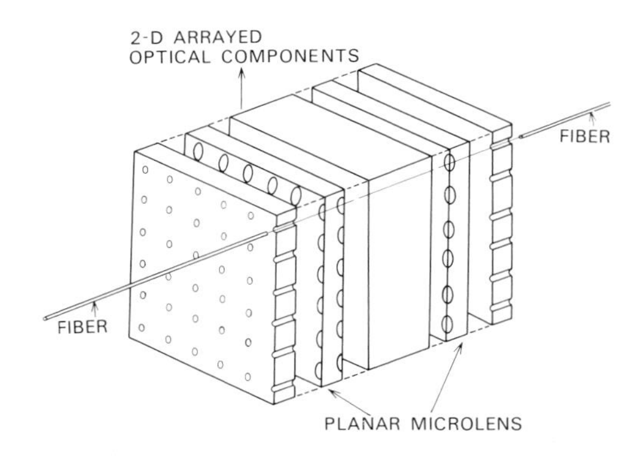

FIG. 28. Stacked planar optics. [From Iga, K., Oikawa, M., Misawa, S., Banno, J., and Kokubun, Y. (1982). *Appl. Opt.* **21,** 3456.]

2. construction of planar optical devices;
3. optical alignment;
4. adhesion;
5. connection of optical fibers in the case of arrayed components; and
6. separation of individual components in the case of discrete components and connection of optical fibers.

Features of stacked planar optics include

1. mass production of standardized optical components of circuits, since the planar devices are fabricated by planar technology,
2. optical alignments, and
3. connection in tandem optical components of different materials such as glass, semiconductors, and electrooptical crystals. This had been

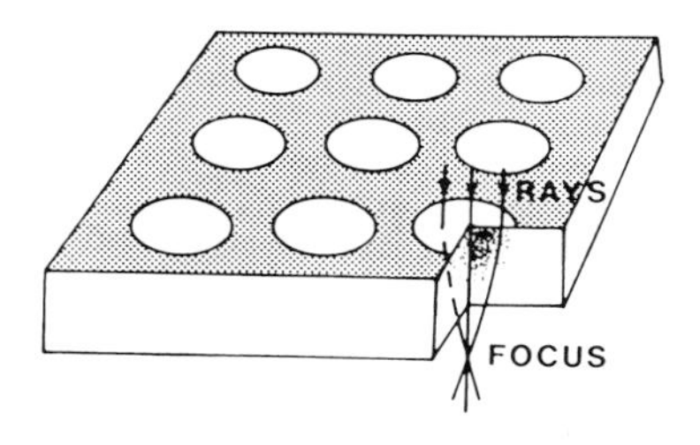

FIG. 29. Distributed-index planar microlens. [From Iga, K., Oikawa, M., Misawa, S., Banno, J., and Kokubun, Y. (1982). *Appl. Opt.* **21,** 3456.]

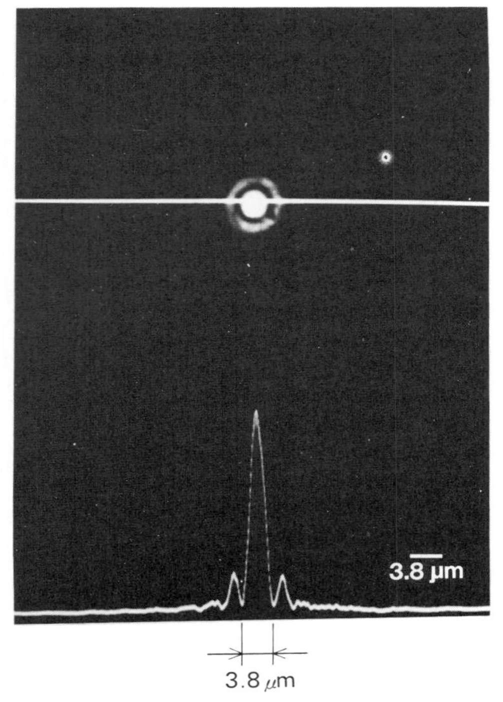

FIG. 30. Focused spot of a planar microlens. [From Iga, K., Oikawa, M., Misawa, S., Banno, J., and Kokubun, Y. (1982). *Appl. Opt.* **21,** 3456.]

thought difficult in integrated optics consisting of planar substrates in which the connection of different components requires optical adjustment, since light is transmitted through a thin waveguide of only a few microns in thickness and width.

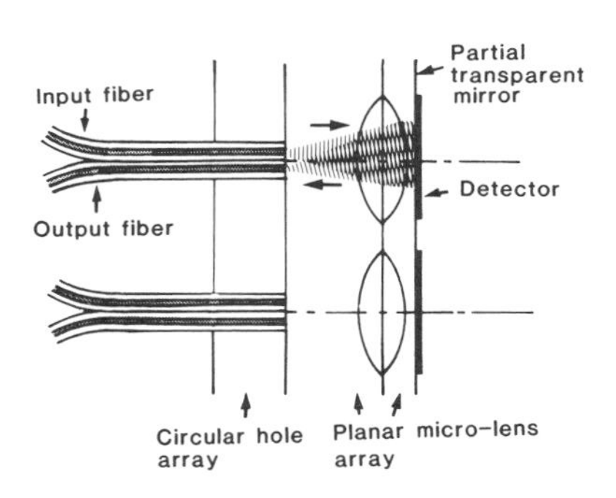

FIG. 31. Optical tap consisting of stacked planar optics. [From Iga, K., Oikawa, M., Misawa, S., Banno, J., and Kokubun, Y. (1982). *Appl. Opt.* **21,** 3456.]

D. Applications of Stacked Planar Optics

Many kinds of optical circuits can be integrated in the form of stacked planar optics, as is summarized in Table V.

We introduce some components as examples of stacked planar optics. The optical tap in Fig. 31 is the component for monitoring the part of the light being transmitted through an optical fiber. The problem of optical tap is that of reducing the scattering and diffraction loss at the component. The light from the input fiber is focused by the use of a partially transparent mirror placed at the back surface of the device. Some of the light is monitored by the detector, which is placed on the back of the mirror. The main beam is again focused by the same lens on the front surface of the output fiber. With this con-

TABLE IV. Design Data of Planar Microlens[a]

	Required	Simple	Pained
NA	0.2–0.5	0.24	0.38
NA_{eff} (aberration free)	0.2	0.18	0.2
Diameter $2a$ (mm)	0.2–1.0	0.86	0.86
Focal distance l (in glass)	0.3–3.8	3.0	1.76
Lens pitch L_p	0.13–1.0	1.0	1.0

[a] From Iga, K., Kokubun, Y., and Oikawa, M. (1984). "Fundamentals of Microoptics." Academic Press, Orlando.

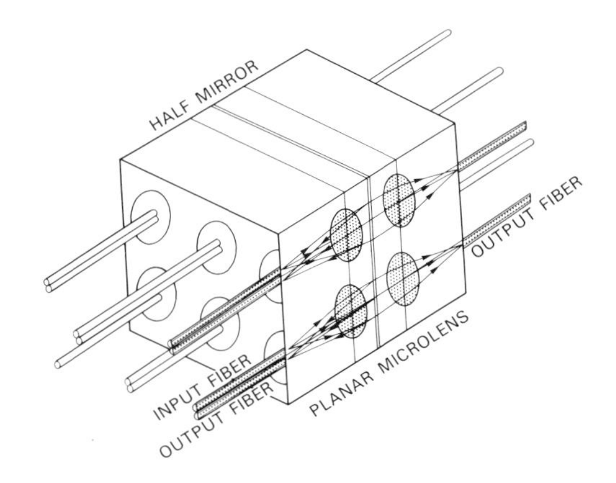

FIG. 32. Branching component made of planar microlenses. [From Oikawa, M., Iga, K., and Misawa, S. (1984). *Dig. Tech. Pap. Top. Meet. Integr. Guided Wave Opt., 1984.*]

figuration we can fabricate many optical taps on the same substrate.

The 2 × 3 branching component has been produced with two pieces of stacked planar microlenses and a half-mirror, as shown in Fig. 32. The light from the input fiber was collimated by the first stacked planar microlens, and a part of the light was reflected by the half-mirror and focused again to output fiber 2. Output fiber 2 was put in contact with input fiber 1. The off-axial value was then 62.5 μm, the fiber radius. The collimated light through the half-mirror was also focused to output fiber 3. In order to put the fibers on each surface, the thickness of the planar microlens was carefully designed and adjusted by using the ray matrix method.

For the moment we are not concerned with the coupling effect among optical components in the stacked planar optical circuit. But we can

TABLE V. Basic Optical Components for Stacked Planar Optics and Optical Circuits[a]

Basic components	Application	Reference
Coaxial imaging components	Coupler[b]	4, 11, 12
Noncoaxial imaging components (transmission-type)	Branching circuit[b]	3, 4, 12, 13
	Directional coupler[b]	13
	Star coupler[c]	12
	Wavelength demultiplexer[b]	4, 12, 13
Noncoaxial imaging components (reflection-type)	Wavelength demultiplexor[b]	12
	Optical tap[b]	12, 13
Collimating components	Branching insertion circuit[b]	3, 4, 12, 13
	Optical switch[c]	12, 13
	Directional coupler[b]	3, 4, 11, 13
	Attenuater[b]	11

[a] From Iga, K., Kokubun, Y., and Oikawa, M. (1984). "Fundamentals of Microoptics." Academic Press, Orlando.
[b] Circuit integrated in a two-dimensional array.
[c] Circuit integrated in a one-dimensional array.

construct a three-dimensional optical circuit that structures the network by allowing coupling among adjacent components.

Since the accumulation of lens aberration may bring about coupling loss, the number of stackings is limited by the aberration of the planar microlenses. The reduction of aberration in the planar microlens is important, therefore, if we apply stacked planar optics to more complex components with a large number of stacks.

Stacked planar optics, a new concept in integrating optical circuits, has been proposed. By using stacked planar optics, we not only make possible the monolithic fabrication of optical circuits, such as the directional coupler and wavelength demultiplexer, but we can also construct three-dimensional optical circuits by allowing coupling among individual components in the array with a suitable design.

BIBLIOGRAPHY

Ezekiel, S., and Arditty, H. J. (eds.) (1982). "Fiber-Optic Rotation Sensors." Springer-Verlag, Berlin.

Iga, K., and Kokubun, Y. (1986). "Optical Fibers." Ohm, Tokyo.

Iga, K., Kokubun, Y., and Oikawa, M. (1984). "Fundamentals of Microoptics." Academic Press, New York.

Suematsu, Y. (1983). *Proc. IEEE* **71,** 692–721.

Suematsu, Y., and Iga, K. (1976). "Introduction to Optical Fiber Communication." Ohm, Tokyo (Wiley, New York, 1980).

Tomlinson, W. J. (1980). *Appl. Opt.* **19,** 1117.

Uchida, T., and Kobayahsi, K. (1982). *Jpn. Annu. Rev. Electron. Comput. Telecommun. Opt. Devices & Fibers* **4,** 172.

Uchida, T., Furukawa, M., Kitano, I., Koizumi, K., and Matsumura, H. (1970). *IEEE J. Quant. Electron.* **QE-6,** 606.

MODE-LOCKING OF LASERS

P. W. Smith
A. M. Weiner *Bell Communications Research Inc.*

Glossary

Active mode-locking: Technique for mode-locking in which an externally-driven modulating element is placed within the laser resonator.

Autocorrelation: Autocorrelation of a pulse is an integral over time of the product of the pulse with a time-delayed replica of itself.

Chirp: Time-varying frequency.

Color-center laser: Laser in which the gain medium is a defect center in an alkali halide crystal.

Diffraction grating: Optical component with many closely spaced grooves that serves to separate light into its component colors or wavelengths.

Dye laser: Laser in which the gain medium consists of an organic dye dissolved in a liquid solvent.

Femtosecond (fsec): One quadrillionth (10^{-15}) of a second. In 1 fsec, light travels ~1% of the thickness of a human hair.

Group velocity dispersion: Variation in pulse-propagation velocity with pulse wavelength.

He–Ne laser: He–Ne laser uses a gas discharge of helium and neon as the gain medium. It was the first laser to emit a continuous output beam.

Ion laser: Laser that uses a gas discharge containing ions as the gain medium.

Laser: Laser is an acronym for *l*ight *a*mplification by *s*timulated *e*mission of *r*adiation.

Laser resonator: Arrangement of open mirrors used to direct light back and forth through the laser gain medium.

Modes: Modes of a resonator are the self-consistent field configurations of the resonator. Longitudinal modes correspond to frequencies for which the resonator length is an integral number of half-wavelengths of light. Transverse modes are modes with different energy distributions in a plane transverse to the light-propagation direction.

Nd : YAG laser: Laser whose gain medium consists of a neodymium-doped yttrium aluminum garnet crystal.

Passive mode-locking: Technique for mode-locking in which a passive nonlinear optical element is place within the laser resonator.

Picosecond (psec): One trillionth (10^{-12}) of a second. In 1 psec, light travels about one third of a millimeter.

Saturable absorber: Absorbing material whose absorption decreases (saturates) with increasing incident light intensity.

Second harmonic generation: Nonlinear optical process in which light at twice the original frequency (half the wavelength) is generated.

Self-phase-modulation: Nonlinear optical process that occurs when an intense optical pulse changes the refractive index of the medium in which it propagates. The change of refractive index in turn modifies the spectrum of the pulse.

A typical laser consists of two essential elements: gain and feedback. A beam of light passing through the gain, or amplifying, medium stimulates it to release its stored energy in the form of additional light that adds to, or amplifies, the beam. Feedback is achieved by placing the gain medium within a resonator (a set of mirrors that reflects the beam back and forth through the gain medium). As a result of this cumulative process, an intense coherent beam of light is produced. The light from such a laser is composed of a number of discrete wavelengths corresponding to different resonant frequencies, or modes, of the resonator.

The total output of such a laser as a function of time will depend on the amplitudes, frequencies, and relative phases of all of these oscillating modes. If there is nothing that fixes these parameters, random fluctuations and nonlinear effects in the laser medium will cause them to change with time, and the output will vary in an uncontrolled way. If the oscillating modes are forced to maintain equal frequency spacings with a fixed phase relationship to each other, however, the output as a function of time will vary in a well-defined manner. The laser is then said to be mode-locked. The form of this output will depend on which laser modes are oscillating and what phase relationship is maintained. The major interest in mode-locking is as a means of generating trains of ultrashort light pulses, although other types of mode-locking leading to FM-modulated or spatially scanning laser beams are also possible.

I. Introduction

A. Brief History of Mode-Locking and Short Optical Pulse Generation

The first theoretical work to describe clearly mode-locking phenomena was published in 1964 and 1965, some five years after the invention of the laser. The first experimental results describing mode-locking of an He–Ne gas laser were published by Hargrove, Fork, and Pollack in 1964. Self-mode-locking was first observed in 1965, as was FM mode-locking. Mode-locking of several different types of lasers was investigated in 1965, and various mode-locking techniques were tried, of which one of the most important was the use of a saturable absorber dye for passive mode-locking. In 1966, mode-locking of an Nd:YAG laser was reported. By now, the mode-locked pulses were becoming too short to be measured even with the fastest detectors, and new measurement methods using autocorrelation techniques had to be developed. Transverse mode-locking techniques producing a spatially scanning beam were studied in the period 1968–1970. [*See* LASERS.]

Optical pulses in the picosecond (10^{-12} sec) range were first generated in 1966. Since then, dramatic progress has been made. Pulses shorter than 1 psec were achieved in 1974 using a linear cavity, passively mode-locked continuous-wave (cw) dye laser, and in 1981, the invention of the ring-cavity, passively mode-locked cw dye laser [commonly called the colliding-pulse mode-locked (CPM) ring dye laser] resulted in pulsewidths below 100 fsec (1 fsec = 10^{-15} sec). Amplified pulses from a CPM laser have now been compressed to durations as short as 8 fsec. This rapid progress in ultrashort optical pulse technology is illustrated by Fig. 1, which shows the shortest pulsewidths achieved as a function of year. Note that an 8-fsec optical pulse contains only about four optical wavelengths. Since an optical pulse can be no shorter than one wavelength, we are fast approaching fundamental limits.

B. Applications of Mode-Locked Lasers

Ultrashort light pulses from mode-locked lasers are finding many applications in science and technology.

1. They provide scientists with a tool for studying ultrafast processes in physics, chemistry, and biology. An analogy is strobe photography. Short bursts of light on the order of tens of microseconds in duration can "freeze" mechanical motions for stop-action photographs. By this method, we can photograph a bullet as it is shot through an apple or see a golf club hit and initially compress a golf ball. Strobe photogra-

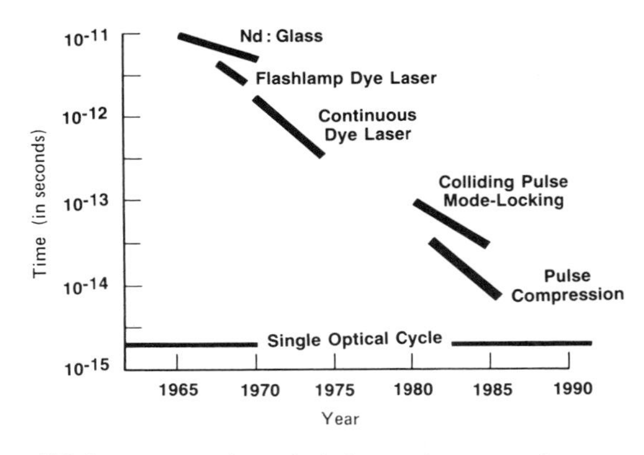

FIG. 1. Progress in optical short-pulse generation using mode-locking and pulse compression techniques. In 1985, pulses of only four optical cycles in duration were produced.

phy works because the light flashes occur so fast that no noticeable mechanical movement occurs during the flash. On a pico- or femtosecond time scale, the microscopic world is alive with motion. Atoms and molecules are vibrating; electrons are colliding with and scattering from the crystal lattice in metals and semiconductors; and rhodopsin molecules in the eye are undergoing complex photochemical changes in the process of converting incident light into a perceived image. Sophisticated measurement techniques using ultrashort light pulses make it possible to study this microscopic world.

2. Ultrashort light pulses are opening up new areas in optoelectronics; for example, they may be used to generate, switch, and sample short electrical pulses. Thus, optical techniques can be used to measure the performance of very fast electronic structures and devices.

3. Ultrashort light pulses are a potential source of very high bit rate pulses for future optical communications or computations systems. The shorter the pulses, the more can be packed into a given time interval. This permits higher data rate transmission or faster computation.

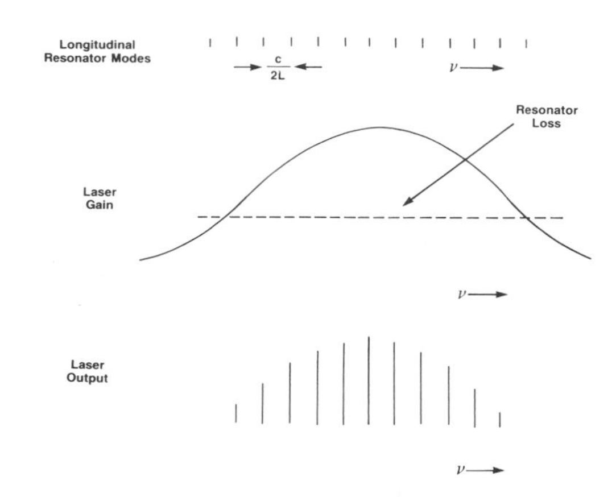

FIG. 2. Laser operation on several longitudinal resonator modes: (a) longitudinal modes of the laser resonator; (b) laser gain versus frequency showing the region where the gain exceeds the cavity losses; (c) laser output—oscillation at resonator mode frequencies for which laser gain exceeds the losses.

II. Modes in Laser Resonators

A mode of a resonator can be defined as a self-consistent field configuration. That is, the optical field distribution reproduces itself after one round trip in the resonator. The modes of an open resonator formed by a pair of coaxial plane or spherical mirrors have been studied in great detail. We can identify a set of longitudinal (or axial) modes that all have the same form of spatial energy distribution in a transverse plane but have different axial distributions corresponding to different numbers of half-wavelengths of light along the axis of the resonator. These longitudinal modes are spaced in frequency by $c/2L$, where c is the velocity of light, and L is the optical path length within the resonator. Figure 2 illustrates the type of laser output we may obtain when a number of longitudinal resonator modes are above threshold for laser operation. Each frequency shown in Fig. 2a corresponds to a different longitudinal mode number, that is, a different number of half-wavelengths of light in the resonator. For most lasers this number is between 10^3 and 10^7. Figure 2b illustrates a typical situation where the laser-amplifying medium exhibits gain exceeding the resonator loss over a range of frequencies. Laser output may then build up in all resonator modes within this range (Fig. 2c).

To obtain laser operation at a single longitudinal mode, it is usually necessary to design a laser resonator sufficiently short so that $c/2L$ is greater than the bandwidth of the gain or to design a complex resonator that has high loss for all modes within the oscillation bandwidth except the favored one.

It is possible to show that for each longitudinal mode number there exists a set of self-consistent solutions for the light energy inside the open resonator that correspond to different energy distributions in a plane transverse to the resonator axis. These are called the transverse modes of the resonator. Figure 3 shows the field distributions for some low-order transverse modes with rectangular symmetry. There exists a complementary set with circular symmetry. The properties of both sets are similar. The cross-sectional amplitude distribution of these modes $A(x, y)$ is given closely by

$$A(x, y) = A_{m,n}\left[H_m\left(\frac{\sqrt{2}x}{w}\right)H_n\left(\frac{\sqrt{2}y}{w}\right)\right] \times \exp\left[\frac{-(x^2 + y^2)}{w^2}\right] \quad (1)$$

where x and y are the transverse coordinates; $A_{m,n}$ is a constant whose value depends on the field strength of the mode; w is the radius of the fundamental mode ($m = 0$, $n = 0$) at $1/e$ of the maximum amplitude (and is often referred to as the fundamental spot size); $H_a(b)$ is the ath-order Hermite polynomial with argument b; and m and n are called the transverse mode numbers. These modes are sometimes referred to as TEM_{mn} modes by analogy with the modes in waveguides. Note that this distribution is independent of the longitudinal mode number. Figure 3 is a plot of A along the y axis, for $m = 0$–5. Note that as the transverse mode number increases, the energy is spread further and further from the axis of the resonator. Figure 4 shows the actual transverse mode patterns found with a 6328-Å He–Ne gas laser.

In order to obtain oscillation in a single transverse mode, it is necessary to use some device that gives high losses to all transverse modes but the desired one. Since higher order modes spread further from the resonator axis, the easiest way to accomplish single-transverse-mode operation is to insert into the laser resonator a circular aperture whose size is such that the fundamental mode experiences little diffraction loss while higher order modes suffer appreciable attenuation.

For resonators that do not have extremely large diffraction losses, the resonant frequency of a mode can be written

$$\nu = \frac{c}{2L}\left[(q+1) + \frac{m+n+1}{\pi} \times \cos^{-1}\sqrt{\left(1-\frac{L}{R_1}\right)\left(1-\frac{L}{R_2}\right)}\right] \quad (2)$$

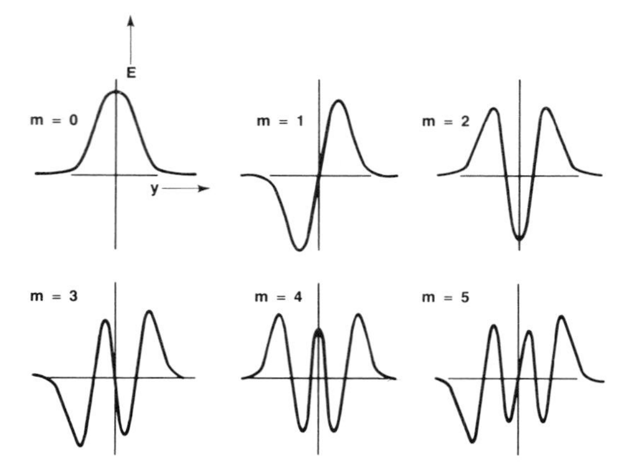

FIG. 3. Plot of electric field amplitude of transverse modes versus the normalized transverse coordinate y/w (w is the spot size of fundamental mode), for $m = 0$–5. Reversal of sign indicates 180° change of phase of the electric field.

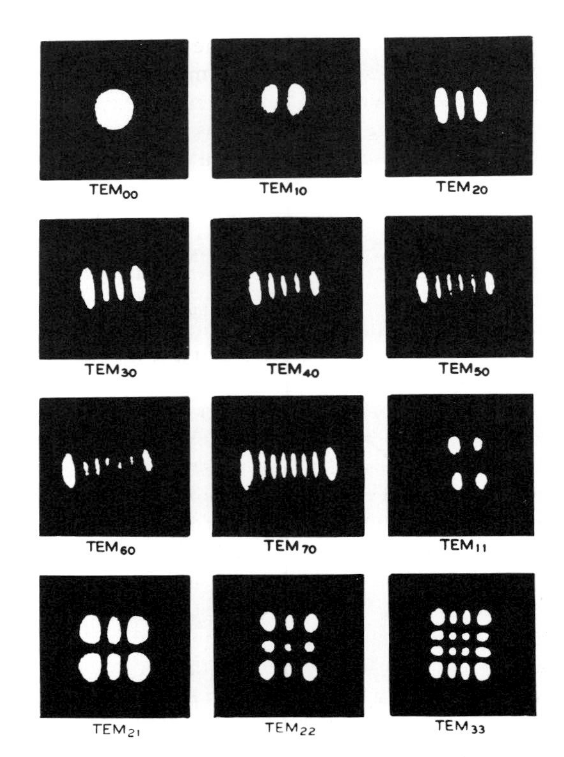

FIG. 4. Experimentally observed transverse mode patterns of a gas laser oscillator with rectangular symmetry.

where $q + 1$ is the number of half-wavelengths of light along the axis of the resonator; m and n are the transverse mode order numbers; and R_1 and R_2 are the radii of the two mirrors making up the laser resonator. Note the following points: (a) for a given transverse mode (given m and n), longitudinal modes with mode number differing by 1 are spaced in frequency by $c/2L$; (b) for a given longitudinal mode number (q), transverse modes with the sum of m and n differing by 1 are spaced in frequency by $(c/2\pi L)\ \cos^{-1}\sqrt{(1 - L/R_1)(1 - L/R_2)}$.

Let us consider the output of a laser operating in a number of longitudinal and transverse modes. The total field is the sum of the individual fields of each of the modes. The optical length L is not the same for all the modes due to the dispersion of the laser material. In general, both the amplitude and the phase of these modes vary with time due to random mechanical fluctuations of the laser resonator length and the nonlinear interaction of these modes in the laser medium. The total field thus varies with time in some uncontrolled way with a characteristic time that is of the order of the inverse of the bandwidth of the oscillating mode frequency spectrum. In Section IV, we shall discuss techniques for fixing the characteristics of the oscillating modes in order to obtain a controlled laser output.

III. Types of Mode-Locking

A. Longitudinal Mode-Locking

Consider a set of oscillating longitudinal modes all corresponding to the same transverse mode, such as those shown in Fig. 2c. If we mode-lock (i.e., fix the frequency spacing, relative phases, and amplitudes of these modes), the laser output will be a well-defined function of time. Consider the nth mode to have amplitude E_n, angular frequency ω_n, and phase ϕ_n. Then, the total laser output field E_T can be written

$$E_t = \sum_n E_n \exp\{i[\omega_n(t - z/c) + \phi_n]\} + \text{c.c.} \quad (3)$$

where c.c. represents the complex conjugate, and we have assumed that the radiation is traveling in the $+z$ direction. If we have equal mode frequency spacing, $\omega_n = \omega_0 + n\Delta$, where $\Delta = 2\pi(c/2L)$, and ω_0 is the optical frequency of the laser output. Thus, we can write

$$E_T = \exp\{i\omega_0(t - z/c)\} \times \sum_n E_n \exp\{i[n\Delta(t - z/c) + \phi_n]\} + \text{c.c.} \quad (4)$$

This corresponds to a carrier wave of frequency ω_0 whose envelope depends on the values of E_n and ϕ_n. Note, however, that (a) the envelope travels with the velocity of light; and (b) the envelope is periodic with period $T = 2\pi/\Delta = 2L/c$. For ϕ_n equal to a constant independent of n, and for a simple mode spectrum, such as that shown in Fig. 2c, this envelope consists of a single pulse in the period T whose width is approximately the reciprocal of the frequency range over which the E_n's have an appreciable value (i.e., the ratio of the pulse spacing to the pulsewidth is approximately the number of oscillating modes). Within the laser resonator this corresponds to a pulse of light traveling back and forth between the resonator mirrors with the velocity of light. Other selections of amplitudes and phases result in different envelopes. For any set of mode amplitudes, the narrowest pulse always results from the phases ϕ_n equal to a constant for all n. (Since origin of time is arbitrary, we shall in subsequent discussion assume that this constant is 0.) The width of this pulse is inversely proportional to the width of the mode spectrum (Fig. 5). We shall refer to the minimum or zero-phase pulsewidth obtainable from a given mode spectrum as the bandwidth-limited pulsewidth.

Figure 6a shows the actual frequency spectrum of the output of a 6328-Å He–Ne laser. The laser is oscillating on a number of longitudinal modes, and the amplitudes of these modes fluctuate with time. Figure 6b shows the same laser output when the modes are mode-locked. The oscillation bandwidth has increased, and the mode amplitudes are stable. Figure 6c shows the corresponding output as a function of time. The output consists of a train of narrow pulses separated by the round-trip time of the light in the optical resonator ($2L/c$). It is also possible to fix mode amplitudes and phases in other ways. By using a phase modulator within the resonator, it has been shown that we can obtain mode amplitudes and phases corresponding to the carrier and sidebands of an FM wave. Under these conditions, the output intensity is constant as a function of time.

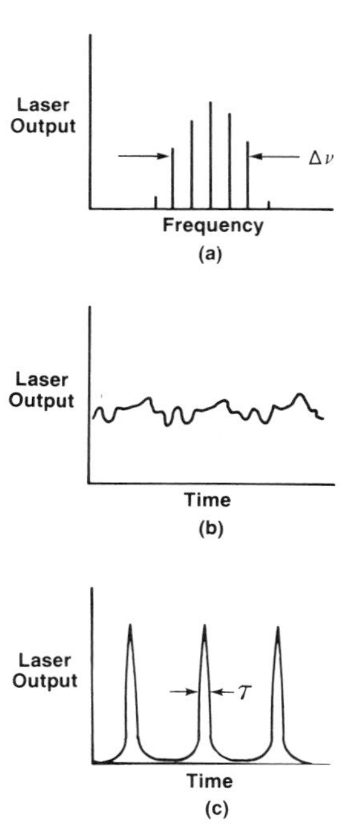

FIG. 5. (a) Output of a laser oscillating in a number of longitudinal modes over a frequency bandwidth $\Delta\nu$. (b) Laser output with randomly phased modes. (c) Mode-locked output with constant mode phases. The minimum pulsewidth (τ) is approximately $1/\Delta\nu$.

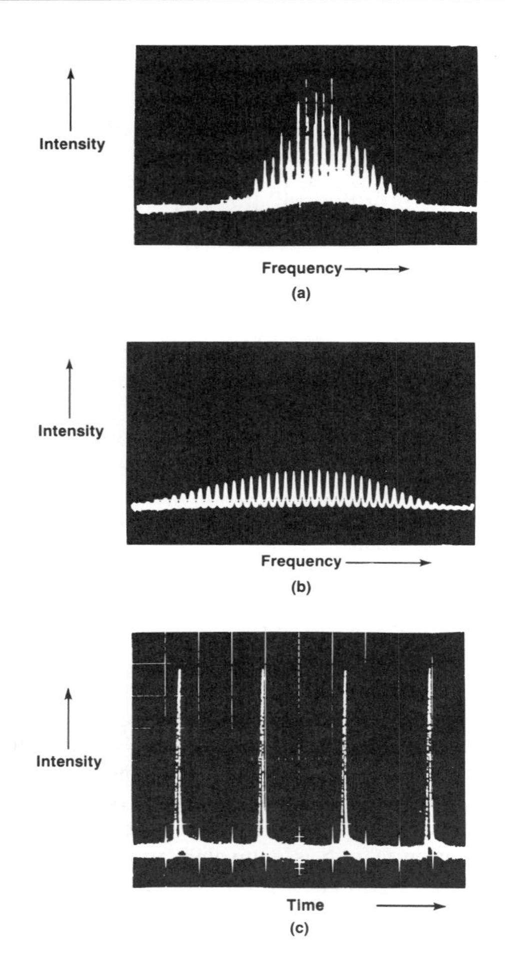

FIG. 6. Experimentally observed laser characteristics: (a) non-mode-locked, output frequency spectrum; (b) mode-locked, output frequency spectrum; (c) mode-locked, output intensity versus time.

In Section IV we discuss the various methods that have been used to fix the mode amplitudes and phases of a set of longitudinal modes so that mode-locked operation can be obtained.

Until now, we have been considering the mode-locking of a set of longitudinal modes all having the same transverse mode distribution. Although the fundamental transverse mode is commonly used, in principle, any set of longitudinal modes having the same transverse mode distribution may be mode-locked.

B. TRANSVERSE MODE-LOCKING

We now consider a set of transverse modes with the same longitudinal mode number but with transverse mode numbers differing by 1 in the y coordinate. These modes are approximately equally spaced in frequency [Eq. (2)]. Consider a set of modes locked with equal frequency spacings and zero phase difference with field amplitudes A_n, where

$$|A_n|^2 = \frac{1}{n!}(\bar{n})^n e^{-\bar{n}} \tag{5}$$

in which $\bar{n}$ is a parameter that determines the number of oscillating transverse modes. Under these conditions, the intensity of the optical field of the laser output is

$$I(\zeta, t) = \frac{1}{\sqrt{\pi}} \exp[-(\zeta - \zeta_0 \cos \Omega t)^2] \tag{6}$$

where $\zeta(=y/w)$ is the transverse coordinate in the y direction normalized with respect to the fundamental spot size; $\zeta_0 = \sqrt{2\bar{n}}$; and Ω is the transverse mode frequency spacing. Equation (6) represents a scanning beam of width equal to the fundamental spot size that moves back and forth in the transverse plane with maximum excursion equal to ζ_0. Figure 7 illustrates this motion over a complete period of $2\pi/\Omega$. The maximum number of resolvable spots is roughly equal to the number of oscillating modes.

It is also possible to consider a set of circularly symmetric transverse modes. In this case it is possible to obtain a beam whose transverse intensity distribution is always Gaussian (i.e., the form of the fundamental transverse mode) but with a periodically time-varying beam radius.

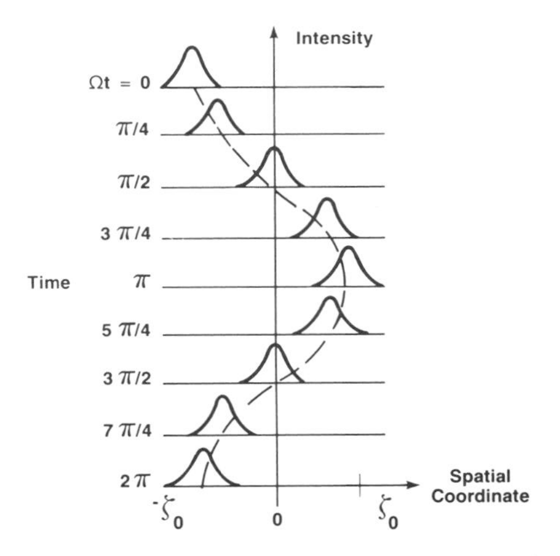

FIG. 7. Plot of spatial intensity distribution as a function of time (in units of $1/\Omega$) for a transverse mode-locked laser.

Simultaneous locking of the modes of a laser oscillating in several longitudinal and transverse modes has also been studied. If a set of transverse modes corresponding to each longitudinal mode number is locked together to form a scanning beam and each of the sets of longitudinal modes is also locked with zero phase difference to form a pulse in the resonator, the light will be confined to a small region of space both in the axial and transverse directions. This "bullet" of light will bounce back and forth in the laser resonator, following the zigzag path to be expected from the laws of geometrical optics.

IV. Techniques of Mode-Locking

To mode-lock a laser, it is necessary to fix the frequency spacing and the relative phases of the laser modes. Although lasers have sometimes been observed to mode-lock spontaneously (self-locking), lasers are usually mode-locked by placing a modulating element, either active or passive, within the laser resonator. The modulating element acts to initiate and maintain the proper relationships between the modes as well as to increase the number of modes that oscillate. In active mode-locking techniques, the modulator is driven by an external power source; in passive mode-locking, the modulator is a nonlinear optical element that is driven by the mode-locked pulses themselves. Table I lists the pulse durations and energies that are obtained with a variety of mode-locked lasers. [*See* NONLINEAR OPTICAL PROCESSES.]

A. Active Techniques

In active mode-locking, an externally driven modulating element is placed within the laser resonator. Loss modulation, gain modulation, and phase modulation have all been employed. When the modulation period is carefully adjusted to match the cavity round-trip time, mode-locking may be achieved. In this case, light incident on a modulator situated at one end of the laser resonator during a certain part of the modulation cycle will again be incident at the same point of the cycle after one round trip in the laser resonator. In the case of loss modulation, light that "sees" loss at one time again sees loss after one round trip. Thus, all the light in the resonator experiences loss except that light which passes through the modulator when the modulator loss is zero. Light tends to build up in narrow pulses in these low-loss time positions. For sufficiently intense modulation, these pulses have a width of the order of the reciprocal of the gain bandwidth.

Mode-locking may also be understood in terms of a frequency domain description. When cavity mode n with frequency ω_n is modulated sinusoidally at frequency Ω, sidebands appear at frequencies $\omega_n \pm \Omega$. Mode-locking occurs when Ω closely matches the mode spacing Δ. (The condition $\Omega = \Delta$ is equivalent to the condition that the modulation period match the cavity round-trip time.) Harmonic mode-locking, where $\Omega = m\Delta$ (m an integer), is also possible but is not considered in the following.

When $\Omega = \Delta$, the sidebands created from

TABLE I. Examples of Mode-Locked Lasers

Mode-locking technique	Laser	Typical pulse duration (psec)	Typical pulse energy (nJ)
Active mode-locking			
Loss modulation	Ion	100	10
	Nd : YAG	100	100
Synchronous pumping	Dye pumped by mode-locked ion or Nd : YAG	1–10	1–10
	Color center pumped by mode-locked Nd : YAG	10	5
	Dye pumped by compressed Nd : YAG	0.3	0.5
Passive mode-locking			
Saturable absorber	Dye	0.5–1	3
CPM	Dye	0.03–0.1	0.1

mode n act as injection signals for modes $n + 1$ and $n - 1$. This has two effects:

1. The modulation tends to ensure that a large number of modes oscillate. For example, starting with only a single mode at frequency ω_0, the modulation produces sidebands that feed modes at $\omega_0 \pm \Delta$. The sidebands generated from these new modes in turn provide injection signals for modes at $\omega_0 \pm 2\Delta$, and so on. The number of modes that will oscillate increases with the strength of the modulation and is ultimately limited by the gain bandwidth (or by a bandwidth-limiting intracavity filter).
2. The result of this injection-locking process is a fixed and stable set of mode amplitudes, frequencies, and phases. This corresponds in the time domain to a train of short pulses, with pulse separation equal to the modulation period and pulsewidth equal to the pulse separation divided by the number of oscillating modes.

Active mode-locking by loss modulation is commonly used to generate pulses from lasers with rather limited gain bandwidths, such as ion lasers or Nd : YAG lasers. The loss modulation is implemented by using an acoustooptic modulator driven by an ultrastable radio frequency (rf) generator. Pulse durations in the range of 80–100 psec are typically achieved.

Although phase modulation has been used in the past to produce mode-locked pulses and to achieve FM mode-locking (Section III), it is no longer used extensively and is not further discussed here.

Mode-locking by gain modulation is similar to mode-locking by loss modulation, except that pulses build up at the times of highest gain rather than those of lowest loss. This type of mode-locking is used with semiconductor diode lasers and organic dye lasers, among others.

In a diode laser, gain is obtained by passing a current through a p–n junction: consequently, gain modulation is conveniently accomplished by adding an rf modulation signal to the direct current (dc) bias current. Actively mode-locked external resonator diode lasers generate pulses a few tens of picoseconds in duration.

A special type of gain modulation, called synchronous mode-locking (or synchronous pumping) has been used to actively mode-lock organic dye lasers and color center lasers. The setup for synchronous mode-locking, shown schematically in Fig. 8, involves two lasers. The first, which is used to pump the amplifying medium of the second, is itself mode-locked, with a pulse

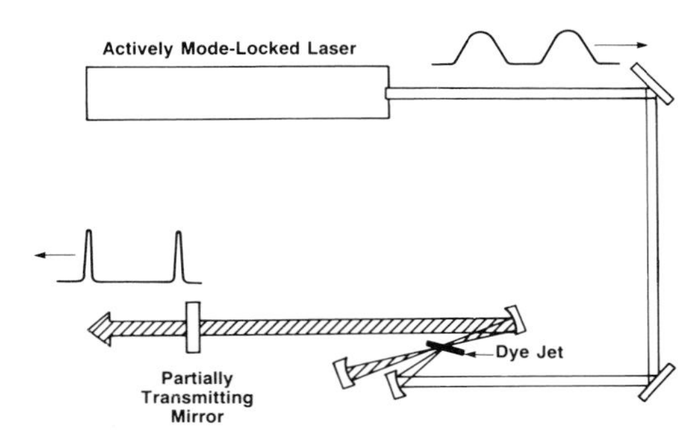

FIG. 8. Schematic of a synchronously pumped mode-locked dye laser.

repetition rate given by $c/2L_1$, where L_1 is the length of the first laser. Thus, the second laser experiences a (nonsinusoidal) gain modulation at the repetition rate of the first laser. Synchronous mode-locking results when the mode spacing $c/2L_2$ of the second laser matches the modulation frequency $c/2L_1$. This is achieved by making the optical lengths of the two lasers equal.

Synchronously mode-locked dye lasers are usually pumped by ion lasers or frequency-doubled Nd : YAG lasers. Despite the relatively long (~100 psec) pump pulses, these dye lasers can generate pulses with durations in the subpicosecond range.

Pulse shaping in a synchronously pumped laser is caused by a combination of gain modulation (due to the pump pulses) and gain saturation. Let us consider a steady-state situation (in a dye laser). The variation of the gain with time is shown in Fig. 9. Before the pump pulse, the gain is zero. After the pump pulse arrives at the dye, the gain builds up and eventually exceeds the loss in the resonator. Subsequently, the dye laser pulse arrives at the gain medium and is amplified: the energy extracted in the process saturates the gain and brings it below the losses, thus turning off the dye laser pulse. The gain now builds up again during the remainder of the pump pulse, but in properly mode-locked operation, the gain will not again exceed the losses until the arrival of the next pump pulse. After the passage of the pump pulse, the gain slowly relaxes back to zero.

For proper mode-locking, the resonator lengths of the two lasers must be extremely well matched; for subpicosecond operation, the tolerance of the length adjustment drops into the submicron range. When L_2 is too short, the dye

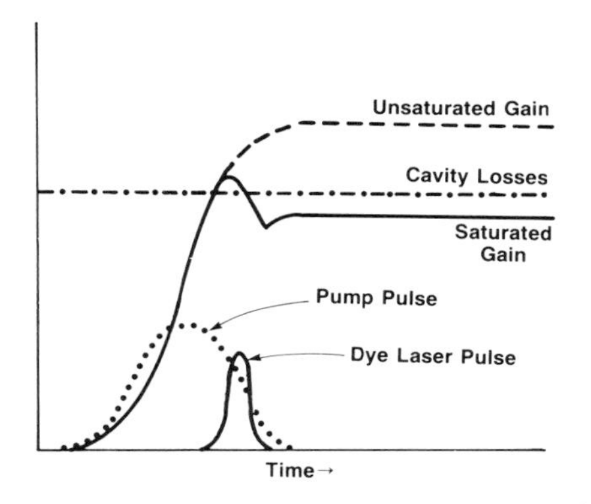

FIG. 9. Dynamics of a synchronously mode-locked dye laser. This diagram illustrates the relationship among the pump pulse, the dye laser pulse, and the gain.

laser pulse arrives too early at the gain medium; this allows the gain to recover after saturation and to exceed the losses for a second time, resulting in the formation of a satellite pulse separated from the main pulse by tens of picoseconds. If L_2 is too long, the pulse arrives late at the gain medium; meanwhile, spontaneous emission is amplified prior to the pulse's arrival, resulting in a broad and noisy pulse shape.

Two extensions of synchronous mode-locking have resulted in pulse durations well into the femtosecond regime: (a) pumping with compressed pulses, and (b) hybrid mode-locking. Frequency-doubled Nd:YAG pulses, compressed by two orders of magnitude down to 0.4 psec (Section V,A), have been used to synchronously pump dye lasers, resulting in dye laser pulses a few hundred femtoseconds in duration. Hybrid mode-locking is a combination of synchronous mode-locking with passive mode-locking (see Section IV,B). By including a saturable absorber in the synchronously pumped dye laser, it is possible to eliminate satellite pulses and, under optimum conditions, generate pulses under 100 fsec in duration.

B. Passive Techniques

The shortest pulses produced directly from a laser have been achieved by passive mode-locking. A special type of passively mode-locked dye laser, called a colliding-pulse-mode-locked (CPM) ring dye laser, has produced pulses as short as 27 fsec.

As the name suggests, a passively mode-locked laser does *not* employ an externally driven intracavity modulator. Instead, a nonlinear optical element (an element whose optical response depends on the light intensity) is placed within the resonator. Generally, the nonlinear element is a saturable absorber—a partially absorbing medium, usually an organic dye solution, whose absorption decreases with increasing light intensity. When a pulse of light traveling within the laser passes through the saturable absorber, the absorber is bleached: the absorption is temporarily modulated. Because the absorption is modulated, the transmitted optical pulse is also modulated, that is, the pulse modulates the absorption, which in turn modulates the pulse. Thus, the nonlinear element in a passively mode-locked laser produces a modulation that is automatically in synchronism with the resonator round-trip time.

Let us now discuss pulse shortening in a passively mode-locked laser. We consider the case of a slow saturable absorber (slow in the sense that the recovery of the absorber, subsequent to optical excitation, occurs on a time scale much longer than the duration of the optical pulse), and we assume that there is but a single pulse propagating within the laser. When the pulse arrives at the saturable absorber, the leading edge of the pulse is absorbed. This bleaches (saturates) the absorber so that the central and trailing edges of the pulse suffer less attenuation. Thus, the leading edge of the pulse is truncated by the saturable absorber (Fig. 10a). Now, the pulse travels to the gain medium. Here, the central portion of the pulse is amplified. In the process the gain is also saturated, so that there is no gain left for the trailing edge of the pulse (Fig. 10b). Gain saturation, together with linear loss (Fig. 10c), causes the trailing edge to be truncated. Thus, through a combination of absorber saturation, gain saturation, and linear loss, the pulse is shortened. Meanwhile, after the passage of the pulse, the gain and absorber media recover slowly to their original state.

When the laser is initially turned on, the optical field builds up from noise. The largest spike in the initial noise structure will be amplified according to the picture described above; after many transits through the cavity, only a single spike will remain. This spike will grow shorter and shorter, until eventually the pulse shortening is balanced by broadening mechanisms within the laser, and an equilibrium pulse duration is established. Broadening is caused by bandwidth-limiting elements (such as an intracavity filter or the gain medium itself) and by group velocity dispersion. Group velocity dis-

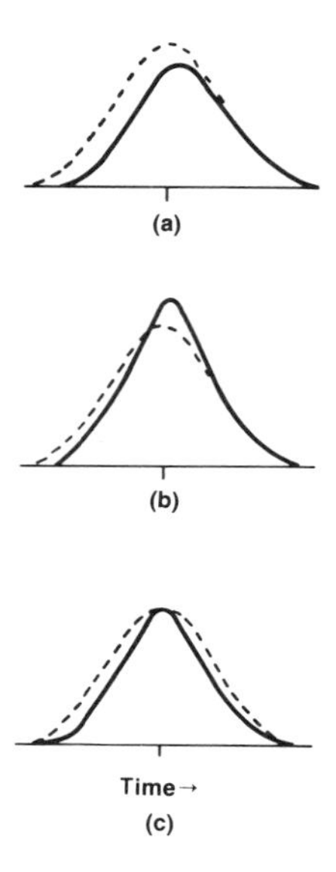

FIG. 10. Pulse shaping in a passively mode-locked laser with a slow saturable absorber. Dashed line: initial pulse shape. Solid line: (a) pulse shape after passage through saturable absorber; (b) pulse shape after passage through gain medium; (c) pulse shape after passage through linear loss.

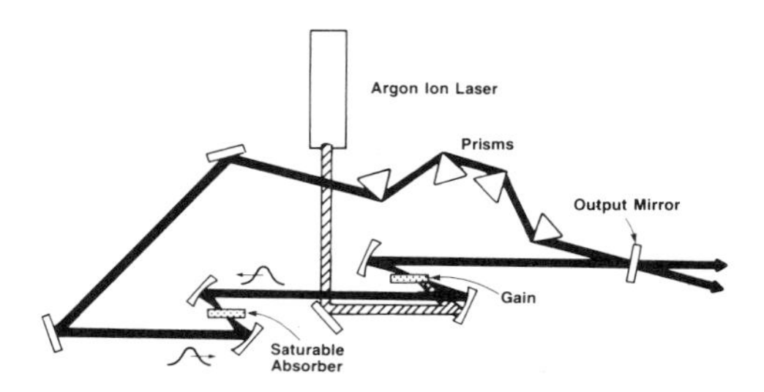

FIG. 11. CPM ring dye laser. Pulses "collide" in the saturable absorber but arrive singly in the gain medium. Pulses as short as 27 fsec are obtained through the partially transmitting output mirror.

persion exists when different wavelengths of light travel at different velocities. A short pulse will broaden as it propagates through materials that exhibit group velocity dispersion, since the different spectral components will disperse in time. Group velocity dispersion can be very important for lasers operating in the femtosecond regime; even reflections from multilayer dielectric mirrors can cause significant pulse spreading.

Mode-locking with a slow saturable absorber has been used and understood for many years. As long ago as 1975, linear cavity passively mode-locked dye lasers produced pulses down to 0.3 psec. The introduction of the ring geometry in 1981 led for the first time to pulses shorter than 100 fsec. Subsequent optimization of the ring resonator laser led to pulse durations as short as 27 fsec.

The layout of a CPM ring dye laser is shown schematically in Fig. 11. The gain is provided by a flowing stream of rhodamine dye dissolved in a suitable solvent, pumped by several watts of light from an argon ion laser. The saturable absorber is a flowing stream of absorbing dye solution. The introduction of the ring geometry leads to an additional mode-locking mechanism that improves the efficiency of the pulse-shortening process. A ring resonator can support two pulses at the same time, one traveling clockwise and the other counterclockwise. It is most favorable energetically for these two pulses to meet, or collide, in the absorber jet. The standing-wave interference pattern generated when the pulses overlap in the absorber minimizes the energy lost because the absorber saturation is greatest where the optical field is most intense and weakest at the nulls of the interference pattern, where there is no optical energy. This colliding-pulse geometry enhances the saturable absorption mechanism described above, leading to shorter pulses and increased stability. The arrangement of four Brewster-angle prisms is used to adjust the sign and amount of group velocity dispersion in the resonator. For the shortest pulses, intracavity pulse compression, due to group velocity dispersion acting in concert with self-phase modulation in the dye, appears to supplement the pulse-shortening mechanisms caused by saturation.

V. Ultrashort Pulses

A. Pulse Compression

The pulses from a mode-locked laser can be made still shorter by using a technique called pulse compression. Although this technique has roots in the chirp radar technology of the 1950s, it has achieved new vitality in the optical field. Relatively long pulses from actively mode-locked lasers may be compressed by a factor of 100 or more down into the subpicosecond regime. Furthermore, the shortest optical pulses achieved to date (only 8 fsec long) have been obtained by applying pulse compression to the already very short pulses produced by a CPM ring dye laser.

A pulse-compression apparatus is pictured schematically in Fig. 12. The input pulse is focused into a length of single-mode optical fiber, where it undergoes a nonlinear optical interaction known as self-phase-modulation. Self-phase-modulation occurs because the refractive index of the glass fiber is modulated by the presence of an intense optical pulse. The change in refractive index in turn affects the optical phase. As a result of this process, the bandwidth of the optical pulse emerging from the fiber is increased dramatically. Since the minimum achievable pulsewidth is limited by the bandwidth, the spectrally broadened pulses have the *potential* to be compressed by a substantial amount.

The additional bandwidth generated by self-phase-modulation takes the form of a "chirp" (a time-varying instantaneous frequency). The central portion of the pulse exhibits a nearly linear up-chirp: the frequency increases linearly with time. It is this linearly chirped portion of the pulse that can be effectively compressed. To achieve compression, the light emerging from the fiber is collimated and then directed through a pair of parallel diffraction gratings. The path length through the grating pair varies linearly with frequency: higher ("blue-shifted") frequencies experience less delay than lower ("red-shifted") ones. When the dispersive delay is properly adjusted (by changing the grating separation), the trailing, blue-shifted spectral components will just catch up to the leading, red-shifted components, and the pulse is compressed.

It is important to note that pulse compression is truly compression and not simply a matter of slicing out a short section from a longer pulse. Most of the original pulse energy is contained in the compressed pulse, so the compressed pulse can be much more intense than the original pulse.

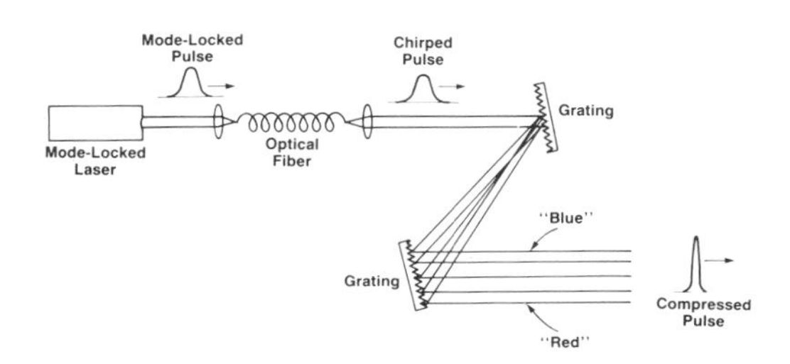

FIG. 12. Apparatus for pulse compression. The pulse is chirped due to self-phase-modulation in a single-mode optical fiber and then compressed to the bandwidth limit by a pair of diffraction gratings.

The description above is somewhat simplified, in that group velocity dispersion in the fiber itself is ignored. In reality, the pulse will broaden temporarally as it broadens spectrally. This usually results in a more linear chirp, however, and thus a better compressed pulse.

The experimental parameters in a pulse-compression experiment vary widely according to the duration, power, and wavelength of the input pulse. Fiber lengths less than 1 cm and greater than 1 km have been used. Table II lists the experimental parameters for a variety of situations that have been studied.

B. Measurement Techniques

Subpicosecond and femtosecond pulses are too short to measure by conventional electronic methods. The fastest sampling oscilloscope currently available has a risetime of 25 psec; fast enough to measure 100-psec pulses from a mode-locked Nd : YAG laser, but far too slow to measure pulses from mode-locked dye lasers. The shortest optical pulses today (8 fsec) are more than three orders of magnitude faster than the fastest oscilloscope.

Since electronic methods are insufficient, all-optical techniques have been developed for picosecond and femtosecond pulse analysis. The idea is that two synchronized ultrafast pulses are used to sample each other. One common apparatus used for pulsewidth measurement is known as a second harmonic generation autocorrelator.

An autocorrelator works as follows. An incoming pulse is first split by a beam splitter (a partially transmitting, partially reflecting mirror) into two replicas of the initial pulse. By causing each pulse to travel a different path length, the relative delay between the two pulses can be adjusted (e.g., a difference of 0.3 mm of path length in air generates a time delay of 1 psec).

The two pulses are focused by a lens to a common spot in a nonlinear crystal, such a potassium dihydrogen phosphate, where second harmonic generation takes place. Second harmonic generation is a nonlinear optical process that produces output light at twice the frequency of the original light. Because the second harmonic intensity is proportional to the *square* of the incident light intensity, more second harmonic light is generated when both pulses are coinci-

TABLE II. Numerical Examples of Pulse Compression

Laser type and wavelength	Pulse duration into fiber	Peak power into fiber	Fiber length	Compressed pulse duration	Compression factor
Nd : YAG laser (1.06 μm)	75 psec	80 W	400 m	0.8 psec	94
Frequency-doubled Nd : YAG laser (0.532 μm)	33 psec	240 W	105 m	0.41 psec	80
Synchronously pumped dye laser (0.59 μm)	5.9 psec	2 kW	3 m	0.2 psec	30
CPM dye laser (0.63 μm)	40 fsec	125 kW	7 mm	8 fsec	5

dent than when the pulses are separated in time. For relative delays that exceed the pulse duration, the amount of second harmonic is reduced. Thus, by measuring the amount of second harmonic light as a function of the delay between the two pulses, we can infer the pulse duration.

An autocorrelator does not directly provide a measurement of the pulse shape. Rather, the dependence of the second harmonic energy on the delay τ yields the intensity autocorrelation function $G(\tau)$ defined by

$$G(\tau) = \frac{\int dt I(t) I(t+\tau)}{\int dt I^2(t)} \tag{7}$$

Here, $I(t)$ is the intensity. In order to determine the pulse duration from the autocorrelation, a particular pulse shape must be assumed. The pulsewidth, defined as the full-width at half-maximum (FWHM) of the intensity, is then obtained by dividing the autocorrelation width (FWHM) by a numerical factor of the order of unity. For pulses with a Gaussian shape, this factor is 1.41.

VI. Current Research

A. Toward Shorter Pulses

Advances in ultrashort optical pulse technology have made possible the generation of ever-shorter pulses at a variety of wavelengths. These advances include the invention of the CPM ring dye laser, the demonstration of pulse compression, and the development of kilohertz repetition rate femtosecond amplifier systems.

The invention of the CPM laser in 1981 made available for the first time optical pulses less than 100 fsec in duration (Fig. 1). Subsequent optimization of the laser parameters led to direct generation of pulses as short as 27 fsec. Meanwhile, the application of pulse-compression techniques to amplified CPM laser pulses led to a series of "world's shortest pulses": 30 fsec in 1982, 16 fsec and then 12 fsec in 1984, and 8 fsec in 1985. The last two results were made possible by the development of kilohertz repetition rate, femtosecond dye amplifiers in 1984. Both copper vapor lasers and mode-locked Nd : YAG lasers have been used to pump these amplifier systems. The copper-vapor-based system generates pulse energies up to several microjoules at repetition rates approaching 10 kHz and with pulse durations as short as 35 fsec.

Pulse compression, using self-phase-modulation in an optical fiber, was first demonstrated in 1982. Since then, pulse-compression technology has developed rapidly. Compression ratios of 100 have been achieved, starting with pulses from actively mode-locked Nd : YAG lasers; and even larger overall compression factors, up to 1000, have been obtained by feeding a compressed optical pulse into a second pulse compressor. This technique has been applied not only at visible wavelengths but also as far in the infrared as 1.3 μm.

A novel type of mode-locked laser operation, using soliton propagation in optical fibers, has been demonstrated at infrared wavelengths longer than 1.3 μm. At these wavelengths, fused silica fibers exhibit negative group velocity dispersion (i.e., the group velocity increases with increasing wavelength). In this case, the fiber dispersion acts to compress rather than disperse the self-phase-modulated pulses, and no external compressor is necessary. The pulses generated by such internal compression are termed solitons. By incorporating a single-mode fiber within the resonator of a synchronously mode-locked color center laser, researchers have demonstrated a "soliton laser," which produces pulses as short as 50 fsec.

Today's shortest pulses contain enormous optical bandwidths; for example, an 8-fsec pulse at 0.63 μm consists of only four optical cycles and has an FWHM bandwidth of 700 Å; roughly one-quarter of the visible spectrum. Working with these bandwidths presents a number of new difficulties. Practically any optical component is capable of broadening such a pulse. Passage through a 1-mm-thick piece of glass will more than double the pulse duration (due to group velocity dispersion). Reflection from certain types of multilayer dielectric mirrors can cause serious distortion of the pulse shape. In order to generate still shorter pulses, a number of challenges must be overcome. For example, pulse compression, used to generate these 8-fsec pulses, is limited because the dispersion contributed by the grating pair does not remain linear over the entire optical bandwidth. Furthermore, measurement techniques such as autocorrelation start to become inaccurate in this very short time regime.

B. Ultrashort-Pulse Tailoring

It has been shown that it is possible to tailor (i.e., to control in detail) the shape of ultrashort laser pulses. Pulses are tailored by using a modified pulse-compressor apparatus, such as that depicted in Fig. 13. The modified compressor incorporates two pairs of diffraction gratings; the first pair separates the various frequency components in space, and the second pair reassembles the frequency components into a well-collimated beam. The temporal dispersion of the two grating pairs is twice that of a single grating pair and is adjusted to provide optimum pulse compression. By inserting a passive, spatial mask into the region between the first and second grating pairs, the amplitude and phase of the individual frequency components may be controlled. Since the temporal pulse shape is the Fourier transform of the frequency spectrum, pulse shapes may be tailored by manipulating the individual spectral components.

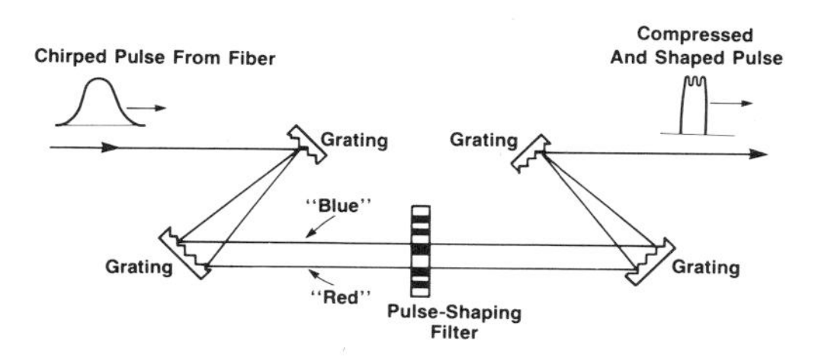

FIG. 13. Apparatus used for ultrashort-pulse tailoring.

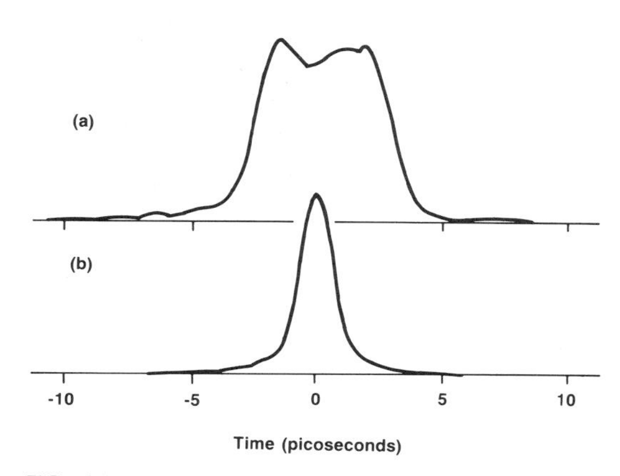

FIG. 14. (a) A phase-coherent picosecond optical square pulse synthesized by using the apparatus shown in Fig. 13. For comparison, the (unshaped) input pulse is shown in (b).

Figure 14 shows a phase coherent, picosecond optical square pulse, synthesized in the manner described above. To produce the square pulse, it was necessary to filter the original frequency spectrum, corresponding to an unshaped compressed pulse, in order to obtain a spectrum of the form

$$E(\omega) = \frac{E_0 T \sin(\omega T/2)}{\omega T/2} \quad (8)$$

where T is the pulse duration. The finite risetime evident in Fig. 14 was due to the limited optical bandwidth (~900 GHz) available for the experiment. It is possible in principle to synthesize any pulse shape, subject only to the limitations of finite bandwidth and spectral resolution.

VII. The Future

As we look toward the future, we can identify a number of areas in which mode-locked laser pulses can be expected to play an important role. There is at present a major worldwide commitment to the installation of optical fiber transmission lines. Because of ever-expanding data-rate requirements, we are already at the point at which great difficulty is experienced in using electronics to modulate and detect signals at the required rates. In order to take advantage of the tremendous bit-rate capacity of modern optical fibers, researchers are devising new all-optical switching and signal-processing elements that do not suffer from the bit-rate limitations of electronics. These devices, driven by ultrashort pulses from mode-locked lasers, may make possible lightwave communications systems that have capacities hundreds of times higher than even today's high-capacity systems. [*See* OPTICAL FIBER COMMUNICATIONS.]

There is much interest at present in the development of digital optical computing systems. Such systems, driven by pulses from mode-locked lasers, may prove useful for a number of special-purpose applications. In particular, researchers are investigating new computer architectures in which optical pulses can be used to make large numbers of operations at the same time. Such computers are likely to find applications in such areas as pattern recognition, image processing, and real-time speech recognition. [*See* OPTICAL CIRCUITRY.]

Mode-locked laser pulses have already had a major impact in the field of scientific measurements. Until the invention of the laser in 1960, the fastest light pulses that had been produced were the strobe light flashes used for "stop-action" photography. Today, we are able to generate compressed pulses from mode-locked lasers that are 100,000 times shorter. These pulses are being used to investigate a new time scale of ultrafast physical phenomena. [*See* ULTRAFAST LASER TECHNOLOGY.]

Another important application of these short pulses is their use to probe the operation of electronic circuits. Researchers are currently demonstrating technologies for measuring currents and voltages in semiconductor integrated devices and circuits with femtosecond precision using ultrashort light pulses.

Thus, we see that ultrashort mode-locked laser pulses are not only being used for probing new frontiers in physics and chemistry, but are also finding use in a new generation of high-speed electrical and optical technology. The future for mode-locked lasers appears bright.

Bibliography

Fleming, G. R., and Siegman, A. E., eds., *Ultrafast Phenomena V* (Berlin, Springer Verlag, 1986). This is a conference proceedings which covers nearly every topic of current ultrashort light pulses research.

Lasers and Applications, Jan. 1985, pp. 79–83 and Feb. 1985, pp. 91–94. Covers the basic principles of active, passive and synchronous mode locking.

Shank, C. V., "Measurement of Ultrafast Phenomena in the Femtosecond Time Domain," *Science*, vol. 219, pp. 1027–1031, March 4, 1983. This reviews the advances in pulse generation technology that occurred ca. 1981-2 following the invention of the CPM laser.

Shank, C. V. and Auston, D. H., "Ultrafast Phenomena in Semiconductor Devices," *Science*, vol. 215, pp. 797–801, Feb. 12, 1982. Discusses application of ultrashort light pulses to measure fast phenomena in semiconductor structures and devices.

Smith, P. W., Duguay, M. A., and Ippen, E. P. "Mode-Locking of Lasers" Pergamon, New York 1973. A rather complete tutorial review of mode-locking work up to early 1973.

NONIMAGING CONCENTRATORS (OPTICS)

R. Winston and J. O'Gallagher *University of Chicago*

GLOSSARY

Čerenkov radiation: Faint light produced by charged particles moving in a medium at a velocity greater than that of light in the medium.

Compound parabolic concentrator: Name given generically to a class of nonimaging collectors with reflecting walls (not necessarily parabolic) that concentrate flux by the theoretical limit.

Dielectric compound parabolic concentrator: Nonimaging collector that operates by total internal reflection.

Edge-ray principle (maximum slope principle): Method for designing optical systems with maximum collecting power.

Étendue: Product of area times projected solid angle.

Flow line concentrator (trumpet): Nonimaging collector in which the reflecting wall follows the lines of vector flux from a Lambertian source.

Nonimaging optics: Optical theory and design that departs from traditional methods and develops techniques for maximizing the collecting power of concentrating elements and systems.

Phase space: Abstract space wherein half the coordinates specify locations and half the direction cosines of light rays.

Nonimaging optics departs from the methods of traditional optical design to develop techniques for maximizing the collecting power of concentrating elements and systems. Designs that exceed the concentration attainable with focusing techniques by factors of 4 or more and approach the theoretical limits are possible. This is accomplished by applying the concepts of Hamiltonian optics, phase space conservation, thermodynamic arguments, and radiative transfer methods.

I. Introduction

A. PURPOSE OF CONCENTRATION

The role of optical concentration in solar collector design is often misunderstood. The simplest and most widely deployed collectors employ no concentration at all and are simply flat panels. This is true for both of the conventional applications, that is, photothermal and photovoltaic conversion systems. For both flat panels and concentrators the amount of energy collected depends on the area of intercepted radiation. The concentrator cannot "intensify the energy" received or "amplify the energy flux," as is sometimes stated in popular articles on the subject. The concentrator does collect energy over some large area A_1 and deliver it to some smaller area A_2. This provides a "geometric concentration ratio"

$$C \equiv A_1/A_2 \tag{1}$$

and in the process sacrifices some angular field of view, which in turn requires, in many cases, that the collector be "tracked" or moved to follow the sun. Only relatively recently, through the use of a class of nonimaging concentrators usually referred to as compound parabolic concentrators or CPCs, has it become possible to relax and sometimes eliminate this tracking requirement.

The motivation for increasing concentration

above unity (corresponding to the flat-panel case) is twofold:

1. Thermal performance: For a solar thermal collector, the heat losses depend on the area of the hot absorber. By reducing the area of the thermal transducer (A_2) relative to the collecting area (A_1), one can achieve respectable efficiencies at higher temperatures than could otherwise be attained. Very high concentrations can be used to generate very high temperatures—in principle approaching that of the surface of the sun.
2. Economic cost effectiveness: If the cost per unit area of the energy transducer (i.e., a solar cell) is very much greater than that of the concentrating optics (concentrating mirrors or lenses), the overall cost per unit energy collected can be dramatically reduced. However, the cost of any required tracking system must be included in the economic analysis.

These are the only two reasons for using concentration. Clearly, only the second applies for photovoltaic applications, while both apply for thermal applications, although the economic benefit is not usually emphasized.

B. Thermodynamic Limit

If one wishes to concentrate radiation according to Eq. (1), it is evident qualitatively that there must be some sacrifice in view angle (the smaller absorbing surface cannot "see" the full field of view comprising the hemisphere visible to the larger aperture). The quantitative relationship between this reduction in view angle and increasing concentration ratio is not intuitively obvious but can be derived in a straightforward manner from thermodynamic arguments. These arguments are based on the fact that in thermodynamic equilibrium the absorber must reradiate back to the environment the same amount of energy as it receives. In particular, for a given acceptance half-angle $\pm\theta_a$ at the collecting aperture, this condition defines a minimum absorber area and corresponding maximum possible geometric concentration given by

$$C \leq 1/\sin\theta_a \tag{2a}$$

if the concentration is done in only two dimensions (troughlike geometries) and

$$C \leq 1/\sin^2\theta_a \tag{2b}$$

if the concentration takes place in three dimensions (conelike geometries). This is referred to as the thermodynamic limit since, if one could make a concentrator that transferred all the incident radiation to an absorber smaller than that given by Eq. (2a) or (2b), it would not have sufficient area to reradiate this energy in thermodynamic equilibrium and its temperature would begin to rise above its surroundings in violation of the second law. Any concentrator system that can attain this limit is referred to as "ideal." Concentrating systems based on imaging or focusing optics fall short of this limit by a factor of 3–4. The CPC and other concentrator shapes determined by the principles of nonimaging optics actually achieve this limit in two dimensions and closely approach it in three dimensions.

Equations (2) can also be derived by application of the principles of étendue or phase space conservation. A bundle of rays propagating in the z direction can be characterized by a distribution of points in a four-dimensional (two position and two directional coordinates) phase space. Phase space conservation then requires that the volume representing an ensemble of rays propagating through an optical system along the z axis must remain constant. If one then considers a distribution uniform in position across an aperture $A_1(z_1)$ and consisting of rays isotropically filling an acceptance angle θ_a with rays of $\theta < \theta_a$ and wants to find the smallest cross-sectional area $A_2(z_2)$ to which these rays can be reduced while allowing the directional distribution to expand to fill the semicircle in directional coordinate space, one is led directly to Eq. (2) in those geometries.

Neither thermodynamic nor phase space conservation arguments tell us actually how to achieve the maximal or ideal concentration; they simply state that these are the limits that no optical system can exceed.

C. Nonimaging versus Imaging Optics

Concentrators designed according to classical imaging optical principles are optimized for paraxial rays and in general fall far short of the limits of Eq. (2). Nonimaging concentrators, on the other hand, are optimized for the extreme angles that are to be included in the acceptance of the optical system and can be maximally concentrating, that is, can approach and in some cases attain the thermodynamic limit. The difference in design techniques can best be explained by reference to typical image-forming optical systems such as camera lenses and slide projector lenses. In these systems there is an axis of symmetry on which the centers of all the lens components lie. Likewise, the object—a scene to be photographed or a slide to be projected—usually has its center or most important point on or near the axis. The designer of the image-forming system then attends first to the region near the axis: the basic properties such as focal length and magnification are determined by the properties of the lens components near the axis, and it follows that the image quality or sharpness must be best near the axis. By adding or changing lens components, the designer can improve the image quality away from the axis; but it inevitably deteriorates gradually until the image is so fuzzy at some distance from the axis that the optical system is unusable beyond that point. This, of course, determines the "field of view."

In nonimaging concentrators, there is a completely opposite emphasis. These systems may have either an axis or a plane of symmetry, or they may even have no symmetry at all; in any case, they are designed to collect entering flux over a certain angular range and larger aperture and concentrate it so that it emerges from a smaller exit aperture or strikes an absorber with a surface smaller than the entry aperture. The design principle to be followed to achieve the maximum concentration is that all rays entering at the extremes of the angular range must emerge just grazing the edge of the exit aperture or absorber surface. This is a concise statement of the "edge-ray principle." The edge-ray principle, together with the laws of refraction and reflection, is the basis of the design of all efficient nonimaging concentrators. It ignores rays from the regions near the axis (if there is an axis) or near the center of the entering angular range; there is no question of sharpness of images anywhere. All that is required is that extreme entering rays meet certain operationally defined conditions—for example, graze the edge of the exit aperture as they emerge—whereas in image-forming systems the extent of the field of view gradually evolves *post hoc* when the center of the field has been taken care of.

A more specific formulation of the ideal concentrator design principle, particularly applied to systems where the active optical surfaces are reflectors, is the "maximum slope principle." It says that each element of the reflector profile curve should be made to assume the maximum possible slope (i.e., rate of increase of the entrance aperture with increasing concentrator depth) consistent with the requirement that rays incident on the aperture at the desired extreme angle (i.e., "edge ray") illuminate the absorber. Given a particular absorber shape and a desired acceptance θ_a, the application of this principle yields a differential equation that allows generation of ideal two-dimensional reflector shapes. Examples of solutions in two dimensions are shown in Fig. 1.

If one considers the two-dimensional geometry defined by two parallel surfaces (A_1 and A_2 in Fig. 1a) perpendicular to a common optical axis, one can show that reflectors connecting these two surfaces with parabolic segments as shown in Fig. 1a actually achieve the limit defined by Eq. (1). The left-hand reflecting surface is a segment of a parabola with its axis tilted by an angle θ_a with respect to the optical axis and positioned so that its focus lies at the lower edge of the right reflector. All rays entering the entrance aperture at an angle θ_a will then strike one reflector and be brought to a focus at the opposite edge of the exit aperture. All rays entering between $\pm\theta_a$ must be reflected to points between the lower mirror edges. This is the simplest manifestation of an ideal concentrator in two dimensions and is the solution that gave rise to the name compound parabolic concentrator, which is now often used to include the other solutions as well. A figure of revolution formed by rotating the shape of Fig. 1a about the concentrator axis does not perform as a true ideal concentrator, essentially

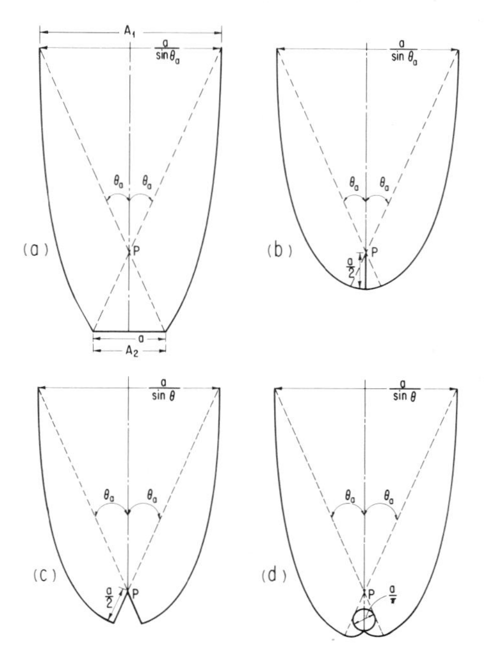

FIG. 1. Cross-sectional profiles of ideal trough concentrators generalized for absorbers of different shapes. In practice the reflectors are usually truncated to about half their full height to save reflector material with only negligible loss of concentration.

because in three dimensions rays that are not in the meridian plane have an additional quantity (angular momentum) that must be conserved as well as the phase space, so the system is overdetermined. However, such three-dimensional cones come very close to ideal performance. Solutions for the other common absorber shapes are shown in Figs. 1b–1d. Three-dimensional versions of these latter profiles do not work as concentrators because angular momentum would not be conserved for skew rays.

The original conical CPCs were first developed in the mid-1960s to maximize the light collection of a system of Čerenkov detectors in a hyperon decay experiment at Argonne National Laboratory. Similar devices were proposed independently around the same time in Germany and the Soviet Union. In 1974 it was determined that a trough with the profile of the Čerenkov light coupler had unique advantages for solar energy concentration. In particular, the wider acceptance angles made possible low and intermediate levels of concentration without continuous tracking. If the long axis of the trough is oriented in an east–west direction and tilted toward the southern sky (in the Northern Hemisphere), a completely stationary concentrator can attain geometric concentrations up to about twofold and, with seasonal adjustments, concentrations up to about 10-fold. Other advantages of nonimaging solar concentrators are greatly relaxed optical tolerances, collection of the circumsolar and a large fraction of the diffuse component of the insolation, and reduced sensitivity to small-angle scattering from dust or scratches.

The application of CPCs for solar thermal collection underwent rapid development in the late 1970s at Argonne National Laboratory and the University of Chicago. In subsequent sections of this article, we survey some of the highlights of this development period. In addition, we present some of the more recent advanced concepts in which nonimaging techniques are used to design two-stage concentrators having an imaging primary and a CPC-like device in the focal zone. Such hybrid devices can be made to approach the maximally concentrating limit and have many interesting applications.

II. Solar Thermal Application

The preferred configurations are those that eliminate heat losses through the back of the absorber, as shown in Figs. 1b and 1d. Note that the geometric concentration is defined relative to the full surface area of the fin absorber (both sides) or tube (full circumference) and that these designs effectively have no back.

In recent years work has centered on the development of CPCs for use with spectrally selective absorbers enclosed in vacuum. Work at Argonne National Laboratory after 1975 was strongly influenced by the emergence of the Dewar-type evacuated absorber. This thermally efficient device, developed by two major U.S. glass manufacturers (General Electric Co. and Owens Illinois) could, when coupled to CPC reflectors, supply heat at higher temperatures than flat-plate collectors while retaining the advantage of a fixed mount. The heat losses associated with such absorbers are so low that the moderate levels of concentration associated with CPCs provide a dramatic improvement in thermal performance and achieve excellent efficiencies up to 300°C. In fact, the gains associated with even higher concentration ratios are marginal and probably negligible when the added complication and expense of active tracking are considered. This is illustrated in Fig. 2, which shows the calculated efficiency relative to total insolation of a collector consisting of evacuated tubes with a selective absorber coating under increasing levels of concentration.

In characterizing the thermal performance of CPCs in general, we describe the thermal collection efficiency η as a function of operating temperature T by

$$\eta(T) = \eta_0 - \frac{U(T - T_a)}{I} - \frac{\sigma\varepsilon(T^4 - T_a^4)}{CI} \quad (3)$$

where C is the geometric concentration ratio, T_a the ambient temperature, I the total insolation, σ the Stefan–Boltzmann constant, ε the absorber surface emittance, and U the linear heat loss coefficient.

The optical efficiency η_0 is the fraction of the incident solar radiation (insolation) actually absorbed by the receiver surface after transmission and reflection losses. The other terms represent parasitic conduction losses and radiation losses from the absorber surface, both of which increase substantially as the collector working fluid temperature is increased. The most important point to be seen from Fig. 2 is that the dramatic reduction in relative thermal losses produced by adding a 1.5× reflector or again effectively tripling the concentration from 1.5× to 5× is not continued as one increases the concentration much beyond 5×. For example, at approximately 300°C ($\Delta T/I \cong 0.3$) the thermal

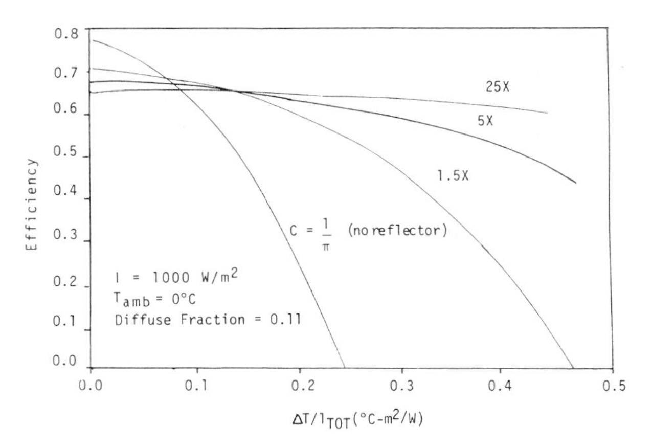

FIG. 2. Calculated thermal performance curves for evacuated tubular absorbers under increasing levels of concentration. Note that the improvement in increasing the concentration above 5× is marginal.

losses for a 5× are already so low that a further factor of 5 reduction corresponds to only a negligible fraction of the operating efficiency. It is not necessary or even desirable to increase the concentration much beyond 5× when using an evacuated selective absorber at these temperatures. Thus the combination of CPCs up to about 5× with these thermally efficient absorber tubes represents a nontracking strategy for practical solar thermal collection up to power generation temperatures with many unique advantages.

The most developed evacuated CPC design is that optimized for totally stationary collection throughout the year, having an acceptance half-angle $\theta = \pm 35°$ corresponding to a maximum ideal concentration of 1.8×. Commercial versions, truncated to net concentrations of about 1.1–1.4×, are now available with typical optical efficiencies between 0.52 and 0.62, depending on whether a cover glass is used. The thermal performance is excellent, with values of $\varepsilon \simeq 0.05$ and $U \simeq 0.5$ W/m^2 K in Eq. (3). Other versions have opened the acceptance angle to ±50° with $C \simeq 1.1\times$ to allow polar orientation and are characterized by $U = 1.3$W/m^2 K (lumping the radiative losses in the linear term). An illustration of the basic configuration is shown in Fig. 3. Several manufacturers introduced collectors of these types for applications ranging from heating to absorption cooling to driving Rankine cycle engines. A number of installations with areas greater than 1000 m^2 have been deployed successfully.

Experimental prototype CPCs have been built in higher concentrations for use with evacuated absorbers. One version with $C = 5.25\times$ is shown in Fig. 4a. This CPC, studied at the University of Chicago, is a large trough coupled to the same glass Dewar-type evacuated tube as used in the 1.5× above. Performance measurements for two modules with different reflecting surfaces are shown in Fig. 4b. The upper curve is for a module with a silver foil reflector. It has been operated at 60% efficiency (relative to a direct beam) at 220°C above ambient. This is to be compared with the measured performance of a fully tracking parabolic trough tested by Sandia Laboratories, as shown by the dashed line in Figure 4b. The performance of the CPC is comparable to that of the parabolic trough at all temperatures tested. The lower curve is for a module with aluminized Mylar reflectors, and even with poorer reflectors it exhibits quite respectable performance. The angular acceptance properties

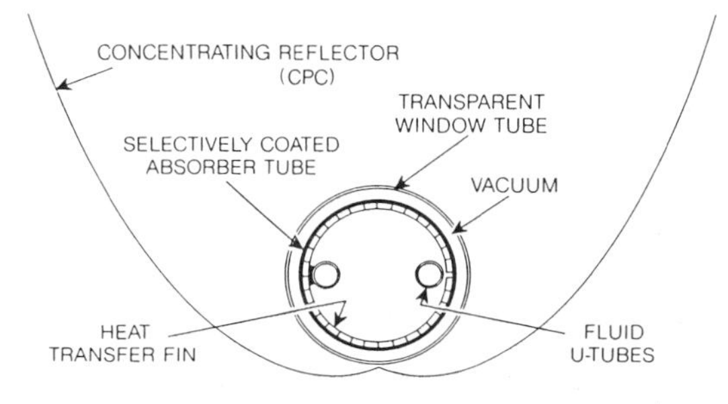

FIG. 3. Cross section of a contemporary commercial evacuated-tube CPC according to the basic design developed by Argonne National Laboratories.

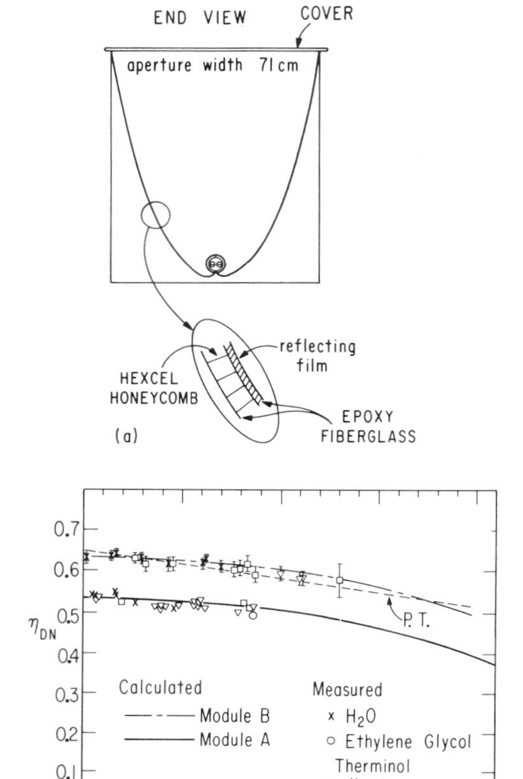

FIG. 4. (a) Profile of an experimental 5× CPC for an evacuated tubular absorber built and tested at the University of Chicago. (b) Measured performance curves for two 5.25× CPC prototype modules. The performance is comparable to that of a commercial parabolic trough, shown by the dashed line.

of the module are in excellent agreement with the design value of ±8°, which allows collection with 12–14 annual tilt adjustments.

An experimental CPC collector under development at the University of Chicago and Argonne National Laboratory that should ultimately lead to the most practical general-purpose solar thermal collector is the integrated stationary evacuated concentrator (ISEC). The optical efficiency of evacuated CPC solar collectors can be significantly improved over that of contemporary commercial versions discussed above by shaping the outer glass envelope of the evacuated tube into the concentrator profile. Improved performance results directly from integrating the reflecting surface and vacuum enclosure into a single unit. This concept is the basis of a new evacuated CPC collector tube that has a substantially higher optical efficiency and a significantly lower rate of exposure-induced degradation than external reflector versions. These performance gains are a consequence of two obvious advantages of the integrated design:

1. Placing the reflecting surface in vacuum eliminates degradation of the mirror's reflectance, thus permitting high-quality (silver or aluminum with reflectance $\rho = 0.91–0.96$) first-surface mirrors to be used instead of anodized aluminum sheet metal or thin-film reflectors ($\rho = 0.80–0.85$) typical of the external reflector designs.
2. The transparent part of the glass vacuum enclosure also functions as an entrance window and thus eliminates the need for an external cover glazing. This increases the initial optical efficiency by a factor of $1/\tau$, where typical transmittances $\tau = 0.88–0.92$.

For the past several years, the solar energy group at the University of Chicago has been developing this concept in collaboration with GTE Laboratories, which fabricated the tubes. Of the 80 prototype built, 45 were assembled into a panel with a net collecting area of about 2 m^2.

The ISEC shown in Fig. 5 is an extended cusp tube CPC matched to a circular absorber of diameter 9.5 mm. The design acceptance half-angle $\theta_a = \pm 35°$ was chosen to permit stationary

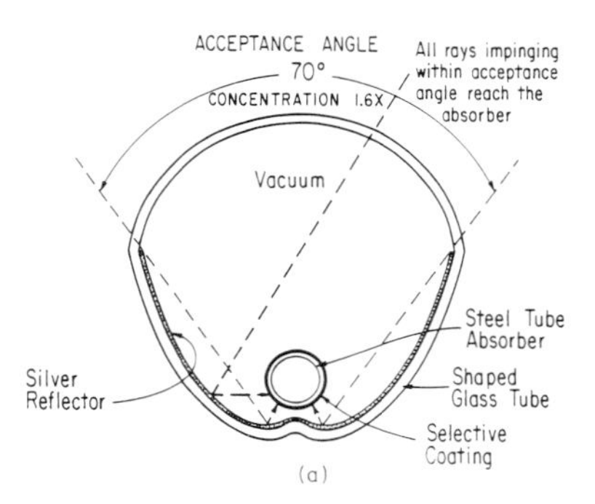

FIG. 5. (a) Details of the actual profile shape and collector design for the integrated stationary evacuated concentrator (ISEC) tube, which has achieved a thermal efficiency of 50% at 200°C. (b) Ray trace diagram showing how essentially all the solar energy incident within ±35° is directed onto the absorber tube. Since the reflector cannot physically touch the absorber, as required for an ideal concentrator, a small fraction is lost in the gap between the reflectors and the absorber.

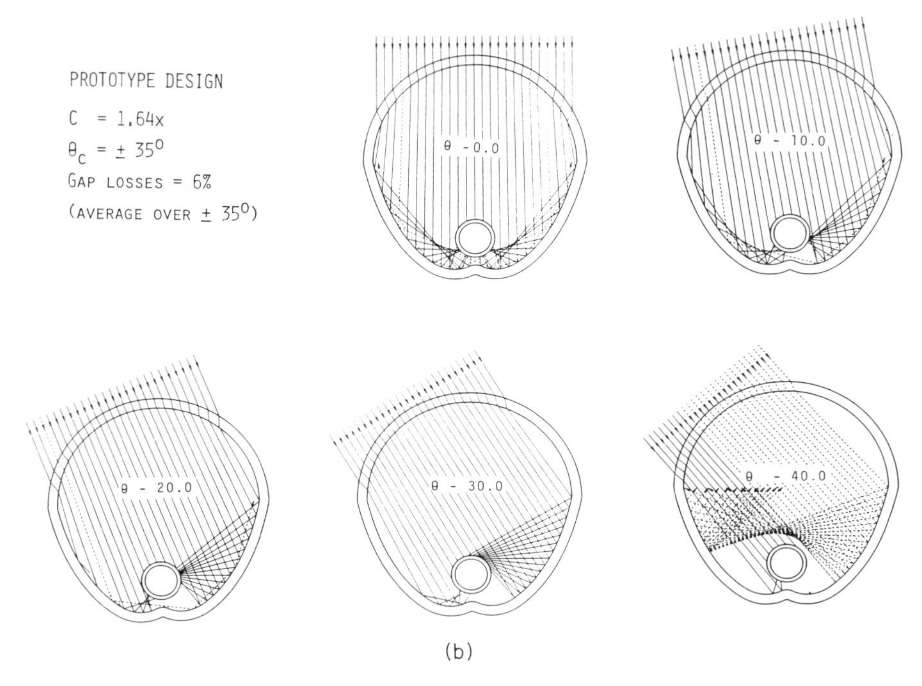

(b)

FIG. 5. *(Continued)*

operation throughout the year. After truncating the CPC, the net concentration was 1.64×. This collector was tested at Chicago for 3 years and routinely achieved the highest high-temperature performance yet measured for a fixed stationary mount collector. Performance curves based on these tests are shown in Fig. 6 along with curves for three other collector types. Note in particular that for temperatures up to about 200°C, the ISEC is comparable to a fully tracking trough and remains respectable up to temperatures approaching 300°C. The relative performance advantages are similar when the comparison is made on an annual energy delivery basis at a variety of locations, as shown for one location in Fig. 7.

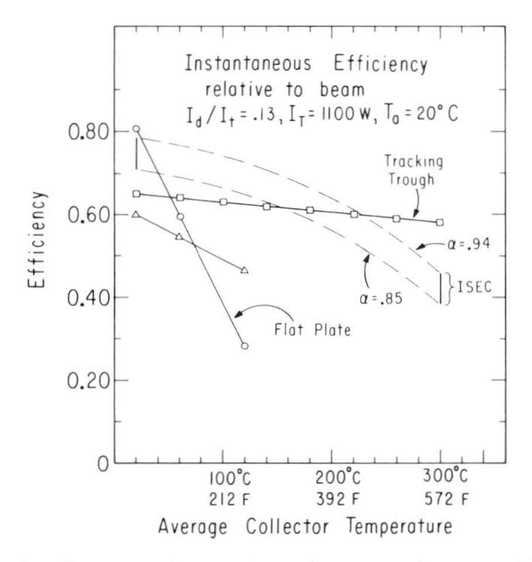

FIG. 6. Comparative peak performance for a tracking parabolic trough, an ISEC, contemporary (external reflector) evacuated CPCs (triangles), and flat plates. The ISEC's superior performance up to temperatures above 200°C, achieved with no moving parts, makes it an extremely flexible solar thermal collector.

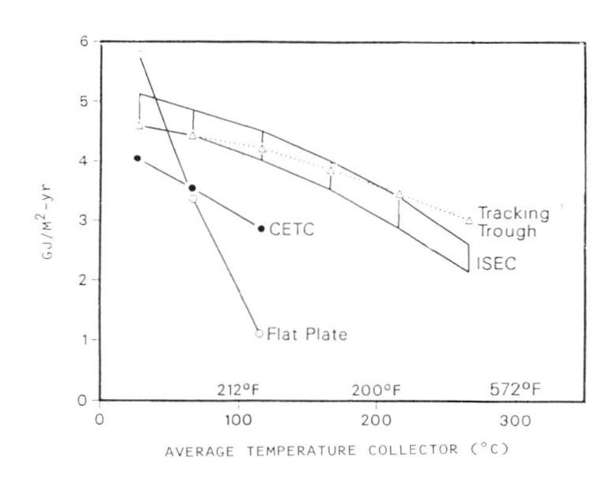

FIG. 7. Comparative annual energy delivery at Phoenix, Arizona, for the collector types in Fig. 6.

The design problems for nonevacuated CPC collectors are entirely different from those for CPCs with evacuated absorbers. The nonevacuated collectors are particularly vulnerable to high heat losses if an improper design is used. One must be careful to minimize or eliminate heat loss via conduction through the reflectors. This can be accomplished by using reflectors whose thickness is negligible compared to the overall dimension (e.g., height, aperture) of the trough, such as metallized plastics or films, or by thermally decoupling the absorber from the reflectors by a small gap maintained by insulating standoffs. Two prototype CPCs with nonevacuated absorbers, a 3× and a 6×, were built and tested extensively as part of the early program at Chicago. The features and performance of these collectors are summarized here. The optical efficiencies and total heat loss coefficients were $\eta_0 = 0.68 \pm 0.01$ and $U = 1.85 \pm 0.1$ W/m^2 °C for the 6× and $\eta_0 = 0.61 \pm 0.03$ and $U = 2.7 \pm 0.02$ W/m^2 °C for the 3×.

The efficiencies of these nonevacuated CPCs are to be compared with typical values for flat plates of $\eta_0 = 0.70$–0.78 and $U = 4.5$–7 W/m^2 K. Despite lower optical efficiencies, the CPCs outperform typical flat-plate collectors above temperatures as low as 10°C above ambient (for the 6×) to about 35°C above ambient (for the 3×). This is particularly important because the 3× should represent a relatively inexpensive collector design. Although a detailed economic analysis cannot be based on the prototype construction methods used here, several unique features contribute to its low cost potential, among them the relatively small absorber cost and very limited insulation requirements.

III. Two-Stage Maximally Concentrating Systems

The principal motivation for employing optical concentration with photovoltaic cells in solar energy is economic. By using what one hopes are relatively inexpensive lenses or mirrors to collect the sun's energy over a large area and redirect it to the expensive but much smaller energy conversion device, the net cost per unit total area of collection can be reduced substantially. Alternatively, to generate electricity through the thermodynamic conversion of solar heat to mechanical energy, concentration is required to achieve the high temperatures necessary to drive a heat engine with reasonable efficiency. In this case the solar flux is directed to an absorber (often a cavity) small enough that the heat losses, even at high temperatures, remain relatively small. It often turns out in both the photovoltaic and thermal conversion cases that the desired concentration is much higher than can be achieved with nontracking CPC-type devices. Conventionally, these higher concentrations are achieved by means of some kind of focusing lens or paraboloidal mirror that is not maximally concentrating.

It is not widely recognized that the nonimaging techniques described in the previous section can be used to design secondary elements that can augment the concentrations of more conventional focusing elements used as primaries, and that such a hybrid optical system can also approach the allowable limit. Applications of such two-stage designs lie in the regime of higher concentration and small angular acceptance, where the geometry of a single-stage CPC becomes impractical. The fundamental advantage is the same as in lower-concentration applications and may be expressed in complementary ways: either significant additional system concentration can be attained (i.e., a smaller, lower-cost absorber) or the angular tolerances and precision can be relaxed while maintaining the same level of concentration.

The limits of achievable levels of solar concentration are represented in Fig. 8 for both line focus and point focus geometries across the range of possible desired angular acceptances from wide angles permitting stationary or seasonal adjusting CPCs down to the angular subtense of the sun (±4.6 mrad). If one could achieve the thermodynamic limit in a point focus geometry (no configuration solution that could accomplish this is known) with no slope or alignment errors, one could reach the thermodynamic limit of 46,000 suns and, in principle, reach the sun's surface temperature of 6000 K. In practice, slope and alignment tolerances, typically about ±0.5°, and the aberrations associated with focusing designs limit the actual values to 30–70× in line focus and 1000–5000× in point focus geometries. Use of a nonimaging secondary can increase these limits (or tolerances) as indicated.

It has been proposed that holographic optical elements (HOEs) could achieve concentrations in the range 10–20× without tracking by stacking individual elements designed to be effective at different times of the day. This is impossible, as indicated by the dashed box in Fig. 8, because it would violate the thermodynamic limit.

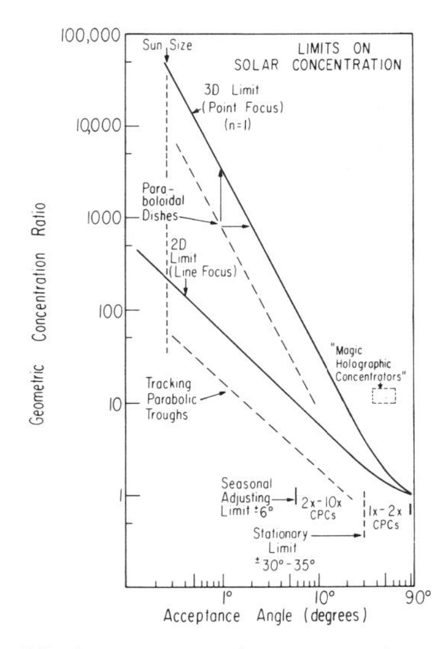

FIG. 8. Maximum geometric concentration ratio corresponding to a given half-angle of incidence permitted by physical conservation laws (the thermodynamic limit) (solid lines). Traditional focusing concentrators fall about a factor of 4 below this limit (dashed lines). Proposed concentrator designs purporting to achieve 10 : 1 ratios with no tracking based on holographic techniques cannot work since they violate this limit.

In this section we describe both photovoltaic applications, where the primary is a lens, and thermal electric applications, where the primary is a paraboloidal mirror. In each case we discuss only the point focus configuration. Schematic drawings of the basic elements for the two cases are shown in Fig. 9.

For the thermal application (Fig. 9a), the primary is characterized by its focal length F and aperture diameter D, which define the rim angle

$$\tan \phi = |2f - 1/8f|^{-1} \tag{4}$$

where $f \equiv F/D$. The secondary is a nonimaging concentrator of either the compound elliptical concentrator (CEC) type, a variant of the more familiar compound parabolic concentrator, or the hyperbolic trumpet type. It is convenient to simplify the analysis by characterizing the primary as having a conical angular field of view of half-angle $\pm\theta_{\rm I}$. This is chosen to accommodate the angular tolerance budget of the primary, including concentrator slope errors, specularity spread, pointing error, and incoming direct sunlight.

The thermodynamic limit in a point focus reflecting geometry is given by Eq. (2b), while geometric arguments show that the geometric concentration of the primary alone must be

$$C_1 < \sin^2 \phi \cos^2 \phi/\sin^2 \phi_{\rm I} \tag{5}$$

to intercept all the energy incident within $\pm\theta_{\rm I}$. The limiting concentration for the secondary is

$$C_2 < 1/\sin^2 \phi \tag{6}$$

Therefore, the maximum combined concentration is

$$C_1 C_2 = \cos^2 \phi/\sin^2 \theta_{\rm I} \tag{7}$$

which approaches the maximum limit for small ϕ (large f).

For the photovoltaic concentrator (Fig. 9b) the nonimaging secondary is formed from a transparent dielectric material with an index of refraction n = 1.3–1.5 in contact with the solar cell. It is usually possible to ensure that total

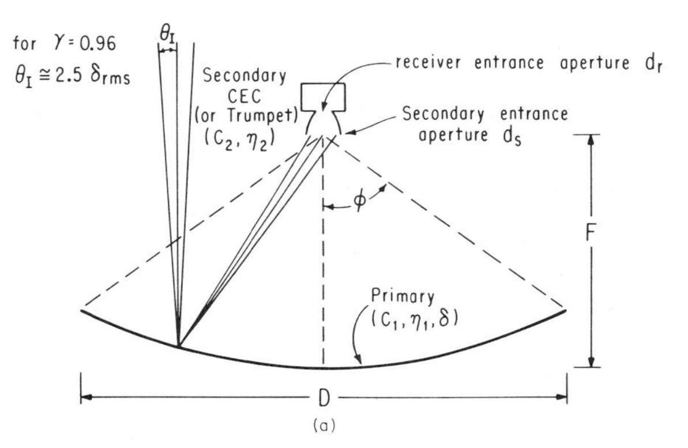

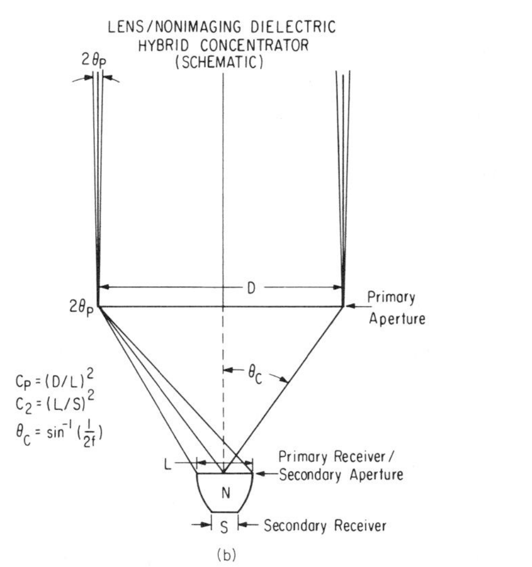

FIG. 9. Geometric and optical parameters characterizing (a) the two-stage point focus reflecting dish and (b) lens concentrators, discussed in the text.

internal reflection (TIR) occurs for all rays accepted by the secondary, thus saving the cost of applying and protecting metallic reflecting surfaces. In such a totally internally reflecting dielectric CPC (DCPC), the refractive power of the dielectric operates in combination with the reflecting profile shape so that secondary concentration ratios are achieved that are a factor n^2 larger than is possible with conventional reflecting secondaries. The primary is a lens (usually a Fresnel lens) of aperture diameter D that focuses normally incident rays at a point that defines the center of the primary receiver. Rays from outer edges of the primary are brought together with a convergence angle $2\theta_c$. If the incident rays subtend an angle of $\pm\theta_p$ on either side of the aperture normal, one can show that the maximum possible geometric concentration C_{max} attainable is

$$C_{max} = n^2/\sin^2 \theta_p \quad (8)$$

Here n is the index of refraction at the final absorber surface relative to that just outside the collecting aperture. Equation (8) for a dielectric secondary corresponds to Eq. (2b) for a reflecting secondary in that it defines the "ideal" limiting concentration allowed by physical conservation laws. For a focusing lens of focal ratio f', where

$$f' \equiv 1/2 \sin \theta_c \quad (9)$$

such that it corresponds to the generalized focal ratio used to express the Abbe sine condition for off-axis imaging, one can show that the *actual* concentration achieved by the lens alone is

$$C_p = 1/4f'^2 \sin^2 \phi_p \quad (10)$$

Comparing Eqs. (10) and (8), we see that for practical lens systems where $f' \gtrsim 1$, C_p falls short of the limit by a factor of $(4f'^2n^2)^{-1}$.

If a DCPC-type secondary with entrance aperture diameter L and exit aperture diameter S is placed at the focal spot of the imaging primary lens, it can achieve an additional geometric concentration $C = L/S$, which is given by Eq. (8) with θ_c replacing θ_p, or

$$C_2 = n^2/\sin^2 \theta_c = 4f'^2n^2 \quad (11)$$

Combining Eqs. (10) and (11) shows that, in principle, the two-stage system can attain an overall geometric concentration equal to the thermodynamic limit. In practice, certain compromises are needed that reduce this somewhat; but typically secondary concentrations in the range 7–10× are readily achievable.

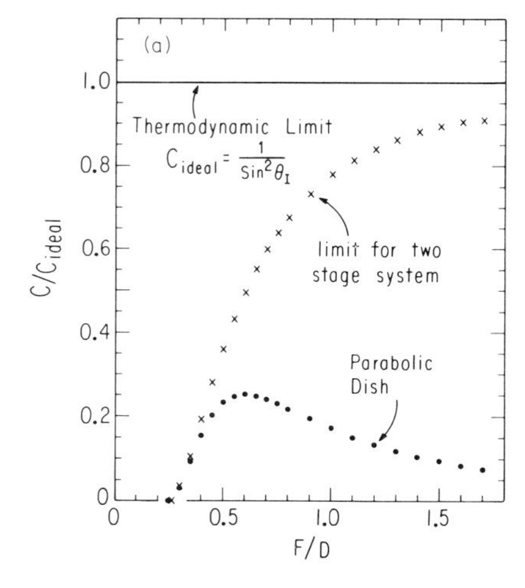

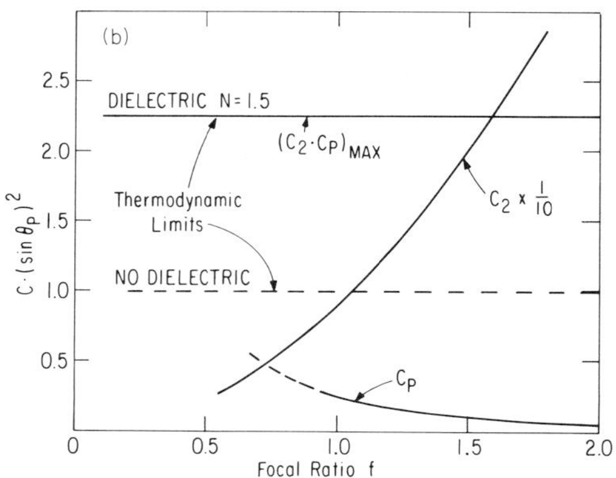

FIG. 10. Maximum geometric concentration achieved by (a) reflecting dish primaries with nonimaging CPC-type secondaries and (b) lens primaries combined with refracting dielectric CPC (DCPC) secondaries.

The limits of concentration for both reflecting and dielectric refracting systems with and without optimized nonimaging concentrators are shown in Fig. 10 as a function of the focal ratio of the primary.

A. Applications

Figure 11 shows examples of practical nonimaging secondaries used to date for both kinds of applications. The flow line or "trumpet"-shaped secondary is a recent development with particular advantages in a retrofit mode, that is, to increase the concentration of a dish that is already designed and built. Figure 11b shows how introducing a small amount of curvature into the front surface of the secondary provides some of the concentration so that the overall height of

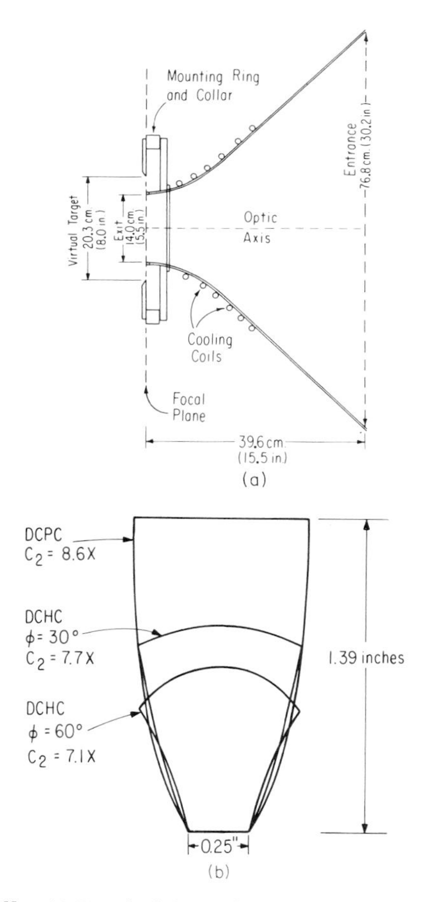

FIG. 11. (a) Practical thermal secondary concentrator referred to as the "trumpet," which has been built and tested at the University of Chicago, and (b) profiles for a DCPC and two dielectric compound hyperbolic concentrators (DCHCs) used for photovoltaic secondaries.

the side-walls can be reduced, yielding substantial savings in material.

For thermal applications with a CPC-type secondary, a particularly attractive option is the use of a primary of longer focal ratio (f-number $\gtrsim$ 1.0) with flat mirror facets, in which case the number of required facets can be reduced by a factor approximately equal to the secondary concentration ratio. For example, with a secondary with $C^2 = 5\times$ the number of flat facets required to achieve $150\times$ can be reduced from nearly 200 to less than 40. Details of such a design are being developed.

A two-stage photovoltaic concentrator presently being developed is shown in Fig. 12. This system has no real analog in contemporary solar concentrator configurations, since conventional

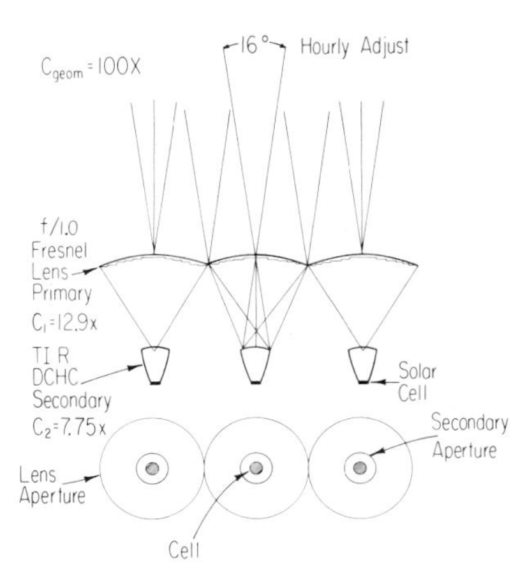

FIG. 12. Illustration of a lens/DCHC combination that could provide 100 : 1 concentration while requiring only approximately hourly adjustment. Prototypes of actual devices based on this concept are under development.

$100\times$ concentrators must track quite accurately ($\lesssim \pm 1°$) and crude tracking devices have much lower concentrations ($\sim$5–$15\times$). Here a good primary is designed to provide a concentration of about $13\times$ with an acceptance angle wide enough to accommodate the sun's movement for close to 1 hr. A TIR secondary provides the additional concentration (here $7.7\times$) required to make the economic savings associated with reduced cell area really worthwhile. No scale is shown in Fig. 12, since a study of the trade-offs associated with size is one of the objectives of the present work. For reference, note that an 0.5-in. cell would correspond to a 5-in.-diameter circular collecting lens. These individual elements could be arranged in a hexagonal close-packed geometry if a high packing density is required, or on a square lattice if not. Finally, note that in this geometry the function of the secondary in redistributing the concentrated sunlight more uniformly on the cell is especially useful.

IV. Summary

Nonimaging optics departs from the methods of traditional optical design in order to develop techniques for maximizing the collecting power of concentrating elements and systems. Designs that exceed the concentration attainable with focusing techniques by factors of 4 or more and approach the thermodynamic limit are possible. This is accomplished by applying the concepts of Hamiltonian optics, phase space conservation, thermodynamic arguments, and radiative transfer methods. Concentrators based on this approach are finding increasing use in solar energy and a variety of other applications and include the now well-known nontracking compound parabolic concentrator (CPC).

Compound parabolic concentrators permit the use of low to moderate levels of concentration for solar thermal collectors without the requirement of diurnal tracking. When used in conjunction with an evacuated absorber with selective surfaces, a fully stationary CPC (designs with concentrations of 1.1–1.4× are now commercially available) has a typical efficiency of about 40% at 150°C (270°F) above ambient with available conventional materials. An experimental 5× CPC requiring approximately 12 tilt adjustments annually, when used with a similar available evacuated absorber, has a measured efficiency of 60% at 220°C above ambient and is capable of efficiencies near 50% at 300°C With such thermally efficient absorbers, higher concentrations are not necessary or desirable.

Argonne National Laboratory and the University of Chicago are working on a research and development program to develop an advanced evacuated-tube collector that will be suitable for mass production by industry, will compete successfully with conventional flat-plate collectors at domestic hot water (DHW) temperatures, and will be suitable for industrial process heat (IPH) and/or cooling applications. The essence of the design concept for these new collectors is the integration of moderate levels of nonimaging concentration inside the evacuated tube itself. This permanently protects the reflecting surfaces and allows the use of highly reflecting front-surface mirrors with reflectances greater than 95%. Fabrication and long-term testing of a proof-of-concept prototype have established the technical success of the concept. Present work is directed toward the development of a manufacturable unit that will be suitable for the widest possible range of applications.

The temperature capabilities of CPCs with nonevacuated absorbers are somewhat more limited. However, with proper design, taking care to reduce parasitic thermal losses through the reflectors, nonevacuated CPCs will outperform the best available flat-plate collectors at temperatures above about 50–70°C.

Nonimaging, near-ideal secondary or terminal concentrators have advantages for any solar concentrating application. In the near term, they are being applied to increase the angular tracking, alignment, and slope definition requirements for the primaries in an effort to reduce system cost and complexity. In future applications, very high geometric concentrations that cannot otherwise be attained can be achieved through the use of these devices.

Acknowledgment

Work described in this review was supported in large part by the U.S. Department of Energy under a number of grants and contracts, including most recently contracts DE-AC02-80-ER10575, DE-AC02-ER10558, DE-AC03-83F11655, and DE-FG02-84CH10201.

Bibliography

O'Gallagher, J., and Winston, R. (1983). Development of compound parabolic concentrators for solar energy. *J. Ambient. Energy* **4,** 171.

O'Gallagher, J., and Winston, R. (1986). Test of a "trumpet" secondary concentrator with a paraboloidal dish primary. *Solar Energy* **36,** 37.

O'Gallagher, J., Welford, W. T., and Winston, R. (1986). Axially symmetric nonimaging flux concentrators with the maximum theoretical concentration ratio. *J. Opt. Soc. Am.* **4,** 66.

Rabl, A., O'Gallagher, J., and Winston, R. (1980). Design and test of non-evacuated solar collectors with Compound Parabolic Concentrators. *Solar Energy* **25,** 335.

Snail, K. A., O'Gallagher, J. J., and Winston, R. (1984). A stationary evacuated collector with integrated concentrator. *Solar Energy* **33,** 441.

Welford, W. T., and Winston, R. (1978). "The Optics of Nonimaging Concentrators." Academic Press, New York.

Welford, W. T., and Winston, R. (1982). Conventional optical systems and the brightness theorem. *Appl. Opt.* **21,** 1531.

NONLINEAR OPTICAL PROCESSES

John F. Reintjes *Naval Research Laboratory*

GLOSSARY

Anti-Stokes shift: Difference in frequency between the pump wave and the generated wave in a stimulated scattering interaction when the generated wave is at a higher frequency than the pump wave.

Brillouin shift: Difference in frequency between the incident wave and the scattered wave in a stimulated Brillouin interaction. It is equal to the sound wave frequency.

Coherence length: Distance required for the phase of a light wave to change by $\pi/2$ relative to that of its driving polarization. Maximum conversion in parametric processes is obtained at L_{coh} when $\Delta k \neq 0$.

Constitutive relations: Set of equations that specify the dependence of induced magnetizations or polarizations on optical fields.

Depletion length: Distance required for significant pump depletion in a frequency-conversion interaction.

Mode-locked laser: Laser that operates on many longitudinal modes, each of which is constrained to be in phase with the others. Mode-locked lasers produce pulses with durations of the order of 0.5–30 psec.

Nonlinear polarization: Electric dipole moment per unit volume that is induced in a material by a light wave and depends on intensity of the light wave raised to a power greater than unity.

Parametric interactions: Interactions between light waves and matter that do not involve transfer of energy to or from the medium.

Phase conjugation: Production of a light wave with the phase variations of its wave front reversed relative to those of a probe wave.

Phase matching: Act of making the wave vector of an optical wave equal to that of the nonlinear polarization that drives it.

Population inversion: Situation in which a higher-energy level in a medium has more population than a lower-energy one.

Q-switched laser: Laser in which energy is stored while the cavity is constrained to have large loss (low Q) and is emitted in an intense short pulse after the cavity Q is switched to a high value reducing the loss to a low value.

Raman shift: Difference in frequency between the incident wave and the scattered wave in a stimulated Raman interaction. It is equal to the frequency of the material excitation involved in the interaction.

Slowly varying envelope approximation: Approximation in which it is assumed that the amplitude of an optical wave varies slowly in space compared to a wavelength and slowly in time compared to the optical frequency.

Stimulated scattering: Nonlinear frequency conversion interaction in which the generated wave has exponential gain and energy is transferred to or from the nonlinear medium.

Stokes shift: Difference in frequency between the pump wave and the generated wave in a stimulated scattering interaction when the generated wave is at lower frequency than is the pump wave.

Susceptibility, *n*th order: Coefficient in a perturbation expansion that gives the ratio of the induced polarization of order n to the nth power of the electric or magnetic fields.

Wave number (cm^{-1}): Number of optical waves contained in 1 cm. It is equal to the reciprocal of the wavelength in centimeters and has the symbol $\tilde{\nu}$. It is commonly used as a measure of the energy between quantum levels of a material system or the energy of the photons in an optical wave and is related to the actual energy by $E = h\tilde{\nu}c$, where h is Planck's constant and c is the speed of light.

Wave vector: Vector in the direction of propagation of an optical wave and having magnitude equal to 2π divided by the wavelength.

Wave-vector mismatch: Difference between the wave vector of an optical wave and that of the nonlinear polarization that drives it. It is also called the phase mismatch.

Zero-point radiation: Minimum energy allowed in optical fields due to quantum-mechanical effects.

Nonlinear optics is a field of study that involves a nonlinear response of a medium to intense electromagnetic radiation. Nonlinear optical interactions can change the propagation or polarization characteristics of the incident waves, or they can involve the generation of new electromagnetic waves, either at frequencies different from those contained in the incident fields or at frequencies that are the same but with the waves otherwise distinguishable from the incident waves—for example, by their direction of polarization or propagation. Nonlinear optical interactions can be used for changing or controlling certain properties of laser radiation, such as wavelength, bandwidth, pulse duration, and beam quality, as well as for high-resolution molecular and atomic spectroscopy, materials studies, modulation of optical beams, information processing, and compensation of distortions caused by imperfect optical materials. Nonlinear optical interactions are generally observed in the spectral range covered by lasers, between the far infrared and the extreme ultraviolet (XUV) (Fig. 1), but some nonlinear interactions have been observed at wavelengths ranging from X rays to microwaves. They generally require very intense optical fields and as a result are usually observed only with radiation from lasers, although some nonlinear interactions, such as saturation of optical transitions in an atomic medium, can occur at lower intensities and were observed before the invention of the laser. Nonlinear optical interactions of various kinds can occur in all types of materials, although some types of interactions are observable only in certain types of materials, and some materials are better suited to specific nonlinear interactions than others.

I. Optical Nonlinearities—Physical Origins

When an electromagnetic wave propagates through a medium, it induces a polarization (electric dipole moment per unit volume) and magnetization (magnetic dipole moment per unit volume) in the medium as a result of the motion of the electrons and nuclei in response to the fields in the incident waves. These induced polarizations and magnetizations oscillate at frequencies determined by a combination of the properties of the material and the frequencies

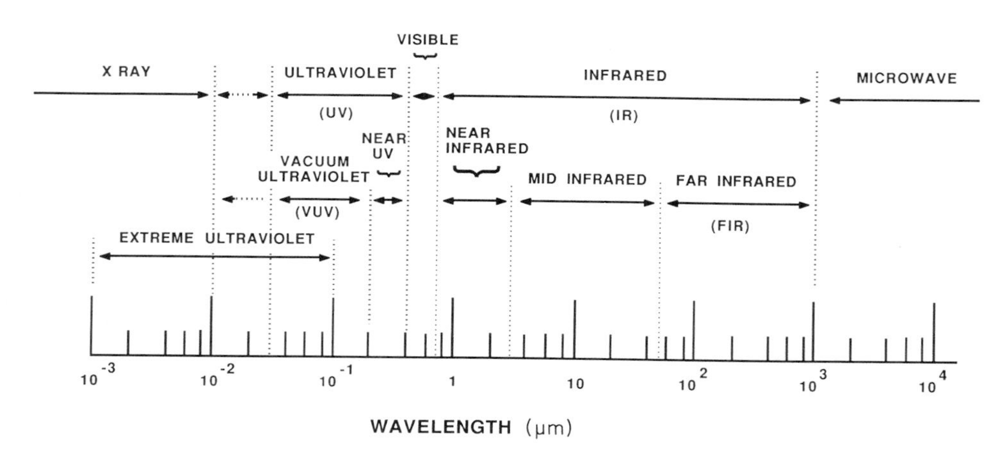

FIG. 1. Designations of various regions of the electromagnetic spectrum between microwaves and X rays. Dashed arrows indicate inexact boundaries between spectral regions.

contained in the incident light waves. The optical properties of the medium and the characteristics of the radiation that is transmitted through it result from interference among the fields radiated by the induced polarizations or magnetizations and the incident fields.

At low optical intensities, the induced polarizations and magnetizations are proportional to the electric or magnetic fields in the incident wave, and the response of the medium is termed linear. Various linear optical interactions can occur, depending on the specific properties of the induced polarizations. Some of the more familiar linear optical effects that result from induced polarizations that oscillate at the same frequency as the incident radiation are refraction, absorption, and elastic scattering (Rayleigh scattering from static density variations, or Tyndall or Mie scattering). Other linear optical processes involve inelastic scattering, in which part of the energy in the incident wave excites an internal motion of the material and the rest is radiated in a new electromagnetic wave at a different frequency. Examples of inelastic scattering processes are Raman scattering, which involves molecular vibrations or rotations, electronic states, lattice vibrations, or electron plasma oscillations; Brillouin scattering, which involves sound waves or ion-acoustic waves in plasmas; and Rayleigh scattering, involving diffusion, orientation, or density variations of molecules. Although these inelastic scattering processes produce electromagnetic waves at frequencies different from those in the incident waves, they are linear optical processes when the intensity of the scattered wave is proportional to the intensity of the incident wave.

When the intensity of the incident radiation is high enough, the response of the medium changes qualitatively from its behavior at low intensities, giving rise to the nonlinear optical effects. Some nonlinear optical interactions arise from the larger motion of the electrons and ions in response to the stronger optical fields. In most materials, the electrons and ions are bound in potential wells that, for small displacements from equilibrium, are approximately harmonic (i.e., have a potential energy that depends on the square of the displacement from equilibrium), but are anharmonic (i.e., have a potential energy that has terms that vary as the third or higher power of the displacement from equilibrium) for larger displacements, as shown in Fig. 2. As long as the optical intensity is low, the electron or ion moves in the harmonic part of the well. In

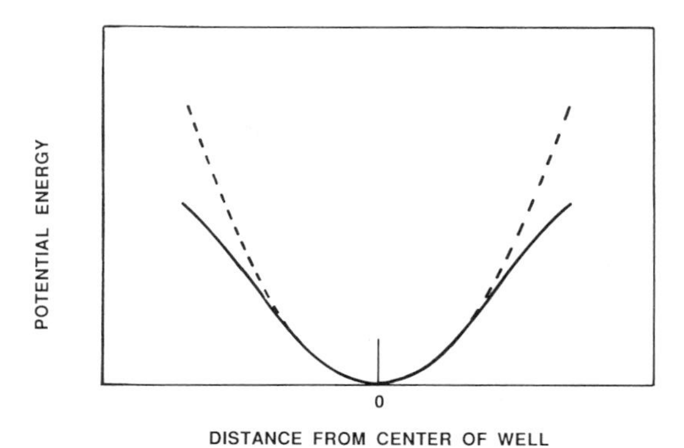

FIG. 2. Schematic illustration of an anharmonic potential well (solid line) that can be responsible for nonlinear optical interactions. A harmonic potential well that does not result in nonlinear interactions is shown for comparison (dashed line).

this regime, the induced polarizations can oscillate only at frequencies that are contained in the incident waves and the response is linear. When the incident intensity is high enough, the charges can be driven into the anharmonic portion of the potential well. The additional terms in the potential introduce terms in the induced polarization that depend on the second, third, or higher powers of the incident fields, giving rise to nonlinear responses. Examples of nonlinear optical processes that occur in this manner are various forms of harmonic generation and parametric frequency mixing.

A second type of nonlinear response results from a change in some property of the medium caused by the optical wave, which in turn affects the propagation of the wave. The optical properties of a material are usually described in the limit of extremely weak optical fields and are therefore considered to be intrinsic properties of the medium. When the optical field is strong enough, however, it can change certain characteristics of the medium, which in turn changes the way the medium affects the optical wave, resulting in a nonlinear optical response. An example of such a response is a change in the refractive index of a medium induced by the optical wave. Such changes can occur, for example, because of orientation of anisotropic molecules along the incident fields, or because of changes in the density of the medium as a result of electrostriction or as a result of a temperature change following absorption of the incident wave. Many of the propagation characteristics of optical waves are determined by the refractive index of the medium. In the situations just described, the refractive index depends on the intensity of the optical wave, and, as a result, many of the propagation characteristics depend on the optical intensity as well.

II. Optical Nonlinearities—Mathematical Description

A. Maxwell's Equations for Nonlinear Materials

The propagation of an optical wave in a medium is described by the Maxwell equation for the optical field, including the effects of the induced polarizations and magnetizations, and a specification of the dependence of the induced polarizations and magnetizations on the optical fields. The Maxwell equation for the electric field of an optical wave in a medium is

$$\nabla^2 E(r, t) - (1/c^2)\, \partial^2 E(r, t)/\partial t^2 = \mu_0\, \partial^2 P(r, t)/\partial t^2 + \partial[\nabla \times M(r, t)]/\partial t \quad (1)$$

Here $E(r, t)$ is the electric field of the optical wave, and the terms on the right are the induced polarizations (P) and magnetizations (M) that describe the effects of any charge distribution that may be present in the propagation path. In vacuum these terms are zero, and in this situation Eq. (1) reduces to the familiar expression for vacuum propagation of electromagnetic radiation.

B. Nonlinear Polarizations

The problem of propagation in a medium is completely specified when the relations between the polarization and the magnetization and the optical fields, termed constitutive relations, are given. In general these relations can be quite complicated. However, for many situations encountered in the laboratory a set of simplifying assumptions can be made that lead to an expansion of the induced polarization in a power series in the electric field of the light wave of the form

$$\mathrm{P} = \varepsilon_0 \chi^{(1)} E + \varepsilon_0 \chi^{(2)} E^2 + \varepsilon_0 \chi^{(3)} E^3 + \cdots + \varepsilon_0 Q^{(1)} \nabla E + \varepsilon_0 Q^{(2)} \nabla E^2 + \cdots \quad (2a)$$

For magnetic materials a similar expansion can be made for the magnetization:

$$M = m^{(1)} H + m^{(2)} H^2 + m^{(3)} H^3 + \cdots \quad (2b)$$

The types of expression in Eqs. (2a) and (2b) are generally valid when the optical fields are weak compared to the electric field that binds the electrons in the material and when the coefficients of the various terms in Eqs. (2a) and (2b) are constant over the range of frequencies contained in the individual incident and generated fields. In addition, the wavelength of the radiation must be long compared to the dimension of the scattering centers (the atoms and molecules of the nonlinear medium), so that the charge distributions can be accounted for with a multipole expansion.

When the first of these conditions is not met, as for example with extremely intense radiation, the perturbation series will not converge and all powers of the incident fields must be used. When the second of the conditions is not met, as, for example, can happen when certain resonance conditions are satisfied, each term in the response of the medium will involve convolutions of the optical fields instead of simply powers of the fields multiplied by constants. In these situations the polarizations induced in the medium must be solved for as dynamical variables along with the optical fields.

The coefficients in the various terms in Eqs. (2a) and (2b) are termed the nth-order susceptibilities. The first-order susceptibilities describe the linear optical effects, while the remaining terms describe the nth order nonlinear optical effects. The coefficients $\chi^{(n)}$ are the nth-order electric dipole susceptibilities, the coefficients $Q^{(n)}$ are the nth-order quadrupole susceptibilities, and so on. Similar terminology is used for the various magnetic susceptibilities. For most nonlinear optical interactions, the electric dipole susceptibilities are the dominant terms because the wavelength of the radiation is usually much longer than the scattering centers. These will be the ones considered primarily from now on.

In some situations, however, the electric quadrupole susceptibilities can make a significant contribution to the response. For example, symmetry restrictions in certain classes of materials can prevent some nonlinear processes from occurring by dipole interactions, leaving the quadrupole interactions as the dominant response. This occurs, for example, with second-harmonic generation in media with inversion symmetry. Other situations in which quadrupole interactions are significant are those in which they are enhanced by resonances with appropriate energy levels or for interactions involving sufficiently short-wavelength X radiation.

The nonlinear susceptibilities are tensors and, as such, relate components of the nonlinear polarization vector to various components of the optical field vectors. For example, the second-order polarization is given by

$$P_i = \varepsilon_0 \chi^{(2)}_{ijk} E_j E_k$$

where i, j, and k refer to the spatial directions x, y, and z. The susceptibility tensors χ have the symmetry properties of the nonlinear medium. As a result they can restrict the combinations of vector components of the various optical fields that can be used effectively. In some situations, such as those involving nonlinear optical processes that change the polarization vector of the optical wave, the tensor properties play a central role in the nonlinear interaction. In other situations, however, the tensor properties are important only in determining which combinations of vector components can, or must, be used for the optical fields and the nonlinear polarizations, and beyond that the nonlinear optical susceptibilities can usually be treated as scalars.

The magnitudes of the nonlinear susceptibilities are usually determined by measurement, although in some cases, most notably for third- and higher-order interactions in atomic gases or second-order interactions in some crystals, they can be calculated using various theories. In some cases they can be estimated with varying degrees of accuracy from products of the linear refractive indices at the various wavelengths involved.

C. Calculation of Optical Fields

In order to calculate the optical fields involved in the nonlinear interactions, we assume that they can be expressed in the form

$$E(r, z, t) = \tfrac{1}{2}[A(r, z, t)e^{i(kz-\omega t)} + \text{complex conjugate}] \quad (3)$$

This expression describes a wave of amplitude A propagating in the z direction with a frequency ω and a wave vector $k = 2\pi n/\lambda$, where λ is the wavelength of the light in vacuum, and n is the linear refractive index, which takes into account the effects of the linear susceptibilities. Many nonlinear effects involve the interaction of two or more optical fields at different wavelengths. In such situations the optical field is written as

$$E(r, z, t) = \frac{1}{2}\sum[A_i(r, z)e^{i(k_i n_i z-\omega_i \tau)} + \text{complex conjugate}] \quad (4)$$

Here the index for the summation extends over all fields in the problem, including those fields that are generated in the nonlinear interaction as well as those that are incident on the medium.

The nonlinear polarization is written as

$$P(r, z, t) = \frac{1}{2}\sum[P_i(r, z)e^{i(k_i^p z-\omega_i t)} + \text{complex conjugate}] \quad (5)$$

Here P_i is the amplitude of the nonlinear polarization with frequency ω_i. It is determined by an appropriate combination of optical fields and nonlinear susceptibilities according to the expansion in Eq. (2a). It serves as a source term for the optical field with frequency ω_i. The wave vector of the nonlinear polarization k_i^p is in general different from the wave vector of the optical field at the same frequency. This difference plays a very important role in determining the effectiveness of many nonlinear optical interactions.

The amplitudes of the various waves can be calculated from the expressions given in Eqs. (1) and (2), using the fields and polarizations with the forms given in Eqs. (4) and (5). In deriving equations for the field amplitudes, a further simplifying assumption, the slowly varying envelope approximation, is also usually made. This approximation assumes that the field envelopes $A_i(r, z, t)$ vary slowly in time compared to the optical frequency and slowly in space compared to the optical wavelength. It is generally valid for all of the interactions that are encountered in the laboratory. It enables us to neglect the second derivatives of the field amplitudes with respect to z and t when Eqs. (4) and (5) are substituted into Eq. (1). When this substitution is made, we obtain expressions of the following form for the various fields involved in a nonlinear interaction:

$$\nabla_\perp^2 A_1 + 2ik_1[\partial A_1/\partial_z + (n_1/c)\,\partial A_1/\partial t] = -\mu_0\omega_1^2 P_1 e^{-i\Delta k_1 z} \quad (6a)$$

$$\nabla_\perp^2 A_2 + 2ik_2[\partial A_2/\partial_z + (n_2/c)\,\partial A_2/\partial t] = -\mu_0\omega_2^2 P_2 e^{-i\Delta k_2 z} \quad (6b)$$

$$\vdots$$

$$\nabla_\perp^2 A_n + 2ik_n[\partial A_n/\partial_z + (n_n/c)\,\partial A_n/\partial t] = -\mu_0\omega_n^2 P_n e^{-i\Delta k_n z} \quad (6c)$$

Here A_i is the amplitude of the ith optical field, which may be either an incident field or a field generated in the interaction, and Δk_i is the wave-vector mismatch between the ith optical field and the polarization that drives it and is given by

$$\Delta k_i = k_i - k_i^p \tag{7}$$

In nonlinear interactions that involve the generation of fields at new frequencies, the nonlinear polarization for the generated field will involve only incident fields. In many cases the interactions will be weak and the incident fields can be taken as constants. In other situations, the generated fields can grow to be sufficiently intense that their nonlinear interaction on the incident fields must be taken into account. In still other nonlinear interactions, such as those involving self action effects, an incident field amplitude can occur in its own nonlinear polarization. The set of Eq. (6) along with the constitutive relations in Eq. (2) can be used to calculate the optical fields in most nonlinear interactions.

D. Classifications of Nonlinear Interactions

Nonlinear interactions can be classified according to whether they are driven by electric or magnetic fields, whether they involve the generation of fields at new frequencies or change the propagation properties of the incident fields, and whether or not they involve a transfer of energy to or from the nonlinear medium. To a certain extent these classifications are overlapping, and examples of each type of interaction can be found in the other categories.

A list of the more common nonlinear interactions is given in Table I. Most of the commonly observed nonlinear optical interactions are driven by the electric fields. They can proceed through multipole interactions of any order, but the dipole interactions are most common unless certain symmetry or resonance conditions are satisfied in the medium. The interactions are classified first according to whether they involve frequency conversion (that is, the generation of a wave at a new frequency) or self-action (that is, whether they affect the propagation of the incident wave). The frequency-conversion interactions are further classified according to whether they are parametric processes or inelastic processes. Parametric processes do not involve a transfer of energy to or from the nonlinear material. The material merely serves as a medium in which the optical waves can exchange energy among themselves. On the other hand, inelastic frequency conversion processes, generally termed stimulated scattering, do involve a transfer of energy to or from the nonlinear medium.

Self-action effects involve changes in the propagation characteristics of a light wave that are caused by its own intensity. They can involve both elastic and inelastic processes and can affect almost any of the propagation characteristics of the light wave, including both absorption and focusing properties. They can also change the spatial, temporal, and spectral distributions of the incident wave, as well as its state of polarization.

Coherent effects involve interactions that occur before the wave functions that describe the excitations in the nonlinear medium have time to get out of phase with one another. They can involve changes in propagation characteristics as well as the generation of new optical signals.

TABLE I. Common Nonlinear Optical Interactions

Nonlinear process	Description	Incident radiation	Generated radiation
Frequency conversion			
Parametric processes			
qth-Harmonic generation	Conversion of radiation from ω to $q\omega$	ω	$q\omega$
Sum- or difference-frequency mixing	Conversion of radiation at two or more frequencies to radiation at a sum- or difference-frequency combination	Three-wave (second order) ω_1, ω_2	$\omega_g = \omega_1 \pm \omega_2$
		Four-wave (third order)	
		ω_1, ω_2	$\omega_g = 2\omega_1 \pm \omega_2$
		$\omega_1, \omega_2, \omega_3$	$\omega_g = \omega_1 \pm \omega_2 \pm \omega_3$

TABLE I. *(Continued)*

Nonlinear process	Description	Incident radiation	Generated radiation
		Six-wave (fifth order) ω_1, ω_2	$\omega_g = 4\omega_1 \pm \omega_2$
Parametric down-conversion	Conversion of radiation at frequency ω_1 to two or more lower frequencies	Three-wave (second order) ω_1	$\omega_{g1} + \omega_{g2} = \omega_1$
		Four-wave (third order) ω_1	$\omega_{g1} + \omega_{g2} = 2\omega_1$
Optical rectification	Conversion of radiation at frequency ω_1 to an electric voltage	ω_1	$\omega_g = 0 = \omega_1 - \omega_1$
Inelastic processes			
Stimulated scattering	Conversion of radiation at frequency ω_1 to radiation at frequency ω_2 with excitation or deexcitation of an internal mode at frequency ω_0	Stokes scattering ω_1, ω_2	$\omega_2 = \omega_1 - \omega_0$
		Anti-Stokes scattering ω_1, ω_2	$\omega_2 = \omega_1 + \omega_0$
Self-action effects			
Self-focusing	Focusing of an optical beam due to a change in the refractive index caused by its own intensity		
Self-defocusing	Defocusing of an optical beam due to a change in the refractive index caused by its own intensity		
Self-phase modulation	Modulation of the phase of an optical wave due to a time-dependent change in the refractive index caused by its own intensity		
Optical Kerr effect	Birefringence in a medium caused by an intensity-induced change in the refractive index		
Ellipse rotation	Intensity-dependent rotation of the polarization ellipse of a light wave due to the nonlinear refractive index		
Raman-induced Kerr effect	Change of refractive index or birefringence of a light wave at one wavelength due to the intensity of a light wave at another wavelength		
Multiphoton absorption	Increase in the absorption of a material at high intensities		
Saturable absorption	Decrease in the absorption of a material at high intensities		
Coherent effects			
Self-induced transparency	High transmission level of a medium at an absorbing transition for a short-duration light pulse with the proper shape and intensity		
Photon echo	Appearance of a third pulse radiated from an absorbing medium following application of two pulses of proper duration, intensity, and separation		
Adiabatic rapid passage	Inversion of population of a medium by application of a light pulse whose frequency is swept quickly through that of an absorbing transition		
Electrooptic effects			
Pockels effect	Birefringence in a medium caused by an electric field and proportional to the electric field strength		
Kerr effect	Birefringence in a medium caused by an electric field and proportional to the square of the electric field strength		
Magnetooptic effects			
Faraday rotation	Rotation of the direction of linear polarization of a light beam resulting from changes in the relative velocities of oppositely circular polarizations caused by a magnetic field		
Cotton–Mouton effect	Birefringence induced in a material caused by a magnetic field		
Miscellaneous			
Optical breakdown	Rapid ionization of a material by avalanche production of electrons caused by an intense optical field		

Electrooptic and magnetooptic effects involve changes in the refractive index of the medium caused by external electric or magnetic fields resulting in changes in the phase of the optical wave or in its state of polarization.

III. Specific Nonlinear Processes

A. Frequency-Conversion Processes

Frequency-conversion processes are those that involve the generation of radiation at wavelengths other than the ones that are contained in the incident radiation. A typical frequency conversion geometry is illustrated in Fig. 3. Pump radiation, consisting of one or more optical waves at one or more frequencies, is incident on a nonlinear medium and interacts with it to generate the new wave. Depending on the interaction, the generated wave can be at a lower or higher frequency than the waves in the pump radiation, or, in some situations, can be an amplified version of one of the incident waves.

Frequency conversion interactions can either be parametric processes or stimulated scattering interactions. In parametric frequency conversion, the incident wave generates a new wave at a frequency that is a multiple, or harmonic, of the incident frequency, or, if more than one frequency is present in the incident radiation, a sum or difference combination of the frequencies of the incident fields. Parametric frequency conversion can also involve the generation of radiation in two or more longer-wavelength fields whose frequencies add up to the frequency of the incident field. In parametric frequency conversion, energy is conserved among the various optical waves, with the increase in energy of the generated wave being offset by a corresponding decrease in energy of the incident waves.

Parametric frequency-conversion interactions can occur in any order of the perturation expansion of Eq. (2). The most commonly observed processes have involved interactions of second and third order, although interactions up to order 11 have been reported. Parametric interactions are characterized by a growth rate for the intensity of the generated wave that depends on a power, or a product of powers, of the intensities of the incident waves, and they are strongly dependent on difference in wave vectors between the nonlinear polarization and the optical fields.

Stimulated scattering generally involves inelastic nonlinear frequency-conversion processes. They are nonlinear counterparts of the various linear inelastic scattering processes that were mentioned earlier. In these processes, an incident wave at frequency ω_{inc} is scattered into a wave with a different frequency, ω_{scat}, with the difference in energy between the incident and scattered photons taken up by excitation or deexcitation of an internal mode of the material. When the internal mode is excited in the nonlinear process, the scattered wavelength is longer than the incident wavelength, while it is shorter than the incident wavelength when the internal mode is deexcited.

Various types of stimulated scattering processes can occur, including stimulated Raman, stimulated Brillouin, and stimulated Rayleigh scattering, each of which involves a different type of internal mode of the medium, as will be described in Section III,A,2. The optical wave generated in stimulated scattering processes is characterized by exponential growth similar to the exponential growth experienced by laser radiation in the stimulated emission process. [*See* LASERS.]

Frequency-conversion interactions are used primarily to generate coherent radiation at wavelengths other than those that can be obtained directly from lasers, or to amplify existing laser radiation. Various frequency-conversion processes have been used to generate coherent

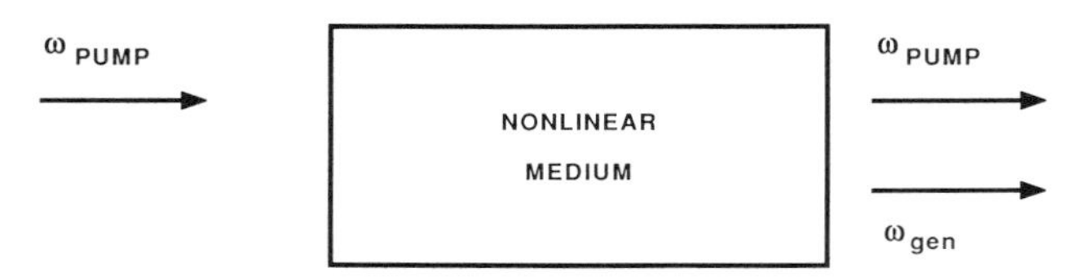

FIG. 3. Schematic illustration of a typical geometry used in frequency conversion interactions. Input radiation is supplied at one or more incident frequencies ω_{pump}. The wave at ω_{gen} is generated in the nonlinear interaction.

radiation at wavelengths ranging from the extreme ultraviolet to the millimeter wave region. The radiation generated through nonlinear frequency-conversion processes has all of the special properties usually associated with laser radiation, such as spatial and temporal coherence, collimation, and narrow bandwidths, and can be tunable if the pump radiation is tunable. The generated radiation can have special properties not possessed by existing laser radiation in a given wavelength range, such as tunability, short pulse duration, or narrow bandwidth, or it can be in a spectral range in which there is no direct laser radiation available. An example of the first is the generation of tunable radiation in spectral regions in the vacuum ultraviolet or the infrared where only fixed-frequency lasers exist. An example of the second is the generation of coherent radiation in the extreme ultraviolet at wavelengths less than 100 nm, a spectral region in which lasers of any kind are just beginning to be developed.

Various frequency-conversion interactions can also be used for spectroscopy, image conversion, control of various properties of laser beams, and certain optical information-processing techniques.

1. Parametric Frequency-Conversion Processes

Parametric frequency-conversion processes include harmonic conversion of various order, various forms of frequency mixing, parametric down-conversion, and optical rectification. They are used primarily for generation of radiation at new wavelengths, although some of the interactions can be used for amplifying existing radiation. Various forms of parametric frequency-conversion interactions have been used to generate coherent radiation at wavelengths ranging from almost the X-ray range at 35.5 nm to the microwave range at 2 mm. Parametric frequency-conversion interactions are usually done in the transparent region of the nonlinear medium. Different interactions, nonlinear materials, and lasers are used to generate radiation in the various wavelength ranges. Some of the more commonly used interactions, types of lasers, and types of nonlinear materials in various wavelength ranges are given in Table II.

a. Second-Order Effects. Second-order effects are primarily parametric in nature. They include second-harmonic generation, three-wave sum- and difference-frequency mixing, parametric down-conversion, parametric oscillation, and optical rectification. The strongest second-order processes proceed through electric dipole interactions. Because of symmetry restrictions, the even-order electric dipole susceptibilities are zero in materials with inversion symmetry. As a result, the second-order nonlinear interactions are observed most commonly only in certain classes of crystals that lack a center of inversion. However, some second-order processes due to electric quadrupole inter-

TABLE II. Wavelength Ranges for Various Nonlinear Parametric Interactions

Nonlinear interaction	Wavelength range	Typical laser	Type of material
Second-harmonic, three-wave sum-frequency mixing	2–5 μm	CO_2, CO	Crystal without inversion center
	Visible	Nd:YAG, Nd:glass	Crystal without inversion center
	Ultraviolet (to 185 nm)	Harmonics of Nd, ruby, dye, argon	Crystal without inversion center
Three-wave difference-frequency mixing	Visible to infrared (25 μm)	CO, CO_2, Nd, dye, Parametric oscillator, spin-flip Raman	Crystal without inversion center
Parametric oscillation	Visible to infrared	Nd, dye, CO_2	Crystal without inversion center
Third harmonic, four wave sum-frequency mixing, four-wave difference-frequency mixing	Infrared (3.3 μm) to XUV (57 nm)	CO_2, CO, Nd, ruby, dye, excimer	Rare gases, molecular gases, metal vapors, cryogenic liquids
Fifth- and higher-order harmonic and frequency mixing	UV (216 nm) XUV (106–35.5 nm)	Nd, harmonics of Nd, excimer	Rare gases, metal vapors

actions have been observed in solids (e.g., sodium chloride) and gases that have a center of inversion. In addition, some second-order processes, such as second-harmonic generation, have also been observed in gases in which the inversion symmetry has been lifted with electric or magnetic fields. In these situations the interactions are actually third-order ones in which the frequency of one of the waves is zero.

The nonlinear polarization for the second order processes can be written as

$$P_{\mathrm{NL}}^{(2)} = 2\varepsilon_0 dE^2 \tag{8}$$

where d is a nonlinear optical susceptibility related to $\chi^{(2)}$ in Eq. (2) by $d = \chi^{(2)}/2$ and E is the total electric field as given in Eq. (4). The right-hand side of Eq. (8) has terms that oscillate at twice the frequency of the incident waves and at sum- and difference-frequency combinations and terms that do not oscillate. Each of these terms gives rise to a different nonlinear process. In general they are all present in the response of the medium. The term (or terms) that forms the dominant response of the medium is usually determined by phase matching, as will be discussed later.

The nonlinear polarization amplitude as defined in Eq. (5) can be evaluated for each of the specific nonlinear processes by substituting the form of the electric field in Eq. (4) into Eq. (8) and identifying each term by the frequency combination it contains. The resulting form of the nonlinear polarization for the sum frequency process $\omega_3 = \omega_1 + \omega_2$ is

$$P_i^{\mathrm{NL}}(\omega_3) = g\varepsilon_0 d_{ijk}(-\omega_3, \omega_1, \omega_2)A_j(\omega_1)A_k(\omega_2) \tag{9}$$

Here i, j, k refer to the crystallographic directions of the nonlinear medium and the vector components of the various fields and polarizations, and g is a degeneracy factor that accounts for the number of distinct permutations of the pump fields.

Similar expressions can be obtained for the other second-order processes with minor changes. For second-harmonic generation, $\omega_1 = \omega_2$, while for difference-frequency mixing processes of the form $\omega_3 = \omega_1 - \omega_2$, ω_2 appears with a minus sign and the complex conjugate of A_2 is used. The degeneracy factor g is equal to 1 for second-harmonic generation and 2 for the second-order mixing processes.

Because the susceptibility tensor d has the same symmetry as the nonlinear medium, certain of its components will be zero, and others will be related to one another, depending on the material involved. As a result, only certain combinations of polarization vectors and propagation directions can be used for the incident and generated waves in a given material. For a specific combination of tensor and field components, the nonlinear polarization amplitude can be written as

$$P^{\mathrm{NL}}(\omega_3) = g\varepsilon_0 d_{\mathrm{eff}}(-\omega_3, \omega_1, \omega_2)A(\omega_1)A(\omega_2) \tag{10}$$

where d_{eff} is an effective susceptibility that takes into account the nonzero values of the d tensor and the projections of the optical field components on the crystallographic directions and $A(\omega_i)$ is the total optical field at ω_i. The nonlinear polarization amplitudes for the other second-order processes have similar expressions with the changes noted above. The forms for the polarization amplitude for the various second-order interactions are given in Table III.

The k vector of the nonlinear polarization is given by

$$k_3^p = k_1 \pm k_2 \tag{11}$$

where the plus sign is used for sum-frequency combinations and the minus sign is used for difference-frequency combinations. The wave-vector mismatch of Eq. (7) is also given in Table III for each nonlinear process. The nonlinear polarization amplitudes and the appropriate wave-vector mismatches as given in Table III can be used in Eq. (6) for the field amplitudes to calculate the intensities of the generated waves in specific interactions.

The second-order frequency-conversion processes are used primarily to produce coherent radiation in wavelength ranges where radiation from direct laser sources is not available or to obtain radiation in a given wavelength range with properties, such as tunability, that are not available with existing direct laser sources. They have been used with various crystals and various types of lasers to generate radiation ranging from 185 nm in the ultraviolet to 2 mm in the microwave region. The properties of some nonlinear crystals used for second-order interactions are given in Table IV, and some specific examples of second-order interactions are given in Table V.

i. SECOND-HARMONIC GENERATION. In second-harmonic generation, radiation at an incident frequency ω_1 is converted to radiation at

TABLE III. Second-Order Nonlinear Polarizations

Nonlinear process	Nonlinear polarization	Wave-vector mismatch	Plane-wave phase-matching condition
Second-harmonic generation	$P(2\omega) = \varepsilon_0 d_{\mathrm{eff}}(-2\omega, \omega, \omega)A^2(\omega)$	$\Delta k = k(2\omega) - 2k(\omega)$[a]	$n(2\omega) = n(\omega)$[a]
		$\Delta k = k(2\omega) - k_1(\omega) - k_2(\omega)$[b]	$n(2\omega) = n_1(\omega)/2 + n_2(\omega)/2$[b]
Three-wave sum-frequency mixing	$P(\omega_3) = 2\varepsilon_0 d_{\mathrm{eff}}(-\omega_3, \omega_1, \omega_2)A(\omega_1)A(\omega_2)$	$\Delta k = k(\omega_3) - k(\omega_1) - k(\omega_2)$	$n(\omega_3)/\lambda_3 = n(\omega_1)/\lambda_1 + n(\omega_2)/\lambda_2$
Three-wave difference-frequency mixing	$P(\omega_3) = 2\varepsilon_0 d_{\mathrm{eff}}(-\omega_3, \omega_1, -\omega_2)A(\omega_1)A^*(\omega_2)$	$\Delta k = k(\omega_3) - k(\omega_1) + k(\omega_2)$	$n(\omega_3)/\lambda_3 = n(\omega_1)/\lambda_1 - n(\omega_2)/\lambda_2$
Parametric down-conversion	$P(\omega_3) = 2\varepsilon_0 d_{\mathrm{eff}}(-\omega_3, \omega_1, -\omega_2)A(\omega_1)A^*(\omega_2)$	$\Delta k = k(\omega_3) + k(\omega_2) - k(\omega_1)$	$n(\omega_1)/\lambda_1 = n(\omega_2)/\lambda_2 + n(\omega_3)/\lambda_3$
	$P(\omega_2) = 2\varepsilon_0 d_{\mathrm{eff}}(-\omega_2, \omega_1, -\omega_3)A(\omega_1)A^*(\omega_3)$		
Optical rectification	$P(0) = 2\varepsilon_0 d_{\mathrm{eff}}(0, \omega, -\omega)A(\omega_1)A^*(\omega_1)$	$\Delta k = k(0) - k(\omega) + k(\omega) = 0$	Automatically satisfied

[a] Type I phase matching.
[b] Type II phase matching.

TABLE IV. Properties of Selected Nonlinear Optical Crystals

Nonlinear material	Symmetry point group	Nonlinear susceptibility d (10^{-12} m/V)	Transparency range (μm)	Effective nonlinearity, $\|d_{eff}\|$: Type I phase matching[a]	Type II phase matching
Ag_3AsS_3 (proustite)	$3m$	$d_{22} = 22$; $d_{15} = 13$	0.6–13	$d_{15} \sin\theta - d_{22} \cos\theta \sin 3\phi$	$d_{22} \cos^2\theta \cos 3\phi$
Te (tellurium)	32	$d_{11} = 649$	4–25	$d_{15} \sin\theta + \cos\theta(d_{11} \cos 3\phi - d_{22} \sin 3\phi)$	$d_{11} \cos^2\theta$
Tl_3AsSe_3 (TAS)		$d_+ = 40$	1.2–18	d_+	
$CdGeAs_2$	$\bar{4}2m$	$d_{36} = d_{14} = d_{25} = 236$	2.4–17	$d_{36} \sin\theta$	$d_{36} \sin 2\theta$
$AgGaS_2$	$\bar{4}2m$	$d_{36} = d_{14} = d_{25} = 12$	0.6–13	$d_{36} \sin\theta$	$d_{36} \sin 2\theta$
$AgGaSe_2$	$\bar{4}2m$	$d_{36} = d_{14} = d_{25} = 40$	0.7–17	$d_{36} \sin\theta$	$d_{36} \sin 2\theta$
GaAs	$\bar{4}3m$	$d_{36} = d_{14} = d_{25} = 90.1$	0.9–17	$d_{36} \sin\theta$	$d_{36} \sin 2\theta$
$LiNbO_3$ (lithium niobate)	$3m$	$d_{15} = 6.25$; $d_{22} = 3.3$	0.35–4.5	$d_{15} \sin\theta - d_{22} \cos\theta \sin 3\phi$	$d_{22} \cos^2\theta \cos 3\phi$
$LiIO_3$ (lithium iodate)	6	$d_{31} = 7.5$	0.31–5.5	$d_{31} \sin\theta$	
$NH_4H_2(PO_4)_2$ (ammonium dihydrogen phosphate, ADP)	$\bar{4}2m$	$d_{36} = d_{14} = d_{25} = 0.57$	0.2–1.2	$d_{36} \sin\theta$	$d_{36} \sin 2\theta$
$KH_2(PO_4)_2$ (potassium dihydrogen phosphate, KDP)	$\bar{4}2m$	$d_{36} = d_{14} = d_{25} = 0.5$	0.2–1.5	$d_{36} \sin\theta$	$d_{36} \sin 2\theta$
$KD_2(PO_4)_2$ (potassium dideuterium phosphate, KD*P)	$\bar{4}2m$	$d_{36} = d_{14} = d_{25} = 0.53$	0.2–1.5	$d_{36} \sin\theta$	$d_{36} \sin 2\theta$
$RbH_2(AsO_4)_2$ (rubidium dihydrogen arsenate, RDA)	$\bar{4}2m$	$d_{36} = d_{14} = d_{25} = 0.47$	0.26–1.46	$d_{36} \sin\theta$	$d_{36} \sin 2\theta$
$RbH_2(PO_4)_2$ (rubidium dihydrogen phosphate, RDP)	$\bar{4}2m$	$d_{36} = d_{14} = d_{25} = 0.48$	0.22–1.4	$d_{36} \sin\theta$	$d_{36} \sin 2\theta$
$NH_4H_2(AsO_4)_2$ (ammonium dihydrogen arsenate, ADA)	$\bar{4}2m$			$d_{36} \sin\theta$	$d_{36} \sin 2\theta$
$KD_2(AsO_4)_2$ (potassium dideuterium arsenate, KD*A)	$\bar{4}2m$	$d_{36} = d_{14} = d_{25} = 0.4$	0.22–1.4	$d_{36} \sin\theta$	$d_{36} \sin 2\theta$
$CsH_2(AsO_4)_2$ (cesium dihydrogen arsenate, CDA)	$\bar{4}2m$	$d_{36} = d_{14} = d_{25} = 0.48$	0.26–1.43	$d_{36} \sin\theta$	$d_{36} \sin 2\theta$
$CsD_2(AsO_4)_2$ (cesium dideuterium arsenate, CD*A)	$\bar{4}2m$	$d_{36} = d_{14} = d_{25} = 0.48$	0.27–1.66	$d_{36} \sin\theta$	$d_{36} \sin 2\theta$
$KTiOPO_4$ (potassium titanyl phosphate, KTP)	$mm2$	$d_{31} = 7$; $d_{32} = 5.4$; $d_{33} = 15$; $d_{24} = 8.1$; $d_{15} = 6.6$	0.35–4.5	$d_{31} \cos 2\theta + d_{32} \sin 2\theta$	
$LiCHO_2 \cdot H_2O$ (lithium formate monohydrate, LFM)	$mm2$	$d_{31} = d_{15} = 0.107$ $d_{32} = d_{24} = 1.25$ $d_{33} = 4.36$	0.23–1.2	$d_{31} \cos 2\theta + d_{32} \sin 2\theta$	
$KB_5O_8 \cdot 4H_2O$ (potassium pentaborate, KB5)	$mm2$	$d_{31} = 0.046$ $d_{32} = 0.003$	0.17–>0.76	$d_{31} \cos 2\theta + d_{32} \sin 2\theta$	
Urea	$\bar{4}2m$	$d_{36} = d_{14} = d_{25} = 1.42$	0.2 –1.43	$d_{36} \sin\theta$	$d_{36} \sin 2\theta$

[a] θ is the direction of propagation with respect to the optic axis, ϕ is the direction of propagation with respect to the x axis.

Damage threshold (10^6 W/cm^2)	Uses
12–40	Harmonic generation, frequency mixing, parametric oscillation in mid infrared
40–60 (at 5 μm)	Harmonic generation, frequency mixing, parametric oscillation in mid infrared
32	Harmonic generation, frequency mixing, parametric oscillation in mid infrared
20–40	Harmonic generation, frequency mixing, parametric oscillation in mid infrared
12–25	Harmonic generation, frequency mixing, parametric oscillation in mid infrared
>10	Harmonic generation, frequency mixing, parametric oscillation in mid infrared
	Harmonic generation, frequency mixing, parametric oscillation in mid infrared; difference frequency generation in far infrared
50–140	Harmonic generation, frequency mixing, parametric oscillation in near and mid infrared; harmonic generation and frequency mixing in visible and near ultraviolet
125	Harmonic generation, frequency mixing, parametric oscillation in near and mid infrared; harmonic generation and frequency mixing in visible and near ultraviolet
500 (60 nsec, 1.064 μm)	Harmonic generation, frequency mixing, parametric oscillation in near infrared; harmonic generation and frequency mixing in visible and near ultraviolet
400 (20 nsec, 694.3 nm) 23,000 (200 psec, 1.064 μm)	Harmonic generation, frequency mixing, parametric oscillation in near infrared; harmonic generation and frequency mixing in visible and near ultraviolet
500 (10 nsec, 1.064 μm) 20,000 (30 psec, 1.064 μm)	Harmonic generation, frequency mixing, parametric oscillation in near and mid infrared; harmonic generation and frequency mixing in visible and near ultraviolet
350 (10 nsec, 694.3 nm)	Harmonic generation, frequency mixing in near ultraviolet
200 (10 nsec, 694.3 nm)	Harmonic generation, frequency mixing in near ultraviolet
	Harmonic generation, frequency mixing in near ultraviolet
	Harmonic generation, frequency mixing in near ultraviolet
500 (10 nsec, 1.064 μm)	Harmonic generation, frequency mixing and parametric oscillation in visible and near infrared
>260 (12 nsec, 1.064 μm)	Harmonic generation, frequency mixing and parametric oscillation in visible and near infrared
160 (20 nsec, 1.064 μm)	Harmonic generation, frequency mixing and parametric oscillation in visible and near infrared
>1000	Harmonic generation and frequency mixing in the ultraviolet
	Harmonic generation and frequency mixing in the ultraviolet
1.4×10^3	Harmonic generation and frequency mixing in the ultraviolet

TABLE V. Performance for Selected Second-Order Frequency Conversion Interactions

Laser	Incident wavelength (μm)	Generated wavelength (μm)	Nonlinear material	Conversion efficiency Power (%)	Conversion efficiency Energy (%)	Conditions, comments
Second harmonic generation						
CO_2	10.6	5.3	Ag_3AsS_3 (proustite)	~1		
	9.6	4.8	Te	5		Limited by multiphoton absorption
			Tl_3AsSe_3	40	25	1 J/cm^2 pump, 100 nsec
			$CdGeAs_2$		27	
			$AgGaSe_2$	60	14	2.1 cm, 75 nsec, 30 mJ, 12 MW/cm^2
			GaAs	2.7		Phase matching with stacked plates
Nd : YAG	1.064	0.532	ADP	23		
			KDP	83		346 J output (with Nd : glass amplifiers)
			KD*P		75	30 psec pulses, 10^{10} W/cm^2 pump
			CDA	60		
			CD*A	57		
			$LiNbO_3$	30		
			KTP	42		
Nd : glass (silicate)	1.059	0.5295	KDP	92		5 psec pulses
Nd : glass (phosphate)	1.052	0.526	KDP		80	
Ruby	0.6943	0.347	ADP	40		
			RDA	40		
Nd : YAG	0.532	0.266	ADP		85	30 psec pulses, 10^{10} W/cm^2 pump intensity
			KDP	70		50 J output (with Nd : glass amplifiers)
			KD*P	75		30 psec, 10^{10} W/cm^2 pump intensity
Dye lasers	0.560–0.620	0.280–0.310	ADP,KDP	8–9		
	0.460–0.600	0.230–0.300	LFM	1–2		
	0.434–0.500	0.217–0.250	KB5	0.2–2		
Ar^+	0.5145	0.2572	ADP,KDP	30		300 mW, intracavity, cw

Three-wave sum-frequency mixing ($\omega_3 = \omega_1 + \omega_2$)					
CO_2	9.6, 4.8	3.53	Tl_3AsSe_3	1.6	1 J/cm^2 pump, overall efficiency from 9.6 μm
Nd : YAG	1.064, 0.532	0.3547	KDP	55	41 J output (with Nd : glass amplifiers)
			RDP	21	
Nd : YAG	1.064, 0.266	0.2128	ADP	60	Measured with respect to power at 266 nm
			KDP		1 kW output, 20 nsec pulses
			KB5	0.002	
Nd : YAG, Dye (second harmonic)	1.064, 0.258–0.300	0.208–0.234	ADP	10	
Nd : YAG, Dye	0.266, 0.745–0.790	0.1966–0.199	KB5	3	40 kW, 5 nsec at 0.1966
Dye (Second harmonic)	0.622–0.652, 0.310–0.326, 0.740–0.920, 0.232–0.268;	0.185–0.2174	KB5	8–12	
Difference-frequency mixing ($\omega_3 = \omega_1 - \omega_2$)					
Nd : YAG, Dye	0.532, 0.575–0.660	2.8–5.65	$LiIO_3$	0.92	Measured with respect to total power
Nd : YAG, Dye	1.064, 0.575–0.640	1.25–1.6	$LiIO_3$	0.7	70 W output
Ruby	0.6943				
Dye	0.84–0.89	3–4	$LiNbO_3$	0.015	6 kW output pulses
	0.80–0.83	4.1–5.2	$LiIO_3$	6×10^{-3}	100 W output pulses
	0.78–0.87	3.2–6.5	Ag_3AsS_3	2.7×10^{-3}	110 W pulses at 5 μm
	0.73–0.75	10–13	Ag_3AsS_3	2×10^{-4}	100 mW at 10 μm
	0.74–0.818	4.6–12	$AgGaS_2$	2×10^{-4}	300 mW at 11 μm
	0.72–0.75	9.5–17	GaSe	0.075	300 W, 20 nsec at 12 μm
Dye	0.81–0.84, 0.81–0.84	50–500	$LiNbO_3$	10^{-5}–10^{-6}	10–100 mW output
Dye	0.440–0.510, 0.570–0.620	1.5–4.8	$LiIO_3$	10^4	400 mW output power

(*continues*)

TABLE V. *(Continued)*

Laser	Incident wavelength (μm)	Generated wavelength (μm)	Nonlinear material	Conversion efficiency Power (%)	Conversion efficiency Energy (%)	Conditions, comments
Dye	0.586, 0.570–0.620	8.7–11.6	$AgGaS_2$	5×10^{-6}		100 μW output power
Dye		300–2 mm	ZnTe, ZnSe			
Argon, Dye	0.5145, 0.560–0.620	2.2–4.2	$LiNbO_3$	10^{-3}		1 μW continuous
CO_2	Various lines between 9.3 and 10.6	70–2 mm	GaAs	10^{-3}		10 W output power
CO_2	10.6	90–111	InSb	4×10^{-8}		2 μW output power
InSb spin-flip	11.7–12					
Parametric Oscillators ($\omega_1 \rightarrow \omega_2 + \omega_3$)						
Nd:YAG	0.266	0.420–0.730	ADP	25		100 kW output power, 30 psec pulses
Nd:YAG	0.472, 0.532 0.579, 0.635	0.55–3.65	$LiNbO_3$	50	30	
Nd:YAG	1.064	1.4–4.0	$AgGaS_2$		16	20 nsec pulses
	1.064	1.4–4.4	$LiNbO_3$	40		
Ruby	0.6943	66–200	$LiNbO_3$	10^{-4}		
Ruby	0.6943	0.77–4	$LiIO_3$	1–10		2–100 kW output
Ruby	0.347	0.415–2.1	$LiNbO_3$	8		10 kW output
Hydrogen fluoride	2.87	4.3–4.5 8.1–8.3	CdSe	10		800 W output, 300 nsec

twice the frequency, and one-half the wavelength, of the incident radiation:

$$\omega_2 = 2\omega_1 \tag{12}$$

For this interaction, the incident frequencies in the nonlinear polarization of Eq. (10) are equal, the generated frequency is ω_2, and $g = 1$. The intensity of the harmonic wave can be calculated for specific configurations from the appropriate equation in Eq. (6) using the nonlinear polarization amplitude $P(2\omega) = \varepsilon_0 d_{\text{eff}}(-2\omega, \omega, \omega)A(\omega)^2$ from Table III. The simplest situation is one in which the incident radiation is in the form of a plane wave and the conversion efficiency to the harmonic is small enough that the incident intensity can be regarded as constant. The harmonic intensity, related to the field amplitude by

$$I(2\omega) = cn_2\varepsilon_0|A(2\omega)|^2/2, \tag{13}$$

where n_2 is the refractive index at ω_2, is then given by

$$I(2\omega) = [8\pi^2 d_{\text{eff}}^2/n_2 n_1^2 c\varepsilon_0\lambda_1^2] \times I_0(\omega)^2 \operatorname{sinc}^2(\Delta k\, L/2) \tag{14}$$

Here $\operatorname{sinc}(x) = (\sin x)/x$, $I_0(\omega)$ is the incident intensity at ω, L is the crystal length, and

$$\Delta k = k(2\omega) - k_1(\omega) - k_2(\omega) \tag{15}$$

where $k(2\omega)$ is the usual wave vector at 2ω, and $k_1(\omega)$ and $k_2(\omega)$ are the wave vectors of the components of the incident wave at ω. If the incident radiation has only one polarization component relative to the principal directions of the crystal (i.e., is either an ordinary or an extraordinary ray), $k_1(\omega) = k_2(\omega) = k(\omega)$ and $\Delta k = k(2\omega) - 2k(\omega)$, whereas if the incident radiation has both ordinary and extraordinary polarization components, $k_1(\omega)$ is in general not equal to $k_2(\omega)$.

In the low conversion regime, the harmonic intensity grows as the square of the incident intensity. This is one of the reasons effective harmonic conversion requires the intense fields that are present in laser radiation. The harmonic intensity is also very sensitive to the value of the wave-vector mismatch Δk. Its dependence on Δk at fixed L and on L for different values of Δk is shown in Fig. 4a,b, respectively. If $\Delta k \neq 0$, the harmonic wave gradually gets out of phase with the polarization that drives it as it propagates through the crystal. As a result the harmonic intensity oscillates with distance, with the harmonic wave first taking energy from the incident wave, and then, after a phase change of π relative to the nonlinear polarization, returning it to the incident wave. The shortest distance at

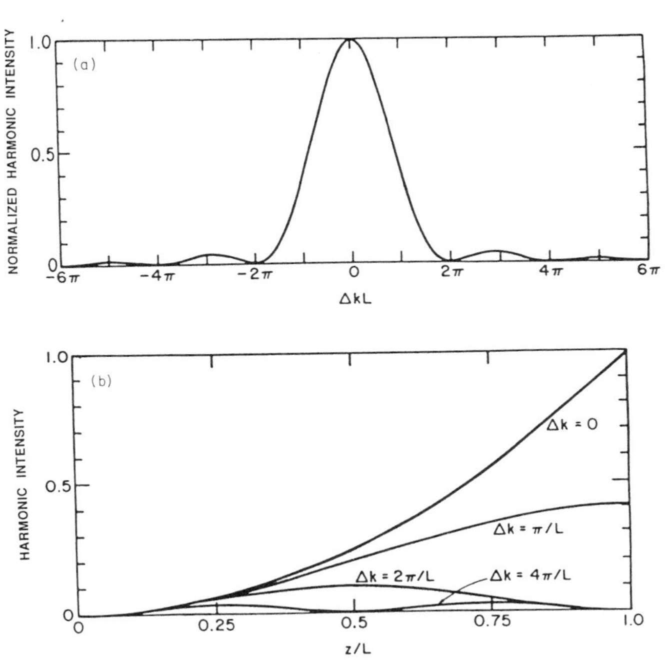

FIG. 4. (a) Variation of second harmonic intensity with Δk at fixed L. (b) Variation of second harmonic intensity with L for various values of Δk.

which the maximum conversion occurs is termed the coherence length and is given by

$$L_{coh} = \pi/\Delta k \tag{16}$$

When $\Delta k \neq 0$, maximum conversion is obtained for crystals whose length is an odd multiple of the coherence length, while no conversion is obtained for crystals whose length is an even multiple of the coherence length.

When

$$\Delta k = 0 \tag{17}$$

the harmonic wave stays in phase with its driving polarization. The process is then said to be phase-matched, and the harmonic intensity grows as the square of the crystal length. Under this condition the harmonic intensity can grow so large that the depletion of the incident radiation must be accounted for. In this case the harmonic and incident intensities are given by

$$I(2\omega_1) = I_0(\omega_1)\tanh^2(L/L_{dep}) \tag{18a}$$

$$I(\omega_1) = I_0(\omega_1)\,\mathrm{sech}^2(L/L_{dep}) \tag{18b}$$

where L_{dep} is a depletion length given by

$$L_{dep} = [n_2 n_1^2 c\varepsilon_0 \lambda_1^2 / 8\pi^2 d_{eff}^2 I_0(\omega_1)]^{1/2} \tag{19}$$

When $L = L_{dep}$, 58% of the fundamental is converted to the harmonic. The harmonic and fundamental intensities are shown as a function of $(L/L_{dep})^2$ in Fig. 5. Although energy is conserved between the optical waves in second-harmonic generation, the number of photons is not, with two photons being lost at the fundamental for every one that is created at the harmonic.

The wave-vector mismatch Δk occurs because of the natural dispersion in the refractive index that is present in all materials. Effective second-harmonic conversion therefore requires that special steps be taken to satisfy the condition of Eq. (17). This process is termed phase matching. For the noncentrosymmetric crystals used for second-order processes, the most common method of phase matching involves use of the birefringence of the crystal to offset the natural dispersion in the refractive indices. In general this means that the harmonic wave will be polarized differently from the incident wave, with the particular combinations being determined by the symmetry properties of the nonlinear crystal. The value of the wave-vector mismatch can be adjusted by varying the direction of propagation of the various waves relative to the optic axis or by varying the temperature of the crystal for a fixed direction of propagation. The angle at which $\Delta k = 0$ is called the phase-matching angle, and the temperature at which $\Delta k = 0$ when $\theta = 90°$ is called the phase-matching temperature.

In the simplest situation, termed type I phase matching, the incident radiation is polarized as an ordinary or an extraordinary ray, depending on the properties of the nonlinear material, and the harmonic radiation is polarized orthogonally to the incident radiation as shown in Fig. 6a. The wave-vector mismatch is then given by

$$\Delta k = k(2\omega) - 2k(\omega) = 4\pi[n(2\omega) - n(\omega)]/\lambda_1 \tag{20}$$

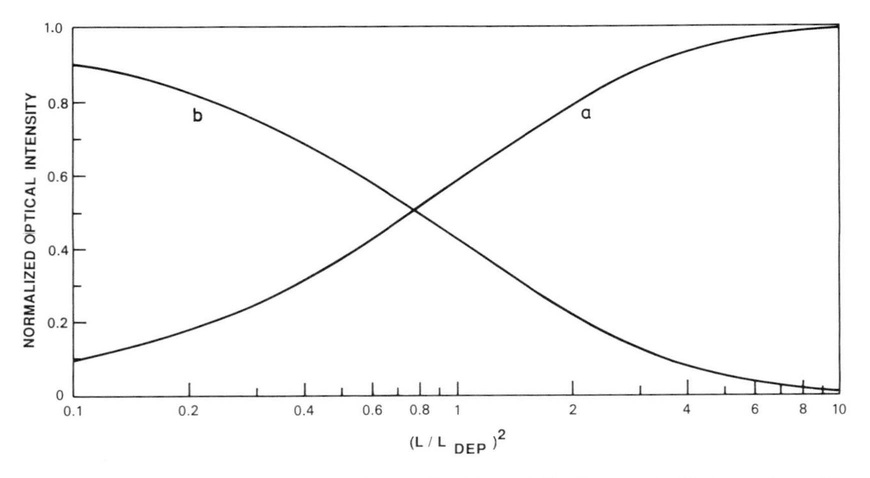

FIG. 5. Variation of the second harmonic (a) and fundamental (b) intensity with $(L/L_{dep})^2$, where L_{dep} is the second-harmonic depletion length, for plane waves at perfect phase matching. [Reproduced from J. Reintjes (1985). Coherent ultraviolet and vacuum ultraviolet sources, *in* "Laser Handbook," Vol. 5 (M. Bass and M. Stitch, eds.). North-Holland, Amsterdam.]

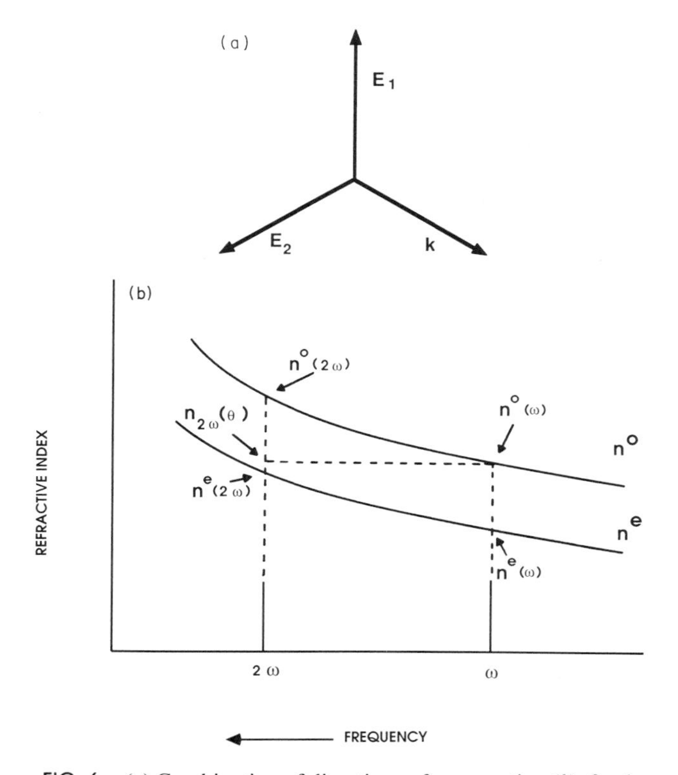

FIG. 6. (a) Combination of directions of propagation (k), fundamental polarization (E_1), and second harmonic polarization (E_2) for type I second-harmonic generation. (b) Typical variation of the ordinary (n^{o}) and the extraordinary (n^{e}) refractive indices with wavelength for a negative uniaxial crystal, showing conditions necessary for type I phase matching of second-harmonic generation. The refractive index of the extraordinary ray at the harmonic varies between n^{o} and n^{e} as the angle of propagation relative to the optic axis is varied. Phase matching occurs when $n_{2\omega}(\theta) = n_{\omega}^{\text{o}}$.

Phase matching entails choosing conditions such that the refractive index for the field at the fundamental wavelength equals the refractive index for the field at the harmonic, that is, $n(2\omega) = n(\omega)$. For example, in a material in which the extraordinary refractive index is less than the ordinary refractive index, the pump wave is polarized as an ordinary ray and the harmonic wave is polarized as an extraordinary ray. The variation of the refractive index of these rays with wavelength is shown in Fig. 6b. The ordinary index is independent of the propagation direction, but the extraordinary index depends on the angle θ between the optic axis and the propagation direction according to the relation

$$[1/n_{2\omega}(\theta)]^2 = (\cos^2\theta)/(n_{2\omega}^{0})^2 + (\sin^2\theta)/(n_{2\omega}^{e})^2 \qquad (21)$$

As θ varies from 0 to 90°, $n_{2\omega}(\theta)$ varies from $n_{2\omega}^{\text{o}}$ to $n_{2\omega}^{e}$. In this situation, phase matching consists of choosing the propagation direction such that $n_{2\omega}(\theta) = n_{\omega}^{\text{o}}$.

In type II phase matching, the incident radiation has both ordinary and extraordinary polarization components, while the harmonic ray is an extraordinary ray (or vice versa, if the nonlinear medium has the property $n^{\circ} < n^{e}$). In type II phase matching, the more general relation among the k vectors of Eq. (15) must be used, and the relationship between the various refractive indexes required for phase matching is more complicated than that given in Eq. (20) for type I, but the propagation direction is again chosen such that Eq. (17) is satisfied.

Second-harmonic generation has been used to generate light at wavelengths ranging from 217 nm (by doubling radiation from a dye laser at 434 nm) to 5 μm, by doubling radiation from a CO_2 laser at 10.6 μm. Examples of specific results are given in Table V. The usefulness of a particular material for second-harmonic generation is determined by a combination of the magnitude of its nonlinear susceptibility, its ability to phase match the process and its degree of transparency to the harmonic and fundamental wavelengths. The short wave limit of second-harmonic conversion is currently set by an inability to phase match the harmonic process at shorter wavelengths, while the infrared limit of 5 μm is determined by increasing absorption at the fundamental.

Second-harmonic conversion is commonly used with high-power pulsed lasers such as Nd:YAG, Nd:glass, ruby, or CO_2 to generate high-power fixed-frequency radiation at selected wavelengths in the infrared, visible, and ultraviolet. It is also a versatile and convenient source of tunable ultraviolet radiation when the pump radiation is obtained from tunable dye lasers in the visible and near infrared. [*See* TUNABLE DYE LASERS.]

Second-harmonic conversion efficiencies range from the order of 10^{-6} for conversion of continuous wave radiation outside the laser cavity to over 90% for high-power pulsed radiation. The laser power required for high efficiency depends on the configuration and nonlinear material. A typical value for 75% conversion of radiation from an Nd:glass laser at 1.06 μm to 530 nm in a 1-cm-long crystal of potassium dihydrogen phosphate (KDP) is 10^8 W/cm^2. Intensities for conversion in other materials and with other lasers are noted in Table V. Conversion of radiation from high-power pulsed lasers is typically done in collimated-beam geometries with angle-tuned crystals chosen for their high damage thresholds. Conversion of radiation from lower-power pulsed lasers or continuous-wave (cw) lasers is often done in focused geometries in materials with large nonlinear susceptibilities. In many situations the maximum conversion efficiency that can be achieved is not determined by the available laser power but by additional processes that affect the conversion process. Some of these competing processes are linear or nonlinear absorption, imperfect phase matching because of a spread of the laser divergence or bandwidth, incomplete phase matching because of intensity-dependent changes in the refractive index (see Section III,B,1), or damage to the nonlinear crystal caused by the intense laser radiation.

ii. THREE-WAVE SUM-FREQUENCY MIXING. Three-wave sum-frequency mixing is used to generate radiation at higher frequencies, and therefore shorter wavelengths, than those in the pump radiation according to the relation

$$\omega_3 = \omega_1 + \omega_2 \tag{22}$$

where ω_3 is the frequency of the radiation generated in the interaction and ω_1 and ω_2 are the frequencies of the pump radiation. Calculation of the generated intensity is similar to that of second-harmonic generation with the appropriate form of the nonlinear polarization as given in Table III used in Eq. (6). At low conversion efficiencies, the generated intensity has the same $\sin^2(\Delta k\, L/2)/(\Delta k\, L/2)^2$ dependence on the wave-vector mismatch as has second-harmonic conversion, but the phase-matching condition is now $\Delta k = k(\omega_3) - k(\omega_1) - k(\omega_2) = 0$, which is equivalent to

$$n(\omega_3)/\lambda_3 = n(\omega_1)/\lambda_1 + n(\omega_2)/\lambda_2 \tag{23}$$

The intensity of the generated radiation grows as the product of the incident pump intensities at low conversion efficiency. In the three-wave sum-frequency mixing interaction, one photon from each of the pump waves is annihilated for each photon created in the generated wave. Complete conversion of the total radiation in both pump waves at perfect phase matching is possible in principle if they start with an equal number of photons. Otherwise the conversion will oscillate with pump intensity or crystal length, even at exact phase matching.

Three-wave sum-frequency mixing is done in the same types of materials as second-harmonic generation. It is used to generate both tunable and fixed-frequency radiation at various wavelengths ranging from the infrared to the ultraviolet. It allows radiation to be generated at shorter wavelengths in the ultraviolet than can be reached with second harmonic conversion if one of the pump wavelengths is also in the ultraviolet.

Examples of specific three-wave sum-frequency mixing interactions are given in Table V. This interaction has been used to produce radiation at wavelengths as short as 185 nm in potassium pentaborate (KB_5), with the cutoff being determined by the limits of phase matching. Three-wave sum-frequency mixing can also be used to improve the efficiency of the generation of tunable radiation, as compared to second-harmonic generation, by allowing radiation from a relatively powerful fixed-frequency laser to be

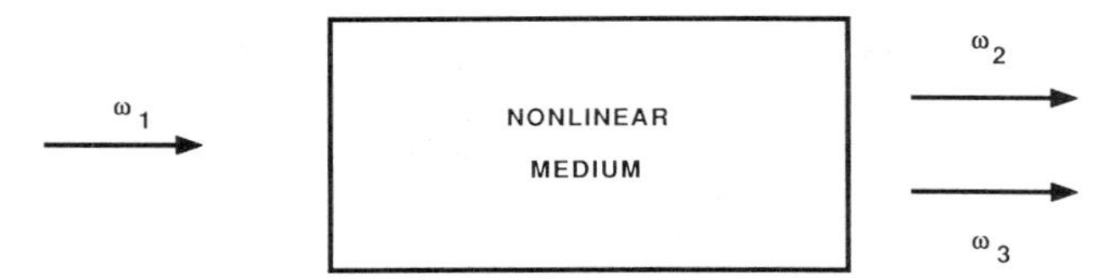

FIG. 7. Schematic illustration of the waves used in parametric down-conversion. Radiation at ω_1 is supplied, and radiation at ω_2 and ω_3 is generated in the nonlinear interaction.

combined with radiation from a relatively weak tunable laser. Another application is in the generation of the third harmonic of certain fixed frequency lasers such as Nd : YAG, Nd : glass, or CO_2 through two second-order processes: second-harmonic conversion of part of the laser fundamental followed by sum-frequency mixing of the unconverted fundamental with the second harmonic. Under certain conditions this process is more efficient than direct third-harmonic conversion (see Section III,A,1,b). Three-wave sum-frequency mixing has also been used for up-conversion of infrared radiation to the visible, where it can be measured by more sensitive photoelectric detectors or photographic film.

iii. THREE - WAVE DIFFERENCE - FREQUENCY MIXING. Three-wave difference-frequency mixing is used to convert radiation from two incident waves at frequencies ω_1 and ω_2 to a third wave at frequency ω_3 according to the relation

$$\omega_3 = \omega_1 - \omega_2 \quad (24)$$

Just as for sum-frequency mixing, the generated intensity at low conversion efficiency grows as the product of the pump intensities at ω_1 and ω_2. In this situation a photon is created at both ω_3 and ω_2 for every photon annihilated at ω_1. The wave-vector mismatch is $\Delta k = k(\omega_3) - [k(\omega_1) - k(\omega_2)]$, and the phase matching condition is

$$n(\omega_3)/\lambda_3 = n(\omega_1)/\lambda_1 - n(\omega_2)/\lambda_2 \quad (25)$$

The materials used for difference-frequency mixing are of the same type as those used for second-harmonic and sum-frequency mixing. Difference-frequency mixing is generally used to produce coherent radiation at longer wavelengths than either of the pump wavelengths. It is used most often to produce tunable radiation in the infrared from pump radiation in the visible or near infrared, although it can also be used to generate radiation in the visible. It has been used to produce radiation at wavelengths as long as 2 mm in GaAs, with the limit being set by a combination of the increasing mismatch in the diffraction of the pump and generated radiation and the increasing absorption of the generated radiation in the nonlinear medium at long wavelengths.

iv. PARAMETRIC DOWN - CONVERSION. Parametric down-conversion is used to convert radiation in an optical wave at frequency ω_1 into two optical waves at lower frequencies ω_2 and ω_3 according to the relation

$$\omega_1 = \omega_2 + \omega_3 \quad (26)$$

This process is illustrated schematically in Fig. 7. The wave at ω_1 is termed the pump wave, while one of the waves at ω_2 or ω_3 is termed the signal and the other the idler. When $\omega_2 = \omega_3$ the process is termed degenerate parametric down-conversion and is the opposite of second-harmonic generation, whereas if $\omega_2 \neq \omega_3$, the process is called nondegenerate parametric down-conversion and is the opposite of sum-frequency generation. The individual values of ω_2 and ω_3 are determined by the phase-matching condition, which for plane-wave interactions is given by

$$\Delta k = k(\omega_1) - k(\omega_2) - k(\omega_3) = 0 \quad (27)$$

and can be varied by changing an appropriate phase-matching parameter of the crystal such as the angle of propagation or the temperature. Off-axis phase matching, as illustrated in Fig. 8, is also possible. The phase-matching condition of Eq. (27) is the same as that for the sum-frequency process $\omega_2 + \omega_3 = \omega_1$. The relative phase of the waves involved determines

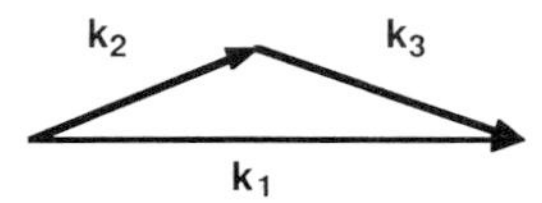

FIG. 8. Off-axis phase-matching diagram for parametric down-conversion. Direction of arrows indicates direction of propagation of the various waves.

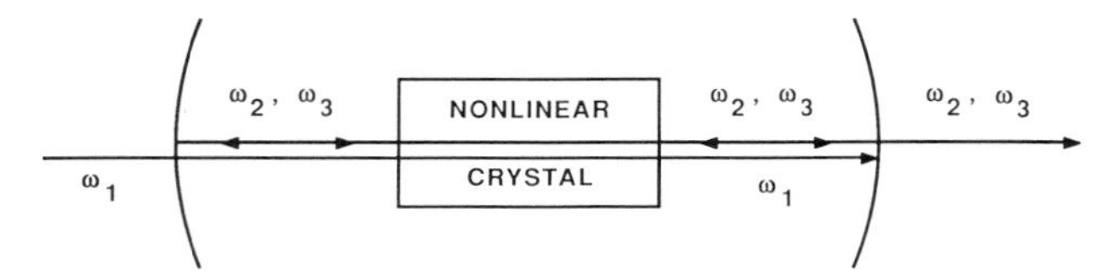

FIG. 9. Illustration of a doubly resonant cavity for a parametric oscillator.

whether the sum process or the parametric down-conversion process will occur.

In the absence of pump depletion, the amplitudes of the waves generated in a parametric down-conversion process are given by

$$A(\omega_2) = A_0(\omega_2) \cosh \kappa z + iA_0^*(\omega_3)(\omega_2 n_3/\omega_3 n_2)^{1/2} \sinh \kappa z \quad (28a)$$

$$A(\omega_3) = A_0(\omega_3) \cosh \kappa z + iA_0^*(\omega_2)(\omega_3 n_2/\omega_2 n_3)^{1/2} \sinh \kappa z \quad (28b)$$

where

$$\kappa^2 = [2\omega_2\omega_3 d_{\text{eff}}^2/n_2 n_3 c^2] I_0(\omega_1) \quad (29)$$

and $A_0(\omega_i)$ and $I_0(\omega_i)$ are the incident field amplitude and intensity, respectively, at ω_i. At low pump intensities the generated field amplitudes grow in proportion to the square root of the pump intensity, while at high pump intensities the growth of the generated waves is exponential.

If there is no incident intensity supplied at ω_2 or ω_3, the process is termed parametric generation or parametric oscillation, depending on the geometry. For this situation the initial intensity for the generated waves arise from the zero-point radiation field with an energy of $h\nu/2$ per mode for each field. If the interaction involves a single pass through the nonlinear medium, as was shown in Fig. 7, the process is termed parametric generation. This geometry is typically used with picosecond pulses for generation of tunable infrared or visible radiation from pump radiation in the ultraviolet, visible, or near infrared. Amplification factors of the order of 10^{10} are typically required for single-pass parametric generation.

Parametric down-conversion can also be used with a resonant cavity that circulates the radiation at either or both of the generated frequencies, as shown in Fig. 9. In this geometry the process is termed parametric oscillation. If only one of the generated waves is circulated in the cavity, it is termed singly resonant, whereas if both waves are circulated, the cavity is termed doubly resonant. In singly resonant cavities the wave that is circulated is termed the signal, while the other generated wave is termed the idler. Optical parametric oscillators are typically used with pump radiation from Q-switched lasers with pulses lasting several tens of nanoseconds, allowing several passes of the generated radiation through the cavity while the pump light is present.

One of the primary uses of parametric down-conversion is the generation of tunable radiation at wavelengths ranging from the visible to the far infrared. Its wavelength range is generally the same as that covered by difference-frequency mixing, although it has not been extended to as long a wavelength. Tuning is done by varying one of the phase-matching parameters such as angle or temperature. A typical tuning curve for a parametric oscillator is shown in Fig. 10. As

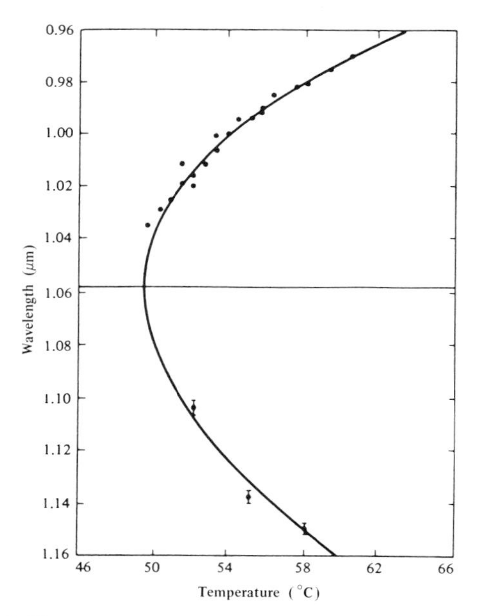

FIG. 10. Tuning curve for a $LiNbO_3$ parametric oscillator pumped by radiation from the second harmonic of a Nd laser at 529 nm. [Reproduced from J. A. Giordmaine and R. C. Miller (1965). *Phys. Rev. Lett.* **14,** 973.]

the phase-matching condition is changed from the degenerate condition, ω_2 increases and ω_3 decreases in such a way as to maintain the relation in Eq. (26). The extent of the tuning range for a given combination of pump wavelength and nonlinear material is generally set by the limits of phase matching, although absorption can be important if ω_3 is too far in the infrared. Radiation in different wavelength ranges can be produced by using different nonlinear materials and different pump sources. Parametric down-conversion has been used to produce radiation at wavelengths ranging from the visible to 25 μm. Some of the combinations of pump sources, nonlinear materials, and tuning ranges are listed in Table V.

Parametric down-conversion can also be used to amplify radiation at ω_2 or ω_3. In this arrangement radiation is supplied at both the pump wavelength and the lower-frequency wave to be amplified, which is termed the signal. The process is similar to difference-frequency mixing, and differs from the difference-frequency process only in that the incident intensity in the signal wave is considerably less than the pump intensity for parametric amplification, whereas the two incident intensities are comparable in difference-frequency mixing. In principle, very high gains can be obtained from parametric amplification, with values up to 10^{10} being possible in some cases.

Optical rectification is a form of difference-frequency mixing in which the generated signal has no carrier frequency. It is produced through the interaction

$$\omega_3 = 0 = \omega_1 - \omega_1 \tag{30}$$

It does not produce an optical wave, but rather an electrical voltage signal with a duration corresponding to the pulse duration of the pump radiation. Optical rectification has been used with picosecond laser pulses to produce electrical voltage pulses with durations in the picosecond range, among the shortest voltage pulses yet produced. These pulses do not propagate in the bulk of the nonlinear crystal as do the optical waves, and observation or use of them requires coupling to an appropriate microwave strip line on the nonlinear crystal.

b. Third- and Higher-Order Processes. Third- and higher-order parametric processes are also used for frequency conversion. They have been used to generate radiation ranging from 35.5 nm, almost in the soft X-ray range, to about 25 μm in the infrared. The most commonly used interactions of this type are given in Table VI, along with the form of the nonlinear polarization, the wavevector mismatch, and the plane-wave phase-matching conditions. These interactions include third- and higher-order harmonic conversion, in which an incident wave at frequency ω_1 is converted to a wave at frequency $q\omega_1$, and various forms of four- and six-wave frequency mixing in which radiation in two or more incident waves is converted to radiation in a wave at an appropriate sum- or difference-frequency combination as indicated in Table VI. Four-wave parametric oscillation, in which radiation at frequency ω_1 is converted to radiation at two other frequencies, ω_2 and ω_3, according to the relation

$$2\omega_1 = \omega_2 + \omega_3 \tag{31}$$

has also been observed.

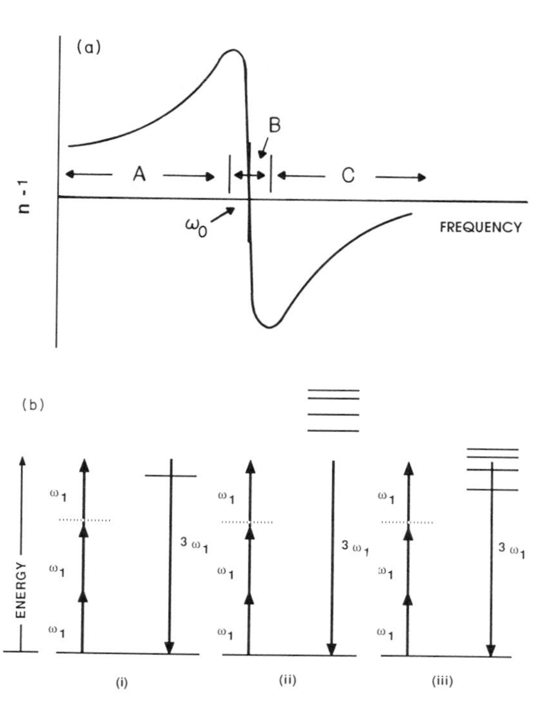

FIG. 11. (a) Variation of the refractive index near an allowed transition showing regions of normal dispersion (A), anomalous dispersion (B), and negative dispersion (C). (b) Energy level diagrams for third-harmonic generation that provide negative dispersion (i) and positive dispersion (ii and iii).

Third-order and higher odd-order processes can be observed with electric dipole interactions in materials with any symmetry. They are used most commonly in materials that have a center

TABLE VI. Nonlinear Polarizations for Third- and Higher-Order Parametric Frequency Conversion Processes

Nonlinear interaction	Process	Nonlinear polarization
qth Harmonic generation	$\omega_q = q\omega_1$	$P(q\omega) = \varepsilon_0\chi(-q\omega, \omega, \omega, \ldots, \omega)A^q(\omega)/2^{(q-1)}$
Four-wave sum-frequency mixing	$\omega_4 = 2\omega_1 + \omega_2$	$P(\omega_4) = 3\varepsilon_0\chi(-\omega_4, \omega_1, \omega_1, \omega_2)A^2(\omega_1)A(\omega_2)/4$
	$\omega_4 = \omega_1 + \omega_2 + \omega_3$	$P(\omega_4) = 3\varepsilon_0\chi(-\omega_4, \omega_1, \omega_2, \omega_3)A(\omega_1)A(\omega_2)A(\omega_3)/2$
Four-wave difference-frequency mixing	$\omega_4 = 2\omega_1 - \omega_2$	$P(\omega_4) = 3\varepsilon_0\chi(-\omega_4, \omega_1, \omega_1, -\omega_2)A^2(\omega_1)A^*(\omega_2)/4$
	$\omega_4 = \omega_1 + \omega_2 - \omega_3$	$P(\omega_4) = 3\varepsilon_0\chi(-\omega_4, \omega_1, \omega_2, -\omega_3)A(\omega_1)A(\omega_2)A^*(\omega_3)/2$
	$\omega_4 = \omega_1 - \omega_2 - \omega_3$	$P(\omega_4) = 3\varepsilon_0\chi(-\omega_4, \omega_1, -\omega_2, -\omega_3)A(\omega_1)A^*(\omega_2)A^*(\omega_3)/2$
Four-wave parametric oscillation	$2\omega_1 \rightarrow \omega_2 + \omega_3$	$P(\omega_2) = 3\varepsilon_0\chi(-\omega_2, \omega_1, \omega_1, -\omega_3)A^2(\omega_1)A^*(\omega_3)/4$
		$P(\omega_3) = 3\varepsilon_0\chi(-\omega_3, \omega_1, \omega_1, -\omega_2)A^2(\omega_1)A^*(\omega_2)/4$
Six-wave sum-frequency mixing	$\omega_6 = 4\omega_1 + \omega_2$	$P(\omega_6) = 5\varepsilon_0\chi(-\omega_6, \omega_1, \omega_1, \omega_1, \omega_1, \omega_2)A^4(\omega_1)A(\omega_2)/16$
Six-wave difference-frequency mixing	$\omega_6 = 4\omega_1 - \omega_2$	$P(\omega_6) = 5\varepsilon_0\chi(-\omega_6, \omega_1, \omega_1, \omega_1, \omega_1, -\omega_2)A^4(\omega_1)A^*(\omega_2)/16$

[a] Phase matching in mixtures or by angle.
[b] Phase optimization in single-component media.
[c] Requirement for optimized conversion.
[d] Positive dispersion allowed but not optimal.

of symmetry, such as gases, liquids, and some solids, since in these materials they are the lowest-order nonzero nonlinearities allowed by electric dipole transitions. Fourth-order and higher even-order processes involving electric dipole interactions are allowed only in crystals with no center of symmetry, and, although they have been observed, they are relatively inefficient and are seldom used for frequency conversion.

The intensity generated in these interactions can be calculated from Eq. (6) using the appropriate nonlinear polarization from Table VI. In the absence of pump depletion, the intensity generated in qth harmonic conversion varies as the qth power of the incident pump intensity, while the intensity generated in the frequency-mixing interactions varies as the product of the incident pump intensities with each raised to a power corresponding to its multiple in the appropriate frequency combination. The generated intensity in the plane wave configuration has the same $[(\sin \Delta k L/2)/(\Delta k L/2)]^2$ dependence on the wave-vector mismatch as do the second-order processes. The plane-wave phase-matching conditions for each of these interactions is the same as for the second-order interactions, namely, $\Delta k = 0$, but the requirements on the individual refractive indexes depend on the particular interaction involved, as indicated in Table VI.

Third- and higher-order frequency conversion are often done with beams that are tightly focused within the nonlinear medium to increase the peak intensity. In this situation, optimal performance can require either a positive or negative value of Δk, depending on the interaction involved. Phase-matching requirements with focused beams for the various interactions are also noted in Table VI.

The isotropic materials used for third- and higher-order parametric processes are not birefringent, and so alternative phase-matching techniques must be used. In gases, phase matching can be accomplished through use of the negative dispersion that occurs near allowed transitions as shown in Fig. 11. Normal dispersion occurs when the refractive index increases with frequency and is encountered in all materials when the optical frequency falls below, or sufficiently far away from, excited energy levels with allowed transitions to the ground state. Anomalous dispersion occurs in narrow regions about the transition frequencies in which the refractive index decreases with frequency, as shown in Fig. 11a. Negative dispersion occurs in a wavelength range above an allowed transition in which the refractive index, although increasing with frequency, is less than it is below the transition frequency. Regions of negative dispersion occur in restricted wavelength ranges above most, but not all, excited levels in many gases. Examples of the energy-level structures that give positive and negative dispersion for third-harmonic generation are shown in Fig. 11b.

Wave-vector mismatch	Plane-wave-matching condition	Dispersion requirement for focused beams in infinitely long media
$\Delta k = k(q\omega) - qk(\omega)$	$n(q\omega) = n(\omega)$	$\Delta k < 0$
$\Delta k = k(\omega_4) - 2k(\omega_1) - k(\omega_2)$	$n(\omega_4)/\lambda_4 = 2n(\omega_1)/\lambda_1 + n(\omega_2)/\lambda_2$	$\Delta k < 0$
$\Delta k = k(\omega_4) - k(\omega_1) - k(\omega_2) - k(\omega_3)$	$n(\omega_4)/\lambda_4 = n(\omega_1)/\lambda_1 + n(\omega_2)/\lambda_2 + n(\omega_3)/\lambda_3$	$\Delta k < 0$
$\Delta k = k(\omega_4) - 2k(\omega_1) + k(\omega_2)$	$n(\omega_4)/\lambda_4 = 2n(\omega_1)/\lambda_1 - n(\omega_2)/\lambda_2$	$\Delta k = 0^a$, $\Delta k \lesseqgtr 0^b$
$\Delta k = k(\omega_4) - k(\omega_1) - k(\omega_2) + k(\omega_3)$	$n(\omega_4)/\lambda_4 = n(\omega_1)/\lambda_1 + n(\omega_2)/\lambda_2 - n(\omega_3)/\lambda_3$	$\Delta k = 0^a$, $\Delta k \lesseqgtr 0^b$
$\Delta k = k(\omega_4) - k(\omega_1) + k(\omega_2) + k(\omega_3)$	$n(\omega_4)/\lambda_4 = n(\omega_1)/\lambda_1 - n(\omega_2)/\lambda_2 - n(\omega_3)/\lambda_3$	$\Delta k > 0$
$\Delta k = k(\omega_2) + k(\omega_3) - 2k(\omega_1)$	$2n(\omega_1)/\lambda_1 = n(\omega_2)/\lambda_2 + n(\omega_3)/\lambda_3$	$\Delta k = 0^a$, $\Delta k \lesseqgtr 0^b$
$\Delta k = k(\omega_6) - 4k(\omega_1) - k(\omega_2)$	$n(\omega_6)/\lambda_6 = 4n(\omega_1)/\lambda_1 + n(\omega_2)/\lambda_2$	$\Delta k < 0$
$\Delta k = k(\omega_6) - 4k(\omega_1) + k(\omega_2)$	$n(\omega_6)/\lambda_6 = 4n(\omega_1)/\lambda_1 - n(\omega_2)/\lambda_2$	$\Delta k < 0^c$, $\Delta k > 0^d$

Phase matching can be accomplished by using a mixture of gases with different signs of dispersion. In this situation each component makes a contribution to the wave-vector mismatch in proportion to its concentration in the mixture. The value of the wave-vector mismatch can be controlled by adjusting the relative concentration of the two gases until the appropriate phase-matching condition is met for either collimated or focused beams, as shown in Fig. 12.

Phase matching can also be done in single-component media with focused beams, provided that the dispersion of the medium is of the correct sign for the interaction and wavelengths involved. With this technique the wave-vector mismatch depends on the density of the gas, and the pressure is adjusted until the proper wave-vector mismatch is achieved. Alternatively, phase matching in a single-component medium can be done by choosing the pump and generated frequencies to lie on either side of the transition frequency so that the phase-matching condition is satisfied. This technique is usually used in gases with plane-wave pump beams.

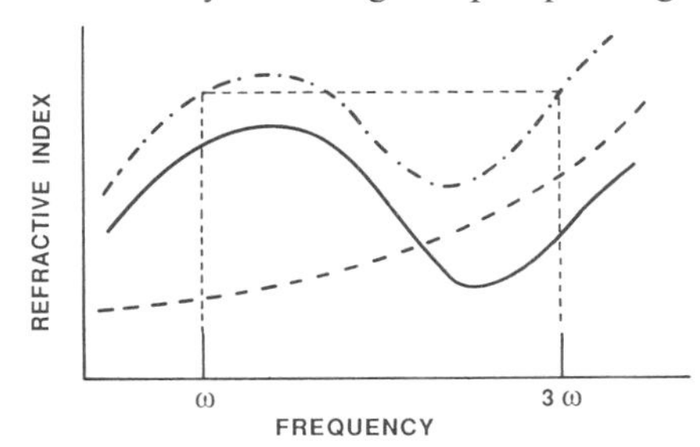

FIG. 12. Illustration of the use of mixtures for phase matching. The solid curve gives the refractive index variation of a negatively dispersive medium, the dashed curve shows a positively dispersive medium, and the chain curve shows the refractive index of a mixture chosen so that the refractive index at ω is equal to that at 3ω.

A fourth method for phase matching is the use of noncollinear waves, as shown in Fig. 13. This technique can be used for sum-frequency processes in media with negative dispersion and for difference-frequency processes in media with positive dispersion. It is commonly used, for example, in liquids for the difference frequency process $\omega_4 = 2\omega_1 - \omega_2$.

The conversion efficiency can be increased significantly if resonances are present between certain energy levels of the medium and the incident and generated frequencies or their sum or difference combinations. This increase in conversion efficiency is similar to the increase in linear absorption or scattering that occurs when the incident wavelength approaches an allowed transition of the medium. For the nonlinear interactions, however, a much greater variety of resonances is possible. Single-photon resonances occur between the incident or generated frequencies and allowed transitions just as with linear optical effects. The effectiveness of these resonances in enhancing nonlinear processes is limited, however, because of the absorption and dispersion that accompanies them.

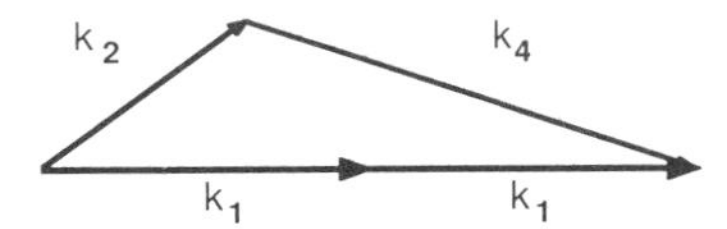

FIG. 13. Off-axis phase-matching diagram for four-wave mixing of the form $\omega_4 = 2\omega_1 - \omega_2$.

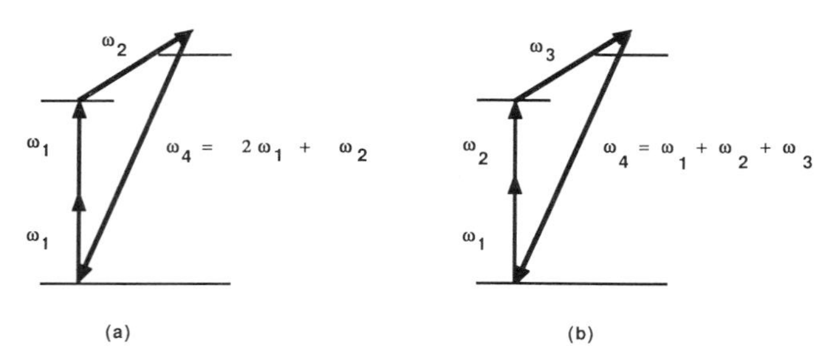

FIG. 14. Level diagrams showing two-photon resonances at (a) $2\omega_1$ and (b) $\omega_1 + \omega_2$ in four-wave sum-frequency mixing.

Resonances in the nonlinear effects also occur when multiples, or sum or difference combinations, of the incident frequencies match the frequency of certain types of transitions. The most commonly used of these resonances is a two-photon resonance involving two successive dipole transitions between levels whose spacing matches twice the value of an incident frequency or a sum or difference of two input frequencies, as indicated in Fig. 14. In single atoms and in centrosymmetric molecules, the energy levels involved in such two-photon resonances are of the same parity, and transitions between them are not observable in linear spectroscopy, which involves single-photon transitions. Near a two-photon resonance, the nonlinear susceptibility can increase by as much as four to eight orders of magnitude, depending on the relative linewidths of the two-photon transition and the input radiation, resulting in a dramatic increase in the generated power as the input frequency is tuned through the two-photon resonance. An example of the increase in efficiency that is observed as the pump frequency is tuned through a two-photon resonance is shown in Fig. 15. Other higher-order resonances are also possible, but they have not been used as commonly as the two-photon resonances. Resonantly enhanced third-harmonic generation and four-wave mixing have proven very useful in allowing effective conversion of tunable radiation from dye lasers to the vacuum ultraviolet to providing high-brightness, narrow-band sources of radiation for high-resolution spectroscopy and other applications.

Some of the applications of third- and higher-order frequency conversion are given in Table VII. The qth harmonic generation is used to pro-

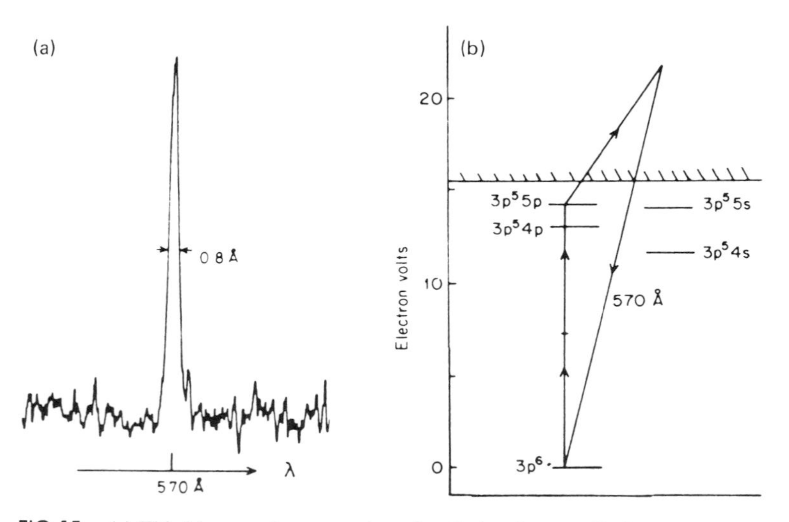

FIG. 15. (a) Third-harmonic conversion of radiation from an Xe laser showing enhancement as the wavelength of the laser is tuned through a two-photon resonance. (b) The resonant-level structure. [Reproduced from M. H. R. Hutchinson *et al.* (1976). *Opt. Commun.* **18**, 203. Copyright North-Holland, Amsterdam.]

TABLE VII. Selected Results for Third- and Higher-Order Frequency Conversion Processes

Interaction	Laser	Pump wavelength (nm)	Generated wavelength (nm)	Nonlinear material	Efficiency (%)
Third harmonic	CO_2	10.6 μm	3.5 μm	CO (liquid), CO (gas), BCl_3, SF_6, NO, DCl	8 (liquid CO)
	Nd:YAG	1.064 μm	354.7	Rb, Na	10
	Nd:YAG	354.7	118.2	Xe	0.2
	Xe_2	170	57	Ar	
	Dye	360–600	120–200	Xe, Kr, Sr, Mg, Zn, Hg	Up to 1
Fifth harmonic	Nd:YAG	266	53.2	He, Ne, Ar, Kr	10^{-3}
	XeCl	308	61.6	He	
	KrF	249	49.8	He	
	ArF	193	38.6	He	
Seventh harmonic	Nd:YAG	266	38	He	10^{-4}
	KrF	249	35.5	He	10^{-9}

duce radiation at a frequency that is q times the incident frequency. The most commonly used interaction of this type is third-harmonic conversion. It has been used to produce radiation at wavelengths ranging from the infrared to the extreme ultraviolet. Third-harmonic conversion of radiation from high power pulsed lasers such as CO_2, Nd:glass, Nd:YAG, ruby, and various rare-gas halide and rare-gas excimer lasers has been used to generate fixed-frequency radiation at various wavelengths ranging from 3.5 μm to 57 nm, as indicated in Table VII. It has also been used with dye lasers to generate radiation tunable in spectral bands between 110 and 200 nm. The extent of the spectral bands generated in this manner is determined by the extent of the negative dispersion region in the nonlinear materials.

Four-wave sum- and difference-frequency mixing interactions of the form

$$\omega_4 = 2\omega_1 \pm \omega_3 \tag{32a}$$

and

$$\omega_4 = \omega_1 + \omega_2 \pm \omega_3 \tag{32b}$$

where ω_1, ω_2, and ω_3 are input frequencies, are also commonly used to produce radiation in wavelength ranges that are inaccessible by other means. These processes can be favored over possible simultaneous third-harmonic generation by the use of opposite circular polarization in the two pump waves, since third-harmonic conversion with circularly polarized pump light is not allowed by symmetry. They have been used to generate radiation over a considerable range of wavelengths in the vacuum ultraviolet, extreme ultraviolet, and the mid infrared. In particular, they have been used to generate tunable radiation over most of the vacuum ultraviolet range from 100 to 200 nm.

These interactions can be used to generate tunable radiation in resonantly enhanced processes, thereby increasing the efficiency of the process. In this situation the pump frequency at ω_1 or the sum combination $\omega_1 + \omega_2$ is adjusted to match a suitable two-photon resonance, while the remaining pump frequency at ω_3 is varied, producing the tunable generated radiation, as illustrated in Fig. 16. The difference frequency processes $\omega_4 = 2\omega_1 - \omega_3$ and $\omega_4 = \omega_1 + \omega_2 - \omega_3$ can be optimized with focused beams in media with either sign of dispersion. As a result, their

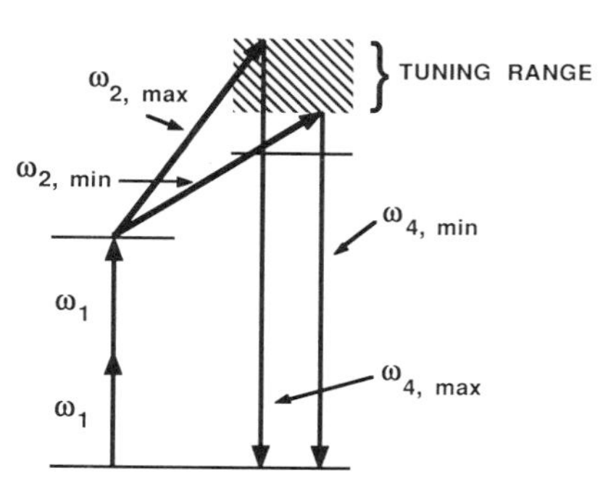

FIG. 16. Level diagram for producing tunable radiation with two-photon resonantly enhanced four-wave sum-frequency mixing. Pump radiation at ω_1 is tuned to the two-photon resonance, pump radiation at ω_2 is tuned over the range $\omega_{2,\min}$ to $\omega_{2,\max}$, and the frequency of the generated radiation tunes over the range $(2\omega_1 + \omega_{2,\min})$ to $(2\omega_1 + \omega_{2,\max})$.

usefulness is not restricted to narrow wavelength ranges above dispersive resonances, and they have been used to generate tunable radiation over extensive ranges in the vacuum ultraviolet between 140 and 200 nm in the rare gases Xe and Kr. Tunable radiation generated in this manner in Xe between 160 and 200 nm is illustrated in Fig. 17.

The difference-frequency processes

$$\omega_4 = 2\omega_1 - \omega_3 \tag{33a}$$

and

$$\omega_4 = \omega_1 \pm \omega_2 - \omega_3 \tag{33b}$$

have also been used to generate tunable radiation in the infrared by using pump radiation from visible and near infrared lasers. In some interactions all the frequencies involved in the mixing processes are supplied externally and in others some of them are generated in a stimulated Raman interaction (see Section III,A,2). Because the gases used for these nonlinear interactions are not absorbing at far-infrared wavelengths, it can be expected that they will ultimately allow more efficient generation of tunable far-infrared radiation using pump radiation in the visible and near infrared than can be achieved in second-order interactions in crystals, although they have not yet been extended to as long wavelengths. Ultimately the limitations on conversion can be expected to arise from difficulties with phase matching and a mismatch between the diffraction of the pump and generated wavelengths. To date, the four-wave difference-frequency mixing interactions have been used to produce coherent radiation at wavelengths out to 25 μm.

Resonances between Raman active molecular vibrations and rotations and the difference frequency combination $\omega_1 - \omega_3$ can also occur. When the four-wave mixing process $2\omega_1 - \omega_3$ or $\omega_1 + \omega_2 - \omega_3$ is used with these resonances it is termed coherent anti-Stokes Raman scattering (CARS). The resonant enhancement that occurs in the generated intensity as the pump frequencies are tuned through the two-photon difference-frequency resonance forms the basis of CARS spectroscopy (see Section IV,A).

Various forms of higher-order interactions are also used for frequency conversion. These consist primarily of harmonic conversion up to order seven and six-wave mixing interactions of the form $\omega_6 = 4\omega_1 \pm \omega_2$, although harmonic generation up to order 11 has been reported. Generally, the conversion efficiency in the higher-order processes is lower than it is in the lower-order ones, and the required pump intensity is higher. As a result, higher-order processes have been used primarily for the generation of radiation in the extreme ultraviolet at wavelengths too short to be reached with lower-order interactions. The pump sources have for

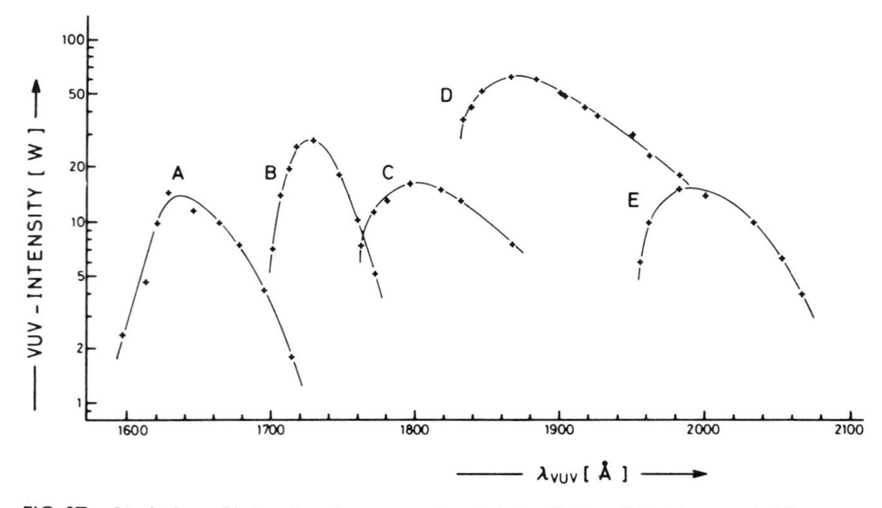

FIG. 17. Variation of intensity of vacuum ultraviolet radiation (VUV) generated in xenon through the process $2\omega_1 - \omega_2$ in the range 160–200 nm. The different regions A–E correspond to pump radiation obtained from different dye lasers. The radiation is continuously variable within each of the regions. [Reproduced from R. Hilbig and R. Wallenstein (1982). *Appl. Opt.* **21**, 913.]

the most part involved mode-locked lasers with pulse durations under 30 psec and peak power levels in excess of several hundred megawatts.

Fifth-harmonic conversion has been used to generate radiation at wavelengths as short as 38.6 nm using radiation from an ArF laser at 193 nm, seventh-harmonic conversion has been used to generate radiation at wavelengths as short as 35.5 nm with radiation from a KrF laser, and ninth-harmonic conversion has been used to generate radiation at 117.7 nm with radiation from a Nd : glass laser at 1.06 μm. Radiation at various wavelengths in the extreme ultraviolet between 38 and 76 nm using fifth- and seventh-harmonic generation and six-wave mixing of radiation from harmonics of a Nd : YAG laser has also been generated.

Observed conversion efficiencies for many of the third- and higher-order processes are noted in Table VII. They range from about 10% for third-harmonic conversion of Nd : YAG laser radiation in rubidium (1.064 μm $\rightarrow$ 354.7 nm) or CO_2 laser radiation in liquid CO (9.6 μm $\rightarrow$ 3.2 μm) to values of the order of 10^{-11} for some of the higher-order processes. The pump intensities used vary between several hundred kilowatts per square centimeter for resonantly enhanced processes to 10^{15} W/cm^2 for nonresonant processes.

The largest conversion efficiencies that can be achieved with third- and higher-order processes are generally less than those that can be obtained with second-order interactions, because competing processes that limit efficiency are more important for the higher-order interactions. Some of the important limiting processes are listed in Table VIII, along with situations for which they are important. As a result the higher-order processes are most useful in generating radiation in spectral regions, such as the vacuum ultraviolet or far infrared, that are inaccessible by the second-order interactions, or for certain applications such as phase conjugation or spectroscopy.

2. Stimulated Scattering Processes

Stimulated scattering processes are nonlinear interactions in which an incident wave at frequency ω_{inc} is converted to a scattered wave at a different frequency ω_{scat}. The difference in photon energy between the incident and scattered frequencies is taken up or supplied by the nonlinear medium, which undergoes a transition be-

TABLE VIII. Competing Processes for Third- and Higher-Order Frequency Conversion[a]

Competing process	Effect on conversion efficiency	Conditions under which competing process can be expected to be important
Linear absorption of generated radiation	Loss of generated power Reduction of improvement from phase matching Limitation on product NL	UV or XUV generation in ionizing continuum of nonlinear medium Generated wavelength close to allowed transition
Nonlinear absorption of pump radiation	Loss of pump power Saturation of susceptibility Disturbance of phase matching conditions Self focusing or self defocusing	Two-photon resonant interactions
Stark shift	Saturation of susceptibility Self-focusing or self defocusing Disturbance of phase-matching conditions	Resonant or near-resonant interactions, with pump intensity close to or greater than the appropriate saturation intensity
Kerr effect	Disturbance of phase-matching conditions Self-focusing or self defocusing	Nonresonant interactions Near-resonant interactions when the pump intensity is much less than the saturation intensity
Dielectric breakdown, multiphoton ionization	Disturbance of phase-matching conditions Saturation of susceptibility Loss of power at pump or generated wavelength	Conversion in low-pressure gases at high intensities Tightly focused geometries

[a] Reproduced from J. Reintjes (1985). Coherent ultraviolet and vacuum ultraviolet sources, *in* "Laser Handbook," Vol. 5 (M. Bass and M. L. Stitch, eds.). North-Holland, Amsterdam.

tween two of its internal energy levels, as illustrated in Fig. 18.

If the medium is initially in its ground state, the scattered wave is at a lower frequency (longer wavelength) than the incident wave, and the medium is excited to one of its internal energy levels during the interaction. In this situation the frequency shift is termed a Stokes shift, in analogy to the shift to lower frequencies that is observed in fluorescence (which was explained by Sir George Stokes), and the scattered wave is termed a Stokes wave. The incident (laser) and scattered (Stokes) frequencies are related by

$$\omega_S = \omega_L - \omega_0 \tag{34}$$

where ω_L and ω_S are the frequencies of the laser and Stokes waves and ω_0 is the frequency of the internal energy level of the medium.

If the medium is initially in an excited state, the scattered wave is at a higher frequency (shorter wavelength) than the incident wave and the medium is deexcited during the interaction, with its energy being given to the scattered wave. In this situation the scattered wave is termed an anti-Stokes wave (shifts to higher frequencies are not possible in fluorescence, as explained by Stokes), and the frequency shift is termed the anti-Stokes shift. The laser and anti-Stokes frequencies are related by

$$\omega_{AS} = \omega_L + \omega_0 \tag{35}$$

Various types of stimulated scattering processes are possible, each involving a different type of internal excitation. Some of the more common ones are listed in Table IX, along with the types of internal excitations and the types of materials in which they are commonly observed. Stimulated Brillouin scattering involves interactions with sound waves in solids, liquids, or gases or ion-acoustic waves in plasmas, and stimulated Rayleigh scattering involves interactions with density or orientational fluctuations of molecules. Various forms of stimulated Raman scattering can involve interaction with molecular vibrations, molecular rotations (rotational Raman scattering), electronic levels of atoms or molecules (electronic Raman scattering), lattice vibrations, polaritons, electron plasma waves, or nondegenerate spin levels in certain semiconductors in magnetic fields. The magnitude of the frequency shifts that occur depends on the combination of nonlinear interaction and the particular material that is used. Orders of magnitude for shifts in different types of materials for the various interactions are also given in Table IX.

Stimulated scattering processes that involve Stokes waves arise from third-order nonlinear interactions with nonlinear polarizations of the form

$$P(\omega_S) = \tfrac{3}{2}\varepsilon_0\chi^{(3)}(-\omega_S, \omega_L, -\omega_L, \omega_S)|A_L|^2 A_S \tag{36}$$

The nonlinear susceptibility involves a two-photon resonance with the difference frequency

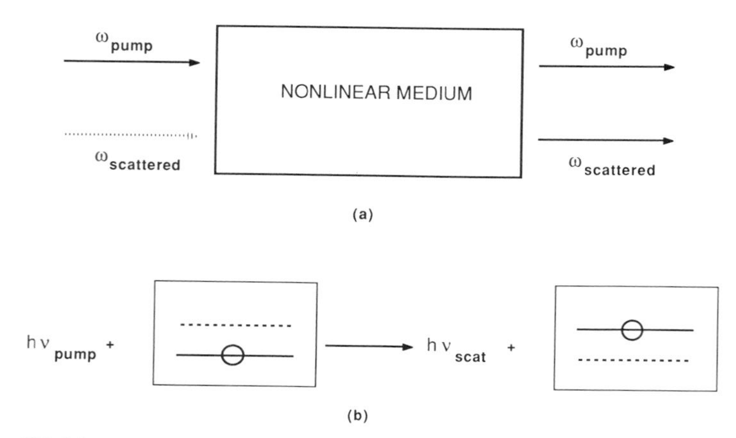

FIG. 18. (a) Schematic illustration of a typical stimulated scattering interaction. The scattered wave can be supplied along with the pump radiation or it can be generated from noise in the interaction. (b) Representation of a stimulated Stokes scattering interaction, illustrating the transitions in the nonlinear medium.

TABLE IX. Stimulated Scattering Interactions

Interaction	Internal mode	Type of material	Typical range of frequency shift $\Delta\nu$ (cm^{-1})
Stimulated Raman scattering	Molecular vibrations	Molecular gases	600–4150
		Molecular liquids	600–3500
	Molecular rotations	Molecular gases	10–400
	Electronic levels	Atomic and molecular gases, semiconductors	7000–20,000
	Lattice vibrations	Crystals	10–400
	Polaritons	Crystals	10–400
	Electron plasma waves	Plasmas	100–10,000
Stimulated Brillouin scattering	High-frequency sound waves	Solids, liquids, gases	100 MHz–10 GHz
	Ion-acoustic waves	Plasmas	0.1–10 MHz
Stimulated Rayleigh scattering	Density fluctuations, orientational fluctuations	Gases, liquids	0.1–1

combination $\omega_L - \omega_S$. As with other two-photon resonances, the energy levels that are involved are of the same parity as those of the ground state in atoms or molecules with a center of inversion. When the two-photon resonance condition is satisfied, the susceptibility for the stimulated scattering interactions is negative and purely imaginary and can be written as

$$\chi^{(3)}(-\omega_S, \omega_L, -\omega_L, \omega_S) = -i\chi''^{(3)}(-\omega_S, \omega_L, -\omega_L, \omega_S) \quad (37)$$

Stimulated scattering processes can be in the forward direction, involving a scattered wave that propagates in the same direction as the incident laser, as shown in Fig. 18, or in the backward direction, involving a scattered wave that propagates in the opposite direction to the incident laser, as shown in Fig. 19. The field amplitude of the wave generated in the forward direction by a susceptibility of the type in Eq. (37) is described by the equation

$$dA_S/dz = (3\omega_S/4n_S c)\chi''|A_L|^2 A_S \quad (38)$$

When pump depletion is negligible, the intensity of the Stokes wave is given by

$$I_S = I_S(0)e^{gIL} \quad (39)$$

where the quantity g is the gain coefficient of the process given by

$$g = (3\omega_S/n_S n_L c^2 \varepsilon_0)\chi'' \quad (40)$$

The waves generated in stimulated scattering processes have exponential growth, in contrast with the power-law dependences of the waves generated in the parametric interactions. The exponential gain is proportional to the propagation distance and to the intensity of the pump radiation. Phase matching is generally not required for stimulated scattering processes, since the phase of the material excitation adjusts itself for maximum gain automatically.

Photon number is conserved in stimulated scattering interactions, with one photon being lost in the pump wave for every one created in the Stokes wave. The energy created in the Stokes wave is smaller than that lost in the pump

FIG. 19. Backward stimulated scattering in which the scattered radiation propagates in the direction opposite to the pump radiation. The scattered wave either can be supplied along with the pump radiation, or can be generated from noise in the interaction.

wave by the ratio ω_S/ω_L, termed the Manly–Rowe ratio, and the difference in photon energy between the pump and Stokes waves represents the energy given to the medium.

When the energy in the Stokes wave becomes comparable to that in the incident laser pump, depletion occurs and the gain is reduced. In principle, every photon in the incident pump wave can be converted to a Stokes photon, giving a maximum theoretical conversion efficiency of

$$\eta_{\max} = [I_S(L)/I_L(0)] = \omega_S/\omega_L \qquad (41)$$

In practice, the efficiency is never as high as indicated in Eq. (41) because of the lower conversion efficiency that is present in the low-intensity spatial and temporal wings of most laser beams. Photon conversion efficiencies of over 90% and energy-conversion efficiencies over 80% have, however, been observed in certain stimulated Raman and Brillouin interactions.

a. Stimulated Raman Scattering. Stimulated Raman scattering can occur in solids, liquids, gases, and plasmas. It involves frequency shifts ranging from several tens of reciprocal centimeters for rotational scattering in molecules to tens of thousands of reciprocal centimeters for electronic scattering in gases. Forward stimulated Raman scattering is commonly used for generation of coherent radiation at the Stokes wavelength, amplification of an incident wave at the Stokes wavelength, reduction of beam aberrations, nonlinear spectroscopy (see Section IV,C), and generation of tunable infrared radiation through polariton scattering. Backward stimulated Raman scattering can be used for wave generation, amplification, pulse compression and phase conjugation (see Section IV,A).

The susceptibility for stimulated Raman scattering in molecules or atoms is given by

$$\chi = -\left[\frac{i}{6\Gamma\hbar^3}\left(1 - \frac{i\Delta}{\Gamma}\right)\right] \times \left[\sum \mu_{0i}\mu_{i2}\left\{\frac{1}{\omega_{i0} - \omega_L} + \frac{1}{\omega_{i0} + \omega_S}\right\}\right]^2 \qquad (42)$$

where $\Delta = \omega_0 - (\omega_L - \omega_S)$ is the detuning from the Raman resonance, Γ the linewidth of the Raman transition, ω_{i0} the frequency of the transition from level i to level 0, and μ_{0i} the dipole moment for the transition between levels 0 and i.

Gain coefficients and frequency shifts for some materials are given in Table X.

Amplification of an incident Stokes wave generally occurs for exponential gains up to about e^8 to e^{10}, corresponding to small signal amplifications of the order of 3000 to 22,000, although under some conditions stable gains up to e^{19} can be obtained. Raman amplifiers of this type are

TABLE X. Stimulated Raman Shifts and Gain Coefficients at 694.3 nm

Material	$\Delta\nu_R$ (cm^{-1})	$g \times 10^3$ (cm/MW)	
Liquids			
Carbon disulfide	656	24	
Acetone	2921	0.9	
Methanol	2837	0.4	
Ethanol	2928	4.0	
Toluene	1002	1.3	
Benzene	992	3	
Nitrobenzene	1345	2.1	
N_2	2326	17	
O_2	1555	16	
Carbon tetrachloride	458	1.1	
Water	3290	0.14	
Gases			
Methane	2916	0.66	(10 atm, 500 nm)
Hydrogen	4155 (vibrational)	1.5	(above 10 atm)
	450 (rotational)	0.5	(above 0.5 atm)
Deuterium	2991 (vibrational)	1.1	(above 10 atm)
N_2	2326	0.071	(10 atm, 500 nm)
O_2	1555	0.016	(10 atm, 500 nm)

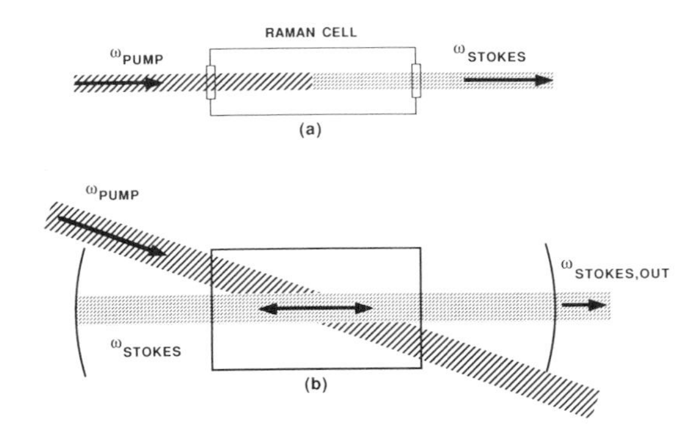

FIG. 20. (a) Generation of a Stokes wave in single-pass stimulated Raman scattering. (b) Generation of a Raman–Stokes wave in a Raman laser oscillator.

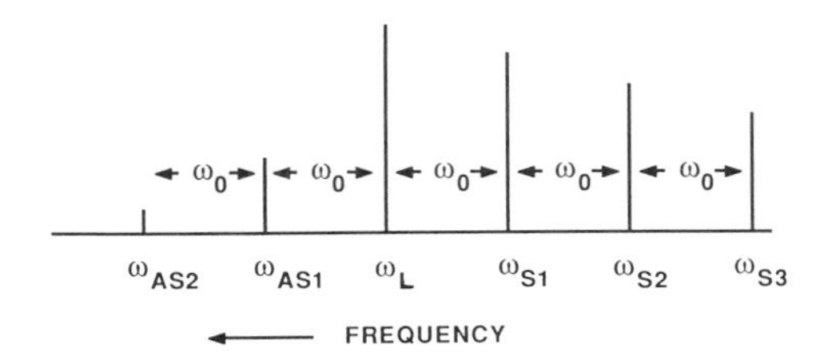

FIG. 21. Schematic illustration of the spectrum produced by multiple Stokes and anti-Stokes scattering.

used to increase the power in the Stokes beam. When the pump beam has a poor spatial quality due to phase or amplitude structure, Raman amplification can be used to transfer the energy of the pump beam to the Stokes beam without transferring the aberrations, thereby increasing the effective brightness of the laser system.

Generation of a new wave at the Stokes frequency can be done in a single pass geometry, as shown in Fig. 20a, or in an oscillator configuration at lower gains, as shown in Fig. 20b. The threshold for single-pass generation depends on the degree of focusing that is used but is generally certain for gains above about e^{23} (10^{10}). Once the threshold for single-pass generation is reached, the process generally proceeds to pump depletion very quickly. As a result, the Stokes frequency that is generated in a Raman oscillator usually involves the Raman-active mode with the highest gain.

If sufficient conversion to the Stokes wave takes place, generation of multiple Stokes waves can occur. In this situation the first Stokes wave serves as a pump for a second Stokes wave that is generated in a second stimulated Raman interaction. The second Stokes wave is shifted in frequency from the first Stokes wave by ω_0. If sufficient intensity is present in the original pump wave, multiple Stokes waves can be generated, each shifted from the preceding one by ω_0 as illustrated in Fig. 21. Stimulated Raman scattering can thus be used as a source of coherent radiation at several wavelengths by utilizing multiple Stokes shifts, different materials, and different pump wavelengths.

Continuously tunable Stokes radiation can be generated by using a tunable laser for the pump radiation or in some situations by using materials with internal modes whose energies can be changed. An example of this type of interaction is the generation of tunable narrow-band infrared radiation from spin-flip Raman lasers in semiconductors, as illustrated in Fig. 22. In these lasers the energy levels of electrons with different spin orientations are split by an external magnetic field. The Raman process involves a transition from one spin state to the other. For a fixed pump wavelength—for example, from a CO_2 laser at 10.6 μm—the wavelength of the Stokes wave can be tuned by varying the magnetic field, which determines the separation of the electron energy levels. Multiple Stokes shifts and anti-Stokes shifts can also be obtained. A list of materials, laser sources, and tuning ranges for various spin-flip lasers is given in Table XI. Radiation that is tunable in bands between 5.2 and 16.2 μm has been generated in

TABLE XI. Spin-Flip Raman Lasers

Pump laser (wavelength, μm)	Material	Tuning range (μm)	Raman order
NH_3 (12.8)	InSb	13.9–16.8	I Stokes
CO_2 (10.6)	InSb	9.0–14.6	I, II, III, Stokes I, Anti-Stokes
CO (5.3)	InSb	5.2–6.2	I, II, III, Stokes I, Anti-Stokes
HF	InAs	3–5	

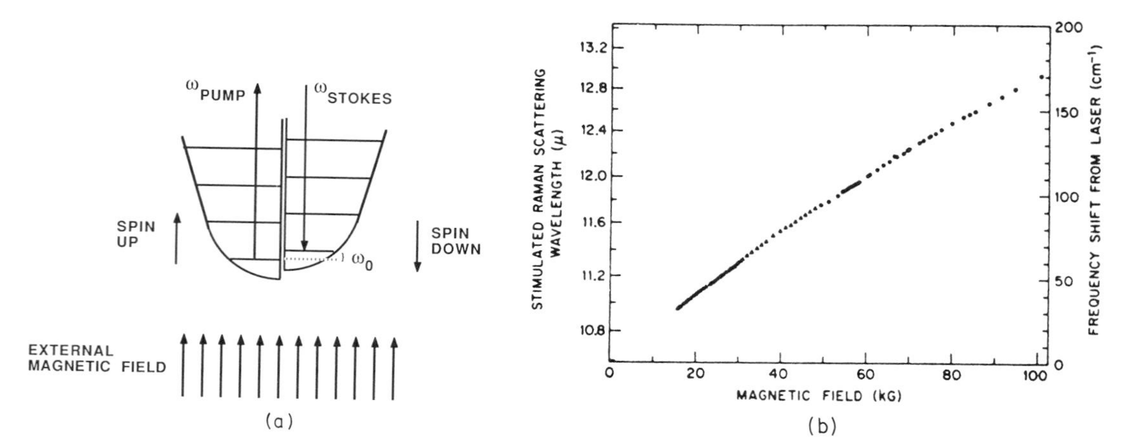

FIG. 22. (a) Level diagram for a spin-flip Raman laser. (b) Typical tuning curve for an InSb spin-flip Raman laser pumped by a CO_2 laser at 10.6 μm. [Part (b) reproduced from Patel, C. K. N., and Shaw, E. D. (1974). *Phys. Rev. B* **3,** 1279.]

this manner with linewidths as narrow as 100 MHz. This is among the narrowest bandwidth infrared radiation available and has been used for high resolution spectroscopy.

Backward stimulated Raman scattering involves the generation or amplification of a Stokes wave that travels in the opposite direction to the pump wave, as was shown in Fig. 19. Backward stimulated Raman scattering requires radiation with a much narrower bandwidth than does forward Raman scattering and is usually observed for laser bandwidths less than about 10 GHz. The backward traveling Stokes wave encounters a fresh undepleted pump wave as it propagates through the nonlinear medium. As a result, the gain does not saturate when the pump wave is depleted, as it does in the forward direction. The peak Stokes intensity can grow to be many times that of the incident laser intensity, while the duration of the Stokes pulse decreases relative to that of the incident pump pulse, allowing conservation of energy to be maintained. This process, illustrated in Fig. 23, is termed pulse compression. Compression ratios of the order of 50 : 1, producing pulses as short as a few nanoseconds, have been observed.

Anti-Stokes Raman scattering involves the generation of radiation at shorter wavelengths than those of the pump wave. Anti-Stokes scattering can occur in one of two ways. The more common method involves a four-wave difference frequency mixing process of the form

$$\omega_{AS} = 2\omega_L - \omega_S \tag{43a}$$

(see Section III,A,1,b) in media without a population inversion. In this interaction the Stokes radiation is generated with exponential gain through the stimulated Raman interaction as described above. The anti-Stokes radiation is generated by the four-wave mixing process, using the Stokes radiation as one of the pump waves. The anti-Stokes radiation that is generated through this interaction grows as part of a mixed mode along with the Stokes radiation. It has the same exponential gain as does the Stokes radia-

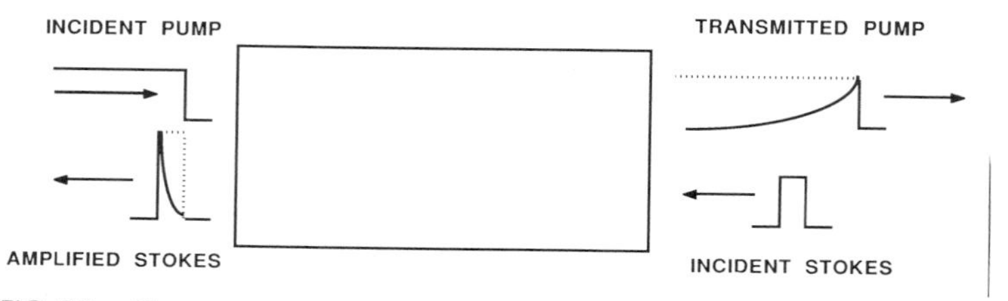

FIG. 23. Illustration of pulse compression with backward stimulated Raman scattering. Backward-traveling Stokes pulse sweeps out the energy of the pump in a short pulse.

tion, but the amplitude of the anti-Stokes radiation depends on the phase mismatch just as for other four-wave mixing interactions. The anti-Stokes generation is strongest for interactions that are nearly phase matched, although neither wave has exponential gain at exact phase matching. For common liquids and gases that have normal dispersion, the anti-Stokes process is not phase matched in the forward direction but is phase matched when the Stokes and anti-Stokes waves propagate at angles to the pump wave, as shown schematically in Fig. 24a. The anti-Stokes radiation is thus generated in cones about the pump radiation in most of these materials. The opening angle of the cone depends on the dispersion in the medium and on the frequency shift. It is of the order of several tens of milliradians for molecular vibrational shifts in liquids and can be considerably smaller for rotational shifts in molecular gases. An example of anti-Stokes emission in H_2 is shown in Fig. 24b. Here the pump radiation was at 532 nm. The anti-Stokes radiation at 435.7 nm was concentrated near the phase-matching direction of about 7 mrad, with a dark band appearing at the exact phase-matching direction and bright emission appearing in narrow bands on either side of phase matching.

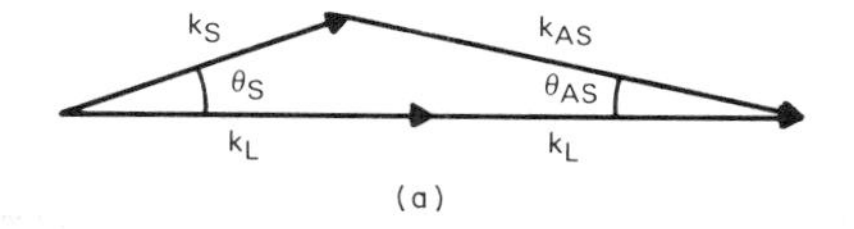

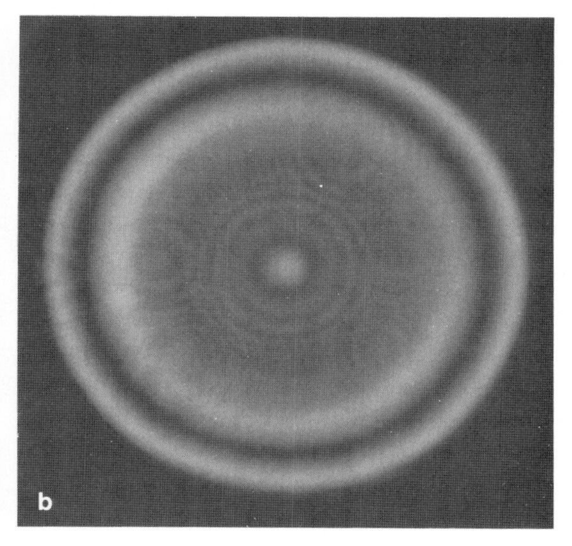

FIG. 24. (a) Off-axis phase-matching diagram for anti-Stokes generation. (b) Anti-Stokes rings produced in stimulated Raman scattering from hydrogen. The dark band in the anti-Stokes ring is caused by suppression of the exponential gain due to the interaction of the Stokes and anti-Stokes waves. [Reproduced from M. O. Duncan *et al.* (1986). *Opt. Lett.* **11,** 803. Copyright © 1986 by the Optical Society of America.]

Multiple anti-Stokes generation can occur through interactions of the form

$$\omega_{AS,n} = \omega_1 + \omega_2 - \omega_3 \tag{43b}$$

where ω_1, ω_2, ω_3 are any of the Stokes, anti-Stokes, or laser frequencies involved in the interaction that satisfy the relations

$$\omega_1 - \omega_3 = \omega_0 \tag{44a}$$

$$\omega_{AS,n} - \omega_2 = \omega_0 \tag{44b}$$

Just as with multiple Stokes generation, the successive anti-Stokes lines are shifted from the preceding one by ω_0 as shown in Fig. 21. Multiple Stokes and anti-Stokes Raman scattering in molecular gases have been used to generate radiation ranging from 138 nm in the ultraviolet to wavelengths in the infrared. Some of the combinations of lasers and materials are listed in Table XII.

In media with a population inversion between the ground state and an excited Raman-active level, radiation at the anti-Stokes wavelength can be produced through a process similar to the one just described for Stokes generation in media that start from the ground state. This combination, illustrated in Fig. 25, is termed an anti-Stokes Raman laser. It has been used to generate radiation at wavelengths ranging from 149 to 378 nm using transitions in atomic gases such as Tl, I, or Br.

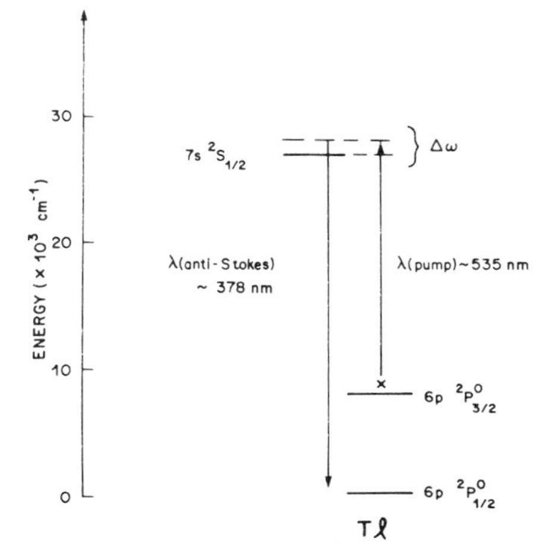

FIG. 25. Level diagram of anti-Stokes Raman laser in Tl. [Reprinted from White, J. C., and Henderson, D. (1982). *Opt. Lett.* **7,** 517. Copyright © 1982 by the Optical Society of America.]

TABLE XII. Wavelength Range (nm) of Tunable UV and VUV Radiation Generated by Stimulated Raman Scattering in H_2 ($\Delta\nu = 4155\ cm^{-1}$) with Dye Lasers[a]

Process (order)	Pump wavelength (nm)					
	600–625 (Rh 101)[b]	570–600 (Rh B)[c]	550–580 (Rh 6G)[b]	300–312.5 (Rh 101, SH)[b]	275–290 (Rh 6G, SH)[b]	548 (Fluorescein 27)[c]
AS (13)						138
AS (12)						146
AS (11)						156
AS (10)						167
AS (9)	185.0–187.3					179
AS (8)	200.4–203.1	196.9–200.4				194
AS (7)	218.6–221.8	214.4–218.6	194.5–198.1			211
AS (6)	240.4–244.3	235.4–240.4	211.6–215.9			231
AS (5)	267.1–271.9	261.0–267.1	256.7–263.0	184.8–189.5		256
AS (4)	300.4–306.6	292.7–300.4	287.3–295.3	200.2–205.7	188.7-195.7	286
AS (3)	343.3–351.3	332.2–343.3	326.3–336.6	218.3–224.9	204.8–213.0	325
AS (2)			377.5–391.4	240.1–248.1	223.8–233.7	376
AS (1)				266.7–276.6	246.8–258.8	
S (1)				342.7–359.1	310.5–329.7	
S (2)					356.5–382.1	

[a] From J. Reintjes (1985). Coherent ultraviolet and vacuum ultraviolet sources, *in* "Laser Handbook," Vol. 5 (M. Bass and M. L. Stitch, eds.), p. 1. North-Holland, Amsterdam.
[b] Data from Wilke and Schmidt (1979). *Appl. Phys.* **18,** 177.
[c] Data from Schomburg *et al.* (1983). *Appl. Phys. B* **30,** 131.

b. Stimulated Brillouin Scattering. Stimulated Brillouin scattering (SBS) involves scattering from high-frequency sound waves. The gain for SBS is usually greatest in the backward direction and is observed most commonly in this geometry, as shown in Fig. 26a. The equations describing backward SBS are

$$dI_B/dz = -g_B I_L I_B \quad (45a)$$

$$dI_L/dz = -g_B(\omega_L/\omega_B) I_L I_B \quad (45b)$$

where the intensity gain coefficient g_B is given by

$$g_B = \frac{\omega_B^2 \rho_0 (\partial\varepsilon/\partial\rho)^2}{4\pi c^3 n v \Gamma_B \varepsilon_0} \quad (45c)$$

where ρ is the density, v the velocity of sound, $\partial\varepsilon/\partial\rho$ the change of the dielectric constant with density, and Γ_B is the linewidth of the Brillouin transition. The wave-vector diagram is shown in Fig. 26b. The incident and generated optical waves are in the opposite directions, and the wave vector of the acoustic wave is determined by the momentum matching condition

$$\boldsymbol{k}_v = \boldsymbol{k}_L - \boldsymbol{k}_S \approx 2\boldsymbol{k}_L \quad (46)$$

where $\boldsymbol{k}_v$ is the $\boldsymbol{k}$ vector of the sound wave and $\boldsymbol{k}_L$ and $\boldsymbol{k}_S$ are the wave vectors of the laser and scattered waves. Because the speed of light is so much greater than the speed of sound, the magnitude of the $\boldsymbol{k}$ vector of the incident wave is almost equal to that of the scattered wave. The corresponding frequency shift of the scattered wave, termed the Brillouin shift, is equal to the frequency of the sound wave generated in the interaction and is given by

$$\Delta\omega_B = 2\omega_L v/c \quad (47)$$

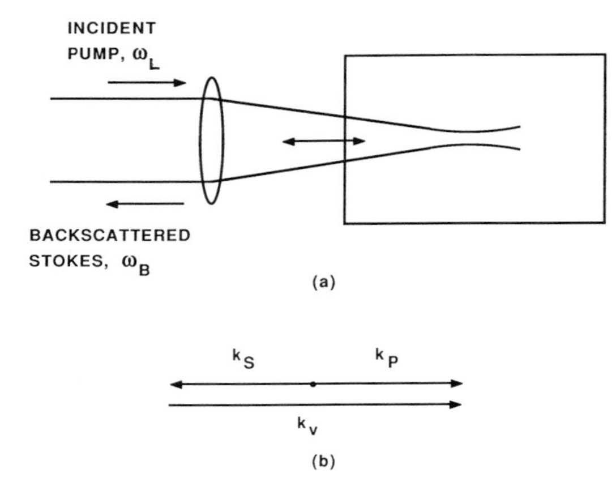

FIG. 26. (a) Schematic diagram of a typical configuration used for stimulated Brillouin scattering. (b) The k-vector diagram for stimulated Brillouin scattering.

where v is the velocity of sound and we have used the approximation that $k_L = -k_S$.

Stimulated Brillouin scattering can be observed in liquids, gases, and solids, and also in plasmas, in which the scattering is from ion-acoustic waves. The SBS shift generally ranges from hundreds of megahertz in gases to several tens of gigahertz in liquids, depending on the particular material and laser wavelength involved. The SBS shifts for some materials are given in Table XIII. The threshold intensity for SBS ranges from 10^8 to 10^{10} W/cm^2, depending on the nonlinear material and the focusing conditions, and maximum reflectivities (the ratio of the intensity of the scattered wave to that of the incident pump wave) commonly exceed 50%. The acoustic waves generated in these interactions are among the most intense high-frequency sound waves generated.

Because the response time of the acoustic phonons is relatively long (of the order of 1 nsec in liquids and up to several hundred nanoseconds in some gases), SBS is observed most commonly with laser radiation with a bandwidth less than 0.1 cm^{-1}. Generally, SBS has a higher gain than stimulated Raman scattering in liquids and usually dominates the interaction for wave generation when the laser radiation consists of narrow-band pulses that are longer than the response time of the acoustic phonon. Stimulated Raman scattering is commonly observed in these materials only for relatively broad-band radiation (for which the SBS interaction is suppressed), for short-duration pulses (for which SBS does not have time to grow), or at the beginning of longer-duration, narrow-band pulses.

In liquids and gases, SBS is used most commonly for phase conjugation (see Section IV,C) and for pulse compression in a manner similar to that described above for stimulated Raman scattering. SBS in solids can also be used for these purposes but is less common because the materials can be easily damaged by the acoustic wave that is generated in the medium.

TABLE XIII. Stimulated Brillouin Shifts and Gain Coefficients at 1.064 μm

Material	$\Delta\nu_B$ (GHz)	g (cm/MW)
Carbon disulfide	3.84	0.13–0.16
Methanol	2.8	0.014
Ethanol	3	0.012
Toluene	3.9	0.013
Benzene	4.26	0.021
Acetone	3	0.019
n-Hexane	2.8	0.023
Cyclohexane	3.66	0.007
Carbon tetrachloride	1.9	0.007
Water	3.75	0.006

B. Self-Action Effects

Self-action effects are those that affect the propagation characteristics of the incident light beam. They are due to nonlinear polarizations that are at the same frequency as that of the incident light wave. Depending on the particular effect, they can change the direction of propagation, the degree of focusing, the state of polarization or the bandwidth of the incident radiation, as was indicated in Table I. Self-action effects can also change the amount of absorption of the incident radiation. Sometimes one of these effects can occur alone, but more commonly two or more of them occur simultaneously.

The most common self-action effects arise from third-order interactions. The nonlinear polarization has the form

$$P(\omega) = \tfrac{3}{4}\varepsilon_0\chi^{(3)}(-\omega, \omega, -\omega, \omega)|A|^2A \quad (48)$$

The various types of self-action effects depend on whether the susceptibility is real or imaginary and on the temporal and spatial distribution of the incident light. Interactions that change the polarization vector of the radiation depend on the components of the polarization vector present in the incident radiation, as well as on the tensor components of the susceptibility.

The real part of the nonlinear susceptibility in Eq. (48) gives rise to the spatial effects of self-focusing and self-defocusing, spectral broadening, and changes in the polarization vector. The imaginary part of the susceptibility causes nonlinear absorption.

1. Spatial Effects

The real part of the third-order susceptibility in Eq. (48) causes a change in the index of refraction of the material according to the relation

$$n = n^L + n_2\langle E^2\rangle = n^L + \tfrac{1}{2}n_2|A|^2 \quad (49)$$

where

$$n_2 = (3/4n^L)\chi' \quad (50)$$

In these equations, $\langle E^2\rangle$ is the time average of the square of the total electric field of Eq. (3), which is proportional to the intensity, n^L is the linear refractive index, χ' is the real part of χ, and n_2 is termed the nonlinear refractive index.

Self-focusing occurs as a result of a combination of a positive value of n_2 and an incident beam that is more intense in the center than at the edge, a common situation that occurs, for example, as a result of the spatial-mode structure of a laser. In this situation the refractive index at the center of the beam is greater than that at its edge and the optical path length for rays at the center is greater than that for rays at the edge. This is the same condition that occurs for propagation through a focusing lens, and as a result the light beam creates its own positive lens in the nonlinear medium. As the beam focuses, the strength of the nonlinear lens increases, causing stronger focusing and increasing the strength of the lens still further. This behavior results in catastrophic focusing, in which the beam collapses to a very intense, small spot, in contrast to the relatively gentle focusing that occurs for normal lenses, as illustrated in Fig. 27.

Self-focusing can occur in any transparent material at sufficiently high intensities and has been observed in a wide range of materials, including glasses, crystals, liquids, gases, and plasmas. The mechanism causing self-focusing varies from one material to another. In solids and some gases the nonlinear index is due to interaction with the electronic energy levels which causes a distortion of the electron cloud, which results in an increase in the refractive index. In materials such as molecular liquids with anisotropic molecules, the nonlinear index arises from orientation of the molecules so that their axis of easy polarization is aligned more closely along the polarization vector of the incident field, as shown in Fig. 28. In such materials the molecules are normally arranged randomly, resulting in an isotropic refractive index. When the molecules line up along the optical field, the polarizability increases in that direction, resulting in both an increase in the refractive index for light polarized in the same direction as the incident field and, because the change in the refractive index is less for light polarized perpendicular to the incident radiation, birefringence. This effect is termed the optical Kerr effect, and the materials in which it occurs are termed Kerr-active. Self-focusing is observed most commonly in these materials.

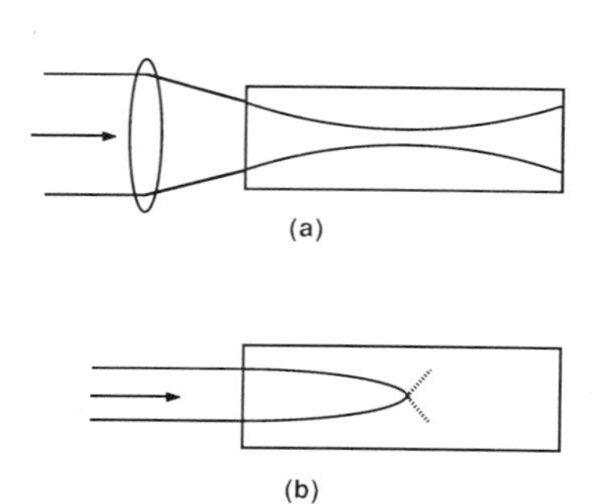

FIG. 27. Schematic of focal region produced by (a) a normal and (b) a self-focusing lens.

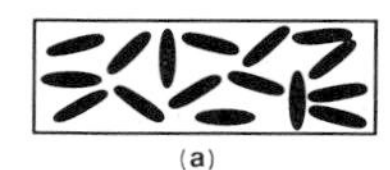

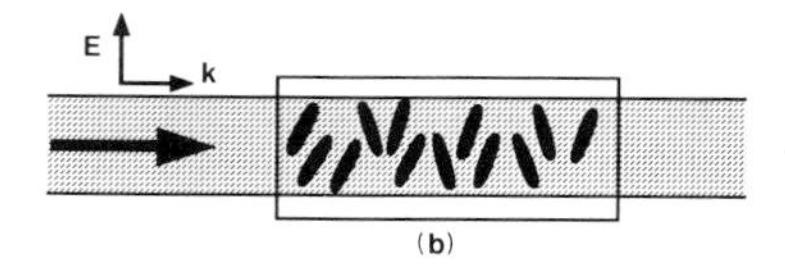

FIG. 28. (a) Random molecular orientation in a liquid with anisotropic molecules produces an isotropic refractive index. (b) Partial alignment with a laser beam produces a nonlinear index and optically induced birefringence.

A nonlinear index can also arise from electrostriction, in which the molecules of the medium move into the most intense regions of the electric field. The resulting increase in density causes an increase in the refractive index near the regions of intense fields. Because of the relatively slow response time of moving molecules, electrostriction has a longer time constant than molecular orientation and is typically important only for pulses that last for several tens to several hundreds of nanoseconds or longer.

Self-focusing in plasmas occurs because of a form of electrostriction in which the electrons move away from the most intense regions of the beam. Because the electrons make a negative contribution to the refractive index, the change in their density distribution results in a positive lens.

In order for a beam to self-focus, the self-focusing force must overcome the tendency of the beam to increase in size due to diffraction. This requirement leads to the existence of a critical power defined by

$$P_c = 0.04\varepsilon_0\lambda^2 c/n_2 \tag{51}$$

For incident powers above the critical power, the self-focusing overcomes diffraction and a beam of radius a focuses at a distance given by

$$z_f = \frac{0.369\ ka^2}{\sqrt{P/P_c} - 0.858} \tag{52}$$

For incident powers below the critical power, the self-focusing cannot overcome the spreading due to diffraction and the beam does not focus, although it spreads more slowly than it would in the absence of the nonlinear index. Values of the nonlinear index and the critical powers for some materials are given in Table XIV.

When the power in the incident beam is just above the critical power, the entire beam focuses as described above in a process termed whole-beam self-focusing. When the incident power exceeds the critical power significantly, the beam usually breaks up into a series of focal regions, each one of which contains one or a small number of critical powers. This behavior is termed beam break-up, and the resulting focusing behavior is termed small-scale self-focusing. If the incident beam contains a regular intensity pattern, the distribution of focal spots can be regular. More commonly, however, the pattern is random and is determined by whatever minor intensity variations are on the incident beam due to mode structure, interference patterns, or from imperfections or dust in or on optical materials through which it has propagated. An example of a regular pattern of self-focused spots developed on a beam with diffraction structure is shown in Fig. 29.

Once the self-focusing process has started, it will continue until catastrophic focus is reached at the distance given in Eq. (52). The minimum size of the focal point is not determined by the third-order nonlinear index but can be determined by higher-order terms in the nonlinear index, which saturate the self-focusing effect. Such saturation has been observed in atomic gases. For self-focusing in liquids, it is thought that other mechanisms, such as nonlinear absorption, stimulated Raman scattering, or multiphoton ionization, place a lower limit on the size of the focal spots. Minimum diameters of self-focal spots are of the order of a few micrometers to a few tens of micrometers, depending on the material involved.

If the end of the nonlinear medium is reached before the catastrophic self-focal distance of Eq. (52), the material forms an intensity-dependent variable-focal-length lens. It can be used in conjunction with a pinhole or aperture to form an optical limiter or power stabilizer.

When the incident beam has a constant intensity in time, the focal spot occurs in one place in the medium. When the incident wave is a pulse that varies in time, the beam focuses to a succession of focal points, each corresponding to a different self-focal distance according to Eq. (52). This gives rise to a moving self-focal point, which, when observed from the side and integrated over the pulse duration, can have the appearance of a continuous track. In special cases the beam can be confined in a region of small diameter for many diffraction lengths in an effect termed self-trapping. This happens, for example, when the nonlinear index is heavily saturated, as, for example, in an atomic transition. When the pulse duration is short compared to the response time of the nonlinear index, which can vary from several picoseconds in common liquids to several hundred picoseconds in liquid crystals, the back end of the pulse can be effectively trapped in the index distribution set up by the front of the pulse. This behavior is termed dynamic self-trapping.

TABLE XIV. Self-Focusing Parameters for Selected Materials

Material	$n_2 \times 10^{-22}$ (MKS units)	Critical power at 1.064 μm (kW)
Carbon disulfide	122	35.9
Methanol	41	106
Ethanol	32	136
Toluene	29	152
Benzene	21	207
Acetone	3.4	1274
n-Hexane	2.6	1645
Cyclohexane	2.3	1880
Carbon tetrachloride	1.8	2194
Water	1.4	3038
Cesium vapor	-2.9×10^{-16} N	

Self-focusing in solids is generally accompanied by damage in the catastrophic focal regions. This results in an upper limit on the intensity that can be used in many optical components and also results in a major limitation on the intensity that can be obtained from some solid-state pulsed lasers.

Because of the tensor nature of the nonlinear susceptibility, the intensity-dependent change in the refractive index is different for light polarized parallel and perpendicular to the laser radiation, resulting in optically induced birefringence in the medium. The birefringence can be used to change linearly polarized light into elliptically polarized light. This effect forms the basis of ultrafast light gates with picosecond time resolution that are used in time-resolved spectroscopy and measurements of the duration of short pulses. The birefringence also results in the rota-

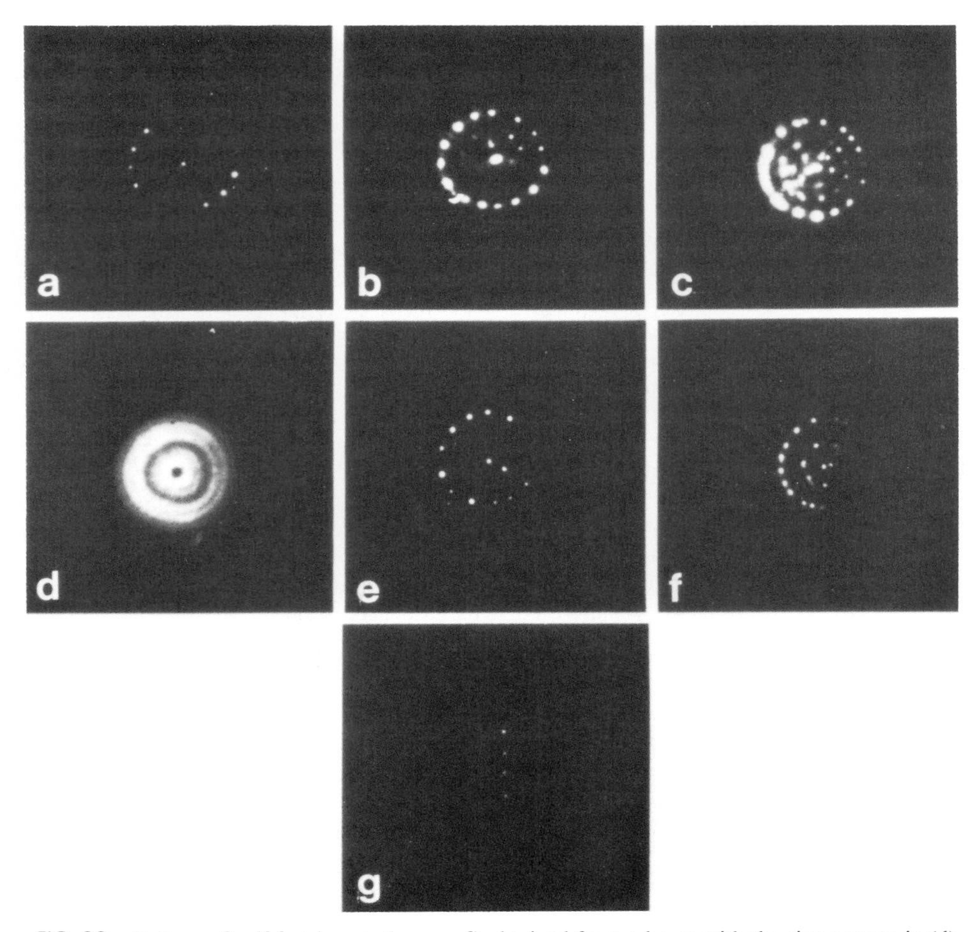

FIG. 29. Pattern of self-focal spots (a–c, e–f) obtained from a beam with the ring pattern in (d). (g) Pattern produced by a beam with a horizontal straight edge. [Reproduced from A. J. Campillo *et al.* (1973). *Appl. Phys. Lett.* **23**, 628.]

tion of the principal axis of elliptically polarized light, an effect used in nonlinear spectroscopy and the measurement of nonlinear optical susceptibilities.

2. Self-Defocusing

Self-defocusing results from a combination of a negative value of n_2 and a beam profile that is more intense at the center than at the edge. In this situation the refractive index is smaller at the center of the beam than at the edges, resulting in a shorter optical path for rays at the center than for those at the edge. This is the same condition that exists for propagation through a negative-focal-length lens, and the beam defocuses.

Negative values of the nonlinear refractive index can occur because of interaction with electronic energy levels of the medium when the laser frequency is just below a single-photon resonance or just above a two-photon resonance. Generally, self-defocusing due to electronic interactions is observed only for resonant interactions in gases and has been observed in gases for both single- and two-photon resonant conditions. A more common source of self-defocusing is thermal self-defocusing or, as it is commonly called, thermal blooming, which occurs in materials that are weakly absorbing. The energy that is absorbed from the light wave heats the medium, reducing its density, and hence its refractive index, in the most intense regions of the beam. When the beam profile is more intense at the center than at the edge, the medium becomes a negative lens and the beam spreads. Thermal blooming can occur in liquids, solids, and gases. It is commonly observed in the propagation of high-power infrared laser beams through the atmosphere and is one of the major limitations on atmospheric propagation of such beams.

3. Self-Phase Modulation

Self-phase modulation results from a combination of a material with a nonlinear refractive index and an incident field amplitude that varies in time. Because the index of refraction depends on the optical intensity, the optical phase, which is given by

$$\phi = kz - \omega t = \frac{2\pi}{\lambda}\,[n^{\mathrm{L}} + \tfrac{1}{2}n_2|A(t)|^2]z - \omega t \quad (53)$$

develops a time dependence that follows the temporal variation of the optical intensity. Just as with other situations involving phase modulation, the laser pulse develops spectral side bands. Typical phase and frequency variations are shown in Fig. 30 for a pulse that has a bell-shaped profile. The phase develops a bell-shaped temporal dependence, and the frequency, which is the time derivative of the phase, undergoes an oscillatory behavior as shown. For a medium with a positive n_2, the down-shifted part of the spectrum is controlled by the leading part of the pulse and the up-

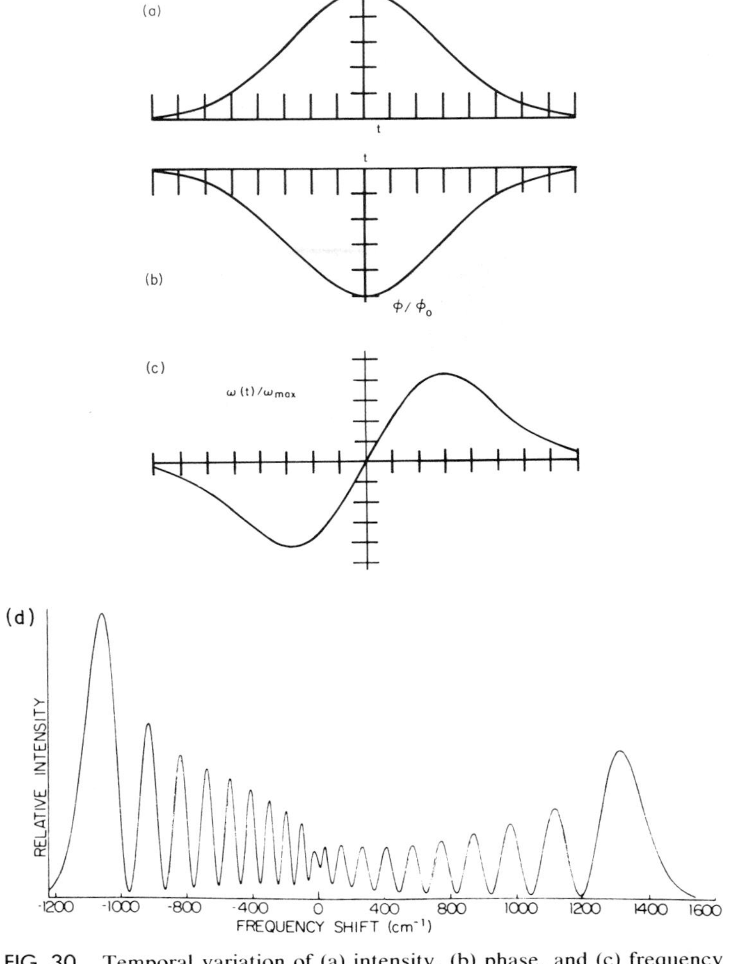

FIG. 30. Temporal variation of (a) intensity, (b) phase, and (c) frequency produced by a medium with a positive nonlinear refractive index. (d) Spectrum of a self phase modulated pulse. The asymmetry results from a rise time that was 1.4 times faster than the full time. [Part (d) reproduced from J. Reintjes (1984). "Nonlinear Optical Parametric Processes in Liquids and Gases." Academic Press, Orlando, Florida.]

shifted part of the spectrum is controlled by the trailing part of the pulse. The number of oscillations in the spectrum is determined by the peak phase modulation, while the extent of the spectrum is determined by the pulse duration and the peak phase excursion. For pulses generated in Q-switched lasers, which last for tens of nanoseconds, the spectral extent is relatively small, usually less than a wave number. For picosecond-duration pulses generated in mode-locked lasers, the spectrum can extend for hundreds of wave numbers and, in some materials such as water, for thousands of wave numbers.

In many instances, self-phase-modulation represents a detrimental effect, such as in applications to high-resolution spectroscopy or in the propagation of a pulse through a dispersive medium when the self-phase-modulation can cause spreading of the pulse envelope. In some instances, however, self-phase-modulation can be desirable. For example, the wide spectra generated from picosecond pulses in materials such as water have been used for time-resolved spectroscopic studies. In other situations, the variation of the frequency in the center of a pulse, as illustrated in Fig. 30c, can be used in conjunction with a dispersive delay line formed by a grating pair to compress phase-modulated pulses in an arrangement as shown in Fig. 31, in a manner similar to pulse compression in chirped radar.

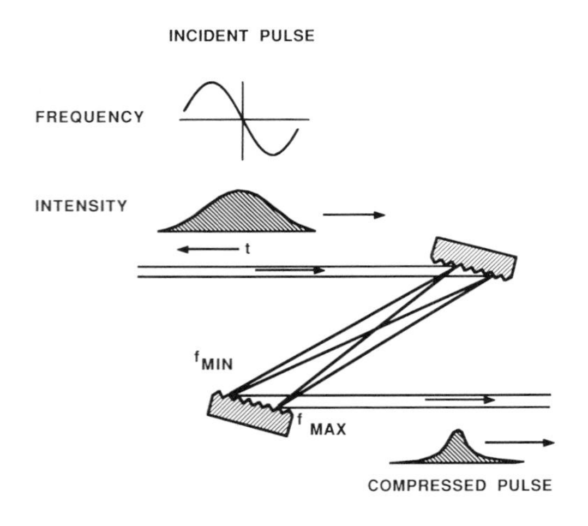

FIG. 31. Pulse compression produced by a self-phase-modulated pulse and a pair of diffraction gratings. The dispersion of the gratings causes the lower-frequency components of the pulse to travel a longer path than that of the higher-frequency components, allowing the back of the pulse to catch up to the front.

Pulse compressions of factors of 10 or more have been achieved with self-phase-modulation of pulses propagated freely in nonlinear media resulting in the generation of subpicosecond pulses from pulses in the picosecond time range. This technique has also been very successful in the compression of pulses that have been phase-modulated in glass fibers. Here the mode structure of the fiber prevents the beam break-up due to self-focusing that can occur for large phase modulation in nonguided propagation. Pulse compressions of over 100 have been achieved in this manner and have resulted in the generation of pulses that, at 8 fsec (8×10^{-15} sec), are the shortest-duration optical pulses yet produced. Self-phase-modulation in fibers, coupled with anomalous dispersion that occurs for infrared wavelengths longer than about 1.4 μm, has also been used to produce solitons, which are pulses that can propagate long distances without spreading because the negative dispersion in the fiber actually causes the phase-modulated pulse to narrow in time. Soliton pulses are useful in the generation of picosecond-duration pulses and in long-distance propagation for optical-fiber communication.

4. Nonlinear Absorption

Self-action effects can also change the transmission of light through a material. Nonlinear effects can cause materials that are strongly absorbing at low intensities to become transparent at high intensities in an effect termed saturable absorption or, conversely, they can cause materials that are transparent at low intensities to become absorbing at high intensities in an effect termed multiphoton absorption.

Multiphoton absorption can occur through absorption of two, three, or more photons. The photons can be of the same or different frequencies. When the frequencies are different, the effect is termed sum-frequency absorption. Multiphoton absorption can occur in liquids, gases, or solids. In gases the transitions can occur between the ground state and excited bound states or between the ground state and the continuum. When the transition is to the continuum, the effect is termed multiphoton ionization. Multiphoton absorption in gases with atoms or symmetric molecules follow selection rules for multiple dipole transitions. Thus two-photon absorption occurs in these materials between levels that have the same parity. These are the same transitions that are allowed in stimulated Raman scattering but are not allowed in a single-photon transitions of linear optics. Multiphoton absorption in solids involves transitions to the conduction band or to discrete states in the band gap. In semiconductors, two- or three-photon absorption can be observed for near-infrared radiation, while for transparent dielectric materials multiphoton absorption is generally observed for visible and ultraviolet radiation. Multiphoton absorption increases with increasing laser intensity and can become quite strong at intensities that can be achieved in pulsed laser beams, often resulting in damage of solids or breakdown in gases. This can be one of the major limitations on the intensity of radiation that can be passed through semiconductors in the infrared or other dielectric materials in the ultraviolet.

The simplest form of multiphoton absorption is two-photon absorption. It is described by an equation of the form

$$dI/dz = -\beta I^2 \tag{54}$$

where β is the two-photon absorption coefficient. This equation has a solution of the form

$$I(L) = I_0/(1 + \beta I_0 L) \tag{55}$$

for the intensity transmitted through a material of length L, where I_0 is the intensity incident on the material at $z = 0$. The form of the solution is indicated graphically in Fig. 32. Note that the transmission as a function of distance is quite different from that encountered for linear absorption, for which the transmitted intensity decreases with distance as $e^{-\alpha L}$. In the limit of large values of $\beta I_0 L$, the transmitted intensity approaches the constant value $1/\beta L$, independent of the incident intensity. Two-photon absorption can thus be used for optical limiting. It can also be used for spectroscopy of atomic and molecular levels that have the same parity as the ground state and are therefore not accessible in linear spectroscopy. Finally, two-photon absorption can be used in Doppler-free spectroscopy in gases to provide spectral resolutions that are less than the Doppler width.

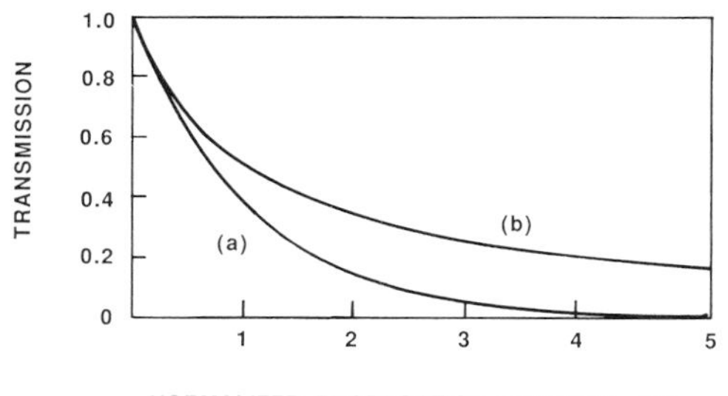

FIG. 32. Dependence of transmitted intensity with distance for (a) linear absorption and (b) two-photon absorption. For (a) $z_0 = \alpha^{-1}$, while for (b) $z_0 = (\beta I_0)^{-1}$.

5. Saturable Absorption

Saturable absorption involves a decrease in absorption at high optical intensities. It can occur in atomic or molecular gases and in various types of liquids and solids, and it is usually observed in materials that are strongly absorbing at low light intensities. Saturable absorption occurs when the upper state of the absorbing transition gains enough population to become filled, preventing the transfer of any more population into it. It generally occurs in materials that have a restricted density of states for the upper level, such as atomic or molecular gases, direct bandgap semiconductors, and certain liquids, such as organic dyes. Saturable absorption in dyes that have relaxation times of the order of several picoseconds have been used to mode-lock solid-state and pulsed-dye lasers, producing optical pulses on the order of several tens of picoseconds or less. These dyes have also been used outside of laser cavities to shorten the duration of laser pulses. Saturable absorbers with longer relaxation times have been used to mode-lock cw dye lasers, producing subpicosecond pulses. Saturable absorption can also be used with four-wave mixing interactions to produce optical phase conjugation. Saturation of the gain in laser amplifiers is similar to saturable absorption, but with a change in sign. It is described by the same equations and determines the amount of energy that can be produced by a particular laser. In a linear laser cavity, the gain can be reduced in a narrow spectral region at the center of a Doppler-broadened line, which forms the basis of a spectroscopic technique known as Lamb dip spectroscopy that has a resolution less than the Doppler width.

C. Coherent Optical Effects

Coherent nonlinear effects involve interactions that occur before the wave functions that describe the excitations of the medium have time to relax or dephase. They occur primarily when the nonlinear interaction involves one- or two-photon resonances, and the duration of the laser pulse is shorter than the dephasing time of the excited state wave functions, a time that is equivalent to the inverse of the linewidth of the appropriate transition. Coherent nonlinear optical interactions generally involve significant population transfer between the states of the medium involved in the resonance. As a result, the nonlinear polarization cannot be described by the simple perturbation expansion given in Eq. (2), which assumed that the population was in the ground state. Rather, it must be solved for as a dynamic variable along with the optical fields.

Virtually any of the nonlinear effects that were described earlier can occur in the regime of coherent interactions. In this regime the saturation of the medium associated with the population transfer generally weakens the nonlinear process involved relative to the strength it would have in the absence of the coherent process.

Coherent interactions can also give rise to new nonlinear optical effects. These are listed in Table XV, along with some of their characteristics and the conditions under which they are likely to occur.

Self-induced transparency is a coherent effect in which a material that is otherwise absorbing becomes transparent to a properly shaped laser pulse. It occurs when laser pulses that are shorter than the dephasing time of the excited-state wave functions propagate through materials with absorbing transitions that are inhomogeneously broadened. In self-induced transparency, the energy at the beginning of the laser pulse is absorbed and is subsequently reemitted at the end of the laser pulse, reproducing the original pulse with a time delay. In order for all the energy that is absorbed from the beginning of the pulse to be reemitted at the end, the pulse field amplitude must have the special temporal profile of a hyperbolic secant. In addition, the pulse must have the correct "area," which is proportional to the product of the transition dipole moment and the time integral of the field amplitude over the pulse duration. The pulse area is represented as an angle that is equivalent to the rotation angle from its initial position of a vector that describes the state of the atom or molecule. In self-induced transparency, the required pulse area is 2π, corresponding to one full rotation of the state vector, indicating that the medium starts and ends in the ground state. If the incident pulse has an area greater than 2π it will break up into multiple pulses, each with area 2π, and any excess energy will eventually be dissipated in the medium. If the initial pulse has an area less than 2π it will eventually be absorbed in the medium.

TABLE XV. Coherent Nonlinear Interactions

Interaction	Conditions for observation
Self-induced transparency	Resonant interaction with inhomogeneously broadened transition; pulse duration less than dephasing time
Photon echoes	Resonant interaction with inhomogeneously broadened transition; pulse duration less than dephasing time; two pulses spaced by echo time τ, with pulse areas of $\pi/2$ and π, respectively
Adiabatic following	Near-resonant interaction; pulse duration less than dephasing time
Adiabatic rapid passage	Near-resonant interaction; pulse duration less than dephasing time; frequency of pulse swept through resonance

Self-induced transparency is different from ordinary saturated absorption. In saturated absorption, the energy that is taken from the pulse to maintain the medium in a partial state of excitation is permanently lost to the radiation field. In self-induced transparency, the energy given to the medium is lost from the radiation field only temporarily and is eventually returned to it.

Photon echoes also occur in materials with inhomogeneously broadened transitions. In producing a photon echo, two pulses are used with a spacing of τ, as shown in Fig. 33. The first pulse has an area of $\pi/2$ and produces an excitation in the medium. The second pulse has an area of π and reverses the phase of the excited-state wave functions after they have had time to dephase. Instead of continuing to get further out of phase as time progresses, the wave functions come back into phase. At a time τ after the second pulse, the wave functions are again all in phase and a third pulse, termed the echo, is produced. Photon echoes are observed most easily when the pulsed spacing τ is larger than the inhomogenous dephasing time caused, for example, by Doppler broadening, but smaller than the

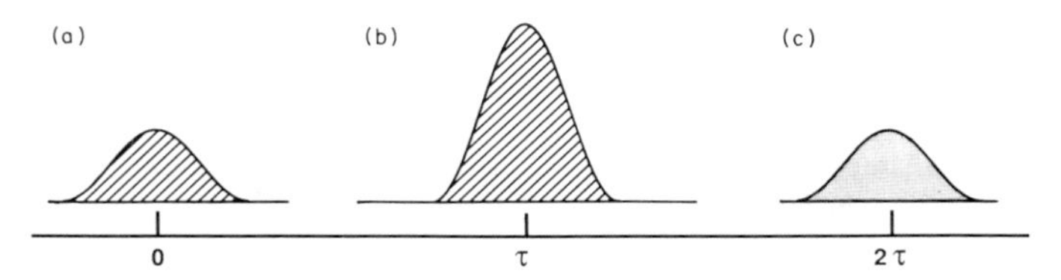

FIG. 33. Arrangement of pulses for generating photon echoes: (a) with an area of $\pi/2$, and (b) with an area of π, are supplied spaced by time τ; and (c) the echo, is generated in the interaction at a time τ after the second pulse.

homogenous dephasing time caused by collisions or population decay.

Other coherent interactions include optical nutation and free induction decay, in which the population oscillates between two levels of the medium, producing oscillations in the optical fields that are radiated, and adiabatic rapid passage, in which a population inversion can be produced between two levels in a medium by sweeping the frequency of a pulse through the absorption frequency in a time short compared to the dephasing time of the upper level.

D. Electrooptic and Magnetooptic Effects

Electrooptic and magnetooptic effects involve changes in the refractive index of a medium caused by an external electric or magnetic field. These are not normally thought of as nonlinear optical effects but are technically nonlinear optical processes in which the frequency of one of the fields is equal to zero. Various electrooptic and magnetooptic effects can occur depending on the situation. Some of these were listed in Table I.

In the Pockels effect, the change in the refractive index is proportional to the external electric field, whereas in the quadratic Kerr effect the change in refractive index is proportional to the square of the electric field. The Pockels effect occurs in solids without inversion centers, the same types that allow second-order nonlinear effects. The quadratic Kerr effect occurs in materials with any symmetry and is commonly used in liquids with anisotropic molecules such as nitrobenzene. The Faraday and Cotton–Mouton effects produce changes in the refractive index that are proportional to the magnetic field strength. Electrooptic and magnetooptic effects generally cause different changes in the refractive indexes in different directions relative to the applied field or the crystal axes, resulting in field-induced birefringence.

Electrooptic and magnetooptic effects are commonly used in light modulators and shutters. The field-dependent changes in the refractive index can be used directly for phase or frequency modulation. This is most commonly done with the Pockels effect, and units with modulation frequencies of up to the order of 1 GHz are commercially available. Field-induced birefringence is used to change the polarization state of a light beam and can be used for both modulation and shuttering. A light shutter can be constructed with either a Pockels cell or a Kerr cell by adjusting the field so that the polarization of the light wave changes by 90°. These are commonly used for producing laser pulses with controlled durations or shapes and for Q-switching of pulsed lasers. The Faraday effect produces a birefringence for circular polarization, resulting in the rotation of the direction of polarization of linearly polarized light. When adjusted for 45° and combined with linear polarizers, it will pass light in only one direction. It is commonly used for isolation of lasers from reflections from optical elements in the beam.

IV. Applications

In the previous sections the basic nonlinear optical interactions have been described, along with some of their properties. In the following sections we shall describe applications of the various nonlinear interactions. Some applications have already been noted in the description given for certain effects. Here we shall describe applications that can be made with a wide variety of interactions.

A. Nonlinear Spectroscopy

Nonlinear spectroscopy involves the use of a nonlinear optical interaction for spectroscopic studies. It makes use of the frequency variation of the nonlinear susceptibility to obtain information about a material in much the same way that variation of the linear susceptibility with frequency provides information in linear spectroscopy. In nonlinear spectroscopy the spectroscopic information is obtained directly from the nonlinear interaction as the frequency of the pump radiation is varied. This can be contrasted with linear spectroscopy that is done with radiation that is generated in a nonlinear interaction and used separately for spectroscopic studies.

Nonlinear spectroscopy can provide information different from that available in linear spectroscopy. For example, it can be used to probe transitions that are forbidden in single-photon interactions and to measure the kinetics of excited states. Nonlinear spectroscopy can allow measurements to be made in a spectral region in which radiation needed for linear spectroscopy would be absorbed or perhaps would not be available. It can also provide increased signal levels or spectral resolution.

Many different types of nonlinear effects can be used for nonlinear spectroscopy, including various forms of parametric frequency conversion (harmonic generation, four-wave sum- and difference-frequency mixing), degenerate four-wave mixing, multiphoton absorption, multiphoton ionization, and stimulated scattering. Some of the effects that have been used for nonlinear spectroscopy are given in Table XVI, along with the information that is provided and the quantities that are varied and detected.

An example of improved spectral resolution obtained through nonlinear spectroscopy is the Doppler-free spectroscopy that can be done with two-photon absorption, as illustrated in Figs. 34 and 35. In this interaction, two light waves with the same frequency are incident on the sample from opposite directions. As the frequency of the incident light is swept through one-half of the transition frequency of a two-photon transition, the light is absorbed and the presence of the absorption is detected through fluorescence from the upper state to a lower one through an allowed transition. Normally, the spectral resolution of absorption measurements is limited by the Doppler broadening caused by the random motion of the atoms. Atoms that move in different directions with different speeds absorb light at slightly different frequencies, and the net result is an absorption profile that is wider than the natural width of the transition. In the nonlinear measurement, however, the atom absorbs one photon from each beam coming from opposite directions. A moving atom sees the frequency of one beam shifted in one direction by the same amount as the frequency of the other beam is shifted in the opposite direction. As a result, each atom sees the same sum of the frequencies of the two beams regardless of its speed or direction of motion, and the absorption profile is not broadened by the Doppler effect. This type of spectroscopy can be used to measure the natural width of absorption lines underneath a much wider Doppler profile. An example of a spectrum obtained with

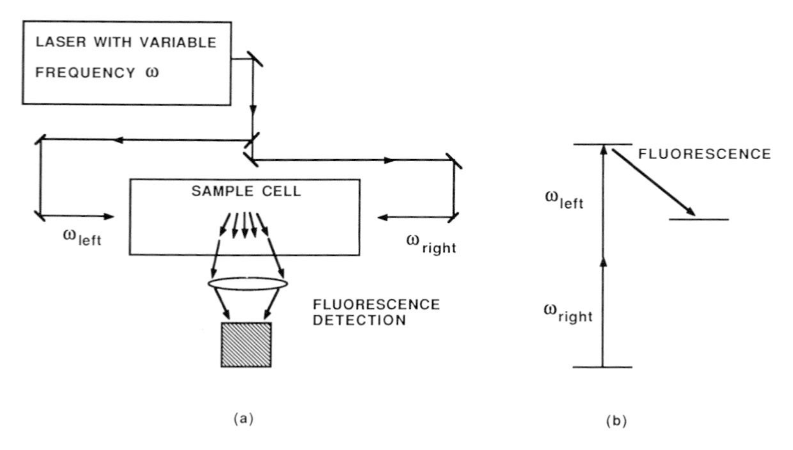

FIG. 34. The use of two-photon absorption with counterpropagating beams for Doppler-free spectroscopy. (a) Experimental arrangement. (b) Level diagram of transitions used.

TABLE XVI. Application of Nonlinear Optics to Spectroscopy[a]

Nonlinear interaction	Quantity varied	Quantity measured	Information obtained
Multiphoton absorption			
Atom + $2\omega_{\text{laser}} \rightarrow$ atom*[b]	ω_{laser}	Fluorescence from excited level	Energy levels of states with same parity as ground state; sub-Doppler spectral resolution
Multiphoton ionization			
Atom + $2\omega_{\text{laser}} + \mathscr{E}^a \rightarrow$ atom$^+$	ω_{laser}	Ionization current	Rydberg energy levels
Atom (molecule) + $(\omega_1 + \omega_2) \rightarrow$ atom$^+$ (molecule$^+$)[c]	ω_2	Ionization current	Even-parity autoionizing levels
Four-wave mixing			
Sum-frequency mixing			
$\omega_4 = 2\omega_1 + \omega_2$	ω_2	Optical power at ω_4	Energy structure of levels near ω_4, e.g., autoionizing levels, matrix elements, line shapes
Third-harmonic generation			
$\omega_3 = 3\omega_1$	ω_1	Optical power at ω_3	Energy levels near $2\omega_1$ with same parity as ground state
Difference-frequency mixing			
$\omega_4 = 2\omega_1 - \omega_2$ (CARS)	ω_2	Optical power at ω_4	Raman energy levels Solids Polaritons Lattice vibrations Gases Measurement of nonlinear susceptibilities Solids Liquids Concentrations of liquids in solutions Concentrations of gases in mixtures Temperature measurements in gases Measurements in flames, combustion diagnostics Interference of nonlinear susceptibilities Time-resolved measurements Lifetimes, dephasing times CARS background suppression
Four-frequency CARS			
$\omega_4 = \omega_1 + \omega_2 - \omega_3$	$\omega_1 - \omega_3$ $\omega_2 - \omega_3$	Optical power at ω_4	Same as CARS, CARS background suppression
Raman-induced Kerr effect			
$\omega_{2,y} = \omega_1 - \omega_1 + \omega_{2,x}$	ω_2	Polarization changes at ω_2	Same as CARS, CARS background suppression
Higher-order Raman processes			
$\omega_4 = 3\omega_1 - 2\omega_2$	ω_2	Optical power at ω_4	Same as CARS
Coherent Stokes–Raman spectroscopy			
$\omega_4 = 2\omega_1 - \omega_2 (\omega_1 < \omega_2)$	ω_2	Optical power at ω_4	Same as CARS
Coherent Stokes scattering			
$\omega_S = \omega_1(t) - \omega_1(t) + \omega_S(t) + \omega_1(t + \Delta t) - \omega_1(t + \Delta t)$	Δt	Optical power at $\omega_S(t + \Delta t)$	Lifetimes, dephasing times, resonant contributions to $\chi^{(3)}$
Raman gain spectroscopy			
$\omega_S = \omega_L - \omega_L + \omega_S$	ω_S	Gain or loss at ω_S	Raman energy levels
Saturation spectroscopy	ω_{laser}	Induced gain or loss	High-resolution spectra

[a] From J. F. Reintjes (1984). "Nonlinear Parametric Processes in Liquids and Gases," pp. 422–423. Academic Press, New York.
[b] An atom or molecule in an excited state is designated by *.
[c] An ionized atom or molecule is designated by $^+$ and $\mathscr{E}^a$ designates ionizing energy supplied by an external electric field.

two-photon Doppler-free spectroscopy is shown in Fig. 35.

Nonlinear spectroscopy can also be used to measure the frequency of states that have the same parity as the ground state and therefore are not accessible through linear spectroscopy. For example, S and D levels in atoms can be probed as two-photon transitions in multiphoton absorption or four-wave mixing spectroscopy. Nonlinear spectroscopy can also be used for spectroscopic studies of energy levels that are in regions of the spectrum in which radiation for

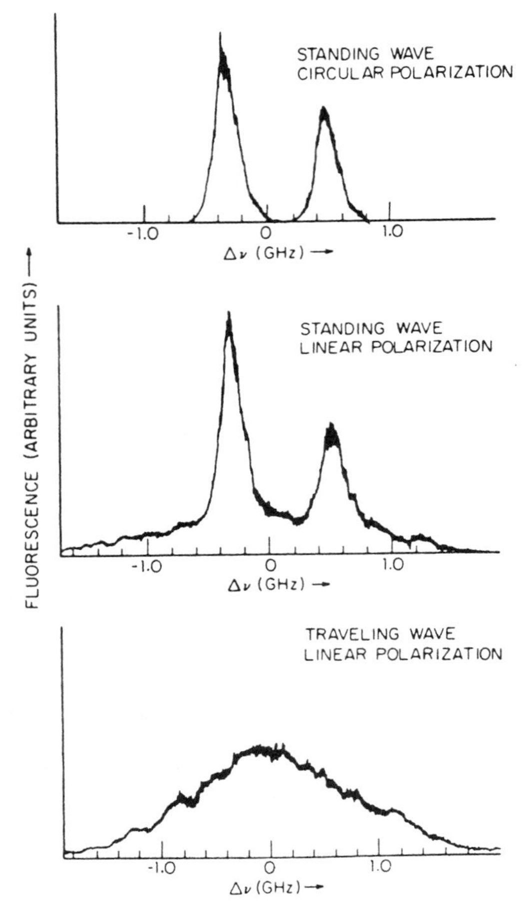

FIG. 35. Example of high-resolution spectrum in sodium vapor obtained with two-photon Doppler-free spectroscopy. [Reproduced from Bloembergen, N., and Levenson, M. D. (1976). Doppler-free two-photon absorption spectroscopy. *In* "High Resolution Laser Spectroscopy" (K. Shimoda, ed.), p. 355. Springer, New York.]

linear spectroscopy either is absorbed or is not available. Examples of such applications are spectroscopy of highly excited levels in the vacuum ultraviolet or extreme ultraviolet or of levels in the far infrared. Nonlinear effects can also be used for spectroscopic studies of excited levels, providing information on their coupling strength to other excited states or to the continuum. Time-resolved measurements can also be made to determine the kinetics of excited states such as lifetimes, dephasing times, and the energy-decay paths.

One of the most extensively used nonlinear processes for spectroscopy is coherent anti-Stokes Raman scattering (CARS). This is a four-wave mixing process of the form $\omega_{AS} = 2\omega_L - \omega_S$, where ω_L and ω_S are the laser and Stokes frequencies that are provided in the incident radiation and ω_{AS} is the anti-Stokes frequency that is generated in the interaction. The process is resonantly enhanced when the difference frequency $\omega_L - \omega_S$ is tuned near a Raman active mode with frequency $\omega_0 = \omega_L - \omega_S$, and the resulting structure in the spectrum provides the desired spectroscopic information. The Raman-active modes can involve transitions between vibrational or rotational levels in molecules or lattice vibrations in solids. Effective use of the CARS technique requires that phase-matching conditions be met. This is usually done by angle phase matching, as was shown in Fig. 24a. Coherent anti-Stokes Raman scattering offers advantages over spontaneous Raman scattering of increased signal levels and signal-to-noise ratios in the spectroscopy of pure materials, because the intensity of the generated radiation depends on the product $(NL)^2$, where N is the density and L is the length of the interaction region as compared with the (NL) dependence of the radiation generated in spontaneous Raman scattering. As a result, spectra can be obtained more quickly and with higher resolution with CARS. Coherent anti-Stokes Raman scattering has been used for measurements of Raman spectra in molecular gases, Raman spectra in flames for combustion diagnostics and temperature measurements, and in spatially resolved measurements for molecular selective microscopy, an example of which is shown in Fig. 36.

Although CARS offers the advantage of increased signal levels in pure materials for strong Raman resonances, it suffers from the presence of a nonresonant background generated by all the other levels in the medium. In order to overcome these limitations, several alternative techniques have been developed involving interactions, such as the Raman induced Kerr effect, in which the sample develops birefringence, causing changes in the polarization of the incident waves as their frequency difference is tuned through a resonance, and four-frequency CARS, in which the variation of the CARS signal near a resonance with one set of frequencies is used to offset the background in the other set.

B. Infrared Up-Conversion

Three-wave and four-wave sum-frequency mixing has been used for conversion of radiation from the infrared to the visible, where photographic film can be used and photoelectric de-

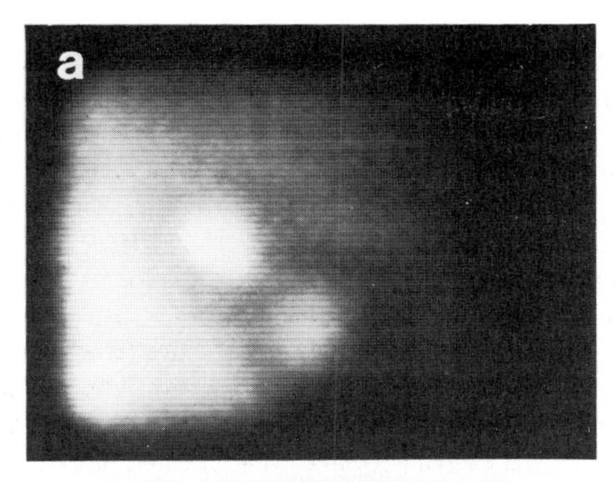

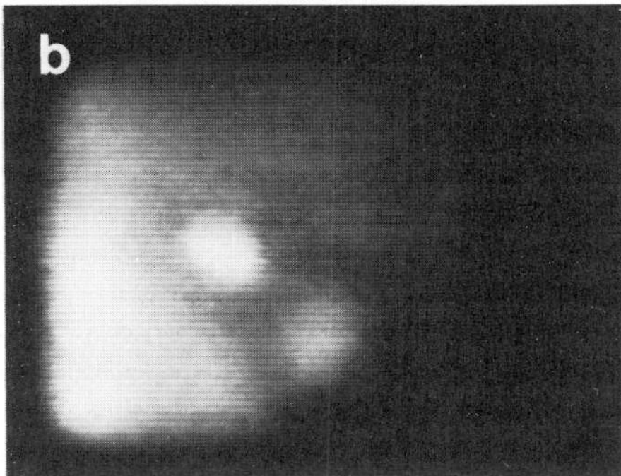

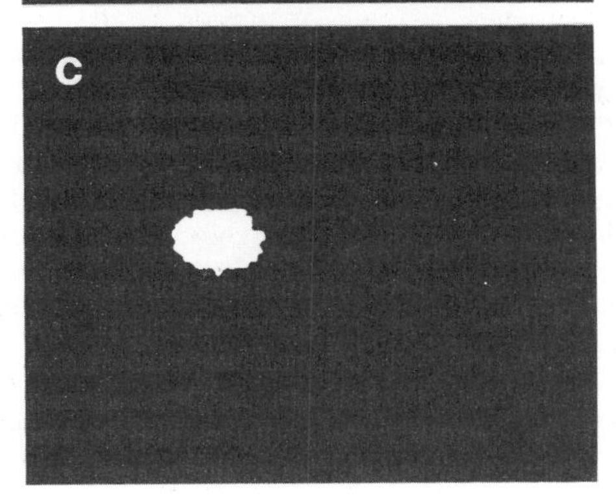

FIG. 36. Molecular selective microscopy using coherent anti-Stokes Raman scattering. (A) On-resonant picture of deuterated and nondeuterated liposomes with two liposomes visible are shown. Only the deuterated liposomes have a Raman resonance for the radiation used. (B) The same picture as that in (A), but the two pump waves have been detuned from the Raman resonance. (C) The nonresonant signal has been subtracted from the resonant one, leaving only the deuterated liposome visible. [Reproduced from M. D. Duncan (1984). *Opt. Comm.* **50,** 307. Copyright © North-Holland, Amsterdam.]

tectors offer the advantages of improved sensitivity and increased temporal resolution. Infrared up-conversion has been used for up-conversion of infrared images to the visible and for time-resolved studies of the rotational spectra of molecules formed in explosions, giving information as to the time dependence of the temperature.

C. Optical Phase Conjugation

Optical phase conjugation, also referred to as time reversal or wavefront reversal, is a technique involving the creation of an optical beam that has the variations in its wavefront, or phase, reversed relative to a reference beam. If the optical field is represented as the product of an amplitude and complex exponential phase,

$$E = Ae^{i\phi} \tag{56}$$

then the process of reversing the sign of the phase is equivalent to forming the complex conjugate of the original field, an identification that gives rise to the name phase conjugation. When optical phase conjugation is combined with a reversal of the propagation direction, it allows for compensation of distortions on an optical beam, which develop as a result of propagation through distorting media, for example, the atmosphere, or imperfect or low-quality optical components such as mirrors, lenses, or windows. Such distortions are familiar to people who have looked through old window glass, through the air above a hot radiator, or in the apparent reflections present on the highway on a hot day. Optical phase conjugation can also be used for holographic imaging and can allow images to be transmitted through multimode fibers without degradation due to the difference in phase velocity among the various modes. It can also be used in various forms of optical signal processing such as correlators and for spectroscopy. [*See* OPTICAL PHASE CONJUGATION.]

The concept of correction of distortions by optical phase conjugation is as follows. When a wave propagates through a distorting medium, its phase contour acquires structure that will eventually diffract, leading to increased spreading, reduced propagation length, reduced focal-spot intensity, and image distortion. The basic idea of compensation of distortions is to prepare at the entrance of the medium a beam whose wave front is distorted in such a way that the distortion introduced by the medium cancels the one that would develop on the beam, resulting in

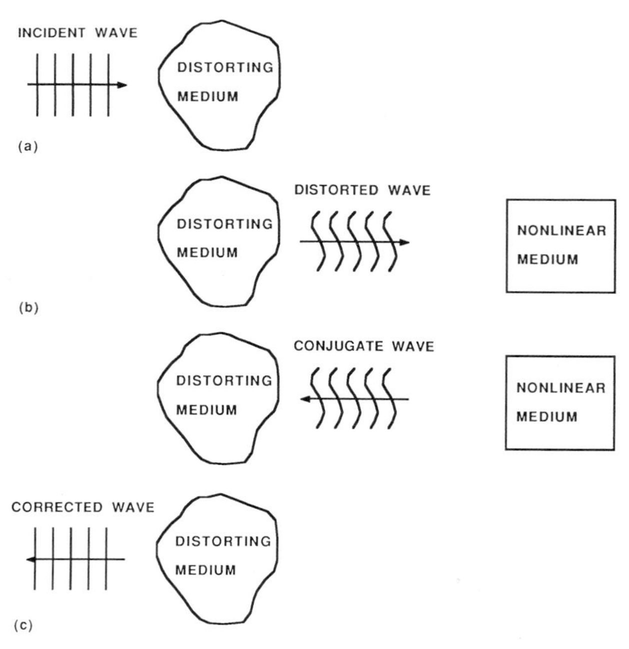

FIG. 37. The use of optical phase conjugation for compensation of distortions. (a) An initially smooth beam propagates from right to left through a distorting medium. (b) After emerging from the distorting medium, the acquired phase variations are conjugated and the direction of propagation is reversed in a suitable nonlinear interaction. (c) After the second pass through the nonlinear medium, the beam emerges with a smooth wavefront.

an undistorted wave front at the exit of the medium.

The wave front required for this compensation to occur is the conjugate of the wave front obtained by propagation of an initially plane wave through the medium. It is usually obtained by propagating a beam through the distorting medium and then into a nonlinear medium that produces a phase-conjugate beam that propagates in the reverse direction, as illustrated in Fig. 37b. Various nonlinear interactions can be used for optical phase conjugation, including degenerate four-wave mixing in transparent, absorbing, and amplifying media and various forms of stimulated scattering such as Raman, Brillouin, and Rayleigh and stimulated emission. The two most widely used techniques are degenerate four-wave mixing and stimulated Brillouin scattering.

Degenerate four-wave mixing configured for phase conjugation is illustrated in Fig. 38. Here two strong waves, A_1 and A_2, are incident on the

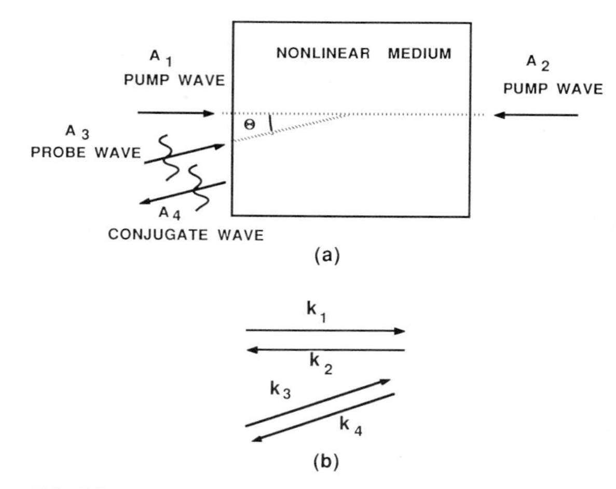

FIG. 38. (a) Configuration for the use of degenerate four-wave mixing for phase conjugation. The frequencies of all waves are equal; the two pump waves, A_1 and A_2, propagate in opposite directions; and the angle θ between the probe wave and the forward pump wave is arbitrary. (b) The k vector diagram is shown.

nonlinear medium from opposite directions. The wave front carrying the information to be conjugated is the probe wave A_3 and the conjugate wave A_4 that is generated in the interaction $\omega_4 = \omega_1 + \omega_2 - \omega_3$ propagates in the direction opposite to A_3. Because the waves are present in counterpropagating pairs and all the waves have the same frequency, the process is automatically phase matched regardless of the angle between the probe wave and the pump waves. Three-wave difference-frequency mixing of the form $\omega_3 = \omega_1 - \omega_2$ can also be used in crystals for phase conjugation, but the phase-matching requirements in nonlinear crystals restrict the angles that can be used, and hence the magnitude of the distortions that can be corrected.

The generation of conjugate waves in nonlinear interactions involves the creation of volume holograms, or diffraction gratings, in the medium through interference of the incident waves. The interference that occurs, for example, between the waves A_1 and A_3 in degenerate four-wave mixing is illustrated in Fig. 39a. The backward wave is created by the scattering of the pump wave A_2 off the diffraction grating created by the interference of waves A_1 and A_3. The information as to the distortions on the incoming wave appears as bending of the contours of the grating. It is transferred to the phase contour of the backward wave as it is created but with its sense reversed. A similar interference occurs between all pairs of the incident waves, and each interference pattern creates a contribution to the backward-propagating phase-conjugate wave.

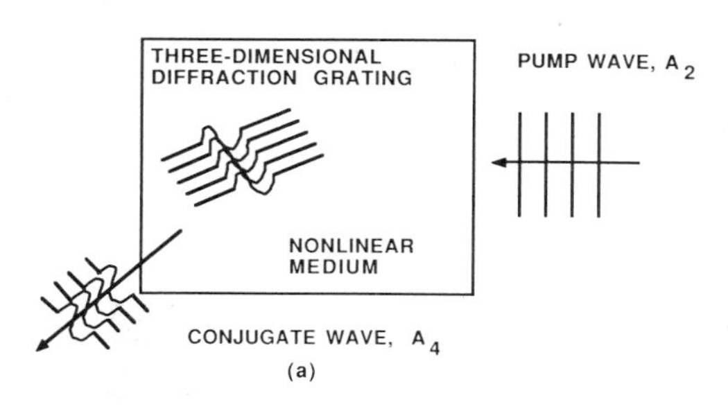

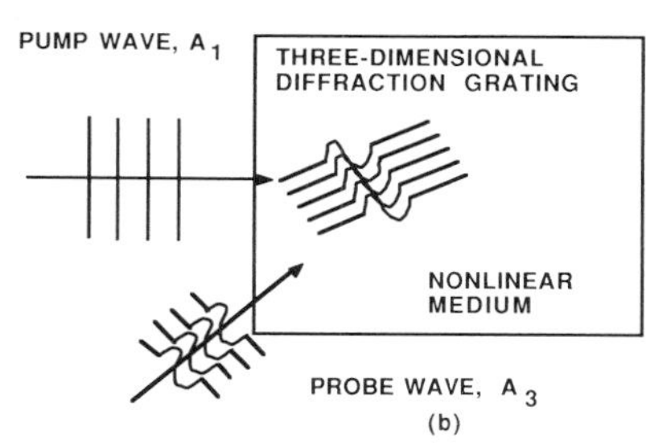

FIG. 39. Illustration of the formation of phase conjugate waves by scattering from the interference patterns formed by pairs of the incident waves. (a) Interference between the probe wave and the forward pump wave forming a contoured grating in the nonlinear medium. (b) Scattering of the backward pump wave from the contoured grating to form the phase-conjugate wave.

In stimulated Brillouin scattering, the incident wave serves as both the pump wave for the nonlinear process and the distorted wave to be conjugated. When the incident wave is highly aberrated, different components of the pump wave interfere in the focus, producing regions of relatively high and low gain. The wave that is ultimately generated in the Brillouin process is the one with the highest average gain. This wave in turn is one that is the phase conjugate of the incident wave, because it is has the most constructive interference in the focus. The interference of the various components of the pump beam in the focus can also be viewed as the creation of diffraction gratings or holograms in a manner similar to that described for degenerate four-wave mixing. The phase-conjugate wave is then produced by scattering of a backward-traveling wave from these diffraction gratings.

Optical phase conjugation has been used for reconstruction of images when viewed through distorting glass windows or other low-quality optical components, for removal of distortions from laser beams, for holographic image reconstruction, and for image up-conversion. An example of image reconstruction using optical phase conjugation is shown in Fig. 40.

D. Optical Bistability

Nonlinear optical effects can also be used to produce bistable devices that are similar in operation to bistable electric circuits. They have potential use in optical memories and the control of optical beams and can perform many of the functions of transistors such as differential gain, limiting, and switching. Bistable optical elements all have a combination of a nonlinear optical component and some form of feedback. A typical all-optical device consisting of a Fabry–Perot (FP) cavity that is filled with a nonlinear medium is illustrated in Fig. 41. The FP cavity has the property that when the medium between its plates is transparent, its transmissivity is high at those optical wavelengths for which an integral number of wavelengths can be contained

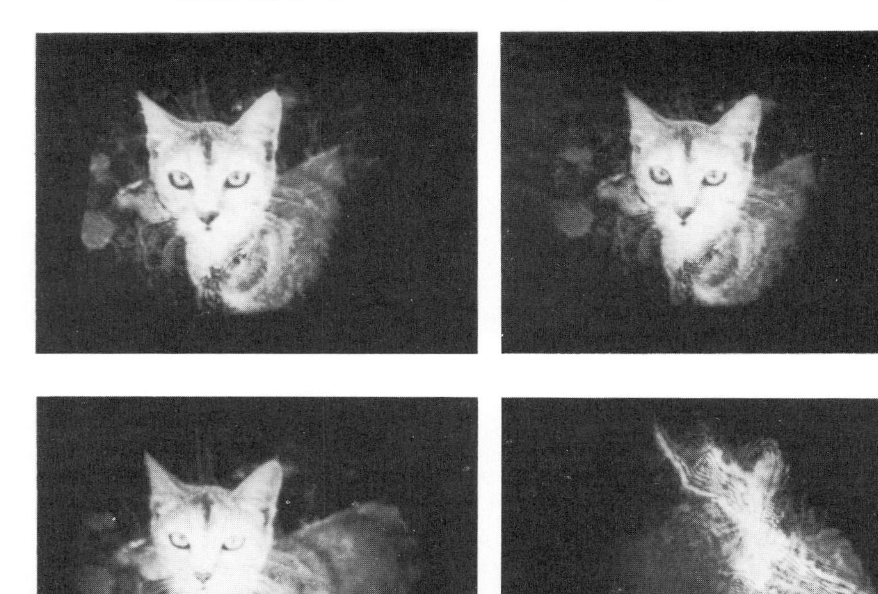

FIG. 40. Example of image reconstruction with phase conjugation. The unaberrated image is shown at the lower left using a plane mirror and at the upper left using a conjugate mirror. The image at the lower right shows the effect of an aberrator (a distorting piece of glass) on the image obtained with a normal mirror, while the image at the upper right shows the corrected image obtained with the aberrator and the conjugate mirror. [From J. Feinberg (1982). *Opt. Lett.* **7,** 488. Copyright © 1982 by the Optical Society of America.]

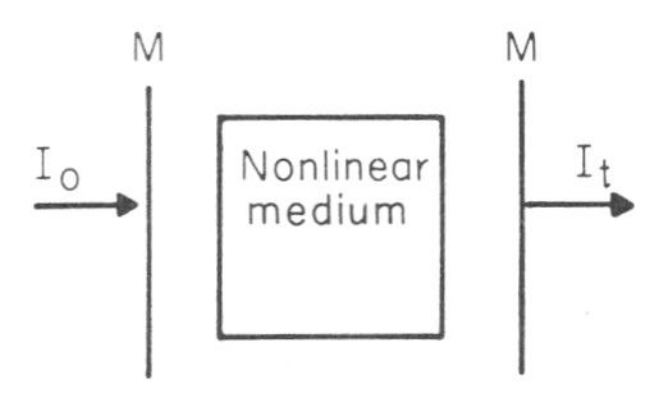

FIG. 41. Illustration of nonlinear Fabry–Perot cavity for optical bistability. The end mirrors are plane and parallel, and the medium in the center exhibits either a nonlinear refractive index or saturable absorption.

between the plates, and its reflection is high for other wavelengths.

The operation of a nonlinear FP cavity can be illustrated by assuming that the nonlinear medium has a refractive index that is a function of the intensity inside the FP cavity. The wavelength of the incident light is chosen to be off resonance so that at low incident intensities the reflectivity is high, the transmission is low, and not much light gets into the FP cavity. As the incident intensity is increased, more light gets

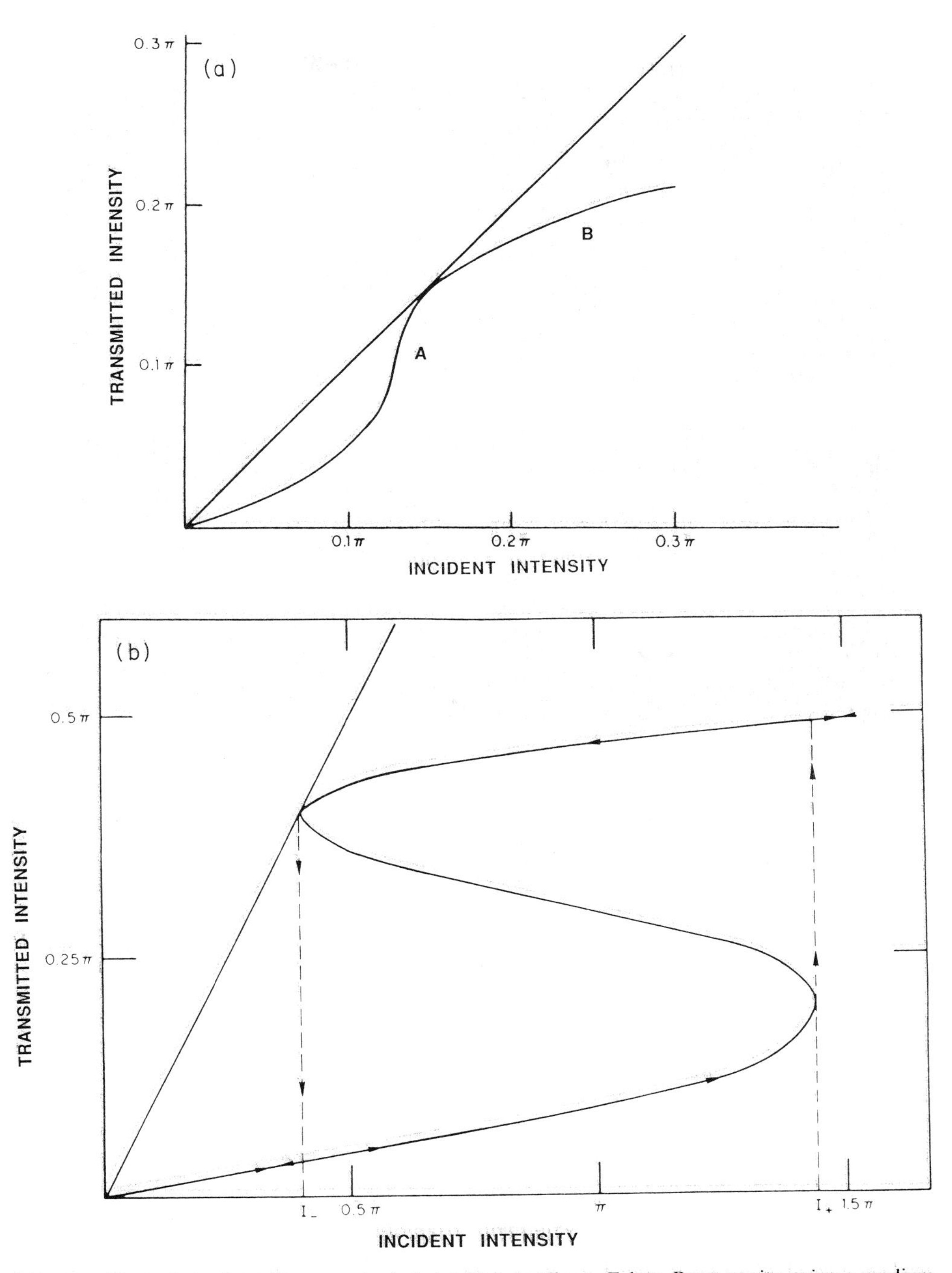

FIG. 42. Illustration of nonlinear transmission with a nonlinear Fabry–Perot cavity using a medium with a nonlinear refractive index. (a) The transmission curve does not show hysteresis and can be used for differential gain (in region A) or optical limiting (in region B). (b) The transmission shows hysteresis and can be used for optical bistability. The difference between the differential gain and bistable operation is determined by the initial detuning from the resonance. [Reproduced from J. Reintjes (1984). "Nonlinear Optical Parametric Processes in Liquids and Gases." Academic Press, Orlando, Florida.]

into the FP cavity, changing the refractive index of the medium, and bringing the cavity closer to resonance. As a result, the fractional transmission increases and the intensity of the light transmitted through the cavity increases faster than does the intensity of the incident light. When the incident intensity is sufficiently high the cavity is brought into exact resonance, allowing full transmission. For further increases in the incident intensity, the cavity moves beyond the resonance and the fractional transmission decreases.

For conditions in which the transmitted intensity increases faster than the incident intensity, the nonlinear FP cavity can be used for differential gain in a manner similar to transistor amplifiers. An example of the transmission curve under these conditions is shown in Fig. 42a. In this situation, a small modulation on the incident light wave can be converted to a larger modulation on the transmitted wave. Alternatively, the modulation can be introduced on a different wave and can be transferred to the forward wave with a larger value. If the incident intensity is held just below the value at which the transmission increases and the cavity is illuminated by another, weaker wave that changes the transmission level, the nonlinear FP cavity can act as a switch. In the range of intensities for which an increase in intensity moves the FP cavity away from resonance, the nonlinear FP acts as an optical limiter. For certain conditions, the level of light inside the cavity is sufficient to maintain the resonance condition once it has been achieved, resulting in hysteresis of the transmission curve with the incident intensity, and providing the conditions for bistability. An example of the transmission curve under this condition is shown in Fig. 42b.

A medium with saturable absorption can also be used for the nonlinear medium. In this case, the resonance condition is always met as far as the wavelength is concerned, but the feedback from the second mirror, which is required for resonant transmission, is not obtained until sufficient light is present inside the cavity to bleach the nonlinear medium. Devices of this type can show differential gain and optical bistability but not optical limiting.

Bistable optical devices can also be constructed in such a way that the transmitted light signal is converted to an electrical voltage that is used to control the refractive index of the material inside the cavity. These hybrid optical–electrical devices can operate with very low levels of light intensity and as such are especially compatible with integrated optical components.

Optical bistability has also been observed in a wide range of optical configurations, such as laser cavities, degenerate four-wave mixing, and second-harmonmic generation. Under some conditions, bistable devices can exhibit chaotic fluctuations in which the light output oscillates rapidly between two extremes.

Bibliography

Bowden, C. M., and Haus, J. (feature eds.) (1989). Nonlinear optical properties of materials. *J. Optical Society of America B, Optical Physics* **6**(4), 562–853.

Fisher, R. (ed.) (1983). "Optical Phase Conjugation." Academic Press, New York.

Kobayashi, T. (1989). "Nonlinear Optics of Organics and Semiconductors," Springer Proceedings in Physics Series, Vol. 36. Springer-Verlag, Berlin and New York.

Levenson, M. D. (1982). "Introduction to Nonlinear Spectroscopy." Academic Press, New York.

Pepper, D. M. (1986). Applications of optical phase conjugation. *Sci. Am.* **254,** (January) (6), 74.

Reintjes, J. (1984). "Nonlinear Optical Parametric Processes in Liquids and Gases." Academic, New York.

Reintjes, J. (1985). Coherent ultraviolet sources. *In* "Laser Handbook" (M. Bass and M. L. Stitch, eds.), Vol. 5, pp 1–202. North-Holland Publ., Amsterdam.

Shen, Y. R. (1984). "The Principles of Nonlinear Optics." Wiley, New York.

Shkunov, V. V., and Zeldovich, B. Y. (1985). Optical phase conjugation. *Sci. Am.* **253,** (December) (6), 54.

NUCLEAR PUMPED LASERS

David A. McArthur *Sandia National Laboratories*

Glossary

Fast burst reactor (FBR): Produces neutron pulses about 0.1–1 msec in width, by rapid mechanical assembly of the critical mass in a neutron-free environment.

Nuclear device: Nuclear explosive which produces a single short pulse of intense gamma and fast-neutron radiation.

Nuclear device pumped laser (NDPL): Laser pumped by the very high radiation fluxes obtainable near an exploding nuclear device.

Range: Mass per unit area of a particular material which is sufficient to absorb all the kinetic energy of a charged particle having a well-defined initial energy (usually expressed in milligrams per square centimeter).

Reactor pumped laser (RPL): Laser pumped by a portion of the leakage neutrons or gamma rays produced by a pulsed or steady-state nuclear reactor.

Substrate: Mechanical support for a thin coating of fissile material in direct contact with the laser medium or nuclear flashlamp medium.

TRIGA: Zirconium-hydride-moderated research reactor with inherently safe operating characteristics which can be operated in a variety of modes (pulsed, steady-state, or multiple pulses).

Volumetric pump rate: The rate of energy deposition per unit volume of laser medium

This work was supported by the U.S. Department of Energy under contract DE-AC04-76PD00789.

Nuclear pumped lasers (NPLs) are optical lasers operating at wavelengths ranging from the ultraviolet through the infrared which are excited directly or indirectly by high-energy charged particles resulting from nuclear reactions. This definition includes optically pumped laser media in which the pumping light source is produced by nuclear reactions, but it excludes lasers in which nuclear radiation is used merely to pre-ionize the laser medium to control a discharge which provides the primary excitation.

Ideas for NPLs were proposed shortly after the invention of the laser. Two broad categories of NPLs have been demonstrated, those pumped by nuclear explosives (NDPL) and those pumped by specialized (usually pulsed) research reactors (RPLs). Table I summarizes basic data on representative NPLs.

I. Examples of Nuclear Pumped Lasers

A. Nuclear Device Pumped Lasers

In 1974 two independent experiments in pumping of laser media by nuclear explosives were reported briefly. Lawrence Livermore Laboratory reported measurements of optical gain and directionally enhanced light emission in high-pressure Xe gas, and Los Alamos Laboratory simultaneously reported a chemical HF laser initiated by radiation. These were the first NPLs reported and demonstrated that nuclear radiation could produce lasing action.

In these studies the laser media were pumped at extremely high volumetric pump rates with an intense short pulse of gamma rays, resulting in laser pulses of about 10 nsec width. The laser cells themselves were small but efficiencies were reasonable. Great care was exercised to minimize radiation damage to the laser optics

TABLE I. Representative Nuclear Pumped Lasers

Laser	Wavelength (10^{-6} m)	Laboratory[a]	Date	Pump source[b]	Target nucleus	Volumetric pump rate (MW/m^3)	Laser energy[b] (J)	Laser efficiency[b] (%)
Xe_2^*	0.17	LLNL	1974	NDPL	gamma	1.5×10^9	N/A	N/A
HF	2.7	LANL	1974	NDPL	gamma	8.0×10^8	2.9	6.5
CO	5.4	Sandia	1975	FBR	^{235}U	1000	0.02	0.5
Xe	3.5	LANL	1975	FBR	^{235}U	200	10^{-6}	10^{-4}
N	0.86, 0.94	UI	1976	TRIGA	^{10}B	3.3	10^{-5}	10^{-4}
Hg^+	0.615	UI	1977	FBR	^{10}B	300	2×10^{-7}	10^{-6}
Xe	2.65	NASA	1980	FBR	$^{235}U/UF_6$	25	10^{-3}	0.05
Ar	1.79	NASA	1981	FBR	^3He	2100	0.12	0.02
Xe	1.73	USSR	1982	N/A	^{235}U	10	2–3	2–5
Cd^+	0.4416	USSR	1982	N/A	^3He	1.9	5×10^{-6}	0.007

[a] LLNL, Lawrence Livermore National Laboratory, Livermore, California; LANL, Los Alamos National Laboratory, Los Alamos, New Mexico; Sandia, Sandia National Laboratory, Albuquerque, New Mexico; UI, University of Illinois, Champaign, Illinois; NASA, NASA Langley Research Center, Hampton, Virginia; USSR, Soviet Union, laboratory not specified.

[b] N/A, not available.

and to construct an optical apparatus which was insensitive to radiation damage. It is likely that the laser apparatus was totally destroyed by the blast or at least made inaccessible by radioactivation. There are no other reports in the nonclassified literature of NDPL experiments, although other countries with nuclear weapons programs and access to underground test facilities may have performed similar experiments.

Although many laser media require such high pump rates that they can only be pumped with the high radiation fluxes from a nuclear device, the difficulty and cost of protecting the laser apparatus from the shock and blast effects of the nuclear explosion may limit NDPLs to single-shot systems. Since much of the cost of a large laser system may be associated with the laser optics, routine destruction of the laser may not be acceptable. The catastrophic effects of an accidental explosion of the nuclear device would also limit deployment of NDPLs to areas far from human activity. In light of these limitations, it is likely that NDPLs could only be deployed in space or in underground locations below the earth's surface. Because of the highly specialized nature of NDPLs and the small amount of published information, further discussion will emphasize the reactor pumped laser.

B. Reactor Pumped Lasers

The more typical NPL reported has been pumped by a pulsed or steady-state nuclear reactor. Reactor pumped lasers (RPLs) are in a sense "laboratory" laser devices, even though in practice there are only a few laboratories with the required reactor facilities. The relatively few workers in the RPL field is probably related to the scarcity of these facilities. In these RPL experiments an intense pulse of neutrons triggers a nuclear reaction, which emits heavy charged particles into a gaseous laser medium. This results in excitation processes qualitatively similar to those produced by an intense electron beam.

The first clear case of reactor pumped lasing was reported by Sandia National Laboratories in 1975 and occurred in CO gas cooled to 77 K (see Table I). Shortly thereafter, Los Alamos Scientific Laboratory reported lasing in He/Xe gas mixtures, and in 1976 the University of Illinois reported lasing in Ne/N_2 gas mixtures at the much lower volumetric pump rates obtainable with their pulsed TRIGA reactor. In all of these feasibility experiments, only a very small fraction of the leakage neutrons from a reactor facility was used to excite the laser media.

Figure 1 shows a representative reactor pumped laser experiment. The laser apparatus consists of a rectangular stainless steel chamber located adjacent to a pulsed reactor (the Sandia National Laboratories SPR-II fast burst reactor). The chamber is fitted with high-reflectivity mirrors which reflect the laser beam in a folded path through the spatial region which contains the highest neutron flux. The chamber is filled

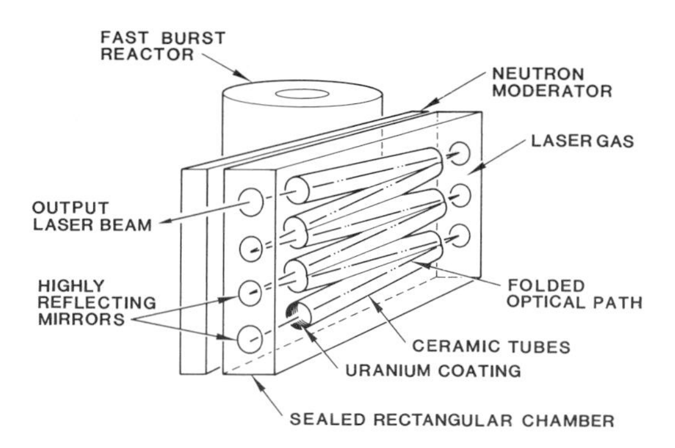

FIG. 1. Reactor pumped laser apparatus for use with a fast burst reactor (FBR).

with laser gas, and the laser gas along the "folded" optical path is excited by fission fragments emitted from the inner surfaces of hollow cylindrical ceramic tubes.

A slab of hydrogen-containing plastic is placed between the reactor and the chamber to slow the fast neutrons produced by the reactor (the process of "neutron moderation"). The slow neutrons are more readily absorbed by the uranium coatings on the inside of the cylinders, triggering the fission reaction. Fission fragments (energetic, highly charged heavy ions) emerge from the coatings to excite the laser gas. The excitation energy is stored in the ^{235}U nucleus, and the function of the neutron is to trigger the release of this fission energy. The energy deposition originates at the spatial location where the neutron is absorbed by the uranium nucleus and extends along the path of the fission fragment until it comes to rest. The volumetric pump rate is easily varied over a wide range by varying the energy released in the reactor pulse.

Figure 2a shows the time dependence of the fission fragment pulse which excites the laser gas, along with the laser pulse from a CO laser gas mixture. The laser action begins at a low excitation power but peaks before the excitation power peaks, and lasing ceases before the end of the excitation pulse. The volumetric pump power is about 1000 W/cm^3, and the total energy deposited is about 0.2 J/cm^3.

Termination of the laser pulse before the end of the excitation pulse is commonly seen for reactor pumped lasers. It may be related to gas heating which reduces the optical gain, to optical distortion of the laser medium by gas heating, or to creation of chemical species which absorb the laser light. In early measurements of optical gain in CO gas, severe focusing of a

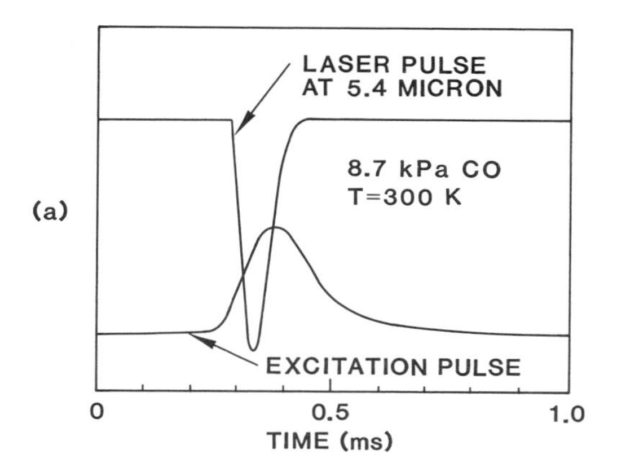

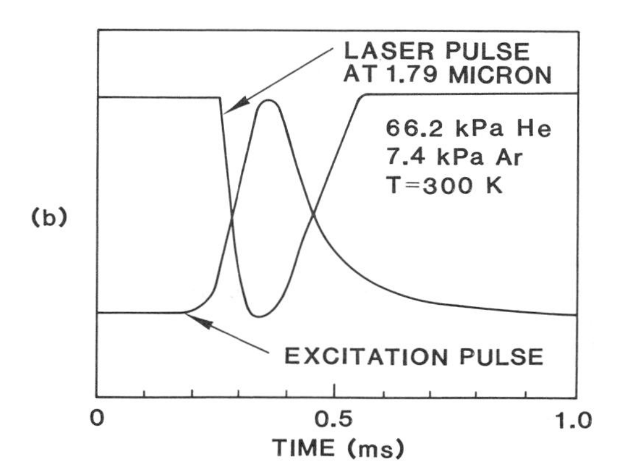

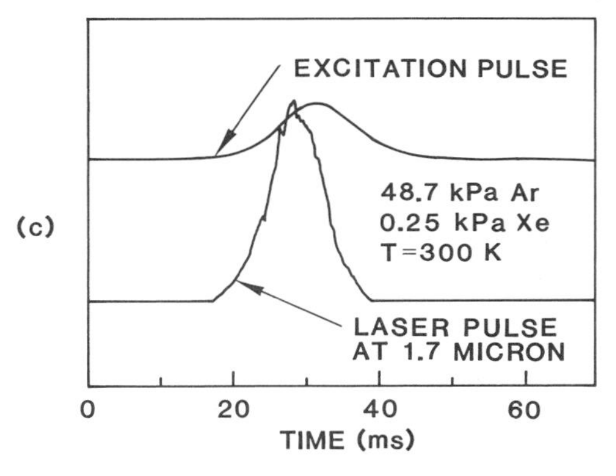

FIG. 2. Representative reactor pumped laser signals: (a) CO RPL excited by Sandia Labs SPR-II (FBR); (b) Ar RPL excited by Sandia Labs SPR-II (FBR); (c) Xe RPL excited by Sandia Labs ACRR (analogous to pulsed TRIGA reactor).

probe laser beam by optical aberrations in the excited medium was observed. More recent measurements have revealed details of this focusing tendency in the excited medium, and this is an active area of current research directed toward improving reactor pumped laser performance. In early experiments, evidence of chemical decomposition of the CO gas was also observed. A yellowish film (probably C_3O_2) accumulated on the laser mirrors. C_3O_2 is a well-known decomposition product observed when CO is irradiated with alpha particles.

Figure 2b shows a laser pulse at about 1.8 μm wavelength which is obtained with the apparatus shown in Fig. 1 when the CO laser mixture is replaced by a He/Ar mixture. The fact that lasing occurs over most of the excitation pulse indicates that laser pulse termination is not a property of the laser apparatus itself but results from differences between the two laser gas mixtures.

Figure 2c shows the excitation pulse and the resulting laser pulse at 1.7 μm wavelength from an Ar/Xe laser gas mixture excited by a reactor analogous to a pulsed TRIGA reactor (the Sandia National Laboratories ACRR). In this case, lasing continues over most of the excitation pulse, which is much longer than the pulse width of the reactor used to obtain the data of Figs. 2a and 2b.

II. Characteristics of Nuclear Pumped Lasers

A. Pumping Mechanisms

There are two types of pump sources: a source of intense gamma radiation or a source of intense neutron flux. The gamma rays can irradiate the laser medium directly and produce excitation. The neutrons ordinarily excite the laser medium indirectly by irradiating other "target" nuclei which undergo nuclear reactions and are the direct agents in exciting the laser medium. At present there are two practical sources: pulsed research reactors and nuclear devices.

The energetic charged particles which excite a nuclear pumped laser medium derive from three main sources: Compton-scattered electrons produced by energetic gamma rays; fission fragments (particles having atomic masses of about 95 and 140 amu, initial energies of 65 to 100 MeV, and initial charges of about +20e); and charged particles such as protons, tritons, and alpha particles produced in exothermic nuclear reactions involving absorption of a neutron.

Table II contains basic data for several nuclear reactions which may be used to pump an RPL. The energy released varies over more than two orders of magnitude, with the largest energy release provided by fissioning of ^{235}U. Among the reactions in Table II, the fission reaction is unique in being able to produce more neutrons to sustain the nuclear reactions.

The charged particles are sufficiently energetic that excitation of a broad range of atomic and molecular states is energetically allowed. There is also, in general, no mechanism for selective excitation of only a few excited states. This selectivity of excitation, which has frequently been used to produce laser action, is not generally available in nuclear pumped lasers.

An important practical limitation is that each laser medium requires a minimum volumetric pump rate to begin lasing and a somewhat greater volumetric pump rate to convert the pumping power to laser light efficiently. One limitation of RPLs is that reactors tend to produce lower volumetric pump rates (no more than 3000 MW/m^3) than can be obtained with some conventional laser excitation methods. This limit is set by temperature limitations of the reactor structure itself. In the case of nuclear explosive devices used as pump sources, much higher pump rates can be obtained, since the survival of the nuclear device and its immediate surroundings is not required.

B. Potential Advantages as Large Lasers

For laser applications that require a very large laser or a laser remote from conventional electrical or chemical power sources, a nuclear pumped laser may be an effective alternative to more conventional lasers. Commercial power reactors routinely operate for months at total fission powers of 3000 MW, with corresponding electrical power outputs of over 1000 MW. Although a large RPL is not expected to resemble a power reactor in detail, conversion of a small fraction of such available fission energy to laser light would produce an extremely powerful, long-lived laser.

Such potential high output powers are possible because of the extremely compact and lightweight energy storage of the fissionable nucleus. Alternatively, the fuel for an electrical power source or the reactants for a powerful chemical laser must be stored in large tanks, in the form of highly corrosive or explosive liquids or gases with an energy storage per unit mass of about 1000 kJ/kg. By contrast, the energy stored in enriched uranium oxide (a dense, refractory ceramic) considerably exceeds 10^{10} kJ/kg. The

TABLE II. Nuclear Reactions for Pumping RPL

Target nucleus	Cross section[a] (10^{-24} cm^2)	Energy release (MeV)	Particles released	Particle energy (MeV)	Range in 1 atm air (cm)
^{3}He	5330	0.76	^{1}H	0.6	1.0
			^{3}H	0.2	0.2
^{10}B	3838	2.79	^{4}He	1.8	1.0
			^{7}Li	1.0	0.5
^{235}U	585	165	^{93}Kr	100	2.3
			^{140}Xe[b]	65	1.8

[a] Thermal neutron energy.
[b] Representative nuclei from a broad spectrum of fission products.

uranium "fuel" of an RPL thus represents a huge, compact, stable, self-sufficient energy source, even allowing for incomplete uranium fuel utilization.

Another potential advantage of the RPL is that the energy is deposited in the laser medium based on its density, and the initial steps of the energy deposition process are largely independent of details of the laser medium condition (such as electron density in the medium or chemical composition of the medium).

The long absorption length of the neutron also allows the excited laser medium to be scaled to volumes of about 100 m^3, while remaining well within the bounds of current reactor technology. This scaling process involves replication of mechanical reactor substructures, such as plates of ceramic neutron-moderator material coated with very thin uranium coatings. Proliferation of complex, relatively fragile high-voltage apparatuses is not required, nor is it necessary to separate the laser gas from adjacent high-vacuum regions with thin, relatively fragile foils. Although the close spacing of the RPL structure presents problems with the extraction of a high-quality optical beam, several potential solutions are being studied. These unusual scaling laws may be of significant advantage in constructing very large lasers.

III. Basic Physics of Reactor Pumped Lasers

A. Nuclear Aspects

1. Reactor Characteristics

A nuclear reactor normally is designed to operate for many years at its maximum output (pulsed or steady-state); so the practical limitation on the lifetime of an RPL is typically related to other factors, such as limitations on coolant supply or degradation of laser optics.

Fast burst reactors (FBRs) have the advantage of producing high pump power in the RPL in a short pulse which does not overheat the laser medium. FBRs are also typically housed in laboratory facilities which allow convenient access to the volume around the reactor for setup of laser apparatuses. However, FBRs have very low pulse repetition rates because of the need to cool a large compact mass of reactor fuel between pulses.

Another relatively common research reactor, the pulsed TRIGA reactor, operates by pulsing from a very low initial power level using control rods. The pulsed TRIGA reactor can be pulsed more frequently because it typically is located at the bottom of a tank of water which acts as both coolant and shield. The neutron spectrum of a TRIGA reactor also has a much lower average energy, which reduces or eliminates the need for neutron-moderator material in the laser apparatus.

The practical laboratory sources of intense neutron fluxes are pulsed or steady-state reactors fueled primarily by uranium. Since each ^{235}U fission event releases about 168 MeV of fission fragment energy and about 2.4 neutrons, the energy cost per neutron created in a reactor is at least 70 MeV/neutron. In fact, because some of the neutrons are absorbed in the reactor fuel to sustain the reaction, the energy cost per neutron is greater than 70 MeV per neutron. Thus the use of low-energy-release reactions from Table II severely reduces the efficiency of a large RPL system. However, for demonstration experiments any of the nonfission reactions in Table II may be used, in some cases with great practical advantages compared to use of the fission reaction.

There is typically some unavoidable waste energy deposited in the reactor structure itself, and the maximum temperature which it can withstand usually limits the obtainable pump rate in RPLs. For typical research reactors, pump rates of perhaps 3000 MW/m^3 can be obtained in pulses with minimum widths ranging from 0.2 to 10 msec. The neutron flux from high-flux steady-state reactors limits the laser pumping power per unit volume to about 10 MW/m^3.

2. Substrate Pumping

The least impact on the laser medium conditions is obtained by separating the target nucleus from the laser medium by using substrate pumping. In substrate pumping the target nuclei are chemically bound into a solid compound that is coated onto the surface of a substrate in contact with the laser gas. From Table II, the maximum path lengths of the charged particles in air at 1 atm pressure are typically a few centimeters. If substrate pumping is used, the pumping of the laser medium will be quite inhomogeneous if the source substrates are too far apart (so that the charged particles do not even reach the center of the laser medium, for example). A typical optimum separation between source substrates is approximately equal to the charged particle range. The energy deposition is also more uniform if the substrates are in the form of large parallel plates.

In all RPL experiments reported to date, a substrate has been used to support the target nuclei (with the exception of ^3He). This method has thus far been limited to excitation of gaseous laser media because of the extremely short range of charged particles in liquids and solids. The target nucleus compound is usually chosen for its chemical and mechanical stability and its adherence to the substrate material. The compound is deposited on the substrate in a thin layer, since the charged particles produced in the coating typically have ranges of a few milligrams per square centimeter of the coating material.

A portion of the energy of the charged particles is deposited in the coating and substrate material. This produces waste heat which does not contribute to laser excitation. For a given coating material and thickness, a constant fraction of the total energy created in the coating is emitted from the coating and is available to excite the laser medium. The output energy per unit area of coating increases as the slow neutron flux increases. For a typical uranium oxide coating of 1 μm thickness, the coating efficiency is about 32%. For neutron fluences at the edge of the core of the Sandia Laboratories ACRR reactor facility, volumetric energy depositions of 1.5 J/cm^3 and volumetric pump rates of 200 W/cm^3 are readily obtained for typical laser gases at about 1 atm pressure.

As the coating thickness increases, the coating efficiency decreases and the flux of energy from the coating approaches a constant value. In experiments it may be useful to maximize this energy flux to obtain the maximum excitation with a given fixed neutron flux. However, in a practical RPL system it would be important to maximize efficiency while maintaining the necessary pumping power and overall reactor characteristics desired.

3. Homogeneous Pumping

Dispersing the target nucleus throughout the laser medium (by choosing a gaseous compound containing the target nucleus and mixing it with the laser gas) normally produces the greatest spatial uniformity of pumping. Homogeneous pumping also uses essentially all the nuclear reaction energy at high pressures of the laser medium, since no energy is lost in a coating. However, the presence of the target nucleus in its host molecule may interfere with the processes which produce gain in the laser medium.

Experiments performed as part of the NASA RPL program have studied homogeneous pumping to the greatest degree, culminating in a 1 kW laser (Table I). Development of a large, efficient RPL system based on ^3He is, however, limited by the very small energy release and the very large neutron absorption cross section of the ^3He reaction (Table II). If the ^3He partial pressure in the laser mixture is several atmospheres, the neutron absorption length shortens to a few centimeters, and the uniformity of pumping will be limited by the neutron absorption length itself.

If a laser medium tolerant of gaseous uranium compounds could be developed, homogeneous pumping would be preferable to any other scheme. The report of lasing at 2.65 μm in Xe in the presence of a small pressure of $^{235}UF_6$ is therefore of considerable interest (Table I). Researchers at Los Alamos National Laboratory have also proposed liquid RPL media containing uranium compounds.

B. Laser Medium Aspects

The charged particles give up their kinetic energy to the laser medium by producing both ionization and excitation of the laser medium. Low-energy secondary electrons produce further ionization and excitation in the medium. The number of ions and excited atoms/molecules can be estimated by using measured properties of the laser gas, such as the average energy required to create an electron–ion pair. However, there is no obvious mechanism to produce highly selective excitation directly. Thus, population inversion relies on selective kinetic and photophysical processes which preferentially populate certain states.

The excitation of high-lying excited states may stimulate high chemical reactivity of the laser medium, making it necessary to replace the laser gas periodically or scrub it.

For substrate pumping, the relatively short range of charged particles requires a separation between the substrates of a few centimeters. This results in large, slablike gain regions separated by opaque barriers. Extraction of a laser beam with good beam quality and high efficiency from this type of system is the most challenging aspect of large-system RPL design. Relatively little work has been done thus far to deal with this optical complexity, which must be weighed against the potential advantages of the simpler RPL excitation structure.

Finally, the relatively low pumping power densities obtainable with reactor pumping requires that the laser medium possess long lifetimes for retaining excitation. The CO laser was initially chosen as a likely candidate for reactor pumping, based on the long lifetimes of the upper energy levels of the laser transitions (~30 msec). The Xe laser may also have a long effective lifetime for excitation because of energy recycling through a high-lying metastable level.

C. Medium Inhomogeneity

For the relatively long energy deposition times involved with typical pulsed reactors, the laser medium can move in response to spatial variations in the energy deposition. This gives rise to nonuniformities in the laser medium density and gradients of the index of refraction which cause distortion of the laser beam (tilt, focusing, defocusing, or higher order aberrations). To produce a very-high-quality laser beam, these distortions must be corrected.

In general, there are two possible causes of inhomogeneous pumping: charged-particle-range effects and effects from absorption of neutrons by the target nucleus. Charged-particle-range effects are a problem with substrate pumping, and neutron-absorption effects are a problem with homogeneous pumping.

For substrate pumping (assuming efficient usage of the fission fragment energy), the spatial scale of the nonuniform energy deposition is of the order of the separation between adjacent substrates. If the substrates are in the form of large parallel plates, a nearly one-dimensional density variation results. Wavefront distortion measurements have been reported for this excitation geometry. For intense pulsed excitation, the density can vary by as much as a factor of two, producing an effective focal length of less than 1 m for a cell only 30 cm long. Such time-dependent focusing can have significant effects on the modal stability of a laser. In fact, resonator stability transitions have been observed in the atomic Xe laser when energy depositions approach 1 J/cm^3.

D. Contrasts with Conventional Lasers

The RPL differs from electrically, optically, or chemically excited lasers in several respects.

1. The maximum excitation power per unit volume in the laser medium is lower than the comparable maximum for electrically excited lasers (assuming that specialized RPL designs based on existing reactor technology are used). However, the energy deposition per unit volume of laser medium is at least comparable to that obtained with electrical or electron-beam excitation.
2. The geometry of the laser excitation region is complex if substrate pumping is used, which demands accurate relative phasing of beams from many independent gain regions to produce a single coherent laser beam.
3. If substrate pumping is used, the excitation process itself gives rise to variations in laser medium density which severely affect beam quality unless the resulting laser beam aberrations are corrected.
4. The laser optics must operate reliably in a

relatively severe radiation environment, which consists primarily of high-energy gamma rays and fast neutrons.

5. The construction of the laser typically involves robust mechanical apparatus rather than electrical or chemical apparatus.

IV. Reactor Pumped Lasers

A. Technological Progress

About 20 RPLs have been discovered since the first report of an RPL in 1975. These are summarized in review articles mentioned in the bibliography.

In 1977 several concepts for constructing large RPL systems were analyzed and reported by Sandia workers. The concepts were based solely on substrate pumping and described specialized nuclear reactor designs which could excite large volumes of laser gas at efficiencies approaching the coating efficiency. These relatively efficient system concepts tended to resemble the TRIGA or ACRR reactors more than the FBR. These initial calculations showed that with a high intrinsic laser efficiency, perhaps 10% of the total reactor energy might be obtained as laser light, using current or near-term reactor technology. However, the problem of extracting a high-quality optical beam from the complex reactor structure was not addressed.

Experimental research directed toward discovery of new RPLs has continued at the University of Illinois. System studies of possible large-scale RPL applications and detailed measurements of radiation-induced absorption in optical materials have been reported. Several comprehensive review articles on RPL research and applications have been published [see Miley (1989) and (1984) in the bibliography].

The NASA program at Langley Research Center was broad and productive: New RPLs were discovered, homogeneous pumping with ^{3}He was developed, the volume limits of RPLs were studied, excitation processes were investigated theoretically, and a reactor pumped amplifier was demonstrated.

In 1979 researchers in the Soviet Union reported near-infrared RPLs using mixtures of noble gases pumped at very low volumetric pump rates, with surprisingly high efficiencies and high energy outputs (Table I). The combination of relatively short wavelength, high efficiency, relatively high pressure, and stable laser gases may make these lasers attractive for RPL applications.

Moderate-efficiency lasing was observed by Russian researchers in Cd vapor at 0.442 to 0.54 μm wavelength using very low volumetric pump rates. Thus, it is possible that high-power visible RPLs with reasonable efficiency can be developed.

Finally, alternate forms of substrate pumping were considered in some detail. The energy escape fraction is largest from small spheres of uranium metal or oxide. Such spheres cannot be incorporated directly into the laser medium because of the optical scattering loss they introduce, even if they were overcoated with a thin reflective coating. However, concepts have been developed to use such reflective aerosols to pump a fluorescer medium, which produces intense incoherent light usable for optically pumping a laser medium. Such nuclear flashlamp pumped RPLs have been investigated in some detail theoretically but have not been demonstrated experimentally.

B. Radiation Effects on Optics

One concern in early research on RPLs was the question of degradation of optics by the intense radiation field of the reactor environment. The radiation environment of the reactor consists of broad spectra of both neutrons and gamma rays ranging in energy up to several MeV.

In early experiments at Sandia in the FBR environment, it was found that no visible optical damage occurred to several convenient window materials, such as optical-grade NaCl, BaF_2, and CaF_2. Metallic mirrors showed almost no measurable change in reflectivity, perhaps because they are already highly conductive. In more recent experiments, dielectric-coated mirrors made of elements with low neutron absorption have also demonstrated very little change in reflectivity in intense pulsed radiation fields. Schlieren-grade fused silica also displays relatively low radiation-induced absorption for visible and near-infrared wavelengths, and is thus a convenient and versatile RPL optical material.

However, materials containing elements such as ^{10}B or ^{6}Li (such as borosilicate glass) should be avoided in optical components because of their large neutron-absorption cross sections. In these nuclei, absorption of a neutron creates tracks of intense ionization in the host material and atomic displacements which can act as defect centers for optical absorptions. Therefore, a basic step in optical design for RPLs is to avoid neutron-absorbing elements in window materials, coating materials, and substrate materials for mirrors.

Optical components can be protected from bombardment by low-energy neutrons by relatively thin sheets of materials such as Cd metal or B_4C suspensions in plastics. Very energetic neutrons cause damage by displacing lattice atoms in elastic collisions. A low-mass recoil atom may have enough energy to create ionization tracks in the material. High-energy neutrons can be moderated to low energies (for absorption in Cd or B) by scattering from shielding materials containing high densities of hydrogen (such as high-density polyethylene).

Gamma rays produce energetic Compton-scattered electrons which may cause significant damage even in optical materials which do not contain neutron-absorbing nuclei. The energetic Compton electrons degrade in energy by producing ionization in the optical material so that a weak transient conductivity is induced by the radiation. Complete shielding of the gamma rays requires thick layers of high-density materials such as Pb, but this is usually not practical in the vicinity of the laser apparatus.

C. Development Toward Applications

The unique characteristics of nuclear pumped lasers include the following: Essentially infinite energy storage; relatively straightforward scaling of the volume of excited laser medium; extremely high total excitation power with current reactor technology; no requirement for pulsed power systems or high-voltage electrical power; and robust mechanical construction.

However, these applications must be realized in the presence of the unique disadvantages of nuclear pumping: Low pump power per unit volume (for RPLs, not NDPLs); intense radiation fields; extraction of the laser beam from the complex reactor structure (for substrate pumping); and the health, safety, and proliferation concerns of nuclear technology in general.

Lasers with moderate total energies have found numerous practical applications. It may be that as applications for very large lasers emerge, the relatively robust "mechanical" nature of NPLs will have important system advantages. NPLs have been considered recently for specialized applications such as for space power, for large industrial chemical excitation, and as weapons. Several articles in the bibliography treat some of these potential applications in greater detail. [See, for example, Miley (1984) and Prelas and Loyalka (1988) in the bibliography.]

Bibliography

Fitaire, M., ed. (1978). *Int. Symp. Nucl. Induced Plasmas Nucl. Pumped Lasers [Pap.], 1st, 1978*. Les Editions de Physique, Orsay, France.

Jalufka, N. W. (1983). "Direct Nuclear-Pumped Lasers," NASA Tech. Pap. No. 2091. National Aeronautics and Space Administration, Washington, D.C.

McArthur, D. A., *et al.* (1988). Recent results on reactor-pumped laser studies at Sandia National Laboratories, *Laser Interact. Relat. Plasma Phenom.* **8,** 75.

Miley, G. H. (1982). Direct nuclear pumped lasers—status and potential applications, *Laser Interact. Relat. Plasma Phenom.* **4A,** 181.

Miley, G. H. (1984). Review of nuclear pumped lasers, *Laser Interact. Relat. Plasma Phenom.* **6,** 47.

Miley, G. H., *et al.* (1989). Fission reactor pumped lasers: History and prospects, *in* "50 Years with Nuclear Fission" (J. W. Behrens and A. D. Carlson, eds.), Vol. 1, p. 333. American Nuclear Society, La Grange Park, Illinois.

Prelas, M. A., and Loyalka, S. K. (1982). A review of the utilization of energetic ions for the production of excited atomic and molecular states and chemical synthesis, *Prog. Nucl. Energy* **8,** 35.

OPTICAL CIRCUITRY

H. M. Gibbs, G. Khitrova, S. Koch, N. Peyghambarian, and
G. I. Stegeman *University of Arizona*
U. J. Gibson *Dartmouth College*
C. T. Seaton *Coherent Radiation*
M. Warren *Sandia National Laboratories*

GLOSSARY

Exciton: Excited electron–hole pair. In a crystal an electron is excited with insufficient energy to place it in the conduction band, but gains that energy minus the binding energy to the hole left in the valence band. Because they remain together in an exciton as they move through the crystal, the electron and hole require less energy than they would were they free of each other. An exciton is analogous to a hydrogen atom or a positronium atom.

Fabry–Perot etalon: Optical interferometer consisting of parallel reflecting surfaces, which cause light to reflect back and forth. When the optical length of the etalon equals an integral number of wavelengths of the light, multiple beams interfere constructively, resulting in high transmission and a greater light intensity inside the etalon than incident to the etalon. Otherwise, the beams interfere destructively, most of the light is reflected by the etalon, and the intensity inside the etalon is less than that incident to it.

Heterostructure: Structure consisting of two or more species of crystals (e.g., GaAs and AlGaAs).

Multiple-quantum-well: Arrangement in which a number of quantum wells are grown on top of each other, for example, the growth of alternating layers of GaAs and AlGaAs.

Nonlinear Fabry–Perot etalon: Fabry–Perot etalon whose transmission properties change with light intensity. The most promising all-optical logic devices so far are based on an intensity-dependent index of refraction.

Nonlinear refractive index: Intensity-dependent index of refraction. Near an optical resonance of the material, the intensity dependence is often complicated. Far from optical resonance, the total index n can be expanded in powers of the intensity (i.e., $n = n_0 + n_2 I + \cdots$ where n_0 is the weak-intensity limit of n).

Nonlinear waveguide: Device that confines light relative to free-space propagation and whose optical properties (e.g., guiding ability and phase shift) depend on the light intensity inside the device.

Optical bistability: Characteristic of a system in which the output intensity can have two values for one input intensity. This phenomenon implies a hysteresis in output versus input, no matter how slowly the input is varied.

Optical nonlinearity: Characteristic of a material in which the index of refraction and/or absorption depend on the intensity of the light beam propagating through the material.

Optical pathlength: Product of a material's index of refraction multiplied by its physical length.

Quantum well: Confinement arising from barriers at the interface between two crystal layers. When a thin layer of GaAs is grown between two AlGaAs layers, carriers and excitons created by the absorption of light in the GaAs layer are confined to that layer. Because AlGaAs has a larger energy gap than that of GaAs, a carrier in the GaAs layer sees a potential barrier at interfaces with the AlGaAs layers, that is, the carrier is trapped in a well. When the GaAs layer's thickness is equal to or less than the separation between the electron and hole in an exciton (250 Å in bulk GaAs), the exciton wave function and binding energy are changed by quantum effects.

Superlattice: Multiple-quantum-well structure with barriers so small that its optical properties cannot be described by a single quantum well. Properties of excitons in a GaAs quantum well can depend on neighboring quantum wells if the AlGaAs barrier is small enough ($\sim$50 Å or less) to allow the wave function to penetrate into one or more adjacent wells.

Optical circuitry is a generic term that includes a wide range of optical devices in which a light signal is used to control another signal (electrical or optical). This definition encompasses practically all optical devices: all-optical devices for logic operations, preprocessing of optical images, and optical computing; optical interconnections between electronic processors and interfaces between electrical and optical arrays; and light-controlled electrical switches for ultrafast oscilloscopes and slow high-voltage or high-current control applications.

I. Introduction

We are in the midst of a communications and information explosion that demands faster and higher capacity signal processing. The development of rapid transportation, instant telecommunication, and high-speed computers has greatly increased the desire for high-volume data transmission. More and more data can be utilized by supercomputers, and they in turn produce information and control decisions at an ever-increasing rate. Most of the expansion has been made possible by the electronic transistor and its descendants, including the latest very-large-scale integrated circuits. But progress in electronics has begun to reach fundamental limitations. Optical circuitry has the potential to surpass the speed of electronics.

The potential of optical circuitry has been discussed for many years, but excitement is growing rapidly in response to the success of optical fibers used for optical transmission; the generation of subpicosecond optical pulses; and the development of promising optical logic elements, namely optical bistable and gate devices. Much research remains undone in efforts to discover the best nonlinear optical materials and fabrication techniques [*See* OPTICAL FIBER COMMUNICATIONS.]

Optical transition energies are on the order of 1 eV, permitting room-temperature and very-high-frequency operation but requiring more energy per bit than Josephson devices. Certainly, the speed and packing density of integrated electronics are impressive. However, in the generation and transmission of picosecond and femtosecond pulses, optics far surpasses electronics. Optical circuitry is much more resistant to electromagnetic interference and to cross-talk than is electronics. If suitable all-optical control and logic devices can be developed, optics should complement electronics for very-high-speed decision making and switching. Recently developed nonlinear optical devices have the desired characteristics, but the nanosecond switching times must be significantly reduced. Switching in the 0.1- to 10-psec regime has been achieved only with optical circuitry; such switching could permit military signal encryption or high-speed multiplexing and demultiplexing to utilize the large bandwidth of optical fibers. [*See* NONLINEAR OPTICAL PROCESSES.]

In this article we emphasize all-optical devices. We examine their functioning potential for use in signal processing (Section II); origins of the light dependence of the index of refraction on which they are based (Section III); the growth of nonlinear thin films and their implementation in all-optical devices (Section IV); and their potential application in nonlinear photonics (Section V). Hybrid optoelectronic devices are not treated here. Devices that control light with electrical signals (e.g., modulators) are especially important for optical communications; control is often achieved by pulsing the laser itself. Devices that use optical sampling techniques to analyze ultrafast signals underscore the fact that ultrashort pulses can be generated and handled much more easily optically.

The optical circuitry devices discussed herein are categorized as waveguide or etalon devices. Waveguide devices confine the light over distances much longer than free-space propagation would permit. Confinement lowers the input power required to obtain the nonlinear effect necessary for a logic operation. Losses in the material must be minimized for this advantage to be realized. The extended interaction length also results in a longer transit time which may not be acceptable in some cases, but which is acceptable for pipeline processing. An optical waveguide is the optical analog of an electrically conducting wire; its guiding dimension must be at least half an optical wavelength in the material. Consequently, guided-wave optical devices are likely to find application in high-speed serial application such as optical communications, data encryption, or convolvers, rather than in computing. Guided-wave interconnections are permanent and shock-resistant, but they do not easily utilize light's unique capability for massive parallelism and interconnectivity. An array can be imaged onto another array using only a lens. Light beams can pass through each other with no interaction; even the direction of a beam can be changed. No doubt waveguiding will play an important role in many optical control devices, but unguided devices that employ propagation of many beams perpendicular to the plane will likely dominate in parallel processing applications. Etalon devices can perform picosecond logic operations on light beams focused onto the etalon. These devices have the potential for massive parallelism and multiple interconnections. In addition, their short transit time makes them ideal for ultrafast computations in which the outputs of one decision plane are needed as inputs to the next decision plane.

Note that both guided-wave and etalon devices employ thin films of nonlinear optical material. The search for better materials and for better growth and fabrication techniques constitutes a significant portion of current research in optical circuitry. [*See* OPTICAL FIBERS, DRAWING AND COATING.]

II. Device Principles and Applications

The propagation and spatial manipulation of light energy is a well-established technology, both in the guided-wave regime with optical fibers and planar waveguides and in free-space propagation through the use of lenses, mirrors, etc. The crucial elements that allow optical circuitry to perform decision making and computational tasks are components whose transmissions depend on some variable parameter of the light incident on them. The property of light used to interact with the components must be easily modified and measured. In addition, the output of one component should be suitable as input to another component. For example, a device whose output phase depends on the input polarization might have some useful functions but would not be a desirable component for building processing systems. Consequently, most of the work in optical circuitry so far has concentrated on intensity-dependent nonlinear materials that can produce an output intensity that is a nonlinear function of the input intensity. [*See* OPTICAL SYSTEMS DESIGN.]

Such nonlinear optical devices are analogous to electronic transfers. Devices have been designed and fabricated for use either in guided-wave applications (see Section II,B) in which the light is confined in a planar or cylindrical structure and/or for applications in which the light is incident perpendicular to the material surface. Most, but not all, of the latter devices are based on the nonlinear Fabry–Perot etalon.

A. Nonlinear Etalons

A nonlinear Fabry–Perot etalon consists of a material with an intensity-dependent refractive index positioned between two partially reflecting mirrors of a Fabry–Perot resonator (Fig. 1a).

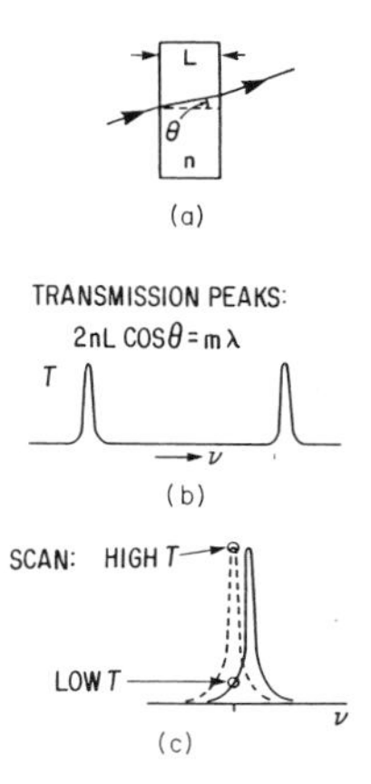

FIG. 1. (a) Nonlinear Fabry–Perot etalon, consisting of a light-intensity-dependent material between two parallel reflecting surfaces. (b) Transmission function of a Fabry–Perot etalon. (c) Shift of Fabry–Perot peak toward laser frequency ν_L with increasing light intensity.

As in an ordinary Fabry–Perot etalon, if the optical length of the cavity is equal to an integral number of half-wavelengths of the incident light, the etalon will be in resonance and highly transmitting (Fig. 1b). The use of nonlinear material allows the optical length of the etalon to be intensity dependent (Fig. 1c). Consequently, the transmission of the etalon can be a highly nonlinear function of the intensity of light incident on it. If the feedback into the nonlinear material from the partially reflecting mirrors is great enough, the device can exhibit optical bistability in which there are two stable output intensity levels possible for a given input intensity. If the output intensity of an optical bistable device is plotted against its input intensity, a hysteresis loop appears (Fig. 2).

Various materials exhibit a nonlinear change in refractive index (see Section III). Semiconductors have received the most attention for etalon applications because they interact efficiently with light over short distances. The semiconductors studied include GaAs, GaAs–AlGaAs multiple quantum wells, InSb, InAs, CuCl, CdS, and CdHgTe. These materials have very large nonlinear indices of refraction arising from electronic mechanisms near their absorption band edge. GaAs devices are the most promising. They operate at room temperature and are very fast, with switch-on times of 1 psec and switch-off times of less than 200 psec. High-finesse GaAs etalons lase easily and have been made into arrays of low-threshold surface-emitting lasers.

Another group of materials exhibits thermally induced changes in both refractive index and physical thickness. In an etalon structure, the effects can combine to produce optical bistability. These materials include semiconductors such as GaAs, ZnS, and ZnSe, as well as color filters and liquid dyes. Of particular interest are the ZnS and ZnSe materials. Devices made from

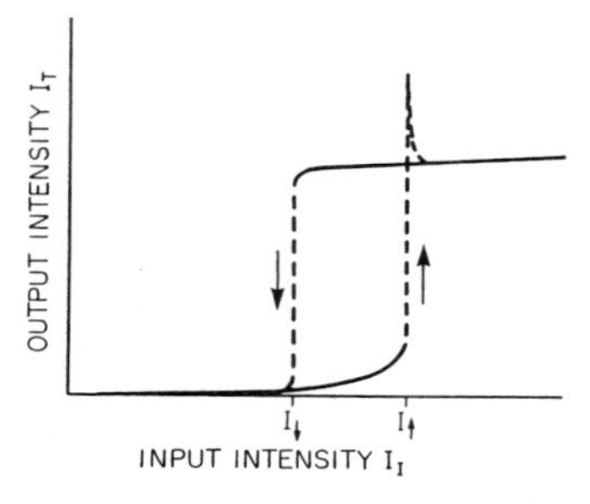

FIG. 2. Characteristic curve for an optical bistable system.

these materials take the form of conventional interference filters, with the semiconductor as the space layer of the filters. Although much slower than the GaAs etalons, ZnS and ZnSe interference filters are more easily fabricated and operate with visible light.

Finally, devices such as the self-electrooptic effect device (SEED) are not based on a nonlinear Fabry–Perot etalon and hence do not require mirrors to provide feedback. The SEED is based on the increase of optical absorption of a GaAs–AlGaAs multiple-quantum-well (MQW) structure arising from shifts caused by the application of an electric field perpendicular to its surface. The MQW layer is fabricated in the device as part of a PIN diode (see Fig. 3), which acts not only as the means of applying a field to the MQW layer but also as a photodetector that allows the device to operate with electrical rather than optical feedback. The device operates at room temperature in the near-infrared region. Through the use of positive or negative feedback modes, the SEED can exhibit either optical bistable operation or linear modulation. Although its construction is more complicated than that of the nonlinear Fabry–Perot etalon, the SEED shows great potential as an optical circuitry component and as an interface between electronics and optics (e.g., as a spatial light modulator).

Nonlinear Fabry–Perot etalons exhibit a variety of operating modes that suggest application as both analog and digital components in optical circuitry. Analog operation of nonlinear etalons includes differential gain if the device is operated on a time scale that is long in comparison to the response time of the optical nonlinearity (Fig. 4). In this case, the device allows one beam to control a more intense beam. In this and in most other applications, the nonlinear etalon is operated as a three-port device, again analogous to a transistor. Another analog-type operating mode is as a limiter, a nonlinear etalon operated in the upper branch of bistability will maintain an almost constant transmitted intensity as the incident intensity is varied. This is because the etalon transmission peak shifts with varying cavity intensity (Fig. 5). Recent investigations have shown that the differential energy gain vanishes when the device speed is reduced to the response time of the nonlinearity. This effect causes problems in application and is currently under extensive study.

Etalon peak shifts can also be used in performing gate operations for digital logic. Zero,

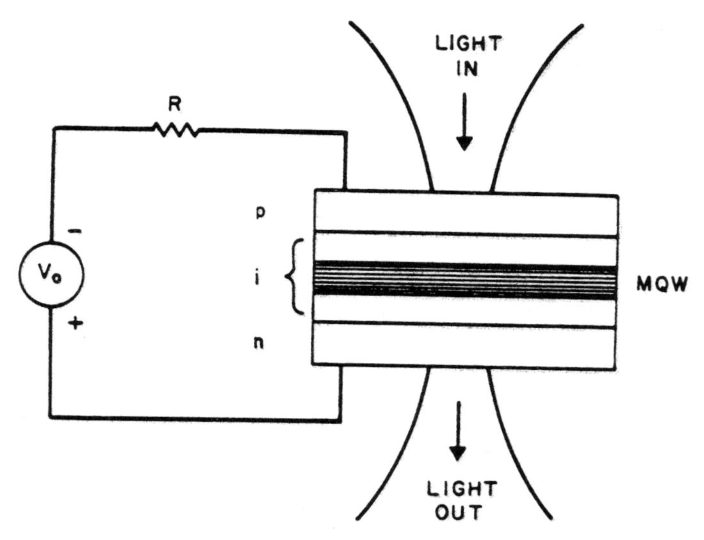

FIG. 3. Schematic of the quantum-well SEED.

one, or two input pulses are simultaneously incident on a device at a wavelength that may be far from the transmission peak (Fig. 6). The pulses serve as the logic inputs for the gate. Another pulse with a wavelength equal to or close to the transmission peak serves as the probe pulse, and its transmission (or reflection) is the output state of the gate. The type of gate (e.g., NOR or AND) is determined by the detuning of the probe pulse from the initial transmission peak of the etalon and from the energies of the short input pulses. The complete set of two-input logic gates has been demonstrated in both dye and GaAs etalons (Fig. 7). High-speed optical logic gates have also been demonstrated using the optical Stark effect in GaAs–AlGaAs etalons. If the input pulse frequency is tuned below the exciton, into the transparency region of the material, the electric field of the laser can be used to switch the etalon. The advantage of switching by the optical Stark effect is that carrier lifetime is no longer a limitation, because carriers are not really excited in this scheme. The disadvantage of such high-speed devices is that higher input powers are required to realize them. A 1-psec recovery using this effect has been demonstrated, showing AND-gate switch-on and switch-off times of ~1 psec (Fig. 8). A disadvantage of this type of logic-gate operation is that the input and output signals have different wavelengths, which complicates the design of a sys-

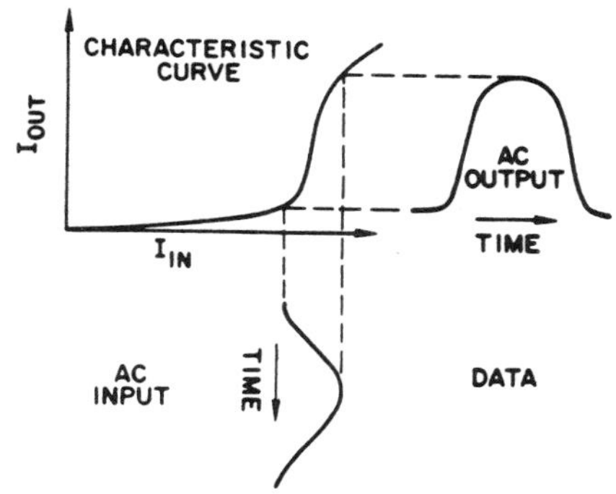

FIG. 4. Differential gain in sodium vapor.

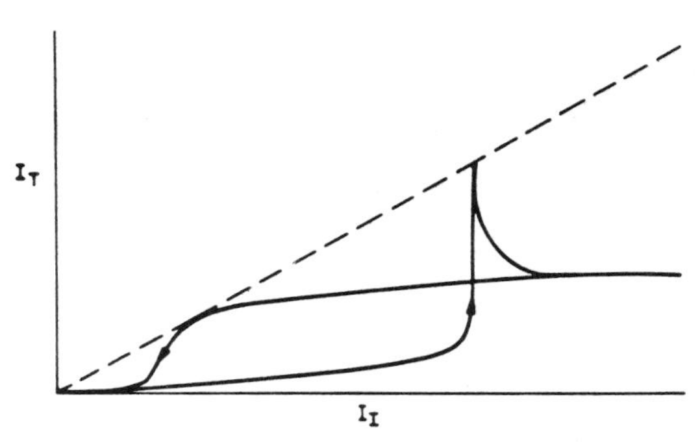

FIG. 5. Optical bistability and limiter action in a 2-mm-long etalon consisting of a polished Corning 3-142 filter with $R = 0.8$. In the on state, the output changes very little for a factor of 4 change in the input. The spike in the transmission occurs as the device turns on and the etalon's frequency of peak transmission is swept through the lase frequency.

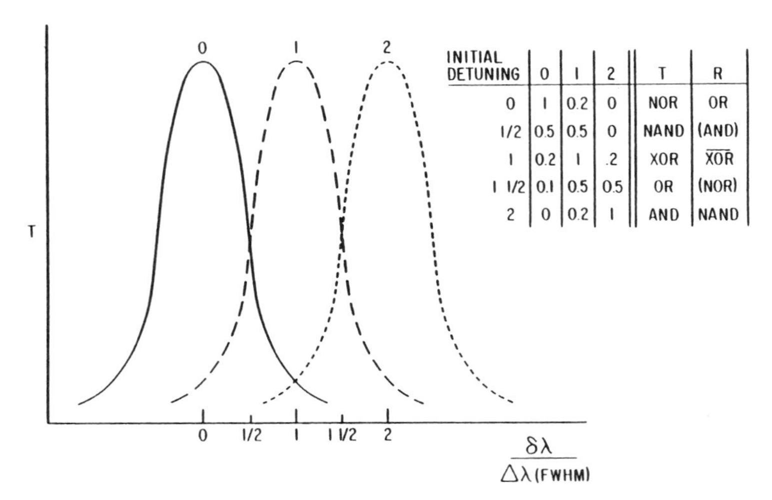

INITIAL DETUNING	0	1	2	T	R
0	1	0.2	0	NOR	OR
1/2	0.5	0.5	0	NAND	(AND)
1	0.2	1	.2	XOR	$\overline{\text{XOR}}$
1 1/2	0.1	0.5	0.5	OR	(NOR)
2	0	0.2	1	AND	NAND

FIG. 6. Use of a nonlinear etalon to perform all of the logic operations. The transmission curves are labeled 0, 1, 2 to indicate the peak position seen by a probe pulse after no (0), one (1), or two (2) input pulses. With the probe wavelength at one of the five labeled values [expressed by the initial detuning in full widths at half maximum (FWHMs) of the transmission peak], the gates in the table are obtained. The fractional values in the columns below 0, 1, and 2 (of inputs) are the approximate transmissions when each input shifts the peak by 1 FWHM. In reflection, the AND and NOR have poor contrast.

tem with sequential logic operations. This disadvantage can be eliminated by bistable operation of a nonlinear etalon with the pulse length much longer than the medium recovery time. A bistable device can operate as an AND gate in transmission ($I_1, I_2 < I_\uparrow$ but $I_1 + I_2 > I_\uparrow$ in Fig. 2) and can also function in steady state as an optical memory device. In both cases, the input and output wavelengths can be identical.

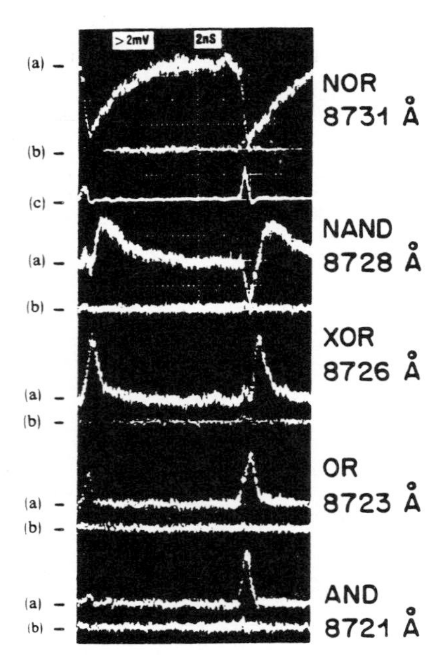

FIG. 7. Logic gate operation using a GaAs etalon. (a) Transmission of a continuous-wave (cw) probe beam. Response on the left side is due to 8 pJ (one input), while that on the right is due to 16 pJ (two inputs). (b) Probe zero line. (c) Input pulses (same for all gates).

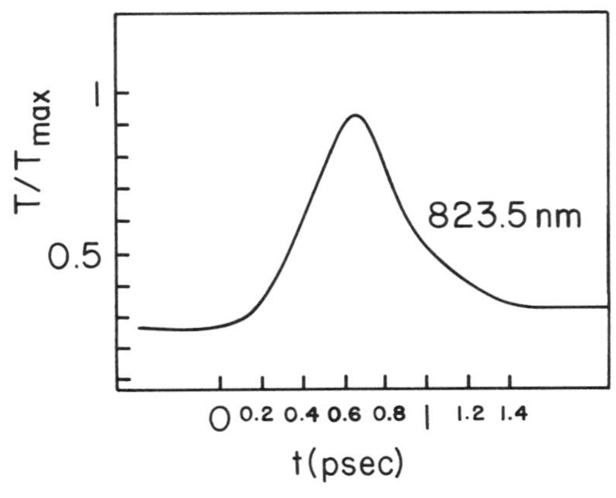

FIG. 8. Picosecond response and recovery of a GaAs AND gate operating on the optical Stark effect.

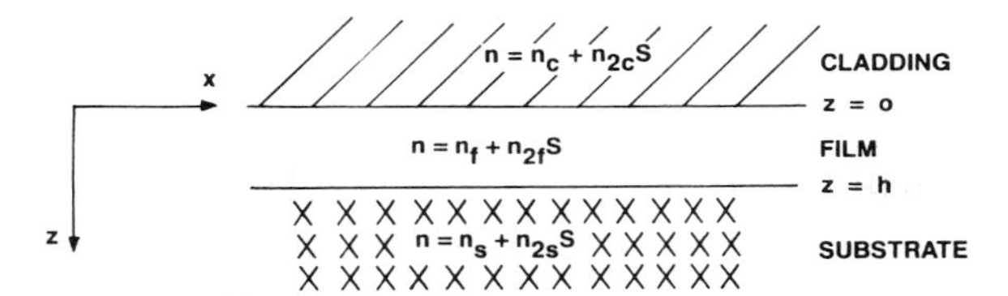

FIG. 9. Schematic of a nonlinear waveguide in which the guide, substrate, and cladding are all nonlinear (S is the intensity).

B. Nonlinear Waveguides

Nonlinear interactions occur whenever the optical fields associated with one or more laser beams are large enough to produce polarization fields proportional to the product of two or more fields. These nonlinear polarization fields radiate in phase with the generated field under optimum conditions (phase matching), growing linearly with propagation distance; hence the key to obtaining efficient (low-power) nonlinear interactions is to maintain high optical intensities over as long a distance as possible.

Optical beams can be confined to an optical wavelength in one dimension by total internal reflection at the boundary of a film whose refractive index is higher than its surroundings. Diffractionless propagation in the confined dimension occurs down the film for centimeter distances limited by absorption, scattering, or both. Nonlinear interactions can take place either in the film or in the neighboring media by means of the evanescent fields that accompany the guided wave (see Fig. 9). Rectangular channels with cross-sectional dimensions of optical wavelengths (rib wavelengths) can also be used to further confine the optical beams.

An impressive number of nonlinear optical devices such as bistable switches and logic gates have been successfully demonstrated in a plane-wave context in a large variety of materials exhibiting some form of intensity-dependent index of refraction (see Section III). The goal of these and, in fact, of any all-optical device is to perform some signal-processing function with a minimum amount of power and device volume. Since it is desirable to operate with input and output beams of the same frequency, the nonlinear interactions are limited to those based on the third-order nonlinearity, for which the pertinent coefficient is usually the intensity-dependent refractive index n_{2i} for the medium i. Most devices require an intensity-dependent phase shift accumulated over some propagation distance, making guided-wave geometries optimum for such interactions. Other waveguide devices are those used on power-dependent field distributions. Experiments reported to date on third-order guided-wave phenomena include degenerate four-wave mixing, coherent anti-Stokes Raman scattering, nonlinear waveguide coupling, optical limiting, and nonlinear coherent coupling. Experimental observations represent the initial stages of this field, and other potential applications have been proposed. Because fabrication techniques are planar-technology oriented, such a waveguide approach to optical signal processing is primarily of interest for serial, rather than parallel, operations.

Degenerate four-wave mixing in waveguides involves the mixing of three input beams of the same frequency to produce a fourth wave of the same frequency. This interaction has potential application to real-time signal processing operations that involve the mixing of two or more waveforms, such as convolution, correlation, or time inversion, as shown in Fig. 10.

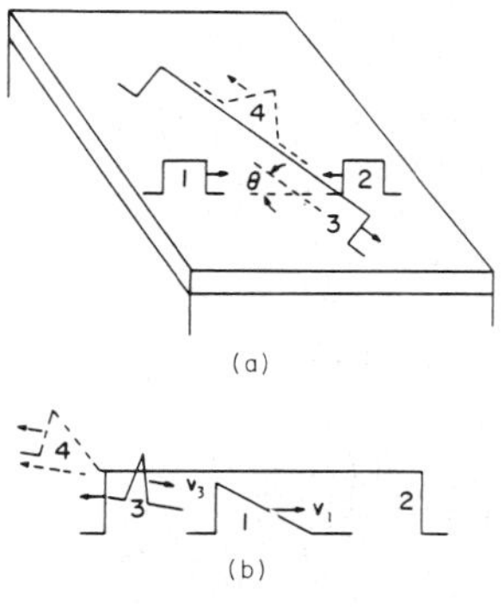

FIG. 10. Schematic diagram of degenerate four-wave mixing in a thin-film optical waveguide. (a) The convolution (waveform 4) of the two input waveforms (1 and 2). (b) Time inversion (waveform 4) of the input waveform 1 by virtue of the δ-function pulse (3) overtaking 1 in the interaction region.

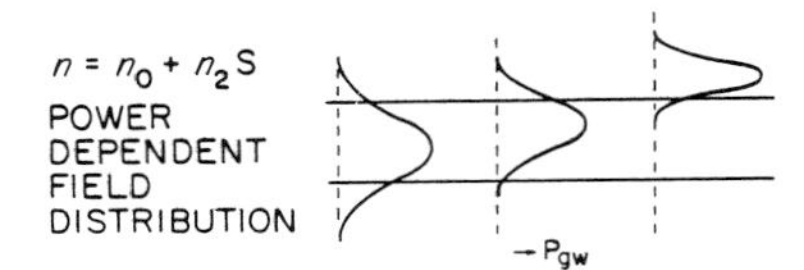

FIG. 11. Schematic of intensity-dependent guiding. At low intensities the light is guided in the central layer, but the self-focusing nonlinearity in the cladding results in guiding in the cladding at high intensities.

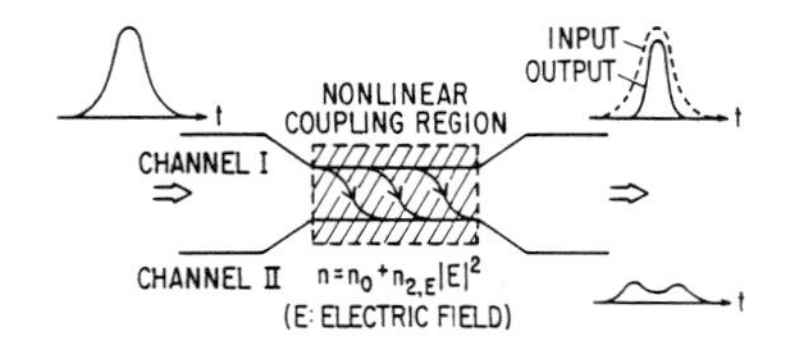

FIG. 12. Schematic of a nonlinear coherent coupler. Schematic of two coupled waveguides with a nonlinear material in the coupling region. At low (high) intensities, the output comes out the lower (upper) guide.

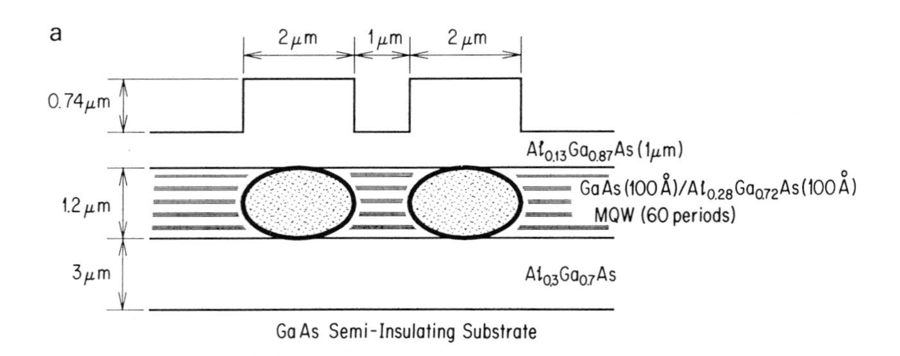

FIG. 13. (a) Structure of an MBE-grown nonlinear directional coupler (NDLC). The total thickness of the top AlGaAs layer is 1 μm. (b) Electron micrograph of a NDLC with 1 μm between 2-μm strips.

For most guided waves, the effect of a nonlinear refractive index is to produce a power dependence of the guided-wave wave vector that can be analyzed with coupled-mode theory. However, if, for example, the cladding medium in a nonlinear waveguide is nonlinear, then the field distributions can also be affected if the optically induced change in the cladding index is comparable to the zero-field index difference between the film and the cladding medium. This leads to an intensity-dependent field distribution, as shown in Fig. 11. Furthermore, self-focusing of the guided wave can occur in a cladding material with a positive n_2, with the light creating its own waveguiding medium in the cladding. This leads to the anomalous variation in guided-wave refractive index with guided-wave power that gives rise to a number of new power-dependent guided-wave phenomena. For example, a cladding medium with positive n_2 in an asymmetric waveguide can cause a device to guide only above a minimum power threshold, which can be set by controlling the waveguide parameters. Alternatively, a limiting device arising from power-dependent control of waveguide cut-off can be obtained by using a cladding medium with negative n_2. When a film is bounded by two nonlinear media, new power-dependent wave solutions occur with the possibility of switching between them.

The other major class of devices based on intensity-dependent refractive indices utilizes a power-dependent wave vector. Such devices generally are a variation of the basic nonlinear coherent coupler (Fig. 12), whose transfer efficiency from one channel to the other depends on the guided-wave power. Mode-locked pulses with 10-psec duration were used to observe all-optical switching in a GaAs nonlinear directional coupler (Fig. 13). At low input energy, nearly complete linear cross-coupling (1 : 3) is observed at 870 nm (Fig. 14). If the input energy is increased to 400 pJ, most of the light remains in the input channel (3 : 1). The recovery time is a few nanoseconds, indicating a carrier-induced nonlinearity. Preliminary measurements using the optical Stark effect have yielded 20% modulation.

Optical control of the transfer in a nonlinear directional coupler may someday be used to produce optical logic operations in devices such as Mach–Zehnder interferometers (Fig. 15) and ring-channel resonators (Fig. 16).

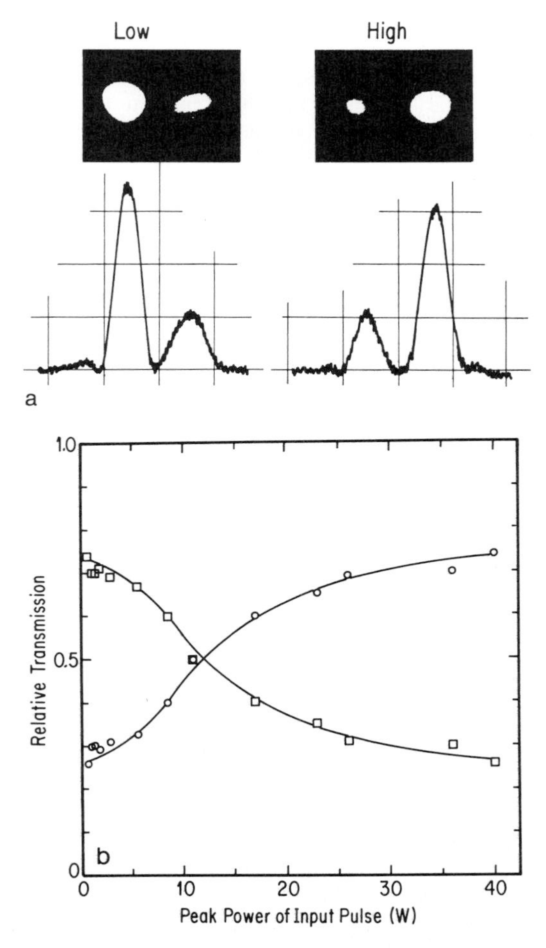

FIG. 14. (a) TV-camera and line-grabber spatial profiles at low and high input energies, with the input focused into the right waveguide. (b) Relative outputs of the two waveguides as a function of the peak power of 10-psec input pulse. The solid lines are guides for the eye.

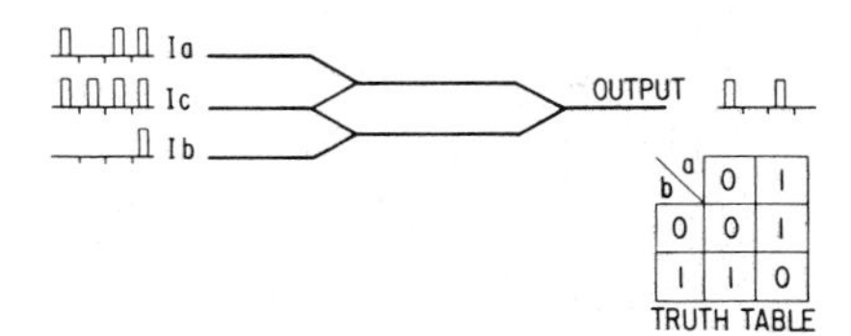

FIG. 15. Schematic of a nonlinear optical Mach–Zehnder interferometer operated as an exclusive OR gate.

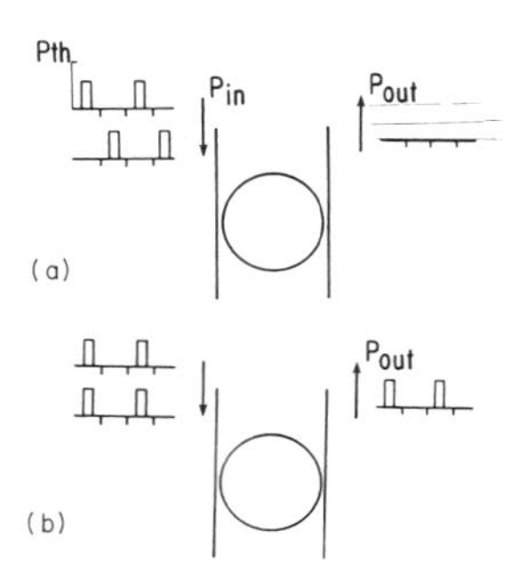

FIG. 16. Schematic of a nonlinear ring-channel resonator operated as an AND gate. (a) The inputs are never coincident, so P_{out} is zero. (b) When the inputs are coincident, their sum exceeds P_{th} and there is output.

Another class of promising nonlinear devices is based on optical tuning of the Bragg condition of a grating embedded in a nonlinear waveguide (Fig. 17). This can lead to bistable switching for guided waves and numerous all-optical logic operations.

All of the applications and devices mentioned here are dependent on materials with a large nonlinearity that can be used in a thin-film format suitable for optical waveguiding. Standard thin-film fabrication techniques (see Section IV) can be used where appropriate, as can techniques such as in-diffusion, ion-exchange, and Langmuir–Blodgett deposition, depending on the materials selected for a given application.

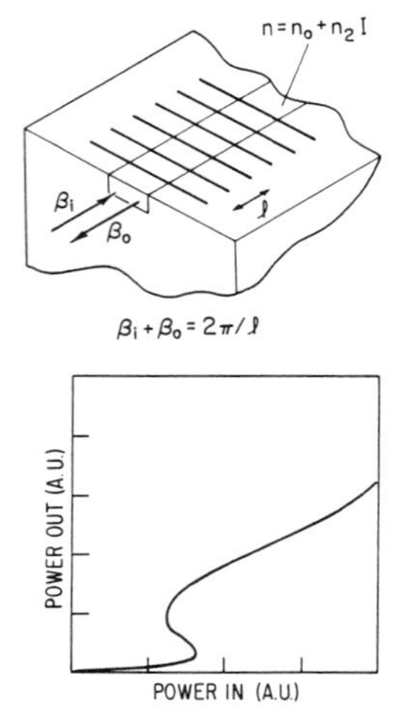

FIG. 17. Schematic of a nonlinear grating operated as a bistable device with calculated transmission function.

III. Nonlinear Refractive Indices

In Section II, various nonlinear optical devices were described. In almost every case, the basic operation mechanism in an intensity-dependent change $\delta(nL)$ in optical pathlength nL or corresponding change $\delta\phi$ in phase shift ϕ, with $\delta\phi = (2\pi/\lambda)\delta(nL)$, where λ is the wavelength, L the sample length, and n the refractive index. The refractive index can be written as $n = n_0 + \Delta n$, where n_0 is the background refractive index and Δn is the nonlinear portion. Under certain conditions, Δn is proportional to the applied intensity ($\Delta n = n_2 I$), but more generally, Δn is a function of carrier density (e.g., band filling and excitonic effect) or electronic temperature (thermal nonlinearities). For comparison of different materials, it is often useful to use the n_2 value, defined as the maximum of the achieved index change, using light of intensity I at a given frequency. The term $\delta(nL)$ may have contributions from a change in physical length δL and from a change in refractive index δn. One of the simplest origins for a $\delta\phi$ is a thermal one (i.e., an intensity-dependent heating of the medium), resulting in both δn and δL. Bistability and logic operations have been achieved through thermal effects, but electronic effects are faster.

The Kramers–Kronig relations express the relationship between the absorption spectrum $\alpha(\omega) \propto \chi''(\omega)$ of a material and its refractive index $n(\omega) \propto \chi'(\omega)$. Consequently, if the absorption spectrum changes, so does the refractive index, as shown in Fig. 18. An absorption resonance gives rise to the well-known anomalous dispersion, with greater refractive index below the resonance and smaller refractive index above. If the absorption profiles can be changed by increasing the light intensity, the refractive index is also intensity-dependent (i.e., nonlinear), and the material can be used for dispersive optical bistability. Contributions to the refractive index by a simple two-level transition in a collection of gas-phase atoms is an illustration of this type of nonlinearity.

Band filling can be envisioned as in Fig. 19. If light of energy exceeding the bandgap is sufficiently intense to fill all of the lower states in the conduction band more quickly than these states decay, the above-bandgap absorption is effectively reduced. This absorption saturation leads to a reduction of the refractive index below the band edge as in Fig. 19. This band filling is the dominant band-edge mechanism in InSb (effec-

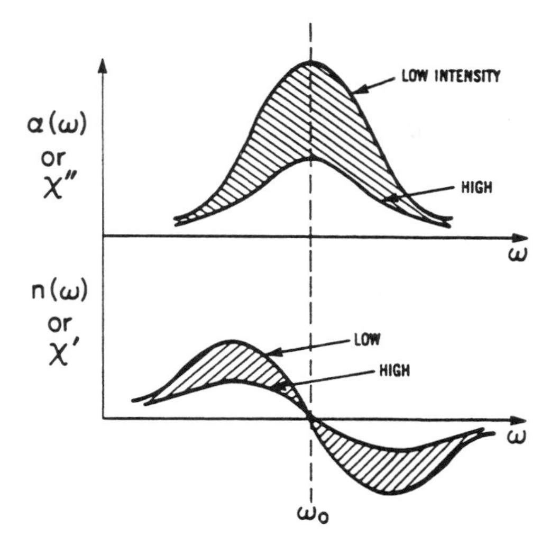

FIG. 18. Saturation of absorption and refraction for a simple absorption line as a function of frequency.

tive $n_2 \cong 1$ cm^2/kW at 5.4 μm and 77 K) and InAs (effective $n_2 \cong 0.03$ cm^2/kW at 3.1 μm and 77 K) and is also very important in room-temperature GaAs.

Exciton nonlinear refraction can be viewed similarly. Less energy is required to create an electron and hole that are bound to each other to form an exciton than to create a free electron and free hole. The exciton resonance appears as an absorption peak at an energy slightly below the band edge. Saturation of the exciton absorption—for example, by creating so many excitons that they overlap and shorten each other's lifetime by screening the Coulomb potential responsible for their binding—changes the refractive index as in Fig. 18. Phase-space filling leads to a large excitonic optical nonlinearity in multiple-quantum-well material (effective $n_2 \cong 0.2$ cm^2/kW for GaAs–AlGaAs at 0.84 μm and 300 K).

The intensity dependence of a two-photon absorption peak arising from biexcitons (i.e., excitonic molecules) likewise gives rise to a nonlinear refractive index, but of the opposite sign, because the absorption increases with increased light intensity (e.g., as in CuCl). Nonlinear refraction can also arise from an electron–hole plasma. Free electrons and holes in semiconductors behave like mobile charge carriers that can be moved by applied electric (optical) fields. Their effective masses are often less than one-tenth of that of a free electron, so they move easily and produce large effects. If one compares the values of the ratio of n_2 to the response time as a measure of the size of the nonlinearity and the speed, one finds that the only mechanisms with values as large as 0.01–10 cm^2 kW μsec are resonant nonlinearities in semiconductors. For this reason, most research directed toward fast practical devices is concentrated on semiconductors.

There are several techniques for measuring n_2. Perhaps the most popular method at present is optical phase conjugation in the form of degenerate four-wave mixing, which has been applied to the measurement of both small and very large n_2 values. Wavefront encoding leading to changes in the far-field profile has also been

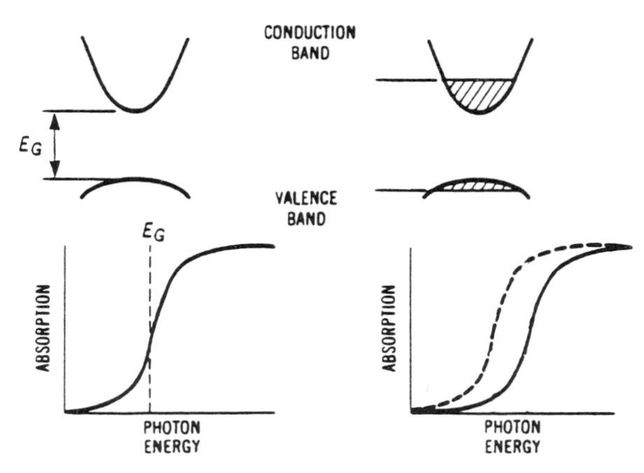

FIG. 19. Schematic illustration of the shift in optical absorption in a direct-gap semiconductor due to band filling (excitonic effects are omitted for simplicity).

used, and various techniques utilize a change in interference fringes.

IV. Growth of Materials and Fabrication of Devices

Exploitation of the physical mechanisms described in the previous section, for the development of optical circuits, requires precise control over the form and quality of the materials involved. This section describes the methods employed to produce high-purity materials and to combine them in forms that allow observation of their nonlinear behavior.

Both classes of devices discussed, the etalon and guided-wave configurations, rely on the production of thin, uniform layers of materials with known properties. The thicknesses and indices of the layers must be accurately controlled, and the chemical purity must be maintained at a high level, particularly for semiconductors. Both vacuum-deposition and bulk-modification methods are employed in the production of these devices.

A. Vapor-Deposition Growth

The etalon configuration and some of the waveguide structures require thin layers with optical thicknesses that vary by less than 1% of the film thicknesses. The best way to achieve this type of uniformity is by deposition from a vapor, either in a vacuum or through a chemical reaction at a heated substrate. A number of variations in each of these methods, applicable to the production of materials for optical circuitry, will be discussed.

1. Vacuum Evaporation

This technique is used for the deposition of thin layers of ZnS, ZnSe, and CuCl, as well as for the deposition of the dielectric multilayer stacks that act as nonabsorbing reflectors in the etalon configuration. A vessel is evacuated to a pressure of about 10^{-6} Torr, and either resistive heating or electron-beam bombardment is used to heat the source material. When the vapor pressure of the material exceeds the base pressure of the system, material travels from the source to the substrate, where the vapor condenses to form a thin layer. With an appropriate choice both of source configuration and source-to-substrate distance, the uniformity of this layer can be excellent. Thickness of the layer may be monitored either with a quartz-crystal micro-balance or by observing the quarterwave fringes optically. Concentration of impurities depends primarily on the purity of the source and on the presence of residual gases.

Films deposited in this manner are, in general, polycrystalline, but use of a single-crystal substrate, low evaporation rates, and heated substrates may result in epitaxial growth. Reduction of the base pressure and use of either elemental or compound effusion ovens as sources gives rise to the technique known as molecular beam epitaxy.

2. Molecular Beam Epitaxy

Molecular beam epitaxy (MBE) is employed extensively in both research and manufacturing environments for the production of high-quality GaAs. In addition to the high purity achieved by the use of extremely low base pressures, slow deposition rates allow fine control over the thicknesses of the layers, as well as the production of abrupt interfaces. These capabilities are crucial for the formation of quantum-well structures. Recent modifications to the MBE process allow the use of some gaseous sources in the chamber as well as the normal solid evaporation sources. These methods, referred to as chemical beam epitaxy, metal organic molecular beam epitaxy, or gas source molecular beam epitaxy, combine some aspects of chemical vapor deposition and MBE. High-quality $Ga_xIn_{1-x}As$–InP materials have been grown by this technique.

The principle of MBE is the same as for conventional vapor deposition, but associated hardware differs. Deposition is performed in a stainless-steel chamber with all-metal seals, which can be baked under vacuum to desorb gases from its walls. This chamber, coupled with a high-capacity pump (often of the entrainment type), allows base pressures in the range of 10^{-12} to 10^{-10} Torr. Sources are temperature-regulated ovens known as Knudsen cells, which maintain a constant flux at fixed temperatures. The cells are enclosed in a liquid-nitrogen-cooled shroud, which keeps the background pressure at the substrate low and prevents cross-talk between sources. For GaAs, separate sources are used for Ga and for As, as the compound evaporates incongruently. Careful control is required over the arrival rates of the constituents to form stoichiometric films, as the two vapors have different sticking coefficients. The use of a hot single-crystal substrate permits epitaxial crystalline growth. Various analysis techniques are used *in vacuo* to assess the quality of the films. X-ray photoelectron or Auger spectroscopy may be used to detect impurities, and electron diffraction to assess the quality of the crystal surface. The quantum wells with GaAs–$Ga_{1-x}AlAs$ layers are made by adding an Al source and opening and closing the shutter as required to create the desired structure.

3. Atomic Layer Evaporation

Atomic layer evaporation (ALE) is a relatively new technique which demonstrates promise for the production of high-quality II–VI semiconductors; it has also been applied to the production of GaAs. ALE has been performed in both ultra-high-vacuum and medium-high-vacuum conditions. It involves exposure of the substrate to alternating vapor bunches from two or more sources. A heated substrate is used to drive a chemical reaction between the two source materials. The deposit is a compound, and the reactant species, as well as any waste products, must have a much higher vapor pressure at the substrate temperature than the desired deposit. Either elements or compounds may be used as source materials. In the case of elemental sources, no waste products would be anticipated. After each vapor pulse, which should contain in excess of the number of atoms or molecules required for a monolayer, there is a time lag before the reactant pulse is sent in. This lag allows the first material to diffuse on the surface and to react at each available site, and allows re-evaporation of the excess material. With this method, the thickness of the deposit is determined by the number of cycles of the two source ovens. This scheme allows deposition with highly controlled thicknesses and possibly superlattice formation. Use of a single-crystal substrate has yielded epitaxial films of CdTe.

4. Chemical Vapor Deposition

Chemical vapor deposition (CVD) is similar to ALE except that there is no temporal separation of the reactive materials. A heated substrate is held beneath the flow of gases containing the desired deposit constituents. The reaction that is then thermally driven results in the deposition of a film of uniform thickness and, usually, low intrinsic stress. Use of organometallics and single-crystal substrates has yielded high-quality epitaxial layers. The equipment for metal–organic chemical vapor deposition (MOCVD) is modestly priced; the gases involved require special handling, however. There has been some controversy about the abruptness of the interfaces formed through use of this technique, but they are of sufficient quality to permit lasing in heterostructures produced by CVD.

B. Other Growth Techniques

1. Liquid-Phase Epitaxy

In liquid-phase epitaxy (LPE), condensation of each new layer of material occurs from a liquid solution above the single-crystal substrate. A graphite tray holds the solute, and the substrate is slid beneath it. Precise cooling of the solute allows growth of an epitaxial layer. If more than one layer, with different compositions, is required, the substrate slides underneath the solutions with different compositions. The technique is economical and satisfactory for the growth of single layers, but interdiffusion at, and characterization of, boundaries can be a significant problem. LPE is applicable to a wide range of materials, including GaAs and compounds thereof.

2. In-Diffusion

In-diffusion is used extensively for the production of waveguides and other devices for integrated optics. Rather than growing a new layer for waveguiding, the index of the outer region of the substrate is altered through controlled introduction of chemical impurities. The two most often used techniques are Ti doping of $LiNbO_3$ and silver-ion exchange in glasses containing alkali. In each case, a controlled heating process, with the desired impurity close to the surface, results in a reproducible increase in the refractive index profile at that surface. Complex structures can be defined by the use of photoresist patterning techniques.

C. Patterning

A long-established need in waveguide devices, and an emerging one in etalons, is for control of the form of the layers within the plane defined by one of the growth techniques discussed above. Most applications require pattern generation on the scale of a few micrometers. This is easily achieved using photoresist techniques adopted from the semiconductor industry.

A solution of polymer containing a photosensitive element is spread in a thin layer (0.5–2.0 μm) on the substrate by spinning at high speeds. This layer is exposed to ultraviolet light through the desired pattern, and a selective developer removes the exposed or unexposed portion, depending on the type of photoresist. The pattern is transferred to the active layer of the device either by etching or in-diffusion. In some devices, such as grating couplers, the pattern is holographically formed, and the resist itself may be used as a grating rather than a transferred copy.

There are several methods for removing material through a mask, including chemical, ion, and reactive-ion etching. Chemical etching is the simplest method to implement and is invaluable in many processes. However, undercutting of the mask and selective directional and defect-etching effects may make the process difficult to control. These factors have led to the increased use of ion-beam techniques.

The most straightforward method of ion-beam etching is to accelerate inert gas ions to energies of the order of 500 eV and to direct them toward the substrate. The ions sputter material away wherever they strike the substrate. The photoresist protects the regions under it, and the pattern is transferred to the substrate. This method can be set up in any existing vacuum system, but the etch rates for some materials of interest are rather slow. This factor has led to the development of reactive-ion etching techniques. Reactive-ion etching is usually performed in a dedicated vacuum system and uses a radio-frequency (RF) plasma discharge rather than a directed beam of ions. The discharge creates fragments that react chemically with the substrate as they bombard it, increasing the etch rate dramatically. Very sharp etch profiles can be achieved if the appropriate reactive gas is chosen.

V. Nonlinear Photonics

The emphasis in this article has been on optical nonlinearities and all-optical devices. Nonlinear optical devices are most often proposed for digital optical computing, optical neural computing, and photonic switching.

In contrast with the strong emphasis on serial communication in waveguide technology, advances in two-dimensional optical circuitry have focused on the development of parallel computing or signal-processing systems that can utilize the inherent parallelism of light. A single image formed by a simple optical system can contain millions of bits of encoded information. This information can be communicated simultaneously through space using a variety of optical systems with ease and without mutual interference.

The basic concepts of the digital optical (parallel-processing) systems are two-dimensional arrays (Fig. 20) of all-optical logic gates, with the outputs of each gate array communicating to the inputs of the next array through free-space reimaging interconnects. The free-space interconnection of the logic-gate arrays could be provided by classical optical components (e.g., prisms and lenses) for systems with simple regular interconnection patterns, as might be used for image processing applications. More complex interconnection patterns may require the use of holographic optical elements.

The impressive success of digital electronic computers and the problems with nonlinear optical devices (cascading, high switching energy, and tolerance) make it doubtful that nonlinear photonics will impact computing anytime soon. If nonlinear photonics does make an impact, it will probably be in the form of an ultrafast preprocessor or as an ultrafast switch that removes some bottleneck. It is also unlikely that nonlinear photonics will contribute to optical neural computing. Some system designs include nonlinear etalons for decision making or for control-

FIG. 20. An array of 9 × 9-μm GaAs pixels defined by reactive etching. Some have been operated as NOR gates.

ling sigmoidal transmission to achieve backward error propagation. Because GaAs etalons and ZnSe interference filters require energy of ≥ 1 kW/cm^2, they are incompatible with efficient photorefractive crystals requiring $\cong 1$ mW/cm^2, unless one is performing whole-image decision making. Unless slow, low-power, all-optical devices are developed, linear electrooptic devices such as the SEED (Fig. 3) may be the only practical solution. Such electrooptic devices may be able to preserve the phase, as is often required. However, no photorefractive crystal works well at the GaAs SEED wavelength. Also, photorefractive crystals have their own problems, so they may be a bottleneck in implementing neural networks optically. In addition, in the neural net area, as in the digital computing area, the electronic competition is quite formidable and is growing rapidly. Perhaps optics should be used only for the vitally important interconnect function.

Finally, nonlinear optics can be applied to photonic switching for multiplexing, time reordering, high-speed routing, sampling, image processing, dynamic interconnects, and other applications. Optics leads electronics by far in short-pulse generation, transmission, and diagnostics. Several switches that open or close in a picosecond have been demonstrated, with recovery times varying from 1 psec to several nanoseconds. Multiple-quantum-wells and optical glass fibers have been used in picosecond switches, and many organic materials exhibit the appropriate response times. Thus, switching is the area of nonlinear photonics in which nonlinear optics can make unique contributions, accomplishing feats not possible at all or as well with electronics. Clearly, the disadvantages of high switching energy and two-wavelength operation are not nearly as serious for a single switch as for an optical computer. Also, in this area, one or a few devices may be essential in the realization of a faster overall system, making photonic switching the area most ripe for nonlinear photonic expansion.

Bibliography

Ghatak, A. K., and Thyagarajan, K. (1989). "Optical Electronics." Cambridge Univ. Press, London.

Gibbs, H. M. (1985). "Optical Bistability: Controlling Light with Light." Academic Press, New York.

Gibbs, H. M., McCall, S. L., and Venkatesan, T. M. C. (1980). Optical bistable devices: The basic components of all-optical systems?, *Opt. Eng.* **19**, 463.

Haug, H., ed. (1989). "Optical Nonlinearities and Instabilities in Semiconductors." Academic Press, Boston.

Koch, S. W., Peyghambarian, N., and Gibbs, H. M. (1988). Band-edge nonlinearities in direct-gap semiconductors and their application to optical bistability and optical computing, *J. Appl. Phys.* **63**, R1.

Macleod, H. A. (1969). "Thin-Film Optical Filters." Adam Hilger, London.

Miller, D. A. B. (1983). Dynamic nonlinear optics in semiconductors: Physics and applications, *Laser Focus* **19**(7), 61.

Stegeman, G. I., Wright, E. M., Finlayson, N., Zanoni, R., and Seaton, C. T. (1988). Third-order nonlinear integrated optics, *J. Light. Tech.* **6**, 953.

OPTICAL FIBER COMMUNICATIONS

B. K. Tariyal *AT&T Technologies, Inc.*
A. H. Cherin *AT&T Bell Laboratories*

GLOSSARY

Attenuation: Decrease of average optical power as light travels along the length of an optical fiber.

Bandwidth: Measure of the information carrying capacity of the fiber. The greater the bandwidth, the greater the information carrying capacity.

Barrier layer: Layer of deposited glass adjacent to the inner tube surface to create a barrier against OH diffusion.

Chemical vapor deposition: Process in which products of a heterogeneous gas–liquid or gas–solid reaction are deposited on the surface of a substrate.

Cladding: Low-refractive-index material that surrounds the fiber core.

Core: Central portion of a fiber through which light is transmitted.

Cut-off wavelength: Wavelength greater than which a particular mode ceases to be a bound mode.

Dispersion: Cause of distortion of the signal due to different propagation characteristics of different modes, leading to bandwidth limitations.

Graded index profile: Any refractive index profile that varies with radius in the core.

Microbending: Sharp curvatures involving local fiber axis displacements of few micrometers and spatial wavelengths of a few millimeters. Microbending causes significant losses.

Mode: Permitted electromagnetic field pattern within an optical fiber.

Numerical aperture: Acceptance angle of the fiber.

Optical repeater: Optoelectric device that receives a signal and amplifies it and retransmits it. In digital systems the signal is regenerated.

Communications using light as a signal carrier and optical fibers as transmission media are termed optical fiber communications. The applications of optical fiber communications have increased at a rapid rate, since the first commercial installation of a fiber-optic system in 1977. Today every major long-distance telecommunication company is spending millions of dollars on optical fiber communication systems. In an optical fiber communication system, voice, video, or data are converted to a coded pulse stream of light using a suitable light source. This pulse stream is carried by optical fibers to a regenerating or receiving station. At the final receiving station the light pulses are converted to electric signals, decoded, and converted into the form of the original information. Optical fiber communications are currently used for telecommunications, data communications, military applications, industrial controls, and medical applications.

I. Introduction

Since ancient times, humans have used light as a vehicle to carry information. Lanterns on ships and smoke signals or flashing mirrors on land are early examples of uses of how humans used light to communicate. It was just over a hundred years ago that Alexander Graham Bell (1880) transmitted a telephone signal a distance greater than 200 m using light as the signal carrier. Bell called his invention a "Photophone" and obtained a patent for it. Bell, however, wisely gave up the photophone in favor of the electric telephone. Photophone at the time of its invention could not be exploited commercially because of two basic drawbacks: (1) the lack of a reliable light source, and (2) the lack of a dependable transmission medium.

The invention of the laser in 1960 gave a new impetus to the idea of lightwave communications (as scientists realized the potential of the dazzling information carrying capacity of these lasers). Much research was undertaken by different laboratories around the world during the early 1960s on optical devices and transmission media. The transmission media, however, remained the main problem, until K. C. Kao and G. A. Hockham in 1966 proposed that glass fibers with sufficiently high-purity core surrounded by a lower-refractive-index cladding could be used for transmitting light over long distances. At the time, available glasses had losses of several thousand decibels per kilometer. In 1970, Robert Maurer of Corning Glass Works was able to produce a fiber with a loss of 20 dB/km. Tremendous progress in the production of low-loss optical fibers has been made since then in the various laboratories in the United States, Japan, and Europe, and today optical fiber communication is one of the fastest growing industries. Optical fiber communication is being used to transmit voice, video, and data over long distance as well as within a local network. [*See* OPTICAL FIBERS, DRAWING AND COATING.]

Fiber optics appears to be the future method of choice for many communications applications. The biggest advantage of a lightwave system is its tremendous information carrying capacity. There are already systems that can carry several thousand simultaneous conversations over a pair of optical fibers thinner than human hair. In addition to this extremely high capacity, the lightguide cables are light weight, they are immune to electromagnetic interference, and they are potentially very inexpensive.

A lightwave communication system (Fig. 1) consists of a transmitter, a transmission medium, and a receiver. The transmitter takes the coded electronic signal (voice, video or data) and converts it to the light signal, which is then carried by the transmission medium (an optical fiber cable) to either a repeater or the receiver. At the receiving end the signal is detected, converted to electrical pulses, and decoded to the proper output. This article provides a brief overview of the different components used in an optical fiber system, along with examples of various applications of optical fiber systems.

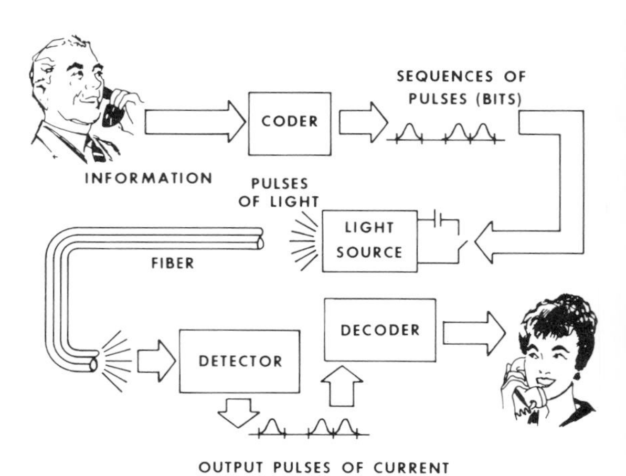

FIG. 1. Schematic diagram of a lightwave communications systems.

II. Classification of Optical Fibers and Attractive Features

Fibers that are used for optical communication are waveguides made of transparent dielectrics whose function is to guide light over long distances. An optical fiber consists of an inner cylinder of glass called the core, surrounded by a cylindrical shell of glass of lower refractive index, called the cladding. Optical fibers (lightguides) may be classified in terms of the refractive index profile of the core and whether one mode (single-mode fiber) or many modes (multimode fiber) are propagating in the guide (Fig. 2). If the core, which is typically made of a high-silica-content glass or a multicomponent glass, has a uniform refractive index n_1, it is called a *step-index fiber*. If the core has a nonuniform refractive index that gradually decreases from the center toward the core–cladding interface, the fiber is called a *graded-index fiber*. The cladding surrounding the core has a uniform refractive index n_2 that is slightly lower than the refractive index of the core region. The cladding of the fiber is made of a high-silica-content glass or a multicomponent glass. Figure

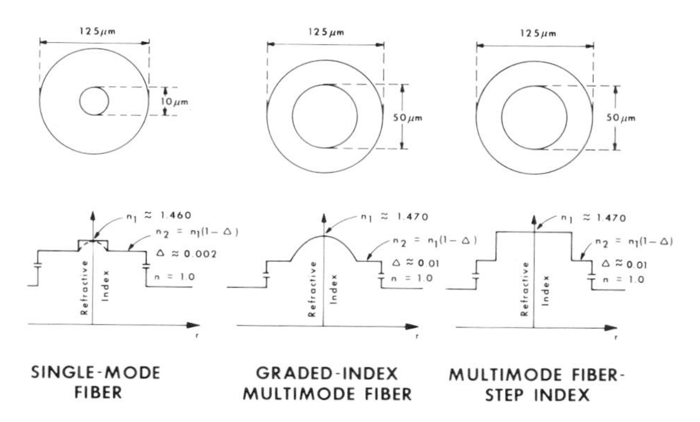

FIG. 2. Geometry of single-mode and multimode fibers.

2 shows the dimensions and refractive indices for commonly used telecommunication fibers. Figure 3 enumerates some of the advantages, constraints, and applications of the different types of fibers. In general, when the transmission medium must have a very high bandwidth—for example, in an undersea or long-distance terrestrial system—a single mode fiber is used. For intermediate system bandwidth requirements between 200 MHz km and 2 GHz km, such as found in intracity trunks between telephone central offices or in local area networks, either a single-mode or graded-index multimode fiber would be the choice. For applications such as short data links where lower bandwidth requirements are placed on the transmission medium, either a graded-index or a step-index multimode fiber may be used.

Because of their low loss and wide bandwidth capabilities, optical fibers have the potential for being used wherever twisted wire pairs or coaxial cables are used as the transmission medium in a communication system. If an engineer were interested in choosing a transmission medium for a given transmission objective, he or she would tabulate the required and desired features of alternate technologies that may be available for use in the applications. With that process in mind, a summary of the attractive features and the advantages of optical fiber transmission will be given. Some of these advantages include (a) low loss and high bandwidth; (b) small size and bending radius; (c) nonconductive, nonradiative, and noninductive; (d) light weight; and (e) providing natural growth capability.

To appreciate the low loss and wide bandwidth capabilities of optical fibers, consider the curves of signal attenuation versus frequency for three different transmission media shown in Fig. 4. Optical fibers have a "flat" transfer function well beyond 100 MHz. When compared with wire pairs of coaxial cables, optical fibers have far less loss for signal frequencies above a few megahertz. This is an important characteristic that strongly influences system economics, since it allows the system designer to increase

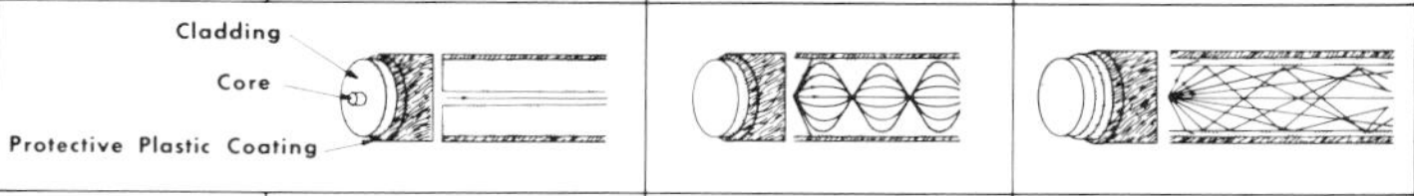

	SINGLE-MODE FIBER	GRADED-INDEX MULTIMODE FIBER	STEP-INDEX MULTIMODE FIBER
Cladding Core Protective Plastic Coating			
SOURCE	LASER PREFERRED	LASER or LED	LASER or LED
BANDWIDTH	VERY VERY LARGE > 3 GHz km	VERY LARGE 200 MHz 3 GHz km	LARGE < 200 MHz km
SPLICING	DIFFICULT DUE TO SMALL CORE	DIFFICULT BUT DOABLE	DIFFICULT BUT DOABLE
EXAMPLE OF APPLICATION	SUBMARINE CABLE SYSTEM	TELEPHONE LOOP DISTRIBUTION SYSTEM	DATA LINKS
COST	LEAST EXPENSIVE	MOST EXPENSIVE	EXPENSIVE

FIG. 3. Applications and characteristics of fiber types.

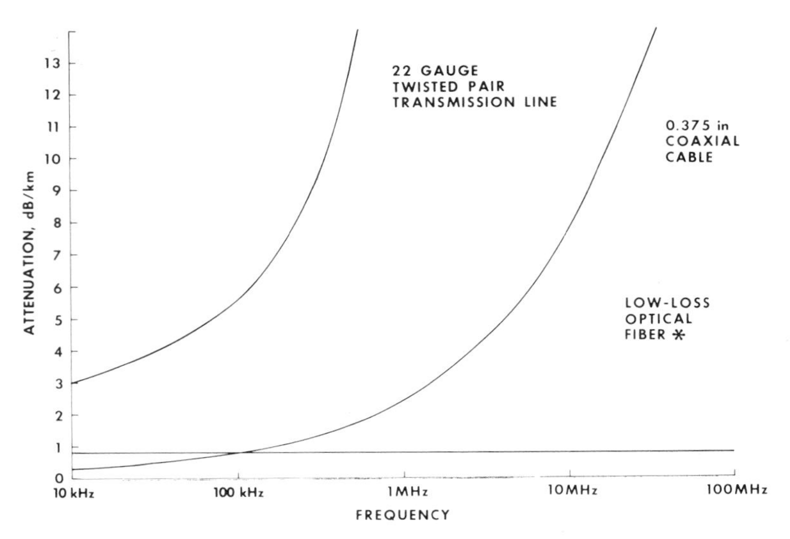

FIG. 4. Attenuation versus frequency for three different transmission media. Asterisk indicates fiber loss at a carrier wavelength of 1.3 μm.

the distance between regenerators (amplifiers) in a communication system.

The small size, small bending radius (a few centimeters), and light weight of optical fibers and cables are very important where space is at a premium, such as in aircraft, on ships, and in crowded ducts under city streets.

Because optical fibers are dielectric waveguides, they avoid many problems such as radiative interference, ground loops, and, when installed in a cable without metal, lightning-induced damage that exists in other transmission media.

Finally, the engineer using optical fibers has a great deal of flexibility. He or she can install an optical fiber cable and use it initially in a low-capacity (low-bit-rate) system. As the system needs grow, the engineer can take advantage of the broadband capabilities of optical fibers and convert to a high-capacity (high-bit-rate) system by simply changing the terminal electronics. A comparison of the growth capability of different transmission media is shown in Table I. For the three digital transmission rates considered (1.544, 6.312, and 44.7 Mbit/sec), the loss of the optical fiber is constant. The loss of the metallic transmission lines, however, increases with increasing transmission rates, thus limiting their use at the higher bit rates. The optical fiber system, on the other hand, could be used at all bit

TABLE I. Growth Capability, Transmission Media Comparisons

	Loss in dB/km at half bit rate frequency (digital transmission rates)		
Transmission media	T1 (1.544 Mbit/sec)	T2 (6.312 Mbit/sec)	T3 (44.736 Mbit/sec)
26-Gauge twisted wire pair	24	48	128
19-Gauge twisted wire pair	10.8	21	56
0.375-in.-Diameter coaxial cable	2.1	4.5	11
Low-loss optical fiber[a]	0.75	0.75	0.75

[a] Fiber loss at a carrier wavelength of 1.3 μm.

rates and can grow naturally to satisfy system needs.

III. Fiber Transmission Characteristics

The proper design and operation of an optical communication system using optical fibers as the transmission medium requires a knowledge of the transmission characteristics of the optical sources, fibers, and interconnection devices (connectors, couplers, and splices) used to join lengths of fibers together. The transmission criteria that affect the choice of the fiber type used in a system are signal attenuation, information transmission capacity (bandwidth), and source coupling and interconnection efficiency. Signal attenuation is due to a number of loss mechanisms within the fiber, as shown in Table II, and due to the losses occurring in splices and connectors. The information transmission capacity of a fiber depends on dispersion, a phenomenon that causes light that is originally concentrated into a short pulse to spread out into a broader pulse as it travels along an optical fiber. Source and interconnection efficiency depends on the fiber's core diameter and its numerical aperture, a measure of the angle over which light is accepted in the fiber.

Absorption and scattering of light traveling through a fiber leads to signal attenuation, the rate of which is measured in decibels per kilometer (dB/km). As can be seen in Fig. 5, for both multimode and single-mode fibers, attenuation depends strongly on wavelength. The decrease in scattering losses with increasing wavelength is offset by an increase in material absorption such that attenuation is lowest near 1.55 μm (1550 nm).

The measured values given in Table III are probably close to the lower bounds for the attenuation of optical fibers. In addition to intrinsic fiber losses, extrinsic loss mechanisms, such as absorption due to impurity ions, and microbending loss due to jacketing and cabling can add loss to a fiber.

TABLE II. Loss Mechanisms

Intrinsic material absorption loss
Ultraviolet absorption tail
Infrared absorption tail
Absorption loss due to impurity ions
Rayleigh scattering loss
Waveguide scattering loss
Microbending loss

TABLE III. Best Attenuation Results (dB/km) in Ge–P–SiO_2-Core Fibers

Wavelength (nm)	$\Delta \approx 0.2\%$ (Single-mode fibers)	$\Delta \approx 1.0\%$ (Graded-index multimode fibers)
850	2.1	2.20
1300	0.27	0.44
1500	0.16	0.23

The bandwidth or information-carrying capacity of a fiber is inversely related to its total dispersion. The total dispersion in a fiber is a combination of three components: intermodal dispersion (modal delay distortion), material dispersion, and waveguide dispersion.

Intermodal dispersion occurs in multimode fibers because rays associated with different modes travel different effective distances through the optical fiber. This causes light in the different modes to spread out temporally as it travels along the fiber. Modal delay distortion can severely limit the bandwidth of a step index multimode fiber to the order of 20 MHz km. To reduce modal delay distortion in multimode fibers, the core is carefully doped to create a graded (approximately parabolic shaped) refractive index profile. By carefully designing this in-

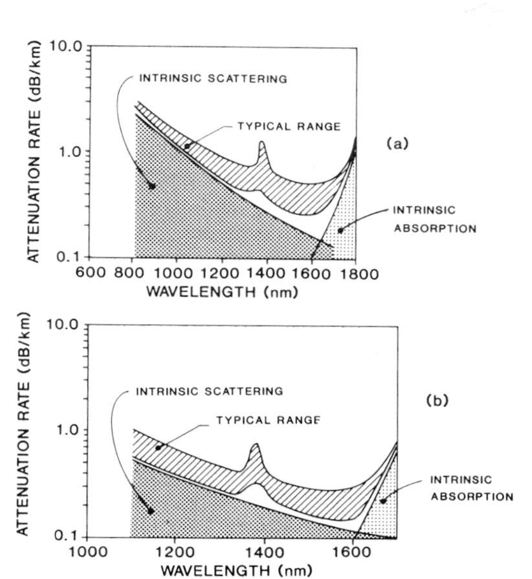

FIG. 5. Spectral attenuation rate. (a) Graded-index multimode fibers. (b) Single-mode fibers.

dex profile, the group velocities of the propagating modes are nearly equalized. Bandwidths of 1.0 GHz km are readily attainable in commercially available graded index multimode fibers. The most effective way of eliminating intermodal dispersion is to use a single-mode fiber. Since only one mode propagates in a single mode fiber, modal delay distortion between modes does not exist and very high bandwidths are possible. The bandwidth of a single-mode fiber, as mentioned previously, is limited by the combination of material and waveguide dispersion. As shown in Fig. 6, both material and waveguide dispersion are dependent on wavelength.

Material dispersion is caused by the variation of the refractive index of the glass with wavelength and the spectral width of the system source. Waveguide dispersion occurs because light travels in both the core and cladding of a single-mode fiber at an effective velocity between that of the core and cladding materials. The waveguide dispersion arises because the effective velocity, the waveguide dispersion, changes with wavelength. The amount of waveguide dispersion depends on the design of the waveguide structure as well as on the fiber materials. Both material and waveguide dispersion are measured in picoseconds (of pulse spreading) per nanometer (of source spectral width) per kilometer (of fiber length), reflecting both the increases in magnitude in source linewidth and the increase in dispersion with fiber length.

Material and waveguide dispersion can have different signs and effectively cancel each other's dispersive effect on the total dispersion in a single-mode fiber. In conventional germanium-doped silica fibers, the "zero-dispersion" wavelength at which the waveguide and material dispersion effects cancel each other out occurs near 1.30 μm. The zero-dispersion wavelength can be shifted to 1.55 μm, or the low-dispersion

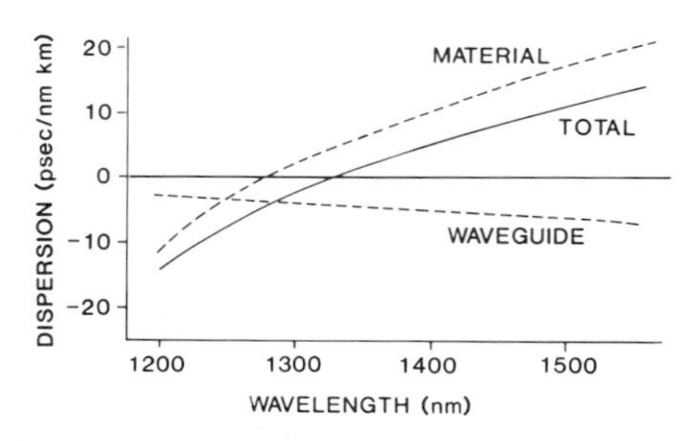

FIG. 6. Single-mode step-index dispersion curve.

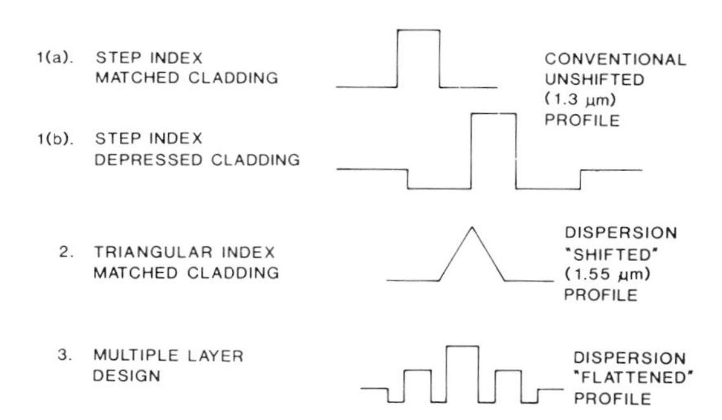

FIG. 7. Single-mode refractive-index profiles.

characteristics of a fiber can be broadened by modifying the refractive-index profile shape of a single-mode fiber. This profile shape modification alters the waveguide dispersion characteristics of the fiber and changes the wavelength region in which waveguide and material dispersion effects cancel each other. Figure 7 illustrates the profile shapes of "conventional," "dispersion-shifted," and "dispersion-flattened" single-mode fibers. Single-mode fibers operating in their zero-dispersion region with system sources of finite spectral width do not have infinite bandwidth but have bandwidths that are high enough to satisfy all current high-capacity system requirements.

IV. Optical Fiber Cable Manufacturing

Optical fiber cables should have low loss and high bandwidth and should maintain these characteristics while in service in extreme environments. In addition, they should be strong enough to survive the stresses encountered during manufacture, installation, and service in a hostile environment. The manufacturing process used to fabricate the optical fiber cables can be divided into four steps: (1) preform fabrication, (2) fiber drawing and coating, (3) fiber measurement, and (4) fiber packaging.

A. Preform Fabrication

The first step in the fabrication of optical fiber is the creation of a glass preform. A preform is a large blank of glass several millimeters in diameter and several centimeters in length. The preform has all the desired properties (e.g., geometrical ratios and chemical composition) necessary to yield a high-quality fiber. The preform is subsequently drawn into a multi-kilometer-long hair-thin fiber. Four different preform manufacturing processes are currently in commercial use.

The most widely used process is the modified chemical vapor deposition (MCVD) process invented at the AT&T Bell Laboratories. Outside vapor deposition process (OVD) is used by Corning Glass Works and some of its joint ventures in Europe. Vapor axial deposition (VAD) process is the process used most widely in Japan. Philips, in Eindhoven, Netherlands, uses a low-temperature plasma chemical-vapor deposition (PCVD) process.

In addition to the above four major processes, other processes are under development in different laboratories. Plasma MCVD is under development at Bell Laboratories, hybrid OVD–VAD processes are being developed in Japan, and Sol-Gel processes are being developed in several laboratories. The first four processes are the established commercial processes and are producing fiber economically. The new processes are aimed at greatly increasing the manufacturing productivity of preforms, and thereby reducing their cost.

All the above processes produce high-silica fibers using different dopants, such as germanium, phosphorous, and fluorine. These dopants modify the refractive index of silica, enabling the production of the proper core refractive-index profile. Purity of the reactants and the control of the refractive-index profile are crucial to the low loss and high bandwidth of the fiber.

1. MCVD Process

In the MCVD process (Fig. 8), a fused-silica tube of extremely high purity and dimensional uniformity is cleaned in an acid solution and degreased. The clean tube is mounted on a glass

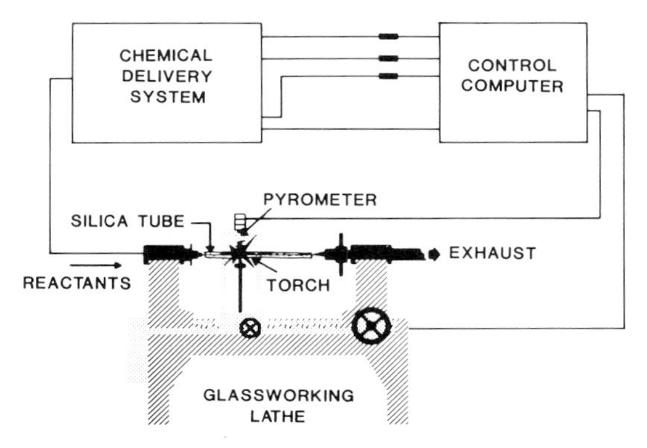

FIG. 8. Schematic diagram of the MCVD process.

working lathe. A mixture of reactants is passed from one end of the tube and exhaust gases are taken out at the other end while the tube is being rotated. A torch travels along the length of the tube in the direction of the reactant flow. The reactants include ultra-high-purity oxygen and a combination of one or more of the halides and oxyhalides ($SiCl_4$, $GeCl_4$, $POCl_3$, BCl_3, BBr_3, SiF_4, CCl_4, CCl_2F_2, Cl_2, SF_6, and $SOCl_2$).

The halides react with the oxygen in the temperature range of 1300–1600°C to form oxide particles, which are driven to the wall of the tube and subsequently consolidated into a glassy layer as the hottest part of the flame passes over. After the completion of one pass, the torch travels back and the next pass is begun. Depending on the type of fiber (i.e., multimode or single-mode), a barrier layer or a cladding consisting of many thin layers is first deposited on the inside surface of the tube. The compositions may include $B_2O_3–P_2O_5–SiO_2$ or $F–P_2O_5–SiO_2$ for barrier layers, and SiO_2, $F–SiO_2$, $F–P_2O_5–SiO_2$, or $F–GeO_2–SiO_2–P_2O_5$ for cladding layers. After the required number of barrier or cladding layers has been deposited, the core is deposited. The core compositions depend on whether the fiber is single-mode, multimode, step-index, or multimode graded index. In the case of graded-index multimode fibers, the dopant level changes with every layer, to provide a refractive index profile that yields the maximum bandwidth.

After the deposition is complete, the reactant flow is stopped except for a small flow of oxygen, and the temperature is raised by reducing the torch speed and increasing the flows of oxygen and hydrogen through the torch. Usually the exhaust end of the tube is closed first and a small positive pressure is maintained inside the deposited tube while the torch travels backward. The higher temperatures cause the glass viscosity to decrease, and the surface tension causes the tube to contract inward. The complete collapse of the tube into a solid preform is achieved in several passes. The speed of the collapse, the rotation of the tube, the temperature of collapse, and the positive pressure of oxygen inside the tube are all accurately controlled to predetermined values in order to produce a straight and bubble-free preform with minimum ovality. The complete preform is then taken off the lathe. After an inspection to assure that the preform is free of defects, the preform is ready to be drawn into a thin fiber.

The control of the refractive-index profile along the cross section of the deposited portion of the preform is achieved through a vapor delivery system. In this system, liquids are vaporized by passing a carrier gas (pure O_2) through the bubblers, made of fused silica. Accurate flows are achieved with precision flow controllers that maintain accurate carrier gas flows and extremely accurate temperatures within the bubblers. Microprocessors are used to automate the complete deposition process, including the torch travel and composition changes throughout the process. Impurities are reduced to very low levels by starting with pure chemicals, and there is further reducing of the impurities with in-house purification of these chemicals. Ultra-pure oxygen and a completely sealed vapor-delivery system are used to avoid chemical contamination. Transition-metal ion impurities of well below 1 ppb and OH^- ion impurities of less than 1 ppm are typically maintained to produce high quality fiber.

2. The PCVD Process

The PCVD process (Fig. 9) also uses a starting tube, and the deposition takes place inside the tube. Here, however, the tube is either stationary or oscillating and the pressure is kept at 10–15 torr. Reactants are fed inside the tube, and the reaction is accomplished by a traveling microwave plasma inside the tube. The entire tube is maintained at approximately 1200°C. The plasma causes the heterogeneous depositions of glass on the tube wall, and the deposition efficiency is very high. After the required depositions of the cladding and core are complete, the tube is taken out and collapsed on a separate equipment. Extreme care is required to prevent impurities from getting into the tube during the transport and collapse procedure. The PCVD process has the advantages of high efficiency, no tube distortion because of the lower temperature, and very accurate profile control because of the large number of layers deposited in a short time. However, going to higher rates of flow presents some difficulties, because of a need to maintain the low pressure.

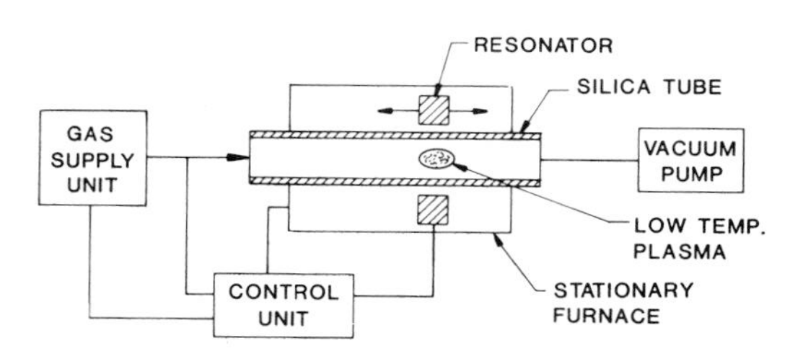

FIG. 9. Schematic diagram of the PCVD process.

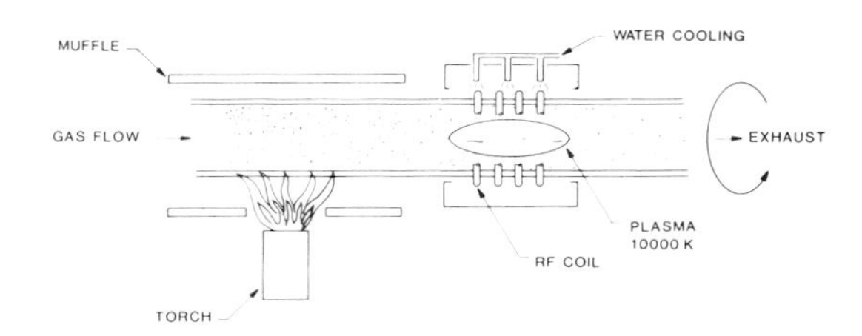

FIG. 10. Schematic diagram of the PMCVD process.

3. The PMCVD Process

The PMCVD is an enhancement of the MCVD process. Very high rates of deposition (up to 10 g/min, compared to 2 g/min for MCVD) are achieved by using larger diameter tubes and an RF plasma for reaction (Fig. 10). Because of the very high temperature of the plasma, water cooling is essential. An oxyhydrogen torch follows the plasma and sinters the deposition. The high rates of deposition are achieved because of very high thermal gradients from the center of the tube to the wall and the resulting high thermophoretic driving force. The PMCVD process is still in the development stage and has not been commercialized.

4. The OVD Process

The OVD process does not use a starting tube; instead, a stream of soot particles of desired composition is deposited on a bait rod (Fig. 11). The soot particles are produced by the reac-

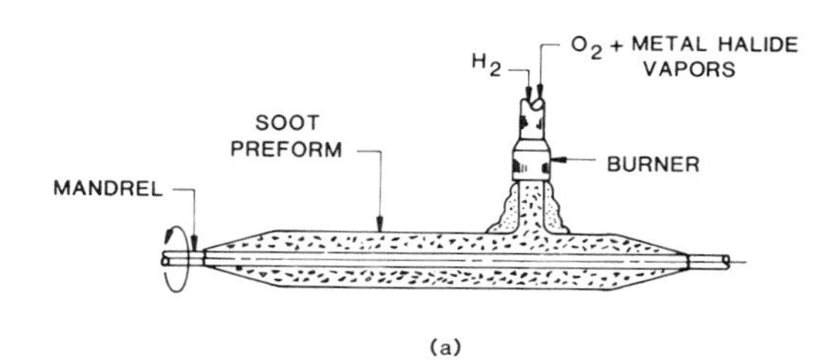

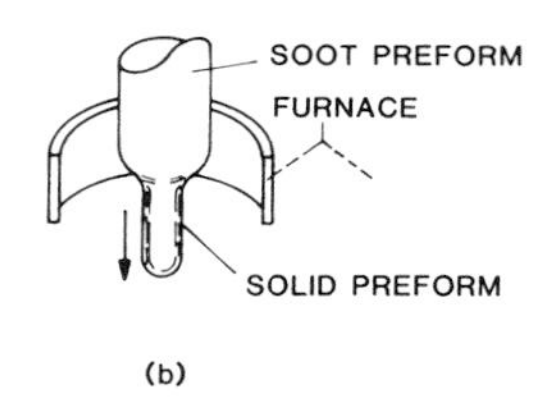

FIG. 11. Schematic diagram of the outside vapor deposition. (a) Soot deposition. (b) Consolidation.

tion of reactants in a fuel gas–oxygen flame. A cylindrical porous soot preform is built layer by layer. After the deposition of the core and cladding is complete, the bait rod is removed. The porous preform is then sintered and dried in a furnace at 1400–1600°C to form a clear bubble-free preform under a controlled environment. The central hold left by the blank may or may not be closed, depending on the type of preform. The preform is now ready for inspection and drawing.

5. The VAD Process

The process is very similar to the OVD process. However, the soot deposition is done axially instead of radially. The soot is deposited at the end of a starting silica-glass rod (Fig. 12). A special torch using several annular holes is used to direct a stream of soot at the deposition surface. The reactant vapors, hydrogen gas, argon gas, and oxygen gas flow through different annular openings. Normally the core is deposited and the rotating speed is gradually withdrawn as the deposition proceeds at the end. The index profile is controlled by the composition of the gases flowing through the torch and the temperature distribution at the deposition surface. The porous preform is consolidated and dehydrated as it passes through a carbon-ring furnace in a controlled environment. $SOCl_2$ and Cl_2 are used to dehydrate the preform. Because of the axial deposition, this process is semicontinuous and is capable of producing very large preforms.

B. Fiber Drawing

After a preform has been inspected for various defects such as bubbles, ovality, and straightness, it is taken to a fiber drawing station. A large-scale fiber drawing process must repeatedly maintain the optical quality of the preform and produce a dimensionally uniform fiber with high strength.

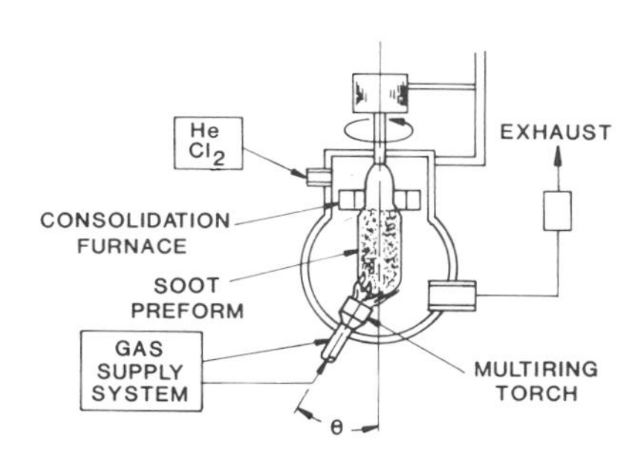

FIG. 12. Schematic diagram of the vapor axial deposition.

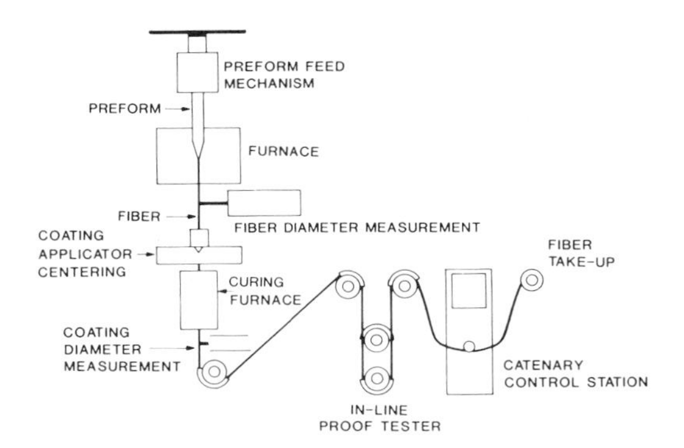

FIG. 13. The fiber drawing process.

1. Draw Process

During fiber drawing, the inspected preform is lowered into a hot zone at a certain feed rate V_p, and the fiber is pulled from the softened neck-down region (Fig. 13) at a rate V_f. At steady state,

$$\pi D_p^2 V_p/4 = \pi D_f^2 V_f/4 \qquad (1)$$

where D_p and D_f are the preform and fiber diameters, respectively. Therefore,

$$V_f = (D_p^2/D_f^2)V_p \qquad (2)$$

A draw machine, therefore, consists of a preform feed mechanism, a heat source, a pulling device, a coating device, and a control system to accurately maintain the fiber diameter and the furnace temperature.

2. Heat Source

The heat source should provide sufficient energy to soften the glass for pulling the fiber without causing excessive tension and without creating turbulence in the neck-down region. A proper heat source will yield a fiber with uniform diameter and high strength. Oxyhydrogen torches, CO_2 lasers, resistance furnaces, and induction furnaces have been used to draw fibers. An oxyhydrogen torch, although a clean source of heat, suffers from turbulence due to flame. A CO_2 laser is too expensive a heat source to be considered for the large-scale manufacture of fibers. Graphite resistance furnaces and zirconia induction furnaces are the most widely used heat sources for fiber drawing. In the graphite resistance furnace, a graphite resistive element produces the required heat. Because graphite reacts with oxygen at high temperatures, an inert environment (e.g., argon) is maintained inside the furnace. The zirconia induction furnace does not require inert environment. It is extremely important that the furnace environment be clean in order to produce high-strength fibers. A zirconia induction furnace, when properly designed and used, has produced very-high-strength long-length fibers (over 2.0 GPa) in lengths of several kilometers.

3. Mechanical Systems

An accurate preform feed mechanism and drive capstan form the basis of fiber speed control. The mechanism allows the preform to be fed at a constant speed into the hot zone, while maintaining the preform at the center of the furnace opening at the top. A centering device is used to position preforms that are not perfectly straight. The preform is usually held with a collet-type chuck mounted in a vertically movable carriage, which is driven by a lead screw. A precision stainless-steel drive capstan is mounted on the shaft of a high-performance dc servomotor. The fiber is taken up on a proper-diameter spool. The fiber is wound on the spool at close to zero tension with the help of a catenary control. In some cases fiber is proof-tested in-line before it is wound on a spool. The proof stress can be set at different levels depending on the application for which the fiber is being manufactured.

4. Fiber Coating System

The glass fiber coming out of the furnace has a highly polished pristine surface, and the theoretical strength of such a fiber is in the range of 15–20 GPa. Strengths in the range of 4.5–5.5 GPa are routinely measured on short fiber lengths. To preserve this high strength, polymeric coatings are applied immediately after the drawing. The coating must be applied without damaging the fiber, it must solidify before reaching the capstan, and it should not cause microbending loss. To satisfy all these requirements, usually two layers of coatings are applied: a soft inner coating adjacent to the fiber to avoid microbending loss, and a hard outer coating to resist abrasion. The coatings are a combination of ultraviolet (UV) curable acrylates, UV-curable silicones, hot melts, heat-curable silicones, and nylons. When dual coatings are applied, the coated fiber diameter is typically 235–250 μm. The nylon-jacketed fiber typically used in Japan has an outside diameter of 900 μm. All coating materials are usually filtered to remove particles that may damage the fiber. Coatings are usually applied by passing the fiber through a coating cup and then curing the coating before the fiber is taken up by the capstan. The method of application, the coating material, the temperature, and the draw speed affect the proper application of a well-centered, bubble-free coating.

Fiber drawing facilities are usually located in a clean room where the air is maintained at class 10,000. The region of the preform and fiber from the coating cup to the top of the preform is maintained at class 100 or better. A class 100 environment means that there are no more than 100 particles of size greater than 0.5 μm in 1 ft^3 of air. A clean environment, proper centering of the preform in the furnace and fiber in the coating cup, and proper alignment of the whole draw tower ensure a scratch-free fiber of a very high tensile strength. A control unit regulates the draw speed, preform feed speed, preform centering, fiber diameter, furnace temperature, and draw tension.

The coated fiber wound on a spool is next taken to the fiber measurement area to assure proper quality control.

5. Proof Testing of Fibers

Mechanical failure is one of the major concerns in the reliability of optical fibers. Fiber drawn in kilometer lengths must be strong enough to survive all of the short- and long-term stresses that it will encounter during the manufacture, installation, and long service life. Glass is an ideal elastic isotropic solid and does not contain dislocations. Hence, the strength is determined mainly by inclusions and surface flaws. Although extreme care is taken to avoid inhomogeneities and surface flaws during fiber manufacture, they cannot be completely eliminated. Since surface flaws can result from various causes, they are statistical in nature and it is impossible to predict the long-length strength of glass fibers. To guarantee a minimum fiber strength, proof testing has been adopted as a manufacturing step. Proof testing can be done in-line immediately after the drawing and coating, or off-line before the fiber is stored.

In proof testing, the entire length of the fiber is subjected to a properly controlled proof stress. The proof stress is based on the stresses likely to be encountered by the fiber during manufacture, storage, installation, and service. The fibers that survive the proof test are stored for further packaging into cables.

Proof testing not only guarantees that the fiber will survive short-term stresses but also guarantees that the fiber will survive a lower residual stress that it may be subjected to during its long service life. It is well known that glass, when used in a humid environment, can fail under a long-term stress well below its instantaneous strength. This phenomenon is called static fatigue. Several models have been proposed to quantitatively describe the relationship between residual stress and the life of optical fibers. Use is made of the most conservative of these models, and the proof stress is determined by a consideration of the maximum possible residual stress in service and the required service life.

C. Fiber Packaging

In order to efficiently use one or more fibers, they need to be packaged so that they can be handled, transported, and installed without damage. Optical fibers can be used in a variety of applications, and hence the way they are packaged or cabled will also vary. There are numerous cable designs that are used by different cable manufacturers. All these designs, however, must meet certain criteria. A primary consideration in a cable design is to assure that the fibers in the cables maintain their optical properties (attenuation and dispersion) during their service life under different environmental conditions. The design, therefore, must minimize microbending effects. This usually means letting the fiber take a minimum energy position at all times in the cable structure. Proper selection of cabling materials so as to minimize differential thermal expansion or contraction during temperature extremes is important in minimizing microbending loss. The cable structure must be such that the fibers carry a load well below the proof-test level at all times, and especially while using conventional installation equipment. The cables must provide adequate protection to the fibers under all adverse environmental conditions during their entire service life, which may be as long as 40 years. Finally, the cable designs should be cost-effective and easily connectorized or spliced.

Five different types (Fig. 14) of basic cable designs are currently in use: (a) loose tube, (b) fluted, (c) ribbon, (d) stranded, and (e) Lightpack Cable. The loose tube design was pioneered by Siemens in Germany. Up to 10 fibers are enclosed in a loose tube, which is filled with a soft filling compound. Since the fibers are relatively free to take the minimum energy configuration, the microbending losses are avoided. Several of these buffered loose tube units are stranded around a central glass–resin support member. Aramid yarns are stranded on the cable core to provide strength members (for pull-

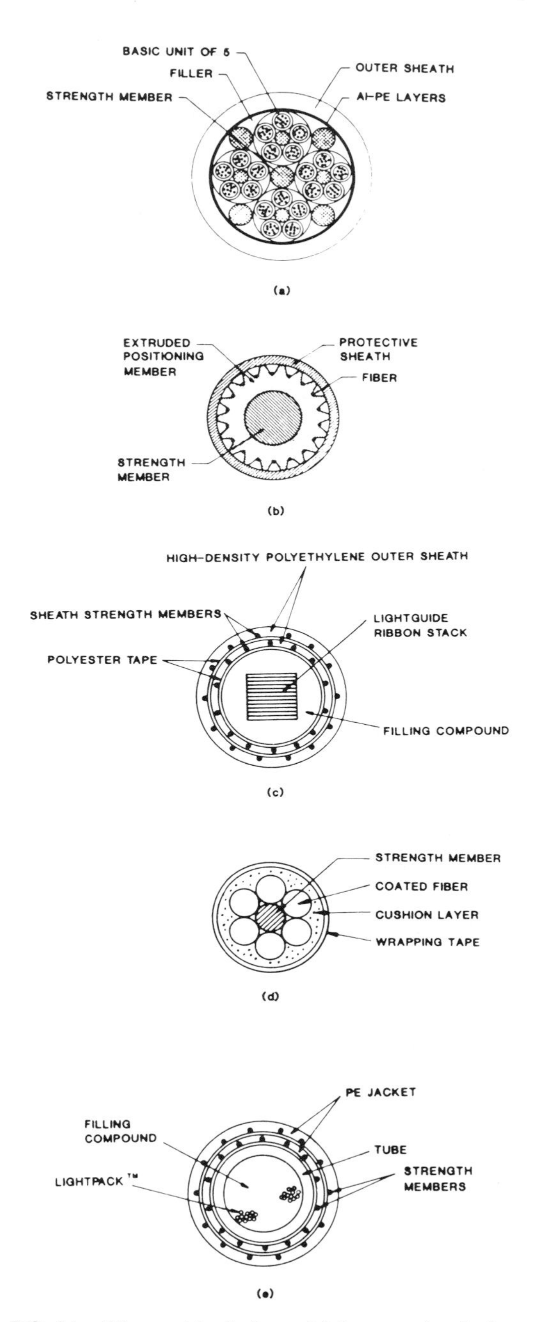

FIG. 14. Fiber cable designs. (a) Loose tube design. (b) Slotted design. (c) Ribbon design. (d) Stranded unit. (e) Lightpack™ Cable design.

ing through ducts), with a final polyethylene sheath on the outside. The stranding lay length and pitch radius are calculated to permit tensile strain on the cable up to the rated force and to permit cooling down to the rated low temperature without affecting the fiber attenuation.

In the fluted designs, fibers are laid in the grooves of plastic central members and are relatively free to move. The shape and size of the grooves vary with the design. The grooved core may also contain a central strength member. A sheath is formed over the grooved core, and this essentially forms a unit. Several units may then be stranded around a central strength member to form a cable core of desired size, over which different types of sheaths are formed. Fluted designs have been pioneered in France and Canada.

The ribbon design was invented at AT&T Bell Laboratories and consists of a linear array of 12 fibers sandwiched between two polyester tapes with pressure-sensitive adhesive on the fiber side. The spacing and the back tension on the fibers is accurately maintained. The ribbons are typically 2.5 mm in width. Up to 12 ribbons can be stacked to give a cable core consisting of 144 fibers. The core is twisted to some lay length and enclosed in a polyethylene tube. Several combinations of protective plastic and metallic layers along with metallic or nonmetallic strength members are then applied around the core to give the final cable its required mechanical and environmental characteristics needed for use in specified conditions. The ribbon design offers the most efficient and economic packaging of fibers for high-fiber-count cables. It also lends the cable to preconnectorization and makes it extremely convenient for installation and splicing.

The tight-bound stranded designs were pioneered by Japanese and are used in the United States for several applications. In this design, several coated fibers are stranded around a central support member. The central support member may also serve as a strength member, and it may be metallic or nonmetallic. The stranded unit can have up to 18 fibers. The unit is contained within a plastic tube filled with a water-blocking compound. The final cable consists of several of these units stranded around a central member and protected on the outside with various sheath combinations.

The Lightpack Cable design, pioneered by AT&T, is one of the simplest designs. Several fibers are held together with a binder to form a unit. One or more units are laid inside a large tube, which is filled with a water-blocking compound. This design has the advantage of the loose tube design in that the fibers are free of strain, but is more compact. The tube-containing units can then be protected with various sheath options and strength members to provide the final cable.

The final step in cabling is the sheathing operation. After the fibers have been made into identifiable units, one or more of the units (as discussed earlier) form a core, which is then covered with a combination of sheathing layers. The number and combination of the sheathing layers depend on the intended use. Typically, a polyethylene sheath is extruded over the filled cable core. In a typical cross-ply design, metallic or nonmetallic strength members are applied over the first sheath layer, followed by another polyethylene sheath, over which another layer of strength members is applied. The direction of lay of the two layers of the strength members is opposite to each other. A final sheath is applied and the cable is ready for the final inspection, preconnectorization, and shipment. Metallic vapor barriers and lightning- and rodent-protection sheath options are also available. Further armoring is applied to cables made for submarine application.

In addition to the above cable designs, there are numerous other cable designs used for specific applications, such as fire-resistant cables, military tactical cables, cables for missile guidance systems, cables for field communications established by air-drop operations, air deployment cables, and cables for industrial controls. All these applications have unique requirements, such as ruggedness, low loss, and repeaterless spans, and the cable designs are accordingly selected. However, all these cable designs still rely on the basic unit designs discussed above.

V. Sources and Detectors

In this section we will review the characteristic of optical sources and detectors that are used in fiber-optic communication systems.

A. Optical Sources

Semiconductor light emitting diodes (LEDs) and injection-laser diodes (ILDs) are attractive as optical carrier sources because they are dimensionally compatible with optical fibers.

They emit at wavelengths (0.8–0.9 μm and 1.3–1.6 μm) corresponding with regions of low optical-fiber loss, their outputs can be rapidly controlled by varying their bias current and therefore they are easy to modulate, and finally they offer solid-state reliability with lifetimes now exceeding 10^6 h. Although LEDs and ILDs exhibit a number of similar characteristics, there are important differences between them that must be understood before one can select a source for a specific fiber-optic communication system.

One major difference between LEDs and ILDs is their spatial and temporal coherence. An ILD radiates a relatively narrow beam of light that has a narrow spectral width. In contrast, LED sources have much wider radiation patterns (beam width) and have moderately large spectral widths. These factors govern the amount of optical power that can be coupled into a fiber and the influence of chromatic dispersion on the bandwidth of the fiber medium. The second difference between ILDs and LEDs is their speed. The stimulated emission from lasers results in intrinsically faster optical rise and fall times in response to changes in drive current than can be realized with LEDs. The third difference between the devices is related to their linearity. LEDs generate light that is almost linearly proportional to the current passing through the device. Lasers, however, are threshold devices, and the lasing output is proportional to the drive current only above threshold. The threshold current of a laser, unfortunately, is not a constant but is a function of the device's temperature and age. Feedback control drive circuitry is therefore required to stabilize a laser's output power. Table IV illustrates typical ILDs and LED characteristics found in fiber-optic communication systems.

TABLE IV. Optical Source Characteristics

	ILDs	LEDs
Output power, mW	1–10	1–10
Power launched into fiber, mW	0.5–5	0.03–0.3
Spectral width (rms value), nm	2–4	15–60
Brightness, W/cm^2 sr	~10^5	10–10^3
Rise time, 10–90%, nsec	<1	2–20
Frequency response (−3 dB), MHz	>500	<200
Voltage drop, V	1.5–2	1.5–2.5
Forward current, mA	10–300	50–300
Threshold current, mA	5–250	Not applicable
Feedback stabilization required	Yes	No

GaAlAs devices (both ILDs and LEDs) emitting in the wavelength region of 0.8–0.9 μm are commercially available and widely used in optical fiber systems. InGaAsP devices, with their emission wavelengths in the region 1.0–1.6 μm, are available for application near 1.3 and 1.6 μm where fiber chromatic dispersion and transmission losses are minimal. The high-radiance Burrus type (surface emitting) LED is well suited for application in systems of low to medium bandwidth (<50 MHz). The power that can be coupled from an LED into a fiber is proportional to the number of modes the fiber can propagate, that is, to its core area times its numerical aperture squared. For simple butt coupling, where the emitting area of the LED is equal to or less than the core area of the fiber, presently available surface emitting GaAlAs and InGaAsP LEDs can launch about 50 μW into a graded-index fiber of numerical aperture (NA) 0.2 and core diameter 50 μm. The spectral width of an LED is a function of the operating wavelength, the active-layer doping concentration, and the junction current density. The rms spectral width of a typical 0.85-μm GaAlAs Burrus-type LED is about 16 nm, while that of a 1.3-μm InGaAsP LED is about 40 nm (spectral width is approximately proportional to λ^2). The modulation bandwidth of an LED depends on the device geometry, its current density, and the doping concentration of its active layer. Higher doping concentration yields higher bandwidth, but only at the expense of lower output power and wider spectral width. Figure 15 shows the tradeoff between output power and modulation bandwidth for a group of Burrus type GaAlSAs LEDs. Typically, a 50-μm-diameter LED that can butt couple 50 μW into a 0.2-NA, 50-μm-core graded-index fiber can be current modulated at rates up to about 50 MHz.

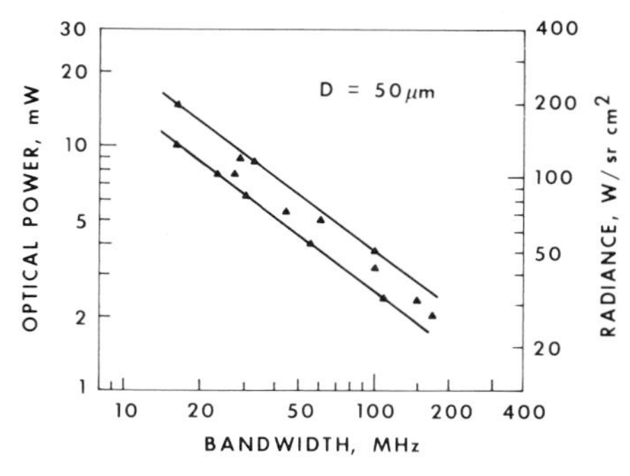

FIG. 15. Output power versus 3-dB bandwidth for Burrus-type LED.

ILDs are well suited for application in medium- to high-bandwidth fiber-optic communication systems. Compared to LEDs, injection lasers offer the advantage of narrower spectral width (<3 nm), larger modulation bandwidth (>500 MHz), and greater launched power (1 mW). ILDs are the sources most compatible for use with single-mode fibers. However, ILDs are not as reliable as LEDs, are more expensive, and require feedback circuity to stabilize their output power against variations due to temperature and aging effects.

B. Optical Receivers–Photodetectors

The basic purpose of an optical receiver is to detect the received light incident on it and to convert it to an electrical signal containing the information impressed on the light at the transmitting end. The receiver is therefore an optical-to-electrical converter or O/E transducer. An optical receiver consists of a photodetector and an associated amplifier along with necessary filtering and processing, as shown in Fig. 16. The function of the photodetector is to detect the incident light signal and convert it to an electrical current. The amplifier converts this current into a usable signal while introducing the minimum amount of additional noise to corrupt the signal. In designing an optical receiver, one tries to minimize the amount of optical power that must reach the receiver in order to achieve a given bit error rate (BER) in digital systems, or a given signal-to-noise ratio (S/N) in an analog system. In this section we will describe the characteristics of the photodetectors used in fiber optic systems. Since the performance of an optical receiver depends not only on the photodetector but also on the components and design chosen for the subsequent amplifier, we will also briefly describe configurations for this amplifier and their associated resulting receiver sensitivities.

In all the installed commercial fiber optic communication systems in existence today, the photodetector used is either a semiconductor *p–i–n* or avalanche photodiode (APD). These devices differ in that the *p–i–n* basically converts one photon to one electron and has a conversion efficiency of less than unity. In an APD carrier, multiplication takes place that results in multiple electrons at the output per incident photon.

The reasons for choosing *p–i–n* or APD photodetector are usually based on cost and on required receiver sensitivity. The avalanching process in the APD has a sharp threshold, which is sensitive to ambient temperature, and may require dynamic control of a relatively high bias voltage. The APD control and driver circuits are more expensive than those for the *p–i–n* detector, and the APD itself is more expensive than the *p–i–n* device. An APD with optimum gain, however, provides about 15 dB more receiver sensitivity than that achieved with a *p–i–n* diode.

An excellent spectral match exists between

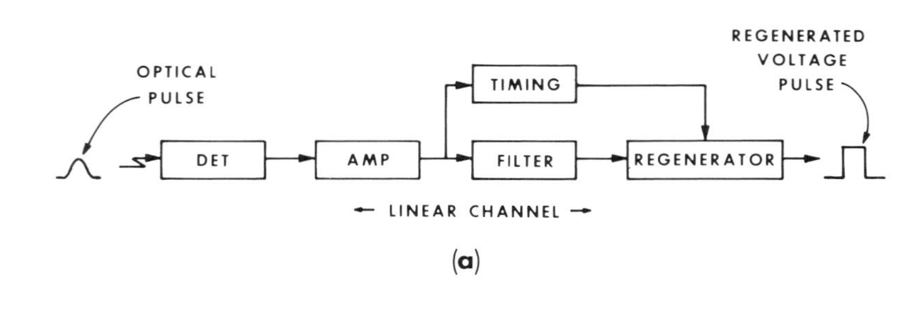

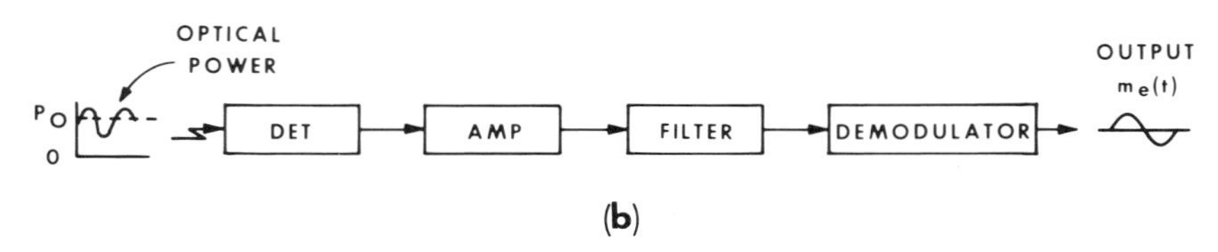

FIG. 16. Schematic of digital receiver. (a) Digital. (b) Analog.

TABLE V. Typical Photodetector Characteristics

Characteristic	Silicon	Germanium	*p–i–n* Diodes InP	Silicon	Germanium
Wavelength range, μm	0.4–1.1	0.5–1.8	1.0–1.6	0.4–1.1	0.5–1.65
Wavelength of peak sensitivity, μm	0.85	1.5	1.26	0.85	1.5
Quantum efficiency, 1%	80	50	70	80	70
Rise time, nsec	0.01	0.3	0.1	0.5	0.25
Bias voltage, V	15	6	10	170	40
Responsitivity, A/W	0.5	0.7	0.4	0.7	0.6
Avalanche gain	1.0	1.0	1.0	80–150	80–150

GaAlAs sources operating in the wavelength range of 0.8–0.9 μm and photodiodes made of silicon (spectral range 0.5–1.1 μm). The silicon *p–i–n* diode having no gain but with low dark current ($<10^{-9}$ A) and large bandwidth (1 GHz) is best suited for applications where receiver sensitivity is not critical. Silicon APDs are preferable in applications that demand high sensitivity, and those employed in presently installed telecommunication systems have current gains of about 100 and primary dark currents in the range of 10^{-10} to 10^{-11} A. Germanium photodiodes are used in the longer-wavelength region (1.3–1.6 μm), since the response of silicon decreases rapidly as λ increases beyond 1.0 μm. Germanium APDs with gain bandwidth products of approximately 60 GHz have been made, but their dark currents are high (10^{-8} to 10^{-7} A) and their excess noise factors are large. InGaAs and InGaAsP diodes are also used in the long-wavelength region. InGaAs *p–i–n* diodes have been made with very low capacitance (<0.3 pf) and acceptably low dark current ($<5 \times 10^{-9}$ A). However, further work is required in the area of long-wavelength APDs to reduce their high excess noise factor. Table V summarizes the characteristics of photodetectors used in fiber-optic communication systems.

To calculate the system margin for a communication system, a knowledge of receiver sensitivity is needed. Receiver sensitivity is determined primarily by the characteristics of the photodetector and the low-noise front-end amplifier, which is optimized for use with the detector. Figure 17 shows two commonly used configurations. To achieve the best receiver sensitivity, the amplifier should have a high input impedance or provide feedback as in a transimpedance amplifier. The first stage can be either a GaAs field-effect transistor (FET) or a silicon bipolar transistor with a suitably adjusted emitter bias.

In the short-wavelength region (0.8–0.9 μm), silicon APDs can provide sufficiently high gain and low excess noise to overcome the input amplifier noise. In this wavelength region, design of the first amplifier stage is not very critical. A conventional silicon bipolar transistor having an emitter capacitance of 10 pF and a current gain of 150 has been used to build a digital receiver that requires only 10 nW average optical input power (−50 dBm) for a bit error rate (BER) of 10^{-9} at 100 Mbit/sec.

Since leakage currents (high noise factor) severely limit the use of avalanche gains as low-noise amplification process in the long-wavelength region (1.3–1.6 μm), *p–i–n* detectors are usually used. With a *p–i–n* diode, the microwave GaAs FET is well suited for use in the first amplifier stage because it has a low gate capacitance and high transconductance. It is typically used in a transimpedance configuration and offers wide bandwidth and good dynamic range. The best receivers have been built using InGaAs *p–i–n* detectors and GaAs FETs and require an

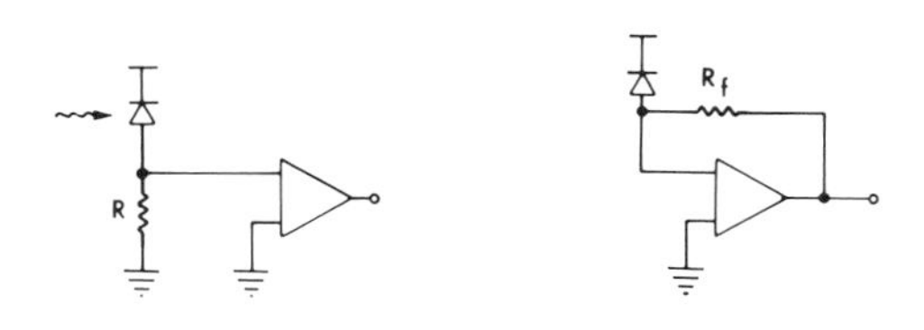

FIG. 17. Schematic of amplifier designs.

average optical input power of 25 nW (−46 dBm) for 10^{-9} BER at 100 Mbit/sec.

Curves for receiver sensitivity as a function of bit rate (bandwidth), along with curves of power available from LED and ILD sources, will allow us to calculate the net transmission loss tolerable between regenerators (system margin) as a function of bit rate (frequency).

Figure 18 shows, for digital systems, the average optical power required at the receiver (for BER of 10^{-9}) as well as the power available from optical sources as a function of bit rate. The lower boundary for receiver performance applies for receivers using silicon APDs at wavelengths of less than 1 μm. The upper receiver curve in Fig. 18 reflects the performance of *p–i–n* FET receivers sensitive in the wavelength range between 1.1 and 1.6 μm. Sources made of GaAlAs and InGaAsP have similar characteristics in terms of modulation speed and power delivered into a fiber. Light emitting diodes are usually restricted to those applications where the modulation bandwidths are less than 50 MHz. The separation between the sources and receiver bands in Fig. 18 is an indication of the gross transmission margin of a system. Practical repeater spans are designed with about 10 dB subtracted from the maximum values given in Fig. 18. This will account for variation in the transmitter and receiver components due to temperature variations and aging. As a general rule of thumb, the required fiber bandwidth in a digital system is equal to or larger than the specified system bit rate. The noise penalty for using this rule is less than 1 dB. Along with this potential noise source, the 10-dB safety margin also allows for signal degradation from various noise sources in the transmitter and receiver. Once the net transmission loss that is tolerable between regenerators is obtained, the distance between regenerators (link length) can be determined from the loss characteristics for the fibers and interconnection devices used in the system.

VI. Source Coupling, Splices, and Connectors

In this section we will investigate the coupling of energy from an optical source into a fiber, and the effects of intrinsic and extrinsic splice-loss parameters on the transmission characteristics of an optical fiber link. In addition, we will give examples of different types of optical fiber connectors and splices.

A. Source Coupling

The factors that effect the coupling efficiency of a source into a fiber can be broadly divided into two categories, as shown in Table VI. The first category, loss due to unintercepted illumination, can be caused by the source's emitting area being larger than the fiber's core area. Even if the source is smaller than the core, one can still have problems with unintercepted illumination if separation and misalignment of the source and fiber axes allow emitted light to miss the core and become lost. Coupling loss due to unintercepted illumination can be eliminated, however, if the source emitting area and the fiber-core area are properly matched and aligned. Figure 19 shows a fiber "pigtail" permanently

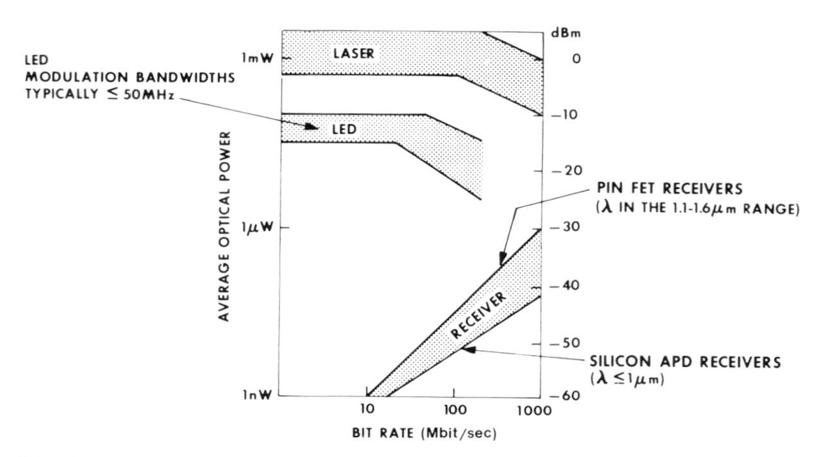

FIG. 18. Transmission margin versus bit rate for digital transmission systems.

TABLE VI. Factors Influencing Source Coupling Efficiency into a Fiber

1. Unintercepted illumination loss
   (a) Area mismatch between source spot size and fiber core area
   (b) Misalignment of source and fiber axis
2. Numerical aperture loss, caused by that part of the source emission profile that radiates outside of the fiber's acceptance cone

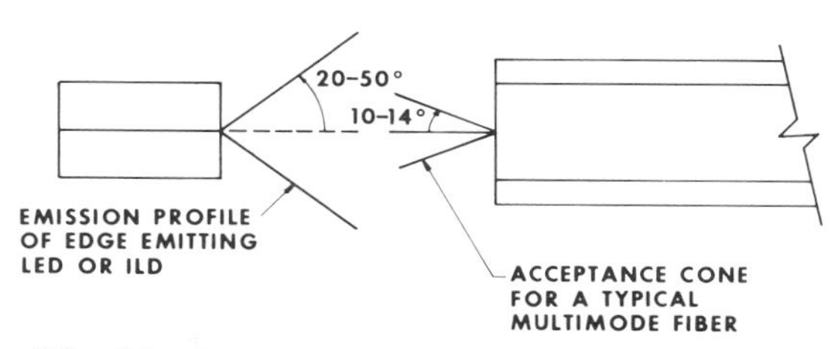

FIG. 20. Schematic showing numerical-aperature mismatch of source and fiber.

mounted and properly aligned on a Burrus-type light emitting diode. This configuration essentially eliminates loss due to unintercepted illumination. The second category of coupling loss that affects efficiency of source coupling into a fiber is due to mismatches between the source beam and fiber numerical apertures. This type of mismatch is shown schematically in Fig. 20. For fiber-optic communication systems, two types of light sources, light emitting diodes (LEDs) and injection laser diodes (ILDs), are typically used. A lambertian source whose radiant intensity varies with the cosine of the angle between a line perpendicular to it and another line drawn to an observation point is often used to model the radiation pattern of a surface emitting LED. Some sources such as edge emitting LEDs and ILDs exhibit radiation patterns that are narrower than a Lambertian source.

The radiation pattern obtained from an edge emitting LED is elliptical in cross section with half-power beam divergence angles of approximately ±60° and ±30°. The radiation pattern obtained from a double heterojunction laser diode is also elliptical in cross section but narrower in beam width than a LED. For example, the typical half-power beam divergence angles of an ILD are ±25° and ±5° perpendicular and parallel to the junction plane, respectively (see Fig. 21).

If we consider the problem of coupling energy from an LED into a multimode graded-index fiber and assume that the LED is a lambertian source in direct contact with the fiber core and covering its entire cross section, we can calculate the source coupling efficiency to be approximately equal to $\frac{1}{2}(\mathrm{NA})^2$. Sources such as edge emitting LEDs and ILDs, which have more directional emitters, will have a higher coupling efficiency than a lambertian source. ILDs with a properly aligned fiber pigtail will have a source coupling efficiency of 10–20%. For the case where the emitting area of the source is smaller than that of the fiber core, imaging optics can

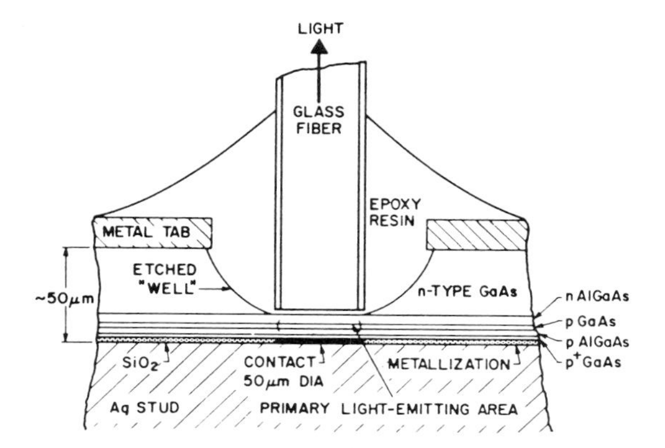

FIG. 19. Burrus-type LED with a fiber pigtail. Small-area, high radiance GaAs–$Al_xGa_{1-x}As$ double-heterostructure surface emitter.

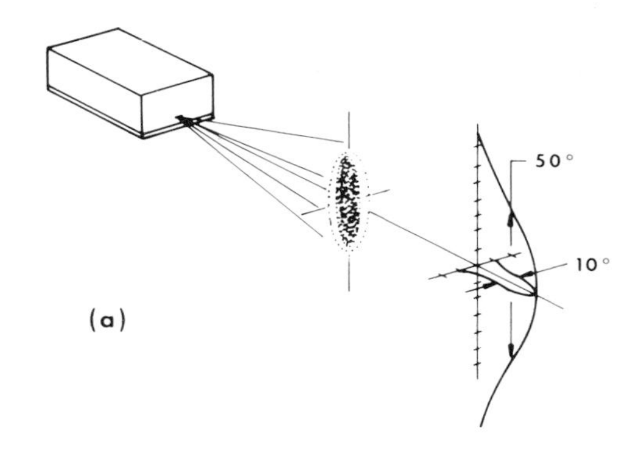

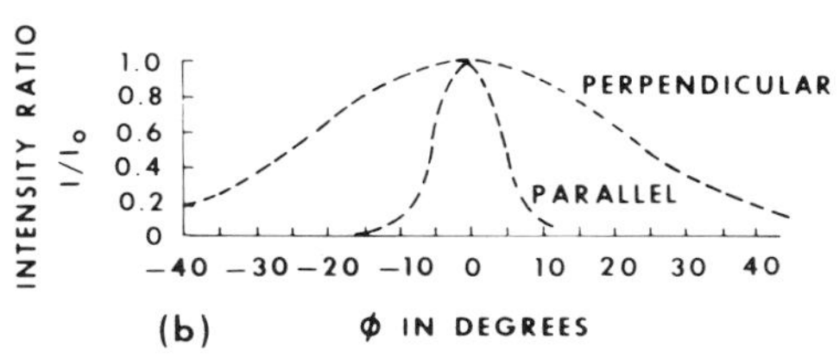

FIG. 21. Radiation pattern of an injection laser diode. (a) Schematic representation of far-field radiation pattern. (b) Far-field intensity pattern measured in planes parallel and perpendicular to the junction.

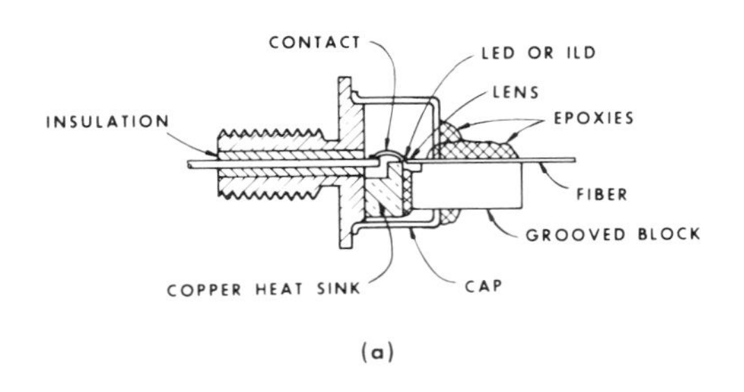

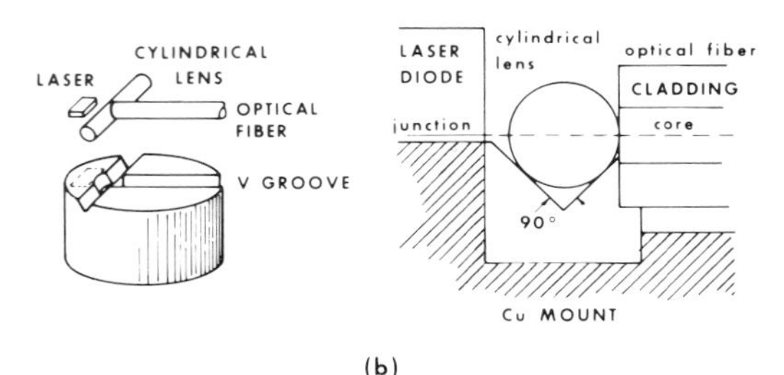

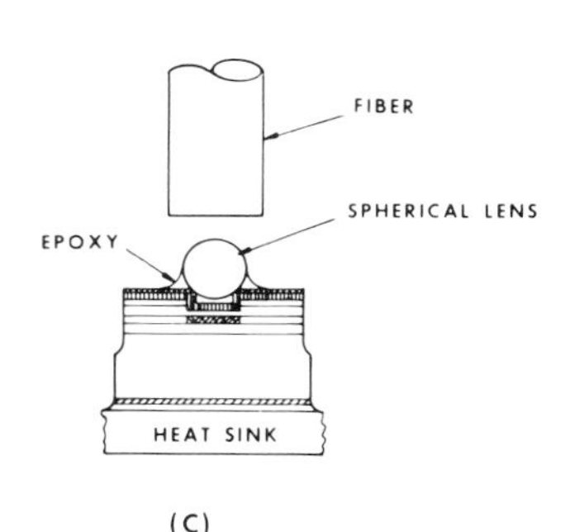

FIG. 22. Lens system to increase source coupling efficiency. (a) LED or ILD with lens incorporated in the source package. (b) ILD with cylindrical lens coupling source energy into graded index fiber ($\eta_c \approx$ 40–50%). (c) LED with spherical lens coupling source energy into fiber ($\eta_c \approx$ 8–10%).

improve the source coupling efficiency. Figure 22 illustrates a number of different lens arrangements that have been used and their associated source coupling efficiencies. Efficiently coupling a light source into a single-mode fiber is a difficult task because of its small core diameter and small difference in the core cladding refractive index. Mechanical tolerances are more stringent for small-core fibers, and careful alignment is needed between the emitting area of the source and the fiber core. Table VII is an estimate for the power launched by surface (SLED) and edge (ELED) emitting LEDs and ILDs into both single-mode and multimode fibers.

TABLE VII. Launched Power in dB min

	Wavelength (μm)				
	0.85	1.3		1.55	
Diode	MMF[a]	MMF	SMF	MMF	SMF
SLED	−15	−17	−34	−18	−35
ELED	−9	−11	−22	−12	−23
ILD	+8	+5	+3	+4	+2

[a] MMF, multimode fiber; SMF, single-mode fiber.

B. Splices and Connectors

The practical implementation of optical fiber communication systems requires the use of interconnection devices such as splices or connectors. A connector, by definition, is a demountable device used where it is necessary or convenient to easily disconnect and reconnect fibers. A splice, on the other hand, is employed to permanently join lengths of fiber together. The losses introduced by splices and connectors are an important factor to be considered in the design of a fiber-optic system, since they can be a significant part of the loss budget of a multikilometer communication link. In this section we will divide losses of splices and connectors into two categories, as shown in Table VIII. The first category of losses is related to the technique used to join fibers and is caused by extrinsic (to the fiber) parameters such as transverse offset between the fiber cores, end separation, axial tilt, and fiber end quality. The second category of losses is related to the properties of the fibers joined and is referred to as intrinsic (to the fibers) splice loss. Intrinsic parameters include variations in fiber diameter (both core and cladding), index profile (α and Δ mismatch), and ellipticity and concentricity of the fiber cores.

TABLE VIII. Splice Loss Factors

Extrinsic
Transverse offset
Longitudinal offset
Axial tilt
Fiber end quality
Intrinsic
Fiber diameter variation
α Mismatch
Δ Mismatch
Ellipticity and concentricity of fiber core

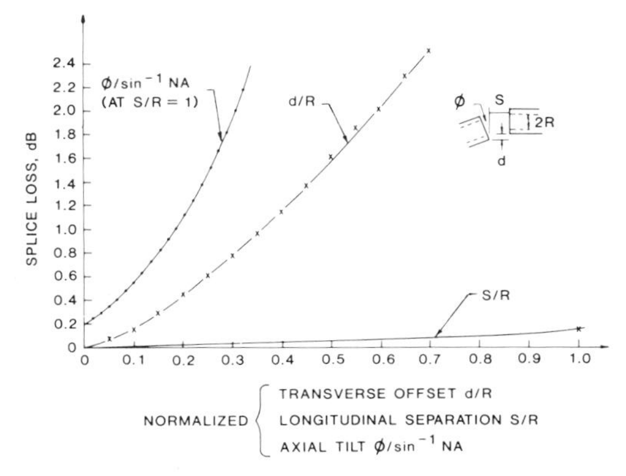

FIG. 23. Splice loss due to extrinsic parameters.

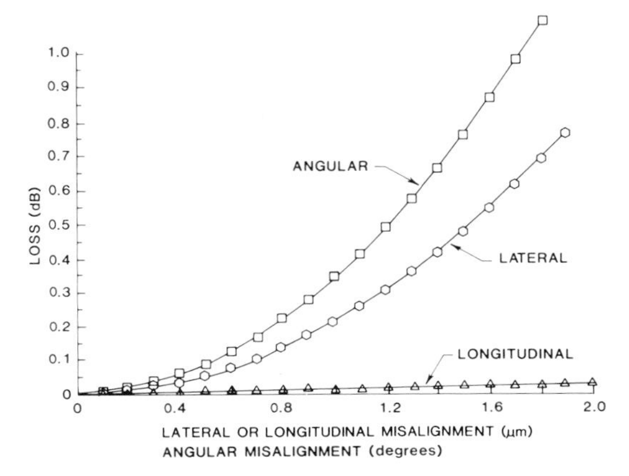

FIG. 24. Extrinsic splice loss of single mode fibers.

Figure 23 compares the relative influence on splice loss of the major extrinsic parameters of transverse offset, end separation, and axial tilt, for multimode graded-index fibers. Splice loss is significantly more sensitive to transverse offset and axial tilt than it is to longitudinal offset. For example, a transverse offset of 0.14 core radii or an axial tilt of 1° (for a fiber with NA = 0.20) will produce a splice loss of 0.25 dB. A longitudinal offset of one core radius will produce a loss of only 0.14 dB. Fiber end quality has a minimal effect on splice loss if proper fracturing or grinding and polishing end-preparation techniques are used in conjunction with an index-matching material. A matching material with a refractive index approximately the same as that of the core is used to reduce the Fresnel reflection loss caused by the glass–air interfaces between the coupled fibers of a joint.

Figure 24 compares, for single-mode fiber, the effects of transverse offset, end separation, and axial tilt on splice loss. Notice that very small (fractions of a micrometer) transverse offsets or axial tilts (fraction of a degree) can cause a significant amount of splice loss (>0.1 dB). The mismatch of intrinsic fiber parameters can also significantly affect the loss of a splice. For multimode graded-index fibers, splice loss is most sensitive to a mismatch of core diameter or delta. A normalized core radius or delta mismatch of 0.1 will produce a splice loss of approximately 0.2 dB. For single-mode fibers, the splice is most sensitive to a mismatch in mode field diameter. A mismatch in the mode diameter of 0.5 μm can cause a splice loss of 0.05 dB. To minimize the effect of intrinsic parameters on splice loss, tight manufacturing tolerances on the fibers used in a low-loss communication system must be maintained.

There is a variety of techniques that have been developed to interconnect fibers. Table IX describes some of the salient features of three of the techniques that will be illustrated in this text.

Figure 25 is a good example of a passively aligned plug and alignment sleeve type of connector. This molded biconical plug connector is widely used, with both single-mode and multimode fibers, as part of a jumper cable for a vari-

TABLE IX. Characteristics of Interconnection Methods

Interconnection method	Alignment technique		Single-mode fiber	Multimode fiber	Connection	
	Active	Passive			Single	Multiple
Plug and Alignment sleeve	X	X	X	X	X	
V-Groove alignment		X	X	X	X	X
Fusion	X	X	X	X	X	

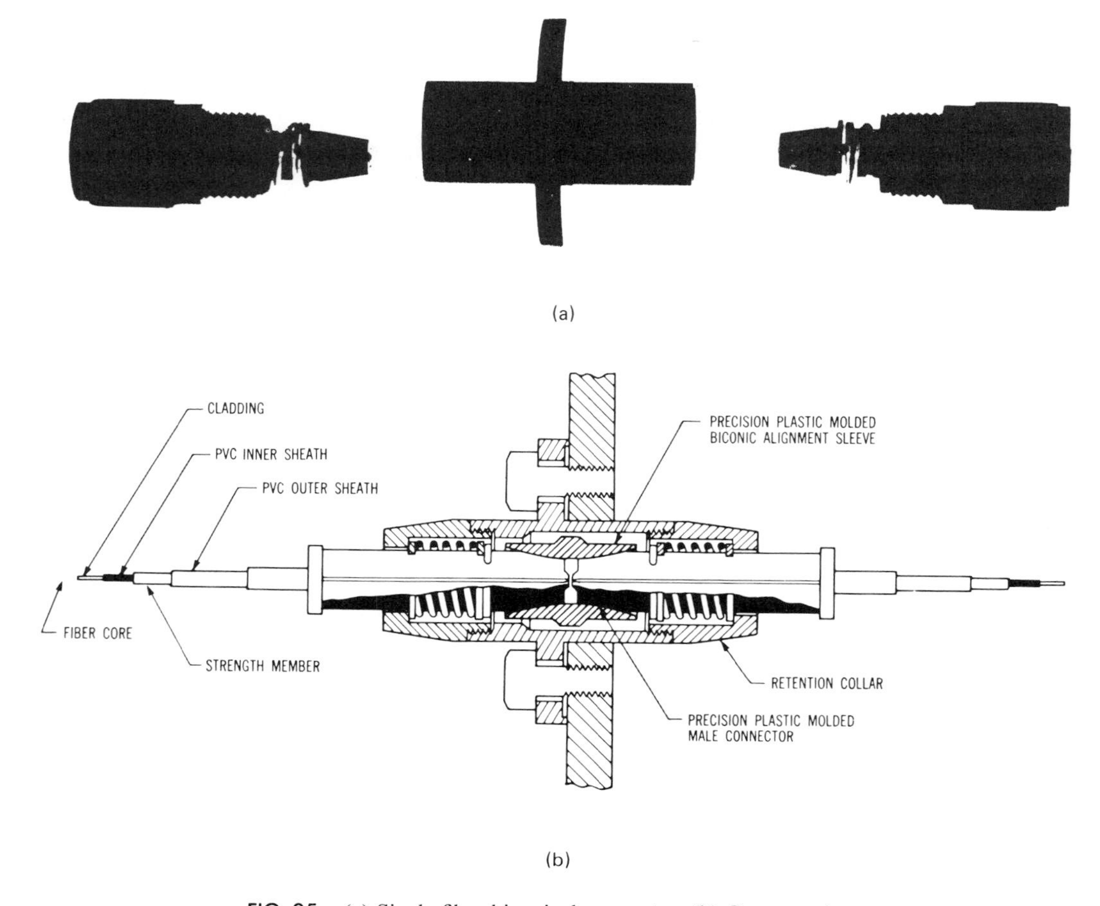

FIG. 25. (a) Single-fiber biconical connector. (b) Cross section.

ety of central-office applications. The heart of this connector is a biconical sleeve that accepts two plugs and aligns the axes of the fiber ends that are centrally located in these plugs. An inherent advantage of the conical alignment configuration is that virtually no abrasive wear occurs between the mating parts until the plug is fully seated within the biconical alignment sleeve. Losses of less than 1.0 dB are usually obtained with this type of connector. Another example of a plug and sleeve splice that illustrates the concept of active alignment of fibers is shown in Fig. 26. For single-mode splices, micrometer-type extrinsic and intrinsic tolerances must be maintained to achieve low-loss connections. To avoid putting very stringent constraints on both the fibers and the parts used to interconnect them, the technique of active alignment of the fiber cores while monitoring a signal related to transmission loss is used. Single-mode fibers mounted in precision glass cylindrical plugs are actively aligned in an eccentric sleeve to produce the very-low-loss (<0.1 dB) rotary splices shown in Fig. 26.

The silicon chip array splice shown in Fig. 27 is a very good example of the use of V grooves as the alignment mechanism for joining fibers.

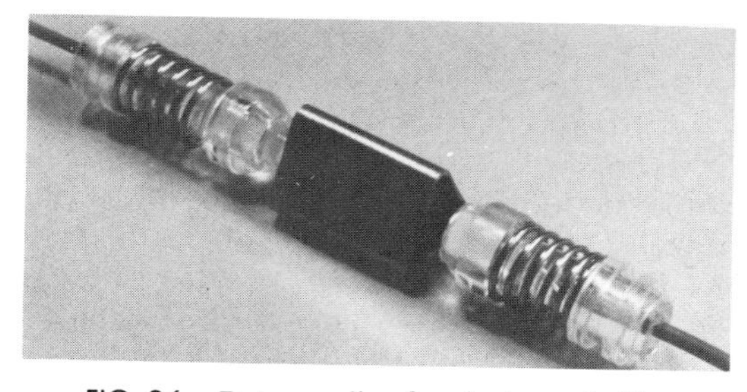

FIG. 26. Rotary splice for single-mode fibers.

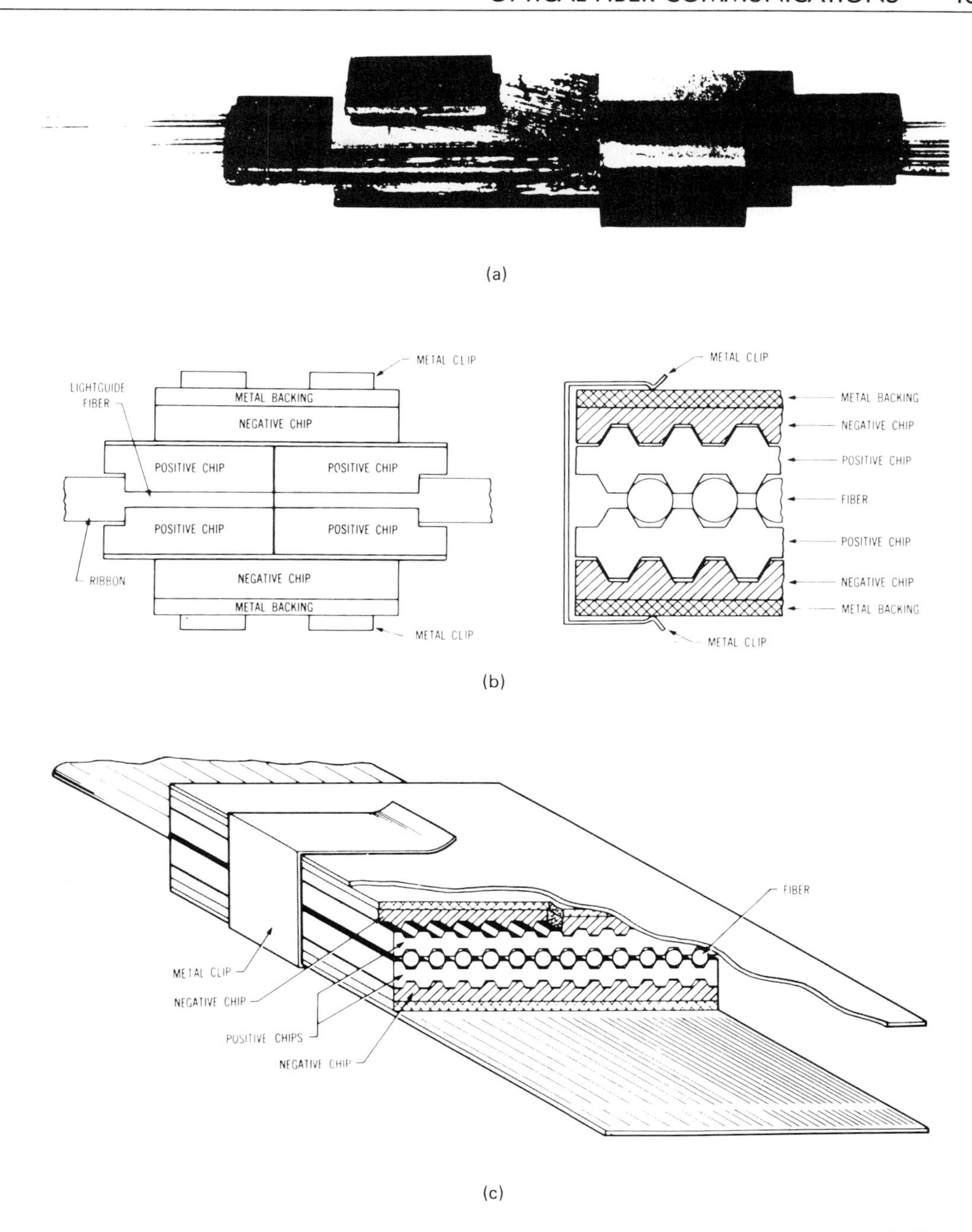

FIG. 27. Multiple-fiber silicon chip array connector. (a) 12-Fiber silicon chip array connector. (b) Side view and partial cross section of array connector. (c) Cross section of array connector.

This splice is used in conjunction with ribbon-type structures to simultaneously splice groups of optical fibers (multiple-fiber splice). This reenterable splice consists of two array halves, two negative chips with metal backing, and two spring clips. An array half is formed by permanently affixing two positive preferentially etched silicon chips to the end of a ribbon. The fiber ends are then simultaneously prepared by grinding and polishing the end of the array. The splice halves are usually assembled on a cable prior to shipment from a factory, although they can be assembled in the field. A craftsperson assembles a splice by simply aligning two array halves with

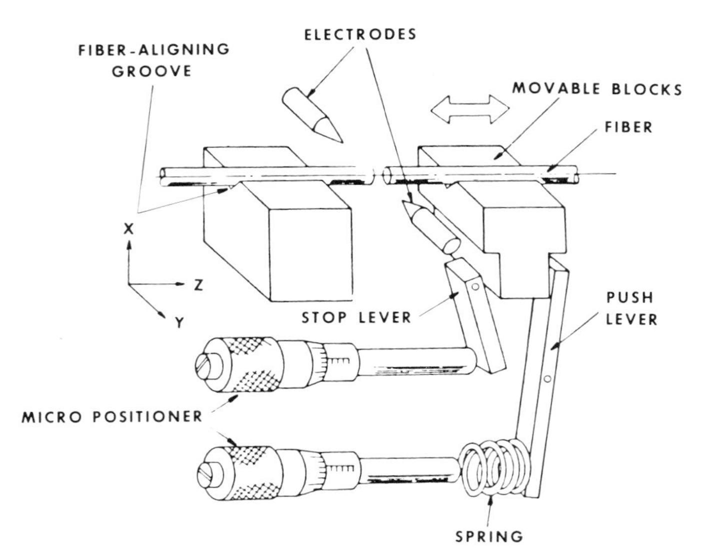

FIG. 28. Schematic showing fusion splicing using an electric arc.

two negative chips. The splice is held together with spring clips, and a matching material is inserted to complete the connection. The reenterable array splice can be disconnected and reconnected in the field, although it was not designed for use where many connect–disconnect operations are required. The average loss of the silicon chip array splice is less than 0.2 dB for multimode fibers and less than 0.6 dB for single mode fibers.

Fusion splicing, illustrated schematically in Fig. 28, is a widely used method for permanently joining individual fibers together. Fusion splicing, or welding as it is sometimes called, is accomplished by applying localized heating at the interface between two butted, realigned fiber ends, causing them to soften and fuse together. Commercially available fusion splicing machines produce splices with average losses of less than 0.3 dB (active alignment) with single-mode fibers and less than 0.2 dB with multimode fibers.

VII. Applications of Optical Fiber Communications

The unique advantages of lightwave communications—the tremendous information-carrying capacity, freedom from electromagnetic interference, the light weight of fibers, and the relative chemical and high-temperature stability of fibers—suggest a communication revolution will occur in the near future. Already the telecommunications market for optical fibers has exploded in the United States, and unprecedented demand for fibers has been experienced. Within a short space of 5 years, from 1980 to 1985, the total U.S. circuit mile capacity installed with optical fibers far exceeded the total existing U.S. circuit mile capacity using copper cables in 1980. So far the field of fiber optics has been technology-driven, but the future will be application-driven. Lightwave application may be broadly classified into telecommunications applications, industrial applications, military application, medical applications, and sensor applications.

A. Telecommunications Applications

By far in terms of product volume and revenues the telecommunications application has been the major market. The need for greater circuit capacity coupled with the problem of congested duct space in the cities led to the initial applications of optical fiber communication for the interoffice trunk in the big cities (Fig. 29). However due to the tremendous progress made in reducing the transmission loss and increasing the bandwidth (information carrying capacity) of the optical fibers, they are now the economic choice for most telecommunication applications. Today the biggest market for fiber is the long-distance application. Japan has already completed its long distance routes using single-mode fibers. In the United States, thousands of

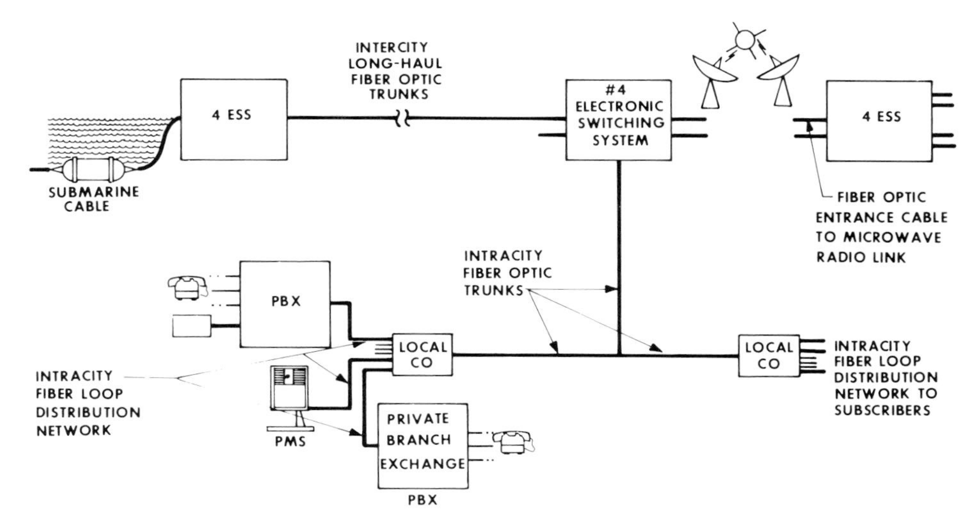

FIG. 29. Optical fiber telecommunications application.

miles of lightguide cables are being installed by AT&T, MCI, GTE, and several other long-distance companies, as well as regional Bell and independent operating companies. The majority of these companies use single-mode fibers that operate at 1300 nm wavelength, with future upgradability to 1550 nm, when reliable sources and electronics are developed. Already commercial systems have been installed with bit rates as high as 565 Mbit/sec, which is equivalent to 7680 two-way conversations over a pair of fibers. Another application for the fibers is in the feeder loop, that is, a cable that is used between a telephone central office and the local distribution interface. Since fibers have very high capacity and can transmit voice, data, and video, efforts are underway to install fibers into individual houses. AT&T has announced its intention to develop and offer a universal information service (UIS) to satisfy all the video, voice, and data communications needs of a subscriber. One of the most attractive features of the universal information system is the use of interactive systems. Systems are already in use that provide continuing interactive service for smoke and heat detectors to automatically alert fire departments of potential or active fire hazards, police alarms, and medical alert alarms to summon aid. In addition to the above, all the other information services can be obtained on demand. The potential for completely automating all the control needs of a household is there. Several trial projects have been implemented in Germany, France, Japan, the United Kingdom, Canada, and the United States. However, the information needs of individual households have not yet reached a level where fibers can be economically justified. However, definite potential exists for an information outlet in each household, which could be used for a multitude of services such as video phone, television, news, banking, and time-share computing.

B. Local Area Networks

Networks using optical fibers to transmit voice, video, and data within a building or within industrial and university campuses have been offered by several vendors. These systems are called local area networks (LANs). For economic reasons, many of today's LANs use a fiber with high (higher than that used for trunking applications) numerical aperture and operate with light emitting diodes working at 1.30 μm. Fiber-optics LANs can improve the communications inside a high-use area, reduce the bulk of copper cables, and in many cases eliminate congestion in computer rooms by remoting peripheral equipment. LANs are likely to be installed in most office buildings, industrial facilities, university campuses, and military bases. This is potentially one of the biggest future markets for fiber-optic applications.

C. Industrial Applications

Industrial applications can be of different types, such as process control, manufacturing automation, and energy management and application. Applications for process control are in nuclear, petrochemical, chemical, and food industries. Applications for manufacturing automation are for numerical control of machines and large data systems. Applications for control exist in airways, highways, shipping, railways, gas and oil transportation, and distribution control. Although the information rates needed for such application are low compared to fiber capacity, fiber-optic control systems have several major advantages over the electronic wire systems. These include freedom from electromagnetic interference, improved signal quality, large potential for upgrading, and safety in a hazardous environment such as oil or gas. Also, remote control can be easily achieved. In the case of power failures, fiber-optic systems can be easily maintained with auxiliary power sources.

D. Computer Applications

Computer applications are the other part of industrial use of fiber optics. Fibers are ideally suited for internal links that require very high data rates, of the order of gigabits per second (Gbit/sec). Auxiliary equipments require lower data rates and hence can be handled both by fibers or copper wires; however, fibers offer the added advantage of longer-distance network and error-free operation, because fiber transmission is unaffected by the electromagnetic noise. The fibers will gradually be used in greater volume as inter- and intracomputer links.

E. Military Applications

Three attributes of fiber optics—security, freedom from electromagnetic interference, and light weight—make fibers very attractive for use in the defense applications. Fibers can be used for data busing in aircrafts with tremendous reduction in weight; thereby increasing the range of these aircraft, along with increased information capacity and more secure and error-free operation. Such aircraft include surveillance aircraft, attack aircraft, space vehicles, and strategic air command bombers. Similar fiber applications exist for inter- and intraship communications, submarine mobile command centers, ship-to-satellite communication links, and all types of missile guidance systems. For military terrestrial applications, all communication within any base could be a local area network and could include missile-center links and ground-to-air and ground-to-ship communications. All the above types of applications exist today, making the miliary applications a large market for fiber optics communications.

F. Medical Applications

In the medical field there are two distinctly different types of fiber-optics systems being offered. The first type takes the form of a LAN that will satisfy all the communications needs for a hospital, from keeping patient records to the energy-management and financial-management systems. The second category of medical fiber-optic systems uses fibers in diagnostic equipment and as sensors. Because of their small size and relatively benign impact upon body elements, much effort is underway to develop fiber-optic sensors for medical applications. The mechanical flexibility and small size allows the measurements of parameters inside the tissue, organs, and blood vessels. Also because of the freedom from electromagnetic interference, these instruments can be used in conjunction with other electronic monitoring systems. Three types of applications for biomedical diagnostics exists: (1) sensors for temperature, pressure, and flow of blood; (2) sensors for monitoring oxygen saturation, pH, and oxygen and carbon dioxide concentrations; and (3) sensors for immunoassay reactions. The sensors consist of fibers through which light is carried to the sample and returned back to analyzing equipment. At the end of the fiber is a suitable miniaturized transducer. The backscattered light is then analyzed to determine the velocity of moving blood cells using the Doppler shift. Spectral analysis can also be performed to measure concentration of blood constituents such as oxyhemoglobin. Concentration of gases in the bloodstream can be measured by coating the fiber end with an appropriate reagent for spectrophotometric or fluorimetric analysis. For monitoring immunoassay reactions, an antibody is attached to the surface of the fiber core from which cladding has been stripped. The reaction to the antigen or reaction product is measured at the fiber output end. [*See* OPTICAL FIBER TECHNIQUES (MEDICINE).]

G. Sensor Applications

Use of fiber optics for sensors has been continuously increasing. As with other applications, the sensor applications also fall into three categories: military, medical, and industrial. The biomedical sensors were discussed in the previous section. Ultrasonic sensors use the phase shift caused in the light propagating through the core through radial stresses. The other sensors used are the pressure and temperature sensors, radiation sensors, depth sensors, acoustic sensors, sensors for color and turbidity; and sensors for underwater bubble detection. In general, the sensor system consists of an electronic control module, sensor head, and fiber-optic cable. The sensor head senses the pressure, temperature, velocity, and reaction and converts it into a change in the optical signal, which is then analyzed to measure the desired change by the electronic control unit. The field of fiber-optics sensors has expanded to the extent that separate conferences are held on fiber-optic sensors.

Bibliography

Barnoski, M. K. (ed.) (1976). "Fundamentals of Optical Fiber Communications." Academic Press, New York.

Bendow, B., and Shashanka, S. M. (ed.) (1979). "Fiber Optics: Advances in Research and Development." Plenum Press. New York.

Cherin, A. H. (1983). "Introduction to Optical Fibers." McGraw Hill. New York.

Li, T. (ed.) (1985). "Optical Fiber Communications." Academic Press. New York.

Midwinter, J. E. (1979). "Optical Fibers for Transmission." John Wiley & Sons. New York.

Miller, S. E., and Chynoweth, A. G. (ed.) (1979). Optical Fiber Telecommunications." Academic Press. New York.

Suematsu, Y., and Ken-ichi, I. (1982). "Introduction to Optical Fiber Communication." John Wiley and Sons. New York.

OPTICAL FIBER TECHNIQUES (MEDICINE)

Abraham Katzir *Tel Aviv University*

GLOSSARY

Acceptance angle: Maximum incident angle for which an optical fiber will transmit light by total internal reflection.

Catheter: Flexible hollow tube normally employed to inject liquids into or to drain fluids from body cavities.

Cladding: Outer part of an optical fiber; has a lower refractive index.

Coherent bundle: Assembly of optical fibers in which the fibers are ordered in exactly the same way at both ends of the bundle.

Core: Inner part of an optical fiber; has a higher refractive index.

Critical angle: Minimum incidence angle in a medium of higher refractive index for which light is totally internally reflected.

Endoscope: Optical instrument used for viewing internal organs. It often contains a fiberscope and ancillary channels for medical instruments and for irrigation or suction.

Fiberscope: Viewing instrument that incorporates a coherent bundle for imaging and a light guide for illumination.

Laser catheter: Catheter that incorporates an optical fiber for the transmission of a laser beam.

Laser endoscope: Endoscope that incorporates an optical fiber for the transmission of a laser beam.

Light guide: Assembly of optical fibers that are bundled but not ordered (noncoherent), used for illumination.

Numerical aperture: Light-gathering power of an optical fiber. It is proportional to the sine of the acceptance angle.

Optical fiber: Thin and transparent fibers through which light can be transmitted by total internal reflection.

Power fiber: Optical fiber that can transmit a laser beam of relatively high intensity.

Total internal reflection: Reflection of light at the interface between media of different refractive indices, when the angle of incidence is larger than a critical angle.

Optical fibers are thin, flexible, and transparent guides through which light can be transmitted from one end to another. An ordered bundle of optical fibers forms a fiberscope and can be used for transmitting images. A medical endoscope incorporates such a fiberscope. Thin, flexible endoscopes enable physicians to get clear images from areas inside the body that were previously inaccessible, such as inside the heart or the bronchial tree. Optical fibers can be used for medical diagnostics, by inserting them inside the body and making physical or chemical measurements through them. Measurements such as blood flow, pressure, and gas content or sugar content in blood can be performed quickly and reliably. Optical power fibers may be used to transmit relatively high laser power to areas inside the body. The laser radiation, in turn, may be used to cut tissues, to coagulate bleeding vessels, to remove tumors, to clean occluded arteries, and to destroy cancer cells. Compound laser endoscopes may include several channels: fiber-optic ones for image transmission, diagnostics, and power transmission, and ancillary channels for injection of liquids or aspiration of debris. It is expected that the laser endoscopes could be fairly thin and flexible. These endoscopes could be used in myriad applications and will undoubtedly revolutionize many fields in medicine.

I. Introduction

Light can be transmitted through a cylinder of transparent material by a series of internal reflections. This phenomenon was probably known by ancient glass blowers, but the earliest demonstration of the effect was given by J. Tyndall in England in 1870. Thin rods that are used for light transmission are called optical fibers. Such fibers appear in nature. For example, tissues of plant seedlings can guide light to coordinate their physiology. There are millions of optical fibers in the retina of the human eye, and fibers may play a role in communication between insects.

An ordered bundle of glass or plastic fibers may be used for the transmission of images. Some of the earliest applications of optical fibers, in the period 1925–1930, involved the use of such bundles in medicine. Only in the early 1950s, with the development of optical fibers that consisted of inner core and outer cladding, did the flexible fiber-optic bundle become a powerful practical tool. At first the medical instruments, endoscopes, that incorporated fiberoptic bundles, enabled physicians to obtain images from organs inside the body. Yet the optical quality of the fibers (and the endoscopes) was not very high in that period.

During the 1970s and 1980s there has been rapid progress in the development of glass fibers with extremely low optical transmission losses. The main challenge is in the use of these fibers in communication systems. Light signals emitted from semiconductor lasers can be transmitted through very long optical fibers and can be detected by semiconductor detectors. The laser light can be modulated in such a way that it may carry with it voice communication, radio and television broadcasting, and computer data. Thus, optical fibers would play a major role in future communication networks. [*See* OPTICAL FIBER COMMUNICATIONS.]

There has been increasing recognition of the enormous possibilities for utilizing the low-loss optical fibers in medical applications. Ultrathin endoscopes with improved optical properties have been developed and utilized. Optical fibers have been used as diagnostic tools. Laser power has been transmitted through optical fibers and used for ablation of tissue or for photomedical applications inside the body.

We review here some of the physical principles involved and some of the revolutionary uses of optical fibers in medicine. [*See* MICROOPTICS.]

II. Optical Fibers

In this section we describe the physical principles of light guiding in optical fibers and in bundles of fibers. We discuss the materials and methods of fabrication of various fibers and bundles.

A. Properties of Optical Fibers

1. Total Internal Reflection

Let us consider two transparent media of refractive indices n_1 and n_2, where $n_1 > n_2$ (Fig. 1). A ray of light propagates at an angle θ_1 with respect to the normal to the interface between the media. At the interface, part of the beam will be reflected back to medium 1, and part will be refracted into medium 2. The reflection is specular, that is, the angle of reflection is equal to the angle of incidence θ_1. The refraction obeys

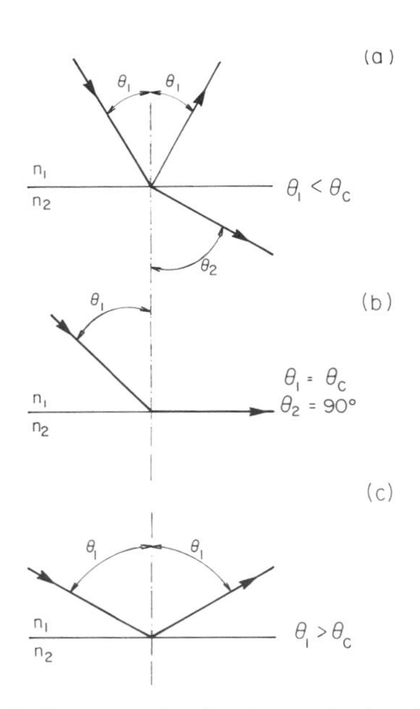

FIG. 1. Reflection and refraction at the interface between two media of refractive indices n_1 and n_2 ($n_1 > n_2$). (a) Incident angle $\theta_1 < \theta_c$. (b) Incident angle $\theta_1 = \theta_c$. (c) Incident angle $\theta_1 > \theta_c$; total internal reflection.

Snell's law, as shown in Fig. 1(a), so that

$$n_1 \sin \theta_1 = n_2 \sin \theta_2 \tag{1}$$

If the angle θ_1 is increased, one may reach some angle $\theta_1 = \theta_c$ for which $\theta_2 = 90°$. Here θ_c is called the critical angle, and for this angle we write

$$n_1 \sin \theta_c = n_2 \tag{2}$$

For every angle of incidence $\theta_1 > \theta_c$ there is no refracted beam. Had there been such a beam, its angle θ_2 should have been given by the equation

$$\sin \theta_2 = \frac{n_1 \sin \theta_1}{n_2} > \frac{n_1 \sin \theta_c}{n_2} = 1$$

Since this of course is impossible, there is only a reflected beam. This phenomenon, shown in Fig. 1(c), is called total internal reflection. If medium 1 is glass with $n_1 = 1.5$ and medium 2 is air with $n_2 = 1$, the critical angle is given by $\sin \theta_c = 1/1.5$. In this case, $\theta_c = 42°$. If medium 2 is soda-lime glass with $n_2 = 1.52$ and medium 1 is flint glass with $n_1 = 1.67$, then $\theta_c = 65°$. It should be mentioned that in practice the total internal reflection is very efficient. In this process the proportion is more than 0.9999 of the energy reflected, in comparison to about 0.95 for good metal mirrors. [*See* GLASS.]

2. Optical Fibers

Consider now a rod of transparent material of refractive index n_1 in air ($n_0 = 1$). Two rays of light incident on one end face of the rod are shown in Fig. 2(a). Ray II will be refracted inside the rod and refracted back in air. Ray I, on the other hand, will be totally internally reflected inside the rod and will emerge from the second face of the rod. This will happen also when the rod is very thin and flexible, and in this case the rod is called an optical fiber. A fiber consisting of a transparent material in air is called unclad fiber. Light will propagate inside the rod (or fiber) by a series of internal reflections even if the rod is not in air, as long as $n_2 < n_1$. In particular, a compound rod may consist of inner part of index n_1 called core and outer part of index n_2 called cladding. If this rod is thin, the optical fiber thus formed is called clad fiber. The cross section of the rod (or the fiber) is shown in Fig. 2(b), with the trajectory of an incident ray of light. Let us assume that the angle of incidence in air is α_0 and that inside the core the beam is refracted at an angle α_1, as shown. We can write now

$$n_0 \sin \alpha_0 = n_1 \sin \alpha_1 = n_1 \cos \theta$$
$$= n_1(1 - \sin^2 \theta)^{1/2}$$

where $n_0 = 1$ is the refractive index in air. The angle θ can assume several values, but its maximum value for total internal reflection is the critical value θ_c given by Eq. (2), $n_1 \sin \theta_c = n_2$. We could calculate α_{0c}, the value of α_0 corresponding to this value of θ:

$$n_0 \sin \alpha_{0c} = n_1(1 - \sin^2 \theta_c)^{1/2}$$
$$= n_1 \left[1 - \left(\frac{n_2}{n_1}\right)^2\right]^{1/2} = (n_1^2 - n_2^2)^{1/2}$$

This value of $n_0 \sin \alpha_{0c}$ is defined as the numerical aperture NA:

$$\mathrm{NA} = n_0 \sin \alpha_{0c}$$
$$= (n_1^2 - n_2^2)^{1/2} \approx \sqrt{2\Delta} \tag{3}$$

where $\Delta = (n_1 - n_2)/n_1$.

Rays of light impinging on the surface at any angle $\alpha_0 \leq \alpha_{0c}$ will be transmitted through the rod (or the optical fiber). All these rays form a cone of angle α_{0c}. For angles of incidence $\alpha_0 > \alpha_{0c}$, the ray will be refracted into the cladding and then into the air. The cone of light rays that could be transmitted by the fiber is a measure for the light-gathering capability of the fiber. The angle of the cone α_{0c} is called the acceptance angle. For an example, if again $n_1 = 1.62$ and $n_2 = 1.52$, then NA = 0.56. If the fiber is in air, $n_0 = 1$ and $\alpha_{0c} = 34°$, but if it is immersed in water, $n_0 = 1.3$ and then $\alpha_{0c} = 25°$. The acceptance angle of the fiber in water is therefore smaller than in air.

In the case of a straight fiber, a ray incident at an angle α will be transmitted by the fiber and emerge with same angle α. For an incident cone

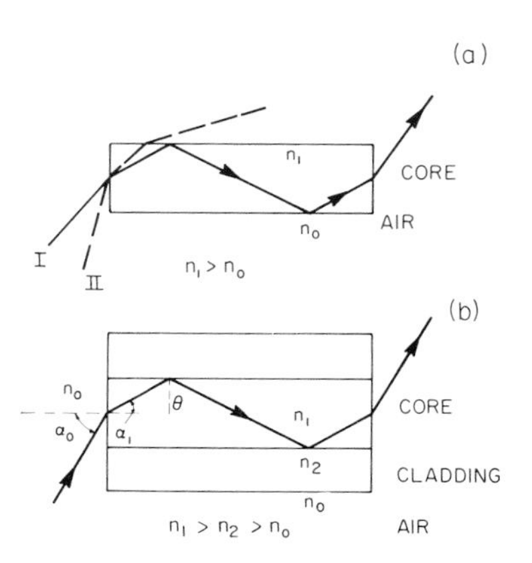

FIG. 2. Trajectory of a ray in a cylindrical rod. (a) A rod of refractive index n_1 in air ($n_0 = 1$). (b) A rod whose core has a refractive index n_1 and cladding index n_2 ($n_2 < n_1$).

of rays, the light emerging from the fiber will also be a cone, with same apex angle.

3. Transmission in Optical Fibers

A beam of light of intensity I_0 impinges on transparent material of thickness L (cm). The beam is transmitted through the material and emerges with intensity I. In many cases the ratio between I and I_0 is given by Beer–Lambert law,

$$I = I_0 \exp(-aL) \qquad (4)$$

where a (cm^{-1}) is the absorption coefficient. Physically, after travelling $L = 1/a$ cm, the intensity I_0 is reduced to $I_0/e \approx I_0/3$, and after travelling $L = 2 \times 1/a$ cm it is reduced to $I_0/e^2 \approx I_0/10$.

In the engineering literature, the transmission loss in the medium A is given in decibels (dB), as defined by

$$A\ (\text{dB}) = -10 \log_{10}(I/I_0) \qquad (5)$$

As an example, for $I = I_0/10$, we get $A = -10 \log_{10} 10^{-1} = 10$ dB. If the light intensity was reduced to I after travelling a distance L in the material, the transmission losses are stated in A/L db/m (or db/km).

A beam of light of wavelength λ and of intensity I_0 may propagate in a clad optical fiber, as shown in Fig. 3. The total transmission through real fibers depends on many factors, and several of them are listed here:

(a) Losses in the coupling of light from a light source into the fiber.
(b) Reflection losses at each end of the fiber. This Fresnel reflection is proportional to $[(n_1 - n_0)/(n_1 + n_0)]^2$.
(c) Absorption in the pure material of the core. This absorption depends on λ.
(d) Absorption by impurities.
(e) Scattering from the inherent inhomogeneous structure of the core (Rayleigh scattering); scattering from small inhomogeneity of refractive index in the core, and in the interface between core and cladding.
(f) If the fiber is bent, some of the light leaks out from the cladding.

In order to ensure high transmission through the fiber, one has to reduce the losses. In unclad fibers the outer surface of the fiber is exposed to mechanical or chemical damage, which gives rise to scattering. Therefore the fiber has to be a clad fiber. The cladding material should be well matched to the core one, and the interface should be of high quality. The core should be very pure, in order to reduce absorption, free of defects, and fairly homogeneous, to reduce scattering. The ray trajectory shown in Fig. 3 is schematic, and as a matter of fact some of the optical power is transmitted through the cladding. Therefore the cladding layer should also be highly transparent. In medical applications, low losses are very important when trying to transmit high optical power through optical fibers. High losses may give rise to excessive heating and to melting of the power-carrying fiber. [*See* OPTICAL FIBERS, DRAWING AND COATING.]

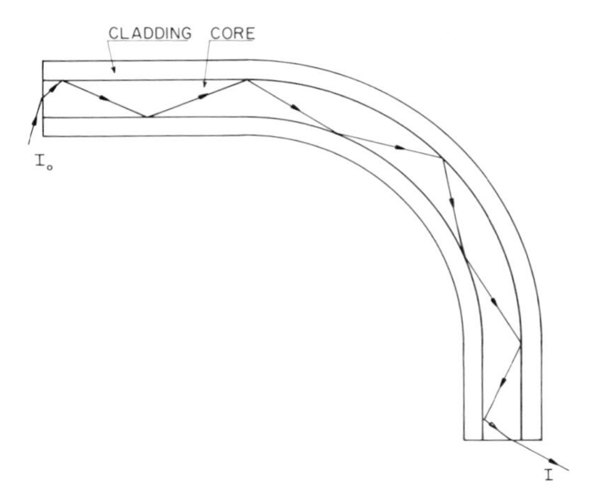

FIG. 3. The transmission of light in a clad fiber.

B. Material and Methods of Fabrication of Optical Fibers

1. Optical Fibers Made of Silica Glass

Optical fibers are usually made of glass, and most often this is an oxide glass based on silica (SiO_2) and some additives. A rod of glass (called preform) is heated to soften, and then a fiber is drawn from the hot glass. This fiber consists only of a core. The preform itself may consist of two glasses, such as shown schematically in Fig. 2(b). In this case a clad fiber is fabricated by drawing. In both cases the outer diameter of the fiber is on the order of 0.1 mm, in order to obtain flexibility.

In the past the preform was prepared by melting, and the glass components were not very pure; also, there were defects in the materials and the interface. Nowadays, exceptionally pure materials are prepared by deposition in a vacuum system, and special efforts are being taken to reduce the number of defects. In the case of optical communications the fibers are coated with a plastic primary coating as they are being drawn, in order to protect them from moisture, scratches, etc. Good-quality optical fibers have transmission losses on the order of 1 dB/km. Such quality is normally not needed for medical applications.

In optics the wavelength λ is normally specified in nanometers (1 nm = 10^{-9} m), or in micrometers (1 μm = 10^{-6} m). The optical spectrum is broadly divided into three spectral regions: ultraviolet (UV), visible, and infrared (IR). The wavelength division is as follows: UV for $\lambda < 400$ nm; visible $\lambda = 400$–700 nm; IR for $\lambda > 700$ nm.

Optical fibers that are based on some mixture of SiO_2, B_2O_3, and Na_2O, transmit well in the visible and near-infrared regions of the optical spectra (0.4 μm $< \lambda <$ 0.7 μm).

2. Special Optical Fibers

Regular glass fibers transmit only in the visible, and they do not transmit in the UV or in the far-infrared. Nor could these glass fibers transmit visible beams of high intensity. For the transmission of UV ($\lambda = 0.2$–0.3 μm), one could use quartz (pure silica) as the core material. Clear plastic (of lower index of refraction) serves as a cladding layer. These quartz fibers also serve as power fibers for the transmission of laser power. For the transmission of the infrared one may use nonoxide glasses such as chalcogenides (e.g., As_2S_3) or fluorides (e.g., ZrF_4–BaF_2–LaF_3). Alternatively, crystalline materials may be used for the infrared. Crystals of silver halides or thallium halide can be extruded through dies. The resulting fibers were found to be transparent in the infrared.

Optical fibers made of plastic materials had been used in the past for the visible range. For example, polystyrene may serve as a core in such fibers, and polymethylmethacrylate as cladding. Such fibers are fairly flexible and less expensive to fabricate, but their optical properties are inferior to those of glass. In particular, plastic fiber transmit well only in the spectral range $\lambda > 450$ nm.

III. Lasers for Fiber-Optic Medical Systems

A. Medical Lasers

There are many regular light sources (e.g., incandescent lamps, high-pressure lamps, etc.) that emit light in some spectral range and that are used in medicine. In all these cases the light consists of many wavelengths; it is emitted from a relatively broad area, and it is spread in many directions. The laser light has several characteristics that make it useful in medicine:

Monochromatic—practically one wavelength.

Collimated beam—the laser light is emitted in a parallel beam.

High intensity—practically all the laser energy is concentrated in a narrow pencil.

Coherent—the laser emission is "ordered" both in space and in time. The spatial coherence helps us to focus the beam to a very small point.

All lasers are based on some active medium that is called the lasing medium. This medium can be a solid crystal (e.g., Nd : YAG crystal), a liquid (e.g., dye in a solution), gas (e.g., Ar, CO_2), or semiconductor (e.g., GaAs). The medium has to be excited (pumped) in order to start the lasing process, and excitation is often provided by electric current. Also, in order to get laser emission the laser should contain two mirrors which reflect the light back and forth through the lasing medium. One of these mirrors is partially transparent, and the laser beam emerges from this mirror. [*See* LASERS.]

There are many lasers that emit light in the UV, visible, and IR. Some lasers emit continuously light of wavelength λ, and the total power emitted is measured in watts. These are called cw (continuous wave) lasers. Other lasers emit short bursts of light, in which case the lasers are called pulsed lasers. The energy emitted from these lasers can be specified by the pulse length (seconds), the number of pulses per second (pps), energy per pulse (joules), and average power (watts). Some of the important lasers which have been used in medicine are given in Table I.

B. The Use of Various Lasers in Medicine

As a laser beam impinges on biological tissue, its energy is attenuated (gradually decreased) in the tissue due to scattering and to absorption. In the absorption process the energy is transformed to another form of energy—often to heat. The attenuation due to absorption is given in general by the Beer–Lambert law [Eq. (4)]. For every tissue and every wavelength λ there is an absorption coefficient a_λ, and one could then write $I_\lambda = I_{0\lambda} \exp(-a_\lambda L)$, for a tissue thickness L. The thickness of tissue in which 99% of the incident energy is absorbed is called the extinc-

TABLE I. Medical Lasers

Lasing medium	Type[a]	Wavelengths (nm)	Operation	Maximum average power	Peak pulse output (J)
Excimer, ArF	G	193	Pulse	10 W	10 nsec at 0.5 J; 100 Hz
Excimer, KrF	G	249	Pulse	25 W	10 nsec at 0.75 J; 100 Hz
He–Cd	G	325, 442	cw	20 mW, 40 mW	
Ar	G	458, 488, 514	cw	20 W (total)	
Kr	G	413, 458, 568, 647	cw	6 W (total)	
Metal vapor, Cu	G	510, 578	Pulse	40 W	10 nsec at 8 mJ
Metal vapor, Au	G	628	Pulse	5 W	10 nsec at 1 mJ
He–Ne	G	633	cw	50 mW	
Dye	L	400–900	cw	1 W	
GaAs/GaAlAs	SC	780–905	cw	0.5 W	
Nd : YAG	S	1064	cw	600 W	
			Pulse	400 W	100 J; 300 Hz
HF	G	2600–3000	cw	150 W	
			Pulse	12 W	200 nsec at 0.6 J; 20 Hz
CO	G	5000–6500	cw	20 W	
CO_2, Sealed	G	10,600	cw	100 W	
CO_2, Waveguide	G	10,600	cw	40 W	

[a] Type: G, gas; L, liquid; S, solid; SC, semiconductor.

tion length l. One can easily see that for wavelength λ, $l_\lambda = 4.6/a_\lambda$ (cm), since $I_\lambda = I_{0\lambda} \exp[-a_\lambda(4.6/a_\lambda)] \approx 0.01 I_{0\lambda}$, so that 99% of the light is absorbed after travelling $4.6/a_\lambda$ cm in the tissue.

Absorption of laser light depends on chromophores such as water, hemoglobin, melanin, keratin, and protein. Water does not absorb in the visible, but the absorption is rather high in the far infrared ($\lambda > 3$ μm). For the CO_2 laser ($\lambda = 10.6$ μm), the extinction length l is less than 0.1 mm. The same is true for the CO laser ($\lambda = 5$ μm). Red pigment in hemoglobin absorbs blue–green laser light, and therefore the extinction length of argon or krypton laser in blood is also small. On the other hand, Nd : YAG laser light ($\lambda = 1.06$ μm) is not absorbed well by tissue, and the extinction length is a few millimeters. In the UV, excimer laser light ($\lambda < 300$ nm) is highly absorbed by tissue.

In the visible or in the infrared, the absorbed laser energy is converted into heat. For low laser intensity (energy per unit area), this gives rise to coagulation of blood. For high intensity, the tissue vaporizes (in most cases it is water that boils away). In practice the laser beam is focused on a certain area, and the vaporization removes tissue from this area. The common medical lasers are bulky and heavy. The output laser beam is often delivered to a desired spot via an articulating arm. This is a system of mirrors, attached to a distal hand piece. This articulating arm enables the user to focus the beam and to move it from place to place. By moving the beam the physician can therefore ablate tissue or perform an incision or an excision. Tissues outside the area of the focal spot are also affected. They heat up to a temperature that is dependent on the thermal diffusivity and the extinction length of the tissue. The heating effect may cause damage to surrounding tissue. The damage is lower if one uses lasers for which the extinction length is small and if one uses pulses of energy rather than a cw beam. The interaction described above is basically photothermal. The interaction between excimer lasers ($\lambda < 300$ nm) and tissue may be different. It involves the absorption of UV light, and the ablation of tissue is probably due to photochemical processes that do not generate heat. This decomposition is under study now, and the excimer laser may have surgical applications. [*See* LASER–MATERIALS INTERACTIONS.]

C. Lasers and Fibers

Laser beams in the visible and near-infrared ($\lambda = 0.3$–3.0 μm) could be easily transmitted by silica-based glass fibers. Glass fibers have been used for transmitting the radiation of argon, copper vapor, and Nd : YAG lasers. Relatively high intensities have been transmitted (both for cw and pulsed lasers) through pure silica (quartz) fibers.

Excimer laser radiation at λ = 250–300 nm could be transmitted only through pure quartz fibers, and even quartz is not very useful at shorter wavelengths. Some better fibers are needed for this spectral range.

The radiation of infrared lasers, and especially that of a CO_2 laser (λ = 10.6 μm), presents a problem. There has been a major effort to find suitable fibers for this radiation. At the moment the best fibers are polycrystalline fibers that are made of halide crystals.

IV. Fiber-Optic Endoscopes

The name endoscope is based on two Greek words: *endon* (within) and *skopein* (view). It is being used to describe optical instruments that facilitate visual inspection and photography of internal organs. Endoscopes may be inserted into the body through natural openings (ear, throat, rectum, etc.) or through a small incision in the skin. The simplest endoscope is an open tube, and it has been used in medicine for thousands of years. The development of modern endoscopes started about a hundred years ago with the addition of artificial illumination (i.e., incandescent lamp) and lenses to the open tube. Rigid endoscopes built in a similar manner are still in use. The nature of endoscopy changed dramatically in the early 1950s when optical-fiber bundles were introduced.

A. Light Guides for Illumination

1. Light Sources

In endoscopy we would like to transmit light through an optical fiber, in order to illuminate an internal organ. The light sources that are suitable for this purpose should deliver enough energy through the fiber to facilitate (later) viewing or photography. In general, high-intensity light sources are needed, such as tungsten lamps, mercury or xenon high-pressure arc lamps, or quartz iodine lamps. In most of these cases, the light has to be focused on a fiber (or a fiber bundle), by means of a lens or a reflector. Special provisions are often required to dissipate the excessive heat generated by the light sources.

Another useful light source is the laser. Some of the lasers that were mentioned earlier are suitable for endoscopic illumination.

2. Light Guides (Nonordered Bundles)

The light from a regular high-intensity source cannot be focused to a spot whose size is equal to the size of a thin optical fiber. Therefore a bundle of fibers is normally used for illumination. This assembly consists of numerous clad fibers of a certain length, which are bundled together but not ordered. In order to increase the light collection efficiency, the fibers are designed to have a relatively high numerical aperture. The diameter of each individual fiber is 20–60 μm, so it is flexible. The ends of the fibers are cemented with epoxy resin or fused together. The remaining fibers are left free, so that the whole bundle can be bent. A thin plastic sleeve is normally used to protect the fibers, and often a metal tube protects the ends of the bundle.

When trying to develop ultrathin fiberscopes, space is at a premium, and the light guide can include only a small number of fibers. In this case it is advantageous to use lasers for illumination. The major problem in using a laser is the nonuniform distribution of light intensity on the illuminated region. This nonuniformity is called speckle, and it stems from interference effects, which appear only with laser light (due to its temporal coherence characteristics). One way of getting rid of speckle is by vibrating the light guide somewhere along its length, keeping the ends fixed.

B. Coherent Bundles for Image Transmission

Optical fibers could be accurately aligned in a bundle, such that the order of the fibers in one end is identical to the order in the other end. Such a systematic optical-fiber assembly is generally called a coherent bundle (not to be confused with the coherent nature of laser light). If a picture is projected and imaged on one end of such a bundle, each individual fiber transmits the light impinging on it, and the ordered array will transmit the whole picture to the other end. This is shown schematically in Fig. 4.

The individual fibers used in a bundle are clad fibers. If unclad fibers were used, light would leak from a fiber to its nearest neighbors (cross talk), and the quality of the image would deteriorate. The optical fibers in a bundle (or at least on each end) are closely packed, as shown schematically in Fig. 5. We define d as diameter of the cladding and c as the diameter of the core.

A coherent (aligned) bundle has to faithfully

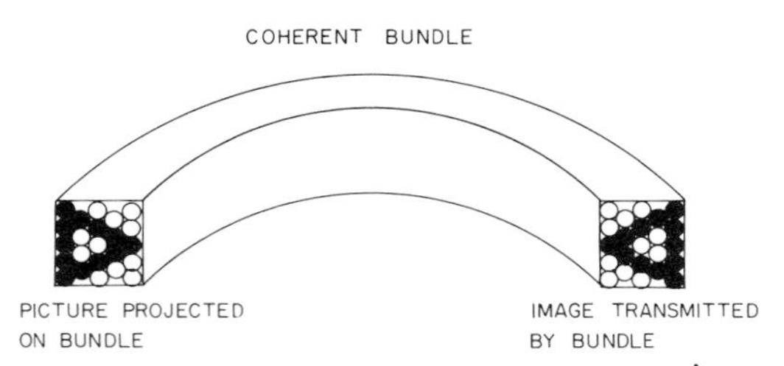

FIG. 4. Image transmission through a coherent (aligned) bundle.

transmit the (colored) image of an illuminated object. In order to get a bright image, as required for photography, each individual fiber should have a high numerical aperture. Toward this goal, one should judiciously choose the materials needed for fabricating core and cladding. The thickness of the cladding layer of each individual fiber $(d - c)/2$ should be of the order of 1.5–2.5 μm, in order to ensure guiding with little loss and to minimize cross talk. On the other hand, in order to get high spatial resolution one has to decrease the diameter c of the fiber core to 10–20 μm. In this case the ratio between the area of the cores to that of the bundle is not high, and the light transmission through the bundle is reduced. The fibers should therefore have low loss in order to obtain a good picture.

The image quality also depends on a variety of factors. Fibers that were broken during fabrication or during use do not transmit light and give rise to dark spots. Fibers that were not packed correctly give rise to dark lines and other defects. The individual fibers should be fabricated uniformly and should have good optical transmission through the whole visible spectral range. Otherwise the picture obtained will not faithfully reproduce color.

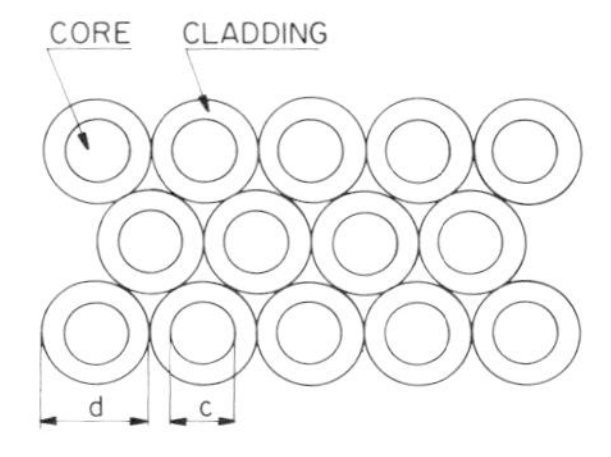

FIG. 5. Close packing of clad optical fibers in a bundle, with c the diameter of the core and d the diameter of the cladding.

C. Fabrication of Fiber-Optic Bundles

A coherent (aligned) bundle may be fabricated by winding a clad optical fiber on a precision machine and continuously winding many layers one on top of another. The fibers are then cut, glued together at each end, and polished. Tens of thousands of fibers, of diameters 10–20 μm, can be bundled in this way.

Alternatively, one starts with an assembly of carefully aligned lengths of clad fibers. The assembly is heated in a furnace, and a compound fiber is drawn—much like the drawing of a single fiber. The resulting fiber is called a multifiber, and it may contain fibers of diameter of about 5 μm. If the number of fibers in a multifiber is small (e.g., 100), it is quite flexible. A few multifibers could then be aligned to form a flexible coherent bundle of very high spatial resolution. If the number of fibers is higher than 1000, the multifiber is rigid. It is then often called an image conduit, and it may serve for image transmission in a rigid endoscope.

Finally, a flexible multifiber may be fabricated by a leaching process. In this case a special clad fiber is fabricated with two cladding layers. The inner cladding layer is made of an acid-resistant glass, and it has a lower refractive index than the core. The outer cladding layer is made of glass that is soluble in an acid. The multifiber is made in the same way described above. Then it is cut to a desired length, and the two ends are protected by some plastic material. The multifiber is then immersed in an acid bath, the outer cladding is leached out, and the individual fibers are separated from each other. A flexible coherent (aligned) bundle may again consist of fibers 10–20 μm in diameter.

D. Fiberscopes and Endoscopes

With the availability of fiberoptic bundles, people were able to design and fabricate a viewing instrument called a fiberscope. The structure of the fiberscope is shown schematically in Fig. 6.

Light from a lamp (or a laser) is focused onto the input end of a flexible light guide. The light is transmitted through this incoherent (nonaligned) bundle and illuminates an object. An objective lens forms an image of the object on the distal end face of a coherent bundle. The image is transmitted through the bundle to the proximal end, and it is viewed through an eyepiece. A photographic camera or a television (video)

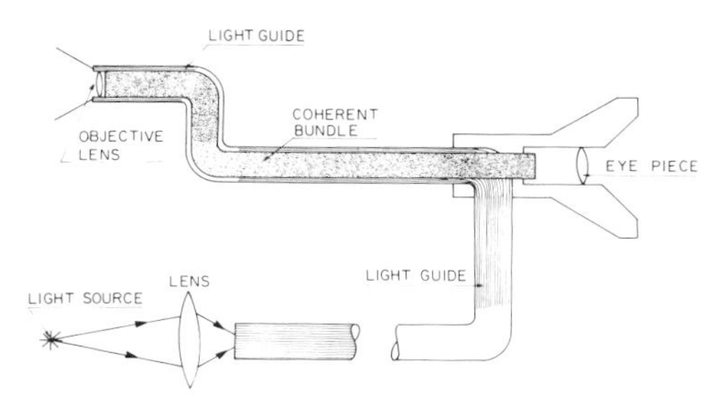

FIG. 6. Schematic diagram of a fiberscope.

camera could also be connected to the proximal end of the bundle with a special optical adaptor. The image-transmitting bundle could be a rigid image conduit or a flexible bundle, resulting in a rigid fiberscope or a flexible one, respectively.

Modern medical endoscopes incorporate fiberscopes that enable the physician to view and to examine organs and tissue inside the body. In addition, an endoscope may include ancillary channels through which the physician can perform some other tasks. A schematic cross section of an endoscope is shown in Fig. 7. One open channel serves for inserting biopsy forceps, snares, and other instruments. Another channel may serve for insufflation with air or for injection of transparent liquids to clear the blood away and to improve visualization. A channel may also serve for aspiration or suction of liquids. Many endoscopes have some means for flexing the distal tip to facilitate better viewing. Some typical data on commercially available endoscopes are given below:

Length:	300–1200 mm
Outer diameter:	2.5–15 mm
Instrumental channel diameter:	0.5–3 mm
Flexion up/down:	180°/60°
Focus depth:	5–100 mm
Field of view	50–70°

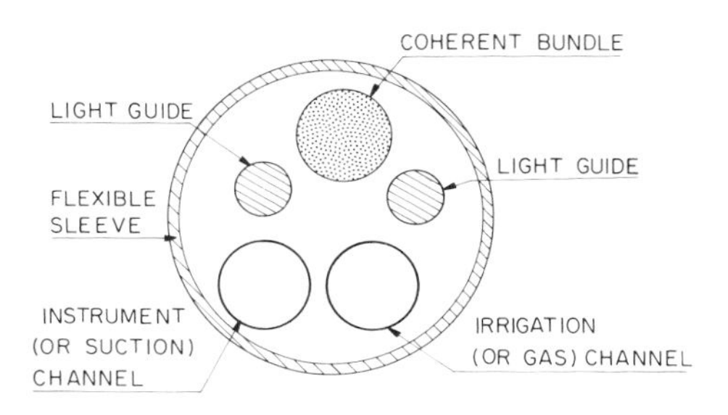

FIG. 7. Schematic cross section of an endoscope.

E. Clinical Applications of Endoscope

Fiber optic endoscopes have been used in a variety of medical applications. A few of these applications will be briefly described here:

1. Regular Endoscopes

a. Gastroscope. The first fiber-optic endoscopes called gastroscopes were developed for viewing the upper part of the gastrointestinal tract. Gastroscopes have been used for examining the stomach, the esophagus, the bile channels, etc. Surgical instruments have been inserted through the ancillary channel and used for removing biliary stones, for example.

b. Colonoscope. This endoscope is used for examining the colon, and it may be utilized for the early detection of carcinoma of the large bowel. Moreover, benign polyps may be detected and removed. For this purpose a special snare is inserted through the instrument channel, it is passed over the polyp, and a high-frequency current is used to remove the polyp (electroresection).

c. Bronchoscope. Most of these endoscopes are fairly thin, as they are used to visualize thin bronchi. Bronchoscopes have also been used for removal of foreign bodies, for early detection of cancer, and for the removal of tumors.

2. Ultrathin Endoscopes

Recently there has been some progress in developing endoscopes of very small diameter. Endoscopes of diameter of about 2 mm may incorporate about 7000 fibers, each of diameter

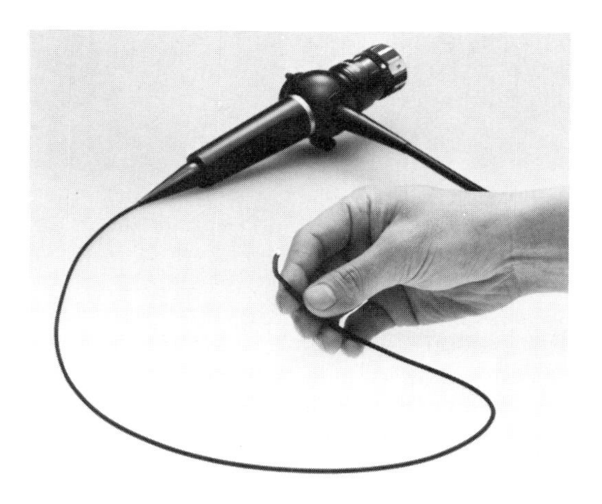

FIG. 8. A picture of an Olympus ultrathin fiberscope.

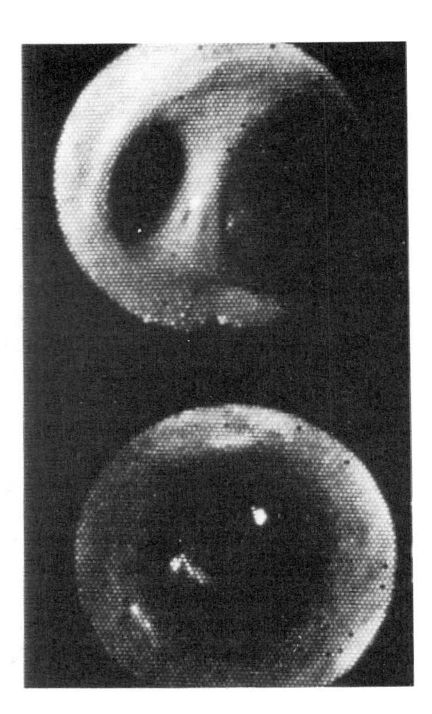

FIG. 9. A picture of the bile duct taken through an ultrathin fiberscope. (Courtesy of Olympus.)

9 μm. The length of the endoscope is about 1 m, and the resolving power is high enough to see a thin polypropylene suture inside a blood vessel. These endoscopes are mainly intended for cardiology, and they are then called angioscopes.

Angioscopes have already been inserted through blood vessels into the heart and used for examining heart valves. They have also been inserted into the coronary arteries and used for viewing atherosclerotic plaque. A picture of an Olympus ultrathin fiberscope is given in Fig. 8, and a picture taken through this fiberscope is given in Fig. 9.

In addition to these endoscopes, there are many types of other endoscopes. Some of the endoscopes are flexible and some are rigid. Some are inserted through natural orifices, and some are inserted via a metal tube through an incision in the skin (percutaneously). Basically all these endoscopes are similar.

V. Fiber Optics for Medical Diagnostics

There is a pressing need to improve some of the diagnostic medical techniques. At present, blood samples, for examples, are sent to a laboratory for analysis. This laboratory may be far in distance from both the patient and the physician, and there are bound to be delays or even unintentional errors in the clinical chemical results. There has been a concentrated effort to use miniaturized electronic devices as sensors that could perform chemical analysis inside the body. Fiber optics offers an alternative method for performing medical diagnostics inside the body of a patient. In principle, this method may turn out to be more sensitive, reliable, and cost-effective.

A. Diagnostic Systems

A typical fiber-optic sensor system that could be used for medical diagnostics is shown schematically in Fig. 10. The laser beam is coupled into the proximal end of an optical fiber and is transmitted to the distal end, which is located in a sampling region. Light is transmitted back through the same fiber (or a different fiber) and is then reflected into an optical instrument for optical analysis. The fiber-optic sensors fall into two categories: direct and indirect, as shown in Fig. 11. When using direct sensors, the distal end of the fiber is bare, and it is simply inserted into the sampling region. Indirect sensors incorporate some transducer at the distal end (sometimes called optrode). The laser light interacts with the transducer, which, in turn, interacts with the sample. Each of these categories should be further divided into two: physical sensors and chemical ones. Physical sensors respond to some physical change in the sample such as temperature or pressure, while the chemical sensors respond to changes that are chemical in nature, such as pH. We will discuss each category sepa-

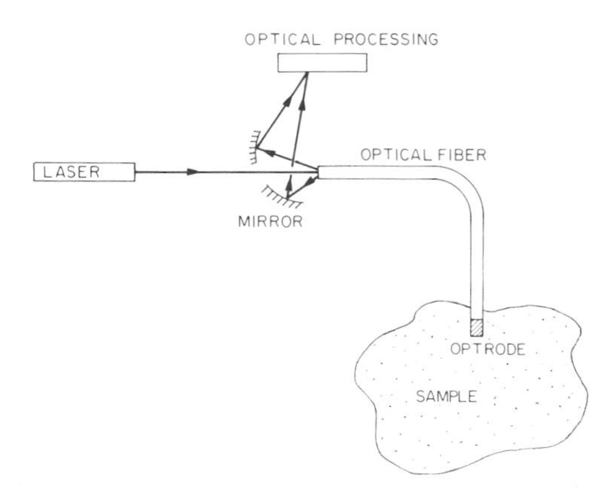

FIG. 10. A fiber-optic sensor for medical diagnostics.

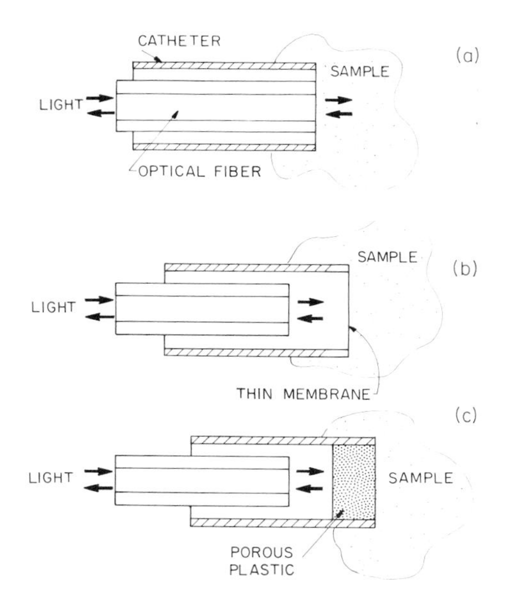

FIG. 11. Fiber-optic sensors. (a) A direct sensor. (b) An indirect physical sensor. (c) An indirect chemical sensor.

rately and illustrate their operation with a few examples.

B. Direct Sensors

1. Physical Sensors

Blood velocity may be measured by using a fiber-optic technique called laser Doppler velocimetry (LDV). A glass fiber is inserted through a catheter (plastic tube) into a blood vessel. An He–Ne laser beam is sent through the fiber, directly into the blood. The light scattered back from the flowing erythrocytes (blood cells) is collected by the same fiber and transmitted back. This scattered light is shifted in frequency (with respect to the regular frequency of the He–Ne) because of the Doppler effect. The frequency shift is proportional to the velocity. Using the LDV technique, instantaneous arterial blood velocity measurements were performed *in vivo* in dogs. Blood-velocity profiles at an arterial stenosis (narrowing) were also studied with the same technique.

Blood flow may also be measured by a dye dilution technique. As an example, sodium fluorescein may be injected into the blood. This dye has a yellow-green luminescence when excited by blue light. An optical fiber may be inserted into an artery, and the same fiber may transmit both the excitation light and the emitted luminescence. The dye dilution in the blood can be deduced from the decrease in luminescence with time. This in turn may be used to calculate blood flow.

2. Chemical Sensors

Oxygen in blood is carried by the red blood cells. Oxygen saturation is the ratio (in percent) between the oxygen content in a blood sample and the maximum carrying capacity. In arteries the blood is normally more than 90% saturated, and in the veins only about 75%. There is a large difference between hemoglobin and oxyhemoglobin in absorption of light at $\lambda = 660$ nm. On the other hand, there is little difference at $\lambda = 805$ nm. Light at 660 nm, emitted from a semiconductor source (or incandescent lamp), is sent through an afferent optical fiber. The light scattered back from blood is transmitted through an efferent fiber, and its intensity I_{660} is measured. Two other fibers are used for monitoring the intensity I_{805} for calibration. The ratio I_{660}/I_{805} is used to determine the oxygen saturation in blood. Measurements have been routinely performed through a thin catheter (of diameter 2 mm) *in vivo*.

C. Indirect Sensors

1. Physical Sensors

In this case a transducer (optrode) is attached to the distal tip of an optical fiber catheter, in order to perform physical measurements. Two measurements are of special importance: temperature and pressure.

There are several schemes for measuring temperature. One scheme is based on the fact that the optical absorption of GaAs crystal near the absorption edge is highly dependent on temperature. A GaAs crystal is attached to the end of the fiber, its absorption is measured using fiber-optic techniques, and the change in absorption is translated into temperature. Another scheme is based on change in the luminescence of a phosphor with temperature. The phosphor powder is attached to the distal end of a plastic clad silica fiber. Ultraviolet light is sent through the fiber and excites the phosphor. The visible luminescence is returned through the same fiber, and its intensity is measured. The change in intensity is translated into temperature. The accuracy of both methods is better than 0.1°C near room temperature.

Pressure may be measured via mechanical

transducers attached to the optical fiber. For example, a reflective surface may be attached to the distal end of a fiber by flexible bellows [Fig. 11(b)]. Light sent through the fiber is reflected back through the same fiber. The light output depends on the shape of the reflective surface, which in turn depends on pressure. Both temperature and pressure have been measured *in vivo* inside blood vessels.

2. Chemical Sensors

In this case the miniature transducers, which are attached to the end of the optical fiber, are sensitive to chemical changes in the sample of interest (e.g., blood). The basic design of an indirect chemical sensor is shown in Fig. 11(c). A special reagent is trapped inside a porous polymer. Light is sent through an afferent fiber, and reflected light or luminescence from the reagent is transmitted back. The reagent is allowed to interact with blood, for example, through diffusion. This, in turn, is manifested as a change in luminescence (or change in reflected light). Sensors for measuring pH are based on a dye indicator. Some dyes luminesce under UV excitation, and the luminescence intensity is determined by the pH of the sample. For other dyes the absorption spectra is determined by the pH. In practice, fiber-optic sensors have been used for measuring pH of blood to include the physiological/pathological range (6.8–7.7) with accuracy better than 0.01 pH units.

A different sensor, of similar construction, is used for monitoring the partial pressure of oxygen pO_2 in blood. In this case one may again use a dye that fluoresces. In some dyes the fluorescence emission decreases with increase in pO_2 (quenching). The fluorescence intensity is therefore a direct measure of pO_2. Similar sensors have also been used for measuring the partial pressure of CO_2, pCO_2, in blood.

A chemical sensor based on slightly different design was successfully used for monitoring the sugar content in blood.

D. The Detection of Cancer Cells

It has been known since the 1940s that certain compounds, called porphyrins, are preferentially concentrated in malignant tumors, with respect to healthy tissue. Porphyrin fluoresces under UV excitation, and therefore by illuminating tissue one could distinguish malignant tumors from benign tissue. In the 1960s a compound called hematoporphyrin derivative (HPD) was found to have even better properties than porphyrin. The fluorescence of HPD is mostly in the red part of the spectrum, with two prominent emission peaks at 630 and 690 nm. This fluorescence can be excited by UV or by blue light, but excitation at around 400 nm gives rise to the highest fluorescence. The fluorescence efficiency (emitted red power divided by excitation UV power) is fairly low, and therefore lasers are required for excitation. A krypton laser emitting about 0.25 W at 413 nm is most suitable for this application.

In practice, HPD may be injected into a patient, and after a few days this dye concentrates only in cancer tissue. If a tissue area is now illuminated with a suitable UV source, malignant tissue will emit characteristic red light. In order to see this red light one has to attach to an imaging system an optical filter that transmits red light at 630 nm and that blocks the exciting light at 413 nm.

This diagnostic method can be readily adapted in endoscopy. A special endoscope could incorporate a quartz fiber for UV illumination, and a red transmitting filter in front of the imaging fiberscope. With such an endoscope, tumors inside the body could be irradiated by krypton laser light, and by the red image formed, one could locate malignant tumors. This may be used, for example, for early detection of lung cancer, in cases where the malignant tumors are small and minimally invasive. Such tumors are too small to be detected by chest X ray or by computed tomography, but they could be detected by fluorescence of HPD.

VI. Fiber-Optic Therapeutic Tools

A. Power Transmission through Optical Fiber

Therapeutic applications of lasers, and in particular surgical applications, call for using relatively high laser power. Typically an average laser power of 10–50 W is required. With the advent of low-loss fibers it is possible now to transmit such power levels through thin optical fibers. Power fibers may replace the cumbersome articulating arms for delivering the beam from the laser to the operating site. These power fibers may also be inserted inside the human body through natural orifices, through flexible endoscopes, or through rigid needlescopes. Laser beams sent through the fibers may then be used for a variety of medical applications: coagulation, ablation, incision, etc.

The three lasers that are dominant in laser surgery are the CO_2, the Nd : YAG, and the Ar-ion laser. These are also the lasers that have been tried first with fiber-optic delivery systems. The argon laser radiation ($\lambda = 514$ nm) is in the visible spectral range, and its radiation is readily transmitted by quartz fibers. The same is true for the Nd : YAG laser, whose radiation is in the near-IR ($\lambda = 1.06$ μm). Quartz fibers of diameters 0.1–0.2 mm and of lengths 1–2 m have been used. The power levels that have been continuously transmitted through the fibers were up to 10 W for the argon-ion laser, and up to 60 W for the Nd : YAG laser. The CO_2 laser radiation is in the far-infrared ($\lambda = 10.6$ μm), and it can be transmitted by polycrystalline halide fibers. Fibers made of the salt TlBrI may be best as far as transmission is concerned, but they are highly toxic and fairly brittle, which may limit their use for medical applications. Fibers made of silver halide may be better suited. In both cases, fibers of diameter of about 1 mm and length of about 1 m have been used. Power levels of tens of watts have been continuously transmitted through infrared-transmitting fibers.

The enormous advantage of using fiberoptic delivery systems for surgical operations inside the body are obvious. Inserting the fibers inside the body is a least-invasive surgical procedure, and the need for major surgical operation may be eliminated in many instances.

Yet there are still many hurdles. Some of the problems involved in using optical fibers for power transmission are as follows.

Damage to the ends of the fibers. The power density (W/cm^2) at each end of the fiber is very high. It may easily reach values of 10^4 W/cm^2. Any defect on the fiber end, a splattered drop of blood or a fingerprint, may increase the absorption of laser beam on the end face. With these high power densities, the end face will be damaged.

Bending. With most fibers today, one cannot bend the fibers beyond a certain value. Bending increases the loss, and again the fiber may be damaged at the bend.

Divergence. The light emitted from fibers is not collimated, like a regular laser beam. It is highly divergent, and the divergence is determined by the numerical aperture. When transmitting the beam through the fiber, the distal end of the fiber has to be kept clean, and it must not touch the tissue. If this end is held 2–3 mm away from the tissue, the power density at the tissue may be too low for incision.

Dry field. Blood will coagulate under laser radiation. In order to prevent coagulation during fiber-optic laser surgery, one has to replace the blood near the distal end of the fiber by saline solution or by transparent blood substitute, or push the blood back using pressurized CO_2 gas.

B. Surgery and Therapy

During the last few years there has been an increasing number of physicians who try to use optical fibers for surgical applications. Most of these experiments and the operations performed on patients were carried out using argon ion or Nd : YAG lasers and quartz fibers. In many experiments the fibers were inserted in catheters, which were then called laser catheters. In some experiments the quartz fibers were incorporated in endoscopes, called laser endoscopes, and the experiments themselves were performed under direct viewing. Few experiments were performed with CO_2 lasers and IR fibers. Some of the important developments will be reviewed.

1. Gastroenterology

One of the first uses of the fiberoptic laser catheter was in treating bleeding ulcers. The laser light could photocoagulate blood and therefore cause hemostasis (cessation of bleeding). Among the three important lasers (CO_2, Nd : YAG, and Ar), some physicians prefer the Nd : YAG laser because it penetrates deep into tissue and its effects are not localized at the surface. Others have used the argon ion laser for controlling gastric hemorrhage (bleeding). In both cases the laser beam was sent through thin quartz fibers. Experiments have been performed *in vivo* on patients with bleeding ulcers in the stomach, the esophagus, or the colon, and the results have been very good.

2. Urology

A thin quartz fiber was inserted in one of the ancillary channels in a laser endoscope. An Nd : YAG laser beam, sent through the fiber, was used for destroying small tumors in the urinary bladder *in vivo*. In other cases larger tumors were electroresected (using the endoscopic snare described earlier) and Nd : YAG radiation was then used for coagulation. Again, the deep penetration of Nd : YAG in tissue may be advantageous in this case. Some other experiments *in vivo* were performed using the Nd:YAG laser power in order to rapidly heat and shatter stones in the bladder.

3. Cardiology

In the cardiovascular system a major role is played by the coronary arteries, which supply blood to the heart muscle, and a myocardial infarction may result if a coronary artery becomes occluded. A common problem with the arteries is a build up of atherosclerotic plaque on the interior walls. The plaque, consisting of fatty material, calcium, etc., starts blocking the coronary arteries, and the blood flow through them is reduced. This results in angina pectoris, a condition that afflicts millions of people.

A common technique for diagnosis is an X-ray study of the arteries called angiography. A thin catheter is inserted through an incision in the groin (or the arm) and is pushed through the arterial system until it reaches the coronary arteries. Then an X-ray-opaque liquid is injected through the catheter, and the shadow of this liquid is examined by X-ray imaging. A blockage in the blood flow can thus be examined. If the coronary artery is only partially blocked, the situation can sometimes be improved by using a method called balloon angioplasty. The more exact name for this method is percutaneous transluminal coronary angioplasty (PTCA). This method makes use of a special catheter that includes a tiny balloon at its end. The tip of the catheter is inserted through the partially constricted artery, the balloon is inflated under high pressure (10 atm), and the blockage is reduced. Unfortunately, PTCA can be successfully used only on a small number of patients. In most cases, if the arteries are blocked, one has to resort to a surgical procedure called coronary artery bypass grafting (CABG). A vein is removed from some other part of the body (usually the leg) and is implanted in parallel to the blocked artery. This operation requires an open-heart procedure, which is traumatic, risky, and expensive.

With the development of optical fibers that could deliver laser energy, the road was paved for a novel method. It was suggested that during cardiac catheterization, when substantial blockage of the coronary is observed, an optical fiber could be inserted through the catheter. A laser beam sent through the fiber could then be used to vaporize the plaque and open a clear channel through which blood flow can then resume. This procedure is called laser angioplasty or vascular recanalization, and a schematic cardiovascular laser catheter and its operation is shown in Fig. 12. Many of the lasers mentioned in Table I could serve as candidates for recanalization procedure, in conjugation with suitable fibers. The major problem with the argon-ion laser is that it is absorbed in the red chromophores and not in water. Therefore the absorption in blood is high, and this laser may be best suited for removing blood clots in arteries, but in plaque the extinction length may not be small. As a result, some of the radiation may be transmitted through plaque and may perforate the blood vessel wall. Yet preliminary experiments have been performed successfully both on animals and on patients. In some experiments, blocked arteries were recanalized during an open-heart "bypass" operation. In other experiments, blocked arteries in the legs of patients were recanalized.

In principle, the Nd:YAG laser is even less suitable for laser angioplasty because of its longer extinction length. On the other hand, Nd:YAG may be operated in a pulsed mode, where energy is delivered in a series of short pulses, which facilitate vaporization of plaque

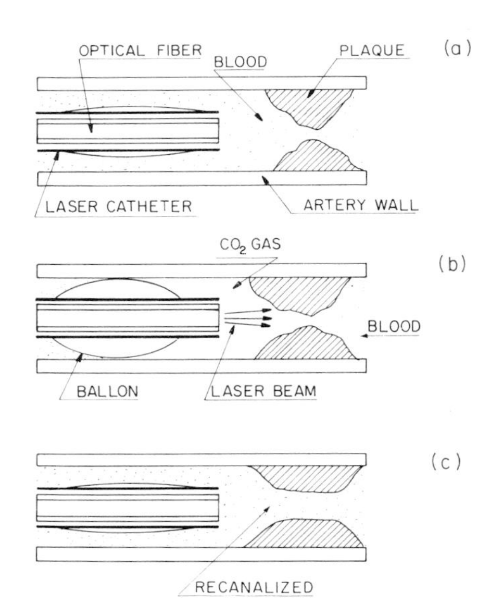

FIG. 12. Laser catheter for recanalization of arteries. (a) A laser catheter inside an artery blocked by plaque. (b) A balloon is inflated, and CO_2 gas (or saline solution) pushes the blood away. Laser beam vaporizes the plaque. (c) Artery recanalized, and blood flow resumed.

without causing excessive damage to the artery wall. Nd : YAG laser radiation was also successfully tried in experiments, both on animals and on patients.

The CO_2 laser radiation (or CO laser radiation) may be most suitable for vaporization of plaque because the extinction length is less than 0.1 mm; therefore, when plaque is vaporized all the energy is spent on vaporization, and the blood vessel wall is not damaged. The major problem with infrared radiation is the delivery system. Suitable optical fibers have only recently been developed and are being tried.

Excimer lasers, in the UV ($\lambda < 200$ nm), do not cause heating. As explained earlier, they may ablate plaque through a photodecomposition process. Excimer lasers are also being evaluated for laser angioplasty. The major problems with using these lasers are threefold: (a) the need for suitable fibers that transmit well in the UV; (b) a possible cancer risk; and (c) the lasers require regular replenishment of dangerous gases.

Years of research may be needed before a "best" procedure will be chosen.

C. Photomedical Applications

The retention of hematoporphyrin derivative (HPD) in malignant tumor cells has already been described. It was found that the use of HPD with lasers can serve not only for diagnosis of cancer but also for therapeutic purposes. If, rather than illuminating a tumor with UV light, one uses red light ($\lambda = 630$ nm) of sufficient energy, the results are strikingly different. HPD absorbs the red light, a series of photochemical reactions occur, and the end result is the release of some photoproduct (probably singlet oxygen), which kills the host malignant tissue. The method itself is called photoradiation therapy (PRT). This is actually photochemotherapy, because it involves the administration of a drug plus optical radiation for triggering the process that cures the disease. Roughly 10–50 mW/cm^2 of red light are needed for this photochemotherapy.

Photoradiation therapy is also adaptable for use in conjunction with fiber-optic systems. A high-intensity red light can be transmitted through a quartz fiber, and delivered directly inside the tumor. This light may then selectively destroy cancer cells. In the past, a cw dye laser emitting at $\lambda = 630$ nm was used for this purpose. More recently, gold-vapor lasers have been utilized for the same application. There has also been progress with the development of miniature semiconductor lasers, which emit continuously more than 0.1 W in the near-infrared. Similar lasers emitting at 630 nm have also suggested for photoradiation therapy.

VII. Novel Fiber-Optic Medical Systems

With the development of new flexible fiberscopes of very high optical quality and the concurrent development of optical fibers for laser power transmission, the road is clear for developing novel endoscopes. Such a laser endoscope is shown schematically in Fig. 13. The endoscope would consist of several channels. One channel would include a complete fiberscope, consisting of an imaging bundle and illumination light guides. Another channel would include optical fibers for diagnostic purposes, and the distal end of these fibers might even include transducers (optrodes). A third channel would include a power fiber, for laser power delivery. A fourth channel could be used for injecting liquids such as drugs, dyes, saline solution, etc. This (or another) channel could be used for pumping out liquids or for the aspiration of gases etc.

Two such catheters may be described in some more detail.

A. Angioplasty Laser Endoscopes

The compound laser endoscope will probably have a diameter of less than 3 mm. The fiberscope (angioscope) in it will be about 1 mm in diameter. The power fiber for transmitting laser

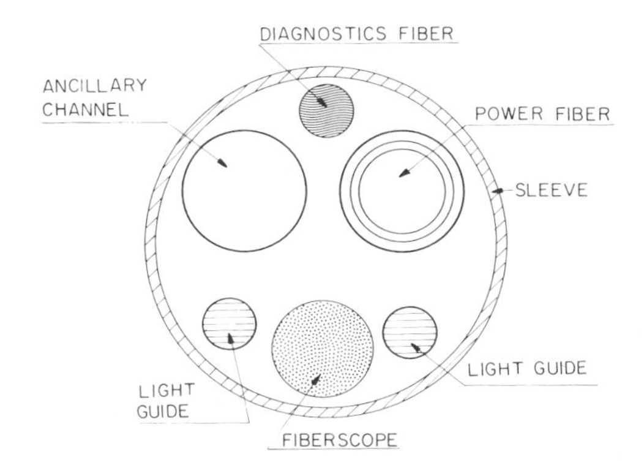

FIG. 13. A compound endoscope, which includes a fiberscope, a fiber-optic diagnostic system, a power fiber, and an ancillary channel for irrigation or suction.

power would have to be compatible with the laser used in angioplasty (UV, visible, or IR). The fiber-optic sensors would have to measure pressure, temperature, and blood flow. In the catheterization laboratory, the endoscope would be inserted into the coronary artery, much like in a regular catheter. The physician would look at the blockage in an artery and measure pressure and blood flow. For angioplasty, the physician will have to block the flow with a balloon and push away the blood from the distal tip, by injecting a clear liquid or pressurized gas through the irrigation channel. The physician would then operate the laser for a predetermined period. Gases released upon vaporization of the plaque would be pumped out through the aspiration channel. The temperature would be monitored during the procedure. After operating the laser, the balloon would be deflated. The blood flow would be measured again, and if the flow were found to be satisfactory, the catheter would be withdrawn and the procedure would be complete.

B. Laser Endoscope for Treatment of Cancer

In this case, the endoscope should include a special fiberscope, a power fiber for transmitting excitation light, and a power fiber for transmitting red light for photoradiation therapy. The endoscope would be inserted through a natural orifice or transcutaneously and brought near a suspected tumor. Krypton laser light would be transmitted through the first fiber, and the fiberscope image would be viewed through a red transmitting filter. Malignant tumors would be identified by their red emission, and then a high-intensity red light ($\lambda = 630$ nm) would be sent through a quartz fiber in order to destroy them. The laser that is most suitable for this purpose may be the gold-vapor laser (see Table I). In case of large tumors, the quartz fiber might be inserted inside the tumor to facilitate better interaction with the HPD inside the tumor. After a few weeks (or after several treatments), the same method might be applied again to ascertain that the tumor growth was arrested.

VIII. Outlook

During the past decade there has been rapid progress in the development of optical fibers and fiberscope systems. Foremost was the development of silica-based glass fibers, which have extremely low transmission losses. These fibers are the major building blocks of future communication systems. Other optical fibers were developed for more specialized applications, such as fibers with good transmission in the far-infrared or in the deep-ultraviolet parts of the spectrum. Among these we may also include power fibers that are capable of transmitting laser beams of fairly high intensities. With the development of the various optical fibers, there has been an increasing recognition of the enormous possibilities for utilizing them in medical applications. Three major uses of optical fibers—endoscopy, diagnostics, and power transmission—have been reviewed here.

In the past, endoscopes were rigid and bulky, and their uses were quite limited. A drastic change occured when fiberscopes (especially the flexible ones) were included in endoscopes. An endoscopist could get a clear image of an internal organ, and even perform surgical operations such as biopsy by inserting medical instruments through ancillary channels. There has been further progress in the development of ultrathin flexible fiberscopes whose outer diameter is less than 2 mm. Ultrathin fiberscope have been inserted into blood vessels and used to visualize atherosclerotic plaque within the coronary arteries or to view the heart valves. With further development, such fiberoptic endoscopes will undoubtedly have a growing role in medicine.

Optical fibers can be used for diagnosis by inserting them into the body and making physical and chemical measurements through them. As direct sensors, the fibers serve simply as light guides for transmitting laser light into the body and back. Such direct sensors have been used for measuring blood flow or for the early detection of cancer. In a different mode of operation, tiny transducers (optrodes) are attached to the fibers to form indirect sensors. With these optrodes, physical measurements, such as blood pressure and temperature, or chemical measurements, such as pH or pO_2, can be carried out. Currently, blood samples are extracted from a patient and sent to the laboratory for chemical analysis. Fiber-optic techniques may bring the laboratory "closer" to the patient. They may enable the performance of fast and repetitive chemical analysis at the patient's bedside or during an operation. The use of optical fibers as sensors may simplify some of the diagnostic techniques and make them more reliable and less costly.

Recently there has also been noticeable progress in the development of power fibers, which could transmit relatively high laser power. Quartz fibers have been used for the transmission of Nd : YAG, Ar-ion, and excimer laser beam, and crystalline infared fibers for the CO_2 laser beam. Power fibers would undoubtedly replace articulating arms in surgical laser systems. Such fibers, inside laser catheters, may be inserted into the body and used to perform surgical operations inside the body, without necessitating a large incision. As an example, we described the laser angioplasty procedure. Power fibers will certainly be used for similar applications in the future.

Finally, one may consider a compound laser endoscope that would contain several channels. One would be an ultrathin fiberscope that will enable the physician to see what he is doing. A second channel would be a fiber optic sensor, for diagnostic purposes. A third channel would be a power fiber for transmitting high intensity laser beams. Other channels would be used for injection of liquids or drugs, for inserting pressurized gases, or for sucking out debris. Similar endoscopes may be used in a multitude of applications. In the future, laser endoscope systems may be inexpensive and simple to use, and they may even be used in small clinics. In large hospitals, high-power lasers could be kept in a central control room. The laser beam would then be delivered to the operating rooms via power fibers. In each operating room, these fibers would be attached to suitable laser endoscopes.

The novel optical-fiber techniques described here may be used in a variety of medical applications. Laser endoscopes could be used in neurosurgery for vaporizing tissue, without causing mechanical damage to surrounding areas. They could be used in gynecology for microsurgery in the fallopian tubes. Laser endoscopes could be used for the removal of tumors on the vocal cords, in the bladder, inside the eyes, or in the brain. There is little doubt that with refinement these novel techniques will have a tremendous impact, and they may revolutionize many fields in medicine.

IX. Recent Developments

During the last five years there has been a rapid advancement in the field of biomedical optics, including lasers and optical fibers in medicine. The basic and preclinical research that was described in Sections I–VIII has led to the development of medical systems and many of the techniques are now used clinically. In this section we discuss some of these developments.

A. Areas of Development

1. Lasers

Medical laser systems have dramatically improved during the last few years. These lasers are now more reliable and more efficient and have made their debut in the clinical setting. Excimer lasers, which cut tissue with little thermal damage, are being tried in ophthalmology and in cardiology. Dye lasers are being tried for stone shattering and in dermatology. Powerful semiconductor lasers are being tested for laser heating of tissue and they may replace metal vapor lasers in photodynamic therapy.

2. Endoscopic Imaging

The quality of fiber-optic endoscopes has markedly increased during the last five years. Several ultrathin endoscopes (see Section IV.E.2) have been tried clinically in cardiology. Novel techniques have been added to endoscopy, such as video endoscopy, where the imaging device is a miniature charge coupled device (CCD) camera. Research is being carried out in new areas such as computerized image processing and holographic endoscopy.

3. Fiber-Optic Sensors

The feasibility of using optical fibers as biomedical sensors has been established, but some technical problems (e.g., response time, calibration, and shelf life) need further attention. Some of the sensors described in Section V have already been introduced into clinical use. A new family of indirect chemical sensors (see Section V.C.2) incorporates biomolecules such as enzymes or antibodies. These sensors can monitor the body's levels of glucose or penicillin and may soon measure metabolic substances, toxins, and microorganisms in the body.

4. Power Transmission

The power handling capabilities of optical fibers has increased as a result of improvements in the properties of the fibers and better properties of the fiber end faces. Many of the problems associated with coupling of high power laser beams into fibers have been solved. As a result, special fibers have been added to medical lasers as standard items for power transmission.

5. Laser Catheters

Many of the new devices are multichannel catheters in which an optical "power" fiber is inserted in one of the channels. Cooling liquid may also be injected into this channel. Other channels are used for irrigation with saline solution or for introducing drugs. A metal "guide wire" is often inserted in one of the channels to facilitate easy insertion of the catheter into the body. Recent progress in such catheters has led to their clinical use in cardiology, gynecology, and orthopedic surgery.

6. Laser Endoscopes

These systems integrate lasers, optical fibers, and endoscopes (see Section VII.A). Some of these systems incorporate novel devices such as miniature ultrasound devices that provide three-dimensional imaging inside blood vessels. Laser endoscopes have reached the stage where they are being used clinically in several medical disciplines such as otolaryngology, gynecology, gastroenterology, and certain areas of brain surgery.

B. Clinical Applications

These developments have already been used in clinical applications in several medical disciplines. Some examples of "least invasive" procedures that replace major operations are discussed in the following sections.

1. Urology

Laser catheters based on dye lasers and Nd : YAG lasers have been used for lithotripsy (shattering of stones). In this case the distal tip of the fiber is brought in contact with the stone. Laser pulses sent through quartz fibers generate a plasma, and the resulting shock waves shatter the stone. This procedure is particularly useful in cases where other noninvasive methods can not be applied. Laser lithotripsy has been successfully applied on hundreds of patients for the removal of urinary stones or biliary stones.

2. Cardiology

One of the most difficult laser procedures is laser angioplasty of the coronary arteries. This has been successfully accomplished with laser catheters based on the excimer laser and on quartz fibers. The tip of the catheter is guided using X-ray fluoroscopy, and the excimer laser is used to vaporize plaque with little thermal damage. Several patients have already been treated by this method. Laser catheters based on other lasers (e.g., Ar or Nd : YAG) and fibers are also being tried. One of the problems that has not yet been fully solved is that of monitoring and control of the laser beam during the procedure. Some of the laser catheters use fiber-optic sensor techniques to distinguish between plaque and normal arterial wall and to prevent perforation of the arteries. At the same time there has been progress in the use of power transmitting fibers in which special tips are attached to the distal ends. In one case the tip is made of metal and the laser power transmitted through the fiber heats the tip. The laser catheter may be introduced into a blocked artery and the "hot tip" used to open the blockage. Laser catheters based on this principle have been used to clear peripheral arteries (e.g., in the legs) in thousands of patients, and they have been used clinically for coronary arteries. Progress in the use of laser catheters for laser angioplasty of the coronary arteries has been slower than hoped, due to the complexity of the procedure.

Laser catheters and laser endoscopes have been tried in other medical disciplines in preclinical and clinical studies. They have been used for endoscopic diagnosis and treatment of cancer. They have been used for endoscopic orthopedic surgery on joints and in neurosurgery for treating brain tumors. There is no doubt now that fiber-optic techniques will replace more traditional procedures in the coming years.

Bibliography

Alfano, R. R., and Doukas, A. G. (ed.) (1984). Lasers in biology and medicine, special issue of *IEEE Journal of Quantum Electronics,* **QE–20,** No. 12.

Atsumi, K. (1984). "New Frontiers in Laser Medicine and Surgery." *Excerpta Medica.* Elsevier. New York.

Carruth, J. A. S., and McKenzie, A. L. (1984). "Medical Lasers—Science and Clinical Practice." Adam Hilger, Bristol and Boston.

Deutch, T. F., and Puliafito, C. A. (eds). (1987). "Lasers in Biology and Medicine," special issue of *IEEE J. Quantum Electron.* **QE-23** (10).

Dixon, J. A. (1984). "Lasers in surgery." *Current Problems in Surgery*. **21** No. 9.

Katzir, A. (ed.) (1985, 1986, 1987, 1988, 1989, 1990). *Proc. SPIE* **576, 713, 906, 1067, 1201.**

Wolf, H. F. (ed.) (1984). "Handbook of Fiber Optics: Theory and Applications." Granada Publishing. Great Britain.

OPTICAL FIBERS, DRAWING AND COATING

L. L. Blyler, Jr. and F. V. DiMarcello *AT&T Bell Laboratories*

Glossary

Cladding: Transparent, low-refractive-index region surrounding the core of an optical fiber, which confines light energy to propagate within the core by total internal reflection.

Core: Highly transparent central region of an optical fiber through which light energy propagates.

Decibel (dB): Unit of relative power; the transmission loss, in decibels, of an optical fiber is expressed as dB = 10 log(power in/power out).

Fire polishing: Act of using a flame to locally melt and smooth the surface of a body, such as a glass preform.

Glass transition: Temperature interval over which the physical properties of a noncrystalline material change from those of a glassy solid to those of a liquid or rubber.

Microbending: Term applied to small-amplitude random bends of the axis of an optical fiber, with periodic components in the millimeter to centimeter range, which give rise to added transmission losses.

Packaging: Act of incorporating one or more optical fibers into a structure, such as a cable or a unit that comprises a portion of a cable.

Preform: Cylindrical glass rod, fabricated with an appropriate radial refractive index profile comprising the core and cladding, from which an optical fiber may be drawn.

Proof test: Test in which a tensile stress or strain is applied to the entire length of an optical fiber by passing it through the test equipment; the test ensures that there are no sites along the fiber having a strength below the tested level.

Silica: Vitreous silicon dioxide.

Transmission loss: Attenuation of light energy propagating in a fiber, which results from absorption and scattering processes, commonly expressed in decibels/kilometer.

Waveguide: Structure that confines and directs electromagnetic radiation along a path.

UV-curable: Ability of a liquid resin to be polymerized into a solid, cross-linked network via initiation of a chemical reaction by exposure to ultraviolet light.

Optical fibers are hair-thin glass filaments consisting of a highly transparent core, surrounded by a cladding of lower refractive index. Light energy, directed on the end face of the fiber, is confined to propagate along the core by total internal reflection. Optical fibers are drawn from preforms that are fabricated to provide the refractive index profile desired in the drawn fiber. The drawing and coating process must be carefully controlled to ensure that important fiber performance characteristics are realized.

I. Fundamentals

A. Performance Properties of Fibers

The fiber drawing and coating process plays a very important role in determining the performance properties of optical fibers used in telecommunication systems. Among the performance properties of concern are strength, dimensional precision, and transmission loss. In addition, the fiber drawing and coating process impacts strongly on manufacturing economy and productivity.

In order to put the above remarks into perspective, it is useful to describe what would be considered high performance. In the case of fiber strength, although the silica glass used to fabicate most optical fibers is inherently strong, its strength is markedly reduced by the presence of surface flaws, which cause stress concentrations. A fiber with a pristine surface should have a tensile strength in excess of 5.5 GN/m^2 (800,000 lb/in.^{2}). If that fiber possessed a surface flaw only 1 μm in size, its tensile strength would be reduced to less than 0.7 GN/m^2 (100,000 lb/in.^{2}). Because it is very difficult to fabricate a fiber that has no surface flaws over its entire length, the term "high strength" is generally applied to fibers that have tensile strengths of 1.4 GN/m^2 (200,000 lb/in.^{2}) or greater over lengths of several kilometers. The tensile strength of a fiber can only be assured by strength testing (called proof or screen testing) the entire fiber. Special testing machines have been developed to carry out this quality control operation as part of the manufacturing process.

Dimensional precision is a very important consideration for connectorization and splicing of optical fibers. Misalignments of the fiber cores by only 1 μm may introduce unacceptable losses, especially for single-mode fibers. Since fiber cores are usually aligned by locating on the outer diameter of the fibers, their diameters must be closely controlled. High precision is therefore taken to mean that the fiber diameter is held to the desired value with a standard deviation of no more than 0.2%. Thus a 125-μm-diameter fiber should have a standard deviation of 0.25 μm or less.

Transmission loss as influenced by the drawing and coating process is most strongly affected by the coating. Typically, the coating influences the microbending losses, which are introduced in the fiber by packaging (cabling) and by subjecting the packaged fiber to different environments. For high performance, the microbending losses due to packaging or to environmental exposure should be less than 0.1 dB/km. [*See* OPTICAL FIBER COMMUNICATIONS.]

B. Description of the Fiber Drawing and Coating Process

The drawing and coating process is displayed schematically in Fig. 1. The important features are listed at the right, while the fiber performance properties affected by each stage of the operation are listed at the left. A glass preform

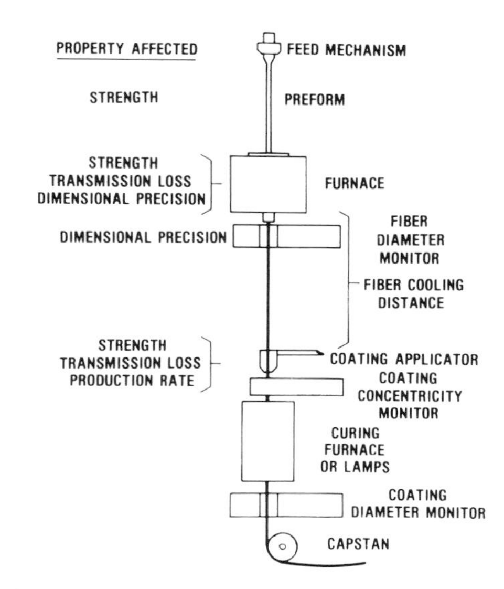

FIG. 1. Schematic diagram of the drawing and coating process.

rod, manufactured in a separate process, is fed at constant rate into the heat source, usually a high-temperature furnace, as depicted in Fig. 2, operated at approximately 2200°C. The fiber is freely drawn from the heated end of the softened preform through the use of a capstan located at the base of the apparatus. Immediately below the heat source, the diameter of the fiber is monitored. A signal based on this measurement is generated for use in a feedback control loop, which adjusts capstan speed to keep the fiber diameter constant.

Because the surface of the glass fiber is extremely susceptible to damage by abrasion, it is necessary to coat the fiber in-line, as it is drawn, before it contacts another surface. Prior to coating, the fiber must first traverse a region where it is allowed to cool. Then it is coated, generally with one or more organic polymers, via a method that avoids damaging the glass. The coating must be solidified very rapidly, before the fiber reaches the capstan. Further, certain coating parameters, such as coating diameter and concentricity, are monitored, and these properties are then controlled by appropriate techniques.

C. Process–Performance Relationships

Figure 3 depicts the interrelationships between fiber performance properties and features of the fiber drawing and coating process. For example, fiber transmission losses are greatly in-

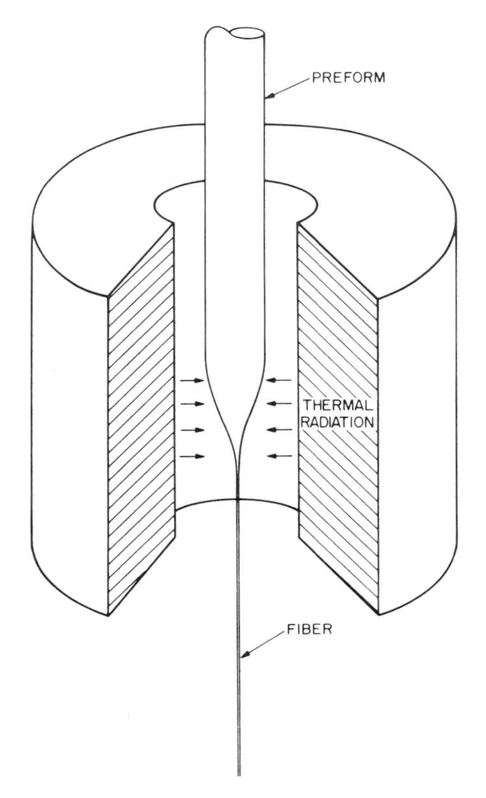

FIG. 2. Schematic diagram of a fiber drawing furnace.

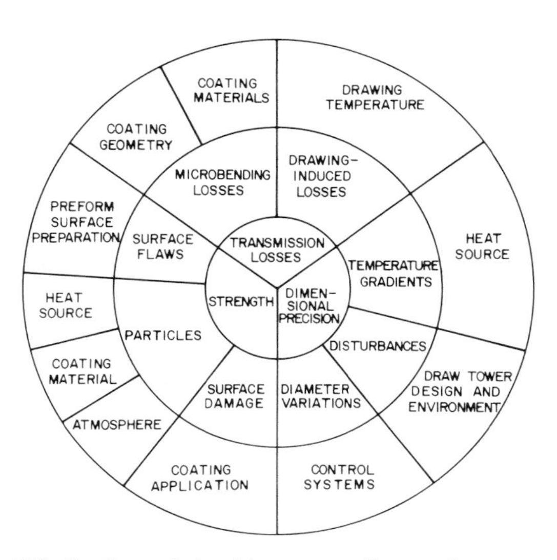

FIG. 3. Interrelationships among fiber performance properties and features of the fiber drawing and coating process.

creased by microbending, a random perturbation of the fiber axis with periodic components in the millimeter range. Such perturbations may result from stresses placed on the fiber by a surrounding cable structure. The geometry and mechanical properties of the coating strongly affect the microbending sensitivity of fibers and, hence, the transmission losses of cabled fibers in the field. Low-modulus polymer coatings, and particularly dual coatings—consisting of a low-modulus primary layer adjacent to the fiber, surrounded by a high-modulus secondary layer—are very effective in isolating the fiber from external stresses that give rise to microbending. In addition, improper heat-source temperatures may contribute to drawing-induced loss mechanisms in the glass material.

Fiber dimensional precision is influenced by temperature gradients and fluctuations in the heat source, as well as vibrations or convective air flow disturbances in the drawing environment. It is important to isolate the process from these effects. Low-frequency fiber diameter variations, arising, for example, from preform diameter variations, must be eliminated by the use of a suitable fiber diameter measurement and control system.

High fiber strengths over lengths of kilometers are required to insure the integrity of fiber transmission paths deployed in the field. Figure 4 exhibits photomicrographs of low-strength fiber fracture surfaces that arise from a variety of sources. Preform surface preparation, for example, is needed to eliminate surface flaws that may be incompletely healed by the drawing process (Fig. 4a). In addition, particles emanating from the heat source, or convected into the heat source from the drawing environment, may become lodged in the molten preform, thereby producing defects (Fig. 4b), which seriously degrade strength. Therefore fibers must be drawn in a highly particle-free environment, and heat sources, such as drawing furnaces, must be scrupulously maintained. Particles in the coating material, which may locate at the coating-glass interface (Fig. 4c), present a similar hazard. Thus coating compounds require filtration in the micrometer range for successful implementation. Finally, improper coating application techniques may result in damage to the fiber surface, and therefore specialized coating methods have been developed to circumvent this difficulty.

The following sections deal with each stage of the fiber drawing and coating process. In particular, the actions and controls that must be un-

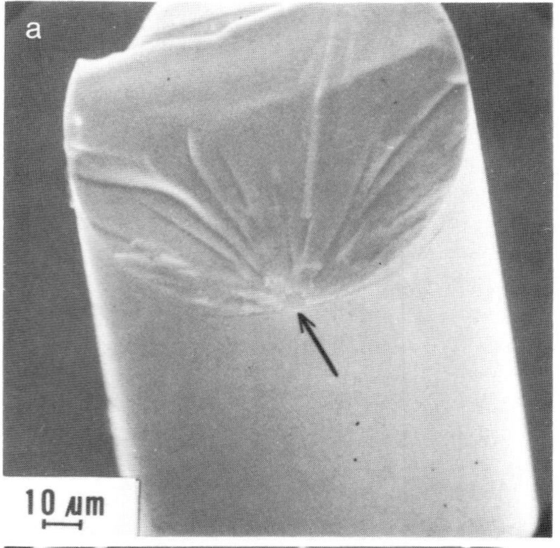

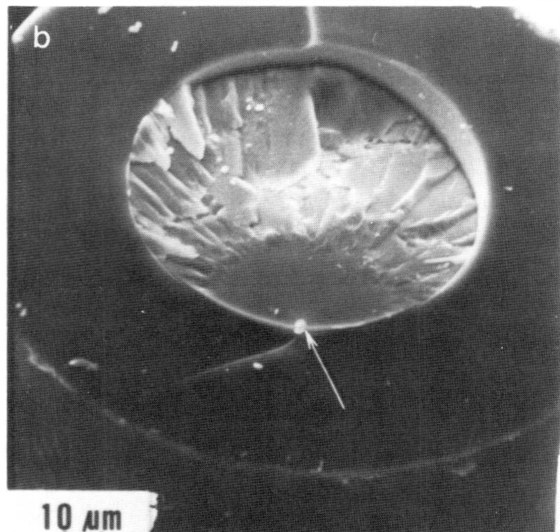

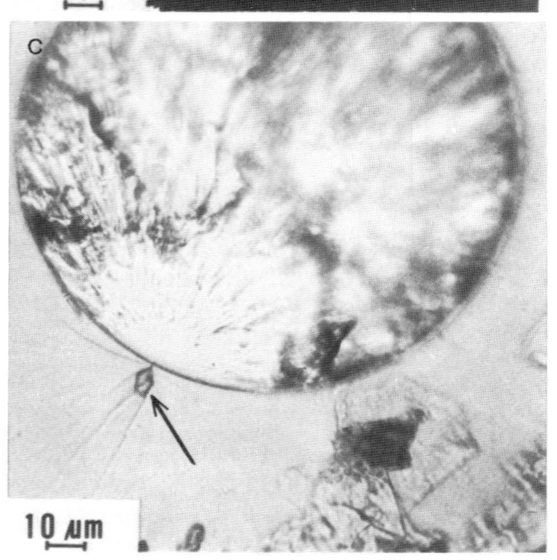

FIG. 4. Photomicrographs of fiber fracture surfaces with failure origins indicated by arrows: (a) surface flaw incompletely healed by the drawing process, (b) foreign particle lodged in the fiber surface, and (c) foreign particle in the coating at the fiber suface.

dertaken at each stage in order to produce high-performance optical fibers are detailed.

II. Features of the Process

A. Preform Surface Preparation

When the fracture surfaces of fibers that have failed in proof testing are examined microscopically, a significant fraction are seen to be surface-initiated failures of nondeterminable origin. Such failures may arise from preform surface flaws that are not healed by the drawing process. Therefore, preform surface preparation is a prerequisite to the attainment of high fiber strengths. Three methods of surface preparation that have been investigated are fire polishing with an oxyhydrogen flame, etching with hydrofluoric acid, and CO_2-laser polishing. Fire polishing has proven to be a highly consistent method for achieving a high-quality preform surface, independent of the original surface condition.

B. Heat Sources for Fiber Drawing

Several types of heat sources have been investigated for the fiber drawing process. CO_2-lasers are very clean heat sources but are currently impractical for drawing large-diameter preforms. Oxyhydrogen torches are similarly clean but tend to produce unacceptable fiber diameter variations due to flame turbulence. The heat sources that are most extensively used in commercial practice are graphite resistance furnaces and zirconia induction furnaces. Both provide uniform heating of the preform by blackbody radiation. The graphite furnace must be operated such that the graphite element is continuously purged with an inert gas, such as argon, so that the element is not oxidized. The zirconia furnace may be operated with no protective atmosphere. However, rapid heating or cooling of the zirconia furnace is to be avoided to prevent cracking of the zirconia susceptor due to thermal shock. Such an event can lead to the evolution of zirconia particles, which may become embedded in the molten preform and thereby produce flaws in the drawn fiber.

C. Fiber Diameter Measurement and Control

Two types of fiber diameter measurement systems are in general use: the laser scanning technique, and the forward light-scattering method. The laser scanning technique involves the scanning of a laser beam over the area through which the fiber passes and detection of the intensity changes due to beam interception. The time between intensity spikes corresponding to the fiber edges is proportional to the fiber diameter.

The light-scattering technique involves directing a laser beam on the fiber and detecting the resultant forward scattering pattern on a photodiode array. The diameter can be determined from the modulation of the scattering pattern. The resolution is 0.1 μm, and a typical measurement rate is 500 Hz.

The measurement system provides a fiber diameter value that can be used in a feedback control system. The value is compared with a desired value, and a control signal is then generated that alters the speed of the drawing capstan so as to drive the fiber diameter toward the desired value. Because of the dynamic response of the process and the time delay that results from the distance between the molten preform neck and the position of fiber measurement, vibrations or other disturbances that cause high-frequency diameter variations must be avoided. Using a suitable feedback control system, the mean diameter of the controlled fiber can be held to within 0.1 μm of the desired value. The standard deviation of the fiber diameter, which chiefly represents the rapid, uncontrollable disturbances in the process, can usually be reduced to 0.25 μm.

D. Fiber Coatings

Two types of fiber coating geometries, depicted in Fig. 5, are in common use. The single-layer high-modulus (or medium-modulus) coating is the simplest structure to produce. It is used in high-strength applications or in packaging designs where microbending loss sensitivity is not a prominent factor. Dual coatings, consisting of a low-modulus primary layer surrounded by a high-modulus secondary layer, are used in cable applications where microbending losses due to cabling or environmental conditions predominate.

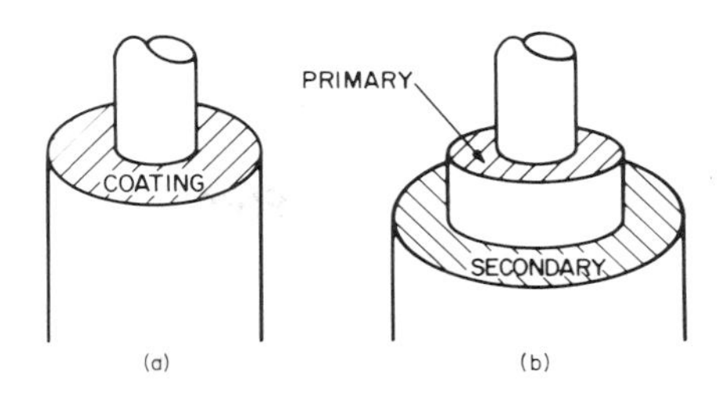

FIG. 5. Schematic drawing of fibers with a (a) single or (b) dual coating.

Microbending losses can be very high, often many times the intrinsic loss of the fiber. They may arise from two sources, as shown in Fig. 6. First, laterally applied stresses encountered in a cable can cause local deformation of the fiber. Second, differential shrinkage of the components of a fiber cable will impose axial compressive loads on the fiber. Such loads may result from the differential thermal contraction of the cable materials or from the recovery of residual strains in the viscoelastic polymers used in the cable construction. Once these axial compressive loads reach a critical value, buckling of the fiber will occur. The dual coating provides microbending resistance to the fiber in both the axial and lateral loading situations. The low-modulus primary layer cushions the fiber against lateral loads and tends to promote long bending periods, outside the microbending range, in the axial loading case.

From a materials standpoint, the primary coating in a dual-coating structure should have a low modulus and a low glass transition (the temperature interval, which may span several degrees, over which a material changes from a rub-

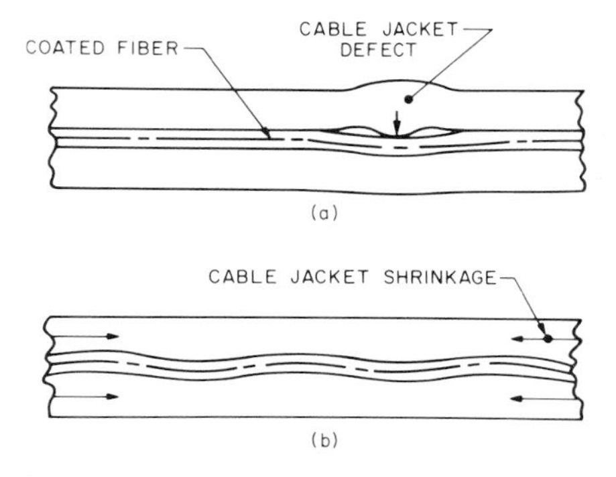

FIG. 6. Schematic depiction of microbending arising from (a) lateral forces caused by a cable defect and (b) longitudinal forces on a fiber in a cable structure.

ber to a glassy solid). The low glass transition ensures that the coating will provide protection from microbending at the lowest temperatures encountered in the field, frequently −40°C or below. The secondary coating should serve as a hard elastic shell; consequently, its glass transition should be well above room temperature. The primary coating should be free of particulate matter of sizes ranging downward to the micrometer level. Such particles could lodge at the coating–glass interface and produce flaws during fiber handling. These particles are removed by fine filtration techniques.

Three types of primary coating materials have been commercially used: thermally curable silicones, ultraviolet-curable (UV-curable) polymers, and thermoplastic rubber compounds. Their modulus–temperature characteristics are displayed in Fig. 7. Thermally curable silicones have been employed because of their exceptionally low glass transition temperatures, ranging from −50°C to −120°C, depending on chemical structure. They can be cured (cross-linked) rapidly at high temperatures. However, the curing chemistry commonly employed with these systems (platinum-catalyzed silane addition to olefinic bonds) provides a mechanism for hydrogen generation from the coating. Because hydrogen can diffuse into silica glass, where it causes increases in optical transmission loss, the use of thermally curable silicones as coatings has fallen into disfavor.

Ultraviolet-radiation-curable primary coatings are now used more extensively than any other type. Their chief advantage is rapid cure, a factor of prime importance for commercial fiber production at reasonable draw rates. Urethane acrylates are most common; however, their glass transition temperatures frequently lie above −40°C. UV-curable formulations based on silicones or polybutadiene have been developed more recently in order to exploit their potential for lower glass transition temperatures.

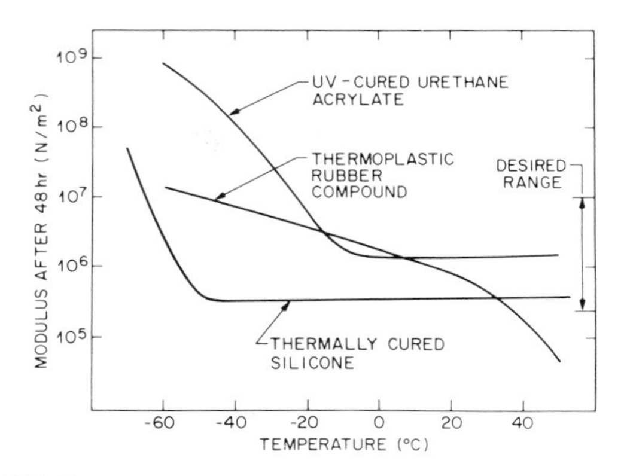

FIG. 7. Modulus–temperature relationships for three representative primary coating materials. The modulus values are taken after allowing stress relaxation for 48 h.

Thermoplastic rubber compounds, applied to the fiber as hot melts (molten, low-viscosity liquids) that solidify rapidly upon cooling, have also been used commercially as primary coatings. Materials with low glass transitions can be formulated, but particle removal is a major problem, and particle formation due to decomposition processes that occur upon prolonged heating of these materials presents a difficult quality-control situation. Consequently, the evolution of improved UV-curable formulations has largely rendered thermoplastic systems obsolete.

Secondary coatings for dual-coating structures are usually extruded thermoplastics, such as nylon, or hard, UV-curable acrylates. Epoxy and urethane acrylates are common. Such UV-curable formulations are also used extensively as single coatings.

Hermetic coatings deserve mention as a special class of materials designed to improve the resistance of glass fibers to moist environments. Silica glass is subject to stress-accelerated crack propagation in the presence of water, termed static fatigue, and hermetic coatings can prevent this failure mechanism. Metal coatings, such as aluminum and indium, as well as ceramic coatings, such as silicon oxynitride and amorphous carbon, have been employed with varying degrees of success. Commonly, the preservation of important performance properties of fibers—for example, strength and transmission loss—when hermetic coatings are employed is a serious problem.

E. Coating Application

Coating application represents an extremely crucial part of the fiber drawing process. It poses the major limitation to achieving high draw rates, and many types of coating defects introduced at this stage adversely affect fiber yield.

Conceptually, the coating process entails passing the fiber through a reservoir containing the liquid coating formulation, as shown in Fig. 8. The liquid wets the moving fiber, which in turn drags the liquid downward into a die that

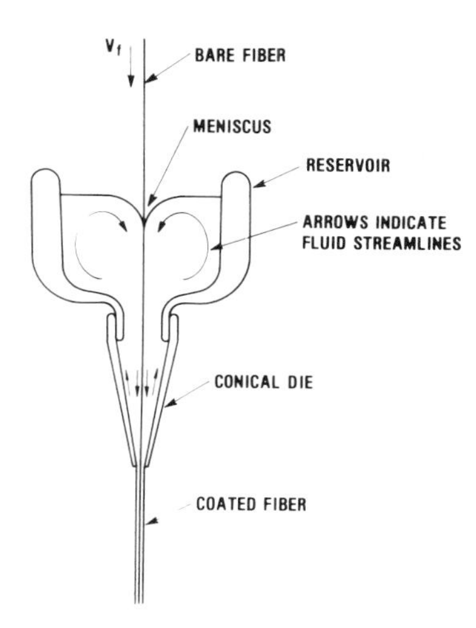

FIG. 8. Schematic diagram of an applicator for coating optical fibers. [From Aloisio, C. J., and Blyler, L. L., Jr. (1982). *Annu. Tech. Conf.—Soc. Plast. Eng., 40th, 1982,* pp. 117–120.]

imparts the appropriate coating thickness. A meniscus is formed at the point of entry of the fiber into the liquid, directed into the reservoir.

A variety of coating defects can arise from the improper implementation of the application process described above. One such defect is coating eccentricity, illustrated in Fig. 9a. Because the fiber cannot be contacted by any solid surface until the coating has solidified, it cannot be guided through the center of the die. However, coating concentricity may be monitored by directing a visible laser beam perpendicular to the fiber just below the coating die. Refraction of the light rays passing through the coating–glass composite produces a forward scattering pattern, the symmetry of which relates to the symmetry of the coating and glass in that plane. By splitting the laser beam into two orthogonal beams that intersect the fiber at 90° to one another, the coating-fiber concentricity may be observed. The coating applicator, mounted on a precision XY stage, can then be moved with micropositioners so that the coating die is concentrically placed around the fiber. The method requires that substantial differences exist between the refractive indices of the glass and coating materials, with 0.04 being a practical minimum. The refractive index of the coating may be either greater than or less than that of the glass.

Air bubbles in the coating, as shown in Fig. 9b, represent another defect that can affect both the strength and loss properties of the fiber. Figure 10 depicts the way in which air bubbles are

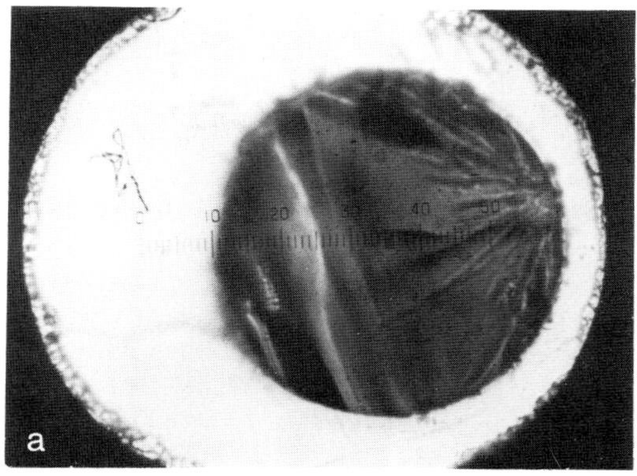

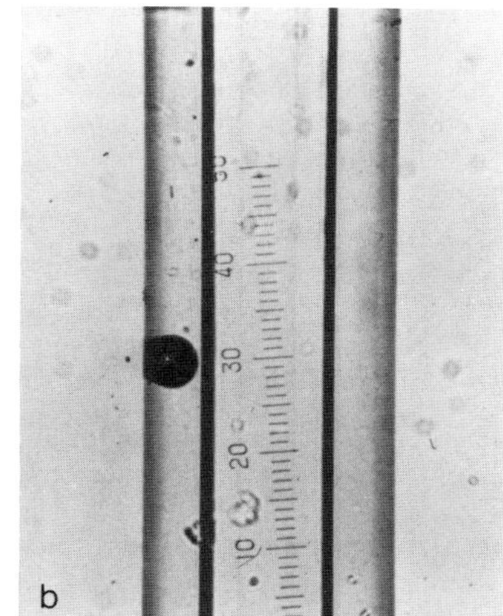

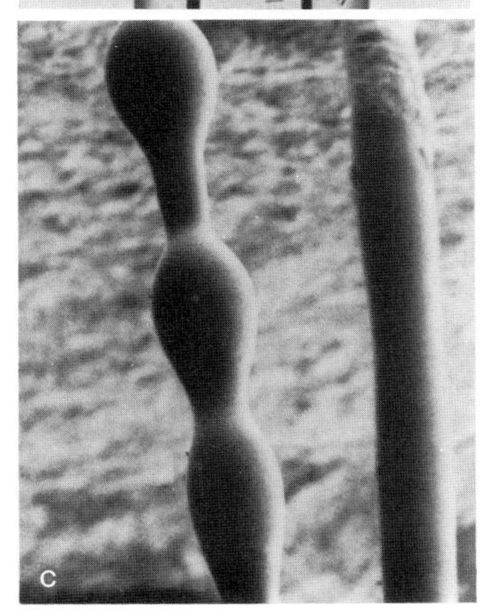

FIG. 9. Photomicrographs of coating defects: (a) eccentric coating, (b) trapped air bubble within the coating, and (c) intermittent coating.

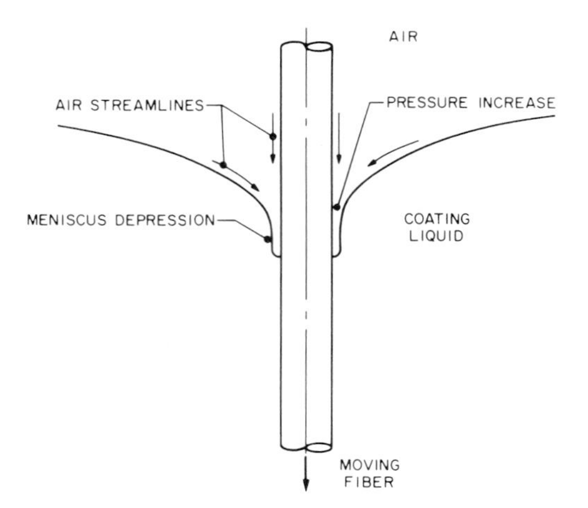

FIG. 10. Schematic diagram of the air streamlines at the fiber and liquid surfaces that produce a depression in the liquid meniscus in a coating applicator.

included in the coating. The fiber entering the coating drags air toward the meniscus. The moving liquid at the top of the reservoir does the same. As a result, there is a pressure increase at the point of entry of the fiber into the liquid. As the fiber speed is increased, the pressure rises and a depression forms in the meniscus. The depression is unstable, and it alternately forms and collapses, trapping air bubbles, which are dragged further into the liquid by the fiber. Some bubbles penetrate the die and are entrained in the fiber coating.

The bubble entrainment problem can be controlled through the use of a pressurized coating applicator. An example is displayed in Fig. 11. In this device, the coating liquid is fed under pressure into a lower chamber. The pressure in this chamber produces a backflow through an internal die to the reservoir section of the applicator, which strips the bubbles from the fiber surface.

The most serious meniscus instability, which represents the major limitation to achieving high draw rates, occurs as the fiber velocity is increased well beyond the point of bubble entrainment. In this instance, the depression of the meniscus propagates as an air column surrounding the fiber (Fig. 12). The air column may extend completely through the applicator and an intermittent coating may result, as shown in Fig. 9c. The condition may be recognized by the disappearance of a circulating flow in the reservoir, since the fiber no longer drags liquid toward the die. The phenomenon is exceptionally sensitive

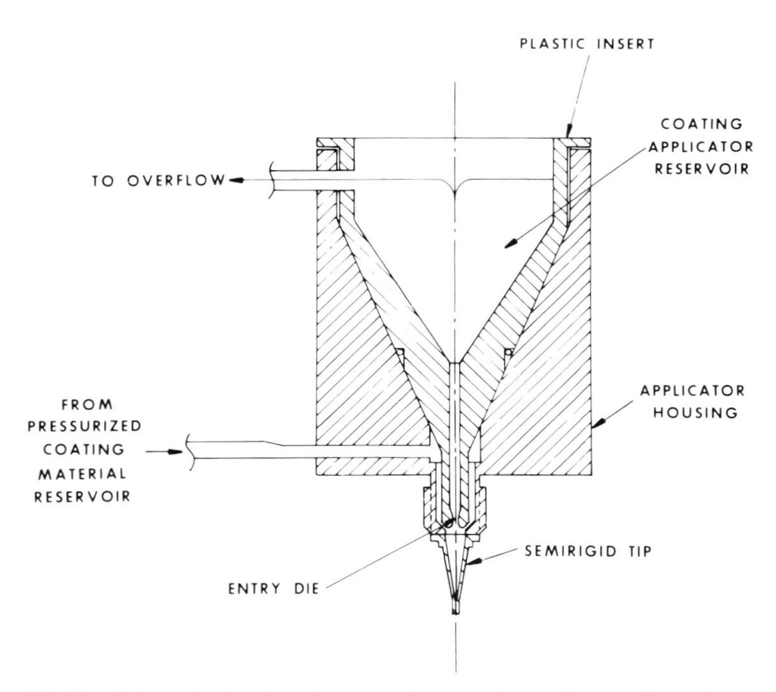

FIG. 11. Schematic drawing of a pressurized coating applicator. [From Lenahan, T. A., and Taylor, C. R. (1982). *Tech. Dig. Top. Meet. Opt. Commun., 41st*.]

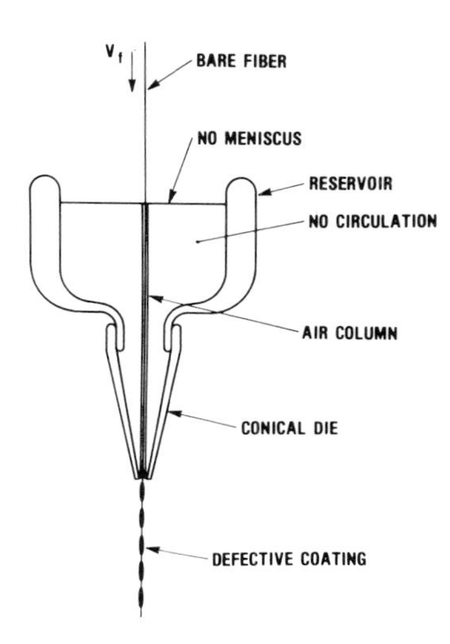

FIG. 12. Schematic drawing of a coating applicator showing the air column surrounding a fiber, which results from a meniscus instability.

to the temperature of the fiber and is frequently encountered when the cooling distance from the heat source to the coating applicator is inadequate. It is for this reason that most draw towers are now constructed so as to allow long distances (3 m or more) for convective cooling of the fiber prior to coating. Thus, tall draw towers, 7 m or more in height, have become commonplace in efforts to achieve high fiber production rates.

F. High-Speed Drawing and Coating

Overwhelmingly, the most serious limitations to the achievement of high fiber drawing speeds (5 m/sec or greater) are imposed by the coating application process. High speeds lead to more frequently encountered coating defects as described above, as well as higher viscous shear forces in the coating die with accompanying higher fiber draw tensions. These shear forces depend strongly on coating liquid viscosity. If the coating viscosity is too high, fiber tensions may become high enough to cause fiber failures during drawing. Typical fiber tension–draw speed plots obtained for fiber drawing with and without coating application are shown in Fig. 13. When no coating is applied, a linear relationship exists between fiber tension and draw speed, and tension levels may be adjusted by increasing or decreasing furnace temperature. When a coating is applied, considerably higher draw tensions may result below the coating applicator, and a linear relationship no longer results. Cur-

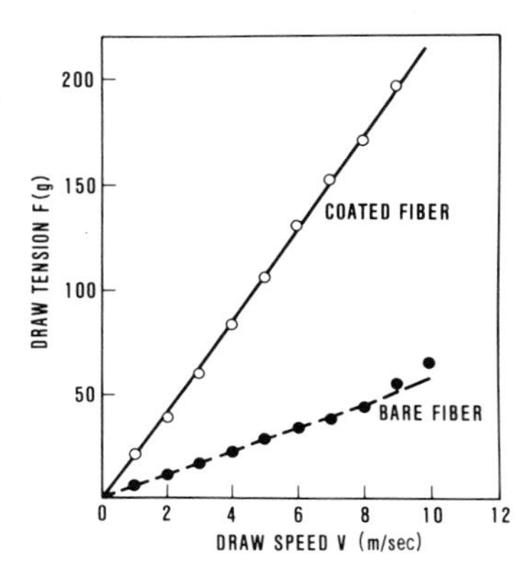

FIG. 13. Relationships between fiber tension and draw speed, with and without coating application. [From Paek, U. C., and Schroeder, C. M. (1984). *Electron. Lett.* **20**(7), 304–305.]

vature in the fiber tension–draw speed plot is associated primarily with viscosity reduction in the coating liquid caused by heat transfer from the fiber, which enters the coating at higher temperatures as draw speed is increased. Figure 14 shows experimental and calculated curves of fiber temperature versus draw speed for various distances between the furnace and the coating applicator, assuming convective cooling as the controlling mechanism. Note that fiber temperatures increase exponentially at higher draw speeds (depending on cooling distance). It is this exponential dependence that limits the draw rate severely. A high fiber temperature causes the liquid viscosity to drop sharply, which in turn causes the coating thickness to decrease. Coat-

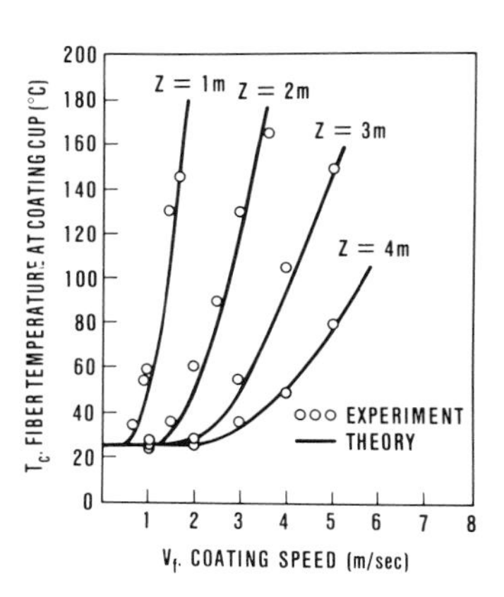

FIG. 14. Relationships between fiber temperature and draw speed for different cooling distances from the furnace. The solid curves are calculated and the circles are measured values. [From Paek, U. C., and Schroeder, C. M. (1981). *Appl. Opt.* **20**(23), 4028–4034.]

ing geometry becomes difficult to maintain within acceptable limits. Even more serious are the coating flow instabilities described earlier that result from the hot fiber entering the cooler coating liquid. These difficulties must be overcome by ensuring that the fiber temperature at the point of coating application remains at an appropriate level (generally not more than 50°C above ambient) throughout the drawing process. This condition may be assured by convective cooling of the fiber through the use of tall draw towers with long cooling distances. In addition, the process may be augmented by flowing a cool gas around the fiber for increased heat transfer. Tall draw towers also allow for the placement of additional UV-curing lamps so that adequate radiation doses for cross-linking the coating are maintained at the higher draw speeds.

The performance properties of fibers drawn at speeds as high as 10 m/sec are indistinguishable from those of fibers drawn at lower speeds. Thus far, no fundamental limitations to the commercial achievement of even higher draw speeds have been encountered.

III. Fiber Performance

Understanding the fundamentals of optical fiber drawing and coating and the relationships between processing techniques and fiber properties has allowed the development of high-performance optical communication systems. As these communication systems become more sophisticated, the drawing and coating process must be improved to accommodate more stringent fiber requirements. Currently, military and undersea transmission systems require the most advanced optical fiber performance. The U.S. military is supporting the development of optical fibers with high tensile strengths to withstand the abuse of a variety of cable deployment techniques and harsh in-service environments. Low-loss fibers with strengths of 1.4–2.1 GN/m^2 (200,000–300,000 lb/in.^{2}) over 10-km lengths have been produced for missile guidance systems, for ruggedized and air-layable cables used in U.S. Army and Air Force tactical communication systems, and for a variety of undersea communication systems for the U.S. Navy.

Commercial undersea cable systems also require optical fibers with high performance properties. Since fibers for these systems must survive the tensile strains encountered during recovery operations from deep ocean waters, strengths of 1.4–2.1 GN/m^2 are necessary. Because extremely long fiber spans are needed between repeaters (30–70 km), consistently flaw-free fibers are required to achieve high yields. Furthermore, drawing- and coating-related transmission losses must be near zero in order to take advantage of the inherent low losses of fibers designed for undersea systems, and highly precise dimensional tolerances are needed to ensure low splice losses.

Undersea lightwave communication systems have been demonstrated in the United States, Europe, and Japan. In 1982, AT&T conducted a sea trial of a system using a 19-km-long cable that contained 12 low-loss optical fibers with proof-tested tensile strengths of 1.4 GN/m^2 (200,000 lb/in.^{2}), joined together using a fusion splicing technique. The trial demonstrated error-free transmission at a rate of 274 Mbits/sec over fibers looped together in lengths greater than 50 km. The mechanical integrity of the fibers and cable was maintained during cable laying and retrieval. The technology developed during the trial has been implemented in a manufacturing setting. Fibers are being produced for use in a trans-Atlantic cable system known as TAT-8, to be installed in 1988. Concurrently, experiments designed to increase the capacity of lightwave systems and to extend unrepeatered transmission distances have established records for high-speed transmission. An unamplified signal has been sent at rates of 420 Mbits/sec over 203 km and 2 Gbits/sec over 130 km.

Results such as those discussed above have demonstrated the feasibility of producing commercial quantities of optical fibers having high strengths, superior optical transmission properties, and precise dimensional tolerances. These high-performance properties have been realized, in part, by the research and development applied to the fiber drawing and coating process described in this article.

Bibliography

Bendow, B., and Shashanka, S. M. (eds.) (1979). "Fiber Optics: Advances in Research and Development." Plenum, New York.

Blyler, L. L., Jr., and Aloisio, C. J. (1985). Polymer coatings for optical fibers. *In* "Applied Polymer Science" (R. W. Tess and G. W. Poehlein, eds.), 2nd ed. American Chemical Society, Symposium Series no. 285, Washington, D.C.

Blyler, L. L., Jr., and DiMarcello, F. V. (1980). Fiber drawing, coating and jacketing. *Proc. IEEE* **68**(10), 1194–1198.

Cherin, A. H. (1983). "Introduction to Optical Fibers." McGraw-Hill, New York.

Li, T. (ed.) (1985). "Advances in Optical Fiber Communications." Academic Press, New York.

Midwinter, J. E. (1979). "Optical Fibers for Transmission." Wiley, New York.

Miller, S. E., and Chynoweth, A. G. (ed.) (1979). "Optical Fiber Telecommunications." Academic Press, New York.

Paek, U. C., and Schroeder, C. M. (1981). High speed coating of optical fibers with UV-curable materials at a rate of greater than 5 meters/second. *Appl. Optics* **20**(23), 4028–4034.

Suematsu, Y., and Iga, K.-I. (1982). "Introduction to Optical Fiber Communication." Wiley, New York.

Watkins, L. S. (1982). Control of fiber manufacturing processes. *Proc. IEEE* **70**(6), 626–634.

OPTICAL INFORMATION PROCESSING

Mir Mojtaba Mirsalehi *The University of Alabama in Huntsville*

GLOSSARY

Bistability: Property of having two stable output states for a given input. In optical bistable devices, the input and the output are beams of light, and the two states are normally low and high transmittance of a media.

Diffraction: Deviation of light rays from rectilinear paths that cannot be interpreted as reflection or refraction.

F number: Ratio between the diameter of a lens and its focal length.

Holography: Technique for recording and reconstructing information about both amplitude and phase of light scattered from an object.

Index of refraction: Ratio of the speed of light in a vacuum to the speed of light in the material.

Monochromatic light: Radiation at a single frequency.

Photorefractive material: Material whose index of refraction can be altered by exposing it to light.

Pupil function: Function that describes the finite extent of a lens or an object.

Space–bandwidth product: Product of the diameter of the input aperture of an optical system and the highest spatial frequency over which that system can operate.

Spatial filtering: Manipulation of the spectrum of an image by a filter plate.

Spatial light modulator: Device that manipulates the amplitude or/and the phase of a beam of light at different points of space.

Optical information processing involves processing a two-dimensional array of data using light. The main advantage of optical processors is their ability to process data in parallel (i.e., an array of data points is processed simultaneously). However, most of the optical systems are analog in nature and thus suffer from lack of accuracy and flexibility. Recently, hybrid (optical analog/digital electronic) systems have been developed to overcome this shortcoming. Significant research is currently underway to develop optical digital computing systems.

I. History of Optical Information Processing

One of the earliest contributions in the field of optical information processing is Abbe's theory of microscope imagery which was introduced in 1873. He showed that the image of an object might lose its detail or obtain false detail if not all parts of the diffracted light are transmitted through the optical system. Porter demonstrated this point by a remarkable experiment in 1906 (Fig. 1), in which lens L_1 collimates the light, and L_2 provides an image of the object at P_i. Porter used a piece of wire gauze as the object. The Fourier transform of the object is formed at the back focal plane (P_f). If all parts of the spectrum are allowed to pass, a faithful image of the gauze is obtained. Porter showed that if a horizontal slit that allows only one row of the spectrum to pass is put at P_f, the vertical structure of the image is lost. Similarly, the horizontal structure is lost if a vertical slit is used (Fig. 2). This

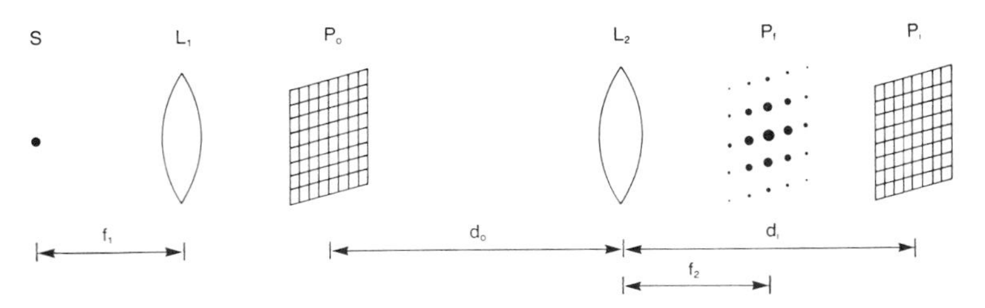

FIG. 1. Porter's experiment, where d_i is the image distance; d_o, object distance; f_1, focal length of L_1; f_2, focal length of L_2; L_1, collimating lens; L_2, imaging lens; P_f, focal plane; P_i, image plane; P_o, object plane; and S, light source.

experiment is the first reported demonstration of spatial filtering.

In 1935, a significant advancement in the field of microscopy was realized with Zernike's phase-contrast microscope. This Nobel prize-winning invention made the observation of transparent objects possible. In 1946, the use of the Fourier integral in analyzing optical systems was introduced by Duffieux. In the early 1950s, Maréchal and his co-workers at Paris University successfully demonstrated a variety of coherent spatial filtering techniques to improve the quality of photographs. Also in that decade, a fruitful exchange of knowledge was begun between scientists working in communications and those in optics. The resemblance of many problems in optics to concepts in communications such as optimum filtering, detection, and estimation was shown by Elias and O'Neill. In the 1960s, coherent optical processing was successfully applied in the field of radar signal processing. Much of this progress was made by Cutrona and his co-workers at the University of Michigan Radar Laboratory. Two other important events in the field of optical information processing that occurred during the 1960s are the inventions of holographic spatial filtering by Vander Lugt in 1963 and computer-generated holograms for spatial filtering by Lohmann and Brown in 1966. In the 1970s, the field of optical processing was broadened by progress in coherent and incoherent processing and the development of hybrid systems. Also, stimulated by advances in technology and the rediscovery of residue arithmetic, the idea of building an optical digital computer was reborn in the mid-1970s. Significant research and development in all of the areas mentioned above is currently being pursued. The area of optical digital computing seems especially promising with new architectures for the future computers. [*See* OPTICAL CIRCUITRY.]

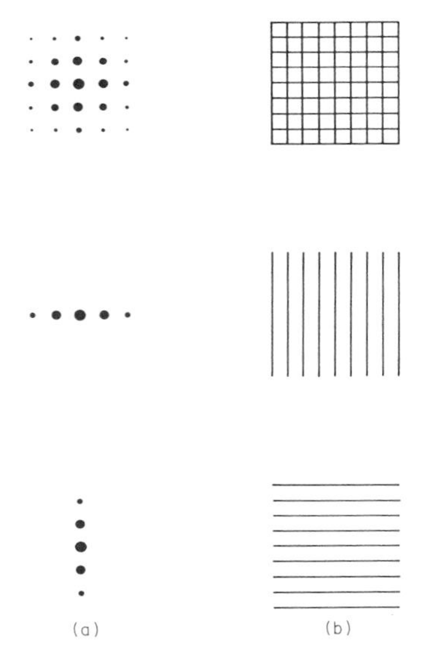

FIG. 2. Results from Porter's experiment: (a) the spectra at the focal plane and (b) the corresponding images at the image plane.

II. Background

The analysis of optical systems requires some background knowledge. The fundamental concepts needed are reviewed in this section. This material is divided into three subsections: mathematics, system theory, and optics. Because of space constraints, the concepts are described briefly. References are cited in the bibliography for further information.

A. Relevant Mathematics

1. Special Functions

One of the functions used in analyzing optical systems is the Dirac delta (impulse) function

$\delta(x)$. This function has the following properties:

$$\delta(x) = 0, \qquad x \neq 0 \tag{1a}$$

$$\int_{x_1}^{x_2} \delta(x)\, dx = 1, \qquad x_1 < 0 < x_2 \tag{1b}$$

A point source of light can be modeled as a two-dimensional delta function, since it is radiation of a finite energy through a small aperture. A periodic array of impulse functions is known as a comb function.

Among other functions used in optics are the rectangle, triangle, Gaussian, and sinc functions. These are, respectively, defined as

$$\text{rect}(x) = \begin{cases} 1, & |x| \leq \frac{1}{2} \\ 0, & \text{otherwise} \end{cases} \tag{2}$$

$$\Delta(x) = \begin{cases} 1 - |x|, & |x| \leq 1 \\ 0, & \text{otherwise} \end{cases} \tag{3}$$

$$\text{Gaus}(x) = \exp(-\pi x^2) \tag{4}$$

$$\text{sinc}(x) = \sin(\pi x)/\pi x \tag{5}$$

These functions can be extended to two dimensions. Another useful two-dimensional function is the circle function, which is defined as

$$\text{circ}(r) = \begin{cases} 1, & r = (x^2 + y^2)^{1/2} \leq 1 \\ 0, & \text{otherwise} \end{cases} \tag{6}$$

The above functions are shown in Fig. 3.

2. Convolution and Correlation Operations

The convolution of two real or complex functions $\mathbf{f}(x)$ and $\mathbf{g}(x)$ may be represented by $\mathbf{f}(x) * \mathbf{g}(x)$ and is defined as

$$\mathbf{f}(x) * \mathbf{g}(x) = \int_{-\infty}^{+\infty} \mathbf{f}(\alpha)\, \mathbf{g}(x - \alpha)\, d\alpha \tag{7}$$

Convolution is a commutative operation [i.e., $\mathbf{f}(x) * \mathbf{g}(x) = \mathbf{g}(x) * \mathbf{f}(x)$]. The result of convolving two functions can be considered as the area under the curve that is obtained by multiplying one function with a reversed (in α) and shifted (by x) version of the other function. Convolution is a useful operation in analyzing linear shift-invariant (LSI) systems as is shown later. Most optical systems can be considered as two-dimensional LSI systems. The two-dimensional version of Eq. (7) is

$$\mathbf{f}(x, y) * \mathbf{g}(x, y) = \iint_{-\infty}^{+\infty} \mathbf{f}(\alpha, \beta) \times \mathbf{g}(x - \alpha, y - \beta)\, d\alpha\, d\beta \tag{8}$$

The complex cross-correlation operation, denoted by ★, is defined as

$$\mathbf{f}(x) \bigstar \mathbf{g}(x) = \int_{-\infty}^{+\infty} \mathbf{f}(\alpha)\, \mathbf{g}^*(\alpha - x)\, d\alpha \tag{9}$$

where $\mathbf{g}^*$ represents the complex conjugate of $\mathbf{g}$. Although the cross-correlation and convolution operations look similar [in fact, $\mathbf{f}(x) \bigstar \mathbf{g}(x) = \mathbf{f}(x) * \mathbf{g}^*(-x)$], they differ in nature. Unlike convolution, cross-correlation is not commutative [i.e., $\mathbf{f}(x) \bigstar \mathbf{g}(x) \neq \mathbf{g}(x) \bigstar \mathbf{f}(x)$]. A special case of cross-correlation is called autocorrelation, when the two functions are the same [i.e., $\mathbf{f}(x) = \mathbf{g}(x)$]. Autocorrelation and cross-correlation are useful operations in pattern recognition.

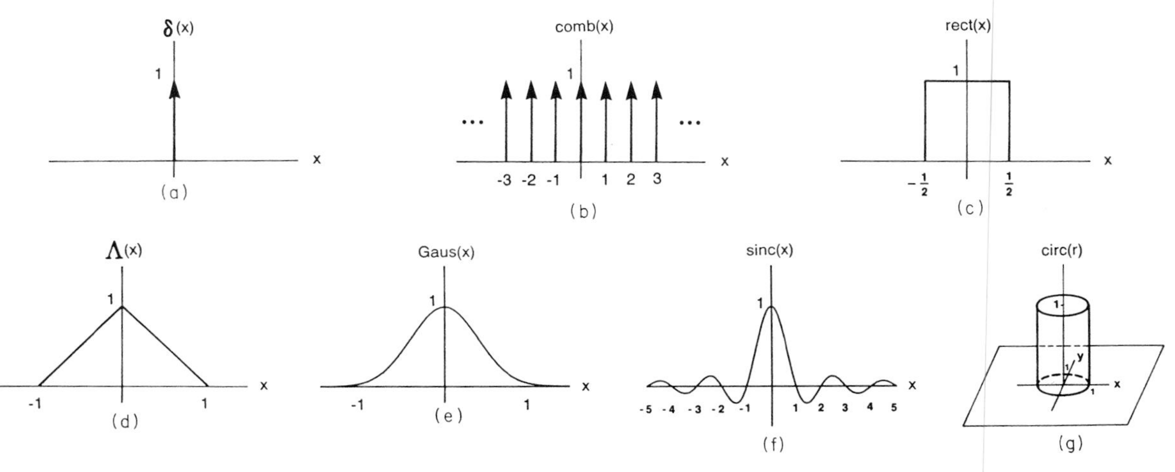

FIG. 3. Some functions frequently used in optics: (a) impulse function, (b) comb function, (c) rectangle function, (d) triangle function, (e) Gaussian function, (f) sinc function, and (g) circle function.

3. Integral Transforms

An integral transform is a linear operation that converts a function $\mathbf{f}(x)$ to another function $\mathbf{F}(u)$ via the following integral:

$$\mathbf{F}(u) = \int_a^b \mathbf{f}(x)\, \mathbf{K}(x, u)\, dx \qquad (10)$$

The function $\mathbf{K}(x, u)$, known as the kernel of the transform, and the limits of the integral are specified for a particular transform. Integral transforms are used to map one domain into another in which the problem is simpler to analyze. For example, the analysis of linear time-invariant systems usually becomes easier if the time domain representation is changed to the frequency domain representation using the Fourier transformation.

A number of integral transforms, each suitable for different problem areas, exist. The most common transform in optics is Fourier transform. The Fourier transform of a one-dimensional function $\mathbf{f}(x)$ is defined as

$$\mathbf{F}(u) = \int_{-\infty}^{+\infty} \mathbf{f}(x) \exp(-j2\pi ux)\, dx \qquad (11)$$

where $j = (-1)^{1/2}$. The corresponding inverse transform is

$$\mathbf{f}(x) = \int_{-\infty}^{+\infty} \mathbf{F}(u) \exp(j2\pi ux)\, du \qquad (12)$$

One of the properties of the Fourier transform is known as the convolution theorem. According to this property, the Fourier transform of the convolution of two functions is the product of their Fourier transforms, that is,

$$\mathcal{F}\{\mathbf{f}(x) * \mathbf{g}(x)\} = \mathcal{F}\{\mathbf{f}(x)\}\, \mathcal{F}\{\mathbf{g}(x)\} \qquad (13)$$

where $\mathcal{F}$ represents the Fourier transform operation. This is a useful tool in analyzing LSI systems.

Most optical systems are two-dimensional, hence, they require two-dimensional transformations. The two-dimensional versions of Eqs. (11) and (12) can be written as

$$\mathbf{F}(u, v) = \iint_{-\infty}^{+\infty} \mathbf{f}(x, y) \exp[-j2\pi(ux + vy)]\, dx\, dy \qquad (14)$$

and

$$\mathbf{f}(x, y) = \iint_{-\infty}^{+\infty} \mathbf{F}(u, v) \exp[j2\pi(ux + vy)]\, du\, dv \qquad (15)$$

It is common to replace u and v in these equations with f_X and f_Y which denote the spatial frequencies along the x and y axes.

Another integral transform of interest in optics is the Mellin transform, which is defined as

$$\mathbf{F}(\mathbf{s}) = \int_0^{\infty} \mathbf{f}(x)\, x^{\mathbf{s}-1}\, dx \qquad (16)$$

where the parameter $\mathbf{s}$ is a complex variable. If for some $k > 0$, the inequality $\int_0^\infty x^{k-1}|\mathbf{f}(x)|\, dx < \infty$ is satisfied, the inverse Mellin transform can be obtained for any $a > k$ using the following equation:

$$\mathbf{f}(x) = \frac{1}{2\pi j} \int_{a-j\infty}^{a+j\infty} \mathbf{F}(\mathbf{s}) x^{-\mathbf{s}}\, d\mathbf{s} \qquad (17)$$

An interesting property of the Mellin transform is that it is scale-invariant. In other words, if the input function is stretched, the shape of the transform remains constant. This property can be stated as $\mathcal{M}\{\mathbf{f}(bx)\} = b^{-\mathbf{s}}\mathcal{M}\{\mathbf{f}(x)\}$, where $\mathcal{M}$ represents the Mellin transform and b is a constant.

4. Boolean Algebra

Boolean algebra was introduced by George Boole in 1847 as a tool for the mathematical analysis of logic. It was not used for practical purposes, however, until the late 1930s, when A. Nakashima and, independently, C. E. Shannon used it for analyzing relay contact networks. After World War II, the switching theory was extended to include sequential systems (i.e., systems whose outputs depend not only on the inputs but also on the previous states of the system). Based on these advances in theory, digital electronic computers and other digital systems were developed. Today, digital electronic systems are widely used, and the corresponding theories are studied in mathematics, computer science, and electrical engineering. The purpose of this subsection is to define only the basic terms used in the analysis of digital systems, so that a reader without a background in this field can understand the material presented in the optical digital computing section.

A digital system is defined as a system whose inputs and outputs are discrete quantities. These quantities are normally represented by binary variables. A binary variable has two possible values: 0 and 1 (also known as false and true). A special case of Boolean algebra that deals with binary variables is known as switching algebra. A function that accepts binary inputs and pro-

TABLE I. A Truth Table

X	Y	Z	F
0	0	0	0
0	0	1	1
0	1	0	1
0	1	1	0
1	0	0	0
1	0	1	0
1	1	0	1
1	1	1	0

vides binary outputs is called a binary function. A binary function can be defined by a corresponding truth table that provides the output for all possible combinations of values of the input variables. Table I shows a truth table that corresponds to a function with three input variables (X, Y, and Z) and one output variable (F).

Another method of defining a binary function is to express each output variable in terms of the input variables using some of the basic logic operations. The basic logic operations include NOT, AND, OR, NAND, NOR, and EXCLUSIVE OR. These are defined by their truth tables in Table II. With this method, the function F of Table I can be expressed as $F = \bar{X}\bar{Y}Z + \bar{X}Y\bar{Z} + XY\bar{Z}$. This form of expression is known as the sum-of-products form, and each term is called a minterm. The number of terms can be minimized by applying logical reduction techniques based on the theorems of Boolean algebra. For example, the above expression can be simplified to $F = \bar{X}\bar{Y}Z + Y\bar{Z}$.

TABLE II. Basic Logic Operations

NOT

X	$\bar{X}$
0	1
1	0

AND

X	Y	XY
0	0	0
0	1	0
1	0	0
1	1	1

OR

X	Y	X + Y
0	0	0
0	1	1
1	0	1
1	1	1

NAND

X	Y	$\overline{XY}$
0	0	1
0	1	1
1	0	1
1	1	0

NOR

X	Y	$\overline{X + Y}$
0	0	1
0	1	0
1	0	0
1	1	0

EXCLUSIVE OR

X	Y	$X \oplus Y$
0	0	0
0	1	1
1	0	1
1	1	0

B. Relevant System Theory

1. Linear Shift-Invariant (LSI) Systems

A system is defined as a mapping of an input function $\mathbf{f}(x)$ into an output function $\mathbf{g}(x)$. This is represented as $\mathbf{g}(x) = S\{\mathbf{f}(x)\}$, where $S\{\mathbf{f}(x)\}$ indicates a general function operating on $\mathbf{f}(x)$. Both the input and the output are functions of x, which could be time, distance, or another parameter. A system is linear if it satisfies the superposition property, defined as

$$S\{\mathbf{k}_1\mathbf{f}_1(x) + \mathbf{k}_2\mathbf{f}_2(x)\} = \mathbf{k}_1 S\{\mathbf{f}_1(x)\} + \mathbf{k}_2 S\{\mathbf{f}_2(x)\} \tag{18}$$

where $\mathbf{f}_1$ and $\mathbf{f}_2$ are two arbitrary input functions, and $\mathbf{k}_1$ and $\mathbf{k}_2$ are two arbitrary complex constants.

A system is called shift-invariant if a shift in the input results only in an equal shift in the output, that is,

$$S\{\mathbf{f}(x - x_0)\} = \mathbf{g}(x - x_0) \tag{19}$$

If the input and output are functions of time, this property is known as time-invariance. In optical systems, the input and output are usually two-dimensional spatial functions, so this property is known as space-invariance.

2. Properties of LSI Systems

Linear shift-invariant systems have interesting properties that make their analysis much simpler than shift-variant systems. One of the important features of an LSI system is that if the response of the system to an impulse function is known, the output of the system can be predicted for any input. This is achieved through the convolution process. The output is given by

$$\mathbf{g}(x) = \int_{-\infty}^{+\infty} \mathbf{f}(\alpha)\, \mathbf{h}(x - \alpha)\, d\alpha \tag{20}$$

where $\mathbf{h}$ represents the impulse response of the system. The analysis of LSI systems usually becomes easier if the Fourier transforms of the functions are used. The above convolution is then replaced by the following simple relation,

$$\mathbf{G}(u) = \mathbf{F}(u)\, \mathbf{H}(u) \tag{21}$$

where $\mathbf{F}(u)$, $\mathbf{G}(u)$, and $\mathbf{H}(u)$ are the Fourier transforms of the input function, output function, and impulse response, respectively. $\mathbf{H}(u)$ is known as the transfer function. The characteristics of an LSI system are completely specified by its transfer function.

C. Relevant Optics

There are five commonly used representations of optics: geometrical optics, wave optics, Fourier optics, quantum optics, and relativistic optics. Although these treatments are on the same general subject, each is suitable for analyzing a particular group of phenomena. To the lowest order of approximation, images obtained by lenses or mirrors are analyzed by the laws of geometrical optics. Aperture diffraction, grating diffraction, and scattering by particles are among the topics that require consideration of the wave nature of light. Fourier optics is a useful tool for studying optical systems. Some phenomena, such as the photoelectric effect, can be explained only by quantum optics, and finally, relativistic optics is based on Einstein's postulates and can explain modern experimental measurements related to the speed of light. Geometrical optics, wave optics, and Fourier optics are all useful in analyzing optical information processing systems. Some aspects of the wave optics and Fourier optics are explained in this section. [*See* DIFFRACTION, OPTICAL.]

1. Basic Concepts

In wave optics, light is treated as an electromagnetic wave. If the light source radiates at a fixed frequency, the light is called monochromatic. If the light is not monochromatic but has a relatively narrow spectrum, it is known as quasi-monochromatic. Light is called polarized if the direction of the electric field is predictable (e.g., the output of a laser) and unpolarized if the orientation of the electric field is random (e.g., the radiation from an incandescent lamp). The polarization could be linear, circular, or elliptical. If the direction of the electric field at a point in space is always along the same line, the light is linearly polarized. If the two components of the electric field along the x and y axes have equal magnitude but differ in phase by 90°, the tip of the electric field vector rotates on a circle, and the light is circularly polarized. Elliptical polarization corresponds to the case where the two components of the field are not equal, and/or they differ in phase by an arbitrary angle. Actually, elliptical polarization is a general case, and linear and circular polarizations can be considered as special cases of it.

A linearly polarized monochromatic wave can be represented by a scalar function $u(x, y, z; t)$ that can be a component of the electric field or the magnetic field. This scalar function can be represented as

$$u(x, y, z; t) = U(x, y, z) \cos[2\pi ft + \phi(x, y, z)] \tag{22}$$

where $U(x, y, z)$ and $\phi(x, y, z)$ are the amplitude and phase of the wave, respectively, and f is the frequency of the light. With complex notation, this equation can be written

$$\begin{aligned} u(x, y, z; t) &= \mathrm{Re}[\mathbf{u}(x, y, z; t)] \\ &= \mathrm{Re}[\mathbf{U}(x, y, z) \exp(j2\pi ft)] \end{aligned} \tag{23}$$

where Re means the real part, $\mathbf{U}(x, y, z) = U(x, y, z) \exp[j\phi(x, y, z)]$ is a complex amplitude, and $\mathbf{u}(x, y, z; t) = \mathbf{U}(x, y, z) \exp(j2\pi ft)$. It is common to drop the time-dependent exponential term and work with the complex amplitude. However, the physical field at any point can be obtained only after the complex amplitude is multiplied by $\exp(j2\pi ft)$ and the real part of the result is taken. The above complex amplitude should satisfy the time-independent wave equation, known as the Helmholtz equation.

Two particular waves of interest are the plane wave and the spherical wave. The names indicate the shapes of the surfaces of constant phase (wavefronts). The complex amplitude of a plane wave can be written

$$\mathbf{U}(x, y, z) = A \exp[jk(\gamma_x x + \gamma_y y + \gamma_z z)] \tag{24}$$

where A is the wave amplitude, and $k = 2\pi/\lambda$ where λ is the wavelength. The parameters γ_x, γ_y, and γ_z are the cosines of the angles between the direction of the field propagation and the x, y, and z axes, respectively. If a spherical wave is generated by a source at $P_0 = (x_0, y_0, z_0)$, the complex amplitude corresponding to a point $P_1 = (x_1, y_1, z_1)$ is given by

$$\mathbf{U}(x, y, z) = \frac{A}{r_{01}} \exp(jkr_{01}) \tag{25}$$

where A is a constant and $r_{01} = [(x_1 - x_0)^2 + (y_1 - y_0)^2 + (z_1 - z_0)^2]^{1/2}$ is the distance between the points P_1 and P_0. If the source and the observation points are restricted to regions near the z axis such that

$$(z_1 - z_0)^3 \gg \frac{\pi}{4\lambda} [(x_1 - x_0)^2 + (y_1 - y_0)^2]^2 \tag{26}$$

then Eq. (25) can be approximated as

$$\mathbf{U}(x, y, z) \simeq \frac{A}{z_1 - z_0} \exp[jk(z_1 - z_0)] \exp\left[\frac{jk[(x_1 - x_0)^2 + (y_1 - y_0)^2]}{2(z_1 - z_0)}\right] \tag{27}$$

This is known as the quadratic approximation of a spherical wave.

In general, radiation is a statistical phenomenon, and light should be described by probability functions. The information about the statistics of light is usually provided by coherence functions. The coherence of a wave field is a measure of its ability to produce observable constructive and destructive interference (i.e., bright and dark regions) when different portions of it are combined. A distinction is often made between temporal coherence and spatial coherence. These are mathematically described by two complex functions. The temporal coherence function $\mathbf{\Gamma}$ is defined as

$$\Gamma(\tau) = \langle \mathbf{u}(x, y; t)\, \mathbf{u}^*(x, y; t - \tau) \rangle \qquad (28)$$

where $\mathbf{u}(x, y; t)$ and $\mathbf{u}(x, y; t - \tau)$ are the analytic functions corresponding to the electric field intensity at point (x, y) and at instants t and $t - \tau$. The angle brackets represent an infinite time average. For a monochromatic light, $\mathbf{u}$ has a single frequency, and $\mathbf{\Gamma}$ has a magnitude that is independent of τ. Such a completely temporal coherent light does not exist. In practice, a coherent light source has a band of frequencies. As a result, the magnitude of the coherent function decreases as τ is increased. This can be experimentally observed as a decrease in the visibility of the interference pattern as the time delay between the two portions of the beam is increased. The coherence time of a light source is defined as the value of τ for which the magnitude of coherence function falls to $1/e$ of its maximum value. Lasers usually have coherence times in the order of milliseconds in comparison with picoseconds coherence times of typical spectral sources. For time intervals greater than the coherence time, the source is considered to be incoherent. Another useful parameter is the complex degree of coherence, $\gamma(\tau)$. This is the normalized form of the temporal coherence function which is obtained by dividing $\Gamma(\tau)$ to its maximum value [i.e., $\Gamma(0)$]. A monochromatic light has a degree of coherence with a magnitude of unity, while a completely incoherent light has a degree of coherence of zero.

The parallel definition for spatial coherence function, $\Gamma(\eta, \xi)$, is

$$\Gamma(\eta, \xi) = \langle \mathbf{u}(x, y; t)\, \mathbf{u}^*(x - \eta, y - \xi; t) \rangle \qquad (29)$$

In this case, the analytic function $\mathbf{u}$ is measured at two points that are located at a plane perpendicular to the light rays. Again, for monochromatic light, $\mathbf{\Gamma}$ has a constant magnitude. In practice, a light source has a coherence length, which is defined as the spatial distance at which the magnitude of $\mathbf{\Gamma}$ falls to $1/e$ of its maximum value at $\eta = \xi = 0$.

Another parameter of interest in wave optics is the light intensity, which is defined as the radiated power per unit area. The intensity of monochromatic light is proportional to the square of its electric field amplitude. Neglecting a constant factor that depends on the unit system used, the equation

$$I(x, y) = |\mathbf{U}(x, y)|^2 \qquad (30)$$

relates the complex amplitude $\mathbf{U}$ to the intensity I of monochromatic light. In general cases, the intensity is defined as

$$I(x, y, z) = 2\, \langle u^2(x, y, z; t) \rangle \qquad (31)$$

where the angle brackets represent an infinite time average.

Inside a material, light travels slower than in a vacuum. The ratio of the speed of light in a vacuum to the speed of light in a material is called the index of refraction of that material. Some materials, known as photorefractive materials, have the property that their index of refraction can be altered by exposing them to light. The mechanism of the photorefractive effect is as follows. When the material is exposed to light, extra charged particles (electrons and holes) are created. These particles then migrate in certain directions and produce regions of excess positive and negative charge inside the material. The electric field induced by this spatial charge distribution changes the index of refraction of the material.

2. Diffraction Phenomena

To the lowest degree of approximation, the spreading of light when it encounters obstacles or apertures is described by ray-tracing rules from geometrical optics. The wave nature of light is ignored in this analysis, and the result becomes inaccurate as the dimensions of the obstacle or aperture are decreased. This departure from the predictions of geometrical optics is representative of diffraction phenomena.

There are two common treatments of diffraction: scalar diffraction theory and vectorial diffraction theory. In the former method, light is treated as a scalar parameter that represents one component of the electric field or magnetic field, and it is assumed that the other components behave similarly. This simplifies the analysis, but it ignores the fact that the components of the electric and magnetic fields are coupled by Maxwell's equations. This coupling is considered in

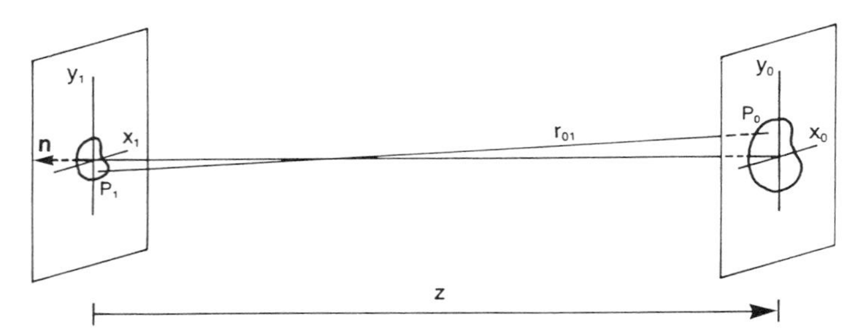

FIG. 4. Diffraction geometry, where the vector **n** is of unit length and is normal to the aperture plane.

the vectorial theory, but the relations obtained, except for few cases, are too complicated to solve. Fortunately, the results from the scalar theory are very accurate if two conditions are satisfied. These are (1) $\lambda/d \ll 1$, where λ is the light wavelength and d the maximum linear dimension of the aperture or obstacle, and (2) $r \gg d$, where r is the distance at which the diffracted light is observed. Both conditions are satisfied in most optical systems of interest here.

A case of great interest in optics is the diffraction by a finite planar aperture (Σ) shown in Fig. 4. In this figure, the vector **n** is an outward vector of unit amplitude that is normal to the plane of aperture. It is assumed that the incident wave is a monochromatic plane wave traveling along the z axis, and the diffracted light is observed at a distance z. Assuming that the complex amplitude corresponding to a component of the incident electric field is known, one seeks the corresponding component at the observation plane. The scalar diffraction theory for this problem was first developed by Kirchhoff in 1882. In his analysis, Kirchhoff made two assumptions about the values of the complex amplitude (**U**) and its normal derivative ($\partial\mathbf{U}/\partial\mathbf{n}$) immediately behind the $z = 0$ plane. These assumptions, known as Kirchhoff's assumptions, are (1) for the points on the aperture [i.e., $(x_1, y_1) \in \Sigma$], **U** and $\partial\mathbf{U}/\partial\mathbf{n}$ are the same as they would be in the absence of the screen and (2) for the points outside the aperture [i.e., $(x_1, y_1) \notin \Sigma$], **U** and $\partial\mathbf{U}/\partial\mathbf{n}$ are zero. Based on these assumptions, Kirchhoff found the following relation for the field at P_0:

$$\mathbf{U}(P_0) = \frac{1}{4\pi} \iint_{\Sigma} \frac{\exp(jkr_{01})}{r_{01}} \times \left[\frac{\partial \mathbf{U}}{\partial \mathbf{n}} - jk\mathbf{U}\cos(\mathbf{n}, \mathbf{r}_{01})\right] ds \qquad (32)$$

where $r_{01} = [(x_0 - x_1)^2 + (y_0 - y_1)^2 + z^2]^{1/2}$ is the magnitude of the vector $\mathbf{r}_{01}$ that joins a point at the observation region P_0 to a point on the aperture P_1, and $(\mathbf{n}, \mathbf{r}_{01})$ is the angle between **n** and $\mathbf{r}_{01}$.

Kirchhoff's assumptions were later shown to be inconsistent. Sommerfeld used another approach to remove this inconsistency. However, the results obtained from Kirchhoff's formula are in remarkably good agreement with experiments, and his theory is still widely used. Sommerfeld's formula for the same system is

$$\mathbf{U}(P_0) = \frac{1}{j\lambda} \iint_{\Sigma} \mathbf{U}(P_1) \frac{\exp(jkr_{01})}{r_{01}} \cos(\mathbf{n}, \mathbf{r}_{01})\, ds \qquad (33)$$

This equation is simplified if it is assumed that the distance between the two planes is much larger than both the dimension of the aperture and the dimension of the region of interest in the observation plane. Based on these assumptions, the cosine term of Eq. (33) can be approximated by unity, and the r_{01} term in the denominator can be replaced by z. A further approximation can be made in the exponent term if z is large enough to satisfy the following condition:

$$z^3 \gg (\pi/4\lambda)\,[(x_0 - x_1)^2 + (y_0 - y_1)^2]^2_{\max} \qquad (34)$$

Under these conditions, Eq. (33) can be approximated as

$$\mathbf{U}(x_0, y_0) = \frac{\exp(jkz)}{j\lambda z} \iint_{-\infty}^{+\infty} \mathbf{U}(x_1, y_1) \exp\{j\frac{k}{2z}[(x_0 - x_1)^2 + (y_1 - y_0)^2]\}\, dx_1\, dy_1 \qquad (35)$$

where the limits of the integral have been changed assuming $\mathbf{U}(x_1, y_1) = 0$ outside the aperture. This formula represents Fresnel diffraction, and the region in which it is valid is referred to as the Fresnel region.

If the distance between the planes satisfies the more restrictive condition,

$$z \gg (k/2)\,(x_1^2 + y_1^2)_{\max} \tag{36}$$

Eq. (35) can be written as

$$\mathbf{U}(x_0, y_0) = \frac{\exp(jkz)}{j\lambda z} \exp\left[\frac{jk(x_0^2 + y_0^2)}{2z}\right] \iint_{-\infty}^{+\infty} \mathbf{U}(x_1, y_1) \exp\left[\frac{-jk(x_0x_1 + y_0y_1)}{z}\right] dx_1\, dy_1 \tag{37}$$

This represents Fraunhofer diffraction, and the region that satisfies condition (36) is called the Fraunhofer region. Because the terms in front of the integral are constants, this formula can be simplified to

$$\mathbf{U}(x_0, y_0) = C \iint_{-\infty}^{+\infty} \mathbf{U}(x_1, y_1) \times \exp\left[\frac{-j2\pi(x_0x_1 + y_0y_1)}{\lambda z}\right] dx_1\, dy_1 \tag{38}$$

where $C = [\exp(jkz)/j\lambda z] \exp[jk(x_0^2 + y_0^2)/2z]$. Equation (38) can be considered as a two-dimensional Fourier transform of the function $\mathbf{U}(x_1, y_1)$ evaluated at spatial frequencies $f_X = x_0/\lambda z$ and $f_Y = y_0/\lambda z$. This result indicates that the shape of the diffraction pattern does not change in the Fraunhofer region. The change in the z value enters only as a scaling factor.

If the aperture is covered by a transparency having a complex transmittance of $\mathbf{t}(x_1, y_1)$, the above formulas are valid provided that $\mathbf{U}(x_1, y_1)$ is replaced by $\mathbf{U}(x_1, y_1) = \mathbf{t}(x_1, y_1)\, \mathbf{U}^-(x_1, y_1)$, where $\mathbf{U}^-(x_1, y_1)$ is the complex amplitude of the incident wave just before reaching the aperture. For the case of an incident plane wave $\mathbf{U}^-(x_1, y_1) = A$, where A is the wave amplitude. The Fraunhofer diffraction then becomes the Fourier transform of $\mathbf{t}(x_1, y_1)$ multiplied by a constant.

3. Fourier Transform Properties of Lenses

Lenses are among the most important components of the optical information processing systems. A lens is composed of an optically dense and transparent material like glass. A lens is called thin if the output light emerges at approximately the same coordinates as the input light. Figure 5 shows a lens and its parameters. The effect of a lens on the incident light can be considered as producing a phase factor delay of

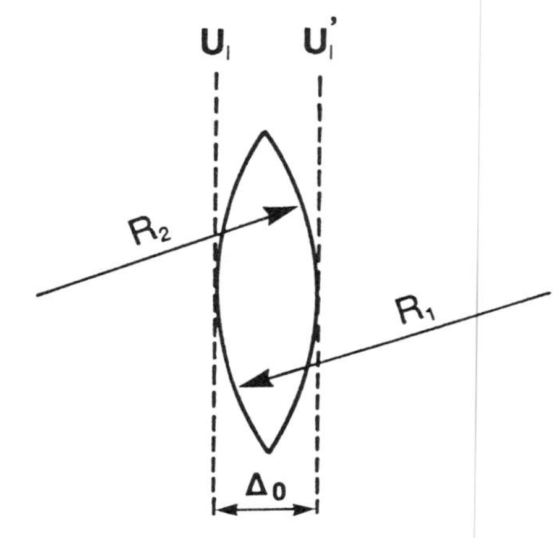

FIG. 5. Thin lens, where R_1 and R_2 are the radii of the lens surfaces; $\mathbf{U}_l$, complex amplitude of the input wave; $\mathbf{U}_l'$, complex amplitude of the output wave; and Δ_0, maximum thickness of the lens.

$$\mathbf{t}(x, y) = \exp(jkn\,\Delta_0) \exp\left[\frac{-jk(x^2 + y^2)}{2f}\right] \tag{39}$$

which, when multiplied by the complex amplitude of the field just in front of the lens ($\mathbf{U}_l$), provides the complex amplitude of the field just behind the lens ($\mathbf{U}_l'$). The parameter f in Eq. (39) is the focal length of the lens, which is given by

$$f = \frac{1}{(n - 1)[(1/R_1) - (1/R_2)]} \tag{40}$$

This formula is valid for all kinds of lenses if the following convention is used to determine the signs of R_1 and R_2: As rays travel from left to right, the radius of each convex surface encountered is considered positive, and the radius of each concave surface is considered negative. Therefore, R_1 in Fig. 5 is positive, while R_2 is negative. Finally, n is the index of refraction.

Practically, Eq. (39) is not valid for all values of x and y, since the lens has a finite aperture. The effective transmittance function is obtained by multiplying $\mathbf{t}(x, y)$, as given in Eq. (39), by a pupil function $P(x, y)$. The pupil function is unity for a point on the lens aperture (Σ) and is zero for a point outside the aperture, that is,

$$P(x, y) = \begin{cases} 1, & (x, y) \in \Sigma \\ 0, & (x, y) \notin \Sigma \end{cases} \tag{41}$$

An important property of a lens is its capability of Fourier transforming. It can be shown that if an object with a complex transmittance of $\mathbf{t}(x, y)$ is placed at a distance d_1 ($d_1 \geq 0$) in front of a lens or at a distance d_2 ($f > d_2 \geq 0$) behind the lens, and an incident monochromatic plane wave illuminates the object, the Fourier transform of $\mathbf{t}(x, y)$ multiplied by a constant and by a phase factor is obtained at the back focal plane. This property can be described as

$$\mathbf{U}_f(x_f, y_f) = (A/j\lambda f)C\,\mathcal{F}\{\mathbf{t}(x, y)\}\Big|_{\substack{f_X = x_f/\lambda f \\ f_Y = y_f/\lambda f}} \quad (42)$$

where x_f and y_f are the coordinates of a point in the focal plane, $\mathbf{U}_f(x_f, y_f)$ is the complex amplitude of the wave at the back focal plane, A the amplitude of the incident wave, and C a phase factor. The constant $j\lambda f$ is usually ignored in the analysis. For the special case that the object is at the front focal plane ($d_1 = f$), the phase factor is unity, and an exact Fourier transform is obtained.

Two types of lenses are widely used in optical information processing: spherical lenses and cylindrical lenses. The surface of a spherical lens has curvature in both directions. This type can perform a two-dimensional Fourier transform as discussed above. A cylindrical lens has curvature in one dimension (Fig. 6); hence, it can perform a one-dimensional Fourier transform. The cylindrical lenses are used for multichannel processing where different data channels are stacked along the axis of the lens.

4. Analysis of Optical Imaging Systems

A simple imaging system that consists of a single positive lens is shown in Fig. 7. From

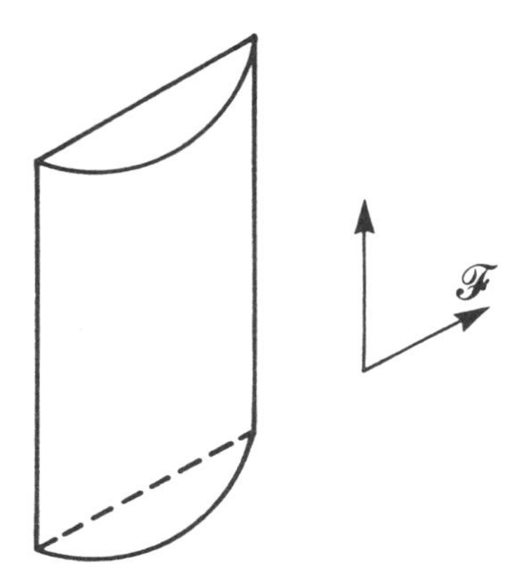

FIG. 6. Cylindrical lens, where the direction along which the Fourier transform is performed is indicated by $\mathcal{F}$.

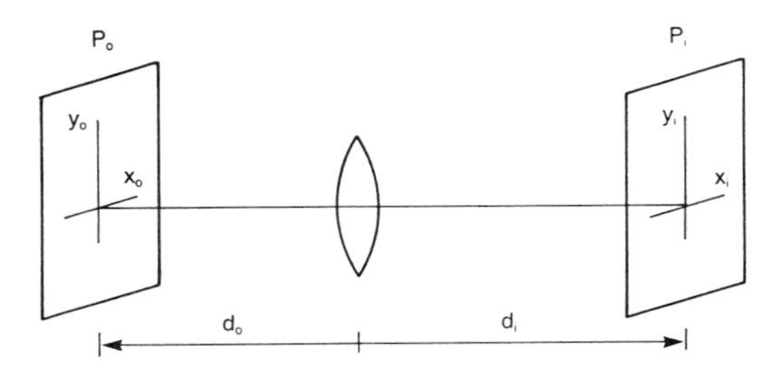

FIG. 7. Simple imaging system, where d_i is the image distance; d_o, object distance; P_i, image plane; and P_o, object plane.

geometrical optics, if the relation

$$\frac{1}{d_o} + \frac{1}{d_i} = \frac{1}{f} \quad (43)$$

is satisfied, an image of linear magnification $m = d_i/d_o$ is obtained. A more rigorous treatment includes the effects of diffraction of light between the two planes. In this analysis, it is assumed that the incident light is monochromatic. The finite aperture of the lens or the physical aperture usually used with the lens is modeled by a pupil function $P(x, y)$, defined in Eq. (41). The result obtained from the diffraction analysis shows that the complex amplitude of the image, $\mathbf{U}_i(x_i, y_i)$, can be obtained from the complex amplitude of the image predicted by the geometrical optics, $\mathbf{U}_g(x_i, y_i)$, using

$$\mathbf{U}_i(x_i, y_i) = C\,\tilde{\mathbf{h}}(x_i, y_i) * \mathbf{U}_g(x_i, y_i) \quad (44)$$

The parameter C is a quadratic phase factor that does not affect the intensity of the image, and it is normally ignored. The function $\tilde{\mathbf{h}}(x_i, y_i)$ is related to the pupil function through

$$\tilde{\mathbf{h}}(x_i, y_i) = \left(\frac{1}{\lambda d_i}\right)^2 \hat{P}\left(\frac{x_i}{\lambda d_i}, \frac{y_i}{\lambda d_i}\right) \quad (45)$$

where $\hat{P}(x_i/\lambda d_i, y_i/\lambda d_i)$ represents the Fourier transform of the pupil function evaluated at the spatial frequencies $f_X = x_i/\lambda d_i$ and $f_Y = y_i/\lambda d_i$.

In general, imaging systems are more complex. They usually include several lenses and apertures. Also, pure monochromatic light is not available. The effects of all apertures in the system can be modeled by one effective aperture located behind the last lens. The pupil function corresponding to this effective aperture can be obtained by projecting all the apertures to the location of the last lens and choosing the common area. The quasi-monochromaticity of the light is measured by the coherence function. In the analysis of imaging systems, illuminations with high degree of coherence are considered coherent, and those with low degree of coherence are considered incoherent.

In a coherently illuminated system, since the phase relation between the different waves that reach a point on the image plane is fixed, the waves add together as complex amplitudes. Therefore, the results obtained for the monochromatic case are valid for coherent systems, with the understanding that **U** now represents the relative amplitude and phase of the light. The analysis of coherent systems becomes simpler if performed in the spatial frequency domain. The convolution Eq. (44) c then be replaced by

$$\hat{\mathbf{U}}_i(f_X, f_Y) = \mathbf{H}(f_X, f_Y)\, \hat{\mathbf{U}}_g(f_X, f_Y) \qquad (46)$$

where $\hat{\mathbf{U}}_i$ and $\hat{\mathbf{U}}_g$ represent the Fourier transforms of $\mathbf{U}_i$ and $\mathbf{U}_g$, respectively, and $\mathbf{H}(f_X, f_Y) = \mathcal{F}\{\tilde{\mathbf{h}}(x_i, y_i)\}$ is called the coherent transfer function of the system. It can be shown that the transfer function is related to the effective pupil function by

$$\mathbf{H}(f_X, f_Y) = P(-\lambda d_i f_X, -\lambda d_i f_Y) \qquad (47)$$

that is, the transfer function is a magnified and inverted version of the effective pupil.

In a noncoherent illuminated system, the waves that reach a point on the image plane are statistically independent. Therefore, they add on a power (or intensity) basis. As a result, the system is linear in intensity, and the image intensity $I_i(x_i, y_i)$ can be obtained from

$$I_i(x_i, y_i) = C\, |\tilde{\mathbf{h}}(x_i, y_i)|^2 * I_g(x_i, y_i) \qquad (48)$$

where C is a constant, $I_g(x_i, y_i)$ the intensity of the image obtained from geometrical optics, and $\tilde{\mathbf{h}}(x_i, y_i)$ is defined in Eq. (45). The equivalent form of the above equation in the frequency domain is

$$\hat{I}_i(f_X, f_Y) = C\, \mathcal{F}\{|\tilde{\mathbf{h}}(x_i, y_i)|^2\}\, \hat{I}_g(f_X, f_Y) \qquad (49)$$

It is common to normalize the parameters in Eq. (49) to their values at $f_X = f_Y = 0$. The normalized form of $\mathcal{F}\{|\hat{\mathbf{h}}(x_i, y_i)|^2\}$ is called optical transfer function (OTF) and is represented by $\mathcal{H}(f_X, f_Y)$. OTF has a maximum value of unity at the origin.

III. Coherent Optical Processing

A. Spectrum Analysis

As shown in Section II,C,3, the Fourier transform of the transmittance of a coherently illuminated object can be obtained by a simple lens. While the Fourier transform is a computationally intensive operation with an electronic digital computer, it is performed in parallel with optics and thus is very fast. The input to an optical spectrum analyzer is a planar transducer with an amplitude transmittance $\mathbf{t}(x, y)$ that represents the function under the investigation. If the Fourier transform of a complex function is desired, the input transducer should be capable of modulating both the amplitude and phase of the light. This can be achieved by a two-layer photographic film. In the case of real functions, a simple black and white film suffices. The spectrum obtained at the back focal plane can be detected visually or electronically, or it can be recorded on another photographic film. Depending on the system configuration, as described later, the Fourier transform of two-dimensional or one-dimensional signals can be obtained.

A characteristic parameter of an optical spectrum analyzer is the space-bandwidth product. This is analogous to the time-bandwidth product of an electronic amplifier. The space-bandwidth product of an optical system is defined as the product of the highest spatial frequency over which it can operate and the size of the input aperture. Comparing two optical systems, the system with the larger space-bandwidth product is capable of analyzing higher spatial frequencies for a given aperture size or larger aperture sizes for a given spatial frequency. Space-bandwidth products of up to 10^5 are achievable with high-precision optical systems. Even with an ordinary 35-mm camera lens, space-bandwidth products greater than 1000 can be obtained.

The only shortcoming of optical spectrum analyzers is their low accuracy. The accuracy of these systems depends on the energy level of the light source, the quality of the optical components, and the noise of the detecting system. The limiting factor is normally the noise of the photodetector. The accuracies of practical systems are typically in the range of 1 to 10%. This limits the practical application of these systems to areas where high speed is required while low accuracy is tolerable.

1. Two-Dimensional Spectrum Analysis

A schematic diagram of a two-dimensional spectrum analyzer is shown in Fig. 8. A beam expander system provides plane-wave illumination from the output of the laser. The system consists of two spherical lenses and a pinhole

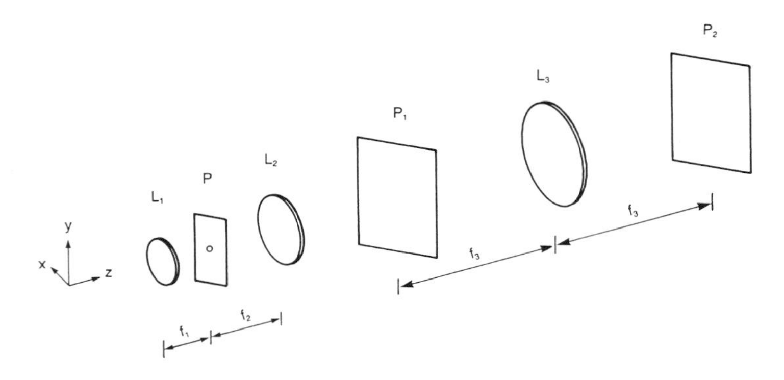

FIG. 8. Schematic diagram of a two-dimensional spectrum analyzer, where L_1 is the condensing lens; L_2, collimating lens; L_3, Fourier transform lens; f_1, f_2, and f_3, focal lengths of lenses L_1, L_2, and L_3, respectively; P, pinhole aperture; P_1, input plane; and P_2, output plane.

aperture. The first lens L_1 acts as a condenser. At the front focal plane of L_1, a pinhole aperture with a diameter of about 10 μm is used to filter out the undesired modes and to provide a point source of light to the second lens L_2. The second lens, located at a focal length from the pinhole, collimates the light and provides a uniform illumination to the input plane. The input and output planes are located at the front and back focal planes of the third lens L_3 to satisfy the condition for an exact Fourier transform. In most practical systems, L_3 is replaced by an optical system that consists of several lenses to improve the quality of the transform. The frequency components of the input function can be obtained by measuring the coordinates of the bright spots at P_2 and using the relations $f_X = x/\lambda f_3$ and $f_Y = y/\lambda f_3$, where λ is the free-space wavelength of the light. Figure 9 shows an example of the input and the output of the spectrum analyzer as recorded on photographic films. The input is a biased one-dimensional sinusoidal function

$$\mathbf{t}(x, y) = b + a \sin(2\pi f_o y) \tag{50}$$

where $b > a$ to ensure a positive real function. The Fourier transform of this function consists of three delta functions, that is,

$$\mathcal{F}\{\mathbf{t}(x, y)\} = b\,\delta(f_X, f_Y) - j\frac{a}{2}\,\delta(f_X, f_Y - f_o) + j\frac{a}{2}\,\delta(f_X, f_Y + f_o) \tag{51}$$

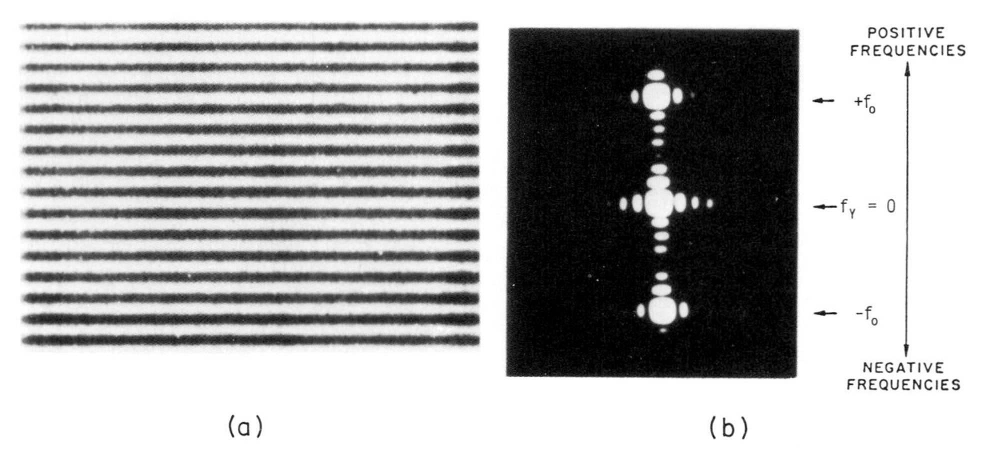

FIG. 9. Example of (a) input and (b) output of an optical spectrum analyzer. [From Preston, Kendall, Jr. (1972). "Coherent Optical Computers," McGraw-Hill, New York.]

These three impulses appear as three bright spots along the f_Y axis in Fig. 9b. The extra spots visible in that figure are due to the finite dimension of the input aperture. The Fourier transform of a rectangular aperture is a two-dimensional sinc function. Since the input function is multiplied by a rectangle function at P_1, the spectrum obtained at P_2 is the convolution of Eq. (51) with a sinc function.

2. Multichannel Spectrum Analysis

The schematic diagram of a multichannel spectrum analyzer is shown in Fig. 10. This is similar to Fig. 8 except that the spherical lens L_3 is replaced by three cylindrical lenses (L_3, L_4, and L_5). The input consists of several channels stacked along the vertical axis (y). The amplitude transmittance of each channel represents a one-dimensional function. The combination of the three cylindrical lenses performs the Fourier transformation along the x axis and images along the y axis. As a result, the Fourier transform of each channel is obtained in a horizontal band located at a particular distance from the x axis. To prevent the spectra of the neighboring channels from affecting each other, spatial guard bands should be considered in the input plane. Assuming 20 channels per millimeter and using an active region of 27 mm in 35-mm diameter optics, one can obtain 540 channels.

3. Folded-Spectrum Technique

As described in the previous section, the spectrum analysis of one-dimensional functions can be performed by multichannel systems. It is sometimes desired to analyze the spectrum of a long one-dimensional signal. This can be done by dividing the signal into a number of segments and providing those segments as the inputs to a multichannel system. In the output plane, the spectra obtained should be integrated at each frequency over all channels. Such an integration can be implemented by a vertical slit and a photomultiplier. The photomultiplier integrates the light that is passed through the slit, and the entire assembly moves along the x axis to detect all the frequencies. The problem with this system is its low space-bandwidth product. Actually, the multichannel capability has been sacrificed to analyze a long signal, but since each segment is treated individually, the space-bandwidth product remains the same. Therefore, the resolution of the system is not increased. To increase the space-bandwidth product, a method was proposed and analyzed by Thomas in 1966 and later implemented and demonstrated by Markevitch in 1969. Thomas suggested that a two-dimensional spectrum analyzer, as shown in Fig. 8, be used for analyzing long signals. The input signal is recorded in a raster format, but the different segments are not isolated as different channels. In fact, since a two-dimensional Fourier transformation is performed, all the segments contribute to the output spectrum in both directions. Thomas showed that the space-bandwidth obtained by this method is the square of the value that is achieved by the multichannel technique. Some of the properties of the spectrum obtained are described here. For a more detailed analysis, the reader is referred to the original paper by Thomas [C. E. Thomas, *Applied Optics* **5** (11) 1782–1790, 1966].

Figure 11a shows the raster scan input format.

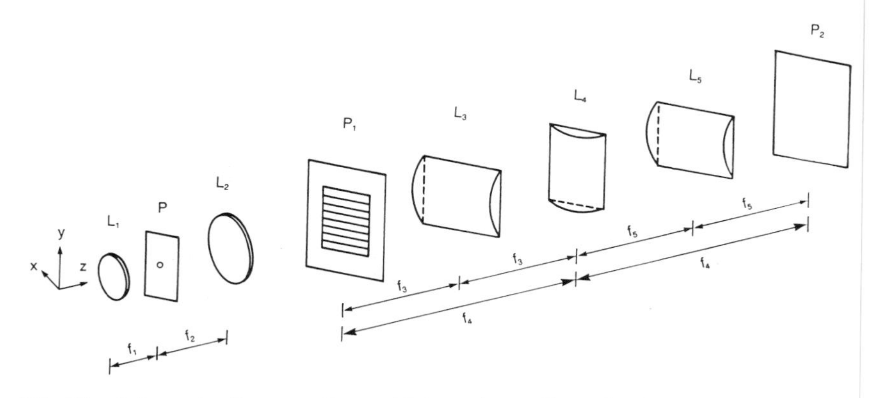

FIG. 10. Schematic diagram of a multichannel one-dimensional spectrum analyzer, where f_3, f_4, and f_5 are the focal lengths of cylindrical lenses L_3, L_4, and L_5, respectively. Other parameters are the same as in Fig. 8.

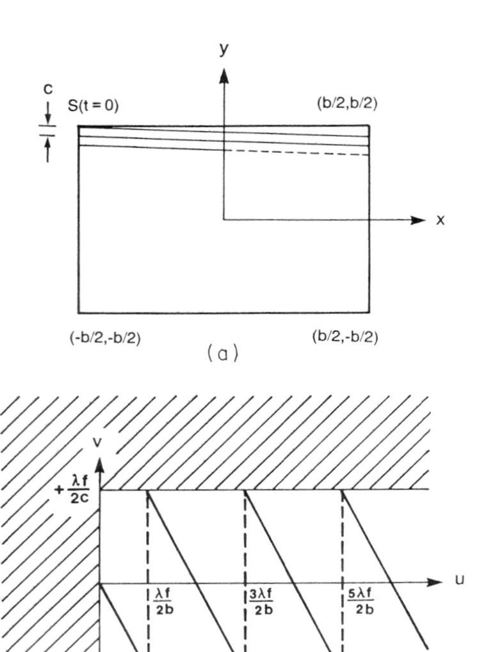

FIG. 11. Raster scan (a) input and (b) output formats in folded spectrum technique, where f is the focal length of the Fourier transforming lens; S, recorded signal; and λ, free-space wavelength of the laser. [From Thomas, C. E. (1966). *Applied Optics* **5** (11), 1782–1790.]

The input signal, which is a time-dependent function, is used to modulate the intensity of an electron beam as it scans the screen. The scan rate is adjusted so that the maximum frequency of the input signal generates a resolvable spatial frequency on the film. This is achieved using a deflection speed of $V_s = f_{max}/p$, where f_{max} is the maximum frequency of the signal and p the number of resolvable line pairs per unit length of the film. The exposed film is then appropriately linearized during the developing process. Therefore, the transmittance of the developed film along the scan lines resembles the amplitude of the signal. The spectrum of the recorded signal has a periodic structure along the v axis that is due to the effective periodic sampling nature of the raster lines. The information about the spectrum can be obtained from one of these periodic regions, such as the one shown in Fig. 11b. Since the amplitude spectrum of a real signal is a symmetric function, the analysis of only positive frequencies is sufficient. If the recorded signal is a sinusoidal wave, its frequency appears as a bright spot in the above region. If the spectra of waveforms with different frequencies are obtained, it can be seen that as the frequency is increased, the bright spot on the output plane scans through the observed region. These output scan lines are tilted with respect to the input lines by 90°.

Thomas has shown that the number of resolution elements along each line of the output spectrum is equal to the number of input scan lines. Conversely, the number of scan lines in the output spectral plane is equal to the number of resolution elements along each scan line of the input data. Thus, the obtained spectrum has as many resolution elements as the input data. Due to the nature of the appearance of this spectrum, it is known as the folded spectrum. Also, the spatial frequency axes u and v are referred to as the coarse frequency and fine frequency axes, respectively. Each raster line in the output plane represents a part of the spectrum with a bandwidth of $\Delta f = 1/T$, where T is the total length of the recorded signal in seconds.

The power of the folded spectrum technique can be seen by way of a numerical example. Using a photographic film with a resolution of 50 line pairs/mm, one can record 200 msec of a 10-MHz bandwidth signal on a 2 cm × 2 cm area. The corresponding folded spectrum consists of 1000 lines with 2000 resolvable spots on each line. The total space-bandwidth product of the system is 2 cm × 10^6, and the system is capable of detecting frequencies up to 10 MHz with a resolution of 5 Hz.

B. Spatial Filter Synthesis

1. System Configuration

In spatial filtering techniques, the Fourier transform of an input function that is obtained by a lens is manipulated by a filter. A second lens performs the Fourier transform operation on the modified spectrum and provides the output.

Figure 12 shows an optical system that is commonly used for spatial filtering analysis. The input plane P_1 is illuminated by a plane wave that propagates along the z axis. The function $\mathbf{f}(x, y)$ is introduced at P_1 as the amplitude transmittance of either a photographic film or a spatial light modulator device. An exact Fourier trans-

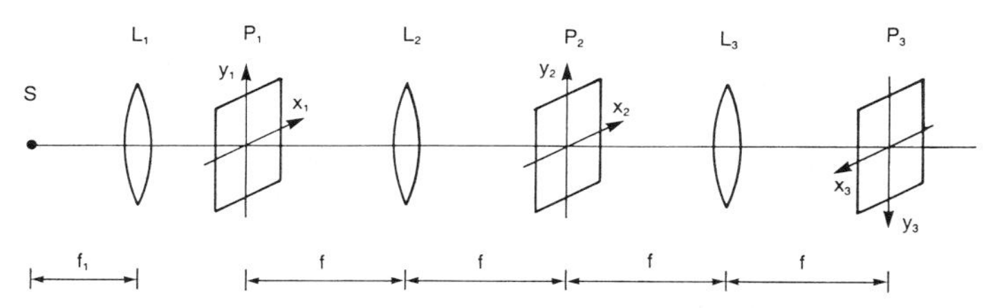

FIG. 12. Configuration of a spatial filtering system, where f_1 is the focal length of lens L_1; f, focal length of lenses L_2 and L_3; L_1, collimating lens; L_2 and L_3, Fourier transforming lenses; P_1, object plane; P_2, focal plane; P_3, output plane; and S, light source.

form of the input function is obtained at the focal plane P_2 as $\mathbf{F}(f_X, f_Y)$, where the spatial frequencies f_X and f_Y are related to the coordinates x_2 and y_2 through the relations $f_X = x_2/\lambda f$ and $f_Y = y_2/\lambda f$. As light passes through the filter plane, the Fourier transform is multiplied by the amplitude transmittance function of this plane. A second Fourier transformation is then performed on the filtered spectrum by L_3. Since two succeeding Fourier transforms result in an inversion of the input function, the axes of the output plane P_3 are usually inverted to simplify the analysis. It is not essential that L_2 and L_3 have the same focal length. However, this is usually assumed, and the system is referred to as the 4-f system. Also, the constants in Eq. (42) are usually ignored. Considering these points, one can write the output function at P_3 as

$$\mathbf{g}(x, y) = \mathbf{f}(x, y) * \mathbf{h}(x, y) \tag{52}$$

where $\mathbf{h}(x, y)$ is the inverse Fourier transform of the function introduced at P_2. In the frequency domain, the equivalent relation to the above equation is

$$\mathbf{G}(f_X, f_Y) = \mathbf{F}(f_X, f_Y)\, \mathbf{H}(f_X, f_Y) \tag{53}$$

where **G**, **F**, and **H** are the Fourier transforms of **g**, **f**, and **h**, respectively.

2. Simple Filters

Simple filters are filters that alter either the amplitude or the phase of the spectrum. An amplitude filter can be obtained by the process of developing an exposed photographic film. A phase filter requires the control of optical path length of the filter plate, for example, by an appropriately varying thickness. A special type of amplitude filter is the binary filter which consists of opaque and transparent regions.

Because of their construction simplicity, these filters were among the first implemented filters. Porter's experiment, described in Section I, is an example of binary filtering. Zernike's phase-contrast microscope was based on introducing an additional phase of 90° to the undiffracted components of light. A useful operation in image processing is to attenuate the low-frequency components by an amplitude filter. Since the average brightness of the image is determined by these components, this filtering results in decreasing the average brightness and increasing the contrast.

3. Holographic Filters

In spatial filtering, it is usually necessary to find the Fourier transform of a desired impulse response and to construct the corresponding filter plate. The Fourier transform of an arbitrary two-dimensional function is computationally intensive, and the result is usually a complex function. Although complex filters can be constructed by cascading an amplitude filter and a phase filter, practical problems in fabricating phase filters make the task difficult. To overcome these difficulties, Vander Lugt introduced holographic filters in 1963.

Figure 13a shows the recording process for a Vander Lugt filter. The desired impulse response $\mathbf{h}(x, y)$ is introduced as the amplitude transmittance of the input transparency. The Fourier transform of the impulse response $\mathbf{H}(f_X, f_Y)$ is obtained optically at the back focal plane P_2 of a lens, where it is interfered with a tilted plane wave. The resultant interference pattern is recorded on a photographic film. Although the developed film consists solely of absorption patterns, it contains the information about both amplitude and phase of the desired transfer function. This can be seen from the intensity pattern at P_2. The tilted wavefront can be represented by $A \exp(-j\, 2\pi y_2 \sin\theta/\lambda)$, where A is the amplitude and θ the angle of the tilted wave. At P_2 this complex amplitude is added to $(1/j\lambda f)\, \mathbf{H}(x_2/\lambda f, y_2/\lambda f)$ which is obtained by lens L_1. The resul-

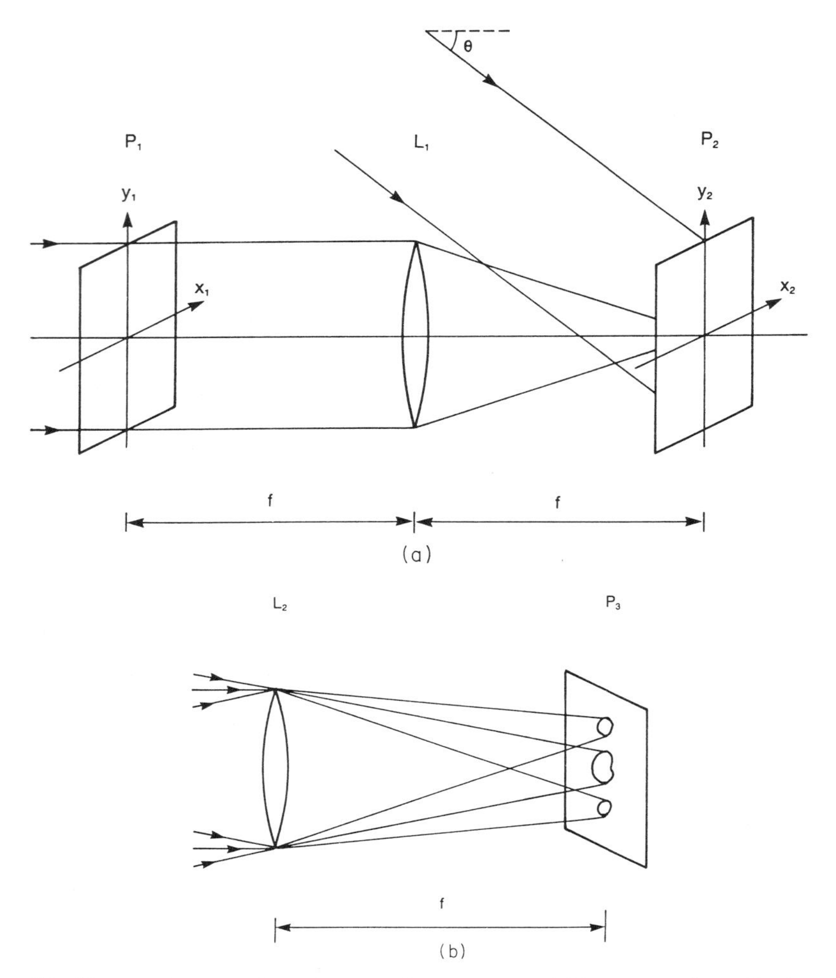

FIG. 13. Vander Lugt filter: (a) recording process and (b) output, where f is the focal length of lenses L_1 and L_2; L_1 and L_2, Fourier transforming lenses; P_1, input plane; P_2, focal plane; and P_3, output plane.

tant intensity, given by

$$I(x, y) = |A \exp(-j 2\pi y_2 \sin\theta/\lambda) + (1/j\lambda f)\, \mathbf{H}(x_2/\lambda f, y_2/\lambda f)|^2 \quad (54)$$

is recorded as a hologram. The recorded hologram is then placed at the filter plane P_2 in the system shown in Fig. 12. The input function $\mathbf{f}(x, y)$ is introduced at P_1, and its Fourier transform is obtained at P_2. At P_2, the light is diffracted by the hologram. This is equivalent to multiplying the Fourier transform of $\mathbf{f}(x, y)$ by the transmittance function of the hologram, which is the same as the above intensity pattern. The diffracted light has four components. Two terms propagate along the z axis and are of no interest. The other two terms propagate at angles $+\theta$ and $-\theta$ with respect to the z axis and provide the convolution and cross-correlation of the input function with the impulse response. This is shown in Fig. 13b. To prevent overlapping of different terms, the condition

$$\theta > (W_f + 1.5\, W_h)/f \quad (55)$$

should be satisfied at the recording process, where W_f and W_h are the maximum dimensions of the functions $\mathbf{f}(x, y)$ and $\mathbf{h}(x, y)$ along the y direction.

The Vander Lugt filter has several advantages. It is capable of implementing complex filters with a single amplitude filter. The Fourier transform of the impulse response is obtained optically, and finally, both convolution and cross-correlation results are provided in the output.

4. Computer-Generated Filters

Although the Vander Lugt filter simplifies the implementation of complex spatial filters, it requires the physical existence of the impulse response. In many cases of interest, the mathematical description of the desired impulse response is known, but it is difficult or even impossible to construct it physically. These cases can be handled using a digital computer. Computer-generated holograms were first introduced by Brown and Lohmann in 1966. One of the advantages of their method was that complex spatial filtering was performed with binary masks. These masks were obtained by photographic reduction of the output of a computer-controlled plotter. Other techniques for making computer-generated holograms have been introduced since Brown and Lohmann's invention. Today, computer-generated holograms are widely used in various applications and are considered to possess outstanding potential for future optical information-processing systems. Because of space constraints, only a brief description of Brown and Lohmann's method is presented here.

The holograms introduced by Brown and Lohmann are known as detour-phase holograms. The name indicates the method used to achieve the required phase variations. It is well known that a periodic array of slits (a grating) diffracts an incident wave in different directions. These directions can be calculated using the relation

$$(\sin \alpha_n - \sin \alpha_0)\, d = n\lambda \tag{56}$$

where n is an integer that indicates the order of the diffracted wave, α_0 the angle of the incident wave, α_n the angle of the nth diffracted beam, d the spacing between the slits, and λ the free-space wavelength of the light. If some of the slits are shifted from their periodic positions, the phase of the diffracted light is affected as shown in Fig 14. This is known as the detour phase and can be used to implement the phase of a complex function.

To make a computer-generated hologram, the complex function that describes the transmittance of the desired filter is sampled in both the x and y directions according to the sampling theorem. That is, the sampling period is chosen to be less than or equal to $1/2f_{\max}$, where $f_{\max}$ is the maximum spatial frequency along the direction of sampling. At each sampled point, the value of the function is represented by an amplitude and a phase, such as $A_{nm} \exp(j\phi_{nm})$, where n and m are the indices of the sampled point along the x and y directions, respectively. A computer-controlled plotter is used to draw an enlarged version of the filter. This plot consists of rectangular cells corresponding to the sampled values. Each cell, as shown in Fig. 15, includes a transparent aperture on an opaque background. The complex value of the wavefront is encoded by the size and position of the aperture. One method is to choose the height of the aperture proportional to the amplitude and its position proportional to the phase. This plot is then photographically reduced to the actual size of the filter, which is usually about 1–2 cm. The size of the apertures after reduction is typically around 10–20 μm. Because of the binary nature of this filter, it can be easily reproduced. Also, film grain noise and film nonlinearities are of little consequence. Computer-generated holograms are usually made as an array of 128 × 128 pixels.

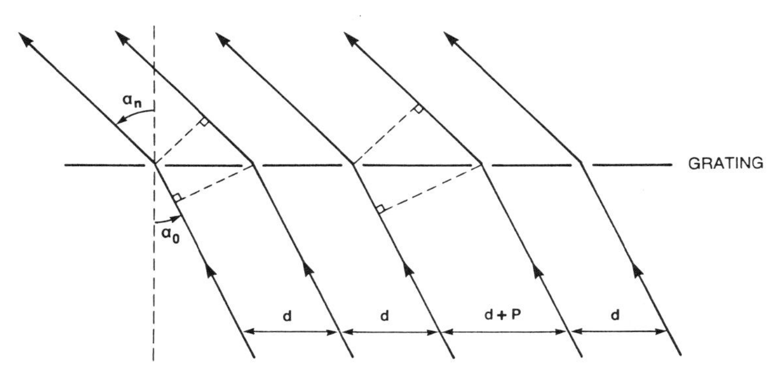

FIG. 14. Detour phase effect, where α_0 is the angle of the incident wave and α_n, angle of the nth diffracted beam. [From Lohmann, A. W., and Paris, D. P. (1968). *Applied Optics* **7** (4), 651–655.]

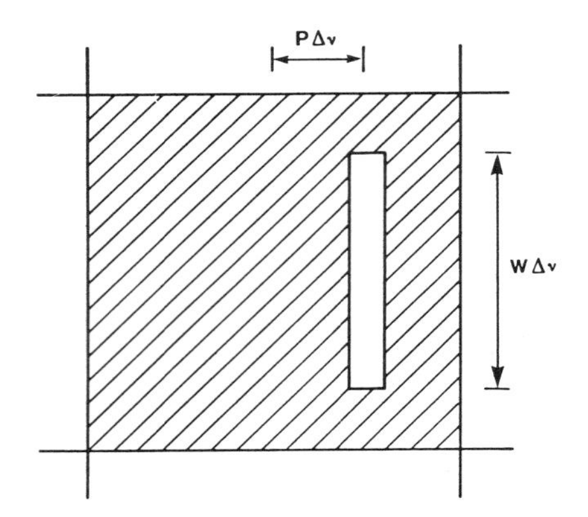

FIG. 15. Example of a sampling cell in a detour phase hologram, where $\Delta\nu$ is the resolution of the plotter and W and P are proportional to the amplitude and phase of the sampled wavefront.

5. Other System Configurations

The system shown in Fig. 12 is not the only processor used for spatial filtering. Figure 16a shows a two-lens processor that is widely used. The distances in this system are such that (1) the point source S is imaged at plane P_2 by lens L_1 and (2) the input plane P_1 is imaged at the output plane P_3 by lens L_2. It can be shown that a Fourier transform of an input function at P_1 is obtained at P_2 where it is affected by the filter. One of the advantages of this system is that the scale of the Fourier transform can be varied by changing the distance between P_1 and P_2. Also, the distances between P_1, L_2, and P_3 can be selected to obtain the desired magnification. Figure 16b shows a one-lens system that can be used for spatial filtering. In this system, the point source S is imaged at plane P_2 by lens L_1, and the input plane P_1 is imaged at the output plane P_3 by the same lens.

Another useful system is the joint transform correlator shown in Fig. 17. This is a 4-f system with two input functions, $\mathbf{g}_1(x, y)$ and $\mathbf{g}_2(x, y)$, that are vertically separated in the input plane by a distance Δ. First, a Fourier hologram is recorded at P_2. Given a linear recording of the light intensity, the transmittance of the resultant hologram is

$$\begin{aligned}\mathbf{t}(f_X, f_Y) &= |\mathbf{G}_1(f_X, f_Y)\exp(j\pi\Delta f_Y) \\ &\quad + \mathbf{G}_2(f_X, f_Y)\exp(-j\pi\Delta f_Y)|^2 \\ &= |\mathbf{G}_1(f_X, f_Y)|^2 + |\mathbf{G}_2(f_X, f_Y)|^2 \\ &\quad + \mathbf{G}_1(f_X, f_Y)\,\mathbf{G}_2^*(f_X, f_Y)\exp(j2\pi\Delta f_Y) \\ &\quad + \mathbf{G}_1^*(f_X, f_Y)\,\mathbf{G}_2(f_X, f_Y)\exp(-j2\pi\Delta f_Y)\end{aligned} \tag{57}$$

where $f_X = x_2/\lambda f$ and $f_Y = y_2/\lambda f$ are the spatial frequencies, and $\mathbf{G}_1$ and $\mathbf{G}_2$ the Fourier transforms of $\mathbf{g}_1$ and $\mathbf{g}_2$, respectively. Then, the input plane is removed, and the Fourier transform of the above function is obtained at P_3. The third and fourth terms of the transmittance expression produce the cross-correlation of the two functions around the ($x_3 = 0$, $y_3 = +\Delta$) and ($x_3 = 0$, $y_3 = -\Delta$) points, respectively. To separate these

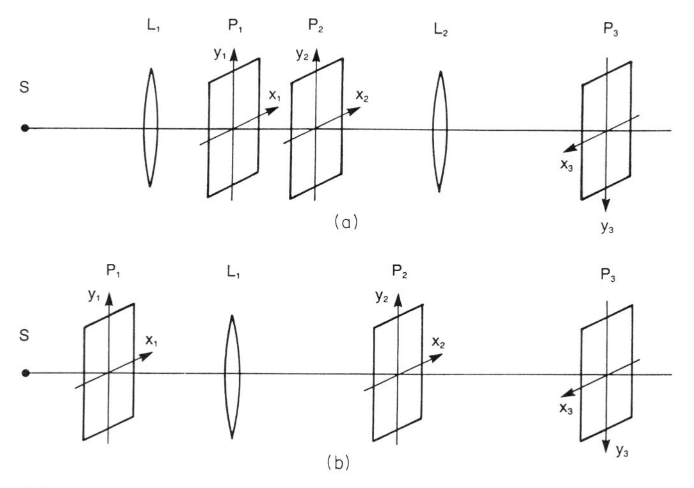

FIG. 16. (a) Two-lens coherent optical processor. (b) One-lens coherent optical processor. Where L_1 and L_2 are lenses; P_1, input plane; P_2, filter plane; P_3, output plane; and S, light source.

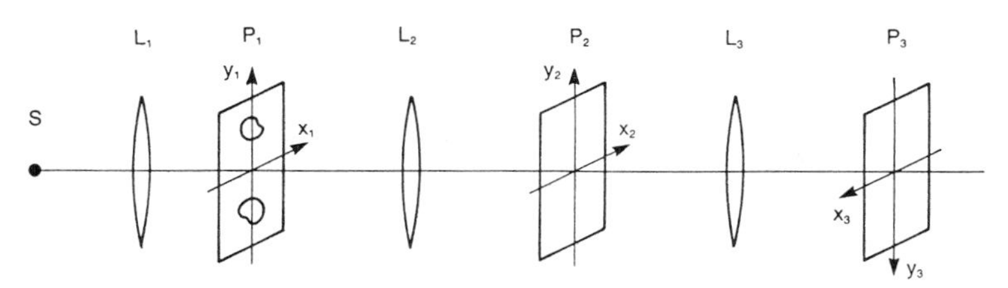

FIG. 17. Joint transform correlator, where L_1 is the collimating lens; L_2 and L_3, Fourier transform lenses; P_1, input plane; P_2, focal plane; P_3, output plane; and S, light source.

two terms from the other two terms that appear near the origin, the relation

$$\Delta \geq 0.5(W_1 + W_2) + \max(W_1, W_2) \quad (58)$$

should be satisfied, where W_1 and W_2 are the dimensions of the two functions along the y axis.

IV. Incoherent Optical Processing

A. Introduction

Coherent optical processing suffers from several problems. Data from commonly used noncoherent sources, such as LEDs and CRTs, cannot be directly processed. Dust particles on the optical components or small defects inside the lenses can produce scattered waves and affect the output significantly. Incoherent systems do not have these problems. To compare the two types of processing, the corresponding systems are shown in Fig. 18. In a coherent system, the entire object is illuminated by a plane wave. This is similar to a one-channel system. If a dust particle exists on one of the lenses, a particular region of the output is affected. In the case of an incoherent system, the light source is extended and can be considered as a number of point sources. Each point of the object plane, the Fourier plane, or the output plane is illuminated by waves that originate from different points of the source. This is similar to a multichannel system. As a result, a dust particle on a lens does not produce a severe effect in any particular region of the output. The effect actually spreads to all regions but becomes more tolerable since each point in the output plane is illuminated by different waves. Therefore, the overall signal-to-noise ratio is slightly decreased without having severe effects in a particular region of the output.

In spite of these advantages, incoherent systems have their own shortcomings. A major problem is that they are inherently capable of handling only nonnegative real functions. This is due to the fact that they work by manipulating the light intensity, not the complex amplitude. Although, as is described later, there are methods that include bipolar and complex functions, they usually require hybrid systems. Also, the total information capacity of incoherent systems, in general, falls below that of coherent systems. However, because of the unique features of incoherent optical processing mentioned above, it is the preferred choice for some applications.

Incoherent optical processors can be divided into two groups: (1) the systems that are based on the diffraction phenomenon, and (2) the systems that can be explained by geometrical optics rules. These are explained in the following two subsections. Methods for including bipolar and complex functions are also discussed and the corresponding problems are addressed.

B. Diffraction-Based Systems

Figure 19 shows a diffraction-based incoherent optical processor. The object can be self-luminous, or it can be illuminated by an extended source. As discussed in Section II,C,4, an incoherent system is linear in intensity. The intensity of the output image, $I_i(x, y)$, can be obtained from

$$I_i(x, y) = C\, s(x, y) * I_g(x, y) \quad (59)$$

where C is a constant and $I_g(x, y)$ the intensity of the image that is predicted by geometrical optics. The function $s(x, y)$ represents the diffraction effects of the system. It is known as the point-spread function and is defined as the intensity distribution that is obtained in the image plane if a point source is located at the origin of the object plane. For the system of Fig. 19, it can be shown that

$$s(x, y) = \left(\frac{1}{\lambda f}\right)^4 \left|\hat{\mathbf{P}}\left(\frac{x}{\lambda f}, \frac{y}{\lambda f}\right)\right|^2 \quad (60)$$

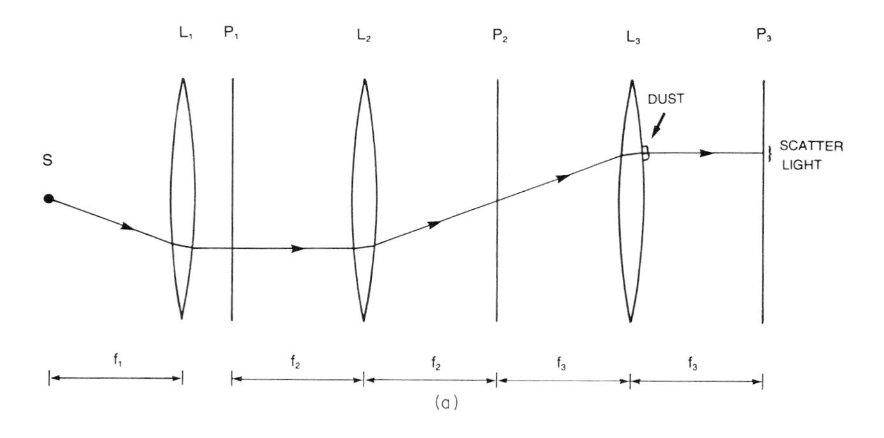

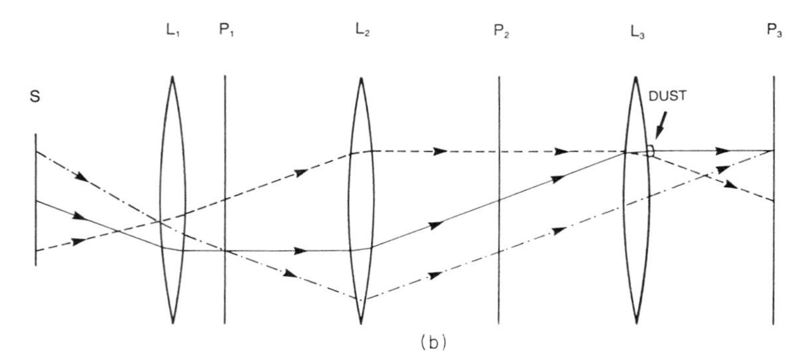

FIG. 18. Effect of dust particles in (a) coherent and (b) incoherent optical systems, where f_1, f_2, and f_3 are the focal lengths of L_1, L_2, and L_3, respectively; L_1, collimating lens; L_2 and L_3, Fourier transform lenses; P_1, object plane; P_2, Fourier plane; P_3, output plane; and S, light source. [From Lee, S. H., ed. (1981). "Optical Information Processing Fundamentals," Springer-Verlag, New York.]

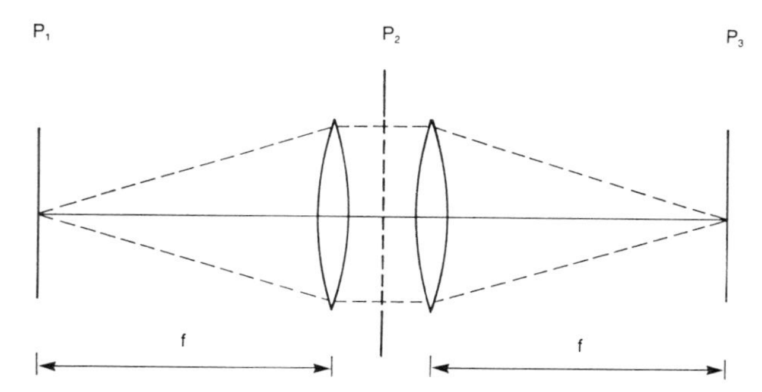

FIG. 19. Diffraction-based incoherent optical processor, where f is the focal length of the lenses; P_1, object plane; P_2, pupil plane; and P_3, output plane. [From Lee, S. H., ed. (1981). "Optical Information Processing Fundamentals," Springer-Verlag, New York.]

where $\hat{\mathbf{P}}(x/\lambda f, y/\lambda f)$ represents the Fourier transform of the complex pupil function (complex transmittance of the aperture) at spatial frequencies $f_X = x/\lambda f$ and $f_Y = y/\lambda f$. It is convenient to use a scaled pupil function defined as $\mathbf{P}'(x, y) = \mathbf{P}(\lambda fx, \lambda fy)$. Equation (60) is then simplified to

$$s(x, y) = |\hat{\mathbf{P}}'(x, y)|^2 \tag{61}$$

Since the image predicted by geometric optics is an inverted version of the object, the relation

$$I_g(x, y) = I_o(-x, -y) \tag{62}$$

holds between the intensity of the object I_o and the intensity of the image I_g. The axes of the image plane are usually inverted to make the two distributions identical. Equation (59) can then be written as

$$I_i(x, y) = C\, s(x, y) * I_o(x, y) \tag{63}$$

The equivalent form of this relation in the Fourier domain is

$$\hat{I}_i(f_X, f_Y) = C\, \hat{s}(f_X, f_Y)\, \hat{I}_o(f_X, f_Y) \tag{64}$$

where ˆ represents the Fourier transform operation. Using Eq. (61), one can find the Fourier transform of the point-spread function as the autocorrelation of the scaled pupil function, that is,

$$\hat{s}(f_X, f_Y) = \mathbf{P}'(f_X, f_Y) \star \mathbf{P}'(f_X, f_Y) \tag{65}$$

The functions in Eq. (64) are usually normalized by their values at the origin. The normalized form of $\hat{s}(f_X, f_Y)$ [i.e., $\hat{s}(f_X, f_Y)/\hat{s}(0,0)$] is known as the optical transfer function and is usually denoted by $\mathcal{H}(f_X, f_Y)$. Using the normalized quantities, one can rewrite Eq. (64) as

$$\hat{I}_i(f_X, f_Y) = \mathcal{H}(f_X, f_Y)\, \hat{I}_o(f_X, f_Y) \tag{66}$$

This equation is the basis for the spatial filtering process using incoherent systems. It shows that the spatial-frequency components of the light intensity at the object plane can be manipulated in the output plane by an appropriate optical transfer function.

The first diffraction-based incoherent optical processor was introduced for character recognition by Armitage and Lohmann in 1965. Their system is shown in Fig. 20. The left part of the system is coherent and provides the Fourier transform of the transmittance of the input plane, $\mathbf{U}_n(x, y)$, at the back focal plane of lens L_1. The intensity distribution at P_1, $I_n(x, y)$, is related to $\mathbf{U}_n(x, y)$ by

$$I_n(x, y) = C_1\, |\hat{\mathbf{U}}_n(x/\lambda f, y/\lambda f)|^2 \tag{67}$$

where C_1 is a constant. A rotating plate made of ground glass is used to make the phase a random variable. The rest of the system works as an incoherent processor. The point-spread function of the incoherent system is $s(x, y) = (1/\lambda f)^4 |\hat{\mathbf{P}}_m(x/\lambda f, y/\lambda f)|^2$, where $\hat{\mathbf{P}}_m$ is the pupil function corresponding to the character of interest. With these relations, the distribution of light intensity in the output plane can be found from Eq. (63) as a convolution integral. Since in the character recognition process, both $\mathbf{U}_n(x, y)$ and $\mathbf{P}_m(x, y)$ are real quantities, their Fourier transforms are symmetric, and convolution becomes similar to the correlation process. The intensity of the output light can be written

$$I_{out}(x, y) = C_2 \iint_{-\infty}^{+\infty} \left|\hat{\mathbf{U}}_n\left(\frac{\xi}{\lambda f}, \frac{\eta}{\lambda f}\right)\right|^2 \times \left|\hat{\mathbf{P}}_m\left(\frac{\xi - x}{\lambda f}, \frac{\eta - y}{\lambda f}\right)\right|^2 d\xi\, d\eta \tag{68}$$

where C_2 is a constant. At the output plane, a pinhole is located at the ($x = 0, y = 0$) point. A detector behind the pinhole measures the inten-

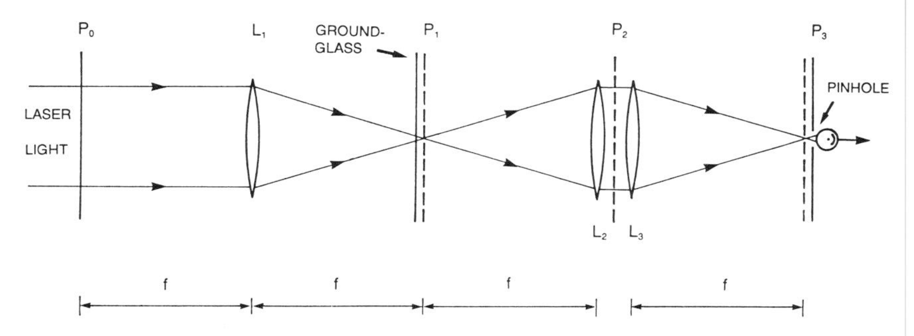

FIG. 20. Optical system for character recognition, where f is the focal length of lenses L_1, L_2, and L_3; P_0, input plane; P_1, focal plane; P_2, pupil plane; and P_3, output plane. [From Armitage, J. D., and Lohmann, A. W. (1965). *Applied Optics* **4**(4), 461–467.]

sity of the output light at the origin given by

$$I_{\text{out}}(0, 0) = C_2 \iint_{-\infty}^{+\infty} \left|\hat{\mathbf{U}}_n\left(\frac{\xi}{\lambda f}, \frac{\eta}{\lambda f}\right)\right|^2 \times \left|\hat{\mathbf{P}}_m\left(\frac{\xi}{\lambda f}, \frac{\eta}{\lambda f}\right)\right|^2 d\xi \, d\eta \quad (69)$$

This quantity is defined as the cross-correlation coefficient between the energy functions of the two characters, that is, $|\hat{\mathbf{U}}_n(\xi/\lambda f, \eta/\lambda f)|^2$ and $|\hat{\mathbf{P}}_m(\xi/\lambda f, \eta/\lambda f)|^2$. If the input character is the same as the reference character, the two energy functions are identical, and the cross-correlation coefficient becomes maximum, indicating a match case. The advantage of this system is that a character can be recognized regardless of a spatial shift. This is due to the fact that the system works based on the intensity functions of Fourier transforms, so a spatial shift of a function does not change the intensity of its Fourier transform. The disadvantage of the system is that some characters, such as *O* and *Q*, have similar energy functions, and therefore the corresponding correlation factors are high. As a result, the probability of an error in detection is increased for these cases.

A significant advance in the field of incoherent spatial filtering occurred in 1968 when Lohmann and, independently, Lowenthal and Werts used holographic techniques to implement the desired filters. This is the incoherent version of the Vander Lugt filtering technique. The recording process is similar to that shown in Fig. 13a, except that a diffuser is put in contact with the input transparency at P_1. As a result, the phase of the light is made a random variable. An input transparency with an intensity transmittance that resembles the desired point-spread function is used, and its Fourier hologram is recorded. This hologram is then used at the pupil plane of the system shown in Fig. 19. The results of the convolution and cross-correlation operations between the object intensity and the point spread function appear in the image plane at different locations along the y axis.

C. Geometrical Optics-Based Systems

Incoherent optical processors that are based on geometrical optics are of two types: imaging systems and shadow casting systems. In the imaging systems, the input function $f_1(x, y)$ is introduced as the intensity transmittance of a transparency at the input plane P_1. This is imaged onto a second transparency that represents the second function $f_2(x, y)$. The output light is demagnified by a lens and spatially integrated by a detector. Several useful operations can be achieved by this scheme. A simple system is shown in Fig. 21. This system is capable of integrating the product of two functions, that is,

$$g = \iint_{-\infty}^{+\infty} f_1(x, y) f_2(x, y) \, dx \, dy \quad (70)$$

where g is the detected power, and an imaging with unit magnification is assumed. With a spatial or temporal scanning technique, convolution and cross-correlation operations can be achieved. For example, if P_2 in Fig. 21 is shifted by x_0 and y_0 along the x and y axes, respectively, the following cross-correlation is obtained at the output

$$g = \iint_{-\infty}^{+\infty} f_1(x, y) f_2(x - x_0, y - y_0) \, dx \, dy \quad (71)$$

To implement the convolution process, one of the transparencies should be inverted with respect to the coordinate axes and then shifted by the desired amount. The system can work as a two-dimensional processor or a multichannel one-dimensional processor. In the latter case,

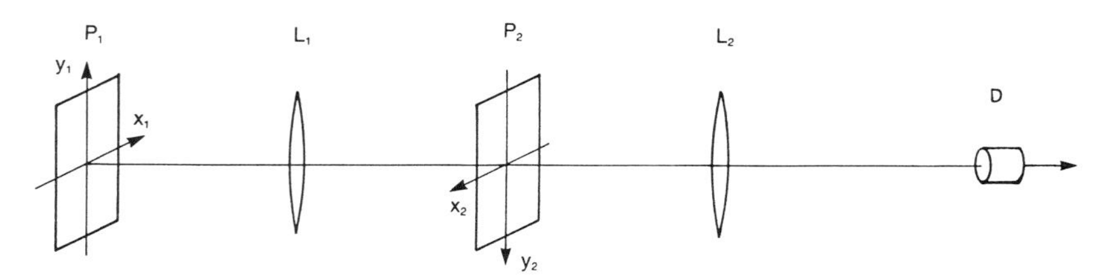

FIG. 21. Incoherent optical system for realizing the integral of a product, where D is the detector; L_1, imaging lens; L_2, integrating lens; P_1, input plane; and P_2, image plane. [From Lee, S. H., ed. (1981). "Optical Information Processing Fundamentals," Springer-Verlag, New York.]

one or both of the transparencies consist of a number of stacked one-dimensional distributions. A cylindrical lens is used as L_2 to integrate only in one dimension, and the output of each channel is detected by a separate detector element.

The systems based on shadow casting can provide convolution and cross-correlation of two functions without requiring mechanical motion. Figure 22 shows a system that performs the cross-correlation process. Again, the two functions are introduced as the intensity transmittance of the transparencies at P_1 and P_2. A point at $(-x_0, -y_0)$ on the source illuminates the first transparency with a tilted plane wave. If diffraction is negligible, the tilted wave propagates between the two planes according to geometrical optics rules and illuminates the second plane. The intensity distribution of the incident light on the second plane is a shifted version of the intensity distribution on the first plane, that is, $f_1[x - (d/f)x_0,\ y - (d/f)y_0]$. This is multiplied by the intensity distribution of the second lens, and the result is focused to (x_0, y_0) on the detector. The power at this point on the detector is

$$I(x_0, y_0) = K \iint_{-\infty}^{+\infty} f_1\left\{x - \left(\frac{d}{f}\right)x_0,\ y - \left(\frac{d}{f}\right)y_0\right\} f_2(x, y)\, dx\, dy \qquad (72)$$

where K is a constant. Therefore, the distribution of the power at the detector plane provides the cross-correlation between the two functions.

The geometrical optics-based processors are simple and powerful. However, they require that diffraction effects be negligible. This means that the transparencies should not have fine structure. As a result, the space-bandwidth products of these systems are lower than the diffraction-based systems, and in terms of information handling they are not efficient.

D. Bipolar Processing

A major limitation of incoherent optical processors is that their point-spread function is a nonnegative real quantity, while many useful operations require systems with bipolar impulse responses. A number of methods have been investigated to implement bipolar incoherent processors. The general idea behind these techniques is to use two pupil functions and get the desired operation by combining the two images at the output. One method, introduced by Rhodes, is the phase-switching technique. In this method, as shown in Fig. 23, the input light is split into two beams and passes through two spatial filters P_1 and P_2. The beams are then recombined, and an output image is constructed. Two images are obtained with the same filters but with different phases between the beams (0° and 180°). This is achieved by moving one of the mirrors shown in Fig. 23. The two output images are subtracted from each other by an electronic system or by a digital computer.

Complex quantities can also be represented in incoherent systems. One technique is to break the real and imaginary components into positive and negative parts and use four separate pupils to manipulate them. A more efficient method, proposed by Burckhardt, is to use three unit vectors at 0°, 120°, and 240° as the basis vectors. Any complex function can then be represented by three nonnegative components.

Although these techniques expand incoherent processing beyond positive impulse response systems, they generally require hybrid systems. Consequently, the physical size of the systems is increased. Also, the two-pupil method does

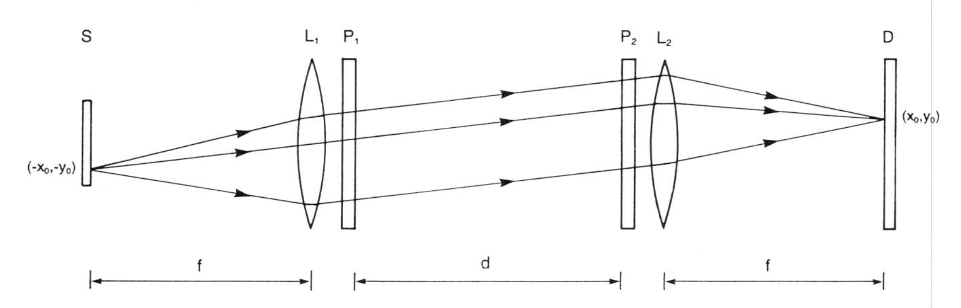

FIG. 22. Incoherent spatial system for realizing the cross-correlation operation, where D is the detector; d, distance between the two input planes; f, focal length of lenses L_1 and L_2; P_1 and P_2, input planes; and S, light source. [From Lee, S. H., ed. (1981). "Optical Information Processing Fundamentals," Springer-Verlag, New York.]

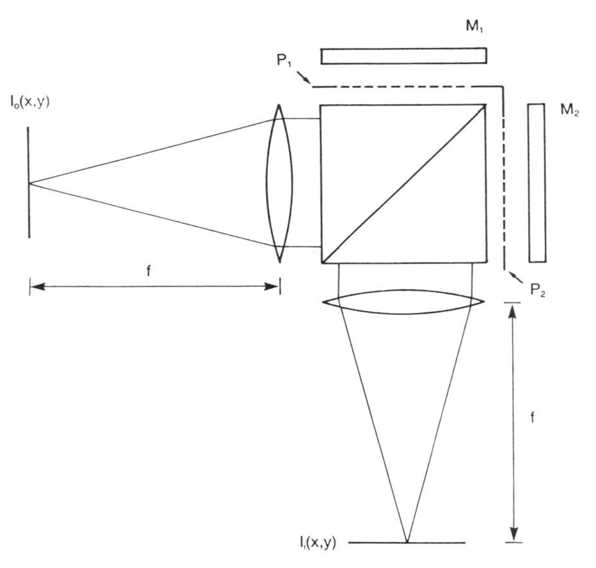

FIG. 23. Two-pupil system based on Michelson interferometer for bipolar incoherent processing, where f is the focal length of the lenses; I_o, intensity of the input object; I_i, intensity of the output image; M_1 and M_2, mirrors; and P_1 and P_2, pupils. [From Lee, S. H., ed. (1981). "Optical Information Processing Fundamentals," Springer-Verlag, New York.]

not have a unique solution for a desired point-spread function, and it is not known what the best choice is.

V. Hybrid Processors

A. Introduction

Optical processors have several important advantages. Among them are parallel processing, high throughput of data, and the ability to perform powerful operations, such as Fourier transformation, with simple systems. However, optical systems have their own shortcomings, one of which is lack of flexibility. Also, most current optical processor designs are inherently analog and cannot provide highly accurate outputs. In contrast, digital electronic systems are both flexible and accurate, and the technology of electronic computers is more advanced. Therefore, it seems reasonable to combine the two systems to utilize the advantages of both. In a hybrid (optical/electronic) processor, the optical system performs a kind of transformation on the input data. The output of the optical system is then analyzed by a digital electronic system.

One of the major design considerations of hybrid systems is the interfacing required to convert the data from one form into the other. Data are entered into the optical system through an input interface. To use the favorite features of the optical system, the input interface should have a high space-bandwidth product and be capable of handling high data rates. Two common types of input devices for incoherent optical systems are TV monitors and arrays of LEDs. Different types of spatial light modulators are used as input devices for coherent systems. The output of the optical system should be converted to an electrical signal, sampled, and digitized for use by the electronic system. These are performed by an output interface. In some hybrid systems, the spectrum of the input signal is manipulated by an interface at the Fourier plane. All three types of interfaces mentioned above are usually connected to a digital computer. Using the flexibility of computer programming, one can control the optical system and analyze the results.

Hybrid processors are made for a variety of applications, each with a special structure suitable for a particular job. A general description of the different types of hybrid systems is provided in the following subsections.

B. Coherent Hybrid Processors Based on Power Spectrum Analysis

A schematic diagram of a hybrid processor that is based on power spectrum analysis is shown in Fig. 24. A two-dimensional pattern is introduced by a spatial light modulator. The Fourier transform of this pattern is detected at the back focal plane of a spherical lens. Since the detectors are sensitive to light intensity, the detected signal is actually the square of the Fourier transform (i.e., the power spectrum). This signal is sampled and digitized. The resultant values are transmitted to an electronic digital computer. The computer uses a programmed algorithm to analyze the sampled values and to provide the desired output.

This type of processor benefits from the data compression property of Fourier transform. To get an intuitive feeling about this property, recall that the value of the Fourier transform of a function at a point in the spatial frequency domain is affected by all points in the function. If the Fourier plane is divided into a number of regions and the power spectrum is integrated inside each region, this set of numbers provides information about the function. To get detailed information, it is required to have sufficiently small regions. However, even if the regions are not small, the values provide information about the general behavior of the function, which could be sufficient for detecting the desired feature. With this technique, the amount of data is significantly reduced and becomes compatible with the capacity of the electronic system.

Different types of detectors can be used for converting the optical output to an electrical signal. A commonly used device is the wedge–ring detector. This is a two-dimensional array fabricated in silicon by planar diffusion. It has a circular shape and typically consists of 32 semicircular rings and 32 wedges; each element can be accessed individually. This structure takes advantage of the inversion symmetry property of the power spectrum. The rings provide information about the amplitudes of spatial frequency components of the spectrum, and the wedges provide information about the orientation of the spatial frequencies. If the output of each segment is represented by a 10-bit number, a total of 640 bits is sufficient to represent all the values. This is significantly lower than the number of bits in the input pattern. If the above detection is too coarse, other detectors such as a TV camera can be used.

Hybrid processors based on power spectrum analysis have found application areas both in the laboratory and in industry. Among the laboratory applications are analysis of aerial imagery, detection of black lung disease from chest X-ray images, and automatic sorting of handwriting. In industry, hybrid systems are used for inspecting products such as medical syringe needlepoints, woven clothes, and printing papers.

C. Coherent Hybrid Processors Based on Spatial Filtering

A schematic diagram of a hybrid processor that is based on spatial filtering is shown in Fig. 25. The optical system is a 4-f configuration as described in Section III,B,1. The electronic digital computer controls the optical system through

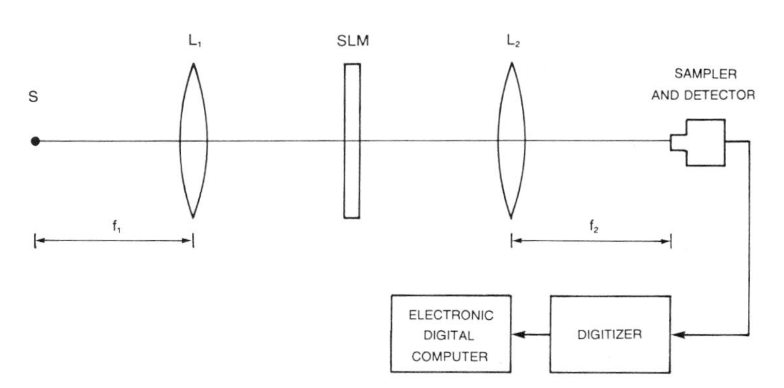

FIG. 24. Coherent hybrid processor based on power spectrum analysis, where f_1 and f_2 are the focal lengths of lenses L_1 and L_2, respectively; L_1, collimating lens; L_2, Fourier transforming lens; S, light source; and SLM, spatial light modulator. [From Stark, Henry, ed. (1982). "Applications of Optical Fourier Transforms," Academic, New York.]

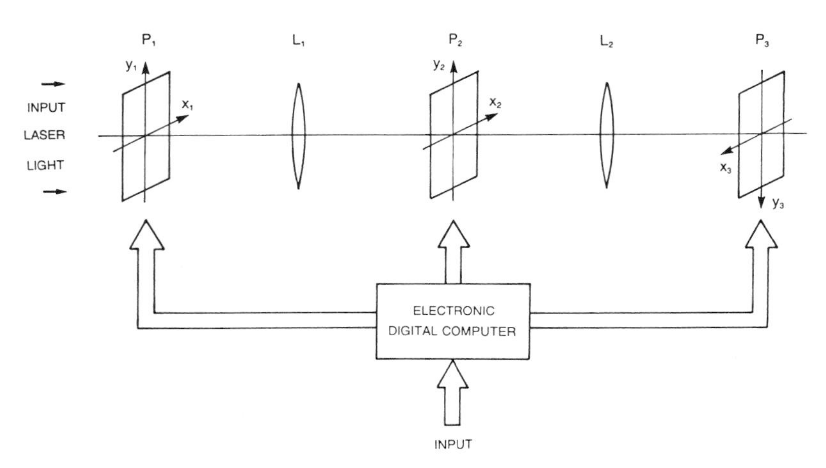

FIG. 25. Schematic diagram of a coherent hybrid processor based on spatial filtering, where L_1 and L_2 are the Fourier transform lenses; P_1, input plane; P_2, Fourier transform plane; and P_3, output plane. [From Lee, S. H., ed. (1981). "Optical Information Processing Fundamentals," Springer-Verlag, New York.]

three interfaces at the input, output, and Fourier transform planes. The input and output interfaces are similar to those described for power spectrum-based processors. The type of the third interface depends on the application of the hybrid system. If it is desired to perform different spatial filtering operations (such as low-pass or high-pass filtering), electronically controlled two-dimensional spatial light modulators are suitable. One example of such devices consists of two layers of liquid crystal. Each layer is divided into a number of regions. The regions can be activated individually by addressing the corresponding electrodes. The electrodes in one layer have ring structure and in the other layer are wedge-shaped. The ring-shaped electrodes are used to block particular spatial frequency bands, while the wedge-shaped electrodes are used to eliminate different orientations. To perform a filtering operation, the required electrodes are activated by an electronic computer. The regions in contact with these electrodes become opaque, and consequently the corresponding frequencies are blocked.

In pattern recognition, it is sometimes required to get the correlation between an input pattern and a number of reference patterns. These reference patterns could be different Vander Lugt filters recorded in a two-dimensional array format. Since the result of the correlation is very sensitive to the position of the Vander Lugt filter, an interfacing system is usually designed to move the desired filter to the required position with sufficient accuracy. These systems may be implemented with stepping motors that are controlled by digital electronic computers. The user specifies the desired filter through the computer terminal. The number of steps is determined by a program, and the specified filter is moved to within a few micrometers of the required position.

Hybrid systems have recently been used for recording high-quality computer-generated holograms. For this purpose, the Fourier transform of the desired function is first calculated and then sampled. During the recording process, the laser beam is deflected to the desired point on the film, and an exposure determined by the sampled value at that point is performed. The process is controlled by a computer for all the sampled values.

Hybrid processors based on spatial filtering are used in numerous applications. Most of these processors function as correlators; they compare the input pattern(s) with one or more reference patterns and detect the matched cases. Among the applications of optical correlators are character recognition, cloud motion analysis, water pollution monitoring, radar processing, and image processing.

D. Incoherent Hybrid Processors

Incoherent processing like coherent processing benefits from the flexibility and accuracy of hybrid systems. In addition to these advantages, incoherent processing can be extended beyond nonnegative real functions by using hybrid systems. As previously described, optical incoherent processors are restricted to nonnegative quantities. Bipolar and complex quantities should be decomposed into a number of nonnegative components (e.g., the positive and negative parts in the case of a bipolar quantity) and then processed separately. The components are recombined at the output to produce the desired result. These tasks can be performed by electronic systems.

As an example of incoherent hybrid processors, the system introduced by Goodman is described here (Fig. 26). This system is capable of performing vector–matrix multiplication in parallel. The values of the input vector are introduced in parallel as intensities of the LEDs. Lenses L_1 and L_2 image the light distribution horizontally and expand it vertically so that all parts of the mask M are illuminated. Lens L_3 is a field lens. The mask M is a two-dimensional array of apertures. The area of the aperture in the mth row and nth column is proportional to the value of the corresponding element in the matrix. As light passes through the mask, different product terms are obtained. The lens combination L_5 integrates the light distribution behind the mask along the horizontal direction and provides the results to a column array of detectors. Lens L_4 is a cylindrical lenslet array that improves the light throughput.

Goodman used this system to perform discrete Fourier transformations (DFTs). DFT is the analogous form of the integral Fourier transform for discrete signals. In the case of one-dimensional signal, it is defined as

$$\mathbf{G}(n) = \sum_{m=0}^{N-1} \mathbf{g}(m) \exp\left(\frac{-j2\pi mn}{N}\right), \quad n = 0, 1, \ldots, N-1 \tag{73}$$

where $\mathbf{g}(m)$ is a discrete function that consists of N values and $\mathbf{G}(n)$ the nth value of the corresponding DFT. Equation (73) is actually a vector–matrix multiplication, where the vector $\mathbf{g}(m)$ is multiplied by an $N \times N$ matrix whose elements are $\exp(-j\pi 2mn/N)$. To include complex quantities, Goodman represented each complex number by three positive components, as described in Section IV,D. This requires an increase in the dimension of the system by a factor of three in both the horizontal and vertical directions.

The intensities of the LEDs in the above processor are controlled by an electronic system. Also, the detected signal is an electrical signal that can be used for another operation or can be digitized and stored. A fixed mask can be used if a particular operation is needed, or a spatial light modulator controlled by a digital computer can be used if a variety of operations are desired. This system is very powerful. It can perform various operations including space-variant and nonlinear processing. Also, since the input data are introduced in parallel, very high throughput processing can be achieved.

VI. Optical Digital Computing

A. Introduction

The optical processors described in previous sections are all analog in nature. Although hybrid systems benefit from digital computation, they actually convert an optical analog signal to an electrical digital signal and then perform digital computation in the electronic system. To utilize the full advantage of the speed and parallelism of optics, it is desirable to perform all the computations optically. Digital computing requires nonlinear processing such as analog-to-digital conversion, logical operations, and bistability. In this section, the general aspects of optical digital computing are described. [*See* NONLINEAR OPTICAL PROCESSES.]

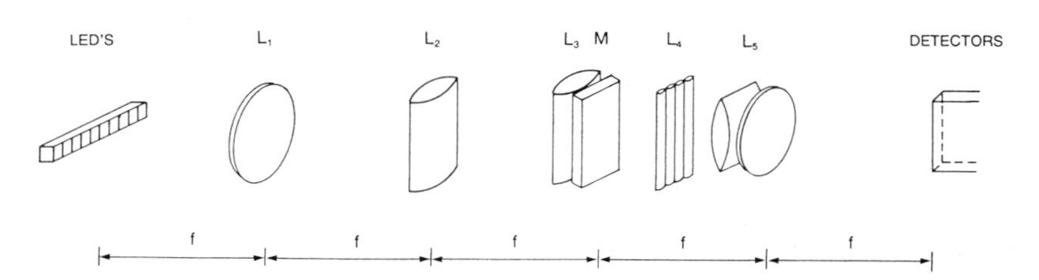

FIG. 26. Schematic diagram of a matrix–vector multiplier, where f is the focal length of lenses L_1, L_2, and L_5; L_1, spherical lens; L_2 and L_3, cylindrical lenses; L_4, cylindrical lenslet array; L_5, integrating lens system; and M, mask. [From Goodman, J. W., Dias, A. R., and Woody, L. M. (1978). *Optics Letters* **2**(1), 1–3.]

A central question is why optical digital computing is needed. The fact is that in spite of recent advances in electronic digital computers, there exist a number of problem areas such as meteorology, aerodynamics, molecular dynamics, fusion energy, and defense early warning systems that demand computing powers well beyond what is currently available. The major difficulty in increasing the computing power of existing electronic computers is the "von Neumann bottleneck." To illustrate this problem, a schematic diagram of a von Neumann type machine is shown in Fig. 27. Electronic computers generally have this structure. The memory addressing mechanism shown in this figure was proposed by von Neumann in the mid-1940s to decrease the number of connections between the central processing unit and the memory. As an example, consider a memory that has 65,536 binary cells (64-Kbit memory). Direct access to all the cells requires 65,536 wire connections. With binary addressing, the number of connections is reduced to $\log_2 65,536 = 16$. In this method, each cell is addressed by a 16-bit binary number. The price that is paid for this reduction is that only one cell can be accessed at a time. This puts a limit on the speed of the processor and it is known as the von Neumann bottleneck. Optical systems do not suffer from such a connection problem. They are inherently two-dimensional, and numerous points can be addressed simultaneously by different beams of light. Therefore, optical digital processors are promising candidates to achieve larger computing capacities.

The idea of building an optical digital computer started in the mid-1960s when reliable lasers became available. The direction persued at that time was to use the nonlinear properties of materials to perform logic operations. Although the feasibility of this technique was demonstrated, the large amount of power required made the approach impractical. Recently, an increasing demand for parallel processing, advances in electrooptic devices, and the rediscovery of residue arithmetic have made the optical digital computing area active again. A number of different optical systems capable of performing logical operations have been introduced during the last decade. The basic principles of residue arithmetic are explained in the next subsection. Various approaches for building an optical digital computer are then described, and the problems associated with them are discussed.

B. Residue Arithmetic

Residue arithmetic is a very old subject. The first recorded work in this field is the verse of the Chinese mathematician Sun-Tsu, published around the first century A.D. In this verse, he gives an algorithm for finding a number whose remainders upon division by 3, 5, and 7 are known. A general theory of remainders (now known as the Chinese Remainder Theorem) was established by the German mathematician K. F. Gauss in the 19th century. The application of residue arithmetic in computers, however, is a relatively recent idea. In 1955, the first investigation on this subject was published by A. Svoboda and M. Valach in Czechoslovakia. Their work stimulated more research on residue arithmetic and its applications. Here, a short introduction to residue arithmetic is presented.

Unlike the commonly used decimal and binary number systems, the residue number system (RNS) is an unweighted system. A number system is said to be weighted if any number X

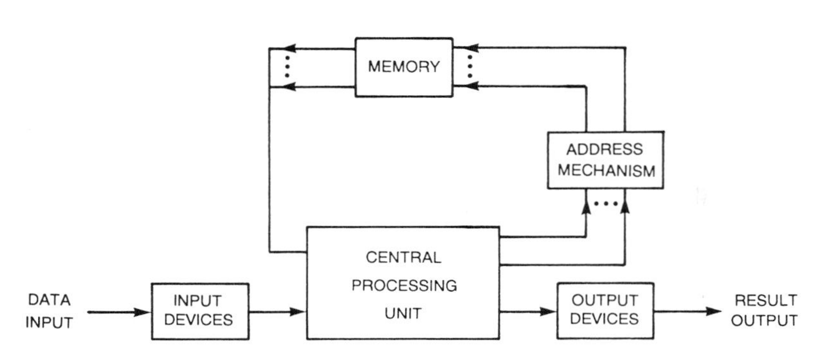

FIG. 27. Schematic diagram of a von Neumann machine.

can be expressed as $X = \Sigma_{i=1}^{N} d_i w_i$, where all w_i are constants, all d_i a set of permissible digits, and N is the number of digits used to represent X. Weighted-number systems have favorable features. However, the dependency that exists between the digits puts a limit on the processing speed when an operation is performed by a computer. For example, the addition of two numbers should be done digit by digit, since the addition of two digits may result in a carry and affect the addition of the next two digits. This is not the case in the RNS.

An RNS is defined by choosing n relatively prime (containing no common factors) numbers $m_1, m_2, \ldots, m_n$ called moduli. Any integer X can then be represented as an n-tuple $(x_1, x_2, \ldots, x_n)$, where $x_i = |X|_{m_i}$ (read X modulo m_i) is the least positive integer remainder that is obtained from the division of X by m_i. For example, if the set of moduli $\{3, 4, 5, 7\}$ is chosen, then the numbers $X = 23$ and $Y = 14$ are represented as $X = (2, 3, 3, 2)$ and $Y = (2, 2, 4, 0)$. In general, given the set of moduli $\{m_1, m_2, \ldots, m_n\}$, a range of $M = \Pi_{i=1}^{n} m_i$ integer numbers can be uniquely represented. This range can be allocated to positive numbers or negative numbers, or it can be partitioned to include both positive and negative numbers.

The important feature of RNS is that basic arithmetic operations can be performed on each digit individually. That is, if $X = (x_1, x_2, \ldots, x_n)$ and $Y = (y_1, y_2, \ldots, y_n)$ are two numbers of the same system, then $Z = X * Y = (z_1, z_2, \ldots, z_n)$, where $z_i = |x_i * y_i|_{m_i}$, for $i = 1, 2, \ldots, n$, and $*$ represents the addition, subtraction, or multiplication operation. Division is possible, but it is difficult, except for the remainder zero case. As an example, the results of performing addition, subtraction, and multiplication on the two numbers mentioned above are $X + Y = (1, 1, 2, 2)$, $X - Y = (0, 1, 4, 2)$, and $X \cdot Y = (1, 2, 2, 0)$. These are the residue representations of the correct answers, that is, 37, 9, and 322, respectively.

In residue arithmetic, there are a number of basic operations that are difficult to perform. These are division, scaling, sign detection, overflow detection, and relative-magnitude determination. In spite of these difficulties, the fact that the calculations associated with different moduli are independent of each other makes RNS suitable for parallel processing. Residue arithmetic is especially well suited for optical computing. One can use the cyclic nature of the phase or polarization of a light beam to implement residue numbers. Also, RNS significantly reduces the storage size required for truth-table look-up processing described in the next section.

C. Realization of Optical Digital Computers

1. Elements of Optical Digital Computers

The fundamental parts of an optical digital computer are central processing unit, memory, input devices, and output devices. The input devices convert the input data to a form of digital signal suitable for the central processing unit. Among the input devices are analog-to-digital converters, residue converters, and encoders. If the input data are not in the format used by the computer, special devices are used to convert the format. Examples of these devices are serial/electronic-to-parallel/optical converters and serial/optical-to-parallel/optical converters. After the data are processed, the output devices convert the result to a form suitable for detection. Among the output devices are decoders, residue converters, and digital-to-analog converters. Depending on the format of the output signal, the detector may be a single element, or it may be a one-dimensional or a two-dimensional array of elements.

Different types of memories might be required in an optical computer. These include scratch pad memory for storing intermediate results, archival storage unit for storing algorithms, and associative memory for truth-table look-up processing. Among the candidates for optical memories are spatial light modulators, optical disks, holographic memories, and bistable devices. One- or two-dimensional spatial light modulators can be used for temporary storage. Optical disks are suitable for cases that require large storage capacities. Today, optical disks with capacities of 10^{11} bits are commercially available. Volume holographic devices are also suitable for large storage capacities. Storage sizes of $(n/\lambda)^3$ bits/m^3 are theoretically predicted for holographic memories, where λ is the free-space wavelength of the light and n the refractive index of the material used. For a 10 mm $\times$ 10 mm $\times$ 2 mm crystal of $LiNbO_3$ with $n = 2.2$, and with a He–Ne laser with wavelength of 632.8×10^{-9} m, the above storage capacity would be 8.4×10^{12} bits. Due to practical problems, the above limit has not yet been reached. Another advantage of holographic memories is that they can function as associative memories. This means that the whole memory is activated by the input and responds with the associated output. Different beams of light can access the memory simultaneously and perform parallel processing. Another technique for realizing optical memories employs bistable devices. A bistable device has two stable states for an input signal. Hence, it can be used as a binary memory cell to store a logic state of 0 or 1. Electronic bistable devices (flip-flops) have been a major part of electronic digital computers since their invention. Optical bistability was first predicted by A. Szoke and his colleagues at the Massachusetts Institute of Technology in 1969. However, due to material problems, they were not successful in showing this effect experimentally. In 1975, H. M. Gibbs and his co-workers at Bell Laboratories demonstrated the effect using sodium vapor as the nonlinear medium. Four years later, the first semiconductor bistable device was built from GaAs by the same group. Today, optical bistability is observed in different materials. Among them, GaAs appears to be the most suitable candidate. With this material, optical bistability has been obtained at room temperature with low powers (<10 mW) and with very high switching times ($\simeq 1$ psec switch-on and ≤ 10 nsec switch-off). Further improvements are expected as the parameters are optimized.

The functional heart of an optical digital computer is the central processing unit which performs various digital arithmetic operations. This unit contains the hardware implementation of the desired operation. It may be made of a single lens to perform Fourier transformation or may include different elements. The techniques for realizing a digital arithmetic operation are described in the following subsection.

2. Methods of Optical Digital Processing

In general, a digital arithmetic operation can be realized in two ways: using logic gates or using truth-table look-up techniques. In the first method, the operation is performed by devices with specific input–output relationships that resemble some arithmetic or logic operations. In the second technique, the result of an operation is read from a memory in which the outputs for all possible input cases are previously stored.

a. Gate-Based Processing. To realize optical gates, several optical quantities such as amplitude, polarization, and spatial coordinates have been proposed for coding. If amplitude coding is used, various logic operations can be obtained by using transparencies that have controllable transmission. A spatial light modulator that is addressed by an electrical or optical signal can be used as a controlled transparency. Logic operations such as AND or OR can then be implemented by putting these transparencies in cascade or in parallel and applying the inputs as the control signals to these planes. If electro-optic materials are used, the polarization of a beam of light can be changed by applying a voltage. By this technique, optical gates for binary logic or residue arithmetic can be implemented. In the spatial coding technique, different values are represented as the presence of light at a particular set of spatial coordinates. This is particularly suitable for coding residue numbers.

b. Truth-Table Look-Up Processing. The use of holography for performing Boolean algebraic operations was first demonstrated by Preston. He recorded Fourier holograms that, upon reconstruction, produced specific phase shifts among the light waves associated with the individual input variables. At the input of this system, the binary states 0 and 1 are modeled as pinhole apertures that introduce phase shifts of 0 and 180° in the laser light, respectively. During processing, the light beams coming out of the input apertures experience specific phase shifts as they are diffracted by the hologram. As a result of the complex amplitude addition of the diffracted waves, the required logic outputs are obtained at the detector plane in the form of bright and dark spots representing 1s and 0s, respectively. Using this technique, Preston performed the logic operation IDENTITY and EXCLUSIVE OR on four pairs of input bits and described how the operation OR could be obtained. Given these, all other logic operations can be constructed.

Preston's idea of using holography for logic operations was expanded by Guest and Gaylord to truth-table look-up processing in their NAND-based processor. Figure 28 shows a simple example of an optical NAND-based processor that implements the multiplication of a pair of two-bit numbers. During the recording process, for each output bit, all the input bit combinations that produce a 1 in that output bit are stored as holograms in a photorefractive crystal such as $LiNbO_3$. During the processing step, the input bits are compared with all the stored reference patterns that correspond to each output bit. If a match is detected, the output bit is consid-

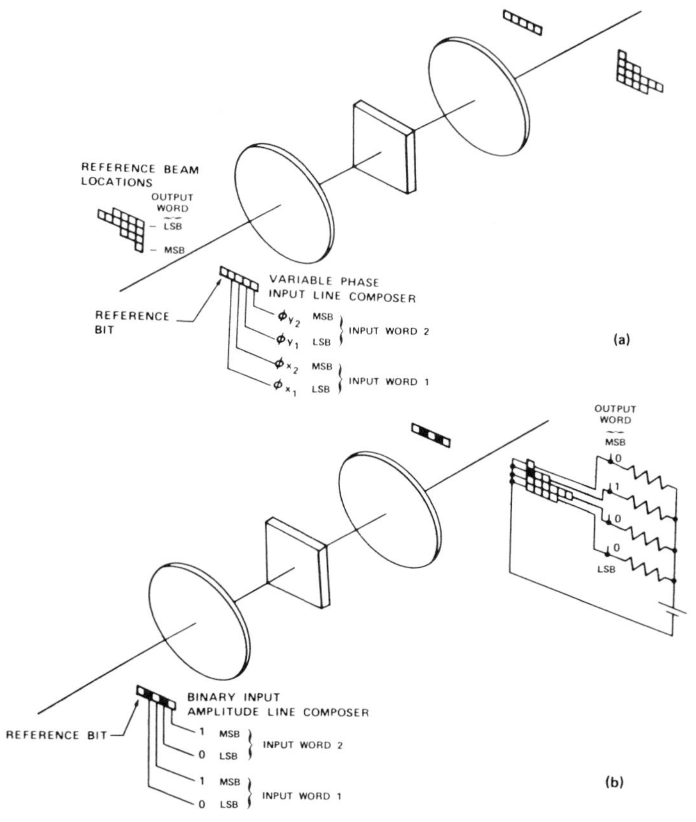

FIG. 28. NAND-based numerical optical processor: (a) recording the truth-table holograms and (b) example of multiplication with the processor, where LSB is the least significant bit; and MSB, most significant bit. [From Guest, C. C., and Gaylord, T. K. (1980). *Applied Optics* **19**(7), 1201–1207.]

ered to be a 1, otherwise it is considered to be a 0. A reference pattern consists of the bits of the two input numbers and an additional bit that is called the reference bit. For example, in the simple case of multiplication of a pair of two-bit numbers, there are four input combinations that produce a 1 in the least significant bit of the output. These are 0101, 0111, 1101, and 1111 which correspond to 01 × 01, 01 × 11, 11 × 01, and 11 × 11, respectively. Similarly, the number of input combinations that produce a 1 in the second, third, and fourth bits of the output are 6, 3, and 1, respectively. Therefore, a total of 4 + 6 + 3 + 1 = 14 reference patterns need to be stored, each consisting of five bits (four input bits and one reference bit). Each reference pattern is recorded as a hologram using a reference laser beam at a particular angle of incidence. The required phase shifts for the recording process are obtained by a phase-shifting line composer.

To use the NAND-based processor, the input bits are coded as transparent and opaque regions representing 1s and 0s, respectively. The location of the reference bit is always made transparent. The above coding can be implemented by a binary amplitude line composer. The light passing through the transparent apertures, upon diffraction by the holograms, reconstructs the reference beam at different spatial locations. Depending on the phase of the recorded bits, these diffracted beams are added to or subtracted from each other. For each output bit, if the input pattern matches one of the reference patterns, the diffracted beams cancel each other at the detector array at the location that corresponds to the matched pattern. As a result, a dark spot is produced at that location. The presence of a dark spot is electronically detected, and it is interpreted as a 1 for that particular output bit.

The NAND-based processor described above

has several advantages. Input patterns can be entered as two-dimensional arrays, and they can access the whole memory simultaneously. Therefore, full parallel processing can be achieved. Another advantage of this processor is its flexibility. Any function that can be represented as a truth table could be implemented by this method. Finally, the device is insensitive to the variations in the amplitude and phase of the laser during the processing step. The only limiting problem is the number of reference patterns that can be holographically recorded. The RNS is very useful in decreasing the required number of reference patterns for a particular operation. Further reduction can be obtained if logical minimization techniques, and multilevel coding techniques are applied. With these reduction techniques, operations of practical interest, such as 32-bit addition and multiplication, can be implemented by this processor.

3. Current Status of Optical Digital Computing

Various methods have been proposed for optical digital computing. However, the experiments that have been performed so far have been limited to simple prototype devices. This is mainly due to the problems that exist in the fabrication of the required components, such as spatial light modulators and integrated optical devices. There is no doubt that the demand for larger capacity computational devices will grow and that optical systems that use full parallel processing are promising candidates to obtain increased computing powers.

To realize optical digital computers, the needed research can be divided into two areas: new computer architectures and progress in the technology of optical devices. As mentioned before, optical systems benefit from parallel processing capability. However, it is not completely known how to use this power. Most of the hardware and software developed for electronic digital computers are based on sequential processing. To take full advantage of optical processing, hardware that can perform parallel processing should be developed, and new software to utilize this parallelism is needed. The progress in fabrication of optical devices is also essential. Among the devices that need to be improved are spatial light modulators, integrated optical devices, and optical bistable devices.

VII. Applications

The applications of optical information processing are so diverse that even brief descriptions of them are not possible here. Some of the areas in which optical processors have been used are antenna pattern analysis, crystal structure analysis, image enhancement, matched filtering, metrology, missile guidance, nondestructive testing, ocean surface analysis, pattern recognition, signal processing, and synthetic aperture radar. Here, we briefly cover two application areas in which optical processing techniques have made significant contributions in their development: synthetic aperture radar and character recognition.

A. Synthetic Aperture Radar

In conventional radar, an electromagnetic pulse is transmitted through the air from a directive antenna. If an airplane is present in the path, the wave is reflected from it and is received by the same antenna. From the time delay between the received pulse and the transmitted pulse, the distance between the antenna and the target can be calculated. Also, from similar measurements at different instants and along different directions, the shape of the target and its velocity can be determined. Theoretically, if the antenna is carried by an airplane, the same method can be used to get information about a target on the ground or to obtain a map of the terrain. However, for high-resolution radar imagery, impractically large antennas are required.

The geometry of an airborne side-looking radar is shown in Fig. 29. It is assumed that the airplane moves with a constant velocity along the x axis. Since the slant-range is determined from the delay time between the transmitted and the received signals, the slant-range resolution can be improved by using shorter pulses. Mathematically, the ground-range resolution corresponding to the slant-range resolution can be expressed as

$$\Delta l = c\tau/(2 \cos \psi) \tag{74}$$

where c is the velocity of light, τ the pulse duration, and ψ the depression angle of the line of sight to the target with respect to the local horizontal axis. The factor of two in the denominator is due to the fact that the electromagnetic wave travels the same path twice. For a pulse of 1-nsec duration, the maximum ground-range resolution corresponding to $\psi = 0°$ is 15 cm. The

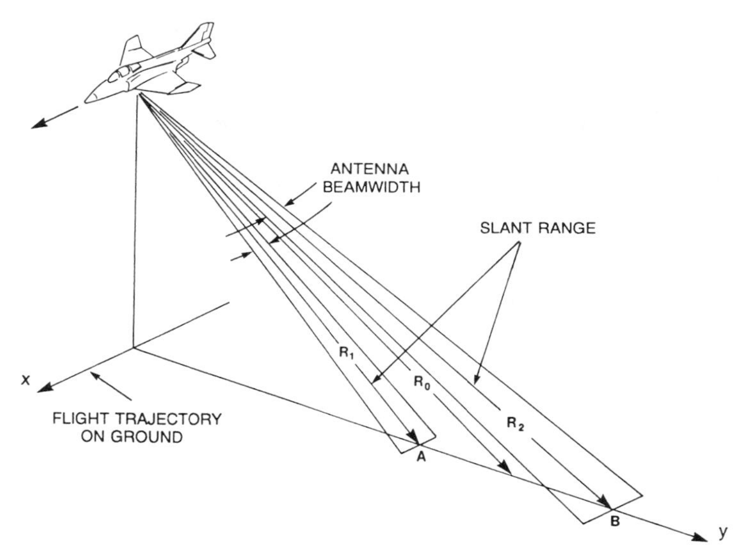

FIG. 29. Synthetic-aperture-radar geometry. [From Preston, Kendall, Jr. (1972). "Coherent Optical Computers," McGraw-Hill, New York.]

along-track or azimuthal resolution depends on the antenna pattern. To get high azimuthal resolution, the half-power angular width of the antenna pattern β should be narrow. This quantity depends on the wavelength of the electromagnetic wave λ and the largest dimension of the aperture of the antenna D. A good approximation is

$$\beta \simeq \lambda/D \tag{75}$$

where λ and D are expressed in the same unit of length and β is in radians. As an illustrative example, if the airplane is 10 km from the target, an azimuthal resolution of 15 cm is obtained with a half-power angular width of $\beta = 0.15/10^4 = 1.5 \times 10^{-5}$ radians. For a wavelength of 3 cm, the required dimension of the aperture antenna is $D = 3 \times 10^{-2}$ m$/1.5 \times 10^{-5} = 2000$ m, which is obviously impractical.

The synthetic-aperture technique was developed in the 1950s to improve the azimuthal resolution of an airborne radar. The idea is to synthesize the effect of a large antenna from the data obtained with a small antenna. To get some intuition about this technique, notice that in Fig. 29, a point at range R_0 is illuminated by the pattern of the antenna while the antenna moves a distance of $L = \beta R_0$. The physical antenna, which has an aperture of dimension D, can be considered as an element of a large antenna of length L. If the pulse rate of the radar is such that the antenna does not move more than D during the period between the succeeding pulses, the response of the long antenna can be determined from the gathered data. This is actually a trade-off between time and space. With this technique, an along-track resolution of D can be achieved by processing the gathered data. The process consists of amplitude weighting, phase-shifting, and adding the results. Also, to realize the ultimate along-track resolution for large values of L, it is required to compensate for the curvature of the reflected wavefronts. This focusing operation is range dependent, and it complicates the process. The amount of information that needs to be stored and the number of computations that are required are so large that real-time analysis could not be achieved even with the most powerful electronic computers. Various methods for storing and analyzing the data were proposed during the early 1950s. Among them, the optical processing approach was found to be the most suitable technique. Finally, in 1957, Cutrona and his colleagues at the University of Michigan successfully demonstrated a synthetic aperture radar using a coherent optical data processor. Their system is briefly described here.

The radar transmits short pulses of microwave energy of duration τ and rate ν_s, as shown in Fig. 30. The delay relation between the outgoing and the incoming pulses is shown in this figure. The incoming signal is demodulated in the receiver using a reference oscillator similar to

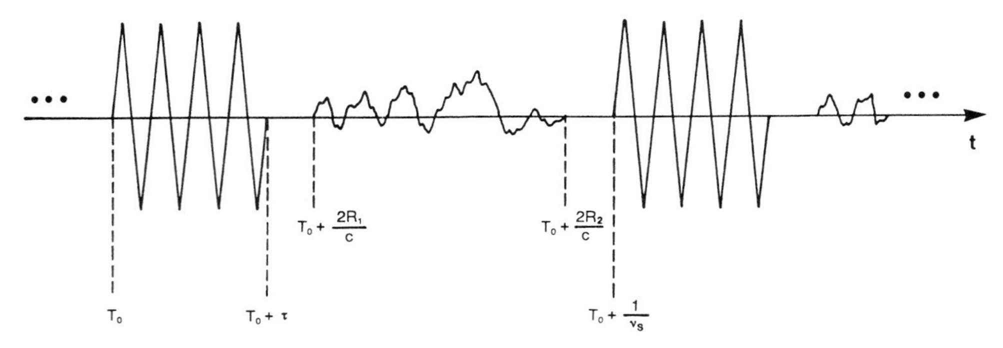

FIG. 30. Outgoing and incoming side-looking radar pulses, where R_1 and R_2 represent the nearest and farthest slant range probed; τ, pulse duration; and ν_s, pulse rate. [From Brown, W. M., and Porcello, L. J. (1969). *IEEE Spectrum* **6,** 52–62.]

the one used for transmission. Since the demodulated signal is bipolar, a dc bias level is added to the signal to make it monopolar. This positive signal is then used to modulate an electron beam created by a cathode-ray tube while the beam is swept along a straight line as shown in Fig. 31. Each sweep corresponds to one return signal and is recorded along a vertical line on a photographic film. The film moves along the horizontal direction, so that different return signals are recorded at different lines. The variable x_1 is therefore a measure of the along-track position while y_1 corresponds to the slant range. The exposed film is developed such that after process, its amplitude transmittance is a linear function of the amplitude of the demodulated signal. This film is then used as the input to the optical system shown in Fig. 32. In this system, a conical lens is used to insert a range-dependent quadratic phase delay to perform focusing operation. The cylindrical and spherical lens combination takes the Fourier transform of the focused signal along the x direction and images the input along the y direction. The output can be observed through a vertical slit. If a photographic film is placed behind the slit, and it is moved with the same speed as the data film, a high-resolution two-dimensional picture of the terrain illuminated by the radar is obtained.

B. Character Recognition

One of the applications of optical information processing that clearly demonstrates the parallel processing capability of optical systems is character recognition. This is a special case of a broader class of processes known as pattern recognition. In pattern recognition, the locations at which a particular pattern appears in an input data page are determined. In character recognition, it is desired to locate a particular character or word in an input text page. To perform this search, one technique is to use matched filters. From communication theory, if the impulse response of a system is denoted by $\mathbf{h}(x)$, a filter whose impulse response is $\mathbf{h}^*(-x)$, where * represents the complex conjugate, is known as the corresponding matched filter. Matched filters

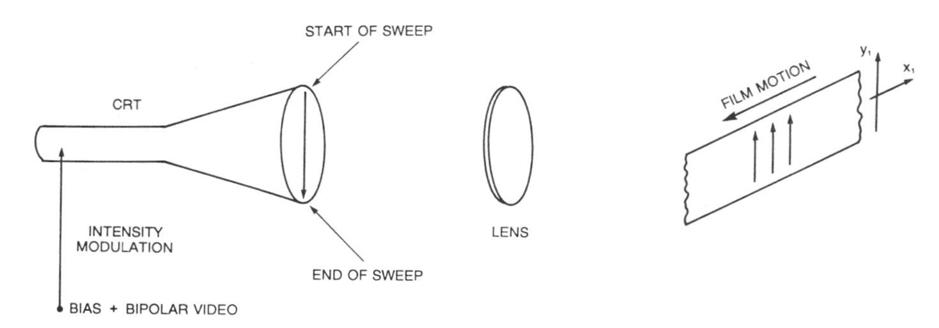

FIG. 31. Recording process in synthetic aperture radar. [From Brown, W. M., and Porcello, L. J. (1969). *IEEE Spectrum* **6,** 52–62.]

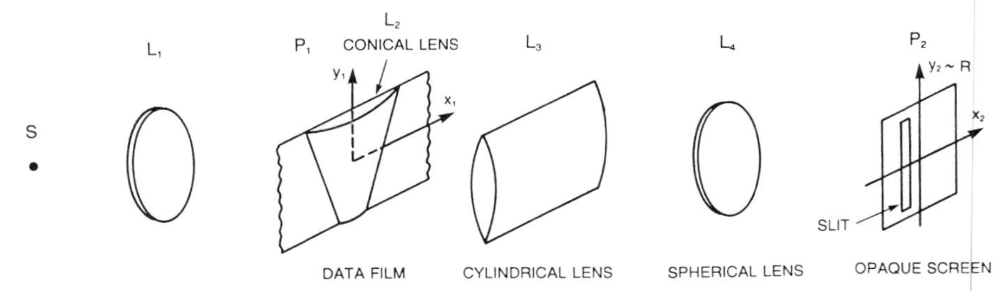

FIG. 32. Optical processor for synthetic aperture radar analysis. [From Brown, W. M., and Porcello, L. J. (1969). *IEEE Spectrum* **6,** 52–62.]

are used for signal detection. It can be shown that if a signal is buried in a noise that has a uniform spectrum (white noise), the matched filter corresponding to that signal provides the best detection. In character recognition, the signal of interest is a particular character, and it is two-dimensional. The unwanted characters are considered to be noise.

An optical system capable of character recognition can be implemented using a Vander Lugt filter. First, a Fourier hologram of the character or word of interest is recorded using the system shown in Fig. 13a. The transmittance of this hologram has the desired term $\mathbf{H}^*(f_X, f_Y)$, which is the Fourier transform of $\mathbf{h}^*(-x, -y)$. The recorded hologram is used in the Fourier plane of Fig. 12. The input text is placed at the P_1 plane, and the output is detected at the P_3 plane. As described in Section III,B,3, the output of a Vander Lugt filter has four terms. In character recognition, the desired term is the cross-correlation of the input text with the character of interest. In the output plane, at the locations that the desired character appears, the cross-correlation becomes an autocorrelation, and bright spots are produced; at the other parts of the page, the light is diffracted, and these regions are less bright. Example of the impulse response and the output of the processor are shown in Fig. 33. The advantage of optical processing for

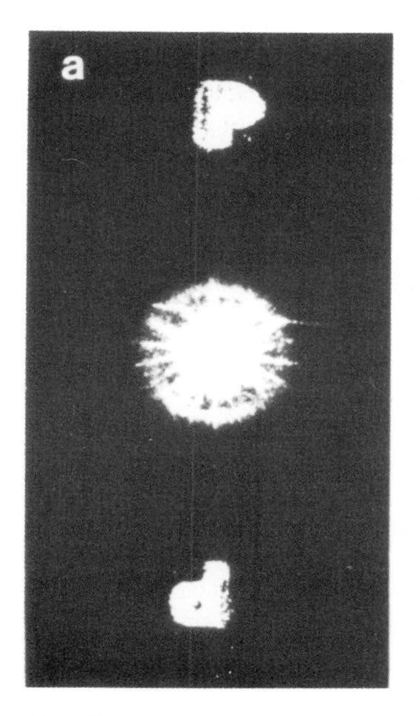

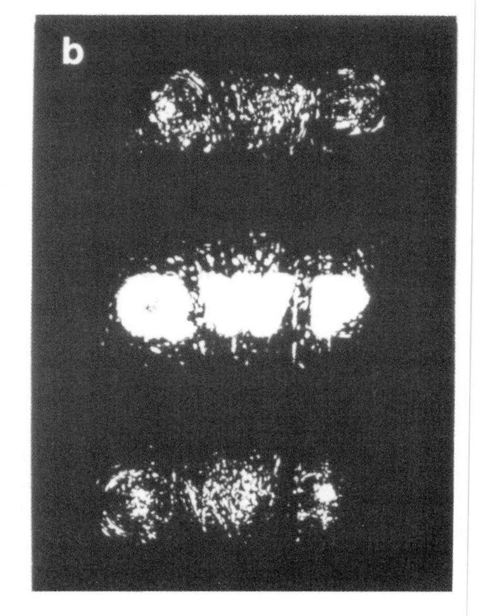

FIG. 33. Photographs of (a) the impulse response of a Vander Lugt filter and (b) the response of the matched-filter portion of the output to letters Q, W, and P. [From Goodman, Joseph W. (1968). "Introduction to Fourier Optics," McGraw-Hill, New York.]

character recognition is evident from the fact that the search in the input data is performed fully in parallel.

BIBLIOGRAPHY

Arrathoon, Raymond, ed. (1986). *Optical Engineering,* special issue on Digital Optical Computing, **25,** (1), January.

Bell, Trudy E. (1986). Optical computing: A field in flux, *IEEE Spectrum* **23,** 34–57, August.

Brown W. M., and Porcello, L. J. (1969). An introduction to synthetic aperture radar, *IEEE Spectrum* **6,** 52–62, September.

Casasent, D., ed. (1978). "Optical Data Processing Applications." Springer-Verlag, New York.

Caulfield, H. J., and Athale, R. A., eds. (1986). *Applied Optics,* special issues on Optical Computing, **25** (10 and 14), May 15 and July 15.

Gibbs, Hyatt M. (1985). "Optical Bistability: Controlling Light with Light." Academic, New York.

Goodman, Joseph W. (1968). "Introduction to Fourier Optics." McGraw-Hill, New York.

Jenkins, F. A., and White, H. E. (1976). "Fundamentals of Optics." McGraw-Hill, New York.

Lee, S. H., ed. (1981). "Optical Information Processing Fundamentals." Springer-Verlag, New York.

Preston, Kendall, Jr. (1972). "Coherent Optical Computers." McGraw-Hill, New York.

Stark, Henry, ed. (1982). "Applications of Optical Fourier Transforms." Academic, New York.

Szabo, N. S., and Tanaka, R. I. (1967). "Residue Arithmetic and Its Applications to Computer Technology." McGraw-Hill, New York.

OPTICAL PHASE CONJUGATION

Robert A. Fisher *R. A. Fisher Associates*

Glossary

Degenerate Kerr-like four-wave mixing: Coupling of three waves to produce a fourth in a medium exhibiting the optical Kerr effect.

Nonlinear optics: Study of the variety of new phenomena that may occur when light passing through a material is made to be very bright.

Optical Kerr effect: Nonlinear optical effect in which the variation in speed of light in a medium is proportional to the intensity of the light wave.

Optical phase conjugation: Use of nonlinear optical effects to produce a phase-conjugate wave. The phase-conjugate wave precisely retraces the path of the incoming beam, even in the presence of arbitrary distorters.

Photorefractive effect: Effect in ferroelectric crystals in which light liberates charge carriers, causing them to concentrate in the darker regions. The resultant voltage modifies the refractive index of the material.

Piston error: Change in the optical path length of an optical system. In double-pass, this change could be compensated by simple translation (pistonlike motion) of the turnaround mirror.

Stimulated Brillouin scattering: Coupling of two light waves in the presence of a sound wave. In backward stimulated Brillouin scattering, the two light waves are counterpropagating.

Optical phase conjugation is a technique that incorporates nonlinear optical effects to reverse precisely the direction of propagation and the overall phase factor for each plane wave in an arbitrary beam of laser light. The process can be regarded as a kind of mirror with very unusual image-transformation properties. A beam reflected by a phase conjugator retraces its original path. This remarkable autoretracing property suggests a wide variety of applications to problems associated with maintaining high optical quality after passing light beams through unavoidably nonuniform or distorting media, to automatic pointing and tracking, and to many other problems.

I. General Picture of Optical Phase Conjugation

As illustrated in Fig. 1, optical phase conjugation produces a reflection that is markedly different from the reflection produced by a conventional mirror. A conventional plane mirror (Fig. 1a) changes the sign of the **k**-vector component normal to the mirror surface while leaving the tangential component unchanged. An incoming light ray can thus be redirected arbitrarily by suitably tilting the mirror. On the other hand, the phase conjugator (Fig. 1b) causes an inversion of the vector quantity **k**, so that the incident ray exactly returns upon itself, independent of the orientation of the conjugator. A simple extension of Fig. 1 indicates that an incident diverging beam would be conjugated to become a converging beam and that an incident converging beam would be conjugated to become a diverging beam.

To understand how optical phase conjugation can be used to "undo" unavoidable distortions,

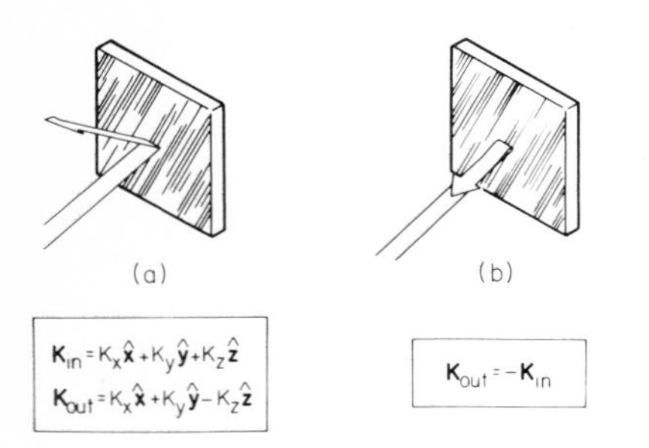

FIG. 1. Comparison of (a) a mirror reflection and (b) a phase-conjugate reflection. The mirror reflection reverses the **k**-vector component normal to its surface, whereas the phase conjugator reverses the vector quantity **k**. Tipping the conventional mirror changes the reflected direction, whereas tipping the phase-conjugating mirror will not. [After Bigio, I. J., *et al.* (1978). In *Proc. Int. Conf. Lasers*, p. 532. STS Press, McLean, Virginia.]

compare the two arrangements depicted in Fig. 2. Here we consider a transmitting prism as the simplest of the class of irregular distorters. In Fig. 2a, a light ray is returned by a phase-conjugate reflector, and the prism is about to be put into the beam. In Fig. 2b, the prism has been pushed up into the beam, thereby slightly redirecting the beam downward on its way to the conjugator. Note that the conjugate beam returns upward to the same spot on the prism and that the return beam is subsequently deflected by the prism to continue retracing the path of the incoming beam. The result is that the properties of the beam returning to the source are exactly the same whether or not the prism is in the path. The conjugator has returned the deflected light to exactly the point of deflection on the prism, and the prism then bends the return light so that it is indeed returning exactly toward the source. The deflecting prism can be of arbitrary orientation and arbitrary angle of deflection as long as all the deflected light is captured and returned by the conjugator.

One can now get a feeling of how this correction works for an arbitrary distorter by treating the distorter as an array of randomly oriented wedges. Naturally, the passage of any high-quality optical beam through any distorter causes a disturbed beam to be produced. If this disturbed beam is reflected by a phase conjugator, the return beam will have impressed upon it all the information necessary so that, as the conjugate beam is returned through the distorter, every disturbance that had been put on the beam in the forward passage is removed from the beam in the backward passage, leaving the incoming beam and the outgoing beam unaffected by the presence of the distorter.

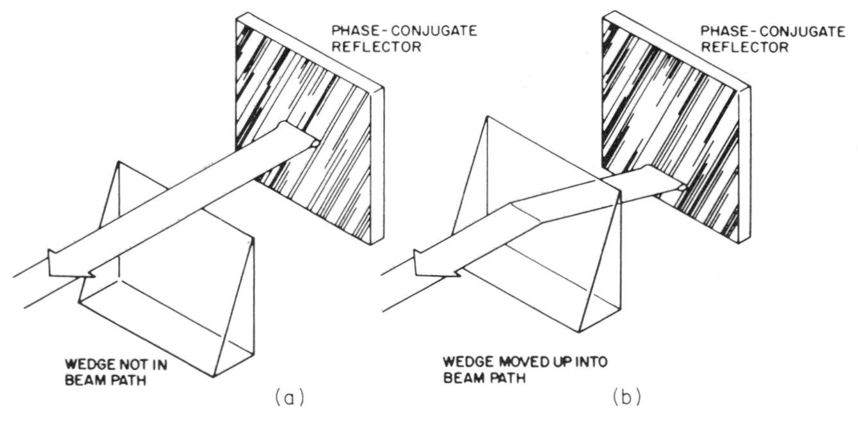

FIG. 2. Pictorial demonstration of simplest correction. In (a), a prism is about to be placed between the conjugator and the light source. In (b), the prism has been moved into the beam, thereby deflecting the incoming beam downward. The conjugate beam is redirected to hit the same spot on the prism and is then deflected by the prism to continue on its route directly back to the light source. After the returning beam has passed through the prism, it is identical to the return beam that would be there if the prism had been absent. Thus the properties of the returning beam are not affected by the presence of the prism as long as the conjugator is large enough to intercept all of the deflected light. [After Yariv, A., and Fisher, R. A. (1983). *In* "Optical Phase Conjugation" (R. A. Fisher, ed.), Chap. 1. Academic Press, New York.]

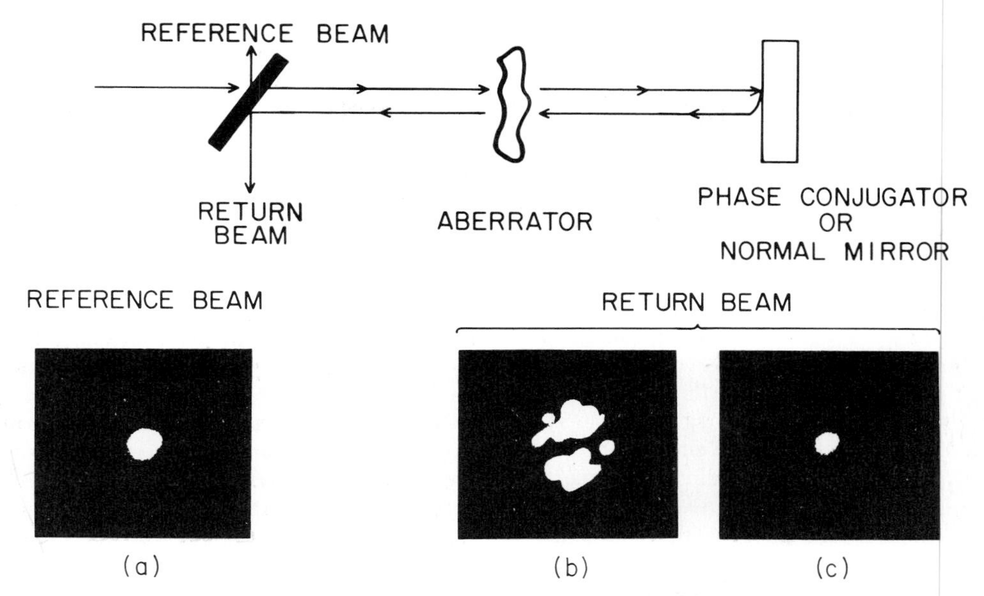

FIG. 3. Experimental demonstration of phase-conjugate reflection: (a) An undistorted laser beam is double-passed through an aberrating material. (b) Conventional reflection for the return trip results in a highly distorted beam, whereas (c) phase-conjugate reflection removes the distortions and only a uniform intensity change is obvious.

Figure 3 is an example of this distortion-undoing property of the optical phase-conjugation process. In this figure, a high-quality beam is brought through a beamsplitter and is subsequently brought through a test aberrator, which for infrared light might be a water-treated salt window or for visible might be an irregular piece of glass. As the beam propagates to the right in the figure, it is then reflected by either a phase conjugator or a normal mirror to return through the aberrator and to be examined as it is deflected downward from the beamsplitter. The spots in the bottom half of the figure are photographs taken of the spatial pattern of input and output beams. The reference beam, as can be seen in (a), is a small compact beam. The return beam is shown for the two cases: (b) the return beam is redirected through the distorter by a conventional mirror and (c) the return beam is redirected through the distorter by a phase conjugator. The beam returned by the conventional mirror is a badly distorted beam because the disturbances put on by the aberrator are doubled after the mirror-reflected beam is returned through the aberrator. However, the beam returned by a phase conjugator has had all its disturbances repaired so that once again it becomes a compact beam of light. This means that when an aberrator disturbs a good beam or a beam with spatial information on it, the beam becomes "bad." The phase conjugator prepares from that bad beam another beam and sends it back to the aberrator. This beam we may call "anti-bad" in the sense that although it has poor optical quality, it has programmed on it all of the information necessary so that when the anti-bad beam strikes the aberrator in the backward passage, all of the influence of the disturbances is undone, restoring the distorted beam to a good beam. This means that a phase-conjugate reflector can allow a high-quality beam to be double-passed through a poor optical system without any loss in optical beam quality. [*See* OPTICAL SYSTEMS DESIGN.]

Let us now look at Fig. 4 to see how such a distortion-undoing property could be utilized. In the case of Fig. 4, we have depicted a large laser gain medium that unfortunately has imperfections in it. These imperfections may be a form of optical machining errors in windows, mirrors, lenses, telescopes, and so on, or they may arise from unavoidable disturbances in the gain medium. An example of these unavoidable disturbances might be nonuniformities of temperature or pressure in the gain medium or sound waves or shock waves that may be propagating throughout the gain medium. Nevertheless, one can use a phase conjugator to allow the efficient

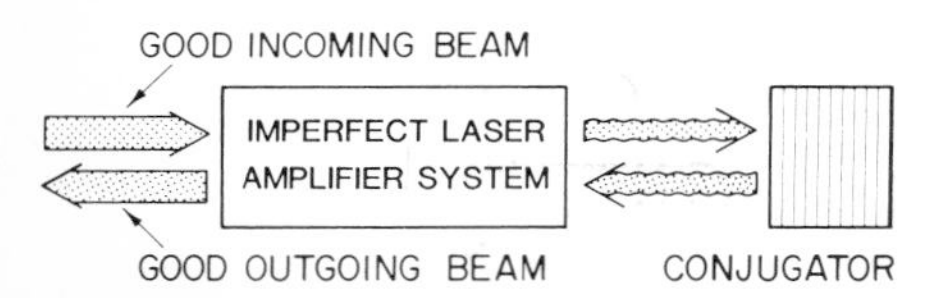

FIG. 4. Pictorial example of the use of optical phase conjugation to double-pass an irregular amplifier medium without loss in beam quality. The distortions that are put on the beam on the first pass are removed on the second pass.

extraction of energy in this stored gain medium without losing beam quality. In Fig. 4 a high-quality beam is brought in through the laser system. It is amplified in its first, or forward, passage through the laser gain medium. However, it becomes badly distorted because of unavoidable nonuniformities in the system. The phase conjugator takes the amplified beam and turns it around, preparing it to undo all the disturbances that had been encountered in the forward passage of the beam. When the backward propagating beam returns through the laser system, all the phase disturbances in the gain medium now straighten out the conjugate beam, allowing a high-quality outgoing beam to emerge from this poor optical quality double-pass amplifier system. This appears to be an extremely useful application for optical phase conjugation.

In addition to the aberration-correcting properties, phase conjugators can be used for automatic pointing and tracking. To visualize pointing applications, consider looking into a phase-conjugating mirror: You would not see your face in the mirror, because any light emanating from a particular point on your face (the chin, for example) would be returned by the conjugator to your chin, not to your eye. The only light seen would be that which had struck the conjugator after emanation as a diffuse reflection of room light scattered from the cornea covering the pupil of either eye. If you increased the illumination of one eye (perhaps with a flashlight), the entire conjugator would appear, to that eye only, to become relatively brighter.

This example shows the pointing and tracking application. A conjugator returns to a target any glint of light that the conjugator can intercept. Should an amplifier be put in front of the conjugator, then a doubly amplified return is redirected at the target. This procedure is remarkably simpler than the conventional means of aiming a laser at a target. Figure 5 compares the two techniques. Part (a) shows a conventional laser system in which an oscillator pulse is intensified in a chain of amplifiers and then directed to a target by an aiming and focusing system (such as a telescope), which is pictorially represented by a simple lens in the figure. The aiming and focusing telescope must be directed to the necessary accuracy by information acquired from a tracking system. This can become a computationally cumbersome task, especially for targets that are apparent only briefly. Part (b) shows the phase-conjugate approach. Light from a low-intensity illumination laser is directed at a target. This illumination beam can be spatially broad and need not be critically aligned. Some of the scattered radiation is gathered by a focusing system and undergoes amplification as it travels through the laser amplifiers. At the far end of the amplifier chain, the radiation is returned by a phase conjugator through the laser chain for further amplification to exceedingly high intensities. Regardless of the optical distortions encountered in the first pass, the phase conjugator automatically redirects the beam back to its source, the target. The amplified beam cannot miss! This technique allows the use of lower-quality optics and eliminates much of the expense of the tracking systems that are usually required.

Such an arrangement also allows for automatic correction of atmospheric distortion between the laser and the target. Thus the use of phase conjugation is an alternative to the conventional adaptive optics techniques for aiming a powerful laser at a small distant target. Clearly, such a simple automatic pointing, tracking, and correcting procedure pertains only to targets whose angular velocity (as seen from the vantage point of the laser) is not so great that the target moves laterally by its own length in the time that it would take light to make a target–conjugator round trip. This would preclude some long distance applications. Similarly, the intervening disturbances cannot vary significantly in the time that it would take to make a distorter–conjugator round trip. Phase conjugation can also correct for random variations in birefringence, a correction that cannot be performed by adaptive optics methods.

Although useful for an introductory description, the simple ray pictures in Figs. 1 and 2 do not completely specify the conjugation process. In addition to reversing each **k** vector, the device complex-conjugates the overall multiplica-

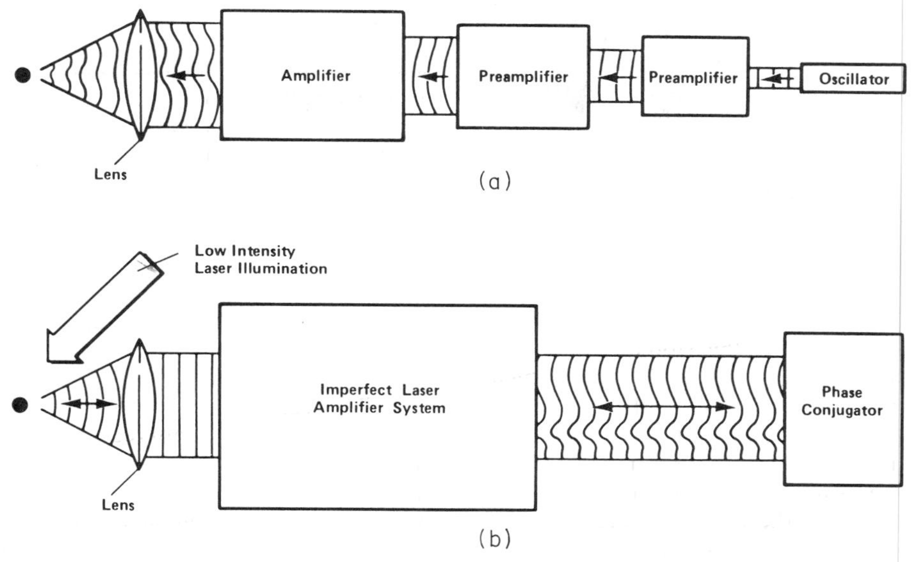

FIG. 5. Comparison of the conventional and phase-conjugate approaches to aiming a powerful laser at a target. In the conventional system (a), one carefully aims the output of the laser systems at the target. The conventional system uses a long chain of laser amplifiers that may gradually introduce distortions in the beam arriving at the fusion target. In the phase-conjugate laser-fusion system (b), a spatially broad, low-intensity laser illuminates the target. A small fraction of this illumination is reflected off the target into the solid angle of the focusing optics and is amplified, phase-conjugate reflected, and further amplified on its return. Because the phase-conjugate beam exactly retraces its path, the amplified beam automatically hits the tiny fusion target. In addition, any phase distortions imparted to the beam by the laser amplifier are removed on the return pass. [From Feldman, B. J., *et al.* (1982). *Los Alamos Sci.* (Fall).]

tive electric field amplitude associated with each plane wave in the beam. The distortion correction property of a phase conjugator requires both of these properties. The phase-conjugate process is formally defined as follows: If an incoming light beam (denoted E_p, where the p stands for probe) is written as a superposition of plane waves, then

$$E_p = \frac{1}{2}\sum_{j=1}^{m} \mathscr{E}_j \exp[-i(\omega t - \mathbf{k}_j \cdot \mathbf{r})] + \text{c.c.} \quad (1)$$

This equation could represent a uniform plane wave (in which there would be only one term in the sum), a TEM_{00} laser mode, a beam with spatial information impressed upon it, or a badly aberrated beam. In any case, E, the phase conjugate of the wave represented by Eq. (1), is given by

$$E_c = \frac{1}{2}\sum_{j=1}^{m} \mathscr{E}_j^* \exp[-i(\omega t + \mathbf{k}_j \cdot \mathbf{r})] + \text{c.c.} \quad (2)$$

Note that the conjugation process involves two changes. Each complex, slowly varying envelope function $\mathscr{E}_i$ is converted into $\mathscr{E}_i^*$, and each $\mathbf{k}_i$ is converted into $-\mathbf{k}_i$. These changes (which differ from those associated with conventional mirror reflections) lead to unusual image transformation properties and to many practical applications for optical phase conjugators.

II. Distortion-Undoing Properties of a Phase-Conjugated Wave

We now present a mathematical proof of the precise distortion-undoing nature of the conjugation process. This proof closely follows that of Yariv. Consider a monochromatic wave propagating through a linear but distorting medium whose arbitrary electric permittivity (dielectric constant) is given by the arbitrary function $\varepsilon(r)$. The real quantity $\varepsilon(r)$ represents the presence in the beam of passive linear components, such as lenses and wedges, or the presence of distorting media, such as turbulent atmosphere. We assume that all components (lenses, distorters, etc.) are lossless, so that no light misses striking the conjugator. The scalar near-forward ($+z$) propagating beam is taken as

$$E_1(r, t) = \tfrac{1}{2}\mathscr{E}_1(r) \exp[-i(\omega t - kz)] + \text{c.c.} \quad (3)$$

In the limit of slow spatial and temporal variations, the scalar wave equation obeyed by Eq. (3) is taken as

$$\nabla^2\mathscr{E}_1 + [\omega^2\mu\varepsilon(r) - k^2]\mathscr{E}_1 + 2ik\frac{\partial\mathscr{E}_1}{\partial z} = 0 \quad (4)$$

Let us next, as a purely mathematical operation, consider the complex conjugate of Eq. (4),

$$\nabla^2\mathscr{E}_1^* + [\omega^2\mu\varepsilon(r) - k^2]\mathscr{E}_1^* - 2ik\frac{\partial\mathscr{E}_1^*}{\partial z} = 0 \quad (5)$$

Equation (5) is the same wave equation but is applied to a wave propagating in the $-z$ direction of the form

$$E_2(r, t) = \tfrac{1}{2}\mathscr{E}_2 \exp[-i(\omega t + kz)] + \text{c.c.}$$

provided we put

$$\mathscr{E}_2(r) = a\mathscr{E}_1^*(r) \quad (6)$$

where a is any constant. Thus a backward-going wave $\mathscr{E}_2$ whose complex amplitude is everywhere the complex conjugate of $\mathscr{E}_1$ (within an arbitrary multiplicative constant) satisfies the same wave equation obeyed by $\mathscr{E}_1$. This means that if, after traversing a distorting medium with beam E_1, we can generate a beam E_2 that is the phase conjugate of E_1 (within a multiplicative constant), then E_2 will propagate backwards and its amplitude $\mathscr{E}_2$ will remain everywhere the complex conjugate of $\mathscr{E}_1$. Therefore, its wave fronts everywhere coincide with those of E_1.

The incident wave E_1 and the reflected conjugate wave E_2 have the same time dependence, namely, $\exp(-i\omega t)$. It is only the spatial part $\mathscr{E}_1(r)\exp(ikz)$ that is complex conjugated. However, the proof given holds only when $\varepsilon(r)$ is real. When the propagation medium is lossy or amplifying, the dielectric function $\varepsilon(r)$ becomes complex, and the proof does not apply. The exception is the case where the loss or gain is independent of r. In this case the overall loss or gain behavior can be harmlessly factored out of the propagation equation, and good correction is guaranteed.

III. Interferometric Recombination of Phase-Conjugate Beams

The interferometric recombination of phase-conjugate waves is a very revealing experiment,

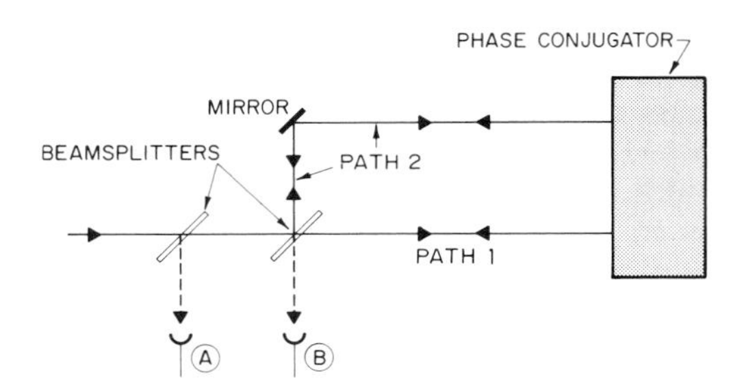

FIG. 6. Interferometric recombination of two separate phase-conjugation events. The right-traveling beam is converted by the beam splitter into two beams that are directed along paths 1 and 2 to the conjugator. The phase-reversal properties of the conjugator allow the return beams to recombine at the beam splitter with constructive interference in the left-going direction and with destructive interference in the down-going direction.

and it shows the remarkable propensity for phase-conjugate waves to return to their origins. Let us examine Fig. 6, in which a beam of light is brought in from the left. The beam first passes through a beamsplitter to allow examination of the backward-going conjugate signal. It is then split into two paths by the subsequent beamsplitter, and both the transmitted light and the split-off light are directed to the conjugator. Here the transmitted light goes through a path of length l_1, and the split-off light traverses a path of length l_2 to the conjugator. The setup is reminiscent of an unequal-path Michelson interferometer, but we shall see that the presence of the conjugators (instead of conventional mirrors) effectively makes this an equal-path interferometer. Along path 1, the transmitted beam, the light wave accumulates a phase of $\exp(ik \cdot l_1)$ in getting to the conjugator, which then complex conjugates the complex amplitude, converting the phase of the returning wave to $\exp(-ik \cdot l_1)$. This return wave then returns a length l_1 to the beamsplitter, being therefore multiplied by $\exp(ik \cdot l_1)$. Thus one takes the complex amplitude of the incoming wave and multiplies it by $\exp(0)$, or unity, to obtain the return wave. Thus the properties of the return wave do not contain any information about the path length l_1. By similar arguments, the split-off wave going its distance l_2 and returning has no information about the distance l_2. Here we have a situation where this unequal arm-length Michelson interferometer can be treated as if the arms were of equal length.

Now let us look at the possibility that return

light might be deflected downward to detector B when the beams try to recombine on the beam-splitter. Independently, each path (l_1 and l_2) would produce a downward-going component; but it turns out that, together, the downward-going contributions are equal and opposite, causing these two contributions to cancel. Therefore there is no output in this downward direction, and all the return light therefore travels to the left after recombining on the beam-splitter. This is a remarkable demonstration of the automatic ability of optical phase conjugation to compensate for variations in optical path difference (OPD), and this can allow the interferometric recombination of light beams that have double-passed a multiple-aperture optical system.

Later in this article, it will be indicated that some optical phase-conjugation techniques exhibit an overall arbitrariness in phase and therefore are not readily adaptable to the multiple aperture problem. In general, any form of optical phase conjugation that uses a pump for a reference is suitable, and those that are self-pumped or self-starting have overall phase ambiguities that prevent compensation for optical path differences and cause the consequent production of downward-going light from the recombining beam splitter of Fig. 6.

IV. Optical Phase-Conjugation Techniques

Since optical phase conjugation uses the nonlinear optical properties of a material to generate a conjugate wave from an incoming wave, we must first understand the origins of the various nonlinear optical effects. For the purposes of this article, the field of nonlinear optics is the study of what changes when light passing through a material becomes very intense. In addition to a material's linear optical properties (those used with mirrors, lenses, microscopes, cameras, projectors, etc.), a large complement of nonlinear effects comes into play at high intensity. If only linear optical effects were operative, then a collection of optical signals would pass through a linear system where each signal would have no effect on the others, and where no new signals would be generated. Nonlinear optical effects (such as Brillouin scattering, Kerr effects, saturation of resonances, photorefractive effects, etc.) allow for the coupling of waves and therefore, the possibility of generating new waves, one of which may be a phase-conjugate signal. [*See* NONLINEAR OPTICAL PROCESSES.]

Stimulated Brillouin scattering, for example, concerns the nonlinear coupling of two light waves in the presence of a sound wave. Here the nonlinearity is associated with the electrostrictive properties (the tendencies of materials to be squeezed into the regions where the optical fields are high) of liquids, solids, and gases. The optical Kerr effect concerns the tendency of cigar-shaped molecules in a liquid to align along the direction of an electric field, thereby increasing the polarizability where the light is intense. Saturable resonances can be absorbing or amplifying media that are driven so hard that their response is nonlinear, giving rise to many nonlinear phenomena including saturable absorption, photon echoes, optical nutation, free-induction decay, and grating echoes. Three-wave mixing concerns the coupling of three waves in a nonlinear crystal that lacks inversion symmetry. Nonlinear surface phenomena concern intensity-dependent (or energy-dependent) changes in absorption, reflectivity, or phase shift for a nonlinear surface. Examples include liquid crystal phenomena, free-electron generation on surfaces of semiconductors, electrostriction at a surface, thermal expansion at a surface, and other forms of optically induced surface modification. Photorefractive effects concern the optically induced migration of charge carriers in insulating ferroelectric materials; these charges produce voltages that modify the material's basic properties. Thermal grating phenomena occur in materials that are slightly absorbing and in which the index of refraction is a function of temperature so that temperature variations are converted into index variations.

A. Kerr-Like Degenerate Four-Wave Mixing

Kerr-like media are characterized by having a speed of light that depends linearly upon the intensity of the light. In contrast, a linear medium has a speed of light that is independent of intensity. Kerr-like materials are often called n_2-materials, because the index of refraction can be written as the sum of the linear index n_0 and the term n_2E^2, where E is the electric field associated with the light wave. The optical Kerr effect is responsible for self-focusing (the collapse of a beam upon itself because of the induced lens), self-phase modulation (the production of a chirp or frequency sweep on a time-varying pulse), and self-steepening (the production of optical shocks on pulses even in the total absence of linear dispersion).

Examples of Kerr-like media are liquids of cigar-shaped molecules and semiconductor materials. The cigar-shaped molecules in liquids are nonlinear because applied light fields induce a torque on each molecule, trying to orient the molecules so that their long axes are parallel to the applied electric field. In semiconductors, nonlinearity arises from many effects, including nonparabolicity of bands and Franz–Keldysh effects.

To use the optical Kerr effect to generate a phase-conjugate return, one sets up the arrangement depicted in Fig. 7. The arrows in the figure denote the directions of the propagation vectors. Here two strong counter-propagating (pump) beams with k-vectors $\mathbf{k}_1$ and $\mathbf{k}_2$ (at frequency ω) are directed to set up a standing wave in a clear material whose index of refraction varies linearly with intensity. This arrangement provides the conditions in which a third (probe) beam with k-vector $\mathbf{k}_3$, also at frequency ω, incident upon the material from any direction would result in a fourth beam with k-vector $\mathbf{k}_4$ being emitted in the sample, precisely retracing the third one. (The term degenerate indicates that all beams have exactly the same frequency.) In this case, phase matching (even in birefringent materials) is obtained independent of the angle between $\mathbf{k}_3$ and $\mathbf{k}_1$. The electric field of the conjugate wave $\mathbf{k}_4$ is given by

$$\mathscr{E}_4 = \mathscr{E}_3^* \tan\left(\frac{2\pi}{\lambda_0}\,\delta n\, l\right) \qquad (7)$$

Here δn is the index change induced by one strong counterpropagating pump wave, λ_0 the free-space optical wavelength, and l the length

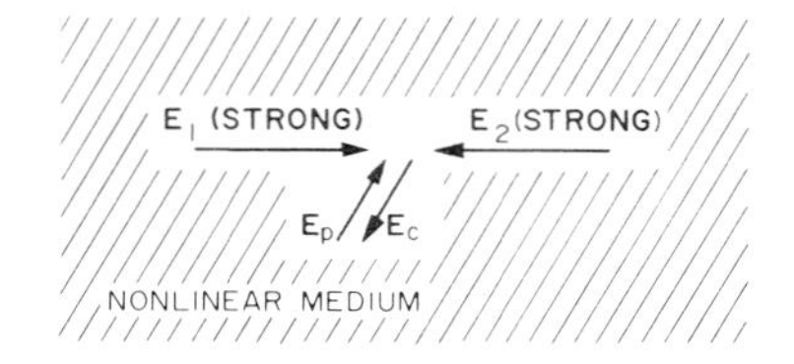

FIG. 7. Configuration for Kerr-like degenerate four-wave mixing. Here two strong counterpropagating pump waves (E_1 and E_2) are brought into a Kerr-like medium (one with an index of refraction that depends linearly upon intensity). The nonlinear interaction couples these two pump waves to any probe wave E_p, producing the radiation of the conjugate wave E_c.

over which the probe beam overlaps the conjugation region. The conjugate reflectivity is defined as the ratio of reflected and incident intensities, which is the square of the above tangent function. The essential feature of phase conjugation is that $\mathscr{E}_4$ is proportional to the complex conjugate of $\mathscr{E}_3$. Although degenerate four-wave mixing is a nonlinear optical effect, it is a linear function of the field one wishes to conjugate. This means that a superposition of $\mathscr{E}_3$s generates a corresponding superposition of $\mathscr{E}_4$s; therefore faithful image reconstruction is possible.

B. Conjugation Using Saturated Resonances

Instead of using a clear material, as outlined above, the same four-wave mixing beam geometry can be set up in an absorbing or amplifying medium partially (or totally) saturated by the pump waves. When the frequency of the light is equal to the resonance frequency of the transition, the induced disturbance corresponds to amplitude gratings that combine the four waves. When the single frequency of the four waves differs slightly from the resonance frequency of the transition, then both amplitude gratings and index gratings become involved. Because of the complex nature of the resonant saturation process, one does not obtain the simple $\tan^2[(2\pi/\lambda_0)\,\delta n\, l]$ expression for the conjugate reflectivity. Instead, this effect is maximized when the intensities of the pump waves are about equal to the intensity that saturates the transition.

C. Stimulated Brillouin Scattering

Earliest demonstrations of optical phase conjugation were performed in the Soviet Union by focusing an intense optical beam into a waveguide containing materials that exhibit stimulated Brillouin scattering (SBS). This effect involves an inelastic process whereby an intense laser beam in a clear material is backscattered by the production of sound waves or acoustic phonons. The production of the sound wave involves the electrostrictive effect, in which the Brillouin scattering medium experiences a force toward regions of high electric field. In the initiation phase of the SBS process, a bright laser beam is focused into an appropriate medium. Additionally, because of zero-point noise in the initial distributions of both sound and light, there is already in the medium the minute presence of appropriate sound wave and backscattered light wave. The coupled process responsible for generating stimulated backward Brillouin scattering can then be understood as a forward-going laser beam being Bragg-scattered off a sound wave to make backward-going light, which is frequency downshifted because the sound wave is moving. At the same time, the forward-going and backward-going light beams interfere to produce an intensity modulation or grating that is moving away from the laser at the sound velocity and beating the frequency of the sound wave. Through the electrostrictive effect this intensity modulation promotes the growth of the sound wave. This doubly coupled effect grows from noise and rapidly depletes the incoming laser beam, thus producing the backward beam and leaving a strong sound wave propagating in the medium. There are actually two backward-wave solutions: a phase-conjugate solution and a non-phase-conjugate solution. If there are sufficient phase disturbances in the beam and if the focused beam is not too far above threshold, then the backward gain for the phase-conjugate wave is twice the gain for the non-phase-conjugate wave, and the former dominates.

Stimulated Brillouin scattering for phase-conjugate double-pass of nonuniform amplifiers has many attractive features. First, because the stimulated Brillouin mirror is not reflective until the incident light is above the threshold intensity, such a system provides automatic optical isolation against unwanted parasitic oscillations. This prevents the buildup of weak light that would be reflected from conventional mirrors if such were used to double-pass the amplifier. Furthermore, the stimulated Brillouin process works alone; there is no requirement for intense high-quality pump waves (as there would be for Kerr-like degenerate four-wave mixing). Third, stimulated Brillouin scattering is normally done in simple liquids or gases that are self-healing in comparison with the solid materials often used in other phase-conjugation techniques. Such solids, once damaged, must be changed before the conjugator can be used again. A fourth attribute is that the energy-handling capability of stimulated Brillouin scattering seems to exceed that of the other common techniques.

For the above reasons, stimulated Brillouin scattering appears to be an attractive choice for controlling the output of large laser systems. As one example, let us assume that in generating the desired output energy of a laser–amplifier system, one finds limitations that prevent the gain medium from becoming arbitrarily large. For this reason, many large-gain media may be assembled in a multiple-aperture configuration to provide the necessary extractible optical energy. In such a multiple-aperture scenario, Fig. 8 identifies the anticipated use of stimulated Brillouin scattering for the coherent recombination of beams. The SBS cells conjugate the light and return it through each segment of gain medium for subsequent recombination. The recombination may take the form of a subsequent interferometer or the process may extend through a multiple-aperture telescope with the anticipated recombination done in the far field. This degree of simplicity offered by SBS is extremely advantageous in large systems. Unfortunately, the ability of SBS to compensate for variations in overall phase shift is not automatic. Random variations in overall phase shift through a double-pass optical system correspond to variations in optical path delay (OPD), and such errors are often called piston errors because a simple translation (pistonlike motion) of the double-pass mirror would provide correction. If the sound wave P_s and the conjugate wave E_c comprise a consistent solution to the SBS equations, then for arbitrary ϕ, the set $e^{-i\phi}P_s$ and $e^{i\phi}E_c$ would equally be solutions. Therefore each independent SBS event may impress an arbitrary overall phase ϕ on the return signal. With no control over the phase of input noise from which the scattered beam grows, simple SBS cannot compensate for piston error. This does not, of course, impair the ability of SBS to compensate for distortions.

As was shown in Fig. 6, one can obtain the interferometric comparison of two independent SBS events. A beam is split into two channels, each of which is subsequently brought to focus in a Brillouin scattering material. Should overall piston error (between the two independent arms) be compensated, then all of the return

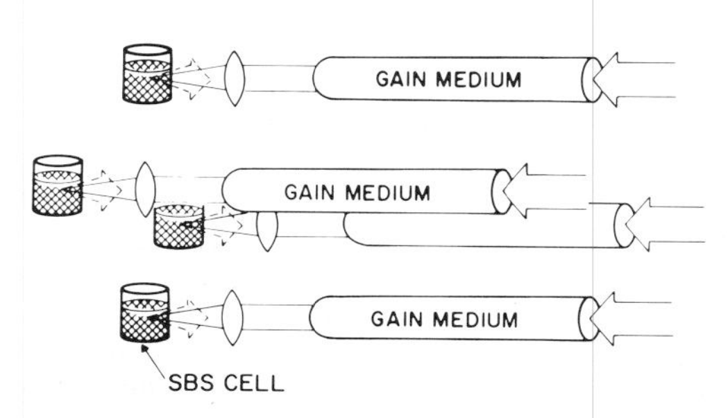

FIG. 8. Use of stimulated Brillouin scattering (SBS) for the phase-conjugate return through a multiple-aperture system.

light would interfere constructively in the "horizontal" path indicated in the figure and would interfere destructively in the undesired downward path.

Because of statistical fluctuations in the Brillouin scattering process, the two return beams start with an arbitrary relative phase and go in and out of phase with each other in a time that is approximately five times the phonon lifetime. Thus the return output oscillates between the desired path and the undesired downward path. The temporal variations of the phase are attributed to spectral line-narrowing considerations in the two independent Brillouin scattering events. For scattering of visible light, typical phonon lifetimes vary from a few nanoseconds to a few tens of nanoseconds, and therefore the constructive interference from two independent SBS events cannot be maintained in the desired channel for very long. This problem is exacerbated in the ultraviolet, where lifetimes of the participating phonons are even shorter by the square of the laser wavelength.

Two approaches have arisen that are likely to eliminate the phase fluctuations associated with SBS control of multiple-aperture systems. The first approach, popular in the Soviet Union, is to focus all of the beams into the same SBS interaction volume, the "common focus approach." Another approach that appears promising is the procedure of backward seeding the SBS process, so that the noise from which the signals grow is prescribed in phase. This involves injecting into the focal volume a backward-traveling SBS shifted signal that initiates the SBS process, thereby specifying the initial conditions.

D. Photorefractive Effects

Photorefractive effects have received a remarkable amount of attention in the last few years because of their applicability to low-power visible lasers. Studies have been performed on $Bi_{12}SiO_{20}$ (BSO), $Bi_{12}GeO_{20}$ (BGO), KTN, SBN, and most importantly, barium titanate. Although most nonlinear optical effects take megawatts of power to produce (SBS, Kerr-like degenerate four-wave mixing, etc.), photorefractive phenomena require only microwatts of power. The nonlinear behavior is spatially and temporally nonlocal, and the effects are independent of intensity; for weaker light, the response just takes more time to develop.

In general, photorefractors are very hard to understand. They are ferroelectric, photorefractive, photoconducting, Pockells, pyroelectric, piezoelectric, and photoacoustic. Typically, the materials are single-ferroelectric domain uniaxial perovskite structures in which the crystal polarization is in the direction of the c axis. To understand the basic nonlinear optical interaction in photorefractors, consider the curves depicted in Fig. 9. Here two waves of identical optical wavelength intersect in the crystal. The top curve denotes the optical intensity associated with the interference of the two beams. It is well-known that the presence of light releases charges in these otherwise insulating materials and that these charges can migrate until they are trapped. If, however, the light is in the form of an interference pattern, then the migrating

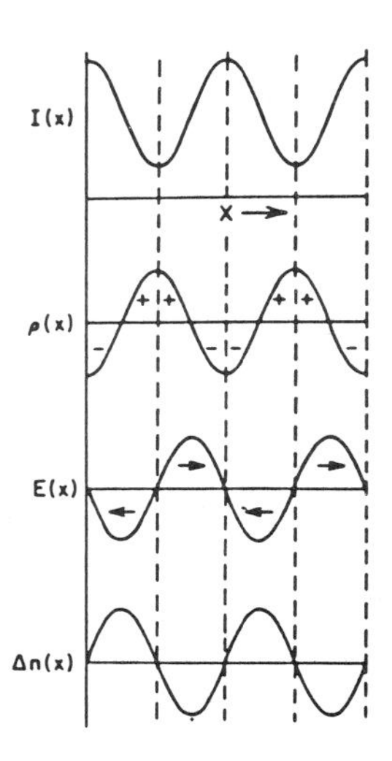

FIG. 9. Pictorial explanation of the formation of photorefractive gratings. The $I(x)$ curve shows the position dependence of an intensity interference pattern when two different plane waves intersect in the medium. The $\rho(x)$ curve shows the position dependence of the charge distribution induced by the light pattern. Note that the majority carriers accumulate in the valleys of the interference pattern, and that their vacancies are most prominent at the peak of the interference pattern. The $E(x)$ curve shows the induced electric field as a function of position. Note that the electric field has maxima and minima at points equidistant between the peaks and valleys of the above curve, and that by symmetry the field is zero at points of maximum or minimum in the charge distribution. The $\Delta n(x)$ curve shows the index change induced by the voltage distribution through the Pockels effect. Most important is that the first and last curves (interference lightwave pattern and induced index change) are out of phase from each other by 90°. [After Feinberg, J. (1983). *In* "Optical Phase Conjugation" (R. A. Fisher, ed.), Chap. 11. Academic Press, New York.]

charges that get trapped in the interference peaks are more likely to be re-released than would those that become trapped in the valleys. This process causes the charges to migrate from regions of high optical intensity to concentrate in the valleys of the interference pattern, as can be seen in the second curve of Fig. 9. The periodic charge separation sets up large periodic electric fields, as can be seen in the third curve of Fig. 9. Note that the peak of the electric field is halfway between the positive and negative charges and that the electric field is (by symmetry) zero at the points of maxima and minima of the charge distribution. For this reason, the electric field grating is 90° out of phase with respect to the initial interference pattern, and this shift leads to the very important principle of two-beam coupling. As a final step in the photorefractive process, the periodically varying electric field, through the Pockels effect, modulates the material's index of refraction. This can be seen in the last curve of Fig. 9, completing the evolution from interfering light beams to a refractive index grating.

The gratings made by the above process are responsible for degenerate four-wave mixing and two-beam coupling. Degenerate four-wave mixing causes the generation of a phase-conjugate beam when a probe beam impinges upon a photorefractor while it is simultaneously exposed to a pair of exactly counter-propagating waves (pump waves). The probe wave and one pump wave produce an optical interference pattern that, through the above mechanism, produces a refractive index grating. The other pump wave is appropriately oriented to Bragg-reflect off that grating to provide optical power into the phase-conjugate direction. Concurrently, the analogous process takes place in which the roles of the two pump waves are reversed. It is important to note that once a photorefractor is producing (by whatever manner) a phase-conjugate beam, then the region through which the probe beam enters and the phase-conjugate beam emerges contains a pair of counterpropagating waves (namely, the probe and the conjugate). This region is therefore a conjugator, and an oscillation can often develop between this region and an arbitrarily placed nearby mirror.

Two-beam coupling, on the other hand, is a unique manifestation of the nonzero phase shift between the optical grating and the index grating. When two beams intersect in a photorefractor, the index grating causes the scattering of light from the beam whose **k**-vector has a component antiparallel to the c axis, and the scattered light increases the intensity of the beam whose **k**-vector has a component parallel to the c axis. Figure 10 shows this two-beam coupling. In Fig. 10a, the beam directed more toward the c axis becomes intensified at the expense of the other beam. In Fig. 10b the beams do not intersect in the crystal, and no coupling is observed. In Fig. 10c the beams are once again made to intersect, but the crystal is rotated to reverse the direction of the c axis. Here, the process is reversed to again provide gain for the beam most directed toward the c axis. Two-beam coupling can lead to the production of beam fanning, in which rays are produced that are directed from the main beam in a direction that is more toward the c axis. Furthermore, the entire beam can bend towards the c axis as it propagates through the crystal.

We now examine the internally reflecting self-pumped conjugator, where a beam is brought in a face parallel to the c axis. As it propagates through the crystal, two-beam coupling allows the growth of auxiliary rays directed toward one corner. After two internal reflections, this light (the auxiliary ray) is returned to reintersect the main beam at another position on the main beam. Here, the two-beam coupling gratings produce light that propagates in the reverse direction along the auxiliary beam, and shortly, a significant amount of optical power is circulating in this auxiliary loop. The loop contains coun-

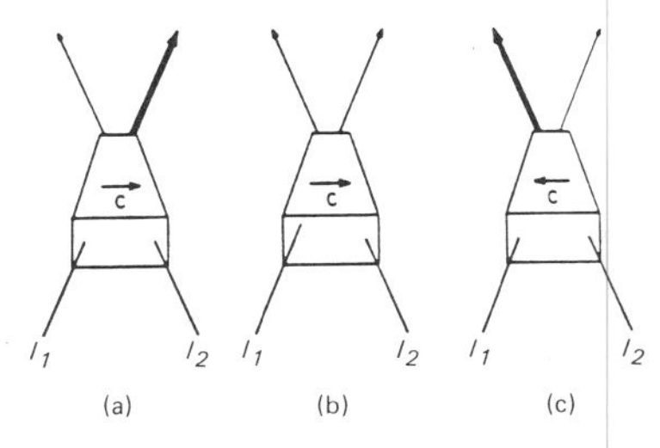

FIG. 10. Demonstration of two-beam coupling in a photorefractive crystal. In all cases the direction of the c axis is indicated. The spots above the figures are recorded laser beam profiles. In (a) the two beams intersect in the crystal, and the one directed more toward the c axis becomes more intense. In (b) the beams miss each other in the crystal, and they therefore do not affect each other. In (c) the crystal has been turned around, and when the beams intersect in the crystal, the power flow is seen to be the opposite of that in (a). [After Feinberg, J., Heiman, D., Tanguay, A. R., Jr., and Hellwarth, R. W. (1980). *J. Appl. Phys.* **51,** 1297.]

terpropagating waves; thus, where it intersects the main beam, the standard degenerate four-wave mixing considerations produce a conjugate of the incoming wave. The process eventually becomes so strong that very little of the incoming light is transmitted through the crystal. A sufficient number of auxiliary loops seem to evolve regardless of the precise orientation of the beam and regardless of the specific shape of the corner. Thus this device is a stand-alone conjugator that needs no pump waves; it builds them out of the main beam using two-beam coupling.

E. Surface Phase Conjugation

Here a strong pump wave is directed normal to a surface that exhibits nonlinear behavior, and the wave to be conjugated impinges at any arbitrary angle. Phenomena that can couple the light waves include electrostriction, heating, damage, phase changes, surface charge production, and liquid crystal effects. In general, the surface phase-conjugation process is equivalent to two-dimensional (thin) holography.

F. Thermal Scattering Effects

Here an optical interference pattern in a slightly absorbing material is converted to an index grating if the material has a temperature-dependent index of refraction. In the four-wave mixing geometry, these thermal gratings can then scatter pump waves into the conjugate wave direction.

V. Applications of Optical Phase Conjugation

Many practical applications of phase conjugators utilize their unusual image-transformation properties. Because the conjugation effect is not impaired by the placement of an aberrating medium into the beam, the effect can be used to repair the damage done to a laser beam by otherwise unavoidable distortions. This technique can be applied to improving the output beam quality of optical systems that contain phase inhomogeneities or imperfect optical components. In a nonuniform laser amplifier system, a phase-conjugate mirror can be used to reflect the beam back through the amplifier, thereby efficiently extracting the stored energy while undoing any aberrations. The optical beam quality would not be degraded by inhomogeneities in the amplifying medium, by deformations or imperfections in the optical elements, windows, mirrors, and so forth, or by accidental misalignment of the optical elements. This then is an example of phase conjugation's ability to allow the high beam quality double-pass through a poor optical quality system.

As a second example, a laser oscillator might have one of its mirrors replaced by a phase-conjugating mirror. Such a phase-conjugate laser oscillator can be configured to produce a high-quality output even in the presence of medium inhomogeneities. Furthermore, there are no unstable modes of such an oscillator; all modes are stable. Thus one might use such an oscillator design to outcouple the energy of a gain medium that is not shaped like a long slender cylinder as is required in lasers having normal mirrors.

Another application might concern the difficult task of aiming a laser beam through an imperfect medium to strike a distant target. This imperfect medium might be the turbulent air, an air–water interface, or the focusing mirror (or lens) in a laser-fusion experiment. Conventional approaches would entail accurately aiming a weak laser beam through a large amplifier chain to strike the distant target. Alternatively, the phase-conjugation approach entails first illuminating the target with a broadly divergent laser. A portion of the glint returning from the target would pass through the imperfect medium, through the laser output telescope, through the laser amplifier system to then strik[illegible] e phase conjugator. The conjugate beam would contain essentially all of the information needed to converge on the target after retraversing the amplifier, the telescope, and the turbulent atmosphere. This automatic pointing and tracking feature would correct for distortions in the laser and in the intervening atmosphere.

Another variation of this technique pertains to correction for images that become blurred in passing through a multimode optical fiber; here a phase-conjugate beam can be generated and sent through an identical length of a second equivalent fiber. If the two fibers are identical, then the image emerging from the second portion is identical to the input image; here the conjugator serves as a midpoint repeater. Also related is the application of phase conjugation to lensless imaging.

The photorefractive effect seems to have some of its own special applications, including image intensification, edge enhancement, convolution, correlation, optical pattern manipulation and comparison, and associative memory. Phase conjugation in resonant absorbers can lead to specialized spectroscopic information about the absorbers. In slightly absorbing conjugators, one can learn about thermal diffusion properties.

Conjugators also have potential applications in the time-domain manipulation of optical pulses. A suitably thin, continuous-wave pumped Kerr-like four-wave mixing device reverses the chirp on a pulse. Just as a conjugator is useful as a repeater in a multimode fiber image transmitter, so a chirp-reversing conjugator could aid in an optical pulse communication network. A fiber-optic pulse communication network is limited in signal bandwidth because of dispersive spreading; if the pulses spread into one another, then the information is lost. A chirp-reversing phase conjugator, however, can send into the subsequent length of fiber the chirp-reversed pulses that are automatically compressed by the very same dispersion originally responsible for the spreading. This time-reversal aspect of phase conjugation could undo the spreading associated with linear dispersion and could therefore increase the possible data rate.

It is also surmised that Kerr-like four-wave mixing conjugators can change the quantum state of the ambient (zero-point noise) light. These new "squeezed" states have markedly different photocount statistics, and the use of such light into the unused channel of an interferometer is expected to increase the sensitivity of quantum-limited detectors (such as gravity-wave detectors) to below what is known as the "naive quantum limit."

Bibliography

Basov, N. G. (ed.) (1986). "Phase Conjugation of Laser Emission," Proceedings of the Lebedev Institute, Vol. 172. (English translation published by Nova Science Publ., Commack, New York.)

Ducloy, M. (1982). Festkoerperprobleme (*Adv. Solid-State Phys.*) **23,** 35–60. Vieweg, Braunschweig.

Feinberg, J. (1982). *Opt. Lett.* **7,** 486.

Feinberg, J. (1988). Photorefractive nonlinear optics, *Phys. Today* **41,** 6, October.

Feldman, B. J., Bigio, I. J., Fisher, R. A., Phipps, C. R., Watkins, D. E., and Thomas, S. J. (1981). Through the looking glass with phase conjugation, *Los Alamos Sci.* **3,** 2.

Fisher, R. A. (1987). Phase conjugation materials, *in* "Handbook of Laser Science and Technology," Vol. V (M. J. Weber, ed.), Part 3, p. 261. CRC Press, Boca Raton, Florida.

Fisher, R. A. (1983). "Optical Phase Conjugation." Academic Press, New York.

Fisher, R. A., and Feldman, B. J. (1980). "1980 McGraw-Hill Yearbook of Science and Technology." McGraw-Hill, New York.

Fisher, R. A. (1986). A Videocassette Course on Optical Phase Conjugation. P.O. Box 3537, Santa Fe, New Mexico, 87501–0537.

Giuliano, C. R. (1981). *Phys. Today,* April.

Gunter, P., and Huignard, J.-P. (1988). "Photorefractive Materials and Applications," Vol. I. Springer-Verlag, Berlin.

Gunter, P., and Huignard, J.-P. (1989). "Photorefractive Materials and Applications," Vol. II. Springer-Verlag, Berlin.

Hoffman, H. J. (1986). *J. Opt. Soc. Am.* **B3,** 253.

Pepper, D. M. (1982). *Opt. Eng.* **21,** 156, and the several articles that follow.

Pepper, D. M. (1986). *Sci. Am.* **254,** 74, January.

Pepper, D. M. (ed.) (1989). Special issue on optical phase conjugation, *IEEE J. Quantum Electron.* **QE-25(3).**

Shkunov, V. V., and Zel'dovich, B. Ya. (1985). *Sci. Am.* **253,** 54, December.

Soviet Academy of Applied Physics (1982). "Wavefront Reversal of Optical Radiation in Nonlinear Media," in Russian. Gorkii.

Yariv, A. (1978). Phase conjugate optics and real-time holography, *IEEE J. Quantum Electron.* **QE-14,** 650.

Yariv, A. (1979). Comments on phase conjugate optics and real-time holography; author's reply, *IEEE J. Quantum Electron.* **QE-15,** 524.

Zel'dovich, B. Ya., Pilipetsky, N. F., and Shkunov, V. V. (1985). "Principles of Phase Conjugation." Springer-Verlag, Berlin.

OPTICAL SYSTEMS DESIGN

Robert E. Fischer *Ernst Leitz Canada Ltd.*

GLOSSARY

Aberration: Geometrical errors in imagery whereby a perfect (or stigmatic) image is not formed. Typical aberrations include spherical aberration, astigmatism, coma, and chromatic aberrations. Lens bendings, locations, powers, materials, and the number of lenses and position of the aperture stop are all used to minimize aberrations.

Aperture stop: Location within a lens system where the chief or principal ray passes through and crosses the optical axis. The presence of a mechanical limiting aperture typically creates the limiting size.

Astigmatism: Aberration in which light in one plane (the "plane of the paper") focuses at a different location from light in the orthogonal plane.

Chromatic aberration: Axial or off-axis aberration whereby different colors have different focus positions, magnifications, spherical aberration, or other aberrations.

Coma: Off-axis aberration whereby the outer periphery of a lens system has a higher (or lower) magnification than the central portion of the lens. The image typically is comet shaped.

Distortion: Aberration that is a change in magnification with field of view. It is typically cubic with the field of view, and makes a square object look like a pincushion (positive distortion) or like a barrel (negative distortion).

Entrance pupil: Position along the optical axis of a lens where the chief ray would intersect the optical axis if it were not redirected by the lens. It is also the image of the aperture stop on the object space side of the lens.

Exit pupil: Position along the optical axis of a lens where the chief ray exiting the system appears to have crossed the optical axis. It is also the image of the aperture stop on the image space side of the lens.

***f*/Number:** Ratio of the focal length to the clear aperture diameter.

Field of view: Angular coverage of the optical system in object space.

Field curvature: Inherent property of imaging optics whereby the image may be formed on a curved surface. Field curvature can be eliminated by proper selection of lens powers, locations, and materials.

Focal length: Distance measured along the optical axis from the image to the plane where the axial imaging cone of light intersects the input light bundle.

Gaussian beam imagery: Imagery using a Gaussian beam intensity profile. This is typical of laser illumination.

Modulation transfer function (MTF): Ratio of the modulation in the image to the modulation in the object for a sinusoidally varying object intensity as a function of spatial frequency. The MTF is affected by geometrical aberrations and diffraction. For a perfect system, the maximum resolvable spatial frequency is $1/(\lambda\ f/\text{number})$, and this is where the MTF goes to zero.

Optical path difference (OPD): Difference between a reference or perfect spherical wavefront centered about the image and a real wavefront. If the OPD is one quarter of a wavelength, then the imagery will be nearly diffraction limited.

Spherical aberration: Axial aberration where rays from the outer periphery of the lens focus closer (or further) from the lens than rays from closer to the axis. Spherical aberration is typically cubic with aperture.

Vignetting: Clipping or truncation of the off-axis ray bundles by elements distant from the aperture stop. Vignetting is usually intentional in visible systems, as elements can be made smaller in diameter and the resulting imagery is better. For typical visible systems, vignetting of 30% or even as much as 50% is not uncommon. Vignetting is usually not acceptable in infrared systems.

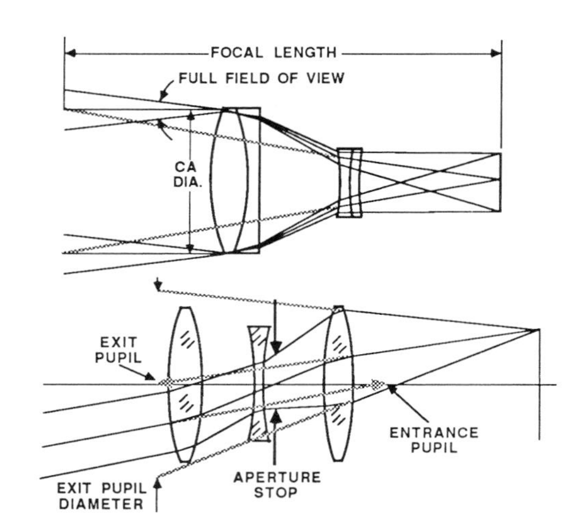

FIG. 1. Optical system first-order parameters.

Optical system design is the application of scientific and engineering principles to the design of image forming optical systems. In order to produce an acceptable optical design for a given application, geometrical image aberrations that otherwise would produce unacceptable image blur must be minimized. These aberrations can affect all wavelengths uniformly, or can be chromatic in nature whereby aberrations change with wavelength. The optical system design is optimized in performance by altering the lens and/or mirror radii, shapes, materials, locations, and number of elements.

I. Introduction

A. Purpose of an Optical System

An optical system as we will define it here has the purpose of forming an image of an object on a detector or sensor, such as photographic film, a charge coupled device (CCD) solid-state detector array, or some other detector medium (or even the human eye). The system imagery covers a desired angular field of view in object space. The resolution of the optical system is typically matched to the smallest detector elemental size so as to resolve a desired minimum sized object. Finally, the optical system clear aperture must be sufficient for the desired sensitivity.

B. First-Order Optics and Optical System Specifications

A generic optical system form is shown in Fig. 1, along with the basic first-order system parameters. Although optical systems come in many forms with different numbers of lens elements and/or mirrors, the basic system parameters described next will always apply:

Focal length. If we extend the rays forming an on axis image forward until they intersect the incoming rays from an infinitely distant object, the distance from this intersection point to the image is the focal length. In some forms of optical systems, such as telescopes or binoculars, the exiting rays are parallel to the incoming rays from infinity. These systems have an infinite focal length and are known as "afocal."

Clear aperture diameter. The limiting diameter of the imaging radiation.

f/Number. The ratio of the focal length to the clear aperture diameter is known as the *f*/ number (i.e., *f*/number = focal length/clear aperture diameter).

Field of view. The angular coverage of the optical system and its detector in object space is the field of view.

Aperture stop. The aperture within the optical system that limits all of the ray bundles. This is usually a mechanical aperture, such as an adjustable iris in a camera lens.

Chief ray (or principal ray). This is the ray passing through the center of the aperture stop from an off-axis object point. The chief ray height from the optical axis on the image surface is usually used as a reference for the image height.

Entrance pupil. The position along the optical axis where the chief ray would intersect the optical axis if it were not redirected by the lens. This is also the image of the aperture stop in object space.

Exit pupil. Tracing back from the image, the position along the optical axis where the chief ray intersects the optical axis or appears to have come from. This is also the image of the aperture stop in image space.

II. Fundamentals of Imagery: Ray Optics and Wave Optics

Optical systems are designed and evaluated by two different yet closely related means: ray optics, and wave optics. Rays represent the direction that the radiation is traveling through the optical system on its way to the final image, whereas the wavefront represents the surface of equal phase of the light (or other electromagnetic radiation) that is propogating through the optical system on its way to the final image. There is a direct correspondence between the two means of considering imagery. A perfectly spherical wavefront on its way to an image will produce a perfect image, and the rays (which are the normals to the wavefront) will converge to a perfect image as well.

Imagery, however, can never be perfect in the geometrical sense, and it is degraded either by *geometrical aberrations,* which are a function of the lens parameters, and/or *diffraction,* which is a physical optics effect, each of which will be discussed next.

A. Geometrical Image Defects (Aberrations)

Ideally, the image of a point object would be a point image. Geometrical aberrations are mathematically predictable errors that create an imperfect or blurred image. These aberrations are minimized or eliminated by the proper selection of the optical system configuration, along with the relative shapes or bendings and locations of elements, the materials properties, and other system parameters that are varied during the design optimization process. Computer programs using damped least squares or other optimization routines are typically used to optimize the performance of a lens system by minimizing the geometrical aberrations.

B. Diffraction

Even if a lens system were perfectly corrected for zero geometrical aberrations, the resulting image would still be degraded by diffraction, which is due to the wave nature of light. The image of a point object is ideally an "Airy disc," which is mathematically the Fourier transform of the pupil shape. [*See* DIFFRACTION, OPTICAL.]

For a system used at wavelength λ, the diameter of the Airy disk is given by

$$\text{physical diameter } d = 2.44\lambda f/\text{number}$$

and

$$\text{angular diameter } d = 2.44\lambda/\text{clear aperture diameter}$$

Figure 2 shows the relationship between the geometrical diameter and the angular diameter of the Airy disk. For high-performing systems, we generally attempt to place approximately 80% of the energy from the Airy disk within the detector elemental size (which defines the system *f*/number), and we select a clear aperture diameter that produces an angular Airy-disk diameter consistent with our desired system angular resolution.

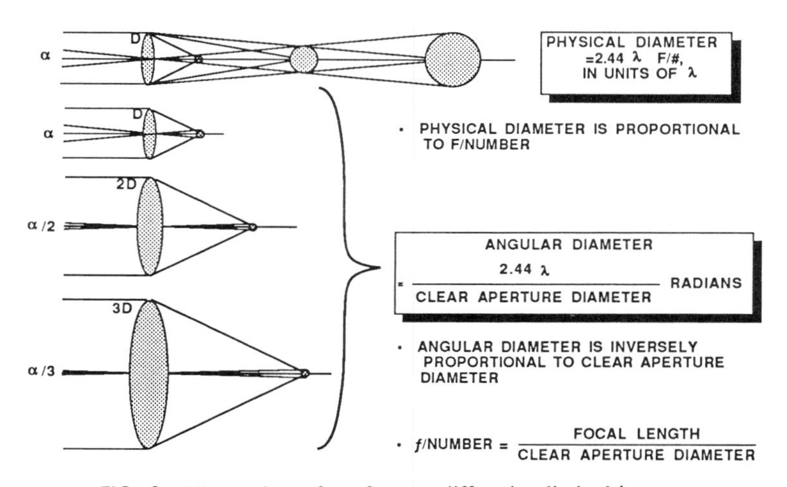

FIG. 2. Illustration of perfect or diffraction-limited imagery.

III. Aberrations and Methods of Elimination

The geometry of focusing light to an image with spherical surfaces is mathematically imperfect, yet spherical surfaces are utilized almost exclusively in optical systems due to their ease of fabrication. Since the shape of a sphere is constant everywhere on the sphere, high-speed production methods can be used where a mating tool and lens surface are worked together. If perfect imagery were possible, the wavefront would converge to an ideal point image, and similarly if we consider ray optics, the rays would focus to a perfect point image. But imagery is not perfect, and the residual image defects are known as "aberrations." The optical design task is to select the proper form, material, and grouping of lens elements and/or mirrors to minimize or eliminate these aberrations.

Aberrations are usually classified as monochromatic or chromatic in form. Monochromatic aberrations are usually computed at a central, or reference, wavelength, where the system has its maximum sensitivity. Chromatic aberrations refer to those aberrations that change with wavelength.

A. Monochromatic Aberrations

If we trace rays rigorously through an optical system using well-established ray-tracing techniques, we can characterize the image defects as to their dependence on aperture and field of view. By tracing "paraxial" rays we can generate an "aberration polynomial," which gives a close approximation of the aberrations. Paraxial rays are traced using ray-trace equations where the sines and tangents are equal to the angles. A typical aberration polynomial will have the following form:

$$\begin{aligned}\Delta y' &= A_1 s^3 \cos\phi + A_2 s^2 h(2 + \cos 2\phi) \\ &\quad + (3A_3 + A_4) s h^2 \cos\phi + A_5 h^3 \\ &\quad + B_1 s^5 \cos\phi + (B_2 + B_3 \cos 2\phi) s^4 h \\ &\quad + (B_4 + B_6 \cos 2\phi) s^3 h^2 \cos\phi \\ &\quad + (B_7 + B_8 \cos 2\phi) s^2 h^3 + B_{10} s h^4 \cos\phi \\ &\quad + B_{12} h^5 + C_1 s^7 \cos\phi + \cdots\end{aligned}$$

where $\Delta y'$ is the ray displacement from the chief ray, ϕ and s are the polar coordinates of the ray in the entrance pupil, A_1 is the spherical aberration, A_2 is the coma, A_3 is the astigmatism, A_4 is the Petzval field curvature, and A_5 is the distortion. The B and C constants are the fifth-order and seventh-order aberrations.

The lowest order of aberration is known as third order, and represents the first term in the aberration polynomial for each aberration type or form. Also, third-order aberrations are the most dominant in a system until they are corrected. We will discuss below the form of each type of image aberration.

1. Spherical Aberration

Spherical aberration results when rays focusing from the outer periphery of a lens focus closer to (or further from) the lens than rays from the center of the lens (the paraxial rays). The aberration is an axial aberration that is constant across the field of view. In Fig. 3a we show a lens with a significant amount of spherical aberration. Spherical aberration measured laterally on the image surface is typically cubic with aperture (third-order spherical aberration). Thus, a lens with a given amount of spherical aberration used at half of its aperture diameter will have one-eighth of the spherical aberration.

In addition to simply reducing the aperture, we can control spherical aberration by changing the lens shape, or "bending" the lens, and/or splitting the optical power into two or more lenses. In Fig. 3b we show a lens of the same focal length as in Fig. 3a; however, we have bent the lens for minimum spherical aberration. Note that there is still a finite image blur residual. Splitting the optical power from, say, one lens to two or more weaker individual lenses will also reduce the spherical aberrations.

2. Coma

Coma is an off-axis aberration in which the outer periphery of a lens has a higher or lower magnification than dictated by the central or chief ray. The resulting image looks like a small comet, as shown in Fig. 4. Coma is quadratic with aperture and linear with field of view.

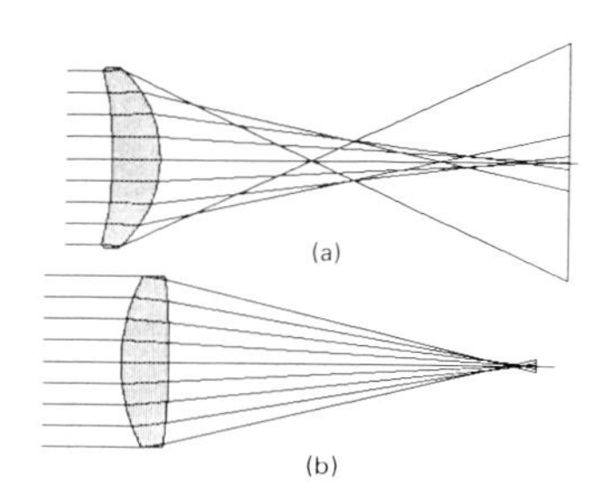

FIG. 3. Illustration of (a) substantial amount of spherical aberration and (b) lens bent for minimum spherical aberration.

FIG. 4. Comatic image of point object.

3. Astigmatism

Astigmatism is also an off-axis aberration, as shown in Fig. 5. It results when light in one plane (the *YZ* plane) focuses at a different position than light in the opposing plane (the *XZ* plane).

4. Field Curvature

In the absence of astigmatism, the image formed by an optical system is formed on a curved surface called the *Petzval surface,* where the curvature of this image surface is given by the following relationship:

(curvature of Petzval image surface)

$$= \left(\frac{1}{\text{radius of image surface}}\right) = -\sum \frac{1}{n/f'}$$

for all optical elements in the system. Thus, for example, for a single element lens of standard BK7 glass, the image is formed on a surface of radius $r = -1.5$ times the focal length, since the refractive index is 1.5.

5. Distortion

Distortion is a change in magnification with field of view, and results when the chief ray at off-axis image positions departs from the location where the paraxial chief ray strikes the image (which would produce an undistorted image). Distortion is basically a deviation from rectilinear imaging. Figure 6 shows the formation of distortion, as well as the two forms of distortion: positive or pincushion distortion, and negative or barrel distortion.

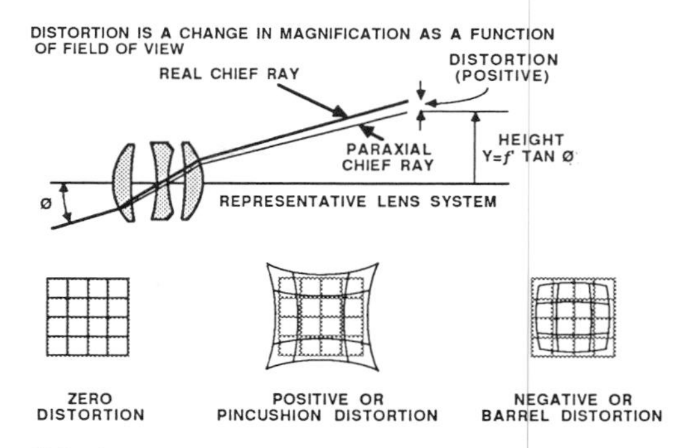

FIG. 6. Formation and illustration of distortion.

6. Other. There are other higher-order aberrations that are sometimes encountered in optical systems; however, those just described are the primary aberrations. In more complex systems, aberrations such as tangential oblique spherical aberration, sagittal oblique spherical aberration, and even aberrations of names like "wings" and "arrows" are possible. These can become important in complex, high-performing optical systems.

The dependence of the third-order aberrations on aperture and field is summarized as follows:

Aberration	Aperture dependence	Field dependence
Spherical aberration	Cubic	—
Coma	Quadratic	Linear
Astigmatism	Linear	Quadratic
Field curvature	Linear	Quadratic
Distortion (percent)	—	Cubic

B. Chromatic Aberration

Due to the fact that the refractive index or bending power of a lens changes with wavelength (a characteristic known as "dispersion"),

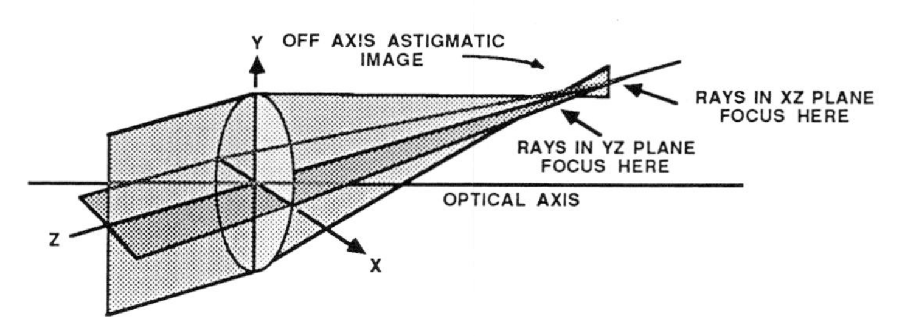

FIG. 5. Formation of astigmatic image.

the aberrations of the preceding section can all change with wavelength. The two basic forms of chromatic aberration are axial color and lateral color; however, variations of each of the discussed monochromatic aberrations with wavelength are sometimes encountered as well. [*See* COLOR, SCIENCE OF.]

1. Axial Color

Due to the dispersive properties of glass, the refractive index increases with shorter wavelengths. In a visible system with a single element, this causes the blue light to focus closer to the lens than the red light, as shown in Fig. 7a. This resulting change in focus position with color is known as *primary axial color*.

By utilizing two materials of different dispersion to form an "achromatic doublet" form of lens, we can bring to a common focus the two outer wavelengths, leaving the central wavelength slightly defocussed towards the lens, as shown in Fig. 7b. The residual aberration is known as *secondary axial color*.

In addition to the primary and secondary color, we can and often do have a variation in spherical aberration with wavelength, which is known as *spherochromatism*.

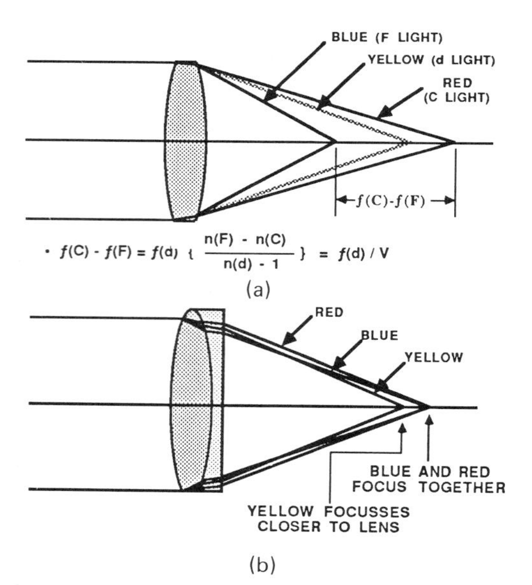

FIG. 7. Axial color for (a) primary axial color of single-element lens and (b) doublet corrected for primary axial color, showing residual secondary color.

2. Lateral Color

Lateral color is an off-axis aberration that in effect is a change in focal length with wavelength. The result is a lateral spreading of the image, which is often seen as color fringing in a radial direction on objects toward the edge of the field of view.

IV. Glass Selection and Correction of Chromatic Aberration

A. Optical Glass

For a visible system we characterize glass by its refractive indices at 0.5876 μm (the sodium d line), 0.4861 μm (the hydrogen blue, or C, line), and 0.6563 μm (the hydrogen red, or F, line). The nominal index is at the d line, and characterize the dispersion by a quantity known as the *Abbe number* or V number, where the Abbe number $= (n_d - 1)/(n_F - n_C)$. Figure 8 shows a typical glass map with the many glasses available plotted against n_d and V.

B. Correction of Chromatic Aberration

It can be shown that the difference in focal length for a thin lens between the F and the C light is

$$f_C - f_F = f_d\{n_F - n_C)/(n_d - 1)\} = f_d/V$$

We also know that for a thin lens,

$$\text{Optical power} = 1/f = (n - 1)\{1/R_1 - 1/R_2\}$$

and for two thin lenses in contact the focal length of the combination f_{12} is given by

$$1/f_{12} = 1/f_1 + 1/f_2$$

By combining the above equations, we can solve for the individual focal lengths of each of the lenses in order to eliminate the primary axial color:

$$f_1 = f_{12}\{V_1 - V_2)/V_1\}$$

$$f_2 = f_{12}\{V_1 - V_2)/V_2\}$$

Solving the above equations produces a doublet combination that has zero primary axial color, and hence the F and the C light come to a common focus. The central-wavelength d light will be defocused toward the lens, as shown in Fig. 7b. This residual can be reduced and sometimes eliminated by the use of "special glasses" that have similar dispersive characteristics.

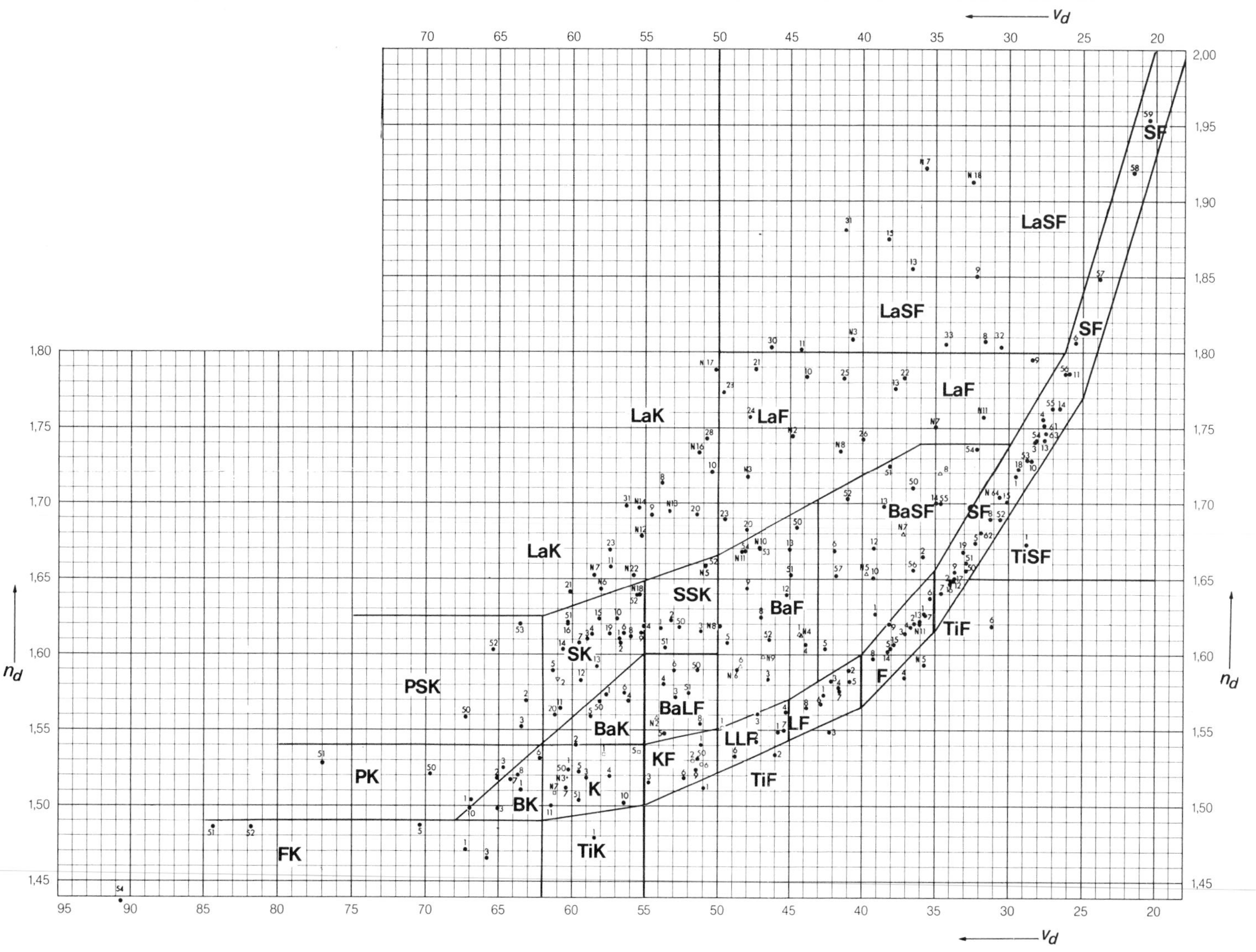

FIG. 8. Diagram of optical glasses: zinc crowns (ZK), □; barium light crowns (BaLk), +; short flints (K_2F), ○; special short flints (K_2FS), △; special long crowns (LgSK), ▽; and other glasses, ●.

V. Optical System Configurations

Optical systems can take on many varied forms, ranging from simple configurations of one or two elements, such as might be used in a simple camera system, to complex configurations of 12–15 or more elements, as might be used in a high-resolution microlithography system. The basic configuration or design form *must* be inherently capable of providing the desired level of system performance (i.e., aberration control).

A. Lens Systems: Representative Design Forms

Shown in Figure 9 is a progression of configurations ranging from a single element suitable only for a magnifying glass or a light collector through multielement lenses providing a high level of aberration correction and performance. The comments regarding performance are only a very rough guideline.

The first lens form in which the all of the primary aberrations can be controlled is the Cooke triplet. With the eight useful variables (six radii and two air spaces) we can control the eight system requirements and primary aberrations (focal length, spherical, coma, astigmatism, field curvature, distortion, primary axial color, and lateral color). The double Gauss lens is a form typical of fast high-performing camera lenses ($f/2$ or faster). Although for most systems the aperture stop is toward the center of the lens for symmetry reasons, the eyepiece is unique in that the aperture stop is actually the pupil of the eye, and as such is remote from the lens. There are obviously many additional forms of lenses, and each will have some unique merit or advantage in performance.

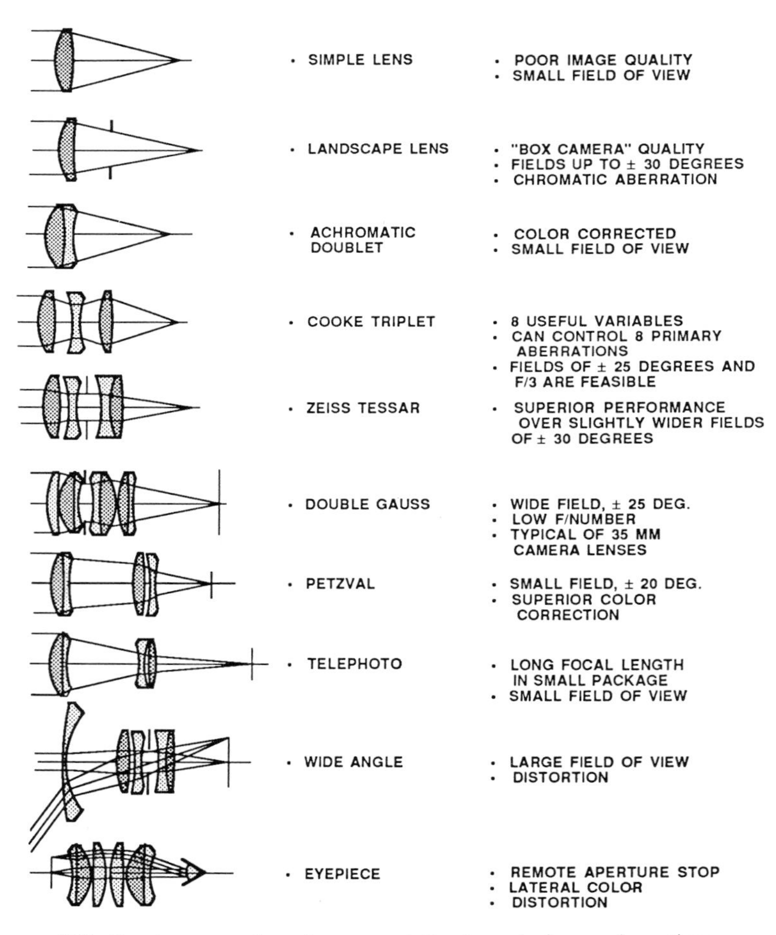

FIG. 9. A progression of representative lens design configurations.

B. Mirror Systems

1. Relative Merits

Optical configurations utilizing mirrors are becoming more popular in recent years, due in part to the advent of contemporary methods of optical fabrication such as single-point diamond turning (see Section IX). The relative merits of refractive or lens systems versus reflective or mirror systems are listed:

Lens systems	Mirror systems
Advantages	Advantages
Straight through, no obscuration	No chromatic aberration
Spherical surfaces	Potentially athermal
Conventional assembly	Shorter length
Disadvantages	Lighter weight
Thermally sensitive	Disadvantages
	Difficult to align
	Aspheric surfaces
	Reduced throughput (due to obscuration)
	Reduced performance (MTF)

One of the major considerations of a reflective system is that mirrors get in each other's way. Thus a mirror system can usually be only two or three surfaces at most, whereas a lens system can be over 15 elements (or about 30 surfaces). As we have shown above, splitting of the optical power is used very effectively to reduce the residual aberrations in a lens system. But in a mirror system the restriction to two or three surfaces necessitates the use of aspheric (nonspherical) surfaces in order to control the aberrations.

2. Representative Design Forms

Similar to lens systems, the system configuration is most important towards achieving a successful design. In Fig. 10 we show a progression of configurations for reflective systems. We define here a reflective system as one in which the majority of the optical power is achieved with the mirrors. In many reflective systems refracting components are used in order to reduce the residual aberrations, as evidenced with the Schmidt telescope, for example.

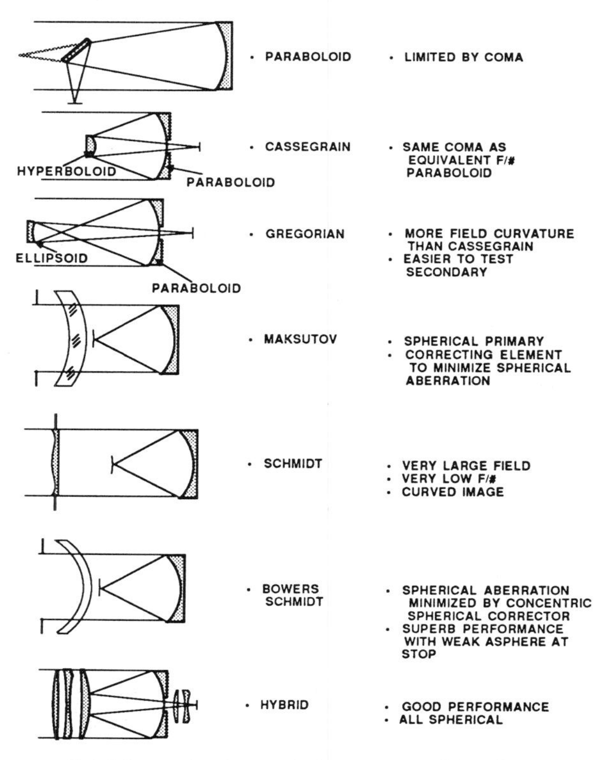

FIG. 10. Reflective optical system configurations.

VI. Design Optimization

A. Optimization Methodology

Specialized computer programs are used for the performance optimization and associated analysis of optical systems. The optimization methodology is that each variable (radius, thickness, index, etc.) is first changed by a small incremental amount and the effect to performance is determined. The result is the partial derivatives

$$\partial P/\partial V_1,\ \partial P/\partial V_2,\ \partial P/\partial V_3,\ \ldots$$

where V_1, V_2, V_3, ... are the individual variables, and P is a measure of performance. The measure of performance is typically a single number or "merit function," which is relatively fast to compute and well represents the imaging characteristics of the system for all wavelengths and fields of view [e.g., the rms (root mean square) blur size over the field of view]. This set of simultaneous equations is solved so as to minimize the residual performance or merit function. System constraints (e.g., focal length, packaging requirements, etc.) can be added into the matrix or can be solved for exactly. Although each computer program differs slightly in its methodology, the basic technique is the same.

B. Vignetting

Although not an optimization routine in the normal sense, a technique often used to improve performance is vignetting, or the selective truncation or clipping of the radiation bundles off-axis. In order to maximize the relative illumination over the field of view, we usually desire to keep the entrance pupil diameter the same at all field angles. However, as we proceed off-axis, the elements separated from the aperture stop will necessarily become large in diameter. Figure 11 illustrates this, where we see that if the full entrance pupil diameter were used off-axis the element diameters would be substantially larger. Further, the elements would have to be thicker, the housing would be larger, and the system would be heavier. Also important is that the rays at positions A and B are often the most severely bent or refracted by the lenses and will produce higher-order aberrations. By truncating the apertures of these outer elements, we can eliminate these aberrated rays and take advantage of a smaller, lighter-weight, and lower-cost system. Visible systems (cameras, projectors, binoculars, etc.) can usually tolerate 30–50% vignetting at the edge of the field of view, and hence we find that many lenses and systems have selective vignetting.

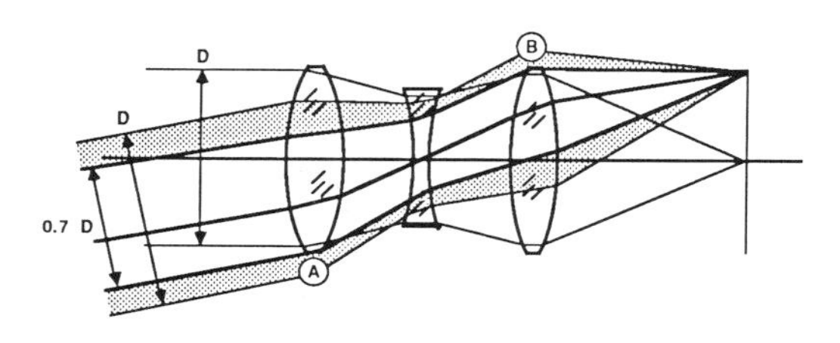

FIG. 11. Illustration of vignetting.

VII. Performance Computation

A. Ray-Trace Analysis

In order to evaluate the performance of an optical system, we require a fast and simple means to describe the imagery. The transverse ray aberration curve as shown in Fig. 12 for coma is used. We trace a fan of rays into the system entrance pupil in two orthogonal planes, the meridional plane (usually drawn as the plane of the paper) and the sagittal plane (in/out of the plane of the paper). The deviation from where each ray intersects the image from the chief ray intercept is plotted as a function of pupil coordinate. This is repeated for each field angle and each wavelength. Through the use of ray-trace curves the designer can gain a great deal of insight into the specific aberrations present in the system.

B. Optical Path Difference

The optical path difference (OPD) from a nearest reference spherical wavefront to the actual system wavefront is also used as a measure of performance. In addition to graphical representations of the wavefront error, the peak-to-valley (P-V), or alternately the rms wavefront error are often used as a specification and/or measure of performance. The P-V wavefront error is the distance between the longest and shortest paths from anywhere on the wavefront leading to a selected focus, whereas the rms wavefront error is root-mean-square wavefront error. The rms is typically one-fifth of the P-V error, although the ratio varies with the nature

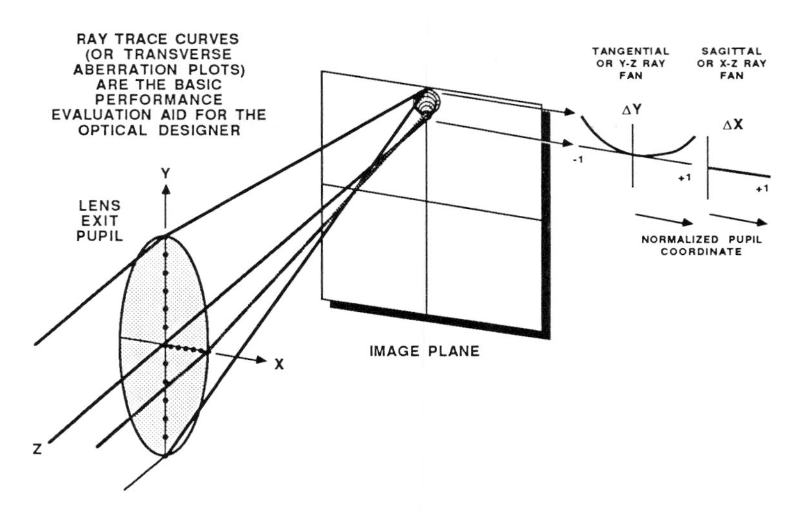

FIG. 12. Formation of comatic image and explanation of ray-trace curves.

of the error. The "Rayleigh criteria" is a rule of thumb that states that if the maximum P-V wavefront error is less than or equal to 0.25 waves, then the imagery will be almost indistinguishable from perfect. Figure 13 shows diffraction-based image point-spread functions for a perfect system and for systems with 0.25, 0.5, and 1.0 waves of spherical aberration, from which we can see that the quarter-wave system is indeed nearly perfect.

C. Modulation Transfer Function

The modulation transfer function (MTF) is a measure of how well the modulation is transferred from the object to the image by the optical system. Figure 14 shows a graphical representation of the MTF. MTF is expressed as a function of spatial frequency, usually at the image, and is in the form of line pairs (lp) per millimeter or cycles per milliradian. This is the analog of fre-

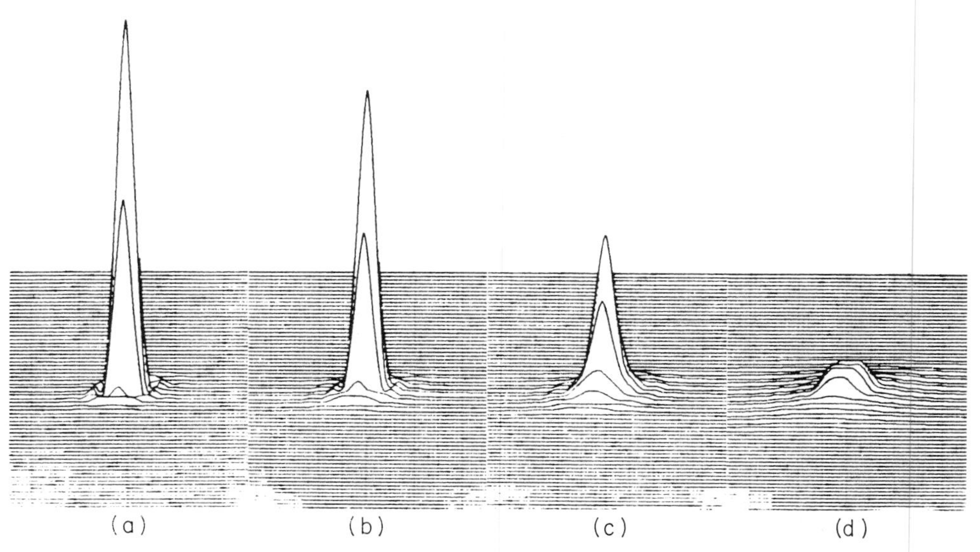

FIG. 13. Imagery of a point object for different amounts of spherical aberrations: (a) perfect system, (b) 0.25 waves spherical aberration, (c) 0.5 waves spherical aberration, and (d) wave spherical aberration.

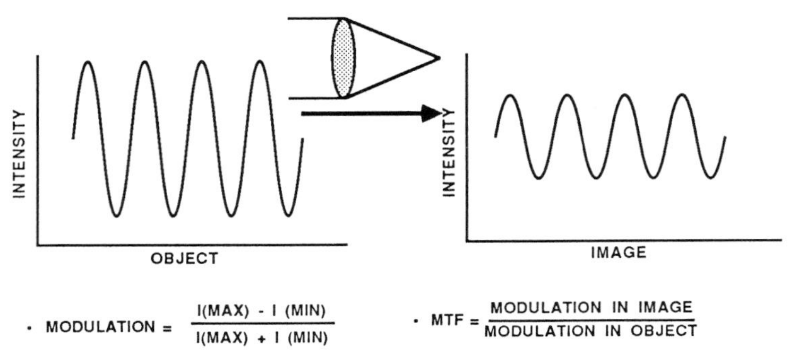

FIG. 14. The concept of the modulation transfer function (MTF).

quency in time-dependent systems. Figure 15 shows a typical MTF curve for a perfect system, a perfect system with a central obscuration (e.g., a reflective Cassegrain telescope), and a representative real system. Note that the theoretical cutoff frequency where the modulation goes to zero is $1/(\lambda f/\text{number})$. Thus for example, an $f/4$ system designed at 0.5876 μm wavelength has a cutoff frequency of 425 lp/mm.

VIII. Optical Design for the Thermal Infrared

The thermal infrared (IR) is typically considered to be the spectral bands at 3–5 μm and the 8–12 μm, as the atmospheric transmission is maximum in these regions. We are in effect "looking at" or sensing thermal radiation or heat, and infrared systems are in wide use. Optical systems for the IR are similar to their visible counterparts in many ways; however, due to the requirements of specialized materials and other factors unique to IR systems, they require special design considerations.

A. Characteristics of Infrared Systems

In the thermal infrared the detector is sensing thermal energy, and in order to maximize its sensitivity the detector is cryogenically cooled to 77 K or lower and enclosed in an evacuated and thermally insulated bottle, or Dewar, as shown in Fig. 16. The "cold finger" that butts up against the detector to keep it cool is typically cooled with liquid nitrogen. A "cold stop" or "cold shield" provides a limiting aperture for the imaging radiation, outside of which there will be no thermal energy sensed by the detector. Finally, an IR-transparent Dewar window seals the assembly.

The IR detectors are typically combinations of special materials such as HgCdTe, InSb, or other materials, and are configured as a single element, a small array of elements, a long linear array, or a full two-dimensional array. When the detector is a full two-dimensional array, then the

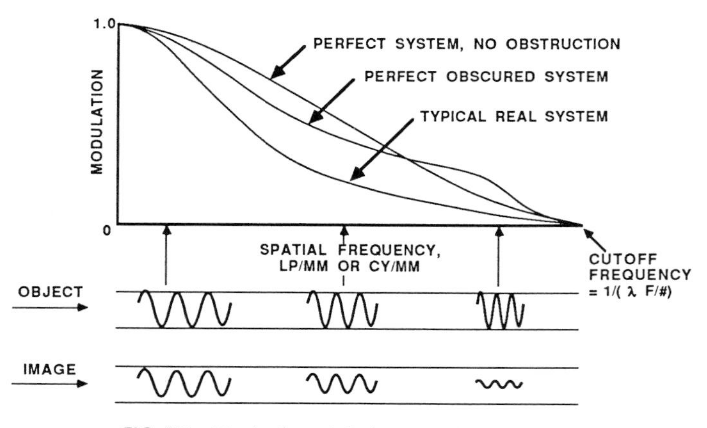

FIG. 15. Typical modulation transfer functions.

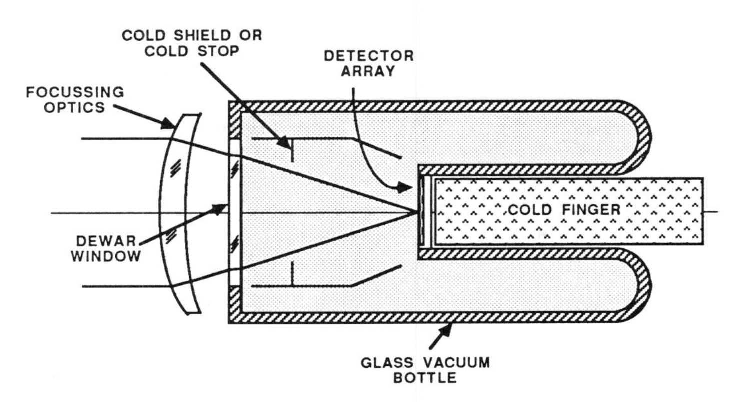

FIG. 16. Typical detector/dewar assembly for infrared imaging system.

full image is recorded at once. However, when we are using smaller arrays we need to scan the image over the array, as shown in Fig. 17. The small array requires a "serial scan" system to sweep the image over the array in two dimensions, whereas the linear array requires a "pushbroom scan" system to sweep the image across the array in one dimension. Figure 18 shows typical configurations for creating the required scan motions.

B. Materials for Use in the Infrared

Lens materials for thermal infrared systems are either dielectric or semiconductor in nature. Although there are many viable materials, only a limited number are typically used. The more common IR materials are listed in Table I. These materials are produced by processes such as chemical vapor deposition or CVD (zinc sulfide and zinc selenide), hot pressing (also used for zinc sulfide and zinc selenide), and crystal growth (germanium and silicon). Although these processes yield a relatively pure material, impurities can lead to polycrystalline grain structure, scattering centers, and other imperfections that may affect imagery. In the case of germanium and silicon, polycrystalline materials are often used, except in cases where very high performance is required.

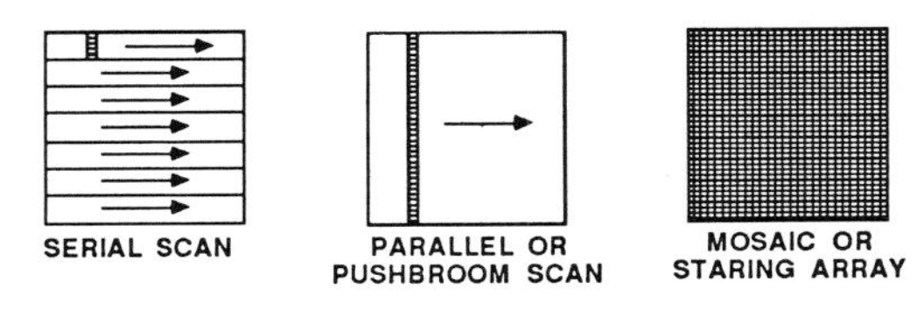

FIG. 17. Scanning and staring for thermal infrared systems.

C. Lens Design and System Configurations for the Infrared

For the most part, IR system configurations are similar to those intended for visible applications. There are several fundamental factors that make the design forms different in the IR.

The refractive index of IR-transmitting materials is typically much higher than for visible-transmitting materials (e.g., the refractive index of germanium is approximately 4.0). This produces inherently lower aberrations, as evidenced by the fact that a single glass element of 1.0 in. diameter at $f/2$ bent for minimum spherical aberration will yield approximately 10 waves rms measured at 0.5876 μm. This reduces to 1.0 wave rms at an index of 4.0. However, this is referenced to a wavelength in the visible, and when we reference to a 10-μm wavelength the residual aberration reduces by a factor of 20 to 0.05 waves rms. This is essentially diffraction

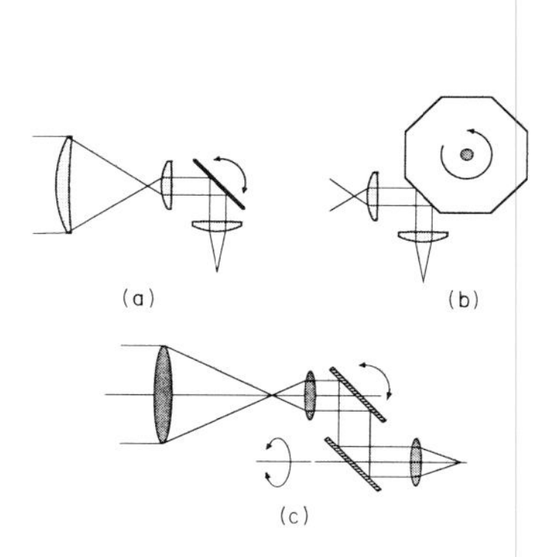

FIG. 18. Optical system configurations for parallel and serial scanning. (a) Oscillating scan mirror for parallel scanning. (b) Rotating polygon mirror for parallel scanning. (c) Two oscillating scan-mirror motions required for serial scanning.

TABLE I. Common Infrared-Transmitting Materials Used for Lens Systems

Material	Refractive index 4 μm	Refractive index 10 μm	dn/dt [(°C)$^{-1}$]	Comments
Germanium	4.00243	4.0032	0.000396	Expensive, large *dn/dt*
Silicon	3.4255	3.4179[b]	0.00015	Large *dn/dt*
Zinc sulfide (CVD)	2.2520	2.2005	0.0000433	
Zinc selenide (CVD)	2.4331	2.4065	0.000060	Expensive, low absorption
AMTIR 1 (Ge/As/Se)	2.5141	2.4976	0.000072	
Magnesium fluoride	1.3526	—[a]	0.000020	Low cost, no coating, high scatter
Sapphire	1.6753	—[a]	0.00001	Hard, low emissivity at high temperature
Arsenic trisulfide	2.4112	2.3816	[c]	
Calcium fluoride	1.4097	—[a]	0.000011	
Barium fluoride	1.4580	—[a]	−0.000016	

[a] Does not transmit
[b] Not recommended due to absorption band.
[c] Not available.

limited, from which we can conclude that fewer elements are typically required in IR systems due to the inherently lower aberrations.

Materials such as germanium and silicon have very low dispersion, which often eliminates the need for chromatic correction. We often find in IR systems using low-dispersion, high-refractive-index materials that only a single element is required at each location dictated by first-order requirements.

D. Unique Problems of IR Systems

As discussed earlier, an IR system is in effect looking at heat sources, and if the detector is permitted to see any hot or warm objects other than the scene (such as housings or other internal system structure) the sensitivity of the system is reduced. More important, however, is that if this unwanted radiation changes over the field of view and/or through scan, we will see image anomalies that will appear as banding or other cosmetically undesirable effects. These effects are:

Narcissus: a change through scan of the reflected radiation from lens and/or mirror surfaces reaching the detector.

Scan noise: modulation through scan from outside the aperture stop and inside the cold shield (this term is often used to label other forms of thermal noise).

Clipping: vignetting or infringement of the radiation bundles by a cell or a housing inside the cold stop through scan.

Beam wander: scan morror geometrically induced lateral beam shift that can cause clipping in poorly aligned systems.

Ghosting: for polygon scanners, the effect where radiation reflected from the adjacent facet from the imaging facet reaches the detector.

Shading: slowly varying exit pupil size variation, due for example to distortion, which alters the scene energy to the detector.

Virtually all of these effects are due to the detector seeing a change in radiation levels from within the system through scan or over the field of view.

IX. Tolerancing and Producibility Issues

A. Error Budget Establishment and Performance Prediction

The performance of an optical system will be degraded from theoretically perfect not only by diffraction and geometrical aberrations but also by fabrication, assembly, and alignment errors. Every optical system must be toleranced so as to assure that the fabricated system meets its performance requirement. This usually requires a computer tolerance analysis in which expected errors are analyzed and summed so as to predict the expected net performance. System parameters including lens radius, thickness, wedge, decentration within its cell, tilt within its cell, air space following the lens, irregularity of the surface profiles (or lack of sphericity), and properties of the lens material all must be taken into account for all components within the system. In certain circumstances a simple-looking lens may have severe sensitivity to tolerances, and the lens may have to be redesigned, possible with more elements, in order to make it more producible.

B. Single-Point Diamond Turning

A relatively new fabrication technique is single-point diamond turning. The process is perhaps better called ultraprecision machining, and it uses a single-crystal diamond tool along with a massive vibration-isolated and thermally controlled machine equipped with air bearings and air slides and, as required, numerical control. Diamond turning can produce optical-quality surfaces on most nonferrous metals, IR-transmitting materials such as germanium and zinc sulfide, and plastics, directly with no conventional grinding or polishing. Thus, flats, spheres, and aspheric surfaces can be efficiently produced. This is especially useful with fast aspheric surfaces that could not be conventionally fabricated. The process shears the surface at the molecular level, and the surface is left intrinsically clean. This has the effect of maximizing reflectivity and minimizing oxidation.

X. Gaussian Beam Imagery

All of the discussion thus far has assumed a uniformly illuminated pupil. Lasers, which are in wide use today in optical systems, produce a Gaussian beam intensity profile across the pupil.

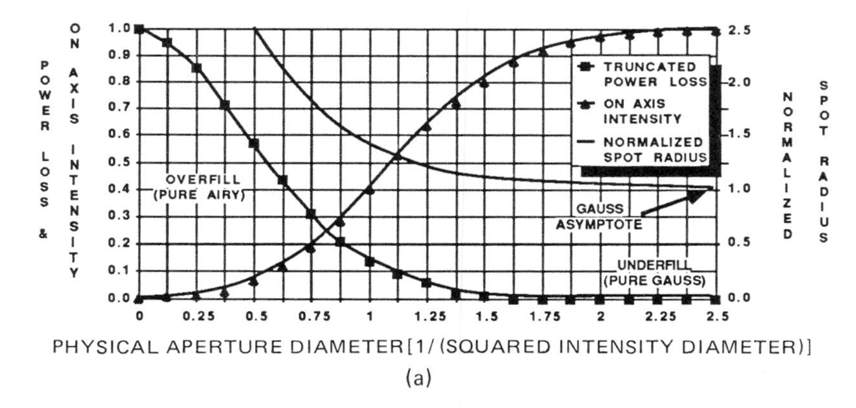

(a)

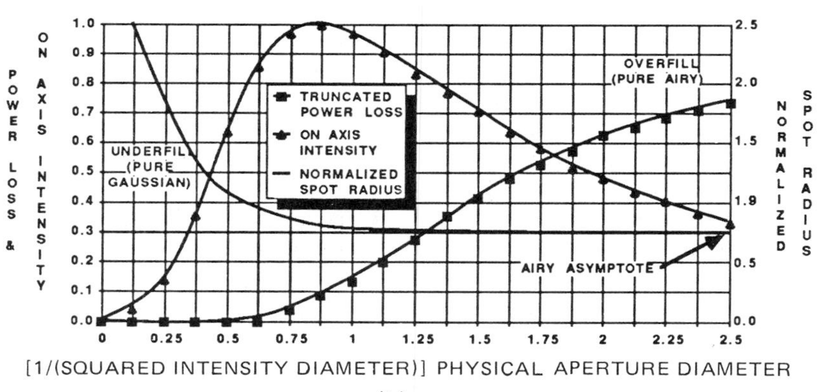

(b)

FIG. 19. Truncated Gaussian beam imagery. (a) Aperture changes relative to a constant $1/e^2$ beam diameter. (b) $1/e^2$ diameter changes relative to a constant aperture diameter.

Although a uniformly illuminated pupil will transform to an Airy-disk image, a Gaussian intensity profile in the entrance pupil will ideally transform to a Gaussian image intensity profile if it is not truncated. In order to produce minimum spot diameters as required in many laser scanning systems, we must understand the effects of truncating the beam. Gaussian beams are normally characterized by their $1/e^2$ intensity diameter (13.5% of the intensity). Figure 19 shows the effect of truncating the beam on spot size, on axis intensity, and on power loss. We show both the effect of the aperture changing relative to a constant $1/e^2$ beam diameter and the $1/e^2$ diameter changing relative to a constant aperture diameter.

Bibliography

Hecht, E., and Zajac, A. (1974). "Optics." Addison-Wesley, Reading, Mass.

Kingslake, R. (1978). "Lens Design Fundamentals." Academic Press, New York.

Kingslake, R., ed. (1965–1980). "Applied Optics and Optical Engineering," Vol. 1–9. Academic Press, New York.

O'Shea, D. (1985). "Elements of Modern Optical Design." Wiley, New York.

Smith, W. (1966). "Modern Optical Engineering." McGraw-Hill, New York.

QUANTUM OPTICS

J. H. Eberly *University of Rochester*
P. W. Milonni *Los Alamos National Laboratory*

GLOSSARY

AC Stark effect: Effective increase of the Bohr transition frequency of a two-level atom which is being excited by a strong laser beam, the amount of increase being the Rabi frequency.

Bloch vector: Fictitious vector whose rotations are equivalent to the time dependence of the wave function or quantum mechanical density matrix associated with a two-level atom.

Coherence time: Limiting time interval between two segments of a light beam beyond which the superposition of the segments will no longer lead to interference fringes.

Coherent state: Quantized state of a light field whose fluctuation properties are Poissonian; it is considered the most classical quantized field state.

Degree of coherence: Normalized measure of the ability of a light beam to form interference fringes.

Optical bistability: Existence of two stable output intensities for a given input intensity of a steady light beam transmitted through a nonlinear optical material.

Optical Bloch equations: Dynamical equations that determine the motion of the Bloch vector; they are a special type of quantum Liouville equation.

Photon echo: Burst of light emitted by a collection of two-level atoms signaling the realignment of their Bloch vectors after initial dephasing; similar to the spin echo of nuclear magnetic resonance.

Rabi frequency: Steady frequency of rotation of the Bloch vector of an atom exposed to a constant laser beam, proportional to the atom's transition dipole moment and the laser's electric field strength.

Superradiance: Spontaneous emission from many atoms exhibiting collective phase-coherence properties, such as radiation intensity proportional to the square of the number of participating atoms.

Two-level atom: Fictitious atom having only two energy levels which is used as a model in theoretical studies of near-resonance interactions of atoms and light, particularly laser light.

Quantum optics is the study of the statistical and dynamical aspects of the interaction of matter and light. It is concerned with phenomena ranging from spontaneous emission and single photon absorption to the highly nonlinear processes induced by laser fields and has connections with laser physics, nonlinear optics, quantum electronics, quantum statistics, and quantum electrodynamics.

I. Introduction

A. Central Issues of Quantum Optics

Planck's quantum, announced to the Prussian Academy on October 19, 1900, as a solution to the blackbody puzzle re-opened the wave–particle question in optics, a question that Fresnel and Young had settled in favor of waves almost two centuries earlier. Planck's quantum could not be confined to light fields. Within three decades, all of particle mechanics had been quantized and rewritten in wave mechanical form, and wave–particle duality was understood to be both universal and probabilistic.

Quantum optics is fundamentally concerned with coherence and interference of both photons and atomic probability amplitudes. For example, it provides one of the main avenues at the present time for detailed study of wave–particle duality. The central issues of quantum optics deal with light itself, with quantum mechanical states of matter excited by light, and with the process of interaction of light and matter.

Questions arising in the description of a single atom and its associated radiation field, as the atom makes a transition between two energy states and either emits or absorbs a photon, are among the most central questions in quantum optics. Observations of individual optical emission and absorption events are possible, and the interpretation of such observations is at the heart of quantum theory.

Various elements of these central considerations are to a degree independent of each other and are understood separately. Among these are (1) the probability that an atomic electron occupies one or another state and the rate at which these occupation probabilities change, (2) the statistical nature of the photons emitted during transitions, (3) correlations between atom and photon states, (4) the characteristic parameters that control the light–matter interaction, and (5) the intrinsically quantum mechanical features of the atom's response to the radiation.

Quantum optics also concerns itself with problems that grow out of these central considerations and whose answers can be expressed within the conceptual framework established by the central problem. Areas related in this way to the core of quantum optics deal, for example, with correlated many-atom light–matter interactions; near-resonant transitions among three and more states of an atom or molecule or solid; optical tests of quantum electrodynamics and measurement theory; multiphoton processes; quantum limits to noise and linewidth; quantum theory of light amplification and laser action; and manifestations of nonlinearity, bistability, and chaos in optical contexts. A wide variety of quantum optical phenomena that bear on one or another of these issues are now known and widely studied. [*See* NONLINEAR OPTICAL PROCESSES.]

B. The *A* and *B* Coefficients of Einstein

The second half of the twentieth century has seen remarkable advances in our understanding of light, of its generation, propagation, and detection. The laser is one manifestation of these advances. Lasers generally depend on the quantum mechanical properties of atoms, molecules, and solids because quantum properties determine the ways that matter absorbs light and emits light. Conversely, the properties of laser beams have made optical studies of quantum mechanics possible in a variety of new ways. It is this interplay that has created the field of quantum optics since about 1960. [*See* LASERS.]

From a different historical perspective, however, quantum optics is much older than the laser and even older than quantum mechanics. The quantum concept first entered physics in 1900 when Planck invented the light quantum to help understand black body radiation. The understanding of other quantum optical phenomena, such as the photoelectric effect, first explained by Einstein in 1905, was well underway almost two decades before a quantum theory of mechanics was properly formulated in 1925 and 1926 by Heisenberg and Schrödinger. Indeed, these early developments in quantum optics played an essential role in the first quantum pictures of atomic matter given by Bohr and others in the period from 1913 to 1923.

Only two parameters are needed to understand the interaction of light with atomic (and molecular) matter, according to Einstein. These two parameters are called A and B coefficients. These coefficients are important because they control the rates of photon emission and absorption processes in atoms, as follows. Let the probability that a given atom is in its nth energy level be written P_n. Suppose there are photons present in the form of radiation with spectral energy density (J/m^3 Hz) denoted by $u(\omega)$. Then the rate at which the probability P_n changes is due to three fundamental processes:

$$(dP_n/dt)_{\text{absorption of light}} = +Bu(\omega)P_m \quad (1a)$$

$$(dP_n/dt)_{\text{spontaneous emission of light}} = -AP_n \quad (1b)$$

$$(dP_n/dt)_{\text{stimulated emission of light}} = -Bu(\omega)P_n \quad (1c)$$

Here P_m is the probability that the atom is in a lower level, the mth, which is related to the nth through the energy relation $E_n - E_m = \hbar\omega$, where $\hbar = h/2\pi$ and h is Planck's famous quantum constant. Einstein's great insight was to include stimulated emission [Eq. (1c)] among the three elementary processes, in effect, to recognize that an atom in an upper energy state could be encouraged by the presence of photons [the existence of $u(\omega)$] to hasten the rate at which it would drop down to a lower state.

The three contributions to the rate of change of P_n shown in Eq. (1a–c) can be added to make an overall single equation for the total rate of change of P_n:

$$dP_n/dt = +Bu(\omega)P_m - AP_n - Bu(\omega)P_n \quad (2a)$$

Einstein applied this equation to an examination of blackbody light. He showed that the steady-state solution

$$P_m/P_n = 1 + A/Bu(\omega) \quad (2b)$$

implies the validity of Planck's formula for $u(\omega)$:

$$u(\omega, T) = \frac{\hbar\omega^3}{\pi^2c^3}\frac{1}{\exp\{\hbar\omega/kT\} - 1} \quad (3)$$

and the value of the prefactor is just the ratio A/B:

$$\frac{A}{B} = \frac{\hbar\omega^3}{\pi^2c^3} \quad (4)$$

where k is Boltzmann's constant and T the temperature in degrees Kelvin. Table I contains the values of physical constants used in evaluating various radiation formulas. For typical optical radiation, the value of the fundamental ratio A/B is approximately 10^{-14} J/m^3 Hz. The corresponding intensity, namely cA/B, is approximately 3×10^{-6} J m^{-2}, or $6\pi \times 10^{-6}$ W m^{-2} per Hz of bandwidth. The value of the spectral intensity of thermal radiation is usually many orders of magnitude lower than this because of the second factor in Eq. (3). At optical wavelengths, the second factor is much smaller than 1 for all temperatures less than about 5000 K.

After the development of a fully quantum mechanical theory of light by Dirac in 1927, it was possible to give expressions for A and B separately:

$$A = \frac{1}{4\pi\varepsilon_0}\frac{4D^2\omega^3}{3\hbar c^3} \quad (5)$$

$$B = \frac{1}{4\pi\varepsilon_0}\frac{4\pi^2D^2}{3\hbar^2} \quad (6)$$

In these formulas we have separated the factor $1/4\pi\varepsilon_0 = 8.9874 \times 10^9$ N m^2/C^2 to display A and B in atomic units as well as SI units, and D denotes the quantum mechanical "dipole matrix element" associated with the $m \rightarrow n$ transition under consideration.

The values of these important coefficients can be obtained for transitions of interest in quantum optics by assuming that the dipole matrix element is approximately equal to the product of the electron's charge and a "typical" electron displacement from the nucleus. Thus, we take $\omega = 2\pi\nu$ and e and h from Table I and $D = er$, with r equal to about 1 to 3 Å ($1\text{–}3 \times 10^{-10}$ m). In this case the values are $A \approx 10^8$ sec^{-1} and $B \approx 10^{22}$ m^2 J^{-1} sec^{-2}, respectively.

The advantage of Einstein's approach, and the reason it still provides one basis for understanding light–matter interactions, is that it breaks the interaction process into its separate elements, as identified above in Eqs. (1a–c). To repeat this important identification, these processes are (a) absorption, (b) spontaneous emission, and (c) stimulated emission.

However, it must be pointed out that Einstein's formulas are not universally valid, and Eqs. (1) and (2) can be seriously misleading in some cases, particularly for laser light. Laser light typically has a very high spectral energy density $u(\omega)$. In this case different formulas and

TABLE I. Physical Constants Used in Evaluating Radiation Formulas

Constant	Value
h (Planck's constant)	6.6×10^{-34} J sec^{-1}
k (Boltzmann constant)	1.38×10^{-23} J K^{-1}
c (Speed of light)	3×10^8 m sec^{-1}
e (Electric charge)	1.6×10^{-19} C
λ (Typical optical wavelength, yellow)	600×10^{-9} m
ν (Typical optical frequency)	5×10^{14} Hz

equations, and even entirely different concepts with their origins in wave mechanics, may be required.

A large body of experimental evidence has accumulated since 1960 showing that many aspects of the interaction between light and matter depend on electric radiation field strength E directly, not only on energy density $u \approx E^2$. Just those aspects of the light–matter interaction that depend directly on E also depend directly on quantum mechanical state amplitudes ψ, not only on their associated probabilities $|\psi|^2$. Issues of coherence and interference of both radiation fields and probability amplitudes are fundamental to these studies. It is principally the experiments and theories that deal with light and matter in this domain that make up the field of quantum optics.

C. Two-State Atom and Maxwell–Bloch Equations

As Einstein's arguments suggest, in quantum optics it is often sufficient to focus attention on just two energy levels of an atom—the two levels that are closest to resonance with the radiation, satisfying the energy condition

$$E_2 - E_1 \approx \hbar\omega$$

where $\omega = 2\pi\nu$ is the angular frequency of the radiation field. This is shown schematically in Fig. 1.

Under these circumstances the wave function of the atom is a sum of the wave functions for the two states

$$\psi(\mathbf{r}, t) = C_1\phi_1(\mathbf{r}) + C_2\phi_2(\mathbf{r}) \tag{7a}$$

For simplicity of description we will assume each level corresponds to a single quantum state and will usually use "level" and "state" synonymously.

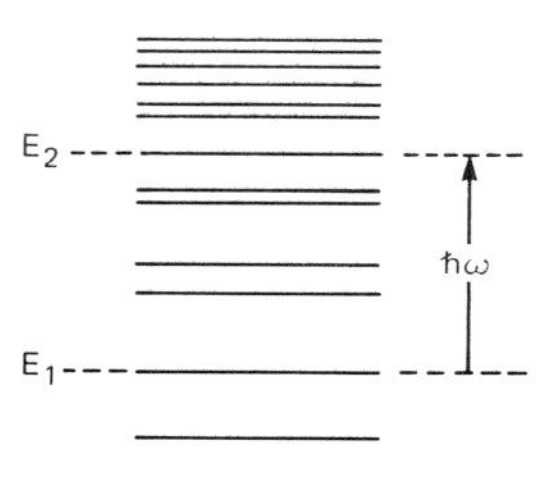

FIG. 1. Schematic energy level diagram showing a two-level subsystem.

The assumption that the electron is certainly in one or the other or a combination of these two levels is expressed mathematically by the equality

$$|C_1|^2 + |C_2|^2 = 1 \tag{7b}$$

Each term in this equation is called a level probability, and all probabilities must add to 1, of course. These probabilities are the quantities labeled by the letter P in the Einstein equations. The C's themselves are not probabilities, and they have no counterpart in classical probability theory. They are called probability amplitudes. It is the remarkable nature of quantum mechanics that the fundamental equation (the Schrödinger equation) governs these amplitudes C, not the probabilities $|C|^2$.

The Schrödinger equation for either one of the amplitudes is

$$i\hbar\, dC_m/dt = E_m C_m + V_{mn} C_n \tag{8}$$

where m and n take either the value 1 or 2, but $m \neq n$. Here $\hbar$ is the usual abbreviation for $h/2\pi$, and V_{mn} is called the interaction matrix element between the atom and the radiation field. In almost all cases of interest in quantum optics, this interaction comes from the potential energy $-\mathbf{d} \cdot \mathbf{E}(\mathbf{r}, t)$ of the atomic dipole in the electric radiation field (a dipole $\mathbf{d} = e\mathbf{r}$ exists because of the separation of the negative electronic charge in its planetary orbit from the position of the positively charged nucleus.

In principle $\mathbf{E}$ is a quantum operator field (see Section III,C). The so-called semiclassical theory of radiation uses its average (or "expectation") value instead. This approximation is usually justified when the field is intense, because quantum fluctuations, which almost always occur at the single photon level, are then negligible. In this section we describe the semiclassical theory and ignore quantum field effects entirely.

The dipole interaction is distributed over the atomic orbitals involved, with the result that

$$\begin{aligned} V_{mn} &= \langle\phi_m| - e\mathbf{r} \cdot \mathbf{E}(\mathbf{r}, t)|\phi_n\rangle \\ &\approx \int d^3r\phi_m^*(\mathbf{r})[-e\mathbf{r} \cdot \hat{\mathbf{e}}E(\mathbf{r}_N, t)]\phi_n(\mathbf{r}) \\ &= -\mathbf{d}_{\mathrm{mn}} \cdot \hat{\mathbf{e}}E(\mathbf{r}_N, t) \end{aligned} \tag{9a}$$

In Eq. (9a) we have written

$$\mathbf{E}(\mathbf{r}, t) = \hat{\mathbf{e}}E(\mathbf{r}, t) \approx \hat{\mathbf{e}}E(\mathbf{r}_N, t)$$

where $\hat{\mathbf{e}}$ is the polarization and $E(\mathbf{r}_N, t)$ is the amplitude of the electric field at the position of the atom (i.e., of the nucleus) $\mathbf{r}_N$ instead of at the position of the electron. This approximation is usually well justified because the electron's orbit, and thus the range of the integral, extends mainly over a region much smaller than an optical wavelength. Thus, over the whole range of the integral $E(\mathbf{r}, t) \approx E(\mathbf{r}_N, t)$ and the only $\mathbf{r}$ dependence comes from the dipole moment $e\mathbf{r}$ itself. This is called the dipole approximation. The integral in Eq. (9a) is called the dipole matrix element, d,

$$d = \hat{\mathbf{e}} \cdot \mathbf{d}_{mn} = \hat{\mathbf{e}} \cdot \int d^3r\phi_m^*(\mathbf{r})e\mathbf{r}\phi_n(\mathbf{r}) \quad (9b)$$

If only one atom is under consideration it is common to put it at the origin of coordinates and write $E(0, t)$ or simply $E(t)$. If several or many atoms are under consideration this is generally not possible [see, e.g., Eq. (20)]. Equation (8) can be written in a simpler form by anticipating an interaction with a quasi-monochromatic radiation field

$$E(t) = \mathscr{E}_0(t)e^{-i\omega t} + \text{c.c.} \quad (10)$$

that is nearly resonant with the atom, where c.c. means complex conjugate. This means that the angular frequency $\omega = 2\pi\nu$ of the radiation field is approximately equal to the angular transition frequency of the atom: $\omega_{21} = (E_2 - E_1)/\hbar$. In this case there is a strong synchronous response of the atom to the radiation, and the equations simplify if one removes the synchronously "driven" component of the atomic response by defining new variables a_n:

$$a_1 = C_1 \quad (11a)$$

$$a_2 = C_2e^{i\omega t} \quad (11b)$$

Note that $|a_1|^2 + |a_2|^2 = |C_1|^2 + |C_2|^2 = 1$, thus, the a's are also probability amplitudes.

If $E(t)$ is sufficiently monochromatic (the field amplitude, i.e., $\mathscr{E}_0$ is practically constant in time; which means $|d\mathscr{E}_0/dt| \ll \omega|\mathscr{E}_0|$), then a rotating wave approximation (RWA) is valid if the anticipated atom field resonance is sufficiently sharp, that is, if $|\omega_{21} - \omega| \ll \omega$. The most important consequence of the RWA is that factors such as $1 + e^{\pm 2i\omega t}$, which appear in the exact equations for the new amplitudes a_1 and a_2, may be replaced by 1 to an excellent approximation. The result is that the frequencies ω_{21} and ω individually pay no further role in the Schrödinger equation, and Eq. (8) takes the extremely compact RWA form:

$$i\, da_1/dt = -\tfrac{1}{2}\chi a_2 \quad (12a)$$

$$i\, da_2/dt = \Delta a_2 - \tfrac{1}{2}\chi a_1 \quad (12b)$$

where

$$\Delta = \omega_{21} - \omega \quad (12c)$$

is called the detuning of the atomic transition frequency from the radiation frequency, and

$$\chi = 2d\mathscr{E}_0/\hbar \quad (12d)$$

is called the Rabi frequency of the interaction. For simplicity, the Rabi frequency χ will be assumed here to be a real number.

For a strictly monochromatic field (time-independent $\mathscr{E}_0$) the solution to Eq. (12a and b) is easily found in terms of $\Omega = \sqrt{\chi^2 + \Delta^2}$, where Ω is called the generalized or detuning-dependent Rabi frequency. In the most important single case the atom is in the lower state at the time the interaction begins, which means that $a_1(0) = 1$ and $a_2(0) = 0$. In this case the solution is

$$a_1(t) = [\cos \Omega t/2 + (i\, \Delta/\Omega)\sin \Omega t/2]e^{-i\Delta t/2} \quad (13a)$$

$$a_2(t) = i[(\chi/\Omega)\sin \Omega t/2]e^{-i\Delta t/2} \quad (13b)$$

and the corresponding probabilities are

$$P_1 = \cos^2(\Omega t/2) + (\Delta/\Omega)^2\sin^2(\Omega t/2) \quad (14a)$$

$$P_2 = (\chi/\Omega)^2\sin^2(\Omega t/2) \quad (14b)$$

These solutions describe continuing oscillation of two-level probability between levels 1 and 2. They have no steady state. Figure 2 shows graphs of $P_2(t)$. One already sees, therefore, that the quantum amplitude equations [Eqs. (12a–d)] make strikingly different predictions from the two-level equations of Einstein [recall the steady-state solution Eq. (2b)].

Within the RWA, the dynamics remain unitary or probability conserving: $P_1 + P_2 = 1$ for

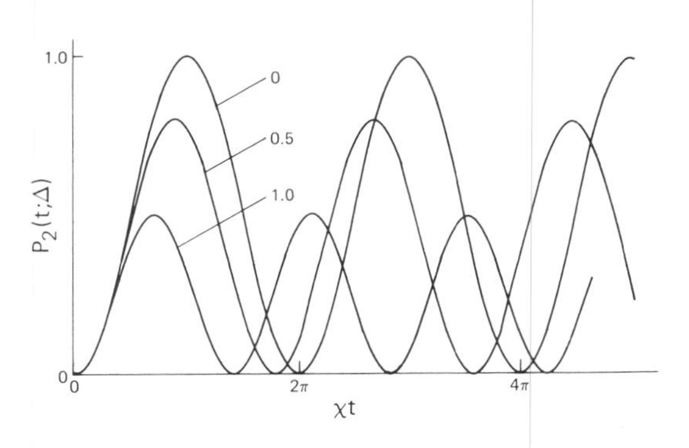

FIG. 2. Rabi oscillations for different values of Δ/χ.

all values of t. As a result, the compact version [Eq. (12)] of Schrödinger's equation remains valid even for very strong interactions between the atom and the radiation. This is important when considering the effects on atoms of very intense laser fields. The limits of validity of Eq. (12) are determined to a large extent by a generalized statement of the limits of the RWA:

$$\omega_{21} \text{ and } \dot{\omega} \gg \sqrt{\Delta^2 + \chi^2} \qquad (15)$$

There are two other real quantities associated with the amplitudes a_1 and a_2 of the radiation–atom interaction. They belong to the atomic dipole's expectation value $\langle \mathbf{d} \rangle$ which, according to Eqs. (7a), (9b), and (11), is

$$\langle \psi(t)|e\mathbf{r}|\psi(t)\rangle = a_1(t)a_2^*(t)e^{i\omega t}\mathbf{d}_{21} + \text{c.c.} \qquad (16)$$

The new quantities are designated by u and v:

$$u = a_1^* a_2 + a_1 a_2^* \qquad (17a)$$

$$v = i(a_1^* a_2 - a_1 a_2^*) \qquad (17b)$$

Along with u and v, a third variable, the atomic inversion $w = P_2 - P_1$ plays an important role. The solutions for u, v, and w corresponding to Eqs. (13) and (14) above are

$$u = (\Delta\chi/\Omega^2)(1 - \cos \Omega t) \qquad (18a)$$

$$v = (\chi/\Omega)\sin \Omega t \qquad (18b)$$

$$w = -(1/\Omega^2)(\Delta^2 + \chi^2 \cos \Omega t) \qquad (18c)$$

which share the oscillatory properties of the probabilities. They also obey the important conservation law:

$$u^2 + v^2 + w^2 = 1 \qquad (18d)$$

which is the same as $|a_1|^2 + |a_2|^2 = 1$.

For many purposes in quantum optics the semiclassical dipole variables u and v, and the atomic inversion w, are the primary atomic variables. They obey equations which are equivalent to Eq. (8) and take the place of Schrödinger's equation for the level's probability amplitudes a_1 and a_2 (or C_1 and C_2):

$$du/dt = -\Delta v \qquad (19a)$$

$$dv/dt = \Delta u + \chi w \qquad (19b)$$

$$dw/dt = -\chi v \qquad (19c)$$

All of these considerations assume that the field is monochromatic or nearly so.

Although the field $\mathbf{E}(\mathbf{r}, t)$ is not considered an operator in the semiclassical formulation, it can still have a dynamical character, which means it is not prescribed in advance, but obeys its own equation of motion. It is naturally taken to obey the Maxwell wave equation, with the usual source term $\mu_0 d^2/dt^2 \mathbf{P}(z, t)$. In the semiclassical theory $\mathbf{P}(z, t) = N\langle \mathbf{d} \rangle$ where N is the density of two-level atoms in the source volume and $\langle \mathbf{d} \rangle$ the average dipole moment of a single atom, already calculated in Eq. (16).

As Eq. (16) indicates, $\langle \mathbf{d} \rangle$ is quasimonochromatic, essentially because a two-level atom is characterized by a single transition frequency. Therefore the $\mathbf{E}$ that it generates may also be regarded as monochromatic or nearly so, as Eq. (10) assumed. This internal consistency is an important consideration in the semiclassical theory. In many cases it is also suitable to regard the field as having a definite direction of propagation, say z, and to neglect its dependence on x and y (plane wave approximation):

$$\mathbf{E}(z, t) = \hat{\mathbf{e}}\mathscr{E}(z, t)e^{-i(\omega t - kz)} + \text{c.c.} \qquad (20)$$

where $k = \omega/c$ and the amplitude or envelope function $\mathscr{E}(z, t)$ is a complex generalization of $\mathscr{E}_0(t)$ in Eq. (10). It obeys a "reduced" wave equation in variables z and t:

$$[\partial/\partial z + \partial/\partial ct]\mathscr{E}(z, t) = i(\pi N d\omega/4\pi\varepsilon_0 c)[u - iv] \qquad (21)$$

The reduced wave equation Eq. (21) and Eqs. (19) for u, v, and w are called the semiclassical coupled *Maxwell–Bloch equations*.

Inspection of the semiclassical Eqs. (19) and (21) reveals their major flaw. One easily sees that the semiclassical approach to radiation theory does not include the process of spontaneous emission. A completely excited atom will not emit a photon in this theory. That is, if the atom is in its excited state, then $a_2 = 1$ and $a_1 = 0$, so $w = 1$ and $u = v = 0$. Thus, according to Eq. (21) no field can be generated. By the same token, if $\mathscr{E} = 0$ then $\chi = 0$ and $dw/dt = 0$, and no evolution toward the ground state can occur. The flaw in the semiclassical Maxwell–Bloch approach arises from the assumption that all of the dynamics can be reduced to a consideration of average values, that is, averages of dipole moment and inversion and field strength. In reality, fluctuations about average values are an important ingredient of quantum theory and essential for spontaneous emission.

Spontaneous emission does not play a dominant role in many quantum optical processes, particularly those involving strong radiation fields. The semiclassical Maxwell–Bloch equations give an entirely satisfactory explanation of these effects, and the next sections describe some of them.

In nature there are no actual two-level atoms, of course. However, the selection rules for allowed optical dipole transitions are sufficiently restrictive, and optical resonances can be sufficiently sharp that very good approximations to two-level atoms can be found in nature. A good example is found in a pair of levels in atomic sodium, and they have been used in recent quantum optical experiments.

In sodium the nucleus has spin $I = 3/2$, and the lowest electronic energy level is $3S_{1/2}$ with hyperfine splitting ($F = 1$ and $F = 2$). The first excited levels $3P_{1/2}$ and $3P_{3/2}$ are responsible for the well-known strong sodium D lines of Fraunhofer in the yellow region of the optical spectrum at wavelengths 589.0 nm and 589.6 nm. The "two-level" transition is between the $m_F = +2$ magnetic sublevel of $F = 2$ of the ground state and the $m_F = +3$ magnetic sublevel of $F = 3$ in the $3P_{3/2}$ excited state. Circularly polarized dye laser light can be tuned within the 15-MHz natural linewidth of the upper level, and the $\Delta m = +1$ selection rule for circular polarization prevents excitation of the other magnetic sublevels.

Spontaneous decay from the upper state, which obeys no resonance condition, is also restricted in this example. The final state of spontaneous decay could, in principle, have $m_F = 4$, 3, 2. However, in sodium there are no $m_F = 4$ or 3 states below the $3P_{3/2}$ state, and only one $m_F =$ 2 state, namely the one from which the excitation process began. Thus, the state of the sodium atom is very effectively constrained to this two-state subset out of the infinitely many quantum states of the atom.

II. Induced Atomic Coherence Effects

The ability of an atomic system to have a coherent dipole moment during an extended interaction with a radiation field is a necessary condition for a wide variety of effects associated with quantum optical resonance. The dipole moment should be coherent in the sense that it retains a stable phase relationship with the radiation field. Long-term phase memory may be difficult to achieve, for example because spontaneous emission and collisions destroy phase memory at a rate that is typically in the range 10^8 sec^{-1} or much greater. Coherence is also lost if the radiation bandwidth is too broad. The importance of dipole coherence effects shows that light–matter interactions do depend most fundamentally on the dipole moment and electric field strength, not on radiation intensity and the B coefficient.

A. π Pulses and Pulse Area

The solution for the level probabilities given in Eq. (14) is the single most important example in quantum optics of the coherent response of an atom to a monochromatic radiation field. "Coherence" in this context has several connected meanings, all associated with the well-phased steady oscillation of the probabilities considered as a function of time. This time dependence was shown in Fig. 2 for several values of the parameters Δ and χ. The significance of the Rabi frequency χ is clear—it is the frequency at which the inversion oscillates when the atom and the radiation are at exact resonance, when $\Delta = 0$, since

$$w(t) = -\cos \chi t \tag{22}$$

Fruitful connections to the spin vector formalism of magnetic resonance physics are obtained by regarding the triplet $[u, v, w]$ as a vector $\mathbf{S}$. Equations (19) for u, v, and w can then be written in compact vector form:

$$d\mathbf{S}/dt = \mathbf{Q} \times \mathbf{S}, \tag{23}$$

where $\mathbf{Q}$ is the vector of length Ω with components $[-\chi, 0, \Delta]$. The vector $\mathbf{Q}$ can be called the torque vector for $\mathbf{S}$, which is variously called the pseudo-spin vector, the atomic coherence vector, and the optical Bloch vector.

This vector formulation in Eq. (23) of Eqs. (19) shows that the evolution of the two-level atom in the presence of radiation is simply a rotation in a three-dimensional space. The space is only mathematical in quantum optics because the components of the optical Bloch vector are not the components of a single real physical vector, whereas in magnetic resonance they are the components of a real magnetic moment. In both cases the nature of the torque equation leads to a useful conservation law: $d/dt(\mathbf{S} \cdot \mathbf{S}) = 0$. That is, the length of the Bloch vector is constant. In the u, v, w notation this means $u^2 + v^2 + w^2 = 1$, and it implies that the vector $[u, v, w]$ traces out a path on a unit sphere as the two-level atom changes its quantum state.

Further consideration of the on-resonant atoms ($\Delta \to 0$ and $\Omega \to \chi$) gives information about the interaction of atoms with (nonmonochromatic) pulsed fields. According to Eq. (19a) $u(t)$ can be neglected if $\Delta = 0$ and a new form of solutions to Eqs. (19b and c) follows immediately:

$$v(t) = -\sin \phi(t) \tag{24a}$$

$$w(t) = -\cos \phi(t) \tag{24b}$$

where $\phi(t)$ is called the "area" of the electric field because it is related to the time integral of the electric field envelope:

$$\phi(t) = \int_0^t \chi(t')\, dt' = (d/\hbar) \int_0^t \mathscr{E}_0(t')\, dt' \tag{25}$$

In the monochromatic limit when $\mathscr{E}_0$ = constant, then $\phi(t) \rightarrow \chi t$.

Recall that the rotating wave approximation (RWA) will not permit $\mathscr{E}_0$ to vary too rapidly. If τ_p is the pulse length, then $\mathscr{E}_0$ is not too rapidly varying if τ_p is long enough, namely if $1/\tau_p \ll \omega$. Pulse lengths in the range 1 nsec $\geq \tau_p \geq$ 10 fsec are of interest and are compatible with the RWA [1 fsec (femtosecond) = 10^{-15} sec].

The significance of $\phi(t)$ is evident in Eq. (24). The term ϕ is just the angle of rotation of the on-resonance Bloch vector $\mathbf{S} = [u, v, w]$ during the passage of the light pulse. A light pulse with $\phi = \pi$ is called a *π pulse*, that is, a pulse that rotates the initial vector $[0, 0, -1]$, which points down, through 180° to the final vector $[0, 0, +1]$, which points up. Thus, a π pulse completely inverts the atomic probability, taking the atom from the ground state to the excited state. A *2π pulse* is one that returns the atom via a 360° rotation of its Bloch vector to its initial state, after passing through the excited state. The remarkable nature of this rotation is not so much that an inversion of the atomic state is possible, but that it can be done fully coherently and without regard for the pulse shape. Only the total integral of $\mathscr{E}_0(t)$ is significant.

The physics behind Bloch vector rotation is essentially the same in magnetic resonance, but perhaps less remarkable since the Bloch vector in that case is a physically "real" magnetic moment. In optical resonance there is no "real" electric moment vector whose Cartesian components can be identified with $[u, v, w]$. Early evidence of the response of the optical Bloch vector to coherent pulses was obtained in experiments of Tang, Gibbs, Slusher, Brewer, and others (see Sections II,B and II,C), and further experiments probing these properties continue to be of interest.

B. Photon Echoes

Photon echoes are an example of spontaneous recovery of a physical property that has been dephased after many relaxation times have elapsed. In the case of echoes the physical property is the macroscopic polarization of a sample of two-level atoms. Recovery of the polarization means recovery of the ability to emit radiation, and the signature of a photon echo is the appearance of a burst of radiation from a long-quiescent sample of atoms. The burst occurs at a precisely predictable time, not randomly, and is due to a hidden long-term memory. The echo principle was discovered and spin echoes were observed by Hahn in 1950 in magnetic resonance experiments. Photon echoes were first observed by Hartmann and co-workers in 1965.

Photon echoes are possible when a sample of atoms is characterized by a broad distribution $g(\Delta)$ of detunings. This may occur in a gas, for example, because of Doppler broadening or in a solid because of crystalline inhomogeneities. For the latter reason it is said that $g(\Delta)$ indicates the presence of *inhomogeneous broadening*. The Maxwellian distribution of velocities in a gas is equivalent to a Maxwellian distribution of detunings since each atom's Doppler shift is proportional to its velocity. If the number of atoms in the sample is N, then the fraction with detuning Δ is given by $Ng(\Delta)\, d\Delta$, where

$$g(\Delta) = \frac{1}{\sqrt{(2\pi)}\delta\omega_D} e^{(-1/2)[(\Delta-\bar{\Delta})^2/(\delta\omega_D)^2]} \tag{26}$$

Here $\bar{\Delta}$ is the average detuning and $\delta\omega_D$ is the Doppler linewidth, $\delta\omega_D = \omega_L(kT/mc^2)^{1/2}$.

The Bloch vector picture is well suited for describing photon echoes. Assume that the Bloch vectors for a collection of N atoms all lie in the equatorial $(u - v)$ plane of the unit sphere along the negative v axis, that is, $\mathbf{S} = [0, -1, 0]$. This arrangement can be accomplished by excitation from the ground state $[0, 0, -1]$ with a strong $\pi/2$ pulse for which $\mathbf{Q} = [-\chi, 0, \Delta] \approx [-\chi, 0, 0]$ if $\chi \gg \Delta$. The total Bloch vector is then $\mathbf{S}_N = [U, V, W] = [0, -N, 0]$, which corresponds to a macroscopic dipole moment of magnitude Nd. After this excitation pulse the sample begins to radiate coherently at a rate appropriate to the dipole moment Nd. However, following the excitation pulse we again have $\chi = 0$ and $\mathbf{Q} = [0, 0, \Delta]$, and according to Eq. (23) the individual Bloch vectors immediately begin to precess freely about the w axis (in the $u - v$ plane) at rates depending on their individual detunings Δ. Specifically, $(u - iv)_t = (u - iv)_0 e^{-i\Delta t}$ for an atom with detuning Δ, if $\chi = 0$.

As a consequence, the total Bloch vector will rapidly shrink to zero in a time $\delta t \approx 1/\delta\omega$, where $\delta\omega$ is the spread in angular velocities in the N atom collection. That is, the coherent sum of N dipoles rapidly dephases and $[0, N, 0] \to [0, 0, 0]$, with the result that the sample quickly stops radiating. This is called *free precession decay*, or free induction decay after the similar effect in magnetic resonance, because the decay is due only to the fact that the individual dipole components $u - iv$ get out of phase with each other due to their different precession speeds, not because any individual dipole moment is decaying.

If the distribution of Δ's is determined by the Doppler effect, as in Eq. (26), then since $(u - iv)_0 = i$, one finds

$$\begin{aligned} U - iV &= iN \int d\Delta\ g(\Delta)e^{-i\Delta t} \\ &= iNe^{-i\bar{\Delta} t} \exp[-\tfrac{1}{2}(\delta\omega_D^2)^2 t] \end{aligned} \tag{27}$$

The decay is very rapid if the Doppler width is large. Typically $\delta\omega_D \approx 10^9$ to 10^{10} sec^{-1}, thus, within a few tenths of a nanosecond $U - iV \to 0$ and radiation ceases.

The echo method consists in applying a second pulse to the collection of atoms after U and V have vanished, that is, at some time $T \gg (\delta\omega_D)^{-1}$. Each single atom with its detuning Δ, after the time interval T, still has $(u - iv)_T = ie^{-i\Delta T}$. Only the *sum* of these u and v values is zero due to their different Δ values. The torque vector describing the second pulse is $\mathbf{Q} = [-\chi, 0, \Delta]$, which can again be approximated by $[-\chi, 0, 0]$ if $\chi \gg \Delta$. The effect of the second pulse is again to rotate the Bloch vectors about the u axis. The ideal second pulse is a π pulse, in which case $u \to u$, $v \to -v$ and $w \to -w$, that is, a rotation by 180°. Thus, for times $t \geq T$ after the π pulse the $u - v$ components are

$$(u - iv)_t = (u - iv)_{T'} e^{-i\Delta(t-T)}$$

where T' signifies the rotated coherence vector immediately following the π pulse at time T:

$$(u - iv)_{T'} = (u + iv)_T = -ie^{i\Delta T}$$

The remarkable feature of a π pulse is that it accomplishes an effective reversal of time. Following it the Bloch vectors do not continue to dephase, but begin to rephase:

$$\begin{aligned} (u - iv)_t &= -ie^{i\Delta T} e^{-i\Delta(t-T)} \\ &= -ie^{-i\Delta(t-2T)} \end{aligned} \tag{28}$$

Thus, at the exact time $t = 2T$, the individual u's and v's all rephase perfectly: $[u, v, w] = [0, 1, 0]$. Their Bloch vectors are merely rotated 180° from their positions after the original $\pi/2$ pulse, and they again constitute a macroscopic dipole moment $\mathbf{S}_N = [0, N, 0]$, and therefore the collection will begin to radiate again.

Because of the timing of this radiation burst, exactly as long (T) after the π pulse as the π pulse was after the original $\pi/2$ pulse, it is natural to call the signal a *photon echo*. Because of the separation by the intervals T and $2T$ from the π and $\pi/2$ excitation pulses, the observation of an echo can be in practice an observation that is very noise free. Following the echo pulse, the coherence vectors again immediately begin to dephase, but they can again be rephased using the same method, and a sequence of echoes can be arranged.

Because of collisions with other atoms, the individual u and v values will actually get smaller during the course of an echo experiment, independent of their Δ values. Thus, the rephased Bloch vector is not quite as large as the original one. One of the possible uses of an echo experiment is to measure the rate of collisional decay of u and v, say as a function of gas pressure, by measuring the echo intensity in a sequence of experiments with different values of T since the echo intensities will get smaller as collisions reduce the length of the rephased Bloch vector. The way in which Eq. (23) is rewritten to account for collisions is taken up in Section II,D.

C. Self-Induced Transparency and Short Pulse Propagation

The polarization of a dielectric medium of two-level atoms, such as any atomic vapor excited near to resonance, is linearly related to the incident electric field strength $\mathscr{E}_0$ at low-light intensities but becomes nonlinear in the strong-field, short-pulse regime. This is most evident in Eq. (18), where the sine and cosine functions contain all powers of $\chi = 2d\mathscr{E}_0/\hbar$. These nonlinearities have striking consequences for optical pulse propagation.

If a 2π pulse is injected into a collection of on-resonant two-level atoms, it cannot give any energy to them because after its passage the atoms have been dynamically forced back into their initial state. Thus, a 2π pulse has a certain energy stability and so does a 4π pulse and every $2n\pi$ pulse for the same reason. However, all other pulses are obviously not stable since they must give up some energy to the atoms if they do not rotate the atomic coherence vectors all the way back to their initial positions.

The effect on the injected pulse due to atomic absorptions is given by the Maxwell equation (21). When there is a broad distribution $g(\Delta)$ of detunings among the atoms, then Eq. (21) leads to a so-called area theorem, a nonlinear propagation equation for a pulse of area ϕ:

$$d\phi(z)/dz = -\tfrac{1}{2}N\sigma \sin \phi(z) \qquad (29)$$

Here $\phi(z)$ means the total pulse area $\int \chi(z, t')dt'$, where the integral extends over the duration of the pulse. The solution is $\tan \phi(z)/2 = e^{-\alpha z/2}$, and the attenuation coefficient is $\alpha = N\sigma$, where N is the density of atoms and σ is the inhomogeneous absorption cross section:

$$\sigma = \int g(\Delta')\sigma_a(\Delta')d\Delta' \qquad (30)$$

where $\sigma_a(\Delta')$ is the single-atom cross section (see Section II,E). For very weak pulses with $\phi \ll \pi$, one can replace $\sin \phi(z)$ by $\phi(z)$ and recover from Eq. (29) the usual linear law for pulse propagation: $d\phi(z)/dz = -(\alpha/2)\phi(z)$, which predicts exponential attenuation of the pulse: $\phi(z) = \phi(0) \exp[-\alpha z/2]$. The factor of $\frac{1}{2}$ arises because ϕ is proportional to the electric field amplitude, not the intensity.

Remarkably, one of the "magic" pulses with $\phi = 2\pi n$, which do not lose energy while propagating, also preserves its shape. This is the 2π pulse, for which there is a constant-shape solution of the reduced Maxwell–Bloch equations:

$$\chi(z, t) = (2/\tau_p)\text{sech}[(t - z/V)/\tau_p] \qquad (31)$$

That is, the entire pulse moves at the constant velocity V, which can be several orders of magnitude slower than the normal light velocity in the medium. In ordinary light propagation this would correspond to an index of refraction $n \approx 1000$. All of these remarkable features were discovered by McCall and Hahn in the 1960s and labeled by the term *self-induced transparency* to indicate that a light pulse could manipulate the atoms in a dielectric in such a way that the atoms cannot absorb any of the light.

Self-induced transparency is an example of soliton behavior. The nonlinearity of the coupled Maxwell–Bloch equations opposes the dispersive character of normal light transmission in a polarizable medium to permit a steady nondispersing solitary wave (or soliton) [Eq. (31)] to propagate unchanged. This happens only for fields sufficiently strong that the 2π pulse condition can be met. The Maxwell–Bloch equations can in many cases be shown to be equivalent to the sine–Gordon soliton equation or generalizations of it.

D. Relaxation

Both photon echoes and self-induced transparency demonstrate the existence of optical phenomena depending on $\chi \approx \mathscr{E}$, and not on $\mathscr{E}^2$, that is, on the Maxwell–Bloch equations and not on the Einstein rate equations. How are these two approaches to light–matter interactions connected? To answer this question it is necessary to extend the scope of the Bloch equations and include the effects of line-broadening and relaxation processes.

The upper levels of any system have a finite lifetime, and so $|a_2|^2$ cannot oscillate indefinitely as Eq. (14b) implies, but must relax to zero. This is most fundamentally due to the possibility of spontaneous emission of a photon, accompanied by a transition in the system to the lower level. Such transitions occur at the rate A, as in Einstein's equation (1b).

Other relaxation processes also occur. For example, collisions with other atoms cause unpredictable changes in the state of a given two-level atom. These collisional changes typically affect the dipole coherence of the two-level atom instead of the level probabilities, that is, they affect u and v instead of w. We suppose that the rate of such processes is γ. Although γ does not appear in Einstein's rate equations, its existence is implied. This will be clarified below.

The fundamental equations of optical resonance, Eqs. (19), can be rewritten to include these relaxations as follows:

$$du/dt = -\Delta v - (\gamma + A/2)u \qquad (32a)$$

$$dv/dt = +\Delta u + \chi w - (\gamma + A/2)v \qquad (32b)$$

$$dw/dt = -\chi v - A(w + 1) \qquad (32c)$$

In the absence of relaxation ($\gamma = A = 0$), the solutions of Eqs. (32) are purely oscillatory [recall Eq. (18)] and are said to be *coherent*. In the absence of the radiation field ($\chi = 0$), the on-resonance solutions are completely nonoscillatory:

$$u = u_0 e^{-(\gamma + A/2)t}$$

$$v = v_0 e^{-(\gamma + A/2)t}$$

$$w = -1 + (w_0 + 1)e^{-At} \qquad \text{all for } \chi = 0$$

These solutions are said to be *incoherent*. In each case "coherence" refers to the existence of oscillations with a well-defined period and phase. In Bloch's notation, the relaxation rates are written $\gamma + A/2 = 1/T_2$, where $A = 1/T_1$, and T_1 and T_2 are called the "longitudinal" and "transverse" rates of relaxation.

Relaxation theory is a part of statistical physics, and in quantum theory statistical properties of atoms and fields are usually discussed with the aid of the quantum mechanical density matrix ρ. The density matrix for a two-level atom has four elements, ρ_{11}, ρ_{12}, ρ_{21}, and ρ_{22}. These are related to u, v, and w by the equations $u = \rho_{12} + \rho_{21}$, $v = -i(\rho_{12} - \rho_{21})$, and $w = \rho_{22} - \rho_{11}$, which have the inverse forms:

$$\rho_{12} = \tfrac{1}{2}(u + iv) = \langle a_1 a_2^* \rangle \quad (33a)$$

$$\rho_{21} = \tfrac{1}{2}(u - iv) = \langle a_2 a_1^* \rangle \quad (33b)$$

$$\rho_{11} = \tfrac{1}{2}(1 - w) = \langle a_1 a_1^* \rangle \quad (33c)$$

$$\rho_{22} = \tfrac{1}{2}(1 + w) = \langle a_2 a_2^* \rangle \quad (33d)$$

Here the brackets $\langle \ldots \rangle$ are understood to refer to an average over an ensemble of parameters and variables inaccessible to direct and deterministic evaluation, such as the inital positions and velocities of all the atoms in a collection that may collide with and disturb a typical two-level atom.

The equations given in Eqs. (19) for u, v, and w can also be obtained from ρ via the Liouville equation of quantum statistical mechanics: $i\hbar\, d\rho/dt = [H, \rho]$. The equations for the density matrix elements (Eqs. (33)] are

$$d\rho_{21}/dt = -(\gamma + A/2 + i\Delta)\rho_{21} - (i\chi/2)(\rho_{22} - \rho_{11}) \quad (34a)$$

$$d\rho_{12}/dt = -(\gamma + A/2 - i\Delta)\rho_{12} + (i\chi/2)(\rho_{22} - \rho_{11}) \quad (34b)$$

$$d\rho_{11}/dt = A\rho_{22} - (i\chi/2)(\rho_{12} - \rho_{21}) \quad (34c)$$

$$d\rho_{22}/dt = -A\rho_{22} + (i\chi/2)(\rho_{12} - \rho_{21}) \quad (34d)$$

We now demonstrate the connection between Einstein's equations and the quantum optical Eqs. (34). Consider the weak-field limit $\chi \ll |\gamma + A/2 + i\Delta|$. In this limit the rate of change of the "off-diagonal" density matrix elements ρ_{21} and ρ_{12} is dominated by the first factor $-(\gamma + A/2 \pm i\Delta)$, and both ρ_{21} and ρ_{12} decay rapidly. If in addition $A \ll |\gamma + A/2 \pm i\Delta|$, then ρ_{22} and ρ_{11} change relatively slowly, and so ρ_{21} and ρ_{12} rapidly adjust themselves to the small quasi-steady values:

$$\rho_{21} \approx -(i/2)\chi[\gamma + A/2 + i\Delta]^{-1}(\rho_{22} - \rho_{11}) \quad (35a)$$

$$\rho_{12} \approx (i/2)\chi[\gamma + A/2 - i\Delta]^{-1}(\rho_{22} - \rho_{11}) \quad (35b)$$

These solutions show that the off-diagonal elements of the density matrix can be determined from constant numerical factors and combinations of the diagonal density matrix elements. They can then be eliminated from Eqs. (34c and d).

This procedure is referred to as *adiabatic elimination* of off-diagonal coherence because the remaining equations for ρ_{11} and ρ_{22} no longer exhibit coherence. That is, they no longer have oscillatory solutions. The term *adiabatic* is appropriate in the sense that ρ_{21} and ρ_{12} are entrained by the slower ρ_{11} and ρ_{22}. The reverse procedure, the elimination of ρ_{11} and ρ_{22} in favor of ρ_{21} and ρ_{12}, is not possible because the reverse inequality $|\gamma + A/2 \pm i\Delta| \ll A$ is not possible.

These adiabatic off-diagonal solutions, once inserted into Eq. (34) lead to an equation for the slowly changing ρ_{22} as follows:

$$\frac{d\rho_{22}}{dt} = -A\rho_{22} - \left[\frac{1}{2}\chi^2 \frac{\gamma + A/2}{\Delta^2 + (\gamma + A/2)^2}\right](\rho_{22} - \rho_{11}) \quad (36)$$

Recall $\rho_{22} = |a_2|^2$ is the probability that the atom is in its upper state, and thus plays the same role as P_n in the Einstein equation (2a). Similarly ρ_{11} plays the role here of P_m there. By comparing Eqs. (2a) and (36) one sees that they are identical in form and content if one identifies the coefficient of ρ_{11} in Eq. (36) with the coefficient of P_m in Eq. (2a). In other words, the density matrix equations [Eqs. (34a–d)] of quantum optics contain Einstein's equation in the weak-field and adiabatic limits $\chi \ll |\gamma + A/2 \pm \Delta|$ and $A \ll |\gamma + A/2 \pm i\Delta|$. The B coefficient can be derived in this limit (see Section II,E) if one properly interprets the factor $\frac{1}{2}\chi^2(\gamma + A/2)/[\Delta^2 + (\gamma + A/2)^2]$.

Relaxation processes affect the Maxwell field as well as the Bloch variables. To determine the form of relaxation to assign to Maxwell equation (21) it is sufficient to consider the conservation of energy. From Eq. (21) and (19c) it follows that

$$[\partial/\partial z + \partial/\partial ct]|\mathscr{E}|^2 = -(N\hbar\omega/4\varepsilon_0 c)[\partial w/\partial t]$$

which is equivalent to an equation for photon flux and level probability

$$[\partial/\partial z + \partial/\partial ct]\Phi(z, t) = -N\,\partial P_2/\partial t \quad (37a)$$

since $\hbar\omega\Phi \equiv I = 2c\varepsilon_0|\mathscr{E}|^2$ and $w = 2P_2 - 1$.

Equation (37a) is Poynting's theorem for a "one-dimensional" medium. It expresses the conservation of photon flux in terms of atomic excitations. However, in the absence of the resonant two-level atoms ($N = 0$), Eqs. (37) predicts $\Phi(z, t) = \Phi_0(z - ct)$, which means that the photon flux has the constant value $\Phi(z_0)$ at every point $z = z_0 + ct$ that travels with the pulse. This is contradictory to ordinary experience in two respects. The medium that is host to the two-level atoms (e.g., other gas atoms, a solvent, or a crystal lattice) always causes both dispersion and absorption. They can be taken into account by modifying Eq. (37) slightly:

$$[\partial/\partial z + \kappa + \partial/\partial v_g t]\Phi(z, t) = -N\, \partial P_2/\partial t \quad (37b)$$

where v_g is the group velocity for light pulses in the medium and κ its linear attenuation coefficient.

This form of the flux equation implies a similar alteration of Maxwell's equation (21):

$$[\partial/\partial z + \kappa/2 + \partial/\partial v_g t]\mathscr{E} = i(Nd\omega/4\varepsilon_0 c)[u - iv] \quad (38)$$

This form of Maxwell's equation is useful in describing the elements of laser theory (see Section II,G).

E. Cross Section and the *B* Coefficient

How does one use quantum optical expressions to obtain basic spectroscopic formulas, such as for the absorption cross section and B coefficient? That is, given expressions derived from Eqs. (34), which are based on the Rabi frequency χ instead of the more familiar radiation intensity I or spectral energy density $u(\omega)$, how does one recover a cross section, for example? Consider the quantum optical derivation of the Einstein formula in Eq. (36). The transition rate (absorption rate or stimulated emission rate) can be identified readily. With the abbreviation $\beta = \gamma + A/2$, one obtains:

$$\text{abs. rate} = \frac{1}{2}\chi^2 \frac{\beta}{\Delta^2 + \beta^2} \quad (39a)$$

The absorption rate is a peaked function of Δ whose value drops to $\frac{1}{2}$ of the maximum value at $\Delta = {}^{\pm}\beta$. Thus, β is called the *half width at half maximum* (HWHM) of the absorption lineshape. This shows that relaxation leads to *line broadening,* and since β applies equally and individually to every atom, it is an example of a *homogeneous linewidth.* Recall (Section I,B) that inhomogeneous broadening is not a characteristic of individual atoms but of a collection of them. From expression (39a) at exact resonance one obtains the relationship:

$$\text{reasonant transition rate} = \chi^2/\text{full linewidth}$$

This is the single most concise relationship between the parameters of incoherent optical physics (transition rate and linewidth) and the central parameter of coherence (Rabi frequency). It holds in situations much more general than the present example and allows rapid and accurate translation of formulas from one domain to the other.

With the use of $\chi = 2d\mathscr{E}/\hbar$ and $I = 2c\varepsilon_0\mathscr{E}^2$, Eq. (39a) becomes

$$\text{abs. rate} = \frac{D^2 I}{3\varepsilon_0 c\hbar^2} \frac{\beta}{\Delta^2 + \beta^2} \quad (39b)$$

The introduction of the new dipole parameter D here is based on the assumption that all orientations of the atomic dipole matrix element $\mathbf{d}_{21}$ are possible (in case, e.g., all magnetic sublevels of the main levels 1 and 2 are degenerate). Then $d^2 \equiv |\mathbf{e} \cdot \mathbf{d}_{21}|^2$ must be replaced by its spherical average, that is, by $D^2/3$, where $D^2 = |\mathbf{d}_{21} \cdot \mathbf{d}_{12}|$. We assume this is appropriate in the remainder of Section II.

The atomic absorption cross section σ_a is, by definition, the ratio of the rate of energy absorption, $\hbar\omega_{21} \times$ (abs. rate), to the energy flux (intensity) I of the photons being absorbed. From Eq. (39b) this ratio is

$$\sigma_a(\omega; \omega_{21}) = \frac{D^2\omega}{3\varepsilon_0\hbar c} \frac{\beta}{(\omega_{21} - \omega)^2 + \beta^2} \quad (40)$$

where $\gamma = \beta - A/2$ is the specifically collisional contribution to the halfwidth. Formula (40) shows that a quantum optical approach to light absorption through the weak field limit of Eqs. (34) gives conventional results of atomic spectroscopy. If Doppler broadening is present, then Eq. (40) must be integrated over the Doppler distribution of detunings Eq. (26) as was done in Eq. (30).

Lineshape plays a key role in understanding the relationship of Eqs. (39) to the Einstein expression (1a) relating absorption rate to the B coefficient. The absorption cross section can be written $\sigma_a = \sigma_t S_a(\omega; \omega_{21})$, where σ_t is the total frequency-integrated cross section:

$$\sigma_t = \pi D^2 \omega_{21}/3\varepsilon_0\hbar c \quad (41a)$$

and S_a is the atomic lineshape

$$S_a(\omega; \omega_{21}) = \frac{\beta/\pi}{(\omega_{21} - \omega)^2 + \beta^2} \quad (41b)$$

which is normalized according to $\int d\omega\, S_a = 1$, and in this case has a Lorentzian shape. A lineshape also exists for the radiation field and is expressed by $u(\omega)$, the spectral energy density function. One connects $u(\omega)$ with I by the frequency integral $c \int u(\omega')d\omega' = I$. In the monochromatic case $cu(\omega')$ takes the idealized singular form $cu(\omega') = I\delta(\omega - \omega')$, and Eq. (39b) is the result for the absorption rate.

In the general nonmonochomatic case, the expression for absorption rate involves the integrated overlap of $u(\omega)$ and the atomic lineshape function:

$$\text{abs. rate} = \frac{c\sigma_t}{\hbar\omega_{21}} \int S_a(\omega'; \omega_{21})u(\omega')d\omega' \quad (42)$$

This has the desired limiting form, involving $u(\omega)$, and not $I = c \int u(\omega)d\omega$, if the spectral width $\delta\omega_L$ of $u(\omega)$ is very broad in comparison to the width β of the absorption lineshape S_a. In this limit, which is implicit in Einstein's discussion, S_a acts in Eq. (42) like a δ-function peaked at $\omega' = \omega_{21}$, and $u(\omega)$ is evaluated at $\omega = \omega_{21}$. From Eq. (42) one can then extract the Einstein B coefficient:

$$B = \frac{\pi D^2}{3\varepsilon_0\hbar^2} \quad (43)$$

A simple relation obviously exists between B and σ_t, namely $\hbar\omega_{21}B = c\sigma_t$.

Another quantity of interest is $\sigma_a(0)$, the on-resonance or peak cross section:

$$\sigma_a(0) = \frac{D^2\omega_{21}}{3\varepsilon_0\hbar c\beta} \quad (44)$$

By definition, $\sigma_a(0) = \sigma_a(\omega = \omega_{21})$; or, conversely, $\sigma_a(\omega; \omega_{21}) = \pi\beta\sigma_a(0)S_a(\omega; \omega_{21})$. Representative values of $\sigma_a(0)$ for an optical resonance transition lie in the range $\sigma_a(0) \approx 10^{-13}$ to 10^{-17} cm^2 for absorption linewidths in the range $\beta \approx 10^8$ to 10^{11} sec^{-1}.

F. Strong Field Criterion and Saturation

The inequality $\chi \ll |\gamma + A/2 \pm i\Delta|$, on which the absorption rate formula (39) and thus the Einstein B coefficient is based, is important in quantum optics and radiation physics generally because it provides a criterion for distinguishing weak radiation fields from strong radiation fields. The inequality implies that there is a critical value $\mathscr{E}_{cr}$ for field strength that gives a universal meaning to the terms *weak field* and *strong field,* namely $\mathscr{E}_0 \ll \mathscr{E}_{cr}$ and $\mathscr{E}_0 \gg \mathscr{E}_{cr}$, respectively, where:

$$\mathscr{E}_{cr} = (\hbar/d)|\gamma + A/2 \pm i\Delta| = (\hbar/d)|\beta \pm i\Delta| \quad (45)$$

However, since the parameters γ, A, d, and Δ may vary by many orders of magnitude from case to case, the numerical value of $\mathscr{E}_{cr}$ may fall anywhere in an extremely wide range. Thus, it is possible that in one experiment a laser with the power level 10^{20} W m^{-2} must be designated "weak," while another laser in a different experiment with the power level 1 W m^{-2} must be considered "strong." This factor of 10^{20} is one indication of the great extent of the domain of quantum optics.

In conventional spectroscopy one sometimes encounters saturation effects. These are of course strongest in the strong field regime and are of interest in quantum optics.

There are two distinct time regimes of saturation phenomena. If $\chi \gg \beta$, there is a range of times $t \ll \delta T \approx \beta^{-1}$ that can still contain many Rabi oscillations since $\delta T \gg \chi^{-1}$. During the time $0 \leqq t \ll \delta T$, the fully coherent undamped formula (14b) can be used for the upper state probability. On average, during this time the probability that the atom is in its upper level is

$$P_2 = \tfrac{1}{2}\chi^2/[\Delta^2 + \chi^2] \qquad \text{(short time average)} \quad (46a)$$

which is a Lorentzian function of $\Delta = \omega_{21} - \omega$ with the power-broadening halfwidth $\delta\omega_p = \chi$. Power broadening is a saturation effect, because if $\chi \gg \Delta$, then $P_2 \to \frac{1}{2}$ on average, which is obviously saturated, that is, unchanged if χ is made still larger.

Another saturation regime exists for long times, $t \gg \delta T \approx \beta^{-1}$. The solution of Eqs. (24) for $P_2 = \rho_{22}$ in this limit is

$$P_2 = \frac{1}{2}\chi^2 \frac{\beta/A}{\Delta^2 + \beta^2 + \chi^2\beta/A} \quad (46b)$$

In this case χ begins to dominate the width when $\chi^2 > \beta A$, which defines the saturation value of χ:

$$\chi_{sat} = \sqrt{(\beta A)} \quad (47)$$

The power-broadening part of the width of Eq. (46b) is different than in Eq. (46a), namely $\delta\omega_p = \chi\sqrt{(\beta/A)}$. Depending on the value of $\gamma \equiv \beta - A/2$, the power width here can be anything between a minimum of $\chi/\sqrt{2}$, if $\gamma = 0$, and a maximum of $\chi\sqrt{(\gamma/A)}$, if $\gamma \gg A$. This distinction between the saturated power-broadened linewidth $\delta\omega_p$ predicted by Eq. (46a) for short times and by Eq. (46b) for asymptotically long times has caused some confusion in the past. Further study indicates that Eq. (46b) breaks down for sufficiently large χ, basically because Bloch-type relaxation, such as assumed in Eqs.

(32), becomes invalid. The first experimental reports of this regime of optical resonance were made by Brewer, DeVoe, Mossberg, and others in the early 1980s.

In the case of asymptotically long times the expression for P_2 can be written in several ways, using Eq. (47) or (40):

$$P_2 = \frac{\frac{1}{2}(\chi/\chi_{\text{sat}})^2}{1 + (\Delta/\chi_{\text{sat}})^2 + (\chi/\chi_{\text{sat}})^2} = \frac{\sigma_a I}{\hbar\omega_{21} A + 2\sigma_a I} = \frac{\Phi/\Phi_{\text{sat}}}{1 + 2\Phi/\Phi_{\text{sat}}} \tag{48}$$

where $\Phi = I/\hbar\omega_{21}$, and we have introduced the saturation flux required for saturation of the transition $\Phi_{\text{sat}} = A/\sigma_a$ (or $= 1/T_1\sigma_a$ in Bloch's notation, which is more appropriate if there are other contributions than spontaneous emission rate A to the level lifetimes). Figure 3 shows the effect of both power broadening and saturation on the steady-state probability P_2.

In common with all two-level saturation formulas, Eq. (48) and Fig. 3 predict $P_2 = \frac{1}{2}$ at most. However, this prediction is valid only for weak fields or for long times. As the solutions in Eq. (14) and in Fig. 2 show, strong monochromatic resonance radiation can repeatedly transfer the electron to the upper level with $P_2 \approx 1$ for times $\ll \beta^{-1}$. Experiments that show $P_2 > \frac{1}{2}$ have been practical only with lasers. Laser pulses are both short and intense, allowing $\chi \gg \beta$ as well as $t \ll \beta^{-1}$.

G. Semiclassical Laser Theory

The coupled Maxwell–Bloch equations can be used as the basis for laser theory. It is a semiclassical theory, but still adequate to illustrate the most important results, such as the roles of inversion and feedback, the existence of threshold, and the presence of frequency pulling in steady-state operation. A fully detailed semiclassical theory of laser operation was already given by Lamb in 1964.

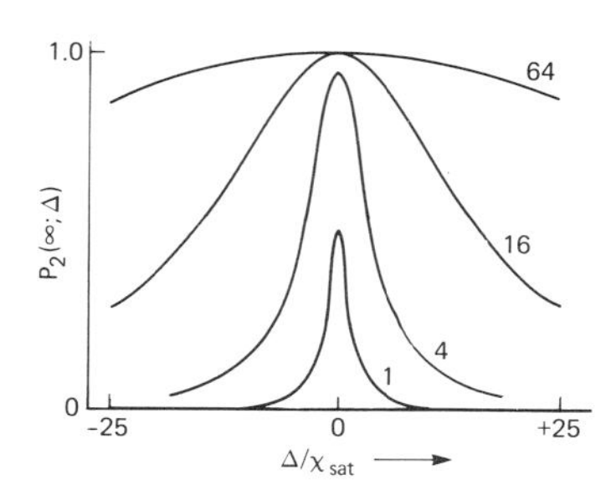

FIG. 3. Population saturation for different values of χ/χ_{sat}.

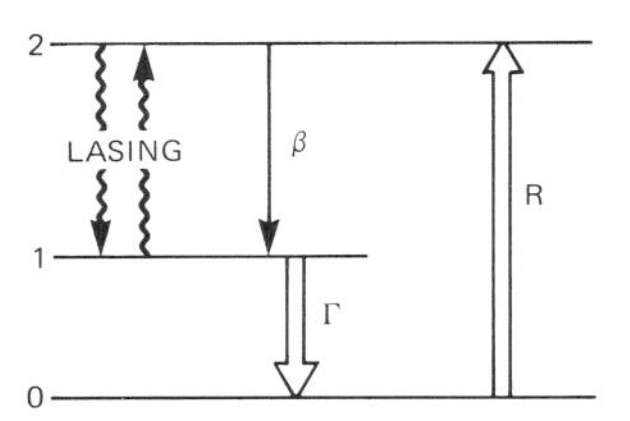

FIG. 4. Schematic diagram of three-level laser system.

The density matrix equations (24) must be modified to allow for pumping of the upper laser level and to allow the lower level to decay to a still lower level labeled 0 and not previously needed. The rates of these new processes are denoted R and Γ, respectively. This is basically the scheme of a so-called three-level laser, as sketched in Figure 4.

The diagonal equations change to

$$d\rho_{11}/dt = -\Gamma\rho_{11} + A\rho_{22} + (i/2)\chi_m(\rho_{12} - \rho_{21}) \tag{49}$$

$$d\rho_{22}/dt = -A\rho_{22} - (i/2)\chi_m(\rho_{12} - \rho_{21}) + R \tag{50}$$

Here the index m or χ_m shows that we are dealing with the electric field of the mth mode of the laser cavity. The laser is easy to operate only if $\Gamma \gg A$ because only then can a large positive inversion be maintained between the two levels with a value of R that is not too high and without seriously depleting the population of level 0. Under other conditions, the equation for the density matrix element ρ_{00} would also have to be included. The off-diagonal density matrix equations (24a and b) are basically unchanged if we interpret γ as including another contribution $\Gamma/2$, where Γ is the decay rate of the lower level probability to level 0 in Fig. 4.

The reduced Maxwell equation (28) is now useful. We will ignore the difference between v_g and c, and write Eq. (38) in terms of χ instead of $\mathscr{E}$, and use $\rho_{21} = \frac{1}{2}(u - iv)$ to find

$$[\partial/\partial z + \kappa/2 + \partial/\partial ct]\chi_m = i(Nd^2\omega/\varepsilon_0\hbar c)\rho_{21} \tag{51}$$

Note that in a cavity κ can arise principally from mirror losses and not from absorption in the cavity volume. The same adiabatic elimination of dipole coherence undertaken in Eqs. (35) provides the value for ρ_{21} to insert into Eq. (51), which in steady state ($\partial\chi_m/\partial t = 0$) becomes:

$$[\partial/\partial z + \kappa/2 - \tfrac{1}{2}(g + i\delta k)]\chi_m = 0 \quad (52)$$

where

$$g + i\delta k = (ND^2\omega/3\varepsilon_0\hbar c)[w_{ss}/(\beta + i\Delta)] \quad (53)$$

and now $\beta = \gamma + (A + \Gamma)/2$. By using Eq. (40) one can obtain

$$g = N\sigma_a(\Delta)w_{ss} \quad (54a)$$

$$c\delta k = -(gc/\beta)(\omega_{21} - \omega) \quad (54b)$$

where Eq. (54a) has been used to simplify Eq. (54b). Here w_{ss} denotes the inversion $\rho_{22} - \rho_{11}$ in steady state.

It is clear that g is the intensity gain coefficient, since if $g > \kappa$, Eq. (52) would predict exponential growth, $|\chi_m|^2 \approx \exp[(g - \kappa)z]$, in an open-ended medium such as a laser amplifier. It is thus clear that $g = \kappa$ is the threshold condition for amplification or laser operation. Also obviously, g is not positive unless w_{ss} is positive. Recall that in an ordinary noninverted medium $w = -1$, and in this case $g = -N\sigma_a = -\alpha$, where α is the ordinary absorption coefficient.

Rather than growing indefinitely, the field in a laser cavity must conform to the spatial period determined by the mirrors. Thus, at steady state $\chi_m(z) \approx \chi_0 \exp[i\Delta k_m z]$, where the phase $\Delta k_m z$ is the difference between the actual phase of steady-state laser operation $k_m z$ and the phase $kz = \omega z/c$ that was assumed initially in defining the field carrier wave and envelope functions in Eq. (20). Since χ_m does not depend on the transverse coordinates x and y, this theory cannot describe transverse mode structure, and $k_m = m\pi/L$, where L is the cavity length and m is the longitudinal mode number. Operating values could be $m \approx 10^6$ and $L \approx 10$ cm, in which case the laser would run at a frequency near to $m\pi c/L \approx 3 \times 10^{15}\ \text{sec}^{-1} \approx 5 \times 10^{14}$ Hz.

In laser operation $\partial/\partial z$ can be replaced in Eq. (52) by $i\Delta k_m$ and the imaginary part of Eq. (52) becomes $\Delta k_m = \tfrac{1}{2}\delta k$, which leads directly to a condition for the operating frequency ω:

$$\omega_m - \omega = -(\kappa c/2\beta)(\omega_{21} - \omega) \quad (55)$$

This requires ω to lie somewhere between the empty cavity frequency $\omega_m = m\pi c/L$ and the natural transition frequency ω_{21} of the atom, and it is said that the two-level laser medium "pulls" the operating frequency away from the cavity frequency $m\pi c/L$. The solution of Eq. (55) for ω is:

$$\omega = [\beta\omega_m + \tfrac{1}{2}\kappa c\omega_{21}]/[\beta + \tfrac{1}{2}\kappa c] \quad (56)$$

H. Optical Bistability

The input–output relationships between light beams injected into and transmitted through an empty optical cavity are linear relationships. However, the situation changes dramatically if the cavity contains atoms. This is obvious in the case of laser action. However, even if the atoms are not pumped, their nonlinearities can be significant.

Consider a laser beam injected into an optical cavity filled with two-level atoms. For simplicity we consider the case where the frequencies of the atoms, the cavity, and the injected laser light are all equal: $\omega = \omega_c = \omega_{21}$. Only the v–w Bloch equation are needed then:

$$dv/dt = -\beta v + (\chi_m + \chi_0)w \quad (57a)$$

$$dw/dt = -A(1 + w) - (\chi_m + \chi_0)v \quad (57b)$$

Here χ_0 and χ_m are the Rabi frequencies associated with the injected field strength and the cavity mode field generated by the atoms, and $\beta = \gamma + A/2$. The Maxwell equation for the internally generated χ_m is

$$(\kappa/2 + \partial/\partial ct)\chi_m = (Nd^2\omega/2\varepsilon_0\hbar c)v \quad (58)$$

Note that there is no term $i\Delta k_m$ from $\partial/\partial z$, as there is in the discussion of laser operation, only because of the three-way resonance assumption.

The question is, how does the presence of the atoms in the cavity affect the transmitted signal? Since the transmitted signal differs only by a factor of mirror transmissivity from the total field in the cavity $\chi_t = \chi_m + \chi_0$, we ask for the relation between χ_0 and χ_t. The dynamical evolution of the system is complicated. Early attention was given to this situation in the 1970s by Szöke, Bonifacio, Lugiato, McCall, and others.

As with the laser, steady state is sufficiently interesting, so we put $dv/dt = dw/dt = d\chi/dt = 0$ and solve for χ_0 or χ_t. They obey a simple but nonlinear relation:

$$\chi_0 = \chi_t + (\alpha/\kappa)(\beta A)[\chi_t/(\beta A + \chi_t^2)] \quad (59)$$

where α is the (on-resonance) two-level medium's absorption coefficient, $\alpha = ND^2\omega/3\varepsilon_0\hbar c\beta$, and $\beta A = \chi_{sat}^2$, the saturation parameter identified in Eq. (47). If both χ_0 and χ_t are normalized with respect to χ_{sat} by defining $\xi = \chi_0/\chi_{sat}$ and $\eta = \chi_t/\chi_{sat}$ then one finds the dimensionless relation:

$$\xi = \eta + (\alpha/\kappa)[\eta/(1 + \eta^2)] \quad (60)$$

Of course the inverse relation $\eta = \eta(\xi)$, that is, the total field strength as a function of the input

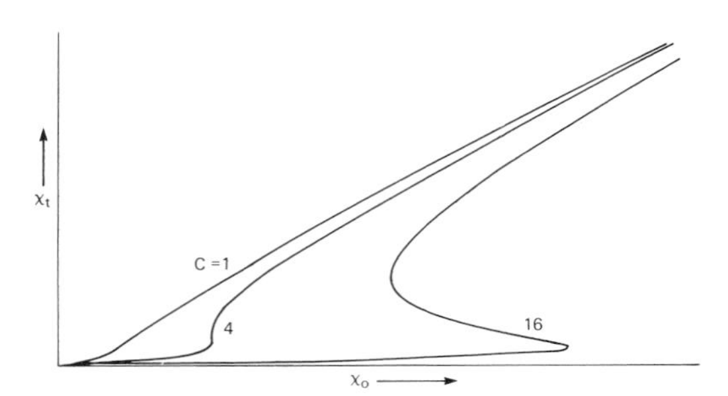

FIG. 5. Bistable input–output curves for different values of $C = \alpha/2\kappa$.

field strength, is more interesting. Figure 5 shows the important features of this relation. It demonstrates that the total field η is double-valued as a function of input field ξ if α/κ is larger than a certain critical value. In this simple model the critical value is $\alpha/\kappa = 8$, and the vertical segment in the central curve is an indication of this. The double-valued nature of the curves for $\alpha/\kappa > 8$ is termed *optical bistability*. In the bistable region of the third curve hysteresis can occur, and a hysteresis loop is shown.

The elements of a primitive optical switch are evident in the bistable behavior shown here. If the input field is held near to the lower turning point, then a very small increase in ξ can lead to a very large jump in η, the transmitted field. The possibility of optical logic circuits and eventually optical computers is clearly suggested even by the simple model described here, and efforts being made around the world to realize these possibilities in a practical way have already achieved limited success.

III. Radiation Coherence and Statistics

A. Coherence of Light

In quantum optics the coherence of light is treated statistically. The need for a statistical description of light, whether quantized or classical, is practically universal since all light beams, even those from well-stabilized lasers, have certain residual random properties that are not uniquely determined by known parameters. These random properties lead to fluctuations of light. A satisfactory description can be based on a scalar electric field:

$$E(\mathbf{r}, t) = E^{(+)}(\mathbf{r}, t) + E^{(-)}(\mathbf{r}, t) \quad (61a)$$

where $E^{(+)}$ is the *positive frequency part* of E. That is, $E^{(+)}$ is the inverse Fourier transform of the positive frequency half of the Fourier transform of E. From this definition, one has $[E^{(-)}(\mathbf{r}, t)]^* \equiv E^{(+)}(\mathbf{r}, t)$. The split-up into positive and negative frequency parts $E^{(\pm)}(\mathbf{r}, t)$ is motivated by the great significance of quasi-monochromatic fields, for which one can write

$$E^{(+)}(\mathbf{r}, t) = \mathscr{E}(\mathbf{r}, t)e^{-i\omega t} \quad (61b)$$

$$E^{(-)}(\mathbf{r}, t) = \mathscr{E}(\mathbf{r}, t)^* e^{i\omega t} \quad (61c)$$

The term *quasi-monochromatic* means that $\mathscr{E}(\mathbf{r}, t)$ is only slowly time dependent, that is, $|d\mathscr{E}/dt| \ll \omega|\mathscr{E}|$.

The intensity of the light beam associated with this electric field is given by:

$$\begin{aligned} I(\mathbf{r}, t) &= c\varepsilon_0 E(\mathbf{r}, t)^2 \\ &= c\varepsilon_0[2|\mathscr{E}(\mathbf{r}, t)|^2 + \mathscr{E}^2(\mathbf{r}, t)e^{-2i\omega t} + \text{c.c.}] \end{aligned} \quad (62)$$

where c.c. means complex conjugate. In principle I is rapidly time dependent, but the factors $e^{\pm 2i\omega t}$ oscillate too rapidly to be observed by any realistic detector. That is, a photodetector can respond only to the average of Eq. (62) over a finite interval, say of length T beginning at t, where $T \gg 2\pi/\omega$. The $e^{\pm 2i\omega t}$ terms average to zero and one obtains:

$$\bar{I}_T(t) = (1/T)\int_0^T I(t + t')dt' = 2c\varepsilon_0|\mathscr{E}(t)|^2 \quad (63)$$

The overbar thus denotes a coarse-grained average value that is not sensitive to variations on the scale of a few optical periods. We have dropped the $\mathbf{r}$ dependence as a simplification. If the beam is steady and the averaging interval T is long enough, then both the length T and the beginning value $I(t)$ are unimportant, and $\bar{I}_T(t)$ is also independent of t. This is the property of a class of fields called *stationary*. Obviously even nonstationary fields can be considered stationary in the sense of a long T average, and we will adopt stationarity as a simplification of our discussion and write $\bar{I}_T(t)$ simply as $\bar{I}$.

Consider now the operation of a *Michelson interferometer* (Fig. 6). An incident beam is split into two beams a and b, and after traveling different path lengths, say ℓ and $\ell + \delta$, the beams are recombined and the beam intensity $I_{a+b} = c\varepsilon_0[E_a(t) + E_b(t)]^2$ is measured. If the beam splitter sends equal beams with field strength $E(t)$ and intensity $\bar{I}$ into each path and there is no absorption during the propagation, then one

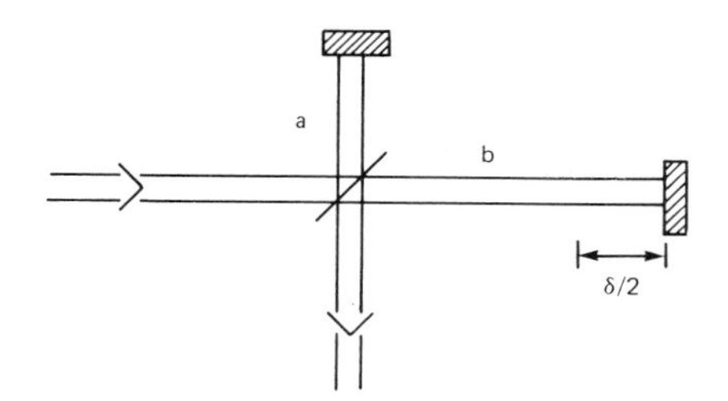

FIG. 6. Sketch of Michelson interferometer.

measures

$$\bar{I}_{a+b} = (T)^{-1} \int_0^T c\varepsilon_0[E_a(t + t') + E_b(t + t')]^2 dt'$$
$$= (T)^{-1} \int_0^T c\varepsilon_0[E(t + t') + E(t + \delta/c + t')]^2 dt'$$
$$= \bar{I} + \bar{I} + 4c\varepsilon_0 \times \mathrm{Re}[\langle \varepsilon^*(t)\varepsilon(t + \delta/c)\rangle e^{-i\omega\delta/c}] \quad (64)$$

where Re means "real part" and the angular brackets indicate the time average. If $\delta = 0$, then $\bar{I}_{a+b} = 4\bar{I}$, because the two beams interfere fully constructively. The quantity $\Gamma(\delta/c) = \langle \varepsilon^*(t)\varepsilon(t + \delta/c)\rangle e^{-i\omega\delta/c}$ is called the *mutual coherence function* of the electric field, and the appearance of fringes at the output plane of the interferometer is due to the variation of Γ with δ. From the factor $e^{-i\omega\delta/c}$ it is clear that a fringe shift (a shift from one maximum of $\bar{I}_{a+b}$ to the next) corresponds to a shift of δ by $2\pi c/\omega = \lambda$.

The mutual coherence function is conveniently normalized to its maximum value which occurs when $\delta = 0$, and the normalized function $\gamma(\delta/c)$ is called the *complex degree of coherence*. That is,

$$\gamma(\delta/c) = \langle \varepsilon(t)\varepsilon^*(t + \delta/c)\rangle e^{+i\omega\delta/c} / \langle \varepsilon(t)\varepsilon^*(t)\rangle \quad (65)$$

and the output intensity can be written:

$$\bar{I}_{a+b} = 2\bar{I}[1 + \mathrm{Re}\,\gamma(\delta/c)] \quad (66)$$

where Re γ must satisfy $-1 \leqq \mathrm{Re}\,\gamma \leqq 1$.

It is common experience that if the path difference δ is made too great in the interferometer, the fringes are lost and the output intensity is simply the sum of the intensities in the two beams. This can be accounted for by introducing a *coherence time* τ for the light by writing

$$\gamma(\delta/c) = \gamma(0)e^{-|\delta|/c\tau}e^{+i\omega\delta/c} \quad (67)$$

This representation for γ has the correct behavior since it vanishes whenever δ is large enough, specifically whenever $\delta \gg c\tau$. For obvious reasons, $c\tau$ is called the *coherence length* of the light. This does not explain the fundamental origin of the coherence time τ, but the lack of such a deep understanding of the light beam can be one reason that a statistical description is necessary in the first place. Typically one adopts Eq. (67) as a convenient empirical relation and interprets τ from it.

The fringe visibility is usually the important quantity that describes an interference pattern, not the absolute level of intensity. The *visibility* is defined by $V = (\bar{I}_{\max} - \bar{I}_{\min})/(\bar{I}_{\max} + \bar{I}_{\min})$, and this is directly related to the complex degree of coherence. Since

$$\mathrm{Re}\,\gamma = |\gamma(0)|e^{-|\delta|/c\tau}\cos(\omega\delta/c + \phi)$$

one has

$$V = \frac{4\bar{I}|\gamma(0)|e^{-|\delta|/c\tau}}{4\bar{I}} = |\gamma(\delta/c)| \quad (68)$$

Thus, the magnitude of the complex degree of coherence is a directly measurable quantity.

One of the foundations of quantum optics was established by Wolf and others in classical coherence theory when it became understood in the 1950s how to describe optical interference effects in terms of measurable autocorrelation functions such as γ. In a sense, the first example of this was provided much earlier by Wiener in 1930 when he showed that the spectrum $S(\omega)$ of a stationary light field is given essentially by the Fourier transform of γ, considered as a function of a time difference τ rather than a path difference δ:

$$S(\omega) = 2\bar{I}\,\mathrm{Re}\int d\tau' e^{-i\omega\tau'}\gamma(\tau') \quad (69)$$

As a specific example, suppose a light beam with carrier frequency ω_L has a Gaussian degree of coherence, with coherence time τ, that is, $\gamma(\tau') = |\gamma(0)|e^{i\omega_L\tau'}\exp[-\frac{1}{2}(\tau'/\tau)^2]$. This is another example, similar to the exponential $\gamma(\delta/c)$ given above, in which a simple analytic function is used to model the normal fact that correlation functions have finite coherence, that is, that $\gamma(\tau) \to 0$ as $\tau \to \infty$. In this example the spectrum is Gaussian:

$$S(\omega) = 2\bar{I}|\gamma(0)|\sqrt{(2\pi\tau^2)}\exp[-\tfrac{1}{2}(\omega - \omega_L)^2\tau^2] \quad (70)$$

The effective bandwidth is the frequency range over which $S(\omega)$ is an appreciable fraction of its peak value. In this case the spectrum is centered at $\omega = \omega_L$, and the bandwidth is given by $\Delta\omega = 2\pi\Delta\nu = 1/\tau$.

This is an example of the general rule that the bandwidth is the inverse of the coherence time. A laser beam with bandwidth $\Delta\nu = 100$ MHz has a coherence time $\tau = 1.6$ nsec and a coherence length $\Delta\delta = c\tau = 0.5$ m, while sunlight with a bandwidth six orders of magnitude broader ($\Delta\nu = 10^{14}$ Hz) has a coherence time $\tau = 1.6 \times 10^{-3}$ psec $= 1.6$ fsec and a coherence length $c\tau = 0.5\ \mu$m that are six orders of magnitude smaller. In the latter case, $\Delta\nu \approx \nu$, and $c\tau \approx \lambda$, so sunlight cannot be called quasi-monochromatic in any sense.

A hierarchy of correlation functions can be defined for statistical fields. One denotes by the *degree of first-order coherence* the normalized first-order correlation of positive and negative frequency parts of the field:

$$g^{(1)}(\mathbf{x}_1, \mathbf{x}_2) = \frac{\langle E^{(+)}(\mathbf{x}_1)E^{(-)}(\mathbf{x}_2)\rangle}{[\langle|E^{(+)}(\mathbf{x}_1)|^2\rangle\langle|E^{(-)}(\mathbf{x}_2)|^2\rangle]^{1/2}} \tag{71}$$

where $\mathbf{x} = (\mathbf{r}, t)$ for short. In the case of a quasi-monochromatic field $g^{(1)}$ is just the same as γ defined above. The terms "first-order coherent," "partially coherent," and "incoherent" refer to light with $|g^{(1)}|$ equal to 1, between 1 and 0, and 0, respectively.

Definition (71) and other average values can be interpreted in an ensemble sense, as well as in a time average sense. In the ensemble interpretation the angular brackets $\langle ... \rangle$ are associated with an average over "realizations," that is, over possible values of the fields at given points $\mathbf{x}$, weighted by appropriate probabilities. Two important examples are (i) so-called chaotic fields, fields that are stationary complex Gaussian random processes, and (ii) constant intensity fields. The chaotic distribution is characteristic of ordinary (thermal) light such as omitted by the sun, flames, light bulbs, etc. The constant intensity distribution is an idealization associated with highly stabilized single-mode laser (coherent) light. The ergodic assumption that time and ensemble averages are equal is usually made. We will write $\langle I \rangle$ and $\bar{I}$ interchangeably.

If $p[\mathscr{E}, t]d^2\mathscr{E}$ is the probability that the complex field amplitude has the value $\mathscr{E}$ within $d^2\mathscr{E} = \mathscr{E}d|\mathscr{E}|d\phi$ in the complex $\mathscr{E}$ plane, then the thermal (chaotic) distribution for the complex amplitude $\mathscr{E}(t)$ is a Gaussian function:

$$p^{\text{th}}[\mathscr{E}] = (2\pi\mathscr{E}_0^2)^{-1/2}\exp[-\tfrac{1}{2}(|\mathscr{E}|^2/\mathscr{E}_0^2)] \quad \text{(thermal)} \tag{72a}$$

and the coherent distribution is a delta function:

$$p^{\text{coh}}[\mathscr{E}] = (|\mathscr{E}_0|/\pi)\delta^2(|\mathscr{E}|^2 - |\mathscr{E}_0|^2) \quad \text{(coherent)} \tag{72b}$$

The spatial and temporal coherence properties of radiation, measured via γ or $g^{(1)}$ or $S(\omega)$, determine the degree to which the fields at two points in space and time are able to interfere. By the use of pinholes, lenses and filters any sample of ordinary light can be made equally monochromatic and directional as laser light. Its spatial and temporal coherence properties can be made equal to those of any laser beam. If the laser light is then suitably attenuated so that it has the same low intensity as the ordinary light, one may ask whether any important differences can remain between the ordinary light and the laser light.

Surprisingly, the answer to this fundamental question is yes. The differences can be found in higher order correlation functions, involving $\mathscr{E}^*\mathscr{E}$ to powers higher than the first, such as $\langle E^2(t)E^2(t + \tau)\rangle$. These can be referred to as intensity correlations. The first measurements of optical intensity correlation functions were made on thermal light by Brown and Twiss in the 1950s before lasers were available.

Discussions of intensity correlations are typically made with the aid of a quantity called the *degree of second-order coherence* and denoted $g^{(2)}$. Higher order degrees of coherence are also defined. If the fields are stationary only the time delays between the measurement points are significant, and one has:

$$g^{(2)}(\tau_1) = \langle I(t)I(t + \tau_1)\rangle/\langle I\rangle^2 \tag{73a}$$

$$g^{(3)}(\tau_1, \tau_2) = \langle I(t)I(t + \tau_1)I(t + \tau_2)\rangle/\langle I\rangle^3 \tag{73b}$$

and so on. Because of stationarity we can put $t = 0$ everywhere. We will deal mostly now with $g^{(2)}$ and no confusion should occur if we omit the index (2) whenever we have a single delay time τ. The assumption that the light is stationary gives $g(\tau) = g(-\tau)$. Since I is an intrinsically positive quantity, it is clear that $g(\tau)$ is positive. There is no upper limit, so g satisfies $\infty > g(\tau) \geqq 0$. In addition, for any distribution of I one has $\langle I^2\rangle \geq \langle I\rangle^2$, so obviously $g(0) \geq 1$; and one can also show that $g(0) \geq g(\tau)$.

With the aid of the thermal and coherent probability distributions given above, several higher order coherence functions can be calculated immediately. For example, if $\tau_1 = \tau_2 = ... = 0$, the nth order moments of the intensity are simply $\langle I^n\rangle \approx \int |\mathscr{E}|^{2n}p[\mathscr{E}]d^2\mathscr{E}$:

$$\langle I^n \rangle = n!\,\langle I \rangle^n = n!\bar{I}^n \qquad \text{(thermal)} \quad (74a)$$

$$\langle I^n \rangle = \langle I \rangle^n = \bar{I}^n \qquad \text{(coherent)} \quad (74b)$$

For large values of n these are obviously quite different, even if $\bar{I}$ is the same for two light beams, one thermal and the other coherent. The difference plays a role in multiphoton ionization experiments with $n = 2$ and larger. This is one example of a fundamental difference between ordinary light and ideal single-mode laser light.

It should be clear that in addition to thermal light and laser light there may be still other forms of light, characterized by distributions other than those in Eq. (72). Quantum theory also predicts that there can even be kinds of light beams for which the underlying probability distribution does not exist in a classical sense, for example, it may be negative over portions of the range of definition. Generalized phase space functions that play the quantum role of classical probability densities were developed by Glauber and Sudarshan in the 1960s. These are still a principal theoretical tool in studies of photon counting.

B. Photon Counting

In photon counting experiments the arrival of an individual photon is registered at a photodetector, which is essentially just a specially designed phototube that gives a signal when an arriving photon ionizes an atom at the phototube surface. A typical experiment consists of many runs of the same length T in each of which the number of photons registered by the photodetector is counted. The counts can be organized into a histogram (Fig. 7), which is interpreted as giving the probability $P_n(T)$ for counting n photons during an interval of length T.

An expression for $P_n(T)$ follows from a consideration of the photodetection process, which begins with the ionization of a single atom by a single photon at the surface of the phototube. In any event, the rate of counting is proportional to the intensity of the light beam, so one writes $\alpha I(t)dt$ for the probability of counting a photon at the time t in the interval dt. The factor α takes account of the atomic variables governing the ionization process as well as the geometry of the phototube. It was first shown by Mandel in the 1950s that $P_n(T)$ is then given by

$$P_n(T) = \langle e^{-\alpha \int_0^T dt' I(t')} [\alpha \textstyle\int_0^T dt' I(t')]^n / n! \rangle \quad (75)$$

where the average is over the variations in intensity during the (relatively long) counting intervals. Alternatively, it can be considered an average over an ensemble of identically prepared runs in which the value of $\bar{I}$ is statistically distributed in some way.

The simplest example occurs if the light intensity does not fluctuate at all, which is characteristic of an ideal single-mode laser (coherent) light beam. In this case Eq. (75) is independent of the t' average and, with $\bar{n} = \alpha \bar{I} T$,

$$P_n = e^{-\bar{n}} (\bar{n})^n / n! \qquad \text{(coherent)} \quad (76)$$

which is the well-known Poisson distribution. It is easily verified that $\bar{n} = \sum nP_n(T)$. That is, as its form indicates, $\bar{n}$ is the average number of photons counted in time T. It is a feature of the Poisson distribution that its dispersion is equal to its mean:

$$\langle (\Delta n)^2 \rangle = \langle n \rangle^2 2 - \langle n \rangle^2 = \bar{n} \qquad \text{(Poisson)} \quad (77)$$

A plot of an ideal Poisson photocount distribution is shown in Fig. 7. The Poisson distribution is also called the coherent or coherent-state distribution because it is predicted by the quantum theory of light to be applicable to a radiation field in a so-called coherent state (see Section III,C). A well-stabilized single-mode laser gives the best realization of a coherent state in practice.

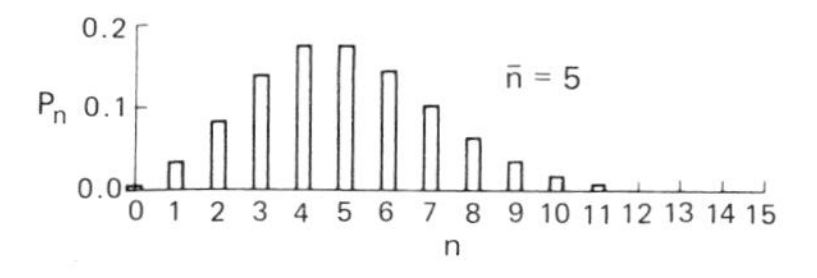

FIG. 7. Ideal Poisson photocount distribution. [Adapted with permission from Loudon, R. (1983). "The Quantum Theory of Light," 2nd ed. Oxford Univ. Press, Oxford. Copyright 1983 Oxford University Press.]

It should be obvious that the same Poisson law will be found even if $\langle I(t') \rangle$ is not constant, so long as T is made great enough that all fluctuations associated with a particular interval are averaged out. The counting fluctuations associated with steady $\langle I \rangle$ are due to the discrete single-photon character of the atomic ionization event that initiates the count and are called particle fluctuations.

Although the Poisson distribution is the result for $P_n(T)$ in the simplest case, constant $\langle I(t') \rangle$, it is not the correct result for ordinary thermal light unless T is very long, in fact $2\pi \Delta\nu T \gg 1$ is necessary. We have posed a question at the end of the last section about differences between laser light and thermal light with equal (very narrow) bandwidths. In that case $\Delta\nu$ is very small, so we cannot automatically assume $2\pi\Delta\nu T \gg 1$. In fact it illustrates the point to assume the reverse. If $2\pi\Delta\nu T \ll 1$, we can assume $\langle I(t')\rangle$ is constant over such a short time T. However, $\langle I(t')\rangle$ can still fluctuate with t', that is, from run to run. In order to evaluate the average over t', which is now essentially an average over runs, we use the thermal distribution Eq. (72a) for $\mathscr{E}$, which is equivalent to the normalized exponential distribution for I

$$p(I) = (1/\bar{I})e^{-I/\bar{I}} \qquad \text{(thermal)} \qquad (78)$$

since $I \approx |\mathscr{E}|^2$. Then Eq. (75) gives

$$P_n(T) = (\bar{n})^n/(1+\bar{n})^{1+n} \qquad \text{(thermal)} \qquad (79)$$

which is variously called the thermal, chaotic, or Bose–Einstein distribution, and $\bar{n}$ is defined as before, $\bar{n} = \alpha\bar{I}T$, with the understanding that here $\bar{I}$ means the average over many runs, all shorter than $1/2\pi\Delta\nu$. The difference between the photocount distributions for thermal light under the two extreme conditions $2\pi\Delta\nu T \ll 1$ and $2\pi\Delta\nu T \gg 1$ is shown in Fig. 8. The nature of the difference between the thermal and coherent probability distributions (76) and (77), and thus of the fundamental difference between natural light and single-mode laser (coherent) light, can show up directly in the record of photocount measurements, as is clear by inspecting Figs. 7 and 8. Other photon count distributions than these two correspond to light that is somehow different from both laser light and thermal light. In quantum optics the most interesting examples are examples of purely quantum mechanical light beams.

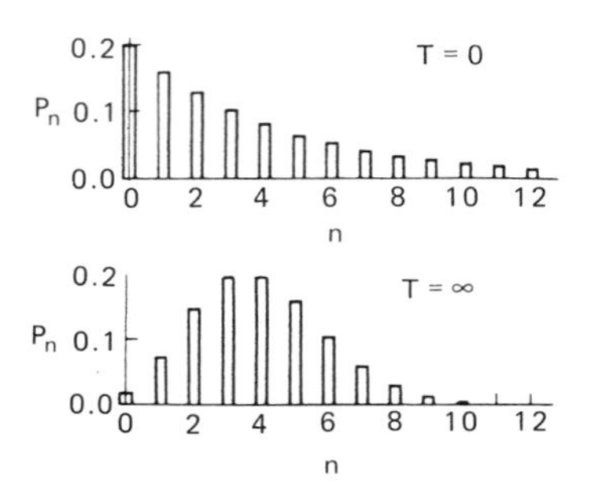

FIG. 8. Thermal photocount records for $T\Delta\nu \gg 1$ and $T\Delta\nu \ll 1$. [Adapted with permission from Loudon, R. (1983). "The Quantum Theory of Light," 2nd ed. Oxford Univ. Press, Oxford. Copyright 1983 Oxford University Press.]

C. Quantum Mechanical States of Light

The quantum theory of light assigns operators in a Hilbert space to the fields of electromagnetism. Any field $F(\mathbf{r})$, quantum or classical, can be written as a sum of plane waves (as a three-dimensional Fourier series or integral):

$$\begin{aligned} F(\mathbf{r}) &= \sum_k f_k \exp[i\mathbf{k}\cdot\mathbf{r}] \\ &= \sum_k \{f_k \exp[i\mathbf{k}\cdot\mathbf{r}] \\ &\quad + f_{-k}\exp[-i\mathbf{k}\cdot\mathbf{r}]\} \qquad k \geqq 0 \end{aligned} \qquad (80)$$

and $f_{-k} = f_k^*$ if $F(\mathbf{r})$ is a real function. If F is operator-valued real (i.e., hermitean), then the expansion coefficients f_k are operators and one writes $f_{-k} = f_k^\dagger$, where the dagger denotes hermitean adjoint in the usual way. In the case of electromagnetic radiation a given plane wave mode has the transverse electric field

$$\mathbf{E}_k(\mathbf{r}) = \hat{\mathbf{e}}N_k[a_k e^{i\mathbf{k}\cdot\mathbf{r}} + a_k^\dagger e^{-i\mathbf{k}\cdot\mathbf{r}}] \qquad (81)$$

where N_k is an appropriate normalization constant. The full field is obtained by summing over k: $\mathbf{E}(\mathbf{r}) = \sum_k \mathbf{E}_k(\mathbf{r})$.

Because a_k and $a_k^\dagger$ are operators rather than numbers, $a_k^\dagger a_k \neq a_k a_k^\dagger$. This is expressed by saying that they obey the "canonical" commutation relations:

$$[a_k, a_k^\dagger] = 1 \qquad (82)$$

where the square bracket means the difference $a_k a_k^\dagger - a_k^\dagger a_k$, as is usual in quantum mechanics. All other commutators among a's and $a^\dagger$'s for this mode, and all commutators of a_k or $a_k^\dagger$ with operators of all other modes vanish, for example, $[a_k, a_\ell] = 0$. The time dependences of these operators, and thus of the electric and magnetic fields, are determined dynamically from Maxwell's equations, which remain exactly valid in quantum theory. In the absence of charges and currents the time dependences are

$$a_k(t) = a_k e^{-i\omega_k t} \quad \text{and} \quad a_k^\dagger = a_k^\dagger e^{i\omega_k t} \qquad (83)$$

where $\omega_k = kc$.

The hermitean operator $a^\dagger a$ (we drop the mode index k temporarily) has integer eigenvalues and is called the photon number operator:

$$a^\dagger a|n\rangle = n|n\rangle, \qquad n = 0, 1, 2, \ldots \qquad (84)$$

and the eigenstate $|n\rangle$ is called a Fock state or *photon number state*. The operators $a^\dagger$ and a are called the photon *creation operator* and *destruc-*

tion operator because of their effect on photon number states:

$$a^\dagger|n\rangle = \sqrt{(n+1)}|n+1\rangle, \quad \text{and}$$
$$a|n\rangle = \sqrt{n}|n-1\rangle \tag{85}$$

That is, $a^\dagger$ and a respectively increase and decrease by 1 the number of photons in the field state. The photon number states are mutually orthonormal: $\langle m|n\rangle = \delta_{mn}$, and they form a complete set for the mode, and all other states can be expressed in terms of them. They have very attractive simple properties. For example, the photon number is exactly determined, that is, the dispersion of the photon number operator $a^\dagger a$ is zero in the state $|n\rangle$ for any n:

$$\begin{aligned}\langle(\Delta n)^2\rangle &= \langle n|(a^\dagger a)^2|n\rangle - \langle n|\, a^\dagger a|\rangle^2 \\ &= n^2 - n^2 = 0.\end{aligned}$$

However, the nice properties of the $|n\rangle$ states are in some respects not well suited to the most common situations in quantum optics. For example, loosely speaking, their phase is completely undefined because their photon number is exactly determined. As a consequence, the expected value of the mode field is exactly zero in a photon number state. That is $\langle n|\mathbf{E}(\mathbf{r})|n\rangle = 0$, because Eq. (85) and the orthogonality property together give $\langle n|a|n\rangle = \langle n|a^\dagger|n\rangle = 0$.

In most laser fields there are so many photons per mode that it is difficult to imagine an experiment to count the exact number of them. Thus, a different kind of state, called a *coherent state,* is usually more appropriate to describe laser fields than the photon number states. These states $|\alpha\rangle$ are right eigenstates of the photon destruction operator a:

$$a|\alpha\rangle = \alpha|\alpha\rangle \tag{86}$$

They are normalized so that $\langle\alpha|\alpha\rangle = 1$. Since a is not hermitean, its eigenvalue α is generally complex. The expected number of photons in the mode in a coherent state is $\langle n\rangle = \langle a^\dagger a\rangle = \langle\alpha|a^\dagger a|\alpha\rangle = |\alpha|^2$. Thus, α can be interpreted loosely as $\sqrt{\langle n\rangle}$, the square root of the mean number of photons in the mode. The number of photons in the mode is not exactly determined in the state $|\alpha\rangle$. This can be seen by computing the dispersion in the number:

$$\begin{aligned}\langle(\Delta n)^2\rangle &= \langle\alpha|(a^\dagger a)^2|\alpha\rangle - \langle\alpha|a^\dagger a|\alpha\rangle^2 \\ &= \langle\alpha|a^\dagger(a^\dagger a - 1)a|\alpha\rangle - \langle\alpha|a^\dagger a|\rangle^2 \\ &= \langle\alpha|a^\dagger a|\alpha\rangle = |\alpha|^2 = \langle n\rangle\end{aligned} \tag{87}$$

Thus, the dispersion is equal to the mean number, a property already noticed for the Poisson distribution (77). If $\langle n\rangle \gg 1$, then the relative dispersion $\langle(\Delta n)^2\rangle/\langle n\rangle^2$ is very small, and the number of photons in the state is well determined in a relative sense. At the same time, a relatively well-defined phase can also be associated with the state.

The clearest interpretation of the amplitude and phase associated with a coherent state $|\alpha\rangle$ is obtained by computing the expected value of $\mathbf{E}$ in the state $|\alpha\rangle$. Since $\langle\alpha|a^\dagger|\alpha\rangle = \langle\alpha|a|\alpha\rangle^* = \alpha^*$, one easily determines that

$$N_k = \sqrt{(\hbar\omega_k/2\varepsilon_0 V)} \tag{88}$$

where V is the mode volume, and then N_k can be interpreted loosely as the electric field amplitude associated with one photon. This shows that $N_k\alpha = \sqrt{(\hbar\omega_k/2\varepsilon_0 V)}\alpha$ is the mean amplitude of the expected electric field, or what is sometimes called the classical field amplitude. The phase of α thus determines the phase of the field described by the state $|\alpha\rangle$.

The coherent states $|\alpha\rangle$ can be expressed in terms of the photon number states:

$$|\alpha\rangle = \exp[-|\alpha|^2/2] \sum_m \left[\frac{a^m}{\sqrt{m!}}\right] |m\rangle \tag{89}$$

where the sum runs over all integers $m \geqq 0$. From Eq. (89) one can compute the probability $p_n(\alpha)$ that a given coherent state $|\alpha\rangle$ contains exactly n photons. According to the principles of quantum theory this is given by $|\langle n|\alpha\rangle|^2$, and we find:

$$p_n(\alpha) = e^{-\langle n\rangle}\langle n\rangle^n/n! \tag{90}$$

which is exactly the "purely random" Poisson probability distribution shown in Fig. 7. This is interpreted as meaning that photon-counting measurements of a radiation field in the coherent state $|\alpha\rangle$ would find exactly n photons with probability (90).

The photon creation and destruction operators are not hermitean and are generally considered not observable, but their real and imaginary parts (essentially the electric and magnetic field strengths, or in other terms, the in-phase and quadrature components of the optical signal) are in principle observable. Thus, one can introduce the definitions

$$a = \tfrac{1}{2}(a_1 - ia_2), \qquad a^\dagger = \tfrac{1}{2}(a_1 + ia_2) \tag{91}$$

and their inverses

$$a_1 = a + a^\dagger, \qquad a_2 = i(a - a^\dagger) \tag{92}$$

What are the quantum limitations on measurement of the hermitean operators a_1 and a_2? They must, of course, obey the Heisenberg *uncertainty relation* $\langle(\Delta a_1)^2\rangle\langle(\Delta a_2)^2\rangle \geqq |\frac{1}{2}\langle[a_1, a_2]\rangle|^2$, where $\langle\ldots\rangle$ indicates expectation value in any given quantum state, and $\Delta a_1 \equiv a_1 - \langle a_1\rangle$. Since $[a, a^\dagger] = 1$, it follows from Eq. (91) that $[a_1, a_2] = -2i$, so the Heisenberg relation for a_1 and a_2 is:

$$\langle(\Delta a_1)^2\rangle\langle(\Delta a_2)^2\rangle \geq 1 \tag{93}$$

A coherent state $|\alpha\rangle$ can be shown to produce the minimum simultaneous uncertainty in a_1 and a_2. That is, $\langle\alpha|(\Delta a_1)^2|\alpha\rangle = 1$ and $\langle\alpha|(\Delta a_2)^2|\alpha\rangle = 1$. Both a_1 and a_2 are therefore said to reach the *quantum limit* of uncertainty in a coherent state.

However, it is only the product of the operator dispersions that is constrained by the Heisenberg uncertainty relation, and there is no fundamental reason why either a_1 and a_2 could not have a dispersion equal to $\mu \ll 1$, so long as the other had a dispersion at least as large as $1/\mu \gg 1$. A quantum state of the radiation field that permits one of the components of the destruction operator to have a dispersion smaller than the quantum limit is said to be a *squeezed state*. Squeezed states could in principle provide the ability to make ultraprecise measurements such as are projected for gravity wave detection. A squeezed state of radiation was first generated and measured by Slusher and others in 1985.

IV. Quantum Interactions and Correlations

It should be remembered that the highly successful semiclassical version of the quantum theory of light (Sections I and II) does not ignore quantum principles, or put $\hbar \to 0$, but it does ignore quantum fluctuations and correlations. It is a theory of coupled quantum expectation values.

In the following sections a number of phenomena depending directly on quantum fluctuations and correlations are described. None of them have been successfully treated by the semiclassical theory or by any other theory than the generally accepted and fully quantized version of the quantum theory of light. For this reason the observation of these and similar quantum optical effects can offer a means of testing the accepted theory.

Such tests are of great interest for two related reasons. Because the quantum theory of light (or quantum electrodynamics) is the most carefully studied quantum theory, and because it serves as a fundamental guide to all field theories, it plays a key role in our present understanding of quantum principles and should be tested as rigorously as possible. Moreover, tests of the effects described below play a special role because they bear on the theory in a different way compared with traditional tests, such as high precision measurements of the Lamb shift, the fine structure constant, the Rydberg, and the electron's anomalous moment.

A. Fully Quantized Interactions

The electric field is mainly responsible for optical interactions of light with matter, and the magnetic field plays a subsidiary role, becoming significant only in situations involving magnetic moments or relativistic velocities. The most important light–matter interaction is the direct coupling of electric dipoles to the radiation field through the interaction energy $-\mathbf{d} \cdot \mathbf{E}$.

A systematic description of the fully quantized interactions of quantum optics begins with the total energy of the atom H_A and the radiation H_R and their energy of interaction:

$$H = H_A + H_R - \mathbf{d} \cdot \mathbf{E} \tag{94}$$

In the quantum theory the atomic, radiation, and interaction energies are given by the Hamiltonian operators

$$H_A = \sum_j E_j|j\rangle\langle j| \tag{95a}$$

$$H_R = \sum_k \hbar\omega_k a_k^\dagger a_k \tag{95b}$$

$$-\mathbf{d} \cdot \mathbf{E} = -\sum_i \sum_j \sum_k \hbar f_{ij}^k\{a_k^\dagger + a_k\}|i\rangle\langle j| \tag{95c}$$

Here E_j and $|j\rangle$ are the quantized energies and eigenstates of the atom including level-shifting and level-splitting due to static fields that give rise to Zeeman and Stark effects, etc. The dipole coupling constant is $\hbar f_{ij}^k = N_k\hat{\mathbf{e}} \cdot \mathbf{d}_{ij}$, in the notation of Eqs. (9) and (39). Also, $a_k^\dagger$ and a_k are the photon creation and destruction operators introduced in Section III,C.

The two-level version of this Hamiltonian is the most used in quantum optics. It is obtained by restricting the sums over i and j to the values 1 and 2. It is not necessary that $|1\rangle$ and $|2\rangle$ be the two lowest energy levels. In photoionization, which is the physical process underlying the operation of photon counters, the upper state is not even a discrete state but lies in the continuum of energies above the ionization threshold (see Section IV,B). In its two-level version H becomes

$$H = E_1|1\rangle\langle 1| + E_2|2\rangle\langle 2| + \sum_k \hbar\omega_k a_k^\dagger a_k - \sum_k \hbar f_{12}^k\{a_k^\dagger + a_k\}\{\sigma + \sigma^\dagger\} \tag{96}$$

where $\sigma = |1\rangle\langle 2|$ and $\sigma^\dagger = |2\rangle\langle 1|$, and f_{12}^k has been taken to be real for simplicity. Note that σ has the effect of a lowering transition when it acts on the two-level atomic state $|\Psi\rangle = C_1|1\rangle + C_2|2\rangle$. That is, $\sigma|\Psi\rangle = C_2|1\rangle$, so σ takes the amplitude C_2 of the upper state $|2\rangle$ and assigns it to the lower state $|1\rangle$. By the same argument $\sigma^\dagger$ causes a raising transition.

The term $a_k^\dagger\sigma^\dagger$ in Eq. (96) is difficult to interpret because it has $\sigma^\dagger$ raising the atom into its upper state together with $a_k^\dagger$ creating a photon. One expects photon creation to be associated only with a lowering of the atomic state. The term $a_k\sigma$ presents similar difficulties. It can be shown, however, that these two terms are the source of the very rapid oscillations $\exp[\pm 2i\omega t]$ which were discussed above Eq. (12) and were eliminated by the rotating wave approximation (RWA). The adoption of the RWA eliminates them here also. With the RWA and the convenient convention that $E_1 = 0$ (and therefore that $E_2 = \hbar\omega_{21}$), the working two-level Hamiltonian is

$$H = \tfrac{1}{2}\hbar\omega_{21}(\sigma_z + 1) + \sum_k \hbar\omega_k a_k a_k^\dagger - \sum_k \hbar f_{12}^k\{a_k^\dagger\sigma + \sigma^\dagger a_k\} \tag{97}$$

The operators σ, $\sigma^\dagger$, and σ_z are closely related to the 2 × 2 Pauli spin matrices:

$$|1\rangle\langle 2| = \sigma \rightarrow \tfrac{1}{2}(\sigma_x - i\sigma_y) = \begin{bmatrix} 0 & 0 \\ 1 & 0 \end{bmatrix} \tag{98a}$$

$$|2\rangle\langle 1| = \sigma^\dagger \rightarrow \tfrac{1}{2}(\sigma_x - i\sigma_y) = \begin{bmatrix} 0 & 1 \\ 0 & 0 \end{bmatrix} \tag{98b}$$

$$|2\rangle\langle 2| - |1\rangle\langle 1| \rightarrow \sigma_z = \begin{bmatrix} 1 & 0 \\ 0 & -1 \end{bmatrix} \tag{98c}$$

One can easily confirm that the matrix representation of σ is a "lowering" operator in the two-dimensional space with basis vectors

$$|2\rangle \rightarrow \begin{bmatrix} 1 \\ 0 \end{bmatrix} \quad \text{and} \quad |1\rangle \rightarrow \begin{bmatrix} 0 \\ 1 \end{bmatrix} \tag{98d}$$

and $\sigma^\dagger$ is represented by the corresponding "raising" operator. The σ's obey the commutator relations:

$$[\sigma, \sigma_z] = 2\sigma, \qquad [\sigma^\dagger, \sigma_z] = -2\sigma^\dagger, \qquad [\sigma^\dagger, \sigma] = \sigma_z \tag{99}$$

Changes in the radiation field occur as a result of emission and absorption of photons during transitions in the atom. One consequence is that a_k obeys an equation obtained from the Heisenberg equation $i\hbar\partial O/\partial t = [O, H]$ that is valid for all operators O:

$$\partial a_k/\partial t = -i\omega_k a_k + if_k\sigma \tag{100}$$

Here we have simplified the notation, rewriting f_{12}^k as f_k. The solution of Eq. (100) is:

$$a_k(t) = a_k(0)e^{-i\omega_k t} + if_k \int_0^t dt'\ e^{-i\omega_k(t-t')}\sigma(t') \tag{101}$$

The first term represents photons that are present in the mode k but not associated with the two-level atoms (e.g., from a distant laser), and the second term represents the photons associated with a transition in the two-level atom.

The σ operators also change in time in the course of emission and absorption processes. Their time dependence is also determined by Heisenberg's equation and one finds the equations:

$$d\sigma/dt = -i\omega_{21}\sigma - i\sum_k f_k a_k \sigma_z \tag{102a}$$

$$d\sigma^\dagger/dt = i\omega_{21}\sigma^\dagger + i\sum_k f_k a_k^\dagger \sigma_z \tag{102b}$$

$$d\sigma_z/dt = 2i\sum_k f_k(\sigma^\dagger a_k - a_k^\dagger\sigma) \tag{102c}$$

These are the operator equations underlying the semiclassical equations for the Bloch vector components given in Eq. (19).

The correspondence between the quantum and semiclassical sets of variables is

$$\sigma \rightarrow \tfrac{1}{2}(u - iv)e^{-i\omega t} \tag{103a}$$

$$\sigma^\dagger \rightarrow \tfrac{1}{2}(u + iv)e^{i\omega t} \tag{103b}$$

$$\sigma_z \rightarrow w \tag{103c}$$

This identification is precise if the radiation field is both intense and classical enough. This means that one retains only one mode and replaces a_k and $a_k^\dagger$ by their coherent state expectation values $\alpha = \alpha_0\ e^{-i\phi}$ and $\alpha_0^* e^{i\phi}$. Then Eqs. (102) are identical with Eq. (19) under the previous assumptions, that is, the field is quasi-monochromatic so $\phi = \omega t$, and $N_k\alpha$ is the slowly varying electric field expectation value, which is interpreted as the classical field amplitude.

Thus, one has

$$2f_k\alpha_0 \to 2d\mathscr{E}_0/\hbar = \chi \qquad (104)$$

The part of a_k that depends on σ in Eq. (100) acts as a radiation reaction field when substituted into Eq. (102). It causes damping and a small frequency shift in the atomic operator equations even if there are no external photons present and thus is associated with spontaneous emission. The damping constant is exactly the correct Einstein A coefficient for the transition, because the A coefficient is a two-level parameter.

The frequency shift is only a primitive two-level version of a more general many level radiative correction such as the Lamb shift. One observes here a natural limitation of any two-level model. It is intrinsically incapable of dealing with any precision with effects, such as radiative level shifts, that depend strongly on the contributions of many levels. However, in the cases of interest described in the remainder of Section IV, the effects of other levels are negligible and the numerical value of the frequency shift is irrelevant. It can be assumed to be included in the definition of ω_{21}.

B. Quantum Light Detection and Statistics

The quantum theory of light detection is based on the quantum theory of photoionization because photon counters are triggered by an ionizing absorption of a photon. Photoionization is a weak field phenomenon because the effective γ is so large [recall Eq. (45)]. Thus, perturbative methods are adequate and one computes the absolute square of the ionization matrix element, in this case given by $\langle F| -e\mathbf{r} \cdot \mathbf{E}(\mathbf{r})|I\rangle$, where $|I\rangle$ and $|F\rangle$ are the initial and final states of the photoionization. The initial state consists of an atom in its ground state, described by the electronic orbital function $\phi_0(\mathbf{r})$, and the initial state of the radiation field $|\Psi\rangle$. The final state consists of the atom in an ionized state described by an electronic orbital function $\phi_f(\mathbf{r})$ appropriate to a free electron with energy above the ionization threshold, and another state of the radiation field $|\Psi'\rangle$. Then the matrix element becomes

$$\begin{aligned}\langle F|-e\mathbf{r} \cdot \mathbf{E}|I\rangle \\ &= -\int \phi_f^*(\mathbf{r})e\mathbf{r} \cdot \langle\Psi'|\mathbf{E}(\mathbf{r})|\Psi\rangle\phi_0(\mathbf{r})d^3r \\ &= -\mathbf{d}_{f0} \cdot \Sigma_k\hat{\mathbf{e}}_k N_k\langle\Psi'|a_k|\Psi\rangle \qquad (105)\end{aligned}$$

where $\mathbf{d}_{f0}$ is the so-called dipole matrix element for the $0 \to f$ transition in the atom. We have also made the "dipole" approximation, in which $\exp[i\mathbf{k} \cdot \mathbf{r}] \approx \exp[i0] \approx 1$ over the entire effective range of the matrix element integral [recall the discussion following Eq. (9)].

Only the part of $\mathbf{E}$ that lowers the photon number of the field, namely the "a" part, is effective in ionization, essentially because ionization is a photon absorption process. In addition we can take a single k value if the incident light is monochromatic. Thus, the ionization rate depends on $f_k^2|\langle\Psi'|a_k|\Psi\rangle|^2$, where $f_k = |\mathbf{d}_{f0} \cdot \hat{\mathbf{e}}_k|N_k$. The actual final states of the atom and of the field are never completely observed, and all the unobserved features must be allowed for, that is, included by summation:

$$\begin{aligned}\text{rate} &\approx f_k^2 \sum_{\Psi'} \langle\Psi|a_k^\dagger|\Psi'\rangle\langle\Psi'|a_k|\Psi\rangle \\ &\approx f_k^2\langle\Psi|a_k^\dagger a_k|\Psi\rangle \approx \langle\Psi|a_k^\dagger a_k|\Psi\rangle\end{aligned}$$

since $\Sigma_{\Psi'}|\Psi'\rangle\langle\Psi'| = 1$ for a complete set of final states. In this expression for the ionization rate we write "$\approx$" instead of "$=$" because we are really interested here only in the effects of field quantization on the rate, not the exact numerical value of the rate. If the radiation field is quantum mechanical we do not know perfectly the properties of the incident light, and these properties must be averaged. This average over the properties of the incident light is the same average discussed from a classical standpoint in Sections III,A and III,B. Thus, we finally have

$$\text{rate} \approx \langle a_k^\dagger a_k\rangle \qquad (106)$$

where the angular brackets now mean an average over the initial field, that is, the quantum mechanical expectation value in state $|\Psi\rangle$.

The significance of Eq. (106) is in the ordering of the field operators. The nature of the photoionization process mandates that they be in the given order and not the reverse. Since $a^\dagger a$ is not equal to $aa^\dagger$ for a quantum field, the order makes a difference. The order given, in which the destruction operator is to the right, is called *normal order*. Photoionization is a normally ordered process by its nature, and therefore so is photodetection and photon counting. This has fundamental consequences for quantum statistical measurements, as we now explain.

Let us consider the degree of second-order coherence $g^{(2)}$ in quantum theory. This was written in Eq. (73a) as an intensity correlation: $g^{(2)}(\tau) = \langle I(t)I(t + \tau)\rangle/\langle I\rangle^2$. Because of the normally ordered character of photoionization, if $g^{(2)}$ is measured with photodetectors as usual, its correct definition according to the quantum theory of ionization is normally ordered:

$$g^{(2)}(\tau) = \frac{\langle E^{(-)}(t)E^{(-)}(t+\tau)E^{(+)}(t+\tau)E^{(+)}(t)\rangle}{\langle E^{(-)}E^{(+)}\rangle\langle E^{(-)}E^{(+)}\rangle} \quad (107)$$

If photon fields were really classical, and these quantum mechanical fine points were unnecessary, then the ordering would make no difference, since the fields $E^{(\pm)}$ would be numbers, not operators, and the original expression for $g^{(2)}$ would be recovered. However, we now exhibit the effects of these quantum differences in a few specific cases.

In the case of a single-mode field, there are no time dependences and $g^{(2)}(\tau) = g^{(2)}(0)$ simplifies to

$$g^{(2)}(0) = \langle a^\dagger a^\dagger aa\rangle/\langle a^\dagger a\rangle^2 \quad (108)$$

which we evaluate in Table II. Among these examples the Fock state is special because its $g^{(2)}$ violates the condition $g^{(2)}(0) \geq 1$, which is one of the classical inequalities given below Eqs. (73). The Fock state is therefore an example of a state of the radiation field for which the quantum and classical theories make strikingly different predictions. It has not yet been possible to study a pure Fock state of more than one photon in the laboratory.

Photon bunching is a term that refers to the fact that photon beams exist in which photons are counted with statistical fluctuations greater than would be expected on the basis of purely random (that is, Poisson) statistics. In fact, almost any ordinary beam (thermal light) will have this property, and this is reflected in that $g^{(2)} > 1$ for thermal light. Photon bunching therefore arises from the Bose–Einstein distribution [Eq. (39)]. A coherent state with its Poisson statistics is purely random and does not exhibit bunching.

A qualitative classical explanation of photon bunching is sometimes made by saying that light from any natural source arises from broadband multimode photon emission by many independent atoms. There are naturally random periods of constructive and destructive interference among the modes, giving rise to large intensity "spikes," or "bunches" of photons, in the light beam. Unbunched light comes from a coherently regulated collection of atoms, such as from a well-stabilized single-mode laser. From this point of view, unbunched coherent light is optimally ordered.

TABLE II. Second-Order Degree of Coherence for Single-Mode Quantum Mechanical Fields in Different States

Quantum field state	Value of $g^{(2)}(0)$
Vacuum state $\lvert 0\rangle$	0
Fock state $\lvert n\rangle$	0 if $n < 2$
	$1 - 1/n$ if $n \geq 2$
Coherent state $\lvert \alpha\rangle$	1
Thermal state	2

However, *photon antibunching* can also occur. There are "antibunched" light beams, in which photons arrive with lower statistical fluctuations than predicted from a purely coherent beam with Poisson statistics. Antibunched light beams have values of $g^{(2)} < 1$, in common with a pure Fock state beam, and are therefore automatically nonclassical light beams.

The first observation of an antibunched beam with $g^{(2)} < 1$ was accomplished by Mandel and others in 1977 in an experiment with two-level atoms undergoing resonance fluorescence. Antibunching occurs in such light for a very simple reason. A two-level atom "regulates" the occurrence of pairs of emitted photons very severely, even more so than the photons are regulated in a single-mode laser. A second fluorescent photon cannot be emitted by the same two-level atom until it has been re-excited to its upper level by the absorption of a photon from the main radiation mode. Thus, a high Rabi frequency χ permits the degree of second-order coherence $g^{(2)}(\tau) = \langle a^\dagger a^\dagger(\tau)a(\tau)a\rangle/\langle a^\dagger a\rangle^2$ to be nonzero after a relatively short value of the time delay τ, but $g^{(2)}$ is strictly zero for $\tau = 0$. A graph showing the experimental observation is given in Fig. 9.

The significance of photon statistics and photon counting techniques in quantum optics and in physics is clear. They permit a direct examination of some of the fundamental distinctions between the quantum mechanical and classical concepts of radiation.

C. Superradiance

Nonclassical photon counting statistics arise from the multiphoton correlations inherent in specific states of the light field. Similarly, multi-atom quantum correlations can give rise to unusual behavior by systems of radiating atoms, as was pointed out by Dicke in 1954. The most dramatic behavior of this kind is called *Dicke superradiance* or *superfluorescence*.

Multi-atom correlations can exist in N-atom systems even if N is as small as $N = 2$. A pair of two-level atoms labeled a and b can have quantum states made of linear combinations of the

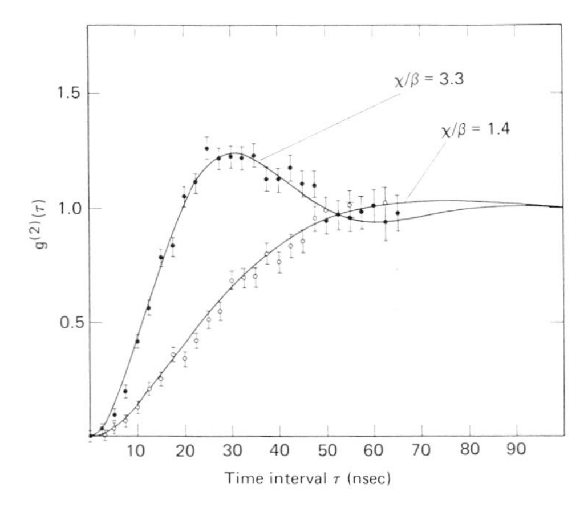

FIG. 9. Curves of the second-order degree of coherence, showing photon antibunching. [Reprinted with permission from Dagenais, M., and Mandel, L. (1978). Investigation of two-time correlations in photon emission from a single atom, *Phys. Rev. A*, **18,** 2217–2228.]

elementary two-atom states

$$|a+\rangle * |b+\rangle \tag{109a}$$

$$|a+\rangle * |b-\rangle \tag{109b}$$

$$|a-\rangle * |b+\rangle \tag{109c}$$

$$|a-\rangle * |b-\rangle \tag{109d}$$

Here the sign $*$ indicates a tensor product of the vector spaces of the two atoms, and + and − designate the upper and lower energy levels, with energies E_2 and E_1, in each of the identical atoms. The two middle states Eq. (109b and c) are degenerate, with total energy $E_1 + E_2 = E_2 + E_1$. It is useful to define the "singlet" and "triplet" states $\|S\rangle$ and $\|T\rangle$ that are linear combinations of the degenerate states as follows:

$$\|S\rangle = (1/\sqrt{2})\{|a\rangle * |b-\rangle - |a-\rangle * |b+\rangle\} \tag{110a}$$

$$\|T\rangle = (1/\sqrt{2})\{|a\rangle * |b-\rangle + |a-\rangle * |b+\rangle\} \tag{110b}$$

in analogy to singlet and triplet combinations of two spin $-\frac{1}{2}$ states.

The two-atom interaction with the radiation field is through $\mathbf{d} \cdot \mathbf{E}$ as in the one-atom case [recall Eq. (95)], and here both atoms contribute a dipole moment:

$$\begin{aligned}\hat{\mathbf{e}} \cdot \mathbf{d} &= \hat{\mathbf{e}} \cdot [\mathbf{d}(a) + \mathbf{d}(b)] \\ &= \hat{\mathbf{e}} \cdot (\mathbf{d}_{21})_a\{|a+\rangle\langle a-| + |a-\rangle\langle a+|\} \\ &\quad + \hat{\mathbf{e}} \cdot (\mathbf{d}_{21})_b\{|b+\rangle\langle b-| + |b-\rangle\langle b+|\} \\ &= d\{|a+\rangle\langle a-| + |a-\rangle\langle a+| \\ &\quad + |b+\rangle\langle b-| + |b-\rangle\langle b+|\}\end{aligned} \tag{111}$$

where we have taken equal matrix elements: $\hat{\mathbf{e}} \cdot (\mathbf{d}_{21})_a = \hat{\mathbf{e}} \cdot (\mathbf{d}_{21})_b = d$.

The main features of superradiance lie in the fact that the two-atom dipole interaction causes transitions between the various states of the system at different rates, and the reason for the difference is the existence of greater internal two-atom coherence in the case of the triplet state. Suppose that the two-atom system is fully excited into the state $\|+2\rangle = |a+\rangle * |b+\rangle$, which has energy $2E_2$, and then emits one photon. The system must drop into a state with energy $E_2 + E_1$. From this state it can decay further by emission of a second photon to the ground state $\|-2\rangle = |a-\rangle * |b-\rangle$ which has energy $2E_1$.

According to the Fermi Golden Rule, the rate of these transitions depends on the square of the interaction matrix element between initial and final states, $\langle F|\mathbf{d} \cdot \mathbf{E}|I\rangle$, summed over all possible final states. We consider the second transition, so there is only one possible final atomic state, namely $\|-2\rangle$. The matrix elements factor into an atomic part and a radiation part. We can use Eq. (81) in the dipole approximation to write

$$\langle F|\mathbf{d} \cdot \mathbf{E}|I\rangle = \textstyle\sum_k \langle F_A|\mathbf{d} \cdot \hat{\mathbf{e}}_k|I_A\rangle\langle F_R|N_k(a_k + a_k^\dagger)|I_R\rangle$$

Since superradiance deals only with spontaneous emission, the initial radiation state is the empty or vacuum state. Thus, the field contribution to the matrix element comes just from $a_k^\dagger$ and is the same in all cases and not interesting.

The various possible atomic matrix elements are

$$\langle -2\|\mathbf{d} \cdot \hat{\mathbf{e}}_k|a+\rangle * |b-\rangle = d \tag{112a}$$

$$\langle -2\|\mathbf{d} \cdot \hat{\mathbf{e}}_k|a-\rangle * |b+\rangle = d \tag{112b}$$

$$\langle -2\|\mathbf{d} \cdot \hat{\mathbf{e}}_k\|0, S\rangle = 0 \tag{112c}$$

$$\langle -2\|\mathbf{d} \cdot \hat{\mathbf{e}}_k\|0, T\rangle = (\sqrt{2})d \tag{112d}$$

so their squares have the relative values 1 : 1 : 0 : 2. That is, the triplet initial state can radiate twice as strongly as either of the original degenerate states, and the singlet state cannot radiate at all. Both the triplet and singlet states are said to be two-body "cooperative" states because they cannot be factored into one-atom states.

It is tempting to interpret these results by saying that the triplet state radiates more rapidly because it has a larger dipole moment than the others and that the singlet state has no dipole moment. Such an interpretation is in the spirit of semiclassical radiation theory, as described in Section I,C, where the expectation values of quantum operators are treated as if they were classical variables. This interpretation has a number of useful features, but must also contain serious flaws, because the observation that it is based on is not true. A calculation of the dipole expectation value shows $\langle\Psi\|\mathbf{d}\|\Psi\rangle = 0$, where $\|\Psi\rangle$ can be any of the two-atom states above, including the rapidly radiating triplet state.

A state with a large dipole moment expectation does exist, namely the factored state

$$\|\Psi_d\rangle = \tfrac{1}{2}\{|a+\rangle + |a-\rangle\} * \{|b+\rangle + |b-\rangle\} \tag{113a}$$

This state is actually an eigenstate of the total dipole operator:

$$\begin{aligned}\hat{\mathbf{e}} \cdot \mathbf{d}\|\Psi_d\rangle &= [\hat{\mathbf{e}} \cdot (\mathbf{d}_{21})_a\{|a+\rangle\langle a-| + |a-\rangle\langle a+|\} \\ &\quad + \hat{\mathbf{e}} \cdot (\mathbf{d}_{21})_b\{|b+\rangle\langle b-| + |b-\rangle\langle b+|\}]\|\Psi\rangle \\ &= 2d\|\Psi\rangle\end{aligned} \tag{113b}$$

so one has $\langle\Psi_d\|\mathbf{d}\cdot\hat{\mathbf{e}}\|\Psi_d\rangle = 2d$. This state is also predicted to radiate strongly.

The extrapolation of these predictions to an N-atom system leads to the prediction of a very large N-atom emission intensity I_N. Related predictions are that the N-atom cooperative process begins with a relatively slow buildup. After a delay of average length $\delta\tau_N$, an ultrashort burst of radiation of duration τ_N occurs. In the ideal case one predicts

$$I_N \approx N^2\hbar\omega_{21}/\tau_1 \tag{114a}$$

$$\tau_N \approx \tau_1/N \tag{114b}$$

$$\delta\tau_N \approx \tau_1 \ln(N) \tag{114c}$$

where τ_1 is the single-atom radiative lifetime: $1/\tau_1 \equiv A$.

If one imagines even small collections of atoms with $N \approx 10^{12}$, then N^2 is impressively very much bigger than N and the term superradiance is indeed apt. If $A \approx 10^8$ sec^{-1} is taken as typical for 2 eV optical transitions, then $N \approx 10^{12}$ suggests that 2×10^{12} eV of energy can be expected at the rate 10^{20} sec^{-1} for a purely spontaneous power output of 2×10^{32} eV sec$^{-1} \approx 3 \times 10^{11}$ W.

Other aspects of superradiance are equally interesting on fundamental grounds. For example, which of the two large dipole states, $\|T\rangle$ or $\|\Psi_d\rangle$, is actually responsible for superradiance? They both predict $I_N \approx N^2\hbar\omega_{21}$, but their correlation properties are completely different, in somewhat the same way that a photon number state $|n\rangle$ and a coherent state $|\alpha\rangle$ have very different correlation properties even if they predict the same mode energy $|\alpha|^2\hbar\omega = n\hbar\omega$. Consider only the fluctuations in **d** itself for the two states. If one calculates the expectation of the dispersion of $\Delta\mathbf{d}^2 \equiv \langle[\mathbf{d}\cdot\hat{\mathbf{e}} - \langle\mathbf{d}\cdot\hat{\mathbf{e}}\rangle]^2\rangle$, one finds:

$$\langle\Psi_d\|\Delta\mathbf{d}^2\|\Psi_d\rangle = 0 \tag{115}$$

$$\langle T\|\Delta\mathbf{d}^2\|T\rangle = d^2 \tag{116}$$

One can infer that radiation from the state $\|T\rangle$ can be expected to exhibit strong fluctuations of a kind completely absent in radiation from the state $\|\Psi_d\rangle$.

The fluctuations predicted from the state $\|T\rangle$ are consistent with the fact that it is exactly the state connected directly to the initial fully excited state $\|+2\rangle$ by the total dipole operator **d**. That is, $\hat{\mathbf{e}}\cdot\mathbf{d}\|+2\rangle = (\sqrt{2})d\|T\rangle$. The fluctuations can be associated with the quantum uncertainty in the emission time of the first photon. Such fluctuations will influence all subsequent evolution, and if $N \gg 1$, they can be regarded as an example of a *macroscopic quantum fluctuation,* that is, a fluctuation with quantum mechanical origins that achieves direct macroscopic observability.

For a period of years, superradiance was a controversial and unobserved phenomenon. The intense and highly directional light beam predicted for the effect suggests that each emitted photon contributes to a spontaneous radiation field, which helps to stimulate the emission of further photons. Such a self-reactive process would provide a feedback analogous to that provided by mirror reflections in a laser cavity. It has been suggested that the physical origins of superradiance and laser emission are in fact the same thing. Important differences exist, however. During laser action, the dipole coherence of an individual atom is interrupted by collisions extremely frequently. The incoherent adiabatic solution for ρ_{21} is a quite satisfactory element of all laser theories (recall Section II,G). By contrast, in ideal Dicke superradiance all N-atom dipole coherence is fully preserved during the entire radiation process.

The experimental observation of superradiance was first achieved in the 1970s by Feld, Haroche, Gibbs, and others. Agreement has been found with the correlated state predictions, particularly with the statistical nature of the delay time fluctuations, and there is no longer any controversy over its existence. However, important questions about quantum and propagation effects on the spatial coherence properties of superradiance remain open.

D. Two-Level Single-Mode Interaction

We have emphasized that beginning with Einstein's reconsideration of Planck's radiation law, the most fundamental interacting system in quantum optics is a single two-level atom coupled to a single mode of the radiation field. This interaction was described semiclassically in Section II,A, and the quantum mechanical origin of the semiclassical equations was explained in Section IV,A. Fully quantum mechanical studies of the quantum coherence properties of this simplest interacting system were initiated by Jaynes in the 1950s. Some of the differences between general quantum and semiclassical theories have been clarified in this context.

When only one mode is significant the Hamiltonian [Eq. (97)] can be reduced to:

$$H_{JC} = \tfrac{1}{2}\hbar\omega_{21}(\sigma_z + 1) + \hbar\omega a^\dagger a + \hbar\lambda(a^\dagger\sigma + \sigma^\dagger a) \qquad (117)$$

Here $\lambda = f^k_{12}$, which is assumed real for simplicity. Remarkably, the effective Hamiltonian (Eq. (117)] for such a truncated version of quantum electrodynamics (the so-called Jaynes–Cummings model) has a number of important properties, and experimental studies by Haroche and Walther and others were begun in the early 1980s. Exact expressions are known for the eigenvalues and eigenvectors of H_{JC}. With $\hbar = 1$ for simplicity and $\Delta = \omega_{21} - \omega$, the eigenvalues are

$$E_{n,\pm} = E_1 + n\omega + \tfrac{1}{2}[\Delta \pm \Omega_n] \qquad (118a)$$

where we have re-inserted $E_1 \neq 0$ for the lower level energy, and the corresponding eigenvectors are given by

$$\|n, +\rangle = \cos\Phi\,|n-1\rangle * |+\rangle + \sin\Phi\,|n\rangle * |-\rangle \qquad (118b)$$

$$\|n, -\rangle = -\sin\Phi\,|n-1\rangle * |+\rangle + \cos\Phi\,|n\rangle * |-\rangle \qquad (118c)$$

where $\cos\Phi = \sqrt{(\Omega_n + \Delta)/2\Omega_n}$ and $\sin\Phi = \sqrt{(\Omega_n - \Delta)/2\Omega_n}$. Here the use of Ω_n for $\sqrt{4\lambda^2 n + \delta^2}$ is a deliberate reminder of Ω defined following Eqs. (12) because they play the same roles in their respective quantum and semiclassical theories. Similarly, one writes $\chi(n)$ as a reminder of χ for the same reason, that is, $\chi(n) = 2\lambda\sqrt{n}$ is the QED equivalent of the Rabi frequency $2d\mathscr{E}_0/\hbar$ [and not to be confused with the χ_m of Eqs. (49) and (50)].

One of the observable quantities of the Jaynes–Cummings model is the atomic energy. Its expectation value can be calculated exactly, without approximation:

$$\langle\sigma_z(t)\rangle = -\Sigma_n\, p_n \cos\chi(n)t \qquad \text{(quantum)} \qquad (119a)$$

Here p_n is the probability that the single mode has exactly n photons. This result can be contrasted with Eq. (18c) in the same limit $\Delta = 0$ and under the same circumstances, namely, with the field intensity $I \approx \chi^2$ distributed with some probability $p(I)$ over a range of values:

$$w(t) = -\int p(I)\cos\chi(I)t\, dI \qquad \text{(semiclassical)} \qquad (119b)$$

The apparent similarity of these results disguises fundamental differences in dynamical behavior between Eqs. (119a and b). These differences arise from the discreteness of the allowed photon numbers in Eq. (119a), which are discrete precisely because the field is quantized, that is, because the mode contains only whole photons and never fractional units of the energy $\hbar\omega$. By contrast, in the semiclassical theory (recall Section I,C) the intensity I and Rabi frequency χ can have any values.

For the quantum photon distribution associated with a coherent state, for which p_n is given by Eq. (90), a plot of $\langle\sigma_z(t)\rangle$ is shown in Fig. 10. The nearly immediate disappearance or "collapse" of the signal shortly after $t = 0$ can be explained on the basis of the interference of many frequencies $\chi(n)$ in the quantum sum of Eq. (119a). Such a collapse can be expected for any broad distribution p_n and would be predicted by the semiclassical Eq. (119b) as well, if $p(I)$ is a broad distribution function. However, the predicted reappearances or "revivals" of the signal are a sign that the field is quantized. They occur, and at regular intervals, only because p_n is a discrete distribution. The semiclassical expression (119b) leads inevitably to an irreversible collapse. Only quantum theory can provide the step-wise discontinuous photon number distribution that is the basis for the revivals.

The revivals and other quantum mechanical predictions implied by the truncated Hamiltonian [Eq. (117)] are of interest because this Hamiltonian is simple enough to permit exact calculations, without further approximations of the kind familiar in most of radiation theory. For

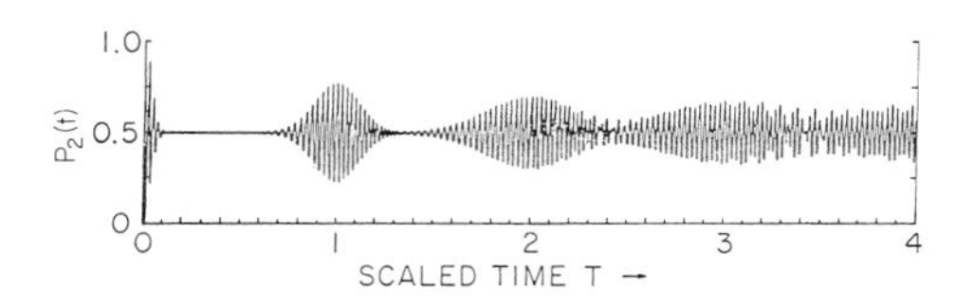

FIG. 10. Quantum collapse and revival of atomic inversion.

example, the expression for quantum inversion in Eq. (119a) has the following unusual properties:

(a) it is not restricted to any finite range of t values;

(b) it holds for all values of the coupling constant λ, which is contained in $\chi(n) = 2\lambda\sqrt{n}$;

(c) it is completely free of decorrelations, such as the commonly used approximation $\langle a^{\dagger}a\sigma_z\rangle \approx \langle a^{\dagger}a\rangle\langle\sigma_z\rangle$;

(d) it is finite even at exact resonance ($\omega_{21} = \omega$) without the aid of ad hoc complex energies;

(e) it is fully quantum mechanical with nontrivial commutators preserved: $[a, a^{\dagger}] = 1$, $[\sigma, \sigma^{\dagger}] = 2\sigma_z$, etc.; and

(f) it is realistically nonlinear (it saturates because the atomic energy cannot exceed E_2).

This combination of properties is unique in atomic radiation theory. They indicate, for example, that a system obeying Hamiltonian [Eq. (117)] would permit some fundamental questions in quantum electrodynamics to be studied independently of the restrictions of the usual perturbation methods that are based on short-time expansions and a small coupling constant. Experimental realization of the model is unlikely in the optical frequency range because of the restriction to a single radiation mode. However, quantum optical techniques, including the detection of single photons, are rapidly being extended to much lower frequencies, where single-mode cavities can be built. Observations of the Jaynes–Cummings model is expected to play a guiding role in microwave single-mode experiments with Rydberg atoms.

E. AC Stark Effect and Resonance Fluorescence

Just as the Bloch vector provides a powerful descriptive framework for a wide variety of quantum optical phenomena, so does the Jaynes–Cummings model. The Hamiltonian [Eq. (117)] can be regarded as a zero-order approximation to a "true" Hamiltonian, in which the atom is allowed more than two levels or the field has more than one mode. It is an unusual zero-order approximation because it includes the interaction Hamiltonian as well as the noninteracting atomic and radiation Hamiltonians.

If the atom interacts resonantly with a single strong mode of the field, then its interactions with other modes, perhaps involving other levels of the atom, can be treated approximately. The energy spectrum of the Jaynes–Cummings Hamiltonian makes it clear how this can be done. In Fig. 11 the RWA energy spectrum is shown in the absence of a strong resonant interaction (i.e., with $\lambda = 0$), and also with $\lambda \neq 0$. The spectrum shows that the state $|n\rangle * |1\rangle$, which corresponds to the atom in its lower level and n photons in the mode when $\lambda = 0$ is pushed down to become the state $\|n, -\rangle$ when $\lambda \neq 0$. That is, for Eq. (118a) we obtain

$$E_{n,-} = E_1 + n\omega - \tfrac{1}{2}[\Omega_n - \Delta] \qquad (120)$$

Since $\Omega_n \geqq \Delta$, the lower level is pushed down. Similarly, the corresponding upper state $|n\rangle * |2\rangle$ is pushed up by the same amount. This shift is called the *AC Stark shift* because it is due to the interaction with an oscillating (alternating) electric field. The size of the AC Stark shift δ_{AC} varies as a function of Δ in the range

$$\tfrac{1}{2}\chi_n \geq \delta_{AC} \geq \chi_n^2/4\Delta \qquad (121)$$

depending on whether the atom and the field mode are near to or far from resonance.

An external probe of the coupled two-level plus single-mode system can reveal these details of its spectrum. For example, the two nearly degenerate states $\|n, +\rangle$ and $\|n, -\rangle$ are split by twice the AC Stark shift. This splitting can be observed by absorption spectroscopy if a weak second radiation field is allowed to induce transitions to a third level in the atom. This was first described and observed in 1955 by Autler and Townes.

A different kind of probe is provided by *resonance fluorescence*, that is, by spontaneous emission into modes other than the main mode.

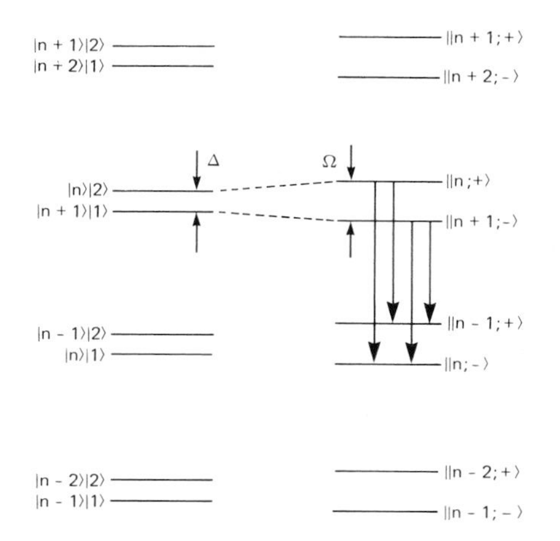

FIG. 11. Jaynes–Cummings RWA energy spectrum.

In this case a strong laser field provides the main mode radiation. Dipole selection rules determine that from the same nearly degenerate states $\|n, +\rangle$ and $\|n, -\rangle$ spontaneous transitions can be made only to the next lower pair of nearly degenerate states, $\|n-1, +\rangle$ and $\|n-1, -\rangle$. There are four separate emission lines predicted, as shown in Fig. 11. The strengths of the four lines are all equal on resonance, but since two of them have the same transition frequency only three lines are actually expected. They have the intensity ratio 1 : 2 : 1, but the side peaks have different widths than the center peak, and the peak height ratio is 1 : 3 : 1. This fluorescence triplet was predicted in the late 1960s, and after a period of controversy about the exact line structure, the predictions mentioned here were verified experimentally in the mid-1970s by Stroud, Walther, Ezekiel, and others. It should be clear from Fig. 11 that the resonance fluorescence peak separation is equal to the Autler–Townes splitting, which is just the quantum Rabi frequency $\chi(n)$ at resonance. The sequence of spectra shown in Fig. 12 illustrates the increased peak separation that accompanies an increased Rabi frequency when the main mode intensity is increased.

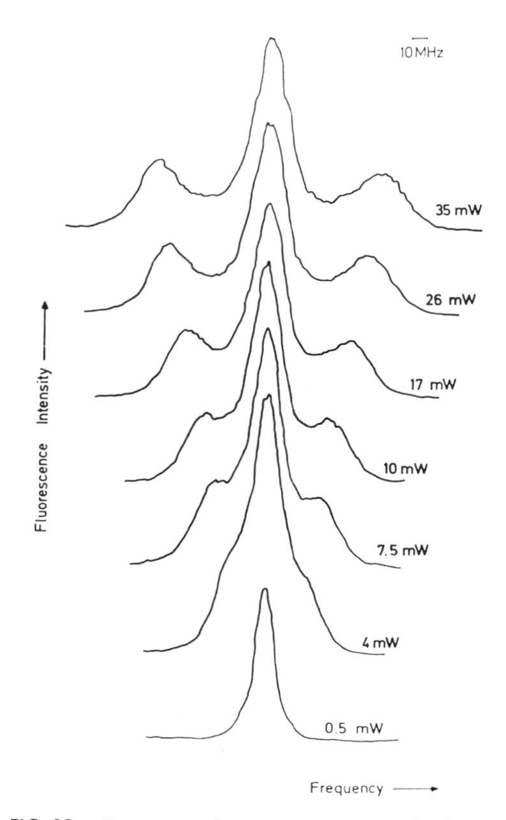

FIG. 12. Resonance fluorescence spectra in the presence of AC Stark splitting. [Reprinted with permission from Hartig, W., Rasmussen, W., Schieder, R., and Walther, H. (1976). Study of the frequency distribution of the fluorescent light induced by monochromatic radiation, *Zeitschrift für Physik,* **A278,** 205–210.]

F. Tests of Quantum Theory

It has long been recognized that quantum theory stands in conflict with naive notions that "physical reality" can be independent of observation. As is well known, the *Heisenberg uncertainty principle* mandates a limit on the mutual precision with which two noncommuting variables may be observed. This curious feature was put into sharp focus by Einstein, Podolsky, and Rosen in 1935, but for nearly half a century it remained mainly of philosophical interest, as experimental tests were difficult to conceive or implement. The situation changed dramatically during the 1970s when it was realized that quantum optical experiments could test different conceptions of "reality."

Einstein, Podolsky, and Rosen (EPR) gave a precise meaning to the concept of reality in this context: "If, without in any way disturbing a system, we can predict with certainty [i.e., with probability equal to unity] the value of a physical quantity, then there exists an element of physical reality corresponding to this physical quantity." It was of primary concern to EPR whether quantum theory can be considered to be a "complete" theory. A necessary condition for completeness of a theory, according to EPR, is that "every element of the physical reality must have a counterpart in the physical theory." Using these definitions of reality and completeness, and the properties of correlated quantum states, EPR concluded that quantum theory does not provide a complete description of physical reality.

An illuminating example due to Bohm is provided by the singlet state of two spin-$\frac{1}{2}$ particles:

$$\|S\rangle = (1/\sqrt{2})[\,|a+, \hat{\mathbf{n}}\rangle * |b-, \hat{\mathbf{n}}\rangle - |a-, \hat{\mathbf{n}}\rangle * |b+, \hat{\mathbf{n}}\rangle] \qquad (122)$$

where $|a\pm, \hat{\mathbf{n}}\rangle$ is the state for which particle a has spin up (+) or down (−) in the direction $\hat{\mathbf{n}}$. The unit vector $\hat{\mathbf{n}}$ can point in any direction. In light of the EPR argument, one notes that the spin of particle b in, say, the $\hat{\mathbf{x}}$ direction, can be predicted with certainty from a measurement of the spin of particle a in that direction. If the spin of particle a is found to be up, then the spin of particle b must, for the singlet state, be down, and vice versa. Thus, the spin of particle b in the $\hat{\mathbf{x}}$ direction can be predicted with certainty "without in any way disturbing" that particle. According to EPR, therefore, the $\hat{\mathbf{x}}$ component of the spin of particle b is an element of physical reality.

Of course, one may choose instead to measure the $\hat{\mathbf{y}}$ component of the spin of particle a, in which case the $\hat{\mathbf{y}}$ component of particle b can be predicted with certainty. It follows therefore that both the $\hat{\mathbf{x}}$ and $\hat{\mathbf{y}}$ components of the spin of particle b (and, of course, particle a) are elements of physical reality. However, according to quantum mechanics, the $\hat{\mathbf{x}}$ and $\hat{\mathbf{y}}$ components of spin cannot have simultaneously predetermined values, because the associated spin operators do not commute. Therefore, quantum theory does not account for these elements of physical reality, and so, according to EPR, it is not a complete theory.

The EPR experiment can be criticized on the ground that "the system" must be understood in its totality and cannot refer to just one of the particles, for instance, of a correlated two-particle system. In the Bohm gedanken experiment just considered, a measurement on particle a does disturb the complete (two-particle) system because the quantum mechanical description of the system is changed by the measurement. One cannot consistently associate an element of physical reality with each spin component of particle b, even though the particles may be arbitrarily far apart and not interacting in any way.

Motivated by the EPR argument, one may ask whether it is possible to formulate a theory in which physical quantities do have objectively real values "out there," independently of whether any measurements are made. These objectively real values may be imagined to be determined by certain *hidden variables* which may themselves be stochastic. One can ask, is it possible, in principle, to construct a hidden variable theory in full agreement with the statistical predictions of quantum mechanics, but which allows for an objective reality in the EPR sense?

In the early 1960s Bell considered the most palatable class of hidden variable theories, the so-called local theories. He demonstrated that such theories cannot fully agree with quantum mechanics. In particular, certain *Bell inequalities* distinguish any local hidden variable theory from quantum mechanics. Another way of stating "Bell's theorem" is that no local realistic theory can be in full agreement with the predictions of quantum theory.

Bell inequalities are not difficult to derive for the Bohm thought experiment. A local hidden variable theory is postulated to give $\pm\frac{1}{2}$ for each spin component, as determined by the hidden variables. The theory is also supposed to account for the spin correlations: if one is up the other must be down. The difference between such a theory and quantum mechanics is that the spins are predetermined (by the hidden variables) before any measurement, that is, they are objectively real. The condition of locality enters through the additional assumption that a measurement of the spin of each particle is not affected by the direction in which the spin of the other particle is measured. This is certainly reasonable if all the spin components are predetermined, since the two particles may be very far apart when a measurement is made.

The question now is whether any measurements can distinguish such a theory from quantum mechanics. Bell considered $E(\hat{\mathbf{m}}, \hat{\mathbf{n}})$, the expectation value of the product of the spin components of particles a and b in the $\hat{\mathbf{m}}$ and $\hat{\mathbf{n}}$ directions, respectively. He obtained the inequality

$$|E(\hat{\mathbf{m}}, \hat{\mathbf{n}}) - E(\hat{\mathbf{m}}, \hat{\mathbf{p}})| \leqq \tfrac{1}{4} + E(\hat{\mathbf{n}}, \hat{\mathbf{p}}) \quad (123)$$

which must be satisfied by the entire class of local hidden variable theories. This inequality is violated by quantum theory, as can be seen from the quantum mechanical prediction $E(\hat{\mathbf{m}}, \hat{\mathbf{n}}) = -\frac{1}{4}\hat{\mathbf{m}} \cdot \hat{\mathbf{n}}$. It is therefore possible to test experimentally the predictions of quantum theory *vis-a-vis* the whole class of plausible "realistic" theories.

Although Bell's theorem promoted philosophical questions about hidden variables to the level of experimental verifiability, it remained difficult to conceive of specific experiments that could be undertaken. In 1969, however, Clauser and coworkers suggested that Bell inequalities could be tested by measuring photon polarization correlations if certain additional but reasonable assumptions about the measurement process were made. The spin considered by Bell is replaced by photon polarization, another two-state phenomenon. Correlated two-photon polarization states are produced in atomic cascade emissions, and efficient polarizers and detectors are available for optical photons.

Consider a $J = 0 \rightarrow 1 \rightarrow 0$ atomic cascade decay, with polarizer–detector systems on the $\pm z$ axes. Linear polarization filters may be employed to distinguish the photons by their energy so that each polarizer–detector system records photons of one frequency but not the other. It may be shown that the two-photon polarization state has the form

$$\|\Psi\rangle = (1/\sqrt{2}[|a, \hat{\mathbf{x}}\rangle * |b, \hat{\mathbf{y}}\rangle + |a, \hat{\mathbf{y}}\rangle * |b, \hat{\mathbf{x}}\rangle] \quad (124)$$

where $|a, \hat{\mathbf{x}}\rangle$ is the single-photon state in which photon a is linearly polarized along the $\hat{\mathbf{x}}$ direction, etc. A similar form applies if a circular polarization basis is used.

The correlated photon state of Eq. (124) is obviously analogous to the spin-$\frac{1}{2}$ correlated state of Eq. (122). A hidden variable theory of such polarization correlations lead to a Bell inequality analogous to Eq. (123):

$$|E(\alpha, \beta) - E(\alpha, \gamma)| \leq 1 - E(\beta, \gamma) \quad (125)$$

where $E(\alpha, \beta)$ now refers to photon polarization components and α and β are the filter orientations with respect to some reference axis. The differences between Eqs. (125) and (123) arise because we are now dealing with spin-one particles and because Eq. (124) describes a positive correlation. The quantum mechanical prediction for $E(\alpha, \beta)$ is simply $\cos 2(\alpha - \beta)$, and Eq. (125) becomes

$$|\cos 2(\alpha - \beta) - \cos 2(\alpha - \gamma)| \leqq 1 - \cos 2(\beta - \gamma) \quad (126)$$

which, in fact, is not satisfied for all angles α, β, and γ. Such violations of Bell inequalities have been observed in independent experiments led by Clauser, Fry, and most recently, Aspect. The results of such experiments are in agreement with quantum theory, and appear to rule out any local hidden variable theory. There are possible loopholes in the interpretation of the experiments, but at the present time none of them seem very plausible. According to Clauser and Shimony, "The conclusions are philosophically unsettling: Either one most totally abandon the realistic philosophy of most working scientists, or dramatically revise our present concept of space–time."

From the viewpoint of quantum optics, the photon polarization correlation experiments measure a second-order field correlation function. Such a correlation function not only distinguishes between classical and quantum radiation theories, but also between quantum theory and local realistic theories. In quantum optics it is often possible to address such questions from essentially first principles and to carry out accurate tests of theory in the laboratory.

Bibliography

Allen, L., and Eberly, J. H. (1975). "Optical Resonance and Two-Level Atoms." Wiley, New York, reprinted by Dover (1987).

Delone, N. B., and Krainov, V. P. (1985). "Atoms In Strong Light Fields." Springer-Verlag, Berlin.

Fontana, P. (1982). "Atomic Radiative Processes." Academic Press, San Diego.

Knight, P. L., and Allen, L. (1983). "Concepts of Quantum Optics." Pergamon, Oxford.

Knight, P. L., and Milonni, P. W. (1980). The Rabi frequency in optical spectra, *In* "Physics Reports," Vol. 66, pp. 21–107. North-Holland, Amsterdam.

Loudon, R. (1983). "The Quantum Theory of Light," 2nd ed. Oxford Univ. Press, Oxford.

Mandel, L. (1976). The case for and against semiclassical radiation theory, *In* "Progress in Optics," (E. Wolf, ed.), Vol. XIII. North-Holland, Amsterdam.

Milonni, P. W. (1976). Semiclassical and quantum-electrodynamical approaches in nonrelativistic radiation theory, *In* "Physics Reports," Vol. 25, pp. 1–81. North-Holland, Amsterdam.

Perina, J. (1984). "Quantum Statistics of Linear and Nonlinear Phenomena." D. Reidel, Dordrecht.

Rosen, H. J., and Gustafson, T. K. (eds.) (1989). Quantum electronic applications and optical studies of high-T_c superconductors, *Quantum Electron.* **25**(11), 2357–2409.

Stenholm, S. (1984). "Foundations of Laser Spectroscopy." John Wiley, New York.

Yoo, H. I., and Eberly, J. H. (1985). Dynamical theory of an atom with two or three levels interacting with quantized cavity fields, *In* "Physics Reports," Vol. 118, pp. 239–337. North-Holland, Amsterdam.

RARE GAS-HALIDE LASERS

Minoru Obara and Fumihiko Kannari *Keio University*

Glossary

Excimer: Molecule that is strongly bound in an excited state but normally has a dissociative ground state. The excited state has a very short lifetime of less than 10 nsec. The term excimer comes from "excited dimer." Excimer also includes exciplex (excited complex).

Pumping: Means by which an inversion is produced in a laser system.

Tunable: Describes a laser having an output wavelength that can be adjusted within a gain bandwidth.

Vacuum ultraviolet (VUV): Region of the spectrum with wavelength less than 200 nm.

Strong laser emission has been obtained from ArF, KrF, XeF, KrCl, XeCl, and XeBr excimers. Strong output wavelength of rare gas-halide lasers spans the spectrum from the near ultraviolet to the vacuum ultraviolet. Rare gas-halide lasers have been operated only in a pulsed mode. Their laser pulse length extends from picoseconds to a microsecond time domain. Rare gas-halide lasers may be pumped in a pulsed mode by electric discharges, intense electron beams, proton beams, or optical sources. These lasers are wavelength tunable over wavelength ranges of several nanometers.

I. Rare Gas-Halide Excimers

A. History

The first demonstration of an excimer laser was made by Basov, Danilychev, and Popov in 1971. They showed stimulated emission of Xe_2 at 172 nm using electron-beam pumping of cryogenic liquid Xe.

The first lasing of a rare gas-halide (RGH) excimer (XeBr) was reported by Searles and Hart in 1975. Shortly thereafter, lasing from XeF was reported by Brau and Ewing. Both XeBr and XeF lasers were pumped by intense electron beams. Subsequently, other RGHs shown in Table I were reported to lase. In addition to electron-beam pumping, researchers have also employed volume-uniform avalanche discharges with X-ray, UV, or corona preionizations, electron-beam controlled discharges, and proton beams successfully to pump a variety of excimer lasers. [*See* LASERS.]

Table I lists RGH lasers. The strong output wavelength of RGH lasers spans the spectrum from the near UV to the vacuum ultraviolet (VUV). Under normal operating conditions, excimer lasing occurs on the wavelength based on the B → X transition described in Section I,B.

In addition to the diatomic RGHs, triatomic RGH excimers can provide tunable coherent photon sources in the visible to the UV region of the spectrum. Table II shows a list of triatomic RGH excimers.

In addition to the triatomic RGHs, a new broad band emission from a fouratomic RGH excimer, Ar_3F at (430 ± 50) nm was reported by Sauerbrey *et al.* in 1986. A fouratomic Ar_3F RGH excimer emission (not lasing) was detected from electron-beam pumped Ar/F_2 and Ar/NF_3 mixtures.

B. Spectroscopic Features of Rare Gas-Halide Excimer Lasers

Figure 1 shows the schematic potential curves for the RGH excimers. The upper laser level is an ionically bound state while the ground state is covalently bound. The upper laser level is

TABLE I. A List of RGH Excimer Lasers

RGH	Wavelength (nm)	Author	Reported year
ArCl	175[a]	Waynant	1977
ArF	193	Hoffman, Hays, and Tisone	1976
KrF	248 (275)[b]	Ewing and Brau	1975
KrCl	222 (240)	Murray and Powell	1976
XeBr	282 (300)	Searles and Hart	1975
XeCl	308 (345)	Ewing and Brau	1975
XeF	351,353 (460)	Brau and Ewing	1975

[a] Wavelength due to B → X transition having the strongest gain coefficient.
[b] Wavelength due to C → A transition in parenthesis.

formed via ionic or neutral reactions. At close internuclear separation, the potential energy curve splits into the $^2\Sigma$ and $^2\Pi$ states as shown in Fig. 1. By convention the $^2\Sigma$ state is referred to as the B state. Under normal high-pressure operating conditions both the B ($^2\Sigma$) and C ($^2\Pi$) state are collisionally mixed.

The ground-state manifold consists of two states, of which the $^2\Sigma$ state has the lowest energy, and is referred to as the X state. This X state is generally nearly flat or weakly bound with the exception of an XeF excimer having a strong bound state with a 1065-cm^{-1} binding energy. The other manifold is the $^2\Pi$ state, which is always repulsive as shown in Fig. 1. This $^2\Pi$ state is referred to as the A state.

Figures 2 and 3 show the potential curves of KrF and XeCl excimers, respectively.

The emission spectrum of RGH excimers consists of two bands such as B ($^2\Sigma$) → X($^2\Sigma$) and C($^2\Pi$) → A($^2\Pi$). The B → X transition has a stronger stimulated-emission cross section than that of the C → A transition, indicating that the B → X transition usually gives intense lasing. The C → A band consists of relatively broad continua, which is attributed to the repulsive structure of the A state.

The stimulated-emission cross section of the B → X bands of RGHs like XeCl and KrF may be written as

$$\sigma = \frac{1}{4\pi c\tau}\sqrt{\frac{ln2}{\pi}}\frac{\lambda^4}{\Delta\lambda}$$

assuming that near the line center the line shape is approximately Gaussian, where c is the velocity of light, λ the wavelength at the line center, τ the lifetime of excimers, and $\Delta\lambda$ the bandwidth of the spectrum.

TABLE II. A List of Triatomic RGH Excimers[a]

RGH	Wavelength (nm)
Ar_2F	285 ± 25
Ar_2Cl	245 ± 15
Kr_2F	420 ± 35
Kr_2Cl	325 ± 15
Kr_2Br	~318
Xe_2F	610 ± 65
Xe_2Cl	490 ± 40
Xe_2Br	440 ± 30

[a] Wavelength and tuning range are based on the data by F. K. Tittel *et al.*

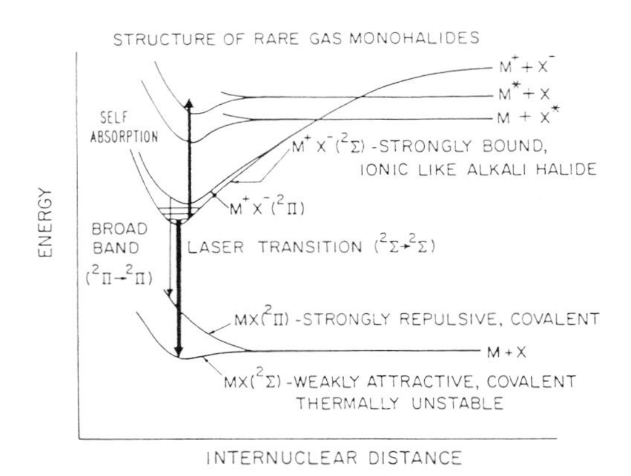

FIG. 1. Schematic potential energy diagram of rare gas halides. [Reproduced with permission from Ch. A. Brau (1984). Rare gas halogen excimers, *in* "Excimer Lasers" (Ch. K. Rhodes, ed.), Springer-Verlag, Berlin and New York.]

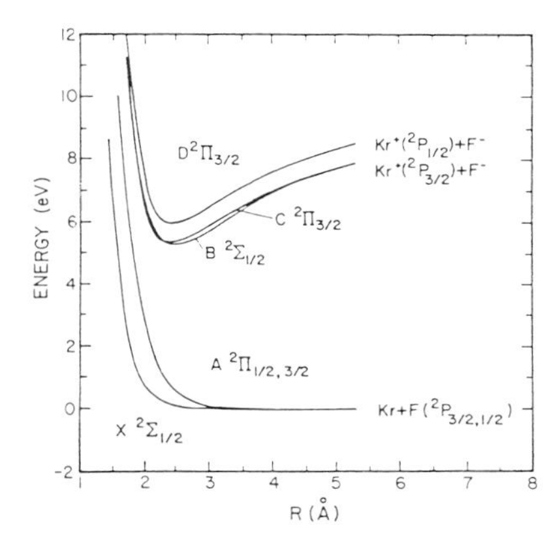

FIG. 2. Potential energy diagram of the KrF excimer. [Reproduced with permission from Ch. A. Brau (1984). Rare gas halogen excimers, *in* "Excimer Lasers" (Ch. K. Rhodes, ed.), Springer-Verlag, Berlin and New York.]

II. Rare Gas-Halide Excimer Laser Kinetics

The kinetic processes involved in a RGH laser are very complicated compared to those of discharge-pumped CO_2 lasers, because neutral reactions and ionic reactions, two-body and three-body reactions, superelastic reactions, and absorption reactions are all responsible for the RGH laser. The kinetic processes involved in the individual RGH lasers are to some extent similar with the exception of those of the XeF laser. Therefore, the kinetic processes for only the discharge-pumped XeCl laser and the e-beam-pumped KrF laser are presented here as an example.

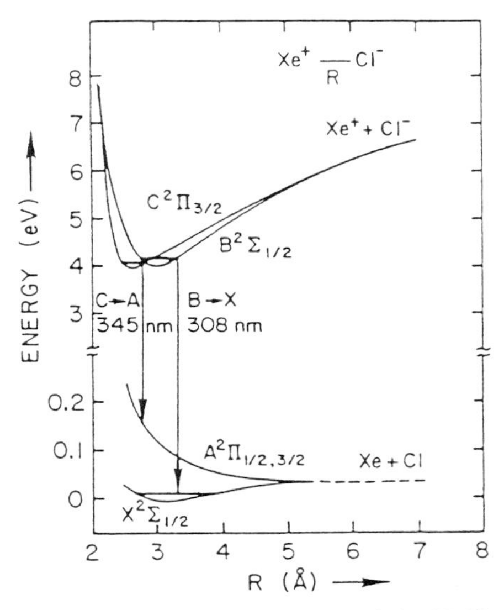

FIG. 3. Potential energy diagram of the XeCl excimer. [Reproduced with permission from D. L. Huestis, G. Marowsky, and F. K. Tittel (1984). Triatomic rare-gas-halide excimers, *in* "Excimer Lasers" (Ch. K. Rhodes, ed.), Springer-Verlag, Berlin and New York.]

A. Discharge-Pumped XeCl Lasers

A typical gas mixture for a self-sustained discharge-pumped XeCl laser is 3-atm mixture of Xe/HCl/Ne = 1/0.1–0.2/balance(%). A list of kinetic reactions responsible for XeCl lasers with their rate constants is shown in Table III. The electron energy distribution in the discharge mixture can be calculated using a Boltzmann equation code, an example of which is shown in Fig. 4. The case is treated of the above mixture pumped by a 100-nsec discharge pulse at an excitation rate of 3 MW/cm^3.

Formation reactions for XeCl(B) are shown in Fig. 5. The percent contribution of the XeCl formation is varied with the HCl concentration. Dominant formation reactions are $Xe^+ + Cl^- \rightarrow$ XeCl(B) (ion recombination reaction) and $NeXe^+ + Cl^- \rightarrow$ XeCl(B) + Ne. A little contribution comes from $Xe^* + HCl(v) \rightarrow XeCl(B) + H$ and $Xe_2^+ + Cl^- \rightarrow XeCl(B) + Xe$. Over 23% of the electrical energy deposited into the discharge can be supplied to form XeCl(B).

Relaxation reactions for XeCl(B) are shown in Fig. 6. About 65% of the formed XeCl(B) con-

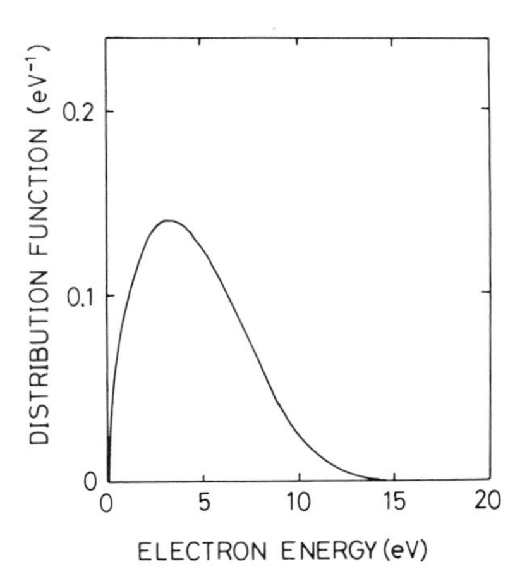

FIG. 4. Electron energy distribution in the 3-atm mixture of Xe/HCl/Ne = 1/0.1/98.9(%) pumped at 3 MW/cm^3. E/N = 2.9×10^{-17} V cm^2.

TABLE III. Kinetic Reactions Involved in the Discharge-Pumped XeCl Laser[a]

Reaction	Rate constant
Secondary electron process	
$e + Ne \rightarrow Ne^* + e$	
$e + Ne \rightarrow Ne^+ + 2e$	
$e + Ne^* \rightarrow Ne^+ + 2e$	
$e + Xe \rightarrow Xe^* + e$	Calculated by Boltzmann Eq.
$e + Xe \rightarrow Xe^+ + 2e$	
$e + Xe^* \rightarrow Xe^+ + 2e$	
$e + HCl \rightarrow HCl(v) + e$	
$e + HCl \rightarrow H + Cl^-$	
$e + HCl(v) \rightarrow H + Cl^-$	
$e + Cl_2 \rightarrow Cl^- + Cl$	1.1(−10) cm³/sec
Electron quenching	
$Ne^* + e \rightarrow e + Ne$	Calculated by Boltzmann Eq.
$Xe^* + e \rightarrow e + Xe$	
$Ne_2^* + e \rightarrow e + 2Ne$	Analogy to Ne*
$Xe_2^* + e \rightarrow e + 2Xe$	Analogy to Xe*
$Cl_2^* + e \rightarrow e + Cl_2$	3.0(−7) cm³/sec
$NeCl^* + e \rightarrow e + Ne + Cl$	2.0(−7)
$XeCl^* + e \rightarrow e + Xe + Cl$	2.0(−7)
HCl(v) quenching	
$HCl(v) + Ne \rightarrow HCl + Ne$	6.2(−17) cm³/sec
$HCl(v) + HCl \rightarrow 2HCl$	2.7(−14)
Neutral reaction	
$Ne^* + 2Ne \rightarrow Ne_2^* + Ne$	4.1(−34) cm⁶/sec
$Xe^* + 2Xe \rightarrow Xe_2^* + Xe$	8.0(−32)
$Xe^* + Xe + Ne \rightarrow Xe_2^* + Ne$	1.6(−32)
$Xe^* + HCl \rightarrow Xe + H + Cl$	5.6(−10) cm³/sec
$Xe^* + HCl(v) \rightarrow Xe + H + Cl$	5.6(−10)
$Xe^* + HCl(v) \rightarrow XeCl^* + H$	2.0(−10)
Penning ionization	
$Ne^* + Xe \rightarrow Xe^+ + Ne + e$	7.5(−11) cm³/sec
$Ne^* + Xe \rightarrow NeXe^+ + e$	1.8(−11)
$Xe^* + Xe^* \rightarrow Xe^+ + Xe + e$	5.0(−10)
$Xe_2^* + Xe_2^* \rightarrow Xe_2^+ + 2Xe + e$	3.5(−10)
Charge transfer	
$Ne^+ + 2Ne \rightarrow Ne_2^+ + Ne$	4.4(−32) cm⁶/sec
$Ne^+ + Xe \rightarrow Xe^+ + Ne$	1.0(−14) cm³/sec
$Ne^+ + Xe + Ne \rightarrow NeXe^+ + Ne$	1.0(−31) cm⁶/sec
$Ne_2^+ + Xe \rightarrow NeXe^+ + Ne$	1.0(−13) cm³/sec
$Ne_2^+ + Xe + Ne \rightarrow Xe^+ + 3Ne$	4.0(−30) cm⁶/sec
$Xe^+ + 2Xe \rightarrow Xe_2^+ + Xe$	3.6(−31)
$Xe^+ + 2Ne \rightarrow NeXe^+ + Ne$	2.5(−31)
$Xe^+ + Xe + Ne \rightarrow Xe_2^+ + Ne$	1.0(−31)
$NeXe^+ + Xe \rightarrow Xe^+ + Ne + Xe$	5.0(−10) cm³/sec
$NeXe^+ + Xe \rightarrow Xe_2^+ + Ne$	5.0(−12)
Ion-electron recombination	
$Ne_2^+ + e \rightarrow Ne^* + Ne$	3.7(−8)$Te^{-0.43}$ cm³/sec
$Xe_2^+ + e \rightarrow Xe^* + Xe$	2.2(−7)$Te^{-0.5}$
$NeXe^+ + e \rightarrow Xe^* + Ne$	2.0(−7)$Te^{-0.5}$
Ion-ion recombination	
$Ne^+ + Cl^- \rightarrow NeCl^*$	} ~2.0(−6) cm³/sec
$Ne_2^+ + Cl^- \rightarrow NeCl^* + Ne$	

(*continues*)

TABLE III. *(Continued)*

Reaction	Rate constant
$Xe^+ + Cl^- \rightarrow XeCl^*$	Pressure dependent rate
$Xe_2^+ + Cl^- \rightarrow XeCl^* + Xe$	$\sim 2.0(-6)$ cm³/sec
$NeXe^+ + Cl^- \rightarrow XeCl^* + Ne$	
$Cl^+ + Cl^- \rightarrow Cl_2^*$	$2.0(-6)$ cm³/sec
Radiation	
$Ne_2^* \rightarrow 2Ne + h\nu$	$3.6(+8)$ sec^{-1}
$Xe_2^* \rightarrow 2Xe + h\nu$	$6.0(+7)$
$Cl_2^* \rightarrow Cl_2 + h\nu$	$5.0(+7)$
$XeCl^* \rightarrow Xe + Cl + h\nu$	$2.5(+7)$
$XeCl^* + h\nu \rightarrow Xe + Cl + h\nu$	$1.25(-16)$ cm²
$Xe_2Cl^* \rightarrow 2Xe + Cl + h\nu$	$7.4(+6)$ sec^{-1}
Predissociation	
$NeCl^* \rightarrow Ne + Cl^+ + e$	$1.0(+10)$ sec^{-1}
$XeCl^*$ quenching	
$XeCl^* + Ne \rightarrow Xe + Cl + Ne$	$1.0(-12)$ cm³/sec
$XeCl^* + Xe \rightarrow 2Xe + Cl$	$3.2(-12)$
$XeCl^* + HCl \rightarrow Xe + Cl + HCl$	$1.7(-9)$
$XeCl^* + HCl(v) \rightarrow Xe + Cl + HCl$	$7.7(-10)$
$XeCl^* + 2Ne \rightarrow Xe + Cl + 2Ne$	$1.0(-33)$ cm⁶/sec
$XeCl^* + 2Xe \rightarrow Xe_2Cl^* + Xe$	$7.3(-31)$
$XeCl^* + Xe + Ne \rightarrow Xe_2Cl^* + Ne$	$1.5(-31)$
Xe_2Cl^* quenching	
$Xe_2Cl^* + Cl_2 \rightarrow 2Xe + Cl + Cl_2$	$2.6(-10)$ cm³/sec
$Xe_2Cl^* + Xe \rightarrow 3Xe + Cl$	$6.0(-15)$
Absorption	
$Ne_2^+ + h\nu \rightarrow Ne^+ + Ne$	$7.4(-18)$ cm²
$Xe_2^+ + h\nu \rightarrow Xe^+ + Xe$	$2.6(-17)$
$NeXe^+ + h\nu \rightarrow Xe^+ + Ne$	$1.0(-19)$
$Xe_2^* + h\nu \rightarrow Xe_2^+ + e$	$1.4(-17)$
$Xe^* + h\nu \rightarrow Xe^+ + e$	$6.0(-20)$
$Xe_2Cl^* + h\nu \rightarrow$ products	$2.6(-17)$
$Cl^- + h\nu \rightarrow Cl + e$	$2.1(-17)$

[a] *Te* is electron temperature.

tributes to the stimulated emission as an intracavity laser flux due to its large stimulated-emission cross section. About 30% of the XeCl(B) is collisionally quenched and spontaneous emission is negligible during lasing. Collisional quenching processes concerned with the XeCl(B) excimer are shown in Fig. 7. At this high excitation rate of 3 MW/cm³, a large fraction is occupied by the discharge electron, which is called a superelastic collision process.

All the intracavity laser flux cannot be extracted because the RGH laser mixture contains many absorbers at the laser wavelength. The percent contribution of the absorption channel is shown in Fig. 8. Main absorbers seem to be Cl^- and Xe_2^+. The photon extraction efficiency, defined as the ratio of the extracted laser energy to the intracavity laser energy, is in excess of 70% here (see Fig. 6). This means that over 70% of the intracavity laser energy can be extracted as a laser output. Here, the maximum extraction efficiency can be written as

$$\eta_{max} \simeq (1 - \sqrt{\alpha/g})^2$$

where g and α are small-signal gain and absorption coefficients, respectively. Therefore, the efficiency is increased with increasing values of g/α.

If the mixing ratio of Xe/HCl/Ne is varied, the electron energy distribution in the discharge plasma changes. As a result, formation of precursors Xe^+, Xe^*, Ne^+, and Ne^* is greatly affected. If helium is used as a diluent gas in place of Ne, the electron temperature also changes, resulting in a different pathway for the XeCl(B) formation and a less effective formation than that of the Ne diluent.

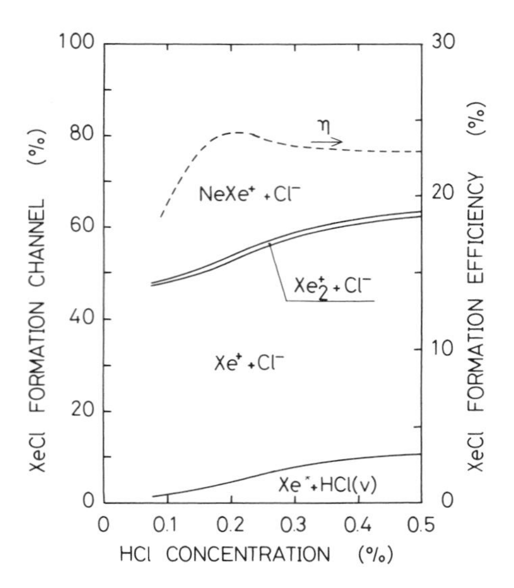

FIG. 5. Variation of percent contribution of the XeCl* formation channels and the XeCl* formation efficiency with the HCl concentration. The gas mixture is X% HCl/1% Xe/(99-X)% Ne. [Reproduced with permission from M. Ohwa and M. Obara (1986). *J. Appl. Phys.* **59**(1), 32.]

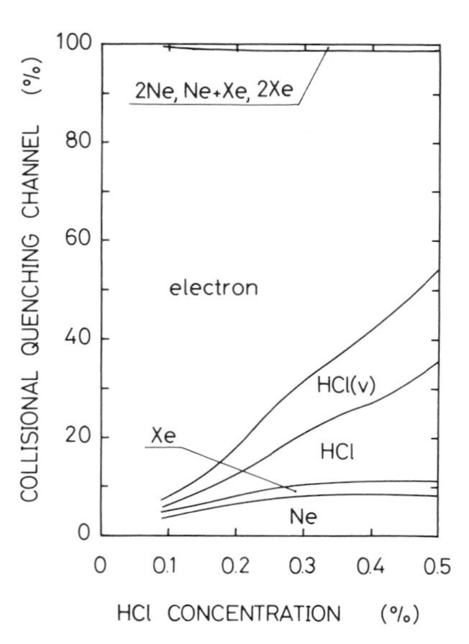

FIG. 7. Variation of percent contribution of the XeCl* collisional quenching channels with the HCl concentration. [Reproduced with permission from M. Ohwa and M. Obara (1986). *J. Appl. Phys.* **59**(1), 32.]

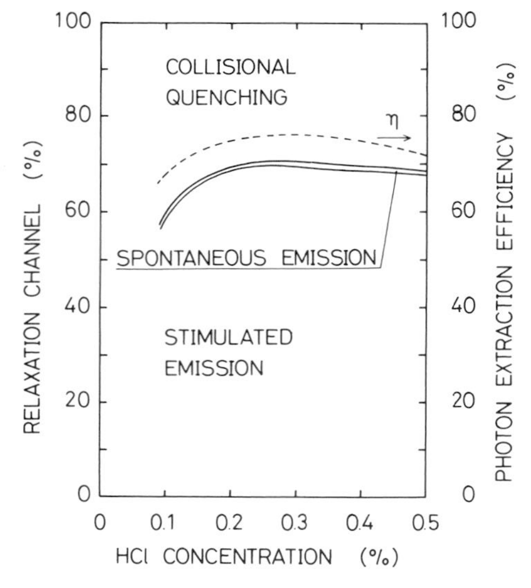

FIG. 6. Variation of percent contribution of the XeCl* relaxation channels and the photon extraction efficiency with the HCl concentration. [Reproduced with permission from M. Ohwa and M. Obara (1986). *J. Appl. Phys.* **59**(1), 32.]

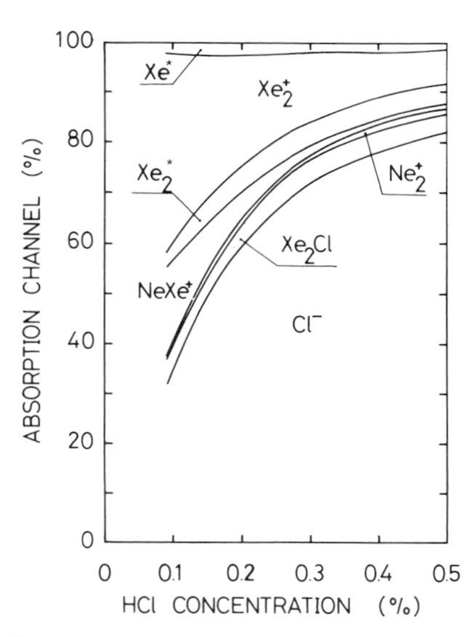

FIG. 8. Variation of percent contribution of the absorption at 308 nm with the HCl concentration. [Reproduced with permission from M. Ohwa and M. Obara (1986). *J. Appl. Phys.* **59**(1), 32.]

B. Electron-Beam Pumped KrF Lasers

A list of kinetic processes concerned with the e-beam-pumped KrF laser using a typical mixture of $Kr/F_2/Ar$ is shown in Table IV.

High-energy electrons of 300 keV to 2 MeV generated from a pulsed relativistic e-beam generator are deposited into the $Kr/F_2/Ar$ mixture to produce the precursors Ar^+, Ar^*, Kr^+, and Kr^*. A typical e-beam current density is from 10 A/cm^2 to 1000 A/cm^2, and its excitation rate ranges from 0.1 MW/cm^3 to 10 MW/cm^3, approximately, depending on the mixture pressure. The high-energy electrons can rapidly be thermalized in He, Ar, and Kr gases in the nanosecond time range, and as a result the secondary electrons reach a thermalized energy of around 1 eV. The initially formed Ar^+ and Ar^* are used to form ArF^*, and then KrF^* is formed via the reaction $ArF^* + Kr \rightarrow KrF^* + Ar$. Here Kr^+ is formed via the charge transfer reaction $Ar^+ + Kr \rightarrow Kr^+ + Ar$, and Kr^+ is also used to form

TABLE IV. KrF* Kinetic Processes Involved in the e-Beam-Pumped KrF Laser[a]

Reaction	Rate constant
Neutral reaction	
$Ar^* + F_2 \rightarrow ArF^* + F$	7.5(−10)
$Ar^* + Kr + Ar \rightarrow ArKr^* + Ar$	1.0(−32) cm^6/sec
$Ar^* + 2Ar \rightarrow Ar_2^* + Ar$	1.14(−32) cm^6/sec
$Ar^* + Kr \rightarrow Ar + Kr^*$	6.2(−12)
$Ar^* + F_2 \rightarrow Ar + 2F$	3.1(−10)
$Ar^{**} + F_2 \rightarrow ArF^* + F$	4.7(−10)
$Ar^{**} + Ar \rightarrow Ar^* + Ar$	1.0(−10)
$Ar^{**} + F_2 \rightarrow Ar + 2F$	3.1(−10)
$Ar_2^* + F_2 \rightarrow Ar_2F^* + F$	2.5(−10)
$Ar_2^* + F \rightarrow ArF^* + Ar$	3.0(−10)
$Ar_2^* + Kr \rightarrow Kr^* + 2Ar$	4.0(−10)
$Ar_2^* + F_2 \rightarrow ArF^* + Ar + F$	3.0(−10)
$Kr^* + F_2 \rightarrow KrF^* + F$	8.1(−10)
$Kr^* + 2Ar \rightarrow ArKr^* + Ar$	1.0(−32) cm^6/sec
$Kr^* + Kr + Ar \rightarrow Kr_2^* + Ar$	2.7(−32) cm^6/sec
$Kr^{**} + F_2 \rightarrow KrF^* + F$	8.1(−10)
$Kr^{**} + Ar \rightarrow Kr^* + Ar$	1.0(−10)
$Kr_2^* + F_2 \rightarrow Kr_2F^* + F$	3.0(−10)
$Kr_2^* + F_2 \rightarrow KrF^* + Kr + F$	3.0(−10)
$Kr_2^* + F \rightarrow KrF^* + Kr$	3.0(−10)
$ArKr^* + F_2 \rightarrow KrF^* + Ar + F$	6.0(−10)
$ArKr^* + Kr \rightarrow Kr_2^* + Ar$	1.0(−10)
$ArKr^* + F_2 \rightarrow ArKrF^* + F$	3.0(−10)
$F + F + M \rightarrow F_2 + M$	1.0(−33) cm^6/sec
ArF* relaxation	
$ArF^* + Kr \rightarrow KrF^* + Ar$	1.6(−09)
$ArF^* + 2Ar \rightarrow Ar_2F^* + Ar$	4.0(−31) cm^6/sec
$ArF^* + F_2 \rightarrow Ar + F + F_2$	1.9(−9)
$ArF^* + Ar \rightarrow 2Ar + F$	9.0(−12)
Ar_2F^* relaxation	
$Ar_2F^* + F_2 \rightarrow 2Ar + F + F_2$	2.0(−10)
$Ar_2F^* + Kr \rightarrow ArKrF^* + Ar$	1.0(−10)
$Ar_2F^* + Kr \rightarrow KrF^* + 2Ar$	1.0(−10)
$Ar_2F^* + Ar \rightarrow 3Ar + F$	1.0(−10)
KrF* relaxation	
$KrF^* + Kr \rightarrow 2Kr + F$	2.0(−12)[b]
$KrF^* + Ar \rightarrow Kr + Ar + F$	1.8(−12)[b]
$KrF^* + 2Ar \rightarrow ArKrF^* + Ar$	7.0(−32) cm^6/sec[b,c]
$KrF^* + Kr + Ar \rightarrow Kr_2F^* + Ar$	6.5(−31) cm^6/sec[b,c]
$KrF^* + F_2 \rightarrow Kr + F + F_2$	7.8(−10)[b]

(*continues*)

TABLE IV. *(Continued)*

Reaction	Rate constant
$KrF^* + 2Kr \rightarrow Kr_2F^* + Kr$	$6.7(-31)$ cm^6/sec[b,c]
$KrF^* + F \rightarrow Kr + 2F$	$7.8(-10)$[b]
Kr_2F^* relaxation	
$Kr_2F^* + F_2 \rightarrow 2Kr + F + F_2$	$1.5(-10)$
$ArKrF^*$ relaxation	
$ArKrF^* + F_2 \rightarrow Ar + Kr + F + F_2$	$1.0(-09)$
$ArKrF^* + Kr \rightarrow Kr_2F^* + Ar$	$2.0(-11)$
$ArKrF^* + Ar \rightarrow Ar_2F^* + Kr$	$2.0(-11)$
Electron excitation, ionization, and attachment	
$Ar + e \leftrightarrow Ar^* + e$	$5.0(-09)Te^{0.74}\exp(-11.56/Te)$
$Ar + e \leftrightarrow Ar^{**} + e$	$1.40(-80)Te^{0.71}\exp(-13.15/Te)$
$Ar^* + e \leftrightarrow Ar^{**} + e$	$8.90(-07)Te^{0.51}\exp(-1.590/Te)$
$Ar + e \rightarrow Ar^+ + 2e$	$2.30(-08)Te^{0.68}\exp(-15.76/Te)$
$Ar^* + e \rightarrow Ar^+ + 2e$	$6.80(-09)Te^{0.67}\exp(-4.20/Te)$
$Ar^{**} + e \rightarrow Ar^+ + 2e$	$1.80(-07)Te^{0.61}\exp(-2.61/Te)$
$Ar_2^* + e \rightarrow Ar_2^+ + 2e$	$9.00(-08)Te^{0.70}\exp(-3.66/Te)$
$Kr + e \leftrightarrow Kr^* + e$	$9.00(-09)Te^{0.72}\exp(-9.96/Te)$
$Kr + e \leftrightarrow Kr^{**} + e$	$2.30(-08)Te^{0.72}\exp(-11.47/Te)$
$Kr^* + e \leftrightarrow Kr^{**} + e$	$9.00(-07)Te^{0.50}\exp(-1.51/Te)$
$Kr + e \rightarrow Kr^+ + 2e$	$2.70(-08)Te^{0.70}\exp(-14.00/Te)$
$Kr^* + e \rightarrow Kr^+ + 2e$	$7.00(-08)Te^{0.68}\exp(-4.04/Te)$
$Kr^{**} + e \rightarrow Kr^+ + 2e$	$2.00(-07)Te^{0.62}\exp(-2.53/Te)$
$Kr_2^* + e \rightarrow Kr_2^+ + 2e$	$9.00(-08)Te^{0.70}\exp(-3.53/Te)$
$ArKr^* + e \rightarrow ArKr^+ + 2e$	$9.00(-08)Te^{0.70}\exp(-3.60/Te)$
$F_2 + e -$ beam $\rightarrow F_2^+ + e -$ beam $+ e$:	32 eV/ion electron
$Ar + e -$ beam $\rightarrow Ar^+ + e -$ beam $+ e$:	26.2 eV/ion electron
$Kr + e -$ beam $\rightarrow Kr^+ + e -$ beam $+ e$:	24.3 eV/ion electron
$Ar + e -$ beam $\rightarrow Ar^* + e -$ beam:	$0.28 \times$ ionization rate
$Kr + e -$ beam $\rightarrow Kr^* + e -$ beam:	$0.28 \times$ ionization rate
$F_2 + e \rightarrow F^- + F$	$4.5(-09)$ ($Te = 1$ eV)
Penning ionization	
$Ar^{**} + Ar^{**} \rightarrow Ar^+ + Ar + e$	$5.0(-10)$
$Ar^* + Ar^* \rightarrow Ar^+ + Ar + e$	$5.0(-10)$
$Ar_2^* + Ar_2^* \rightarrow Ar_2^+ + 2Ar + e$	$5.0(-10)$
$Kr_2^* + Kr_2^* \rightarrow Kr_2^+ + 2Kr + e$	$5.0(-10)$
$Kr^* + Kr^* \rightarrow Kr^+ + Kr + e$	$5.0(-10)$
Charge transfer reaction	
$Ar_2^* + Kr \rightarrow Kr^+ + 2Ar$	$7.5(-10)$
$Ar^+ + Kr \rightarrow Kr^+ + Ar$	$3.0(-11)$
$Ar^+ + 2Ar \rightarrow Ar_2^+ + Ar$	$2.5(-31)$ cm^6/sec
$Ar^+ + Kr + Ar \rightarrow ArKr^+ + Ar$	$1.0(-31)$ cm^6/sec
$Kr^+ + 2Ar \rightarrow ArKr^+ + Ar$	$1.0(-31)$ cm^6/sec
$ArKr^+ + Kr \rightarrow Kr_2^+ + Ar$	$3.2(-10)$
$Kr^+ + 2Kr \rightarrow Kr_2^+ + Kr$	$2.5(-31)$ cm^6/sec
$Kr^+ + Ar + Kr \rightarrow Kr_2^+ + Ar$	$2.5(-31)$ cm^6/sec
Three-body ion–ion recombination reaction	
$Ar^+ + F^- + (Ar, Kr) \rightarrow ArF^* + (Ar, Kr)$	$\sim 2.0(-6)$ cm^3/sec (effective two-body collision rate)
$Ar_2^+ + F^- + (Ar, Kr) \rightarrow ArF^* + Ar + (Ar, Kr)$	
$ArKr^+ + F^- + (Ar, Kr) \rightarrow KrF^* + (Ar, Kr)$	
$ArKr^+F^- + (Ar, Kr) \rightarrow ArKrF^* + (Ar, Kr)$	
$Kr^+ + F^- + (Ar, Kr) \rightarrow KrF^* + (Ar, Kr)$	
$Kr_2^+ + F^- + (Ar, Kr) \rightarrow KrF^* + Kr + (Ar, Kr)$	
$F_2^+ + F^- \rightarrow 3F$	$4.0(-8)$

(continues)

TABLE IV. (*Continued*)

Reaction	Rate constant
Desociative recombination reaction	
$Ar_2^+ + e \rightarrow Ar^* + Ar$	$0.6(-6)[Te(K)/300]^{-0.66}$
$Ar_2^+ + e \rightarrow Ar^{**} + Ar$	1.1(−07)
$Kr_2^+ + e \rightarrow Kr^* + Kr$	$1.2(-6)[Te(K)/300]^{-0.55}$
$Kr_2^+ + e \rightarrow Kr^{**} + Kr$	1.9(−07)
$ArKr^+ + e \rightarrow Kr^{**} + Ar$	$1.2(-6)[Te(K)/300]^{-0.55}$
Superelastic reaction	
$ArF^* + e \rightarrow Ar + F + e$	1.6(−07)
$Ar_2F^* + e \rightarrow 2Ar + F + e$	1.0(−07)
$Kr_2F^* + e \rightarrow 2Kr + F + e$	1.0(−07)
$KrF^* + e \rightarrow Kr + F + e$	2.0(−07)
$ArKr^* + e \rightarrow Ar + Kr + e$	1.0(−07)
$ArKrF^* + e \rightarrow Ar + Kr + F + e$	1.0(−07)
$Ar_2^* + e \rightarrow 2Ar + e$	1.0(−07)
$Kr_2^* + e \rightarrow 2Kr + e$	1.0(−07)
Radiative lifetime	
$KrF^*(B) \rightarrow Kr + F + h\nu$	1.43(08) sec^{-1}
$KrF^*(C) \rightarrow Kr + F + h\nu$	1.33(07) sec^{-1}
$ArF^*(B) \rightarrow Ar + F + h\nu$	2.50(08) sec^{-1}
$ArF^*(C) \rightarrow Ar + F + h\nu$	2.27(07) sec^{-1}
$Kr_2^* \rightarrow 2Kr + h\nu$	8.0(07) sec^{-1}
$Ar_2^* \rightarrow 2Ar + h\nu$	6.0(07) sec^{-1}
$Kr_2F^* \rightarrow 2Kr + F + h\nu$	5.6(06) sec^{-1}
$Ar_2F^* \rightarrow 2Ar + F + h\nu$	5.4(06) sec^{-1}
$ArKr^* \rightarrow Ar + Kr + h\nu$	8.0(07) sec^{-1}
$ArKrF^* \rightarrow Ar + Kr + F + h\nu$	5.0(06) sec^{-1}
248-nm absorption cross section	(cm^2)
$F_2 + h\nu \rightarrow 2F$	1.2(−20)
$F^- + h\nu \rightarrow F + e$	5.6(−18)
$Kr_2^+ + h\nu \rightarrow Kr^+ + Kr$	1.6(−18)
$Ar_2^+ + h\nu \rightarrow Ar^+ + Ar$	1.3(−17)
$Ar^{**} + h\nu \rightarrow Ar^+ + e$	6.0(−18)
$Kr^{**} + h\nu \rightarrow Kr^+ + e$	6.0(−18)
$Ar^* + h\nu \rightarrow Ar^+ + e$	4.5(−18)
$Kr^* + h\nu \rightarrow Kr^+ + e$	1.0(−19)
$Kr_2F^* + h\nu \rightarrow 2Kr + F$	6.0(−18)
$Ar_2F^* + h\nu \rightarrow 2Ar + F$	1.0(−18)
$KrF^* + h\nu \rightarrow Kr + F + 2h\nu$	2.6(−16)
$ArKr^+ + h\nu \rightarrow Ar^+ + Kr + h\nu$	1.5(−17)
$Kr_2^* + h\nu \rightarrow Kr_2^+ + e$	1.8(−18)

[a] Units are in cm^3/sec unless otherwise noted. Te is electron temperature.

[b] These quenching rate constants were experimentally evaluated from the product of quenching rate constant k and radiative lifetime of $RgX^*(B)$ state $\tau(B)$. Considering the mixing process of $RgX^*(B)/(C)$ at a finite rate, these quenching rate constants were modified using the effective radiative lifetime, $\tau(B/C)$. Reduction factors for ArF^* quenching rate constants and KrF^* quenching rate constants are 0.418 and 0.433, respectively.

[c] Expressed for 300 K gas temperature.

KrF* via $Kr^+ + F^-$. The negative ion F^- is formed via an electron attachment process with the thermalized slow electron $e + F_2 \rightarrow F^- + F$.

The percent contribution to the KrF* formation as a function of excitation rate is shown in Fig. 9. The Kr/F_2/Ar mixture of 1.5 atm is treated as a typical system. The KrF* excimer is mainly formed through $Kr^+ + F^-$, ArF* + Kr, and Kr* + F_2. The KrF* formation efficiency reaches roughly 20%

The percent contributions of the KrF* relaxation channels are shown in Fig. 10. Sixty to eighty percent of the KrF* excimers can contribute to the stimulated emission as an intracavity laser flux, depending on the excitation rate. Other relaxation processes are by a slow electron, F_2, Kr, and Ar. In these collisional relaxation reactions the reaction KrF* + Kr forms the Kr_2F^* trimer, and the reaction KrF* + Ar forms the ArKrF* trimer.

About 85% of the intracavity laser flux can be extracted as a laser output energy, as shown in Fig. 11. The rest of the energy is lost in the

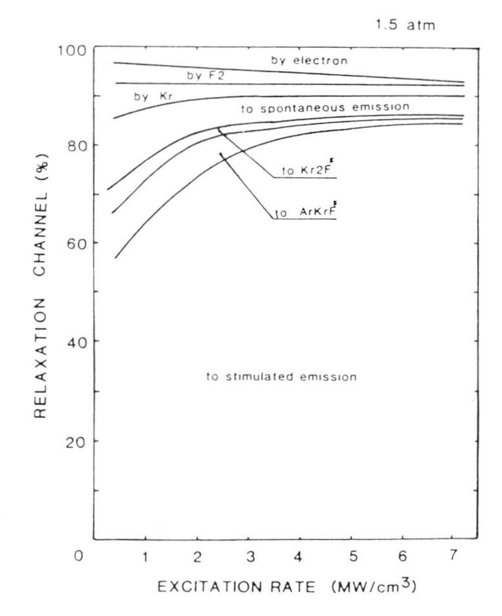

FIG. 10. Variation of percent contribution of the KrF* relaxation channels with the excitation rate for the 1.5-atm mixture. [Reproduced with permission from F. Kannari *et al.* (1983). *IEEE J. Quantum Electron.* **QE-19**(2). Copyright © 1983 IEEE.]

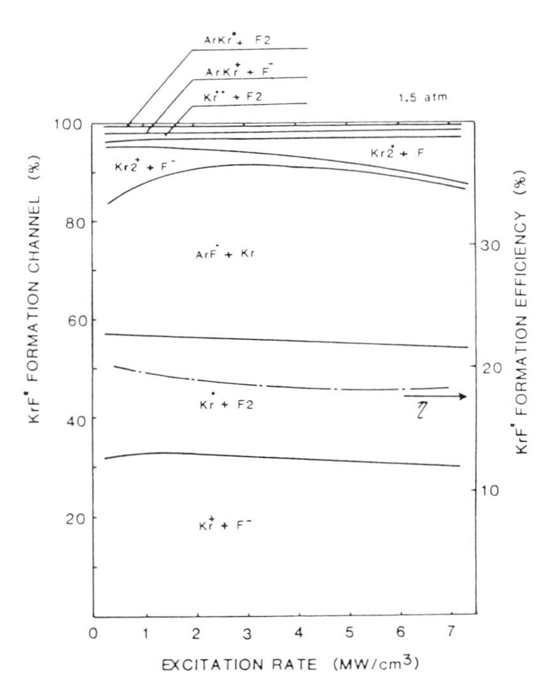

FIG. 9. Variation of percent contribution of the KrF* formation channels and the KrF* formation efficiency with the excitation rate for the 1.5-atm mixture with argon diluent pumped by a 70-nsec e-beam. [Reproduced with permission from F. Kannari *et al.* (1983). *IEEE J. Quantum Electron.* **QE-19**(2). Copyright © 1983 IEEE.]

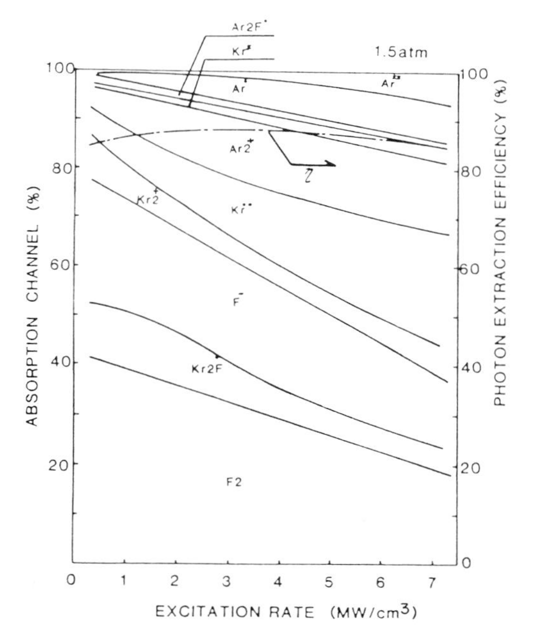

FIG. 11. Variation of percent contribution of the 248-nm absorption channels and the photon extraction efficiency with the excitation rate for the 1.5-atm mixture. [Reproduced with permission from F. Kannari *et al.* (1983). *IEEE J. Quantum Electron.* **QE-19**(2), 232. Copyright © 1983 IEEE.]

cavity by absorptions due to F_2, F^-, Ar_2^+, Kr_2^+, and Kr_2F^*, etc.

III. Rare Gas-Halide Excimer Laser Technology

Typical gas mixtures of rare gas-halide lasers are high-pressure (approximately 1–5 atm) rare gas mixtures containing a small amount of halogen donor. High excitation-rate pumping of several 100 kW/cm^3 to several MW/cm^3 is necessary to efficiently produce high gain of the excimer laser of interest in relation to the absorption. For this purpose, pulsed e-beam pumping and self-sustained discharge pumping have been mainly employed, while proton-beam pumping, nuclear pumping, e-beam sustained discharge pumping, and rf discharge pumping have also been tried.

Shown in Fig. 12 is a plot of the laser output energy as a function of the pulse repetition rate for various reported RGH lasers. The oblique lines show the average RGH laser output power (laser energy per pulse multiplied by pulse repetition rate). It is understood that the recent RGH laser power level lies in the power region less than 1 kW, although the laser power level is rapidly being increased. Discharge-pumped RGH lasers capable of producing laser output up to 100 W are now commercially available. Region A in Fig. 12 shows the regime in which the average laser power is gained not by pulse energy but by pulse repetition rate. These lasers are predominantly discharge pumped. Region B indicates the regime in which the average power is increased by both pulse energy and pulse repetition rate. Finally, Region C shows the regime in which low-repetition-rate or single-pulse lasers are operated with high-energy laser output.

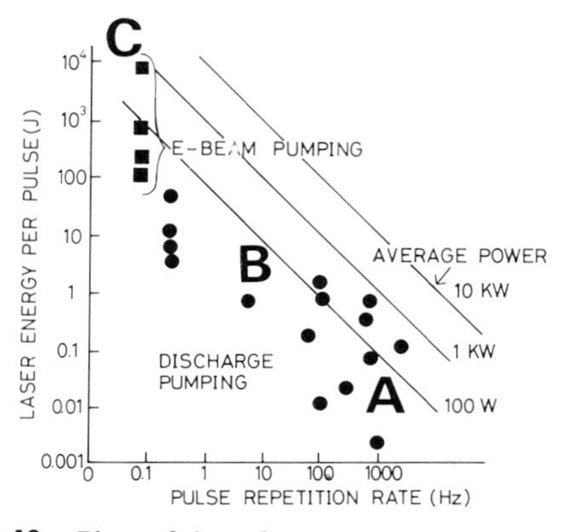

FIG. 12. Plots of the RGH laser energy as a function of pulse repetition rate for various RGH lasers reported.

These lasers are mainly pumped by intense relativistic e-beams. The high-repetition-rate technology for these lasers is not a current issue, but energy-scaling technology for single pulse operation is currently a rather urgent issue. The lasers concerned in this region are e-beam-pumped high-energy KrF and XeF lasers for inertial confinement fusion applications and for military applications, respectively. The world's largest KrF laser reported to date is the system called LAM (large aperture module), developed at Los Alamos National Laboratory, and is used as an inertial confinement fusion driver. In 1985 this laser could produce ~10 kJ per pulse output. The highest-energy XeCl laser of 60-J level output pumped by discharges was developed at the US Naval Research Laboratory in 1985.

A. Electron-Beam Pumping

The pulsed-power technology involved in the efficient and spatially uniform generation of intense (high current-density) relativistic e-beams from relatively large aperture diodes has advanced recently for single shot operation. The e-beam generation technology is so scalable in energy that high-energy RGH excimer lasers are more readily realizable than discharge-pumped lasers.

The relevant components of a cold cathode-type e-beam diode suited for RGH laser pumping are shown schematically in Fig. 13. The e-beam diode consists of a cold cathode to which the pulsed high voltage (negative polarity) of 30 to 1500 nsec is applied, a thin foil anode, and a pressure foil separating the high-pressure laser gas mixture from the vacuum diode chamber. These foils are made of a half-mil to three-mil-thick titanium foil or aluminized polyimide (e.g., KAPTON®) film. The support structure is placed between the foils like a Hibachi assembly

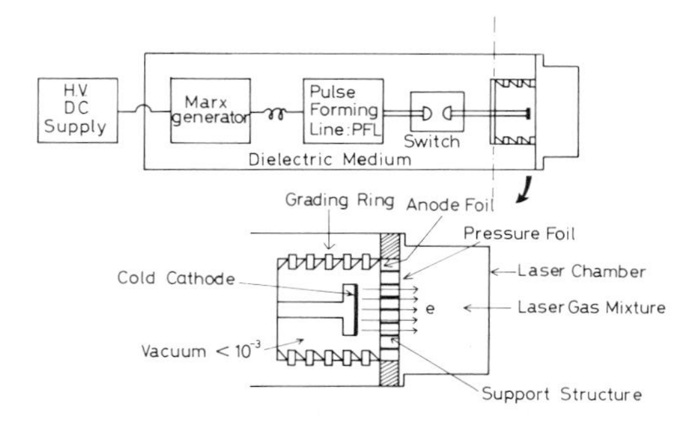

FIG. 13. Schematic diagram of a typical e-beam pumping system for excimer lasers.

to withstand the pressure of the laser gas mixture. The diode chamber is typically evacuated to low pressures less than 10^{-3} torr. An intense relativistic e-beam with a current density of 10 to 1000 A/cm² and an electron energy of 0.3 to 2 MeV has been used to pump RGH lasers. This high current-density e-beam can be generated only from a cold cathode-type diode made of carbon felt, or multi-blades of tantalum or titanium foils, which can enhance the local electric field strength up to typically 1 MV/cm. The operational characteristics of the cold cathode-type diode approximately obey the Child–Langmuir law (space-charge-limited electron flow). Therefore, the e-beam current density can be approximately written as

$$J_{eb} = \frac{\sqrt{2}}{9\pi}\left(\frac{e}{m_0}\right)^{1/2}\frac{V^{3/2}}{d^2} = 2.3 \times 10^3 \frac{V^{3/2}}{d^2} \text{ A/cm}^2$$

where J_{eb} is the current density in A/cm², V the applied voltage in MV, and d is the anode–cathode spacing in cm. Here e and m_0 are the electron charge and mass, respectively. This expression is accurately the nonrelativistic expression. In the region of $V > \sim 0.5$ MV some relativistic correction is required. Thermionic electron emitters (hot cathode) are not used for direct e-beam pumping of RGH lasers (except for indirect e-beam pumping, such as an e-beam-sustained discharge pumping), due to their inherent characteristic of low current-density e-beam generation.

The axial magnetic field is in some cases applied to ensure uniform e-beam generation if total e-beam current is larger than the critical current for self pinching, and also to improve the e-beam utilization.

A variety of e-beam-pumped RGH laser layouts have been successfully tried. Four major geometries are shown schematically in Fig. 14. The merits and demerits of these layouts should be discussed in terms of the energy scaling, the aspect ratio of the pumped region, uniform pumping, e-beam utilization, and output performance of an available pulsed high-voltage generator. To improve the e-beam energy deposition into the laser mixture, the external magnetic field is applied in some cases as shown in Fig. 14. The depth of the pumped region is determined by the electron penetration depth at a given accelerating voltage of electrons. The variation of the electron range in one atmospheric

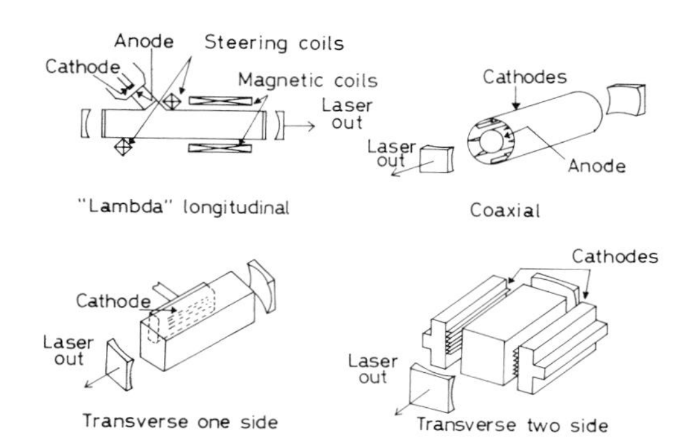

FIG. 14. Schematic of four major layouts for e-beam pumped RGH lasers.

pressure of various rare gases commonly used for RGH lasers is shown in Fig. 15. As can be seen in Fig. 15, a higher-Z rare gas can deposit more e-beam energy because of its shorter propagation distance. Actually, a typical KrF mixture is a 2–3 atm mixture of $Kr/F_2/Ar$ = 10/1/89(%). In 1983, low pressurization of the KrF laser mixture was proposed and successfully tried by increasing a higher Z krypton content than argon or by using an argonless mixture of Kr/F_2 to keep the e-beam deposition constant.

Under the high-excitation-rate pumping, the precursor for a RGH excimer is a rare gas ion, as mentioned previously. Some typical values of the average energy (eW_i) to produce an ion pair of rare gas by an e-beam are shown in Table V along with their ionization potentials. If the energy deposited into the rare gas mixture is calculated, or measured, the net rate of ion-pair pro-

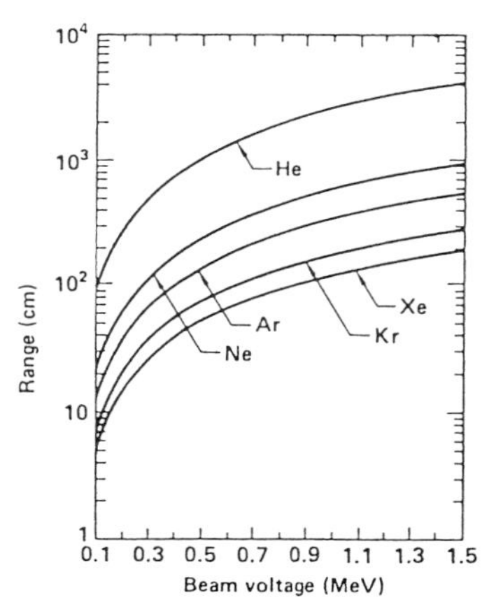

FIG. 15. Berger and Seltzer range for commonly used rare gases at 1 atm. [Reproduced with permission from J. J. Ewing (1979). Excimer lasers, *in* "Laser Handbook" Vol. 3 (M. L. Stitch, ed.), North-Holland Publ., Amsterdam.]

TABLE V. Comparison of Average Energy to Ionize Rare Gases and Their Ionization Potentials

Rare gas	eW_i (eV)	Ionization potential (eV)
He	42	24.6
Ne	36	21.6
Ar	26	15.8
Kr	24	14.0
Xe	22	12.1

duction can be calculated. The secondary electrons can rapidly cool down to around 1 eV in pure rare gases such as He, Ar, Kr, etc. In Ne gas the electrons cannot be thermalized rapidly. The e-beam energy deposited into the rare gases can be measured by a calorimetric and pressure-jump method. This energy measurement is required to estimate the intrinsic efficiency for e-beam-pumped lasers, defined as a ratio of the laser output energy to the deposition energy. Experimentally, intrinsic efficiencies for KrF and XeCl lasers are typically 6–12% and around 5%, respectively. As a single shot, or low-repetition-rate device several companies can supply complete e-beam accelerators for RGH laser pumping.

The world-wide KrF lasers pumped by e-beams are shown in Fig. 16. It is understood

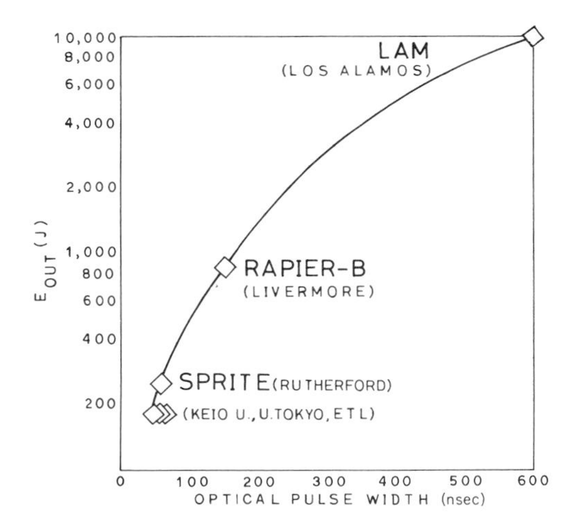

FIG. 16. World-wide KrF lasers pumped by e-beams. [Adapted with permission from L. A. Rosocha (1985). A short-pulse multikilijoule KrF inertial fusion laser system, presented at CLEO'85 (Conference on Lasers and Electro-Optics), May, LA-UR-85-1506.]

that the scaling-up of the laser energy is done experimentally by increasing laser pulse width (in other words, pumping pulse width). These large-scale KrF laser systems have all been constructed to be used as a power amplifier in a fusion laser system. The SPRITE laser employed a four-sided transverse excitation scheme, while RAPIER and LAM employed two-sided transverse excitation schemes as shown in Fig. 14. The size of LAM in Los Alamos National Laboratory is as large as $2 \times 1 \times 1$ m^3. The repetitive operation of the e-beam-pumped RGH laser is mainly limited by the heating of the anode foil. By specially cooling the anode foil, the repetitive operation is realizable at thermal loadings of up to several hundreds of watts per square centimeter onto the anode foil.

B. Discharge Pumping

The discharge pumping technology is well suited to excite high-repetition-rated RGH lasers, which can operate so far with laser output energies of several milijoules per pulse to a joule level at a repetition rate up to several kHz.

Applying an X-ray preionization to a high-pressure mixture of Xe/HCl/Ne, an energy scaling study of discharge-pumped XeCl lasers has also been done since 1979 and in 1985 a high-energy laser of 60-J level output has been successfully demonstrated as a single pulse device at the U.S. Naval Research Laboratory.

A pumping rate on the order of 1 GW per liter of discharge volume is necessary to efficiently produce RGH laser gains. The discharge resistance of the RGH discharge load is typically around 0.2 ohms. Therefore, the typical voltage of 20 kV gives discharge current as high as 100 kA. This high current cannot be switched by thyratron switches due to the heavy loadings of the switch. Hence, the primary low power and long pulse is produced in a primary circuit and then this pulse is compressed in the secondary circuit into the secondary high power and short pulse, which can efficiently pump the RGH lasers. Discharge pumping circuits developed so far are mainly classified into capacitor transfer circuit, pulse forming line (PFL) circuit, magnetic pulse compressor circuit, and spiker sustainer circuit, which are shown schematically in Fig. 17.

1. Pumping Circuits

a. Charge Transfer Circuit. The charge transfer circuit is a circuit that is widely used in rela-

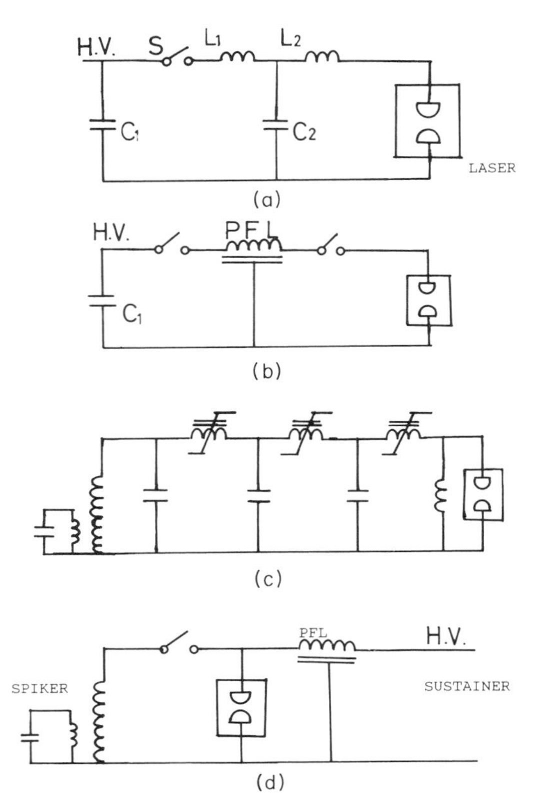

FIG. 17. Schematic of widely used RGH laser excitation circuits. (a) Charge transfer circuit, (b) pulse forming line (PFL) circuit, (c) magnetic pulse compressor circuit, and (d) spiker sustainer circuit.

tively small-scale repetition-rated RGH lasers and commercially available RGH lasers. Among this type of excitation circuit, the resonant charge transfer mode ($C_1 = C_2$ in Fig. 17) is mostly employed, because the charge transfer efficiency from C_1 to C_2 is maximized. Figure 18 shows a set of typical values for this type of excitation circuit used to pump RGH lasers. When operating the circuit shown in Fig. 18, the operating characteristics of the XeCl laser can be numerically analyzed by computer simulation, the results of which are shown in Fig. 19. A 4-atm mixture of Xe/HCl/Ne = 1.3/0.1/98.6(%) is assumed as an optimum gas mixture. Peak value of the primary current I_1 is less than 5 kA, which is within the current ratings of thyratrons widely used for RGH lasers. While peak value of the secondary current I_2 is increased up to about 18 kA. This increase is attributed to the fact that L_2 is much less than L_1 as shown in Fig. 18. The corresponding excitation rate is ~1.6 MW/cm^3,

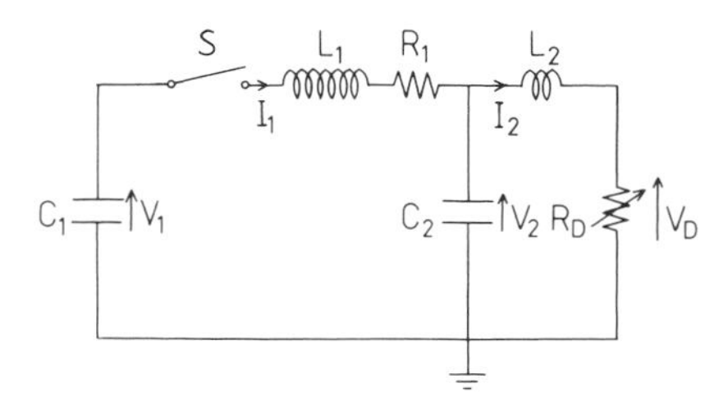

FIG. 18. Equivalent circuit of the charge transfer circuit. $C_1 = 60$ nF, $L_1 = 200$ nH, $R_1 = 0.2\ \Omega$, $C_2 = 60$ nF, $L_2 = 5$ nH, and V_D = laser discharge voltage. Discharge volume is 2.0(H) × 1.0(W) × 6.0(L) = 120 cm^3.

which gives a specific laser energy of 3 J/liter. The excessive excitation rate will result in the low laser efficiency mainly due to the excimer deactivation by discharge electrons.

Using this type of excitation circuit, a maximum laser efficiency of nearly 3% for both XeCl and KrF lasers was obtained with output energies of around 300 mJ, while only a 1.3% laser efficiency was obtained for the ArF laser. It is noted that laser efficiency and output energy are experimentally traded off. This is attributed to the fact that the breakdown voltage is determined not by the charging voltage V_1 but by the laser gas mixture. This laser is mostly equipped

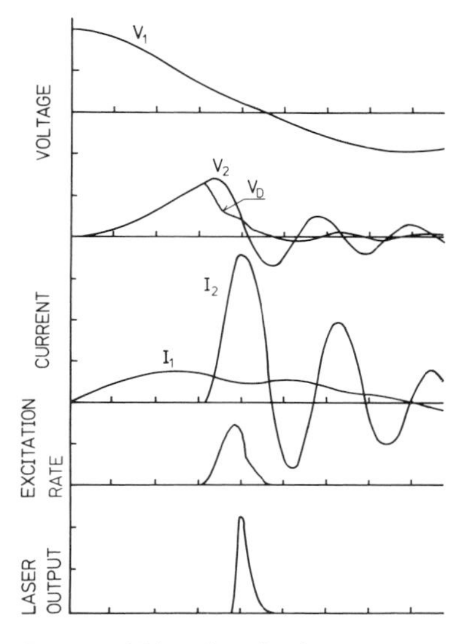

FIG. 19. Temporal histories of voltage, current, excitation rate, and XeCl laser output at a charging voltage of 20 kV. The inductance L_2 of the secondary loop is 5 nH. Each ordinate is as follows: voltage; 10 kV/div.; current, 10 kA/div.; excitation rate, 1 MW/cm^3/div.; and laser output; 5 MW/div.

with the UV spark array preionization, which is directly and simultaneously driven by the main discharge circuit because of its simplicity.

b. Pulse Forming Line (PFL) Circuit. The PFL circuit uses a low-impedance (typically less than 1 ohm) PFL consisting of solid or liquid dielectric materials in place of capacitor C_2 in the capacitor transfer circuit in Fig. 17. Solid dielectric materials include Mylar® (polyamide) and epoxy sheets, while liquid materials are deionized water or deionized ethylene glycol. Coaxial-type, parallel-plate-type, and Blumlein-type PFLs were used for the RGH laser excitation.

The advantage of this circuit is that it makes it possible to inject a quasi-rectangular waveform pulse into a discharge load. The pulse duration and output impedance are simply selected by changing the length and geometry of the PFL, respectively. This pumping system is well suited for high-energy XeCl lasers of in excess of several joules per pulse.

Figure 20 shows a schematic of a PFL discharge-pumped XeCl laser together with measured and theoretical time histories of discharge voltage and current. The scaling-up of the PFL circuit is readily realizable. However, establishment of large-volume uniform self-sustained discharge in a RGH laser mixture is a state of the art technology. As an output switch in this scheme, rail gaps, UV-laser-triggered rail gaps, or magnetic switches, which will be described below, are used instead of thyratrons.

In 1985 a maximum output of 60 J in an XeCl laser was obtained with X-ray preionization at the U.S. Naval Research Laboratory. An X-ray preionized XeCl laser of 100 W at 100 Hz is commercially available that is pumped by a PFL discharge.

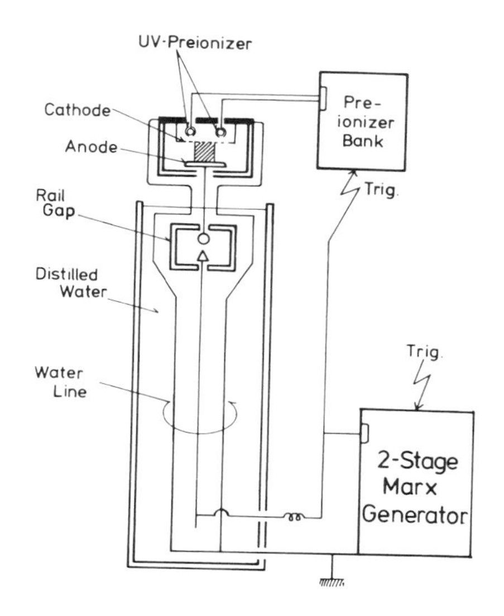

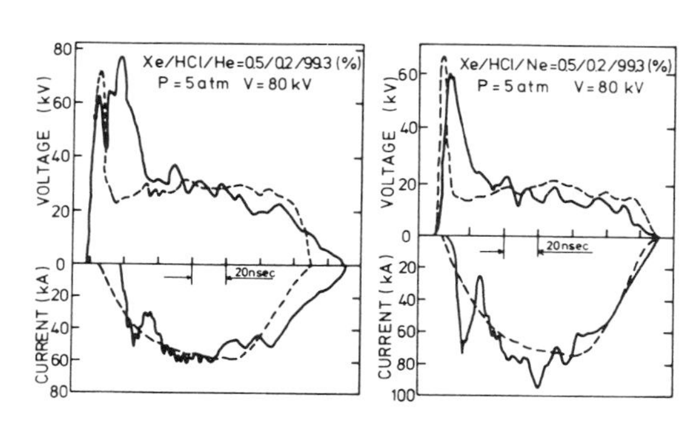

FIG. 20. Schematic of a PFL discharge pumped XeCl laser (top) together with measured (solid line) and theoretical (dashed line) time histories of the discharge current and voltage (bottom). [Reproduced with permission from H. Hokazono *et al.* (1984). *J. Appl. Phys.* **56**(3), 680.]

c. Magnetic Pulse Compressor (MPC) Circuit. A magnetic switch (MS) is a new switch consisting of a magnetic core made of ferromagnetic materials, being completely different from a gas-discharge switch such as thyratron. A magnetic core is completely saturated in an ON phase. If these MSs are connected in series, the primary pulse is successively compressed only by decreasing the saturated inductance of the magnetic coils. A three-stage MPC is schematically shown in Fig. 17. A typical MPC circuit is shown in Fig. 21, together with each voltage and current. An 8-μsec pulse is compressed to a 100 nsec pulse. The MS is a long-life or endless-solid-state switch because it experiences no erosion, and it can act reliably at high-repetition frequency. An XeCl laser operating at 150 W in average laser output at 500 Hz is commercially available.

d. Spiker Sustainer Circuit. The spiker sustainer circuit is an advanced excitation circuit for RGH lasers. The low-impedance PFL sees initially the discharge load being an open load so that the reflection of the voltage pulse may occur due to the impedance mismatch. To eliminate this unfavorable voltage reflection, the high-voltage high-impedance pulser initially breaks down the laser gas mixture and then the other

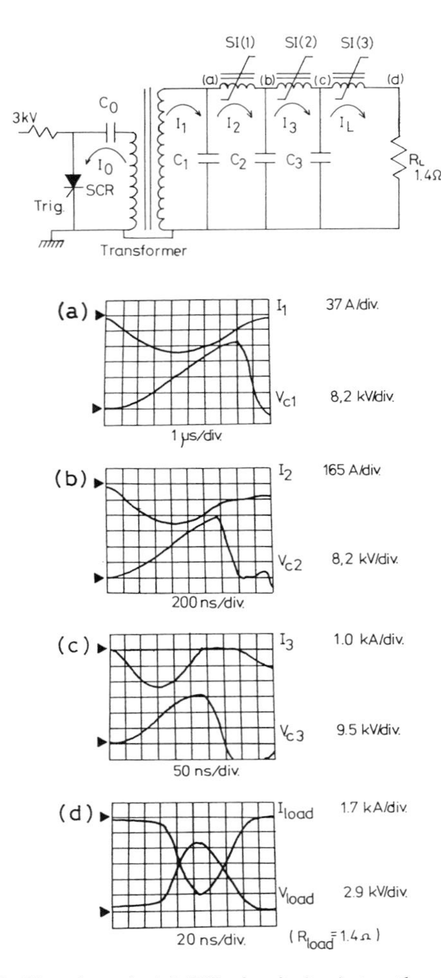

FIG. 21. A typical MPC circuit (top) together with each voltage and current (bottom). $C_0 = 2.5\ \mu F$ and $C_1 = C_2 = C_3 = 13.6$ nF. (a) First stage, (b) second stage, (c) third stage, and (d) load. [Courtesy of Tsutomu Shimada.]

power supply maintains the discharge plasma to condition the electron energy distribution for efficient pumping of the RGH laser. In this scheme, the former is called a spiker, the latter being a sustainer. This scheme is more complicated than the PFL circuit alone, but higher overall electrical efficiency is expected, because of sufficient impedance matching. Using this scheme, an XeCl laser of 4.2 J has been realized with a high efficiency of 4.2%.

2. Preionization Technology

To initiate a volumetrically uniform avalanche discharge in the 2–4 atm rare gas–halogen mixture, preionization of the high-pressure mixture prior to the initiation of the main discharge is indispensable. Spatial uniformity of the preionization in the rare gas-halogen mixture is the most important issue. Especially, its uniformity perpendicular to the discharge electric field is of importance. The preionization electron number density on the order 10^6 to 10^{12} #/cm^3 is experimentally utilized, and is dependent on the preionization strength employed and gas mixtures.

A variety of preionization technologies developed so far are shown in Fig. 22. The simplest and most convenient preionization technology is a UV photo-preionization using a photo-electron emission process. The UV photons are generated by the use of a pin-arc discharge and a dielectric surface discharge, both of which are induced in the laser gas mixture, and UV RGH laser beams. UV preionization via pin-spark discharge is widely used in commercially available RGH lasers.

X-ray preionization technology was successfully applied in 1978 to high-pressure RGH lasers, and has been used preferably in large-scale RGH laser devices. A high-energy XeCl RGH laser of 60-J level output employed the X-ray preionization. By masking the X-ray beams by Pb plates as shown in Fig. 23 to determine the preionized volume, only the preionized (X-ray irradiated) volume can be discharge pumped with less effect of the nonuniformity of the electric field induced by main electrode geometries than that with noncollimated UV preionizers.

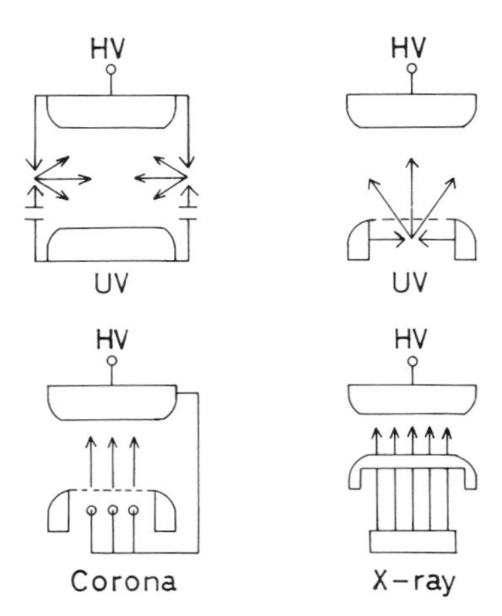

FIG. 22. A variety of preionization technologies developed to date. UV and corona preionization are suitable for high repetition rate and relatively small lasers. The X-ray preionization is suitable for large-scale discharge lasers.

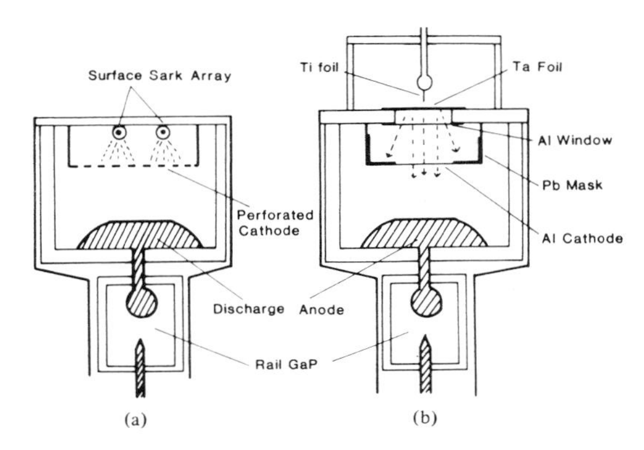

FIG. 23. Comparison of (a) noncollimated UV and (b) collimated X-ray preionizers. [Reproduced with permission from K. Midorikawa *et al.* (1984). *IEEE J. Quantum Electron.* **QE-20**(3), 198. Copyright © 1984 IEEE.]

The effect of the preionization electron number density is large and it is experimentally revealed that the RGH laser energy and laser pulse width are increased in logarithmic form with increasing initial electron number density. An example for the KrF laser is shown in Fig. 24.

3. Long Pulse Operation

Long pulse operation of the RGH lasers pumped by a self-sustained discharge is interesting in view of gaseous electronics and ultrashort pulse generation via mode-locking technology. Many RGH lasers have so far been operated in a short pulse (10–50 nsec) regime. In 1985, long pulse oscillations of 1.5-μsec XeCl lasers were demonstrated by conditioning the discharge in a sophisticated manner. Long pulse

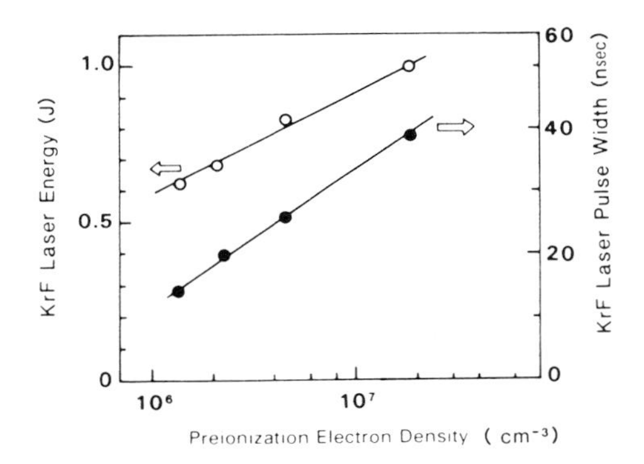

FIG. 24. Effect of preionization electron number density on KrF laser energy and pulse width. [Reproduced with permission from K. Midorikawa *et al.* (1984). *IEEE J. Quantum Electron.* **QE-20**(3), 198. Copyright © 1984 IEEE.]

RGH lasers with reduced intensity can efficiently deliver their energies through quartz fibers so that they can be used in photomedicine such as laser-angioplasty and laser surgery.

4. High-Repetition-Rate Operation

High-repetition-rate operation of RGH lasers is desirable for high average-power generation. For this purpose thyratrons have been commonly used as a switching element. The operational performance of the used switch is one of the repetition-rate limiting issues.

The relationship of high-repetition-rate RGH lasers to the switching elements used is shown in Table VI. Because of rapid progress in both the pulsed power technology involved in the modulator and laser gas purification, operations at repetition rates of up to 2 kHz and an average laser power of up to 300 W have been demonstrated separately. Using a single thyratron an average laser power of up to 55 W is obtained and an average laser power of in excess of 120 W is obtained using multi-thyratrons in parallel, a thyratron with a magnetic assist, or spark gaps. The magnetic assist means that using a saturable inductor in series with a thyratron switch the current-rise rate through the thyratron is reduced so as to decrease the energy dissipated in

TABLE VI. Relationship of High Repetition Rate RGH Lasers and the Switching Elements Used

Year reported	RGH	Repetition rate (Hz)	Average power (W)	Switch[a]
1976	KrF	20	16×10^{-3}	G
1977	XeF	200	50×10^{-3}	T
	KrF	1000	40	T
1978	KrF	1000	10	T
	XeF	500	1.5	T
1979	KrF	1000	55	T
	XeF	2000	24	$4 \times$ T
1980	KrF	1000	200	$2 \times$ G
	ArF	250	6	$2 \times$ T
	KrF	400	28	$2 \times$ T
	XeCl	600	45	$2 \times$ T
	XeF	250	5	$2 \times$ T
1982	XeCl	750	3.5	T + M
1983	XeCl	1500	130	T + M
	XeCl	300	120	$2 \times$ T
	XeCl	400	180	G
1984	XeCl	500	150	T + M
	XeCl	250	40	S + M
1986	XeCl	500	300	T + M

[a] G: Spark Gap, T: Thyratron, M: Magnetic Switch, S: Semiconductor Switch.

the thyratron. The allowable rise rate of the current for thyratron is less than 10^{11} A/sec, while that of a gas-insulated spark gap reaches 10^{13} A/sec. However, their operational lifetime becomes extremely short under the heavily loaded conditions required for efficient RGH laser excitation.

For attaining the nearly endless lifetime of a RGH laser exciter, an all-solid-state circuit is the state of the art and appears to be promising. At present, commercially available high-power semiconductor switches have been well developed, but maximum ratings such as hold-off voltage, peak current, and current-rise rate cannot fulfill the switching requirements necessary for efficient RGH laser excitation. Therefore, additional voltage transformer circuits and a magnetic pulse compression circuit are needed, as shown previously in Fig. 21.

In addition to the high-repetition-rate exciter, gas purification and aerodynamic technologies such as a fast gas circulation system and an acoustic damper at repetition rates exceeding the multi-kilohertz range are required to realize long-life high-repetition-rated operation of the RGH lasers. A long operational shot life of up to 10^8 shots has been demonstrated to date.

C. Control of Laser Properties

1. Spectral Properties

The fluorescence from the rare gas-halides with bound lower states (e.g., XeF and XeCl) typically contains line structure originating from the transitions from the lowest vibrational level of the upper laser state to various vibrational levels of the ground electronic state. When these lasers are operated without wavelength tuning elements in the resonator, the laser emission consists of radiation on the two or three strongest vibrational transitions. For the rare gas-halides with unbound lower electronic states (e.g., KrF and ArF), the fluorescence shows a weakly structured band. The untuned laser emission typically consists of a band approximately 100 cm^{-1} wide near the peak of the fluorescence.

The linewidth of the RGH lasers can be reduced and the tunability of the RGH lasers can be realized using gratings, prisms, etalons, or some combination of these elements in the laser resonator. The lowest level of frequency selection is usually obtained with a grating used in a Littrow geometry at one end of the resonator or with one or more prisms for the tuning element. This technique provides tunability over the emission band of the RGH lasers and realizes selection of single vibrational transition in lasers such as XeF or XeCl. Narrow linewidths of 80 cm^{-1} for KrF and 5 cm^{-1} for ArF have been demonstrated with two prisms in the resonator.

Narrower linewidths of RGH lasers are needed especially for the photolithography of the ultra-large-scale integration (ULSI) devices. With intracavity etalon, either alone or with a grating, narrower linewidths are achievable. A combination of three etalons has been used in a XeCl laser to produce a linewidth of 220 MHz and single-mode operation with a linewidth of 30 MHz. Narrower linewidths can also be obtained by expanding the intracavity laser beam cross section to cover the large number of grooves on the grating, which is used especially in dye lasers. This configuration includes grazing-incidence gratings and Littrow gratings with a prism beam expander. Linewidths of 0.3 cm^{-1} for KrF and 0.5 cm^{-1} for ArF and XeCl have been achieved by the use of a grazing-incidence grating.

2. Spatial Properties

Because of the high gain and wide gain bandwidth of RGH lasers, high-Fresnel-number stable resonators, which are commonly used to efficiently extract the available laser energy, produce a laser output with high spatial divergence. A low divergence laser beam from RGH lasers can be obtained by using a confocal unstable resonator with a large magnification or by simply using a low-Fresnel-number stable resonator. The use of an unstable resonator can efficiently extract laser output with high beam quality, while the use of low-Fresnel-number stable resonators results in a substantial decrease in

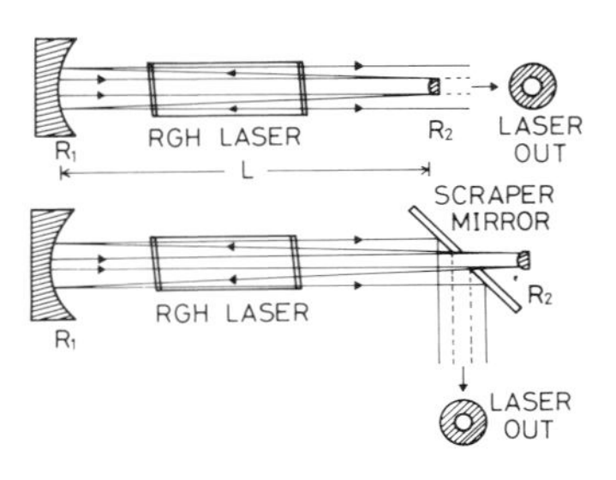

FIG. 25. Typical output coupling methods for confocal unstable resonators.

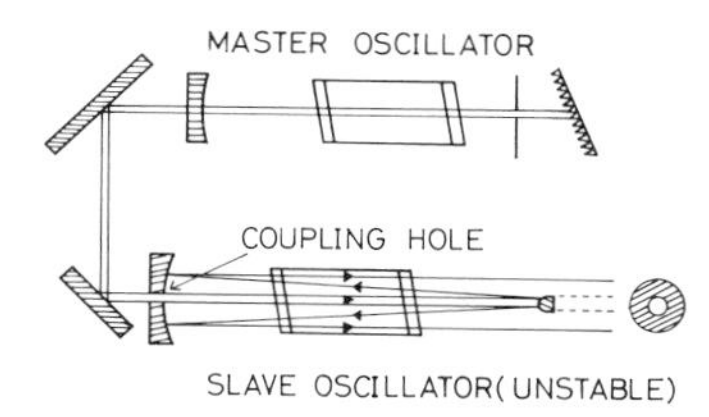

FIG. 26. Schematic of an injection-locked resonator (unstable).

the available laser energy. Rare gas-halide laser beams with divergencies within a factor of one or two of their diffraction limits are obtainable without any difficulty by the use of high-magnification confocal unstable resonators. Typical output coupling methods of confocal unstable resonators are schematically shown in Fig. 25.

3. Injection Locking

For some applications, it is necessary to produce RGH lasers at a reasonable output power level with low spatial divergence and narrow bandwidth. To fulfill these requirements, use of either a master oscillator/power amplifier system or an injection-locked resonator is the most convenient method. Figure 26 shows a schematic of an injection-locked resonator. In this scheme the slave oscillator is an unstable resonator that can control the output energy and spatial mode quality. Input pulse can be made by a master oscillator typically consisting of a cavity with a line-narrowing function. An input power level on the order of 10^{-3} or less of the output laser level is effective. When a phase-conjugate mirror employing an effective stimulated Brillouin scattering (SBS) material in the UV wavelength (such as C_6H_{14} or SF_6) is used as a back mirror of an amplifier, phase front distortion induced on the laser beam in the amplifier can be eliminated, and then the highly controlled properties of the input laser can be realized on the amplified beam. High-power laser extraction is more readily achievable as a by-product with injection locking than without injection because the injected seed pulse assists the build-up of the intracavity flux and reduces energy loss. This effect is more remarkable in shorter duration lasers.

In injection-locked RGH lasers, an input pulse can be made by an entirely different kind of laser. When radiation generated from a visible dye laser chain pumped by a flashlamp, an Ar ion laser, a frequency-doubled YAG laser, or an RGH laser is injected into the RGH amplifier after one or more stages of frequency conversion, a highly controlled laser pulse with desirable spectral, spatial, and temporal beam quality can be amplified to high energy level. The use of a cw single-mode dye laser provides a narrow band seed signal whose wavelength is tunable over the bandwidth of the RGH laser. A system of this type has also been adapted for an e-beam pumped XeF(C-A) laser, and the extremely wide bandwidth of 80 nm was continuously tuned with a linewidth of less than 0.001 nm and an intrinsic efficiency of 1.5%. Beam divergencies within a factor of three of the diffraction limit were achieved.

D. Ultrashort Pulse Amplification

The broad bandwidth afforded by RGH lasers makes these systems promising candidates for constructing ultrashort pulse laser systems. Although direct generation of ultrashort RGH laser pulses by using either active or passive mode locking is limited to its pulsewidth on the order of or slightly shorter than 1 nsec due to the short gain duration, successful ultrashort pulse generation using visible dye lasers and their wavelength conversion technologies have made it possible to generate high-power subpicosecond UV lasers using RGH gain media for an amplifier. The shortest pulses and highest powers achieved by various RGH lasers are listed in Table VII. The highest laser peak power of 4

TABLE VII. Ultrashort Pulse RGH Lasers

Excimer	Bandwidth (cm^{-1})	Bandwidth-limited pulsewidth (fsec)	Saturation energy (mJ)	Present status: Output energy	Present status: Pulsewidth
XeCl	120	88	1.35	300 mJ	310 fsec
			(<3 psec)	12 mJ	160 fsec
KrF	325	32	2.0	1.5 J	390 fsec
				4 mJ	45 fsec
ArF	400	26	3.0	30 mJ	10 psec

TW in a 390 fsec pulse has been obtained by amplifying a frequency tripled short pulse dye laser generated in a synchronously pumped dye laser stage by three discharge pumped KrF amplifiers and an e-beam pumped KrF amplifier. The shortest RGH laser pulse of 45 fsec was obtained by amplifying a frequency doubled short pulse dye laser which was generated in a distributed feedback dye laser cavity pumped by a cavity quenched dye laser pulse.

Since the storage times of RGH lasers are typically 1–2 nsec, saturation energy is typically 1–2 mJ/cm^2, which is much lower than the values of other high-power lasers. Therefore, an amplifier with large aperture is required to amplify a laser pulse to very high energy levels. In such a system, amplified spontaneous emission (ASE) generated from the final large aperture amplifier tends to reduce the aspect ratio of the output laser pulse. At 248 nm wavelength, acridine dye dissolved in ethanol acts as an effective saturable absorber.

When focusing a subpicosecond RGH laser pulse generated in these systems, intensity reaches well over 10^{17} W/cm^2. High-order harmonic generation of these short pulse RGH lasers can decrease their wavelength down to the XUV wavelength region. A 14.6 nm pulse was obtained by 17th-harmonic generation with a picosecond KrF laser. An ultrashort UV probe continua was also generated through self-phase modulation (SPM) induced in high peak power RGH lasers focused into high-pressure gases. A TW-level RGH laser will be able to generate continuum pulse in the XUV spectral range.

Bibliography

Bass, M. and Stitch, M. L. (1985). "Laser Handbook," Vol. 5, North-Holland, Amsterdam.

Kannari, F., Obara, M., and Fujioka, T. (1985). An advanced kinetic model of electron-beam-excited KrF lasers including the vibrational relaxation in KrF*(B) and collisional mixing of KrF*(B,C), *J. Appl. Phys.* **57**(9), 4309–4322.

Midorikawa, K., Obara, M., and Fujioka, T. (1984). X-ray preionization of rare-gas-halide lasers, *IEEE J. Quantum Electron.* **QE-20**(3), 198–205.

Ohwa, M., and Obara, M. (1986). Theoretical analysis of efficiency scaling laws for a self-sustained discharge pumped XeCl laser, *J. Appl. Phys.* **59**(1), 32–34.

Rhodes, C. K., Egger, H., and Pummer, H. (1983). "Excimer Lasers—1983" American Institute of Physics, Vol. 100, New York.

Rhodes, C. K. (1984). "Excimer Lasers," 2nd ed., Springer-Verlag, Berlin and New York.

Stitch, M. L. (1979). "Laser Handbook," Vol. 3, North-Holland, Amsterdam.

Suda, A., Obara, M., and Noguchi, A. (1985). Properties of a KrF laser with atmospheric-pressure Kr-rich mixture pumped by an electron beam, *J. Appl. Phys.* **58**(3), 1129–1134.

SEMICONDUCTOR INJECTION LASERS

Peter J. Delfyett and Chang-Hee Lee *Bell Communications Research*

GLOSSARY

Auger recombination: Three-body collision involving two electrons and a hole (or two holes and an electron). The energy released by the recombination of an electron and a hole is immediately absorbed by another electron (or hole), which then dissipates the energy by emitting phonons.

Band gap energy: Energy difference between the conduction and valence bands in a semiconductor material, which also corresponds to the energy required to promote an electron from the valence band to the conduction band.

Conduction band: Band of energy levels above the valence band which can support the conduction of electrons.

Degenerate doping: Amount of doping required to bring the Fermi energy level to a level comparable to the conduction band energy for electrons and the valence band energy for holes.

Direct band gap semiconductor: Semiconductor in which the valence band maximum and the conduction band minimum occur at the same position in momentum space.

Dopant: Material which is incorporated into a semiconductor material which adds excess electrons (*n*-type dopant) or excess holes (*p*-type dopant).

Electron-hole recombination: Radiative (or nonradiative) process in which an electron in the conduction band recombines with a hole in the valence band. In the radiative recombination process, photons are generated, while in the nonradiative recombination process, phonons are generated.

Fermi level: Energy level which represents a 50% chance of finding a state occupied by an electron.

Indirect band gap semiconductor: Semiconductor in which the conduction band minimum and the valence band maximum do not occur at the same place in momentum space. In these semiconductors, electrons in the conduction band minimum require a change in momentum in order to recombine with a hole at the valence band maximum. The recombination process in this case usually requires a phonon.

Nonlinear gain: Nonlinear part of the gain in semiconductor lasers, which manifests itself in the gain versus photon density curve. In this case, the gain decreases with an increasing photon density, due to finite intraband scattering and dynamic carrier heating.

***p–n* junction:** Region which joins two materials of opposite doping. This occurs when *n*-type and *p*-type materials are joined to form a continuous crystal.

Population inversion: Necessary condition which ust exist in order for the stimulated emission process to occur. This condition exists when the population of the upper lasing level exceeds the population of the lower lasing level.

Q-switching: Method for producing a high power pulse from a laser system. This is usually accomplished by inserting an optical shutter in the cavity to control resonant Q or quality factor.

Quantum well: Material structure which can spatially confine electronic charges to spatial dimensions on the order of tens of angstroms (10^{-10} m). These dimensions are comparable to the de Broglie wavelength of the electron/hole.

Schawlow–Townes equation: Equation which theoretically predicts the linewidth of the lasing transition of a single mode laser. The linewidth predicted from this equation is inversely proportional to the output power and the cavity quality factor (Q-factor).

Valence band: Energy band corresponding to the valence electrons of a semiconductor crystal. This energy band is normally filled and as a result does not allow the conduction of electrical current.

Window structure: Modification to the standard laser facets of a semiconductor injection laser. The modification involves processing the laser facets so that the energy gap of the laser facet material is larger than the emitted photon energy. This reduces optical absorption at the laser facets and prevents irreversible facet damage.

Semiconductor injection lasers are highly efficient laser light emitting devices which are extremely small, with typical linear dimensions being on the order of a few hundred micrometers ($1\ \mu m = 10^{-6}$ m). These lasers belong to a specific class of solid state lasers which are constructed from semiconductor materials, as opposed to conventional solid state lasers which are made from insulating crystals doped with active ions. The semiconductor injection laser derives its input power from an electrical current which is directly passed or injected through the device. The typical threshold currents for initiating lasing in these devices is on the order of a few milliamperes, with conversion efficiencies between the injected electrons and the generated photons exceeding 90%. Other methods of excitation are possible, such as optical pumping and electron beam pumping. However, these methods of pumping are considered to be less attractive because they do not take advantage of the compactness of the diode laser, due to the relatively large size of the pumping sources, and the pumping scheme is normally less efficient. Semiconductor injection lasers are normally constructed from several semiconductor material systems, most notably the gallium arsenide/aluminum gallium arsenide system and the indium phosphide/indium gallium arsenide phosphide system. The material system used is dependent on the desired emisson wavelength of the laser. The field of semiconductor injection lasers began in 1962, just three years after the first laser was invented. These early devices were very crude; however, during the past 29 years, tremendous advances have been made in the development of these devices. Now, many scientific, industrial, and commercial applications rely on the existence of these devices. The present article reviews the basic principles of light generation in semiconductor injection lasers, surveys several important semiconductor laser structures and their applications in the scientific and commercial communities, and contemplates the future directions and trends in this rapidly growing and exciting field of research.

I. Brief Historical Overview

The first semiconductor laser devices were made from chips of gallium arsenide (see Fig. 1). The gallium arsenide was grown such that a *p–n* junction, or diode, was formed inside the crystal. The chip had a metallic base with a wire contact attached to the top to allow the injection of the electrical current. Smooth end faces were formed on the diode and acted as mirrors to provide the optical feedback necessary to attain laser oscillation, while the side walls of the laser chip were roughened to prevent laser oscillation in the direction perpendicular to the desired direction of lasing. These devices had very high threshold currents and could only be operated at very low temperatures. The light output characteristics from these laser chips were far from what their solid state and gas laser counterparts could produce. The laser emission from the chips suffered from a lack of coherence due to the wide spectral bandwidth inherent in all semiconductor light emitting devices, and the output beam emission was contained in a very broad far-field pattern. Despite these initial drawbacks, it was apparent that semiconductor lasers would have a very promising future. Twenty-eight years later, semiconductor lasers have been developed to the point where they have easily overcome the drawbacks which plagued them in the early days. [*See* SOLID-STATE LASERS.]

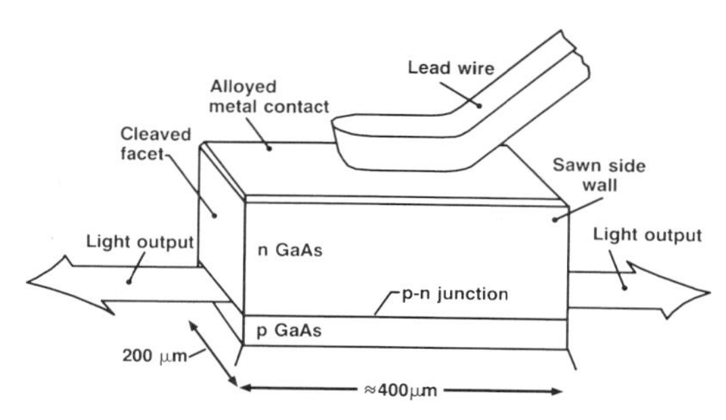

FIG. 1. Broad contact semiconductor injection laser mounted on a heat sink, with a wire contact. [From Thompson, G. H. B. (1980). "Physics of Semiconductor Laser Devices." Wiley, New York.]

The techniques used today for developing semiconductor lasers are identical to the technology which is used for manufacturing electronic devices. This fact means that semiconductor lasers can be mass-produced with a comparable reliability as standard electronic components. As a result of this link in processing technology, electronic devices can be integrated with semiconductor lasers on the same wafer. This feature has had a tremendous impact on the fields of integrated optoelectronics, optical communications, and optical data storage, and it is what makes semiconductor lasers very attractive for technological and commercial applications.

II. Basic Principles

The basic principles involved in the light generation process of semiconductor injection lasers can most easily be described by examining the energy level diagram of a p–n junction made from a direct band gap semiconductor. Indirect band gap semiconductors are not used for semiconductor lasers due to the nonradiative processes which are the dominant mechanism for electron-hole recombination. These processes generate lattice vibrations, or phonons, which are mechanisms for generating sound and heat and are not useful for light generation. In Fig. 2, an energy level diagram of a p–n junction is shown. Two semiconductor layers with opposite doping are grown in contact with each other. The n-doped side has an excess number of electrons; the p-doped side has an excess number of holes. The doping levels are typically 10^{18} cm^{-3} and assure that the Fermi level, shown as the dotted line in Fig. 2a, lies within the upper level or conduction band on the n-doped side and the lower level or valence band lies on the p-doped side. When this occurs, the doping is said to be degenerate.

In thermal equilibrium, electrons and holes can not recombine with each other due to the potential barriers which exists between the p–n junction. As a forward bias voltage is applied across the junction, the potential barrier is lowered. If the forward bias voltage is increased to a level which is nearly equal to the band gap energy, as in Fig. 2b, that is,

$$V_{\text{fb}} \leq \frac{E_{\text{bg}}}{e}, \tag{1}$$

then both electrons and holes are injected into the active region of the p–n junction. In the above equation, V_{fb} is the forward bias voltage, E_{bg} the band gap energy of the host semiconductor in units of electron volts, and e the electron charge. This biasing condition produces a population inversion in the active region, which is a necessary condition for the lasing process. The electrons and holes are now allowed to recombine with each other directly, emitting photons

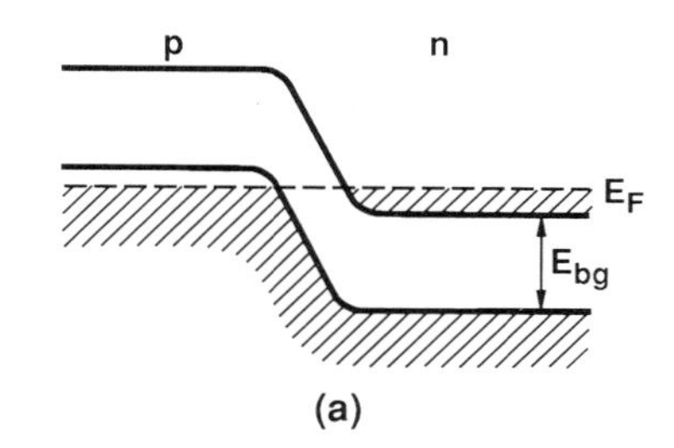

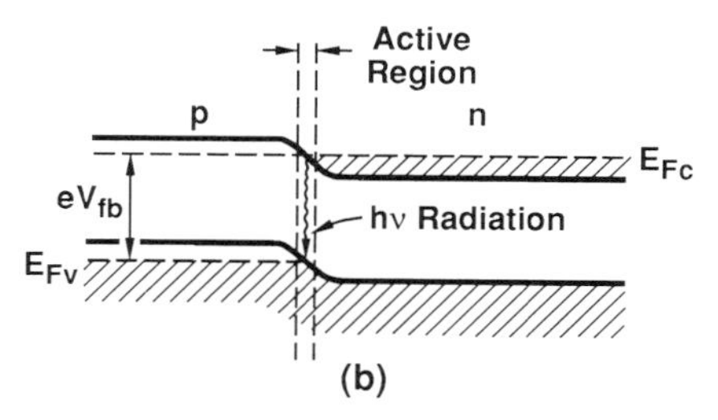

FIG. 2. (a) Degenerate p–n junction at zero bias. (b) Junction with a forward bias voltage comparable to the band gap energy. [From Yariv, A. (1975). "Quantum Electronics," 2nd ed. Wiley, New York.]

which experience a gain that satisfies the relation

$$E_{bg} < h\nu < E_{Fc} - E_{Fv} \tag{2}$$

In the above equation, E_{Fc} and E_{Fv} are the quasi-Fermi levels of the conduction and valence bands, respectively, h is Planck's constant, and ν is the optical frequency of the laser emission. Once the light has been generated and amplified, the optical feedback is provided by the cleaved end facets of the semiconductor. The cleaved end facets act as mirrors which reflect the light back and forth inside the semiconductor, initiating laser oscillation. The gain inside the semiconductor laser can approach 100× in a single pass. Due to this large gain factor, relatively low reflectives from the cleaved facets are sufficient to initiate lasing. The typical reflectivity of the cleaved facets is approximately 30% as compared to 60–99% for solid state and dye laser mirrors.

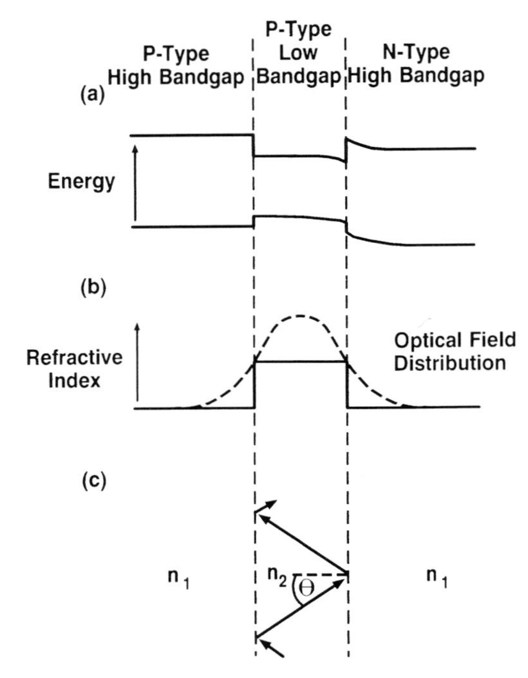

FIG. 3. Schematic of a double heterostructure laser. (a) Band diagram under forward bias. (b) Refractive index profile and optical field distribution. (c) Wave-guiding effect produced by total internal reflection and the index profile shown in (b). [From Kapon, E. (1989). *In* "Handbook of Solid State Lasers" (P. K. Cheo, ed.), Marcel Dekker, New York.]

A. Optical and Electronic Confinement: Transverse Confinement

The first semiconductor injection lasers developed were called homostructures. This is because the *p–n* junction was made from one type of semiconductor material, for example GaAs (gallium arsenide). In these lasers, there is no mechanism to confine the injected carriers or created photons. As a result, these lasers suffered from high threshold currents and a poor quality output beam. These problems, however, were overcome with the development of heterostructure lasers, that is, a semiconductor laser made from different semiconductor materials, such as gallium arsenide (GaAs) and aluminum gallium arsenide ($Al_xGa_{1-x}As$). In this notation, x is the fractional content of aluminum and is typically 30%.

The most common type of heterostructure laser is the double heterostructure laser, illustrated in Fig. 3. This laser is made from a combination of semiconductor materials which have different band gap energies for current confinement and different optical indices of refraction for optical confinement. The double heterostructure laser is made of three types of semiconductor materials; a layer of *p*-type low band gap material, such as GaAs, sandwiched between a *p*-type and an *n*-type material with higher band gap, such as $Al_xGa_{1-x}As$. The value of aluminum content is chosen such that the AlGaAs regions will have a higher band gap energy and a lower optical index of refraction than the GaAs active region.

With the design of the band gaps of the double heterostructure laser, as depicted in Fig. 3a, it is possible to understand the current confinement by considering the injected carriers being trapped inside a potential well. Electrons and holes are injected into the active region from the high band gap material, AlGaAs. The electrons and holes become trapped in the potential well created by the low band gap material, GaAs. This potential well acts to confine the carriers in the active region of the laser. As a result, the electrons and holes cannot diffuse out of the active layer and are forced to recombine with each other in the GaAs material, contributing to the light generation.

The optical beam confinement is provided by the waveguiding properties of the AlGaAs/GaAs/AlGaAs material structure depicted in Figs. 3b and 3c. The AlGaAs layers have a lower optical index of refraction than that of the GaAs

(Fig. 3b). This type of spatial index profile leads to the confinement of an optical beam inside the higher index material. This is because of the total internal reflection experienced by an optical beam inside the GaAs material (Fig. 3c). The minimum angle of incidence which an optical beam can have and still undergo total internal reflection is given by

$$\sin\theta = \frac{n_1}{n_2}, \tag{3}$$

where θ is the critical angle of incidence and n_1 and n_2 are the indices of refraction of the AlGaAs and GaAs materials, respectively.

III. Laser Structures

The optical and electrical confinement just discussed provides confinement of the injected carriers and the optical beam in a direction which is parallel to the growth direction of the layered structure. Optical and electrical confinement are also needed in the lateral direction, that is, in a direction parallel to the layers. This can be provided by two means: (1) gain guiding and (2) index guiding. The physics describing the operation of these two general classes of semiconductor injections lasers can best be understood by considering some specific laser structures as described in the next section.

A. Gain-Guided and Index-Guided Laser Structures: Lateral Confinement

Gain guiding is one method of providing electrical and optical confinement of the injected carriers and the generated photons in a direction which is perpendicular to the direction of growth of the layered structure. It is most easily accomplished by fabricating a narrow electrical opening in an otherwise normally insulating region, forming a stripe on the top of the laser structure. The carriers are injected and confined into the active layer directly under the contact stripe. This provides optical gain only under the stripe contact, thus giving the name gain-guided laser. Some specific laser structures which utilize the gain-guiding principle are the oxide stripe laser and the proton implanted laser. These lasers are depicted schematically in Fig. 4.

The oxide strip laser in Fig. 4a has a narrow opening on the order of several micrometers which is created in an electrically isolating oxide layer. A metallization layer is then deposited on

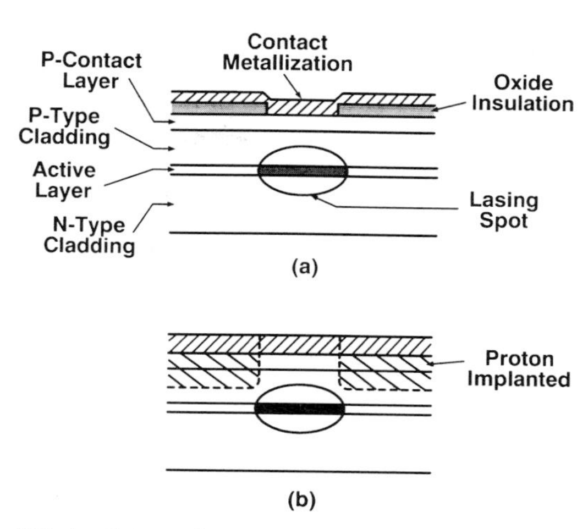

FIG. 4. Schematic cross sections of two types of gain-guided laser structures. (a) The oxide stripe laser. (b) The proton implanted laser. [From Kapon, E. (1989). *In* "Handbook of Solid State Lasers" (P. K. Cheo, ed.), Marcel Dekker, New York.]

top of the oxide layer to create the stripe contact. The layers located above the active region are made sufficiently thin so that the injected carriers are confined in a narrow stripe in the active region.

The proton implanted laser in Fig. 4b has a contact stripe which is created by implanting protons into the metallization layer, leaving only a small stripe which has not been implanted. The implanted regions become highly resistive to current injection, while the unimplanted region has a low resistivity which serves as the electrical stripe contact. Using this method, it is possible to control the amount of current confinement by varying the depth of proton implantation.

In gain-guided structures, the unpumped regions on both sides of the stripe are lossy at the lasing wavelength. This leads to a quasi-gaussian lateral gain profile. The peak of the gain profile is located directly under the stripe, while the wings of the gain profile extend into the lossy regions. This leads to guidance of the optical field in a direction along the laser stripe. The features of the gain-guided optical field are not only determined by the gain distribution. Additional guiding is naturally provided by the index change due to local lateral temperature gradients and carrier density gradients in the active layer.

Alternative methods of providing lateral electrical and optical beam confinement can be seen by examining index-guided structures. Index-guided structures provide optical lateral confinement by fabricating a refractive index distribution which is parallel to the laser's active layer. This, in combination with the double heterostructure configuration, gives a two-dimensional confinement of the optical field. The electrical confinement can be provided by a small opening in an electrically isolating layer, as in the gain-guided structures, or by employing reversed biased p–n junctions which sandwich the lasing region in the lateral direction. Some types of index-guided structures are the ridge wave guide laser and the buried heterostructure (BH) laser. These laser structures are depicted schematically in Fig. 5.

In the ridge waveguide structure (Fig. 5a), the top of the laser device is etched down to be very close to the active layer. Only a small part of the diode is not etched. This produces a type of plateau above the active layer, which ultimately becomes the lasing region. The evanescent field of the propagation wave extends into the plateau region and is efficiently guided in this process. In this structure, the electrical confinement is provided by the opening in the insulating oxide layer.

The buried heterostructure laser in Fig. 5b is designed so that the active region is completely surrounded by material which has a lower optical index and also a higher energy band gap. As a result of this two-dimensional index profile, the optical beam is tightly confined in both transverse and lateral directions to the direction of

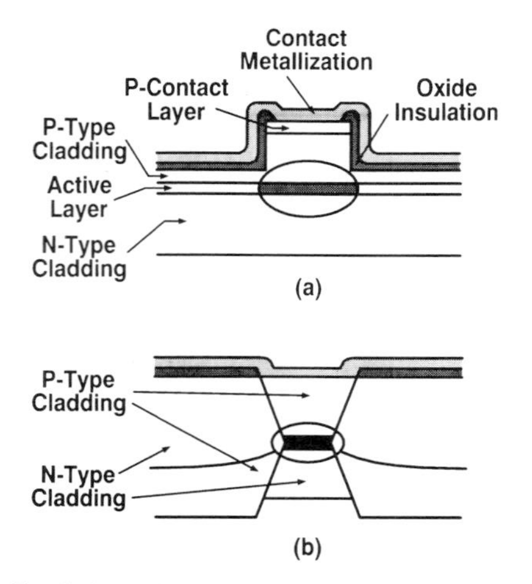

FIG. 5. Schematic cross sections of two types of index-guided lasers. (a) The ridge waveguide laser. (b) The buried heterostructure laser. [From Kapon, E. (1989). *In* "Handbook of Solid State Lasers" (P. K. Cheo, ed.), Marcel Dekker, New York.]

propagation. This structure employs two reversed biased p–n junctions so that the current also becomes laterally confined to the lasing region. With this type of optical and current confinement, threshold currents of less than 1 mA have been achieved.

As we have seen, there are several methods for confining both the injected electrical current and the generated optical beam. The gain-guided structures are easier to fabricate than the index-guided structures, however, gain-guided structures have higher threshold currents and an inferior output beam as compared to index-guided lasers. As a general rule, by properly designing the optical and electronic properties of the semiconductor laser, the injected carriers and generated photons can be confined into a well-defined region, yielding the maximum interaction, which leads to lower threshold currents and a high quality output beam.

B. High Power Semiconductor Lasers

Typical optical power output levels in standard single-stripe diode lasers are on the order of several milliwatts. These power levels are adequate for many commercial applications which use diode lasers, for example, fiber communications and optical disks. However, for many other applications, such as optical pumping, optical time domain reflectrometry, laser radar, and nonlinear optics, these power levels are not sufficient.

The most obvious way of increasing the output power of a semiconductor laser is to increase the driving current of the laser. This method works up to the point where the output power damages the facets of the laser or until the heat generated in the active region starts to degrade the performance of the laser diode.

Other methods of increasing the output power of the laser diode rely on increasing the volume of the active region. This is typically done by either increasing the width or the thickness of the active area. Increasing the length of the laser only reduces the threshold current. The output power in this case is still limited by the catastrophic facet damage.

By increasing the active region stripe width, one produces what is termed a broad area laser. These devices have very wide active areas typically ranging from 10 μm to over 250 μm. These lasers are capable of providing large output powers on the order of a few watts with several amperes of driving current. These lasers can be used as an optical pump in solid state laser systems or in other applications which require a compact, high output power laser. The problems associated with these lasers are that they usually require large biasing currents. This leads to potentially damaging thermal effects in the active region, ultimately leading to device failure. Another problem associated with broad area lasers is that the laser tends to operate in a multispatial mode pattern. The poor beam quality which is produced makes it difficult to focus the beam to a small spot, making the device unsuitable for fiberoptic applications. Another serious limitation to broad area lasers is due to beam filamentation. The filamentation is caused by a nonlinear optical effect called self-focusing. Self-focusing is caused by an increase of the optical index of refraction which is proportional to the optical intensity of a laser beam. An optical beam which has spatial nonuniformities in the intensity of the transverse beam profile causes a spatial distribution of varying refractive index. This causes the beam to collapse or self-focus into a filament which causes damage to the laser and, ultimately, device failure.

To overcome these difficulties, new device structures were developed. The problem of high driving currents was addressed by the development of quantum well lasers. These lasers rely on properties of the electron which occur when the electron is confined to small region in space. The active area thickness of typical quantum well lasers is approximately 100 Å. Broad area lasers which utilize a quantum well as the active area can have threshold currents as low as 80 mA with a stripe width of 100 μm, and have output powers as high as 1 W with 1 A of driving current.

The problem of beam filamentation has been addressed by the development of laser arrays. A typical laser array structure is shown in Fig. 6. Several laser stripes are formed by proton implantation of an initially wide stripe. This is shown by the shaded areas of the p^+ GaAs region. The disordered regions become highly resistive to current flow, while the unimplanted regions become the lasing stripes. The stripes which define the width of the active region are kept small, typically a few micrometers. This forces each individual laser to operate in the single fundamental transverse mode, which eliminates the problem of the beam filamentation. The individual stripes are positioned close enough to the neighboring stripes to allow the evanescent fields of the individual stripes to overlap and interact. This interaction allows coupling of the spatial modes of the individual stripes, forcing the laser array structure to operate coherently. Utilizing a structure with 10 single emitters with 10-μm wide stripes in an external cavity configuration, output powers in excess of 700 mW have been produced in a single lobed beam.

These one-dimensional array structures can be stacked on top of each other to produce a

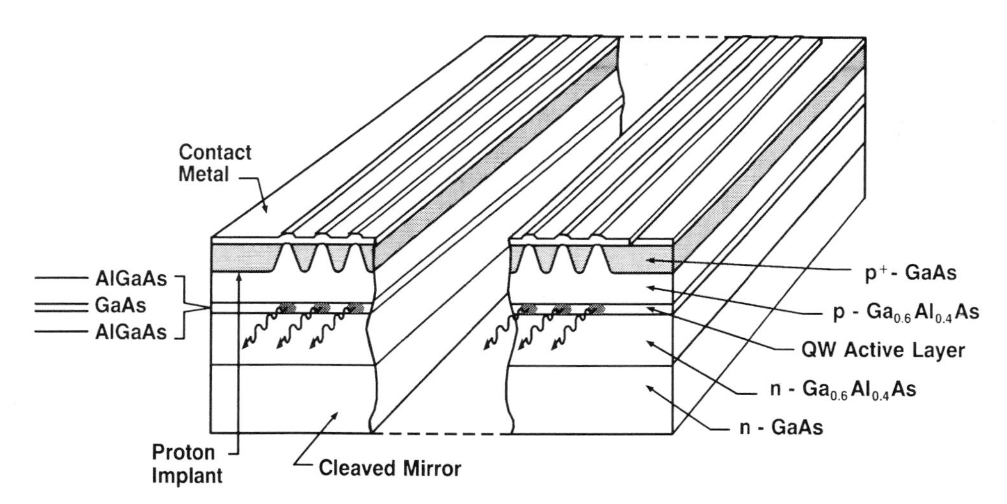

FIG. 6. Schematic cross section of a multiple laser array. [From Streifer, W., Scifres, D., Harnagel, G., Welch, D., Berger, J., and Sakamoto, M. (1988). Advances in diode laser pumps, *IEEE J. Quantum Electron.* **QE-24,** 883.]

two-dimensional laser array. However, at the present stage of development, it is difficult to achieve coupling of all of the individual stripes without the use of external optics. The highest power produced from this type of two-dimensional laser array structure has exceeded 100 W. Although the emitted energy is not totally coherent, due to the inability to couple all of the individual emitters, the output power is sufficient to allow the device to be used as an optical pump to excite a solid state laser material, such as Nd:YAG. This type of application is very promising for semiconductor laser array structures.

C. High Modulation Frequency Lasers

The direct modulation property of semiconductor injection lasers is one of its most unique characteristics, and this feature provides the potential for many scientific and commercial applications. Unlike other solid state and liquid lasers, the optical output of the semiconductor injection laser can be controlled directly by modulating the injection current.

The modulation dynamics of semiconductor injection lasers can be described by the rate equations for the laser, which have been derived from the Maxwell–Bloch equations after adiabatically eliminating the atomic polarization. These equations are

$$\frac{dN}{dt} = \frac{I}{eV} - \frac{N}{\tau_s} - g(1 - \varepsilon S)(N - N_a)S \quad (4a)$$

and

$$\frac{dS}{dt} = g(1 - \varepsilon S)(N - N_a)S - \frac{S}{\tau_p} + \beta\frac{N}{\tau_s} \quad (4b)$$

In these equations, N is the electron density in the active region, S the photon density, V the active region volume, τ_s and τ_p the electron lifetime and photon lifetime, respectively, N_a the carrier density at transparency, β the spontaneous emission coupling coefficient, g the differential gain coefficient, ε the gain saturation coefficient, and e the electron charge.

By employing a small signal analysis of the laser rate equations above the lasing threshold, one can derive the resonance frequency f_r and the damping constant α_d for the modulation response of the photon density to the injection current. This yields

$$f_r = \frac{1}{2\pi}\sqrt{\frac{gS_0}{\tau_p}} \quad (5)$$

for the resonance frequency and

$$\alpha_d = \frac{1}{\tau_s}(1 + gS_0\tau_s) + \frac{\tau_s}{\tau_p}\left[\beta\frac{1 + g\tau_p N_a}{g\tau_s S_0}\frac{1}{\tau_s} + \varepsilon g S_0\right] \quad (6)$$

for the damping constant, where S_0 is the steady state photon density.

In Fig. 7, typical modulation response curves which have been experimentally measured for several values of output power are shown. The circuit diagram shows how the measurement is performed. The modulating signal $i_1(\omega)$ is applied to the laser diode and varied from a low modulation frequency of ~200 MHz to a high modulation frequency of ~10 GHz. In this notation, ω denotes the applied frequency. The current $i_2(\omega)$ is measured from a wide bandwidth photodiode, and the ratio of the output current to the input current is plotted on a logarithmic scale. These curves show three salient features which are characteristic of the modulation response of semiconductor laser diodes. For quasi-static modulation, the modulation frequency applied to the laser is much less than the laser resonance frequency. In this region, the optical output of the laser follows the modulating current, and the modulation efficiency is equal to the slope efficiency of the steady state light output versus dc current input curve of the laser diode. At the high frequency limit, the coupled electron–photon system can not follow the modulation current, and the modulation efficiency decreases rapidly. The interesting feature, that is, the resonant enhancement of the modulation response, is observed when the modulating current is at or near the resonance frequency of the laser. This resonance peak in the modulation response is indicative of a damped oscillation of the photon density after being perturbed from the steady state.

The peak in the modulation response can be described by a strong two-way interaction between the populations of injected carriers and photons. The stored energy of the system can swing between the two populations with a natural resonance frequency which depends on particular circumstances but is normally in the vicinity of a few gigahertz. Little damping is supplied by the optical resonator since under lasing conditions its Q-factor is very large and the main contribution to damping comes from the spontaneous recombination time of the carriers, giving a decay time on the order of 5 nsec.

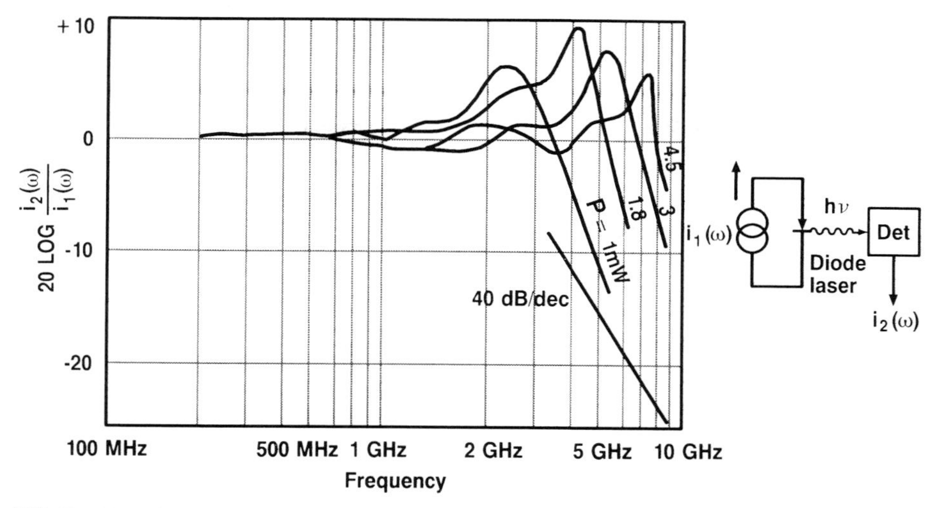

FIG. 7. Experimentally measured continuous wave output power versus current characteristic of a laser of length 120 micrometers. The inset shows the method of measurement. [From Lau, K., Bar-Chaim, N., Ury, I., and Yariv, A. (1983). Direct amplitude modulation of semiconductor GaAs lasers up to X-band frequencies, *Appl. Phys. Lett.* **43,** 11.]

The resonance manifests itself as a transient oscillation during laser switching and also as an enhancement of the modulation response to a small sinuosoidal current in the relevant frequency range.

The modulation bandwidth of the semiconductor injection laser, given by Eq. (5), is determined by the gain coefficient, photon lifetime, and the steady state photon density. These parameters are influenced by the structure of the laser, the operating temperature, and the laser diode material system. For a given laser, the bandwidth can be increased by increasing the injection current as shown in Fig. 7. However, the maximum current is limited by the catastrophic facet damage for AlGaAs lasers and by nonlinear gain, Auger recombination, and current leakage in InGaAsP lasers. By fabricating a window structure near the facets in AlGaAs lasers, the damage threshold can be increased due to the decrease in the optical absorption occurring at the facets. These lasers have been demonstrated to have modulation bandwidths in excess of 11 GHz at room temperature.

Another method of increasing the modulation bandwidth is to decrease the photon lifetime. This is most easily accomplished by decreasing the laser dioide cavity length. This, however, increases the laser threshold current level and therefore lasers with extremely low threshold current levels are required for this method. Utilizing a 40-μm-long AlGaAs multiple quantum well laser, a modulation bandwidth of 24 GHz has been achieved.

In addition to excessive current pumping and shortening the laser diode cavity length as a means to increase the modulation bandwidth, the differential gain coefficient can be increased. This can be achieved by lowering the operation temperature of the laser diode. For example, the bandwidth of a GaAs buried heterostructure laser which is 7 GHz at room temperature (25°C) can be extended to 9.5 GHz at −50°C for a 175-μm-long cavity length. Similarly, the bandwidth of an InP constricted mesa laser can be increased from 16 GHz at 20°C to 26.5 GHz at −60°C. The drawback with this method is that there is a decrease in the modulation efficiency due to an increase in the series resistance of the diode which occurs at low temperatures.

Higher differential gain coefficients can also be obtained by utilizing semiconductor quantum well structures. Due to the confinement of carriers in a direction of the quantum well growth, the density of states becomes stepwise and increases the differential gain coefficient. This staircase density of states distribution increases the modulation bandwidth by a factor of 2. It is expected that another factor of 2 increase in the modulation bandwidth will be obtainable with quantum wire structures, which have a quasi-discrete density of states distribution.

Other phenomena which influence the modulation dynamics of semiconductor lasers include the spontaneous emission, gain saturation, and the external electrical connections to the device. These effects normally manifest themselves as a suppression of the relaxation oscillations which reduces the peak of the modulation response and also reduces the modulation bandwidth.

D. Narrow Linewidth Lasers

The spectral linewidth of standard Fabry–Perot semiconductor lasers is typically on the order of 100 MHz. For many applications, such as holography, spectroscopy, and coherent optical communication systems, these broad linewidths are unsatisfactory. The main difficulty of obtaining narrower linewidths in semiconductor lasers is that there is a modulation of the index of refraction due to fluctuations of the excited population caused by spontaneous transitions. This has an effect of broadening the spectral linewidth of the lasing transition which is predicted by the Schawlow–Townes formula by a factor of $(1 + \alpha_l^2)$. The parameter α_l is called the linewidth enhancement factor and is defined as

$$\alpha_l = \frac{\chi_r^{(3)}}{\chi_i^{(3)}} = \frac{\partial \chi_r / \partial N}{\partial \chi_i / \partial N} \tag{7}$$

where $\chi_{r/i}^{(3)}$ is the real/imaginary part of the third-order nonlinear susceptibility, χ the linear plus nonlinear components of the susceptibility, and N the population density. In most gas and solid state lasers, the lasing gain spectrum is symmetric due to a single atomic transition. This implies that α_l is zero for these laser systems, and the linewidth is correctly predicted by the Schawlow–Townes equation. In semiconductor injection lasers, the gain spectrum is not symmetric due to the density of states and the Fermi–Dirac statistics of electrons and holes. This leads to a nonzero value for α_l in semiconductor lasers and thus a broader spectral linewidth of the lasing transition. The typical value of α_l is ~5 for bulk semiconductor injection lasers and ~2.5 for quantum well and semiconductor injection lasers.

Because of this difficulty in obtaining inherently narrow spectral linewidths, there has been great interest in developing new laser structures and techniques which can yield linewidths less than a few megahertz. The most promising method for reducing the linewidth of a semiconductor laser incorporates a grating structure inside the semiconductor chip itself or utilizes a grating in an external cavity configuration.

The two most popular laser structures which utilize gratings as an integral part of the laser structure are (1) the distributed feedback (DFB) laser structure and (2) the distributed Bragg reflector (DBR) laser structure. In these laser structures, part of the semiconductor waveguide is fabricated to include a corrugated region. This region acts as a diffraction grating which couples a narrow spectral portion of the total spontaneous emission back into the laser diode. The relationship between the corrugation period and the laser emission wavelength must satisfy the relation.

$$m\Lambda \cong \frac{\lambda}{2n} \tag{8}$$

where Λ is the grating period, λ the laser wavelength, n is the index of refraction, and m an integer.

In the DFB laser, the grating structure is distributed along the entire length of the laser diode, just slightly above the gain region. This is shown schematically in Fig. 8. Due to the frequency selective feedback provided by the grating, lasing occurs at a single longitudinal frequency defined by the laser cavity. The power contained in the longitudinal modes that have frequencies which do not correspond to the Bragg condition [(Eq. (8)] is lower by 1000 times that in the lasing mode. Another attractive feature of the DFB laser is that the cavity is defined

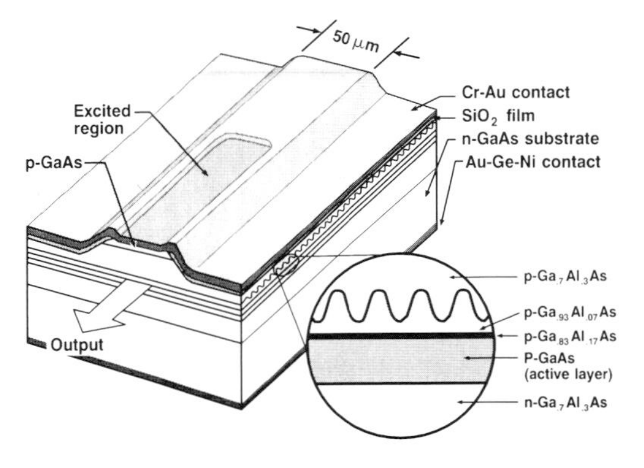

FIG. 8. Schematic of a GaAs–GaAlAs distributed feedback laser. The inset gives a detailed illustration of the laser layers and the corrugated structure. [From Yariv, A. (1988). "Quantum Electronics," 3rd ed. Wiley, New York.]

by the grating structure, not by cleaved facets. This allows the DFB laser to be incorporated directly into an integrated optoelectronic device.

The distributed Bragg reflector laser is similar to the DFB laser. Both lasers incorporate grating structures directly into the waveguiding path of the laser mode. The difference between these two structures in that the DBR laser utilizes a passive waveguiding region to contain the corrugated structure. The main advantage of this type of structure is that it does not require epitaxial growth over the grating structure and, hence, is easier to fabricate. The disadvantage is that the coupling efficiency between the grating structure and the active region is lower, due to the waveguide discontinuity between the active and passive regions. Both of these laser structures provide lasing emission linewidths on the order of 1 MHz, which is two orders of magnitude narrower than the conventional Fabry–Perot laser structure.

Narrower linewidths can be obtained by utilizing an external cavity. Typical cavities include an objective lens for collecting and collimating the laser emission and a reflecting diffraction grating or etalon for frequency selection. When the optical feedback from the external cavity is strong, an antireflection coating is required on the laser facet which is inside the external cavity and faces the frequency selective element. The antireflection coating is necessary because it eliminates the longitudinal modes of the Fabry–Perot structures. If the antireflection coating is not employed, there will be competition between the longitudinal modes defined by the external cavity and the cleaved facets. With external cavity semiconductor lasers employing a diffraction grating, linewidths as narrow as 1 kHz have been achieved. The disadvantage of external cavity lasers is due to the relatively large dimensions, typically on the order of several centimeters.

IV. Applications

Since the development of semiconductor lasers there has been an increasing desire for their utilization in many applications. This is due to their compact size, efficient electrical-to-optical conversion, and extremely low cost. In this section, several scientific and commercial applications of semiconductor injection lasers are reviewed.

A. Diode-Pumped Solid State Lasers

The advances of semiconductor laser arrays has recently caused a resurgence of research activity in the solid state laser community. This is mainly because diode laser arrays have the potential to be used as an efficient optical pump for these laser systems. The advantages of a semiconductor laser array pump as compared to a conventional flashlamp pumped system are numerous. The semiconductor laser arrays are compact, efficient, robust, and potentially inexpensive. The emission wavelength of the laser arrays can be controlled by varying the material composition during the wafer growth. As a result, the emission wavelength can be made to conicide precisely with the peak absorption lines of the solid state laser crystal, thus making the laser array a much more efficient optical pump as compared to a broadband flashlamp. In addition, diode pumps produce less undesirable heating of solid state laser crystal and less potential for damage than a flashlamp pumped system, due to the high energy photons present in a flashlamp source. Another attractive feature of a semiconductor laser array pump is that the light emitted is partially coherent, enabling the light to be efficiently focused or mode-matched to the solid state laser system.

Specific examples of a semiconductor laser array pump would include devices made from the AlGaAs material system. The emission wavelength can be varied from 700 to 900 nm by varying the Ga/Al composition. These emission wavelengths can be made to coincide with absorption bands of several important solid state lasing ions, such as neodymium, chromium, holmium, erbium, and promithium.

Solid state lasers can be made to operate in various manners and geometric configurations, all of which can utilize a diode array as the pump source. The solid state laser can be an optical fiber, a slab, or a standard cylindrical rod. The solid state laser may operate as a continuous wave (CW), quasi-CW or pulsed, or Q-switched device, and the emission may be single- or multispatial and longitudinal mode. Each operating condition and geometry places different requirements on the pump source. Longitudinal or end pumping is most efficient for optical fiber and rod active mediums, whereas transverse or side pumping is particularly suitable for slab active mediums.

Presently, the diode laser arrays which are

used as the pump source in a longitudinally pumped solid state laser system are typically 100–200 μm wide and rated to operate ~5 years with 1 W of CW output. The pumping is normally coupled by optical fibers or bulk optics so that there is mode matching between the diode pump and the solid state laser. The advantages of this geometry are the long absorption path of the pump light and the efficient utilization of the pump light resulting from a mode-matched system.

Figure 9 illustrates a typical longitudinally, diode-pumped solid state laser system. The diode array is made from AlGaAs material system and is mounted on a thermoelectric cooler which controls the operating temperature and the emission wavelength of the laser array. The coupling optics collect the light emission from the laser array and spatially manipulate the beam shape so that the transverse spatial mode patterns of the diode pump and the Nd : YAG laser cavity are matched. The laser rod is an Nd : YAG crystal which has high reflecting (HR) and antireflecting (AR) coatings to allow the reflection and transmission of the desired wavelengths. The 95% reflecting curved mirror serves as the output coupler of the complete Nd : YAG laser. This configuration has been used to produce over 450 mW of CW output power with approximately a 10% electrical-to-optical conversion efficiency.

The longitudinally pumped scheme is limited in output power by the diode output power per unit width. Higher power can be obtained by overdriving the laser arrays, which ultimately reduces the operating lifetime of the array. One simple solution is to use several arrays, each coupled into an optical fiber. The fiber is then combined into a bundle and used as the pumping source. Utilizing this method, ~1 W of TEM_{00} power was produced from an Nd : YAG laser emitting at 1.06 μm. Intracavity frequency doubling was also employed, which produced ~200 mW of light at 530 nm.

Transverse pumping can be utilized in both rod and slab geometries. The advantage of this method is the large area available for pump light illumination, with the disadvantage being inferior mode-matching capabilities. Utilizing a quasi-CW pumping scheme in Nd : YAG, a 0.4 cm^2, 13-bar, two-dimensional stack laser array was employed, emitting 90 W of pumping power into the laser crystal. Particularly noteworthy are the intracavity frequency doubling results, which produced ~3 W of green light at 532 nm with this pumping configuration.

The field of semiconductor laser arrays is still a new and developing technology. With future advancement in achieving higher output power and higher efficiencies, their use in increasingly larger solid state laser systems will be unavoidable. In addition, as manufacturing volumes and yields increase, the cost of these pump sources will decrease with an improvement of diode quality. Even with the present state of technology, it is apparent that semiconductor laser arrays will be an important player in the field of compact solid state laser systems.

B. Optical Communications

Semiconductor lasers have a tremendous potential for use in optical communication systems, due to their high efficiency, small size, and direct modulation capability. Because of their small size, they become the ideal source for op-

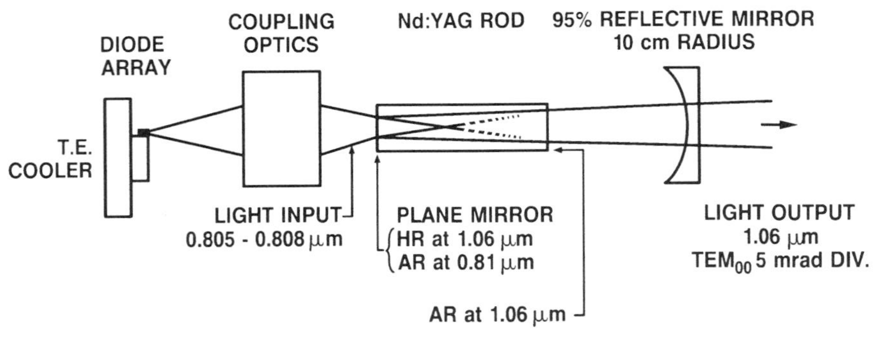

FIG. 9. Schematic of a longitudinally, diode-pumped Nd : YAG laser. [From Streifer, W., Scifres, D., Harnagel, G., Welch, D., Berger, J., and Sakamoto, M. (1988). Advances in diode laser pumps, *IEEE J. Quantum Electron.* **QE-24,** 883.]

tical radiation to be utilized in an optical fiber transmission system. The reason for developing an optical communication system is that these systems can take advantage of the large bandwidth that optics has to offer. For example, in a conventional copper wire communication system, the bandwidth of the transmission channel (i.e., the coaxial cable) is limited to approximately 300 MHz. In a normal telephone conversation, frequency components up to 3 kHz are required for the listener to understand the conversation. By multiplexing many telephone conversations on different carrier frequencies, approximately 100,000 conversations can be sent through a single conventional coaxial cable without interference between conversations. However, in an optical fiber communication system, the bandwidth of the fiber is typically several terahertz. Thus by multiplexing the conversations with a suitable optical carrier signal, more than one billion conversations can be sent over a single optical fiber. This example clearly shows the potential advantage of utilizing an optically based communication system.

Communication systems have three main components: the transmitter, the receiver, and the transmission path. Each component has certain characteristics which ultimately define how the communication system is to be designed. An example of this is seen clearly in the early designs of an optical communication system. These systems were based on the GaAs semiconductor laser wavelengths. The main reason for this was due to the availability of these devices. However, due to the dispersion and absorption losses of the transmission channel (i.e., the optical fiber), this wavelength region was considered to be appropriate only for short distance communication networks. Long distance communication networks which were based at this wavelength would be costly, due to the losses of the transmission channel and the necessity of a short repeater spacing. With the development of longer wavelength semiconductor injection lasers from the InP/InGaAsP material systems, communication systems can take advantage of special properties which are characteristic of glass optical fibers.

In most optical communication systems, the information is encoded onto a beam of light and transmitted through an optical fiber. In an ideal system, the information is undistorted by the transmission path. However, in a real optical fiber, the power level of the signal is decreased as it is transmitted, due to absorption and scattering losses in the fiber. As a result, the signal needs to be regenerated along the transmission path in order for the signal to be detected at the receiver side. Another detrimental effect caused by the fiber is due to the chromatic or group velocity dispersion. This has an effect of broadening bits of information, causing them to overlap in time. These detrimental effects are minimized by utilizing specific wavelengths which take advantage of special physical properties of optical fibers. For example, optical fibers have a minimum loss window at 1.55 μm, typically 0.2 dB/km, and a minimum dispersion window at 1.3 μm, typically 3 psec/km/nm. By utilizing semiconductor lasers which emit in these wavelength regions, the problems can be minimized. The InP/InGaAsP semiconductor lasers emit in these wavelength regions and have been developed sufficiently so that long distance communication systems are being designed to take advantage of these optical windows.

The most basic type of optical communication system utilizes an intensity modulation scheme to encode information onto an optical beam. In these schemes, the coherent properties of the laser are not employed. For detection, the detector simply counts the number of incident photons during a given amount of time in order to recover the modulation signal. Although the intensity modulation scheme does not utilize the coherent properties of the semiconductor laser output, many applications still require single longitudinal mode lasers in order to reduce the effects of chromatic dispersion in an optical fiber. These effects are most important for long distance fiber communications.

Many of the advantages of an optical communication system exploit the coherent properties of the laser output. As in a conventional radio communication system, coherent optical communication systems have a local oscillator at the receiver side. By employing a local oscillator, the detected signal sensitivity is increased by an order of magnitude. This ultimately translates into a longer repeater spacing. This feature is very important for ultralong distance communications (i.e., transoceanic optical communication links). Figure 10 shows a typical coherent optical communication system. The transmitter consists of a semiconductor injection laser, an external modulator, and an amplifier which amplifies the modulated signal before it is launched into the transmission channel. The

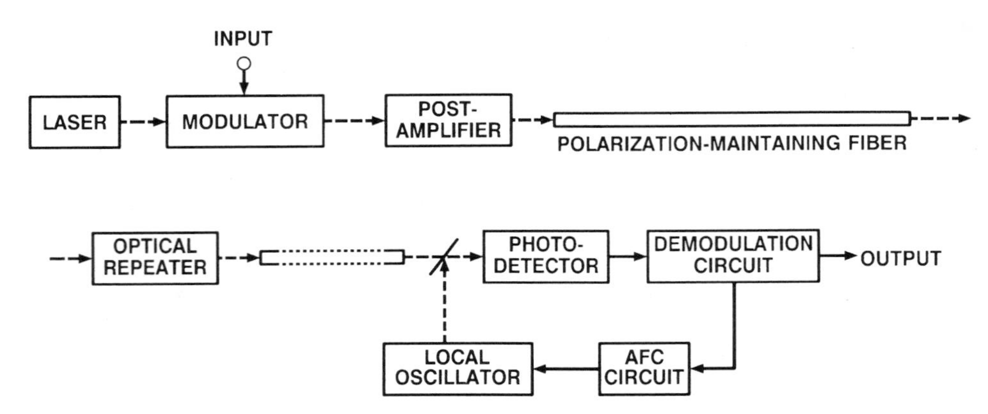

FIG. 10. Schematic of a coherent optical communication system.

transmission channel consists of an optical fiber and an optical repeater, which regenerates the transmitted signal. On the receiver end there is a photodetector, which detects the transmitted signal along with the local oscillator signal. The demodulation circuit demodulates the signal from the photodetector to produce the desired received information. The demodulation circuit also provides a feedback signal to the automatic frequency control circuit. This circuit controls the frequency of the local oscillator so that it can track the fluctuation of the input frequency.

There are numerous modulation schemes that depend on which parameter of the electric field is modulated, that is, amplitude shift keying (ASK), frequency shift keying (FSK), and phase shift keying (PSK). Since the coherent properties of the laser output are utilized in these systems, the bandwidth of the transmitted signal can be reduced to the bandwidth of the modulation signal. Thus by appropriately multiplexing many signals together, whether time division multiplexing (TDM) or frequency division multiplexing (FDM), the communication system can take full advantage of the channel capacity that the optical fiber has to offer.

V. Directions and Trends

The field of semiconductor lasers is only 29 years old. Since its development, it has revolutionized the science of laser optics and given birth to the field of integrated optoelectronics. The field is growing at a tremendous rate, with new laser structures yielding lower threshold currents and higher output power being reported in the technical journals each month. With advances such as these, it is apparent that the future will be even more exciting than the fields' development up to the present. It is difficult to predict the future; however, recent trends point to areas in which tremendous progress will be made. Next, a few areas of semiconductor research which will play an important part in the future development of the field are highlighted.

New semiconductor laser structures will rely on the physics of quantum wire and quantum box effects. This will enable the devices to operate with only a few microamperes of injection current. In addition, these devices will have tunable bandwidths spanning several tens of nanometers.

Two-dimensional surface emitting lasers are another area which will contribute to future semiconductor laser development. These devices have the potential to impact fields of high power semiconductor laser arrays, optical storage, and optical computing.

Semiconductor lasers have certainly advanced to the stage where, in the near future, they will replace the more common solid state and gas lasers which have been the workhorses in both the scientific and industrial arenas. Semiconductor lasers will be successfully mode locked and the resultant ultrashort optical pulses will be amplified to peak power levels approaching the kilowatt region. This will have a tremendous impact on the ultrafast laser community, by providing an inexpensive, efficient, and compact source for ultrafast nonlinear optical studies. In addition, real-time optical signal processing and optical computing will take one step closer to reality with this advancement.

New wavelengths will become available by utilizing strained layer semiconductor material. The key difficulty to overcome in this area is to make the devices thermodynamically stable on a time scale equivalent to the human life span. These new material structures will not only provide wavelengths previously unattainable from semiconductor injection lasers, but will undoubtedly uncover new physics resulting from the combined interaction of quantum well structures and strained layer materials.

Visible diode lasers are now becoming available with wavelengths in the 600-nm regime. Utilizing the wide band gap II–VI material systems, such as ZnSeTe, ZnCdTe, and CdSSe, wavelengths extending to 480 nm will be available. The major hurdle in this field is that p–n junctions of sufficiently high quality are difficult to fabricate, due to the out-diffusion process of the p-type dopant. However, present research indicates that this problem will soon be overcome.

Semiconductor injection lasers are also having a major impact in the area of optical amplification. By simply modifying the device structure, a semiconductor injection laser can be transformed into an amplifying device. This type of device is of paramount importance in integrated optical receivers, optical repeaters, and high power amplifiers for semiconductor mode-locked lasers.

Terabit communication systems will undoubtedly be a part of the future communication networks. This will come about with the continued development of multisegmented semiconductor injection lasers. In these devices, the laser, a modulator, and a frequency selective element are integrated on a single chip. With the continued advancements in high speed electronics, these devices will directly emit pulses on the order of a trillionth of a second, with repetition rates exceeding 100 GHz.

From this brief look into the future, it is apparent that semiconductor injection lasers will have a tremendous impact on the telecommunications, medical, scientific, and industrial communities. The advances in device technology have far exceeded the expectations of the early researchers. We can only assume that the past advances will continue into the future, bringing devices and technologies, which at this time can only be dreamed of, into reality. How these new technologies will shape the development of the world and what impact they will have on society remains to be seen.

Acknowledgments

The authors wish to thank A. M. Weiner and E. Kapon for reading the manuscript.

Bibliography

Kapon, E. (1989). *In* "Handbook of Solid State Lasers" (P. K. Cheo, ed.). Marcel Dekker, New York.

Lau, K., Bar-Chaim, N., Ury, I., and Yariv, A. (1983). Direct amplitude modulation of semiconductor GaAs lasers up to X-band frequencies, *Appl. Phys. Lett.* **43,** 11.

Streifer, W., Scifres, D., Harnagel, G., Welch, D., Berger, J., and Sakamoto, M. (1988). Advances in diode laser pumps, *IEEE J. Quantum Electron.* **QE-24,** 883.

Thompson, G. H. B. (1980). "Physics of Semiconductor Laser Devices," Wiley, New York.

Yariv, A. (1975). "Quantum Electronics," 2nd ed. Wiley, New York.

Yariv, A. (1988). "Quantum Electronics," 3rd ed. Wiley, New York.

SOLID-STATE LASERS

Stephen A. Payne
Georg F. Albrecht *Lawrence Livermore National Laboratory*

GLOSSARY

Active medium or active ion: Material or dopant ion responsible for the lasing.

Energy storage: Capability of a laser material to store energy in the inverted population.

Flash lamps: High-current plasma discharges contained in a fused silica envelope whose light output pumps a laser.

Fusion driver: Laser system of sufficient power to initiate nuclear fusion on a laboratory scale.

Host material: Crystal or glass which is host to the ion whose energy transitions make the lasing possible.

Inversion: Amount of population density by which the upper laser level population exceeds that of the lower laser level.

Laser transition: Electronic change in the state of an active ion that gives rise to the emission of a photon at the lasing wavelength.

Optical switch: Device used to control the pulse format generated by a laser.

Pumping: Process intended to drive a laser medium to or above the threshold inversion density.

Sensitizer ion: Impurity ion used to enhance the absorption of pump light by the laser medium and then transfer this energy to the active ions.

Tunable laser: Laser for which the output wavelength may be tuned to different values.

A solid-state laser is a device in which the active medium is based on a solid material. While this material can be either an insulator or a semiconductor, semiconductor lasers are covered elsewhere in this encyclopedia. Solid-state lasers based on insulators include both materials doped with, or stoichiometric in, the laser ions and materials which contain intrinsic defect laser species, known as F-centers. F-center lasers, however, are considered elsewhere in this encyclopedia as well. Thus, the designation solid-state lasers is intended to denote those laser systems which are based on crystalline or glassy insulating media, in which a stoichiometric component or extrinsic dopants incorporated into the material serve as the laser species.

The physics and engineering of solid-state lasers is both a mature field and an area burgeoning with new activity. While there are many concepts and laser designs that have been established, each year continues to bring remarkable discoveries that open new avenues of research. This article is not only intended to provide an accounting of the known physics of solid-state laser sources but also to convey a sense of the enormity of the field and the likelihood that many new laser materials and architectures will be discovered during the next decade.

I. Introduction

The principle of laser action was first demonstrated in 1960 by T. H. Maiman. This first system was a solid-state laser; a ruby crystal served as the active element and it was pumped with a flash lamp. With this report of laser action, the

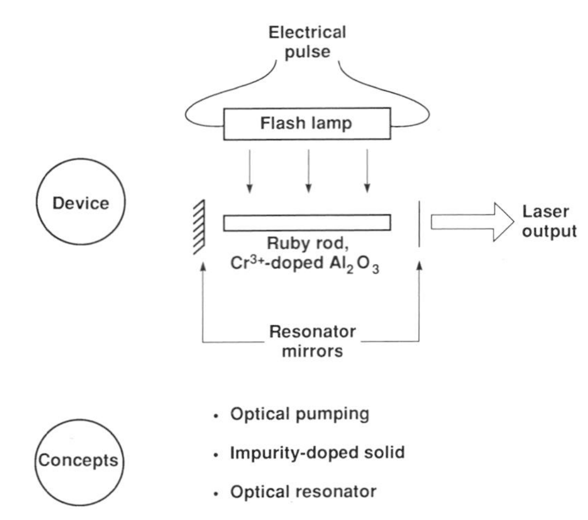

FIG. 1. Schematic desorption of the ruby system for which laser action was first reported and a listing of the fundamental concepts that were introduced.

main concepts on which solid-state lasers are based became established (see Fig. 1). The idea of optically pumping the laser rod was realized, as well as the use of an impurity-doped solid as the laser medium. Lastly, the concept of a laser resonator, as adapted from the work of Townes and Schalow, was experimentally demonstrated. Much of this article is essentially an exposition of the extensive technical progress which has occurred in each of these three areas. Optical pumping has evolved considerably by way of optimization of flash lamps and through the additional use of laser-pumping techniques. The number of impurity-doped solids that have now been lased stands at over 200. Optical resonators have also become remarkably sophisticated in terms of the manipulation of the spatial, temporal, and spectral properties of the laser beam. [*See* LASERS.]

II. Solid-State Laser Materials

A. Laser-Active Ions in Solids

Solid-state lasers are based on a wide variety of materials. All of these materials are conceptually similar, however, in that a laser-active impurity ion is incorporated into the solid material, referred to as the host. In nearly all cases of interest to us, the host is an ionic solid (e.g., MgO), and the impurity carries a positive charge (e.g., Ni^{2+}). As a simple illustration of this situation, a two-dimensional view of the MgO : Ni^{2+} system is pictured in Fig. 2. Here, a small fraction of the Mg^{2+} sites are substituted by Ni^{2+} ions. While the pure MgO crystal is clear, the NiO doping leads to green coloration. It is the impurity ions that are responsible for the laser action. The host medium nevertheless profoundly affects the electronic structure of the impurity and is, of course, responsible for the bulk optical, thermal, and mechanical properties of the laser medium.

1. Ions in Crystals and Glasses

On the basis of the introduction and a cursory glance at the periodic table of the elements, it may seem that the number of potential impurity–host systems is virtually limitless. The reported laser ions are indicated in Table I, along with a very abbreviated list of host crystals. While 22 laser ions are reported in Table I, it should be noted that these ions lase with varying degrees of proficiency, and generally possess both advantages and disadvantages. For example, Ti^{3+} has a wide tuning range but lacks the ability to store energy, Pm^{3+} is predicted to lase quite efficiently but is radioactive, Sm^{2+} lacks good chemical stability, and so on.

Each of the laser host materials possesses certain attributes and handicaps. The ease of crystal growth is important, as is the "meltability" of glassy materials. It is also noted that only certain impurity ions will be compatible with a particular host. Generally, the size and charge of the substitutional host metal ion must be similar

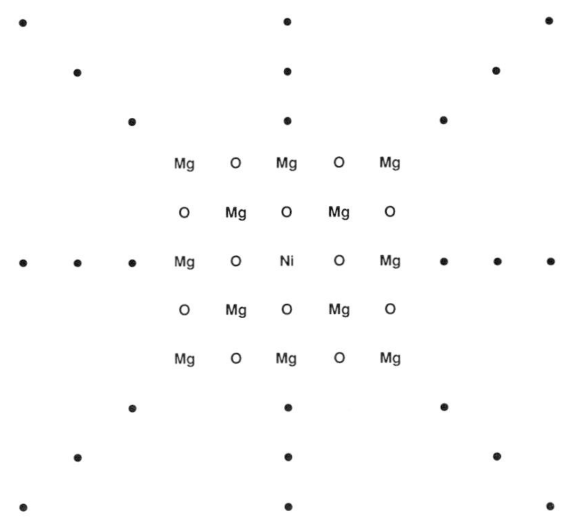

FIG. 2. Example of an impurity-doped host, MgO : Ni^{2+}.

TABLE I. Laser Ions and Abbreviated Listing of Host Materials

Laser ions

Transition metal ions: Ti^{3+}, V^{2+}, Cr^{3+}, Cr^{4+}, Co^{2+}, Ni^{2+}

Trivalent rare earth ions: Ce^{3+}, Pr^{3+}, Nd^{3+}, Pm^{3+}, Sm^{3+}, Eu^{3+}, Gd^{3+}, Tb^{3+}, Dy^{3+}, Ho^{3+}, Er^{3+}, Tm^{3+}, Yb^{3+}

Divalent rare earth ions: Sm^{2+}, Dy^{2+}, Tm^{2+}

Actinide ion: U^{3+}

Examples of laser hosts

Fluoride crystals: MgF_2, CaF_2, LaF_3, $LiYF_4$, $LiCaAlF_6$

Oxide crystals: MgO, Al_2O_3, Y_2O_3, $BeAl_2O_4$, $YAlO_3$, $CaWO_4$, YVO_4, $Y_3Al_5O_{12}$, $Gd_3Sc_2Ga_3O_{12}$, $LiNdP_4O_{12}$

Glasses: ZrF_4–BaF_2–LaF_3–AlF_3, SiO_2–Li_2O–CaO–Al_2O_3, P_2O_5–K_2O–BaO–Al_2O_3

to that of the impurity ion. For example, Ni^{2+} can be incorporated into the Mg^{2+} sites of MgF_2 and MgO, and Nd^{3+} into the Y^{3+} sites of $LiYF_4$, Y_2O_3, $YAlO_3$, and $Y_3Al_5O_{12}$.

2. Energy Levels

The nature of the energy levels and dynamics of the impurity–host system determine the character and effectiveness of the laser. A generic representation of the impurity energy levels and the energy flow appears in Fig. 3. An ideal laser crystal or glass would efficiently absorb the light from the pump source (step 1). The energy then relaxes to the lowest excited state (step 2). This level typically has a lifetime that is long enough to "store" energy. Gain occurs in the next step, as the impurity undergoes a transition between

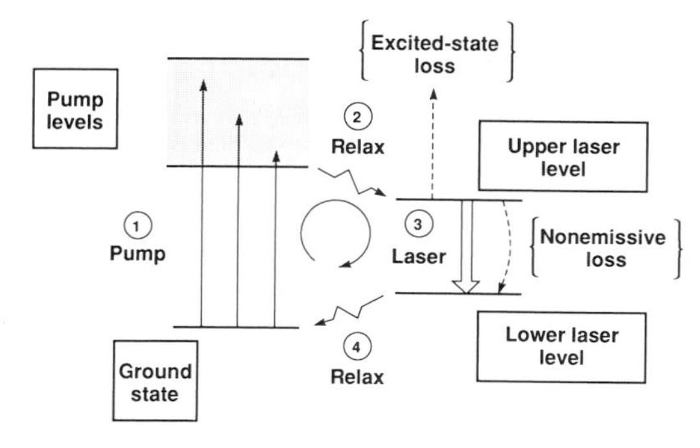

FIG. 3. Generic representation of the energy levels and energy flow (steps 1–4) for an idealized impurity laser ion. The dashed arrows illustrate two fundamental loss mechanisms.

the upper and lower laser levels. While there are many potential loss mechanisms in a laser material, two that are fundamental in nature are noted in Fig. 3. One is excited-state absorption, where the upper level "sees" some amount of loss, rather than gain, due to the excitation to a higher lying excited state. A second loss involves the degradation of the energy of the upper laser level into heat, resulting in nonemissive loss. Step 4 shows the lower laser level relaxing back to the ground state. This last step is critically important since a system in which this process is absent (known as a 3-level laser) requires considerably more energy in order to lase. The advantage of the 4-level scheme depicted in Fig. 3 is that the lower laser level is unoccupied and therefore cannot absorb the laser light (thereby introducing ground state absorption loss into the system).

B. Host Materials

The host materials that are utilized in laser systems must exhibit adequate transparency, mechanical strength, and thermal properties. In addition, the material must be able to sustain a precise optical polish and be cast or grown adequately within reasonable economic and time constraints. The host must afford the impurity ions the type of spectroscopic properties that are appropriate for good laser performance. As a result of the numerous requirements, not many materials turn out to be useful in practical circumstances. Next, the nature of glasses and crystals is discussed, and the important physical properties are briefly outlined.

1. Glasses

Most laser glasses fall into one of several categories, including silicates, phosphates, and fluorides. These glasses may also be mixed, yielding fluorophosphates, silicophosphates, etc. In all cases, the glass is imagined to consist of two components: the network former and the modifiers. The network is a covalently bonded three-dimensional system, while the modifiers are ionically bonded and are imagined to disrupt the network structure. The silicate glasses provide the simplest description of the interplay between the network and the modifiers. First, consider crystalline quartz, or SiO_2, as illustrated in Fig. 4. Here, every oxygen bridges between two silicons. Fused silica is similar although it is glassy, meaning that the highly ordered nature of the

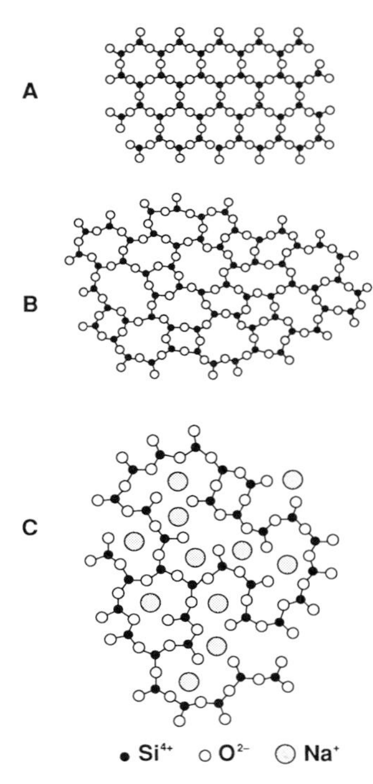

FIG. 4. Two-dimensional view of the structure of SiO_4 tetrahedrons in (A) quartz crystal, (B) fused silica, and (C) sodium silicate glass.

system has been eliminated. If modifiers such as Na_2O are added, some of the oxygens become "nonbridging." There are some favorable features afforded to the glass by the modifier ions: the melt acquires a much lower viscosity and may be easily poured and cast, and the glass is able to dissolve rare earth ions much more effectively than fused silica.

A similar situation exists for other types of glasses as well. For example, the P_2O_5 in phosphate glasses forms the network, and alkali and alkaline earth oxide compounds are added as modifiers. For the case of fluoride glasses, ZrF_4, ThF_4, or BeF_2 may serve as the network former.

2. Crystals

The growth of most crystals turns out to be considerably more difficult than is the case for melting and casting glassy materials. Crystals provide important advantages, however, since a precisely defined site is available to the laser ion rather than the broad distribution of sites that characterize a glass. Crystals often have more favorable thermal and mechanical properties as well. For example, the thermal conductivity tends to be much higher and oxide crystals tend to be very strong mechanically compared to glasses. As a result, it is often advantageous to generate the crystalline media.

Crystals may be grown in many different ways, two examples of which are shown in Fig. 5. The Bridgman method typically involves slowly lowering a crucible through a zone in which the temperature abruptly drops from above to below the melting point of the crystal. A seed crystal is sometimes placed at the bottom of the crucible to initiate the growth. Also shown is the Czochralski method, in which a seed is dipped into the melt and then slowly raised as it is rotated. Most crystals are grown at the rate of 0.1–10 mm/hr. It is important to note that there are many other methods of crystal growth that have not been discussed here (solution and flux growth, flame-fusion, etc.). All methods are based on the concept of slowly enlarging the seed crystal.

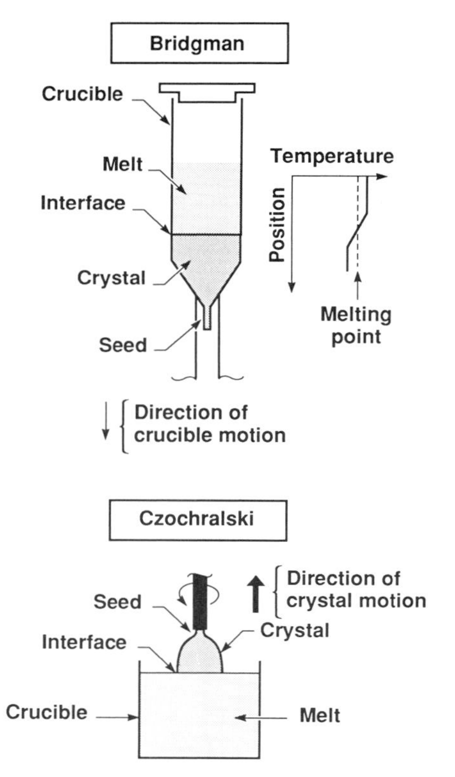

FIG. 5. Schematic drawing of two common crystal growth methods.

TABLE II. Summary of Some Thermal, Mechanical, and Optical Properties for Several Laser Hosts

	Thermal		Mechanical		Optical	
Material	Melting (softening) points, T (°C)	Thermal conductivity, κ (W/m °C)	Young's modulus, E (GPa)	Fracture toughness, K_{IC} (MPa/m$^{1/2}$)	Index change, dn/dT (10^{-6}/°C)	Nonlinear index, n_2 (10^{-13} esu)
Glasses:						
ED-2 (silicate)	(590)	1.36	92	0.83	+3.8	1.4
LG-750 (phosphate)	(545)	0.62	52	0.45	−5.1	1.1
Crystals:						
Al_2O_3	2040	34	405	2.2	+1.6	1.3
$Y_3Al_5O_{12}$	1970	13	282	1.4	+8.9	2.7
CaF_2	1360	9.7	110	0.27	−11.5	0.4
$LiYF_4$	820	5.8, 7.2	75	0.31	−2.0, −4.3	0.6

3. Physical Properties

The physical properties of the host material, in part, stipulate the architecture of the laser system. The properties of several types of host materials, including glasses and crystals, are outlined in Table II. The qualitative trends evidenced in the table can be enumerated as follows: (a) The high-melting oxides tend to provide the highest mechanical strengths, as noted by the magnitudes of Young's modulus (the "stiffness") and the fracture toughness (a measure of the material's resistance to breaking). (b) The thermal conductivity of crystals is much higher than glasses. This permits the rapid cooling of the laser material when deployed in a system. (c) The temperature difference between the surface and center of a laser material gives rise to a thermal lens. The magnitude and sign of the change of refractive index with temperature, dn/dT, partially determines the extent of thermal lensing, which is typically smallest for fluorides and largest for oxides. (d) The nonlinear refractive index, n_2, is smallest for fluorides. The value of n_2 determines the power at which self-focusing occurs. In this process the power of the beam itself creates a lens in the material, which may result in catastrophic self-focusing of the light.

C. Rare Earth Ion Lasers

As mentioned above, 13 of the rare earth (RE) ions have been lased. In passing from Ce^{3+} to Yb^{3+} the $4f$ shell becomes filled with electrons, $4f^1$ to $4f^{13}$. It is the states that arise from the $4f^n$ shell that give rise to nearly all of the RE laser transitions. In this section a few specific systems will be discussed in detail in order to provide representative examples of the nature of RE lasers.

1. Nd^{3+} Lasers

All of the relevant absorption and emission features of Nd^{3+} are due to $4f \rightarrow 4f$ transitions. Dozens of electronic states arise from the $4f^3$ electronic configuration of Nd^{3+}, many of which are indicated in Fig. 6. The ground states consist of the ^4I_J manifold, where the ^4I designation describes the spin ($S = 3/2$) and orbital ($L = 6$) angular momenta, and the $J = 9/2$, $11/2$, $13/2$, $15/2$ indicate the coupling between these two momenta. Absorption transitions occur from the $^4I_{9/2}$ ground state to the indicated energy levels, and each of these transitions are manifested in the absorption spectrum. Figure 7 contains the absorption spectrum of Nd^{3+} in a phosphate glass known as LG-750. The appearance of the spectrum for Nd^{3+} is similar in any host, since it is primarily due to the free-ion properties of Nd^{3+}. The main effect of the host glass (crystal) is to split and broaden the atomic transitions into bands (lines).

Following excitation into any of the absorption bands shown in Figs. 6 and 7, the energy rapidly relaxes to the metastable $^4F_{3/2}$ excited state and generates significant heat during this process. The $^4F_{3/2}$ state, on the other hand, decays in a radiative manner to all the states of the

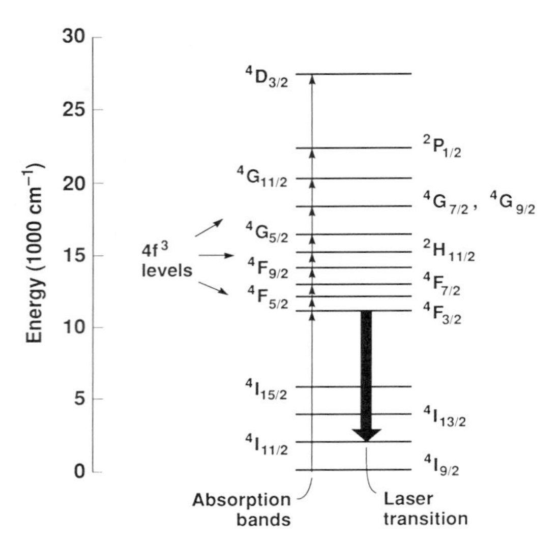

FIG. 6. Energy levels of Nd^{3+} showing the transitions responsible for the absorption bands and the laser action. Note that for clarity of presentation only the major states involved in the transitions have been indicated.

^4I manifold. The $^4F_{3/2} \rightarrow {}^4I_{11/2}$ transition produces an emission band which peaks at 1054 nm (see Figs. 6 and 7). It is this transition that provides the strongest gain for Nd^{3+}-doped laser glasses and crystals. Nd^{3+} lasers are by far the most technologically important type of lasers. The three main materials that are routinely utilized with Nd^{3+} are $Y_3Al_5O_{12}$ and $LiYF_4$ (known as Nd : YAG and Nd : YLF, respectively) and several kinds of glass.

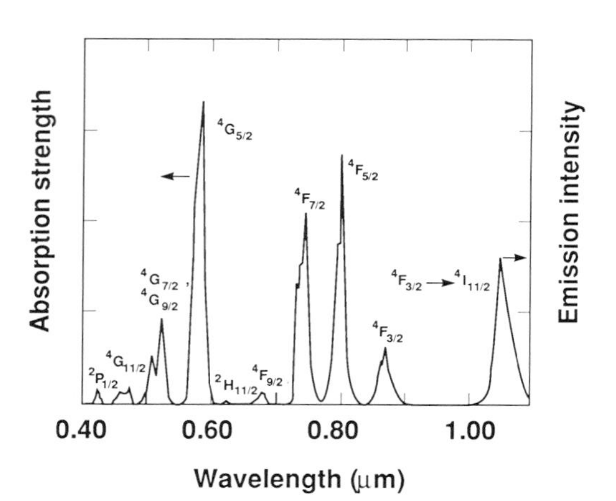

FIG. 7. Absorption and emission spectra of Nd^{3+} in phosphate glass (LG-750). The final states of the absorption transitions are indicated.

2. Er^{3+} Lasers

The Er^{3+} ion has been lased in many crystals and glasses, and this ion has in fact been reported to lase on 13 different transitions. The relevant energy levels and the wavelengths of the laser transitions are shown in Fig. 8. Of all the laser lines, however, only the $^4I_{13/2} \rightarrow {}^4I_{15/2}$ transition is utilized in a commercial system. The Er^{3+} laser ion illustrates the wide gap between demonstrated laser transitions and those that are suitable for technological application. This disparity may be taken as an indication of the progress in laser physics and design that is likely to occur in the coming years.

One question that may come to the reader's mind is "why does all the energy relax to the $^4F_{3/2}$ metastable level for Nd^{3+} irrespective of which state is excited, while so many of the higher lying states of Er^{3+} are found to emit sufficiently to allow for laser action?" The answer to this question lies in the "energy gap law." Simply put, it states that the rate of nonemissive decay between two energy levels of a rare earth ion is solely dependent on the energy gap between them. Since the energy gaps are generally

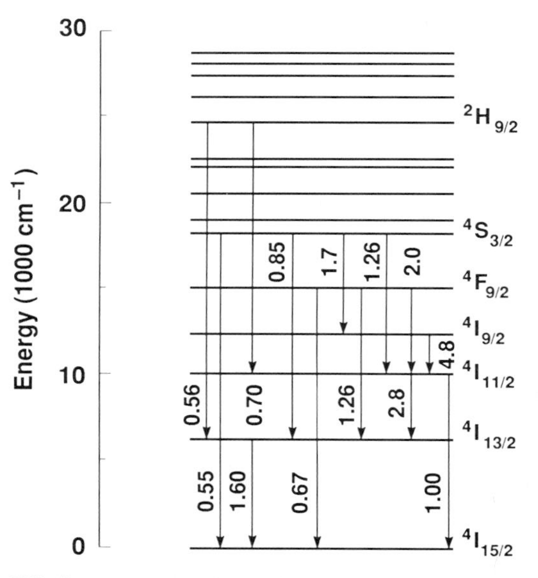

FIG. 8. Energy levels and demonstrated laser transitions of Er^{3+}-doped materials. The wavelengths of the laser action are indicated in micrometers.

larger for Er^{3+} than for Nd^{3+} (compare Figs. 8 and 6), more of the energy levels are able to emit in the case of Er^{3+} and therefore may potentially be able to lase.

3. Other Rare Earth Lasers

As noted above, there is a great disparity between the number of RE ions that have been lased and the relatively small number of materials that are actually used in practical applications. Nevertheless, great potential exists for the future. A summary of all the wavelengths that have been generated by RE ions appears in Fig. 9. (Note that the particular transitions involved have not been indicated on the figure.) The wavelengths span the range from 0.18 to 5.2 μm, and both divalent and trivalent RE ions are represented. The efficiencies of many of these systems are extremely low, however, and may

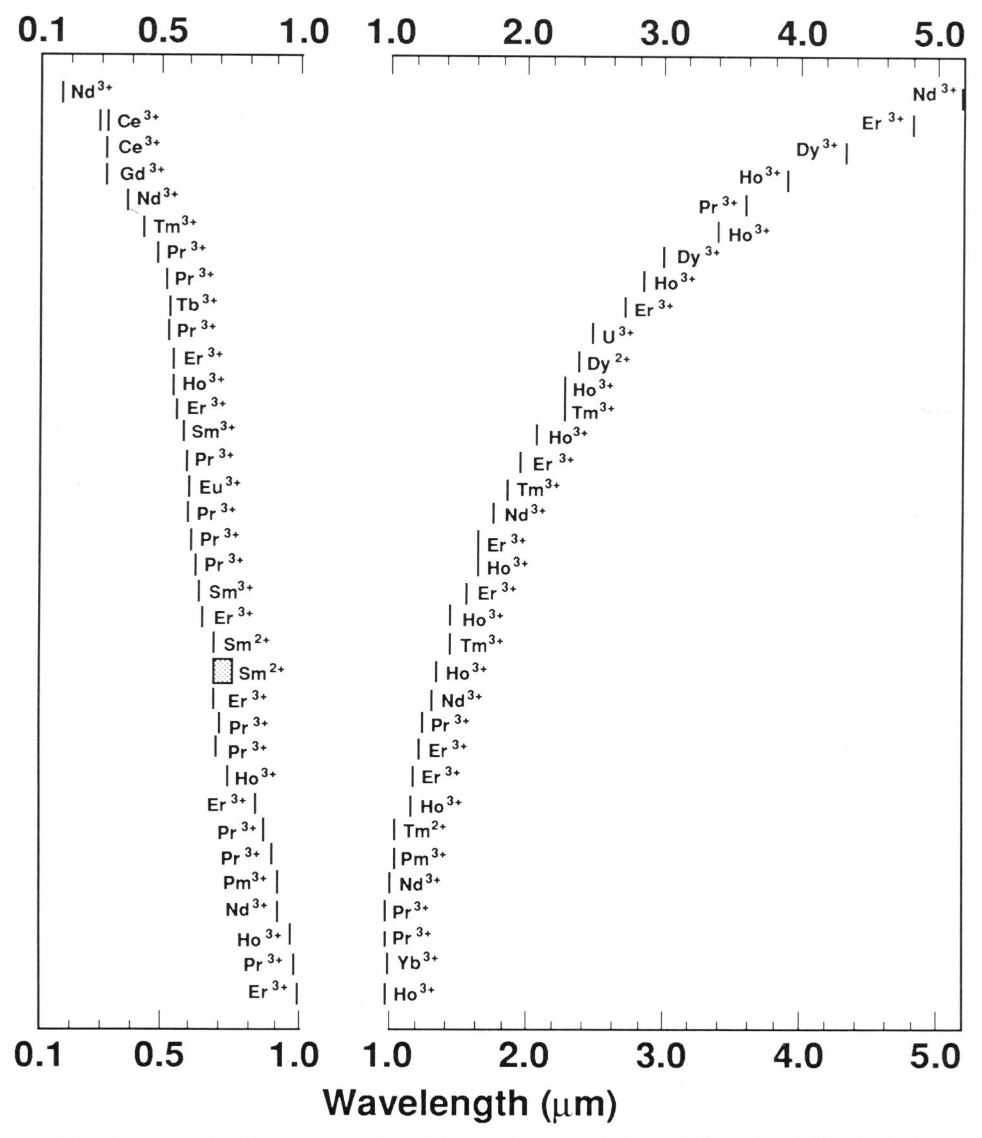

FIG. 9. Summary of all the wavelengths that may be generated by RE ions. Each line indicates a particular transition for a RE ion that may have been lased in numerous host materials.

never prove to be useful in practical circumstances.

D. Transition Metal Ion Lasers

1. Ti^{3+} Lasers

The optical properties of transition metal ions are fundamentally different from the rare earth species. This is primarily becase the $3d \rightarrow 3d$ electronic transitions that are responsible for the absorption and emission features of transition metal ions interact strongly with the host, in contrast to relative insensitivity of the rare earth $4f \rightarrow 4f$ transitions. The type of situation that arises is depicted in Fig. 10 in terms of a "configuration coordinate" model. The Ti^{3+} ion in Al_2O_3 has been selected for illustrative purposes because its valence shell contains only a single d electron, $3d^1$. The $3d$ electron is split into two states, the ^2E and $^2T_2$, by the six nearest-neighbor oxygen anions surrounding the Ti^{3+} ion. (The splitting is much larger than that experienced by the $4f$ electrons of RE ions.) As depicted in Fig. 10, the average Ti–O distance is slightly larger in the ^2E state than in the $^2T_2$ state. This difference is particularly important because it produces wide absorption and emission features.

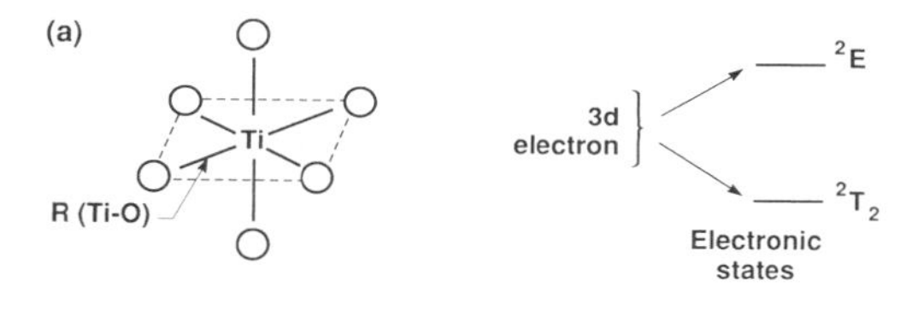

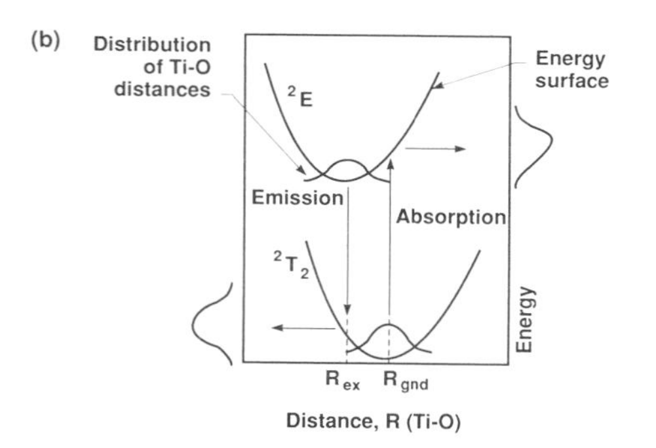

FIG. 10. (a) Splitting of the $3d$ electron of Ti^{3+} into the ^2E and $^2T_2$ states due to the interaction with the six nearest-neighbor oxygen anions of Al_2O_3. (b) Configuration coordinate model of the Ti^{3+} impurity depicting how the displacement between the $^2T_2$ and ^2E states results in broad absorption and emission features.

The configuration coordinate diagram of Fig. 10 explains how the different Ti–O distances in the ground and excited states give rise to broad spectral features. The gaussian curve drawn on the ground state potential energy surface indicates the probabilistic distribution of Ti–O distances that occurs. Since the electronic transition to the excited state occurs rapidly compared to the motion of the Ti–O atoms, this ground state Ti–O distance distribution is simply "reflected" off the rising side of the upper state energy surface, thereby producing a broad absorption feature. A similar argument applies to the emission process.

The actual absorption and emission spectra of $Al_2O_3 : Ti^{3+}$ are shown in Fig. 11. This material (known as Ti : sapphire) may be optically pumped with a doubled Nd : YAG laser, an Ar^+ laser, or a flash lamp. The output of the laser can be tuned from 0.66 to 1.2 μm. Ti : sapphire operates efficiently and is not adversely impacted by the detrimental loss mechanisms indicated in Fig. 3.

2. Cr^{3+} Lasers

Cr^{3+} lasers are similar to Ti^{3+} lasers in that these crystals exhibit broad spectral features. Cr^{3+}-doped systems possess an important advantage, however, in that they have three absorption bands rather than one and therefore absorb flash lamp light more efficiently. Furthermore, because the trivalent oxidation state is very stable, Cr^{3+} may be incorporated into a wide variety of hosts. A summary of the tuning ranges achieved by Cr^{3+}-doped materials appears in Fig. 12, where it is seen that wavelengths from 0.70 to 1.25 μm can be covered with different host materials. It is crucial to emphasize, however, that many of the reported Cr^{3+}-lasers are flawed in various ways, such as by having low efficiency or perhaps by permanently coloring under the influence of ultraviolet flash lamp light. The two most promising lasers are Cr^{3+}-doped $BeAl_2O_4$ and $LiCaAlF_6$ (known as alexandrite and Cr : LiCAF, respectively).

Another important distinction between Ti^{3+} and Cr^{3+} lasers pertains to the lifetime of the metastable excited state, which is typically near 1–10 μsec for Ti^{3+} and 100–300 μsec for Cr^{3+}. As a result, Cr^{3+} lasers can be arranged to "store" more energy than Ti^{3+} lasers.

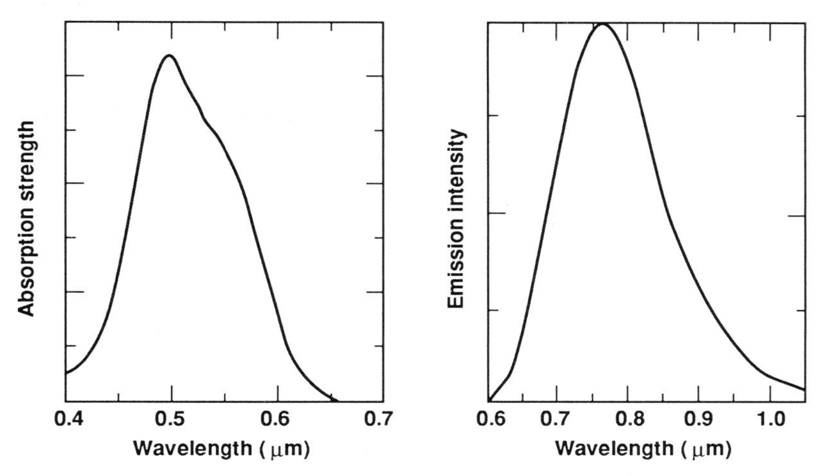

FIG. 11. Absorption and emission spectra of Ti^{3+}-doped Al_2O_3 at 300 K.

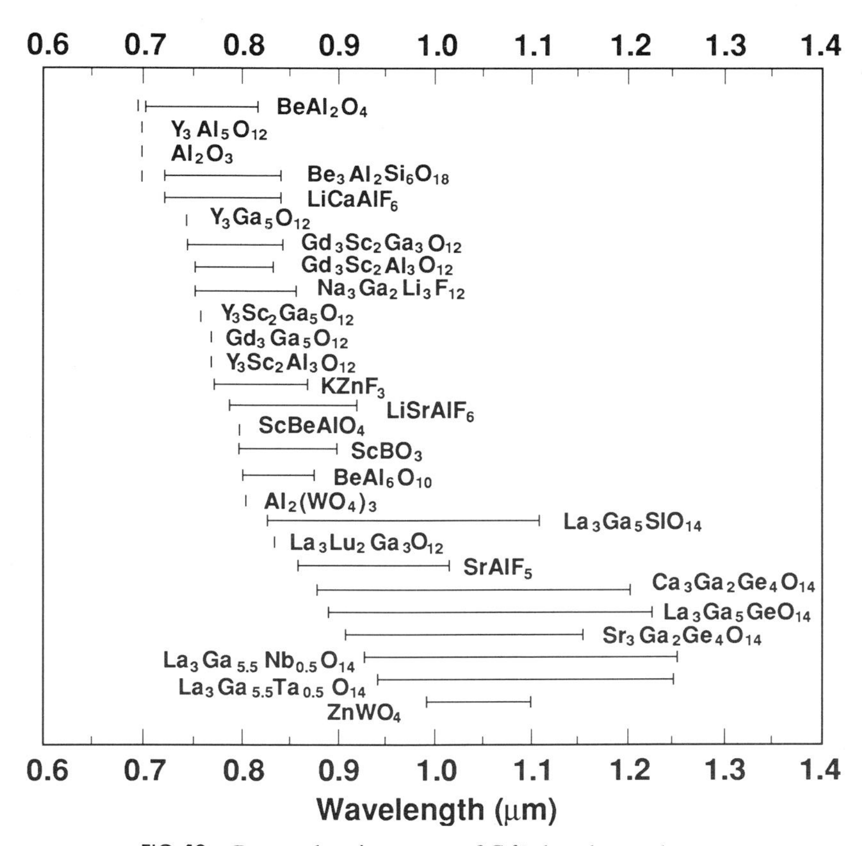

FIG. 12. Reported tuning ranges of Cr^{3+}-doped crystals.

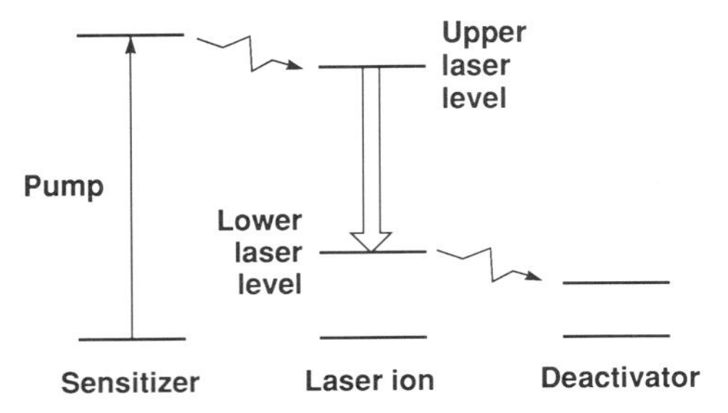

FIG. 13. Illustration showing the role of a sensitizer to excite the upper laser level or that of a deactivator to depopulate the lower laser level.

3. Other Transition Metal Ion Lasers

There are, in total, 30 transition metal ions, including the first, second, and third row ions. In addition, several oxidation states exist for each of these ions. In spite of this diversity, only four ions other than Ti^{3+} and Cr^{3+} have been lased, they are V^{2+}, Cr^{4+}, Co^{2+}, and Ni^{2+}. These ions exhibit laser output in the 1–2.3 μm region and, of these materials, $MgF_2:Co^{2+}$ crystals seem to have the most promise.

The limited number of transition metal ion lasers is related to a combination of factors. First, many of the ions do not have stable oxidation states and tend to vaporize rather than dissolve in the host material. Second, many ions turn out either to be nonemissive or to have serious excited state absorption losses, rendering them useless as laser ions.

E. Energy Transfer in Laser Materials

1. Background

The performance of a laser material may be enhanced by the presence of additional impurity ions. Figure 13 illustrates two possible roles for these extra ions. A sensitizer increases the level of pump-light absorption by the laser material, by absorbing in spectral regions where the laser ion normally does not absorb. The sensitizer then efficiently transfers energy to the upper level of the laser ion.

Another role that the additional ions may play is that of "deactivating" the lower laser level. As discussed above, a four-level laser has im-

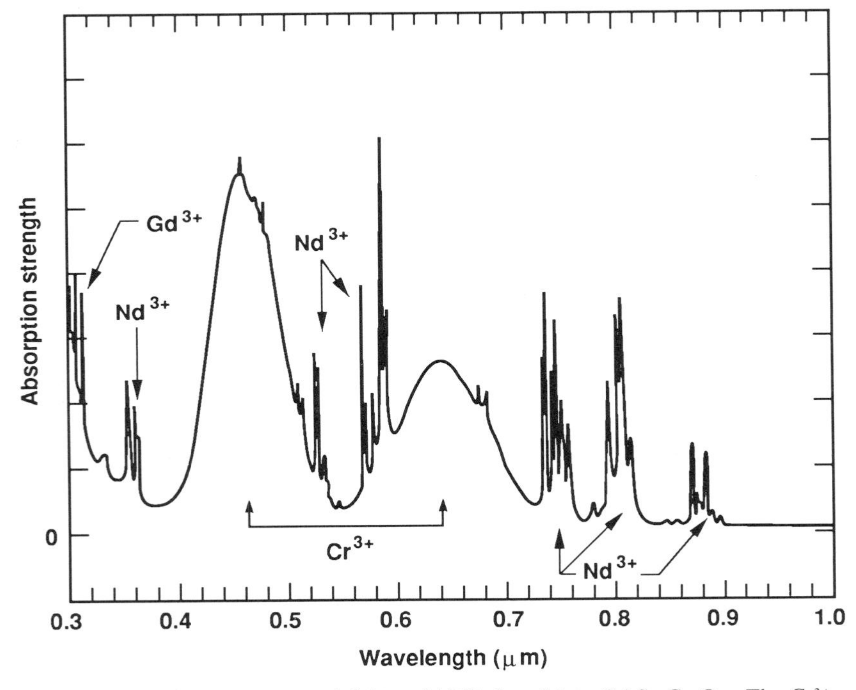

FIG. 14. Absorption spectrum of Cr^{3+} and Nd^{3+} doped into $Gd_3Sc_2Ga_3O_{12}$. The Cr^{3+} sensitizers provide for more efficient flash lamp absorption and then rapidly transfer their energy to the Nd^{3+} laser ions.

portant advantages since the lower laser level remains unpopulated. If this level does not rapidly drain following the laser action, it will become populated. The deactivator serves to funnel this energy away from the laser ion.

2. Examples of Sensitized Laser Materials

There are several sensitized laser materials that have proved to be useful. As an example, consider the case of Cr^{3+} and Nd^{3+} doped together into $Gd_3Sc_2Ga_3O_{12}$ (GSGG), where Nd^{3+} serves as the laser ion and Cr^{3+} as the sensitizer. The absorption spectrum of GSGG : Cr^{3+}, Nd^{3+} is shown in Fig. 14. Since the sharp features are due to the Nd^{3+} ion and the broad bands to Cr^{3+}, it is clear that the Cr^{3+} ion provides greatly enhanced absorption for the flash-lamp-pumped system. (The flash lamp output is essentially a quasi-continuum throughout the ultraviolet–visible–infrared regions.) It is also crucial to note that the Cr → Nd energy transfer is extremely efficient (>90%). GSGG : Cr^{3+}, Nd^{3+} and related systems have provided the highest flash-lamp-pumped efficiencies measured to date.

The energy transfer of the Cr^{3+}, Tm^{3+}, Ho^{3+} : YAG crystal is a striking example of an elegant new laser system. The Cr^{3+} ions efficiently absorb the flash lamp light. The energy is then transferred to the Tm^{3+} ions, as shown in Fig. 15. The Tm^{3+} ions are doped at high concentration to allow for efficient cross-relaxation in which two Tm^{3+} ions are generated in the $^3F_4$ excited state for each Tm^{3+} ion initially in the

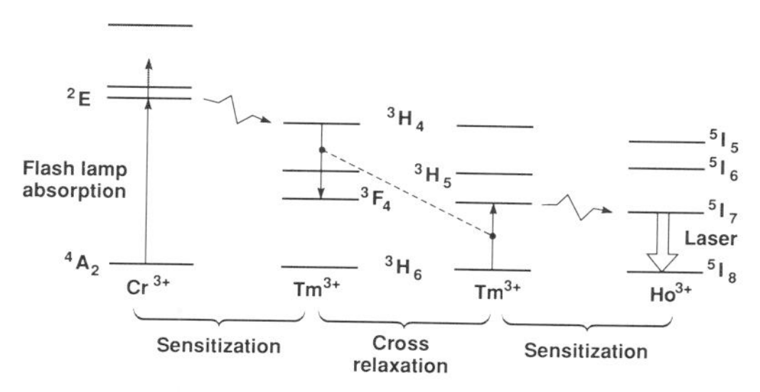

FIG. 15. Energy-transfer dynamics of the Cr^{3+}, Tm^{3+}, Ho^{3+} : YAG laser crystal. The Cr^{3+} impurities initially absorb the light and transfer the energy to Tm^{3+}. The Tm^{3+} ions then cross-relax to produce two excited states for each ion initially excited. Lastly the energy is transferred to the Ho^{3+} laser ions.

$^3H_4$ state. Lastly, the energy is transferred to the Ho^{3+} ions, which exhibit gain near 2.1 μm. Each of the concentrations of the ions must be chosen carefully to optimize the energy transfer steps.

F. Practical Laser Materials

A myriad of issues impact the effectiveness of a laser material. The issues range from the laser parameters to the thermal-optical-mechanical factors to the raw materials costs and the ease of fabrication. A listing of some viable solid-state laser materials appears in Table III. Nd : YAG is by far the most widely used material. It offers high gain and has a long storage time of 240 μsec. The R_T value, a measure of the material's resistance to thermal stress-induced fracture,

TABLE III. Properties of Some Common Laser Materials

Name	Material	Laser wavelength, λ (μm)	Upper level lifetime, τ (μsec)	Thermal stress resistance, R_T ($W/m^{1/2}$)	Type of pumping[a]
Nd : YAG	$Y_3Al_5O_{12}$: Nd^{3+}	1.064	240	1100	FL, DL
Nd : YLF	$LiYF_4$: Nd^{3+}	1.047, 1.053	480	140	FL, DL
LG-750	Phosphate glass : Nd^{3+}	1.054	350	70	FL
Cr, Nd : GSGG	$Gd_3Sc_2Ga_3O_{12}$: Cr^{3+}, Nd^{3+}	1.061	250	660	FL
Ti : sapphire	Al_2O_3 : Ti^{3+}	0.66–1.2	3.2	3400	ArL, D-YAG
Alexandrite	$BeAl_2O_4$: Cr^{3+}	0.70–0.82	250	2350	FL
Cr, Tm, Ho : YAG	$Y_3Al_5O_{12}$: Cr^{3+} Tm^{3+}, Ho^{3+}	2.1	8000	1100	FL
Er : glass	Er^{3+}-doped silicate	1.54	10,000	200	FL

[a] FL, flash lamps; DL, diode laser; ArL, argon ion laser; D-YAG, double neodymium YAG laser.

shows that YAG is a strong crystal. Nd : YLF is mechanically weaker than YAG (low R_T value) because it is a fluoride rather than an oxide. The main advantages of Nd : YLF are that it generates significantly weaker thermal lensing and that it stores energy about twice as long. The main advantage of Nd : glass is that it can be fabricated in large sizes with excellent optical quality. By contrast, the size of crystals is usually much more limited. The sensitization of Nd^{3+} by Cr^{3+}, as has been accomplished for the GSGG host, results in a material that gives flash-lamp-pumped efficiencies that are about twice as large as can be obtained for Nd : YAG.

While Nd^{3+} is probably the most important laser ion, the Ti^{3+}, Cr^{3+}, Ho^{3+}, and Er^{3+} ions are also useful dopants for laser materials (see Table III). The Ti : sapphire and alexandrite crystals allow for broadly tunable output from the laser, while Ho^{3+} and Er^{3+} are capable of generating infrared light.

III. Solid-State Laser Architectures

The many different laser materials described in the previous section are complemented by a similarly large variety of oscillator and amplifier configurations. On one end of the extreme are the minilasers, which can be about the size of a sugar cube and deliver microwatts of power. On the other end of the scale is the Nova laser at the Lawrence Livermore National Laboratory with a light output pulse that delivers terrawatts of peak power, more than the total power consumption of the entire United States at any one instant (see Fig. 25). Solid-state lasers are used in a large variety of applications, and the following examples are only an indication of the possibilities. The smallest lasers are used for memory repair in integrated circuits and for a multitude of alignment tasks. As the power of the laser increases, applications such as ranging and wind velocity measurements become important. Somewhat larger lasers enable activities involving marking, medical, and military applications. Some of the most powerful systems are designed for cutting, drilling, and welding at high rates of throughput. A modern automotive production line now includes many robot-controlled solid-state lasers. The largest lasers are quite unique and serve special research purposes. On the top end of the list is the fusion driver Nova, which, as the name suggests, generates enough energy and power to initiate the same process of thermonuclear fusion which powers the stars.

Regardless of the size of the laser, one always has to start out with an active medium which is pumped by either some lamp or another laser. This creates an inversion which is then extracted by amplification of a signal or spontaneous emission. This process takes place in a resonator cavity which consists of two or more mirrors and which may contain optical switches as well. Such an assembly is referred to as an oscillator, and the basic processes are discussed by W. T. Silfvast elsewhere in this encyclopedia. There are a multitude of ways to arrange the optical switches in a cavity to change the temporal format of the laser output; several of these will be discussed in this section. One significant difference between solid-state lasers and other types is that they are storage lasers. The word *storage* refers to the fact that the typical fluorescent lifetime of an inversion is very much longer (several microseconds to milliseconds) than the time it takes for the lasing process to extract the inversion (microseconds to nanoseconds). Hence, as the active medium is excited by the pump, the inversion can be accumulated over time (it can be stored) and, by use of an electro-optic switch, can be extracted at a chosen instant. To get increased amounts of power or energy one can, up to a point, make larger oscillators. Eventually, however, one has to build external amplifiers to further increase the output of a system. We will discuss amplification toward the end of this section.

A. Methods of Pumping

1. Flash Lamp Pumping

There are three methods with which an inversion can be created in a solid state laser. The cheapest and most common method uses flash lamps (Fig. 16). A flash lamp essentially consists of a fused silica tube of suitable diameter and length with an electrode at each end. Once triggered with a short high-voltage spike, a plasma discharge occurs between the electrodes and converts the supplied electrical power with high efficiency to power radiated as light in the infrared, visible, and ultraviolet. Some of this light is absorbed by the active ion with which the host medium is doped and, by virtue of its energy levels and decay dynamics, an inversion is created. Pulsed lasers operated in a storage-type

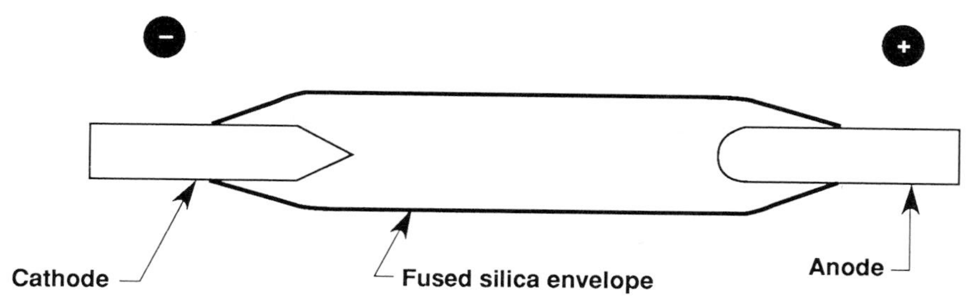

FIG. 16. The basic elements of a flash lamp. The envelope is typically made from fused silica and the electrodes from a tungsten alloy. The shape of the cathode helps to increase lamp life.

mode use different kinds of flash lamps than continuous wave (CW) or quasi-CW systems (pulse duration is long compared to the fluorescent lifetime of the laser ion). The flash lamp used for a CW Nd : YAG laser is filled with several atmospheres of Kr gas. In this case the plasma radiates around 800 nm, which is a wavelength readily absorbed by the neodymium ion. If one wants to pump a pulsed laser, the gas of choice tends to be Xe. In that case the electrical pulse applied to the lamp is on the order of or shorter than the lifetime of the upper laser level. The energy delivered to the lamp electrodes can be in the range of many kilojoules in a pulsewidth of several hundred microseconds. The plasma radiation pumping the laser is approximately described with a blackbody spectrum which extends from the ultraviolet to the infrared spectral regions. As a result, only a small fraction of it is actually absorbed by the active medium. Nevertheless, the peak power densities possible with such pulsed Xe flash lamps are sufficiently intense that high-power pulsed lasers can be constructed in this way. Excessive loading of the flash lamp will eventually result in explosive failure of the component.

2. Diode Pumping

The pumping of small solid-state lasers with laser diodes has attracted great interest within the last few years. This method allows for very efficient solid-state lasers, because the diode lasers themselves convert electrical power to radiated power very efficiently (see the article "Semiconductor Injection Lasers" in this encyclopedia), and the light output by the diode laser can be accurately tuned to the absorption line of the active medium of the solid-state laser. Although diode pumping is at present far more expensive than flash lamp pumping, it is the method of choice where efficiency is a premium, as is the case for most military applications. Diode-pumped lasers are, nevertheless, likely to become cheaper in the future through increased volume of production. Since laser diodes are limited by their peak power, the total energy output necessarily decreases with a shorter pump pulse duration. The single shot intensity of a two-dimensional diode array is on the order of several kilowatts per square centimeter. As a consequence, laser ions with a long fluorescent lifetime are easier to diode pump efficiently than ions with a short fluorescent lifetime. Presently, the most common device involves the use of GaAs diode lasers to pump the Nd^{3+} absorption band near 810 nm in various hosts.

3. Laser Pumping

It is also possible (and the diode pumping mentioned above is a case in point), to pump a laser with another laser. With the exception of diode pumping, this is mostly done for scientific applications. A typical example is the pumping of Ti : sapphire with frequency-doubled Nd : YAG lasers. Here, the upper level lifetime of Ti : sapphire (3.2 μsec) is too short to easily apply conventional flash lamp pumping techniques, whereas the wavelength of frequency-doubled Nd : YAG ideally matches the absorption band of Ti : sapphire and has a suitably short pulsewidth. Furthermore, when compared to flash lamp pumping, a pump laser can deliver considerably higher fluence at the absorption line of the active ion than flash lamps.

A quantity known as the saturation fluence F is given by

$$F = h\nu/\sigma \tag{1}$$

and is an important characteristic parameter for various aspects of laser behavior. In Eq. (1), h is

Planck's constant, ν is the frequency of the light, σ is, in this example, the absorption cross section of the active ion. This last item is the enabling factor in so-called "bleach-pumped" solid-state lasers. As a specific example, an alexandrite laser pulse may be used to pump the 745 nm absorption line of Nd^{3+} in Y_2SiO_5 with a sufficient fluence to put nearly all of the Nd^{3+} ions present in the host crystal into the upper laser level. With essentially no Nd^{3+} ions left in the ground state, the active medium becomes transparent to further pump radiation, hence the name bleach pumping. This method of pumping makes it possible to achieve efficient laser action on transitions on which the active ion would otherwise not lase due to ground state absorption. Although a new development, the number of useful wavelengths for solid-state lasers is greatly extended by this method, and outputs of 0.5 J at several hertz repetition rates have been demonstrated on transitions which could not be lased at such output levels with conventional flash lamp pumping methods.

B. Optical Switches

The aligned mirrors that surround the active medium and permit repeated passes of laser light through the inversion form the resonator cavity. The basic resonator physics is described elsewhere in this encyclopedia (see "Lasers"). Aside from the active medium one can also place a variety of electro-optic switches inside the cavity, including Pockels cells, acousto-optic switches, and saturable dyes. It is these switches which contribute to the great versatility of solid-state lasers and the different output pulse formats that they can generate.

1. Pockels Cell

Figure 17 explains the basic functioning of a very commonly used device known as the Pockels cell. Such a cell often consists of an electro-optic crystal called KD*P (deuterated potassium dihydrogen phosphate) fitted with electrodes to create an electric field within the crystal. The presence of a strong electric field (typically about 3 kV) changes the refractive index of the KD*P crystal for a particular direction of polarization. Since voltages can be switched rapidly, one can also abruptly switch the polarizing properties of the KD*P crystal. For the case of a setup with zero voltage applied to the crystal, the incident beam with vertical polarization passes through the KD*P switch unaffected; this beam will also pass through an appropriately oriented polarizer on the exit side of the crystal. With a voltage applied, however, the KD*P crystal will rotate the direction of polarization of the incident beam by 90°, so that the transmitted beam will now reflect off the polarizer (see Fig. 17). Although the intrinsic response time of the crystal is in the range of tens of femtoseconds, practical rise and fall times are limited by the circuitry to switching times on the order of nanoseconds. Hence a Pockels cell can be used as a very fast polarization switch with a good degree of synchronization between the applied high-voltage pulse and the switched light beam. A switching contrast of 1000 : 1 with a single Pockels cell and good polarizers is readily achievable.

2. Acousto-Optic Switch

Another comon switch, based on the acousto-optic effect, is sketched in Fig. 18. A piezoelectric transducer driven by a radio-frequency (typically tens of megahertz), low-voltage signal launches an acoustic wave into a block of fused silica, where the acoustic wave sets up a sinusoidal refractive index grating in accordance with the photoelastic effect. This refractive index grating then scatters the incident beam out of its original direction with good efficiency, introducing a corresponding loss in the transmitted beam. There are two typical applications of this technique. For a traveling acoustic wave of a given duration, the transmitted beam is switched off while the acoustic wave is present in the beam aperture. Turn-on and turn-off speeds are equal to the time it takes the leading or falling edge of the radio-frequency (RF) train to travel across the beam diameter (a few microseconds for typical intracavity applications). For the case of Q-switching, it is important that the acoustic wave be terminated without reflection from the surface of the quartz block in order to achieve fast switching speeds, (see Section III, C, 3). In comparing this technique to switching an optical beam with a Pockels cell, one finds that the Pockels cell method is much faster and provides better contrast but requires considerably higher voltages and also has more severe limits with respect to the maximum repetition rate that can be obtained.

The other typical application of acousto-optic

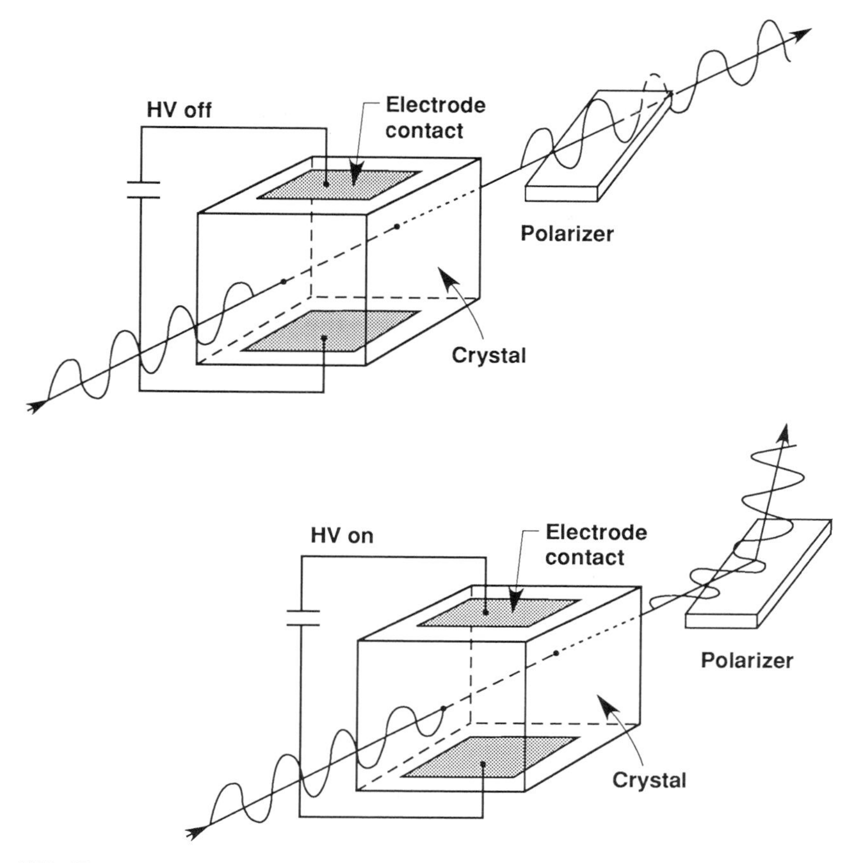

FIG. 17. The basic operation of the Pockels cell is to rotate the polarization of a transmitted beam on application of a high-voltage signal. In conjunction with one or two polarizers, Pockels cells are widely used inside of resonator cavities as Q-switches or cavity dumpers and outside of resonators as fast switchout systems.

switching is based on generating a standing (or stationary) wave in a fused silica block which is explicitly configured to form a resonant cavity for the acoustic wave. This resonant acoustic wave will then create a refractive index grating which oscillates at twice the RF drive frequency, thus periodically scattering the beam from its original direction. An acousto-optic device utilized this way inside of a laser cavity is called a mode-locker (see Section III, C, 5)

3. Saturable Absorber

The last switching method to be discussed involves the use of a saturable absorber. The optical beam enters a cell which contains a material that absorbs at the wavelength of the incident light pulse. If the beam enters above several saturation fluences for the absorption transition [see Eq. (1)], the beginning of the pulse will bleach through the cell, making it transparent for the rest of the pulse. Hence such a saturable absorber solution acts more like an intensity threshold filter than an actual switch. The cell can contain a solution of an appropriate dye, or it can be a crystal containing suitable F-centers. Switching light by this method is very popular as a simple and cheap Q-switching and mode-locking technique in the laboratory. But the statistics of the bleach process, the lack of timing control, and the chemical stability of the dyes make it unsuited for lasers which either need to be switched precisely and with good repeatability or need to be maintenance free.

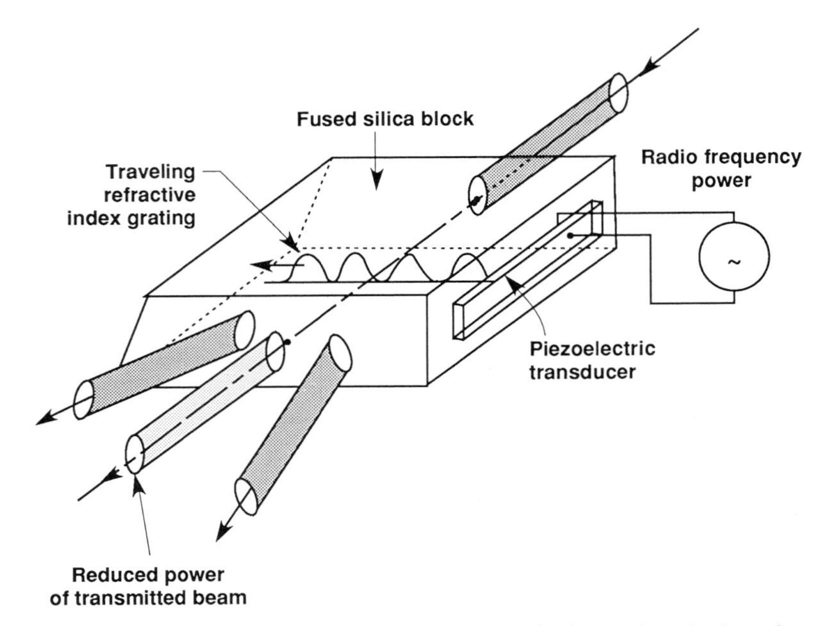

FIG. 18. The basic operation of an acousto-optic device is to reduce the intensity of a transmitted beam by diffraction from a refractive index grating which is generated, via the photoelastic effect, by the RF power supplied to the transducer. The device depicted is used for Q-switching and cavity dumping. If configured as a resonator for the acoustic wave, it serves as a mode-locker.

C. Modes of Operation

The different switches may be configured in various ways in the resonator cavity such that many modes of operating a solid-state laser become possible. A wide variety of output pulse durations and formats can be achieved, ranging in duration from picoseconds to a continuous mode of operation, although some pulse durations are easier to produce than others. The availability of different pulse durations, in conjunction with variable output energies and wavelengths, contributes to the versatility of solid-state lasers. Other intracavity elements include wavelength-tuning or wavelength-narrowing devices such as prisms, birefringent tuners and etalons, and apertures to control the transverse mode behavior. In addition, frequency converters outside the resonator cavity can double the optical frequency of output pulses having sufficient peak power. For example, a large fraction of the output of a 1.06-μm Nd^{3+} laser may be shifted to the visible wavelength of 0.53 μm. Medical applications provide an example of the diversity of the wavelength and pulse duration requirements that arise. For example, a CW output may be used at one particular wavelength to cut tissue and at another wavelength to coagulate blood, while long pulses may be used to break apart kidney stones and short pulses to spot-weld loose retinas in the eye.

1. CW Operation

The simplest mode of operating a laser is with no switches in the cavity, so that the resonator only contains the active medium. For a continuous pump, the laser operates in a CW mode. If no transverse or longitudinal mode control is implemented, the output will fluctuate due to the complex ways in which the longitudinal and transverse modes beat and couple to each other via the active medium. Other output fluctuations will come from fluctuations in the power supply. A laser with full mode control will operate only on a single longitudinal and transverse mode. After stabilizing the power supply, such lasers can have output fluctuations of less than 1%. Continuous wave lasers have many applications, depending on the degree to which they are stabilized and their output power. These range from small alignment lasers (milliwatt range) to medical lasers used for surgery (watt range) to lasers used for cutting steel (kW range).

2. Free Running

The term *free running* is generally used to describe a laser which runs CW for times that are long compared to the storage time of the laser, which is typically greater than several milliseconds. Figure 19 shows the pump pulse, the gain and loss in the active medium, and the output power. As the active medium is pumped sufficiently to exceed threshold, lasing starts with a few output spikes after which it settles down to the CW output level. The output of lasers is often temporally shaped by tailoring the current pulse to the flash lamps. The most frequent application of such lasers is for materials working, industrial drilling, and welding applications.

3. Q-Switching

This mode of operation requires the addition of a Pockels cell and a polarizer into the resonator as sketched in Fig. 17. Figure 20 shows the timing sequence of the pump pulse from the flash lamp, the impact of the Pockels cell voltage on the transmission of the resonator cavity, and the laser output. At first the Pockels cell blocks the light from traveling back and forth between the mirrors. As the active medium is pumped with flash lamp light, the inversion in the upper laser level accumulates, thereby storing the inversion until the pump pulse is nearly over. Note that far larger inversions are built up in this

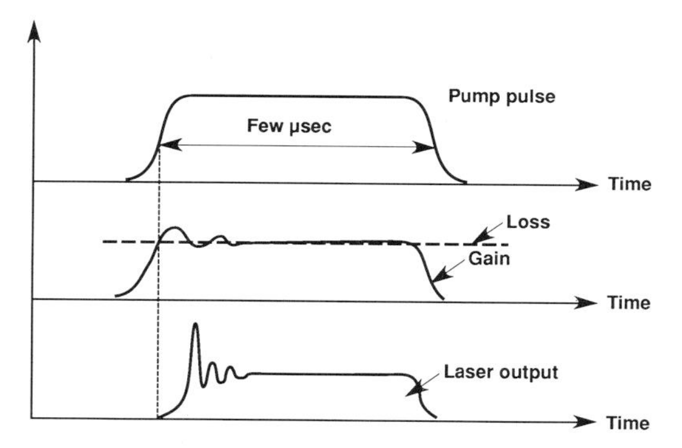

FIG. 19. Timing diagram for a free-running, pulsed oscillator. Depending on how fast the pump pulse causes the gain to rise above the loss initially, the gain and laser output will react with overshoots which will damp out. The loss level includes that of the active medium and the transmission of the resonator mirror, through which the laser light exits the resonator cavity. In CW and quasi-CW lasers the steady-state gain equals the loss, and the excess pump power is converted to laser output.

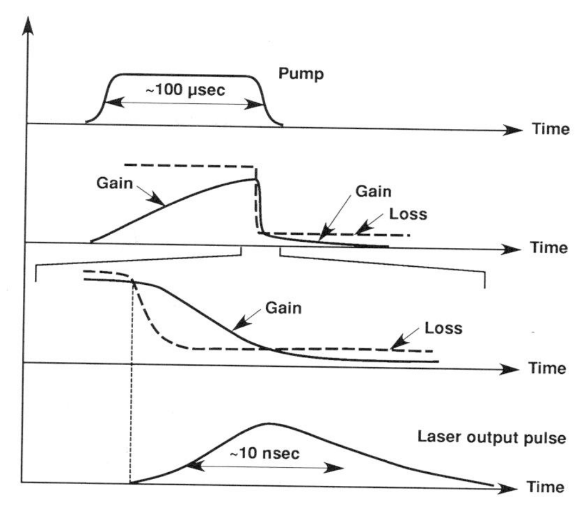

FIG. 20. Timing diagram for a Q-switched oscillator. Initially, the Pockels cell causes a high loss in the cavity so that lasing is inhibited and the pump power is integrated as inversion in the upper laser level. When the pump pulse is over, the Pockels cell loss is switched off, lowering the loss in the resonator cavity to a value corresponding again to the transmission of the mirror through which the laser light leaves the cavity. The laser pulse will build up rapidly (note the much expanded time scale) and reach the peak at the time where the gain is equal to the loss.

way than are possible for the free-running case. Once the peak inversion density is reached, the Pockels cell is switched to transmission. The net gain in the resonator cavity suddenly becomes very high so that the energy of the output pulse rapidly builds up and the inversion is extracted efficiently. Note that the pulse duration depends on the amount of inversion just before switching and that the generation of short pulses requires that a large initial inversion be stored. Typical Q-switched oscillators produce 100-mJ pulses in tens of nanoseconds.

4. Cavity Dumping

A mode of operation closely related to Q-switching is cavity dumping. The essential architecture is the same as that for Q-switching; the timing diagram is given in Fig. 21. The first phase of cavity-dumped operation is similar to that of Q-switching in that energy is stored. Both of the resonator mirrors are selected to be 100% reflective, such that the amplified light remains trapped within the cavity. As the peak intracavity intensity is reached, the Pockels cell rapidly switches the cavity transmission off again. This ejects the light circulating in the cavity by reflec-

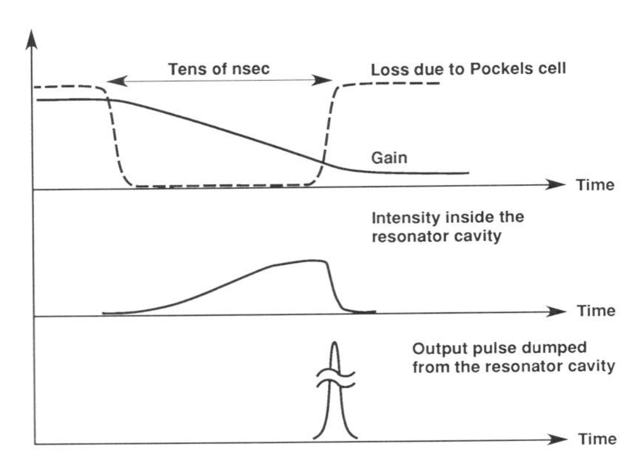

FIG. 21. In the pump phase, the timing diagram for the cavity-dumped case is identical to the Q-switched case. After switching the Pockels cell to enable lasing, the cavity loss now is very small since, for a cavity-dumped architecture, the reflectivity of both resonator mirrors is 100%. Once the intensity in the cavity has reached the maximum value, the Pockels cell is again switched and ejects the intracavity intensity from the laser in a pulse equal to the round-trip time of the resonator cavity.

tion off the polarizer in a pulse whose duration is equal to two cavity passes. This technique is used to produce pulses of a few nanoseconds duration, since the pulsewidth now depends on the length of the resonator cavity and not on the amount of inversion stored before switching. The peak power output is significantly limited in this method since the light circulating inside the resonator can become intense enough to destroy critical components of the laser.

An architecture very closely related to the cavity-dumped oscillator is the regenerative amplifier. The principal difference is that the laser oscillation in a regenerative amplifier does not build up from spontaneous emission but is initiated by a signal injected into the resonator from the outside as the Pockels cell is switched to transmission. This injected signal is then trapped in the cavity and amplified until it has reached maximum intensity, at which point it is ejected (dumped) from the cavity.

5. Mode-Locking

By inserting an acousto-optic mode-locker in the cavity it is possible to produce very short pulses in an oscillator. With a mode-locker in the cavity and the transverse modes suitably constrained, the oscillator can be operated quasi-CW to produce a steady stream of short pulses. It is also possible to add a Pockels cell and a polarizer to produce an output which has the pulse envelope of a Q-switched pulse but which is composed of individual short pulses from the mode-locking process. External to the cavity one can then pick out a single pulse by placing an additional Pockels cell between two polarizers and applying a short high-voltage pulse at just the instant when the desired pulse is at the Pockels cell. Such an arrangement is commonly called a *single-pulse switchout*. This is a standard way to produce individual pulses with durations from 100 psec to 1 nsec and energies on the order of 100 μJ. [*See* MODE-LOCKING OF LASERS.]

D. Types of Oscillators

In this section representative examples of frequently used solid-state lasers are described, starting with the smallest lasers and proceeding to lasers of increasing power and energy. These lasers are practical embodiments of the pumping and operating schemes described above.

1. Minilasers

Figure 22 is a sketch of a generic diode-pumped minilaser. The diode or array of diodes is focused with suitable optics through one of the resonator mirrors into the active medium (typically Nd:YAG). This "end-pumping" method allows the precise matching of the pumping volume to the lasing volume of the lowest order transverse mode, thereby contributing to the good efficiency of minilasers. Output powers are typically in the milliwatt range. The mode of operation ranges from CW to repetitively Q-switched (requiring the addition of a small Pockels cell to the resonator cavity). These lasers offer a stability of a few percent and are extremely rugged; the entire device fits comfortably in the palm of the hand.

2. Watt/Joule Level Laser

This heading encompasses lasers with outputs in the range of a few watts, and with pulsed energies of 0.1 to 1 J. Continuous wave systems as well as pulsed systems are characterized by a great variety of resonator configurations and make up the majority of solid-state lasers sold today. These lasers typically contain a rod with a diameter on the order of 6 mm and a length of 10 cm, which is pumped with one or two flash lamps. The resonator cavity is generally 30 cm to 1.5 m long, depending on whether there is a mode-locker, a Pockels cell, or another combi-

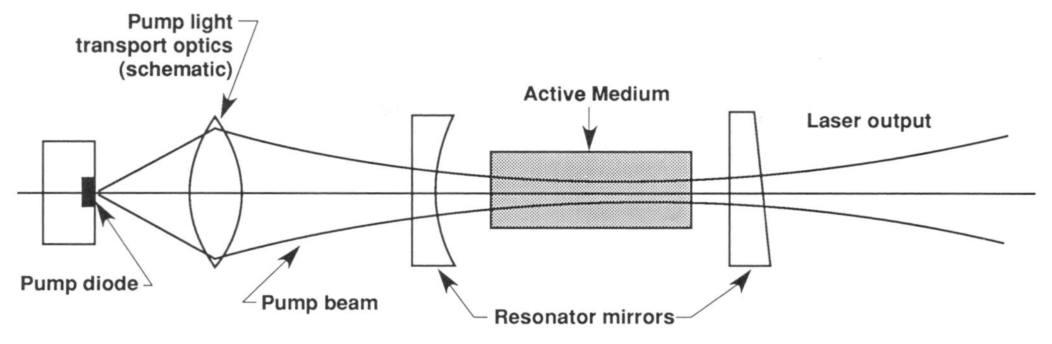

FIG. 22. The essential elements of a diode-pumped minilaser as explained in the text. In practice, the pump light transport optics is a sophisticated array of lenses to change the spatial characteristics of the diode output to those optimally suited to pump the minilaser.

nation of switches. The cavity may be made from low thermal expansion materials for more demanding applications. The power supply operates in the kilowatt range and typically is a floor unit occupying a few cubic feet of space. A photograph of a representative 200-W system appears in Fig. 23. Various CW versions are used for medical applications, such as cutting tissue and coagulating blood, and for marking and scribing in industrial settings. The CW mode-locked laser is an important component of many ultrafast pulse setups. Free-running, pulsed, and Q-switched systems have other uses in the medical field (e.g., breaking up kidney stones without intrusive surgery). Industrial applications include semiconductor processing and

FIG. 23. A 250-W Nd : YAG laser used in the semiconductor industry. (Courtesy of Quantronix Corp.)

working thin metal films. Many lasers for scientific applications operate in this range as well. Commercial laser companies offer a variety of resonators and modes of operation for the scientific market, and further variations can be seen in the laboratory.

3. Materials Working Lasers

These lasers operate mostly in the CW or free-running pulsed mode and deliver output powers up to and exceeding 1 kW. They are configured as oscillators but often have two or three laser rods in the resonator cavity. The resonator cavity is still approximately 1 m long, although the power supply may be significantly larger. Furthermore, these lasers, as well as some of the smaller lasers described above, often incorporate sophisticated beam delivery systems which allow the use of one central laser system on several work stations. Larger systems are used for demanding welding and drilling application (e.g., components for jet engines).

A more recent development is the slab laser (Fig. 24). As opposed to the rod-shaped active media typically used in conventional solid-state lasers, the active medium of slab lasers is in the form of a thin plate or slab. The beam enters and exits through tilted surfaces and zigzags through the active medium in such a way that the optical distortions, which a beam experiences by passing through the material, can largely be eliminated. A slab laser can generate increased output power by employing a larger volume of the active medium, and the beam that is delivered can be brought to a tighter focus. Slab lasers involve a number of complex technological issues but have experienced a large development effort in recent years. A slab laser with 1-kW output has been demonstrated by General Electric Corporation.

E. Laser Scaling

We have thus far primarily described laser oscillators. To achieve higher energies, amplification of the pulse from the oscillator is required. The regenerative amplifier was already mentioned in the section on cavity dumping. In comparison to a single-pass amplifier, the regenerative amplifier is a far more complex setup, although it provides for very efficient extraction of energy from the medium and is favored for short pulse amplifiers in the laboratory. To get to higher energy, further amplification stages are required. One would, of course, like to make the most efficient use of the inversion stored in such an amplifier, while avoiding the potentially destructive consequences of propagating a high peak power pulse through the amplification medium. Consider a pulse making a single pass through an amplifier in which an inversion is stored. The key parameter to describe the amplification process is again the saturation fluence, of Eq. (1), where σ is now the stimulated emission cross section (rather than the absorption cross section, as noted earlier). If the input

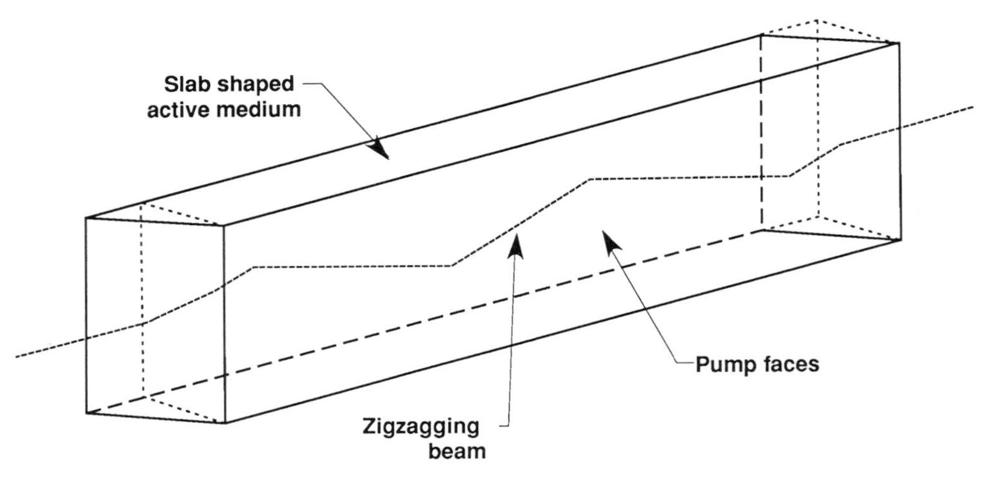

FIG. 24. The basic idea behind a slab laser is to zigzag the beam through the active medium and thus greatly reduce the thermal distortions of the beam. The dotted lines are meant to indicate where perpendicular edges would lie. The actual slab laser has entrance and exit surfaces which are inclined with respect to the incident beam.

fluence to the amplifier remains small compared to the saturation fluence of the transition (called the small signal gain regime), the amplification G proceeds in an exponential manner according to:

$$G = \exp(N_i \sigma x) \qquad (2)$$

where N_i is the inversion density and x is the pathlength in the amplifier. In this regime the amplification factor G is high but the extraction of the inversion is consequently low, since saturation of the inversion does not occur. The other extreme is that the input fluence is much larger than the saturation fluence. In this case extraction of the inversion will approach 100%, and the energy stored in the amplifier is simply added to the energy in the input beam. Generally, the best place to operate is in between the two extremes, where there is a desirable degree of signal amplification at an acceptable extraction efficiency of the energy originally stored in the amplifier. The basic equation which describes this amplification process is the Frantz–Nodvik equation.

There are a variety of effects that limit the energy which may be generated in a large-scale amplifier chain. First, the amplifier medium cannot be made arbitrarily large since it will lose its storage capability. Spontaneous emission, generated by fluorescence decay of the inverted ion population inside the amplifier, can be trapped inside a large amplifier module and deplete the inversion before the extraction beam arrives. This process is called *parasitic oscillation*. Second, as the peak power in the amplified pulse increases its intensity, it becomes large enough that the electric field of the amplified light wave itself changes the refractive index of the medium through which it travels. This can induce a process called *self-focusing*, where the amplified beam destroys the active medium by creating tracks and bubbles inside the amplifier. Finally, amplifier surfaces and other optics are often covered with dielectric coatings to increase, or decrease, their reflectivities. These coatings can be destroyed by a light pulse which has too high a peak power, often at power levels below that at which self-focusing sets in. The resistance of such coatings to this effect is described by a *damage threshold* which characterizes the fluence the coating can withstand at a given pulse duration.

FIG. 25. Nova, the biggest laser in existence. Note the people and automobile shown for scale.

As the ultimate example of a laser which can only be built if all of these types of processes are well understood, Fig. 25 shows the Nova laser at the Lawrence Livermore National Laboratory. It has delivered pulses with 120,000 J of energy in durations of several nanoseconds.

IV. Future Directions

There are three clear areas where solid-state laser architectures will advance significantly in the next few years. The first is to take diode-pumped lasers from the minilaser scale to the level of tens of watts. This requires development of high-average power, two-dimensional diode arrays, which is now well underway. The second is to utilize slab laser technology to operate lasers at the 1–10 J/pulse level at repetition rates of about 100 Hz. Finally, large fusion lasers may ultimately operate at several hertz, and produce average output powers in the megawatt regime.

The new laser materials that are developed will be deployed in systems to generate new wavelengths, operate more efficiently, be produced at lower cost, and have optical properties that are tailored to meet specific technical objectives. Solid-state lasers that operate efficiently in the ultraviolet-blue region are likely to be developed. Many new transition metal lasers having optical properties that are useful may be discovered in the next decade. One change that may occur will involve the advent of tailoring some types of laser materials for a specific application. (This has already occurred in the case of Nd-doped glasses for fusion lasers.) This process of designing, rather than discovering, laser materials will be enhanced by a better understanding of the physics and chemistry of solid-state media.

Acknowledgment

This research was performed under the auspices of the Division of Materials Sciences of the Office of Basic Energy Sciences, U.S. Department of Energy, and the Lawrence Livermore National Laboratory under Contract No. W-7405-ENG-48.

Bibliography

Adair, R., Chase, L. L., and Payne, S. A. (1989). Nonlinear refractive index of optical crystals, *Phys. Rev. B.* **39,** 3337.

Albrecht, G. F. (1990). Average power slab lasers with garnet crystals as the active medium, *in* "YAG and Other Garnet Lasers" (De Shazer, ed.). Wiley, New York.

Ballhausen, C. J. (1962). "The Theory of Transition-Metal Ions." McGraw-Hill, New York.

Belforte, D., and Levitt, M. (1989). "The Industrial Laser Annual Handbook." PennWell, Tulsa, Oklahoma.

Berger, J., Welch, D. F., Scifres, D. R., Streifer, W., and Cross, P. S. (1987). High power, high efficiency Nd : YAG laser end pumped by a laser diode array, *Appl. Phys. Lett.* **51,** 1212.

Brown, D. C. (1981). "High Peak Power Nd : Glass Laser Systems," Springer Series in Optical Sciences, Vol. 25. Springer-Verlag, New York.

Dieke, G. M. (1968). "Spectra and Energy Levels of Rare Earth Ions in Crystals." Wiley (Interscience), New York.

Izumitani, T. S. (1986). "Optical Glass." American Institute of Physics, New York.

Kaminskii, A. A. (1981). "Laser Crystals." Springer-Verlag, New York.

Koechner, W. (1988). "Solid State Laser Engineering," Springer Series in Optical Sciences, Vol. 1. Springer-Verlag, New York.

Krupke, W. F., Shinn, M. D., Marion, J. E., Caird, J. A., and Stokowski, S. E. (1986). Spectroscopic, optical and thermo-mechanical properties of neodymium- and chromium-doped gadolinium scandium gallium garnet, *J. Opt. Soc. Am. B* **3,** 102.

Moulton, P. F. (1986). Spectroscopic and laser characteristics of Ti : Al_2O_3, *J. Opt. Soc. Am. B* **3,** 4.

Payne, S. A., Chase, L. L., Newkirk, H. W., Smith, L. K., and Krupke, W. F. (1988). $LiCaAlF_6 : Cr^{3+}$: a promising new solid-state laser material, *IEEE J. Quantum Electron.* **QE-24,** 2243.

Pfaender, H. G., and Schroeder, H. (1983). "Schott Guide to Glass." Van Nostrand Reinhold, New York.

Siegman, A. E. (1986). "Lasers." University Science Books, Mill Valley, California.

Walling, J. C., Heller, D. F., Samelson, H., Harter, D. J. Pete, J. A., and Morris, R. C. (1985). Tunable alexandrite lasers: Development and performance, *IEEE J. Quantum Electron.* **QE-21,** 1568.

Weber, M. J., ed. (1982). "Handbook of Laser Science and Technology." CRC Press, Boca Raton, Florida.

Yariv, A. (1975). "Quantum Electronics." Wiley, New York.

SPECKLE INTERFEROMETRY

Harold A. McAlister *Georgia State University*

GLOSSARY

Airy pattern: Diffraction pattern produced when light from a point source passed through a circular aperture shows a bright central peak surrounded by a series of decreasingly bright concentric rings. The central bright peak is known as the Airy disk.

Aperture: Diameter of the mirror or lens that first collects light in a telescope and forms an image of the object under observation.

Coherence: Property possessed by two or more beams of light when their fluctuations are highly correlated.

Diffraction: Change in direction of a ray of light when passing an obstacle or a change that occurs when passing through some aperture because of the wave nature of light.

Dispersion: Differential bending of light of different wavelengths by a refracting medium. Atmospheric dispersion occurs when light from a star is observed at angles other than directly overhead.

Interference: Combination of two or more coherent beams of light that leads to the reinforcement of intensity where wave crests overlap and the cancellation of intensity at locations where wave crests overlap with wave troughs.

Isoplanatic angle: Angular size of a single cell of atmospheric coherence is determined by the physical size of the cell and its elevation.

Photoelectric effect: Emission of electrons from certain materials that occurs when light of a wavelength less than some critical value strikes the material.

Pixel: Picture element or the smallest resolution element within an image. It may be a light-sensitive grain in a photographic emulsion or a small rectangular element in an electronic detector.

Rayleigh limit: Theoretical diffraction limit to angular resolution that occurs when the bright central peak of the Airy pattern of a point source is located in the first dark ring of the Airy pattern from a second point source. This limit in radians is given by $1.22l/D$ where l is the wavelength and D is the aperture.

Seeing: Quality related to the blurring of star images by atmospheric turbulence usually expressed in the angular size of a blurred image.

Troposphere: Lowest region of Earth's atmosphere extending to an elevation of about 11 km and in which weather phenomena occur.

Wave front: Surface that is at all points equidistant from the source of light and from which light rays are directed in a perpendicular manner.

Wavelength: Distance between successive wave crests or any other repetition in phase in a beam of light. Visible wavelengths are from about 400 to 650 nm, while infrared wavelengths begin beyond 650 nm.

Galileo first used the telescope for astronomical purposes nearly four centuries ago, and the subsequent history of astronomy is closely tied to the quest for telescopes of ever-increasing power. The fundamental capability of a telescope is determined by its aperture. The aperture area determines the amount of light that a telescope can collect and hence sets an effective limit to the faintest observable objects. Light gathering power has been the primary motivation behind the construction of large telescopes as astronomers seek to understand the most distant and faintest components of the universe. The largest telescopes now in existence can collect millions of times the amount of light in comparison with the human eye, and extensive engineering studies have shown ways to produce affordable giant telescopes that will significantly enhance light collecting ability. The second capability tied to aperture is angular resolution, the detection of fine structural detail in images. Whereas impressive gains in light gathering power have been achieved since Galileo, blurring produced by Earth's atmosphere has set a limit to resolution equivalent to that achievable by an aperture of only about 10 in. Not until 1970 was a method discovered that has allowed astronomers to attain the full theoretical resolution of large-aperture telescopes. This method is known as speckle interferometry. Speckle techniques are providing a wealth of information for wide classes of astronomical objects, including the Supernova 1987A that exploded in the Large Magellanic Cloud.

I. Astronomical Seeing and Resolution

Earth's atmosphere places serious limitations upon astronomical observations for two primary reasons. First, the atmosphere is not equally transparent at all wavelengths, and certain wavelength regimes are completely inaccessible from the ground. Observations from telescopes orbiting above the atmosphere have to a certain extent circumvented this obstacle. Second, turbulence within the atmosphere produces image blurring that seriously degrades the ability of telescopes to resolve detail in images. Beginning in the late nineteenth century, astronomers realized the importance of locating observatories at sites with exceptionally stable air in order to achieve the best possible seeing conditions. It is fortunate that the properties that lend favorable astronomical seeing are also consistent with transparency, and modern observatories are typically located at relatively high elevations in very dry climates. [*See* TELESCOPES, OPTICAL.]

The intrinsic limiting ability of a telescope to resolve fine angular detail is set by the diffraction properties of light. For a telescope such as the 4-m aperture Mayall reflector on Kitt Peak in southern Arizona observing at a wavelength of 550 nm, the center of the visible region of the spectrum, the Rayleigh limit is approximately 0.035 arcseconds, an angle equivalent to that subtended by a nickel seen from a distance of 75 miles.

Unfortunately, the atmosphere thwarts the realization of such resolution and imposes an effective limiting resolution from 1 to 2 arcseconds, a degradation in resolution by a factor of roughly 50. Some locations on Earth offer seeing that is occasionally as good as 0.2 to 0.3 arcseconds, but even these rare and superb seeing conditions are an order of magnitude worse than what would be obtained under ideal circumstances. One obvious option is to put telescopes into orbit above the atmosphere. Indeed, a primary justification for the 2.5-m Hubble Space Telescope (HST) has been its ability to completely avoid atmospheric blurring. For the foreseeable future, ground-based telescopes will continue to be built with apertures significantly larger than their far more expensive space-borne counterparts. The 10-m Keck telescope now under construction on Mauna Kea in Hawaii and an entire family of 8-m-aperture telescopes being planned by several institutions throughout the world offer a significant gain in resolution over the HST. A great deal of interest therefor exists among astronomers to find ways to circumvent the atmospheric blurring of images. In 1970, the French astronomer and optical physicist Antoine Labeyrie pointed out an elegantly simple method for circumventing atmospheric seeing conditions to achieve diffraction limited resolution. Labeyrie's method, which he named speckle interferometry, takes advantage of the detailed manner in which the blurring occurs in order to cancel out the seeing induced effects.

The atmosphere is a turbulent medium with scales of turbulence ranging from perhaps hundreds of meters down to turbulent eddies as small as a few centimeters. The turbulence arises from the dynamics of the atmosphere as driven by Earth's rotation and the absorption of solar radiant energy, which is converted into the thermal energy content of the atmosphere. Turbulence alone does not induce "bad seeing," rather it is the variation in density from one turbulence region to another that causes rays of light to be refracted from otherwise straight paths to the telescope.

Light from a star spreads out in all directions to fill a spherical volume of space. The distances to stars are so great that a typical star can be envisioned as a point source illuminating a spherical surface on which the telescope is located. Because the radius of this imaginary sphere is so enormously large in comparison with the telescope aperture, the light entering the telescope at any instant can be pictured as a series of parallel and plane wave fronts. Equivalently, all rays of light from the star to the telescope can be considered as parallel rays perpendicular to the incoming wave fronts. To this simple picture, we must add the effects of the atmosphere.

A telescope accepts the light from a star passing through a cylindrical column of air pointing to the source and having a diameter equal to the telescope's aperture. If the column of air were perfectly uniform, the incoming wave fronts would remain flat, and a perfect Airy pattern could be formed. The density variations accompanying turbulence exist at elevations throughout the troposphere. The cumulative effect of these fluctuations can be modeled as being equivalent to patches across the telescope entrance aperture, or pupil, such that within one such patch, rays of light remain roughly parallel (or alternately, the wave front remains nearly flat). A given patch, or coherence cell, will produce some net tilt of the parallel bundle of rays and will retard the entire bundle by some amount referred to as a piston error. A telescope whose aperture is stopped down to match the diameter of these cells, a quantity commonly referred to as r_0, would produce an instantaneous image in the form of an Airy pattern. From one instant to the next, a given r_0-size cell moves because of winds and will even dissolve on a slightly longer time scale because of the dynamics of turbulence.

A telescope with an aperture larger than r_0 will at any time contain $(D/r_0)^2$ coherence cells, each of which produces some random tilt and piston deviations on its bundle of rays from the mean of these deviations. At any instant, there will be some fraction of these deviations arising from points distributed randomly throughout the aperture that have nearly identical tilt and piston errors. The light from these coherent subapertures undergoes interference to produce a fringe pattern that shows regions of brightness and darkness. These bright regions are called speckles, and each speckle is in essence a distorted or noisy version of an Airy pattern. The entire distribution of speckles at any instant fills a region whose size corresponds to the Airy disk of a single r_0-size aperture. Under typical seeing conditions, the coherence cell size r_0 is from 8 to 15 cm, with seeing conditions degrading as r_0 becomes smaller. When $r_0 = 15$ cm, the seeing disk diameter will be about 1 arcsecond. The twinkling of starlight as observed by the unaided eye is produced by the rapid passage of individual coherence cells, each of which is significantly larger than the pupil of the eye, across the line of sight with the resultant apparent rapid motion of the star arising from the random tilts from each successive cell. Planets do not appear to twinkle because their disks are sufficiently extended in angular size to average out the tilts from a number of cells at any instant.

The rapid motion and dissolution of seeing cells requires the use of short exposure times in order to record a speckle pattern. For exposures longer than the atmospheric redistribution time t_0, speckle patterns will blur into the classic long-exposure image of a star in which the image profile intensity drops off in a Gaussian-like manner to fill the arcsecond-scale seeing disk. Experience has shown that exposure times no longer than about 0.01 sec are typically sufficient to freeze the speckles. Atmospheric conditions vary considerably from place to place and from time to time, and values of t_0 less than 0.001 sec have been encountered. Exposures on such a short time scale will permit the detection of so few photons, even from a bright object with a large telescope, that speckles cannot effectively be recorded. Fortunately, such rapid seeing conditions are rare.

II. Speckle Camera Requirements and Technology

The first step in Labeyrie's method of speckle interferometry is to record the speckle images from large telescopes using specially designed cameras. Several factors tend to reduce the amount of light available to a speckle camera. Because speckles result from the interference of overlapping wave fronts of light, their sharpness or contrast is related to the size of the wavelength region over which the interference occurs. The production of useful speckle requires that the recorded wavelengths be restricted to ranges of no more than a few tens of nanometers, the remaining light transmitted by the telescope being rejected by filters that transmit only in some preselected wavelength region. Therefore, as much as 95% of the otherwise available light from the object must be filtered out prior to recording speckles. The necessity for exposure times shorter than t_0 provides a weak level of illumination for speckle imagery in comparison with classical astronomical imagery where exposures of many minutes or even hours are commonly used to integrate light. Finally, because each speckle must be resolved by the speckle camera system, a very high magnification is required in order to sample each speckle with more than a single pixel on the camera detector. The high magnification leaves few photons available per speckle during a sub-t_0 exposure time. Clearly, speckle cameras must employ very sensitive detectors in order to record speckle patterns from any but the very brightest stars.

A speckle camera system is schematically illustrated in Fig. 1. Light from the telescope comes to a

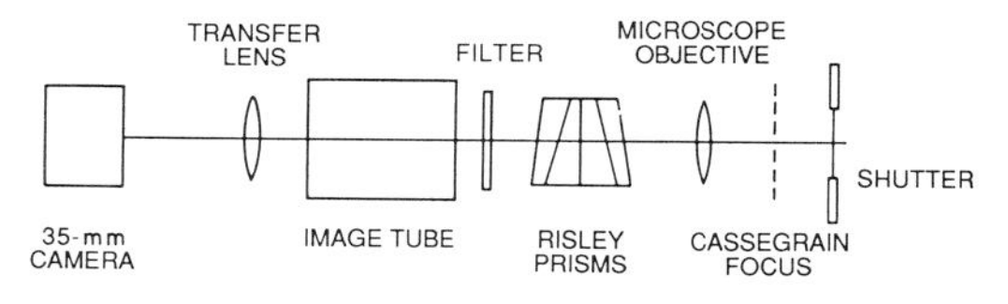

FIG. 1. A schematic view of a simple speckle camera system.

focus just in front of a microscope objective that serves the purpose of increasing the effective magnification of the telescope by a factor of perhaps 20 to 30. The beam from the microscope objective is very slowly converging or can be made collimated to be brought to a focus later by a lens. An interference filter provides the required spectral filtering. For telescope apertures larger than about 1.5 m, the Airy disk size is small enough so that spectral dispersion from the atmosphere produces noticeably elongated speckles, even over the rather narrow spectral regions used. Risley prisms provide a useful means for introducing dispersion that can be adjusted to the appropriate amount in the direction opposite to that arising from the atmosphere to cancel out this effect. The magnified, spectrally filtered, and dispersion-compensated beam then passes through a shutter before striking the detector.

A speckle camera detector is often a composite of several elements, the first of which is an electronic device for intensifying the light from the star to a level sufficiently bright for the detection stage to record. The input side of an image intensifier tube converts the arriving photons into electrons by means of the photoelectric effect. These photoelectrons are then accelerated down the tube by a high-strength electric field to strike a phosphor and produce a glowing spot of light that may be thousands or even a million times brighter than the original group of photons causing their production. The phosphor may then be imaged by a video-type device or by any of a variety of electronic detectors that can convert the phosphor output into a digital record of the (x, y) locations of all arriving photons and their intensities. The intensifier tube may itself be electronically gated, eliminating the need for a mechanical shutter. Only with such sophisticated image intensifiers has speckle interferometry been able to play a useful role in astronomy. A speckle picture of a bright star is shown in Fig. 2. Had Labeyrie's idea occurred fully to the previous generation of astronomers who had only photographic emulsions at their disposal, they would have been frustrated by the lack of technology that would enable speckle interferometry to be carried out beyond a demonstration stage.

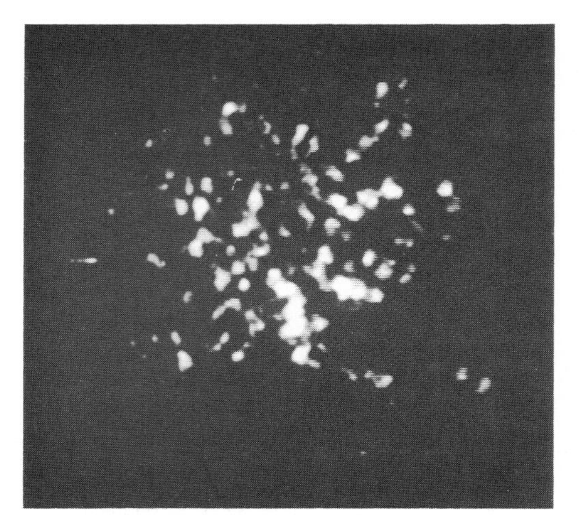

FIG. 2. A speckle image of a single bright star shows the discrete speckles, each of which is a representation of an Airy disk corresponding to the 4-m aperture of the telescope at which this picture was taken.

III. Application to High-Resolution Imaging

Labeyrie showed mathematically that the production of speckles by a telescope and the atmosphere is mathematically equivalent to the convolution of the point-spread function (PSF) of the atmosphere with the object intensity distribution on the sky. The so-called convolution theorem states that the Fourier transform of a convolution is equal to the multiplicative product of the individual Fourier transforms of the two quantities to be convolved. This fact is the essence of speckle interferometry. If, for example, speckle images are obtained of one of the handful of the highly evolved and rare supergiant stars that are near enough to the Sun so that their angular diameters are larger than the Airy disk of a telescope, then the convolution theorem shows that the star diameter can be deconvolved from the speckle data by dividing the Fourier transform of speckle images of the supergiant star by the transform of speckle images from an unresolved star to cancel out the effects of the atmospheric PSF.

Whereas nature provides only a few supergiant stars resolvable by speckle methods with the largest existing telescopes, there are thousands of binary star systems suited to the technique. The two stars comprising a binary system are bound in orbit around a common center of mass by their mutual gravity and may be so close together or the system may be so far from the Sun that the angular separation of the components is smaller than the seeing disk or even smaller than the Airy disk. For binaries with angular separations in the regime of 0.03 to 0.3 arcseconds, speckle interferometry currently provides the best method for accurately measuring their orbital motions. Such measurements lead to the determination of stellar masses, quantities that are relatively rarely known and yet play a vital role in our theoretical understanding of the origin and evolution of stars.

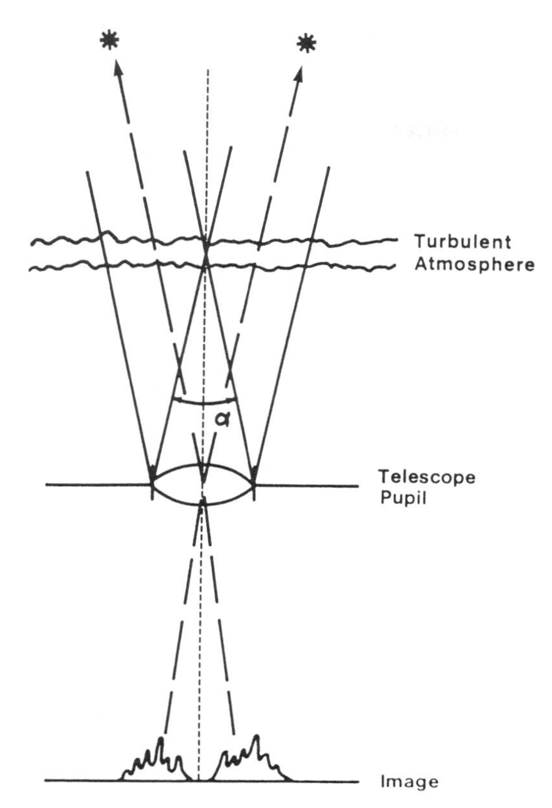

FIG. 3. The condition of isoplanicity is shown for a pair of stars separated by the angle at which the separate beams of light are just no longer passing through any common atmospheric turbulence. The speckle patterns in this case would be uncorrelated or nonisoplanatic.

The two speckle patterns arising from two stars in a binary system are highly correlated as long as the light from the two stars passes through the same collection of coherence cells, a condition known as isoplanicity and illustrated in Fig. 3. The two stars then give rise to speckle patterns that are very nearly identical but are displaced from each other by an amount equal to the angular separation of the binary. The isoplanatic angle is typically a few arcseconds so that binary stars with angular separations less than this amount will produce speckle patterns with very high point to point correlation. Figure 4 is a speckle image of a widely separated binary star in which the high degree of correlation between the speckle patterns from the two stars in the system is obvious. For systems closer in separation than that in Fig. 4, the two speckle patterns will merge together and overlap in such a way that every speckle will be doubled, an example of which is shown in Fig. 5. Thus, a speckle image of a binary star produced at a 4-m-aperture

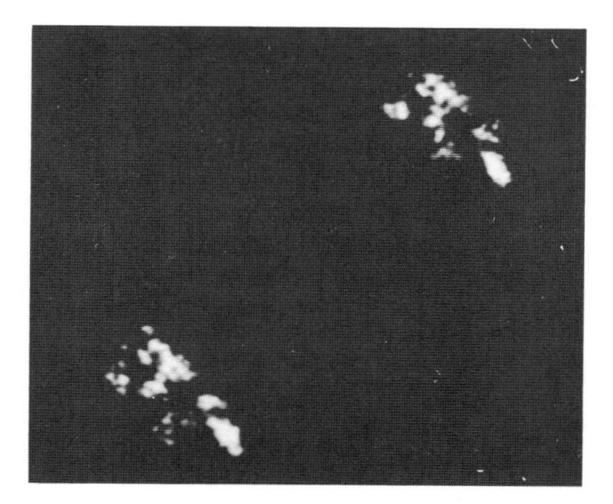

FIG. 4. This 2-arcsecond-separation binary shows the very high correlation between the separate speckle patterns of the two stars indicative of isoplanicity.

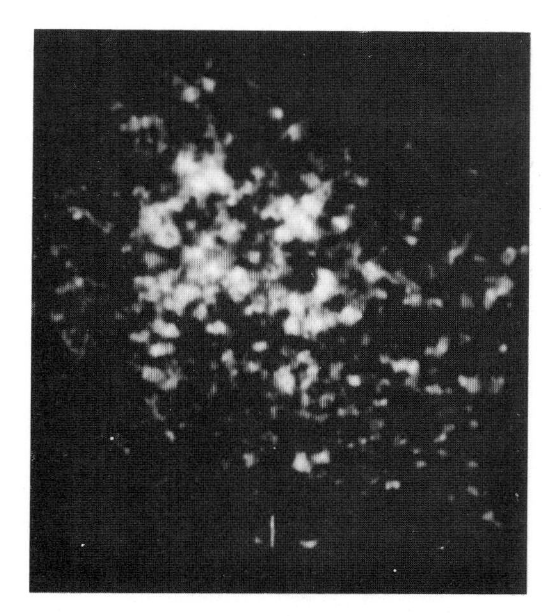

FIG. 5. A speckle photograph of a binary star having an angular separation of 0.25 arcsecond shows an apparently single speckle pattern caused by the complete overlap of the patterns from the two stars.

telescope will contain hundreds of individual representations of the binary star geometry.

A simple method of analysis of binary star speckle data is provided by the method of vector-autocorrelation. A vector-autocorrelogram (VAC) can be produced by plotting all the pairings among speckles that occur within a speckle image. Because the pairing corresponding to the actual binary star pair occurs very frequently compared with every other random pairing among any two speckles in an image, the binary star geometry stands out in a VAC. The individual VACs from many speckle images of a binary can be added together to further increase the relative contribution from the binary star pairing. A VAC possesses a bright central peak because of the pairing of every speckle with itself. This central peak is accompanied by two identical peaks on opposite sides, one of which is produced by the sum of all speckle pairings of star 1 with respect to correlated speckles from star 2, while the other results from pairings of star 2 with respect to star 1. This simple algorithm can operate sufficiently rapidly in a computer so that data can be reduced in real time as they are produced by a digital speckle camera system. A vector-autocorrelogram of approximately 1800 speckle images of the same binary as is exemplified in Fig. 5 is shown in Fig. 6. The binary peaks in a VAC are superimposed upon a background that gradually slopes away from the central peak. This background results from all the uncorrelated speckle pairings and has a radius determined by seeing. It can be subtracted using a VAC produced under the same seeing conditions but for a single star.

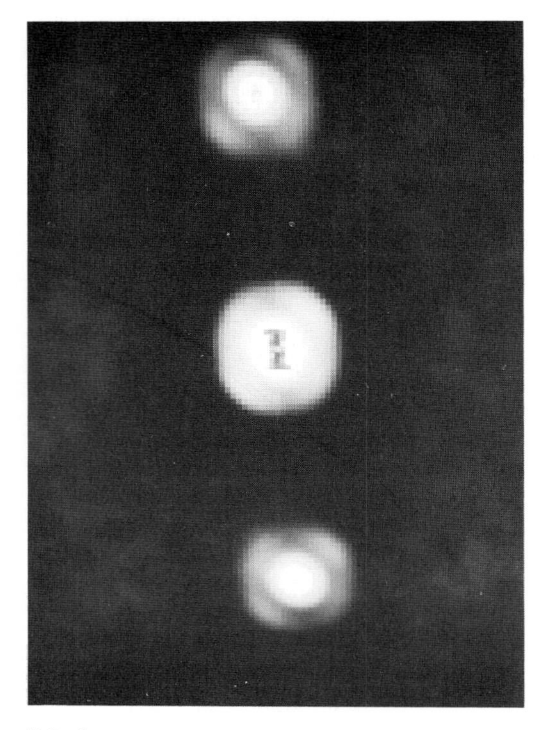

FIG. 6. The computer-generated vector-autocorrelogram of approximately 1800 speckle images of the same system as shown in Fig. 5 shows the characteristic central peak accompanied by identical peaks arising from the binary star geometry.

Deconvolution methods, such as the one described previously, provide an effective means for studying objects that have simple geometries, such as single stars whose diameters are resolved or binary star systems describable only by the angular spacing of the two components and their relative orientation. However, for more complex objects, it is necessary that actual images of these objects with diffraction limited detail be reconstructed from the speckle data. A fundamental problem in such reconstructions is that the atmosphere randomly distorts the relative phase of the wave front to such an extent that the real phase information is lost in a speckle image. To produce a real image requires the incorporation of these missing phases and the amplitudes of the incoming waves. Algorithms have been developed that work with phases or amplitudes, typically in an iterative approach, to attempt image reconstruction, and a number of interesting examples exist to show the promise of these methods.

IV. Astronomical Results from Speckle Interferometry

A. Stellar Angular Diameters

The largest existing telescopes have apertures that just begin to yield resolutions capable of resolving the diameters of stars, and even so, only in the case of the nearest supergiants. Thus, speckle interferometry has been able to measure such stars as Betelgeuse in the constellation Orion and the half dozen or so other supergiants closest to the Sun. Speckle image reconstructions for Betelgeuse have shown this star, whose surface would extend beyond the orbit of Mars if Betelgeuse were to replace the Sun, to be nonuniformly luminous and to be surrounded by a very extended shell of gas and dust. One and possibly two faint companions have been discovered orbiting about Betelgeuse by speckle interferometry. Although such diameter measures are rare, an important result is that speckle measures have been consistent with results obtained by a variety of other techniques.

B. Binary Stars

As a class, binary stars are ideally suited to speckle interferometry. Suitable candidates for measurement exist in almost unlimited supply, and speckle methods are providing hundreds of new discoveries while measuring previously known systems with greatly enhanced accuracy over classical methods. There currently exists approximately 10,000 speckle measurements of some 2000 binary star systems, many of which have been measured through sufficient orbital motions to permit the calculation of the parameters describing these motions. These orbital elements lead to the determination of the masses of the components, quantities that can then be used to confirm or improve theoretical models of stellar structure, formation, and evolution. Because of the increased resolution of speckle interferometry over other methods, speckle surveys for new binaries are able to penetrate into separation regimes not previously detectable. These surveys, although rather limited in extent because of the strong competition for time on the largest telescopes, are supporting already existing evidence that the majority of stars in our galaxy exist in binary or multiple star systems of higher complexity.

C. Infrared Speckle Interferometry

A particularly promising area for speckle interferometry and image restoration is the application of these methods at infrared (IR) wavelengths, where the atmosphere is more benign than at optical wavelengths and where there exist classes of objects of moderate complexity that are ideal candidates for high-resolution imaging. As wavelengths increase, both t_0 and r_0 increase in size, and the observational requirements are relaxed. The first IR speckle observations were made with single pixel detectors across which were scanned at high speeds the images of objects to be analyzed. This approach has been used to measure the heated dust shells surrounding such supergiants as Betelgeuse and Antares and other hot and highly evolved stars. The sizes of protostellar objects from which normal stars will eventually evolve have been measured. These objects are typically enshrouded in dense dust clouds that obscure the visible radiation while radiating at IR wavelengths because of heating from the central hot-star-forming gas. IR sources discovered by standard methods have been found to be highly complex, and very faint and cool companions have been found in orbit around a number of stars. The star T Tauri, the prototype of a class of stars thought to represent the transition between protostars and normal hydrogen burning stars, has been found to have a companion. This provides a rare opportunity to study the circumstances surrounding the formation of a binary star system.

IR speckle methods are now entering into a period of potentially great productivity with the advent of extremely sensitive solid-state detectors with full two-dimensional pixel coverage. These powerful new devices combined with the wealth of objects to which they can be applied make IR speckle interferometry one of the most exciting areas of contemporary astronomy.

D. Supernova 1987A

On February 24, 1987, a very massive star in the nearby irregular galaxy known as the Large Magellanic Cloud (LMC) collapsed and exploded in a split second to initiate a dramatic event known as a supernova. The LMC supernova, named SN 1987A, was the closest such occurrence in over four centuries and has provided astronomers with an unparalleled opportunity to study this dramatic phenomenon. Following the collapse and explosion, the majority of the mass of the star was expelled in all directions at very high velocity. It was realized that within a few months following the explosion this expanding material would grow to fill a volume sufficiently large to be resolved by speckle interferometry. The measurement of this expanding volume provided important constraints upon supernova models and even provided a means for directly measuring the distance to SN 1987A and thus to the LMC. Observations were made and will be continued until the event has faded to such a low level of brightness as to be undetectable. One particularly intriguing aspect of these observations was the discovery by one team of observers of a "mystery spot" of light separated from the supernova itself but close enough to it so that it must somehow be associated with the event. A possible explanation is that the spot represents some sort of material near the supernova that has been illuminated by the brilliant radiation from the event. The exact nature of this spot remains to be established.

V. The Future of High-Resolution Astronomy

The successes of speckle interferometry have firmly established high-resolution astronomy as an important scientific enterprise. New, large telescopes are paying special attention to their interferometric applications as well as to their light collecting potential. Whereas it was not long ago considered necessary to go into space to achieve significant gains in resolution, there are now numerous plans to further extend the boundaries of resolution from the ground using single giant telescopes or arrays of telescopes.

The 1990s will see the inauguration of several large telescopes with apertures in the 8- to 10-m range. A consortium of European countries is committed to the construction of a facility known as the

Very Large Telescope (VLT), which will consist of four separately mounted 8-m telescopes capable of combining their beams to yield the equivalent light collecting power of a 16-m telescope. Each of these telescopes can be used for speckle interferometry at visible and IR wavelengths. To dramatically improve the resolution of the VLT, methods are being developed to combine the separate beams of light from the telescope interferometrically to provide resolution corresponding to an aperture of approximately 100 m.

No single telescope can be constructed with such an enormous aperture, but scientific problems are driving optical astronomers to synthesize these large apertures in ways that have already been carried out at much longer wavelengths by radio astronomers. Arrays of telescopes, each of which may have a relatively small aperture, can be distributed along the ground to effectively synthesize an aperture hundreds of meters across. Optical wavelengths are far more challenging to this approach than are the much longer radio wavelengths, and it has only been in recent years that technology suited to multiple telescope arrays has matured sufficiently. Interferometer arrays have been built in Australia, Europe, and the United States. Experience from speckle interferometry with single telescopes has gone far to improve our knowledge of how the atmosphere will affect such arrays as we strive for a gain of a factor of 100 over the resolution now provided by speckle methods. Interferometry from space has now been designated a primary scientific goal for NASA during the next 25 yr, as resolutions corresponding to synthetic apertures many kilometers across are envisioned.

High angular resolution astronomy will provide a revolutionary approach to viewing the universe. This is remarkable progress in the less than two decades since Labeyrie invented speckle interferometry.

Bibliography

Bates, R. H. T. (1982). Astronomical speckle imaging. *Phys. Rep.* **90,** 203.

Dainty, J. C. (1984). Stellar speckle interferometry. *Topics Appl. Phys.* **9,** 255.

Labeyrie, A. (1976). High-resolution techniques in optical astronomy. *Prog. Opt.* **14,** 47.

Labeyrie, A. (1978). High-resolution techniques in optical astronomy. *Annu. Rev. Astron. Astrophys.* **16,** 77.

McAlister, H. A. (1985). High angular resolution measurements of stellar properties. *Annu. Rev. Astron. Astrophys.* **23,** 59.

McAlister, H. A. (1988). Seeing stars with speckle interferometry. *Am. Sci.* **76**(2), 166–173.

SWITCHING, OPTICAL

S. C. Gratze *GEC Research UK*

Glossary

Nonblocking switch: Switch in which a signal from any input channel can be routed to any free output without disturbing the transmission of any other signals passing through the switch.

Optical channel: Usually the individual physical communications medium (e.g., a particular fiber) carrying an individual optical signal; however, it can be a particular time slot in a time division multiplex frame or a specific wavelength that carries an individual signal.

Optical signal: Data or other information modulated onto an optical carrier (normally in the 0.5 to 2 μm wavelength range).

Switch size: An $n \times m$ switch is one switching between n input channels and m output channels.

Optical switching comprises the techniques and technologies used to switch or transfer optical signals from one transmission channel to another while the signals remain in optical form. Optical switching is characterized by its high "data transparency," meaning that the switching function is equally well performed regardless of what type or rate of data modulation is applied to the optical carrier. A wide variety of practical optical switches have been developed using many different physical effects; the main types are covered in this article. Work in this area is closely linked to and supports developments in fiber optical communications and integrated optoelectronics and is particularly applicable to very high bit rate and coherent telecommunications systems.

I. Introduction to Optical Switching

A. Major Applications and Advantages

In many communication systems, it is necessary to switch signals between transmission channels or circuits so that different destinations can be selected by any information source. This is what happens, for example, in a telephone exchange when a number is dialed. [The opposite situation exists in a broadcast system when each subscriber receives all signals and selects the one(s) required, but here forms of switching can be employed in the receiver as part of the selection process.] Switching is also used for other purposes, such as reconfiguring the communication network. In a conventional optical fiber transmission system exchange, the incoming signals are detected (converted from optical to electrical form), amplified and regenerated, switched electronically, and then reconverted back to optical signals for transmission on the selected channel. In addition to telecommunications systems, optical transmission is being applied to computer networking, TV distribution, microwave signal routing, sensor systems, and many other uses. [*See* OPTICAL FIBER COMMUNICATIONS.]

As the data rates transmitted on optical fiber continue to increase, it becomes increasingly difficult and costly for the electronic components used in conventional switching to handle the bit rates required. In addition, electronic switching destroys the phase and frequency (wavelength) characteristics of the optical signal unless these are specially recovered. These characteristics are vital for coherent communication links, which offer much greater bandwidth/transmission distances than the standard incoherent amplitude modulation approach.

In optical switches, the signal remains in optical form and can be switched or routed without change. The phase and frequency of the optical carrier is therefore preserved and the operation of the switch is independent of the type and bit rate of the modulation employed. Optical switches are generally bidirectional, that is, light can propagate through them equally well in either direction. Integrated optical switches in particular can also be reconfigured at very high speeds (nanoseconds and faster), so the routing through switches can be changed at the data bit (or byte) rate or used to time multiplex a number of lower rate optical channels.

Most, but not all, of the optical switches so far developed are *space switches,* typically connecting any input fiber to the selected output fiber or fibers. *Time switches* are also commonly used in telecommunications, where the time slot into which data from a particular source are multiplexed is changed by the switch, which eventually results in it being routed to the selected destination. In addition, optical switches can be produced that change the polarization state of the optical carrier or selectively route particular optical wavelength, offering new systems possibilities.

B. Main Technologies Utilized

The conceptually simplest type of switch is mechanical, using the movement of a fiber or the deflection of a free-space optical beam to connect the signal from a given input fiber to a selected output fiber. The advantages of this approach are low optical loss (for simple switches) and polarization and wavelength independence. Disadvantages are slow switching speed (milliseconds), the need for finely toleranced construction to achieve low loss, and the considerable development needed to attain reliable operation.

A wide variety of small, nonmechanical switches, which may be classified as microoptic, have been developed employing electrooptic, acoustooptic, and other nonmechanical means to deflect optical beams propagating within the switch. They generally offer microsecond to millisecond switching and fabrication methods that are better suited to the fine tolerances required; however, in some cases, the wavelength and polarization independence is lost or optical losses are increased in comparison with mechanical switches.

An attractive option for providing very low loss is to arrange the switching to take place without leaving the optical fiber. These all-fiber switches generally employ a fiber interferometer arrangement in which the phase in one arm can be varied, and they are suitable for use in single-mode fiber systems. Typically, polarization independent and offering moderate switching speeds, these devices are likely to be used as bypass or wavelength selection switches.

Integrated optical switches that use optical guiding in the surface of a planar substrate have been subject to extensive research and development. They offer direct compatibility with single-mode fiber, high switching speeds (nanoseconds or less), functional flexibility, the capability of integrating quite complex switching matrices, and monolithic construction. Their potential for integration with other optoelectronic components, for example, semiconductor lasers and photodetectors, is high, and this represents one of the most likely directions for the future development of optical switching.

When really large switching matrices are required, that is, > 100 inputs and outputs, all of the previously mentioned switch types become cumbersome, and it is essential to consider using the two-dimensional parallel processing capability of optics. Experimental demonstrations have shown the possibility of using dynamic holograms and programmable diffraction gratings to route light between two-dimensional (e.g., square) arrays of input and output fibers. These types of devices not only have the potential for use in large optical exchanges but also may find application in optical and optoelectronic computers. The main attributes of the various types of switches are summarized in Table I.

II. Mechanical Switches

A. Moving Fiber

Two basic types of moving fiber switch have been developed—end-fire coupled and remote connection. A typical end-fire switch is shown in Fig. 1, employing piezoelectric elements to move the input fibers relative to the output fibers when a control voltage is applied to bend the piezoelectric elements. By using piezoelectric elements to support both the input and output fibers, the control voltage is halved, and together with the use of mechanical "stoppers" to limit the lateral movement, this permits high stability to be obtained. Good long-term reliability can also be achieved.

An alternative end-fire coupled switch uses a funnel arrangement to direct individual fibers that are electromechanically activated into the correct position for coupling with a fixed fiber (Fig. 2). This

TABLE I. Typical Attributes of Optical Switches

Switch type	Size[a]	Loss[b] (dB)	Crosstalk[b] (dB)	Speed[c] (sec)	Bandwidth[d] (nm)	Input type[e]
Mechanical moving fiber	~ 1 1×8	1	<-30	$\sim 10^{-3}$	>1000	M
Moving beam	~ 50 8×8	2	<-30	$\sim 10^{-2}$	>1000	M
Microoptic liquid crystal	~ 1 4×4	4	-30	10^{-3}	~ 100?	M
Liquid dielectric	~ 1 2×2	1	<-30	10^{-2}	~ 1000	M
Guided wave all fiber	~ 5 2×2	0.2	<-25	10^{-4}	~ 100?	S
Lithium niobate	$\sim 10^{-2}$ [f] 8×8	0.5	30	10^{-9}	~ 50	S
III–V	$\sim 10^{-2}$ [f] 2×2			10^{-9}	~ 50?	S
Holographic[g]	10×10	15→20		10^{-2}	1→10	M

[a] Size of 2×2 unit in cm^3 and maximum complexity of typical current unit.
[b] Typical value for 2×2 unit.
[c] Speed of switching or reconfiguration.
[d] Values with ? indicate estimated values.
[e] For listed performance: M, multimode fiber; S, single-mode fiber.
[f] Ex packaging, based on chip real estate; for packaged size see Fig. 11.
[g] Data on holographic switches may not be representative of performance demonstrated in an individual unit.

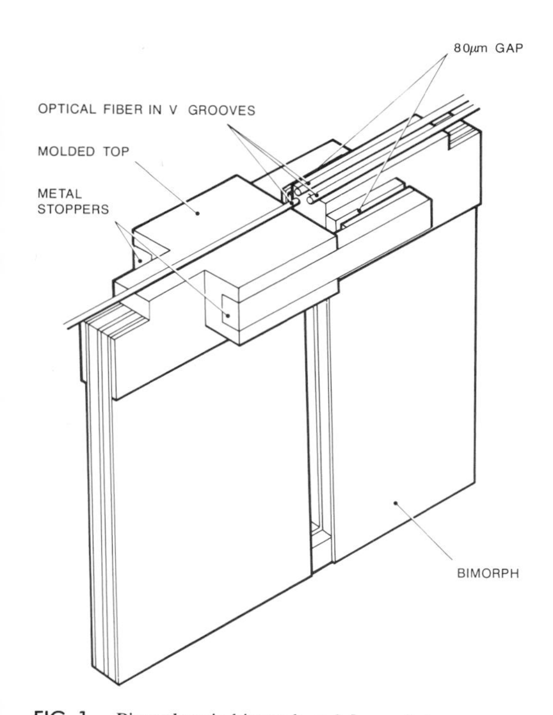

FIG. 1. Piezoelectric bimorph end-fire switch.

switch type is obviously well suited for N to 1 switching, and low-cost, miniature activation mechanisms, such as those used in dot matrix printers, can be employed.

Insertion losses with multimode fiber are low (1 dB or less), but with monomode fiber, adequate tolerances can be difficult to obtain. The switches are inherently polarization and wavelength independent. The major limitation is in the switching speed, which is typically a few milliseconds.

The remote connection arrangement uses two arrays of fibers and lenses facing each other and employs piezoelectric elements to deflect the input fibers, so changing the angle at which the collimated light is launched by the lens. This directs it toward the required output lens, where the position of the output fiber is similarly deflected to provide optimum output coupling. These types of devices have the capability of connecting fairly large numbers of fibers, but the optical losses increase with increasing switch size and assemblies tend to be bulky.

The main applications of these mechanical switches are seen as being in multimode switches of relatively low complexity, for example, computer terminal bypass switches for local area networks.

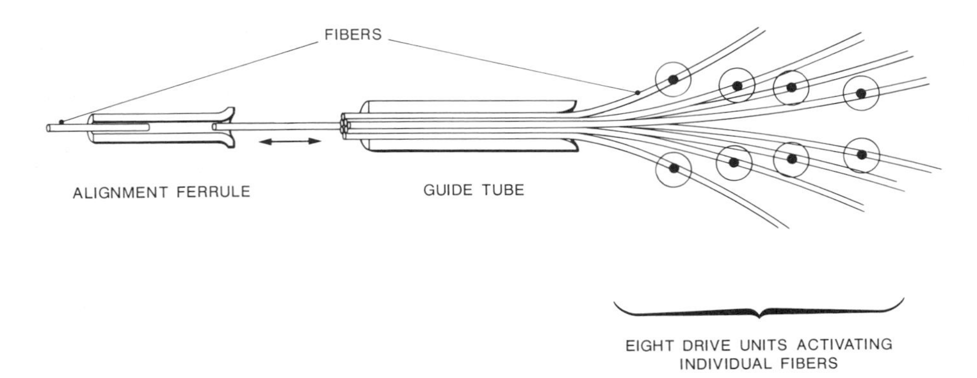

FIG. 2. Funneled fiber switch.

B. Moving Beam

Moving beam switches employ a substantial free-space optical path between fibers, with lenses to collimate and focus the beams, and prisms or mirrors to selectively deflect the light, so defining the switch routing. The input and output fibers and associated lenses can be arranged as arrays along two sides of a rectangular base that houses prisms or mirrors that are electromechanically inserted into the beams (Fig. 3).

The optical losses of this type of switch depend on the quality of the optical elements used for collimation and focusing. If large distances are required for the free-space beams, the lens apertures must be increased, which in turn increases the size of the switch. To provide reliable and accurate beam deflection also necessitates precision electromechanical elements and a rigid base, which tends to lead to bulky and expensive switches. The geometry is, however, well suited to nonblocking $N \times N$ switch matrices, and the performance attributes are similar to those of end-fire switches.

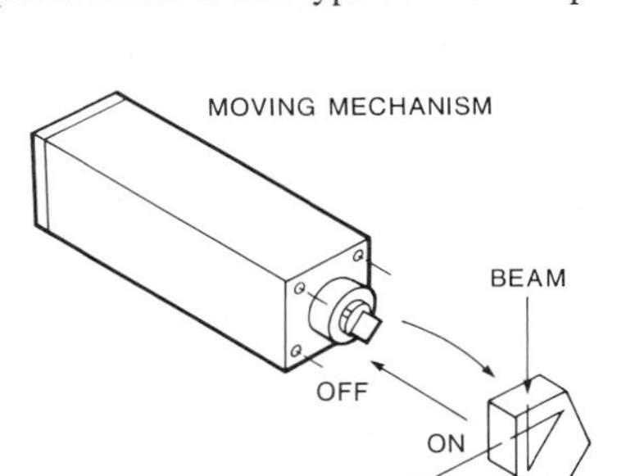

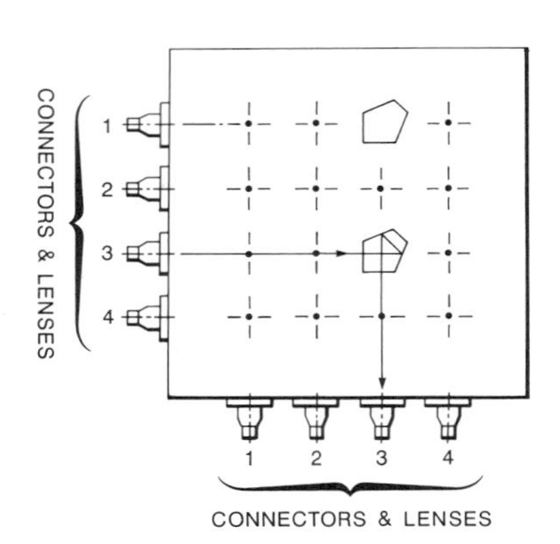

FIG. 3. Prismatic beam deflection switch.

III. Microoptic Switches

A. Liquid Crystal Switches

Microoptic switches is a term used to cover a range of nonmechanical switch technologies, excluding those employing guided wave optics and the holographic techniques covered in later sections. The best developed type employs liquid crystal materials in a structure broadly similar to those used for display applications. As shown in Fig. 4, the liquid crystal layer is sandwiched between glass supports together with transparent conductors (usually indium tin oxide) to apply the control voltage and alignment layers that control the direction of the molecules in the liquid crystal. When a voltage is applied, the molecules in the layer tend to rotate and align along the applied field direction. This causes a change in the refractive index seen by light passing through the layer. For switching, the mode of operation usually employed is to arrange the refractive indices of the glass and liquid crystal and the angle of the incoming beam so that with no applied voltage total internal reflection occurs at the glass–liquid crystal interface, while with an applied field, the indices match and light is transmitted. (The alignment layers and electrodes are

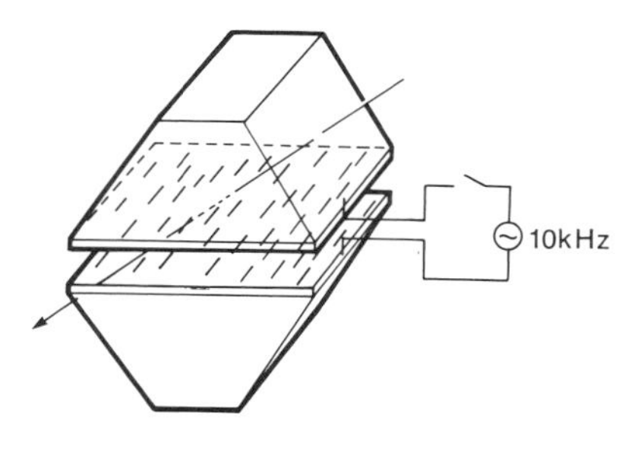

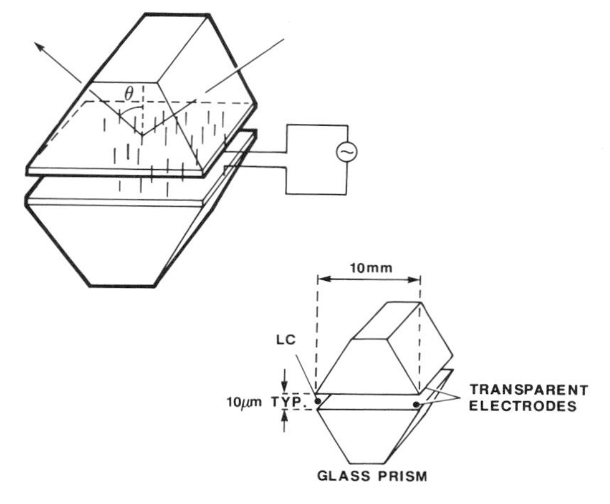

FIG. 4. Liquid crystal switch schematic.

thin compared with the optical wavelength.) Because the liquid crystal is birefringent, this switch requires a polarizer on the input, but the loss resulting from rejecting one polarization can be overcome to some extent in modified designs arranged to separately switch the two polarization components. In the case of multimode fibers, which support a statistically large number of modal polarizations, the added complexity for perhaps a 1-dB loss reduction is not usually accepted, but for a monomode fiber (where the single mode being in the wrong polarization could cause a complete fade), dual polarization capability (or polarization maintaining fiber) is essential. The basic switch offers a loss of typically 4 dB (3 dB of which is from polarization filtering), an isolation of 30 dB, a fairly wide wavelength range (~ 100 nm), control voltages of 30 V, and switching speeds of ~ 1 msec.

By using an array of control electrodes and reflecting the light within the glass blocks, more complex switch arrays can be constructed; a number of ingenious switch architectures and structures have been produced to do this with minimum size and complexity. The operation of the switch and the collection of the deflected light depend on reasonable col-

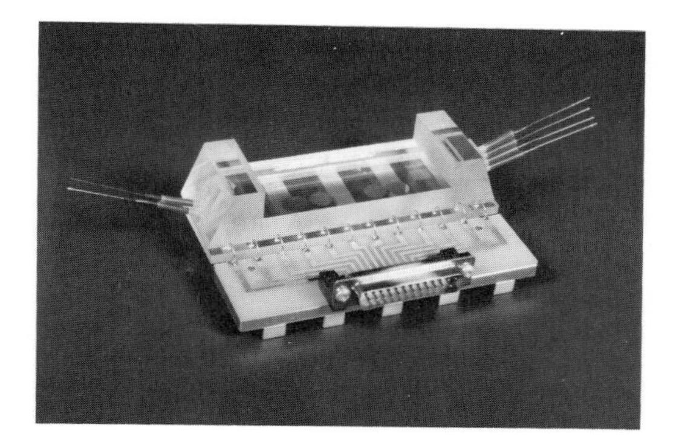

FIG. 5. A 4×4 liquid crystal switch.

limation of the optical beams, and graded index (GRIN) lenses are normally employed at the fiber interfaces to provide this. The degree of collimation obtained depends on the type of fiber and GRIN lenses used, and this dominates the total loss of switch matrices for multimode fiber use. For single-mode fiber, positional accuracy is the most important factor. In either case, switch complexity is currently limited to about 8×8 for acceptable loss (~ 16 dB). A complete 4×4 liquid crystal switch is shown in Fig. 5.

B. Other Microoptic Switches

A total internal reflection switch structure similar to that just presented can be made using a moving (liquid) dielectric (Fig. 6). In this switch, a high dielectric constant liquid is drawn in or out of the region through which the light propagates by electrocapillarity effects when a bias is applied to the control electrodes. The liquid is chosen to have the same refractive index as the glass, so when the liquid is in the beam path, transmission occurs. When it is not, the choice of incidence angle gives total internal reflection. Losses within the switch can be well below 1 dB and isolation better than 30 dB. The device is polarization independent, has a wide wavelength range, and can be designed to be "fail safe"; however, switching times are relatively long (10–100 msec).

Microoptic switches can also be produced using acoustooptic effects in which high-frequency acoustic waves in a crystal are used to deflect an optical beam at an angle dependent on the rf frequency applied to the electroacoustic transducer (Fig. 7). This type of device is also used for acoustooptic, Bragg-cell laser beam deflectors and modulators. Suitable optics focus the light onto the output fiber determined by the deflection angle. A number of individual acoustic channels can be provided in one device, and

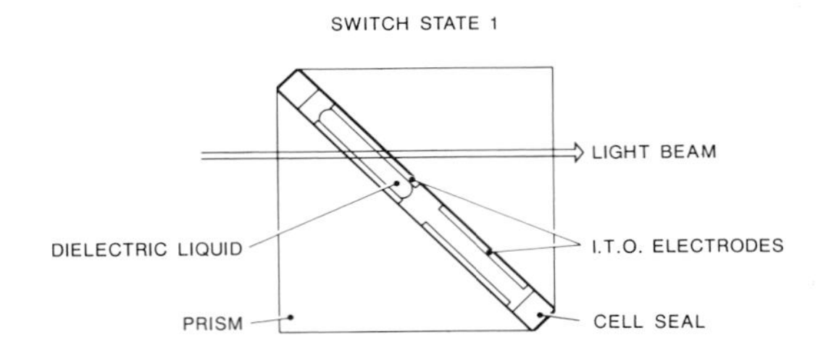

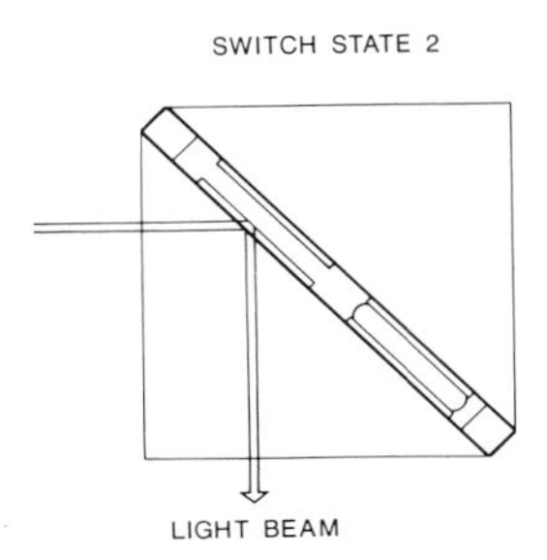

FIG. 6. Moving dielectric switch.

an 8 × 8 switch has been produced. The losses in the Bragg cell can be quite low, and switching times of a few microseconds obtained; however, the device is polarization and wavelength sensitive, and for the more complex devices, the associated optics are critical and can give substantially increased losses.

In general, microoptic switches can provide switching speeds down to a few microseconds, low loss, compatibility with multimode fiber, polarization independence, and capacities up to 8 × 8. One particular type cannot, however, provide all these, and the technological choice depends upon the application. Their exploitation is clearly dependent on the needs for switches with moderate speed and complexities and, in particular, the extent of the use of multimode fiber for which they are particularly well suited.

IV. All-Fiber Switches

A. Interferometric Switches

The structure of a basic all-fiber interferometric switch is shown in Fig. 8. The device comprises a Mach Zehnder interferometer made from single-mode fiber in which the path length of one arm can be varied electrically via a control element, which can be a resistive film deposited on the fiber to provide heating, a piezoelectric device to stretch or bend the fiber, or other forms of activation. When the relative phases of the signals from each arm are varied, the light can be made to switch between the two output fibers. Arrays of such elements can be assembled to make more complex switches. The transfer characteristics of the switch vary sinusoidally with the inverse optical wavelength, and by suitable design, devices with good wavelength discrimination can be produced; alternatively, broad spectral responses can be obtained. The devices are basically simple, insertion losses are very low (0.2 dB), cross talk is better than 25 dB, and the device is polarization independent. Switching speed and stability are dependent on the types of activation employed, but submillisecond response is possible. An immediate application of this type of device should be as a bypass switch

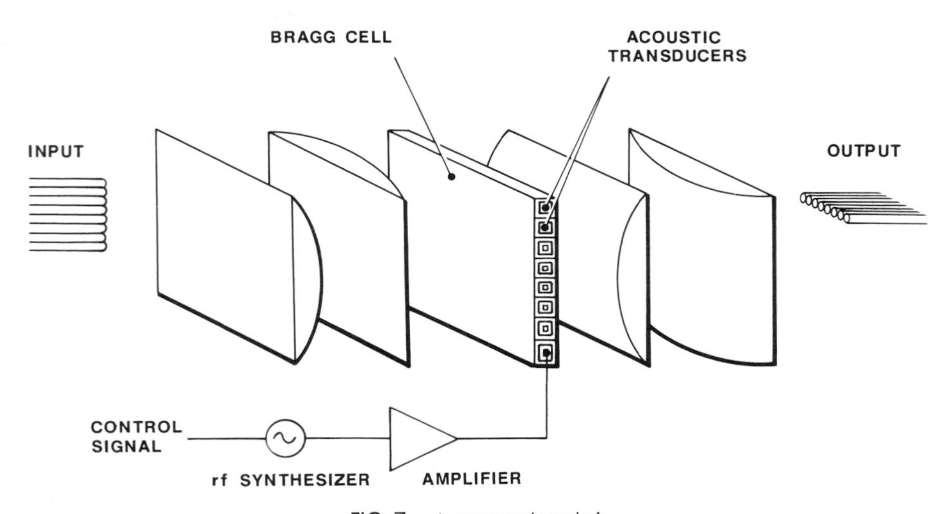

FIG. 7. Acoustooptic switch.

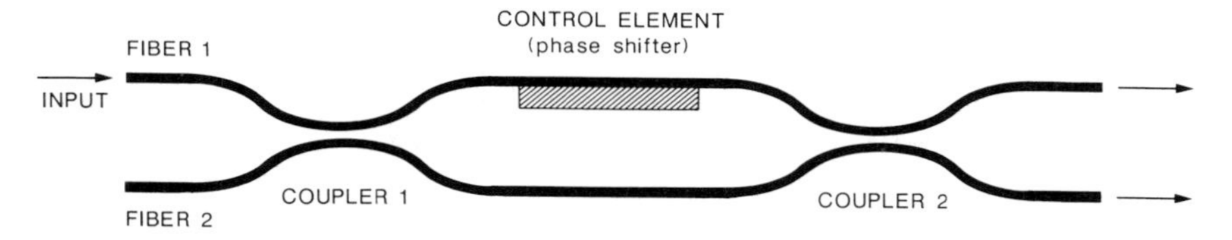

FIG. 8. All-fiber switch.

for optically connected computer workstations and the like.

B. Coupler Switches

If the cladding from two fibers is removed locally and the cores are brought close together, light can couple from one fiber to the other. By providing a thin layer of electrooptic material and suitable control electrodes in the gap between the cores, the coupling can be electrically controlled to provide switching between the output fibers. Periodic structures within the gap can be used to increase the wavelength selectivity of the device. Fabrication techniques exist to make many such switch elements in parallel in one unit. The major advantages of the switch are its direct compatibility with single-mode fiber and potential for fairly high-speed (microseconds) operations; however, substantial development is required to make reliable, high-performance devices at low cost.

V. Integrated Optical Switches

A. Lithium Niobate Switches

Some of the most complex and high-performance optical switches have been made in lithium niobate, and this is the classical material for integrated optics. The technology employed is to diffuse a patterned layer of thin film titanium into the polished surface of a lithium niobate substrate to produce channels of raised refractive index which form integrated optical waveguides. Light coupled into the device from optical fibers can then be routed around the optical circuit. Since lithium niobate is an electrooptic material, application of an electric field via surface electrodes changes its refractive index and hence the velocity of light in the waveguides. This ability to manipulate the light propagation in single-moded waveguides forms the basis of several types of optical switches and a range of other important integrated optical devices (modulators, polarization controllers, filters) with which they can be integrated. The technology used to form the devices is very similar to that used to produce electronic integrated circuits (deposition, photolithography, diffusion, etc.) and therefore has excellent potential for low cost and reliability.

The most extensively developed type of integrated optical switch is the stepped $\Delta\beta$ direction coupler switch (Fig. 9). In this device, two monomode waveguides are brought together, and evanescent coupling between the waveguide modes causes energy to couple back and forth periodically between the guides along the length of the coupling region. By applying an electric field, the coupling period can be changed, so the guide in which optical energy remains at the end of the device can be reliably controlled. The element acts as a cross bar switch having a cross (X) and a bar (=) state. Usually, this is achieved by changing the relative velocities or effective index ($\Delta\beta$) in the guides, but electrode geometries that change the interguide coupling coefficient can also be used ($\Delta\kappa$). To improve performance and ease manufacturing tolerances, the electrodes for the $\Delta\beta$ switch are usually split in two: the first section of the switch adjusts the optical power to be equally split between the two guides, and the second switches from that well-defined state to the required output distribution. These switches typically offer 0.5-dB loss per switch, cross talk better than 30 dB, and useful wavelength ranges of 10 to 100 nm (by design). The control voltages required are inversely proportional to device coupling length, but a typical 5-mm long switch would require 50 V for 1.3 μm operation. Photorefractive effects in lithium niobate limit the power that can be handled at wavelengths below 1 μm unless special processes are used. Switching times are typically in nanoseconds, but special structures can considerably reduce this. Because quite complex switch matrices are being integrated into single chips, there is pressure to reduce device lengths without unacceptably increasing the control voltage. Considerable effort has been put into design refinements that reduce the voltage length products, and novel geometries have been demonstrated that, by using a reflector at one end of the chip, effectively

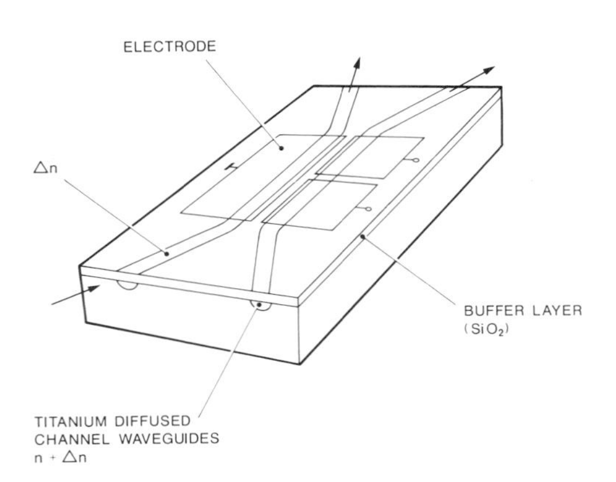

FIG. 9. Stepped $\Delta\beta$ directional coupler switch.

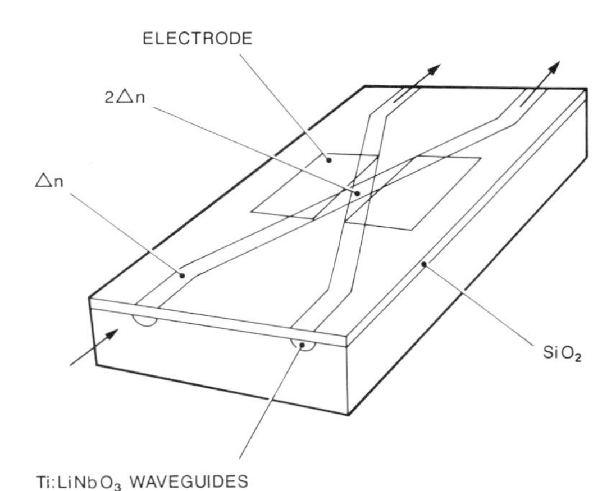

FIG. 11. Integrated optic X switch.

double the useful length. Figure 10 shows a stepped $\Delta\beta$ switch servicing 4 input and 4 output fibers; 8 $\times$ 8 devices have also been produced, and 16 $\times$ 16 arrays are possible.

These performances are for single-polarization devices; but polarization-independent switches have also been demonstrated, although their voltage length products are higher and isolation is lower. Their performance is, however, adequate for many potential single-switch applications, such as bypass switches for computer terminals. Addition of periodic structures in the coupling region can enhance the spectral selectivity of $\Delta\beta$ switches, permitting wavelength selective switches and taps to be produced where required. Tailoring of the gap width along the coupling region provides control of the shape of the spectral response.

As well as the various directional coupler switches, a number of other types have been developed, including the X switch (Fig. 11) and designs intermediate between the two. A common version of the X switch operates by arranging the waveguide in the crossover region to support a second as well as the fundamental mode. The two modes propagate with a velocity difference determined by the control voltage applied, with the result that the combined field distribution at the output split can be made to match the position of the (single) mode of either of the output guides to effect switching. Other variants have an extended region of double-width guide between the crossover region or guide geometries which make their operation more similar to the total internal reflection switches described earlier. Active Y junctions, which form 1 into 2 switches, have also been described.

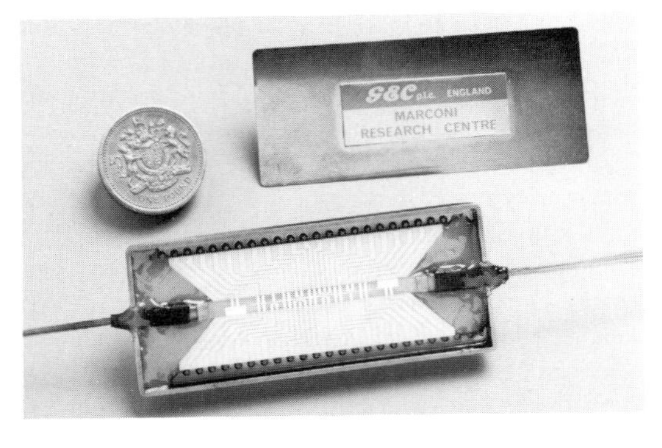

FIG. 10. A 4 $\times$ 4 lithium niobate integrated optic switch.

B. Aternative Integrated Optic Switches

Although lithium niobate has some excellent optical and electrooptic properties and titanium diffused waveguides have a mode size well matched to single-mode fiber yielding low coupling losses, alternative integrated optic switch technologies have been pursued for two reasons. The first objective is to use materials, such as glass or plastic, which offer large-area, low-cost substrates; the second is to use semiconductor materials in which electronic and optoelectronic components can also be integrated, particularly indium-phosphide-based quaternary compounds that offer laser integration.

Techniques have been developed for producing waveguides in glass (ion exchange) and plastics (photopolymerization), but since glass and most plastics are not electrooptic, an active layer must be deposited over the waveguides. Since only a small part of the light will travel in the active layer, to be attractive this must have much higher electrooptic coefficients than lithium niobate, and research has concentrated on novel organic crystals and polymers and on liquid crystals. At present, solid organics exist with coefficients a few times higher than lithium niobate, while liquid crystals, although having effective coefficients orders of magnitude higher, suffer from surface "pinning" effects causing the layers in which the light travels to be relatively inactive. The solution is to develop materials and technologies to provide low-loss optical waveguides in the active organic layers, and many advances are expected in this area from worldwide research.

Waveguides in III–V semiconductor materials with propagation losses as low as those in lithium niobate have already been demonstrated. Although the electrooptic coefficients are lower, the higher index difference and smaller optical mode size permit voltage-length products comparable to lithium niobate to be obtained and the same types of switches to be demonstrated. As yet, however, guiding structures offering both low loss and low switching voltage have not been demonstrated—the low-loss guides are not very amenable to the tight bends needed to form switch matrices, none of the guides have a good modal match to fiber, and the work on integration with optoelectronic and electronic components is only just starting. In addition, basic material size, repeatability, and availability are greatly inferior to that of lithium niobate. However, the major investments being made in III–V technology will change this picture; and monolithic integration to provide signal amplification and regeneration and built-in control circuitry will be a major area for optical switching research and development in the future.

VI. Holographic Switches

The optical switches described so far have been "small switches," individual devices connecting 2 × 2 to 16 × 16 fiber arrays, whereas telecommunication systems often require switching of 100 × 100s or 1000 × 1000s of channels. Although a number of optical switching devices can be interconnected, for example, with optical fiber to form larger switches, the techniques to do this effectively are still being developed. The data rates optical switches can handle are orders of magnitude greater than their electronic counterparts but the technology to exploit this is still experimental. Much larger optical switches that exploit the parallel processing capability inherent in optics are therefore being researched. In these switches, the input fibers are arranged in a regular, for example, square, array and their outputs collimated. The optical beams pass through a programmable deflection unit and are refocused onto the output fiber array. Although these switches have, at present, relatively high insertion losses, these losses are fixed independent of the connection unlike some other switch types.

Programmable beam deflectors have been constructed using photorefractive materials, such as bismuth silicon oxide (BSO), in which an array of holographic patterns is set up to deflect the optical beams from the fibers by interfering two beams from a laser operating at a ("write") wavelength at which BSO is sensitive. By adjusting both the wavelength and spatial distribution of the "write" beams, the holograms required to make the wanted connections can be written. Such systems, as shown schematically in Fig. 12, are, at present, large and power consuming, requiring a tunable dye laser (and argon ion laser pump) and electro- or acoustooptic beam deflectors to write the holograms. Switch sizes similar to those from other technologies have currently been demonstrated, but extrapolating to 100 × 100 capacity and solid-state technologies to produce the write beams indicates this technique may be viable in the future.

An alternative type of beam deflector currently being investigated uses programmable diffraction gratings to deflect the spatially separated collimated input beams. Programmable gratings can be produced using optically written spatial light modulators (SLM) with inbuilt memory. In such a device (Fig. 13), the area into which a grating is to be written is illuminated with a grating produced by an electronically driven interferometer, which has the correct pitch and alignment to diffract the input light in the desired direction. Photocarriers generated in the semiconductor layer replicate the optical write pattern as a voltage pattern across the latching (ferroelectric) liquid crystal layer where it is stored as a phase grating (index change) while other areas of the SLM are programmed. An interesting alternative is to use the VGM (variable grating mode) liquid crystal effect in which gratings can be formed with a pitch determined directly by the applied voltage.

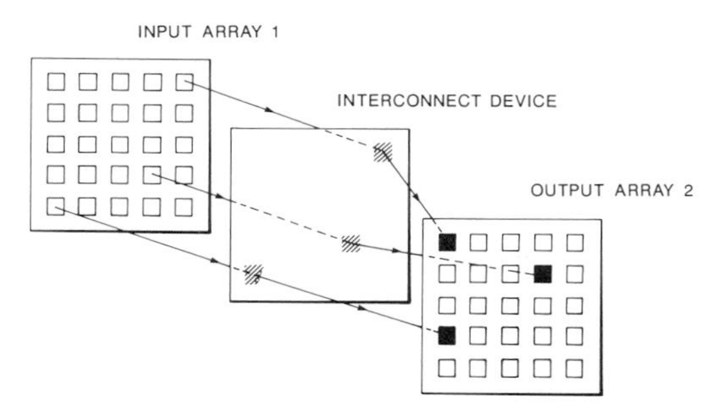

FIG. 12. Principle of operation of the holographic switch.

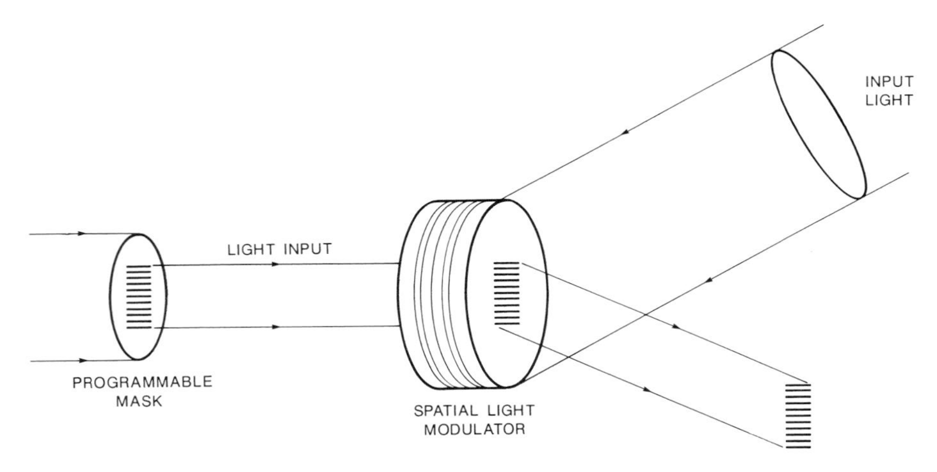

FIG. 13. Spatial light modulator schematic.

These devices and systems are the subject of current research, and representative performance figures cannot yet be quoted. However, they are typically wavelength and polarization sensitive and require precision optical assemblies but offer the potential for extension to quite large switch sizes. For these reasons, the interest in them is based as much on optical computing as it is on telecommunications.

VII. Other Switches

The majority of optical switches so far produced have been space switches, although many can be modified for wavelength switching as previously described. Conventional telecommunication systems also require time switches, and one way of providing these is to use space switches to select appropriate lengths of optical fiber to act as delay lines. Considerable research effort is, however, being directed into developing monolithic optical time switches based on bistable laser diodes, which can be triggered by an incoming optical pulse and can store this information bit until reset. In parallel, work is proceeding on all-optical switches and logic devices in which an optical input switches the state of the optical output signal (derived from an optical "power supply" or pump) from state 0 to state 1 using the intrinsic nonlinear optical properties of the device material. These new types of switch, combined with increased optoelectronic integration, are expected to be of increasing importance in the future.

VIII. Summary and Future Prospects

Some of the telecommunication applications and the related attributes of optical switches, particularly their data transparency and wave preserving characteristics, have been described and their use in rf signal processing and optical computing indicated. The wide range of mechanical, microoptic, all-fiber, integrated optic, and holographic switches has been presented together with their performance capabilities (see Table I). Indications of the expected applications of the various switch types have been given together with some of the determining factors, for example, the extent that multimode fiber will continue to be used.

The main directions for future development are twofold. First, established technologies such as lithium niobate integrated optics will be used to produce prototype switches tailored to special requirements to allow new system applications to be explored, and, where their particular properties are well suited, these switches will be developed further and exploited. Second, new technologies, such as III–V semiconductor optoelectronic integration, organic materials, and possibly holographic techniques, will be researched in parallel with new system applications leading to complex optical processing functions for application in communications and computing. Because the field of optoelectronic research is evolving rapidly, it is impossible to predict in detail the advances that will occur but optoelectronic integration and a blurring of the distinction between optics and electronics is almost inevitable.

Bibliography

Alferness, R. C. (1981). Guided wave devices for optical communications, *IEEE J. Quantum Electron.* **QE17(b),** 946–959.

Granestrand, P., Stoltz, B., Thylen, L., Bergnall, K., Döldissen, W., Heinrich, H., and Hoffman, D. Strictly non-blocking 8 × 8 integrated-optic switch matrix, *Electron. Lett.* **22,** 816–817.

Le Rouzic, J. (1983). Optical switching in the subscriber network, *9th Eur. Conf. Opt. Commun.*, Geneva.

"Proceedings of Topical Meeting on Photonic Switching." (1987). Las Vegas, Nevada. Springer-Verlag, Berlin and New York.

Smith, P. W. (1987). On the role of photonic switching in future communications systems, *IEEE Circuits and Devices,* May 1987.

TELESCOPES, OPTICAL

L. D. Barr *National Optical Astronomy Observatories**

Glossary

Airy disc: Central portion of the diffracted image formed by a circular aperture. Contains 84% of the total energy in the diffracted image formed by an unobstructed aperture. Angular diameter = 2.44 λ/D, where λ is wavelength and D the unobstructed aperture diameter. First determined by G. B. Airy in 1835.

Aperture stop: Physical element, usually circular, that limits the light bundle or cone of radiation that an optical system will accept on-axis from the object.

Coherency: Condition existing between two beams of light when their fluctuations are closely correlated.

Diameter-to-thickness ratio: Diameter of the mirror divided by its thickness. Term is generally used to denote the relative stiffness of a mirror blank: 6 : 1 is considered stiff, and greater than 15 : 1 is regarded as flexible.

Diamond turning: Precision-machining process used to shape surfaces in a manner similar to lathe turning. Material is removed from the surface with a shaped diamond tool, hence the name. Accuracies to one microinch ($\frac{1}{40}$th μm) are achievable. Size is limited by the machine, currently about 2 m.

Diffraction-limited: Term applied to a telescope when the size of the Airy disc formed by the telescope exceeds the limit of seeing imposed by the atmosphere or the apparent size of the object itself.

Effective focal length: Product of the aperture diameter and the focal ratio of the converging light beam at the focal position. For a single optic, the effective focal length (EFL) and the focal length are the same.

Field of view: Widest angular span measured on the sky that can be imaged distinctly by the optics.

Focal ratio (f/ratio): In a converging light beam, the reciprocal of the convergence angle expressed in radians. The focal length of the focusing optic divided by its aperture size, usually its diameter.

Image quality: Apparent central core size of the observed image, often expressed as an angular image diameter that contains a given percentage of the available energy. Sometimes taken to be the full width at half maximum (FWHM) value of the intensity versus angular radius function. A complete definition of image quality would include measures of all image distortions present, not just its size, but this is frequently difficult to do, hence the approximations.

Infrared: For purposes of this article, wavelength region from about 0.8 to 40 μm.

Optical path distance: Distance traveled by light passing through an optical system between two points along the optical path.

Seeing: Measure of disturbance in the image seen through the atmosphere. Ordinarily expressed as the angular size, in arc seconds, of a point source (a distant star) seen through the atmosphere, that is, the angular size of the blurred source.

* Operated by the Association of Universities for Research in Astronomy, Inc., under contract with the National Science Foundation.

Ultraviolet: For purposes of ground-based telescopes, the wavelength region from about 3000 to 4000 Å.

Optical telescopes were devised by European spectacle makers around the year 1608. Within two years, Gallileo's prominent usage of the telescope marked the beginning of a new era for astronomy and a proliferation of increasingly powerful telescopes that continues unabated today. Because it extends what the human eye can see, the optical telescope in its most restricted sense is an artificial eye. However, telescopes are not subject to the size limitation, wavelength sensitivities, or storage capabilities of the human eye and have been extended vastly beyond what even the most sensitive eye can accomplish. Properties of astronomical telescopes operating on the ground in the optical/IR spectral wavelength range from 3000 to about 40,000 Å (0.3–40 μm) will be considered. The atmosphere transmits radiation throughout much of this range. Telescopes designed for shorter wavelengths are either UV or X-ray telescopes and for longer wavelengths are in the radio-telescope category. Emphasis is placed on technical aspects of present-day telescopes rather than history.

I. Telescope Size Considerations and Light-Gathering Power

An astronomical telescope works by capturing a sample of light emitted or reflected from a distant source and then converging that light by means of optical elements into an image resembling the original source, but appropriately sized to fit onto a light-sensitive detector (e.g., the human eye, a photographic plate, or a phototube). Figure 1 illustrates the basic telescope elements. It is customary to assume that light from a distant object on the optical axis arrives

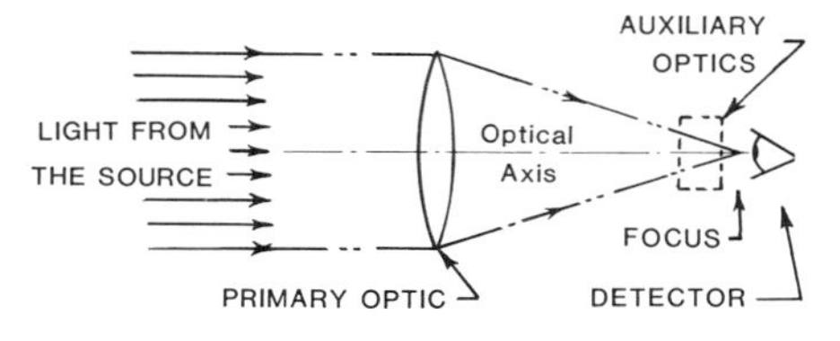

FIG. 1. Basic telescope elements. Refractive lens could be replaced with a curved reflective mirror.

as a beam of parallel rays sufficiently large to fill the telescope entrance, as shown.

The primary light collector can be a lens, as in Fig. 1, or a curved mirror, in which case the light would be shown arriving from the opposite direction and converging after reflection. The auxiliary optics may take the form of eyepieces or additional lenses and mirrors designed to correct the image or modify the light beam. The nature and arrangement of the optical elements set limits on how efficiently the light is preserved and how faithfully the image resembles the source, both being issues of prime concern for telescope designers.

The sampled light may have traveled at light speed for a short time or for billions of years after leaving the source, which makes the telescope a unique tool for studying how the universe was in both the recent and the distant past. Images may be studied to reveal what the light source looked like, its chemistry, location, relative motion, temperature, mass, and other properties. Collecting light and forming images is usually regarded as a telescope function. Analyzing the images is then done by various instruments designed for that purpose and attached to the telescope. Detectors are normally part of the instrumentation. The following discussion deals with telescopes.

A. Telescope Size and Its Effect on Images

The size of a telescope ordinarily refers to the diameter, or its approximate equivalent, for the area of the first (primary) image-forming optical element surface illuminated by the source. Thus, a 4M telescope usually signifies one with a 4-m diameter primary optic. This diameter sets a maximum limit on the instantaneous photon flux passing through the image-forming optical train. Some telescopes use flat mirrors to direct light into the telescope (e.g., solar heliostats); however, it is the size of the illuminated portion of the primary imaging optic that sets the size.

The size of a telescope determines its ability to resolve small objects. The Airy disc diameter, generally taken to be the resolution limit for images produced by a telescope, varies inversely with size. The Airy disc also increases linearly with wavelength, which means that one must use larger-sized telescopes to obtain equivalent imaging resolution at longer wavelengths. This is a concern for astronomers wishing to observe objects at infrared (IR) wavelengths and also explains in part why radio telescopes, operating at even longer wavelengths, are so much larger than optical telescopes. (Radio telescopes are more easily built larger because radio wavelengths are much longer and tolerances on the "optics" are easier to meet.)

Telescopes may be used in an interferometric mode to form interference fringes from different portions of the incoming light beam. Considerable information about the source can be derived from these fringes. The separation between portions of the primary optic forming the image (fringes) is referred to as the baseline and sets a limit on fringe resolution. For a telescope with a single, round primary optic, size and maximum baseline are the same. For two telescopes directing their beams together to form a coherent image, the maximum baseline is equal to the maximum distance between light-collecting areas on the two primaries. More commonly, the center-to-center distance would be defined as the baseline, but the distance between any two image-forming areas is also a baseline. Thus, multiple-aperture systems have many baselines.

As telescope size D increases so does the physical size of the image, unless the final focal ratio F_f in the converging beam can be reduced proportionately, that is,

$$\text{final image size} = F_f \cdot D \cdot \theta$$

where θ is the angular size of the source measured on the sky in radians. With large telescopes this can be a matter of importance when trying to match the image to a particular detector or instrument. Even for small optics, achieving focal ratios below about f/1.0 is difficult, which sets a practical limit on image size reduction for a given situation.

Another size-related effect is that larger telescopes look through wider patches of the atmosphere which usually contain light-perturbing turbulent regions that effectively set limits on seeing. Scintillation (twinkling) and image motion are caused by the turbulence. However, within a turbulent region, slowly varying isotropic subregions (also called isoplanatic patches) exist that affect the light more or less uniformly. When a telescope is sized about the same as, or smaller than, a subregion and looks through such a subregion, the instantaneous image improves because it is not affected by turbulence outside the subregion. As the subregions sweep through the telescope's field of view, the image changes in shape and position. Larger telescopes looking through many subregions integrate or combine the effects, which enlarges the combined image and effectively worsens the seeing. However, these effects diminish with increasing wavelength, which means that larger telescopes observing at IR wavelengths may have better seeing than smaller ones observing in the visible region.

Studies of atmospheric turbulence effects have given rise to the development of special devices to make optical corrections. These are sometimes called rubber mirrors or adaptive optics. An image formed from incoming light is sensed and analyzed for its apparent distortion. That information is used to control an optical element (usually a mirror) that produces an offsetting image distortion in the image-forming optical train. By controlling on a star in the isoplanatic patch with the object to be observed (so that both experience similar turbulence effects), one can, in principle, form corrected images with a large ground-based telescope that are limited only by the telescope, not the atmosphere. In practice, low light levels from stars and the relatively small isoplanatic patch sizes (typically a few arc seconds across) have hampered usage of adaptive optics on stellar telescopes.

B. Telescope Characteristics Related to Size

At least three general, overlapping categories related to telescope size may be defined:

1. Telescopes small enough to be portable. Sizes usually less than 1 m.
2. Mounted telescopes with monolithic primary optics. Sizes presently range up to 6 m. Virtually all ground-based telescopes used by professional astronomers are in this category.
3. Very large telescopes with multielement primary optics. Proposed sizes range up to 25 m for ground-based telescopes. Only a few multielement telescopes have actually been built.

A fourth category could include telescopes small enough to be launched into earth orbit, but the possibility of an in-space assembly of components makes this distinction unimportant.

One cannot, in a short space, describe all of the telescope styles and features. Nevertheless, as one considers larger and larger telescopes, differences become apparent; and a few generalizations can be postulated.

In the category of small telescopes, less than 1 m, one finds an almost unlimited variety of telescope configurations. There are few major size limitations on materials for optics. Polishing of optics can often be done manually or with the aid of simple machinery. Mechanical requirements for strength or stiffness are easily met. Adjustments and pointing can be manually performed or motorized. Weights are modest. Opportunities for uniqueness abound and are often highly prized. Single-focus operation is typical. Most the telescopes used by amateur and professional astronomers are in this size range. Figure 2 illustrates a 40-cm telescope used by professional astronomers.

In the 1–2 m size range, a number of differences and limitations arise. Obtaining high-quality refractive optics is expensive in this range and not practical beyond. Simple three-point mechanical supports no longer suffice for the optics. The greater resolving-power potential demands higher quality optics and good star-tracking precision. Telescope components are typically produced on large machine tools. Instrumentation is likely to be used at more than one focus position. Because of cost, the domain of the professional astronomer has been reached.

As size goes above 2 m, new issues arise. The need to compensate for self-weight deflections of the telescope becomes increasingly important to maintaining optical alignment. Flexure in the structure may affect the bearings and drive gears. Bearing journals become large enough to require special bearing designs, often of the hydrostatic oil variety. The observer may now be supported by the telescope instead of the other way around. Support of the primary optics is more complex, and obtaining primary mirror blanks becomes a special, expensive task. Automated operation is typical at several focal positions. Star-tracking automatic guiders may be used to control the telescope drives, augmented by computer-based pointing correction tables. Figure 3 illustrates a 4-m telescope with all of these features.

At 5 m, the Hale Telescope on Mount Palomar is regarded as near the practical limit for equatorial-style mountings (see Section V). Altitude–azimuth (alt–az) mountings are better suited for bearing heavy rotating loads and are more compact. With computers the variable drive speeds required with an alt–az telescope can be managed. Mounting size and the length of the telescope are basic factors in setting the size of the enclosing building. For technical reasons and lower cost, the present trend in large telescopes is toward shorter primary focal lengths and alt–az mounts. This trend is evident from Table I, which lists the telescopes 3 m in size or larger that have been built since about 1950. Also listed are the major large telescopes proposed for construction in the late 1980s and the 1990s, which will be discussed in the next section. The largest

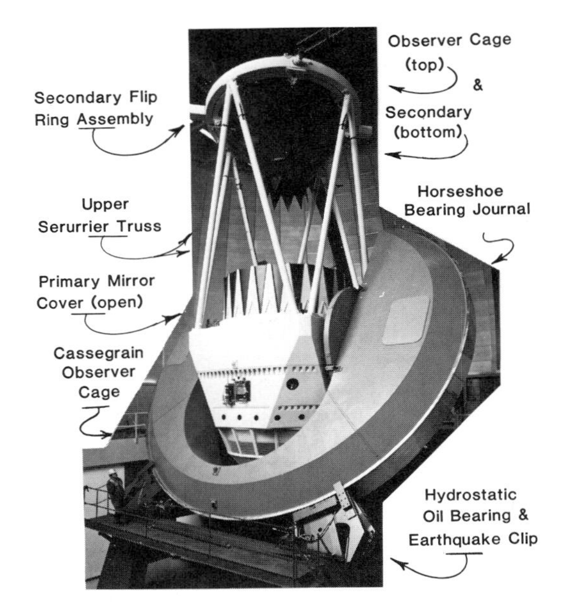

FIG. 3. Mayall 4-m telescope, with equatorial horseshoe yoke mountings. [Courtesy National Optical Astronomy Observatories, Kitt Peak.]

FIG. 2. 40-cm telescope on an off-axis equatorial mount. [Courtesy National Optical Astronomy Observatories, Kitt Peak.]

TABLE I. Telescopes 3 Meters or Larger Built Since 1950

Date completed	Telescope and/or institution	Primary mirror size (m)	Primary focal ratio	Mounting style
1950	Hale telescope, Palomar Observatory	5	3.3	Equatorial horseshoe yoke
1959	Lick Observatory	3	5.0	Equatorial fork
1973	Mayall telescope, Kitt Peak National Observatory	4.0	2.7	Equatorial horseshoe yoke
1974	Cerro Tololo	4.0	2.7	Equatorial horseshoe yoke
1975	Anglo-Australian telescope	3.9	3.3	Equatorial horseshoe yoke
1976	European Southern Observatory (ESO)	3.6	3.0	Equatorial horseshoe yoke
1976	Soviet Special Astrophysical Observatory	6.0	4.0	Alt–az
1979	Infrared Telescope Facility (IRTF)	3.0	2.5	Equatorial English yoke
1979	Canada-France-Hawaii Telescope (CFHT), Hawaii	3.6	3.8	Equatorial horseshoe yoke
1979	Infrared telescope (UKIRT), United Kingdom	3.8	2.5	Equatorial English yoke
1979	Multiple mirror telescope (MMT) Observatory, Mt. Hopkins	4.5[a]	Six 1.8-m[b]	Alt–az at f/2.7
1983	German–Spanish Astronomical Center, Calar Alto	3.5	3.5	Equatorial horseshoe fork
(1986)	Wm. Herschel telescope, La Palma	4.2	2.5	Alt–az
(1987)	European Southern Observatory	3.5	2.2	Alt–az
Proposed	University of Washington, Chicago, Princeton	3.5	1.75	Alt–az
Proposed	University of Texas	7.6	1.8	Alt–az
Proposed	Japanese National Telescope	7.6	2.0	Alt–az
Proposed	Array of separate telescopes, European Southern Observatory	16[a]	Four 8.0-m[b] at f/2	Alt–az
(1991)	Segmented parabolic primary with 36 hexagons, Keck Observatory	10[a]	1.75	Alt–az
Proposed	National New Technology Telescope (NNTT), MMT style	15[a]	Four 7.5-m[b] at f/1.8	Alt–az
Proposed	MMT style, Royal Greenwich Observatory	18[a]	Six 8.0-m[b] at f/1.8	Alt–az
Proposed	Segmented spherical primary with 400 hexagons, USSR	25[a]	2.7	Alt–az

[a] Equivalent circular mirror diameter with equal area.
[b] Number, size, and f/ratio of individual primary mirror.

optical telescope in operation today is the Soviet 6 m, which incorporates a solid, relatively thick (650 mm) primary mirror that had to be made three times in borosilicate glass and finally in a low-expansion material before it was successful. Such difficulty indicates that 6 m may be a practical limit for that style of mirror. New approaches are needed to go beyond.

C. The New Giant Telescopes

The desire for greater light-gathering power and image resolution, especially at IR wavelengths, continues to press astronomers to build telescopes with larger effective apertures. Costs for a given telescope style and imaging performance have historically risen nearly as the primary aperture diameter to the 2.5 power. These factors have given impetus to a number of new technology telescope designs (see Proposed Projects in Table I) that are based on one or more of the approaches discussed in the following. Computer technology plays a strong part in all of these approaches.

1. Extending the Techniques for Making Lightweight Monolithic Mirror Blanks

Sizes up to about 8 m are considered feasible, although the Soviet 6 m is the largest telescope mirror produced before 1985. Further discussion on blank fabrication methods is provided in Section III. Supporting such large mirrors to form good images will be difficult without some active control of the surface figure and thermal conditions in the mirror blank.

Several American universities and Japan are planning telescopes of this variety.

2. Making a Large Mirror from Smaller Segments

Also known as segmented mirror telescope, or SMT. In principle, no limit exists for the size of a mosaic of mirror segments that functions optically as a close approximation to a monolithic mirror. For coherency each segment must be precisely and continuously positioned with respect to its neighbors by means of position sensors and actuators built into its support. The segments may be hexagonal, wedge shaped, or other to avoid large gaps between segments. Practical limits arise from support structure resonances and cumulative errors of the segment positioning system. Manufacturing and testing the segments require special methods since each is likely to be a different off-axis optic that lacks a local axis of symmetry but must have a common focus with all the other segments.

The University of California and the California Institute of Technology have adopted this approach for their Keck Observatory ten-meter telescope, expected to be completed in 1991, which will look similar to the SMT in Fig. 4. The USSR also has announced plans for a 25-m SMT utilizing a spherical primary to avoid the problems of making aspheric segments.

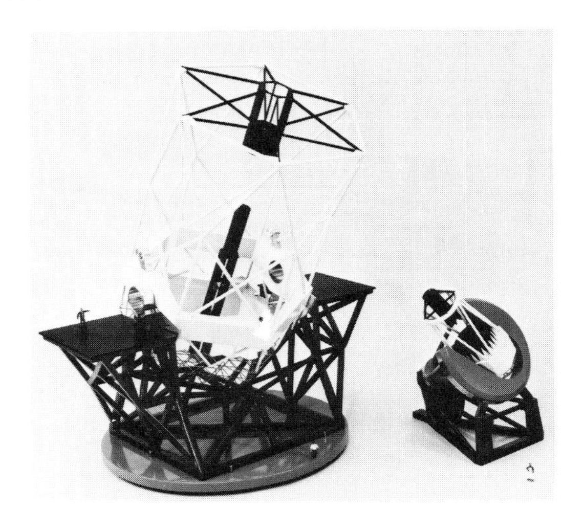

FIG. 4. Segmented mirror telescope (SMT) concept adopted for the Ten-Meter Telescope (TMT) at Keck Observatory (Mauna Kea, Hawaii) and considered for the 15-m National New Technology Telescope (NNTT). The NNTT model is shown with a model of the Mayall 4-m telescope to the same scale. [Courtesy of the National Optical Astronomy Observatories, Kitt Peak.]

3. Combining the Light from an Array of Telescopes

Several methods may be considered:

1. Electronic combination after the light has been received by detectors at separate telescopes. Image properties will be those due to the separate telescopes, and coherent combining is not presently possible. Strictly speaking, this is an instrumental technique and will not be considered further.

2. Optical combination at a single, final focus of light received at separately mounted telescopes. To maintain coherency between separate light beams, one must equalize the optical path distance (OPD) between the source and the final focus for all telescopes, a difficult condition to meet if telescopes are widely separated.

3. Placing the array of telescopes on a common mounting with a means for optically combining the separate light beams. All OPDs can be equal (theoretically), thus requiring only modest error correction to obtain coherency between telescopes. This approach is known as the multiple mirror telescope (MMT).

The simplest array of separately mounted individual telescopes is an arrangement of two on a northsouth (NS) baseline with an adjustable, combined focus between them (the OPD changes occur slowly with this arrangement when observing at or near the meridian). Labeyrie pioneered this design in the 1970s at Centre D'Etudes et de Recherches Géodynamiques et Astronomiques (CERGA) in France, where he used two 25-cm telescopes on a NS variable baseline of up to 35 m to measure successfully numerous stellar diameters and binary star separations, thereby showing that coherent beam combination and the angular resolution corresponding to a long telescope baseline could be obtained. Other schemes for using arrays of separate telescopes on different baselines all require movable optics in the optical path between the telescopes to satisfy the coherency conditions, and so far none has been successfully built. However, the European Southern Observatory is considering construction of an in-line array of four 8-m telescopes with an "optical trombone" arrangement for equalizing OPDs.

The MMT configuration was first used by the Smithsonian Astrophysical Observatory (SAO) and the University of Arizona (UA). The SAO/UA MMT on Mount Hopkins uses six 1.8-m image-forming telescopes arrayed in a circle around a central axis. Six images are brought to a central combined focus on the central axis, where they may be incoherently stacked, coherently combined, or used separately. The effective baseline for angular resolution (i.e., the maximum separation between reflecting areas) is 6.9 m, and the combined light-gathering power is equivalent to a single 4.5-m diameter mirror. Figure 5 shows a version of the MMT that has been adopted for the 15-m National New Technology Telescope (NNTT), where it is planned to use four 7.5-m mirrors with an angular resolution baseline of about 21 m. The NNTT design includes exchangeable secondary optics modules that enable switching from the combined beam mode to a mode allowing usage of the four telescopes for individual instrumentation.

FIG. 5. Planned 15-m National New Technology Telescope based on the multiple mirror telescope (MMT) concept. Four 7.5-m image-forming telescopes will be operated separately or as a square coherent array with a 21-m diagonal baseline and light-collecting area equivalent to a 15-m diameter mirror. Primary mirrors will be lightweight honeycombs, and the top end optics are in modules that can be rapidly interchanged. The combined beam configuration is shown. [Courtesy of the National Optical Astronomy Observatories, Kitt Peak.]

II. Optical Configurations

Light entering the telescope is redirected at each optical element surface until it reaches the focal region where the images are most distinct. The light-sensitive detector is customarily located in an instrument mounted at the focal region. By interchanging optics, one can create more than one focus condition; this is commonly done in large telescopes to provide places to mount additional instruments or to produce different image scales. The arrangement of optics and focal positions largely determines the required mechanical support configuration and how the telescope will be used. [*See* OPTICAL SYSTEMS DESIGN.]

The early telescopes depended solely upon the refractive power of curved transparent glass lenses to redirect the light. In general, these telescopes were plagued by chromatic aberration (rainbow images) until the invention in 1752 of achromatic lenses, which are still used today in improved forms. Curved reflective surfaces (i.e., mirrors) were developed after refractors but were not as useful until highly reflective metal coatings could be applied onto glass substrates. Today, mirrors are more widely used than lenses and can generally be used to produce the same optical effects; they can be made in larger sizes, and they are without chromatic aberration. These are still the only two means used to form images in optical telescopes. Accordingly, telescopes may be refractive, reflective, or catadiotropic, which is the combination of both. [*See* DIFFRACTION, OPTICAL.]

A. Basic Optical Configurations: Single and Multielement

The telescope designer must specify the type, number, and location of the optical elements needed to form the desired image. The basic choices involve material selections and the shapes of the optical element surfaces. Commonly used surfaces are flats, spheres, paraboloids, ellipsoids, hyperboloids, and toroidal figures of revolution.

The optical axis is the imaginary axis around which the optical figures of revolution are rotated. Light entering the telescope parallel to

TYPE	PRIMARY OPTIC	SECONDARY OPTIC	CONFIGURATION 1-PRIMARY 2-SECONDARY 3-EYEPIECES/CORRECTORS 4-FOCUS
KEPLERIAN GALILEAN (if refractive)	SPHERE or PARABOLA	NONE	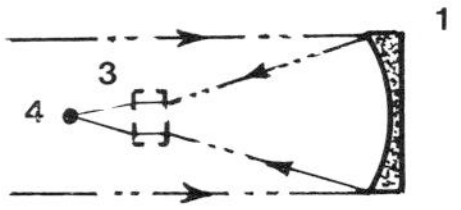
HERSCHELIAN	OFF-AXIS PARABOLA	NONE	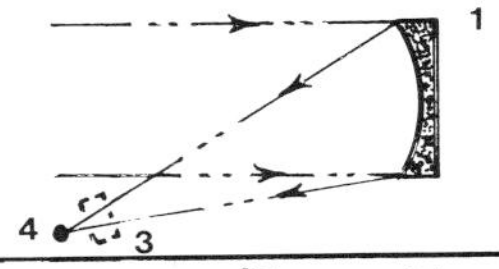
NEWTONIAN	PARABOLA	DIAGONAL FLAT	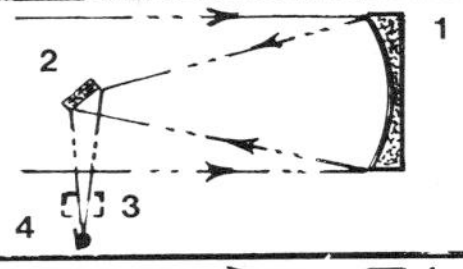
GREGORIAN	PARABOLA	ELLIPSE	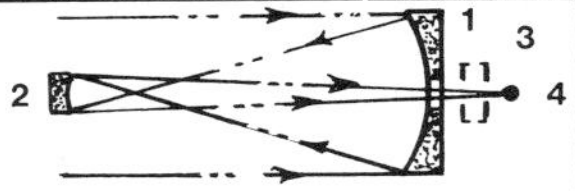
MERSENNE	PARABOLA	PARABOLA	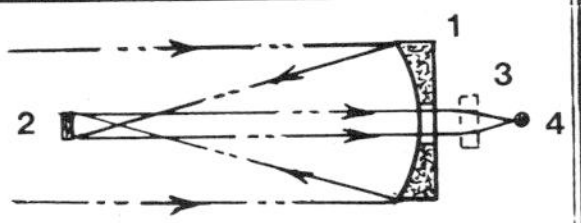
CASSEGRAIN	PARABOLA	HYPERBOLA	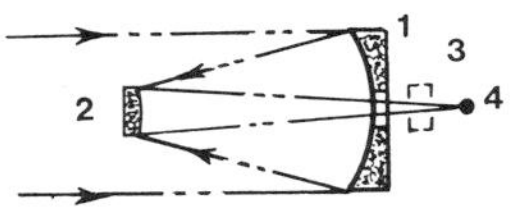
RITCHEY- CHRÉTIEN	MODIFIED PARABOLA	MODIFIED HYPERBOLA	
DALL-KIRKHAM	ELLIPSE	SPHERE	
SCHMIDT	ASPHERIC REFRACTOR	SPHERE	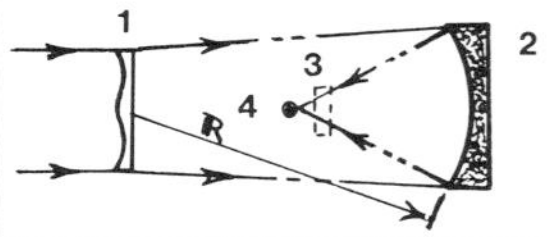
BOUWERS- MAKSUTOV	REFRACTIVE MENISCUS	SPHERE	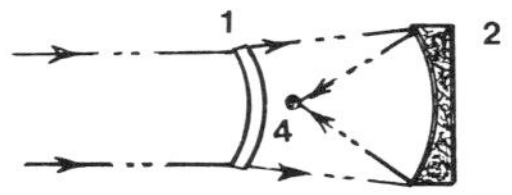

FIG. 6. Basic optical configurations for telescopes.

this axis forms the on-axis (or zero-field) image directly on the optical axis at the focal region. The field of view (FOV) for the telescope is the widest angular span measured on the sky that can be imaged distinctly by the optics.

In principle, a telescope can operate with just one image-forming optic (i.e., at prime focus), but without additional corrector optics, the FOV is quite restricted. If the telescope is a one-element reflector, the prime focus and hence the instrument/observer are in the line of sight. For large telescopes (i.e., >3 m) this may be used to advantage, but more commonly the light beam is diverted to one side (Newtonian) or is reflected back along the line of sight by means of a secondary optic to a more convenient focus position. Figure 6 illustrates the optical configurations most commonly used in reflector telescopes. In principle, the reflectors shown in Fig. 6 could be replaced with refractors to produce the same optical effects. However, the physical arrangement of optics would have to be changed.

In practice, one tries to make the large optics as simple as possible and to form good images with the fewest elements. Other factors influencing the configuration include the following:

1. Simplifying optical fabrication. Spherical surfaces are generally the easiest to make and test. Nonsymmetric aspherics are the opposite extreme.
2. Element-to-element position control, which the telescope structure must provide. Tolerances become tighter as the focal ratio goes down.
3. Access to the focal region for viewing or mounting instrumentation. Trapped foci (e.g., the Schmidt) are more difficult to reach.
4. Compactness, which generally aids mechanical stiffness.
5. Reducing the number of surfaces to minimize light absorption and scattering losses.

Analyzing telescope optical systems requires a choice of method. The geometric optics method treats the incoming light as a bundle of rays that pass through the system while being governed by the laws of refraction and reflection. Ray-tracing methods based on geometric optics are commonly used to generate spot diagrams of the ray positions in the final image, as illustrated in Fig. 7a. More rigorous analysis based on diffraction theory is done by treating the incoming light as a continuous wave and examining its interaction with the optical system.

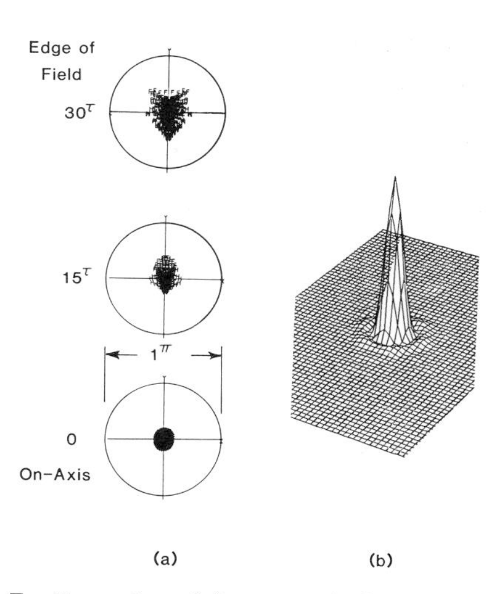

FIG. 7. Examples of image analysis methods. (a) Typical spot diagrams of well-corrected images in the focal plane of a 1 deg field. Each image represents the ray bundle at that field location. (b) Computer-generated 3-dimensional plot of intensity in a perfect image that has been diffracted by a circular aperture.

Figure 7b is a computer-generated plot of light intensity in a diffracted image formed by a circular aperture. The central peak represents the Airy disc. For further detail on the use of these methods the reader is referred to texts on optics design.

B. Wide Field Considerations

Modern ground-based telescopes are usually designed to resolve images in the 0.25–1.0-arcsec range and to have FOV from a few arc minutes to about one degree. In general, distortions or aberrations due to the telescope optics exist in the images and are worse for larger field angles. Table II lists the basic types. Space-based telescopes (e.g., the Hubble telescope) can be built to resolve images in the 0.1 arcsec region because atmospheric seeing effects are absent, but compensation must still be made for optical aberrations. Aberrations can also arise from nonideal placement of the optical surfaces (i.e., position errors) and from nonuniform conditions in the line of sight.

To correct distortions produced by optics, the designer frequently tries to cancel aberrations produced at one surface by those produced at another. Extra optical surfaces may be introduced for just this purpose. Ingenious optical corrector designs involving both refractors and reflectors have resulted from this practice.

TABLE II. Basic Image Aberrations Occurring in Telescopes

Type	Condition
Spherical aberration[a]	Light focuses at different places along the optical axis as a function of radial position in the aperture.
Coma[a]	Image size (magnification) varies with radial position in the focal region. Off-axis flaring.
Field curvature[a]	Off-axis images are not focused on the ideal surface, usually a plane.
Astigmatism[a]	Light focuses at different places along the optical axis as a function of angular position in the aperture.
Distortion[a]	Focused off-axis image is closer to or further from the optical axis than intended.
Chromatic aberration	Shift in the focused image position as a function of wavelength.

[a] Also known as Seidel aberrations.

Many of these designs are described in texts on optics under the originators' names (e.g., Ross, Baker, Wynne, Shulte, and Meinel). It is possible, however, to cancel some field aberrations by modifying the principal optical surfaces or by using basic shapes in special combination. For example, the Ritchey–Crétian telescope design for a wide FOV reduces coma by modifying the primary and secondary surfaces of a Cassegrain telescope. The Mersenne telescope cancels aberrations from the primary (a parabola) with the secondary (another parabola).

A spherical mirror with the aperture stop set at its center of curvature has no specific optical axis and forms equally good images everywhere in the field. Images of distant objects have spherical aberrations, however, and the focal region is curved. The Schmidt telescope compensates for most of the spherical aberration by means of an aspheric refractor at the center of curvature. The Maksutov telescope introduces an offsetting spherical aberration by means of a spherical meniscus refractor. Numerous variations of this approach have been devised yielding well-corrected images in field sizes of 10 deg and more. However, large fields may have other problems.

During a long observing period, images formed by a telescope with a very large FOV (i.e., ~1 deg or more) are affected differently across the field by differential refraction effects caused by the atmosphere. Color dispersion effects (i.e., chromatic blurring) effectively enlarge the images as the telescope looks through an increasing amount of atmosphere. Differential image motion also occurs, varying as a function of position in the FOV, length of observation, and telescope pointing angle. Partial chromatic correction can be made by inserting a pair of separately rotatable prisms, called Risley prisms, ahead of the focal position. Even with chromatic correction, however, images are noticeably elongated (>0.5 arcsec) at the edge of a 5-deg field compared with the on-axis image after a continuous observing period of a few hours.

C. Instrumental Considerations: Detector Matching and Baffling

At any focal position, the distance measured along the optical axis within which the images remain acceptably defined is referred to as the depth of focus. The detector should be adjusted to the most sharply focused position in this region; however, it is more common to adjust the focus (e.g., by moving the secondary) to sharpen the images at the detector. Certain instruments containing reimaging optics (e.g., spectrographs) require only that the image be formed at the entrance to the instrument (e.g., at a slit or aperture plate).

To view all or part of the focused FOV, it is common to insert a field mirror into the beam at 45 deg to divert the desired portion of the field out to an eyepiece or a TV monitoring camera. The undiverted light to be observed passes on to the instrument. The diverted light can be used for guiding purposes or to make different observations.

The angular size of the focused image (i.e., the image scale) and its sensitivity to defocusing changes is governed by the effective focal length (EFL) of the optical system that formed the image:

$$\text{image scale} = \frac{1}{\text{EFL}} = \frac{\text{radians on the sky}}{\text{length in the focal plane}}$$

Large EFL values produce comfortable focal depths, but the image sizes are relatively large. The physical size of the detector may place a limit on the FOV that can be accommodated for a given scale, hence a potential need for a different focal position or reimaging optics.

Light baffles and aperture stops are important but frequently neglected aspects of telescope design. It is common to place a light baffle just ahead of the primary to block out unwanted edge effects, in which case the baffle may become the aperture stop. In some cases, the primary surface may be reimaged further along the optical path where a light baffle can be located. This may be done either to reduce the size of the required baffle or to locate it advantageously in a controlled environment (e.g., at cryogenic temperature to reduce thermal radiation effects). In other cases, the aperture may be set by undersizing one of the optical elements that is further along the optical path. An example is an IR-optimized telescope, where the secondary is made undersized to ensure that the detector cannot see past the edge of the primary mirror. The effective light-gathering power of the telescope can be significantly reduced under these conditions.

Obstructions in the light path further reduce light-gathering power. The most common obstructions are the secondary and auxiliary optics along with the mechanical struts that support them. It is common for 10–20% of the aperture to be obstructed in this manner. Depending upon the telescope style and the instrument location, it may be necessary to use a light baffle to prevent the detector from seeing unwanted radiation. For example, a detector at the Cassegrain focus of a two-mirror telescope can see unfocused light from stars directly past the perimeter of the secondary mirror unless obstructing baffles are provided. The size and location of these baffles are generally determined empirically by ray-tracing, and it is a fact that larger FOV requires larger baffles that mean more obstruction. Thus, one cannot truly determine telescope light-gathering power until the usage is considered.

III. Telescope Optics

Telescope performance depends fundamentally upon the quality of optics, especially the surface figures. That quality in most telescopes is a compromise between what the optical designer has specified, what the glassmakers and opticians can make, and how well the mechanical supporting structures perform. Especially for larger sizes, one should specify the optics and their supports at the same time. How the optics are to be tested should also be considered since the final figure corrections are almost always guided by test results. Optical figuring methods today range from simple manual lapping processes to sophisticated computer-controlled polishers (CCP) and direct machining using diamond tools. The remarks to follow are only a summary of a complex technical field.

A. Refractive Optics

Light passing through the surface of a refractive optic is changed in direction according to Snell's law of refraction:

$$\eta_1 \sin \theta_1 = \eta_2 \sin \theta_2$$

where η_1 and η_2 are the refractive indices of the materials on either side of the surface and θ_1 and θ_2 the angles with respect to the surface normal of incidence and refraction. Producing a satisfactory refractor is, therefore, done by obtaining transparent glass or another transmitting material with acceptable physical uniformity and then accurately shaping the surfaces through which light passes. Sometimes the optician can alter the surface to compensate for nonuniform refractive properties. Losses ordinarily occur at the surface due to scattering and also to the change in refractive index. Antireflection coatings can be applied to reduce surface transmission losses, but at the cost of restricting transmission to a specific wavelength range. Further losses occur internally by additional scattering and absorption.

Manufacturing methods for refractive optics are similar to those for reflectors.

B. Reflective Optics

Light reflecting off a surface is governed by the law of reflection:

$$\theta_1 = -\theta_2$$

where θ_1 and θ_2 are the angles of incidence and reflection, respectively. Controlling the slope at all points on the reflecting surface is, therefore, the means for controlling where the light is directed. Furthermore, any part of the surface that is out of its proper position, measured along the light path, introduces a change at that part of the reflected wavefront (i.e., a phase change) equal to twice the magnitude of the position error.

Surface accuracy is thus the prime consideration in making reflective optics. Achieving high reflectivity after that is usually done by applying a reflective coating.

The most common manufacturing method is to rough-machine or grind the mirror blank surface and then progressively refine the surface with abrasive laps. Certain kinds of mirrors, especially metal ones, may be diamond turned. In this case, the accuracy of the optical surface may be governed by the turning machine, whereas ordinarily the accuracy limits are imposed principally by the optical test methods and the skill of the optician. High-quality telescope mirrors are typically polished so that most of the reflected light (>80%) is concentrated in an image that is equivalent to the Airy disc or the seeing limit, whichever is larger.

C. Materials for Optics

Essential material requirements for all optical elements are related to their surfaces. One must be able to polish or machine the surfaces accurately, and afterward the blank should not distort uncontrollably. Residual stresses and unstable alloys are sources of dimensional instability to be avoided. Stresses also cause birefringence in refractors.

1. Refractors

Refractors should transmit light efficiently and uniformly throughout the operating wavelength region. However, there is no single material that transmits efficiently from 0.3 to 40 μm. One typically chooses different materials for the UV, near-IR (to ≈ 2 μm), and far-IR regions. Glassmakers can control the index of refraction to about one part in 10^6, but only in relatively small blanks (<50-cm diameter). In larger sizes, index variations and inclusions limit the availability of good-quality refractor blanks to sizes less than 2-m diameter.

2. Reflectors

The working part of a reflector optic is usually the thin metallic layer, 1000–2000 Å thick, that reflects the light. An evaporated layer of aluminum, silver, or gold is most commonly used for this purpose, and it obviously must be uniform and adhere well to the substrate. Most of the work in making a reflector, however, is in producing the uncoated substrate or mirror blank. The choice of material and the substrate configuration are critically important to the ultimate reflector performance.

Reflector blanks, especially large ones, require special measures to maintain dimensional (surface) stability. The blank must be adequately supported to retain its shape under varying gravitational loads. It must also be stable during normal temperature cycles, and it should not heat the air in front of the mirror because that causes thermal turbulence and worsens the seeing. A mirror-to-air temperature difference less than about 0.5°C is usually acceptable. The support problem is a mechanical design consideration. Thermal stability may be approached in any of several ways:

1. Use materials with low coefficients of thermal expansion (CTE).
2. Lightweight the blank to reduce the mass and enhance its ability to reach thermal equilibrium quickly (i.e., by using thin sections, pocketed blanks, etc.). Machineability or formability of the material is important for this purpose.
3. Use materials with high thermal conductivity (e.g., aluminum, copper, or steel).
4. Use active controls for temperature or to correct for thermally induced distortion. Elastic materials with repeatable flexure characteristics are desirable. Provisions in the blank for good ventilation may be necessary.

Materials with low CTE values include borosilicate glass, fused silica, quartz, and ceramic composites. Multiple-phase (also called binary) materials have been developed that exhibit near-zero CTE values, obtained by offsetting the positive CTE contribution of one phase with the negative CTE contribution from another. Zerodur® made by Schott Glaswerke, ULE® by Corning Glass Works, and epoxy–carbon fiber composites are examples of multiple-phase materials having near-zero CTE over some range of operating temperature. Fiber composites usually require a fiber-free overlayer that can be polished satisfactorily.

One usually considers a lightweight mirror blank to reduce costs or to improve thermal control. This is particularly true for large telescopes. Reducing mirror weight often produces a net savings in overall telescope cost even if the lightweighted mirror blank is more expensive than a corresponding solid blank. The important initial step is to make the reflecting substrate or faceplate as thin as possible, allowing for polishing tool pressures and other external forces. One then has three hypothetical design options:

1. Devise a way to support just the thin monolithic faceplate. This approach works best

for small blanks. Large, thin blanks require complex supports and, possibly, a means to monitor the surface shape for active control purposes.

2. Divide the faceplate into small, relatively rigid segments and devise a support for each segment. Segment position sensing and control is required: a sophisticated technical task.

3. Reinforce the faceplate with a gridwork of ribs or struts, possibly connected to a backplate, to create a sandwichlike structure. One may create this kind of structure by fusing or bonding smaller pieces, by casting into a mold, or by machining away material from a solid block.

All of these approaches have been used to make lightweight reflector blanks. Making thin glass faceplates up to about 8 m is considered feasible by fusing together smaller pieces. Titanium silicate, fused silica, quartz, and borosilicate glass are candidate materials. Castings of borosilicate glass up to 6 m (e.g., the Palomar 5-m mirror) have been produced; 8 m is considered feasible. Structured (i.e., ribbed) fused silica and titanium silicate mirrors up to 2.3 m have been produced (e.g., the Hubble Space Telescope mirror); up to 4 m is considered feasible.

Metals may also be used for lightweight mirrors but have not been widely used for large telescopes because of long-term dimensional (surface) changes. The advent of active surface control technology may alter this situation in the future. Most metals polish poorly, but this can be overcome by depositing a nickel layer on the surface to be polished. Most nonferrous metals can also be figured on diamond-turning machines; however, size is limited presently to about 2 m.

IV. Spectral Region Optimization: Ground-Based Telescopes

The optical/IR window of the atmosphere from 0.3 to 35 μm is sufficiently broad that special telescope features are needed for good performance in certain wavelength regions. Notably, these are needed for the UV region, less than about 0.4 μm and the thermal IR region centered around 10 μm. In the UV, it is difficult to maintain high efficiency because of absorption losses in the optics. When observing in the IR region, one must cope with blackbody radiation emitted at IR wavelengths by parts of the telescope in the light path, as well as by the atmosphere. Distinguishing a faint distant IR source from this nearby unwanted background radiation requires special techniques. Using such techniques, it is common for astronomers to use ground-based telescopes to observe IR sources that are more than a million times fainter than the IR emission of the atmosphere through which the source must be discerned.

A. UV Region Optimization

The obvious optimizing step for the UV region is to put the telescope into space. If the telescope is ground-based, however, one good defense against UV light losses is to use freshly coated aluminum mirror surfaces. Reflectivity values in excess of 90% can be obtained from freshly coated aluminum, but this value rapidly diminishes as the surface oxide layer develops. Protective coatings such as sapphire (Al_2O_3) or magnesium fluoride (MgF_2) can be used to inhibit oxidations. Also, multilayer coatings can be applied to the surfaces of all optics to maximize UV throughput. However, these coatings greatly diminish throughput at longer wavelengths, which leads to the tactic of mounting two or more sets of optics on turrets, each set being coated for a particular wavelength region. The desired set is rotated into place when needed. This tactic obviously works best for small optics, not the primary.

Many refractive optics materials absorb strongly in the UV. Fused silica is good low-absorbing material. If optics are cemented together, the spectral transmission of the cement should be tested. Balsam cements are to be avoided.

B. IR Region Optimizations

One cardinal rule for IR optimization is to minimize the number and sizes of emitting sources that can be seen by the detector. This includes mechanical hardware such as secondary support structures and baffles, as well as seemingly empty spaces such as the central hole in the primary mirror. All of these emit black body radiation corresponding to their temperatures.

Since the detector obviously must see the optical surfaces, another cardinal rule is to reduce the emissivity of these surfaces with a highly reflective coating. If a coating is 98% reflective, it emits only 2% of the blackbody radiation that would otherwise occur if the surface were totally nonreflective. If possible, the detector should see only reflective surfaces, and these should be receiving radiation only from the sky or other reflective surfaces in the optical train. Objects that must remain in the line of sight can also be advantageously reflective provided that they are not looking at other IR-emitting objects that could send the reflected radiation into the main beam.

Achieving these goals may require one or more of the following special telescope features:

1. Using an exchangeable secondary support structure. This enables elimination of oversize secondaries and baffles that might be needed for other kinds of observation.
2. Using the secondary mirror as the aperture stop and making it sufficiently undersized that it cannot see past the rim of the primary mirror.
3. Putting all of the secondary support (except the struts) behind the mirror so that none of the hardware is visible to the detector.
4. Placing a specially shaped (e.g., conical) reflective plug at the center of the secondary to disperse radiation emitted from the central hole region of the primary mirror.

A basic technique for ground-based IR observing is that of background subtraction. This involves alternating the pointing direction of the telescope between the object (thus generating an object-plus-background signal from the detector) and a nearby patch of sky that has no object (thus generating a background-only signal). Signal subtraction then eliminates the background signal common to both sky regions. Methods for alternating between positions include (1) driving the telescope between the two positions, usually at rates below 0.1 Hz, (2) wobbling the secondary mirror at rates between 10 and 50 Hz (called chopping), and (3) using focal plane modulators such as rotating aperture plates (called focal plane chopping).

V. Mechanical Configurations

The mechanical portion of a mounted, ground-based telescope must support the optics to the required precision, point to and track the object being observed, and support the instrumentation in accessible positions. It is customary to distinguish between the telescope (or tube), which usually supports the imaging optics, and the mounting, which points the telescope and includes the drives and bearings. Tracking motion is usually accomplished by the mounting.

Most telescope mountings incorporate two, and occasionally three, axes of rotation to enable pointing the telescope at the object to be observed and then tracking it to keep it centered steadily in the FOV. In general, the rotating mass is carefully balanced around each axis to minimize driving forces and the location of each rotation axis is chosen to minimize the need for extra counterweights.

Figure 8 illustrates the basic mounting styles that are discussed in the next section.

A. Mounting Designs

A hand-held telescope is supported and pointed by the user. The user is the mounting in this case. The mechanical mounting for a telescope performs essentially the same function, except that a mechanical mounting can support heavier loads and track the object more smoothly. As a rule of thumb, short-term tracking errors in high-quality telescopes are less than 10% of the smallest resolved object that can be observed with the telescope. Smoothness of rotation is important for long-term observations (i.e., no sudden movements) which mandates the use of high-quality bearings. Pressurized oil-film bearings (hydrostatics) are used in large telescopes for this reason.

Telescope drives range widely in style. The chief requirements are smoothness, accuracy, and the ability to move the telescope rapidly for pointing purposes (i.e., slewing) or slowly for

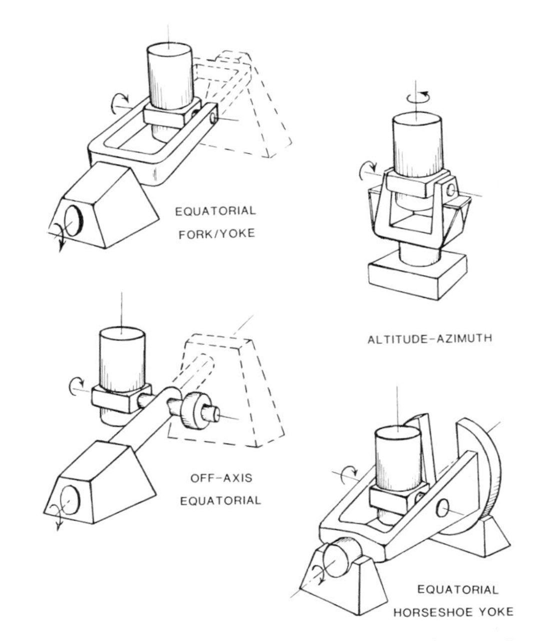

FIG. 8. Basic telescope mounting styles in popular use. Numerous variants on each style are in existence.

tracking (i.e., at one revolution/day or less). Electric motor driven traction rollers, worm gears, or variants on spur gears are most commonly used. Position measuring devices (encoders) are often used to sense telescope pointing and to provide input data for automatic drive controls. Adjustments in tracking rates or pointing are accomplished either by manual control from the observer or, possibly, by star-tracking automatic guiding devices. Pointing corrections may also be based on data stored in a computer from mounting flexure and driving-error calibrations done at an earlier time. Telescope pointing accuracies to about 1 arcsec are currently possible with such corrections. Once the object is located in the FOV, the ability to track accurately is the most important consideration.

1. Equatorial Mounts

Astronomical telescopes ordinarily are used to observe stars and other objects at such great distances that they would appear stationary during an observation period if the earth did not rotate. Accordingly, the simplest telescope tracking motion is one that offsets the earth's rotation with respect to "fixed" stars (i.e., sidereal rate) and is done about a single axis parallel to the earth's north–south (N–S) polar axis. Equatorial mountings are those that have one axis of rotation (i.e., the polar or right ascension axis) set parallel to the earth's N–S axis. This axis is tilted toward the local horizontal plane (i.e., the ground) at an angle equal to local latitude. A rotatable cross-axis (also called declination axis) is needed for initial pointing and guiding corrections, but the telescope does not rotate continuously around this axis while tracking.

The varieties of equatorial mountings are limited only by the designer's imagination. The basic varieties, however, are the following:

1. Those that mount the tube to one side of the polar axle and use a counterweight on the opposite side to maintain balance. For an example, see Fig. 2. These are sometimes called off-axis or asymmetric mounts. It is also possible to mount a second telescope in place of the counterweight.
2. Those that support the tube on two sides in a balanced way to eliminate the need for a heavy counterweight. Yokes and forks are most commonly used, especially for larger telescopes. These are sometimes called symmetric mounts. For an example, see Fig. 3 which shows a horseshoe yoke mount.

2. Other Mounting Styles

The alt–az mounting is configured around a vertical (azimuth) axis of rotation and a horizontal cross-axis (the altitude or elevation axis). The altitude–altitude (alt–alt) mounting, not widely used, operates around a horizontal axis and a cross-axis that is horizontal when the telescope points at the meridian and is tilted otherwise (similar to an English yoke with its polar axis made horizontal). With either of these styles, because neither axis is parallel to the earth's rotation, it is necessary to drive both axes at variable rates to track a distant object. Furthermore, the FOV appears to rotate at the focal region, which often necessitates a derotating instrument mounting mechanism, also moving at a variable rate. These factors inhibited the use of these mountings, except for manually guided telescopes, until the advent of computers on telescopes. Computers enable second-to-second calculation of the drive rates, which is required for accurate tracking. The alt–az configuration cannot track an object passing through the local zenith because the azimuth drive rate theoretically becomes infinite at that point. In practice, alt–az telescopes are operated to within about 1 deg of zenith.

The famous Herschel 20-ft telescope, built in England in 1783, was the first large alt–az telescope. Very few were built after that, but the trend today is toward alt–az mounts (see Section I,B and Table I). The ability to support the main azimuth bearing with a solid horizontal foundation is advantageous, as is the fact that the altitude axis bearings do not change in gravity orientation. These are important considerations when bearing loads of hundreds of tons must be accommodated. The alt–alt mounting is not as suitable for carrying heavy loads because the cross-axis is usually tilted with respect to gravity.

B. Telescope Tubes and Instrument Considerations

Design of the tube begins with the optical configuration. Tube structures are designed to maintain the optics in alignment, either by being stiff enough to prevent excessive deflections or by deflecting in ways that maintain the optics in the correct relative position. The well-known Serrurier truss first used on the Palomar 5-m telescope is a much-copied example of the latter (see Fig. 9). The tube structure is normally used to support the instruments at the focal positions, sometimes along with automatic guiders, field

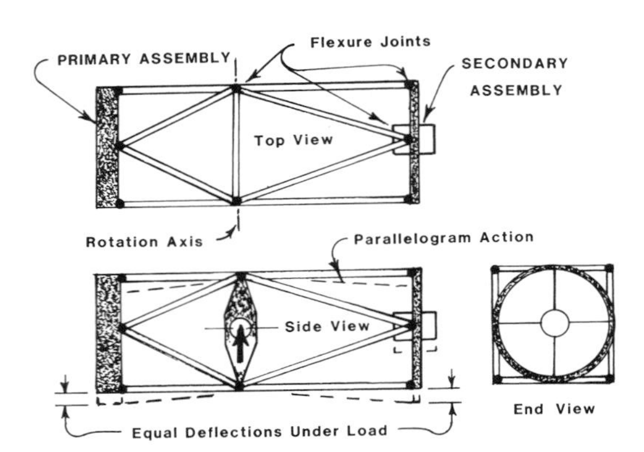

FIG. 9. Serurrier truss used to maintain primary-to-secondary alignment as the tube rotates. Equal deflections and parallelogram action at both ends keep the optics parallel and equidistant from the original optical axis. Similar flexure is designed into most large telescope tube structures.

viewing TV monitors, calibration devices, and field de-rotators.

Focal positions (i.e., instrument mount locations) on the tube obviously move as the telescope points and tracks, which can be a problem for instruments at those locations that work poorly in a varying gravity environment. In those cases, one can divert the optical beam out of the telescope tube along the cross-axis to a position on the mounting or even outside the mounting. Flat mirrors are normally used for this purpose. To reach a constant-gravity position with an equatorial mounting, one must use several mirrors to bring the converging beam out: first along the declination axis, then the polar axis, and finally to a focus off the mounting. This is known as the coudé focus and is commonly used to bring light to spectrographs that are too large to mount on the telescope tube.

One can reach a constant-gravity focus (instrument location) on an alt–az telescope by simply diverting the beam along the altitude cross-axis to the mounting structure that supports the tube. This is called the Nasmyth focus after its Scottish inventor. The instrument rides the mounting as it rotates in azimuth but does not experience a change in gravity direction.

VI. Considerations of Usage and Location

Considering the precision built into most optical telescopes, one would expect them to be sheltered carefully. In practice, most telescopes must operate on high mountains, in the dark, and in unheated enclosures opened wide to the night sky and the prevailing wind. Under these conditions, it is not unusual to find dust or dew on the optics, a certain amount of wind-induced telescope oscillation, and insects crawling into the equipment. Certain insects flying through the light path can produce a noticeable amount of IR radiation. Observer comforts at the telescope at minimal.

In designing a telescope, one should consider its usage and its environment. A few general remarks in this direction are provided in the following sections.

A. Seeing Conditions

The seeing allowed by the atmosphere above the telescope is beyond ordinary control. Compensation may be possible as discussed in Section I,A. but the choice of site largely determines how good the imaging is. Seeing conditions in the region of 0.25 arcsec or less have been measured at certain locations, but more typically, good seeing is in the 0.5–1.0 arcsec range. Beyond 2–3 arcsec, seeing is considered poor. To the extent possible, one should build the telescope to produce images equal to or better than the best anticipated seeing conditions.

Locating the telescope at high altitudes usually reduces the amount of atmosphere and water vapor that is in the line of sight (important for IR astronomy), however, the number of clear nights and the locally produced thermal turbulence should also be considered. In many locations, a cool air layer forms at night near the ground which can be disturbed by the wind and blown through the line of sight. In other cases, warm air from nearby sources can be blown through the line of sight. In either case, telescope seeing is worsened.

Other seeing disturbances can originate inside the telescope enclosure. Any source of heat (including observers) is a potential seeing disturbance. Also, any surface that looks at the night sky, and hence is cooled by radiative exchange, may be a source of cooled air that can disturb seeing if it falls through the line of sight. If possible, it is desirable to allow the telescope enclosure to be flushed out by the wind to eliminate layers and pockets of air of different temperatures. Some telescope buildings have been equipped with air blowers to aid in the process, but dumping the air well away from the building has not always been possible even though it should be done.

The study of atmospheric seeing has become a relatively advanced science, and the telescope builder is well advised to consult the experts in choosing a site or designing an enclosure. Having chosen a site, one may be guided by the truism that seeing seldom improves by disturbing Mother Nature.

B. Nighttime versus Daytime Usage

With the advent of IR astronomy, optical telescopes began to be used both day and night because the sky radiation background is only slightly worse at IR wavelengths during the day compared with night. During the day it is much harder to find guide stars, and the telescope must often point blindly (and hence, more accurately) at the objects to be observed; but much useful data can be obtained. Some problems arise from this practice, however.

A major purpose of the telescope enclosure, other than windscreening, is to keep the telescope as close as possible to the nighttime temperature during the day so that it can equalize more rapidly to the nighttime temperature at the outset of the next night's observing. Obviously, this cannot be done if the telescope enclosure has been open during the day for observation. The condition is worsened if sunlight has been allowed to fall on the telescope during the day. Accordingly, optical/IR telescopes should be designed for rapid thermal adjustment.

Some of the design options in thermal control are (1) to insulate heavy masses that cannot equalize quickly, (2) to reduce weights and masses, (3) to provide good ventilation (i.e., avoiding dead air spaces that act as insulators), (4) to make surfaces reflective so that radiative coupling to the cold night sky is minimized, and (5) to isolate or eliminate heat sources. One should also consider using parts made from materials with low thermal expansion, but these have limited value if their heating effects are allowed to spoil the telescope seeing.

C. Remote Observing

The traditional stereotype of an astronomer is a person perched on a high stool or platform, peering through the eyepiece and carefully guiding the telescope. The modern reality is likely to be quite different. Sophisticated electronic detectors replace the eye. Automatic star-tracking guiders take over the guidance chore. The astronomer sits in a control room sometimes far away from the telescope. A TV monitor shows the FOV or, at least, that part of the field not falling on the detector. A computer logs the data and telescope conditions. The telescope is not even seen by the astronomer: It can be in the next room or even a continent away if the communication link is properly established.

The advent of space-based astronomy clearly marked the time when the astronomer and the telescope were separated. The same separation is taking place in ground-based astronomy, albeit less dramatically. Numerous demonstrations have occurred during the 1970s and 1980s in which astronomers conducted observing runs on telescopes located at distant sites. In one case, the astronomer was in Edinburgh, Scotland and the telescope was on Hawaii. The connection was through a communications satellite. This trend is likely to accelerate as the cost for such connections reduces and the data transmission rates increase.

The future stereotype astronomer is likely to be perched at a computer terminal, not a telescope. For some, the romance of astronomy will be gone; for most, the gain in capability will far outweigh the loss. Remotely located telescopes, attended by highly skilled operators, with automated instrumentation linked by computer to the astronomer, are seen by this author as the forefront astronomical equipment of the future. Galileo would be pleased.

Bibliography

Bell, L. (1981). "The Telescope." Dover, New York.

Burbidge, G., and Hewitt, A., eds. (1981). "Telescopes for the 1980s." Annual Reviews, Palo Alto, CA.

Driscoll, W. G., and Vaughan, W., eds. (1978). "Handbook of Optics." McGraw-Hill, New York.

King, H. C. (1979). "The History of The Telescope." Dover, New York.

Kingslake, R. (1983). "Optical System Design." Academic, Orlando, Florida.

Kuiper, G., and Middlehurst, B., eds. (1960). "Telescopes," Stars and Stellar Systems, Vol. 1. University of Chicago Press, Chicago.

Learner, R. (1981). "Astronomy Through the Telescope." Van Nostrand Reinhold, New York.

Marx, S., and Pfau, W. (1982). "Observatories of the World." Van Nostrand Reinhold, New York.

TUNABLE DYE LASERS

T. F. Johnston, Jr. *Coherent, Inc.*

GLOSSARY

Amplified spontaneous emission (ASE): Background light from dye fluorescence, which in high-gain short-pulse dye lasers limits the achievable gain and must be suppressed between amplifier stages to leave a spectrally pure laser output.

Autocorrelation, background-free: Method of measuring mode-locked laser pulse lengths, in which the pulse train is split into two beams, one of which is delayed, and in which the two beams are focused to overlap in a crystal whose phase-matching conditions permit frequency summing only when one photon is taken from each beam. The summed or doubled power, versus path delay, is proportional to the autocorrelation in time of the pulse with itself and so gives the pulse length, if the functional form of the pulse shape is known.

Brewster's angle: That singular angle of incidence (the angle between the ray direction and the normal to the surface of a dielectric plate of index n) given by $\tan \theta = n$ for which there is 100% transmission (no reflection) of light whose electric field vibrates in the plane of incidence. This low-loss property makes Brewster angle windows and entrance faces very common in intracavity laser optics.

Cavity dumper: Device for coupling single pulses out of the resonator of a mode-locked laser, consisting of an acousto–optic cell situated at a beam focus in the cavity. A transducer bonded to the cell receives a short burst of radiofrequency (rf) energy timed to the pulse, setting up sound waves that diffract the pulse at a slight angle to dump it out of the cavity. This is used both to reduce the pulse repetition rate to a manageable value and increase the energy per pulse.

Chromophore: Series of alternating single and double bonds in a dye molecule (be it linear, branched, or cyclic linkages), which through chemical resonance give rise to laser states, strong visible absorption and color, and delocalized mobile electrons, making the dye molecule a good "antenna" for visible light.

Etalon: Fabry–Perot or parallel-plate interferometer of fixed plate separation (often the coated, parallel sides of a plate of glass), which when tipped at an incidence angle greater than the beam divergence acts in transmission as an intracavity frequency (or longitudinal mode) selecting filter.

Free spectral range (FSR): Frequency spacing between adjacent transmission peaks of a parallel-plate interferometer and, by extension, between orders of other multiply ordered frequency filters.

Jitter: Average residual frequency excursion or linewidth of a single-mode laser, due to residual environmental perturbations of the cavity length.

Mode locking: Formation of a high-peak-power, repetitive train of short-output pulses from a laser of cavity length d by the imposition of constant or "locked" phases for the difference or beat frequencies (nominally $c/2d$) between adjacent longitudinal modes. Here c is the speed of light, and a modulator (mode locker) or saturable absorber placed in the cavity forces the phase locking. The Fourier transform to the time domain of this locked-mode spectrum gives a pulse of period $2d/c$, the round-trip transit time for a circulating pulse in the cavity, and pulse length equal to the inverse of the locked bandwidth.

Modes, longitudinal and transverse: Terms for the discrete frequencies and associated intensity distributions supported by a laser resonator. At a mode frequency, an integral number of wavelengths fits in the path distance of a round trip in the resonator, for then the multiply reflected waves from many transits add up in phase, and oscillation will be fed back there. The integral number of on-axis wavelengths defines the *longitudinal mode* number q, and there are intensity distributions possible peaking off-axis, adding path length, altering the frequency, and producing interference nodes in the two directions transverse to the axis, whose integer numbers m, n label the *transverse mode*. As light waves are transverse electromagnetic waves, by analogy with the designations for modes in microwave cavities, the full designation for a laser cavity mode is TEM_{mnq}, although by convention the q is often suppressed, the lowest-order transverse mode (a single spot of Gaussian intensity profile) being designated the TEM_{00} mode.

Phase matching: Process of adjusting the extraordinary refractive index relative to the ordinary refractive index of a nonlinear crystal either by changing the temperature (temperature-tuned matching) or the direction of propagation of a laser beam relative to the crystal's optic axis (angle-tuned matching), to make the indices of the fundamental beam and its orthogonally polarized harmonic precisely match, within a few parts per million. Then the harmonic frequencies radiated from the microscopic volume elements along the path of the fundamental beam all add up in phase in the forward direction to produce a macroscopic harmonic beam.

Saturation intensity: Intensity producing a sufficiently rapid stimulated transition rate to reduce the level population difference to half its initial value (and thus reduce the absorption or gain on the transition to half its initial value); the scale parameter in saturated absorption or gain saturation.

Singlet and triplet states: Terms referring to the net or total electronic spin of a molecular state, a singlet being a spin of zero, and a triplet a spin of one quantum unit of angular momentum. (In a magnetic field the unity spin can be oriented parallel, antiparallel, or perpendicular to the field, giving rise to the term "triplet.") The distinction is important in dyes that have a singlet ground state, because the optical selection rules forbid strong transitions where the spin changes (spin is conserved), making the excited triplet state metastable and troublesome in designing dye lasers.

Spatial hole burning: Gain saturation by a single-frequency standing wave field proceeds primarily locally at the antinodes of the standing wave, reducing the gain there or "burning spatial holes," a process of importance in mode selection in single-frequency lasers.

Spectral condensation: Efficient spectral narrowing that occurs when a wavelength-selective filter is added to a dye laser cavity; the oscillating linewidth may be reduced many orders of magnitude, while the output power or energy is reduced typically by less than half.

Synchronous pumping: Method of mode-locking a laser (here the dye laser) by gain modulation or repetitive pulsed pumping, where the round-trip transit time in the cavity of the pumped laser is adjusted to be precisely equal to (synchronous with) the period of a mode-locked pumping laser (here, usually an argon laser or doubled Nd : YAG laser).

Wave number: The number $1/\lambda$ of wavelengths in a centimeter (units of cm^{-1}), a convenient spectroscopic unit of frequency because the free spectral range of an interferometer of plate spacing t (cm) is just $1/2t$ (cm^{-1}). To convert the wave number of a photon to its energy in joules, multiply by $hc = 1.99 \times 10^{-23}$ (h is Planck's constant, c the speed of light).

Dye lasers are the light energy convertors of the scientific world, converting input pump light from flashlamps or fixed-wavelength pump lasers into tunable-output wavelengths, broad pumping bandwidths into narrow-linewidth outputs, and short pumping pulses into ultrashort-output pulses. To make these conversions they use optically excited molecular antennas—the conjugated double bonds of the strongly absorbing and fluorescing organic substances commonly called dyes, the same substances that color clothes, photographic emulsions, and Easter eggs. Among the widely variable and highly desirable properties of the output beams from dye lasers are the broadest range of wavelengths from any laser (ultraviolet to infrared), the narrowest linewidths of common spectroscopic sources (10^2 Hz at center frequencies near 10^{15} Hz), and the shortest electrical pulses directly generated in a human-made device (10^{-14} sec).

I. Introduction

A. Dye Laser Types

Laser action in dyes takes place between the broad, diffuse energy bands characteristic of complex molecules in condensed liquid solvents. The tunability of the dye laser—the ability to smoothly vary the output wavelength over a broad range, much as the receiving wavelength of a radio may be continuously tuned—is a result of these broadened electronic states. A large number of organic dyes exist (more than 200 are actively in use as laser dyes), each fluorescing in a specific wavelength region, giving overlapping tuning ranges from which the laser's broad overall spectral coverage is constructed. The tunability of the dye laser is the basic attribute from which its other desirable properties are derived.

The diffuseness of the energy bands is evidence of the short lifetimes of the laser states, due both to the strongly allowed nature of the laser electronic transition and to the frequent collisions of each dye molecule with its neighbors. To invert such states takes large pumping rates, so far only achieved by optical means, with input light from a high-intensity flashlamp or a second, pumping laser. Pump lasers occur most commonly in scientific applications (the focus of this article), since with them the widest range of dye laser output properties are provided; flashlamps were important in the development of the dye laser and are still the most economical pump sources. [*See* LASERS.]

The three main branches of the family of scientific dye lasers are distinguished by the types of their pump sources and by the applications appropriate to their resultant outputs. Pulsed dye lasers take advantage of the active laser medium being a liquid, to scale to large pumped volumes readily cooled by flowing the dye. From pulsed pump laser inputs, they produce high-output energies per pulse, typically tens of millijoules in a 10-nsec pulse, in a beam a fraction of a centimeter across. The resultant peak intensities, exceeding a megawatt per square centimeter, are large enough that the nonlinear terms may dominate in the response function of a dielectric sample placed in the beam. The intense electric field time modulates the index of refraction of the sample, to produce new light frequencies or other effects in the interaction revealing the structure of matter, and making up the body of applications termed nonlinear optics. [*See* NONLINEAR OPTICAL PROCESSES.]

Continuous-wave (cw) dye lasers, pumped by continuous output lasers, take advantage of the efficient spectral narrowing that occurs in a dye laser when a dispersive element is added to the feedback cavity for wavelength control. In the first observation of this effect, a diffraction grating was substituted for an end mirror in an early pulsed dye laser. The term "spectral condensation" was coined to describe the result, that the bandwidth of the output beam shrank to 1% of the original bandwidth, while retaining 70% of the original output energy. The spectral filters for today's cw dye lasers are designed to select and allow oscillation only on a single mode of the dye laser cavity. The narrowness of the spectrum of output frequencies is then determined by how stable is the frequency of that cavity mode, and frequency stabilization servos are employed to produce condensation factors smaller than 10^{-8}. These output bandwidths are less than 10^{-3} of the Doppler broadening width of a typical spectral line in a gas. When incident on a gaseous sample, this light will produce a velocity-selected, "saturated" absorption—an absorption only by those atoms with the proper velocity component to be Doppler-shifted into resonance with the frequency of the laser beam. The selected absorption can be read with a second, probe laser beam. By looking inside the Doppler width, spectral resolutions are reached that are finer by factors of 10^{-1} to 10^{-3} than was possible before, making up the body of applications termed high-resolution laser spectroscopy.

Mode-locked dye lasers take advantage of the broad spectrum over which the dye molecules can lase, to assemble a short pulse of light from oscillations at many dye cavity mode frequencies (tens of thousands of modes are made to oscillate with their phases locked together). The dye cavity or feedback resonator nominally supports oscillations only at the cavity mode frequencies $q(c/2d)$ (where q is a large integer of order 10^6, c is the speed of light, and d is the optical path distance between the cavity end mirrors). These are just the frequencies where the multiply reflected waves, generated in multiple transits of the cavity, reinforce and add up in phase. The frequency differences between adjacent cavity modes are all nominally $c/2d$, but ordinarily the phases of these difference (or mode-beating) frequencies are random. If the phases can be made constant, or *locked,* the Fourier sum of the phase-locked frequencies will produce a repetitive, high-peak-power spike of duration the inverse of the locked bandwidth, and of period of $2d/c$, (the cavity round-trip transit time) corresponding to a short pulse of light circulating back and forth in the cavity. Phase- or mode-locking is produced in two different ways. In passive mode-locking (which uses a cw pump laser), an absorber is introduced into the dye cavity (generally, this is a second dye), whose absorption lessens (saturates) at high input intensities. This favors a form of oscillation that maximizes the peak power incident on the absorber, to produce the greatest saturation, and pulse formation and mode-locking ensue. In active mode-locking, most commonly today of the form called synchronous mode-locking, the dye laser is pumped with a mode-locked pump laser. The optical cavity length of the dye laser is adjusted to precisely equal that of the pump laser, matching the transit times in the two cavities. This favors the formation of that short pulse in the dye cavity that optimally saturates the repetitive, pulsed gain from the pump pulse, and again mode-locking ensues. With the short pulses assembled in this way, it is possible (by saturation and probe techniques) to measure molecular reorientation times and trace energy flow paths in the photochemistry of molecules of biological or chemical interest. This body of applications is termed time-domain laser spectroscopy. [*See* MODE-LOCKING OF LASERS; ULTRAFAST LASER TECHNOLOGY.]

The dye laser has well fulfilled the spectroscopist's wish, expressed soon after the laser was invented, for an almost infinitely malleable laser whose wavelength, bandwidth, or pulse length could be varied as desired. Since its invention, the dye laser has evolved with or caused each advance in pump laser technology, in a 20-year record of increasing malleability. Practical, commercial dye laser systems convert 20–50% of the input light energy into variable, controlled-output light energy—an efficiency record that does not leave much room for other competing nonlinear optical processes to displace the use of dyes. As long as this historical trend of coevolution persists, it is likely that the dye laser will remain the premier instrument bringing a gleam to the spectroscopists' eye.

B. Historical Overview

The evolution of the scientific dye laser family tree is outlined in Fig. 1, which shows the year of introduction of each of the major ideas to be discussed in this article. The figure may serve as a guide in further literature study, to help find an early paper where the idea (then new) is thoroughly discussed. On the margins, indications are given of the pump lasers predominately in use, and on the right edge of each branch the performance specifications available by year are listed. For the linewidths (in hertz) shown for the cw branch, and pulse lengths (in seconds) shown for the mode-locked branch, these specification represent the best performance published through that year, as researchers generally could reproduce these results once the details were published. The average power specification (in watts) given for the pulsed branch is more problematical. For some experiments the energy per pulse is most important, in others it is the pulse repetition rate, and researchers (whose lasers may excel in only one regime, and not the other) do not always report both. Therefore this specification axis gives the average power available from commercial pulsed lasers by year. Flashlamp pumped dye lasers are obviously pulsed lasers, but are shown as a transition leading into the column to which they gave rise, the cw branch. The years 1966–1970, when the three main types were differentiated, are discussed here. A brief sketch is then given of the trends each branch subsequently followed, with more details in later sections.

Several publications in the first years of the 1960s suggested that lasing might be possible in optically excited organic dyes. The first experimental attempt to demonstrate a dye laser was in 1964, where a dye cell positioned between two cavity mirrors and filled with a solution of

FIG. 1. The historical development of the three main types of dye lasers used as scientific instruments today. The year of introduction in the literature is shown of ideas discussed in this text, along with an indication of the types of pump lasers predominately in use and the output specifications available from each type of dye laser.

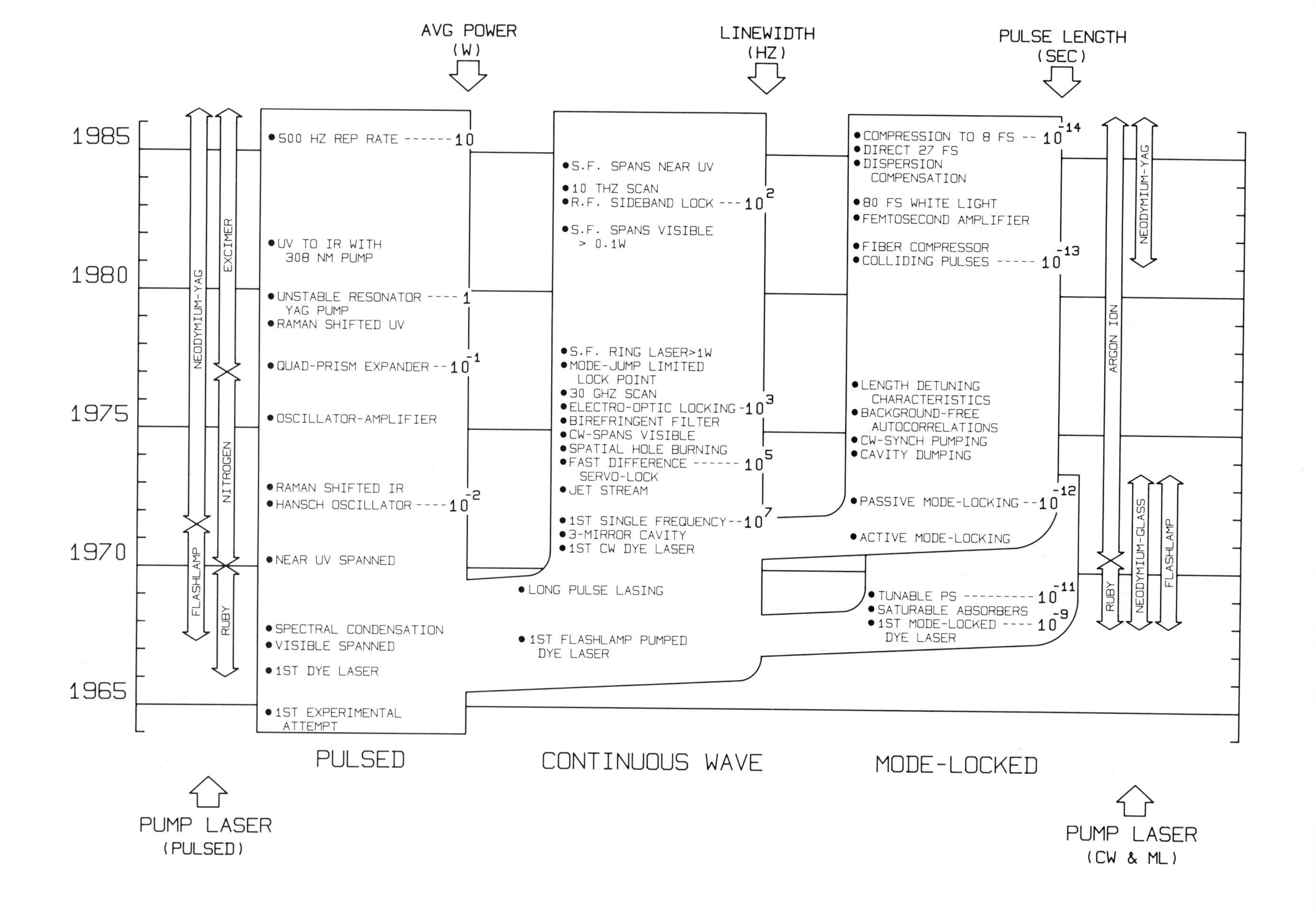

AVG POWER (W)
LINEWIDTH (HZ)
PULSE LENGTH (SEC)
1985
1980
1975
1970
1965
NEODYMIUM-YAG
EXCIMER
NITROGEN
FLASHLAMP
RUBY
500 HZ REP RATE ------ 10
UV TO IR WITH 308 NM PUMP
UNSTABLE RESONATOR ---- 1
YAG PUMP
RAMAN SHIFTED UV
QUAD-PRISM EXPANDER -- 10^-1
OSCILLATOR-AMPLIFIER
RAMAN SHIFTED IR
HANSCH OSCILLATOR ---- 10^-2
NEAR UV SPANNED
SPECTRAL CONDENSATION
VISIBLE SPANNED
1ST DYE LASER
1ST EXPERIMENTAL ATTEMPT
S.F. SPANS NEAR UV
10 THZ SCAN
R.F. SIDEBAND LOCK --- 10^2
S.F. SPANS VISIBLE > 0.1W
S.F. RING LASER>1W
MODE-JUMP LIMITED LOCK POINT
30 GHZ SCAN
ELECTRO-OPTIC LOCKING -10^3
BIREFRINGENT FILTER
CW-SPANS VISIBLE
SPATIAL HOLE BURNING
FAST DIFFERENCE ------- 10^5
SERVO-LOCK
JET STREAM
1ST SINGLE FREQUENCY -- 10^7
3-MIRROR CAVITY
1ST CW DYE LASER
LONG PULSE LASING
1ST FLASHLAMP PUMPED DYE LASER
COMPRESSION TO 8 FS -- 10^-14
DIRECT 27 FS
DISPERSION COMPENSATION
80 FS WHITE LIGHT
FEMTOSECOND AMPLIFIER
FIBER COMPRESSOR
COLLIDING PULSES ----- 10^-13
LENGTH DETUNING CHARACTERISTICS
BACKGROUND-FREE AUTOCORRELATIONS
CW-SYNCH PUMPING
CAVITY DUMPING
PASSIVE MODE-LOCKING -- 10^-12
ACTIVE MODE-LOCKING
TUNABLE PS --------- 10^-11
SATURABLE ABSORBERS
1ST MODE-LOCKED ---- 10^-9
DYE LASER
NEODYMIUM-YAG
ARGON ION
RUBY
NEODYMIUM-GLASS
FLASHLAMP
PULSED
CONTINUOUS WAVE
MODE-LOCKED
PUMP LASER (PULSED)
PUMP LASER (CW & ML)

the dye perylene was illuminated by a flashlamp. This dye is now known to have high losses from excited-state absorptions, and this attempt was ended when there was no clearly positive result.

The first certain dye laser action was in 1966 in the work of Sorokin and Lankard at the IBM Research Center, Yorktown Heights, New York. They were interested in the spectral properties of the phthalocyanine dyes, a class of dyes then in use in *Q*-switch solutions for ruby lasers. (The normal temporally spiking output from a ruby laser may be converted to a single giant pulse, by adding a saturable absorber dye cell, or "*Q*-switch," to the ruby cavity.) When the dye chloroaluminum phthalocyanine was excited by the output of a giant-pulse ruby laser at 694 nm, Sorokin and Lankard observed an anomalous darkening of their spectral plate at 755 nm. This, they suspected, was due to stimulated emission as this wavelength coincided with a peak in the dye fluorescence curve, and they quickly confirmed this by adding resonator mirrors around the dye cell to generate an intense, well-collimated beam. By the end of the year, the same discovery had been made independently three more times, again in the United States by Spaeth *et al.*, in Germany by Schafer *et al.*, and in the Soviet Union by Stepanov *et al.*, all workers also studying *Q*-switch dyes stimulated by ruby laser giant pulses.

Dyes fluoresce and lase only at wavelengths longer than their excitation wavelength; thus, to extend the infrared emission of these first dye lasers to the visible required a shorter-wavelength pump source than the direct ruby laser. This was obtained by doubling the output frequency of ruby or of neodymium–glass (Nd : glass) lasers in the nonlinear crystals potassium dihydrogen phosphate (KDP) or ammonium dihydrogen phosphate (ADP), to give 347-nm (near ultraviolet) or 532-nm (green) pumping beams. Nearly two dozen new lasing dyes were quickly demonstrated with these new pump sources, spanning the visible spectrum with pulsed dye outputs by the end of 1967. These dye lasers were tuned by varying the concentration of the dye in the cell.

The spectral condensation effect was discovered that same year. A diffraction grating in the Littrow orientation replaced one mirror of a two-mirror cavity, to reflect a dispersed fan of color back along the general direction of the incident light. Only a small band of wavelengths lay in the solid angle of the modes of the dye cavity defined by the grating and the second mirror. Not only was most of the original energy retained in the resulting narrow band output, but the wavelength of this output could be tuned (over seven times the previous range) by merely rotating the grating, a vast improvement in ease of tuning.

The requirements for successful flashlamp pumping of a dye could be determined from measurements on laser-pumped dye lasers, and this significant step followed in 1967 as well, making dye lasers accessible to many more scientists. What had not been sufficiently appreciated in 1964 was that the absorption overlapping the lasing wavelengths, arising from the excited, metastable triplet state of the dye molecule, could be largely avoided if the rise time of the flashlamp pulse were fast enough (less than a few tenths of a microsecond instead of the typical 10 μsec). Then lasing could occur before the build-up of an appreciable triplet population. Perylene was made to lase in 1972 with 20-nsec pump pulses from a *Q*-switched, doubled ruby laser.

The large-gain bandwidths demonstrated by the tuning ranges of these dyes prompted efforts to mode-lock the dye laser to produce tunable ultrashort pulses. Mode-locking was achieved in 1968 first by synchronously pumping with a mode-locked, doubled, Nd : glass laser, and then by introducing an absorber dye cell in the cavity of a flashlamp-pumped dye laser. The pulse lengths of these first systems were reported as being "detector limited," less than a half nanosecond, the rise time of a fast oscilloscope. By the end of that year the pulses had been shown by the two-photon fluorescence technique to be 10^{-11} sec in length, and the substitution of the diffraction-grating end reflector had shown these short pulses to be tunable.

There also was interest in generating longer pulses from dye lasers to approach cw operation, by removing the limits imposed by the absorbing triplet states. Careful measurements on long-pulse flashlamp-pumped dye lasers led to the conclusion that in some dyes, notably the rhodamines, the triplet population could be sufficiently quenched (returned to the ground state) by chemical interaction with oxygen from the air dissolved in the dye mix, that the residual triplet absorption did not in principle prevent cw operation. (Other chemical triplet-state quenchers were subsequently discovered.) What terminated the lasing of the dye before the end of the pump pulse, it was then realized, were the losses due to optical inhomogeneities in the heated dye. Peterson, Tuccio, and Snavely, at the Kodak Research Laboratories (Rochester, N.Y.), designed a dye cell and pumping geometry to deal with these thermal problems after carefully evaluating the pump power that would be required to exceed cw laser threshold. They used a 1-W, 514-nm argon laser pump beam critically focused into a near-hemispherical dye cavity 4.5 mm long with a minimum transverse mode diameter, or waist, of only 11 μm. Water was used to dissolve the dye, because of its excellent thermo-optic properties, and the solution was circulated in a fast flow of 4 m sec^{-1} velocity across the pumped spot. Each dye molecule effectively saw a flash, since the transit time through the pumped volume was only 3 μsec. With this apparatus they successfully operated in 1970 the first cw dye laser, in rhodamine 6G dye.

Thus by the end of 1970 the three main branches existed. Developments in the following years carried the pulsed-branch lasers through a sequence of adaptions to ever more versatile and powerful pump lasers. First the nitrogen gas laser, with a 337-nm pumping wavelength, gave access to dyes lasing from the near ultraviolet (UV), through the blue and into the red, with one pump laser. This was then supplanted by the doubled and tripled Nd : YAG pump laser, giving higher pump energies per pulse (hundreds instead of a few millijoules) and the excimer pump laser, giving higher pulse repetition frequencies (hundreds instead of tens of pulses per second). In cw dye lasers, the technology first became easier with the introduction of the three-mirror cavity and free-flowing dye jet stream, which eliminated the critically toleranced and damage-prone dye cell, and then became harder again as frequency servo and control electronics of increasing sophistication were brought in to scan the single-frequency oscillation of ever narrower residual linewidths, over ever broader scan widths. Mode-locked lasers adopted the cw-type pumping format once this became available. The steady conditions allowed for a better characterization and definition of the regime where the mode-locked circulating pulse compresses on each transit of the cavity, and shorter pulses were produced. In this regime, the brevity of the pulse length is limited by the residual pulse-stretching effects in all of the optical elements the pulse sees, and the subsequent history is one of finding these and eliminating them ever since.

It is clear from Fig. 1 that the dye laser is still in a state of rapid evolution. What is also important in the figure, but is not perhaps immediately evident, is that the history of the dye laser gives a splendid example of the way a rich technology evolves. A new idea brings additional new ideas by extrapolation (and laser-pumped dye lasers led to flashlamp-pumped ones, which ushered in cw dye lasers). Improvements make the end device more useful, which leads to a component development allowing further improvements (excimer lasers, argon lasers, and Nd : YAG lasers have all experienced a decade of improving specifications through the demands made for better dye pump lasers). Performance only gets better, as to be adopted, each new idea must produce results that surpass the old. After grasping more of the details of Fig. 1 (in what follows), it is worth looking again on the figure from these points of view.

II. Quantum Chemistry and Physics of Dyes

A. Conjugated Double Bonds as Chromophores

The fundamental properties of a dye laser derive directly from the molecular structure of the dyes. Figure 2 shows the chemical structure in solution of the chloride salt of the most common laser dye, rhodamine 6G. Two equally probable forms [Figs. 2(a) and 2(b)] are shown, differing only in the location of the trivalent nitrogen atom. One form is obtained from the other by transferring an electron from one side of the molecule to the other, along the connecting chain of conjugated carbon bonds (alternating single and double bonds) in the ring structures. To maintain the correct +4 valence of each carbon atom in this transfer, the locations of four of the six double bonds in the three connecting rings must also be changed. In classical theories of dyes, the chain of shifted bonds was called the chromophore, and the light absorption and color of the dye were attributed to this oscillating dipole formed by the transfer of charge—the chromophore was a "molecular antenna." These theories arose from the empirical fact that essentially all organic dyes were found to contain conjugated double bonds, or, equivalently, were represented by a chemical "resonance" between nearly identical structures, differing only in the placement of the double bonds.

The important features of this type of bonding are best grasped by building up to a complicated

RHODAMINE 6G WAVE FUNCTIONS

$$\psi_g = \frac{1}{\sqrt{2}}(\phi_1 + \phi_2)$$
$$\psi_e = \frac{1}{\sqrt{2}}(\phi_1 - \phi_2)$$

(a) where ϕ_1 =   (b) and ϕ_2 =

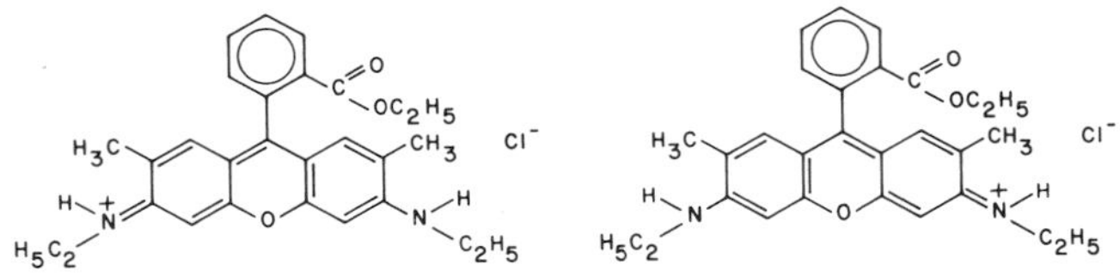

FIG. 2. Chemical symbols for two "resonant" forms of the structure of the rhodamine 6G dye molecule and the wave functions to which this resonance gives rise.

dye structure like Fig. 2, through the simpler molecules shown in Fig. 3. Ethylene, $H_2C{=}CH_2$, illustrates the character of the double bond in carbon (Fig. 3a). To "prepare" the carbon for bonding, three of the four $2s^22p^2$ outer electrons are hybridized into three sp^2 trigonal orbitals, having lobes of the wave function lying 120° apart in a plane (like a clover leaf). The fourth electron remains as a p electron with a wave-function node at the nucleus and the positive- and negative-phase lobes of the p orbital projecting perpendicularly from this plane. The first bond of the double bond is formed by overlapping one sp^2 orbital from each carbon atom, making a bond rotationally symmetric about the nuclear axis, called the σ bond. The remaining four sp^2 orbitals in ethylene are filled by making σ bonds to hydrogen atoms. The second bond in the double bond is formed by overlapping the two remaining p orbitals. The molecular π bond thus formed retains the node, giving a pair of sausage-shaped lobes (of opposite phases) floating above and below a nodal plane containing all the nuclei and the center lines of all the σ bonds (Fig. 3a). The binding energy of the π bond is greatest when the atomic p orbitals overlap most (when their lobe axes are parallel). This explains the absence of free rotation about the double bond, and this and the sp^2 hybridization explain why the characteristic forms of dye molecules are planar structures built up from zigzag chains and hexagonal rings with 120° bond angles.

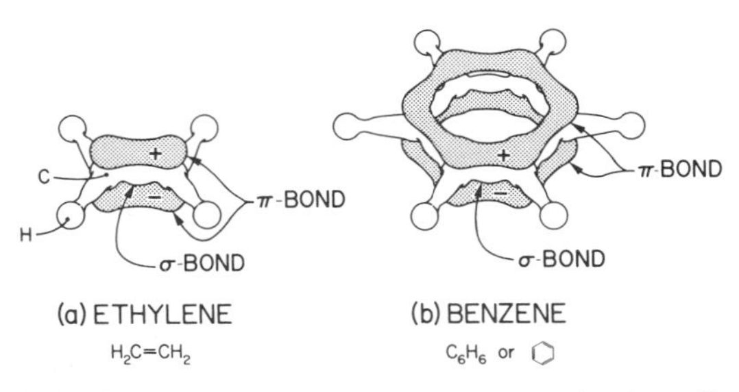

FIG. 3. Chemical symbols and schematic three-dimensional representations of the electron density distributions, for (a) a double bond system, ethylene, and (b) a conjugated double bond system, benzene. The shapes show the surfaces of constant electron density except for distortions necessary for clarity in depicting the overlapping σ-bond and π-bond electron clouds.

As a prototype conjugated molecule, consider benzene, C_6H_6 (Fig. 3b), six sp^2 carbons joined in a plane. X-Ray data shows that all the carbon bond lengths in benzene are equal, and of intermediate length (0.139 nm) between a single (0.154 nm) and double (0.134 nm) carbon bond. By definition, there are only enough p-orbital electrons (one per conjugated atom) to make a second bond over half of the σ-bond carbon linkages. What happens, by symmetry, is that the six available p electrons are *shared equally* by the six carbons, linking them by six "half bonds." They form two ring-shaped lobes or π-bond electron clouds in which the π electrons freely move, floating above and below the molecular skeleton of σ bonds lying in the nodal plane as shown in Fig. 3b. The characteristic of the conjugated bonding found in dyes is that it gives delocalized, mobile π-orbital electrons.

In quantum mechanics this resonance or sharing between possible structures is described by taking for the wave function of the actual structure a linear combination of the wave functions for the resonant possibilities. The energy of the system is calculated with the combined wave function, and the linear coefficients are determined as those that they give stationary values for the system's energy. For example, in combining the equally likely structures in Figs. 2a and 2b, this produces from the old wave functions Φ_1, Φ_2, each associated with the energy E, the two new sum and difference wave functions Ψ_g, Ψ_e of the figure, one associated with an energy less than E and one greater than E. The square of the absolute value of the wave function gives the spatial electron density in the molecule. In the additive combination $\Phi_1 + \Phi_2$, there results an averaging of the positions of the double bonds of the resonant forms, essentially placing an extra half bond everywhere along the chromophore. This reduced spatial specificity, by the uncertainty principle, implies a lower electronic kinetic energy, and the additive combination Ψ_g is the lower-energy state. The amount the energy is lowered over the nonresonant state is called the delocalization energy. Since the molecular energy is lowered, this combination it is termed a *bonding* state, or π state. In the subtractive combination $\Phi_1 - \Phi_2$, the nodes and antinodes of the wave function along the chromophore are emphasized, corresponding to localized electron density peaks and higher kinetic energy for the π electrons, and this is the higher-energy (excited) state. Since the molecular energy is increased, this is termed an *antibonding* state, or π^* state. Electronic states of stable molecules usually correspond to having more electrons in bonding states than in antibonding states.

By similar quantum-mechanical rules, starting with N conjugated atoms in a dye chromophore, each contributing one electron to the π system, it can be shown there will result N separate energy states for π electrons, each capable of holding up to two electrons (in a filled state the electrons must be paired due to the Pauli exclusion principle, with antiparallel spins). Half of these states will be bonding states and the other half antibonding states. The π states in large unsaturated molecules are usually the highest-lying bonding levels and the π^* states the lowest-lying antibonding levels. Thus the normal electronic ground state for a dye uses the available N electrons, to just fill the $\frac{1}{2}N$ bonding π states with two paired electrons each. This gives a resultant zero spin to the system, and the ground state in dyes is a singlet state. The lowest-lying empty state is the first antibonding π^* state, the $\frac{1}{2}N + 1$ level of the system. The characteristic absorption and fluorescence of dyes, which is the working electronic transition in the dye laser is this $\pi(\frac{1}{2}N) \rightarrow \pi^*(\frac{1}{2}N + 1)$ transition.

The energies of these levels, and thus the center wavelength for that dye in a laser, may be calculated successfully with a surprisingly simple model, called the free-electron model. This assumes that the σ-bond framework of the conjugated chain provides a line or confining box of constant electrostatic potential along which the π electrons move freely. The important variables are the size of the box (number of conjugated atoms) and its topology (whether the bonding network is linear, branched, or ring structures). At the ends of the chromophore, the constant potential gives way to a steeply rising Coulomb potential, so the π electrons have the properties found in the quantum-mechanical problem of "a particle in a one-dimensional box." Their wave functions are sinusoidal standing waves with an integral number of half wavelengths fitting into the length of the box, with their energies increasing quadratically with their quantum number (the number of half wavelengths). The electrons with more antinodes in their wave functions thus have larger kinetic energies. From these solutions, correct first-order values are obtained not only for the wavelength of the peak of the dye absorption (which increases with the length of the chromophore in a linear molecule) but also for many other properties. Some of these are the oscillator strength of the transition (which is large, of order unity), the polarization dependence of absorption (which peaks for the incident electric field aligned along the chromophore), and the shift of peak absorption wavelength with atomic substitution in the molecule, or with change of solvent.

It was shown already that the ground state for a dye molecule has zero spin and is a singlet. In the first excited state there are two unpaired electrons (one in the $\frac{1}{2}N$ bonding level, and the other in the $\frac{1}{2}N + 1$ antibonding level). The Pauli exclusion principle does not restrict the spins of these electrons to be antiparallel, since they are in different levels, and a triplet state (parallel spins, for a total spin of unity) is possible in the excited electronic state. The free-electron wave functions can be used to show that due to spatial correlation effects between these unpaired electrons, the energy of the excited triplet state will lie below that of the singlet level. This gives rise to the problem of accumulation of dye molecules after optical pumping in the metastable triplet level, which was of such great importance in the historical development of dye lasers. (The triplet state is metastable or requires a collision to deactivate it, because a change in spin is not allowed in a radiative transition).

To summarize this, the quantum theory substantiates the classical view of mobile electrons, free to move along the chain of conjugated atoms in dyes, giving rise to oscillating dipoles and strong absorption and fluorescence. These are shown to be π-bond electrons, arising from the "resonance" or ambiguity in placement of the double bonds in the chain, which splits the electronic energy into a set of delocalized bonding levels and a set of localized antibonding levels. Laser operation and the strong color of the dye are due to the transition between the highest member of the low-lying set and the lowest member of the high-lying set, and are seen to be phenomena deeply imbedded in the quantum properties of matter.

B. Dye Absorption and Fluorescence Spectra Derived from Potential-Energy Curves

The basic spacing of the levels discussed above is about 20,000 cm^{-1} in rhodamine 6G dye. These levels are broadened by molecular vibration, rotation, and collisions with solvent molecules. Vibration produces sublevels, spaced by up to several hundreds of wave numbers, which in some dyes are sufficiently resolved to give secondary peaks in the absorption spectrum. Rotational and collisional (homogeneous) broadening smears these sublevels into overlapping bands to make the tunable continuum.

Vibrational broadening in molecular spectra is interpreted with a potential energy diagram such as is shown, somewhat schematically, for rhodamine 6G in Fig. 4a. A potential diagram has a clear interpretation for a simple diatomic molecule, where the structural coordinate R (the horizontal axis) is the separation between the two atomic nuclei. The characteristic shape of the potential energy curve is a dip or potential well, as the nuclei approach from infinite separation, by the amount of the binding energy (reached at separation R_0) before rising steeply as the nuclear Coulomb repulsion takes over at smaller separations. The ground-state separation R_0, at the potential minimum, fixes the normal linear dimension in a diatomic molecule. In a complex dye molecule, there are many nuclear separations that change in various directions with different modes of vibration, and the concept of a potential energy depending on a structural coordinate must be generalized. The ideas are the same as in the diatomic case, but the diagram becomes somewhat schematic. For Fig. 4a, the dimension for the structure coordinate axis was chosen (arbitrarily) as the distance between the nitrogen atoms in rhodamine 6G. The measured spectra (for wavelengths less than 700 nm) for rhodamine 6G are plotted in Fig. 4(c), and the schematic nature of the potential diagram is unimportant in this discussion since it was constructed to be consistent with this spectral data.

Each electronic state (ground state S_0, first excited singlet S_1, triplet T_1, and so forth) has a separate potential curve $E(R)$. As the molecular dimensions generally swell in the excited states, larger values for the structural coordinate R are found at the potential minima of the higher curves (a fact having important consequences for laser operation).

Vibrational energy in a molecule is quantized, giving a set of discrete levels shown in Fig. 4a as horizontal lines drawn between the boundaries of the potential wells. The square of the wave function for each of these levels gives the probability the nuclei will be found at a given value of R; three of these, for energies E_0, E_2, and E_3, are shown. These probabilities peak near the potential boundaries that correspond to the classical turning points in the vibratory motion, where the nuclear velocity is near zero and reverses direction. The intermediate points are passed through relatively quckly; the vibrating nuclei are likely to be found near their maximum or minimum positions.

Light absorption or emission is governed by the Franck–Condon principle, which states that the change in electronic energy occurs so much more rapidly than the vibrational motion that neither the position nor the momenta of the nuclei can change appreciably. Thus, on the potential-well diagram, the absorption (or emission) of a photon is represented by a vertical line, which, by the weighting of coordinate values, begins and ends near a potential-curve boundary for the strong transitions. (Transitions to endpoints well away from the curve boundaries are prevented by the second part of the Franck–Condon principle—there can be no discontinuous changes in nuclear kinetic energy.) The length of the vertical line is given by the photon energy (which matches the difference in electronic energy)

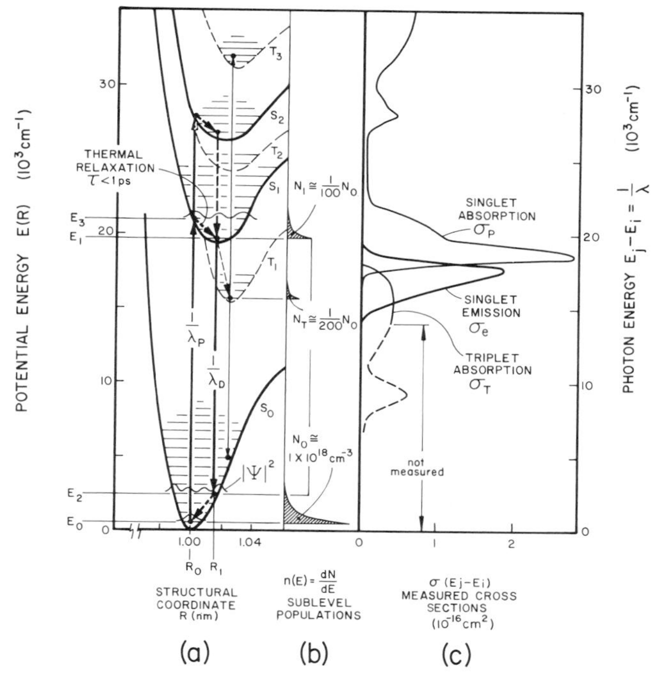

FIG. 4. Schematic potential-energy wells constructed to be consistent with the measured spectra for the rhodamine 6G dye molecule. (a) The potential energy versus generalized structural coordinate (nuclear separation) for the first three singlet and first three triplet energy bands. Laser light absorption and emission (at pump and dye laser wavelengths) are indicated by heavy arrows, and nonradiative (collision) processes by dashed arrows. (b) Thermal equilibrium sublevel population distributions in these six bands (only the lowest three are appreciably populated). (c) Absorption and emission cross sections measured for this dye (except that the triplet absorption between 5000 and 14,000 cm^{-1} is inferred from spectra on similar dyes).

hc/λ, where h is Planck's constant, c the velocity of light, and λ the photon wavelength, or, if energy is measured in wave numbers as in Fig. 4, the length is directly $1/\lambda$. The line strength for the transition is given by quantum mechanics as an overlap integral, calculated with the two end-point vibrational wave functions.

The intensity of the transition is also weighted by the sublevel population density $n(E)$, the number of molecules per unit volume in the initial vibrational sublevel at energy E. The collisional and rotational broadening at room temperature spans the vibrational level spacing, so this sub-level population density has the form appropriate for Boltzmann (thermal) equilibrium (Fig. 4b) or $n(E)$ decreases sharply (falling exponentially) with energy above the bottom of the potential well. The product of this Boltzmann factor and the quantum-mechanical line strength is taken and plotted versus the energy difference of the potential curves, to produce the dye absorption and fluorescent emission curves (Fig. 4c).

These spectra are presented as cross sections for molecular absorption σ_p, or its inverse, stimulated emission σ_e. These are defined by Beer's law, which for example in the case of absorption of pump light from the ground state is

$$I_p(z) = I_p(0)\exp(-N_0\sigma_p z) \qquad (1)$$

Here a plane light wave of uniform intensity I_p (watts per square centimeter) and wavelength λ_p (nanometers) has been assumed, incident on a sample of dye of thickness z (centimeters) with a ground-state dye concentration of N_0 (molecules per cubic centimeter). A cross section has units of centimeters squared and represents in the case of σ_p, for example, the absorbing area per molecule in the ground state. Because these cross sections are defined by an average over the Boltzmann population distribution in the initial state, they are temperature and solvent-dependent.

The total molecular population density N is the sum over all the populated states, which is for Fig. 4 where only the ground state, first singlet, and first triplet are appreciably populated:

$$N = N_0 + N_1 + N_T \quad (2)$$

(Each of these state populations is the integral of the sublevel population density $n(E)$, over the energy range of several kT that the level populations are spread, k being Boltzmann's constant, T the absolute temperature, and $kT = 200\ \text{cm}^{-1}$ at room temperature. In Fig. 4b the energy spread is exaggerated to about twice this range, to allow the exponential fall-off to show). The assumption of a wave of uniform intensity here and in what follows simplifies the equations, and real devices with nonuniform waves can be treated later by integrating these results over the intensity variation within the active volume.

The two most striking features of the curves of absorption and emission versus photon energy are their asymmetric shapes and the fact that these are nearly mirror images of each other (Fig. 4c). These are consequences of the offset $R_0 < R_1$ between the potential minima of the S_0 and S_1 singlet states. The absorption from the ground state (labeled $1/\lambda_p$ in Fig. 4a) proceeds from the populated sublevels near the bottom of that well, to high-lying sublevels at points of minimum R along the steep inner boundary of the upper potential well. Hence, the σ_p spectrum shows a sharp rise, then falls with a more gradual tail, with increasing photon energy. Conversely, the fluorescence decay (5 nsec time constant) or stimulated transition (labeled $1/\lambda_D$ in Fig. 4a) arises from populated sublevels near the bottom of the upper well. These transitions go to the high-lying sublevels at points of maximum R along the descending outer boundary of the ground-state potential curve. Hence the σ_e spectrum rises sharply, then falls with a more gradual tail, with decreasing photon energy.

Other processes shown in Fig. 4a may be listed that compete with the main dye laser cycle of absorption at λ_p and emission at λ_D. Higher-energy pump photons can excite π electrons to the second singlet state directly ($S_0 \rightarrow S_2$, depicted as an extension of the arrow for the $1/\lambda_p$ transition, and as the second, higher-energy peak of σ_p). This excitation relaxes to S_1 by fast *internal conversion* processes (shown as dashed lines; these are collision-induced energy transfers on a picosecond time scale, without radiation). Internal conversion on $S_2 \rightarrow S_1$ allows pumping of the rhodamines and other visible dyes by ultraviolet pump lines such as the 308-nm line from the xenon chloride excimer laser. Internal conversion on $S_1 \rightarrow S_0$, which reduces laser efficiency, is also possible.

Some collisions produce *intersystem crossing*, converting the excited population in S_1 to that in the triplet state T_1 (indicated by a light dashed arrow). The triplet is a long-lived metastable state, as the radiative decay to the ground state $T_1 \rightarrow S_0$ is forbidden by selection rules that do not allow a change in spin in an electronic transition. Interactions with solvent molecules and torsional bending of the dye molecule produces a weak mixing of spin states, allowing a weak, delayed phosphorescence on this transition (thin solid arrow downward in Fig. 4a). Because this state is long-lived (microsecond to millisecond lifetimes), even a slow pump rate from S_1 due to collisions can result in an appreciable T_1 population, and thus also $T_1 \rightarrow T_3$ absorption, which overlaps the lasing wavelengths (thin solid arrow upward in Fig. 4a and absorption spectrum σ_T in Fig. 4c). Estimates of the triplet population can be made by various means, among them by noting the power increase in a dye laser by the addition of the liquid cyclooctatetrane (COT) to the dye mixture. This is a molecule with acceptor states lying below the first triplet level of many visible dyes, which by collisions relaxes (or quenches) the dye triplet population back to the ground state, reducing the deleterious absorption. From a typical cw laser output power increase of +30% in rhodamine 6G, this method estimates the T_1 state population to be half that of S_1. Another well-known triplet-state quencher is oxygen (O_2) dissolved from the air into the dye solution. These processes are described by a set of quantum efficiencies (branching ratios) φ, φ_T, φ_{IC} whose sum is unity, and that give the probabilities for a molecule put into S_1 to decay by each of the three channels. For good laser dyes, the fluorescence efficiency $\varphi = 1 - \varphi_T - \varphi_{IC}$ will exceed 85%, making these competing pathways unimportant (so long as the triplet absorption is quenched). With these definitions, if τ is the spontaneous emission lifetime for fluorescent

$S_1 \rightarrow S_0$ decay, the total decay rate out of the first singlet state is $1/\tau\varphi$.

C. Spectral Condensation, Gain Saturation, and Laser Output Power

Referring to Fig. 4a, dye molecules optically pumped from the ground state to E_3 above the singlet potential minimum thermally relax to E_1 by collisions giving up heat energy to solvent molecules in a time on the order of a picosecond. This relaxation is faster than the optical absorption or stimulated emission rates in actual dye lasers. It is this thermal equilibrium process that, by maintaining the population distribution among the sublevels of each state at the Boltzmann values despite the perturbing optical transitions, allows the pumped dye molecules to exhibit gain and work as a laser medium. The fluorescent decays labeled $1/\lambda_D$ into the ground state are into high-lying depopulated vibrational sublevels (near E_2), which remain so because of the ground-state Boltzmann equilibrium. This greatly reduces the pumping rate needed to establish gain in the dye. A pump rate sufficient to produce a singlet population N_1 of only 1% of that of the ground state N_0 is all that is needed to establish a population inversion between sublevels $n(E_1) > n(E_2)$, and gain in rhodamine 6G dye (see Fig. 4b). The thermal equilibrium process works in the laser for the pump absorption labeled in $1/\lambda_p$ as well, by rapidly removing population from the terminal sublevel E_3 to prevent stimulated emission at λ_p back to the ground state. This makes it possible to totally invert the working states in a dye laser (i.e., put all of the dye molecules into the excited state, $N_1 \gg N_0 \approx 0$). The Boltzmann equilibrium creates a classical four-level laser system out of the two broadened states.

It is also responsible for the spectral condensation effect. When broadband, nonselective feedback is provided in the dye laser, stimulated emission occurs from all occupied sublevels in the upper band S_1. The higher-lying sublevels (in the high-energy tail of the population distribution) emit primarily to higher sublevels in the ground state, (i.e., at longer wavelengths) than do those molecules near the potential minimum E_1. In this case the spectral bandwidth of the emission is broad, 10–20 nm in a pulsed dye laser. When a spectrally selective element is added to the dye feedback cavity, the threshold is raised for all emissions but those in the narrow selected wavelength band, say, the wavelengths emitted from level E_1, which are all that lase. The population in the entire upper state S_1 drains through this channel as the thermal collisions maintaining the Boltzmann equilibrium attempt (successfully) to fill the hole in the depleted sublevel population distribution. The same argument holds if the selective feedback is tuned for lasing at shorter or longer wavelengths, but the higher energy of the draining sublevel permits access only to the higher-energy tail of the Boltzmann distribution, and the process is not as efficient.

How much less efficient may be seen by calculating the steady-state laser output power for the case of a spectrally narrow tunable filter with center wavelength λ_D selecting the oscillation wavelength. The idea of gain saturation is the basis for an output power calculation. Assume uniform plane waves once again, with later integrations over the active volume accounting for intensity spatial variations in real devices. Initially, before lasing begins, the population inversion and the (small signal) gain per pass $k_0 L$ are at a maximum. Here $k_0 L$ is the exponential gain constant in Beer's law, and L is the dye optical thickness along the resonator axis.

This initial gain must decrease, or saturate, because as the lasing intensity I_e builds up, it causes stimulated transitions at rates comparable to the spontaneous fluorescent transition rate $1/\tau\varphi$, and the population inversion is depleted. Thus the gain constant $k(I_e)L$ is a decreasing function of I_e. In the steady state after the buildup transient, the laser intensity inside the resonator is the intensity that decreases the gain per pass to just balance the cavity losses, as then the laser intensity suffers no net change in a cavity round trip. For a ring (or traveling-wave) laser, the dye gain is traversed once per round trip of the resonator (of mean reflectivity r), so that balance is achieved if $re^{kL} = 1$, or

$$k(I_e)L = \ln(1/r) \tag{3}$$

This is the implicit equation for the steady-state laser intensity. Once it is solved for I_e, the output power P_{out} is given by

$$P_{out} = t(\pi w_D^2)I_e \tag{4}$$

Here πw_D^2 is the area of the laser beam of (uniform) radius w_D in the dye medium. The resonator losses consist of the output coupling loss t (transmission of the output mirror) and the dissipative losses a due to light absorption and scatter in the intracavity optics, and the insertion losses of the wavelength selective filter. As t and

a are usually small, $\ln(1/r) = \ln[1 - a - t]^{-1} = a + t$ in Eq. (3).

The form of the gain decrease may be found by solving rate equations for the level populations, which balance in the steady state the rate of transitions into S_1 from pump absorptions ($S_0 \rightarrow S_1$) against the rate out of S_1 due to stimulated laser transitions and spontaneous fluorescent decays. The broadening of the laser states is included by the Boltzmann average in the cross-section definitions, so the rate equations deal with the total state populations N_1 and N_0. (Care must be exercised in the definitions and averaging to arrive at a consistent set of rate equations. Additional terms describing intersystem crossing, triplet and excited singlet state absorptions, and so forth, are not included here. In particular, it is assumed that the triplet population is quenched, $N_T = 0$.)

The essential physics is illustrated with just these four terms, in the time-independent, plane-wave, longitudinal pumping case. (The pump light propagates along the same axis as the dye laser beam.) The gain constant is given by

$$\begin{aligned} k(I_e)L &= (N_1\sigma_e - N_0\sigma_a)L \\ &= [N_1(\sigma_e + \sigma_a) - N\sigma_a]L \end{aligned} \tag{5}$$

since $N = N_1 + N_0$ by Eq. (2). Here σ_a is the singlet absorption cross section, evaluated at the lasing wavelength λ_D. The steady-state balance of the transition rates into and out of the upper laser state gives a value $N_1(I_e)$ for that population density that is dependent on the laser intensity I_e. When used in Eq. (5), there results a saturated gain equation of the form expected for a homogeneously broadened laser,

$$k(I_e)L = \frac{k_0 L}{1 + I_e/I_s} \tag{6}$$

where the initial or small signal gain is

$$k_0 L = NL\left(\frac{\sigma_e\sigma_p I_p - \sigma_a(hc/\lambda_p)(1/\tau\varphi)}{\sigma_p I_p + (hc/\lambda_p)(1/\tau\varphi)}\right) \tag{7}$$

[Equation (7) corrects an error in the corresponding equation in the Johnston article.] The saturation intensity, which by Eq. (6) is the value of I_e dropping the gain to half its initial value, is

$$I_s = \frac{\lambda_p}{\lambda_D}\left(\frac{\sigma_p I_p + (hc/\lambda_p)(1/\tau\varphi)}{\sigma_e + \sigma_a}\right) \tag{8}$$

Setting the saturated gain equal to the resonator losses (asumed small) as in Eq. (3), solving Eq. (6) for I_e, and putting this into Eq. (4) gives the expected form for the output power (valid above threshold, when the expression in the last bracket is positive):

$$P_{out} = t(\pi w_D^2)I_s\left(\frac{k_0 L}{a + t} - 1\right) \tag{9}$$

The output increases linearly with the excess above unity or threshold value of the small-signal gain-to-loss ratio. Since the denominator of Eq. (7) is the same as the numerator of Eq. (8), it is convenient to consider the form the output power takes when the laser is pumped well above threshold (where these factors cancel in the product $I_s k_0 L$), and as a further simplification, at dye wavelengths in the middle of the tuning range or longer where the ground-state reabsorption may be neglected.

Then

$$P_{out} \approx \left(\frac{t}{a + t}\right)\frac{\lambda_p}{\lambda_D}(NL\sigma_p)\rho P_{in} \tag{10}$$

for $k_0 L/(a + t) \gg 1$ and $\sigma_a = 0$. Here $\rho = w_D^2/w_p^2$, the ratio of dye beam area to pumped area, is introduced and $\pi w_p^2 I_p$ is recognized as the input power in the longitudinal pumping case.

As a first use of these results, note the direct connection they establish between the experimental behavior of a dye laser and the underlying molecular physics. In using the laser, the ease of operation is greatest when well above threshold, and differences in behavior at opposite ends of the tuning range of a dye are immediately apparent. On the short-wavelength end of the range, there is a wide tolerance of output to mirror misalignment, dirty optics, etc., which persists even as the short-wavelength oscillation limit is approached and the output abruptly dies. The opposite is true as the long-wavelength limit is approached, where cavity alignment and losses are critical, and there always seems to be some output available at even longer wavelengths if only the critical adjustments can be maintained. The excess gain may be thought of as a measure of the tolerance to cavity losses, while by Eq. (9) the saturation intensity is a weighting factor, giving the value of this gain in determining output power. The behavior just described is thus due to the mirror-image and asymmetry properties of the absorption and emission cross sections. The short-wavelength end of the range is near the peak where the emission cross section σ_e is slowly varying, and the tuning limit is determined by the sharp rise in absorption cross section σ_a with decreasing λ_D. The rise in σ_a both decreases the saturation intensity [Eq. (8)] and adds a cavity loss term due to laser photon reabsorption in the dye itself [Eq. (7)]. The long-wavelength end of the range is determined by the asymmetric slow falloff of σ_e with increasing λ_D. This drops the gain towards the threshold (resonator) loss value, there is negligible reabsorption ($\sigma_a = 0$), and the rise in I_s [with $1/\sigma_e$, Eq. (8)] allows the output power to persist. An example where the k_0L and I_s parameters were measured as a function of λ_D in a cw dye laser is given in Section IV.

This linking of the macroscopic laser properties with the microscopic dye ones may be enhanced by further consideration of these results. In the gain expressions Eqs. (5) and (7), neglect the reabsorption term in σ_a, as is permissible in the middle of the tuning range. Then the small signal gain per pass of the dye medium is

$$k_0L = N_1L\sigma_e = NL\sigma_e\left(\frac{\sigma_pI_p}{\sigma_pI_p + (hc/\lambda_p)(1/\tau\varphi)}\right) \tag{11}$$

which is the product of the gain a uniform beam traversing the medium would see if all the dye molecules were in the upper state, at total inversion (NL being the molecular density in the beam cross section, σ_e the emitting area per molecule), times the fraction of molecules in the upper state (bracketed term). The interpretation of the bracketed term comes from Eq. (11), or directly from the rate equation and particle conservation equation from which it is derived.

The term $(hc/\lambda_p)(1/\tau\varphi)$ is the pump power dissipated (by all spontaneous decay processes) per molecule in the upper laser S_1. It is useful to estimate from Eq. (11) the minimum pump intensity needed to reach threshold in a dye laser. This will be when the pump power density absorbed by all ground-state molecules, $N_0\sigma_pI_p$, balances against the power density dissipated by upper-state decays, $N_1(hc/\lambda_p)(1/\tau\varphi)$, to give that excited-state fraction producing gain equal to the resonator losses.

From Eq. (11), this is:

$$I_p^{th} = \frac{(1/\sigma_p)(hc/\lambda_p)(1/\tau\varphi)}{[NL\sigma_e/(a + t)] - 1} \tag{12}$$

For typical values $\sigma_p = \sigma_e = 1 \times 10^{-16}$ cm^2, $\lambda_p = 500$ nm, $\tau\varphi = 5 \times 10^{-9}$ sec, $N = 1 \times 10^{18}$ cm^{-3}, $(a + t)/L = 10$ cm^{-1}, this gives a threshold pump intensity of 90 kW cm^{-2}, clearly showing the need for bright flashlamps or a laser to pump a pulsed dye laser, and a coherent pump beam focusable to a small spot to pump a cw one. Note that in the example given, the pump power absorbed per ground-state molecule, σ_pI_p, is one-ninth the spontaneously dissipated power per molecule in the upper state. Gains and pump intensities 10 times the threshold value or more are realized in practical dye lasers.

The interpretation of the saturation intensity result, Eq. (8), contains a subtlety. In the conservative two-state system under discussion, a molecule removed from the upper state by laser-stimulated emission at the rate $\sigma_eI_e/(hc/\lambda_D)$ per molecule must appear in the lower state. There it immediately is subjected to a pump rate (per molecule) of $\sigma_pI_p/(hc/\lambda_p)$ returning it to the upper state. Thus for stimulated emission to produce a reduction of the small-signal upper-state population by half, it must be at a transition rate per molecule equal to the sum of the spontaneous decay rate plus the return rate, yielding Eq. (8). This makes the saturation intensity a linear function of the pump intensity; at high pump rates, the dye bleaches, the small-signal gain saturates at the total inversion value $NL\sigma_e$, and the output power increases with pump rate solely through the I_s term in Eq. (9).

Turning now to Eq. (10), the first term is recognized as the coupling efficiency $\eta_C = t/(a + t)$ found in all lasers, expressing the fact that a finite circulating power must be left in the resonator to stimulate emission (not all of the power represented by the initial population inversion can be coupled out of the resonator). High-gain dye lasers with optimized output transmission t have $\eta_C \approx 0.8$. The next term is the Stokes efficiency $\eta_S = \lambda_p/\lambda_D$, accounting for the power lost (for a quantum efficiency of unity) due to the dye photon being of lower energy than the pump photon. The term $NL\sigma_p$ is the fraction of the pump light absorbed in the dye in this uniform plane-wave case, where this factor must be small to maintain the assumption of uniformity. When the pump intensity is allowed to vary along the axis of propagation, this becomes the fraction of pump intensity absorbed in the transit of the dye medium, $\eta_A = [1 - \exp(-N\sigma_pL)]$ by Eq. (1). The ratio ρ of lasing to pumped areas accounts for any pumped volume left unutilized and is replaced by an average value sometimes called the filling factor η_F, equal to a volume integral of this ratio over the pump and dye beam intensity variations in describing actual devices. The threshold term may be restored to Eq. (10), so that in its general form this may be rewritten to give the slope efficiency or ratio for conversion of pump power in excess of threshold to tunable dye laser output power:

$$P_{out}/(P_{in} - P_{th}) = \eta_C \eta_S \eta_A \eta_F \quad (13)$$

Typical values, in the case of the rhodamine 6G cw single-frequency ring laser at the peak of the tuning curve and 6 W of input power, for the last three efficiency factors are $\eta_S = 0.85$, $\eta_A = 0.80$, and $\eta_F = 0.80$, giving a theoretical slope efficiency of 44%. Experimental values for this case reach 35%.

The question of the efficiency of the dye laser convertor and of spectral condensation in particular at wavelengths away from the maximum emission cross section is answered by Eq. (13). As long as the dye is pumped hard enough to create a gain well above threshold, the lasing conversion process occurs at fixed slope efficiency essentially independent of center wavelength or lasing bandwidth.

The most serious omission in the treatment given so far is the neglect of the triplet-state terms in Eqs. (7)–(13). The power loss due to triplet-state absorptions is treated by Peterson, who shows that the laser types are usefully separated by a demarcation time of $(K_{ST})^{-1} \approx 100$ nsec in rhodamine 6G, the inverse of the intersystem crossing rate filling the triplet state from the first excited singlet. This order of time is required to build up the equilibrium triplet population. For short pulse dye lasers (laser-pumped, pulsed lasers) the pump pulses are typically 10 nsec in length and the triplet absorption terms can be neglected. The largest effect is for the cw type, where, for example, in the rhodamine 6G ring laser considered earlier, inclusion of triplet losses would reduce the theoretical small signal gain by 34%, increase the saturation intensity by 17%, and reduce the output power by 30%. To compute these changes requires knowledge of the triplet absorption cross section σ_T, which is known in this case but has not been measured for most other dyes.

In short-pulse dye lasers there is insufficient time for triplet-state buildup and the results above are usable without correction. In flashlamp-pumped dye lasers, not only are triplet terms significant, but often the full time-dependent rate equations are needed, requiring numerical integrations in modeling their behavior. In mode-locked dye lasers the resonator losses are deliberately time-dependent, requiring further discussion (see Section V).

The dye lifetime, or stability against photo decomposition, is another important dye property not in these equations. All dye molecules eventually decompose under repeated pump excitations. In the more stable molecules this occurs after 10^6–10^7 photon absorptions, corresponding to about 10^4 J absorbed energy per cubic centimeter of dye solution, with perhaps a minimum of 1% of this lifetime required for a dye to be usable. In practice, a large volume of a liter or more of dye is circulated through a laser to give several hundred hours of operation between dye changes. Dye concentrations are adjusted to give the desired pump beam absorption depth, in matching the pumped volume to the dye beam diameter (fixed by the dye resonator). Dye concentrations are generally limited to less than 10^{-2} mol/liter, where the limit of solubility for some dyes occurs, and where aggregation (dimer formation) in others alters the dye spectra.

The main intent of the remainder of this article is to give a picture of the art of dye laser design in the three main scientific instrument categories, at the 1986 stage of their evolution.

III. Short-Pulse Dye Lasers

A. Status of Flashlamp-Pumped Dye Lasers

The high threshold pump intensity given by Eq. (12) means that flashlamps for dye laser pumping must be driven hard, blackbody temperatures above 6000 K being required, which results in short-lived, frequently replaced lamps. Much of the lamp radiation is in a broad ultraviolet band at these plasma temperatures, leading to relatively rapid photodecomposition of the dye if the lamp light is unfiltered, and inefficient operation if it is filtered. Lamp pulse lengths of 1–50 μsec are long enough that triplet-state quenching and thermal lensing of the dye must be dealt with. These pulses are too long (compared to the 5- to 30-nsec pulse lengths from pump lasers) to reach the peak dye powers most useful in nonlinear optics and available from laser-pumped dyes. These are the disadvantages of flashlamp-pumped dye lasers.

Nevertheless, because the flashlamp laser has the economy of eliminating the pump laser, many special-purpose lamp designs and dye lasers to go with them have been developed to fill special-purpose needs. There is also a considerable history of contributions to the dye laser art made with flashlamp pumping, ranging from the first demonstration of lasing in a large number of dyes, to clarifying the mechanisms of pulse formation in mode-locked dye lasers. But because of the disadvantages already listed, there are today few flashlamp pumped scientific dye lasers, and no widespread general example of this type in the same sense that these exist for the three types represented as the three main branches (Fig. 1) of the dye laser family tree. The review article by Peterson expertly and extensively covers many flashlamp systems. Here only the short-pulse category of pulsed dye lasers will be discussed.

B. Short-Pulse Oscillators

Dye fluorescent lifetimes are too short (4–8 nsec) to provide energy storage, and for a high small-signal gain, a high-intensity pump pulse (but not a high energy one) is required. Short pump pulses (less than 100 nsec duration) offer the necessary intensity without adding extensive cooling demands and have the advantages of being too fast for triplet-state absorption or thermal lens distortions to develop in the dye. (The index of refraction of a liquid is dominated by its density, and a relatively high change of refractive index with temperature is a consequence of a relatively high rate of thermal expansion—which has a time constant on the order of 10 μsec, however, to develop.) The dye in a short-pulse system is simply circulated or stirred fast enough (about 1 m/sec) to remove energy-deposition temperature gradients in the few milliseconds between shots.

In the early development of pulsed dye lasers these reasons made the nitrogen (N_2) lasers then available attractive as dye pumps. The nitrogen laser has 4- to 10-nsec pulse lengths, 100-kW peak powers, a near-UV wavelength of 337 nm, and a repetition rate of 100 Hz, high enough to complete spectroscopic experiments in hours, (unlike the alternative doubled ruby laser pump at 1 Hz). The design principles for a narrowband spectroscopic short-pulse dye laser were first worked out by Hänsch in 1972 with this pump laser (Fig. 5a). The dye laser market thus established stimulated development of the more powerful Nd : YAG and excimer lasers for pumping use, which were to later displace the nitrogen laser (Fig. 1).

Hänsch realized that the low ratio of pump pulse length to round-trip transit time in the laser cavity dictated the design of the short-pulse oscillator. In a cavity long enough (here 40 cm) to accommodate a tuning device, a circulating,

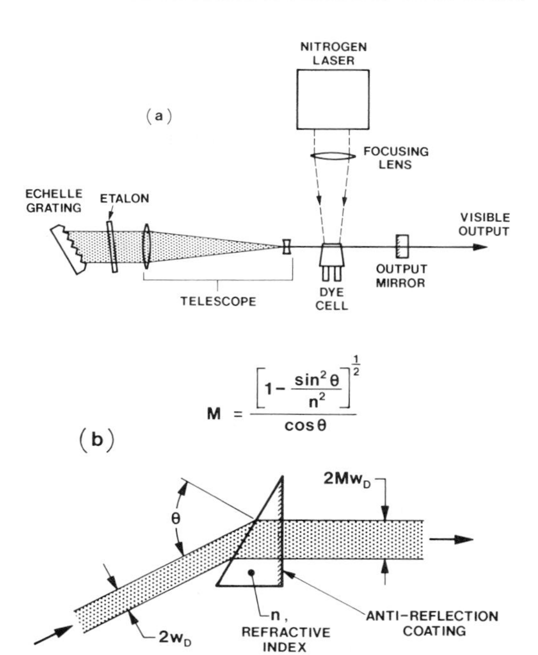

FIG. 5. The original short-pulse tunable dye oscillator. (a) The layout by Hänsch, with an intracavity telescope to increase the resolution of the grating. (b) One-dimensional beam expansion in a prism with high incidence angle on the input face and normal incidence on the exit face. A series of four prisms, arranged for no net dispersion, makes up the "quad-prism expander" that has replaced the Hänsch telescope. [Adapted with permission from Klauminzer, G. K. (1977). *IEEE J. Quantum Electronics* **QE-13,** 103. © 1977 IEEE.]

spontaneously emitted photon has only a few round trips during the excitation to replicate in an avalanche to $\sim 10^{15}$ photons (0.3 mJ at 600 nm), an output representing about 0.5 quantum efficiency of conversion of input photons. This necessitates a high gain in the dye, but nitrogen laser pumping permits this, as unsaturated gains of 10^3 in a few millimeters path had already been demonstrated. (Such gains correspond to nearly total inversion in Eq. (11). To achieve them, Hänsch transversely pumped an $L = 1$ cm long dye cell transversely (from the side) as shown in Fig. 5a. The nitrogen laser beam was focused into a narrow line of 0.15 mm width along the inner cell wall, with a dye concentration adjusted for a symmetrical 0.15 mm absorption depth). The output linewidth in such an avalanche is only a little smaller than the narrowest filter in the cavity, but very narrow filters (even though inevitably lossy) could be used because of the high gain, which also gives a high transmission for the output coupler. This widens the cavity mode widths to be comparable to their spacing, eliminating discrete longitudinal mode structure and allowing the laser to be smoothly tuned without servo control of the cavity length. The laser's frequency stability is determined just by that of the narrowest spectral filter. This short-pulse oscillator is essentially a few-pass amplifier of filtered spontaneous emission.

An appropriate filter set in this design is a grating in Littrow mounting (equal incidence and diffraction angles) plus a high-finesse etalon. An etalon is a pair of parallel reflecting surfaces (often the two sides of a plate of glass), which makes a wavelength filter by interference of the multiply reflected beams. The relative narrowness or half-transmission width of the filter ratioed to the spacing of adjacent orders, called the finesse, usually is determined by and increases with the reflectivity of the two surfaces. The grating resolution is determined by the number of lines illuminated, times the grating order. To increase the number of lines (or equivalently, to decrease the beam divergence on the grating), the cavity contains a beam-expanding telescope of magnification 25 (Fig. 5a). This also reduces the intensity incident on the grating to avoid damaging its aluminum coating, and reduces the etalon losses that are due to walk-off or incomplete overlap of the interfering beams. A coarsely ruled or echelle grating (600 lines/mm) is actually an advantage in that it allows the grating to work over a large surface area at high incidence angles, while keeping only one dye tuning range per grating order. With this design, Hänsch achieved 5-nsec pulses of 0.1 mJ energy and 0.1 cm^{-1} spectral width without the etalon, and 0.01-mJ, 0.01-cm^{-1} pulses with the etalon, tunable with several dyes over the visible spectrum.

The basic Hänsch design is used today with one major change—the telescope is replaced by a set of half-prism beam expanders (Fig. 5b). (As the exit face is used at normal incidence, this is "half" of a normal prism, used with symmetrical entrance and exit angles.) To increase resolution, expansion is required only in the dispersion plane of the grating; by expanding in one dimension only with prisms, the grating tilt in the orthogonal plane is made noncritical. Typically four prisms are used in pairs with substractive dispersions to give an in-line achromatic arrangement, constituting a quad-prism expander (see Fig. 6 in the next section). The magnifica-

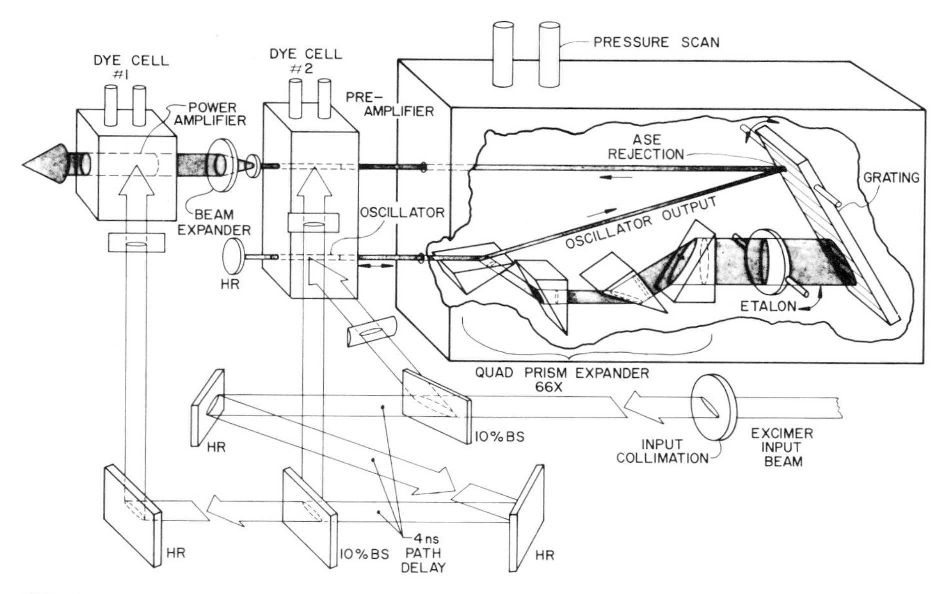

FIG. 6. Schematic diagram of a narrow-linewidth short-pulse dye laser of current design, for pumping by an excimer laser. Two stages of amplification and features increasing resolution and rejecting amplified spontaneous emission are shown. Highly reflecting mirrors are labeled as HR and beamsplitters as BS. (Placement of dye cells, optics, and angles between beams are altered for purposes of clarity.) (Courtesy of Lambda Physik, Göttingen, West Germany.)

tion given by a single prism (Fig. 5b) is large only at very high incidence angles, creating a large entrance-face reflection loss. By using several prisms, the incidence angle can be brought nearer to Brewster's angle to drastically reduce this loss; to produce the 66 times expansion of Fig. 6 with four symmetrical prisms, the incidence angle would be about 74° and the transmission 70% (through all four prisms) for the favored polarization direction lying in the incidence plane. The prisms automatically polarize the dye laser beam and are more economical than an achromatic telescope. Most importantly, large beam expansions are possible in a shorter length, allowing more transits during the excitation pulse and more efficient extraction of the oscillation input energy.

C. Excimer Laser-Pumped Dye Lasers and Oscillator/Amplifier Configurations

As more powerful short-pulse pump lasers were developed, the dye-laser designs evolved to make use of them. The limit on the small-signal gain per pass in a short-pulse oscillator is due to the growth of amplified spontaneous emission (ASE). This will be greatest for wavelengths near the peak of the emission cross section σ_e, and the intensity reached in a single pass must be kept well below one saturation intensity I_s [which by Eq. (8) varies inversely with σ_e] to prevent depletion of the gain at the desired, fed back wavelength λ_D. Calculation of the growth of ASE for a cylindrical geometry shows this will be true if a conservative rule of thumb is followed, that the gain per pass G = $(\exp k_0L)$ be kept below $(L/w_p)^2$, the square of twice the aspect ratio of the pumped volume (w_p is the pump beam radius). (The spectral separation between the ASE and λ_D at the ends of the dye tuning range means the ASE can be spectrally filtered outside the resonator here; in mid tuning range, the heavy gain saturation at λ_D reduces the ASE.)

High gains in one pass thus require a long, thin pumped volume, leading to small total volumes and ineffective use of a powerful pump laser. A better route to high-output energies in a tunable narrowband source is to use an oscillator–multiple-stage amplifier configuration. Here the inefficiency of the oscillator (from lossy spectral filters and the cavity length to contain them) can be ignored, as most of the pump energy is put into the following amplifier cells. By spectral filtering between stages, and timing (with path delays) the arrival of the pump pulse at the amplifier cell, ASE can be minimized. The amplifiers work in the partially saturated regime, which smooths out intensity fluctuations. The beam can be expanded between stages (with the dye concentration adjusted in each cell) to achieve a large active volume and a balance between amplifier extraction efficiency and absorption losses from the excited singlet state, a problem in high-gain systems with a large percentage of inverted molecules.

These features are evident in the modern design for a short-pulse, narrow-linewidth dye laser shown in Fig. 6. (Other designs also in use are the Littmann grating mount, a high-incidence grating with back-up mirror, and sometimes three amplifier stages.) This shows an excimer laser-pumped laser, but the designs for doubled Nd : YAG laser pumping are very similar. The 308-nm line from the xenon chloride excimer is used to pump some 20 dyes covering 332–970 nm. In the longer-wavelength dyes, this is by absorption to higher singlet states, which nonradiatively relax to the upper laser state (see Section II,B). (There is usually reduced photochemical stability with UV pumping of low Stokes efficiency dyes, however.) Pump beams of 100 W average power (0.4 J energy in 28-nsec pulses at 250 Hz rate) are available.

An excimer is a molecular species that is stable only in the excited state. In a XeCl excimer laser, a high-current pulsed discharge in a mixture of xenon and chlorine gas makes Xe^+ and Cl^-, which attract to form the excited $XeCl^*$ molecule, the upper laser state. After radiative decay, the molecule dissociates back to Xe and Cl, so there is no laser lower-state bottleneck to be concerned with in this laser. The discharge channel in an excimer laser is rectangular, giving an oblong cross section to the pumping beam (entering from the right in Fig. 6), making transverse pumping the most convenient. Ten percent of this beam is split off and focused into the dye cell (number 2) of the Hänsch type oscillator whose resonator mirrors are the grating on the right and the high reflector to the left of the dye cell. In the design shown, the first prism of the 66× quad-prism expander is placed at a higher incidence angle to make the reflection off its input surface serve as the oscillator output coupler. This output beam is directed at the grating at a small incidence angle, smaller than the angular width of the diffraction lobe for the unexpanded beam. This second diffraction off the grating thus acts as a spectral filter (that tracks automatically), reducing the amount of oscillator ASE that enters the preamplifier stage. An intracavity etalon with a finesse of 30 and order spacing of 1 cm^{-1} gives an 0.03-cm^{-1} linewidth, scannable either by microprocessor-controlled tilting of the grating and etalon (30 cm^{-1} range before resetting the etalon order) or by changing the gas pressure in the sealed oscillator housing. The 4-nsec path delay of the pump beam, before the 10% split off to pump the preamplifier dye, allows an appropriate oscillator buildup time before gating "on" the amplification of the next stage. The remaining 80% of the pump beam is weakly focused into the enlarged active volume of the final power amplifier stage, which is filled by the appropriately enlarged dye beam (about eight diameters expansion). This dye cell has a separate dye circulation system to allow the concentration to be adjusted for a longer absorption depth.

At 90-W average 308-nm input power, this system has generated, at a 250-Hz pulse rate with fast flows of coumarin 102 dye to avoid burning the dye cells, an average output power of 16 W at 475 nm. This is at 0.2 cm^{-1} linewidth (no etalon), with less than 1% ASE and a beam divergence of twice the diffraction limit. At 22 nsec pulse length this is 2.9 MW peak power. With an etalon giving 0.04 cm^{-1} linewidth, the average power was 13 W. At the time this article is being written, only a few dyes have been characterized at this high an input power, an example of a point in the evolutionary race between short-pulse pump and dye lasers where the pump laser is a little ahead.

D. Dye Lasers Pumped and Mixed With Harmonics of the Nd: YAG Laser

For the highest peak powers (at lower pulse repetition rates and average powers), the dye laser is pumped (and mixed) with harmonics of the *Q*-switched neodymium–YAG laser. Higher peak powers result from a more efficient pump wavelength (green, 532 nm) for rhodamine dyes, coupled with a shorter pump pulse length (of 5–9 nsec).

Pump energy is stored in the optically excited solid-state YAG laser rod over the upper-state lifetime of 200 μsec, and an intracavity electro-optic *Q*-switch dumps this stored energy at a fixed repetition rate, usually 10 Hz, to form these short pulses. (The fixed rate allows the thermal lens in the thermally loaded rod to be accounted for in optimizing the YAG resonator design.) High-power YAG systems consist of an oscillator rod followed by a larger-diameter amplifier rod. For dye pumping, the fundamental wavelength of 1.064 μm must be doubled in a phase-matched nonlinear crystal to the green wavelength. For shorter-wavelength spectral coverage, the green beam is sum-frequency mixed with the unused portion of the fundamental in a second nonlinear crystal to generate a 355-nm ultraviolet pump beam, the third harmonic. Mixing is like doubling, but with photons of unequal energies. Amplitude noise on the fundamental is carefully controlled as the noise level multiplies in these nonlinear conversions.

The same processes of doubling and of mixing with the 1.06-μm fundamental beam, are performed with the strong outputs of the green pumped dyes as well as to achieve spectral coverage into the ultraviolet. These nonlinear conversions are most efficient for beams in the lowest-order transverse intensity profile, or TEM_{00} Gaussian mode. It was the development of low-order-mode YAG lasers, based on unstable resonators with diffraction or polarization output coupling, that made pumping dyes with YAG lasers practical. This is also the reason for the main difference in dye laser design for use with YAGs, namely, that for the green-pumped dyes the final power-amplifier stage is pumped longitudinally, with the round, Gaussian profile pump beam folded in and overlapped along a common axis or at a small angle with the dye beam. The dye laser then replicates the low-order mode of the pump beam, enhancing the subsequent nonlinear conversions.

A 10-Hz YAG–dye system with 532-nm pump pulses of 360 mJ energy produces 120-mJ output pulses at 560 nm with rhodamine 6G dye, in a 5-nsec pulse of 0.3 cm^{-1} linewidth (no etalon), and less than 1% ASE. That is a 33% conversion efficiency, a 24-MW peak power, and 1.2 W average power with a beam divergence of four times the diffraction limit. With an etalon, 90-mJ pulses are produced at 0.05 cm^{-1} linewidth. Linewidths increase in the nonlinear conversions with the number and bandwidths of the photons involved.

E. Short-Pulse Dye Laser Tuning Spectrum

Dye output tuning curves for the two pump sources are compared in Fig. 7. This shows the output energy per pulse versus dye and output wave number, with only that portion shown of a dye's tuning range where the output exceeds

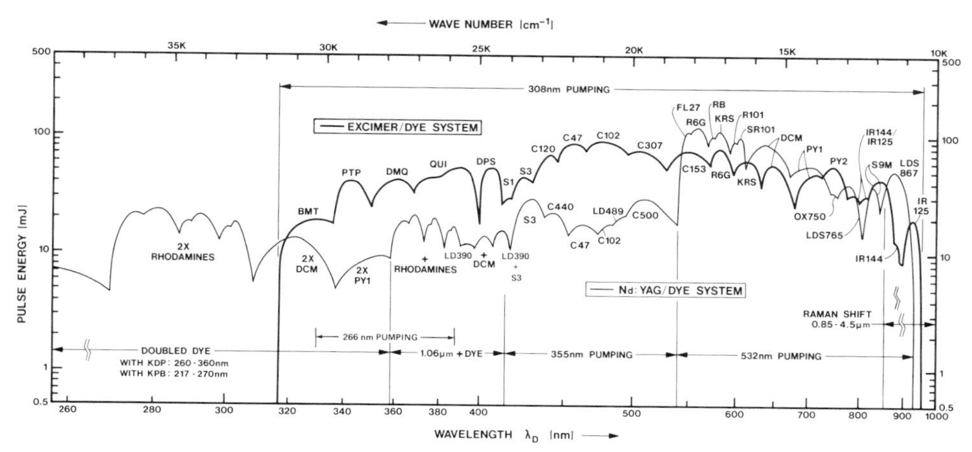

FIG. 7. Output tuning curves in energy per pulse at low repetition rates for excimer laser-pumped (heavy line) and Nd : YAG laser-pumped (light line) short-pulse dye lasers. Only the upper portion of a dye's tuning curve (above that of the adjacent dyes) is shown. Alternative pumping schemes to those shown are also possible for the YAG system, with the 266-nm fourth harmonic to pump dyes with outputs of 334–387 nm and with the 1.06-μm fundamental to pump dyes with outputs of 1.09–1.32 μm. (Excimer data, courtesy of Lambda Physik, Göttingen, West Germany; Nd:YAG data, courtesy of Quanta-Ray, Mountain View, Calif.)

that of the adjacent dyes. To obtain these results, the pump systems are configured to maximize the energy per pulse. This is at low repetition rates (<10 Hz), giving 0.5 J energy per 308-nm pump pulse for the XeCl system, and for the YAG system, 0.36 J per 532-nm pump pulse, 0.15 J per 355-nm pump pulse, and the undoubled portion of the initially 0.8-J, 1.06-μm mixing pulse. At higher rates both systems show a decreasing energy per pulse (but an increasing average power) as the lasers must dissipate more heat. The YAG system reaches a plateau of constant average power at about 15 Hz and is not designed to be used above 30 Hz. The excimer system pulse energy decreases more slowly, to 80% of the single pulse energy at a 250-Hz repetition rate, giving the excimer system the capability for nearly an order of magnitude higher average power. The difference in efficiency of the pump wavelength and in the shorter pulse width gives the YAG system the capability in the rhodamine dye region for nearly an order of magnitude higher peak power, and in this sense the two pumping systems are complementary.

Other features of Fig. 7 are also readily related to differences in the pump sources. For the strong green-absorbing dyes, the lower conversion efficiency of the excimer system is nearly accounted for by the lower Stokes efficiency η_S, which makes the 0.5-J UV input equivalent to 0.29 J of green input. Conversely, the UV excimer efficiently pumps the dyes lasing in the blue, where conversion to the third harmonic cuts the YAG pump power to less than half. In the red output region where there are common dyes, notice the shift to longer wavelengths for the excimer system. This is due to the overlapping σ_a absorption at the higher dye concentration, adjusted for a short transverely pumped absorption length and a more weakly absorbed pump wavelength. (The dye codes either are listed in the next section or are cross-referenced in laser dye manufacturers' literature.)

The YAG system propagates its strength in the red lasing dyes (using appropriate crystals and steering optics) by sum-frequency mixing these outputs with the YAG fundamental to cover the 358-to-413-nm span and doubling these dyes for the 270-to-358-nm span. Thus the characteristic pattern of the red dye tuning peaks appears three times in the YAG/dye spectrum. In both systems, doubling the coumarin and stilbene dye outputs reaches shorter UV wavelengths (only the YAG results are shown in Fig. 7), to 217 nm with two crystals. Output wavelengths are extended into the infrared to about 4.5 μm with both systems by directing the dye beam into a Raman shifter, a long (1 m) high-pressure (30 atm) cell filled with hydrogen gas. The hydrogen molecules are driven in coherent vibrations by the high-intensity input light scattered in the Raman process, and they radiate coherent beams shifted in wave number from the input beam in increments of the hydrogen vibrational level spacing of 4155 cm^{-1} at conversion efficiencies exceeding 20% for the first Stokes wave (light frequencies downshifted by one vibrational spacing) and 2% for the first anti-Stokes wave (upshifted frequencies, used to extend outputs deeper into the ultraviolet). In the days before lasers, the Raman effect was considered a weak process, as hours of exposure time with a several-kilowatt mercury arc source were required to register a Raman-shifted line on a spectrographic plate. With little more electricity taken from the wall than required by the old arc source, the short-pulse dye laser system produces, by stimulated Raman scattering, Raman-shifted beams of a megawatt peak power. Thus the wavelength extension techniques employed with short-pulse dye lasers serve quite well as examples of the nonlinear optics applications for the beams these systems generate.

IV. Continuous-Wave, Single-Frequency Dye Lasers

A. Dye Cells, Dye Jets, and Focusing Cavities

The first continuously operated dye laser focused the pump light from an argon ion laser into a cell of flowing dye to a spot diameter of around 10 μm, an active cross-sectional area of only 10^{-6} cm^2. By Eq. (12), this brought down the pump power needed to be comfortably above threshold to less than a watt, and the dye flow took care of cooling and triplet-state relaxation by exchanging the dye in this tiny active volume every few microseconds.

Focusing the coherent pump beam to this diameter was no problem, but building a stable dye laser resonator with a minimum area or beam waist diameter this small to overlap the pump spot was another matter. The necessity for matching the two beam diameters comes from a calculation of the filling factor η_F of Eq. (13) for the cw case of longitudinal pumping. A Gaussian TEM_{00} mode pump beam of $1/e^2$-radius w_p is assumed, whose axis and waist location overlaps that of a TEM_{00} mode dye beam of radius w_D. As previously defined, the ratio of beam areas is $\rho_0 = w_D^2/w_p^2$, where in this case the subscript zero has been added to indicate that the beam radii are to be taken as their minimum values, at their waists. The results are that the incremental gain constant for the dye beam is reduced from the plane wave result of Eq. (7) (where I_p is taken as the peak pump intensity at the center of the Gaussian profile) by a factor of $1/(1 + \rho_0)$, the threshold pump power in Eq. (13) is increased by a factor $(1 + \rho_0)$, and the filling factor η_F in Eq. (13) becomes $\rho_0/(1 + \rho_0^2)^{1/2}$. The output power is still a linear function of the input power above threshold, as given by Eq. (13), but the slope of this conversion increases only asymptotically with ρ_0, while the threshold power subtracted from the pump power increases linearly. Consequently there is in practice an optimum value of $\rho_0 \approx 1.2$–1.3 (giving $\eta_F \approx 0.8$), namely, that the dye beam diameter should be only slightly larger than the pump beam diameter.

The necessary small beam diameter was achieved with a resonator consisting of a flat and curved mirror (radius R), whose mirror separation d was critically adjusted around $d \approx R$ (but with $d < R$) to be near the stability limit. A resonator is said to be stable when multiply reflected paraxial rays stay within the structure; for this case, as d approaches R, a multiply reflected ray makes increasingly steep angles with the axis as it propagates, until at $d > R$ the ray walks out of the structure, as a ray tracing in the unfolded resonator (an equivalent sequence of lenses) will readily show. As this unstable condition is approached, the beam diameter at the flat mirror shrinks toward vanishing, and at the curved mirror it blows up to exceed the mirror aperture diameter. The first cw dye laser used $R = 4.5$ mm, and the spacing difference $R - d$ was set to 20 μm, to within ± 0.3 μm to keep the dye beam area within $\pm 20\%$ of the desired size. If the curved-mirror radius were made larger in such a resonator (to accommodate a wavelength tuning device), the tolerance on the required mirror spacing would become proportionally smaller. Clearly, a less critical assembly than a two-mirror resonator is needed for a cw dye laser.

A three-mirror or focusing resonator consisting of two curved mirrors and a flat provides this (Fig. 8b). This has a long arm of length $d \approx 0.6$ m where the beam is collimated and of a relatively large diameter $2(2d\lambda/\pi)^{1/2} \approx 1$ mm. The beam is then brought to a small focus of diameter $R(\lambda/2\pi d)^{1/2} \approx 40$ μm by the internal focusing mirror of radius $R \approx 0.1$ m. A rhodamine 6G wavelength of $\lambda = 0.6$ μm has been assumed here.

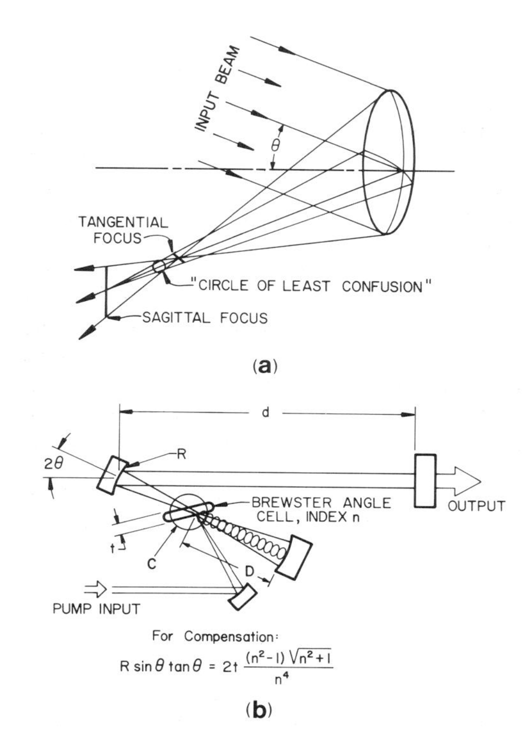

FIG. 8. (a) Astigmatism of an off-axis spherical mirror and (b) its compensation by a Brewster plate in a three-mirror focusing cavity, as originally used for standing-wave cw dye lasers.

(These dimensions are appropriate to the cw laser discussed below, which is pumped with a 6- to 24-W input and so works with a proportionally larger focal area than the earlier cw laser.) A new problem is introduced with this solution, which is fortunately readily solved. A spherical mirror used off-axis like the internal folding mirror in Fig. 8b generates an astigmatic focus, as shown in Fig. 8a. The rays lying in the incidence plane see the mirror's curvature foreshortened into a tighter arc and come to an out-of-plane line focus (the tangential focus) ahead of the on-axis focal length; rays lying in a plane orthogonal to the incidence plane see no foreshortening and reach a line focus (the sagittal focus) lying in the incidence plane beyond the on-axis focal length. In between the line foci, the beam shape becomes round (the "circle of least confusion") but at a considerably larger focal area.

Fortunately, an astigmatism of the opposite sense results when a flat plate is placed in the focused beam in the same plane of incidence but at an angle off normal incidence. In the dye laser this will be the dye cell, and it can compensate the fold-mirror astigmatism and restore the small (round) focus. The incidence angle chosen for the plate will be Brewster's angle θ_B (where $\tan \theta_B = n$, the index of refraction of the plate) for low loss. [Minimizing the nonuseful resonator losses a in Eq. (9) is very important in cw dye lasers that operate at low gain; this accounts for the abundance of Brewster angle faces found on cw intracavity wavelength control devices—which then serve to polarize the beam as well.] For a dye cell of thickness t, the correct fold angle 2θ between the collimated and focused beams that achieves compensation is calculated in Fig. 8b; for a 3-mm-thick glass cell in the example above, the result is a 9° fold angle.

The stability limits for a three-mirror resonator are computed as in the two-mirror case after forming from the combination of the output and fold mirror an optically equivalent single mirror, which has a small radius of curvature $R' = R^2/2(2d - R) = 4.5$ mm in the example. In aligning the dye laser, as the spacing between the two curved mirrors of Fig 8b is varied plus or minus half this distance off the central focus position, the dye spots on the curved mirrors blow up to fill the mirror aperture as the resonator reaches its limits of stable focus. The spot at the focus, in the dye, correspondingly shrinks toward vanishing, which gives a way to adjust ρ_0 downward and experimentally find the optimum value. Since the resonator for a laser must be stable in both the tangential and sagittal planes, if the astigmatism compensation is not correct, this range of stable focus adjustment is reduced by the amount of residual astigmatism (distance between the line foci of Fig. 8a), and the output beam is not round (except for the one adjustment where the circle of least confusion lies in the plane of the dye). For the resonator of the example, assuming perfect compensation, the adjustment range over which the dye beam area stays within ±20% of the central focus size is ±1.5 mm, a manageable tolerance and a vast improvement over the two mirror case. Note in Fig. 8b that the pump beam enters the dye cell at a slight angle to the dye beam (called noncollinear pumping), which is permissible as by the Beer's law absorption profile a good overlap of the two beams is only essential in the first third of the path length in the dye, where over half of the pump beam is absorbed (for a total absorption of 80–85% as normally employed).

The improved ease of operation made by the focusing resonator soon led from milliwatt-level to watt-level cw output powers (at higher input powers), accentuating another problem. No matter how clean or carefully filtered the dye

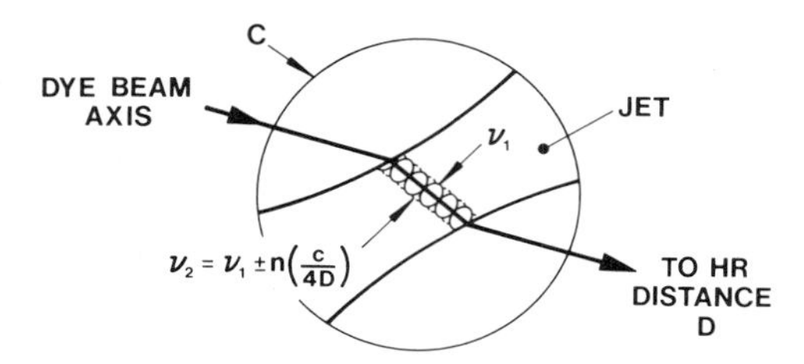

FIG. 9. Enlarged view of the jet, replacing the dye cell inside the circle C of the three-mirror cavity of Fig. 8b, showing the standing waves of the first frequency to lase locally depleting the gain or "burning spatial holes." This leaves gain regions unaddressed and allows a second frequency to lase.

solution was made, something that would burn and stick always seemed to pass through the focused pump spot after a few hours of operation, forcing shutdown and realignment of the system on a clean position on the dye cell. The answer to this was to eliminate the dye cell! It was realized that good optical quality in the region of the dye was needed over an area only a few times larger than the small area of the focus. Experimentation with free jets of dye dissolved in a viscous solvent like ethylene glycol, squirted at 10 m sec^{-1} from a nozzle with a slotted tip and caught in an open tube returning the solution to a fluid pump, showed this jet stream to give adequate surface flatness, stability, and dye velocity. (The flat jet actually necks down near the middle of its cross section as shown in Fig. 9, and is translated during set-up to find the mid position where the two surfaces are parallel.) A jet and catch tube are shown in Fig. 10. In addition to solving the burning problem, nozzles are much more economical than optical quality cells. Astigmatism can be so reduced by closing up the fold angle (possible because of the small jet dimensions, only 4 mm wide by 0.1 mm thick in section at the incidence plane) that compensation is no longer necessary.

B. Spatial Hole Burning and Ring Cavities

To fulfill its promise as a light source for high-resolution spectroscopy, the cw dye laser had to be operated with a single-frequency output (i.e., in a single longitudinal and transverse mode), which led to the development of tunable intracavity selective filters or "filter stacks" capable of supressing oscillation in all but one such resonator mode. These are considered in the next section; here the phenomenon of spatial hole burning is discussed. This limits the single-frequency power obtainable in the resonator of Fig. 8b to small values.

The three-mirror cavity shown is called a standing-wave or linear resonator, because the

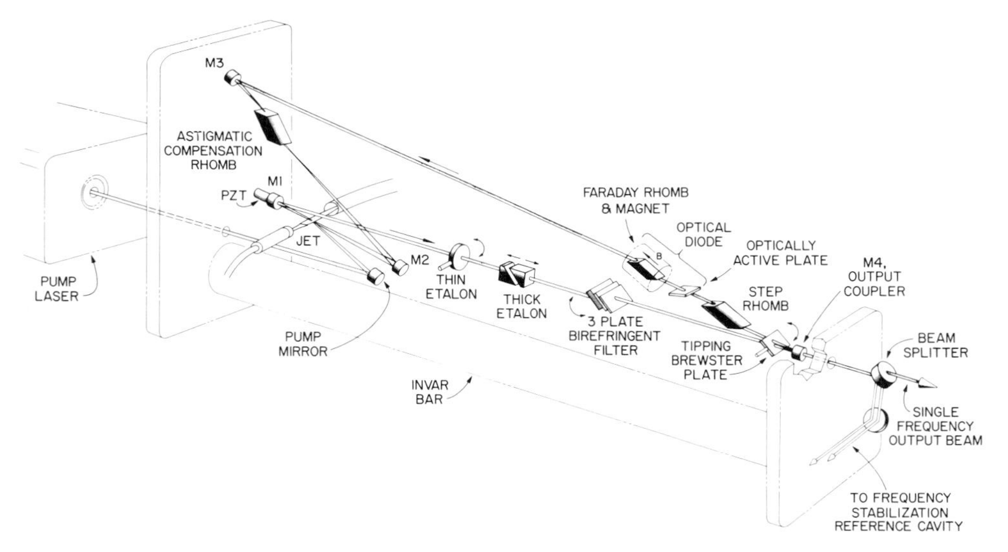

FIG. 10. Tunable single-frequency ring (traveling-wave) dye laser of current design. Spherical folding mirrors M1 to M3 and flat output coupler M4 fold the beam in a figure-eight path, which is traversed in the direction shown by arrows due to the biasing action of the optical diode element. Elements in the horizontal arm lying between M1 and M4 constitute the tunable filter stack, and the output frequency is controlled and the linewidth narrowed by a frequency stabilization servo (whose elements are not shown). (Courtesy of Coherent, Inc., Palo Alto, Calif.)

right-going and left-going dye beams in the resonator coincide and transform into each other at the normal incidence reflections at the two end mirrors. The superposition of these two traveling waves produces standing waves in the cavity that have nodes at the reflecting end surfaces. The allowed resonant frequencies or longitudinal modes of the cavity are precisely those that fit an integer number of half wavelengths into the optical path between the two end mirrors, as this condition gives constructive interference of the multiply reflected waves and results in the standing waves shown over length D in Fig. 8b. The oscillation in this resonator is at a single frequency, say the frequency ν_1. By counting half wavelengths from the nearest end mirror (the curved mirror at distance D from the jet), it is seen that inside the jet (see the magnified view of Fig. 9, inside the circle C of Fig. 8b) the standing waves occupy only part of the pumped volume. A second cavity mode with a frequency near $\nu_2 = \nu_1 \pm n(c/4D)$, where n is an integer, will have antinodes where the first mode has nodes and hence will not be suppressed by the gain saturation due to the first mode—the two modes in the same jet interact with spatially separated dye molecules. The first mode is said to "burn spatial holes" in saturating the gain medium, and the second mode utilizes the gain left between the holes made by the first. When a standing-wave, initially single-frequency dye laser is pumped well above threshold in an attempt to increase the single-frequency output power, at a fairly low input power a second frequency ν_2 begins to oscillate and destroys the single-frequency operation.

This limit is removed by going to the contemporary form of single-frequency dye laser in a traveling-wave or "ring" resonator (Fig. 10). The optical elements of the laser are shown here in their actual positions, with their support structure (invar bar and end plates) indicated in phantom lines. The dye beam is folded in a loop by mirrors M1 to M4, avoiding normal incidence end reflections and standing waves. (Actually, the folded path is more of a figure eight, to keep the fold angles on spherical mirrors M1 to M3 and their astigmatism small and compensable by the thick Brewster plate, or compensating rhomb, between M2 and M3.) Since the pumped volume is wholly utilized by a traveling wave, the first direction to oscillate (right- or left-going) will suppress the other possible one, and a single-frequency (s.f.) ring laser runs stably and naturally in one direction only. But cw dye lasers are subject to having their oscillation very briefly but often interrupted and restarted (due to microscopic particles or bubbles in the jet stream passing through the focal volume). A means to bias favorably one direction is required, or the output beam switches randomly and uselessly in exit angle from the output coupler.

The biasing device is called an "optical diode" and consists of a Faraday rhomb (a glass plate in a magnetic field) and an optically active plate. Considerations of time-reversal symmetry in Maxwell's equations dictate that any one-way or nonreciprocal light wave device be based on the Faraday effect, or the rotation of the direction of a linear polarization in transit of a material in a longitudinal magnetic field. The sense of the rotation depends only on the sign of the magnetic field B; for instance, for B pointing away from M4 in Fig. 10, the electric field of *either* a right-going or left-going beam is tipped away from the observer in a transit of the Faraday rhomb. When this is combined with a *reciprocal* polarization rotator (the Brewster incidence crystal quartz plate cut with optic axis along the ray propagation direction, to show optical activity), the forward or favored wave suffers substractive rotations, resulting in no tip of polarization out of the incidence plane of the remaining Brewster plates in the cavity and no reflection loss there. The backward wave suffers additive rotations and subsequent reflection losses; experimentally, a small differential loss of only 0.4% between these waves was found to be sufficient to always select the forward wave. The Faraday and optically active rotations are made to match over the desired spectral region by appropriate choice of magnetic field, rhomb material, and quartz plate thickness.

In addition to giving an order of magnitude increase in single-frequency output power, ring cavities enhance the intracavity second-harmonic generation process. By having all of the fundamental intracavity power traveling in one direction, the power of the doubled output, which extends single-frequency spectroscopy into the near ultraviolet, is increased an automatic factor of four.

C. Tunable Filter Stacks for Single-Frequency Lasers

Most of the remaining intracavity elements of Fig. 10 comprise the single frequency filter stack, a series of frequency filters of progressively narrower bandwidths, with overlaid center frequencies that tune together, designed to allow oscillation in but a single longitudinal mode of the ring cavity. (The transverse TEM_{00} dye mode is selected by the overlap with a TEM_{00} mode pump beam; this is called "gain aperturing.") Aside from the coarsest element (the lasing bandwidth of the dye itself), these filters all use the constructive interference between a first beam, and a second beam split off and delayed by a path length of an integer number of wavelengths, to generate a transmission peak. Each integer labels an order number of the multiply peaked filter, the spacing in frequency between orders is called the free spectral range (FSR), and the bandwidth of a transmission peak is given as the full width at half-maximum (FWHM). To minimize the intracavity losses [a in Eq. (9)], the FWHM of each element in the stack is made just narrow enough to select a single order of the next finer element of the stack, and spectral condensation plays a considerable role in this selection process.

The filter stack must be designed to span an enormous range. The two parts of Fig. 11 showing the filter stack at low and high resolution are therefore drawn with the expansion of the horizontal scale of 300 times between them directly indicated to help visualize this. The parameters of the figure are appropriate for rodamine 6G dye at its peak 580-nm wavelength. The gain bandwidth of this dye, determined largely by the emission cross section $\sigma_e(\lambda_D)$ derived in Fig. 4c, is about 3×10^3 cm^{-1}, making the curvature of

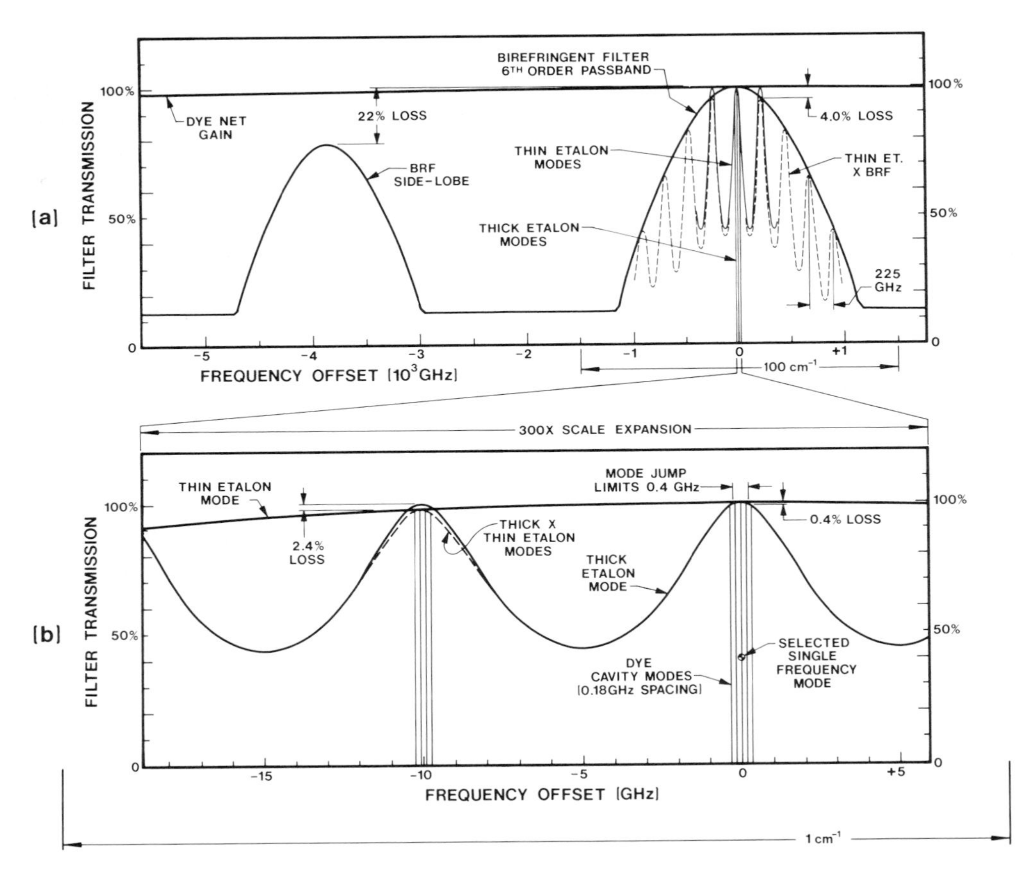

FIG. 11. Transmission functions of the filters in the tunable single frequency ring laser filter stack (a) at low resolution and (b) at 300 times higher resolution. Solid lines are for a filter element used alone, and dashed lines are for the composite stack. The curvature of the dye tuning curve is barely noticeable in (a) and the separation of the selected mode to the next adjacent cavity mode is barely visible in (b), emphasizing the enormous span of frequency over which the filter stack must work to select one mode out of 400,000. (Courtesy of Coherent, Inc., Palo Alto, Calif.)

the gain curve (scaled up here slightly to fit a 0–100% axis) just noticeable on the 250-cm^{-1} span of Fig. 11a.

The FSR of the birefringent filter or BRF, the next finer element in the stack, is chosen to be 3 × 10^3 cm^{-1} so that only the sixth order falls within the region of strong lasing of this dye. Birefringent crystal quartz plates oriented at Brewster's incidence are each cut with their optic axis lying in the face of the plate but rotated about 45° out of the incidence plane. The linearly polarized dye beam is split upon entry into the crystal into orthogonally polarized components, the extraordinary and ordinary rays. These recombine at the exit face back into light linearly polarized in the incidence plane for the central pass frequency, but for other frequencies into elliptically polarized light (which is partially reflected at intracavity Brewster surfaces). Like an etalon where $2nt$ is the second beam path delay (n is the index, t the physical thickness) producing a multipeaked filter of FSR = $1/2nt$ wavenumbers, this arrangement may be thought of as a two-beam interferometer with an $(n_0 - n_e)t$ path delay. In crystal quartz the ordinary–extraordinary index difference is about 0.01, so the birefringent filter acts like an etalon of thickness 1% of the physical thickness and is a convenient way to make a low-resolution interference filter. (An etalon of thickness of only 1 μm would be required to give the same FSR.) Three plates with precisely aligned axes and thicknesses in the exact ratios 1 : 4 : 16 produce an intracavity FWHM of 56 cm^{-1}, about half the width the same filter would produce extracavity between full polarizers. Gain saturation and spectral condensation are very effective in a cw dye laser, and this filter alone in the cavity ("broadband operation") will give an output spectral width of 0.03 cm^{-1}. The filter is tuned by rotation of the three plates together about their face normal, which varies the extraordinary index (and second beam path delay) while keeping Brewster's low-loss angle of incidence. Side lobes or frequency leaks in the filter transmission function correspond to polarizations that emerge from one or more (but not all) of the plates near the low-loss linear polarization. A BRF is not as positive as a grating for rejecting frequencies, because of these side lobes, but it is much lower in loss for the selected frequency (0.1% instead of 2–10%), which is what is important in cw dye lasers. It was the introduction of the low-loss BRF tuner in 1974 that allowed the cw dye laser to first span the visible with the low cw UV power then available to pump the blue lasing dyes.

A pair of low-reflectivity etalons (finesse of 1.8) reduces the BRF oscillating bandwidth to that of a single longitudinal cavity mode, the "thin" etalon of 7.5 cm^{-1} FSR or $\frac{1}{2}$ mm physical thickness, and the "thick" etalon of 0.33 cm^{-1} FSR and 10 mm physical thickness. Two etalons of low finesse give less loss for the selected mode than a single high-finesse etalon, due to a more complete overlap of the interfering beams (less "walk-off" loss). In Fig. 11a the thin-etalon transmission functions for the three central orders are shown as solid lines, and the products of the BRF and thin etalon functions (the composite filter) for several more orders are shown as dashed. The BRF suppresses the thin-etalon orders adjacent to the selected one by adding an extra 4% loss at the frequency offset of these orders. In Fig. 11b, the frequency scale is expanded 300 times to show the selection of a thick-etalon mode by the selected mode of the thin etalon. Here the curvature of the thin-etalon function is just noticeable, and a central thick etalon mode is selected by an extra 2.4% loss at adjacent orders. Finally, the thick etalon selects a single mode of the ring cavity to complete the filter stack, by inserting an additional 0.4% loss at adjacent cavity modes split off from the central one by a c/P = 0.18 GHz frequency offset (P is the path length around the ring perimeter). The total absorption and scatter loss per round trip at the selected mode frequency for the whole filter stack of Fig. 11, the a of Eq. (9), is 2.5%, with the walk-off loss of the thick etalon at 0.7% the largest component.

The center frequencies of this stack of filter elements must track together during scanning to tolerances that maintain proper mode selection. An offset of a coarse element, no greater than one-third of the FSR of the higher-resolution element whose order it selects, is acceptable. The basic continuous scan width of the system is 1 cm^{-1} (30 GHz); longer scans are pieced together from 1-cm^{-1} segments, and over this width the tracking tolerances and means of achieving them are as follows. The BRF need not be tuned at all but is stepped between segments. The thin etalon is tilt tuned with a rotational galvanometer and must stay within 3 GHz (or 10% of the scan range) of the linearly scanning frequency. This is done with a preprogrammed square-root drive to the galvo (the path-length change and hence center-frequency offset of a tilted etalon, goes as the square of the tilt angle away from normal incidence). The thick etalon must track to 60 MHz or 0.2% of the scan range, requiring a servo to meet this tolerance. The thick etalon is made up of two Brewster-surfaced prisms (essentially a solid etalon with a thin slice of air in the middle as shown in Fig. 10), so that its thickness may be tuned by piezoelectric translation. A small dither at a rate of a few kilohertz is added to the thick etalon drive voltage, generating a small-amplitude modulation of the output beam. This is phase-sensitively detected, giving a discriminant signal to lock the thick etalon center frequency to the scanning cavity mode frequency (standard dither-and-track circuit). To scan the cavity mode, a Brewster plate (on another rotational galvanometer) is tipped slightly ($\pm 2°$) off Brewster's angle. This introduces only slight reflection losses ($<0.4\%$) and linearly varies the path length through the plate by 120 μm, enough to scan the cavity frequency over the desired 1-cm^{-1} range. To avoid misaligning the ring cavity, this tipping plate is mounted near the beam vertex at the output coupler M4 and is traversed by both beams, in the design of Fig. 10. It was the introduction of the tipping Brewster plate as the cavity-mode scanning element that first allowed a 1-cm^{-1} or 30-GHz continuous scan range, due to the improvement in linearity and repeatability of this method over the older use of a mirror mounted on a piezoelectric element for the long-range scan.

D. The Narrowest Linewidths–Frequency Stabilization Servomechanisms

The intrinsic linewidth of the oscillation at the frequency of the selected single dye cavity mode, limited by spontaneous emission noise, can be found by the well-known Schawlow–Townes formula to be something less than 10^{-3} Hz. The observed linewidth is much larger and is determined not by this intrinsic width but by the stability of the cavity mode frequency given by $\nu_q = q(c/P)$ (where q is an integer of the order 3×10^6). The optical path P around the ring is subject to random environmental perturbations, a one-wavelength or 0.6-μm change in P, changing the cavity frequency by c/P or 0.18 GHz. The effective linewidth or frequency jitter produced by these perturbations will be (without stabilization) about 15 MHz rms, due to environmental noises, vibrations, and particularly due to the jet stream, whose thickness varies with pressure surges in the dye fluid circuit. This linewidth is reduced with a frequency stabilization servo that locks the dye cavity mode frequency to that of a stable reference interferometer, transferring the frequency stability properties of the output beam from that of the generating cavity, to that of the reference. The reference cavity, which can be compact and need not enclose a jet, is built to be stable by employing low-expansion materials in a temperature-stabilized housing, isolating the structure from vibrations, and sealing it against air-pressure changes.

A frequency servo consists of three parts: a *discriminator,* or error signal that changes sign when the perturbed output frequency passes through that of the reference; gain to boost this signal; and path-length transducers in the dye cavity driven by the amplified signal to bring the output frequency back to the reference value. The responses of the various transducers in this loop are included in the (complex) loop gain $M(f)$, which varies with the servo response rate or Fourier frequency f (the inverse of servo response time). The loop gain can be measured by breaking the loop, terminating the ends with the same impedances that were there before the gap, injecting a sinusoidal signal at frequency f into one end, and measuring the signal [$M(f)$ times as large] returned by the loop to the other end. Circuit analysis of the servo shows that a free-running or open-loop frequency deviation of the dye beam is reduced in magnitude when the loop is closed by the factor $1/[1 - M(f)]$. It is thus desirable to keep $M(f)$ large at high response rates f to correct excursions in P that occur even at rapid rates, to most reduce the output frequency jitter.

In doing this, however, the servo design must manage the phase shifts from all elements in the loop to keep the servo stable. The phase of $M(f)$ must be controlled over the full servo bandwidth where the magnitude of M is greater than unity, in order to keep the feedback negative and corrective and avoid oscillations of the loop, which correspond to a zero in the denominator of the feedback factor. All stable servos thus roll off the loop gain with increasing Fourier frequency to get and keep M less than 1 at high rates where phase shifts become uncontrollable. The two sources of phase shift of main concern are transducer resonances and the roll-off rate of M with f.

All path-length transducers have some form of inertia, and thus, when driven at a sufficiently high Fourier frequency f, they will resonate or reach large excursions at a drive frequency where the phase of their response is 90° relative to their low-f response. This resonance may span a narrow range of response rates, be highly nonlinear, and be critically dependent on small dimensions in the transducer (like the wall thickness of a piezoelectric element). Thus an inverse electrical filter (or "trap") added to the loop can generally only null out a transducer resonance to first order, and a servo designer's rule of thumb is to use gain roll-off to provide attenuation and a stability margin at the resonance by keeping the unity-gain frequency considerably smaller than the frequency of the lowest transducer resonance (no larger than one-third).

This upper bound on the unity-gain frequency is especially restrictive when phase shift due to the smooth, controlled roll-off of the loop gain itself is taken into account. This phase shift may be derived by interpreting mathematically the requirement of causality, that the response of the loop not precede its cause. The simplest case (most often used in practice) is for a constant roll-off slope S on a plot of log M versus log f, where S is constant over a minimum seven-octave span around unity gain (i.e., over frequencies from one-tenth to 10 times the unity-gain frequency). Then servo theory gives that this phase shift is proportional to S and reaches 180° (instability) at a slope of 12 dB/octave. A second designer's rule of thumb is thus to keep the roll-off rate less than 10 dB/octave, a painfully slow rate that limits the loop gain over much of the working servo bandwidth at Fourier frequencies below the unity-gain frequency, but such is the price for stability.

To summarize the basic design of a servo loop; the unity-gain frequency is usually dictated by the lowest resonant frequency of the path length transducers; the loop gain is made to rise at 10 dB/octave from unity gain toward lower response rates, reaching a maximum loop gain at a corner frequency determined by the drift rate of the reference [where generally $M(f)$ is leveled off, since it is pointless to provide more gain, and a tighter lock, good only at response rates where the reference is in error].

Usually, it is desirable to provide several length-control elements of different response rates in the dye cavity because the slow transducers will have larger correction ranges than the fast ones, and there is a corresponding correlation of speed with range among the forces perturbing the cavity length. The system of Fig. 10 connects the dc to 100-Hz components of the error signal to the galvo of the tipping Brewster plate, the slow (800 Hz first resonance) but long-range (30 GHz) transducer, to correct for room-temperature and air-pressure changes. With a crossover network it connects the 0.1-Hz to 10-kHz error information to the mirror M1 mounted on the piezoelectric transducer (PZT), the fast (30–60 kHz first resonance) but limited-range (1 GHz) element, to correct for bubbles or pressure fluctuations in the dye jet. The crossover network allows the first design rule to be followed in both the slow loop and the fast loop, to keep the servo stable. A convenient check that the servo is operating in the ring laser of Fig. 10 is to lightly push on a mirror end plate with a finger; when the servo is connected, the reflected spot off the tipping Brewster plate will reveal a proportional plate rotation, countering the direction of push and keeping P constant.

Frequency servos designed for dye lasers must address the fact that the dye oscillation is subject to brief (but frequent) momentary interruptions, or frequency excursions too fast for the servo transducers to follow due to bubbles in the jet stream. If the frequency servo can be made to automatically and rapidly relock the laser frequency to the original lock point after such an interruption, then the effect of the brief excursion can be made insignificant in the use of the laser, since this is generally to acquire time-averaged spectroscopic signals. The problem posed for the servo designer, since the laser oscillation may jump to a new dye cavity mode upon interruption (a "mode jump"), is how to unambiguously steer this arbitrary new frequency back to that of the original reference cavity mode.

Fortunately, the highest-resolution filter in the filter stack (the thick etalon) operates through a "mode-jump cycling" effect at optical rates as an optical frequency limiter and allows a relocking servo to be built. Figure 12 illustrates the relocking concept used for the laser and filter stack of Fig. 10 and 11. A discriminator is generated here by the "fast differencing" method (also called a "fringe-side" discriminator). This, one of the simpliest discriminators, was the earliest type to be widely used, and is found today in commercial s.f. ring systems. A beamsplitter picks off from the output beam of Fig. 10 two low-intensity beams, one of which is transmitted through the reference cavity, the other attenuated an adjustable amount, and then the beams are detected with photocells connected to the input terminals of a fast differencing amplifier. The attenuation is adjusted to put the zero point

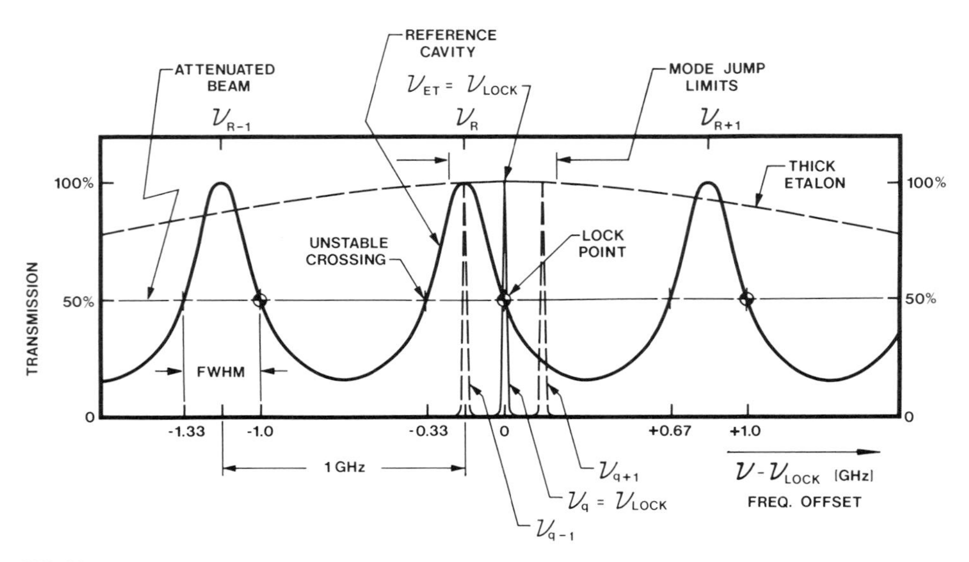

FIG. 12. Design for an automatic relocking servo with the "fringe-side" (fast difference) discriminator used with the ring laser of Fig. 10. The mode-jump limiting frequencies defined around the lock point (the zero-crossing frequency with a negative slope) must lie within that point's acquisition range (the frequency span between positive-slope zero crossings) to insure relocking to the original reference cavity mode after a momentary interruption of lasing. (Courtesy of Coherent, Inc., Palo Alto, Calif.)

of this difference signal at about the half-transmission point of a reference interferometer transmission peak (or half-way up the "side of the fringe"), as shown as a horizontal line labeled "attenuated beam" in Fig. 12. This level signal (independent of optical frequency) functions to prevent the mapping of intensity fluctuations into the difference, or frequency, error signal. The difference between this line, and the reference cavity transmission signal (at the negative slope zero crossing at 0 GHz frequency offset) is negative for frequencies above the zero crossing and positive below that lock point; taking the difference has generated a discriminator. There are additional potential lock points (negative-slope zero crossings) at -1.0 GHz, $+1.0$ GHz, and so forth, corresponding to adjacent reference cavity modes. These are separated by positive-slope, unstable zero crossings (at -1.33 GHz, -0.33 GHz, $+0.67$ GHz, and so forth) representing the boundaries or limits of the acquisition range of each stable lock point. The "mode-jump cycling" effect is used to contain the oscillating dye frequency within the acquisition range of the original lock point (of the reference cavity mode labeled ν_R) despite the interruption. Then when the perturbation has subsided, since there is only the original stable lock point (and no crossings to adjacent lock points) within the frequency range accessible to the dye oscillation, the servo will automatically relock as desired.

To understand mode-jump cycling, consider Fig. 12 and imagine the dye cavity modes ν_{q-1}, ν_q, ν_{q+1}, etc., being swept rapidly up in frequency by a fast perturbation, so rapidly that the thick-etalon tracking circuit cannot respond and the etalon peak is frozen in place. Originally the oscillating qth mode (at ν_q) is at the frequency of the etalon peak, but when this mode reaches an offset from the peak of $+\frac{1}{2}(c/P)$, it suffers a loss due to the etalon equal to that of the (nonoscillating) $(q - 1)^{st}$ mode at $-\frac{1}{2}(c/P)$ offset, and a mode-jump of the oscillation backward by $-c/P$ in frequency becomes possible. Due to the hysteresis introduced by gain saturation, the original oscillation will generally persist out to offsets greater than $\pm\frac{1}{2}(c/P)$ (both signs occur if downsweeps are considered as well as upsweeps). Nevertheless, observations have shown there are definite limits for the actual optical frequency change after the fast perturbation has subsided, regardless of the number of wavelengths of path change the perturbation contained. Measurements for the ring laser of Fig. 10 over the visible dyes (and over the gain variation within the tuning range of each dye) give a maximum mode-jump limit span of less than $3(c/P)$ around the etalon peak (see Fig. 12). Once these limits are known, a relocking servo is constructed by choosing the parameters of the reference cavity such that when the system is normally locked and the thick-etalon frequency is aligned with the lock frequency, only one lock point is accessible from within this mode-jump limit span. For the fast-differencing discriminator of Fig. 12, the nearest unstable crossing point is at one fringe width (or FWHM, the full width at half maximum of the reference cavity peak) lower in frequency than the lock point. For symmetrical mode-jump limiting frequencies about the lock point, a relocking design is thus achieved with a reference cavity for which the FWHM is greater than half the mode-jump limit span.

A logical problem may exist in this relocking design, in that at the first turn-on of the servo, before the thick etalon frequency and reference cavity lock points are aligned, there may be no lock point within the mode-jump limit span. (For example, the etalon frequency could be at +0.33 GHz upon turn-on in Fig. 12). What happens then is that the fast length control element (the PZT mounted mirror) is driven to a limit of its range by mode-jump cycling and saturates, or ceases momentarily to function (until a lock is acquired, and the charge is bled from the capacitance of the PZT drive circuit). The servo response rate thus momentarily drops to that of the slow loop (the galvo response rate), which by design is comparable to the response rate of the thick-etalon dither-and-track circuit. Thus, the thick-etalon frequency can move on initial turn-on to allow acquisition of a lock point.

With other discriminators the details are different but the ideas are the same. A logical trail must be built, such that recovery from a momentary interruption leads only into the acquisition range of the original reference cavity mode. For servos with unity-gain frequencies in the megahertz range (attained with electro-optic path-length transducers), the bubble interruption may last several servo response times. This makes it necessary to clamp various transducer drive voltages within certain bounds and wait out the interruption. The response times and sequencing of elements during the recovery from the clamped state must then be tailored, to stay within the desired acquisition range.

Two other methods are also commonly employed for generating frequency discriminators. The first, due to Hänsch and Couillaud, generates a symmetrical, dispersion-shaped discriminator with relatively large signal amplitudes at large offsets from the lock point, which results in prompt relocking. It derives from a frequency-dependent elliptical polarization induced in the beam reflected from a reference cavity containing a Brewster plate with an incidence plane inclined to the plane of the linear input polarization. The second method, developed for lasers by Drever, Hall, Bjorklund, and their co-workers, is termed the "rf sideband lock," because the input beam to the reference cavity is modulated at a radio frequency with an electro-optic phase modulator. The upper and lower out-of-phase rf sidebands that this generates are split off from the central carrier by the modulation frequency, chosen to be greater than the width of the reference cavity resonance. When the laser (carrier) frequency is on resonance with a reference cavity mode, there is a buildup inside the cavity of standing-wave energy at the carrier frequency but not at the sideband frequencies. A discriminator is formed by heterodyning in a fast detector the fields reflected from the cavity, and phase-sensitively detecting the beat frequencies against the rf modulation source. Analysis of this signal shows it to be the optical analog of the Pound or "magic-tee" stabilizer for microwave sources, and to have the property of acting as an optical phase detector on resonance due to the phase memory of the buildup field leaking out of the cavity.

The linewidths produced by these systems are a function of the speed of the path-length transducers and the resultant unity-gain frequency or Fourier bandwidth of the servo. The commercial system of Fig. 10 with a tipping Brewster plate and a PZT mounted mirror as transducers, using a fast-differencing discriminator, achieves a 5- to 10-kHz servo bandwidth and an optical frequency jitter of 200–500 kHz rms measured over a 1-sec sampling time with a 10-kHz detector bandwidth. The laboratory system with the best reported performance adds an intracavity AD*P electrooptic phase modulator to the list of path-length transducers (a crystal whose index of refraction depends on an applied high voltage) and, using an rf sideband lock with a 40 MHz rf frequency, achieves a servo bandwidth of 3–4 MHz. The resulting jitter was so small that special techniques were required to measure it. A helium–neon laser was similarly rf sideband locked to an adjacent mode of the same reference cavity, so that problems with drift and isolation of the reference would not mask the measurement of the locked frequency jitter. The beat note between these two independently locked lasers was examined and found to be less than 100 Hz in width—only 2 parts in 10^{13} of the dye laser's output frequency. Such systems offer enormous opportunities for precision measurements in physics and other sciences.

The ability of a relocking system to rapidly and automatically recover from a momentary perturbation is what makes electronic control of an s.f. dye laser effective. This ability is put to good use in the most recently evolved version of this instrument, where the laser is coupled to a wavemeter and computer-driven over a 10-THz or 333-cm^{-1} scan width. Such a long scan is done by piecing together many shorter scans stored in memory, where the endpoints of the short scans are abutted by the computer to the wavelength precision of the coupled wavemeter to present seamless output spectra. All of the filter stack elements must be moved to cover this long scan, and reset each in turn after each has been scanned over its free spectral range. To do this, the computer is simply programmed to introduce a "momentary perturbation" of its own by clamping the drive signal for each element at its nominal value, opening the loop, resetting the appropriate element, and again closing the loop. The system then reacquires the original lock point just as though this had been an external perturbation.

E. Single-Frequency Dye Laser Output Tuning Spectrum

The argon and krypton ion pump lasers for the s.f. cw dye laser offer a wide span of pump wavelengths, and the dye laser performs best where the absorption maxima of efficient dyes can be matched to strong pump lines. The overlapping tuning ranges of the 11 combinations giving the highest reported output powers over the fundamental tuning range (the visible spectrum) in a single-frequency laser are shown in Fig. 13. Additional data for this figure are listed in Table I. All of the tuning curves are for the ring laser of Fig. 10, with the exceptions of the short-wavelength half of the stilbene 1 curve and all of the IR 140 curve. For these two ends of the fundamental spectrum, the laser was set up in a standing-wave (three-mirror) configuration, as a ring configuration was not needed for these low-gain combinations and the elimination of the optical diode losses gave better performance there.

Table I gives the pump powers used for these

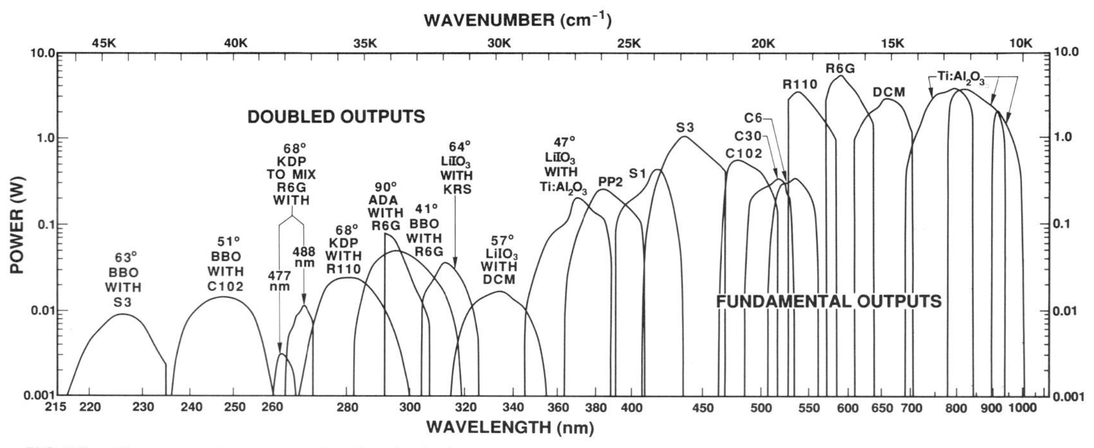

FIG. 13. Output tuning curves for the single frequency laser, pumped with the ion laser input listed in Table 1. The fundamental outputs are extended into the ultraviolet by adding nonlinear crystals to the cavity to produce doubling and mixing. For discussion of the updated results in the media PP2, S1, S3, Ti : Al_2O_3, and doubling with BBO, see the discussion of recent developments in Section IX at the end of the article. [UV data and Ti : Al_2O_3 results courtesy of Coherent, Inc., Palo Alto, California, and visible data adapted with permission from Johnston, T. F., Jr., Brady, R. H., and Proffitt, W., (1982), *Appl. Optics* **21,** 2312.]

TABLE Ia. Fundamental Single-Frequency Outputs

Gain medium symbol	Molecular weight (amu)	Common name	Peak output λ_D (nm)	Peak output P_{max} (W)	$n\pi w_D^2$ (10^{-6} cm^2)	Optimum coupling t_0 (%)	Gain k_0L (%)	I_s (MW/cm^2)	Pump laser P_{in} (W)	Pump laser λ_p[a]
PP2	542	Polyphenyl 2	383	0.25	5.7	1.9	8	2.2	3.4	SUV
S1	569	Stilbene 1[b]	415	0.42	6.2	2.7	11	1.8	6.0	UV
S3	562	Stilbene 3	435	1.0	6.5	3.9	18	2.0	7.0	UV
C102	255	Coumarin 102	477	0.58	7.1	2.9	12	2.4	4.8	V
C30	347	Coumarin 30	518	0.38	7.7	2.4	10	2.1	4.6	V
C6	350	Coumarin 6[c]	535	0.35	8.0	3.8	16	0.8	6.0	B
R110	367	Rhodamine 110[c,d]	540	3.6	14	10	60	0.6	23	BG
R6G	479	Rhodamine 6G[c,d]	593	5.6	16	11	70	0.8	24	BG
DCM	303	Dicyano-methylene[c,d]	661	2.9	17	5.1	23	1.6	20	BG
Ti : Al_2O_3	—	Titanium-sapphire single crystal	800	3.6	70	5.3	22	0.7	20	BG

TABLE Ib. Mixed and Doubled Single-Frequency Outputs

Crystal symbol	Match angle	Crystal name	Peak output λ_{UV} (nm)	Peak output P_{max} (mW)	Gain medium	Ion laser mixing line (nm)	Pump laser P_{in} (W)	Pump laser λ_p[a]
BBO	62°	Beta-barium borate	228	9	S3	—	7.1	UV
BBO	53°		248	15	C102	—	5.5	V
KDP	80°[e]	Potassium dihydrogen phosphate	262	3	R6G	477	7.5	BG
KDP	74°[e]		268	10	R6G	488	7.5	BG
KDP	68°		280	24	R110[c]	—	6.8	G
BBO	41°	Beta-barium borate	295	48	R6G[c]	—	8.5	G
ADA	90°	Ammonium dihydrogen arsenate	292	77	R6G	—	6.0	G
$LiIO_3$	65°	Lithium iodate	313	36	Kiton Red S	—	7.5	G
$LiIO_3$	57°		332	16	DCM	—	7.5	BG
$LiIO_3$	47°		370	200	Ti : Al_2O_3	—	20	BG

[a] Codes for the pump wavelengths are explained in the text, except for SUV, which is the 300–336 nm pump band.
[b] The standing-wave laser configuration used for a portion of the S1 curve in the earlier edition is here replaced by ring laser data.
[c] The triplet-state quencher COT was added here to the dye mix.
[d] Special water-based solvent AMX used (see the text).
[e] The crystal for R110 dye (cut for a match angle of 68° at Brewster incidence) was used here at large tip angle.

curves. The pump wavelengths or bands (groups of lines that lase simultaneously) from the argon ion laser are identified as UV for the ultraviolet (333–368 nm) pump band, BG for the blue-green (455–514 nm) band, B for the blue (488 nm) wavelength, and G for the green (514 nm) wavelength; these last two are the two strongest individual lines and are often used singly. Similarly, for the krypton ion laser, V indicates the violet (407–415 nm) band, R the red (647–676 nm) band, and IR the infrared (753–799 nm) band. The prime example of a strong combination is the high-gain dye rhodamine 6G, which is efficiently pumped either by the blue-green argon band or single-line (for the best dye transverse mode control, important in intracavity doubling) with the 514-nm line, the highest-power line of all ion laser outputs. A single-frequency stabilized output of 5.6 W has been attained with this dye.

The dye laser was fully optimized for these fundamental tuning curves, making the available input power be the limit to the output reached. With the weaker pump lines, the radius of the fold mirror (M1 of Fig. 10) was decreased to tighten the dye and matching pump focal areas and keep the gain well above the 2.5% s.f. cavity losses. (The 10-cm radius for the red dyes became 7.5 cm for the stilbenes, coumarins, and S9M, and 5.0 cm for IR140). The transmission of the output coupler was also optimized across the span of each dye (the breaks on several of the tuning curves are places where a different output coupler was substituted). For the BG pumped dyes, a special dye solvent (called AMX by Johnston) consisting of three parts ammonyx LO and one part ethylene glycol, chilled to 10°C, replaced the normal ethylene glycol jet. This reduced the optical distortion of the jet from thermal lensing, which otherwise would limit the output at these input powers of 20 W or more. The ammonyx LO is mostly water, which has a low refractive index change with temperature and a high specific heat, yet due to the remaining component (a soap, lauryl amine oxide) has a high viscosity at this temperature and can form a stable jet. The additive COT (cyclooctatetraene), which relaxes the triplet state of the dye and enhances output, was used with this AMX solvent and in three other cases identified by footnote *c* in Table I. Small amounts of dissolving agents permitted dye concentrations sufficient for 80–90% pump beam absorption in the 0.1-mm jet thickness. For all of the Fig. 13 dye recipes, the observed locked jitter was less than 0.5 MHz rms, although when COT is added it must be fresh (stored in a vacuum ampule or nitrogen-purged container to avoid contact with oxygen, which turns it oily and immiscible). The vertical drops at the ends of some of the tuning curves are birefringent filter "break points," where the oscillation jumps from the main BRF transmission lobe to a side lobe closer to the peak of the dye gain curve. Tuning can be extended slightly beyond these breaks, by substituting a cavity optic coated to have a high transmission at the side-lobe wavelength.

Just as in the short-pulse dye laser case, these fundamental outputs are extended into the ultraviolet by doubling and mixing. In the short-pulse case, because of the high peak powers, high conversion efficiencies for the nonlinear processes are reached easily and the problem there is one of limiting the peak powers to less than the damage thresholds of the nonlinear crystals. The converse problem exists in the s.f. cw case, that of how to raise the fundamental power incident on the crystal sufficiently to reach a usable conversion efficiency. This is solved by placing the nonlinear crystal inside the dye laser cavity and replacing the output coupler with a high reflector, to reach circulating fundamental powers in the range 20–60 W and s.f. ultraviolet outputs of several milliwatts. The circulating intensity I_e at the focus in the dye jet is proportional (neglecting the threshold term) to the inverse of the cavity dissipative losses a, as shown by solving Eqs. (4) and (9) for I_e and setting $t = 0$, to give $I_e = I_s k_0 L/a$. The crystal absorption and scatter losses contribute to a and must be quite low (less than 1%) to reach the desired circulating powers. Therefore, high-quality, well polished crystals are used in the cavity at Brewster's incidence angle.

The high-power dyes R110 to LD700, for which doubling is attractive, span a broad fundamental wavelength range, best covered without gaps by angle-tuned phase matching. Angle matching produces a lower conversion efficiency to the ultraviolet than the alternative, temperature-tuned matching (with a fixed matching angle of 90°). But unlike this alternative, angle matching is insensitive to temperature gradients in the crystal and requires no slow, high-hysteresis temperature servo. (Temperature-tuned matching was the first method used for intracavity doubling in s.f. ring lasers, and it is still employed to double the long-wavelength half of the rhodamine 6G spectrum. Matching this half spectrum is done by heating an ADA crystal substituted for the astigmatic compensation rhomb of Fig. 10, from room temperature to ~100°C, producing the results in Fig. 13 and Table I).

The angle-matched conversion efficiency is lower because at input beam propagation directions other than 90° (or 0°) to the optic axis the harmonic beam is doubly refracted, or separates ("walks off") at a slight angle from the fundamental beam. This separation that develops between the beams effectively limits the interaction length, or distance over which the harmonic radiation is most efficiently produced, to the distance it takes for harmonic light generated at the crystal input face to be displaced by a beam diameter from the fundamental. This interaction length is the beam diameter divided by the walk-off angle. The conversion efficiency still rises with lengths longer than this, but the rate of increase slows and most of the additional harmonic power goes to make the UV beam progressively more elliptical and distorted in transverse mode profile.

Thus the angle-matched nonlinear crystals used with the dye laser, which must generate a clean beam profile to be usable in spectroscopy, are chosen to have a length of one interaction distance. This choice has the interesting consequence that the nonlinear crystal need not be placed in a beam focus in the cavity. If the crystal is placed in the collimated arm of the cavity where the fundamental beam is of large diameter (0.7 mm), there is no loss in conversion efficiency if the crystal length is scaled with the beam diameter to still be one interaction distance. This amounts to 1-cm-long $LiIO_3$ crystals and 4-cm-long KDP crystals for placement between M3 and the optical diode in the upper arm of the cavity of Fig. 10. This placement is chosen because of its mechanical convenience and because in a parallel beam the crystal may be made to produce negligible interference with the single-frequency filter stack. The doubled beam is coupled out of the resonator with a Brewster-angle dichroic beam splitter.

For most angular orientations in the cavity, the nonlinear crystal would act as a fourth plate of the birefringent filter, and when angle-tuned would surely disrupt the s.f. filter stack. This is avoided if the dye beam sees only one (pure) refractive index, which for the two crystals above is the pure ordinary index. The crystal is oriented so that the electric field of the refracted dye beam is perpendicular to the crystal optic axis, then the dye beam is "resolved" as a pure ordinary wave, and the crystal acts then as would a plate of glass instead of as a doubly refractive material. Setting up this perpendicular condition uses one of a total of two angular degrees of freedom; orienting the crystal for Brewster incidence (to maintain the high circulating fundamental beam power) uses the other. One of these conditions must be relaxed to permit angle-phase matching; for this adjustment the crystal is rotated about an axis parallel to the electric field direction, maintaining the "no-disruption" perpendicular condition. This gives smooth tuning (normal action of the s.f. filter stack) but produces cavity reflection losses that increase from zero as the square of the crystal tip angle away from Brewster's incidence. Several different cuts of crystal are used to minimize these reflection losses, each cut putting the Brewster incidence no-loss point at a wavelength in mid-tuning range of each dye to be doubled, with the results shown in Fig. 13 and Table I. These ultraviolet doubled-dye outputs scan over twice the range that the fundamental is scanned and have twice the jitter.

Tunable ultraviolet s.f. outputs extending to even shorter wavelengths, which still take advantage of the strong red dyes, can be generated by nonlinear mixing of the dye beam with a shorter-wavelength, single-frequency beam from an ion laser, to make the sum frequency. Mixing is just like doubling, but with unequal energies for the input photons; this merely requires slightly different phase-matching angles. For the results of Fig. 13, an ion laser cavity was folded to intersect and include in the ion beam cavity, the 68° KDP crystal in the dye cavity. The collimated-arm crystal placement permits this; the ion and dye beams must be collinear inside the crystal for the summing interaction, but at the Brewster angle faces of the crystal the two wavelengths are refracted differentially and separate with an angle between them of five times the beam divergence. The throw distance to the nearest dye cavity mirror in the collimated arm is large enough for the ion beam to clear the edge of the dye mirror, and on the other side of the crystal for the dye beam to clear the edge of an inserted ion cavity end mirror. The only common element in the two cavities is the KDP crystal. The ion laser is operated in a standing-wave cavity as the phase-matching condition with the traveling wave dye beam picks out the desired running wave ion beam component in the summing interaction. The mixed and doubled dye outputs of Fig. 13 allow single-frequency, high-resolution spectroscopy over the range 260–400 nm. Sum-frequency mixing has been used to generate even shorter UV tunable s.f. outputs, down to 194 nm, in scattered wavelength bands appropriate for particular experiments. The beam at the sum frequency has the scan range of the dye laser and a jitter equal to the sum of the ion laser and dye laser jitter.

F. Gains and Saturation Intensities for Common Dyes

The tuning curve data of Table I may be analyzed to give gains and saturation intensities for these dyes, under the listed pumping conditions. This permits an experimental check on the uniform plane-wave theoretical analysis of Section II.

The analysis uses the optimum output coupling values $t = t_0$ determined experimentally with the tuning curves and the estimate of the single-frequency cavity dissipative losses a = 2.5% for the ring laser of Fig. 10. Differentiating Eq. (9) with respect to t to find the output mirror transmission that maximizes the output power gives

$$t_0 = a[(k_0L/a)^{1/2} - 1] \qquad (14)$$

from which

$$k_0L = a[(t_0/a) + 1]^2 \quad (15)$$

The gain values computed from Eq. (15) are listed in Table Ia for all the dyes (except IR 140, where lasing was too close to threshold to determine an optimum coupling). The output power with this optimum coupling is

$$P_{max} = n\pi w_D^2 I_s(t_0^2/a) \quad (16)$$

which shows that the maximum output is proportional to the inverse of the cavity losses a. The factor of $n = 1.4$, the index of refraction of the dye jet (at λ_D), did not appear in Eq. (9) for the plane-wave, longitudinally pumped case. It is appropriate in expressions like Eqs. (9) or (16) containing a focal area in a Brewster angle dye medium (the present case), because a beam of cross-sectional area πw_D^2 (normal to the beam direction) is expanded in area upon refraction at the Brewster surface to $n\pi w_D^2$. The prism expanders of Figs. 5 and 6 illustrate this expansion. [Note also that this factor of n was erroneously omitted in Eq. (6) of Johnston *et al.*, which resulted in saturation intensities reported there being 1.4 times too large.] Table Ia lists the correct beam focal areas. When the output power was measured using an optimum output coupler, inverting Eq. (16) gives the saturation intensity:

$$I_s = \frac{P_{max}}{n\pi w_D^2}\left(\frac{a}{t_0^2}\right) \quad (17)$$

and if a nonoptimum transmission t was used, Eq. (9) (including the factor of n) is solved for I_s, yielding the values in Table Ia.

The same analysis can be applied over a dye's full tuning range, to show experimentally the wavelength dependence of the gain and saturation intensity. This was done for the high-power tuning curve of rhodamine 6G, by determining the optimum transmission at seven points over the tuning range (Fig. 14b). The Fig. 14a tuning curves show as well the high conversion efficiency from broadband output (~2 GHz linewidth without etalons) to single-frequency operation of a ring laser. These curves were taken with a single output coupler to produce smooth full-range tuning data (this coupler gave a slightly lower peak output than the one that gave the peak power listed in Table Ia). Also, the t_0 values plotted were chosen (within the range of experimental error) to give smooth curves for k_0L, I_s [labeled EXPT in Fig. 14c and d] when put into Eq. (15) or Eq. (17).

These experimental values measure the response of the dye medium to changes in resonator loss, and as such represent the average over the active volume of the microscopic inversion variations in the dye. A direct comparison with theory would thus involve a (numerical) integration of the plane-wave equations over the focal volume, a procedure more complex than is warranted in view of the moderate accuracy of the measured t_0 values (about ±10% of their means). A simpler approximate treatment is to first establish upper bounds for I_s and k_0L by putting into the plane-wave Eqs. (8) and (11) the

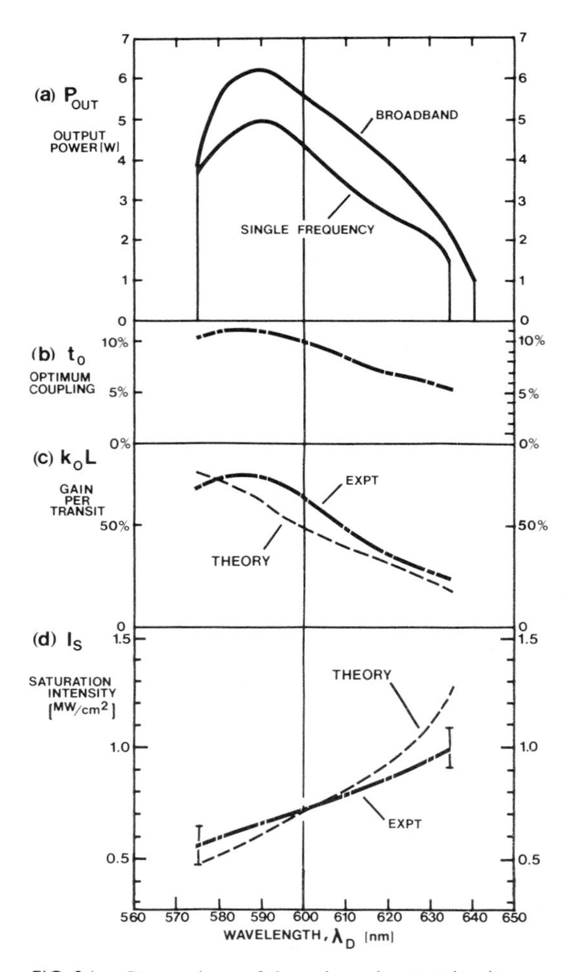

FIG. 14. Comparison of the gain and saturation intensities derived from measurements of optimum coupling (EXPT), with the theoretical plane-wave expressions using spectroscopic cross sections (THEORY) for rhodamine 6G dye in AMX solvent at 24 W pumping. [Adapted with corrections and permission from Johnston, T. F., Jr., Brady, R. H., and Proffitt, W. (1982). *Appl. Optics* **21**, 2311.]

maximum pump intensity reached in the experiment (the intensity at the center of the Gaussian mode profile, on the input side of the jet). This saturation intensity upper bound is then reduced by an estimate of the ratio of the average pump intensity to this maximum value. This gain upper bound is then reduced by the factor $1/(1 + \rho_0)$ quoted earlier (in Section IV,A) derived from the overlap integral at the focus in longitudinal pumping of a dye beam of Gaussian waist radius w_D and pump beam of Gaussian waist radius w_p, where $\rho_0 = w_D^2/w_p^2$. The upper bounds are evaluated for $\lambda_D = 590$ nm, at the peak of the experimental gain curve.

The stimulated pumping rate $(\lambda_p/hc)\sigma_p I_p$ (transitions per second) appearing in Eq. (8) is replaced for a multiline pump beam by the sum of such terms over the wavelengths present. When weighted by the appropriate pump cross sections (Fig. 4), the 24-W BG pump is found to be equivalent to 15 W of 514-nm pumping. To convert power to intensity, note that a TEM_{00} mode beam of total power P and Gaussian $1/e^2$-radius w has a maximum intensity on axis of $2P/\pi w^2$ and an average intensity of *half* this, if the beam cross-sectional area is taken to be πw^2. The calculated pump focal area in the jet was $n\pi w_p^2 = 6.7 \times 10^{-6}$ cm², giving $\rho_0 = 2.3$, and a maximum pump intensity of $I_p = 4.4$ MW/cm². The spontaneous decay rate for rhodamine 6G is $1/\tau\varphi = 0.2$ GHz; the green-equivalent pump intensity above yields a maximum pumping rate of 1.6 GHz, and Eq. (8) gives an upper-bound saturation intensity of $I_s = 4.0$ MW/cm². Because this pump rate is much larger than the spontaneous decay rate, the saturation intensity is proportional to pump intensity and the upper bound can be cut in half to account for the radial, off-axis falloff of the pump intensity. It is cut again by a factor ranging from 0.32 to 0.39, the ratio of the maximum to the average on-axis intensity, to account for exponential absorption along the axis to a 5–10% transmission through the jet. This gives theoretical I_s estimates for the cw laser case of 0.6–0.8 MW/cm². These agree satisfactorily (within the error estimates) with both experimental values at this wavelength, 0.8 MW/cm² from the data of Fig. 13 and 0.65 MW/cm² from the data of Fig. 14.

The gain upper bound, for $N = 1.3 \times 10^{18}$ cm^{-3} (from the estimated dye concentration of 2×10^{-3} M), $L = 0.012$ cm, and a maximum inversion fraction (calculated from the relative transition rates above) of 0.88, is given by Eq. (11) as 210% per pass. Applying the Gaussian beam reduction factor of 0.30 for the 2.3 focal area ratio gives a theoretical gain estimate of 64% per pass, which agrees satisfactorily with the two experimental results of 70% and 73%.

The large inversion fraction implies that the dye jet was bleached on the pump input side, violating the Beer's law absorption profile assumed in deriving both reduction factors and showing the roughness of the present approximations. Thus, the best test of the plane-wave theory is probably the comparisons of the predicted and observed wavelength dependencies of Fig. 14(c) and (d), which involve fewer assumptions. Here the theoretical curves are scaled from the $\lambda_D = 590$ nm results according to Eqs. (7) and (8) by the measured cross sections of Fig. 4. Using the smaller of the two theoretical saturation intensity values gave the best overlap with the experimental curve. The plane-wave analysis is seen to predict satisfactorily the variation of I_s and $k_0 L$ over the tuning range observed in an actual cw single-frequency dye laser.

Figure 15 is an example of sub-Doppler saturated absorption spectroscopy, the basic application of the scanning single-frequency dye laser. This example was chosen to display the connection of laser spectroscopy with the older, Doppler-limited spectroscopy, and conse-

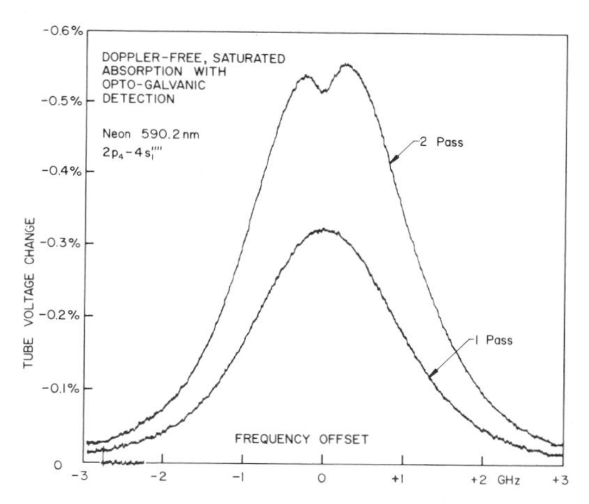

FIG. 15. Doppler-free and Doppler-limited laser spectroscopy compared on a neon absorption line, using opto-galvanic detection. The change in voltage across a helium–neon discharge tube was recorded as the single frequency of a coaxial dye laser beam (passed either once or twice through the discharge) and was scanned through the 590.2-nm resonance. [Adapted with permission from Johnston, T. F., Jr. (1978). *Laser Focus* **Mar.,** 58–63.]

quently is at a low resolution (not at all near the 1-MHz limit imposed by the frequency jitter of this stabilized laser). The laser beam was passed (through end windows) along the axis of a narrow-bore discharge tube run at constant current in a mixture of helium and neon gas. As the frequency of the dye laser scanned through the frequencies of neon spectral lines, power was absorbed from the laser beam, and proportional changes in the discharge power were recorded. This has been appropriately named optogalvanic (current-producing light) detection and is a detection scheme with a wide dynamic range and a good signal-to-noise ratio that is often used with spectral calibration lines for dye laser wavemeters.

The lower trace of Fig. 15 shows the absorption signal when the beam was passed once through the tube and shows a classical Doppler line profile, obtainable before the existence of lasers by high-resolution interferometry. This Gaussian line shape reveals the velocity distribution of the moving atoms in the gas and not the underlying homogeneous width of the spectral line (due to the finite lifetimes of the absorbing states). For laser frequencies below line center (negative offsets in Fig. 15), atoms with an on-axis velocity component along the laser beam (but in an opposed direction) see the laser frequency up-shifted into resonance and absorb the light. Since the atoms in the gas are moving in all directions with a finite average speed, the most likely axial velocity component in the distribution is zero, and the absorption increases (according to the Gaussian weighting function) as the laser frequency moves toward line center. The process is reversed as the laser frequency scans above line center, where the laser light is absorbed by a decreasing number of atoms seeing a down-shifted laser frequency.

If now the transmitted beam is retroreflected to make a second pass in the opposite direction through the tube, the absorption is doubled as expected for off-line center laser frequencies (as shown in the upper trace of Fig. 15), as both a right-going velocity group and a left-going velocity group of atoms have the correct Doppler shifts for absorption from one of the beams. But when the laser scans to line center, the two distinct velocity groups coalesce into a degenerate "zero-axial-velocity" group and there is a saturated absorption dip of width equal to the homogeneous width, marking the center of the line. The appearance of the dip requires sufficient laser power to perturb the velocity profile; a saturation intensity for absorption on the spectral line must be reached. Gas-phase spectral lines are typically 10^{-5} times narrower than the tuning range of a dye and the saturation intensities are correspondingly smaller than those of dyes, so the required intensities are readily reached (40 mW was the input here) even in an unfocused dye laser output beam. The improvement in ability to locate the central position of the line in Fig. 15 is only about an order of magnitude but is still enough to show, by the asymmetry of the dip, the presence of an isotope shift for the 9% of ^{22}Ne in the natural neon gas fill, to the high-frequency side of the line center of the main (91%)^{20}Ne component. For molecular absorption lines, the improvement in resolution reaches a factor of 10^3. This ability to see inside the classical Doppler limit throughout the dye tuning spectrum has revolutionized modern spectroscopy.

V. Ultrashort-Pulse (Mode-Locked) Dye Lasers

Of the three scientific dye laser types, the mode-locked branch is evolving most rapidly at present, as shown by the density of accomplishments in recent years in Fig. 1. This section covers the basics of the mode-locking of dye lasers. [*See* MODE-LOCKING OF LASERS.]

A. Pulse Formation by Phase-Locking of Cavity Modes

In TEM_{00} transverse mode operation, the mode frequencies ν_q of a standing-wave dye laser with resonator optical path length d are given by the longitudinal-mode values $\nu = q(c/2d)$. Here the integer q (of order 10^6) differs by 1 for adjacent modes, making the beat or difference frequencies between nearest neighbors in multimode oscillation all nominally $c/2d$. The beats are generally not precisely at this value, because of mode pulling and pushing effects. Associated with a gain peak at some center frequency there is an index of refraction variation given by the Kramers–Kronig relations tending to pull the mode frequencies slightly toward the center frequency; mode frequencies are slightly repelled from an absorption dip. The homogeneous-width "holes" in the gain curve caused by the competition between modes in a multimode laser thus normally cause a slight random shifting of nearest-neighbor beat frequencies away from the nominal value. This situation is best described by saying that the phases of the beat frequencies are normally random and uncorrelated.

By Fourier's theorem, however, if the phases could all be made constant (or "locked"), then the sum of the many mode amplitudes spaced regularly in frequency by differences of $c/2d$ would produce a periodic function, of period $T = 2d/c$. If initially these locked-mode amplitudes were phased for maximum constructive interference at some point in the cavity (producing a high peak power there), at time intervals of T they would return to this state at that location. The laser's steady multimode output would be converted to a repetitive train of pulses, with an interpulse spacing of T, corresponding to a circulating pulse striking the output coupler at intervals of the cavity round-trip transit time T. The width τ_D (FWHM) of this dye output pulse by Fourier's theorem would be given by a constant divided by the width $\Delta\nu$ (FWHM) of the spectral distribution of amplitudes of the locked modes, the constant being always of order unity but dependent on the form of the pulse shape.

To produce this locking of phases requires a small gain (or loss) term sufficient to overcome the random mode pulling effects; to establish the constructive initial phase distribution, this term should favor a high peak power pulse. Such a term is provided in the argon ion and Nd : YAG pump lasers by placing at one end of their resonators an acoustooptic loss modulator, or mode-locker. A sinusoidal radiofrequency voltage drives an acoustic wave in the mode-locker cell, giving a diffraction loss of the laser beam varying in time at a frequency precisely adjusted to equal the mode spacing frequency of the resonator. In frequency-domain terms, the sidebands introduced on a longitudinal mode carrier frequency by this amplitude modulation produce zero beats with the adjacent mode frequencies to lock the modes. In time-domain terms, the same result occurs because the formation of a short pulse allows oscillation with the least loss, as the pulse can pass through the loss modulator at the times of zero diffraction loss. Because input energy (other than that of the laser beam itself) is required to drive the mode-locker, this is called "active" mode-locking. Typically, a long pump laser is used giving a mode spacing of 76 MHz and a period of 13 nsec. The green pumping pulses from a mode-locked argon laser are about 120 psec long and from a mode-locked, doubled YAG laser are about 70 psec long, the difference reflecting the different gain bandwidths for these lasers.

The dye laser is similarly mode-locked by pumping synchronously (with the cavity round-trip transit times matched as will be described) with a mode-locked pump laser, producing repetitive gain modulation. Alternatively, mode-locking is produced (with constant cw pumping) by adding a second dye jet to the cavity, containing a saturable absorber dye. This adds a loss term that is least (saturates the most) for a high peak-power circulating pulse. This method is called "passive" mode-locking since the absorber is driven by the laser energy itself. Both mode-locking methods may be used together in a "hybrid" mode-locked laser as shown in Fig. 16.

To match the dye cavity length to that of the long pump laser, while keeping the dye laser compact, the beam in Fig. 16 is folded in a zigzag path by the two flat mirrors. The focal diameters produced in the gain jet by the mirrors M1 and M2 and in the absorber jet by mirrors M3 and M4 are given as in the cw case by the three-mirror analysis of Fig. 8. A translation stage (not shown) on the output end of the laser allows precise length adjustment. In the figure, a cavity-dumper assembly (which can be translated) is shown in place of an output mirror. This device is a time-gated output coupler, driven in synchronism with the mode-locker of the pump laser, and is often used to reduce the repetition rate of the train of output pulses while increasing the energy per pulse. An acoustic wave is turned on to generate a diffraction grating in the Brewster cell at the M5 and M6 focus for several nanoseconds at the time of passage of the light pulse. The interference of the diffracted and direct beams returned by M6 is phased in the second diffraction to couple nearly 100% of the pulse energy into the diffracted beam direction, where it is directed out of the laser by the take-out prism. Compared to a transmitting mirror, the use of all high reflectors and this large coupling factor in cavity-dumped operation gives 10 or more times the output energy per pulse, at one-twentieth the repetition rate.

B. Pulse-Shaping by Saturable Gain and Absorption

The number of modes phased together in the dye laser is so large (10^4 in a picosecond pulse) that the process is most conveniently described in the time domain as the shaping of the pulse in passage through the elements of the cavity. A picosecond pulse is spatially only 0.3 mm long and interacts sequentially with each element. In

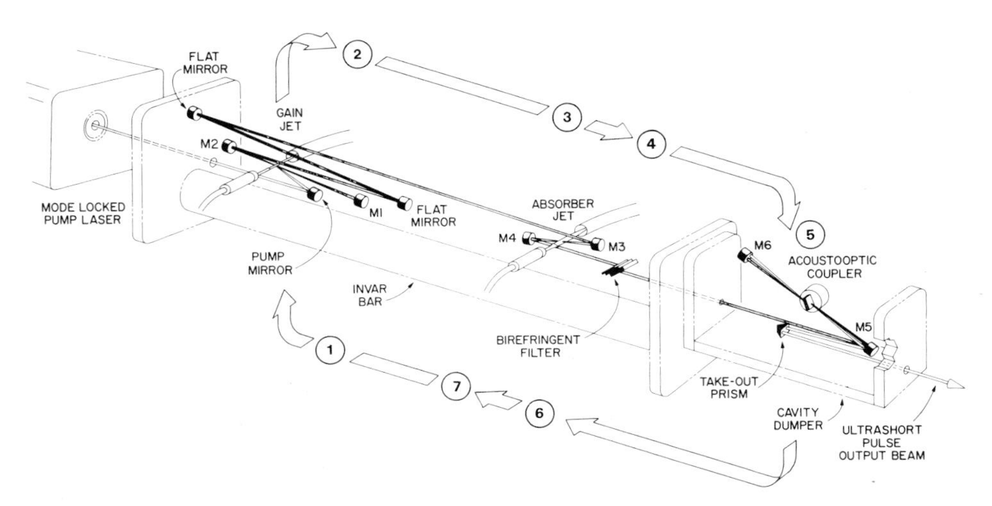

FIG. 16. Ultrashort pulse dye laser of current design for "hybrid" mode-locking (employing both synchronous pumping and a saturable absorber). Shown in place of an output mirror is a cavity dumper assembly, which is driven in synchronism with the circulating pulse to couple out a reduced number of pulses, at higher energy per pulse. (Courtesy of Coherent, Inc., Palo Alto, Calif.)

Fig. 17, the circulating pulse is moving to the right (time increases to the left), and the effect of each (numbered) passage in Fig. 16 is shown schematically.

In passage through the gain jet (1, 2) the leading edge of the pulse is amplified, which increasingly depletes the gain over the remainder of the pulse, since on the time scale of the dye pulse the rates of pumping transitions and spontaneous decays are slow. This steepens the leading edge and advances the timing of the peak of the pulse. For good mode-locking, the focal area in the absorber jet is made smaller than in the gain jet (note the shorter-focal-length mirrors in Fig. 16) to ensure faster saturation in the absorber (3, 7) to further steepen the leading edge (and slightly retard the peak). The uniform, linear loss of an output coupler (5) leaves the pulse shape unchanged, but the attenuation of the trailing edge here, in combination with the growing middle and attenuated leading edge, means a net pulse shortening in passage through these three elements. What keeps the pulse width from collapsing to zero is the finite spectral bandwidth of the frequency filter used in the cavity to tune the center wavelength of the pulse. In Fig. 16, this is shown as a three-plate birefringent filter (whose transmission function is shown in Fig. 11), but broader two- and one-plate filters are also used for shorter pulses. By Fourier's theorem, the clipping by the filter of the highest and lowest frequencies in the pulse

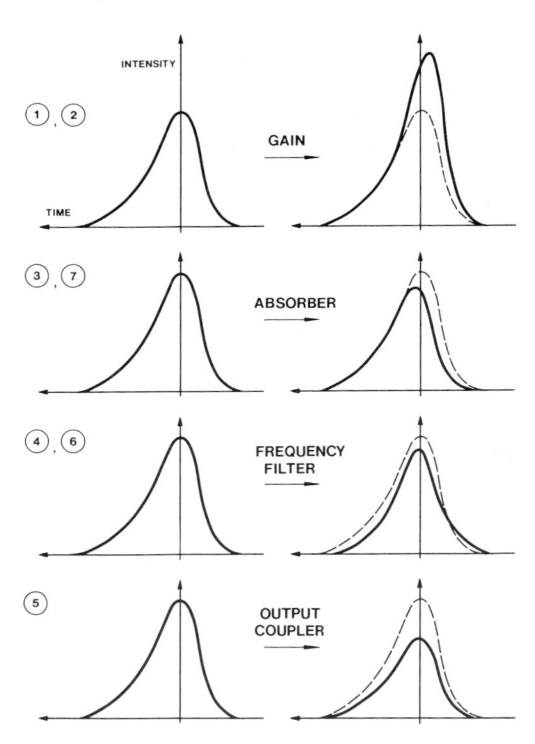

FIG. 17. Schematic diagram of the shaping of the pulse that occurs in the transit of the cavity elements (indicated by the circled numbers) of the laser of the preceding figure. The peak of the pulse is advanced by the interaction with the gain jet and retarded by the absorber; the pulse width is increased in the filter, and left unchanged (though the amplitude is attenuated) at the output coupler.

spreads the pulse shape and balances the compression.

The frequency filter functions in the mode-locking process to clean up the frequency noise and give transform-limited pulses. Pulse formation starts from noise bursts, spontaneous amplitude fluctuations that are limited in rate only by the dye gain bandwidth and that can be quite rapid. These are amplified and filtered in the "good mode-locking" regime, but can persist as temporal substructure in a nonoptimum pulse. The quality of a mode-locked pulse train is judged by measuring both its temporal and spectral widths and by checking that the product of these is close to the Fourier transform constant (implying a consistent pulse shape).

The pulse-shaping process just described applies directly to passive mode-locking, where the gain gradually recovers with cw pumping over a cavity transit time, and the gain seen by the dye pulse does not depend critically on its arrival time back at the gain jet. By contrast, in synchronous mode-locking, the gain rises as the integral of the pump pulse over the ~100-psec pulse width, and the shaping of the dye pulse critically depends on its arrival time (relative to the pump pulse) at the gain jet. This is discussed in Section V,C, after the method of measuring pulse shape is explained. To conclude here note that the smaller linear losses in cavity-dumped operation during the pulse shaping process accounts for output pulses 1.5–2 times greater in pulse width observed in that mode.

To measure the shape of the dye pulse requires optical techniques, since the ~10^{14}-Hz bandwidths involved far exceed the capabilities of conventional circuitry. While a variety of techniques has been developed, the standard today is "background-free autocorrelation"—an average (over interference terms) of the overlap of the pulse train with a delayed sample of itself is recorded as a function of the path delay. The setup is a modified Michelson interferometer where the incoming pulse train is divided at a beamsplitter into two beams, which follow different paths down the two arms of the interferometer and emerge as parallel separate beams. The length of one arm can be varied in a precise manner to give a calibrated path delay. A lens focuses the parallel beams to cross at the same spot on a thin nonlinear crystal oriented to phase match for second harmonic generation only when one photon is taken from each of the two crossing beams. An aperture can be centered to pass only the doubled light as this beam emerges between the two crossing beams, and together with a UV transmitting filter over the photomultiplier detector gives a "background-free" signal. Mathematically this is proportional to the autocorrelation of the pulse shape with itself. By its nature, an autocorrelation signal must be symmetric about zero path delay, and to deconvolve this signal trace to give the width of the original pulse, a prior knowledge of the functional form of the pulse shape is required. Consequently, the autocorrelation width itself is often quoted, which is adequate for many experiments. Table II gives the relationship of the pulse width τ_D, to the width of the autocorrelation trace τ_C (both FWHM) for several functional forms. The squared hyperbolic secant form ($sech^2$) is the shape predicted for dye laser pulses in first-order, linearized mode-locking theories. The Fourier transform "pulse width–spectral width" product $\tau_D \Delta\nu$ used to check for incomplete mode-locking is also listed. Another indication of partial mode-locking is the appearance of a "coherence spike" at zero time delay on the autocorrelation trace, where the noisy substructure in a pulse will have its maximum

TABLE II. Autocorrelation Widths and Spectral Bandwidths for Several Transform-Limited Pulse Shapes

Functional form of pulse shape	Ratio of pulse to autocorrelation width, τ_D/τ_C	Fourier transform product, $\tau_D \Delta\nu$
Square	1	0.886
Gaussian	0.707	0.441
$sech^2$	0.648	0.315
Lorentzian	0.500	0.221
One-sided exponential	0.500	0.110
Symmetric two-sided exponential	0.413	0.142

self overlap. For dye pulses of a few picoseconds or longer, where some dispersion in the autocorrelator apparatus can be tolerated, a convenient instrument is the "rapid scanning" type using a rotating glass block to generate a repetitive path delay scan at 60 Hz, permitting display of the trace on an oscilloscope. For subpicosecond pulse lengths, a stepping motor-driven translation stage with micrometer resolution gives the calibrated path delay and a thin doubling crystal (100 μm KDP) gives adequate detection bandwidth.

C. Pulse-Shape Dependence on Cavity Length in Synchronous Mode-Locking

In synchronous pumping, mismatching the two cavity lengths varies the relative arrival times at the gain jet of the dye and pump pulses, which has a strong pulse-shaping effect. The nature of this effect was revealed in a clever experiment by Firgo, Daly, and Mahr, using what they called an "optical up-conversion light gate." They used rhodamine 6G dye with 514-nm argon laser pumping in a three-mirror cavity equivalent to Fig. 16 without an absorber jet or cavity dumper. Using a lens that collimated fluorescent light from the gain jet, they collected a sample of both the dye and pump beams reflected from the same point at the jet and sent this "signal" beam into one arm of an autocorrelator, with the dye output pulse train incident along the other, variable path-delay arm. After being collinearly combined at a beamsplitter, all of these beams were focused into an angle-matched ADP crystal, and the sum-frequency ("up-converted") UV light from any pair of one of the three signals and the dye beam was detected through a filter monochrometer. This cross correlation of the dye pulse with either the argon pump pulse or the fluorescence signal became in this way a waveform sampling apparatus with high time resolution. All three signals originated at the same point on the jet, and the spatial arrangement of the experiment preserved the relative signal timing as shown in Fig. 18. (This is a fine example of time-domain spectroscopy, the main application of the mode-locked dye laser.)

The chronology of the pulse-forming process demonstrated here is as follows. The gain (and fluorescence) is expected to rise as the time integral of the pump pulse, since the dye's spontaneous decay time far exceeds the time scale of the figure. Indeed, the fluorescence signal closely

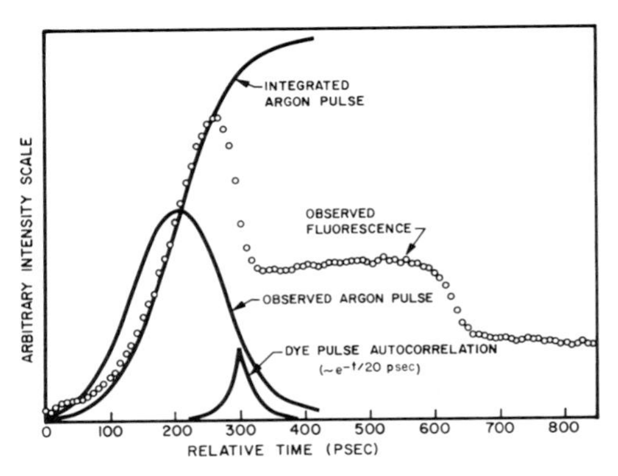

FIG. 18. Results of the optical up-conversion light gate experiment of Mahr *et al.* that measured the arrival time of the dye pulse relative to the pump pulse in synchronous pumping and demonstrated that synchronism is maintained by pulse shaping over the mode-locking range of cavity length. [Reprinted with permission from Frigo, N. J., Daly, T., and Mahr, H. (1977). *IEEE J. Quantum Electronics* **QE-13,** 103. © 1977 IEEE.]

follows the integrated argon pulse (computed from the measured pump pulse shape) except in three places. First, there is a sharp drop in fluorescence coinciding with the time the dye pulse strikes the jet at 300 psec, due to the gain saturation (loss of upper-state population) in amplifying the dye pulse. The dye pulse here was traveling in a direction equivalent to that from M2 to M1 [passage (1)] in Fig. 16. Second, a fluorescence dip occurs 300 psec later, the propagation time from the jet to M1 and back, so this drop is due to the (reduced) amplification of the dye pulse on passage (2). Third, at zero time there is a residual fluorescence signal, due to incomplete spontaneous decay of the inversion produced 10 nsec earlier by the preceding pump pulse and left after the two passes of the preceding dye pulse.

The effect on these signals of mismatching the two cavity lengths was then studied. It was found that the optimum dye cavity length (that giving the shortest pulses) was slightly less than the length setting of Fig. 18. For this optimum length the dye pulse moved some 80 psec closer to the peak of the pump pulse, close enough that there was some gain recovery (rise in observed fluorescence) after the first passage of the dye pulse. This gain rise followed the shape of the integrated argon pulse. The cavity length change that caused this 80-psec time shift was equivalent to only a 0.25-psec cavity transit time change. The circulating dye pulse satisfies two conditions: that its round-trip transit time is precisely equal to the period of pumping pulses (after accumulated advance and retardation adjustments due to pulse shaping), and the "steady-state" condition, that the shape of the pulse is reproduced at each location in the cavity in each transit. To accommodate the shorter cavity length, the pulse position moved down the integrated fluorescence curve to a position of less gain. The pulse shape evolved into a shorter shape having 0.25 psec less net advance per round trip in the cavity. For a further 0.25-psec shortening of the transit time, the dye pulse position moved another 30 psec down the gain curve, and at this position the gain recovery was large enough after the first passage that a second dye pulse began to form.

Examples of this typical pulse-shape behavior for three cavity length settings are shown in Fig. 19, taken with the laser of Fig. 16 in 3.8-MHz pulse repetition rate (cavity-dumped) operation, with rhodamine 6G dye, a three-plate birefringent filter, and 1 W of doubled YAG pumping. The vertical axis is the signal from a rapid scanning autocorrelator, and the horizontal is the calibrated path delay. A display like this is used in setting up the laser to locate the optimum cavity length setting. In this laser, the absorber jet would then be turned on, the three-plate filter exchanged for a one-plate, and the new optimum cavity length setting would be found, yielding shorter pulses of 1 psec autocorrelation trace width, 18 nJ energy per pulse, and 18 kW peak dye power.

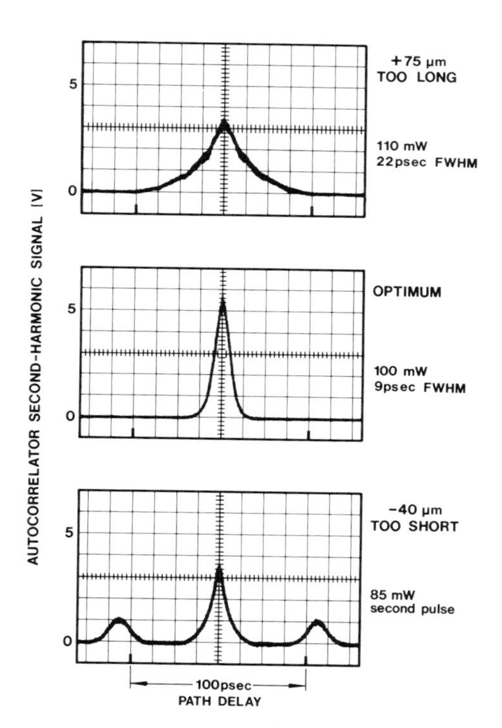

FIG. 19. Autocorrelation traces taken during setup of the cavity-dumped laser of Fig. 16, showing changes in the pulses with cavity length adjustment. A three-plate birefringent filter and a 1-W 532-nm pump beam were used without an absorber jet for this data. (Courtesy of Coherent, Inc., Palo Alto, Calif.)

D. The Shortest Pulses—Colliding-Pulse Ring Lasers and Pulse Compression

This picture of pulse formation in mode-locked dye lasers was sufficient for the generation of subpicosecond pulses and stood as the state of the art until the early 1980s, when two ideas were introduced, one that was new and one that then became widely appreciated, to permit the generation of even shorter pulses. The new idea was that of colliding-pulse mode-locking, due to Fork, Greene, and Shank. Here cw pumping of a ring laser was used, with a thin saturable absorber jet position at one-quarter the perimeter of the ring away from the gain jet. Two pulses circulate in this cavity, one clockwise and one counterclockwise, and collide in the absorber jet to produce interference fringes and a high degree of saturation there. The spacing of the two jets in the cavity ensures that each beam sees the gain jet at half a transit time after the last passage of a dye pulse, to give equal-intensity beams for maximum interference in the absorber. This laser produced 90-fsec pulse widths—a pulse only 0.03 mm long.

The second idea that awoke the scientists' imagination was the demonstration by Nikolaus and Grischkowsky of the compression by a factor of 12 of a 5-psec pulse to 450 fsec with a fiber-grating pulse compressor (Fig. 20). These experimentors realized that the spatially uniform (non-Gaussian) intensity distribution that results when a light beam is coupled into a single-mode optical fiber would result in spatial uniformity of the nonlinear optical effect known as self phase modulation (SPM). The refractive index of the fiber material is time-modulated by the large electric field amplitude of the cavity-dumped dye pulse coupled into the fiber. This phase modulation increases the bandwidth of the pulse. (In fact, SPM has been used to generate a femtosecond white-light continuum by the

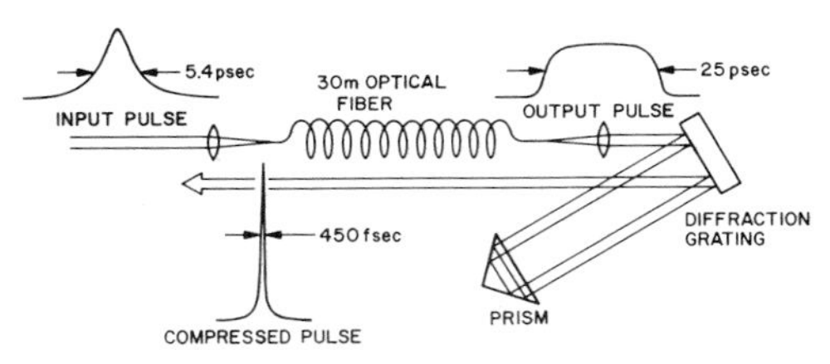

FIG. 20. The pulse compression experiment of Grischkowsky *et al.*, in which self phase modulation and group velocity dispersion of a pulse in an optical fiber are balanced to produce a linear frequency chirp in the output pulse. The two passes off the diffraction grating constitute a dispersive delay time, which compresses this pulse to one-twelfth the width of the input pulse. [Reprinted with permission from Nikolaus, B., and Grischkowsky, D. (1983). *Appl. Phys. Lett.* **42,** 1.]

focusing of ultrashort dye pulses into dielectric media.) In traveling through the fiber, the ordinary group velocity dispersion for this broadened spectral content linearly chirps the pulse—the redder wavelengths emerge first, followed linearly in time (in a longer pulse) across the pulse spectrum by the bluer wavelengths. A dispersive delay line (like the two passes off the grating of Fig. 20) provides a path length proportional to the wavelength and compensates this chirp to produce a compressed pulse.

These results showed that femtosecond-width pulses were possible in mode-locked dye lasers, revitalized the search for the processes limiting the pulse lengths, and led to the present evolutionary burst in this field. Dispersion-compensating delay lines were added intracavity in the dye laser. This looks like the quad-prism expander of Fig. 6 when placed in-line in an arm of a ring laser (only with "full" Brewster prisms instead of the "half" prisms in that figure), or equivalently is a pair of prisms placed in front of an end mirror of a linear laser. One prism is laterally displaced to adjust the relative spectral delay to compensate linear chirp from other intracavity elements. To date, pulses have been generated of 55-fsec width from a linear hybrid mode-locked laser, 27-fsec width from a colliding-pulse ring laser, and the shortest of all (as of 1985), a train of pulses of 8-fsec width was produced by compression from a 40-fsec source.

The molecular relaxation times measured with these ultrashort pulses in time-domain spectroscopy are as diagnostic of the system studied to a chemist or biologist as are atomic energy levels to a physicist. By modeling the chemical interaction and matching the calculated decay times to the observed ones, the actual molecular structure is revealed, and these new laser tools should keep these fields of study active for many years to come.

VI. Specialized Dye Lasers

The majority (in numbers) of dye lasers in existence is represented by the three scientific types already discussed, but these by no means represent the diversity of ways that lasers with a dye gain medium have been built. In this concluding section, four additional dye lasers are briefly discussed as a better indication of the breadth of this laser type and of its potential impact on future technology.

Vapor-phase dye lasers were initially investigated in the hope that the efficiency of the dye laser (plus pumping system) could be improved with excitation by a discharge directly in the dye vapor. This proved not to be the case. While molecular densities comparable to solution dye lasers were reached in the vapor state (by heating to ~400°C nonionic dyes sealed in silica glass cells), and spectra comparable to dye spectra in solution were observed (with slight broadening and a shift to shorter wavelengths), the inevitable result of running a discharge in the vapor was the rapid decomposition of the dye molecules. Several vapor-phase dyes were made to lase as short-pulse oscillators by optical pumping with a pulsed nitrogen laser, with results comparable to the same dyes in solution. The initial goal, however, is still well worth emphasizing. The applications of most lasers are limited by their high consumption of input power (their inefficiency), and the dye laser is one of relatively few lasers with an intrinsically efficient inversion mechanism (described in Section II). Thus, means to improve efficiency by eliminating another laser as the pump source, as in these vapor-phase experiments, are well worth seeking.

The distributed feedback dye laser first demonstrated the tunable, narrow-band output from a very compact structure that would be desired in a light source for integrated optical devices. The optical feedback in these mirrorless lasers is provided by Bragg scattering from a periodic spatial variation of the refractive index or the gain itself in a thin-film gain medium. The first such laser used a holographic phase grating, which was exposed and developed in a dichromated gelatin film on a glass substrate, before the film was dyed by soaking in a solution of rhodamine 6G. This was transversely pumped with a nitrogen laser to produce a 630-nm beam (of 0.05 nm linewidth) from the $0.1 \times 0.1 \times 10$-mm total laser volume. To show a distributed feedback dye laser that was externally tunable, a dye cell was side-pumped with intersecting, interfering, 347-nm beams (from a beam-split, doubled ruby laser) to produce fringes and a periodic gain variation in the sheet of dye. The rhodamine 6G dye output wavelength tuned 70 nm for a 48–56° change in incidence angle for each of the two pump beams. In the tiny volumes of these lasers, if the dye molecules are immobilized in gelatin or plastic, the small number of molecules are recycled so frequently that photodecomposition limits the dye lifetime to a few seconds or less. The actual distributed feedback lasers that are candidate sources for integrated optical circuits, the inheritors of these dye laser results, will be analogs fabricated in semiconductor materials.

On the other end of the size and output power scale are the specialized dye laser oscillator and amplifier devices built for the atomic vapor laser isotope separation (AVLIS) process planned as the prime means in the United States for the enrichment of uranium-235 for nuclear reactor fuel rods near the turn of this century. In this process, the uranium raw stock is vaporized in a vacuum with an electron beam, and the streaming atomic vapor is exposed to dye laser radiations tuned to selectively photoionize (in three absorption steps) only ^{235}U atoms. These charged ions are deflected out of the stream by electrostatic fields to achieve separation. (Unlike the ^{20}Ne to ^{22}Ne isotope shift of Fig. 15, the ^{235}U to ^{238}U isotope shift for spectral lines of interest is greater than a Doppler width.) The general scale and specifications of the three laser beams (at three visible wavelengths) required in the separator are about the same as the design goal reported for the selective step, namely, ~1 J/pulse, ~10^4 Hz pulse repetition rate (or ~10 kW average power), in about a 1-GHz bandwidth locked onto the center of the uranium line. This high average power will be reached with massive banks of copper-vapor lasers (a short-pulse laser like the nitrogen laser but with emissions in the green and yellow) as pumps for short-pulse dye oscillator and amplifier chains. The copper-vapor lasers have a few percent electrical-to-optical efficiency, which is important in the economics of the process, and the yellow-green pump lines are efficient both in pumping red dyes and in giving long dye photodegradation times. Thus, dye lasers in about two decades may be responsible for a sizable fraction of the world's electric power generation.

Continuous-wave dye lasers have also been the basis in experimental investigations for the cure of cancer (Fig. 21). In this work, termed photodynamic therapy, the patient with a tumor is first injected with a drug (hermatoporphyrin derivative, HpD) that has the property of being selectively retained in rapidly dividing tissue (the tumor) and of photodegrading to release free oxygen when exposed to a measured dose of light of the proper color. After a waiting period of 2–3 days for the drug to clear from healthy tissue, the patient is given the light dose through fiber optic guides slipped into needles implanted into the tumor. The released oxygen burns up the tumor but has no effect (and leaves a sharp line of demarcation) on the healthy tissue. In principle, the drug can be activated by ordinary light sources, filtered to give the proper

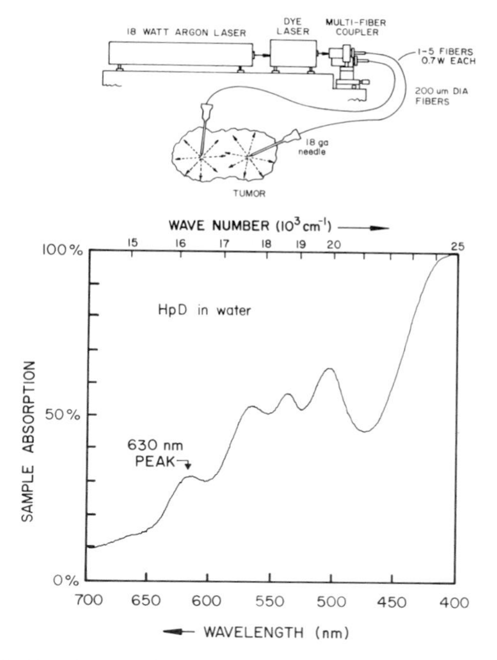

FIG. 21. Schematic diagram of the destruction of cancerous tumors by the activation of the experimental drug HpD (hermatoporphyrin derivative). A dose of dye laser light, tuned to a specific absorption band of the drug, is administered through fiber optic guides to photodecompose the drug and kill the tumor. The absorption spectrum shown is for an early form of the drug in water; the peaks shift somewhat in tissue and with the newer forms of the drug. (Courtesy of Coherent, Inc., Palo Alto, Calif.)

wavelengths (corresponding to an appropriate absorption peak). In practice, the only activating wavelength with enough penetration depth for coverage of a few cubic centimeters volume per implant in most bodily tissue is the one in the red, which requires a 630 ± 5 nm wavelength. A laser source is dictated by the convenience of delivery through fibers (a coherent source focuses efficiently into the fiber's input end) and by the large total dosage of ~40 J cm^{-3} needed in this narrow bandwidth. The cw dye laser is the laser of choice for this application. Experimental medicine using the tunability of the dye laser to selectively absorb in different tissues without drugs is also being done in photocoagulation inside the eye and in other laser surgical procedures. These examples show that the future importance of the dye laser can be expected to extend much beyond today's useage as a research tool.

VII. Recent Developments

Two new crystals have recently become available in high optical quality form to the benefit of the tunable lasers discussed in this article. The first, beta-barium borate (BBO), is a doubling crystal with transparency to slightly below 200 nm that has extended tunable outputs down into this wavelength region. The second, titanium : sapphire (Ti : Al_2O_3), is an optically pumped crystal which has been substituted for the dye medium in single-frequency and mode-locked laser cavities to produce tunable outputs at higher power in the near infrared than previously available. Increased powers from the pump lasers in all three types of scientific dye lasers have boosted tunable outputs as well, particularly in the case of the UV pump lines from argon ion lasers. The pump powers of the standard UV-line mix now are twice as great, and shorter UV pump lines have become available, making possible cw tunable lasing in the UV directly from new dyes. In ultrafast lasers the shortest-pulse record (done by extracavity pulse compression) is now 6 fsec; trains of pulses of 29 fsec duration have been generated directly in the hybrid mode-locking geometry by the addition of prism pairs to the laser cavity to compensate the intracavity group-velocity-dispersion; and commercial systems now attain and maintain sub-100 fsec pulse lengths by incorporating servo systems in their design.

A. Single-Frequency and Short-Pulse Lasers

The resulting changes in wavelength coverage for the single-frequency lasers have been illustrated in the updated tuning spectrum of Fig. 13 of the text. Beginning on the long wavelength end of the spectrum, the solid-state crystal titanium-doped sapphire has produced the highest reported single-frequency output over the 690–1010 nm fundamental tuning range, with a peak of 3.6 W at 800 nm (replacing the 2.0 W peak at 740 nm of the dye LD700). This is a crystalline optically pumped gain medium and not a dye, but due to the homogeneous line broadening at the active Ti^{3+} ion sites, titanium : sapphire shows spectral condensation and tunability like a dye and is used as an alternate gain medium in the same laser as shown in Fig. 10 (now termed a Ti : dye laser). The dye laser cavity is modified for Ti : sapphire by changing the curved mirror M1 to a flat mirror and changing the radius of curvature of M2. The 2 cm long crystal replaces the astigmatic compensation rhomb in the cavity (just as this position was used earlier for the temperature-tuned ADA doubling crystal, see Section IV.E), and the dye jet is turned off. The argon ion laser pump beam is folded to enter the cavity through the mirror M2, after passing through a lens placed behind M2, to focus into the crystal in the colinear pumping geometry needed with a long gain medium. The same single-frequency filter stack and frequency servo system used with a dye produce with the crystal <500 kHz rms frequency jitter as described in the text. Titanium : sapphire is pumped by the blue-green band of lines from the argon ion laser. By clamping the crystal in a water-cooled block it withstands a full 20 W pump power, which accounts for the high output power (the LD700 dye used a red pump beam from a krypton ion laser, for which only 5 W pump power is available). The tunability of this crystal extends beyond 1000 nm, replacing as well the next two longer wavelength dyes, styryl 9M and infrared dye 140.

Improvements in argon ion lasers have raised the UV-band pump power to 7 W, and added a short-ultraviolet wavelength (300–336 nm) pump band (SUV in Table 1) of about 3 W power. The first improvement has increased by a factor of three the blue wavelength outputs from the stilbene 1 and stilbene 3 dye lasers to 0.42 W and 1.0 W, and the second has permitted single-frequency ultraviolet lasing from the dye polyphenyl 2 over the range 364–408 nm with a 0.25 W peak output at 383 nm. Other UV-lasing dyes are under development. An output comparable to the directly UV-lasing dyes was produced by doubling titanium : sapphire radiation intracavity in an angle-matched lithium iodate crystal as described in Section IV.E. This high output (0.20 W at 370 nm with 345–388 nm tunability) was due to the low crystal absorption of <1% at the fundamental wavelengths and to the ability of this crystal to sustain high pump powers.

Advantage could be taken of the improved output powers in the blue dyes as the angle-matched crystal beta-barium borate for doubling this region became available at the same time. When used in the cavity (Fig. 10) of the stilbene 3 and coumarin 102 dye lasers, this extended the single-frequency tuning spectrum over most of the range from 215–260 nm. The small 235–238 nm gap of Fig. 13 is expected to be closed by doubling in stilbene 3 using the technique explained in Section IV.E for pushing the tuning range past a tuning curve break point. In R6G dye, BBO doubling has produced five times more second harmonic power than the earlier angle-matched result with KDP, and it is anticipated that the tuning curves remaining in the figure between the doubled R6G and C102 results will be superceded by BBO doubling results when this crystal is tried in the intervening dyes. To be useful for cw intracavity doubling a crystal must have an absorption of <1% over the length of the crystal (one aperture length of 7 mm here); the excellent results of Fig. 13 demonstrate that the crystals available now are of this high quality. This crystal also has a high damage threshold, permitting it to be used in doubling the blue wavelength outputs from short-pulse dye lasers, which has been done to produce ten times the previously available deep UV pulse energies.

B. Ultrashort-Pulse Lasers

An important recent result in the area of ultrasfast lasers was the demonstration of pulses of 29 fsec duration directly generated in a hybrid mode-locked linear dye cavity (like Fig. 16 of the text), to which a set of four prisms were added and arranged to compensate group-velocity-dispersion as discussed in Section V.D. This nearly matches the shortest pulse (27 fsec) so far generated directly in a colliding pulse mode-locked cavity and favors the use of hybrid mode-locking which has produced considerably higher output power (350 mW versus 50 mW) and considerably greater tunability. But since the pulsewidth, shape, and noise of the pulse train from a synchronously mode-locked cavity are strong functions of the cavity length as discussed in Section V.C, for reproducible sub-100 fsec pulses the cavity length must be held stable to within about 100 nm. Holding this tolerance requires a cavity length servo. From among several possibilities, commercial systems use a length-discriminant based on the output power of the dye laser, which rises linearly through the cavity length region that gives the shortest pulses. With an autocorrelator to identify the desired pulse width and shape, the optimum cavity length and corresponding output power is found and a proportional reference voltage subtracted from the detected power to generate the discriminant. The advantages of this error signal are that it is reproducible over the whole tuning range and in different dyes, and its use simultaneously stabilizes the output power. However, it presumes that the only source of power fluctuation is the dye cavity length. In order to make this assumption true, a fast amplitude noise reduction servo based on acoustooptic diffraction (called a "noise eater") is used on the pump beam, and the pointing fluctuations of the pump beam are stabilized by sensing its positional error at the dye laser and feeding back adjustments to steer the pump laser output mirror. Routine optimization of these interacting servo systems is done by microprocessor control. In generating sub-100 fsec pulses a preference has come about for the doubled Nd : YLF (neodymium–yttrium–fluoride) laser as a pump source instead of the doubled Nd : YAG laser, due primarily to the lower noise characteristics of the former host crystal.

The new shortest pulse record of 6 fsec (compared to the former 8 fsec result) was produced from a train of 50 fsec pulsewidth amplified pulses, linearly chirped in a 0.9 cm length of single-mode quartz fiber, then compressed in a dispersion compensating delay line similar to that discussed in Section V.D except that the two pairs of prisms were preceded by two pairs of gratings in a similar folded geometry. The extra degree of freedom given by adjustment of both gratings and prisms permitted compensation of both the square and cubic terms in the Taylor expansion of the optical phase versus optical frequency offset from the center of the pulse spectrum, with the shorter compressed pulse being the result. The ultrashort pulse (ultrafast) laser area continues to be one of rapid evolution with the results discussed in this section merely suggestive of the types of many such changes occurring in this field.

Bibliography

Bradley, D. J., and New, G. H. C. (1974). Ultrashort pulse measurements. *Proc. IEEE* **62,** 313–345.

Couillaud, B., and Fossati-Bellani, V. (1985). Mode-locked lasers and ultrashort pulses. *Lasers and Applications* **Jan.,** 79–83, **Feb.,** 91–94.

Drever, R. W. P., Hall, J. L., Kowalski, F. V., Hough, J., Ford, G. M., Munley, A. J., and Ward, H. (1983). Laser phase and frequency stabilization using an optical resonator. *Appl. Phys. B* **31,** 97–105.

Fork, R. L., Shank, C. V., Yen, R., and Hirlimann, C. A. (1983). Femtosecond Optical Pulses. *IEEE J. Quantum Electronics* **QE-19,** 500–505.

Hänsch, T. W. (1972). Repetitively pulsed tunable dye laser for high resolution spectroscopy. *Appl. Optics* **11,** 895–898.

Ippen, E. P., and Shank, C. V. (1978). Sub-picosecond spectroscopy. *Phys. Today* **May,** 41–47.

Johnston, T. F., Jr., Brady, R. H., and Proffitt, W. (1982). Powerful single-frequency ring dye laser spanning the visible spectrum. *Appl. Optics* **21,** 2307–2316.

Mason, S. F. (1970). Color and the electronic states of organic molecules. *In* "The Chemistry of Synthetic Dyes," Vol. III (K. Venkataraman, ed.), pp. 169–221. Academic Press, New York.

Peterson, O. G. (1979). Dye lasers. *In* "Methods of Experimental Physics," Vol. 15A (L. Marton, ed.), pp. 251–359. Academic Press, New York.

Ryan, J. P., Goldberg, L. S., and Bradley, D. J. (1978). Comparison of synchronous pumping and passive mode-locking of CW dye lasers for the generation of picosecond and subpicosecond pulses. *Optics Commun.* **27,** 127–132.

Scavennec, A. (1976). Mismatch effects in synchronous pumping of the continuously operated mode-locked dye laser. *Optics Commun.* **17,** 14–17.

Schafer, F. P. (ed.) (1977). "Dye Lasers," Vol. 1 of "Topics in Applied Physics," 2nd revised ed. Springer-Verlag, New York.

Shank, C. V. (1975). Physics of dye lasers. *Rev. Modern Phys.* **47,** 649–657.

Sorokin, P. (1969). Organic lasers. *Sci. Am.* Feb., 30–40.

Valdmanis, J. A., and Fork, R. L. (1986). Design considerations for a femtosecond pulse laser balancing self phase modulation, group velocity dispersion, saturable absorption, and saturable gain. *IEEE J. Quantum Electronics* **QE-22,** 112–118.

ULTRAFAST LASER TECHNOLOGY

Peter J. Delfyett, S. K. Gayen, and R. R. Alfano *City College of New York*

GLOSSARY

Bandwidth: Measure of the spectral range of an optical pulse. It is the difference in frequency between the two points where the intensity of the pulse is half its peak value.

Chirp: Frequency change in time.

Dispersion: Frequency (or wavelength) dependence of the index of refraction of a material medium. Such a medium is called a dispersive medium.

Gain saturation: Reduction of amplification (or gain) due to depletion of population in the excited state of the gain media.

Group velocity dispersion: Frequency (or wavelength) dependence of the group velocity.

Mode: Self-consistent optical field configuration of a laser resonator that reproduces itself after one round trip in the cavity.

Mode locking: Coupling of a set of oscillating modes in the laser resonator and forcing them to oscillate with fixed frequency spacing, relative phases, and amplitudes.

Pulse compression: Shortening of pulse width by sending a frequency-swept pulse through a negative dispersive delay line.

Pulse width: Measure of the temporal duration of an optical pulse. It is the time separation between the two points where the intensity falls to half the peak value.

Saturable absorber: Absorber (usually an organic dye) whose absorption properties are intensity dependent. Below some threshold intensity, the absorption is linear and above it the absorption starts to reduce and ultimately the absorber becomes transparent. This type of absorption is known as saturable absorption.

Self-phase modulation: Modulation of the instantaneous phase of an intense ultrashort laser pulse by the intensity-dependent index of refraction of the medium through which the pulse propagates. This leads to a spectral broadening of the pulse.

Time scale: Millisecond (msec) = 10^{-3} sec, microsecond (μsec) = 10^{-6} sec, nanosecond (nsec) = 10^{-9} sec, picosecond (psec) = 10^{-12} sec, femtosecond (fsec) = 10^{-15} sec.

Ultrafast: Term referring to a laser or a spectroscopic technique. A laser that generates light pulses with temporal duration in the 10^{-12} to 10^{-15}-sec range is called an ultrafast laser. Laser pulses having such temporal duration are referred to as ultrashort pulses. A spectroscopic technique with time resolution in the 10^{-12} to 10^{-15}-sec range is an ultrafast technique.

Ultrafast laser technology encompasses the generation and diagnostic techniques of picosecond (10^{-12}-sec) and femtosecond (10^{-15}-sec) laser pulses and their applications in direct temporal studies of extremely rapid phenomena in nature. This new frontier in optics emerged with the discovery of high-power picosecond light pulses in 1966, and guided by the ever-enthusiastic search for increasingly higher time resolution, it has evolved over the past two decades to the point where pulses as short as 8 fsec have been generated and investigations of ultrafast phenomena on the femtosecond time scale have been accomplished. The present article surveys the methods of ultrafast pulse generation and detection, reviews their use in spectroscopic techniques, and contemplates the future directions and trends in this rapidly growing, immensely exciting area of research.

I. Brief Historical Overview

Many fundamental processes in the submicroscopic world are extremely rapid, with typical temporal durations ranging from a few trillionths of a second to a few tenths and even hundredths of a trillionth of a second. From such primary processes as transfer of excitation in the visual pigment that initiates vision to such exotic events as electron-momentum relaxation in semiconductors, there is a plethora of these ultrafast phenomena in various disciplines of science and engineering. The relevant time scales of a number of these processes from many diverse fields are presented in Table I.

Just 20 years ago, it was not possible to clock such ultrafast events directly, because the temporal resolution of available experimental techniques was less than about a nanosecond (a billionth of a second). Knowledge about these extremely rapid events used to be derived indirectly from an analysis of the characteristic frequencies of the radiation they caused to be absorbed or emitted.

A major breakthrough in clocking fast events in the time domain occurred in 1966 with the discovery of mode-locked glass lasers, which produce light pulses several picoseconds in duration. The advent of high-power picosecond pulses in turn revolutionized the field of nonlinear laser spectroscopy. Many nonlinear effects in condensed matter, unobserved in the past, were discovered. These in turn helped the generation of even shorter pulses. One particularly important example that has shaped the growth of ultrafast laser technology is the discovery in 1970 of self-phase modulation (SPM), which is responsible for the generation of ultrafast white light or supercontinuum pulses. Self-phase modulation has turned out to be a key process for femtosecond pulse generation. Methods of mode locking developed rapidly, and by 1972 continuously mode-locked dye lasers were producing pulses on the order of a picosecond. This type of laser was improved further, leading to the generation of pulses shorter than a picosecond in 1974. The development of the colliding pulse mode-locking technique in 1981 pushed the limit of attainable pulse widths even shorter, to less than 100 fsec. Novel shaping and pulse compression techniques have led to the shortest reported optical pulse width of 8 fsec in 1985. [*See* LASERS; MODE-LOCKING OF LASERS.]

With the advent of even shorter and shorter pulses, the field of ultrafast laser spectroscopy developed to clock extremely fast events directly in the time domain. Over the past two decades an impressive amount of new information on rapid phenomena in matter, including molecular dynamics in liquids, relaxation and transfer of excitation in physical and biological systems, and kinetics of chemical reactions has been obtained by using sophisticated and novel optical techniques. While the early picosecond research could provide qualitative information

TABLE I. Relevant Time Scales of Typical Fast Processes

Event	Time (sec)
Biology	
Vision	$\sim 10^{-8}$–10^{-10}
Photosynthesis, relaxation processes	$\sim 10^{-9}$–10^{-12}
Electron transport steps in reaction centers	$\sim 10^{-12}$
Chemistry	
Singlet–triplet nonradiative transition	$\sim 10^{-10}$–10^{-13}
Molecular reorientation in solvents	$\sim 10^{-12}$
Photodissociation, photoionization	$\sim 10^{-12}$–10^{-13}
Solvent caging, H bonding	$\sim 10^{-13}$–10^{-14}
Physics	
Thermalization of hot electrons	$\geq 10^{-13}$
Vibrational dephasing in excited molecules	$\geq 10^{-13}$
Fluorescence rise time	$\sim 3 \times 10^{-14}$
Relaxation of electron–momentum distribution	$\sim 2 \times 10^{-14}$
Electron–electron and electron–hole scattering	$\sim 10^{-14}$
Electronic cloud deformation	$\sim 10^{-15}$

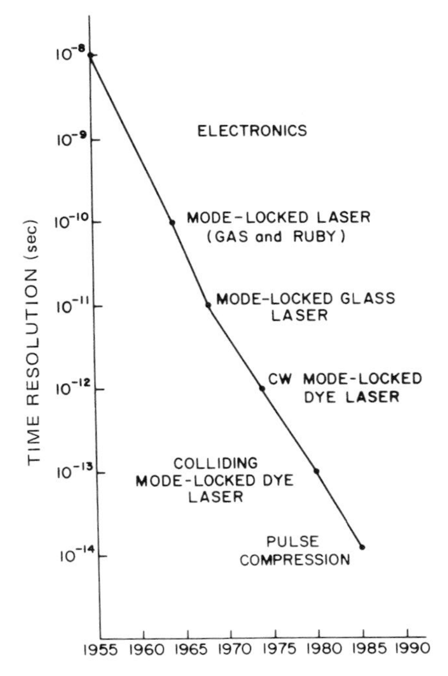

FIG. 1. Progress in time resolution as a function of time. [From Alfano, R. R. (1986), *Proc. Int. Conf. Lasers '85* (C. P. Wang, ed., STS Press, McLean, Virginia.]

and set limits for some of these processes, higher-resolution and better techniques today make detailed quantitative studies possible. A history of the progress achieved in the capability of measuring faster events with increasing temporal resolution is presented in Fig. 1.

New frontiers in ultrafast laser spectroscopy will undoubtedly arise with the discovery of new phenomena. It has the potential to revolutionize the testing of high-speed electronic devices, to refine our understanding of quantum and optics theories, to increase data transmission capacity in optical communication, and to open new fields in optical switching and computing. In the following sections we attempt to introduce the reader to this exciting and rapidly growing new frontier of optics and physics.

II. Generating Ultrashort Light Pulses

In the two decades following the discovery of mode-locked glass lasers in 1966, techniques for generation of shorter pulses have advanced dramatically. These advances have been supported largely by a sound understanding of the physical mechanisms by which ultrashort light pulses evolve from the initial fluorescence of lasing materials and by the simultaneous development of pulse measurement techniques.

A. Solid-State Laser

The first laser to generate ultrashort pulses was a mode-locked solid-state laser. It still is the workhorse of ultrafast laser spectroscopy. A typical mode-locked solid-state laser is shown schematically in Fig. 2.

The laser oscillator consists of a laser rod surrounded by a flash lamp and a saturable absorber dye between two high-reflectivity mirrors. The material of the laser rod may be ruby, Nd : glass, Nd : YAG, alexandrite, or any other crystal with a broadband fluorescence spectrum necessary to support short-pulse production. The ends of the rod are cut at Brewster's angle to prevent subcavities, feedback, and reflection losses from the surface. The flash lamp surrounding the laser rod provides the optical energy necessary to excite the laser-active ions in the rod. The end mirrors are dielectric coated, the one at the back being 100% reflecting and the other in the front 50%–70%, reflecting at the emission band of the laser medium. All surfaces are wedged by 30′ to 2°.

After a laser-active ion in the rod has been raised to an excited state by absorption of radiation from the flash lamp, the excited ion can relax back to the ground state either by spontaneous or, if it is in an electromagnetic field, by stimulated emission of radiation. It is the stimulated emission of radiation that leads to the basic operation of the laser. The frequency, the phase, and the direction of emitted radiation are that of the electromagnetic field.

Depending on the energy-level structure of the lasing material, it is possible to have more

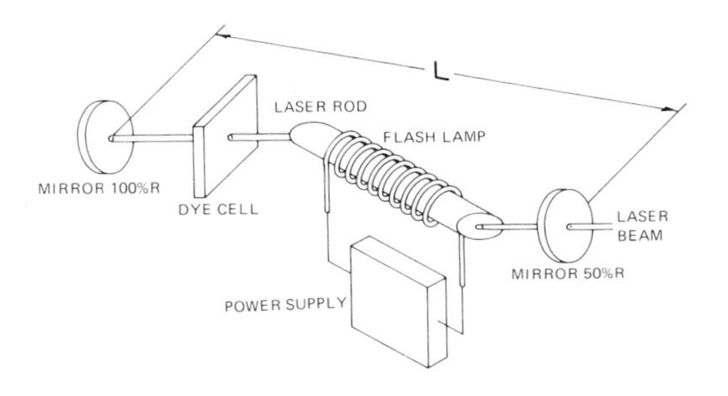

FIG. 2. Schematic diagram of a mode-locked solid-state laser. [From Alfano, R. R., ed. (1982). "Biological Events Probed by Ultrafast Laser Spectroscopy." Academic Press, New York.]

atoms in the excited state than in the terminal state (which may or may not be the ground state), a condition known as population inversion. In solid-state laser materials, the absorption of a sufficient amount of flash-lamp light causes population inversion. The excited atoms decay, emitting radiation in the characteristic fluorescence band of the laser medium. As these photons bounce back and forth through the laser cavity, they stimulate more and more atoms in the laser rod to emit radiation in the same frequency, phase, and direction. Since the excited state population is higher than the terminal state, there is a net amplification of the light until the population of the excited state is depleted.

The characteristics of laser output are determined by the fluorescence spectrum and the quantum yield of the lasing medium and the configuration of the laser cavity. Of all the possible frequencies of the fluorescence spectrum of the laser material, only those frequencies resonant in the laser cavity for which the unsaturated gain is larger than the intracavity losses are allowed to oscillate by the laser cavity. Such a self-consistent optical field configuration that reproduces itself after one round trip in the laser cavity is called a mode. Two types of resonator modes are possible: longitudinal and transverse. Longitudinal (or axial) modes are characterized by the same form of spatial energy distribution in a transverse plane (parallel to the end mirrors) but have different axial distributions corresponding to different number of half-wavelengths of light along the resonator axis. The allowed longitudinal modes in the laser cavity are determined by the condition that the waves in the cavity form a standing wave pattern, satisfying the condition

$$\nu_q = (c/2L)q \tag{1}$$

where c is the speed of light in free space, L the optical length of the cavity, and ν_q the frequency of the qth longitudinal mode. The longitudinal modes are equally spaced in frequency, the frequency spacing being $f = c/2L$. In a typical laser resonator the number of longitudinal modes is of the order 10^6. Figure 3 shows a hypothetical fluorescence profile superimposed on the axial resonator modes. The transverse modes, on the other hand, differ both in frequency and in spatial energy distribution in a transverse plane. Generally, there are a number of transverse modes corresponding to each longitudinal mode. Only those longitudinal modes that are within the threshold limits $\Delta f'$ of the gain envelope get amplified. In a free-running laser both longitudinal and transverse modes oscillate simultaneously with random phase. The pulse width of such a laser closely follows the duration of the flash-lamp pulse, typically milliseconds, while the intensity has the characteristics of thermal noise. To generate an ultrafast pulse of high intensity, many modes of the cavity need to be coupled together and forced to oscillate with fixed frequency spacing, relative phases, and amplitudes. The process is called mode locking or phase locking. The pulse width and bandwidth are related by the relation

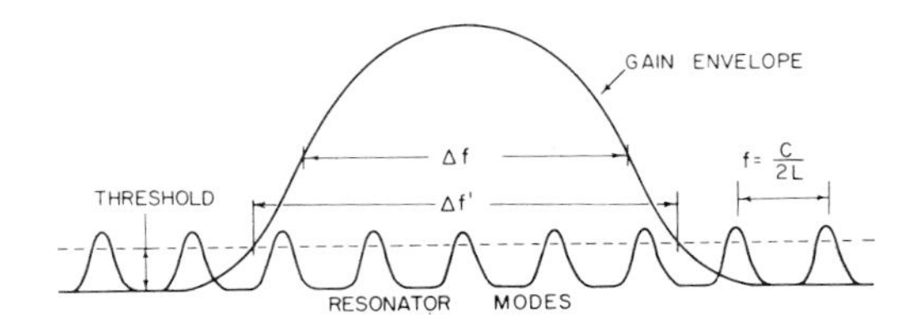

FIG. 3. The gain of the laser line superimposed on the allowed modes of a resonator. [From Alfano, R. R., ed. (1982). "Biological Events Probed by Ultrafast Laser Spectroscopy." Academic Press, New York.]

$$\Delta t \, \Delta f \geq K \tag{2}$$

where Δt is the pulse width [full width at half-maximum height (FWHM)], Δf the bandwidth (FWHM), and K a constant of the order of unity, whose value depends upon the shapes of the intensity profiles. Values of K for some typical pulse shapes are presented in Table II. Since $\Delta f = (c/2L)N = fN$, N being the number of modes within Δf,

$$\Delta t \geq K/Nf \tag{3}$$

Equation (3) shows that the larger the number of

TABLE II. Lower Limits of the Time–Bandwidth Product K for Different Pulse Shapes

Pulse	K
Gaussian	
$\exp[-4 \ln 2t^2/(\Delta t)^2]$	0.4413
Hyperbolic sech	
$\mathrm{sech}^2(1.7627t/\Delta t)$	0.3148
One-sided exponential	
$\exp(t \ln 2/\Delta t)$, $t \geq 0$	0.1103
Symmetric two-sided exponential	
$\exp[-(2\|t\| \ln 2/\Delta t)]$	0.1420
Lorentzian	
$[1 + (2t/\Delta t)^2]^{-1}$	0.2206

From van der Ziel, J. P. (1985). *In* "Semiconductors and Semimetals," Vol. 22B, (W. T. Trang, ed.). Academic Press, Orlando, Florida.

modes that oscillate in phase, the shorter the duration of the pulse.

It is clear that mode locking is a key process in generating ultrashort pulses. It is customary to distinguish between active and passive mode locking. Mode locking of a laser requires the insertion, in the cavity, of an external element that initiates and maintains the proper coupling between the axial modes. If the mode-locking device is to be driven by an external energy source, the method is called active mode locking. If no energy source other than the laser oscillator itself is required, the method is passive mode locking. However, under certain conditions, the nonlinear effects of the laser medium itself may cause a fixed phase relationship to be maintained between the oscillating modes, a phenomenon known as self-locking. We will mainly consider passive mode locking by dyes in this section. It should be noted that only longitudinal mode locking has been mentioned. One of the conditions for perfect longitudinal mode locking is to have only one transverse mode oscillating. This is accomplished by inserting into the laser cavity a circular aperture whose dimension is such that the TEM_{00} mode experiences little diffraction loss, while the higher-order modes are attenuated significantly. Typically, the aperture diameter is on the order of a millimeter.

One of the most successful methods of passively mode locking a solid-state laser is to use a saturable absorber (an organic dye) in the cavity. The absorption characteristics of the dye are such that at low-incident intensity the absorption is constant, but as the intensity increases and exceeds a certain critical value (~50 MW/cm^2), the absorption decreases. It is the combination of this nonlinear absorption in the saturable absorber and the lasing medium that produces ultrashort pulses. [*See* TUNABLE DYE LASERS.]

An intense pulse of light can induce bleaching in a saturable absorber, usually by exciting the absorber's ground state S_0 electrons into its first excited singlet state S_1, thus depleting the ground state. When the pulse enters the absorbing medium, the center of the pulse is not completely absorbed, due to bleaching. The leading edge is more completely absorbed, shortening the pulse. The saturable absorber also discriminates against weaker pulses in the cavity, helping to eliminate the formation of satellite pulses. The pulse becomes shortened to about the characteristic recovery time of the saturable absorber. Saturable absorption alone is sufficient for pulse formation in solid-state lasers, where gain depletion is not an efficient pulse-shortening process.

Initially there is only spontaneous fluorescence in the cavity. Its intensity profile resembles that of thermal noise. The formation of a train of ultrashort laser pulses from noise is illustrated by the computer simulation in Fig. 4. In the first stage of pulse generation [Fig. 4 a–c] called the linear amplification stage, both the amplification in the laser medium and the absorption in the saturable absorber are linear. As the gain increases above threshold, the noiselike signal is amplified and a natural mode selection takes place because the frequency-dependent gain favors cavity modes in the center of the fluorescence line. This spectral narrowing leads to a smoothening and broadening of the amplitude fluctuations. In the second phase of pulse evolution, the gain is still linear, but the absorption is nonlinear because intensity peaks in cavity approach the critical value. The dye preferentially transmits the most intense fluctuation and effectively suppresses the smaller ones. The width of the surviving pulse also decreases in time because the wings of the pulse are more strongly absorbed than the peak [Fig. 4d and e]. The second phase ends when the absorber is completely saturated and the dye becomes transparent. It is essential that the recovery time of the dye be shorter than the round trip time of the pulse in the cavity. Typically, the dye recovery time is on the order of 3 to 10 psec. In the final phase, pulse intensity grows rapidly, typically to several gigawatts per square centimeter within 50 to 60 round trips. As illustrated in Fig. 4f, background pulses are almost completely suppressed. The bouncing single pulse gives rise to an output train of pulses, as a fraction of it escapes from the cavity through the partially reflecting mirror at each round trip. Finally, the population inversion depletes and the pulse decays. Since in the process the dye also performs the act of Q switching, the envelope of the pulse train becomes a giant pulse. This rather simplified description may create the impression that the pulses would tend to become infinitely narrow, but due to the limit of bandwidth which the laser medium can support, the pulse duration is restricted to either $\sim 1/\Delta f$ or the recovery time of the dye, whichever is longer. The time evolution of mode-locked pulses has been experimentally verified in ruby and glass lasers.

In Table III we list the solid-state laser systems that have been successfully mode locked using saturable absorbers.

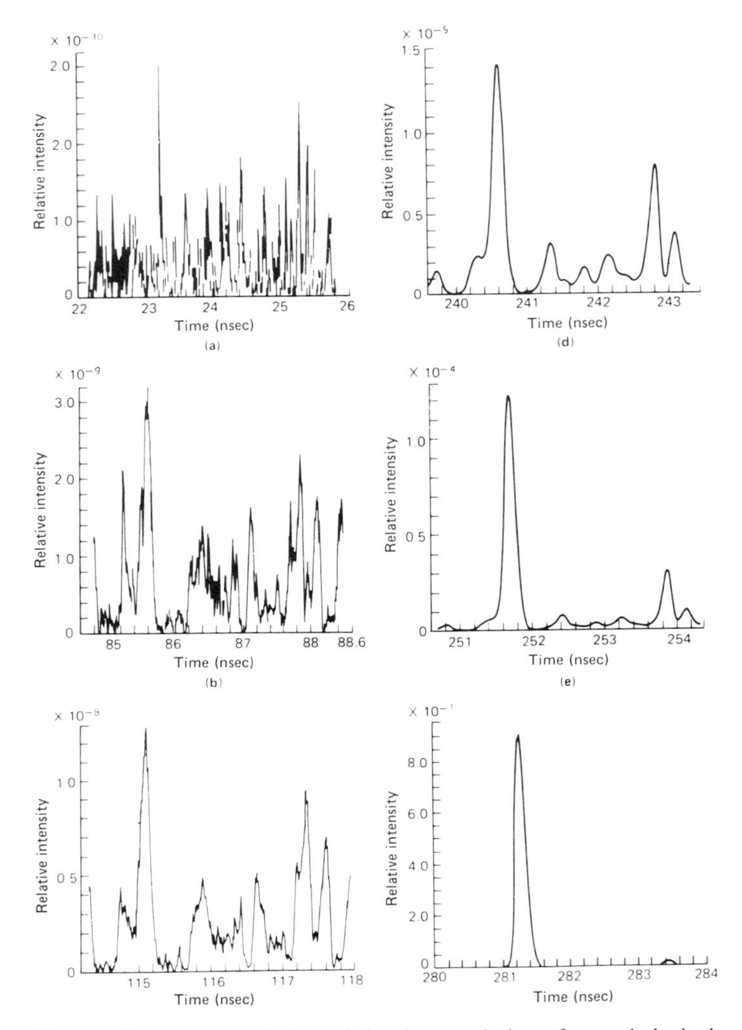

FIG. 4. Computer simulation of the time evolution of a mode-locked pulse from noise. [Reprinted with permission from Fleck, J. A. (1970). *Phys. Rev. B* 1:84.]

TABLE III. Passively Mode-Locked Solid-State Lasers[a,b]

Laser	Wavelength (nm)	Dye	Pulsewidth (psec)
Ruby	694.3	DDI in methanol	10–30
Nd:YAG	1064	Kodak 9860, 9740 in DCE	30
Nd:glass (silicate)	1060	Kodak 9860 in DCE	8–10
Nd:glass (phosphate)	1054	Kodak 9860 in DCE	6–7
		Kodak 5 in DCE	2–4
Alexandrite	760	DDI in methanol	8

[a] Adapted from Alfano, R. R. (Ed.) (1982). "Biological Events Probed by Ultrafast Laser Spectroscopy." Academic Press, New York.

[b] Abbreviations:
DDI, diethyl dicarbocyanine iodide; DCE, dichloroethane.

B. Synchronously Pumped Dye Laser Systems

An important method to generate ultrafast laser pulses is synchronous mode locking. In this technique, the gain medium (usually a dye) is pumped with a continuous train of ultrashort pulses. This modulates the gain of the medium. If the round-trip time of the dye laser cavity is equal to, or a submultiple of, that of the pumping laser, the gain of the dye laser is impulsively driven in synchronism with the round-trip repetition rate, and a train of ultrashort light pulses is generated from the dye laser. It has been shown that the temporal duration of the output laser pulses is proportional to the square root of the ratio of the pump pulse duration to the intracavity bandwidth. Therefore, to produce short pulses the pulse duration of the pumping source is the most crucial parameter of this type of laser system. Another is matching the cavity lengths with submicrometer accuracy.

Over 16 years ago synchronous mode locking utilized the pulse train from high-power mode-locked ruby lasers and from the second harmonic of a mode-locked Nd : glass laser and Nd : YAG laser. The shorter pulses were produced by the Nd : glass (~7 psec) and Nd : YAG systems (~20 psec). Recently, mode-locked argon-ion and YAG lasers have become the primary pumping source for synchronously pumped dye lasers because of their high repetition rate (10^8 Hz) and because of the short pulse durations they produce from the dye lasers.

The continuous-wave (CW) mode-locked Nd : YAG laser has been used by many researchers. Advantages of this type of pump source as compared to the conventional argon and krypton sources are (1) shorter pumping pulse duration (~50 psec as compared with ~150 psec), (2) ease and low cost of replacing the lamps in a CW mode-locked YAG laser as compared to replacing the laser tube, and (3) the potential for externally synchronized laser amplification once the ultrashort pulse has been generated.

One such laser system is depicted in Fig. 5. A CW mode-locked YAG laser is externally frequency doubled via a $Ba_2NaNb_5O_{15}$ doubling crystal. This produces a train of ultrashort pulses at 532 nm. The pulse duration, repetition rate, and average power are 35 psec, 100 MHz, and 500 mW, respectively. The output is focused into a folded, astigmatically compensated laser cavity containing rhodamine 6G (Rh6G). This produces 30 mW of average power at 600 nm with a repetition rate and pulse duration of 100 MHz and 5 psec, respectively. By adding a saturable absorbing liquid such as DODCI into the dye jet, to help enhance the pulse shortening and shaping mechanisms, pulses 70 fsec in duration have been generated.

To produce shorter pulses the pump pulses are compressed. A specific design of this type uses a Spectra Physics CW mode-locked Nd : YAG laser to generate a train of 80-psec pulses. The laser has an average output power of 8 W and a repetition rate of 80 MHz. This train is then passed through a single-mode optical fiber, thus increasing the bandwidth of the laser pulses via self-phase modulation and group velocity dispersion (GVD). The pulses are then compressed by using a grating pair (see Section II, H for details), in the double-pass geometry. This technique enables the YAG pulses to be shortened to 5 psec with an average power of about 3 W. These pulses are then frequency doubled to produce shorter pulses of 3.5 psec with an average power of 1.25 W. This output train is then used to pump a standard dye laser, using a $10^{-3}M$ solution of Rh6G as the gain medium. This technique has produced as much as 360

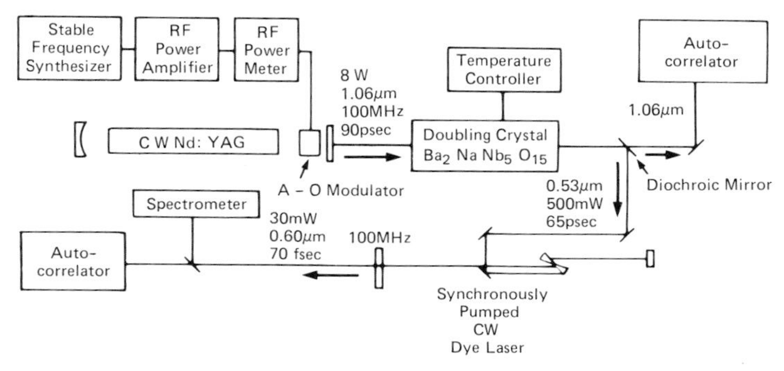

FIG. 5. Schematic diagram of a synchronously pumped dye laser system. [Reprinted with permission from Sizer, T., II, *et al.* (1983). *IEEE J. Quantum Electron.* **QE-19:**506. © 1983 IEEE.]

mW of average power at 584 nm, with pulse durations as short as 220 fsec.

A similar technique is to frequency double the output of a CW mode-locked YAG laser before pulse compression. This enables the pumping pulse to be shortened by pulse compression to 460 fsec as opposed to 3.5 psec. This technique has produced pulse trains from dye lasers with an average power of 40 mW and pulses 210 fsec in duration.

C. Colliding Pulse Mode Locking

The most successful method for generating the shortest pulses to date is ring-colliding pulse mode locking (CPM). This technique has produced pulses as short as 27 fsec directly, that is, without employing any external pulse shaping or pulse-compressing scheme.

The configuration of a colliding pulse mode-locked laser cavity is displayed in Fig. 6. A series of mirrors form a ring cavity whose central elements are a saturable gain dye and a saturable absorber dye. The choice of gain medium depends on the wavelength of interest. A particular design employs rhodamine 6G as the gain medium and diethyloxacarbocyanin iodide as the saturable absorber. The two dyes are placed at two focal points in the cavity. A 5-W CW argon-ion laser operating at 514.5 nm is used to pump the gain medium. As the lasing process builds up, two counterpropagating beams are formed in the cavity.

The pulse-shortening mechanism in the CPM scheme is similar to that in the passively mode-locked dye lasers, with further additions and improvisations. Picosecond and femtosecond pulses in passively mode-locked dye lasers are generated by both gain depletion in the lasing medium and bleaching in the saturable absorber medium. In these lasers, the process of pulse shortening involves two dyes. A slowly relaxing saturable absorber acts to steepen only the leading edge of the pulse. Saturation of the gain of the laser medium in combination with linear loss discriminates against the trailing edge. The absorber should saturate more easily than the gain. This provides a net gain at the peak of the pulse and a loss on either side. For a Rh6G/DODCI laser, spot size in the ratio of 2 : 1 results in a stable pulse formation. Shortening continues until dispersion effects prevent the pulse from becoming narrower.

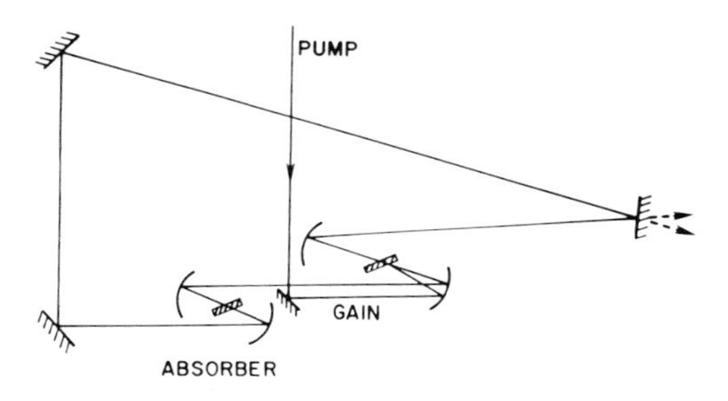

FIG. 6. Schematic diagram of a colliding pulse mode-locked laser cavity [with permission from Shanke, C. V. (1982). *Science* **219:** 1027. Copyright 1982 by the AAAS.]

Gain depletion and saturable absorption are modified by at least three interactions in CPM if pulses cross at either the gain or absorption media. One interaction is simply the doubling of the intensity in the medium. The second, a standing wave interaction, is necessary to explain the shortening of pulses in CPM operation together with a coherent four-wave mixing process. This forms an interference pattern that effectively couples the pulses together. This produces a nonlinear diffraction grating.

If the two pulses meet in a medium within a coherence distance from their centers, the pulses will be in phase within the medium. The standing-wave intensity will be quadruple the intensity of the individual pulses as a result of the constructive interference. However, where the two pulses meet outside their coherence distance, the temporal average in the medium will be only twice the individual intensities. If the two pulses meet in the saturable absorber, the bleaching within the coherence distance of the center is at least twice that of the bleaching just outside the center, thus shortening the pulse by dechirping.

If the pulses meet in the lasing medium, the gain is depleted more in the center of the pulse than at the edges. Alternatively, if the pulses meet even near the gain medium, the first pulse is preferentially amplified with respect to the second pulse. Therefore, counterpropagating subpicosecond pulse trains are stabilized if they meet in the saturable absorber and destabilized if they meet at or near the gain medium.

The separation between the saturable absorber dye and the saturable gain dye and their proper positioning in the cavity are thus crucial to the stable operation of a CPM ring dye laser. A separation of $L/4$, where L is the optical length of the cavity, turns out to be a good choice. In this arrangement, the two counterpropagating pulses see the same gain and collide in the absorber with equivalent power. This $L/4$ spacing also reduces the formation of extra pulses in the cavity.

The standing-wave interaction works only when the centers of both pulses coincide. When the centers do not coincide, but the pulses only overlap in the medium, the saturable absorber will not selectively absorb the tail of each pulse. If the pulses only overlap, there is no standing-wave interaction and the finite lifetime of the saturable absorber broadens the pulse. Therefore, if the jet stream is thicker than each pulse, permitting such an overlap, the CPM mechanism will not be efficient.

The simple CPM ring dye laser cavity that contains a saturable absorber dye and a saturable gain dye as the only two essential elements has produced pulses as short as 65 fsec. Both theoretical and experimental studies have been made to recognize the basic mechanisms that shape and shorten the pulses in the CPM ring cavity and to optimize their effects to achieve improved performance. It has been shown that intracavity phase modulation arising from GVD of nonresonant media like glass path, dye solvent and mirrors in the cavity, as well as from saturation and phase memory of the resonant media, plays a crucial role in the shaping of pulses in CPM dye laser cavities.

The phase modulation introduced by GVD of the mirrors in the cavity has been minimized by careful design of the mirrors. Such mirrors consist of homogeneous dielectric layers with alternating refractive index (high, low, high). Each layer has an optical thickness of $\lambda_0/4$, where λ_0 is the resonance wavelength of the mirror. Number of layers typically range from 10 to 20. In a particular design using a stack of 19 dielectric layers in the mirror design, dispersion was substantially reduced to the point where the main source of phase modulation and dispersion seemed to be the dyes.

In another novel CPM ring laser cavity design, a sequence of four prisms, each cut for minimum deviation operation at Brewster's angle, is incorporated in the cavity. The prism sequence enables intracavity GVD to be adjusted continuously from positive through zero to negative values. In addition to compensating for GVD, the prisms also make compensation of self-phase modulation possible. Using such a cavity and carefully balancing the four basic pulse-shaping mechanisms—self-phase modulation, group velocity dispersion, saturable absorption and saturable gain within a single resonator—pulses as short as 27 fsec have been generated. These are the shortest reported pulses obtained directly from a laser oscillator.

Even shorter optical pulses may be generated by further compressing the output pulses from a CPM laser using various pulse compression techniques. The best effort so far has produced an 8-fsec burst of light, which is just four cycles of visible light! Pulse compression will be discussed in Section II, H in more detail.

The peak power of output pulses from a CPM laser oscillator is typically in the kilowatt range. However, this can be amplified several millionfold to generate femtosecond light pulses at gigawatt power levels. A schematic diagram of a four-stage femtosecond pulse amplifier is shown in Fig. 7. The amplifier stages are pumped by the 530-nm second harmonic light from a nanosecond Nd : YAG laser synchronized to the CPM dye laser. The amplified pulse from each amplifier stage is passed through a saturable absorber dye before it enters the next stage (not shown in Fig. 7). This effectively clips the low-energy wings of the pulse and any accompanying spontaneously emitted light from amplifier dye cell.

Considerable temporal broadening of the pulse to 400 fsec occurs during the amplification process due to group velocity dispersion in the solvent of the amplifying dye and other optical components. This, however, may be compensated by employing a pulse compressor after the final amplifier stage, producing 90-fsec pulses.

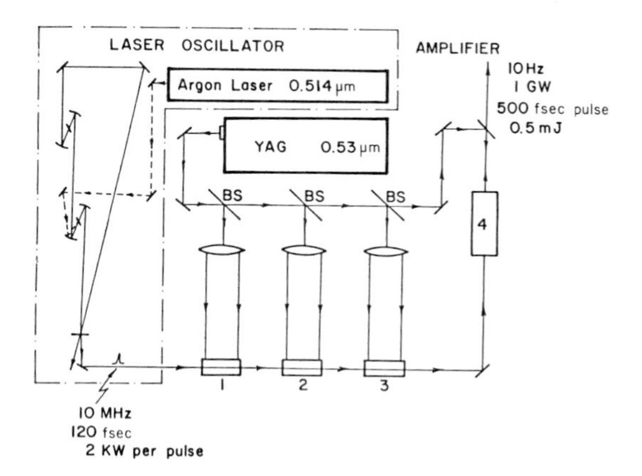

FIG. 7. Schematic diagram of a four-stage amplifier system used for amplifying femtosecond laser pulses. [From Ho, P. P. et al. (1985). *Opt. Commun.* **54,** 57. © Elsevier Science Publishers B.V.]

D. Mode-Locked Semiconductor Lasers

Ultrafast light pulse generation with semiconductor lasers is of great technological interest, especially for such applications as fast data processing and high-bit-rate optical communications. There are many advantages of semiconductor lasers over their solid-state and dye counterparts, which readily lend themselves to the high-repetition-rate optical information systems. These are small size, high efficiency, sturdy construction, wavelength tunability, and ease of integration into the system. Semiconductor lasers, however, have some inherent drawbacks as well. These include broad far-field diffraction pattern and lack of coherence, performance degradation at room-temperature operation, aging effects, and dispersion effects due to the large refractive index of these materials. Technological advances are expected to minimize these problems.

Semiconductor lasers are ideal for mode-locked operation because of their wide gain bandwidth, on the order of 100 cm^{-1}. These devices can, in principle, generate subpicosecond pulses. Techniques for ultrashort optical pulse generation include active mode locking, passive mode locking, and hybrid mode locking (a combination of active and passive mode locking). A brief history of the development of ultrafast semiconductor lasers is presented in Table IV.

Active mode locking of semiconductor lasers is accomplished by modulating the gain of the semiconductor either by a radio frequency (rf) signal or by ultrashort electrical pulses (gain switching). The rf modulation scheme is identical to that employed for active mode locking of solid-state lasers with the essential difference that in the semiconductor laser it is the gain that is modulated as opposed to the cavity losses, which are modulated in solid-state lasers. In this scheme a sinusoidal voltage with frequency modulation proportional to the cavity round-trip time is applied to modulate the gain of the semiconductor. This modulation locks the cavity modes in phase leading to shortening of pulses. A schematic diagram of an actively mode-locked semiconductor laser is presented in Fig. 8.

TABLE IV. Progress in the Generation of Ultrafast Pulses from Semiconductor Diode Lasers

Laser type	Method of pulse generation	Wavelength (nm)	Pulse duration (psec)	Repetition rate (GHz)	Year
GaAlAs, double heterostructure, stripe geometry	Active mode locking	810	23	3	1978
GaAlAs, Zn-diffused, double heterostructure, stripe junction	Gain switching	840	30	150–300	1979
GaAlAs, modified strip, buried double heterostructure	Passive mode locking	830	5.1	0.85	1980
GaAlAs, oxide isolated stripe, double heterostructure	Gain switching	840	15–60	1	1981
GaAlAs, buried optical guide, double heterostructure	Colliding pulse mode locking	830	0.56	0.625	1981
GaAlAs, Brewster angle cut in external cavity	Active mode locking	840–855	9.5	0.326	1983
GaAlAs, buried, double heterostructure	Synchronous mode locking	840	30	0.08	1984
GaAlAs, buried, double heterostructure	Active mode locking	840	12	17.5	1984
GaAlAs, V-groove	Active mode locking	840	4	0.01	1984
InGaAsP, ridge guide	Gain switching	1550	11–16	—	1985
InGaAsP, distributed feedback	Gain switching	1550	24–30	0.082	1986

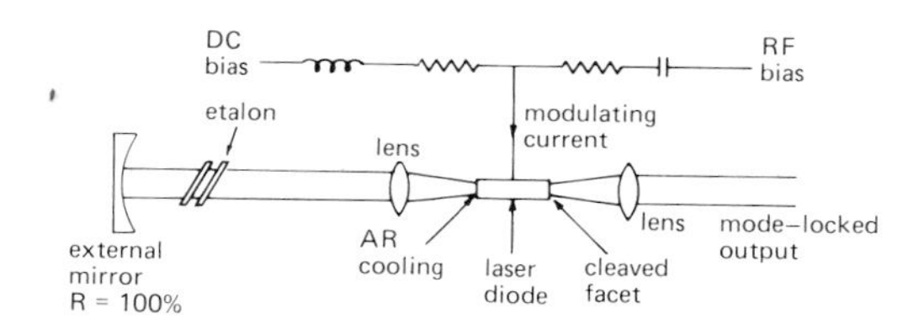

FIG. 8. Schematic diagram of an actively mode-locked semiconductor injection laser. [Reprinted with permission from Demokan, M. S. (1982). "Mode-Locking in Solid-State and Semiconductor Lasers." Wiley, New York. Copyright 1982 John Wiley & Sons.]

The gain-switching scheme uses ultrashort electrical pulses to pump the laser and modulate the gain. No external cavity is used. The laser is pumped above the threshold only momentarily. The cavity round-trip time of semiconductor diode lasers is of the order of tens of picoseconds. So even during the momentary excitation, oscillations can build up and the output is a single light pulse, whose duration depends on the magnitude and to some extent on the duration of electrical pump pulse and on the length of the cavity. Low-threshold, short-cavity, and short-duration high-amplitude pump pulses produce short output pulses. The repetition rate of output light pulses is equal to the applied modulation frequency.

In the passive mode-locking scheme, the semiconductor laser materials are doped with impurities or are irradiated to create defect centers. These defects or impurities act as absorption centers, thereby producing a region of saturable absorption in the gain medium. A microstructure layer at the end of the semiconductor can also be used as a saturable element. These regions produce the same effect as an external saturable absorption cell or dye jet in solid-state and dye lasers.

Mode locking of semiconductor lasers is an active area of current research. Efforts have been made to synchronously mode lock the semiconductor laser by using a picosecond optoelectronic switch for synchronous excitation. This technique has produced 30-psec pulses at 841 nm from a GaAs–GaAlAs double-heterostructure laser. Active–passive mode locking of semiconductor lasers has produced pulses as short as 560 fsec. Passive and active mode locking of semiconductor lasers without an external cavity has been attempted with moderate success. By actively mode locking a Brewster-angled GaAlAs diode laser, 9.5-psec pulses tunable over the 840- to 855-nm range have been obtained.

Application of ultrafast semiconductor lasers is another active research area of technological importance. We note some of those novel applications below. A picosecond laser radar that utilizes the same semiconductor laser both as the ultrafast optical pulse source and as an optical preamplifier for the reflected signal has been developed. This system has shown a maximum optical gain of 17 dB with a dynamic range of 33 dB and a spatial resolution on the order of a millimeter.

Another major application area of the ultrafast semiconductor laser is ultrafast information coding. One type of modulation scheme is based on the gain-switching technique. Ultrashort electrical pulses are used to generate optical pulses that are suitable for transmission over optical fibers. This type of pulse code modulation has been used successfully to generate data at 4 Gbit/sec. Other experimental data have shown that information rates of 25–30 Gbit/sec may be possible in the near future.

Most of the semiconductor lasers that are being developed emit radiation in the near infrared, the reason being that these wavelengths correspond to the minimal dispersion region in today's optical fibers. However, with electron beam pumping of II–IV semiconductors, generation of picosecond pulses in the blue-green region of the spectrum will be possible, with potential applications in underwater communications.

E. Excimer Lasers

Considerable effort has been devoted toward generating ultrafast laser pulses in the ultraviolet (UV) region. Significant progress can be made in the fields of nonlinear optics, photochemistry, X-ray sources, and dye lasers if such sources existed. The large gain bandwidth of excimers in the UV region ($\sim$160 cm^{-1}) makes them prime candidates for ultrashort pulse generation. However, the shortest excimer laser pulse reported so far has a duration of 300 psec and is obtained by simultaneous use of active and passive mode locking in a XeCl excimer oscillator. This clearly shows that the large bandwidth of the excimer media has not yet been fully exploited for the generation of ultrashort pulses. The reason is the relatively short duration of the gain provided by these lasers. So, the number of round trips available for the narrowing of mode-locked pulses is relatively small in an excimer medium. Furthermore, stable mode-locking saturable absorber dyes are not presently available in the UV region.

The main use of excimer media for ultrafast pulse generation has been for amplification of ultrashort UV pulses from another laser. In this technique, as illustrated in Fig. 9, ultrafast pulses to be amplified are generated from another laser system, usually mode-locked dye or solid-state lasers. The wavelength of the pulses from the primary source is then converted to a wavelength that matches the gain bandwidth of the excimer medium. The wavelength conversion occurs via standard nonlinear optical mixing techniques, such as harmonic generation or parametric mixing. The converted pulse is passed through the excimer amplifier producing a high-power ultrashort light pulse in the UV region. The state of the art using this approach has yielded pulses with 2 mJ of energy and a pulse duration of 300 fsec. Table V highlights the recent progress in this area.

An alternate method that does not involve a separate mode-locked oscillator uses a distributed feedback dye laser system. A XeCl oscillator is used to pump a subnanosecond dye laser and several amplifier stages. The output from this first stage then pumps a picosecond dye laser and several amplifier stages. The output is converted by harmonic generation to a wavelength that matches the gain bandwidth of the XeCl media and is amplified. This system provides pulses with an energy of 10 mJ, a temporal duration of 5 psec, and a repetition rate of 2 Hz and is tunable over the bandwidth of the XeCl gain module.

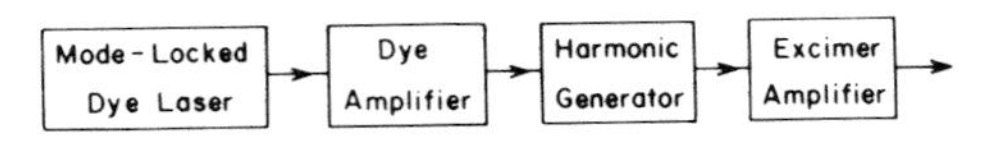

FIG. 9. A block diagram of a typical excimer amplifier system for ultrashort pulse amplification. [Reprinted with permission from Alcock, A. J. (1984). *Proc. S.P.I.E.* **46**. Copyright 1984 SPIE.]

F. HIGH-POWER INFRARED LASER

Previously described methods of ultrafast laser pulse generation use gain–loss modulation to generate a train of mode-locked pulses. An alternative method for generating an ultrafast laser pulse is to gate, or slice out, a portion of a longer-duration laser pulse. This low-power portion is then injected into a regenerative oscillator, where it is amplified many times to a reasonable intensity. This gating technique has been used to produce pulses as short as 0.5 psec at 10 μm using a multiatmosphere carbon dioxide amplifier system. [*See* GAS LASERS.]

The technique is shown schematically in Fig. 10. It is based on the optical dynamics associated with electron–hole (e–h) plasma generation in semiconductors. The e–h plasma modulates the refractive index and absorption, which in turn affects the reflection and transmission characteristics of the materials. A 10-μm pulse of 30 nsec duration illuminates a semiconductor slab set at Brewster's angle. The energy gap of this semiconductor is chosen to be greater than the laser-photon energy. This allows the laser to pass through the semiconductor. An intense ultrashort laser pulse at 620 nm (control pulse) strikes the semiconductor. The photon energy of this laser pulse is greater than the energy gap of the semiconductor. This creates a high-density

TABLE V. Recent Progress in Ultraviolet Ultrafast Pulse Generation[a]

Primary ultrafast pulse source	Excimer gain media	Pulse duration (psec)	Pulse energy (mJ)
Mode-locked Nd : glass laser	XeF	200	7
Mode-locked dye laser, flashlamp pump	XeCl	3.5	2.5
Mode-locked dye laser, synchronously pumped with Nd : YAG laser	KrF	20	20
Mode-locked dye laser synchronously pumped by argon ion laser	XeCl	2	40
Mode-locked dye laser synchronously pumped by argon-ion laser	ArF	10	40
Distributed feedback dye laser	XeCl	10	10
Mode-locked dye laser synchronously pumped by argon ion laser combined with fiber pulse compressor	XeCl	0.35	10

[a] Modified and extended with permission from Alcock, A. J. (1984). Proc. SPIE **46**. Copyright 1984 SPIE.

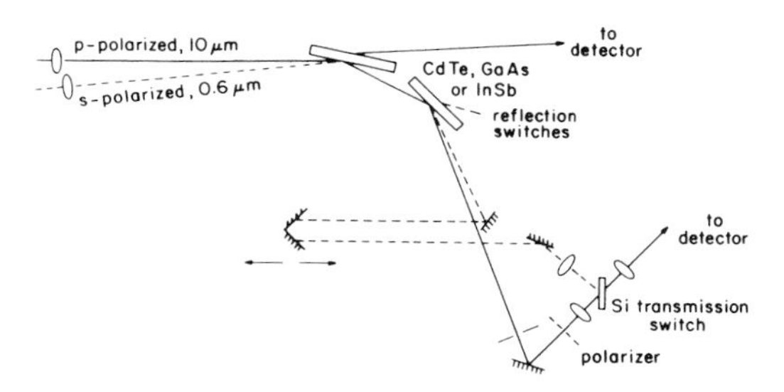

FIG. 10. Experimental configuration used to generate 10-μm picosecond pulses. [Modified with permission from Corkum, P. B. (1985). *IEEE J. Quantum Electron.* **QE-21:**216. © 1985 IEEE.]

e–h plasma, which reflects a small portion of the 10-μm beam. A second semiconductor set at Brewster's angle, with similar band-gap characteristics, is inserted into the optical path of the reflected beam. The 620-nm pulse is then delayed and used again as a control pulse. It is recombined with the reflected 10-μm beam onto the second semiconductor. As a result, the 10-μm beam is transmitted through the second semiconductor until the control pulse hits the semiconductor and generates an e–h plasma. The 10-μm beam then becomes attenuated, or turned off, by the free carriers generated by the second control pulse. This produces a low-energy ultrafast laser pulse. Its temporal duration can be varied from 2 to 40 psec by adjusting the delay time between the two control pulses. The resulting ultrafast pulse is regeneratively amplified (Fig. 11), by a millionfold, producing a train of mode-locked pulses in the infrared.

G. Ultrafast Supercontinuum Laser Source

Ultrafast supercontinuum refers to an ultrashort burst of white light. This burst of white

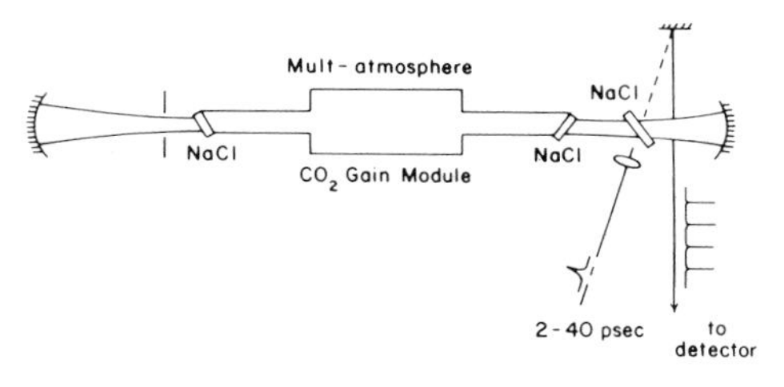

FIG. 11. Experimental configuration used to amplify 10-μm wavelength picosecond pulses. [Reprinted with permission from Corkum, P. B. (1985). *IEEE J. Quantum Electron.* **QE-21:**216. © 1985 IEEE.]

light is generated by passing an intense ultrafast laser pulse through certain material media. The input ultrashort laser pulse undergoes extreme frequency broadening through nonlinear interactions in the material medium, generating the supercontinuum pulse. The principal physical mechanism responsible for the white light generation is self-phase modulation (SPM). Other nonlinear processes such as four-photon parametric mixing, self-focusing, and avalanche ionization occurring simultaneously may also contribute to the spectral broadening of the incident laser pulse. Self-phase modulation is also the key mechanism that supports ultrashort pulse generation by optical pulse compression techniques. It provides the necessary bandwidth broadening and frequency sweep crucial for pulse compression (see the following section).

The modulation of phase is mediated by the intensity-dependent index of refraction. An intense optical pulse traveling through a medium can distort the atomic configuration of the material, resulting in a change in the refractive index via the nonlinear coefficients. The electric field of the laser beam in the time domain after traveling a distance z in the material is given by

$$E(t) = \tfrac{1}{2}E_0(t) \times \exp\,[-i(\omega_L t - n(t)z\omega_L/c)] + \text{cc} \tag{4}$$

where $E_0(t)$ is the envelope of the pulse, ω_L the laser angular frequency, and n the total index of refraction. For a medium having inversion symmetry the index of refraction n becomes intensity dependent:

$$n = n_0 + n_2E^2 \tag{5}$$

The intensity-dependent term in the index of refraction modulates the spectral intensity by modulating the instantaneous phase. This leads to a broadening of the pulse both at the Stokes and anti-Stokes frequencies. The broadening $\Delta\omega$ is given by

$$\Delta\omega(t) = -(\omega_L z/c)/[\partial n_2(t)/\partial t] \tag{6}$$

The pulse modifies its own spectra through a change in phase and envelope. These processes are SPM and self-steepening, respectively. The shape, fine structure and spectral range of ultrafast supercontinuum laser source (USLS) depend on the nonlinear index of refraction of the medium; the shape, wavelength, duration, intensity, and phase modulation of the input laser pulse; and the interaction length of the pulse in the medium. The optical properties such as coherence and pulse duration are determined by the incident laser pulse. The maximum frequency spread is given by

$$|\Delta\omega_{\max}| \approx \frac{\omega_L n_2}{cT_p} E_0^2 z \quad (7)$$

where T_p is the input pulse duration (FWHM) and z the interaction length in the medium.

A variety of materials, both solids and liquids and recently gaseous media, have been used to generate ultrashort supercontinuum by SPM. The typical broadening for picosecond input pulses is over several thousand wave numbers. For femtosecond input pulses, output frequency typically ranges over 10,000 cm^{-1} on either side of the pump laser frequency. Table VI presents a brief history of the development of ultrafast supercontinuum laser source.

Pulse compression is not the only technique that relies on spectral broadening by SPM. The ultrafast supercontinuum source is a versatile tool for time-resolved absorption spectroscopy, excite-and-probe measurements, and nonlinear optical studies such as soliton generation. Its broad spectral range allows the measurement of time delay and relative intensity for many different wavelengths at the same time. Recently, new uses of USLS in ranging, three-dimensional imaging, atmospheric remote sensing, and optical fiber have been suggested. [*See* NONLINEAR OPTICAL PROCESSES.]

H. Pulse Compression

The methods of ultrashort light pulse generation described so far use sophisticated mode-locking or pulse-slicing techniques. Another very essential technique to reduce the pulse width of ultrashort lasers down to the bandwidth-limited region is pulse compression. An additional advantage of pulse compression is that it increases the attainable pulse peak power simultaneously as it narrows down the pulse.

The ultimate duration of a laser pulse is determined by the pulse width–bandwidth uncertainty product $\Delta t \, \Delta f \geq K$, illustrated in Eq. (2) and Table II. Generation of increasingly shorter pulses requires corresponding spectral broadening. However, the bandwidth of a laser is determined by the gain bandwidth of the lasing medium and the design of the optical cavity. Pulse compression, on the other hand, depends on nonlinear optical properties of materials operative from ultraviolet to infrared. It thus has a significant advantage over direct pulse generation in a laser oscillator, especially for the generation of pulses in the femtosecond time domain.

The working principle behind most of the optical pulse compression schemes developed to date owes its origin to chirp radar. The idea is to send a chirped pulse (a pulse in which the frequency of the carrier varies continuously throughout the pulse) through a linearly dispersive delay line. The group velocity of light being dependent on its instantaneous frequency, different spectral portions of the pulse travel at different speeds through the delay line. In principle, the length of the delay line may be adjusted such that the leading edge of the pulse is delayed by just the right amount to coincide with the trailing edge at the output of the delay line. The width of the resulting output pulse may be as short as the inverse of the bandwidth of the frequency sweep.

Optical pulse compression thus involves two phases. In the first phase, a chirp or frequency sweep is impressed on the pulse. If the frequency sweep is such that the lower frequencies lead the higher frequencies, the pulse is called positively chirped. The pulse is negatively chirped if the higher frequencies lead the lower frequencies. The second phase involves com-

TABLE VI. Progress in Ultrafast Supercontinuum Generation

Year	Material	Input wavelength (nm)	Input pulse duration (psec)	Spectral region	Frequency spread (cm^{-1})
1969–1973	Liquids and solids (e.g., CCl_4, water, glasses, quartz)	530 or 1060	8	Visible and near IR	12,000
1974–1976	Fibers	530	30	Visible	1,000
1983	Glycerol	620	0.1	UV, visible, and near IR	16,000
1985	Semiconductors (e.g., GaAs, ZnSe)	10,000	6	IR	10,000

pressing the chirped pulse by propagating it through a dispersive delay line.

Ultrashort optical pulses can be suitably chirped by the nonlinear process of self-phase modulation in a nonlinear optical material. Self-phase modulation has been described in detail in the previous section. It supports pulse compression in two ways. First, it broadens the bandwidth of the input ultrashort laser pulse. Next, it produces the necessary frequency sweep essential for subsequent compression of the pulse. To first order, the magnitude of the frequency sweep induced by SPM is proportional to the first time derivative of the optical pulse shape. As a result, only the central portion of the pulse has the approximate linear chirp suitable for compression. The wings of the pulse have opposite frequency sweep, which leads to temporal broadening. Compression of a (hyperbolic secant)^2 pulse is illustrated in Fig. 12. The

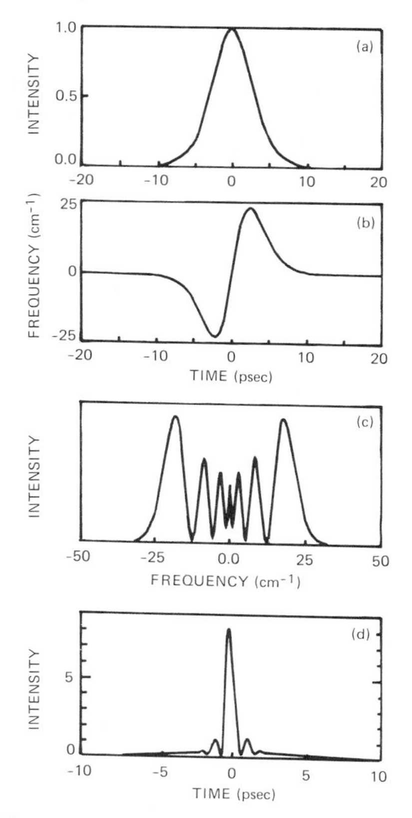

FIG. 12. (a) (Hyperbolic secant)^2 pulse; (b) frequency modulation by SPM; (c) Fourier transform of the pulse; (d) calculated compressed pulse. [Reprinted with permission from Grischkowsky, D., and Balant, A. C. (1982). *Appl. Phys. Lett.* **44**:2.]

spatial intensity distribution of the laser pulse is not uniform, but varies with the distance from the center of the pulse. Since self-phase modulation is an intensity-dependent process, the magnitude of the frequency sweep also varies with the distance from the center. This inhibits uniform compression of the entire profile. However, by using a single-mode optical fiber as the nonlinear medium, these problems as well as the problem of self-focusing associated with SPM in the bulk media may be greatly reduced. The fiber controls mode structure and enhances nonlinearity through confinement and long interaction length. By proper choice of pulse intensity and fiber length, essentially the entire output pulse can be linearly chirped. Single-mode propagation ensures that the entire output beam has the same chirp independent of the transverse spatial position. In addition, the fibers add the factor of group velocity dispersion to SPM. Depending on the location of wavelength near an absorption band, the GVD can be either positive or negative.

The frequency-swept pulse generated by SPM is next sent through a dispersive medium (compressor) for compression. For pulses chirped by SPM in a bulk medium, only the linearly chirped central portion gets compressed by anomalous dispersion in the compressor. Leading and trailing edges of the pulse have different frequency sweep (Fig. 12) and are not compressed. They give rise to wings on compressed pulses.

Pulses chirped by SPM in single-mode optical fibers may be compressed in two different ways, depending on whether the group velocity dispersion is positive or negative. In the wavelength region of negative (or anomalous) GVD, the low-frequency leading edge of the frequency-swept pulse travels slower than the high-frequency trailing edge, and the fiber acts as its own compressor. Alternate short sections of the fiber in this scheme may be looked upon as chirping and compressing the pulse. This mode of pulse compression is called soliton compression. This is simpler since no external compressor is required, and narrower pulses are produced. However, the quality of the compressed pulse is much poorer than that produced by SPM and positive GVD followed by external compression.

By far the most successful method of generating high-quality compressed pulses is chirping by the combined action of SPM and positive (or normal) GVD, also called dispersive self-phase modulation (DSPM), followed by subsequent compression in an external compressor. In

DSPM the leading edge of the pulse is frequency downshifted by SPM and normal GVD advances the arrival, while the trailing edge is frequency upshifted by SPM and dispersion delays arrival. The net effect of positive GVD is thus to broaden the pulse, limiting the bandwidth imposed by the frequency sweep. However, it also linearizes the chirp over most of the pulse length, so that almost all the power of the input pulse appears in the compressed pulse. Thus the quality of the compressed pulse is much better, but the price one pays is that the pulse is not as narrow as would be obtained in the absence of GVD.

The next step is the compression of the DSPM pulse by a pulse compressor. A compressor is simply a delay line in which the delay depends on frequency. The action of the compressor on a pulse with Fourier transform,

$$V(z, \omega) = A(\omega) \exp[i\Phi(\omega)] \tag{8}$$

where $A(\omega)$ and $\Phi(\omega)$ are real functions, is to introduce a phase function $\Phi_c(\omega)$ so that the Fourier transform of the compressed pulse is given by

$$V_c(z, \omega) = A(\omega) \exp\{i[\Phi(\omega) + \Phi_c(\omega)]\} \tag{9}$$

For an ideal compressor $\Phi_c(\omega) = -\Phi(\omega)$, such that at $t = 0$, all the frequency components of the pulse are in phase and the peak amplitude has the maximum possible value. Such an ideal compressor is expected to generate the shortest possible compressed pulse, although the pulse quality may not be the best.

In practice the DSPM pulse has been successfully compressed by a sodium vapor compressor and a grating pair compressor. Since it can be used in any wavelength region, a matched pair of gratings has turned out to be a versatile compressor and has been used in a variety of pulse compression applications. The grating compressor consists of a pair of plane-ruled diffraction gratings arranged with their faces and rulings parallel. Each frequency propagating through the grating pair is diffracted in a different direction and follows a different route, giving rise to a time delay that is a decreasing function of frequency. The grating spacing can be adjusted properly to provide the right amount of group delay, generating the compressed pulse. The variation of the group delay with wavelength for a grating pair compressor is given by

$$\frac{\Delta\tau}{\Delta\lambda} = \frac{b(\lambda/d)}{cd[1 - (\lambda/d - \sin r)^2]} \tag{10}$$

where b is the slant distance, d the grating constant, c the speed of light, and r the angle of incidence.

In terms of the phase function $\Phi_c(\omega)$ the effect of the grating pair compressor is to generate a phase function, which is approximately of the form

$$\Phi_c(\omega) = \Phi_0 - a\omega^2 \tag{11}$$

where a is the experimentally adjustable compressor constant. Because of the quadratic dependence of $\Phi_c(\omega)$ on ω, the grating pair compressor is also called a quadratic compressor. For a linearly chirped pulse a grating pair compressor is the ideal compressor. If the frequency sweep on the pulse does not deviate too much from linearity, a quadratic compressor can still produce reasonable compression, but for high nonlinearity it is not suitable.

The optimal pulse compression using DSPM followed by a grating-pair compressor depends critically on the fiber length, grating separation, and input pulse intensity.

Pulse compression techniques described so far have made it possible to obtain femtosecond pulses from a variety of picosecond laser sources and to compress femtosecond sources to generate pulses as short as a few optical cycles. Typical compression factors realized in practice from a single stage vary from 3 to 10 for femtosecond input pulses, 10 to 30 for picosecond input pulses, and as high as 80 for 33-psec, 532-nm pulses.

The output pulse from a grating compressor can be used as the input pulse for another fiber and compressor. Such a two-stage compressor has been used to compress 9-psec pulses at 1.06 μm down to 200 fsec, an overall compression factor of 450. However, the shortest pulse yet has been produced by DSPM in a fiber followed by a grating pair compressor. A typical arrangement for such a pulse compression scheme is presented in Fig. 13. Input pulses 40 fsec in duration were compressed down to 8 fsec (four cycles of visible light!) with a fiber 3 to 8 mm in length.

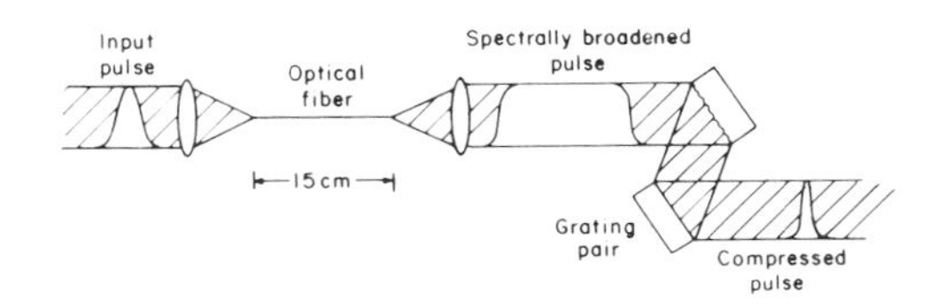

FIG. 13. Experimental arrangement for compressing a positively chirped laser pulse. [Reprinted with permission from Shank, C. V. (1982). *Science* **219:**1027. Copyright 1982 by the AAAS.]

III. Ultrafast Spectroscopic Techniques

In order to clock extremely rapid phenomena in nature several general ultrafast interrogation techniques have been developed and successfully used. These are the streak camera, optical Kerr gate, up-conversion gate, excite-and-probe technique, and transient grating method, to name a few. In this section we present a brief description of these techniques.

A. The Streak Camera

The technique most widely used for ultrafast luminescence measurements incorporates a streak camera. In this technique, shown in Fig. 14, light emitted from a sample photoexcited by an ultrafast laser pulse is focused on to a photocathode. Photoelectrons are released by the photocathode. The flux of electrons emitted is proportional to the light intensity hitting the photocathode. These electrons are accelerated and then deflected by an applied voltage that sweeps the electrons across the phosphor screen. The electrons released at different times from the photocathode strike the phosphor screen at different positions. This causes a track, or streak, which has a spatial intensity profile directly proportional to the incident temporal intensity profile of the luminescence. This phosphorescent streak is then analyzed electronically by a video system. In one shot, a complete fluorescence profile in time can be measured. The temporal resolution of streak cameras commercially available approaches 1 psec with UV or IR spectral sensitivity.

To obtain both temporal and spectral characteristics of luminescence, streak cameras can be connected in tandem with a spectrometer and a two-dimensional detector array. The ultimate resolution in this scheme, as in any simultaneous time–frequency measurement scheme, is determined by the uncertainty relation. Typically, one can measure 10-Å bandwidth with 10-psec resolution.

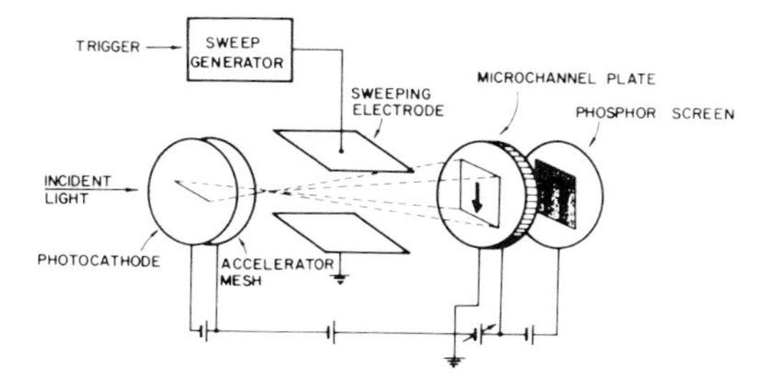

FIG. 14. Schematic diagram of the operation of a streak tube. (Reprinted with permission from Alfano R. R., ed. (1984). "Semiconductors Probed by Ultrafast Laser Spectroscopy." Academic Press, New York.)

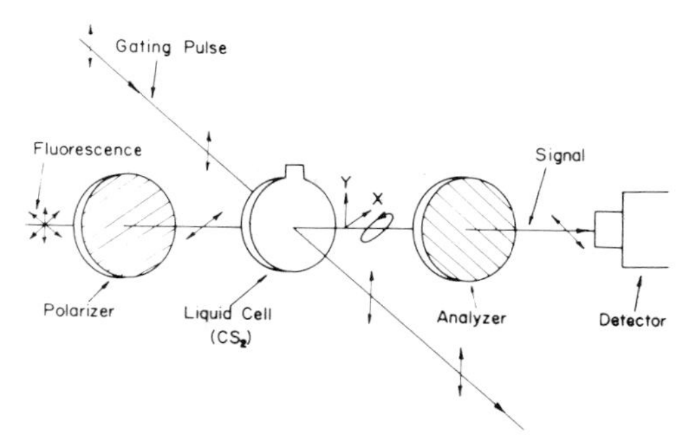

FIG. 15. Schematic diagram of an optical Kerr gate. The gating pulse is polarized at 45° with respect to the two polarizations. (Reprinted with permission from Alfano R. R., ed. (1984). "Semiconductors Probed by Ultrafast Laser Spectroscopy." Academic Press, New York.)

B. Optical Kerr Gate

The earliest ultrafast luminescence studies utilized the optical Kerr gate, which consists of a Kerr active liquid, such as carbon disulfide, situated between two crossed polarizers (see Fig. 15). Because of the crossed polarizers, the gate is naturally closed. Under the action of an intense electric field associated with an ultrafast laser pulse, the molecules of the liquid experience a short-lived induced birefringence. This causes light, which happens to be passing through the Kerr shutter at the same time as the laser pulse, to become elliptically polarized. A portion of the elliptically polarized light is then able to pass through the Kerr gate. Thus, light can only pass through the gate when it temporally coincides with the intense gating laser pulse. The intense laser pulse can be used to carve out successive portions of the temporal profile of the emitted luminescence by varying the delay time of the gating pulse, using a movable prism (see Fig. 16). Typically, this takes 100 measurements to form an intensity profile in time. A wavelength spectra, at a given delay, is obtained by using a spectrometer and video system.

The resolution of the optical Kerr gate depends upon the duration of the gate pulse and/or the reorientation time of the molecules of the

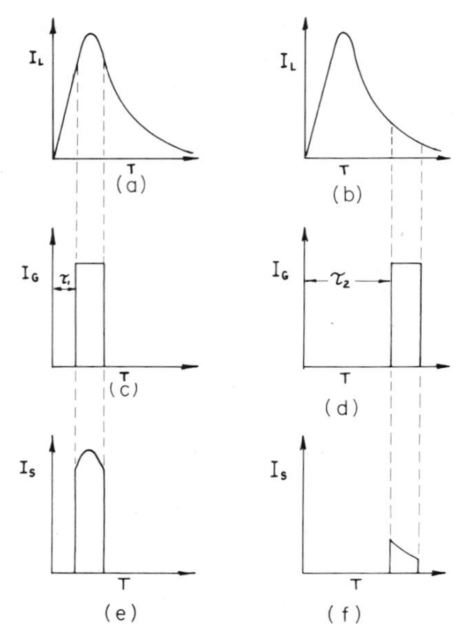

FIG. 16. Principle behind the operation of the Kerr gate and parametric up-conversion gate. The gating pulses at two different delays (c and d) carve out two segments from the luminescence (a and b). By varying the delay time the luminescence profile can be reconstructed (e and f). [Reprinted with permission from Alfano, R. R. (Ed.) (1984). "Semiconductors Probed by Ultrafast Laser Spectroscopy." Academic Press New York.]

active Kerr medium. For CS_2, one of the very fast Kerr active media available, the reorientation time is 2 psec. In the past, the pulse duration of the solid-state glass laser had been the limiting factor of the resolution of the Kerr gate (6 psec). The subpicosecond dye laser requires even faster Kerr active media based on electronic distortion rather than induced orientation of the molecules. Using polymers as Kerr active media, such a gate can attain a resolution of 100 fsec.

C. Parametric Up-Conversion Gate

The process of parametric up-conversion has been used as a mechanism in ultrafast optical shutters. Figure 17 shows the up-conversion technique. After an ultrashort pumping pulse excites a sample, the luminescence from the sample is collected, collimated, and combined with part of the excitation pulse in a noncentrosymmetric crystal (such as KDP, KNO_3, $LiIO_3$). The angle of the crystal is set in order to phase match the frequency of the gating pulse with a selected frequency of luminescence. A signal whose frequency is the sum of the laser and luminescence frequencies is generated by the crystal and detected by a photomultiplier tube. By varying the delay of the gating pulse and measuring the sum frequency signal, the temporal profile of the luminescence is obtained with no background luminescence. To obtain the spectra at a given time, the angle of crystal is usually tuned to the phase-matched sum of the gating pulse frequency and selected luminescence frequency. Since up-conversion involves virtual electronic transitions, this gate has a response time on a femtosecond time scale equal to the pulse width of the pumping laser. Dye lasers are ideal for up-conversion gates because they are wavelength tunable and can generate CW subpicosecond pulses.

D. Excite-and-Probe Technique

The excite-and-probe technique is a sensitive spectroscopic technique in the kinetics study of the excited states of matter. In excite-and-probe measurements, two incident ultrashort light pulses spatially overlap on the sample under investigation. The intense pump pulse excites the sample. The changes initiated in the sample by the pump pulse are monitored by the weaker probe pulse. The time evolution of the excited state is investigated by varying the time delay between the pump and the probe pulses. The probe pulse may be used to monitor such processes as induced absorption, bleaching, and induced Raman scattering.

The excite-and-probe technique has been widely used to study the time behavior of excited states of a variety of biological, chemical, and physical systems. Although the basic principle is the same, there is wide variation in the experimental arrangement. A typical arrangement is shown in Fig. 18. The pump and the

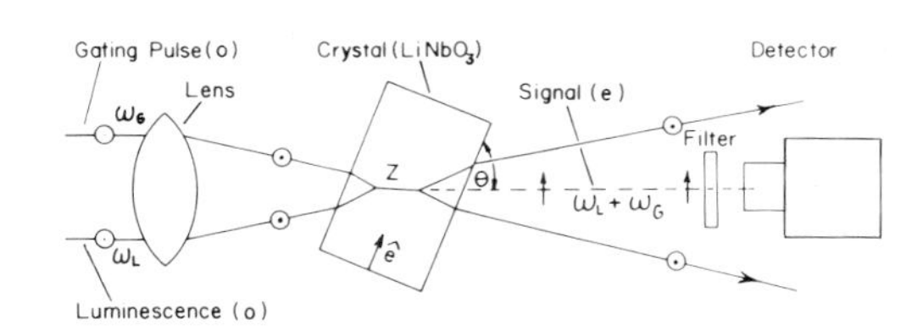

FIG. 17. Schematic diagram of the parametric up-conversion gate. The gating and the luminescence pulses are combined inside the crystal to produce a sum frequency wave. [Reprinted with permission from Alfano, R. R. (Ed.) (1984). "Semiconductors Probed by Ultrafast Laser Spectroscopy." Academic Press, New York.]

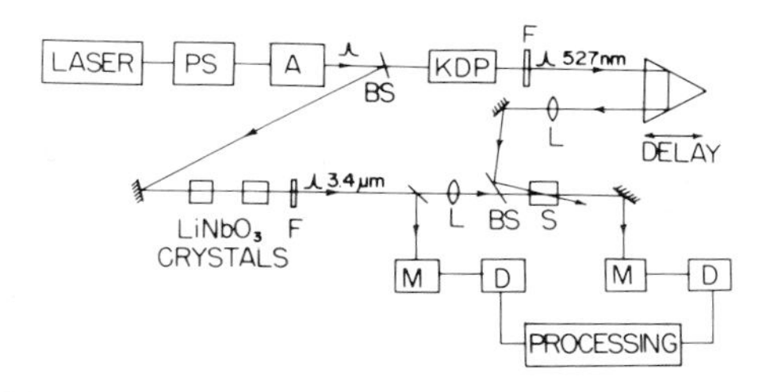

FIG. 18. A typical arrangement for excite-and-probe measurement. (A, amplifier; BS, beam splitter; D, infrared detector; F, filter; L, lens; M, monochromator; PS, single-pulse selector, S, sample.) [Reprinted with permission from Gayen, S. K. *et al.* (1985). *Appl. Phys. Lett.* **47,** 455.]

probe pulses may, in principle, be derived from two different lasers, but for picosecond or femtosecond applications it is extremely difficult to synchronize the timing of the two pulses to below 5 psec. So, for ultrashort time resolution both the pump and probe pulses are generally derived from the same primary laser pulse. Nonlinear frequency-shifting techniques such as harmonic generation, sum and difference frequency generation, stimulated Raman scattering, and ultrafast supercontinuum are used to generate the pump and probe pulses of desired frequency. If time and wavelength resolved absorption spectra of excited states are required, the probe pulse may be frequency broadened by self-phase modulation in a suitable media. The excited state is interrogated by the ultrafast supercontinuum probe pulse and a spectrograph or an optical multichannel analyzer is used for recording the absorption spectrum. A simultaneous three-dimensional recording of intensity, time, and wavelength is thus obtained.

The time delay between the pump and the probe pulses may be varied continuously by using an optical delay line. An optical path difference of 1 mm is equivalent to a time delay of 3.3 psec. The time resolution of this technique is determined by the temporal duration of the pump and the probe pulses. The sensitivity depends on the sensitivity of the probe pulse detection scheme and signal averaging.

E. Transient Grating Method (Four-Wave Mixing)

The transient grating method, sometimes referred to as dynamic holography, is very similar to the excite-and-probe technique. Instead of a single intense laser pulse exciting a sample, two (or more) laser pulses, temporally and spatially coincident, enter the sample at a small angle (see Fig. 19). The coherent superposition of these laser pump pulses ("write" pulse) results in an interference pattern similar to a hologram. The beams that "write" the hologram effectively create a grating or a periodic spatial distribution of the index of refraction by modifying the optical properties of the sample. This modification of the sample's optical properties may occur via any of the following mechanisms: molecular orientation in liquids, electron–hole generation in semiconductors, resonant interactions in two-level atomic systems, deformation of electronic orbitals, and other multiphoton processes.

Once this grating has been formed, a third weak probe beam enters the sample and is scattered in a similar fashion as a third beam is used to read a hologram. By monitoring the intensity of the scattered wave versus the time delay between the writing (pump) beams and the reading (probe) beam, the dynamics of the various processes causing the spatial index variation can be clocked.

A few of the ultrafast spectroscopic clocking techniques that rely on the transient grating method include time-resolved degenerate and nondegenerate four-wave mixing, time-resolved coherent anti-Stokes Raman spectroscopy (CARS), phase conjugation, and time-resolved Raman-induced phase conjugation spectroscopy (RIPCS) (see next section). In these techniques, as in the excite-and-probe technique, the temporal resolution is limited to the excitation pulse widths.

F. Ultrafast Phase Conjugation Spectroscopy

Phase conjugation is another method that can be utilized to measure the ultrafast time response of materials. Four beams are involved in the phase conjugation process: three are input laser beams and the fourth is the generated

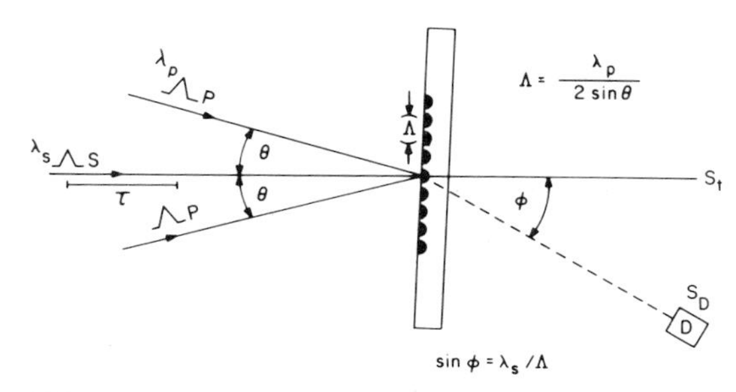

FIG. 19. Principle behind the transient grating method.

phase conjugate signal beam. There are two approaches in utilizing phase conjugation. The first is degenerate four-wave mixing, where the input laser beams are of the same frequency. The second is nondegenerate four-wave mixing, where the input frequencies differ and may be tuned to vibrational or electronic resonances.

In the phase conjugation technique, two pump beams enter the nonlinear material (semiconductors, polymers, dielectrics, and liquids) propagating in opposite directions. A probe beam enters the nonlinear material at a small arbitrary angle and interacts with the two counterpropagating pump beams. The optical mixing of these three waves in the media produces a phase conjugate signal beam that leaves the medium in a direction exactly opposite to the probe beam. The spatial distribution of the conjugate wave is exactly complex conjugated to the probe beam, which is analogous to a "time reversal" of the probe beam. It is this fact that gives the conjugate wave the aberration-correcting properties that are normally associated with phase conjugation.

The production of the conjugate wave can be understood easily by considering the wave mixing as a dynamic quasi-holographic process. The forward pump and probe beams interact in the sample to form an interference pattern or grating. The backward pump beam "sees" this hologram and is Bragg scattered in a direction opposite to the probe beam to form the conjugate wave. Simultaneously, the backward-propagating pump and the probe beams interact to form their own hologram. The forward pump beam sees this hologram and is Bragg scattered in a direction opposite to the probe beam to form the conjugate wave. In addition, the interaction of the two counterpropagating pump beams temporally modulates the optical properties of the medium (temporal grating) as opposed to the spatial modulations produced by a pump and probe beam. This temporal modulation causes the probe beam to be reflected back upon itself to produce the conjugate wave. In actuality, it is the coherent superposition of these processes that generates the phase conjugate beam. The nonlinear polarization produced by this four-wave mixing process is given by

$$P_{\mathrm{NL}} = (E_{\mathrm{f}} \cdot E_{\mathrm{p}}^{*})E_{\mathrm{b}} + (E_{\mathrm{b}} \cdot E_{\mathrm{p}}^{*})E_{\mathrm{f}} + (E_{\mathrm{f}} \cdot E_{\mathrm{b}})E_{\mathrm{p}}^{*} + \mathrm{cc} \qquad (12)$$

The first two terms correspond to the spatial gratings and the third term corresponds to the temporally modulated grating. The symbols E_{f}, E_{b}, and E_{p} correspond to the forward, backward, and probe beams, respectively. The physical mechanisms that produce these interference patterns or holograms are identical to those that produce transient gratings arising from molecular orientation, Kerr effect, and resonant interactions.

When ultrashort laser pulses are used, these gratings can be recorded in the material on an ultrafast time scale. One can perform an excite-and-probe experiment using the three beams, two to create the grating and a third to scatter from the modulation arising from the buildup of elementary excitations. The relaxation time of the gratings can be caused by diffusion, excitation lifetime, and orientational relaxation.

To measure relaxation times of the grating arising from different excitation modes, it is imperative that the correct beam be delayed. By delaying the backward pump beam (which is traveling opposite to the forward pump and probe beams) relative to the coincidence of the other two beams, one obtains the relaxation time of the grating. In this geometry, the effect of diffusion is decreased and the dominant relaxation mechanism of the elementary excitation is measured. On the other hand, if the forward pump beam is delayed relative to the backward pump and probe beams, diffusion will usually dominate the decay. If the probe beam is delayed with respect to both counterpropagating pump beams, the coherence time of the input laser pulses is measured.

Degenerate four-wave mixing has been used extensively to measure the magnitude and response time of the nonlinear index of refraction n_2. Carbon disulfide is commonly used as a reference material.

Nondegenerate phase conjugation techniques have a great potential for measuring dynamics of elementary excitations in materials, where the difference in frequency of the beams can be tuned to generate and interrogate the modes. An example of this is the ultrafast Raman-Induced phase conjugation (RIPC) technique, which is displayed schematically in Fig. 20.

In the RIPC technique, an ultrafast supercontinuum "white light" laser source (see Section II,G) is used as the backward pump beam, while the forward pump and probe beams are ultrashort "monochromatic" laser pulses. In this way, the beating of the white light pulse and the probe beam couple many of the vibration modes, from ~15 cm^{-1} to 3000 cm^{-1}, under study. A transform-limited 1-psec (13-cm^{-1})

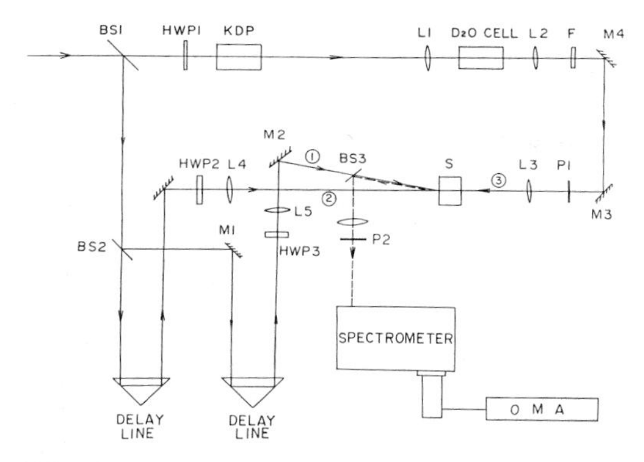

FIG. 20. A schematic diagram illustrating the Raman-induced phase conjugation technique. (BS, beam splitter; HWP, half-wave plate; M, mirror; L, lens; S, sample; F, filter; P, polarizer; OMA, optical multichannel analyzer.)

pulse is necessary to obtain both temporal and spectral information. The lifetimes of the molecular vibrations and optical phonons may be clocked directly by delaying the forward pump beam and monitoring the decay of the phase conjugate signal pulse. Over the next few years, we expect that time resolved RIPC will be used to directly measure the optical phonons in bulk and microstructure semiconductors. It will complement the commonly used coherent anti-Stokes Raman spectroscopy and stimulated Raman scattering methods.

IV. Directions and Trends

Ultrafast laser technology is only 20 years old. Since its inception it has revolutionized our understanding of the world of the ultrafast. The field is growing at an accelerated pace, new resolution limits are being attained, new phenomena are being discovered, and new techniques are being developed. The future is even more exciting.

Promising research areas include generation of even shorter pulses and using them to study transient phenomena in the submicroscopic world. Generating the "ultimate shortest pulse" may be the most challenging problem of the day. Improved stability, tunability over the entire visible, near ultraviolet and near infrared, amplification of pulses to the terawatt range, pulse repetition rates well above the gigahertz range are at the frontier of femtosecond light source research.

Improving the experimental capability of using ultrashort pulses in basic research is of great interest. Areas of research are widely varied: chemical dynamics, quantum-well structures, electron–phonon and electron–electron interactions in solids, hot carrier momentum and energy relaxation, intravalley and intervalley scattering times, coherent excitation, fluorescence rise time, and time-resolved studies of phase transition, to name a few. Femtosecond pulses are shorter than, or comparable to, the period of vibration of elementary excitations. This enables one to monitor directly the time evolution of the interaction of radiation with matter starting from coherent vibration and electronic excitation through its interaction with the dynamic system to the eventual relaxation. New physical phenomena may be induced and/or monitored by femtosecond pulses. For example, the quantization of particle motion in quantum wells introduces new physical phenomena, which may be probed by ultrafast techniques. Also ultrafast processes in ultrasmall domains are expected to yield interesting new effects.

Detection and diagnostic equipment have not kept pace with ultrafast light sources. Detection systems with high sensitivity and high temporal resolution are urgently needed. Streak and framing cameras with femtosecond time resolution, nonlinear optical methods in the femtosecond regime, medical diagnostics and imaging using femtosecond light pulses, and remote sensing using ultrashort light pulses are areas where significant progress needs to be and can be made.

In the area of more practical applications, materials characterization, high-speed electronic measurements, characterization of semiconductor and electro-optical devices of micrometer and submicrometer sizes are being radically transformed by ultrafast laser technology. However, more exciting and far-reaching future applications of ultrashort pulses are expected to be in the fields of optical communication and optical computation. Data transmission at a rate of 4 Gbit/sec over 117 km of fiber has been accomplished. The limiting factors are dispersion in the fiber and detector response time, not the duration of pulses. Even with the pulses currently available, data transmission rates may be increased three or more orders of magnitude. Optical computing based on ultrafast logic units has the potential for revolutionizing the field of computers.

Three centuries ago, two discoveries in optics—the telescope and the microscope—enhanced the spatial resolution several orders of magnitude and brought the vast expanse of the heavens and the unseen microscopic world closer to human experience. Rapid expansion in scientific knowledge in many different disciplines followed those discoveries and continues to date. With the advent of ultrashort light pulses, temporal resolution has increased several orders of magnitude to investigate the nonequilibrium state of matter. How that will shape the human understanding of the physical universe and what impact it will have on other sciences remain to be seen.

Acknowledgments

The research at the Institute for Ultrafast Spectroscopy and Lasers is supported by the Army Research Office, the Air Force Office of Scientific Research, the Office of Naval Research, the National Aeronautics and Space Administration, the National Science Foundation, the Board of Higher Education—Professional Staff Council, and Hamamatsu Photonics KK.

Bibliography

Alcock, A. J., and Bourne, O. L. (1984). *Proc. SPIE* **476**.

Alfano, R. R. (Ed.) (1982). "Biological Events Probed by Ultrafast Laser Spectroscopy." Academic Press, New York.

Alfano, R. R. (Ed.) (1984). "Semiconductors Probed by Ultrafast Laser Spectroscopy." Academic Press, New York.

Alfano, R. R., and Shapiro, S. L. (1970). Emission in the region 4000 to 7000 Å via four photon coupling in glass, *Phys. Rev. Lett.* **24**:584.

Alfano, R. R., and Shapiro, S. L. (1970). Observation of self-phase modulation and small scale filaments in crystals and glasses, *Phys. Rev. Lett.* **24:**592.

Burnett, K., and Hutchinson, M. H. R. (eds.) (1989). Multiphoton physics, *J. Mod. Opt.* **36**(7).

Corkum, P. B. (1985). Amplification of picosecond 10-μm pulses in Multiatmosphere CO_2 lasers, *IEEE J. Quantum Electron.* **QE-21**:216.

Demokan, M. S. (1982). "Mode-Locking in Solid-State and Semiconductor Lasers." Wiley, New York.

Fujimoto, J. G. (1986). "Femtosecond Pulse Generation and Measurement." Laser Institute of America, Toledo, Ohio.

Johnson, A. M., and Simpson, W. M. (1985). Tunable femtosecond dye laser synchronously pumped by the compressed second harmonic of Nd: YAG, *J. Opt. Soc. Am.* **B2**:619.

Knox, W. H., Fork, R. L., Downer, M. C., Stolen, R. H., Shank, C. V., and Valdmanis, J. A. (1985). Optical pulse compression to 8 fs at a 5 kHz repetition rate, *Appl. Phys. Lett.* **46,** 1120.

Levenson, M. D., and Kano, S. S. (1988). "Introduction to Nonlinear Laser Spectroscopy," revised ed. Academic Press, Boston.

Miller, A., and Sibbett, W. (eds.) (1988). Ultrafast phenomena, *J. Mod. Opt.* **35**(12).

Shank, C. V. (1982). Measurement of ultrafast phenomena in the femtosecond time domain, *Science* **219**:1027.

Shapiro, S. L. (Ed.) (1984). "Ultrashort Light Pulses," 2nd ed., Springer Verlag, New York.

Shen, Y. R. (1985). "Principles of Nonlinear Optics." Academic Press, Orlando, Fla.

Sizer, T., II, Kafka, J. D., Duling, I. N., III, Gabel, C. W., and Mourou, G. A. (1983). Synchronous amplification of subpicosecond pulses, *IEEE J. Quantum Electron.* **QE-19**:506.

Tomlinson, W. J., Stolen, R. H., and Shank, C. V. (1984). Compression of optical pulses chirped by self-phase modulation in fibers, *J. Opt. Soc. Am.* **B1**:139.

van der Ziel, J. P. (1985). In "Semiconductors and Semimetals," Vol. 22B, (W. T. Tsang, Ed.). Academic Press, Orlando, Fla.

ULTRASHORT LASER PULSE CHEMISTRY AND SPECTROSCOPY

Joseph L. Knee *Wesleyan University*

GLOSSARY

CPM: Colliding pulse mode-locked dye laser. Dye laser capable of producing optical pulses of ~ 50 fsec duration.

CW: Continuous wave: Describes a laser that operates continuously as opposed to an intermittent mode. In some cases, the output may be pulsed, but the laser is still described as cw because the active medium is continuously pumped.

Femtosecond: Unit of time that is equal to 10^{-15} sec.

LIF: Laser induced fluorescence: spectroscopic technique in which an atom or molecule is excited by a laser pulse, and subsequently this excitation event is detected by emitted photons from the excited state.

Mode-locked: Condition that is created by constraining the longitudinal modes of a laser to have a particular phase relationship. The result is that the laser will produce a short-duration pulsed output.

Nd:YAG: Type of laser named for the active medium that is a yttrium aluminum garnet crystal that is doped with Nd^{3+} ions. This laser produces an output at a wavelength of 1.06 μm.

Photomultiplier: Device that detects photons. For each photon striking the active surface of the device, 10^6 electrons are produced, creating an electrical signal that can easily be detected and processed.

Picosecond: Unit of time that is equal to 10^{-12} sec.

TCSPC: Time correlated single photon counting: picosecond measuring technique that times events by monitoring emitted photons as a function of time after excitation by a picosecond laser pulse.

Ultrafast: General term to describe both picosecond and femtosecond time domains.

Ultrashort laser pulse chemistry and spectroscopy is a field of study wherein the unique properties of modern pulsed lasers are used to measure the rates of fundamental processes in chemistry and biology. Compared to the time scale for macroscopic phenomena that are encountered in everyday life, the rates for microscopic molecular level processes are enormously fast. The processes being investigated by ultrafast techniques can be divided into three broad categories: (1) electron and proton transfer between or within molecules; (2) structural rearrangements, such as chemical isomerization, or solvent sphere rearrangement; and (3) true chemical reactions that involve the making and breaking of chemical bonds. The time scale of these events can range from tens of femtoseconds to hundreds of picoseconds or longer. To make measurements on these time scales, the measuring device must have at least this time resolution. Pulsed lasers currently are the only tools that have the necessary time resolution and so have been applied widely in the investigation of time-dependent phenomena. In 1988, great progress was made in extending the time scale of chemical measurements even further down to the tens and hundreds of femtosecond time regime. In one experiment, 6-fsec laser pulses were used to measure the response of bacteriorhodopsin to the absorption of a photon, which is the initiation step in the vision process. Chemical reactions involving the breaking of bonds were measured in the gas phase with observed reactions times as fast as 150 fsec. In the future, one can expect the limits of time resolution to be pushed in a number of different areas. Also, as this technology begins to be disseminated to the general scientific community, the areas of research can be expected to broaden and yield interesting results that cannot be anticipated at present.

I. Ultrafast Laser Technology

A. Picosecond Lasers

Studies of ultrafast chemical phenomena are driven by developments in laser technology. The production of picosecond width pulses is now quite well established and there are a number of systems commercially available. A summary of typical lasers and their specifications is given in Table I. The lasers can be classified into two main types although this distinction is becoming blurred with time as the limits of performance are pushed. This first are high-repetition-rate low-energy per pulse systems. The common examples of these are the cw mode-locked Ar^+ and Nd:YAG lasers. The gain media in these lasers are excited continuously, but the gain of the cavity is modulated rapidly using an acousto-optic device to produce a mode-locked pulse train. The second class of lasers are Q-switched and mode-locked lasers of which Nd:YAG and Nd:GLASS are the premier examples. In this case, the gain medium is excited transiently, but to a much greater degree than the cw laser case, and a single large picosecond pulse is extracted. Typical operating specifications for each type of laser are given in Table I. [*See* ULTRAFAST LASER TECHNOLOGY.]

These lasers can be used directly in chemical experiments, but often they are not suitable because the output wavelength is fixed. In general, studies of ultrafast dynamics are made by depositing energy in a molecular system via a laser pulse and then following in time the resulting behavior of the system. It is highly desirable (and often necessary) that the amount of energy be controlled by varying the wavelength of the exciting laser pulse. To accomplish this, a dye laser is used. A dye laser consists of an organic dye as an active medium that is excited (pumped) by the picosecond pulses from the lasers listed previously. The picosecond pulse structure of the pump lasers is transferred to the dye laser pulse with the now added feature of tunability, albeit with an overall loss in laser intensity. The directly accessible tuning range is 550–900 nm, but this can be extended by nonlinear frequency generation techniques to cover the 250–3000 nm range with varying degrees of efficiency across this range.

B. Femtosecond Lasers

To obtain the very shortest pulses possible, an entirely different type of laser is used. It is called a colliding pulse mode-locked (CPM) cw dye laser. To date, these lasers have produced the shortest optical pulses with direct outputs being ~50 fsec and pulsed compressed outputs as short as 6 fsec. Pulse compression is a technique for pulse shortening that is applied to the pulse external to the laser cavity. The CPM laser uses rhodamine 6G (an organic dye) as the active medium and is pumped by the cw output of an Ar^+ laser. Pulse formation occurs because of the presence of a saturable absorber (DODCI, another organic dye) within the laser cavity. The output is at a high repetition rate with low energy per pulse.

TABLE I. Summary of Ultrafast Lasers[a]

Laser type	Pulse width (psec)	Wavelength (nm)	Repetition rate (Hz)	Energy/pulse (mJ)
Argon ion	100	514.5	10^8	10^{-6}
CW Nd:YAG				
Fundamental	70	1064	10^8	10^{-5}
2nd harmonic	50	532	10^8	10^{-6}
Q-switched Nd:YAG				
Fundamental	100	1064	20	100
2nd harmonic	70	532	20	50
3rd harmonic	50	355	20	35
4th harmonic	30	266	20	25
Synch-pumped dye laser[b]	1	570–900	10^8	10^{-7}
CPM dye laser	0.050	620	10^8	10^{-7}
Amplified dye laser	1	570–900	20	1
Amplified CPM	0.050	620	20	0.1

[a] Typical values; actual operating parameters vary widely.
[b] Pumped by 514.5-nm argon ion or 532-nm CW Nd:YAG lasers.

The major drawback of this type of laser is that it is not tunable, the output being at 620 nm. Furthermore, the linewidth is quite large, which is unavoidable by virtue of the Heisenberg uncertainty principle, which requires the pulse width in time and the linewidth in wavelength to be inversely related. As an example, a 100 fsec pulse at 620 nm has a bandwidth of ~16 nm. In experiments on chemical systems, this uncertainty in wavelength (energy) is often much too great to make the measurements useful. Therefore, it is desirable to have the time resolution that is required for a particular experiment but to not go very far beyond what is needed.

Both the CPM laser and the high-repetition-rate picosecond dye laser have energies per pulse that are often insufficient for experiments of interest. In this case, these pulses can be amplified to much higher powers using a pulsed dye amplifier with an organic dye as the active medium and a separate laser as the pump source. In the process of amplification, the repetition rate of the dye laser pulses is dramatically reduced. See Table I for amplification specifications.

II. Experimental Techniques

To measure the duration of an event at least two measurements must be made that determine when the start and end of the event occurred. If information is desired as to the progress of an event, then many measurements must be made with the number being related to the degree of detail of the event that is desired. In the ultrafast measurement of chemical processes, there are a large number of techniques for making these timing measurements. However, the major techniques can be classified into three schemes: (1) time-correlated single-photon counting, (2) streak camera measurements, and (3) optical pump–probe techniques.

A. Time-Correlated Single-Photon Counting

Time-correlated single-photon counting (TCSPC) is the most widely applied technique for measuring picosecond molecular dynamics. In this technique, the sample of interest is excited by a single picosecond laser pulse. Following this excitation, photon emission from the sample is monitored by a photomultiplier that is specifically designed for maximized time response. The time between the excitation by the laser pulse and the detection of the emitted photon is recorded. This measurement is repeated many times until a statistical correlation between the time of excitation and emission is recorded. Typical TCSPC data for a simple kinetic process is shown in Fig. 1. The form of the decay represented is typical of first-order kinetics and follows the following relationship:

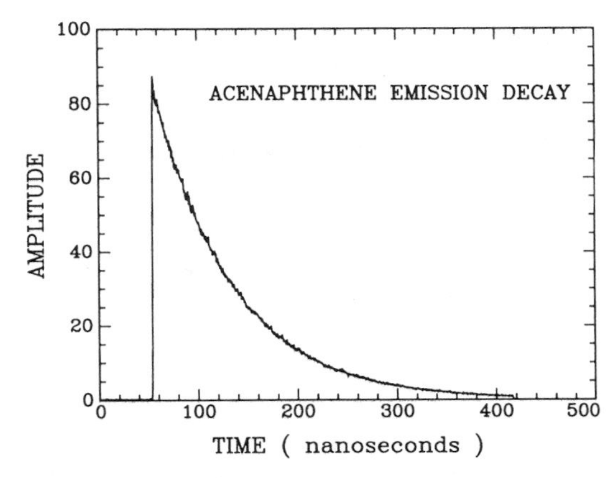

FIG. 1. Transient decay of the total fluorescence of acenaphthene obtained using the TCSPC method. The sample was excited with a 318-nm, 5-psec laser pulse. The decay lifetime, τ (see text), is 80 nsec.

$$I(t) = I_0 e^{-(t-t_0)/\tau}$$

where $I(t)$ is the time-dependent observed emission intensity, I_0 is the intensity at t_0 (time of laser excitation), and τ is the intrinsic kinetic parameter and is equal to the excited state lifetime. In general, one obtains an experimental decay curve such as in Fig. 1 and then fits it to the functional form above to obtain τ, the characteristic lifetime. Often, the data is more complicated (see Fig. 2), in which case more sophisticated models must be applied that typically have more than one fitting parameter.

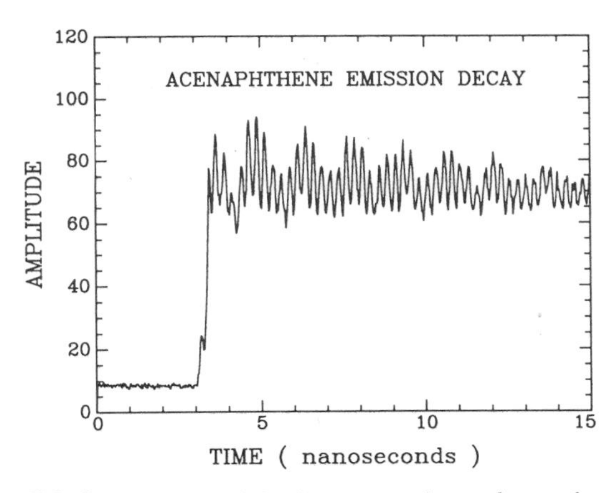

FIG. 2. A portion of the fluorescence decay of acenaphthene obtained by gating on a small part of the emission spectrum so as to monitor only a particular excited state vibrational mode. The fast oscillations are ~100 psec, which demonstrates the timing capabilities of TCSPC.

There are a number of limitations to this technique. First, it is confined to samples that emit radiation after excitation by a picosecond laser pulse, a situation that occurs quite frequently. Second, the time resolution of the technique is limited to at best 50 psec. This limitation is imposed not by the duration of the laser pulses involved but by the characteristics of the photomultiplier detectors.

B. Streak Camera Measurements

A streak camera is a device that can record very fast optical transient events. In essence, it takes a snapshot in time of an event with a resolution on the order of several picoseconds. In a similar way to TCSPC, it is used to measure transient fluorescence decays that result from excitation by a picosecond laser pulse. A streak camera is different in that it can record an entire decay using only a single picosecond laser pulse. The much improved time resolution of several picoseconds is also a benefit. In practice, streak cameras are now not widely used because of their expense and the difficulty of operation.

C. Plump–Probe Techniques

This classification covers a number of experimental schemes. What is common is that the sample is excited by one ultrafast pulse (picosecond or femtosecond), and then the progress of the event at some time later is probed by a second ultrafast pulse. How this second pulse monitors the progress of the event is what distinguishes the different types of pump–probe experiments. The important distinction from TCSPC is that the time resolution is determined by the laser pulse width, not any particular detector characteristics. A specific example of a unimolecular dissociation reaction illustrates the technique. The reaction is

$$\mathrm{NCNO} + h\nu \rightarrow \mathrm{CN} + \mathrm{NO}$$

where $h\nu$ represents the laser pulse that is in this case at 580 nm. The NCNO molecule absorbs the laser photon. This energy causes the molecule to fragment and form the molecular products CN and NO. The rate of this reaction is determined by measuring the time required for the appearance of the products after the absorption of the initiating laser pulse, the pump. The appearance of the products is monitored by a second picosecond laser pulse, the probe (388 nm), using laser induced fluorescence (LIF) of the CN product. The LIF signal is proportional to the amount of CN present and therefore reveals the extent of the reaction. In this way, the development of the reaction as a function of time can be followed by monitoring the LIF signal as a function of the pump and probe laser pulse delay. This is shown for the reaction in Fig. 3.

Other examples of probing techniques are transient absorption, photoionization, and fluorescence up and down conversion using nonlinear optical materials. The important point is that the time resolution is only limited by the duration of the laser pulse and therefore is appropriate for state-of-the-art femtosecond measurements.

III. Electron and Proton Transfer Reactions

Electron and proton transfer are processes that are of central importance in chemistry and biology and have been the focus of much study by ultrafast laser techniques. Electron transfer is a process in which an electron migrates from one position in a molecule to either a different site on the same molecule (intramolecular transfer) or to a different molecule altogether (intermolecular transfer). This is similar for proton transfer. The driving force for these reactions is some change in the chemical environment that normally occurs because of solvent-induced geometrical rearrangements or the creation of excited states by photon absorption. Although proton transfer might be considered in the general classification of bond-breaking chemical reactions, the low mass and hence high mobility and tunneling ability of hydrogen lead to quite unique behavior that is often more similar to electron transfer. Although there has been some work in gas-phase electron and proton transfer, the majority of work has been in the liquid phase. In general, it has been shown that the solvent plays a major role in mediating these types of reactions, and its effects must be considered in detail. In fact, these processes are sensitive probes of solvent–solute interactions.

Ultrafast techniques in general are best suited to studying excited state processes in which an ultrafast pulse is used to initiate the reaction at a well-defined point in time by optically creating the reactive excited state. After creation of this excited state, the electron or proton transfer process takes place with an intrinsic rate. The course of the reaction is monitored by measuring the appearance of the reaction products or by the disappearance of the reactants. One possible monitoring technique is TCSPC, as described previously. This requires that the reactant or products fluoresce and is limited to a time resolution of ~ 50 psec. Pump–probe techniques can also be used with the probe having to be sensitive to the re-

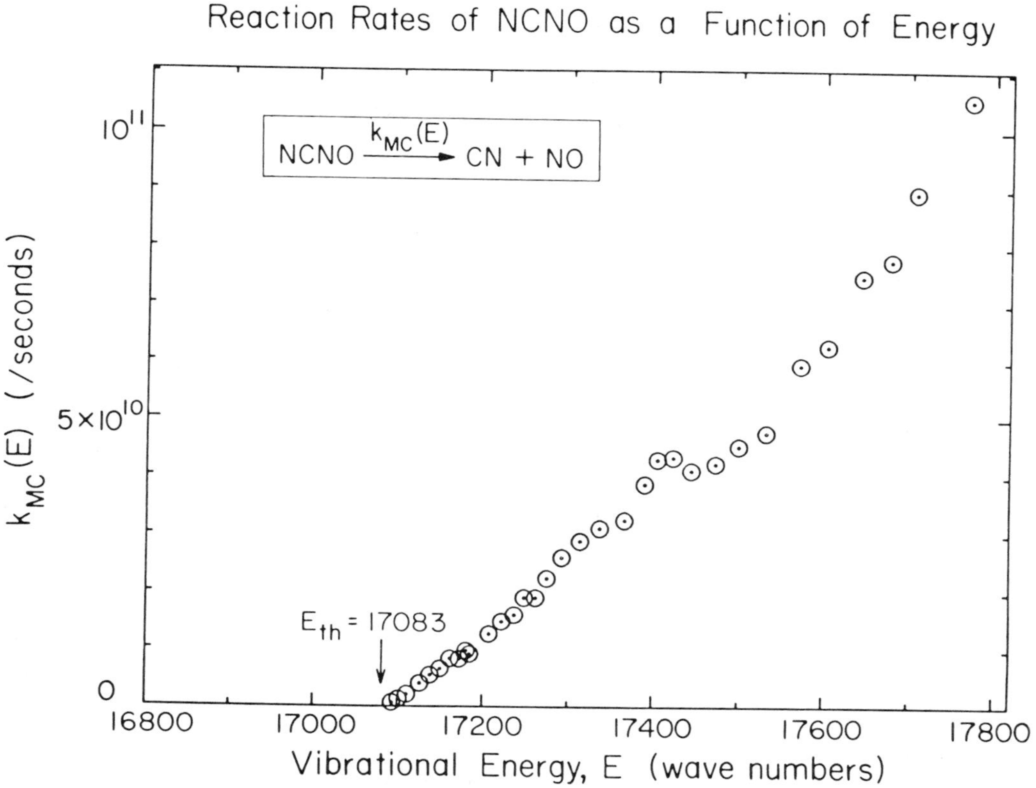

FIG. 3. Rates of the NCNO → CN + NO reaction obtained by a picosecond pump–probe technique. The rates of reaction are seen to increase with increasing pump energy until the limit of the experimental resolution is reached at 10 psec. ($k = 1/\tau$).

actant or product concentrations. The most widely used technique in these solution studies is transient absorption. In this case, the reaction is monitored by measuring the absorption of the probe by the reactant, the product, or perhaps an intermediate that exists only transiently as the reaction proceeds.

An example of research in this area is the study of the primary process of photosynthesis. It has been shown that the initial process is an electron transfer to form a charge separated species consisting of a chlorophyll pair. This charge separation has been measured to occur in less than 100 fsec, and the problem is still under intensive investigation. Other examples of electron transfer show transfer rates on the order of tens or hundreds of picoseconds. In these slower cases, it is assumed that the electron transfer is a two-step process. The first step is a molecular and solvent rearrangement to an appropriate configuration from which a very rapid electron transfer takes place. The measured rates then correspond to this rearrangement process, which is the rate-limiting step. Data on electron transfer rates obtained using ultrafast measurement techniques have been important in testing theories of electron transfer.

IV. Molecular and Solvent Rearrangement

A. Vibrational Energy Redistribution

Chemical and solvent rearrangements, including bond-breaking and -forming reactions, occur because sufficient energy is deposited in a molecule or system to move away from one geometry to another stable position. Generally, this energy is in the form of heat, which, on the molecular level, is motion of the nuclei. These motions can be classified as molecular vibration, rotation, and translation. If this energy is channeled in a particular way, then a chemical rearrangement can take place. How this energy is distributed and the rate of transfer is a question that ultrafast spectroscopy can address. This problem has been investigated in both the liquid and gas phase.

Intramolecular vibrational energy redistribution (IVR) is the process of energy flowing between different vibrational modes of a molecule that is isolated from interaction with other molecules. This is a measure of how energy deposited in a particular vibrational motion will evolve to other motions under only the influences of the potential energy surface of the molecule in question. This has been studied extensively in large aromatic molecules using primarily TCSPC. Data for the acenaphthene molecule is shown in Fig. 2. In this example, a picosecond laser pulse excites a particular nuclear vibration in acenaphthene (in an excited electronic state). Wavelength selective emission is monitored so that only one particular vibration is observed. The oscillating emission measures the flow of energy into and out of the initially prepared state. If a greater amount of energy is deposited in the system, then the energy flow becomes irreversible and a simple nonoscillatory decay is observed.

The result of this type of energy redistribution is particularly evident in molecules such as *trans*-stilbene. In this case, when the vibrational energy is sufficient, the molecule undergoes an isomerization to the *cis* form.

trans- *cis-*

This reaction has been measured in the gas phase and, as expected, the rate depends on the amount of energy in the molecule in excess of the reaction barrier.

In solution, intermolecular interactions greatly influence the observed dynamics. When *trans*-stilbene is placed in solution, it is found that the rate for *trans–cis* isomerization is solvent dependent. In fact, there is a correlation of the reaction rate with solvent viscosity. Molecular motions and interactions in solutions is a very active area of current research and ultrafast measurements are making significant contributions by revealing the important interactions that occur on the time scale of chemical reactions.

V. Chemical Reactions

Ultrafast spectroscopy has been applied to the study of chemical reactions, but the types of reactions have been limited mainly to unimolecular reactions, such as the NCNO example given previously. The reason for this is that most bimolecular or more complicated reactions involve several steps. The rates for each individual step are difficult to deconvolute from the experimental observables. In particular, in a bimolecular reaction one can separate the process into two steps. The first step is for the two reactants to find each other, either in solution or the gas phase. The second step is the reaction process itself. Although these can be successfully sorted out, a detailed determination of the reaction steps is difficult. On the other hand, unimolecular reactions begin to occur immediately upon being energized (which can be performed by an ultrafast laser pulse), making the study of the fundamental steps of a reaction more tractable.

A. Measurement of Reaction Rates

Again, the measurement of reaction rates in the gas phase and solution involve quite different considerations. In the gas phase, the reaction is determined solely by a single molecular potential energy surface. A simple example is the following:

$$\mathrm{ICN} + h\nu \rightarrow \mathrm{ICN}^* \rightarrow \mathrm{I} + \mathrm{CN}$$

In this example, ICN is excited by a 100-fsec pulse to an excited electronic state that is repulsive (see Fig. 4). On the excited surface, there is a strong re-

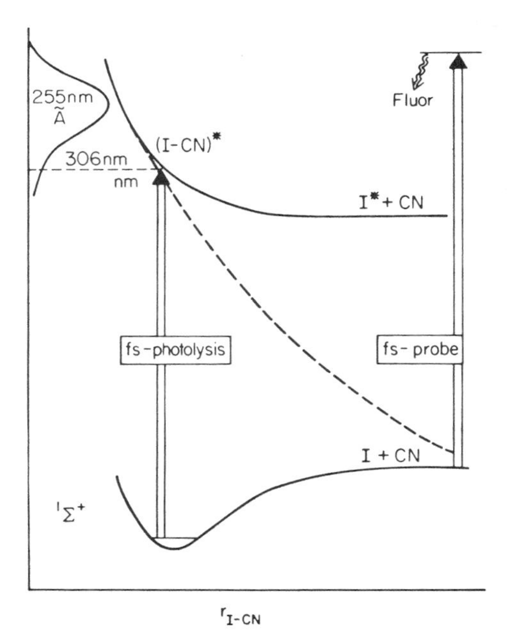

FIG. 4. Schematic representation of the potential energy levels of the ICN molecule and how the unimolecular reaction is initiated and monitored.

pulsive force between the I and the CN that drives the reaction as indicated. The progress of the reaction is followed by monitoring the CN fragment by laser-induced fluorescence, using an ultrafast laser pulse as the probe. By monitoring the CN population as a function of time, the reaction was found to have a fundamental lifetime of ~200 fsec. From the measurement of gas-phase reactions such as this, useful information can be obtained about the potential surfaces involved. Similar studies are now underway on a number of related systems.

Similar experiments can be performed in solution, but there are many more interactions present in such systems. While this makes it more difficult to understand, there is certainly more to learn from such systems, particularly information that relates to more conventional solution-phase chemistry. An example is the unimolecular dissociation of I_2 in a number of solvents. In this case, the intramolecular potential surfaces are similar to the ICN example, but now complex solvent interactions are present. The problem can be envisioned as a unimolecular reaction occurring in a cage of solvent molecules. Ultrafast solution-phase experiments are used to investigate the course of the reaction. In particular, there is interest in geminate recombination, which is the process whereby the two dissociated fragments (in this case I atoms) recombine to form the original molecule. It has been shown to take on the order of tens of picoseconds for the iodine atoms to dissociate and recombine. This apparently simple process is composed of several steps, including vibrational energy relaxation, which occurs as the iodine molecule is formed. This heat of formation is then dissipated to the solvent. Extremely short (~ 50 fsec) pulses are being used to look at solvent interactions that are caused by individual collisional events between solute and solvent molecules. These collisions occur approximately every 500 fsec, so these processes are certainly within experimental limits.

B. Chemical Control Using Ultrafast Laser Pulses

An early goal of chemists using lasers was to be able to control chemical reactions by depositing energy into specific bonds with a laser pulse, thereby weakening the bond and increasing its reactivity with respect to other bonds in the molecule. Although extensive efforts have been directed in this area, they have not been fruitful. The reason for this has been determined by the many other ultrafast experiments, such as those mentioned previously. It is found that energy does not remain localized in one particular vibrational mode of a molecule but instead very rapidly randomizes to all other modes of the system within the limits of energy conservation. The result is that laser excitation behaves very similar to thermal heating at the molecular level. Presently, more sophisticated approaches to laser-assisted chemistry are being investigated in which the effects of energy randomization are considered and can hopefully be overcome by careful control of the laser pulse properties including pulse width and bandwidth.

Bibliography

Austin, D. H., and Eisenthal, K. B., eds. (1984). "Ultrafast Phenomena IV." Springer-Verlag, Berlin and New York.

Demtroder, W. (1982). "Laser Spectroscopy: Basic Concepts and Instrumentation." Springer-Verlag, Berlin and New York.

Fleming, G. R. (1986). Subpicosecond spectroscopy. *Annu. Rev. Phys. Chem.* **37,** 81–104.

Fleming, G. R. (1986). "Chemical Applications of Ultrafast Spectroscopy." Oxford Univ. Press, London and New York.

Mathies, R. A., Cruz, C. H. B., Pollard, W. T., and Shank, C. V. (1988). Direct observation of the femtosecond excited-state *cis-trans* isomerization in bacteriorhodopsin. *Science* **240,** 777–779.

Index